Elastomer Technology Handbook

Elastomer Technology Handbook

Edited by

Nicholas P. Cheremisinoff, Ph.D.

SciTech Technical Services
Morganville, New Jersey

CRC Press
Taylor & Francis Group
Boca Raton London New York

CRC Press is an imprint of the
Taylor & Francis Group, an **informa** business

CRC Press
Taylor & Francis Group
6000 Broken Sound Parkway NW, Suite 300
Boca Raton, FL 33487-2742

© 1993 by Taylor & Francis Group, LLC
CRC Press is an imprint of Taylor & Francis Group, an Informa business

First issued in paperback 2019

No claim to original U.S. Government works

ISBN-13: 978-0-367-44988-9 (pbk)
ISBN-13: 978-0-8493-4401-5 (hbk)

Visit the Taylor & Francis Web site at
http://www.taylorandfrancis.com

and the CRC Press Web site at
http://www.crcpress.com

Library of Congress Card Number 92-37546

Library of Congress Cataloging-in-Publication Data

Elastomer technology handbook / edited by Nicholas P. Cheremisinoff.
    p. cm.
  Includes bibliographical references and index.
  ISBN 0-8493-4401-8
  1. Elastomers--Handbooks, manuals, etc. 2. Polymers--Handbooks,
manuals, etc. I. Cheremisinoff, Nicholas P.
TA455.E4E43 1993
678′.7—dc20

                                                      92-37546
                                                      CIP

PREFACE

The first recorded use of rubber dates back to the early Mayan civilization in Mexico, sometime around 500 B.C., where local inhabitants had a court game similar to basketball. Excavated ruins include stone rings on the walls, revealing basketball-like courts. Spanish explorers in the 15th and 16th centuries recorded bouncing balls, waterproof clothing, and shoes from Mexico and Brazil, but it wasn't until the 1700s that modern day terminology was first introduced. The French Academy of Science gave the name for natural rubber in the 1700s, calling it caoutchouc, or "weeping wood". The scientist Joseph Priestley coined the word "rubber" in 1770 because a piece of the gummy material *rubbed out* lead pencil marks.

Since Priestley's time, scientists have categorized over 400 different vegetations that produce natural rubber latex, including dandelion, quayule, chicle, and gutta percha. Cultivation of rubber trees began in Brazil in the 19th century. Seedlings were transplanted to Southeast Asia, Mexico, Africa, and Central America. In fact, the rubber tree became the focus of cultivation and processing equipment was designed around the properties of the Hevea rubber to facilitate its use.

Perhaps mankind's greatest attribute is its endeavor to copy and outdo nature. In 1826 Michael Faraday identified the latex produced by rubber trees as a polymer of isoprene. Researchers in the mid-1900s duplicated the structure to make synthetic polyisoprene. Earlier, C. Greville Williams isolated isoprene from the distillation of rubber in 1860. In 1879, French scientists A. and G. Bouchardat made a rubber gum from an isoprene monomer by thermally cracking the rubber and, in 1892, W. Tilden prepared the first isoprene synthesized from a non-rubber material. In Tilden's experiments a gum from isoprene derived from pine stump turpentine was prepared.

The synthetic rubber industry has experienced rapid growth during this century. Because of natural rubber shortages, Germany produced methyl rubber during World War I. In 1921, the Soviets improved the technology for manufacturing butadiene, which is the basic monomer for most synthetic rubber. It took just two short years beyond that for emulsion polymerization to be developed in both the United States and Germany.

In the 1930s, polychloroprene and polysulfide elastomers were developed. These materials created new markets for rubbers because of their high oil and solvent resistance. In 1935, general-purpose styrene-butadiene rubber, Buna S, was first introduced by Germany. Shortly afterward, Buna N (Perbunan) was introduced as an oil-resistant nitrile-butadiene elastomer. In 1940 the Standard Oil Company of New Jersey (known today as the Exxon Corporation) and I. G. Farbenindustrie GmbH entered into a patent exchange agreement, thus launching synthetic rubbers in North America.

Between 1940 and 1955 there was an almost explosive introduction of new elastomers. By 1951, the capacity for general-purpose SBR (styrene-butadiene rubber) had reached 850,000 tons. The vast majority of this consumption was in the tire industry.

By the late 1960s, thermoplastic elastomers began to appear on the market. Many automobile bumpers were manufactured from thermoplastic polyurethanes. In the 1970s, the consumer began to see shoe soles made from styrenic block copolymers. Olefinic TPE wire and cable jacketing was also introduced at about the same time. High-grade copolyester TPEs, first marketed in the mid-1970s, are widely used today in the automotive and mining markets.

By the early 1980s, higher grade olefinics began to be marketed as thermoplastic vulcanizates because they contained cured EPDM elastomer imbedded in the thermoplastic matrix. In 1984 a metal processible rubber was introduced that found applications in modified rubber equipment. We also saw, in the 1980s, polyamide TPEs used in a variety of other products including sporting goods.

Since the mid-1980s, major advancements in elastomers have largely centered on the modification of existing polymers rather than on the invention of new materials. However, this is not to say that experimentation in catalyst systems is dormant or that new product development has ceased.

This historical perspective enables one to appreciate that elastomers have enjoyed a long and diverse development. It is a subject that, in fact, is quite fragmented, largely because the development of synthetic elastomers has been market-driven, rather than created out of scientific curiosity. In this regard the subject of elastomeric materials cannot be appreciated or fully comprehended unless there is sensitization to the needs of the consumer. This determines the types of applications for which materials are sought. For this reason, the user will find that many of the chapters in this handbook emphasize applications and property attributes on a commercial basis. Unlike the more conventional handbook format of tabulations of property data, this volume provides a more fundamental treatment of concepts and general principles in the design, fabrication, and uses of synthetic elastomers. As a point of clarification, the terms rubber and elastomer are used interchangeably. Furthermore, because many rubbers are used either in combination with or approach properties akin to plastics, the reader will find some discussions and overlap with plastics technology. It is felt that this volume will be useful to both the researcher and the practitioner. Polymer scientists/chemists, product development specialists, and polymer fabricators should find this a valuable reference.

It should be noted that this handbook represents the contributions of over forty specialists. It also represents the efforts of scores of engineers and associates who provided advice and consultation in the organization of this volume. A heartfelt gratitude is extended to these individuals for making this work possible. Additionally, CRC Press is to be commended on the fine production of this work.

Nicholas P. Cheremisinoff, Ph.D.

ABOUT THE EDITOR

Nicholas P. Cheremisinoff, Ph.D., is president of SciTech Technical Services, specializing in environmental and engineering consulting practices. Dr. Cheremisinoff has had nearly 20 years of industrial and applied research experience in petroleum, petrochemicals, synthetic fuels, elastomers and plastics, and the environmental field. He is internationally recognized as an expert in heat transfer operations and fluid mechanics, having authored/co-authored and edited nearly 100 textbooks on these subjects. He received his B.S., M.S., and Ph.D. degrees in chemical engineering from Clarkson College of Technology.

CONTRIBUTORS

Bhola Nath Avasthi
Department of Chemistry
Indian Institute of Technology
Kharagpur - 721 302, India

Susmita Bhattacharjee
Rubber Technology Centre
Indian Institute of Technology
Kharagpur - 721 302, India

Anil K. Bhowmick
Rubber Technology Centre
Indian Institute of Technology
Kharagpur - 721 302, India

Randi Boyko
Cycle Chem
217 South First Street
Elizabeth, New Jersey 07206

Petra Budrugeac
ICPE Research Institute for Electrical
 Engineering
Bd. T. Vladimirescu, Nr.45-47
Bucharest 79623, Romania

Franklin M. C. Chen, Ph.D.
Kimberly-Clark Corporation
2100 Winchester Road
Neenah, Wisconsin 54956

Ue-De Chen, Ph.D.
Aero Industry Development Center
Chung-San Institute of Science
 Technology
10 Fl.-3. 384# Hwa-Mei Street
Taichung, Taiwan 403
Republic of China

Nicholas P. Cheremisinoff, Ph.D.
SciTech Technical Services
457 Highway 79
Morganville, New Jersey 07751

John D. Culter, Ph.D.
Advanced Materials Engineering, Inc.
7116 Gleason Road
Edina, Minnesota 55439

S. K. De, Ph.D.
Rubber Technology Centre
Indian Institute of Technology
Kharagpur - 721 302, India

Wim C. Endstra
Akzo Chemicals bv
Zutphenseweg 10
P. O. Box 10
7400 AA Deventer
The Netherlands

Gerhard Martin
Hermann Berstorff
Maschinenbau Gmb H
Ander Breiten Wiese 315
D-3000 Hannover, Germany

Eddy I. Garcia-Meitin
Materials Science Group
Texas Polymer Center
Dow Chemical U.S.A.
2301 N. Brazosport Boulevard
Freeport, Texas 77541

B. R. Gupta, Ph.D.
Rubber Technology Centre
Indian Institute of Technology
Kharagpur - 721 302, India

Bruce Hartmann, Ph.D.
Materials Science and Technology Branch
Naval Surface Warfare Center
Silver Spring, Maryland 20903-5640

Iwakazu Hattori, Dr. of Chemistry
Elastomer Laboratory
Technical Center
Japan Synthetic Rubber Company, Ltd.
100, Kawajiri-Cho
Yokkaichi, Mie, Japan

A. B. Hunter
Boeing Defense and Space Group
P. O. Box 3999, MS 8Y-57
Seattle, Washington 98124

R. Hussein
State University of New York
Syracuse, New York 13210

Takashi Inoue, Dr.
Department of Organic and Polymeric
  Materials
Tokyo Institute of Technology
Ookayama, Meguro-ku
Tokyo 152, Japan

Daniel Klempner, Ph.D.
Polymer Institute/Polymer Technologies,
  Inc.
University of Detroit Mercy
4001 West McNichols Road
Detroit, Michigan 48219

Jack L. Koenig, Ph.D.
Department of Macromolecular Science
Case Western Reserve University
Cleveland, Ohio 44106

Michael R. Krejsa, Ph.D.
The Chemical Group of Monsanto
730 Worcester Street
Springfield, Massachusetts 01151

Yoichiro Kubo
Nippon Zeon Company, Ltd.
Research and Development Center
1-2-1 Yako, Kawasaki-ku
Kawasaki, 210, Japan

Laura Leidy
SciTech Technical Services
457 Highway 79
Morganville, New Jersey 07751

Kenya Makino, Dr. of Chemistry
Elastomer Laboratory
Technical Center
Japan Synthetic Rubber Company, Ltd.
100, Kawajiri-Cho
Yokkaichi, Mie, Japan

George Mares
ICPE Research Institute for Electrical
  Engineering
Bd. T. Vladimirescu, Nr. 45-47
Sector 5, Bucharest 79623, Romania

Richard L. Markham, M.S.
Polymer Process/Product Development
Batelle Memorial Institute
505 King Avenue
Columbus, Ohio 43201

Cleopatra V. Oprea, Ph.D.
Department of Organic and
Macromolecular Chemistry and
  Technology
Polytechnic Institute of Jassy
Bulevardul M. Eminescu Nr. 11
6600 Jassy, Romania

Toshiaki Ougizawa, Dr.
National Institute of Materials and
  Chemical Research
Tsukuba, Ibaraki 305, Japan

Marc T. Payne, B.S.
Advanced Elasstomer Systems, L. P.
260 Springside Drive
P. O. Box 5584
Akron, Ohio 44334

Dale M. Pickelman, B. S.
Advanced Polymeric Systems Laboratory
Central Research
The Dow Chemical Company
Midland, Michigan 48674

Marcel Popa, Ph.D.
Department of Organic and
Macromolecular Chemistry and
  Technology
Polytechnic Institute of Jassy
Bulevardul M. Eminescu Nr. 11
6600 Jassy, Romania

Charles P. Rader, Ph.D.
Advanced Elastomer Systems, L. P.
260 Springside Drive
P.O. Box 5584
Akron, Ohio 44334

Sanjoy Roy
Rubber Technology Centre
Indian Institute of Technology
Kharagpur, 721 302, India

Rudy J. School, B.S.
R. T. Vanderbilt Company, Inc.
30 Winfield Street
Norwalk, Connecticut 06856

Werner Schuler
Hermann Berstorff
Maschinenbau Gmb H
Ander Breiten Weise 315
D-3000 Hannover, Germany

Michael P. Sepe
Dickten and Masch Manufacturing
 Company
N44 W33341 Watertown Plank Road
P. O. Box 112
Nashotah, Wisconsin 53058

H. So, Ph.D.
Department of Mechanical Engineering
National Taiwan University
Taipei, Taiwan 10617
Republic of China

Daniel Sophiea, Ph.D.
Department of Acoustic Materials
Dexter Automotive Materials
1200 Harmon Road
Auburn Hills, Michigan 48326

David Sudduth
Materials Science and Technology Branch
Naval Surface Warfare Center
Silver Spring, Maryland 20903-5640

Hung-Jue Sue, Ph.D.
Materials Science Group
Texas Polymer Center
The Dow Chemical Company
2301 N. Brazosport Boulevard
Freeport, Texas 77541

Joachim Sunder, Dr.Ing.
Werner & Pfleiderer Gummitechnik
 Gmb H
Asdorfer Strasse 60
Freudenberg, Germany

Carel T. J. Wreesmann, Ph.D.
Akzo Chemicals bv
Zutphenseweg 10
P. O. Box 10
7400 AA Deventer
The Netherlands

TABLE OF CONTENTS

Chapter 1

ANALYTICAL TEST METHODS FOR POLYMER CHARACTERIZATION

Nicholas P. Cheremisinoff, Randi Boyko, and Laura Leidy

TABLE OF CONTENTS

0-8493-4401-8/93/$0.00 + $.50

© 1993 by CRC Press, Inc.

1

I. CHROMATOGRAPHIC TECHNIQUES

A. CHROMATOGRAPHY FOR ANALYTICAL ANALYSES

Chromatography may be defined as the separation of molecular mixtures by distribution between two or more phases, one phase being essentially two dimensional (a surface) and the remaining phase or being a bulk phase brought into contact in a countercurrent fashion with the two-dimensional phase. Various types of physical states of chromatography are possible, depending on the phases involved.

Chromatography is divided into two main branches. One branch is gas chromatography; the other is liquid chromatography. Liquid chromatography can be further subdivided as shown in Figure 1.

The sequence of chromatographic separation is as follows: a sample is placed at the top of a column where its components are sorbed and desorbed by a carrier. This partitioning process occurs repeatedly as the sample moves toward the outlet of the column. Each solute travels at its own rate through the column; consequently, a band representing each solute will form on the column. A detector attached to the column's outlet responds to each band. The output of detector response vs. time is called a chromatogram. The time of emergence identifies the component and the peak area defines its concentration, based on calibration with known compounds.

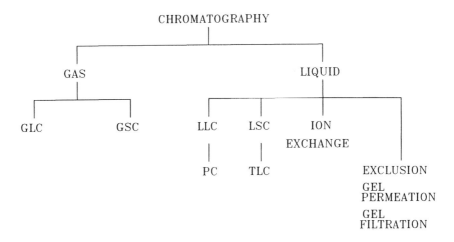

FIGURE 1. Types of chromatographic operations.

B. GAS CHROMATOGRAPHY
1. General
If the moving phase is a gas, the technique is called gas chromatography (GC). In GC the sample is usually injected at high temperature to ensure vaporization. Obviously, only materials volatile at this temperature can be analyzed.

2. Types of GC
If the stationary phase is a solid, the technique is referred to as gas-solid chromatography. The separation mechanism is principally one of adsorption. Those components more strongly adsorbed are held up longer than those which are not.

If the stationary phase is a liquid, the technique is referred to as gas-liquid chromatography and the separation mechanism is principally one of partition (solubilization of the liquid phase).

GC has developed into one of the most powerful analytical tools available to the organic chemist. The technique allows separation of extremely small quantities of material (10^{-6} g).

The characterization and quantitation of complex mixtures can be accomplished with this process. The introduction of long columns, both megabore and capillary, produces a greater number of theoretical plates increasing the efficiency of separation beyond that of any other available technique. The technique is applicable over a wide range of temperatures (-40 to $350°C$) making it possible to chromatograph materials covering a wide range of volatilities. The laboratory uses packed columns along with megabore and capillary. In this way the broadest range of chromatographic problems can be addressed.

The detector used to sense and quantify the effluent provides the specificity and sensitivity for the analytical procedure. Table 1 summarizes significant detector characteristics.

C. LIQUID CHROMATOGRAPHY
1. General
If the moving phase is a liquid, the technique is called liquid chromatography (LC). In LC the sample is first dissolved in the moving phase and injected at ambient temperature. Thus, there is no volatility requirement for samples. However, the sample must dissolve in the moving phase. Note that LC has an important advantage over GC: the solubility requirement can usually be met by changing the moving phase. The volatility requirement is not so easily overcome.

TABLE 1
Summary of Detector Characteristics

Detector	Principle of operation	Selectivity	Sensitivity	Linear range	MDQ[a]	Stability
Thermal conductivity	Measures thermal conductivity of gas	Universal	6×10^{-10}	10^4	10^{-5} g of CH_4 per volume of detector effluent	Good
Flame ionization	H_2-O_2 flame	Responds to organic compounds, not to H_2O or fixed gases	9×10^{-3} for alkane	10^7	2×10^{-11} g for alkane	Excellent
Electron capture	$N_2 + B \rightarrow e^-$ $e^- +$ sample $\rightarrow$	Responds to electron adsorbing compounds, e.g., halogen	2×10^{-14} for CCl_4	10^5	—	Good
Hall electrolytic conductivity detector		In halogen mode responds to halogens		10^0	1×10^{13} g cl/s	Poor

[a]   Minimum detectable quantity.

2. Types of LC

There are four kinds of LC, depending on the nature of the stationary phase and the separation mechanism.

3. Liquid/Liquid Chromatography (LLC)

LLC is partition chromatography or solution chromatography. The sample is retained by partitioning between mobile liquid and stationary liquid. The mobile liquid cannot be a solvent for the stationary liquid. As a subgroup of LLC there is paper chromatography.

4. Liquid/Solid Chromatography (LSC)

LSC is adsorption chromatography. Adsorbents such as alumina and silica gel are packed in a column and the sample components are displaced by a mobile phase. Thin layer chromatography and most open column chromatography are considered LSC.

5. Ion-Exchange Chromatography

Ion-exchange chromatography employs zeolites and synthetic organic and inorganic resins to perform chromatographic separation by an exchange of ions between the sample and the resins. Compounds which have ions with different affinities for the resin can be separated.

6. Exclusion Chromatography

Exclusion chromatography is another form of LC. In the process a uniform nonionic gel is used to separate materials according to their molecular size. The small molecules get into the polymer network and are retarded, whereas larger molecules cannot enter the polymer network and will be swept out of the column. The elution order is the largest molecules first, medium next, and the smallest sized molecules last. The term "gel permeation chromatography" has been coined for separations polymers which swell in organic solvent.

The trend in LC has tended to move away from open column toward what is called high-pressure liquid chromatography (HPLC) for analytical as well as preparative work. The change in technique is due to the development of high-sensitivity, low dead volume detectors. The result is high-resolution, high-speed, and better-sensitivity LC.

7. Type of Information Obtained
a. Form
The output of a chromatographic instrument can be of two types:

- A plot of area retention time vs. detector response. The peak areas represent the amount of each component present in the mixture.
- A computer printout giving names of components and the concentration of each in the sample.

b. Units
The units of concentration are reported in several ways:

- Weight percent of parts per million by weight; volume percent or parts per million by volume.
- Mole percent.

c. Sample
Size — A few milligrams is usually enough for either GC or LC.

State — (1) For GC, the sample can be gas, liquid, or solid. Solid samples are usually dissolved in a suitable solvent; both liquid or solid samples must volatilize at the operating temperature. (2) For LC, samples can be liquid or solid. Either must be soluble in moving phase.

d. Advantages
GC —

- Moderately fast quantitative analyses (0.5 to 1.5 h per sample).
- Excellent resolution of various organic compounds.
- Not limited by sample solubility.
- Good sensitivity.
- Specificity.

LC —

- Separation of high boiling compounds.
- Not limited by sample volatility.
- Moving phase allows additional control over separation.

e. Disadvantages
GC —

- Limited by sample volatility.

LC —

- Less sensitive than GC.
- Detectors may respond to solvent carrier, as well as to sample.

f. Interferences
Interferences in chromatography can generally be overcome by finding the right conditions to give separation. However, this might be costly, since development of separations is largely a trial-and-error process.

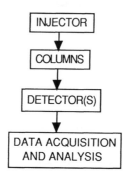

FIGURE 2.   Basic components of GPC/DRI.

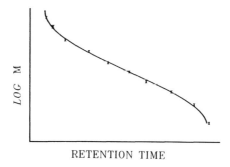

FIGURE 3.   Typical calibration curve using a polystyrene (PS) standard.

D. GPC/DRI

For common linear homopolymers, such as PIB, PE, PS . . . , GPC analysis can be performed with a single DRI detector. Figure 2 shows the basic component of a GPC/DRI system. Most often, a PS calibration curve is generated from narrow molecular weight PS standards (see Figure 3), which can then be converted to the desired polymer (i.e., PIB, EP) if the appropriate calibration constants are available. These constants, known as the Mark-Houwink parameters or k and α, are used to calculate the intrinsic viscosity of the polymer as a function of molecular weight (which is needed to relate the size of one type of polymer to another). If the Mark-Houwink parameters are not available, the molecular weights can be used for relative comparison but will not be correct on an absolute basis. If the sample is branched, the molecular weights will be biased low, and a secondary detector [low-angle laser light scattering (LALLS) or viscometer (VIS)] is needed for accurate results.

E. GEL PERMEATION CHROMATOGRAPHY

Gel permeation chromatography (GPC), also known as size exclusion chromatography (SEC), is a technique used to determine the average molecular weight distribution of a polymer sample. Using the appropriate detectors and analysis procedure it is also possible to obtain qualitative information on long-chain branching or determine the composition distribution of copolymers.

As the name implies, GPC or SEC separates the polymer according to size or hydro-dynamic radius. This is accomplished by injecting a small amount (100 to 400 μl) of polymer

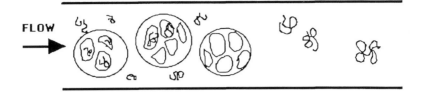

FIGURE 4. Polymer solution flow through GPC column.

solution (0.01 to 0.6%) into a set of columns that are packed with porous beads. Smaller molecules can penetrate the pores and are therefore retained to a greater extent than the larger molecules which continue down the columns and elute faster. This process is illustrated in Figure 4.

One or more detectors is attached to the output of the columns. For routine analysis of linear homopolymers, this is most often a differential refractive index (DRI) or a UV detector. For branched or copolymers, however, it is necessary to have at least two sequential detectors to determine molecular weight accurately. Branched polymers can be analyzed using a DRI detector coupled with a "molecular weight sensitive" detector such as a VIS or a LALLS detector. The compositional distribution of copolymers, i.e., average composition as a function of molecular size, can be determined using a DRI detector coupled with a selective detector such as UV or FTIR. It is important to consider the type of polymer and information that is desired before submitting a sample. The following outline describes each instrument that is currently available.

F. GPC/DRI/LALLS

One can use two instruments with sequential LALLS and DRI detectors. The unit is operated using TCB at 135C and is used to analyze PE, EP, and PP samples. The other, operating at 60 c, is for butyl-type polymers which dissolve in TCB at lower temperatures. The data consists of two chromatograms, plots of detector millivolt signal (LALLS and DRI) vs. retention time. The DRI trace corresponds to the concentration profile, whereas the LALLS signal is proportional to concentration *M, resulting in more sensitivity at the high molecular weight end. An example of the output is shown in Figure 5 for polyethylene NBS 1476. The LALLS trace shows a peak at the high molecular weight end (low retention time) which is barely noticeable on the DRI trace. This suggests a very small amount of high molecular weight, highly branched material. This type of bimodal peak in the DRI trace is often seen in branched EP (ethylene-propylene polymers) or LDPE (low-density polyethylene) samples. The report consists of two result pages, one from the DRI calibration curve as described above, and the second from the LALLS data. An example of a report page is shown in Figure 6. At the top of the page should be a file name and date of analysis. The header also includes a description of the method and detector type, which in this case is the DRI detector and EP calibration curve. Following the header are the parameters integration (i.e., start end times for integration and baseline) and a slice report (i.e., cumulative weight percent and molecular weight as a function of retention time). This section gives details about the distribution such as the range of molecular weights for the sample and the fraction of polymer above a particular molecular weight. At the bottom of the page is a summary of the average molecular weights, whereas Z denotes the Z average molecular weight or Mz, etc.

For a linear polymer (if all the calibration constants are known), the molecular weights from both pages should agree within 10%. A LALLS report that gives higher molecular weights than the DRI suggests that the sample is branched and the values from the LALLS report should be used (again, assuming that the calibration constants are correct).

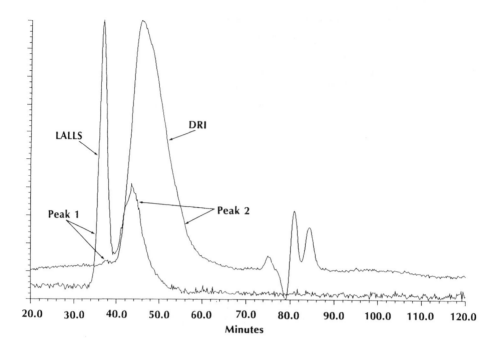

FIGURE 5. Example of output from an LALLS detector.

Occasionally, some of the sample, gel, or insolubles is filtered out during the sample preparation and analysis. The percentage should be indicated on the report.

G. GPD/DRI/VIS

GPC with an on-line VIS can be used instead of a LALLS detector to analyze branched polymers. In this case the intrinsic viscosity is measured so that the Mark-Houwink parameters are not needed. It is complementary to the LALLS instrument in intrinsic viscosities.

H. GPC/DRI/UV

The UV detector is used to analyze chromophores. Its most common use is for graft or block copolymers containing PS or PMS. The data from this instrument consist of two chromatograms, the UV and DRI traces. An example is shown in Figure 7 for an EP-g-PS copolymer (peak 1) with residual PS homopolymer (peak 2). The UV absorption relative to the DRI signal corresponds to the copolymer composition, which is why the relative UV absorption is higher for the pure PS in peak 2. The results report consists of two pages. One is the molecular weight report from the DRI calibration curve as described above. Note that the molecular weights are reported as if the sample is a homopolymer, not copolymer. The other page using the UV data gives an effective extinction coefficient E' which is the UV/DRI ratio. A higher E' indicates a higher composition of the UV active chromophore (e.g., more PS in the graft copolymer). This technique is also used to determine the compositional distribution of ENB (ethylidene norbornene) in EPDM, i.e., whether the ENB is evenly distributed across the molecular weight distribution or concentrated in the low or high molecular weight end. The GPC/DRI/UV instrument can be used to analyze samples that dissolve in THF at 30 to 45°C.

I. GPC/DRI/FTIR

The GPC/DRI/FTIR instrument is complementary to the UV detector for compositional distribution. It runs at 135°C in TCB and can be used for EP analysis. Typical applications include ethylene content as a function of molecular weight, maleic anhydride content in maleated EP, or PCL content in caprolactone-g-EP copolymers. The FTIR detector is off-line so that five to ten fractions of the eluant are collected on KBr plates and analyzed. This procedure gives calibration of IR absorption bands. This method is much more labor intensive than the other techniques and should be used with discretion.

1. Submitting Samples

Samples should be weighed out (typically 30 to 120 μg) in bottles. The submitter should check which is the appropriate amount for a particular test. The sample should be labeled with the contents, exact amount of polymer, and test type. Any other information, such as expected molecular weight range, ENB or other monomer content, dissolution temperature, etc. is helpful for optimizing the analysis. Typically, a single GPC run takes approximately $2^{1}/_{2}$ h, except for GPC/FTIR which can take 5 h for the fractionation and additional time for the FTIR data acquisition.

II. THERMAL ANALYSIS

A. GENERAL PRINCIPLES OF OPERATION

Thermal analysis refers to a variety of techniques in which a property of a sample is continuously measured as the sample is programmed through a predetermined temperature profile. Among the most common techniques are thermal gravimetric analysis (TGA) and differential scanning calorimetry (DSC).

In thermal analysis (TA) the mass loss vs. increasing temperature of the sample is recorded. The basic instrumental requirements are simple: a precision balance, a programmable furnace, and a recorder (Figure 8). Modern instruments, however, tend to be automated and include software for data reduction. In addition, provisions are made for surrounding the sample with an air, nitrogen, or an oxygen atmosphere.

In a DSC experiment the difference in energy input to a sample and a reference material is measured while the sample and reference are subjected to a controlled temperature program. DSC requires two cells equipped with thermocouples in addition to a programmable furnace, recorder and gas controller. Automation is even more extensive than in TA due to the more complicated nature of the instrumentation and calculations.

A thermal analysis curve is interpreted by relating the measured property vs. temperature data to chemical and physical events occurring in the sample. It is frequently a qualitative or comparative technique.

In TA the mass loss can be due to such events as the volatilization of liquids and the decomposition and evolution of gases from solids. The onset of volatilization is proportional to the boiling point of the liquid. The residue remaining at high temperature represents the percent ash content of the sample. Figure 9 shows the TA spectrum of calcium oxalate as an example.

In DSC the measured energy differential corresponds to the heat content (enthalpy) or the specific heat of the sample. DSC is often used in conjunction with TA to determine if a reaction is endothermic, such as melting, vaporization, and sublimation of exothermic, such as oxidative degradation. It is also used to determine the glass transition temperature of polymers. Liquids and solids can be analyzed by both methods of thermal analysis. The sample size is usually limited to 10 to 20 mg.

TA can be used to characterize the physical and chemical properties of a system under conditions that simulate real world applications. It is not simply a sample composition technique.

Much of the data interpretation is empirical in nature and more than one thermal method

029:1 NBS-1476-8   30 Mar 1990   1:24:03   Page 1

Analyzed 3 Apr 1990 10:57:15

High Speed GPC - Calibration Curve 206   INST 3 EP TCB
Detector - DRI

Peak Parameters - Time, min	Mvolts	Baseline Height	Time
Start 36.00	5.0	4.96	35.00
Max 46.00	63.5		
Finish 66.00	6.3	6.26	67.00

Peak Area, mv-sec = 33373.   Molecular Weight at Peak Max   79442

Time, min	Height, mv	dWM/dLM	Cum Wt Pct	Mol Pct	Cum Mol Pct	Mol Wt
36.00	0.000	0.00	0.00	0.00	0.00	5050122
37.20	1.149	0.01	0.15	0.00	0.00	2699310
38.40	0.800	0.01	0.36	0.00	0.00	1501013
39.60	0.934	0.01	0.52	0.00	0.01	866453
40.80	4.760	0.03	1.09	0.02	0.03	518072
42.00	16.159	0.11	3.44	0.15	0.18	320161
43.20	31.938	0.27	8.82	0.53	0.70	204054
44.40	49.795	0.50	18.04	1.38	2.08	133833
45.60	57.835	0.70	30.00	2.65	4.73	90134
46.80	56.523	0.74	42.42	4.01	8.74	62195
48.00	53.013	0.78	54.20	5.43	14.16	43878
49.20	46.600	0.75	64.87	6.87	21.04	31579
50.40	38.402	0.64	73.90	7.98	29.01	23134
51.60	30.066	0.56	81.14	8.63	37.65	17214
52.80	22.703	0.45	86.67	8.80	46.44	12982
54.00	17.144	0.35	90.85	8.75	55.19	9900
55.20	12.479	0.24	93.97	8.52	63.71	7619
56.40	8.370	0.15	96.12	7.59	71.30	5904
57.60	5.380	0.10	97.54	6.43	77.73	4597
58.80	3.529	0.05	98.47	5.39	83.11	3588
60.00	2.208	0.05	99.05	4.38	87.49	2801
61.20	1.595	0.04	99.44	3.69	91.17	2183
62.40	1.085	0.02	99.70	3.25	94.43	1694
63.60	0.517	0.01	99.87	2.54	96.97	1307

```
64.80     0.360     0.01      99.95     1.84      98.82     1000
66.00     0.053     0.00      100.00    1.18      100.00    757


Average Mol Wts          Ratios of Averages          Time Int Std Peak, min

(Z+1) =  2080106         (Z+1)/Z   =  4.417              Expected  85.20
    Z =   470912         (Z+1)/WT  = 23.598                Actual  84.20
   WT =    88146             Z/WT  =  5.342
  VIS =    73154           WT/VIS  =  1.205
 AVIS =    0.726           VIS/NO  =  3.144
   NO =    23266           WT/NO   =  3.789

                         Intrinsic Viscosity   0.993
```

FIGURE 6. Example of a report page.

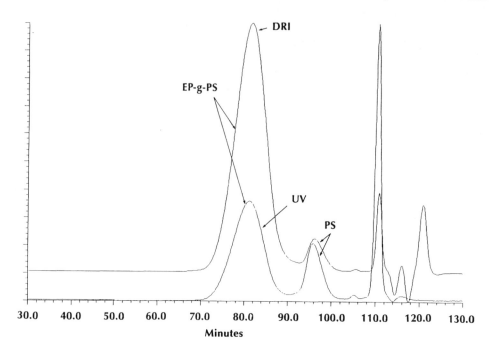

FIGURE 7. Example of UV detector output.

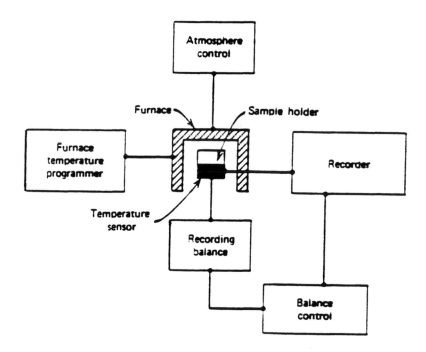

FIGURE 8. Typical components of a TA instrument.

may be required to fully understand the chemical and physical reactions occurring in a sample.

Condensation of volatile reaction products on the sample support system of a TA can give rise to anomalous weight changes.

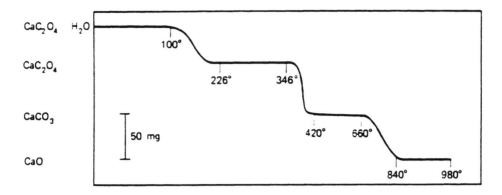

FIGURE 9. TA spectrum of calcium oxalate.

B. THERMAL ANALYSIS OF POLYMERS

A simple example of the relationship between "structure" and "properties" is the effect of increasing molecular weight of a polymer on its physical (mechanical) state: a progression from an oily liquid, to a soft viscoelastic solid, to a hard, glassy elastic solid. Even seemingly minor rearrangements of atomic structure can have dramatic effects as, for example, the atactic and rearrangements of atomic structure can have dramatic effects as, for example, the atactic and syndiotactic stereoisomers of polypropylene — the first being a viscoelastic amorphous polymer at room temperature while the second is a strong, fairly rigid plastic with a melting point above 160°C. At high thermal energies conformational changes via bond rotations are frequent on the time scale of typical processing operations and the polymer behaves as a liquid (melt). At lower temperatures the chains solidifies by either of two mechanisms: by ordered molecular packing in a crystal lattice — *crystallization* — or by a gradual freezing out of long-range molecular motions — *vitrification*. These transformations, which define the principal rheological regimes of mechanical behavior: the melt, the rubbery state, the semicrystalline, and the glassy amorphous solids, are accompanied by transitions in thermodynamic properties at the glass transition, the crystalline melting, and the crystallization temperatures.

TA techniques are designed to measure the above-mentioned transitions both by measurements of heat capacity and mechanical modulus (stiffness).

1. Differential Scanning Calorimetry (DSC)

The DSC measures the power (heat energy per unit time) differential between a small weighed sample of polymer (approximately 10 mg) in a sealed aluminum pan referenced to an empty pan in order to maintain a zero temperature differential between them during programmed heating and cooling temperature scans. The technique is most often used for characterizing the T_g, T_m T_c, and heat of fusion of polymers (see Figure 10). The technique can also be used for studying the kinetics of chemical reactions, e.g., oxidation and decomposition. The conversion of a measured heat of fusion can be converted to a percent crystallinity provided, of course, the heat of fusion for the 100% crystalline polymer is known.

2. Thermogravimetric Analysis (TGA)

TGA makes a continuous weighing of a small sample (approximately 10 mg) in a controlled atmosphere (e.g., air or nitrogen) as the temperature is increased at a programmed linear rate. The thermogram shown in Figure 11 illustrates weight losses due to desorption of gases (e.g., moisture) or decomposition (e.g., HBr loss from halobutyl, CO_2 from calcium carbonate filler). TA is a very simple technique for quantitatively analyzing for filler content of a polymer compound (e.g., carbon black decomposed in air but not nitrogen). While oil

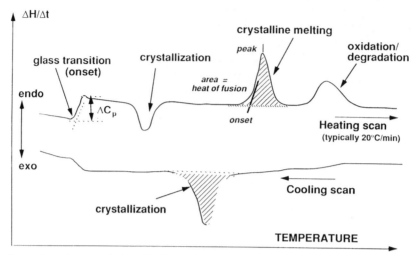

FIGURE 10. Typical polymer DSC thermograms.

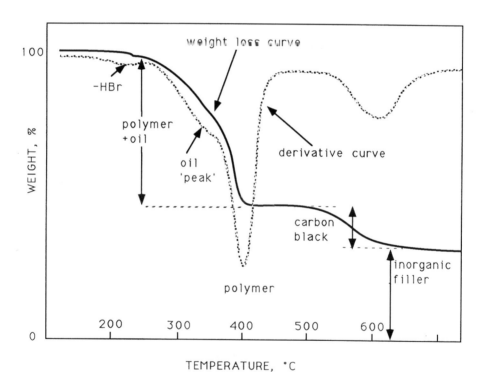

FIGURE 11. TA thermogram of an elastomer compound.

can be readily detected in the thermogram it almost always overlaps with the temperature range of hydrocarbon polymer degradation. The curves cannot be reliably deconvoluted since the actual decomposition range of a polymer in a polymer blend can be affected by the sample morphology.

3. Thermomechanical Analysis (TMA)

TMA consists of a quartz probe which rests on top of a flat sample (a few millimeters square) in a temperature-controlled chamber. When setup in neutral buoyancy (with "flat probe") then as the temperature is increased the probe rises in direct response to the expansion of the sample yielding thermal expansion coefficient vs. temperature scans. Alternatively, with the "penetration probe" under dead loading a thermal-softening profile is obtained (penetration distance vs. temperature). Although this is a simple and versatile experiment it gives only a semiquantitative indication of mechanical modulus vs. temperature. The dynamic mechanical thermal analysis (DMTA) described below gives an absolute modulus measurement.

4. Dynamic Mechanical Thermal Analysis (DMTA)

DMTA is a measurement of the dynamic moduli (in phase and out of phase) in an oscillatory mechanical deformation experiment during a programmed temperature scan at controlled frequency. Thermograms are usually plotted to show elastic modulus, E, and tan δ vs. temperature (Figure 12). The peak of the tan δ is a particularly discriminatory measure of T_g, although this is the center of the relaxation, whereas in the DSC experiment the onset temperature of the T_g relaxation is usually reported. In such a case the DSC T_g will be lower than that for DMTA by an amount that varies with the specific polymer. There is, in addition, a frequency effect which puts the mechanical (c.~1 Hz) T_g about 17°C higher than that for a DSC measurement (c.~0.0001 Hz) for an assumed activation energy of 400 kJ/mol (typical for polymer T_g). The DMTA has a frequency multiplexing capability which can be used for calculating activation energies using time-temperature superposition software.

The temperature range of the DMTA is from −150 to 300°C and frequencies from 0.033 to 90 Hz. The sample size for the usual flexural test mode is $1 \times 10 \times 40$ mm; slightly less sample is required in the parallel plate shear mode.

III. MICROSCOPY FOR POLYMER CHARACTERIZATION

A. GENERAL INFORMATION

This section provides general information on the use of microscopic techniques for polymer characterization. For polymer blends a minimum domain size of 1 μm can be examined in the optical microscope using one or more of the following techniques. A schematic of a typical optical microscope is shown in Figure 13.

1. Phase contrast — thin sections (100 to 200 nm) in thickness (and having refractive indices which differ by approximately 0.005) are supported on glass slides and examined "as is" or with oil to remove microtoming artifacts, e.g., determination of the number of layers in coextruded films, dispersion of fillers, and polymer domain size (Figures 14 and 15).
2. Polarized light — is used if one of the polymer phases is crystalline or for agglomeration of inorganic filters, e.g., nylon/EP blends, fillers such as talc (Figure 16).
3. Incident — is used to examine surfaces of bulk samples, e.g., carbon black dispersion in rubber compounds (Figure 17).
4. Bright field — mainly used to examine thin sections of carbon black-loaded samples, e.g., carbon black dispersion in thin films of rubber compounds.

When the domain size is in the range of <1 μm to 10 nm, scanning electron microscopy (SEM) and/or transmission electron microscopy (TEM) are necessary. A schematic of a scanning electron microscope is shown in Figure 18.

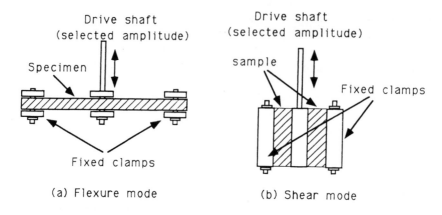

(a) Flexure mode   (b) Shear mode

Dynamic Mechanical Thermal Analyzer

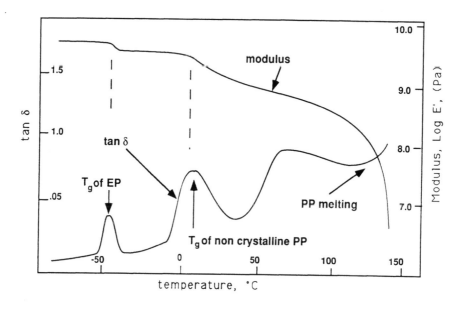

FIGURE 12.   Details of DMTA. The DMTA plot is of an EP/PP blend.

Samples in the SEM can be examined "as is" for general morphology, as freeze-fractured surfaces, or as microtome blocks of solid bulk samples. Contrast is achieved by any one or combination of the following methods:

1.   Solvent etching — when there exists a *large* solubility difference in a particular solvent of the polymers being studied, e.g., PP/EP blends.
2. O_sO_4 staining — there exists at least 5% unsaturation in the polymers being investigated, e.g., NR/EPDM, BIIR/neoprene (Figure 19).
3. RuO_4 staining — when there is no solubility difference or unsaturation this possibility is explored, e.g., knit explored line between two DVAs (dynamic vulcanized alloys) (Figure 20).

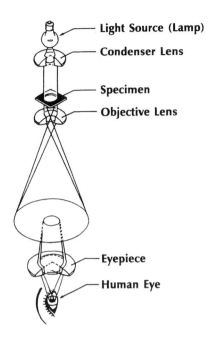

- Light Source (Lamp)
- Condenser Lens
- Specimen
- Objective Lens
- Eyepiece
- Human Eye

Optical Microscope

FIGURE 13.   Schematic of an optical microscope.

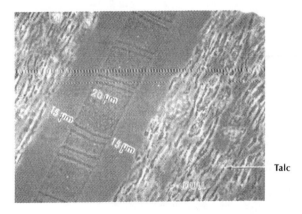

Talc

FIGURE 14.   Light microscopy phase contrast films.

In addition, the SEM can be used to study liquids or temperature-sensitive polymers on a Cryostage.

The SEM is also used in X-ray/elemental analysis. This technique is qualitative. X-ray analysis and mapping of the particular elements present are useful for the identification of inorganic fillers and their dispersion in compounds as well as inorganic impurities in gels or on surfaces and curatives, e.g., aluminum, silicon, or sulfur, in rubber compounds, and Cl and Br in halobutyl blends (Figure 21).

TEM (schematic shown in Figure 22) is used whenever a more in-depth study (when domain sizes are less than 1 μm or so) is required on polymer phase morphologies such as

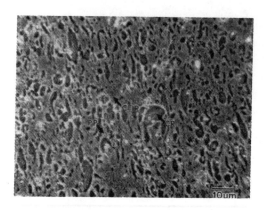

CIIR = Grey Areas

PP = White Areas

Neoprene (CR) = Dark Areas

FIGURE 15. Light microscopy phase contrast polymer domains: chlorobutyl/polypropylene/neoprene blend (CIIR/PP/CR).

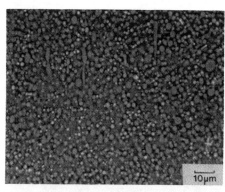

EP is Light,
Dispersed Phase

Nylon is Dark
Matrix

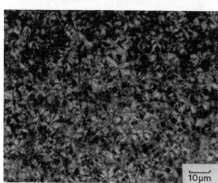

Polarized Light
(Shows Spherultic Structures)

FIGURE 16. Light microscopy phase contrast: nylon/EP blends.

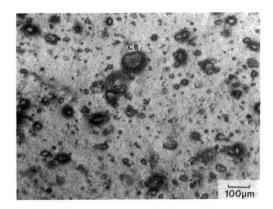

FIGURE 17. Light microscopy incident: light carbon black dispersion in rubber.

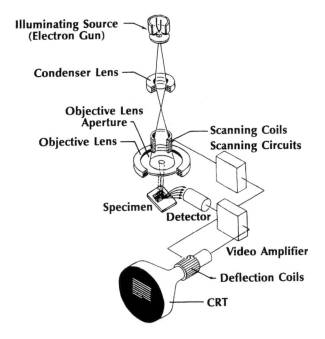

FIGURE 18. Schematic of a scanning electron microscope (SEM).

dynamically vulcanized alloys (Figure 23) and nylon/EP (Figure 24) filler location as in carbon black in rubber compounds (Figure 25) and also in the morphology of block copolymers (Figure 26). Thin sections are required and take anywhere from 1 h to 1 d per sample depending on the nature of the sample. They must be ~100 nm in thickness and are prepared usually by microtoming with a diamond knife at near liquid nitrogen temperatures ($-150°C$). The same contrasting media for SEM apply to TEM. In addition, PIB backbone polymers scission and evaporate in the TEM which helps locate these polymers domains in blends.

B. NON-ROUTINE TECHNIQUES

- Solvent casting when microtoming is not desirable as a method of sample preparation.
- STEM — used for elemental composition study in thin films when resolution better than X-ray analysis is required in the SEM on bulk samples.

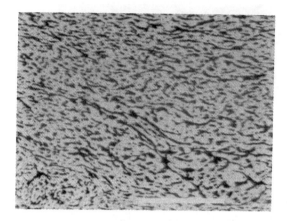

Neoprene is the Light Phase

FIGURE 19.   Enhancing SEM contrast in blends by osmium tetroxide staining: bromobutyl/neoprene blend.

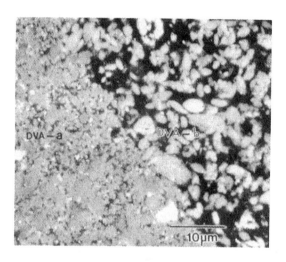

FIGURE 20.   SEM-ruthenium tetroxide stained knit line between two DVAs.

- Cryostage-SEM — to study liquid samples at low temperatures, e.g., butyl slurry.
- Fluorescence microscope — useful in examining polymer/asphalt blends or any sample which is fluorescent.
- OM/hot stage — to observe melting point of either an impurity or other moiety in a compound.

IV. ELEMENTAL AND STRUCTURAL CHARACTERIZATION TESTS

A. ATOMIC ABSORPTION SPECTROSCOPY

In atomic absorption (AA) spectrometry the sample is vaporized and the element of interest atomized at high temperatures. The element concentration is determined based on the attenuation or absorption by the analyte atoms of a characteristic wavelength emitted

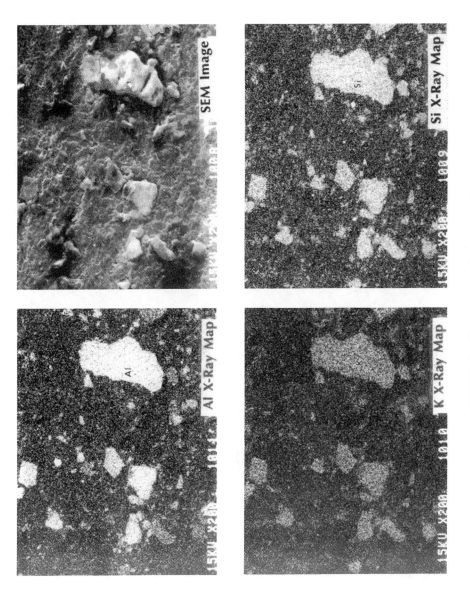

FIGURE 21. X-ray mapping of surface impurity.

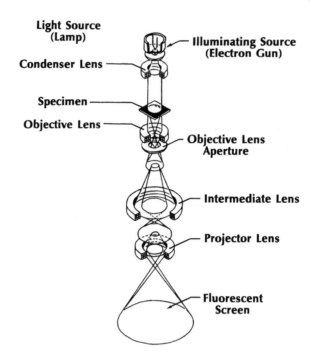

FIGURE 22.   Schematic of a transmission electron microscope (TEM).

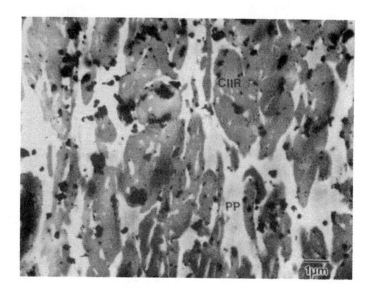

FIGURE 23.   TEM-ruthenium tetroxide-stained polymer domains in a DVA compound.

from a light source. The light source is typically a hollow cathode lamp containing the element to be measured. Separate lamps are needed for each element. The detector is usually a photomultiplier tube. A monochromator is used to separate the element line and the light source is modulated to reduce the amount of unwanted radiation reaching the detector.

Conventional AA instruments (Figure 27) use a flame atomization system for liquid sample vaporization. An air-acetylene flame (2300°C) is used for most elements. A higher

FIGURE 24.   Phase morphology in a nylon/EP-MA blend by TEM.

temperature nitrous oxide-acetylene flame (2900°C) is used for more refractory oxide-forming elements. Electrothermal atomization techniques such as a graphite furnace can be used for the direct analysis of solid samples.

AA is used for the determination of parts per million levels of metals. It is not normally used for the analysis of the light elements such as H, C, N, O, P, S, halogens, and noble gases. Higher concentrations can be determined by prior dilution of the sample. AA is not recommended if a large number of elements are to be measured in a single sample.

Although AA is a very capable technique and is used worldwide, its use in recent years has declined in favor of ICP and XRF methods of analysis. The most common application of AA is for the determination of boron and magnesium in oils.

Conventional AA instruments will analyze liquid samples only. Dilute acid and xylene

FIGURE 25. Location of carbon black in a blend of chlorbutyl and natural rubber and EPDM.

solutions are common. The volume of solution needed is dependent on the number of elements to be determined.

AA offers excellent sensitivity for most elements with limited interferences. For some elements sensitivity can be extended into the subparts per billion range using flameless methods. The AA instruments are easy to operate with "cookbook methods" available for most elements.

Conventional AA uses a liquid sample. The determination of several elements per sample is slow and requires larger volumes of solution due to the sequential nature of the method. Chemical and ionization interferences must be corrected by modification of the sample solution.

Chemical interferences arise from the formation of thermally stable compounds such as oxides in the flame. The use of electrothermal atomization, a hotter nitrous oxide-acetylene flame, or the addition of a releasing agent such as lanthanum can help reduce the interference.

Flame atomization produces ions as well as atoms. Since only atoms are detected, it is important that the ratio of atoms to ions remains constant for the element being analyzed. This ratio is affected by the presence of other elements in the sample matrix. The addition

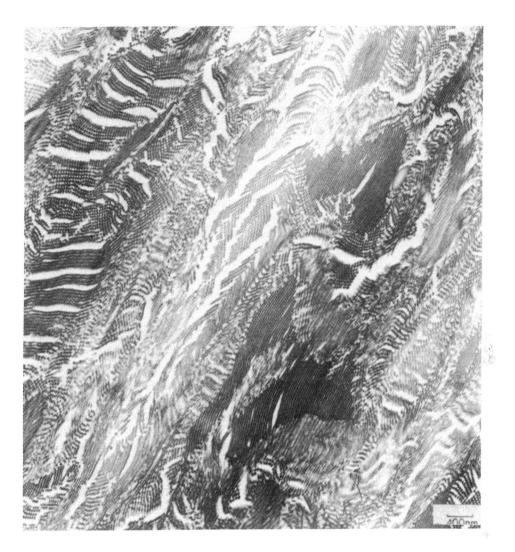

FIGURE 26. REM-ruthenium tetroxide stained graft copolymer.

of large amounts of an easily ionized element such as potassium to both the sample and standards helps mask the ionization interference.

The capabilities of flame AA can be extended by employing the following modifications:

1. A cold quartz tube for containing mercury vapor (for mercury determination).
2. A heated quartz tube for decomposing metallic hydride vapors for As, Se, Sb, Pb, Te, Sn, and Bi determination.
3. A graphite furnace for decomposing involatile compounds of metals, with extremely high sensitivity.

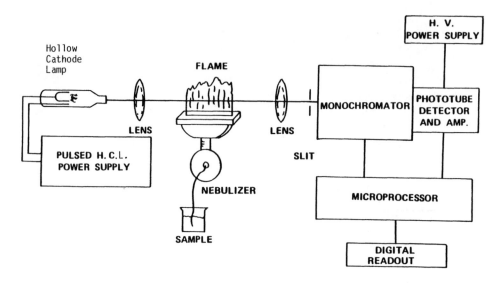

FIGURE 27.   Major components of atomic absorption spectrometer.

B. INDUCTIVELY COUPLED PLASMA ATOMIC EMISSION SPECTROSCOPY

In inductively coupled plasma atomic emission spectroscopy (ICP) the sample is vaporized and the element of interest atomized in an extremely high temperature (~7000°C) argon plasma, generated, and maintained by radiofrequency coupling. The atoms collide with energetically excited argon species and emit characteristic atomic and ionic spectra that are detected with a photomultiplier tube. Separation of spectral lines can be accomplished in two ways. In a sequential or scanning ICP (Figure 28), a scanning monochromator with a movable grating is used to bring the light from the wavelength of interest to a single detector. In a simultaneous or direct reader ICP (Figure 29), a polychromator with a diffraction grating is used to disperse the light into its components wavelength. Detectors for the elements of interest are set by the vendor during manufacture. Occasionally a scanning channel is added to a direct reader to allow measurement of an element not included in the main polychromator.

ICP is used for the determination of parts per million levels of metals in liquid samples. It is not suitable for the noble gases, halogens, or light elements such as H, C, N, and O. Sulfur requires a vacuum monochromator. A direct reader ICP excels at the rapid analysis of multielement samples.

Common sample types analyzed by ICP include trace elements in polymers, wear metals in oils, and numerous one-of-a-kind catalysts.

ICP instruments are limited to the analysis of liquids only. Solid samples require some sort of dissolution procedure prior to analysis. The final volume of solution should be at least 25 ml. The solvent can be either water, usually containing 10% acid, or a suitable organic solvent such as xylene.

ICP offers good detection limits and a wide linear range for most elements. With a direct reading instrument multielement analysis is extremely fast. Chemical and ionization interferences frequently found in AA spectroscopy are suppressed in ICP analysis. Since all samples are converted to simple aqueous or organic matrices prior to analysis, the need for standards matched to the matrix of the original sample is eliminated.

The requirement that the sample presented to the instrument must be a solution necessitates extensive sample preparation facilities and methods. More than one sample preparation method may be necessary per sample depending on the range of elements requested. Spectra

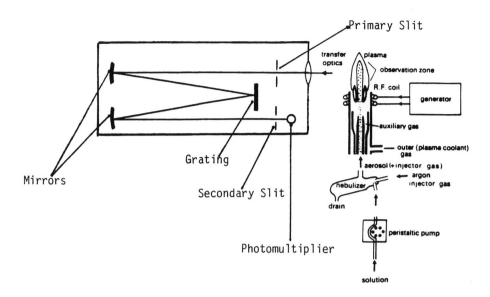

FIGURE 28. Scanning ICP.

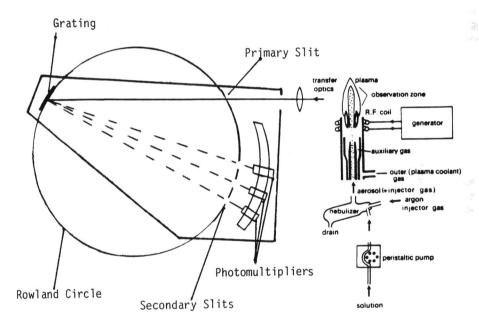

FIGURE 29. Direct reader ICP.

interferences can complicate the determination of trace elements in the presence of other major metals. ICP instruments are not rugged. Constant attention by a trained operator, especially to the sample introduction and torch systems, is essential.

Spectral interferences, such as line overlaps, are prevalent and must be corrected for accurate quantitative analysis. With a scanning instrument it may be possible to move to an interference-free line. With a direct reader, sophisticated computer programs apply mathematical corrections based on factors previously determined on multielement standards.

C. ION CHROMATOGRAPHY (IC)

Commercial ion chromatograph instruments have become available since early 1976. Ion chromatography (IC) is a combination of ion-exchange chromatography, eluent suppression, and conductimetric detection. For anion analysis, a low-capacity anion exchange resin is used in the separator column and a strong cation exchange resin in the H^+ form is used in the suppressor column. A dilute mixture of Na_2CO_3 and $NaHCO_3$ is used as the eluent, because carbonate and bicarbonate are conveniently neutralized to low conductivity species and the different combinations of carbonate-bicarbonate give variable buffered pH values. This allows the ions of interest in a large range of affinity to be separated. The anions are eluted through the separating column in the background of carbonate-bicarbonate and conveniently detected based on electrical conductivity. The reactions taking place on these two columns are, for an anion X:

1. Separator column

$$Resin - N^+HCO_3^- + NaX^+ \rightleftarrows Resin - N^+X^- + Na^+HCO_3^-$$

2. Suppressor

$$Resin - SO_3^- + Na^+HCO_3^- \rightleftarrows Resin - SO_3^-Na^+ + H_2CO_3$$

$$Resin - SO_3^-H^+ + Na^+X^- \rightleftarrows Resin - SO_3^-Na^+ + H^+X^-$$

As a result of these reactions in the suppressor column, the sample ions are presented to the conductivity detector as H^+X^-, not in the highly conducting background of carbonate-bicarbonate, but in the low conducting background of H_2CO_3.

Figure 30 shows a schematic representation of the ion chromatography system. Dilute aqueous sample is injected at the head of the separator column. The anion exchange resin selectively causes the various sample anions of different types to migrate through the bed at different respective rates, thus effecting the separation. The effluent from the separator column then passes to the suppressor column where the H^+ form cation exchange resin absorbs the cations in the eluent stream. Finally, the suppressor column effluent passes through a conductivity cell. The highly conductive anions in a low background conductance of H_2CO_3 are detected at high sensitivity by the conductivity detector. The nonspecific nature of the conductimetric detection allows several ions to be sequentially determined in the same sample. The conductimetric detection is highly specific and relatively free from interferences. Different stable valance states of the same element can be determined.

On the other hand, because of the nonspecific nature of the conductivity detector, the chromatograph peaks are identified only by their retention times. Thus, the two ions having the same or close retention times will be detected as one broad peak giving erroneous results.

Figure 31 shows a typical chromatogram for the standard common anions F^-, Cl^-, NO_2^-, PO_4^{-3}, Br^-, NO_3, and SO_4^{-2}. Numerous applications of ion chromatography have been illustrated in the literature for a variety of complex matrices.

The advantages of ion chromatography are

1. Sequential multi-anion capability; eliminates individual determinations of anion by diverse technique.
2. Small sample size (<1 ml).
3. Rapid analysis (~10 min for approximately seven anions).

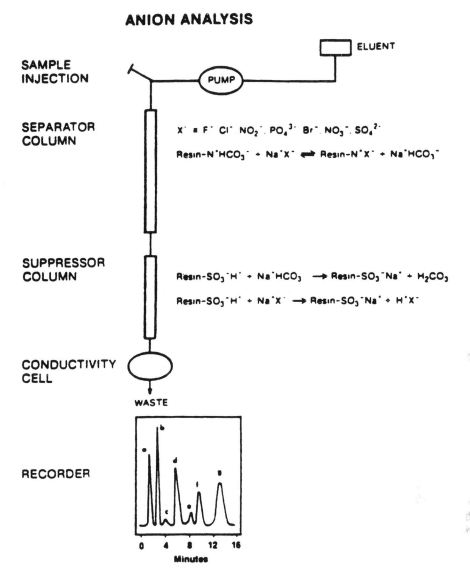

FIGURE 30. Ion chromatography (IC) flow scheme.

4. Large dynamic range over 4 decades of concentration.
5. Speciation can be determined.

The principle disadvantages of IC are

1. Interferences possible if two anions have similar retention times.
2. Determination difficult in the presence of an ion present in very large excess over others.
3. Sample has to be in aqueous solution.
4. Method not suitable for anions with PKa of <7.

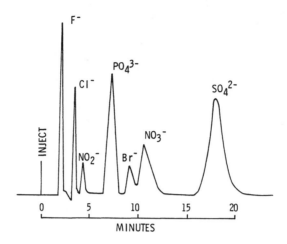

FIGURE 31. Analysis of standard inorganic anions by ion chromatography.

In addition to the common inorganic anions analyzed by IC, a number of other species can also be determined by using appropriate accessories. Some of these applications include

Technique	Species
Ion chromatography	Carboxylic acids
Chemistry — IC	Formaldehyde; borate
Mobile phase IC	Ammonia; fatty acids; ethanol amines
Electrochemical detection	Phenols, CN^-, Br^-, I^-, S^{-2}, etc.

D. ION SELECTIVE ELECTRODES (ISE)

ISE measures the ion activities or the thermodynamically effective free ion concentrations. ISE has a membrane construction that serves to block the interfering ions and only permit the passage of ions for which it was designed. However, this rejection is not perfect, and hence some interferences from other ions occur. The electrode calibration curves are good over 4 to 6 decades of concentration. The typical time per analysis is about a minute, though some electrodes need 15 min for adequate response. The response time is faster as more concentrated solutions are analyzed. Although a single element technique, many elements can be determined sequentially by changing electrodes, provided calibration curves are prepared for all ions. Also, the instrument is portable and is thus useful for field studies. Samples volumes needed are typically about 5 ml, although 300 μl or less can be measured with special modifications. An accuracy of 2 to 5% is achieved. The ISE measures the activity of the ions in solution. This activity is related to concentration and thus, in effect, measures the concentration. However, if an ion such as fluoride which complexes with some metals — Fe or Al — is to be measured, it must be decomplexed from these cations by the addition of a reagent such as citric acid or EDTA. ISEs for at least 22 ionic species are commercially available.

An example is described here for the measurement of fluoride ions in solution. The fluoride electrode uses a LaF_3 single crystal membrane and an internal reference bonded into an epoxy body. The crystal is an ionic conductor in which only fluoride ions are mobile. When the membrane is in contact with a fluoride solution, an electrode potential develops across the membrane. This potential, which depends on the level of free fluoride ions in solution, is measured against an external constant reference potential with a digital pH/mv meter or specific ion meter. The measured potential corresponding to the level of fluoride ions in solution is described by the Nernst equation:

$$E = E_0 - S \log A$$

where E = measured electrode potential, E_0 = reference potential (a constant), A = fluoride level in solution, and S = electrode slope.

The level of fluoride, A, is the activity or "effective concentration" of free fluoride ions in solution. The total fluoride concentration, C, may include some bound or complexed ions as well as free ions. The electrode responds only to the free ions, whose concentration is

$$C_f = C_t - C_b$$

where C_b is the concentration of fluoride ions in all bound or complexed forms.

The fluoride activity is related to free fluoride concentration by the activity coefficient r:

$$A = rC_f$$

Ionic activity coefficients are variable and largely depend on total ionic strength. Ionic strength is defined as

$$\text{Ionic Strength} = \frac{1}{2} \sum C_i Z_i^2$$

where C_i = concentration of ion i and Z_i = charge of ion i.

If the background ionic strength is high and constant relative to the sensed ion concentration, the activity coefficient is constant and activity is directly proportional to concentration. Since the electrode potentials are affected by temperature changes, the sampl6 and standard solutions should be close to the same temperature. At the 20-ppm level a 1°C change in temperature gives a 2% error. The slope of the fluoride electrode response also changes with temperature. The electrode can be used at temperatures from 0 to 100°C, provided that the temperature has equilibrated, which may take as long as an hour. In general, it is best to operate near room temperature.

ISEs are subject to two types of interferences: method and electrode. In the first type, some property of the sample prevents the electrode from sensing the ion of interest; for example in acid solution fluoride forms complexes with H^+ and the fluoride. ISE cannot detect the masked fluoride ions. In the electrode interference, the electrode responds to ions in solution other than the one being measured; for example, bromide ion poses severe interference in using chloride ISE. The extent of interference depends on the relative concentration of analyze to interfering ions. The interfering ions can be complexed by changing pH or adding a reagent to precipitate them. However, finding the right chemistry is not always easy.

Going back to the example of fluoride determination, the fluoride forms complexes with aluminum, silicon, iron, and other polyvalent cations as well as hydrogen. These complexes must be destroyed in order to measure total fluoride, since the electrode will not detect complexed fluoride. This is achieved by adding a total ionic strength adjustment buffer which contains the reagent CDTA (cyclohexylene dinitrilo tetraacetic acid) which preferentially complexes the cations and releases the fluoride ions. The carbonate and bicarbonate anions interfere by making the electrode response slow; hence, these ions are eliminated by heating the solution with acid until all CO_2 is removed. At a pH above 7, hydroxyl ions interfere, while at a pH less than 5 the H^+ ions form complexes such as HF_2 thus producing low fluoride results. Addition of TISAB to both samples and standards and further adjustment of pH to between 5.0 and 5.5 is necessary to eliminate the hydroxide interference and the formation of hydrogen-fluoride complexes. Other common anions such as other halides, sulfate, nitrate, phosphate, or acetate do not interfere in the fluoride measurement.

The advantages of this technique are

1.   Inexpensive and simple to use instrument.
2.   Rapid analysis: about a minute per sample.
3.   Portable instrument; can be used in the field.
4.   Large dynamic range over 4 to 6 decades of concentration.

E. MASS SPECTROMETRY (MS AND GC/MS)

Most of the spectroscopic and physical methods employed by the chemist in structure determination are concerned only with the physics of molecules; mass spectroscopy deals with both the chemistry and the physics of molecules, particularly with gaseous ions. In conventional mass spectrometry, the ions of interest are positively charged ions. The mass spectrometer has three functions:

1.   To produce ions from the molecules under investigation.
2.   To separate these ions according to their mass-to-charge ratio.
3.   To measure the relative abundances of each ion.

In the 1950s, Benyon, Biemann, and McLafferty clearly demonstrated the chemistry of functional groups in directing fragmentation, and the power of mass spectrometry for organic structure determination began to develop.

Today, mass spectrometry has achieved status as one of the primary spectroscopic methods to which a chemist faced with a structural problem turns. The great advantage of the method is found in the extensive structural information which can be obtained from sub-microgram quantities of material.

The methodology of mass separation is governed by both the kinetic energy of the ion and the ion's trajectory in an electromagnetic field. There exists a balance between the centripetal and centrifugal forces which the ion experiences. Centripetal forces are caused by the kinetic energy, and centrifugal forces by the electromagnetic field. We may express this force balance as follows, Figure 32 or Figure 33 with GC:

$$\frac{mU^2}{r} = qUB$$

where: m = ion's mass, U = ion's velocity, r = radius of ion trajectory in the magnetic field, q = ion's charge, and B = magnetic field strength. The right-hand side is the centripetal force and the left-hand side is the centrifugal force.

Solving for mass-to-charge ratio yields

$$\frac{m}{q} = \frac{Br}{U}$$

The kinetic energy of the ion is given by

$$aV = \tfrac{1}{2}\, mU^2$$

where q = charge of the ion, V = accelerating potential, m = ion's mass, and U = ion's velocity.

Solving for U yields

$$U = \frac{2qV^{1/2}}{m}$$

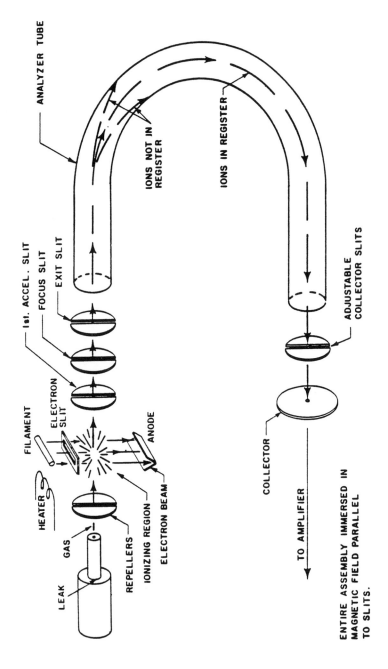

FIGURE 32. 21-104 mass spectrometer analyzer.

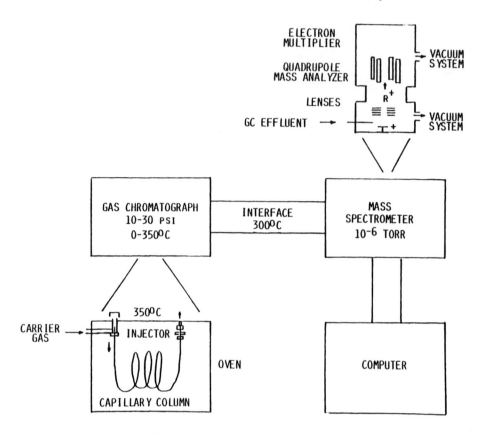

FIGURE 33.   Schematic of a GC/MS instrument.

By substituting for U in the second expression, we obtain

$$m/q = Br/(2qV^{1/2}/m)$$

Squaring each side of the equation yields

$$\frac{m}{q} = \frac{B^2e^2}{2V}$$

Thus, the mass-to-charge ratio can be determined if one knows B, e, and V. Since e is constant for a mass passing through the two slivers, scanning a spectrum is achieved by varying either B or V, keeping the other constant.

Mass spectrometers provide a wealth of information concerning the structure of organic compounds, their elemental composition, and compound types in complex mixtures. A detailed interpretation of the mass spectrum frequently allows the positions of the functional groups to be determined. Moreover, mass spectrometry is used to investigate reaction mechanisms, kinetics, and is also used in tracer work.

The mass spectrum may be either in analog form (chart paper) or digital form (printed paper). Analyses are calculated to give mole percent, weight percent, or volume percent. Either individual components, compound types by carbon number, or total compound type are reported. This is determined by the nature of the sample and the requirements of the submitter.

The characteristics of the sample submitted for an MS test are

1. Size: 1 to 1000 mg.
2. State: gas, liquid, solid, but only the portion vaporizable at about 300°C is analyzed.
3. Phases: if sample has more than one phase, each phase can generally be analyzed separately.
4. Composition limitations: essentially no limits to composition, simple mixtures, and complex mixtures can be handled.
5. Temperature range: samples should be at room temperature and should be thermally stable up to 300°C for bath introduction and may be involatile for field desorption work.

A wide variety of materials from gases to solids and from simple to complex mixtures can be analyzed. The molecular weight and atomic composition are generally determined. Only a very small amount of sample is required. Most calibration coefficients can be used for long periods of time.

Some compounds such as long-chain esters and polyethers decompose in the inlet system and the spectrum obtained is not that of the initial substance. Calibration coefficients are required for quantitative analyses. The sample introduced to the instrument cannot usually be recovered.

Some classes of compounds, such as olefins and naphthenes, give very similar spectra and cannot be distinguished except by analysis before and after hydrogenation or dehydrogenation.

F. NUCLEAR MAGNETIC RESONANCE SPECTROMETER

Nuclear magnetic resonance (NMR) is a spectrometric technique for determining chemical structures. When an atomic nucleus with a magnetic moment is placed in a magnetic field, it tends to align with the applied field. The energy required to reverse this alignment depends on the strength of the magnetic field and to a minor extent on the environment of the nucleus, i.e., the nature of the chemical bonds between the atom of interest and its immediate vicinity in the molecule. This reversal is a resonant process and occurs only under select conditions. By determining the energy levels of transition for all of the atoms in a molecule, it is possible to determine many important features of its structure. The energy levels can be expressed in terms of frequency of electromagnetic radiation and typically fall in the range of 5 to 600 MHz for high magnetic fields. The minor spectral shifts due to chemical environment are the essential features for interpreting structure and are normally expressed in terms of parts per million shifts from the reference frequency of a standard such as tetramethyl silane.

The most common nuclei examined by NMR are ^1H and ^{13}C, as these are the NMR-sensitive nuclei of the most abundant elements in organic materials. ^1H represents over 99% of all hydrogen atoms, while ^{13}C is only just over 1% of all carbon atoms; further, ^1H is much more sensitive than ^{13}C on an equal nuclei basis. Until fairly recently, instruments did not have sufficient sensitivity for routine ^{13}C NMR, and ^1H was the only practical technique. Most of the time it is solutions that are characterized by NMR, although ^{13}C NMR is possible for some solids, but at substantially lower resolution than for solutions.

In general, the resonant frequencies can be used to determine molecular structures. ^1H resonances are fairly specific for the types of carbon they are attached to and to a lesser extent to the adjacent carbons. These resonances may be split into multiples, as hydrogen nuclei can couple to other nearby hydrogen nuclei. The magnitude of the splittings, and the multiplicity, can be used to better determine the chemical structure in the vicinity of a given hydrogen. When all of the resonances observed are similarly analyzed, it is possible to determine the structure of the molecule. However, as only hydrogen is observed, any skeletal feature without an attached hydrogen can only be inferred. Complications can arise if the molecule is very complex, because then the resonances can overlap severely and become difficult or impossible to resolve.

^{13}C resonances can be used to directly determine the skeleton of an organic molecule. The resonance lines are narrow and the chemical shift range (in parts per million) is much larger than for ^1H resonances. Furthermore, the shift is dependent on the structure of the molecule for up to three bonds in all directions from the site of interest. Therefore, each shift becomes quite specific, and the structure can be easily assigned, frequently without any ambiguity, even for complex molecules.

Very commonly, however, the sample of interest is not a pure compound, but is a complex mixture such as a coal liquid. As a result, a specific structure determination for each molecular type is not practical, although it is possible to determine an average chemical structure. Features which may be determined include the hydrogen distribution between saturate, benzylic, olefinic, and aromatic sites. The carbon distribution is usually split into saturate, heterosubstituted saturate, aromatic + olefinic, carboxyl, and carbonyl types. More details are possible, but depend greatly on the nature of the sample, and what information is desired.

Any gas, liquid or solid sample that can be dissolved in solvents, such as CCl_4, CH $\sim$ Cl, acetone, or DMSO to the 1% level or greater can be analyzed by this technique. Samples of ~ 0.1 g or larger of pure material are sufficient. Solids can also be analyzed as solid. However, special arrangements need to be made. In either case, the analysis is nondestructive so that samples can be recovered for further analysis if necessary.

The NMR experiment can be conducted in a temperature range from liquid nitrogen ($-209°C$) to $+150°C$. This gives the experimenter the ability to slow down rapid molecular motions to observable rates or to speed up very slow or viscous motions to measurable rates.

NMR is a very powerful tool. It often provides the best characterization of compound structure, and may provide absolute identification of specific isomers in simple mixtures. It may also provide a general characterization by functional groups which cannot be obtained by any other technique. As is typical with many spectroscopic methods, adding data from other techniques (such as mass or infrared spectrometry) can often provide greatly improved characterizations.

The following are general notes and comments concerning the use of NMR specifically for common rubber characterization problems. A schematic of a Fourier transform NMR spectrometer is given in Figure 34.

Sample preparation — Samples are analyzed by proton (^1H) and/or carbon ^{13}C NMR. Sample requirements are $\sim^1/_2$ g for ^1H and ~ 1 g for ^{13}C.

For butyl-based polymers dissolving is performed in dewatered chloroform at ambient temperature. EP (ethylene-propylene) can be dissolved in deuterated *O*-dichlorobenzene at 140°C.

Wet samples cannot be accurately analyzed. Opaqueness in a rubber sample is generally an indication of moisture in the sample.

Established Methods by ^1H NMR —

Sample Type	Analyzed for
EPDM	ENB, hexdiene, ethylene, DCDP (wt %)
SBB	Isoprene unsaturated (mol %); KR01 (wt %)

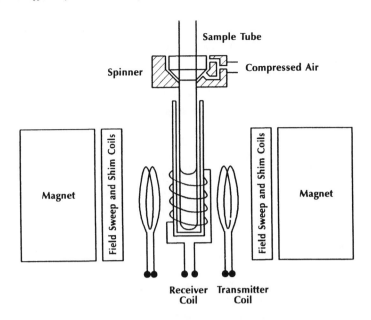

Block diagram of a high-resolution NMR spectrometer and the arrangement of the sample in the probe (cross coil configuration).

FIGURE 34. Schematic of a Fourier transform NMR spectrometer.

Sample Type	**Analyzed for**
Butyl	Isoprene unsaturated (mol %)
Bromobutyl	
Chlorobutyl	Type I, II III (%)

I

$$CH_3$$
$$|$$
$$-CH_2-C=CH-CH_2-$$

II

$$CH_2$$
$$\|$$
$$-CH_2-C-CH-CH_2-$$
$$|$$
$$X$$

III

$$CH_2X$$
$$|$$
$$-CH_2-C=CH-CH_2-$$

Butyl Isoprene unsaturated (mol %)

$$CH_3$$
$$|$$
$$-CH_2-C=CH-CH_2-$$ 1,4 Isoprene

$$CH_3$$
$$|$$
$$-CH_2-C-$$ 1,2 Isoprene
$$|$$
$$CH$$
$$\|$$
$$CH_2$$

where X = Cl or Br.

NMR results are quantitative. Analysis of a ^{13}C or ^1H spectrum would reveal the different types of functionalities, as well as their contents in the sample. For example, Figure 35 shows the ^1H NMR spectrum of the diene (ENB) in an EPDM polymer (ethylene-propylene diene monomer).

Figure 36 shows a ^{13}C NMR spectrum of EP used to determine the sequence distribution. The detectability for ^1H NMR is typically 0.01 mol % and for ^{13}C NMR is 0.1 mol %.

G. FOURIER TRANSFORM INFRARED (FTIR)

Generally, this technique is used to analyze samples that are available either in small quantity or a small entity. Gels within a rubber sample have to be microtomed (i.e., cut into very thin slices) and mounted in KBr plates in a Microscopy Lab. Samples contaminated with inorganic components are usually analyzed by both X-ray and FTIR microscope. Sample size ~20 μm can be analyzed by the FTIR microscope.

FTIR (refer to Figure 37) is commonly used for qualitative identification of various functionalities. For quantitative analysis, FTIR requires the use of well-characterized standards. NMR spectroscopy is typically used to characterize a set of samples which are then used as standards for the FTIR calibration.

Figure 38 shows the FTIR spectrum of a butyl sample. This sample contains BHT, ESBO, stearic acid, and calcium stearate. The contents of all these components can be determined from this single spectrum. In some cases, the assigned peak absorbance is relatively small; therefore, a thick film, ~0.6 mm, is used.

H. ULTRAVIOLET, VISIBLE, AND INFRARED SPECTROMETRY (UV, Vis, IR)

When electromagnetic radiation passes through a sample, some wavelengths are absorbed by the molecules of the sample. Energy is transferred from the radiation to the sample, and the molecules of the sample are said to be elevated to an excited energy state. The total energy state of the ensemble of molecules may be regarded as the sum of the four kinds of energy: electronic, vibrational, rotational, and transnational. Transnational energy is associated with an elevation of the temperature of the sample. Rotational energy comes about by the absorption of very high wavelengths of infrared radiation (25 to 500 μm) and is manifested by an increase in the rotational energy of the sample molecules. Vibrational energy arises when radiation in the mid-infrared region is absorbed (2 to 25 μm) and is manifested by an increase in the vibrational energies of functional groups within the sample molecule. Electronic energy is gained by an ensemble of molecules when an electron is promoted to a higher molecular orbital by absorption in the ultraviolet and visible regions of the spectrum (0.2 to 0.8 μm).

Pure transitions between rotational states represent very small energy changes (high wavelength). Absorption spectra observed in the far infrared are generally "pure" in the sense that the energy absorbed by the molecule is entirely converted into pure rotational motion. This is not the case in the other regions of the spectrum. Thus, when higher amounts of energy are absorbed by molecules, the vibrational motions generated are not restricted to those for which the rotational properties of the molecule remain constant. The absorption band, therefore, will represent a composite of vibrational motions, each occurring in molecules of different rotational levels. The same is true for electronic absorptions, where both rotational and vibrational properties of the molecules are impressed on the electronic transitions.

Another complication arises in the interpretation of absorption spectra. If a molecule vibrates with pure harmonic motion and the dipole moment is a linear function of the

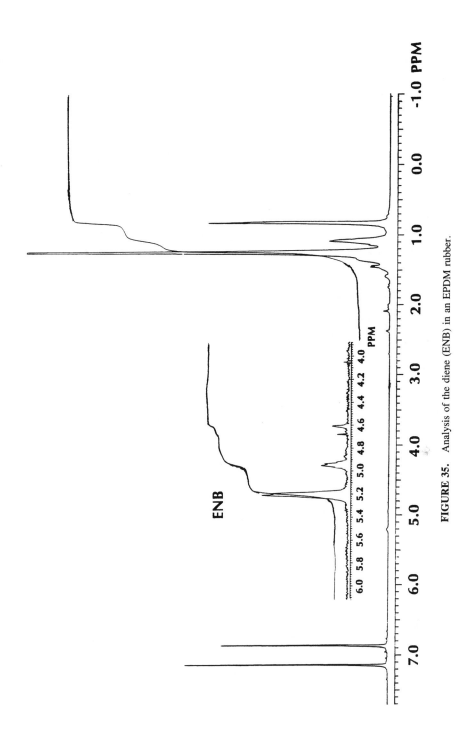

FIGURE 35. Analysis of the diene (ENB) in an EPDM rubber.

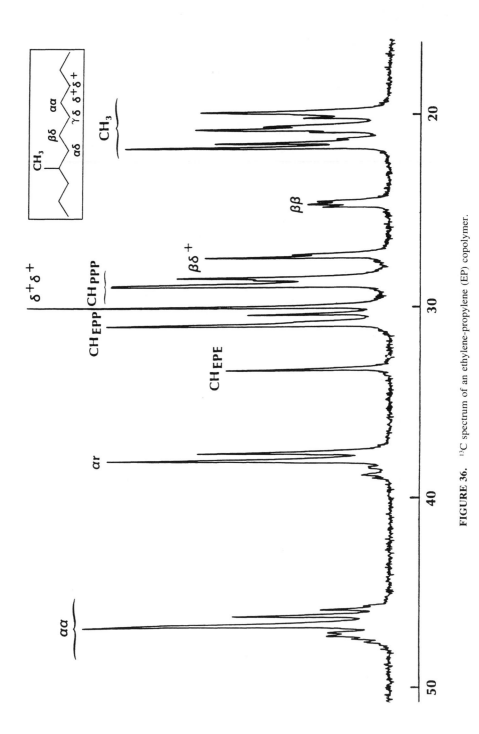

FIGURE 36. ¹³C spectrum of an ethylene-propylene (EP) copolymer.

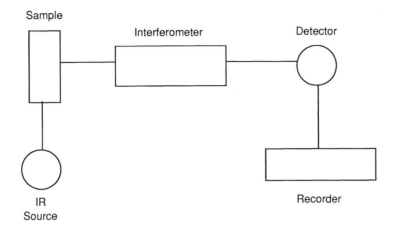

FIGURE 37. Basic components of an FTIR.

displacement, the absorption spectrum will consist of fundamental transitions only. If either of these conditions is not met, as is usually the case, the spectrum will contain overtones (multiples of the fundamental) and combination bands (sums and differences). Most of these overtones and combination bands occur in the near infrared (0.8 to 2.0 μm).

Not all vibrations and rotations are infrared active. If there is no change in dipole moment, there is no oscillating electric field in the motion and there is no mechanism by which absorption of electromagnetic radiation can take place. An oscillation, or vibration, about a center of symmetry, therefore, will not be observed in the infrared spectrum (absorption) but can be observed in the Raman spectrum (scattering).

In summary, therefore, there are five regions of the electromagnetic spectrum of interest:

m	
0.2–0.4	Ultraviolet (electronic)
0.4–0.8	Visible (electronic)
0.8–2.0	Near IR (overtones)
2.0–25.0	Mid IR (vibrational)
25.0–500.0	Far IR (rotational)

Electronic transitions (UV, visible spectra) generally give information about unsaturated groups in the sample molecules. Olefins absorb near 0.22 μm, aromatics near 0.26 to 0.28 μm, carbonyls near 0.20 to 0.27 μm, polynuclear aromatics near 0.26 to 0.50 μm, and conjugated C = S groups near 0.62 μm. Any material which is colored will generally show absorption in the visible region. The intensity of the absorption is proportional to the number of chromophores giving rise to the absorption band.

Overtones (near IR) are useful for studying the presence of groups containing hydrogen. Fundamentals involving hydrogen vibration tend to congregate near the same frequencies in the mid IR, but are easier to distinguish and study in the overtone region.

Vibrational transitions (mid IR) are the most useful of all to study. These give information about the presence or absence of specific functional groups in a sample. Practically all functional groups (that have an infrared-active fundamental) display that fundamental over a very narrow range of wavelength in the mid-IR region. Moreover, the whole spectrum, containing fundamentals, overtones, and combination bands, constitutes a fingerprint of the sample. This means that, although we might not know what a sample is, we will always know it later if it occurs again. Finally, the absorption intensity of any band, whether

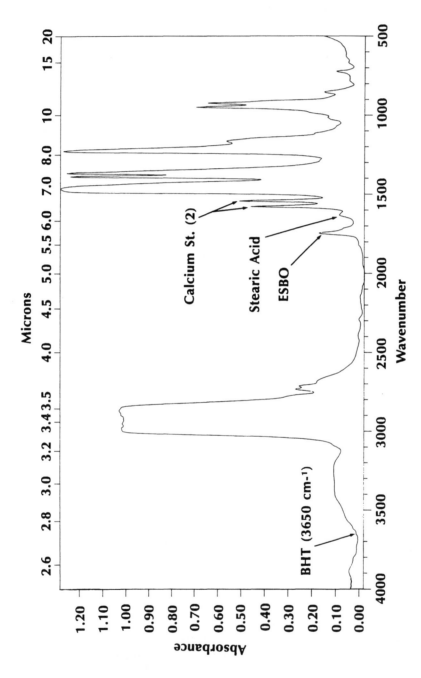

FIGURE 38. Example of an FTIR spectrum for butyl rubber.

fundamental or overtone, is proportional to the number of functional groups giving rise to the signal.

The characteristics of the sample used are

1.   For gases, we generally need about 250 cm^3 at 1 atm to obtain a spectrum.
2.   For liquids, we generally need about 0.25 cm^3 to obtain a spectrum.
3.   For solids, we generally need about 1 mg to obtain a spectrum.
4.   Trace analyses within samples will, of course, increase the sample requirements proportionally.

The advantages of this technique are

1.   Faster and cheaper than most other techniques.
2.   Very specific for certain functional groups.
3.   Very sensitive for certain functional groups.
4.   Fingerprint capability.

The disadvantages of this technique are

1.   Requires special cells, NaCl, KBr, quartz, etc.
2.   Usually requires solubility of sample.
3.   Very difficult to get good quantitation in solids.
4.   Must calibrate all signals.
5.   Water interferes.

Measurement interferences can occur from

1.   Water interferes with practically all IR work.
2.   Solvents generally interfere and must be selected carefully.
3.   Multicomponent samples generally have mutually interfering species. Separations are often required. Sometimes changing the spectral region helps.
4.   Optical components interfere to different extents in different regions. Thus, quartz is good for UV/Vis/Near-IR, but bad for mid-IR/far-IR. KBr is good for mid-IR, bad for far-IR.

FTIR Spectrometry is a special technique. In dispersive spectrometry, the wavelength components of light are physically separated in space (dispersed) by a prism of grating (Figure 39). Modern dispersive spectrometers divide the incident beam into two beams: one beam goes through the sample and the other goes through a suitable reference material. The intensity of both beams are monitored by a suitable detector and final data output can be displayed in either transmittance or absorbance:

$$\text{transmittance} = I_S/I_R$$

$$\text{absorbance} = -\log I_S/I_R$$

where I_S, I_R refer to the intensities in the sample beam and reference beam, respectively. This rationing occurs at each wavelength element, and the final plot is a graphical display with transmittance or absorbance on the Y axis and wavelength or frequency on the X axis.

In Fourier transform spectrometry, the wavelength components of light are not physically separated. Instead, the light is analyzed in the time frame of reference (the time domain)

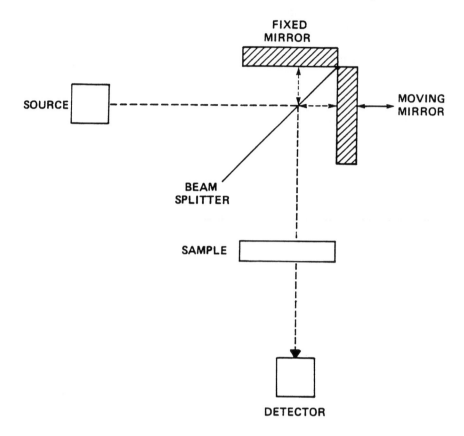

FIGURE 39.   Michelson interferometer.

by passing it through a Michelson interferometer. The Michelson interferometer is so constructed that light is separated into two beams by a beamsplitter. One beam strikes a stationary mirror and is reflected back to the beamsplitter.

The other beam strikes a moving mirror and is reflected back to the beamsplitter. The two beams are recombined at the beamsplitter and proceed on to the sample and the detector. Note that, upon combination, the two beams will interfere constructively or destructively, depending upon whether the difference in pathlength of the two beams is an integral multiple of the wavelength. The difference in pathlength is called the retardation and, when the retardation is an integral multiple of the wavelength, the interference is maximally constructive. If we plot retardation vs. intensity measured at the detector, we have, in effect, a time domain function of the intensity, and this can be transformed by Fourier transform mathematical techniques into a frequency function of the intensity.

Ultimately, then, we get the same information by both techniques, but we get it much faster, and more precisely, by the Fourier transform technique.

I. X-RAY FLUORESCENCE SPECTROMETRY

X-ray fluorescence spectrometry (XRF) is a nondestructive method of elemental analysis. XRF is based on the principle that each element emits its own characteristic X-ray line spectrum. When an X-ray beam impinges on a target element, orbital electrons are ejected. The resulting vacancies or holes in the inner shells are filled by outer shell electrons. During this process, energy is released in the form of secondary X-rays known as fluorescence. The energy of the emitted X-ray photon is dependent upon the distribution of electrons in the excited atom. Since every element has a unique electron distribution, every element

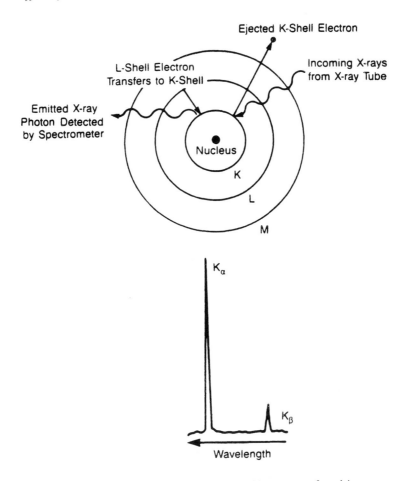

FIGURE 40. XRF excitation process and resulting spectrum for calcium.

produces a unique secondary X-ray spectrum, whose intensity is proportional to the concentration of the element in the sample. The excitation process and resulting X-ray spectrum are illustrated for calcium in Figure 40.

X-ray fluorescence instrumentation is divided into two types, wavelength-dispersive (WDXRF) and energy-dispersive (EDXRF) spectrometer, both of which are often automated with extensive computer systems for unattended operations that include data collection, reduction, and presentation. In a WDXRF (Figure 41), radiation emitted from the sample impinges on an analyzing crystal. The crystal diffracts the radiation according to Bragg's Law and passes it on to a detector which is positioned to collect a particular X-ray wavelength. Most spectrometers have two detectors and up to six crystals to allow optimization of instrument conditions for each element.

In an EDXRF (Figure 42), the emitted X-ray radiation from the sample impinges directly on a solid-state lithium-drifted silicon detector. This detector is capable of collecting and resolving a range of X-ray energies at one time. Therefore, the elements in the entire periodic table or in a selected portion can be analyzed simultaneously. Optimization for specific elements is accomplished through the use of secondary targets and/or filters. In some cases, radioisotope sources are used in place of X-ray tubes in instruments designed for limited element applications.

XRF offers a unique approach for rapid, nondestructive elemental analysis of liquids, powders, and solids. Although the first row transition elements is the most sensitive, elements

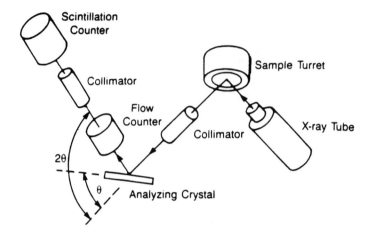

FIGURE 41. WDXRF spectrometer.

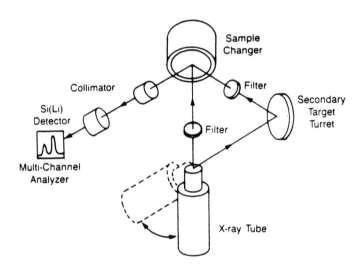

FIGURE 42. EDXRF spectrometer.

from atomic number 12 (magnesium) and greater can be measured over a dynamic range from trace (parts per million) to major (percent) element concentrations. EDXRF is well suited for qualitative elemental identification of unique samples, while WDXRF excels at high precision quantitative analysis.

Whenever quantitative analysis is desired, care must be taken to use proper standards and account for interelement matrix effects since the inherent sensitivity of the method varies greatly between elements. Methods to account for matrix effects include standard addition, internal standard, and matrix dilution techniques as well as numerous mathematical correction models. Computer software is also available to provide semiquantitative analysis of materials for which well-matched *standards are not available.*

XRF is used by both research and plant laboratories to sole a wide variety of elemental analysis problems. Among the most common are the quantitative analysis of additive elements (Ba, Ca, Zn, P, and S) in additives and lubricating oils, lead and sulfur in gasoline, sulfur in crudes and fuel oils, and halogens in polymers. The high analytical precision of WDXRF has enabled the development of methods for the precious metal assay of fresh reforming catalyst that rival the precision of classical wet chemical methods. Percent active metal such as Pt, Ir, Re, or Ru as well as trace contaminants such as S and Cl have also been determined on numerous catalyst types.

The elemental composition of unknown materials such as engine deposits can be determined qualitatively and the information used to develop dissolution methods prior to analysis by inductively coupled plasma atomic emission spectroscopy (ICPAES). Alternatively, a semiquantitative analysis can be provided by XRF alone, especially important when only a limited quantity of sample is available and needed for subsequent tests. The deposit does not even have to be removed from the piston since large objects can be placed directly inside an EDXRF spectrometer.

Aqueous and organic liquids, powders, polymers, papers, and fabricated solids can all be analyzed directly by XRF. The method is nondestructive, so, unless dilution is required, the original sample is returned to the submitter. Although the method can be applied to the analysis of materials ranging in size from milligram quantities to bulk parts such as engine pistons, a minimum of 5 g of sample is usually required for accurate quantitative analysis.

One of the greatest advantages of XRF over other methods of elemental analysis is that it is nondestructive and requires minimal sample preparation. WDXRF, when properly calibrated, offers precision and accuracy comparable to wet chemical methods of analysis. EDXRF offers rapid qualitative analysis of total unknowns.

Due to numerous interelemental matrix effects, matrix-matched standards including a blank are necessary for accurate quantitative analysis. The detection limits for XRF are not as low as other spectrometric methods and a noticeable drop-off in sensitivity is noted for light elements such as magnesium.

The most common interferences are absorption and/or enhancement of the element of interest by other elements in the matrix. Line overlaps may also occur. In the analysis of solids, particle size and geological effects can be important. Computer programs are available to correct for all of these interferences.

J. POTENTIOMETRIC TITRATIONS

A titration is a technique for determining the concentration of a material in solution by measuring the volume of a standard solution that is required to react with the sample. One of the most common titrations is a the acid-base titration in which the concentration of a base can be determined by adding a standard solution of an acid to the sample until the base is exactly neutralized. The exact neutralization point is found by the use of an indicator that changes color when the end point is reached.

There are, however, various cases where the visual method of detecting the end point cannot be used. For example, the solution may be dark or no appropriate indicator is available. In such cases, physicochemical techniques can be employed. A potentiometric titration is one of this type wherein the end point is detected by an abrupt change in voltage (between an electrode and the body of the solution) that may be observed as titrant is added to the solution.

As the solution is being titrated, the potential difference that exists between the indicating electrode and the solution may be continuously monitored with a voltage-measuring device such as a recorder. This affords an objective means of determining the end point which occurs when a very slight excess of titrant (as little as 0.25 ml) causes a sharp voltage change. Automatic recorders are now in use which not only plot the titration curve but also electronically determine the end point and can (if so programmed) complete the calculation producing the final result.

Results are reported in any of the following units: weight percent, milligrams per liter, normality, milliequivalents per 100 gm.

In some cases a plot of milliliter of titrant vs. voltage can be provided. Such plots are useful to determine whether there is more than one titratable material in the sample and to learn something about the character of the material titrated.

Titratable functional groups such as chloride, sulfide, mercaptide, weak or strong acids, weak or strong bases, and certain amines may be determined by this technique.

The sample characteristics are as follows:

- Size — the amount of sample to be submitted depends strictly upon the concentration of the functional group sought. For samples whose functional group is expected to be in the parts per million range, it is advisable to use 100 ml of sample.
- State — sample must be a solid or liquid.
- Composition limitation — sample must be soluble in water or in one of the several special nonaqueous titration solvents available that can accommodate most petroleum fractions (except some of the heavy ends).
- Temperature — titrations are conducted at room temperature.
- Concentration — a wide range of concentrations can be accommodated by varying the amount of sample dissolved in the titration solvent. In many cases concentrations as low as a few parts per million can be reported.

Titrations are relatively simple and rapid. They provide information concerning chemically reactive functional groups that would be difficult to obtain by other techniques.

Generally, the potentiometric titration technique is not good for qualitative purposes. One must indicate a priori what functional group he wants determined. The voltage at which the end point occurs does provide a clue to the material being titrated, but the "voltage spectrum" is too compressed to provide identification by voltage only. For example, strong acids can be differentiated from weak acids but the identity of the acid must be ascertained by other chemical or physical means.

The time required to titrate a single sample may vary somewhat depending upon the nature of the sample and what determinations are requested. An average of about 1 h is needed to run a single sample; however, the time per sample may be less when a series of similar samples is run consecutively. If special solutions have to be prepared, an additional 2 h may be required for the first sample in the series.

K. NEUTRON ACTIVATION ANALYSIS

Neutron activation analysis is a method of elemental analysis in which nonradioactive elements are converted to radioactive ones by neutron bombardment, and the elements of interest are determined from resulting radioactivity (Figure 43). High-energy (14-MeV) neutrons are generated by the reaction of medium-energy deuterium ions with titrium. For oxygen analysis, the carefully weighed sample is irradiated for 15 s to convert a small amount of the oxygen-16 to nitrogen-16, which emits α rays with a half-life of 7.4 s. The irradiated sample is transferred to a scintillation detector where the α rays are counted for 30 s to insure that all usable radioactivity has been counted and that no significant radioactivity remains in the sample. The system is calibrated with standards of known oxygen content.

The raw data consists of counts per 30 s from a digital counter, which can be converted to weight percent of the element of interest.

The characteristics of a proper sample are

- Size — container is plastic cylinder 9 mm I.D. × 20 mm deep. Holds ~1 cc. Sample should fill this container for best accuracy.
- State and phases — solid or liquid, reasonably homogeneous.

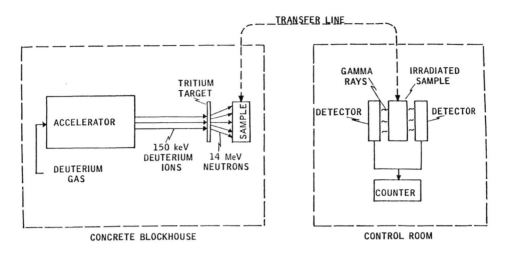

FIGURE 43.   Neutron activation analysis system.

- Composition limitations — must be moisture free if true sample oxygen is desired.
- Temperature range — room temperature only.
- Concentration of oxygen which can be determined — 0.01 to 60%.

The advantages of this technique are

1. Principle method for determining total oxygen directly.
2. Fast (about 10 min per analysis). Repeat analysis on weighed sample requires only 1 minute.
3. Nondestructive.
4. Moderate sensitivity.

V. RHEOMETRY

Although there are numerous rheometric techniques used, this section will only describe three common systems heavily employed for polymer characterization.

A. RHEOMETRICS SYSTEM IV

A schematic of the system is illustrated in Figure 44. For dynamic frequency sweeps (refer to Figure 45), the polymer is strained sinusoidally and the stress is measured as a function of the frequency. The strain amplitude is kept small enough to evoke only a linear response. The advantage of this test is that it separates the moduli into an elastic one, the dynamic storage modulus (G'), and into a viscous one, the dynamic loss modulus (G''). From these measurements one can determine fundamental properties such as:

- Zero shear viscosity (which can be related to weight average molecular weight and long-chain branching).
- tan delta (which is related to the damping properties).
- Plateau modulus (which can indicate the extent and "tightness" of crosslinking).
- Complex viscosity (which can be related to the steady shear viscosity).

Differences in G' and G'' and hence in the properties mentioned above will be found if there are differences in molecular weight, molecular weight distribution (MWD), or long-

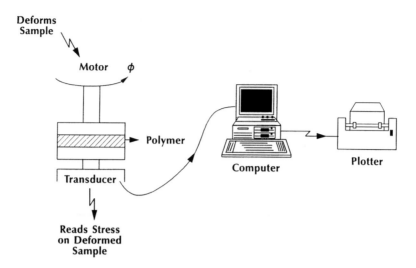

FIGURE 44. Details of Rheometrics system IV.

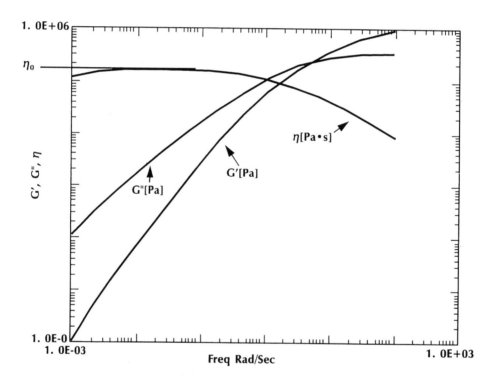

FIGURE 45. Typical frequency sweep.

chain branching. For example, if the MWD is primarily in the high molecular weight end, the value of G' will be higher.

To obtain measurements during oscillatory shear, the drive motor causes the fixture to oscillate from high to low shear rates deforming the sample. The transducer detects the periodic stress which is generated by the deformation. The magnitude of the stress is converted into dynamic shear moduli.

The application of stress relaxation is shown in Figure 46. The relaxation modulus (G) is determined after a step strain as a function of time. A step strain is applied to the sample

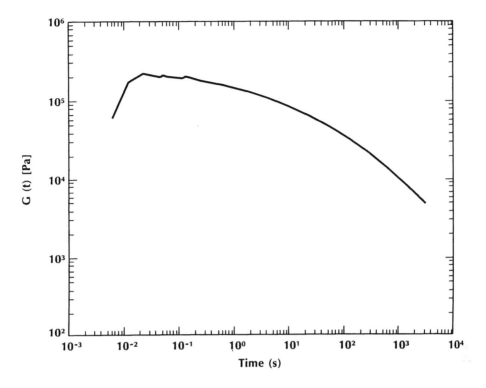

FIGURE 46. Typical stress relaxation.

causing a stress. The modulus is measured as the stress relaxes. The stress relaxation modulus shows how molecular weight affects the relaxation process as a function of time as depicted in Figure 47.

To perform a temperature sweep, the sample is deformed at constant frequency over a temperature range. The modulus is then obtained as a function of temperature at constant frequency. This measurement gives the temperature dependence of the rheological properties as well as a good indication of the thermal stability.

B. CAPILLARY RHEOMETRY

Steady shear viscosities can be measured with two different instruments. The System IV can measure polymer viscosities from about 0.001 to 10 s^{-1} while the Gottfert Capillary Rheometer is capable of obtaining viscosities from 0.1 to 100,000 1/s. In steady shear, the strains are very large as opposed to the dynamic measurements that impose small strains. In the capillary rheometer, the polymer is forced through a capillary die at a continuously faster rate. The resulting stress and viscosity are measured by a transducer mounted adjacent to the die. A schematic of the system is illustrated in Figure 48.

The shear viscosity can be used for relating the polymer flow properties to the processing behavior, extruder design, and many other high shear rate applications. Elongational viscosity, die swell measurements, as well as residence time effects can be estimated. Typical data are shown in Figure 49.

C. TORQUE RHEOMETRY

Torque rheometers are multipurpose instruments well suited for formulating multicomponent polymer systems, studying flow behavior, thermal sensitivity, shear sensitivity, batch

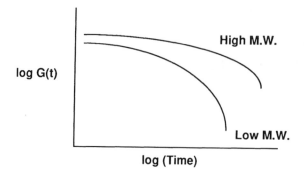

FIGURE 47. How the stress relaxation modulus is related to molecular weight.

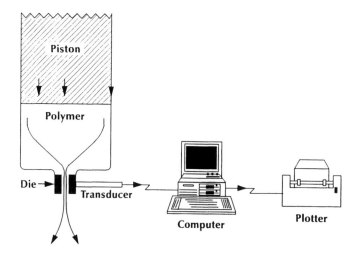

FIGURE 48. Details of the capillary rheometer.

compounding, and so on. The instrument is applicable to thermoplastics, rubber (compounding, cure, scorch tests), thermoset materials, and liquid materials.

When the rheometer is retrofitted with a single-screw extruder one can measure rheological properties and extrusion processing characteristics to differentiate lot-to-lot variance of polymer stocks. It also enables the process engineer to simulate a production line in the laboratory and to develop processing guidelines.

The torque rheometer with a twin-screw extruder is considered a scaled-down continuous compounder. It allows the compounding engineer to develop polymer compound and alloys. It also permits the formulation engineer to assure that the formulation is optimum.

The torque rheometer is essentially an instrument that measures viscosity-related torque caused by the resistance of the material to the shearing action of the plasticating process.

Torque can be defined as the effectiveness of a force to produce rotation. It is the product of the force and the perpendicular distance from its line of action to the instantaneous center of rotation.

The prevalent design is a microprocessor-controlled torque rheometer. The system consists of two basic units: an electromechanical drive unit and a microprocessor unit. The basic system is shown in Figure 50.

Most plastics and elastometric products are not pure materials but rather mixtures of the basic polymer with a variety of additives such as pigments, lubricants, stabilizers,

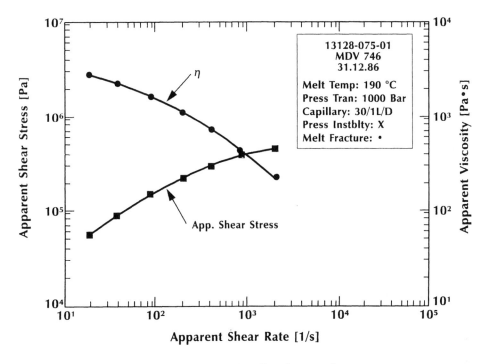

FIGURE 49. Typical capillary rheometer data.

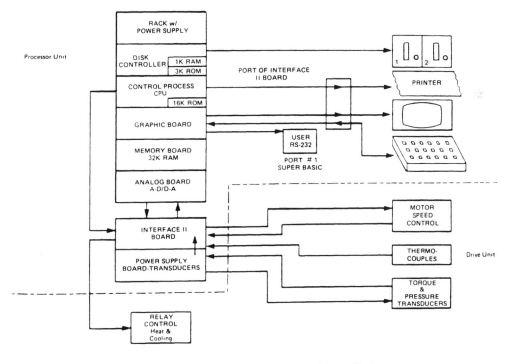

FIGURE 50. Schematic diagram of Haake-Buechler System 40 microprocessor.

antioxidants, flame retardants, antiblock agents, cross-linking agents, fillers, reinforcement agents, plasticizers, UV absorbants, foaming agents, and others. All these additives must be incorporated into the polymer prior to fabrication. Some of the additives take a significant portion of the mixture; others, only minute amounts. Some are compatible; others are not. Depending on the quality of resin and additives and homogenization of the mixtures, the quality of the final product will be varied. Therefore, developing a quality resin and additives that meet with desired physical and mechanical properties of the product and quality control associated with them play an important role in the plastic industry. In the development of formulations and applications for new polymers, the torque rheometer is an invaluable instrument. Common practice is to equip the torque rheometer with a miniaturized internal mixer (MIM) to simulate large-scale production at the bench. The mixer generally consists of a mixing chamber shaped like a figure eight with a spiral-lobed rotor in each chamber. A totally enclosed mixing chamber contains two fluted mixing rotors that revolve in opposite directions and at different speeds to achieve a shear action similar to a two-roll mill.

In the chamber, the rotors rotate in order to effect a shearing action on the material mostly by shearing the material repeatedly against the walls of the mixing chamber. This is illustrated conceptually in Figure 51. The rotors have chevrons (helical projections) which perform additional mixing functions by churning the material and moving it back and forth through the mixing chamber. The mixture is fed to the mixing chamber through a vertical chute with a ram. The lower face of the ram is part of the mixing chamber. There is usually a small clearance between the rotors, which rotate at different speeds at the chamber wall. In these clearances, dispersive mixing takes place. The shape of the rotors and the motion of the ram during operation ensure that all particles undergo high-intensive shearing flow in the clearances.

There are three sets of interchangeable rotors available on the market. They are roller, cam, and sigma rotors designs, although there are many other designs as illustrated in Figure 52. Normally, roller rotors are used for thermoplastics and thermosets, cam rotors are for rubber and elastomers, and sigma rotors are for liquid materials. Banbury rotors are used with miniaturized Banbury mixer for rubber compounding formulation.

Plastic materials with two or more components being processed should be well mixed so that the compounded material provides the best physical properties for the final product. There are distinctive types of mixing processes. The first involves the spreading of particles over position in space (called distributive mixing). The second type involves shearing and spreading of the available energy of a system between the particles themselves (dispersive mixing). In other words, distributive mixing is used for any operation employed to increase the randomness of the spatial distribution of particles without reducing their sizes. This mixing depends on the flow and the total strain, which is the product of shear rate and residence time or time duration. Therefore, the more random arrangement of the flow pattern, the higher the shear rate; the longer the residence time, the better the mixing will be.

A dispersive mixing process is similar to that of a simple mixing process, except that the nature and magnitude of forces required to rupture the particles to an ultimate size must be considered. Essential to intensive mixing is the incorporation of pigments, fillers, and other minor components into the matrix polymer. This mixing is a function of shear stress, which is calculated as a product of shear rate and material viscosity. Breaking up of an agglomerate will occur only when the shear stress exceeds the strength of the particle.

An important aspect of mixing studies on the torque rheometer is the temperature dependency of the mixing process. In general, the viscosity of a polymer decreases as temperature increases and vice versa. Properties of the material also change depending on temperature. Hence, knowledge of the temperature dependency of viscosity is important.

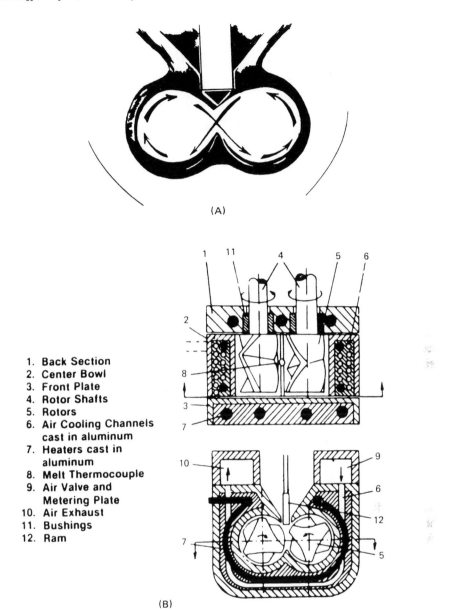

1. **Back Section**
2. **Center Bowl**
3. **Front Plate**
4. **Rotor Shafts**
5. **Rotors**
6. **Air Cooling Channels cast in aluminum**
7. **Heaters cast in aluminum**
8. **Melt Thermocouple**
9. **Air Valve and Metering Plate**
10. **Air Exhaust**
11. **Bushings**
12. **Ram**

FIGURE 51. Mixing action in an MIM.

Large variations of viscosity for a certain range of temperature means that the material is thermally unstable (i.e., requires large activation energy). This kind of material has to be processed with accurate temperature control. Figure 53 shows the torque vs. temperature curve obtained from the microprocessor-controlled torque rheometer to see the temperature dependency of viscosity-related torque on an EPDM material.

The MIM along with the torque rheometer can be used to simulate and optimize a variety of processing applications/problems. Some typical but by no means inclusive examples are in studying fusion characteristics, examining stability and processability, color, or thermal stability testing, examining the gelation of plastisols in developing criteria for the selection

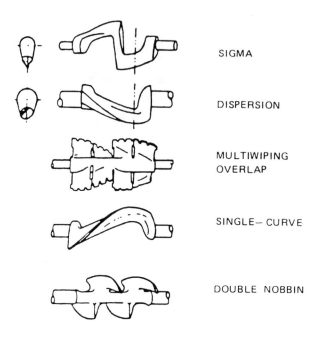

SIGMA

DISPERSION

MULTIWIPING
OVERLAP

SINGLE— CURVE

DOUBLE NOBBIN

FIGURE 52.   Types of mixer rotor configurations.

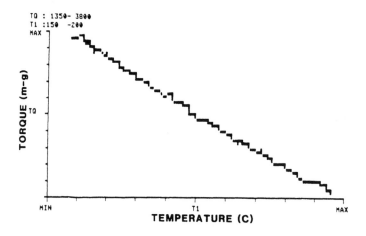

FIGURE 53.   Cross plot of torque and temperature from rheocord.

of blowing agents for foam products, compound formulation optimization, studying the scorch, and cure characteristics of rubber compounds, and in studying the cure characteristics of thermosets. An example is given below.

Lubricants play an important role in processing and the properties of the final product. The lubricants also effect the fusion of the polymer materials, that is, internal lubricants reduce melt viscosity, while external lubricants reduce friction between the melt and the hot melt parts of the processing equipment and prevent sticking, controlling the fusion of the resin. Figure 54 illustrates the results of an experiment aimed at studying the fusion characteristics of PVC. The level of external lubricant used in the formulation affects the fusing

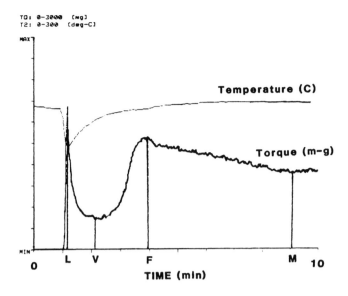

TQ: 0-3000 (mg)
T2: 0-300 (deg-C)

FIGURE 54. Results of PVC fusion study.

time between points L and F on the curve. The higher the level of external lubricant in the formulation, the longer the fusion time will be.

If an unnecessarily high level of external lubricant is used in the formulation, it will take a longer period of time to melt the material in processing, which results in reducing production, increasing energy consumption, and poor products. Meanwhile, if too low a level of external lubricant is used, the material will melt too early in the processing equipment, which may result in degradation in the final product. Therefore, selecting the optimum amount of external lubricant is a must for improvement of processing and for good-quality products.

One of the most frequently applied tests is in the study of additive incorporation and compounding. All of the additives used in a formulation must be incorporated in the major component, and the components should be in a stable molecular arrangement. Figure 55 illustrates a test result for incorporation of minor components to the major component as well as homogeneous compound after the additives are incorporated. The test was performed with an EPDM rubber and reblended additives. The EPDM was loaded into the mixer and mixed for 30 s. Torque values immediately dropped sharply and increased as the additives incorporated. When the ingredients were fully incorporated, a second torque peak was observed and finally stabilized when the material was homogeneously compounded. The second peak is called the "incorporation peak". If hard fillers are added to the polymer, torque increases sharply and generates the second peak. This can be seen when carbon black is incorporated. The time from the addition of the minor components resulting in the incorporation peak is referred to as the "incorporation time" and is critical to standard batch compounding operations with rubbers.

To this point, emphasis has been on applications testing where the torque rheometer has been retrofitted with a MIM. Another common practice is to incorporate a screw extruder. Solids conveying, melting, mixing, and pumping are the major functions of polymer processing extruders. The single-screw extruder is the most widely used machine to perform

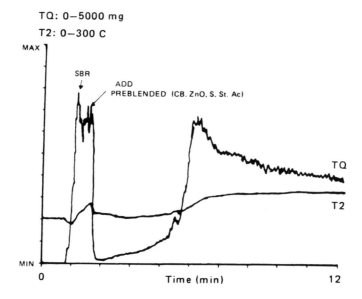

TQ: 0–5000 mg
T2: 0–300 C

FIGURE 55.   Development of incorporation time.

these functions. The plasticating extruder has three distinct regions: solids conveying zone, transition (melting) zone, and pumping zone.

The unit can be fed polymer in the particulate solids form or as strips, as in the case of rubber extrusion. The solids (usually in pellet or powder form) in the hopper flow by gravity into the screw channel, where they are conveyed through the solids conveying section. They are compressed by a drag-induced mechanism in the transition section. In other words, melting is accomplished by heat transfer from the heated barrel surface and by mechanical shear heating.

Simulation of the extrusion process in the laboratory is one of the most important applications of the torque rheometer in conjunction with single-screw extruders. Figure 56 illustrates simulations of widely used extrusion processes in the industries.

It is important for the process engineer to know the rheological properties of a material since the properties dominate the flow of the material in extrusion processes and also dominate the physical and mechanical properties of the extrudates. Therefore, it is also important to measure the properties utilizing a similar miniaturized extruder in the laboratory so that a process engineer knows the flow properties in the system by simulating the production line. Also, it is desirable to know the flow properties of a material to be processed in the range of shear rates of equipments to be used.

VI. CHEMICAL ANALYSIS OF POLYMERS

Chemical analysis of polymer materials is difficult because of the large number and types of such materials and because of modification and compounding practices of conventional polymers.

For exact identification of polymers it is important for the samples to be in the form of pure products without incorporated additives such as plasticizers, fillers, or stabilizers. One must separate additives by extraction or reprecipitation before identification. The solvents

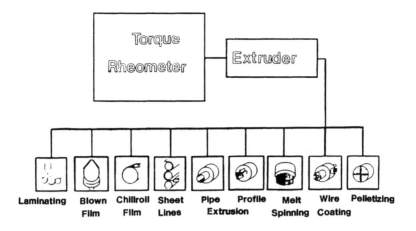

FIGURE 56. Application of torque rheometry to studying extrusion operations.

or mixtures of solvent and precipitant are substance specific and should be chosen separately for each case.

In order to determine the quantitative composition of a polymer, following sequence of operations is followed:

- Comminution of the polymer sample
- Separation of additives
- Qualitative and quantitative investigation of the additives
- Identification and quantitative analysis of isolated polymer samples

A. COMMUNICATION, SEPARATION, AND IDENTIFICATION

Mechanical comminution of the sample is performed because the composition of polymers frequently shows inhomogeneities despite good processing. Furthermore, some tests (e.g., monomer content and H_2O content) depend on the sample.

Products are comminuted with cutting tools such as shears, knives, or razor blades. Drilling, milling, etc. are also suitable. A smaller particle diameter can also be obtained by grinding precooled samples in mills. Depending on the elasticity characteristics of the sample, either dry ice or liquid nitrogen can be effectively used.

After comminution, samples must be conditioned over phosphorus pentoxide at room temperature in a desiccator. In all mechanical loads and also in low-temperature treatment, decomposition processes which may affect polymoleallarity and sample composition must be accounted for. Since the analytical result may be affected, attention must be paid to the reproducibility and uniformity of the comminution processes.

Plasticizers can be separated by extraction with diethyl ether. Stabilizers based on pure organic or organometallic compounds may only be partially separated. Extraction time depends on particle size and on the amount of plasticizer in the sample.

For the quantitative determination of the mass of plasticizer about 1 to 2 g of the comminuted sample are weighed and extracted with anhydrous diethyl ether in a Soxhlet apparatus. After distilling off the ether and drying the extract at 105°C to constant weight, the amount of ether-soluble components is calculated from the difference in weight of the extraction flask before and after the extraction. Preparative separation of the plasticizer before identification of the polymer is performed in an analogous manner.

Plasticizers include the esters of a few aliphatic and aromatic mono- and dicarboxylic acids, aliphatic and aromatic phosphorus acid esters, ethers, alcohols, ketones, amines,

TABLE 2
Reagents for the Detection of Plasticizers in TLC

Type of Plasticizer Reagent

Phthalates, adipates	Ethanolic resorcinol solution (20%) with 1% zinc chloride (10 min at 100°C); subsequently 4 N H$_2$SO$_4$ (20 min at 120°C)
Phosphates	Ammonium molybdate solution made from 3 g of ammonium molybdate in 20 ml of perchloric acid (40%) and 5 ml of concentrated hydrochloric acid in 200 ml of water (10 min 100°C); subsequently saturated hydrazine sulfate solution (20 min 110°C)
Citrates	Ethanolic vanillin solution (20%) 10 min 80°C); subsequently 4 N H$_2$SO$_4$ (10 to 20 min 110°C)

amides, and nonpolar and chlorinated hydrocarbons. These additives are used in various mixtures. For their separation and qualitative detection, thin-layer chromatography (TLC) is preferred. Usually Kieselgur plates, 0.25 mm thick, activated at 110°C for 30 min, in the saturated vapor are used. Methylene chloride and mixtures of diisopropyl ether/petether at temperatures between 40 to 60°C have been successfully used as the mobile phase.

Under selected conditions, polymer plasticizers remain at the starting point. Although variation of the mobile and stationary phases and detection reactions allows selectivity of TLC to be increased, methods such as GC must be used for more complex plasticizer mixtures.

The gas-chromatographic separation of plasticizers can be effected directly or after conversion to low boiling point compounds. This is achieved by a transesterification reaction with methanol or diazomethane. After separation of the plasticizers mixtures with LC, identification by spectroscopic methods is possible (Table 2).

Inorganic fillers are, in general, insoluble in organic solvents. They can be quantitatively separated from the soluble polymers by centrifugation using solutions of 5% by weight in suitable solvents and subsequent pouring off of the liquid. Dissolution has to be accelerated by shaking or stirring. The conditions of centrifugation depend on the density and particle size of the fillers. Carbon black cannot be separated completely, even under the most severe centrifugation and special methods are available for its quantitative separation. The same applies to the fillers in insoluble polymers, e.g., in thermosets and unsaturated polyesters. If required the fillers may be identified using the normal methods of qualitative inorganic analysis.

For the quantitative determination of fillers, a small amount of the polymer material is weighed and dissolved in solvent. After centrifuging the solution is pipetted off from the centrifugate. The residue is suspended once more in the same solvent, centrifuged again, and after isolation is washed once more with the solvent and then several times with methanol. The residue is dried to constant weight and then the solid matter is determined gravimetrically. Table 3 reports the solubility of selected polymers.

B. STABILIZER IDENTIFICATION

Light stabilizers (also known as UV absorbers) and antioxidants are indicative of the great number of compounds which have gained importance. Among the most important heat stabilizers are basic salts of heavy metals, metal salts of organic acids, and nitrogenous organic compounds. Common antioxidants are phenols, aromatic amines, and benzimidazoles. UV absorbers are substances which absorb strongly in the short-wavelength range but are transparent at wave numbers <25000 cm^{-1} so that the stabilized material does not show any coloration. The hydroxybenzophenone derivatives, salicyl esters, and benzotriazoles are examples.

Identification of stabilizers is complex because of their great number and the small amounts usually present. Added to this is the difficulty that stabilizers take part in transfer

TABLE 3
Solubility of Selected Polymers

Polymer	Soluble	Insoluble
Polyvinyl chloride	Dimethyl formamide, tetrahydrofuran, cyclohexanone	Alcohols, hydrocarbons, butyl acetate
Polyvinylidene chloride	Ketones, tetrahydrofuran, dioxan, butyl acetate	Alcohols, hydrocarbons
Polyamides	Phenols, formic acid, trifluoroethanol	Alcohols, hydrocarbons, esters
Polyethylene	Tetralin, decalin, xylene, dichloroethylene at temperatures >100°C	Alcohols, petrol, esters
Polystyrene	Ethyl acetate, benzene, acetone, chloroform, methylene dichloride	Alcohols, water
Polyvinyl acetate	Aromatic hydrocarbons, ketones, chlorinated hydrocarbons, alcohols	Petrol
Polyvinyl alcohol	Formamide, water	Ether, alcohols, petrol, benzene, esters, ketones, hydrocarbons
Polyurethanes	Dimethyl formamide	Ether, alcohols, petrol, benzene, water
Polyacrylonitrile	Dimethyl formamide, nitrophenols	Esters, alcohols, ketones, hydrocarbons
Polyester	Phenols, nitrated hydrocarbons, acetone, benzyl alcohol	Esters, alcohols, hydrocarbons
Aminoplastics	Benzylamine (160°C), ammonia	Organic solvents
Phenoplastics	Benzylamine (200°C)	
Cellulose, regenerated	Schweizer's reagent	Organic solvents
Cellulose ester	Esters, ketones	Aliphatic hydrocarbons
Polybutadiene	Benzene	Alcohols, petrol, esters
Polyisoprene		Ketones
Polyisobutylene	Ether, petrol	Alcohols, esters
Polymethacrylic acid	Aromatic hydrocarbons, esters, ketones, chlorinated hydrocarbons	Ether, alcohols, aliphatic hydrocarbons

or rearrangement reactions during molding processes so that only a portion of the stabilizer is found unchanged in the finished product. Because of the intense toxicity of some of these decomposition products, their detection is of particular importance.

Stabilizers are identified after separation by solid-liquid extraction or after the removal of the polymer by precipitation from the diluted solution. Some extraction solvents for the most important stabilizers and polymers are given in Table 4.

TLC is the preferred separation method because of its high separation efficiency, rapidity, and large variety of detection possibilities. Usually 0.5-mm thick silica-gel-G-plates are used, activated at 120°C for 30 min in a supersaturated atmosphere. Well known polytechniques such as multiple separation in opposite or parallel direction allow the selectivities to be further increased. The selection of an appropriate mobile phase determines the efficiency of separation. Advantage is taken of specific interactions and also of reactivity with the stabilizers under investigation.

C. POLYMER IDENTIFICATION

Polymer identification starts with a series of preliminary tests. In contrast to low molecular weight organic compounds, which are frequently satisfactorily identified simply by their melting or boiling point, molecular weight, and elementary composition, precise identification of polymers is difficult by the presence of copolymers, the statistical character of the composition, macromolecular properties, and by potential polymeric analogous reactions. Exact classification of polymers is not usually possible from a few preliminary tests. Further physical data must be measured and specific reactions must be carried out in order to make

TABLE 4
Extraction Agents for Stabilizers

Polymer	Stabilizers to be extracted	Extracting agent
Polyvinyl chloride	Organotin stabilizers	Heptane: glacial acetic acid 1:1
Polyvinyl chloride	N-containing organic stabilizers	Methanol or diethyl ether
Polyethylene	Antioxidants	Chloroform
Polyoxymethylene	Phenolic antioxidants	Chloroform
Rubbers	Stabilizers; accelarators	Boiling acetone
		Boiling water

a reliable classification. The efficiency of physical methods such as IR and NMR spectroscopy as well as pyrolysis GC makes them particularly important.

One method of analysis is pyrolysis, which is the application of thermal energy, causing covalent bonds to be broken. Fragments are produced whose chain length and structure are dependent upon the temperature on the one hand and on the type of bonds on the other. If oxygen is present, oxidation occurs at the same time which may lead to ignition. Since pyrolysis itself is an endothermic reaction, the required energy must be supplied by an external heat source. The subsequent oxidation process is exothermic. If sufficient energy is released, the sample will burn spontaneously. If the temperature drops after removal of the flame, the polymer will be self-extinguishing. Behavior in an open flame can easily be observed by holding about 0.1 to 0.2 g of sample with a suitable implement in the outer edge of a small Bunsen flame.

To test behavior during dry heating, about 0.1 g of the sample is carefully heated in a 60-mm long glow tube with a diameter of 6 mm over a small flame. If heating is too vigorous, the characteristic phenomena can no longer be observed.

Depolymerization is a special case of thermal degradation. It can be observed particularly in polymers based on a, a'-disubstituted monomers. In these, degradation is a reversal of the synthesis process. It is a chain reaction during which the monomers are regenerated by an unzipping mechanism. This is due to the low polymerization enthalpy of these polymers. For the thermal fission of polymers with secondary and tertiary C-atoms, higher energies are required. In these cases elimination reactions occur. This can be seen very clearly in PVC and PVAC.

Depolymerization is in functional correlation with the molecular weight distribution and with the type of terminal groups which are formed in chain initiation and termination.

Depolymerization, elimination, and statistical chain-scission reactions can be used for polymer analysis. When the monomer is the main degradation product obtained, it can easily be identified by boiling point and refractive index.

Elimination and chain-scission reactions provide characteristics pyrograms which can often be identified by GC or IR spectroscopy.

For testing depolymerization behavior, about 0.2 to 0.3 g of the polymeric substance is carefully and gently heated to a maximum of 500°C in a small distillation flask. The distillate is collected in a receiver and its boiling point and refractive index are determined.

Appendix A

ABBREVIATIONS OF POLYMERS

ABS	Acrylonitrile-butadiene-styrene
AN	Acrylonitrile
CA	Cellulose acetate
CAB	Cellulose acetate butyrate
CAP	Cellulose acetate propionate
CN	Cellulose nitrate
CP	Cellulose propionate
CPE	Chlorinated polyethylene
CPVC	Chlorinated polyvinyl chloride
CTFE	Chlorotrifluoroethylene
DAP	Diallyl phthalate
EC	Ethyl cellulose
ECTFE	Poly(ethylene-chlorotrifluoroethylene)
EP	Epoxy
EPDM	Ethylene-propylene-diene monomer
EPR	Ethylene propylene rubber
EPS	Expanded polystyrene
ETFE	Ethylene/tetrafluoroethylene copolymer
EVA	Ethylene-vinyl acetate
FEP	Perfluoro (ethylene-propylene) copolymer
FRP	Fiberglass-reinforced polyester
HDPE	High-density polyethylene
HIPS	High-impact polystyrene
HMWPE	High-molecular-weight polyethylene
LDPE	Low-density polyethylene
MF	Melamine-formaldehyde
PA	Polyamide
PAPI	Polymethylene polyphenyl isocyanate
PB	Polybutylene
PBT	Polybutylene terephthalate (thermoplastic polyester)
PC	Polycarbonate
PE	Polyethylene
PES	Polyether sulfone
PET	Polyethylene terephthalate
PF	Phenol-formaldehyde
PFA	Polyfluoro alkoxy
PI	Polyimide
PMMA	Polymethyl methacrylate
PP	Polypropylene
PPO	Polyphenylene oxide
PS	Polystyrene
PSO	Polysulfone
PTFE	Polytetrafluoroethylene
PTMT	Polytetramethylene terephthalate (thermoplastic polyester)
PU	Polyurethane
PVA	Polyvinyl alcohol
PVAC	Polyvinyl acetate

PVC	Polyvinyl chloride
PVDC	Polyvinylidene chloride
PVDF	Polyvinylidene floride
PVF	Polyvinyl fluoride
TFE	Polytelrafluoroethylene
SAN	Styrene-acrylonitrile
SI	Silicone
TP	Thermoplastic elastomers
TPX	Polymethylpentene
UF	Urea formaldehyde
UHMWPE	Ultrahigh-molecular-weight polyethylene
UPVC	Unplasticized polyvinyl chloride

Appendix B

GLOSSARY OF POLYMERS AND TESTING

Accelerated aging: Test in which conditions are intensified in order to reduce the time required to obtain a deteriorating effect similar to one resulting from normal service conditions.

Accelerated weathering: Test in which the normal weathering conditions are accelerated by means of a device.

Aging: Process of exposing plastics to natural or artificial environmental conditions for prolonged period of time.

Arc resistance: Ability of plastic to resist the action of a high-voltage electrical arc, usually in terms of time required to render the material electrically conductive.

Apparent density: (Bulk density) weight of unit volume of material including voids (air) inherent in the material.

Abrasion resistance: Ability of material to withstand mechanical action such as rubbing, scraping, or erosion that tends to progressively remove material from its surface.

Amorphous polymers: Polymeric materials that have no definite order or crystallinity. Polymer molecules are arranged in completely random fashion.

Bulk factor: Ratio of volume of any given quantity of the loose plastic material to the volume of the same quantity of the material after molding or forming. It is a measure of volume change that may be expected in fabrication.

Brittle failure: Failure resulting from inability of material to absorb energy, resulting in instant fracture upon mechanical loading.

Birefringence (double refraction): The difference between index of refraction of light in two directions of vibration.

Brittleness temperature: Temperature at which plastics and elastomers exhibit brittle failure under impact conditions.

Burst strength: The internal pressure required to break a pressure vessel such as a pipe or fitting. The pressure (and therefore the burst strength) varies with the rate of pressure build-up and the time during which the pressure is held.

Capillary rheometer: Instrument for measuring the flow properties of polymer melts. Comprised of a capillary tube of specified diameter and length, means for applying desired pressures to force molten polymer through the capillary, means for maintaining the desired temperature of the apparatus, and means for measuring differential pressures and flow rates.

Charpy impact test: A destructive test of impact resistance, consisting of placing the specimen in a horizontal position between two supports then striking the specimen with a pendulum striker swung from a fixed height. The magnitude of the blow is increased until specimen breaks. The result is expressed in inch-pound or foot-pound of energy.

Chalking: A whitish, powdery residue on the surface of a material caused by material degradation (usually from weather).

CIE (Commission Internationale de l'Eclairage): International commission on illuminants responsible for establishing standard illuminants.

Coefficient of thermal expansion: Fractional change in length or volume of a material for unit change in temperature.

Compressive strength: Maximum load sustained by a test specimen in a compressive test divided by original cross-section area of the specimen.

Conditioning: Subjecting a material to standard environmental and/or stress history prior to testing.

Colorimeter: Instrument for matching colors with results approximately the same as those of visual inspection, but more consistently.

Continuous use temperature: Maximum temperature at which material may be subjected to continuous use without fear of premature thermal degradation.

Crazing: Undesirable defect in plastic articles, characterized by distinct surface cracks or minute frostlike internal cracks, resulting from stresses within the article. Such stresses result from molding shrinkage, machining, flexing, impact shocks, temperature changes, or action of solvents.

Crystallinity: State of molecular structure attributed to existence of solid crystals with a definite geometric form. Such structures are characterized by uniformity and compactness.

Crosslinking: The setting up of chemical links between the molecular chains. When extensive, as in most thermosetting resins, crosslinking makes one infusible supermolecule of all the chains. Crosslinking can be achieved by irradiation with high-energy electron beams or by chemical crosslinking agents.

Creep: Due to viscoelastic nature, a plastic subjected to a load for a period of time tends to deform more than it would from the same load released immediately after application. The degree of this deformation is dependent on the load duration. Creep is the permanent deformation resulting from prolonged application of stress below the elastic limit. Creep at room temperature is called cold flow.

Creep modulus (apparent modulus): Ratio of initial applied stress to creep strain.

Creep rupture strength: Stress required to cause fracture in a creep test.

Cup flow test: Test for measuring the flow properties of thermosetting materials. A standard mold is charged with preweighed material and the mold is closed using sufficient pressure to form a required cup. Minimum pressures required to mold a standard cup and the time required to close the mold fully are determined.

Cup viscosity test: Test for making flow comparisons under strictly comparable conditions. The cup viscosity test employs a cup-shaped gravity device that permits the timed flow of a known volume of liquid passing through an orifice located at the bottom of the cup.

Density: Weight per unit volume of a material expressed in grams per cubic centimeter, pounds per cubic foot, etc.

Dielectric strength: Electric voltage gradient at which an insulating material is broken down or "arced through" in volts per millimeter of thickness.

Dielectric constant (permititivity): Ratio of the capacitance of a given configuration of electrodes with a material as dielectric to the capacitance of the same electrode configuration with a vacuum (or air for most practical purposes) as the dielectric.

Dissipation factor: Ratio of the conductance of a capacitor in which the material is dielectric to its susceptance, or the ratio of its parallel reactance to its parallel resistance. Most plastics have a low dissipation factor, a desirable property because it minimizes the waste of electrical energy as heat.

Dimensional stability: Ability to retain the precise shape in which it was molded, fabricated, or cast.

Drop impact test: Impact resistance test in which a predetermined weight is allowed to fall freely onto the specimen from varying heights. The energy absorbed by the specimen is measured and expressed in inch-pounds or foot pounds.

Differential scanning calorimetry (DSC): Thermal analysis technique that measures the quantity of energy absorbed or evolved (given by a specimen in calories) as its temperature is changed.

Ductility: Extent to which a material can sustain plastic deformation without fracturing.

Durometer hardness: Measure of the indentation hardness of plastics. It is the extent to which a spring-loaded steel indentor protrudes beyond the pressure foot into the material.

Elongation: The increase in length of a test specimen produced by a tensile load. Higher elongation indicates higher ductility.

Embrittlement: Reduction in ductility due to physical or chemical changes.

Environmental stress cracking: The susceptibility of a thermoplastic article to crack or craze formation under the influence of certain chemicals and stress.

Extensometer: Instrument for measuring changes in linear dimensions (also called strain gauge).

Extrusion plastometer (rheometer): A type of viscometer used for determining the melt index of a polymer. Comprised of a vertical cylinder with two longitudinal bored holes (one for measuring temperature and one for containing the specimen, the latter having an orifice of stipulated diameter at the bottom and a plungering from the top). The cylinder is heated by external bands and weight is placed on the plunger to force the polymer specimen through the orifice. The result is reported in grams per 10 min.

Fatigue failure: The failure or rupture of a plastic under repeated cyclic stress, at a point below the normal static breaking strength.

Fatigue limit: The stress below which a material can be stressed cyclically for an infinite number of times without failure.

Fatigue strength: The maximum cyclic stress a material can withstand for a given number of cycles before failure.

Fadometer: An apparatus for determining the resistance of materials to fading by exposing them to ultraviolet rays of approximately the same wavelength as those found in sunlight.

Flammability: Measure of the extent to which a material will support combustion.

Flexural modulus: Ratio of the applied stress on a test specimen in flexure to the corresponding strain in the outermost fiber of the specimen. Flexural modulus is the measure of relative stiffness.

Flexural strength: The maximum stress in the outer fiber at the moment of crack or break.

Foamed plastics (cellular plastics): Plastics with numerous cells disposed throughout its mass. Cells are formed by a blowing agent or by the reaction of the constituents.

Gel point: The stage at which liquid begins to gel, that is, exhibits pseudoelastic properties.

Gel permeation chromatography (GPC): Column chromatography technique employing a series of columns containing closely packed rigid gel particles. The polymer to be analyzed is introduced at the top of the column and then is eluted with a solvent. The polymer molecules diffuse through the gel at rates depending on their molecular size. As they emerge from the columns, they are detected by differential refractometer coupled to a chart recorder, on which a molecular weight distribution curve is plotted.

Hardness: The resistance of plastic materials to compression and indentation. Brinnel hardness and shore hardness are major methods of testing this property.

Haze: The cloudy or turbid aspect of appearance of an otherwise transparent specimen caused by light scattered from within the specimen or from its surface.

Hooke's law: Stress is directly proportional to strain.

Hoop stress: The circumferential stress in a material of cylindrical form subjected to internal or external pressure.

Hysteresis: The cyclic noncoincidence of the elastic loading and the unloading curves under cyclic stressing. The area of the resulting elliptical hysteresis loop is equal to the heat generated in the system.

Hygroscopic: Material having the tendency to absorb moisture from air. Plastics, such as nylons and ABS, are hygroscopic and must be dried prior to molding.

Impact strength: Energy required to fracture a specimen subjected to shock.

Impact test: Method of determining the behavior of material subjected to shock loading in bending or tension. The quantity usually measured is the energy absorbed in fracturing the specimen in a single blow.

Indentation hardness: Resistance of a material to surface penetration by an indentor. The hardness of a material as determined by the size of an indentation made by an indenting tool under a fixed load or the load necessary to produce penetration of the indentor to a predetermined depth.

Index of refraction: Ratio of velocity of light in vacuum (or air) to its velocity in a transparent medium.

Infrared analysis: Technique used for polymer identification. An infrared spectrometer directs infrared radiation through a film or layer of specimen and measures the relative amount of energy absorbed by the specimen as a function of wavelength or frequency of infrared radiation. The chart produced is compared with correlation charts for known substances to identify the specimen.

Inherent viscosity: In dilute solution viscosity measurements, inherent viscosity is the ratio of the natural logarithm of the relative viscosity to the concentration of the polymer in grams per 100 ml of solvent.

Intrinsic viscosity: In dilute solution viscosity measurements, intrinsic viscosity is the limit of the reduced and inherent viscosities as the concentration of the polymeric solute approaches zero and represents the capacity of the polymer to increase viscosity.

ISO: Abbreviation for the International Standards Organization.

Isochronous (equal time) stress-strain curve: A stress-strain curve obtained by plotting the stress vs. corresponding strain at a specific time of loading pertinent to a particular application.

Izod impact test: Method for determining the behavior of materials subjected to shock loading. Specimen supported as a cantilever beam is struck by a weight at the end of a pendulum. Impact strength is determined from the amount of energy required to fracture the specimen. The specimen may be notched or unnotched.

Melt index test: Melt index test measures the rate of extrusion of a thermoplastic material through an orifice of specific length and diameter under prescribed conditions of temperature and pressure. Value is reported in grams per 10 min for specific condition.

Modulus of elasticity (elastic modulus; Young's modulus): The ratio of stress to corresponding strain below the elastic limit of a material.

Monomer: (monomer single-unit) A relatively simple compound that can react to form a polymer (multiunit) by combination with itself or with other similar molecules or compounds.

Molecular weight: The sum of the atomic weights of all atoms in a molecule. In high polymers, the molecular weight of individual molecules varies widely; therefore, they are expressed as weight average or number average molecular weight.

Molecular weight distribution: The relative amount of polymers of different molecular weights that comprise a given specimen of a polymer.

Necking: The localized reduction in cross section that may occur in a material under stress. Necking usually occurs in a test bar during a tensile test.

Notch sensitivity: Measure of reduction in load-carrying ability caused by stress concentration in a specimen. Brittle plastics are more notch sensitive than ductile plastics.

Orientation: The alignment of the crystalline structure in polymeric materials so as to produce a highly uniform structure.

Oxygen index: The minimum concentration of oxygen expressed as a volume percent in a mixture of oxygen and nitrogen that will just support flaming combustion of a material initially at room temperature under the specified conditions.

Peak exothermic temperature: The maximum temperature reached by reacting thermosetting plastic composition is called peak exothermic temperature.

Photoelasticity: Experimental technique for the measurement of stresses and strains in material objects by means of the phenomenon of mechanical birefringence.

Poisson's ratio: Ratio of lateral strain to axial strain in an axial-loaded specimen. It is a constant that relates the modulus of rigidity to Young's modulus.

Polarizer: A medium or a device used to polarize the incoherent light.

Polarized light: Polarized electromagnetic radiation whose frequency is in the optical region.

Polymerization: A chemical reaction in which the molecules of monomers are linked together to form polymers.

Proportional limit: The greatest stress that a material is capable of sustaining without deviation from proportionality of stress and strain (Hooke's Law).

Relative viscosity: Ratio of kinematic viscosity of a specified solution of the polymer to the kinematic viscosity of the pure solvent.

Rheology: The science dealing with the study of material flow.

Rockwell hardness: Index of indentation hardness measured by a steel ball indentor.

Secant modulus: The ratio of total stress to corresponding strain at any specific point on the stress-strain curve.

Shear strength: The maximum load required to shear a specimen in such a volume manner that the resulting pieces are completely clear of each other.

Shear stress: The stress developing in a polymer melt when the layers in a cross section are gliding along each other or along the wall of the channel (in laminar flow).

Shear rate: The overall velocity over the cross section of a channel with which molten or fluid layers are gliding along each other or along the wall in laminar flow.

SPE: Abbreviation for Society of Plastics Engineers.

Specific gravity: The ratio of the weight of the given volume of a material to that of an equal volume of water at a stated temperature.

Spectrophotometer: An instrument that measures transmission or apparent reflectance of visible light as a function of wavelength, permitting accurate analysis of color or accurate comparison of luminous intensities of two sources of specific wavelengths.

Specular gloss: The relative luminous reflectance factor of a specimen at the specular direction.

SPI: Abbreviation for Society of Plastics Industry.

Spiral flow test: A method for determining the flow properties of a plastic material based on the distance it will flow under controlled conditions of pressure and temperature along the path of a spiral cavity using a controlled charge mass.

Strain: The change in length per unit of original length, usually expressed in percent.

Stress: The ratio of applied load of the original cross-sectional area expressed in pounds per square inch.

Stress-strain diagram: Graph of stress as a function of strain. It is constructed from the data obtained in any mechanical test where a load is applied to a material and continuous measurements of stress and strain are made simultaneously.

Stress optical sensitivity: The ability of materials to exhibit double refraction of light when placed under stress.

Stress concentration: The magnification of the level of applied stress in the region of a notch, crack, void, inclusion, or other stress risers.

Stress relaxation: The gradual decrease in stress with time under a constant deformation (strain).

Tensile strength: Ultimate strength of a material subjected to tensile loading.

Tensile impact energy: The energy required to break a plastic specimen in tension by a single swing of a calibrated pendulum.

Thermogravimetric analysis (TGA): A testing procedure in which changes in the weight of a specimen are recorded as the specimen is progressively heated.

Thermoplastic: A class of plastic material that is capable of being repeatedly softened by heating and hardened by cooling. ABS, PVC, polystyrene, polyethylene, etc. are thermoplastic materials.

Thermosetting plastics: A class of plastic materials that will undergo a chemical reaction by the action of heat, pressure, catalysts, etc. leading to a relatively infusible, nonreversible state. Phenolics, epoxies, and alkyds are examples of typical thermosetting plastics.

Thermal conductivity: The ability of a material to conduct heat. The coefficient of thermal conductivity is expressed as the quantity of heat that passes through a unit cube of the substance in a given unit of time when the difference in temperature of the two faces is $1°$.

Thermomechanical analysis (TMA): A thermal analysis technique consisting of measuring physical expansion or contraction of a material or changes in its modulus or viscosity as a function of temperature.

Torsion: Stress caused by twisting a material.

Torsion pendulum: Equipment used for determining dynamic mechanical plastics.

Toughness: The extent to which a material absorbs energy without fracture. The area under a stress-strain diagram is also a measure of toughness of a material.

Tristimulus colorimeter: The instrument for color measurement based on spectral tristimulus values. Such an instrument measures color in terms of three primary colors: red, green, and blue.

Ultraviolet: The region of the electromagnetic spectrum between the violet end of visible light and the X-ray region, including wavelengths from 100 to 3900 Å. Photons of radiation in the UV area have sufficient energy to initiate some chemical reactions and to degrade some plastics.

Ultrasonic testing: A nondestructive testing technique for detecting flaws in material and measuring thickness based on the use of ultrasonic frequencies.

Vicat softening point: The temperature at which a flat-ended needle of 1 mm^2 circular or square cross section will penetrate a thermoplastic specimen to a depth of 1 mm under a specified load using a uniform rate of temperature rise.

Viscosity: A measure of resistance of flow due to internal friction when one layer of fluid is caused to move in relationship to another layer.

Viscometer: An instrument used for measuring the viscosity and flow properties of fluids.

Water absorption: The amount of water absorbed by a polymer when immersed in water for a stipulated period of time.

Weatherometer: An instrument used for studying the effect of weather on plastics in an accelerated manner using artificial light sources and simulated weather conditions.

Weathering: A term encompassing exposure of polymers to solar or ultraviolet light, temperature, oxygen, humidity, snow, wind, pollution, etc.

Yellowness index: Measure of the tendency of plastics to turn yellow upon long-term exposure to light.

Yield point: Stress at which strain increases without an accompanying increase in stress.

Yield strength: The stress at which a material exhibits a specified limiting deviation from the proportionality of stress to strain. Unless otherwise specified, this stress will be the stress at the yield point.

Young's modulus: The ratio of tensile stress to tensile strain below the proportional limit.

Appendix C

DESCRIPTION OF PROFESSIONAL AND TESTING ORGANIZATIONS

AMERICAN NATIONAL STANDARDS INSTITUTE (ANSI)

ANSI is a federation of standards competents from commerce and industry, professional, trade, consumer, and labor organizations and government. ANSI helps to perform the following:

- Identifies the needs for standards and sets priorities for their completion.
- Assigns development work to competent and willing organizations.
- Sees to it that public interests, including those of the consumer, are protected and represented.
- Supplies standards writing organizations with effective procedures and management services to ensure efficient use of their manpower and financial resources and timely development of standards.
- Follows up to assure that needed standards are developed on time.

Another role is to approve standards as American National Standards when they meet consensus requirements. It approves a standard only when it has verified evidence presented by a standards developer that those affected by the standard have reached substantial agreement on its provisions. ANSI's other major roles are to represent U.S. interest in nongovernmental international standards work, to make national and international standards available, and to inform the public.

AMERICAN SOCIETY FOR TESTING AND MATERIALS (ASTM)

ASTM is a scientific and technical organization formed for "the development of standards on characteristics and performance of materials, products, systems and services and the promotion of related knowledge." ASTM is the world's largest source of voluntary consensus standards. The society operates through more than 135 main technical committees with 1550 subcommittees. These committees function in prescribed fields under regulations that ensure balanced representation among producers, users, and general interest participants. The society currently has 28,000 active members, of whom approximately 17,000 serve as technical experts on committees, representing 76,200 units of participation.

Membership in the society is open to all concerned with the fields in which ASTM is active. An ASTM standard represents a common viewpoint of those parties concerned with its provisions, namely, producers, users, and general interest groups. It is intended to aid industry, government agencies, and the general public. The use of an ASTM standard is voluntary. It is recognized that for certain work, ASTM specifications may be either more or less restrictive than needed. The existence of an ASTM standard does not preclude anyone from manufacturing, marketing, or purchasing products or using products, processes, or procedures not conforming to the standard. Because ASTM standards are subject to periodic reviews and revision, it is recommended that all serious users obtain the latest revision. A new edition of the Book of Standards is issued annually. On the average about 30% of each part is new or revised.

FOOD AND DRUG ADMINISTRATION (FDA)

The FDA is a U.S. government agency of the Department of Health and Human Services. The FDA's activities are directed toward protecting the health of the nation against impure and unsafe foods, drugs, cosmetics, and other potential hazards.

The plastics industry is mainly concerned with the Bureau of Foods which conducts research and develops standards on the composition, quality, nutrition, and safety of foods, food additives, colors, and cosmetics and conducts research designed to improve the detection, prevention, and control of contamination. The FDA is concerned about indirect additives. Indirect additives are those substances capable of migrating into food from contacting plastic materials. Extensive tests are carried out by the FDA before issuing safety clearance to any plastic material that is to be used in food contact applications. Plastics used in medical devices are tested with extreme caution by the FDA's Bureau of Medical Devices which develops FDA policy regarding safety and effectiveness of medical devices.

NATIONAL BUREAU OF STANDARDS (NBS)

The overall goal of the NBS is to strengthen and advance the nation's science and technology and to facilitate their effective application of public benefit.

The bureau conducts research and provides a basis for the nation's physical measurement system, scientific and technological services for industry and government, a technical basis for increasing productivity and innovation, promoting international competitiveness in American industry, maintaining equity in trade, and technical services, promoting public safety. The Bureau's technical work is performed by the National Measurement Laboratory, the National Engineering Laboratory, and the Institute for Computer Sciences and Technology.

NATIONAL ELECTRICAL MANUFACTURERS ASSOCIATION (NEMA)

NEMA consists of manufacturers of equipment and apparatus for the generation, transmission, distribution, and utilization of electric power. The membership is limited to corporations, firms, and individuals actively engaged in the manufacture of products included within the product scope of NEMA product subdivisions.

NEMA develops product standards covering such matters as nomenclature, ratings, performance, testing, and dimensions. NEMA is also actively involved in developing National Electrical Safety Codes and advocating their acceptance by state and local authorities. Along with a monthly news bulletin, NEMA also publishes manuals, guidebooks, and other material on wiring, installation of equipment, lighting, and standards. The majority of NEMA standardization activity is in cooperation with other national organizations. The manufacturers of wires and cables, insulating materials, conduits, ducts, and fittings are required to adhere to NEMA standards by state and local authorities.

NATIONAL FIRE PROTECTION ASSOCIATION (NFPA)

The NFPA has the objective of developing, publishing, and disseminating standards intended to minimize the possibility and effect of fire and explosion. NFPA's membership consists of individuals from business and industry, fire service, health care, insurance, educational, and government institutions. NFPA conducts fire safety education programs for the general public and provides information on fire protection and prevention. Also provided by the association is the field service by specialists on flammable liquids, electricity, gases, and marine problems.

Each year, statistics on causes and occupancies of fires and deaths resulting from fire are compiled and published. NFPA sponsors seminars on the Life Safety Codes, National Electrical Code, industrial fire protection, hazardous materials, transportation emergencies, and other related topics. NFPA also conducts research programs on delivery systems for public fire protection, arson, residential fire sprinkler systems, and other subjects. NFPA publications include *National Fire Codes Annual*, *Fire Protection Handbook*, *Fire Journal*, and *Fire Technology*.

NATIONAL SANITATION FOUNDATION (NSF)

The NSF is an independent, nonprofit environmental organization of scientists, engineers, technicians, educators, and analysts. NSF frequently serves as a trusted neutral agency for government, industry, and consumers, helping them to resolve differences and unite in achieving solutions to problems of the environment.

At NSF, a great deal of work is done on the development and implementation of NSF standards and criteria for health-related equipment. The majority of NSF standards relate to water treatment and purification equipment, products for swimming pool applications, plastic pipe for potable water as well as drain, waste, and vent (DWV) uses, plumbing components for mobil homes and recreational vehicles, laboratory furniture, hospital cabinets, polyethylene refuse bags and containers, aerobic waste treatment plants, and other products related to environmental quality.

Manufacturers of equipment, materials, and products that conform to NSF standards are included in official listings and these producers are authorized to place the NSF seal on their products. Representatives from NSF regularly visit the plants of manufacturers to make certain that products bearing the NSF seal do fulfill applicable NSF standards.

PLASTICS TECHNICAL EVALUATION CENTER (PLASTEC)

PLASTEC is one of 20 information analysis centers sponsored by the Department of Defense to provide the defense community with a variety of technical information services applicable to plastics, adhesives, and organic matrix composites. For the last 21 years, PLASTEC has served the defense community with authoritative information and advice in such forms as engineering assistance, responses to technical inquiries, special investigations, field trouble shooting, failure analysis, literature searches, state-of-the-art reports, data compilations, and handbooks. PLASTEC has also been heavily involved in standardization activities. In recent years, PLASTEC has been permitted to serve private industry.

The significant difference between a library and technical evaluation center is the quality of the information provided to the user. PLASTEC uses its database library as a means to an end to provide succinct and timely information which has been carefully evaluated and analyzed. Examples of the activity include recommendation of materials, counseling on designs, and performing trade-off studies between various materials, performance requirements, and costs. Applications are examined consistent with current manufacturing capabilities, and the market availability of new and old materials alike is considered. PLASTEC specialists can reduce raw data to the user's specifications and supplement them with unpublished information that updates and refines published data. PLASTEC works to spin-off the results of government-sponsored R & D to industry and similarly to utilize commercial advancements to the government's goal of highly sought technology transfer. PLASTEC has a highly specialized library to serve the varied needs of their own staff and customers.

PLASTEC offers a great deal of information and assistance to the design engineer in the area of specifications and standards on plastics. PLASTEC has a complete visual search microfilm file and can display and print the latest issues of specifications, test methods, and standards from Great Britain, Germany, Japan, U.S., and International Standards Organization. Military and Federal specifications and standards and industry standards such as ASTM, NEMA, and UL are on file and can be quickly retrieved.

SOCIETY OF PLASTICS ENGINEERS (SPE)

The SPE promotes scientific and engineering knowledge relating to plastics. SPE is a professional society of plastics scientists, engineers, educators, students, and others interested in the design, development, production, and utilization of plastics materials, products, and equipment. SPE currently has over 22,000 members scattered among its 80 sections. The individual sections as well as the SPE main body arrange and conduct monthly meetings, conferences, educational seminars, and plant tours throughout the year. SPE also publishes *Plastics Engineering, Polymer Engineering and Science, Plastics Composites,* and the *Journal of Vinyl Technology.* The society presents a number of awards each year encompassing all levels of the organization: section, division, committee, and international. SPE divisions of interest are color and appearance, injection molding, extrusion, electrical and electronics, thermoforming, engineering properties and structure, vinyl plastics, blow molding, medical plastics, plastics in building, decorating, mold making, and mold design.

SOCIETY OF PLASTICS INDUSTRY (SPI)

The SPI is a major society, whose membership consists of manufacturers and processors of plastics materials and equipment. The society has four major operating units consisting of the Eastern Section, the Midwest Section, the New England Section, and the Western Section. SPI's Public Affairs Committee concentrates on coordinating and managing the response of the plastics industry to issues like toxicology, combustibility, solid waste, and energy. The Plastic Pipe Institute is one of the most active divisions, promoting the proper use of plastic pipes by establishing standards, test procedures, and specifications. Epoxy Resin Formulators Division has published over 30 test procedures and technical specifications. Risk management, safety standards, productivity, and quality are a few of the major programs undertaken by the machinery division. SPI's other divisions include Expanded Polystyrene Division, Fluoropolymers Division, Furniture Division, International Division, Plastic Bottle Institute, Machinery Division. Molders Division, Mold Makers Division, Plastic Beverage Container Division, Plastic Packaging Strategy Group, Polymeric Materials Producers Division, Polyurethane Division, Reinforced Plastic/Composites Institute, Structural Foam Division, Vinyl Siding Institute, and Vinyl Formulators Division.

The National Plastics Exposition and Conference, held every 3 years by the Society of Plastic Industry, is one of the largest plastic shows in the world.

UNDERWRITERS LABORATORIES (UL)

UL is a not-for-profit organization whose goods are to establish, maintain, and operate laboratories for the investigation of materials, devices, products equipment, constructions, methods, and systems with respect to hazards affecting life and property.

There are five testing facilities in the U.S. and over 200 inspection centers. More than 700 engineers and 500 inspectors conduct tests and follow-up investigations to insure that potential hazards are evaluated and proper safeguards provided. UL has six basic services it offers to manufacturers, inspection authorities, or government officials. These are product listing service, classification, service, component recognition service, certificate service, inspection service, and fact finding and research.

UL's Electrical Department is in charge of evaluating individual plastics and other products using plastics as components. The Electrical Department evaluates consumer products such as TV sets, power tools, appliances, and industrial and commercial electrical equipment and components. In order for a plastic material to be recognized by UL it must pass a variety of UL tests including the UL 94 flammability test and the UL 746 series, and short- and long-term property evaluation tests. When a plastic material is granted Recognized Component Status, a yellow card is issued. The card contains precise identification of the material including supplier, product designation, color, and its UL 94 flammability classification at one or more thickness. Also included are many of the property values such as temperature index, hot wire ignition, high-current arc ignition, and arc resistance. These data also appear in the recognized component directory.

UL publishes the names of the companies who have demonstrated the ability to provide a product conforming to the established requirements, upon successful completion of the investigation and after agreement of the terms and conditions of the listing and follow-up service. Listing signifies that production samples of the product have been found to comply with the requirements and that the manufacturer is authorized to use the UL's listing mark on the listed products which comply with the requirements.

UL's consumer advisory council was formed to advise UL in establishing levels of safety for consumer products to provide UL with additional user field experience and failure information in the field of product safety and to aid in educating the general public in the limitations and safe use of specific consumer products.

Chapter 2

CHEMICAL CHARACTERIZATION IN POLYMER ANALYSIS

A. B. Hunter

TABLE OF CONTENTS

0-8493-4401-8/93/$0.00 + $.50

I. INTRODUCTION

The use of chemical characterization (CC) techniques is becoming more prevalent throughout industry. They are referenced in Mil-Specifications as well as Test Standards literature. Its popularity stems from the many various practical applications associated with CC techniques. The methodology can be used to fingerprint formulations or, as needed, fully define the formulation. This would include processing as well as stoichiometric information. Other uses would include cure development, cure optimization, cure kinetics, failure analysis, and life cycle kinetics.

CC uses methodology that simply identifies qualitatively and quantitatively the material by its chemistry. This is in contrast with physical characterization where one is evaluating physical properties. Physical characterization includes such methods as tensile strength, compression, and shear strength, whereas CC would include molecular morphology, molecular weight, degree of crystalinity, and other such properties. In theory, the CC data should correlate with physical property data, but in many cases this is difficult to prove. However, some work has been done in this area that is very useful in material analysis and the control of properties.[1] Many different techniques are used in CC; the most commonly used will be discussed later.

II. SURVEY OF CHEMICAL CHARACTERIZATION METHODOLOGY

Figure 1 shows a general schematic used for CC methodology. Although other methods are used, the ones presented seem to be the most popular.

Thermal analysis identifies the methods with the following acronyms:

- FTS: Fourier transform spectroscopy — although this particular method is also identified under Spectroscopy, it is used here in conjunction with thermal scanning to characterize transitions which may occur at identifiable temperatures.
- DSC: differential scanning calorimeter
- TGA: thermal gravimetric analysis
- DMA: dynamic mechanical analysis

Spectroscopy identifies the methods with the following acronyms:

- FTIR: Fourier transform infrared analysis
- RAMAN: Raman spectroscopy
- GC/MS: Gas chromatography interfaced with mass spectroscopy

Inorganic Analysis identifies the methods with the following acronyms:

- X-ray: X-ray diffraction analysis
- MP: microprobe analysis
- EDX: energy-dispersive X-ray
- ES: emission spectroscopy

Microscopy would include the use of heated cells as well as polarized light. Solubility evaluation involves trying various solvents in a systematic way with the insolubles being evaluated by the inorganic analysis methods, and the solubles by chromatography methods. The components are then collected and evaluated by various means such as FTIR and MS.

There are several operational levels at which CC may operate. The most intense level

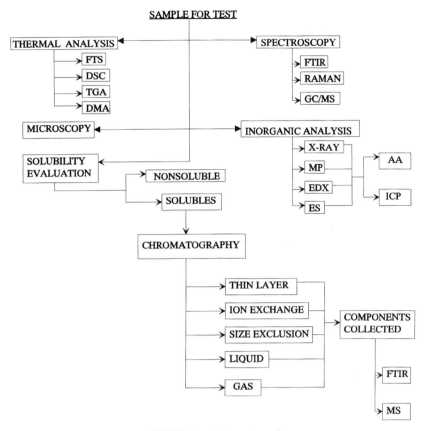

FIGURE 1. Testing schematic.

would be a full characterization which would involve complete identification of all components in the material both from a qualitative and quantitative aspect. This would include a measurement of any degree of staging that may be present in the polymer. A reduced level would involve fingerprinting the material only and using the fingerprint as a standard reference. In reality, because of costs involved, the usual method is to use a combination of both of the above methods. This includes some kind of judgment factor of identifying the key components that should be fully identified. A recent implementation includes the use of statistical process control techniques to monitor the data output for quality assurance purposes.

CC may be used for many different applications. The following are a few of the most common:

1. Quality assurance purposes in product control
2. Understanding reaction mechanisms that may be taking place
3. Cure monitoring and optimization
4. Efficiency of a manufacturing process
5. Efficiency of a scaling-up operation for manufacturing
6. Understanding degradation mechanisms that may be taking place in the material
7. Developing reaction rates
8. Developing degradation rates of the polymers

Each of these could be a full treatise in itself, but because of limitations will only be mentioned here.

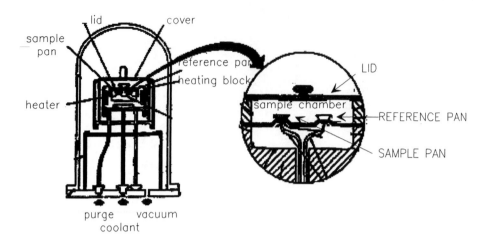

FIGURE 2.   Differential scanning calorimeter cell. (Courtesy of TA, Inc.)

III. TECHNIQUES DESCRIBED

A. THERMAL ANALYSIS

Thermal Analysis covers techniques that measure some property through a temperature range. There are many good textbooks available that cover the subject in depth.[2,3] These should be consulted as needed, since it is beyond the scope of this chapter to cover that kind of detail. All of these methods have in common a cell for the sample that is capable of isothermal or programmable temperature control. In addition, some characteristic is measured throughout the temperature excursion. Referring to Figure 1, FTS operating in this capacity has a controlled temperature sample chamber. Some chambers have pressure control as well. The chamber may be designed in such a way that it has the ability to perform in either transmittance or reflectance mode. Infrared scans are taken periodically through the temperature scan. As the temperature is scanned, it is best to leave the sample in the chamber to allow the subsequent infrared scans to be compared quantitatively. This assures the analysis to be performed on the same location of the sample for all runs. These comparisons can be made fairly accurately as long as the sample thickness remains constant and is not affected by flow with the heat. This analysis is usually done with a film on a salt block or a pellet. If it is done in solution, one need not be concerned about flow or other similar problems. Whether it is done in solution or as a film on a salt block, the data are most useful when done in a relative quantitative manner. Another concern is to make sure the salt block does not affect the reaction kinetics of the material being measured. One can easily make wrong assumptions if the data are used to project how the material in question will react under manufacturing or laboratory conditions when the substrate used in the analysis has any effect whatsoever on the material. By making quantitative measurements of the material throughout the temperature range of interest, one can determine reactant consumptions and perform kinetic studies. In addition, this becomes a tool to aid the interpretation of transitions detected by other thermal analytical tools, such as differential scanning calorimetry. The most benefit is realized from these methods as they work in a symbiotic manner.

DSC and differential thermal analysis (DTA) are very similar. The main features of a DSC cell are shown in Figure 2. The basic features of the two methods are the same. Both methods detect thermal transitions that may be taking place with the sample as compared to an inert reference material. In the case of DSC the sample is placed in the sample pan and an inert substance placed in the reference pan. Because of cell configuration, the DSC

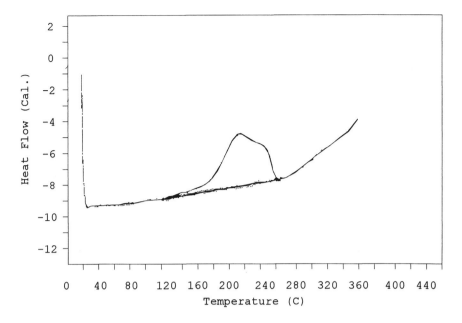

FIGURE 3. DSC analysis of a typical epoxy resin (heat of reaction 182 J/G).

is often used in a quantitative manner, whereas the DTA is used to obtain an accurate measurement of transition temperatures. These measurements can be made from liquid nitrogen temperatures up to about 2000°C for DTA and in the case of DSC the upper temperature is about 800°C. The transition being measured in both cases is the temperature at which there is any abrupt change in the specific heat of the material. In the case of DSC, if the transition is exothermic or endothermic, the measurement can be calculated in terms of calories per mole; for DTA the usual readout would be the temperature or time of transition. Figure 3 is a typical scan of an epoxy resin showing an exotherm beginning at about 120°C and ending at 260°C. This was done at a heat rate of 10°C/min. Most of the present-day equipment has software to calculate the "heat of polymerization" for this type of run. If software is not available, the details of the calculation are available in several good reference books,[2,3] as well as the manufacturers' literature. In the example given the heat of polymerization calculates out to be 182 J/g. (Note: Most software has the capability of calculating out in either calories per mole, or joules per gram.) Because of the quantitative capability of DSC it is probably the most popular. Some of the applications this method is used in are

Melting point
Glass transition
Degree of cure
Reaction kinetics
Oxidation characteristics
Heat of fusion
Heat of polymerization
Degree of crystallinity
Degree of cure
Specific heat
Thermal conductivity

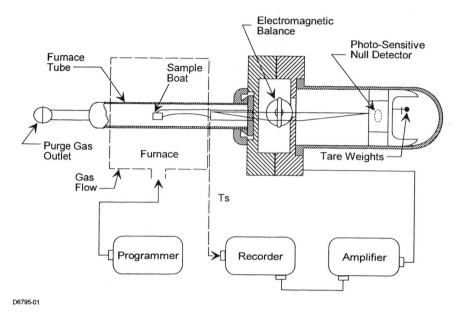

FIGURE 4. Thermal gravimetric analysis cell. (Courtesy of TA, Inc.)

The analysis can also be run under various pressures or vacuum, which enables one to duplicate production techniques.

Several precautions need to be heeded by the thermal analyst. (1) The reference sample must be inert over the temperature range of investigation. (2) Cleanliness of the sample and reference pan is a must in order to prevent false transitions from occurring. (3) The sample size used is so small that any contaminant introduced into the sample may cause a calculated result to be totally erroneous. (4) If the sample is a composite, one must realize the source of the transition that is occurring in order to make proper calculations.

TGA measures sample weight change that occurs with temperature or time. This may be a gain or loss in weight. A typical range of operation is from room temperature to 1000°C, although other ranges are available from instrument manufacturers or by user modification of the equipment. Figure 4 shows a schematic of TGA instrumentation. Instrumentation is available to run simultaneous TGA and DSC. Also, present-day software is capable of determining the derivative of the TGA curve as well as the weight loss curve itself. Typical applications for this type of analysis are

> Compositional analysis
> Thermal stability
> Environmental effect analysis
> Degradation kinetics
> Reaction analysis
> Oxidative analysis

In addition, the TGA equipment can be directly interfaced with a mass spectrometer or an infrared analyzer for identification of the outgassing or reaction products coming off the sample during analysis.

DMA submits the sample to some kind of cyclic deformation over a temperature or time scan. Various deformations may be used such as compressive, flexure, tension, torsion, or bending. The literature adequately describes these various methods.[4-9] The frequency of deformation may be fixed or resonant. Some instrumentation is designed to operate under either condition. A typical commercially available instrument is described in Figure 5. This

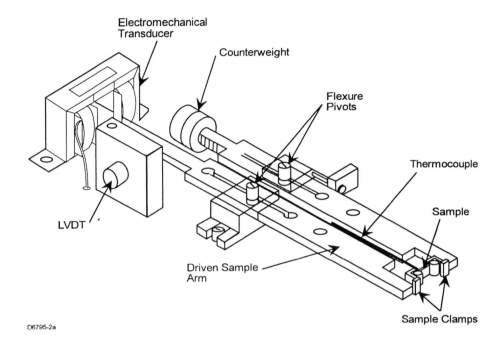

FIGURE 5. Dynamic mechanical analysis cell. (Courtesy of TA, Inc.)

technique places the sample under a sinusoidal load which results in some sinusoidal deformation occurring in the material. Calculations can then be made to determine various modulus properties. This technique is very sensitive to transitions which may be occurring in the material through the temperature or time scan. For example, the glass transition temperature of material can often be detected by DMA, whereas it may not be discernable by DSC. Software is available from the various instrument manufacturers for determining creep analysis and other viscoelastic properties of materials. A typical scan of an epoxy preimpregnated material (Figure 6) identifies its processing range and cure window. The mechanical behavior of this type of material is usually described as viscoelastic. A sinusoidal stress introduced into the material results in some sinusoidal deformation response. This output has two aspects, one referred to as dynamic storage modulus, the other as dynamic loss modulus. The dynamic storage modulus is an indication of the energy stored and recovered during each cycle. It gives a good indication of the stiffness of the material. For example, in Figure 6, although the output has not been calculated in these terms, the storage modulus curve would be similar to the frequency curve. A degradation in the storage modulus curve becomes a good indicator of mechanical property loss occurring in the system with temperature. The other factor, loss modulus (sometimes referred to as damping factor or loss factor), is related to energy dissipation in the system per cycle. It is a good indicator of impact strength or the ability of the system to absorb energy. The damping curve is similar to the loss modulus curve of Figure 6. The instrumentation giving the output for Figure 6 operates in the mode of shear moduli or flexure moduli. The shear loss moduli is usually represented by G'', whereas the shear storage moduli would be G'. Another factor called the loss angle is reported as tan δ and is determined from G''/G'. Other instrumentation may operate in other modes, but the principle remains the same.

B. SPECTROSCOPY

Whereas thermal analysis techniques identify various transitions the material may experience with time and temperature, spectroscopy techniques are used to characterize

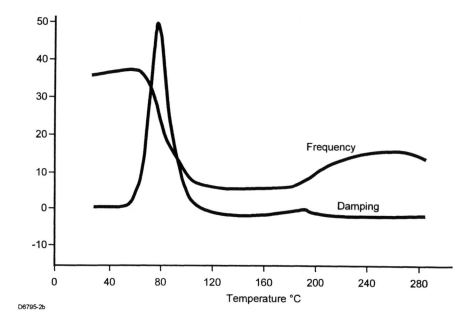

FIGURE 6. Typical DMA scan of epoxy prepreg.

qualitatively and quantitatively the organic portion of the material. The operation may require separation from fillers and inorganic components before analysis. This may be accomplished by solvent action, centrifugation, or a combination of both. The prime spectroscopy candidates used for CC are infrared analysis (IR), Raman, and GC/MS. Although other techniques are used, the above-listed ones are the most common.

IR (in some cases referred to as FTIR) is a good technique for identifying organic materials. Although, it can also be used to characterize inorganic structures, it is not as definitive for inorganics as it is for organics. Many good reference works are available to aid the analyst in sample preparation as well as identification of materials.[10-13] IR measures the ability of material to transmit, absorb, or reflect infrared energy. An FTIR instrument has some advantage over a conventional grating or prism-type infrared instrument. The practical result is faster analysis time with increased sensitivity over conventional instrumentation. The infrared spectrum is normally recorded on chart paper as a graph with transmittance or absorbance intensity on the y-axis and wavelength or wavenumber on the x-axis. Not all materials are infrared active. To interact with infrared energy the material must have some functional groups that absorb infrared energy by rotating or vibrating about an axis. The sample may be in the form of solid, liquid, or gas. Sample cells are available from the various manufacturers for the various types of analysis. One must make sure that the sample blocks or windows that are used are not opaque to the infrared wavelength range used in the analysis. Here again, manufacturers' literature usually are very descriptive of the ranges used with the various windows. The normal range of infrared energy of operation is what is called the middle range of 4000 to 200 wave numbers. Some work has been done in the far infrared region, 200 to 10 wave numbers, and the near infrared region of 12,800 to 4000. Some functional groups are active in the near and far region, but the preponderance of activity is in the middle range. Identification of wavelength activity with functional groups is listed in several good sources.[10,11]

Raman spectroscopy is also useful, although not nearly as popular as infrared spectroscopy. In Raman spectroscopy, measurements are made of the scattering effect of photons colliding with the molecules of the sample. The vibrational or rotational energy of the molecule is changed by some increment in the collision. This increment may be a gain or loss in energy. In either case the change gives rise to the lines that make up the Raman spectrum. Infrared and Raman spectroscopy are not duplicates of each other, even though both involve vibrational and rotational energy levels at the molecular level. They do however compliment each other because the intensity of the bands may be completely different from one method to the other, depending on how the energy of impact transfers at the molecular level. Some bands that are barely discernible by one method may be quite pronounced by the other. Therein lies Ramans usefulness along with infrared spectroscopy.

GC/MS is a combined method that has become a workhorse of the laboratory. Laboratories usually do not receive pure samples, but often mixtures of materials, for analysis. Mass spectrometry is probably one of the best laboratory tools for qualitative analysis. The spectrums produced however can be very complex and difficult to interpret if the sample is very complex. The advent of the computer for doing library searches has been a real boon to the mass spectroscopist in saving man hours of laboriously going over spectrums and identifying samples. Various literature is available for the analyst who still desires to do this kind of work or for the one who wants to understand the rules of interpretation.[15-17] In addition to the computer interface the other advancement in mass spectroscopy is the interfacing of gas chromatography with mass spectroscopy. The chromatograph has allowed the analyst to use the instrumentation that does the best job separating materials into its component parts in conjunction with the mass spectrometer for identifying compounds. Recent advances have even done better by interfacing liquid chromatographs with the mass spectrometer. The use of the liquid chromatograph has expanded the range of materials that can be analyzed by this combined technique considerably. Another recent development has allowed interfacing TGA with mass spectrometry for even further expansion of the capability. These combined techniques will be discussed later in this chapter.

C. MICROSCOPY

Microscopy has become a useful tool in CC for several reasons. The most obvious, of course, is just observing the material in question under magnification. This is especially true for composite-type materials or materials that may have a certain cell structure that may be of interest. Another valuable application is the use of polarized light to identify different components or phases of the sample. One of the most valuable uses though is the controlled heated hot stage. Being able to visually observe the material as it goes through a temperature regime sometimes gives valuable information for interpreting data obtained by thermal analysis. For some materials, crystal formation has been observed during cool down of the sample. Although microscopy is sometime the forgotten tool of characterization, it can provide a wealth of information, many times without a large expenditure of manpower.

D. INORGANIC ANALYSIS

Many of the materials presented for CC have inorganic fillers. Not only is the identification of these fillers important, but also are the size, shape, and distribution. Microscopic methods can aid in these determinations. The actual identification is best done by instrumentation such as X-ray diffraction, microprobe analysis (MP), energy dispersive X-ray (EDX), and emission spectroscopy (ES). Of these methods, MP, EDX, and ES are used for positive identification of the elements present. They do not identify the form that is present. This analysis may be performed on the material as is or on the nonsolubles from the solubility evaluation. Once the elements are identified a quantitative determination may be made by atomic absorption (AA), inductively coupled plasma (ICP), or some other such tool. Although quantitative analysis can be done by MP, EDX, and ES with varying degree of success and difficulty, the most accurate would be AA or ICP. Once the elements are identified quantitatively, the sample may be submitted for X-ray diffraction analysis (X-ray) for positive identification of the compound present. X-ray will only identify the crystalline portion. Although it can detect the presence of amorphous material, it will not identify the compound. Another useful part of this analysis is that it can identify phases and forms as well. One problem with this technique is that, unless some preparation is used to separate the components, the diffraction pattern will represent a composite of all the crystalline components present. This confuses the computerized library searches that are available to aid in identification. The library searches may be time savers on pure compounds but at the present time there is no substitute for an experienced diffractionist in the laboratory.

E. SOLUBILITY EVALUATION

Solubility evaluation is not only a means of preparing the sample for other analytical tests, but by itself gives important characteristics concerning the polymer. The obvious information, of course, is resistance of the polymer to various solvents or liquids. A not-so-obvious characteristic is the rate of solvation. The rate may be evaluated at different temperatures. The rate of solvation gives comparative information concerning molecular size and morphology. The nonsoluble portion should be submitted to other analysis; the soluble portion is often submitted to chromatography analysis.

Thin layer chromatography, although not as popular as other chromatography techniques, does give important information that sometimes is not available by other means.[21] One advantage of thin layer chromatography is that exploratory separations can be performed that are very similar to liquid chromatography without endangering an expensive column. Reverse-phase or absorption chromatography can be evaluated fairly fast with minimum expense. Once solvents and methods are optimized the method can be transferred to the liquid chromatograph for fine tuning. Samples may be collected from the thin layer chromatograph plates for analysis by FTIR or other techniques. Commercially available equipment can be purchased from most scientific supply houses that enable one to use various ingenious techniques for making simultaneous multiruns as well as sampling the fractions.

Ion-exchange chromatography is commonly used to characterize polymer systems. This methodology quantitatively analyzes anion or cation content. In order for this analysis to be successful the detectable components must be in ionic form in the mobile phase. Many times this is not the case for polymeric materials. However, these materials may still be analyzed by destroying the sample in a Parr bomb, for example, which enables one to convert the components of interest into ionic form. In doing this one must realize the method does not differentiate between the different forms that may be present. For example, if chlorine is present in different forms in the original sample, after the Parr bomb activity it will all be seen as chloride ion. One must be careful to work out the chemistry of what has happened before making too many assumptions concerning the data output.

Size exclusion chromatography is an extremely important but often underused characterization tool. Its importance lies in the fact that one obtains molecular weight distributions. There are several good texts on this subject[18,19,22,23] which one can review. It is beyond the scope of this chapter to go into depth. A brief review of its significance is appropriate though. The molecular weight distribution is important because it is a distribution and not a single number. The synthesizing process of making polymers produces some statistical distribution of monomers. This statistical distribution has a tremendous effect on properties. Therefore, it is important that the distribution remains as constant as possible to assure predictable and constant properties of the final product. Size exclusion chromatography, sometimes referred to as gel permeation chromatography, gives a measure of this distribution. It is important to note, however, that with this method, even though the calculations are in terms of molecular weight, the chromatography method actually separates by molecular size. The reason for this is the column used has a certain pore size. As the sample in solution moves through the column, the various size molecules will move in and out of the pore structure. Thus, the larger molecules will move through the column at a faster rate than the smaller ones, and elute sooner, simply because it does not have as many "open spaces" to explore as it moves through the column. In calculating the molecular weights by this method, one usually runs molecular weight standards and makes the determination by using various methods of comparing the sample distribution to the known molecular weight.[19] One obvious problem with this methodology is that, in order to have accurate numbers, one must use standards of the same morphology. Standard molecular weight materials are often not available that would duplicate the material being tested. Therefore, the numbers that are calculated using this method are relative and not absolute. There are several molecular weight numbers that may be determined. For example, there is number average molecular weight (M_n), weight average molecular weight (M_w), viscosity average molecular weight (M_v), and Z-average molecular weight (M_z). The difference in these numbers lies essentially in the manner in which the molecules are counted. For example, M_n is determined by literally counting the number of molecules of each size that are present. Any method that would determine this, such as end group analysis or vapor pressure osmometry, is capable of calculating M_n. To put this in other terms, the M_n in grams would contain Avogadro's number of molecules. M_w on the other hand takes into account the actual weight of each molecule present. Light-scattering techniques, for example, can be used to determine M_w. Vapor pressure osmometry and light-scattering techniques are primary methods of determining molecular weights, since the numbers determined are absolute and not relative to some reference comparison. A comparison of the molecular weight numbers would be as follows (see Figure 7):

$$M_n < M_v < M_w < M_z$$

Various properties are dependent on these molecular weight numbers. For example, many thermodynamic properties are related to M_n. There is also an effect on some mechanical properties such as tensile strength. Bulk properties such as melt viscosity are effected by M_w. Many mechanical properties are drastically effected by M_w. Viscoelastic properties are often effected by M_z. Another consideration is the molecular weight distribution. The distribution is graphically displayed by the chromatograph data. The distribution number is actually a ratio, calculated by M_w/M_n or M_z/M_w. Of these two ratios the M_w/M_n is the most commonly used. A general rule is that narrow distributions give improved mechanical properties but rather poor processing properties. Wider distributions usually have a good effect on processing properties.

Viscosity molecular weight is another important parameter of polymers.[24-26] Recent advances allow this to be calculated from on-line detectors for the liquid chromatograph. This method has an additional advantage in that one can calculate polymer branching from this information as well. Another recent method development is field-flow fractionation (FFF).[27-29] FFF does have certain advantages over normal column chromatography, in that it does not use a packing material. FFF is carried out in an open tube or channel. There are no support particles in the channel and consequently the flow is uniform. However, because of frictional drag at surface boundaries, a difference in velocity occurs across the cross-section of the channel. Externally generated fields are applied across the channel perpendicular to the flow. The applied field is of such a nature that it interacts with the components of the test sample, which is injected in a normal fashion to that of liquid chromatography, causing the components to be driven toward one of the surfaces. The interaction of the components with the applied force field and the channel surface is the basis for the separation.

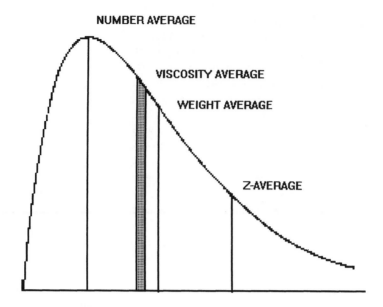

FIGURE 7. Molecular weight distribution.

Liquid chromatography involves several different modes, all of which are described in the literature.[18,20,30-32] Some of the more common modes are liquid-liquid chromatography (LLC), liquid-solid chromatography (LSC), and ion-pairing chromatography. LLC, sometimes referred to as partition chromatography, depends on the partitioning of the solute between two immiscible components, one of which is the stationary phase, the other being the mobile phase. In normal-phase LLC the stationary bed is strongly polar and the mobile phase is nonpolar. Thus, the components of the sample that are more polar are retained on the column longer than the less-polar ones. Reverse-phase LLC is the reverse mechanism, the stationary bed is nonpolar, and mobile phase is more polar. In this case the more nonpolar materials are retained on the column longer. LSC, on the other hand, depends on the mechanism of the stationary phase being an absorbent and the separation depending on repeated adsorption-desorption steps. LSC is sometimes referred to as adsorption chromatography. LLC is commonly used for polymer analysis, whereas LSC is a good separator of isomers. Because of the nature of column used in size exclusion chromatography, as discussed previously, there may be some crossover effect depending on the solvent used. For example, if one uses a gel permeation chromatograph column to determine molecular weight size, there may be a partitioning effect taking place. Alternatively, if one uses a silica type-column, one can obtain an adsorption effect. This explains some strange effects that one sees at times when performing molecular weight distribution analysis. Sometimes elution times for components can be reversed, depending on the solvent used. One needs to minimize this effect as much as possible when performing size exclusion or gel permeation chromatography. Ion-pairing chromatography is a combination of ionic interaction and partition mechanism.

GC[17,33-35] is commonly used in the laboratory for the analysis of solvents, gases, or other materials that are volatile below 150°C. This method is very similar to liquid chromatography except that here the sample moves through the column in a gaseous phase rather than a liquid. Like liquid chromatography, GC utilizes various columns and detectors. With proper standards it becomes a good quantitative technique. In itself it has no way of identifying the components eluting except by time of elution. Therefore, when combined with the mass spectrometer it becomes a most powerful tool for analyzing both qualitatively and quantitatively various samples. It can also be interfaced with an infrared analyzer. In essence the infrared analyzer or mass spectrometer simply becomes a detector for the gas chromatograph. Similarly, the liquid chromatograph can be interfaced with a mass spectrometer or infrared analyzer. These combined techniques supply the laboratory with powerful tools to qualitatively or quantitatively fingerprint materials.

Other miscellaneous methods used include

1. Ultraviolet-visible spectroscopy (UV) is useful to determine the photodegradation of polymers. Photodegradation is often caused by dissociation of bonds within the polymer from interaction with sunlight, mainly the UV portion. Therefore, UV spectroscopy becomes a natural for measuring that effect. Kinetic rates can be developed from this type of experimentation. This technique may also be useful in evaluating certain additives in the polymer which by their very nature are responsive to sunlight. The UV method can be used qualitatively or quantitatively.

2. Nuclear magnetic resonance (NMR) spectroscopy also has value in characterizing polymers. Recent advances in the instrumentation such as ''magic spin technology'' have made it useful for evaluating solid samples. NMR provides a means of evaluating molecular structure at detailed levels. Although solvent limitations and cost of equipment have been a hindrance to wide usage, this methodology has tremendous capability when molecular configuration information is needed. This would include reaction mechanisms, sequences of configuration, chain branching, and other similar type determinations.

3. Electron spin resonance (ESR) spectroscopy is a technique used often to study degradation reactions of polymer. This is especially true if the degradation reaction involves the production of free radicals.

4. Thermal analysis/mass spectroscopy (TA/MS) can be used to characterize reaction product or volatiles that come off materials during processing or curing. Some of this equipment is commercially available such as interfacing a thermal gravimetric analyzer with a mass spectrometer. Other interfaces can be developed in the laboratory. One such device developed in this laboratory is an interface between a high-pressure differential scanning calorimeter and a mass spectrometer. This is useful to evaluate reactions under manufacturing conditions. The thermal gravimetric analyzer enables one to use different environments, whereas the differential scanning calorimeter enables one to look at the effect of varying pressures.

5. Dielectrometry is often used to characterize cure cycles. Different types of instrumentation are available. All of them operate on the same principle, that is, the sample simply becomes a part of a capacitance circuit. It may be operated at a single frequency or sweeping frequencies. Usually for polymeric-type materials the frequency range would be 1000 Hz on down. The capacitance bridge is in a controlled heat environment allowing a temperature scan comparable to manufacturing usage. As the temperature increases, the sample viscosity drops until the material gels, at which point the viscosity increases. This change in viscosity affects the dipole movement as well as ionic movement in the bridge. Thus, electrical measurements, such as dissipation factor, or conductance, change drastically with time and temperature. The output information can be used to determine maximum resin flow windows and completion of cure.

6. Other methods that have limited special applications include scanning electron microscopy (SEM), scanning tunneling microscopy (STM), X-ray photoelectron (XPS), and Auger electron spectroscopy (AES). The chemist needs to be aware of these methods but they will not be covered in any detail here. For further information on these and other methods, references should be consulted[22-24] such as the annual reviews published by the American Chemical Society.

IV. METHODOLOGY APPLIED TO RUBBER TECHNOLOGY

A. GENERAL APPLICATION

The term "rubber" covers a broad spectrum of material. ASTM D 1566[50] gives a fairly comprehensive definition: a material that is capable of recovering from large deformations quickly and forcibly, and can be, or already is, modified to a state in which it is essentially insoluble (but can swell) in boiling solvent, such as benzene, methyl ethyl ketone, and ethanoltoluene azeotrope." ASTM D 1418[45] classifies the basic rubbers from their chemical composition. This covers both dry and latex forms. In addition ASTM D 3853[52] gives the terminology for chemicals other than polymers, fillers, and pigments that are used in the compounding of rubber products. Other ASTM procedures[46-49,51,53] describe various test procedures that may be useful in full characterization of rubber products. Rubber chemistry which is a subsection of macromolecular chemistry is amply described in the literature.[36-44] Important properties that characterize rubber and make it so useful are elasticity, toughness, and solvent resistance. Theoretical considerations of these properties are amply described in the above-referenced literature. An understanding of this theory is important in selecting the chemical characterization methods to define the material. Generally, rubber consists of polymers that have weak interaction forces between them. When some force is applied, such as tension, the polymeric chains become extended. When the force is released, the chains return to their original configuration because the interaction forces are not strong enough to hold them in place. Rubber that is vulcanized has crosslinks introduced between the polymeric chains and, if the crosslink density is high enough, a rigid network is formed that causes the rubber to increase in hardness. Certain rubbers also develop crystallites when stretched which in turn drastically affect properties. A knowledge of the chemistry of fillers is also important in selecting proper CC techniques. Fillers could include antioxidants, antiozonants, plasticizers, softeners, extenders, carbon black, and nonblack-compounding ingredients. A good discussion of these fillers, their purpose, and how they interact with rubber matrix is given in Reference 43. Another aspect of rubber chemistry the technician must be aware of is vulcanization and the use of accelerators. All of these factors should be considered in developing a characterization scheme for the samples in question. As a general rule, the more one knows about the chemistry of the product in question, the better one can develop an adequate characterization test plan. If the product is a complete unknown, much testing must be done to assure that a satisfactory characterization is accomplished. If one has the resources and has done a complete characterization, a design of experiments can be performed to determine variation effect on properties. This would normally include collaboration with the manufacturer to vary his process and determine properties of the finished product. A characterization being done at each level of experiment gives good information on the efficiency of the process and its effect on final properties.

One of the more important properties of polymeric material is the glass transition temperature (T_g). The T_g is basically the temperature at which the material has obtained enough energy to allow some kind of segmental motion to take place in the molecule. Understanding this it is easily seen that a material may have more than one T_g, depending on what part of the molecule has obtained its threshold energy to have freedom of movement. Without this energy the segment is locked in place and does not have this freedom. Some materials by their very nature will not have a T_g; others may have two or even three. These transitions are identified by Greek letters with the highest temperature being the α transition. Some work has been done on materials that have both an α and β T_g that indicate the β transition is related to toughening in the polymer.[8] Many materials will not have a T_g; for others the T_g will occur at the same temperature as that of decomposition. A somewhat oversimplified definition of T_g would be the temperature, or temperature range, that the polymer is in a glassy state below, and a rubbery state above.

In some cases when one performs dynamic mechanical analysis, there will be secondary damping peaks different from the T_g, the T_g being the major transition. Those peaks occurring below the T_g are often associated with motion of short segments on the polymer backbone. In some cases there will be transitions occurring above the T_g, especially if the material is partially crystalline. Although there is much discussion on the origin of these transitions, a predominant school of thought associates them with either motion of the folded chains or some lattice defects. These secondary transitions, below the T_g, often relate to multiple segmental motion around the backbone. When this occurs the mechanism provides a means of energy absorption, thus relating to toughness.

For polymers that go through a cure process, the cure temperature is usually above the T_g to allow molecular motion to take place for the reaction to occur satisfactorily. The T_g is an indicator of amorphosicity, whereas the melt (T_m) is an indicator of crystallinity. A very practical illustration of T_g is what happens to a rubber garden hose left out in the winter. During the cold winter months the hose is very brittle and can be easily shattered by stomping on it. Conversely, during the summer months the hose is very flexible. In one case it is below the T_g; in the other case it is above it.

Another use of characterization techniques is to monitor degradation of the rubber with time. For example, the material may lose plasticizer with time which can be easily detected using characterization techniques. Loss or migration of other additives can also be monitored by these same techniques. Any loss or migration of additives or plasticizers can have a catastrophic effect on properties.

Environmental degradation can also be determined. This may include, not only oxidative degradation resulting in weight loss, but also property loss. Kinetic models can be developed from this information for predictive purposes. One must be extremely careful though to make their predictions within the bounds of the experiment. The property being predicted must respond in a linear fashion over the range of interest.

B. SPECIFIC APPLICATION

Typical DMA scans of neoprene rubber and silicone type rubber are illustrated in Figure 8 to 11. Figure 8 is a scan of frequency and damping. The scan was performed on a Dupont DMA 951, using resonant frequency. This configuration clamps the sample as illustrated in Figure 5, and then a cyclic, sinusoidal type stress is applied. The output shown in Figure 8 indicates a damping peak occurring at approximately $-33°C$. This same information is shown in Figure 9, but calculated out as shear storage modulus and shear loss modulus. The similarity of the scan is obvious. Figure 10 is the same information calculated out as tan delta, or loss tangent. Various transitions can easily be detected from these scans. The T_g is one of the more obvious transitions. Some technicians would determine it as the temperature of the damping peak, others determine it as the temperature of deviation from base line of the frequency or loss tangent. Whichever the technician chooses to use, it must be done consistently to be able to compare materials. It is also obvious from the shear modulus curves of Figure 9 that other changes are taking place in the material before it reaches the T_g. The increase in shear storage modulus, for example, indicates some kind of stiffening occurring in the matrix. In this particular case, this phenomena may be due to some response of a filler or plasticizer in the matrix as the temperature is increasing. This material is highly filled. These scans were made at 5°C per minute in order to allow for temperature equilibrium of the sample to take place. One must be cautious to assure or at least recognize there may be a temperature lag in the sample compared to the heating rate. Figure 11 is the same type of analysis performed on silicone rubber to compare the differences in properties of the two materials. In comparing the damping peak of the two materials, the silicone rubber has a lower temperature limit of use than the neoprene type.

Figures 12 and 13 illustrate DSC scans of two different rubber formulations. Figure 12

DMA

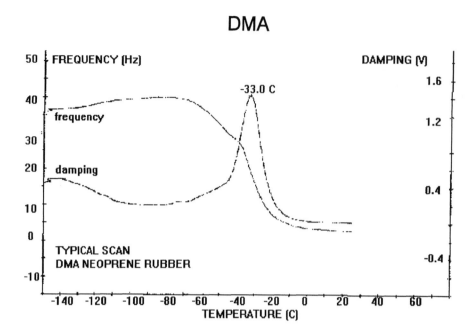

FIGURE 8.   Typical scan of dynamic mechanical analysis of neoprene rubber.

is that of neoprene and Figure 13 of the silicone type. Both of these are different formulations than used in the DMA curves of Figures 9 to 11. Here, once again, the T_g is noted by the change of base line occurring at, for neoprene, $-57.3°C$, and for silicone, $-64.9°C$. The DSC is measuring the result of a change occurring in the specific heat, this being one of the many properties that change at the T_g. These particular scans were run in an air environment. Other options may include the use of an inert atmosphere or, at times, a reactive atmosphere. The effect one is evaluating will affect the choice of atmosphere. For example, to study the effect of oxidation on the material, one may choose to use an inert atmosphere such as nitrogen first, and then repeat using air. Another option is to use open sample pans or sealed pans. One must be careful that good contact of the pan to the detecting surface is maintained throughout the experiment. Sometimes sealed pans have a tendency to bulge because of internal pressure being developed which, in turn, causes the contact to deteriorate.

Figures 14 and 15 illustrate TGA curves of neoprene rubber (Figure 14), and silicone rubber (Figure 15). Both of these scans were accomplished in flowing air. Once again to study oxidative effects, scans can be performed in an inert gas such as nitrogen first, and then repeated in air. These are typical scans for rubber products. Initially, plasticizers, extenders, and other lower molecular weight modifers volatilize off. Then rubber pyrolizate products come off as the polymer thermally degrades. The remaining residue represents inorganic filler. The complete characterization would include a GC/MS identification of the lower boiling products, an IR analysis of the pyrolizate, and an x-ray diffraction analysis of the filler.

Figure 17 illustrates a typical analysis of Pyrolysis Gas Chromatography. The rubber compound is represented by three sections; "A" representing the volatile component, "B" representing the polymer portion, and "C" representing inorganics. Using pyrolysis GC techniques, fingerprints are obtained of the "A" and "B" portions. Figure 16 illustrates some typical fingerprints. In each case the upper single peak portion represents "A"; the

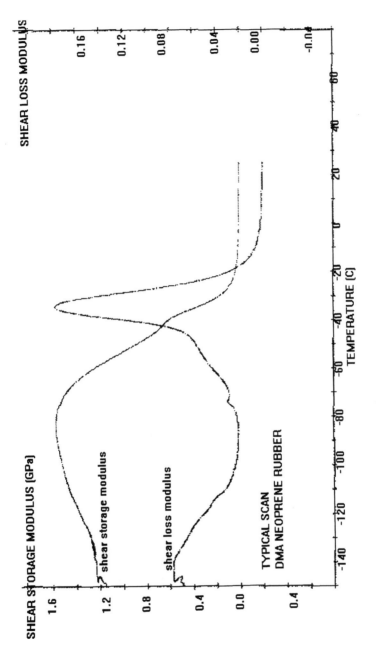

FIGURE 9. Dynamic mechanical analysis of neoprene rubber.

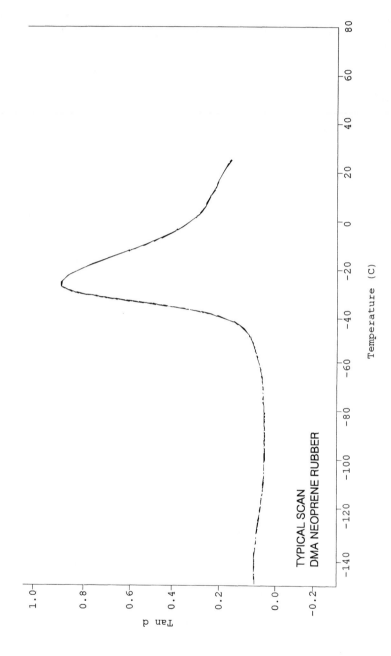

FIGURE 10. Dynamic mechanical analysis of neoprene rubber: tan d.

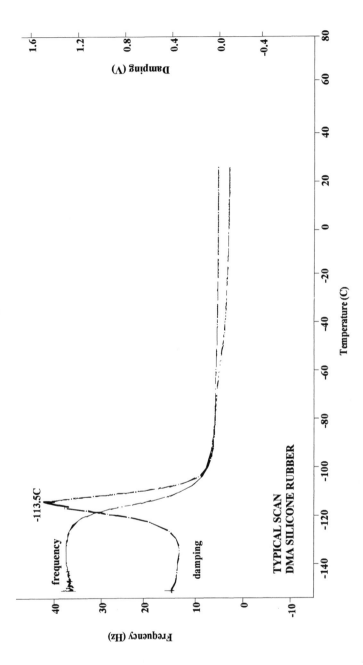

FIGURE 11. Dynamic mechanical analysis: silicone rubber typical scan.

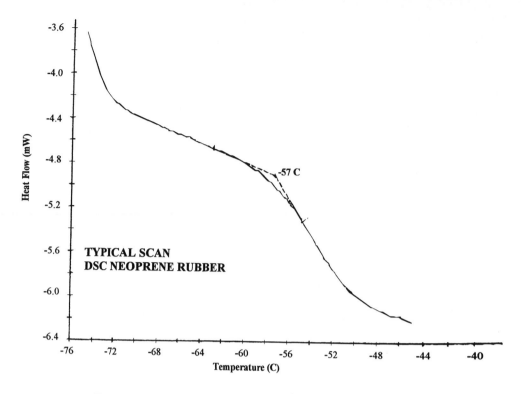

FIGURE 12.   Differential scanning calorimeter: neoprene rubber typical scan.

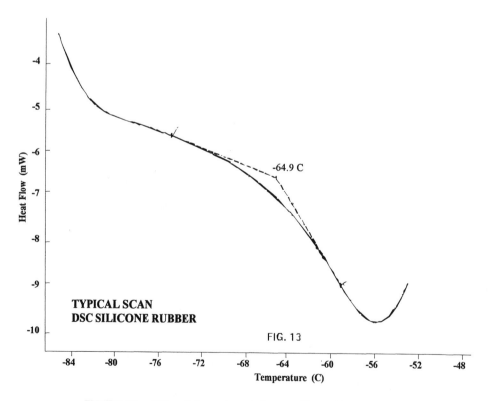

FIGURE 13.   Differential scanning calorimeter: silicon rubber typical scan.

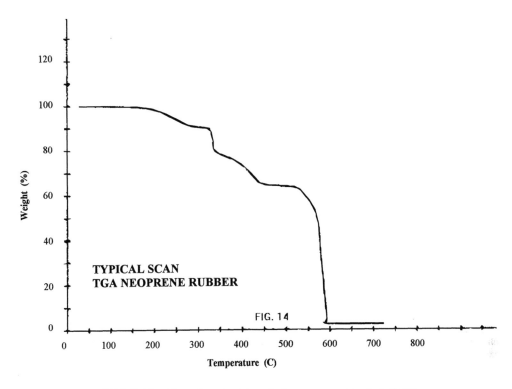

FIGURE 14. Thermal gravimetric analysis; neoprene rubber typical scan.

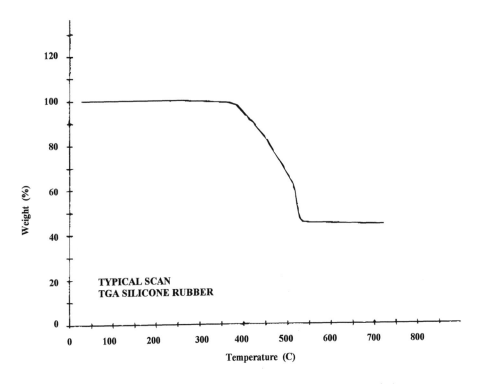

FIGURE 15. Thermal gravimetric analysis; silicon rubber typical scan.

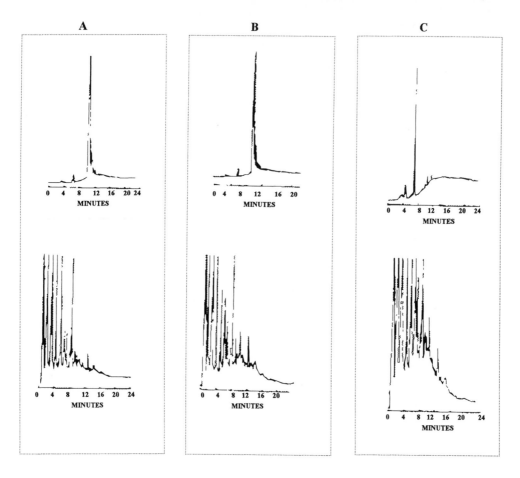

FIGURE 16.   Pyrolysis gas chromatography. Materials: (A) MIL-R-2785, Huntington; (B) MIL-R-6855; (C) Paracril C.

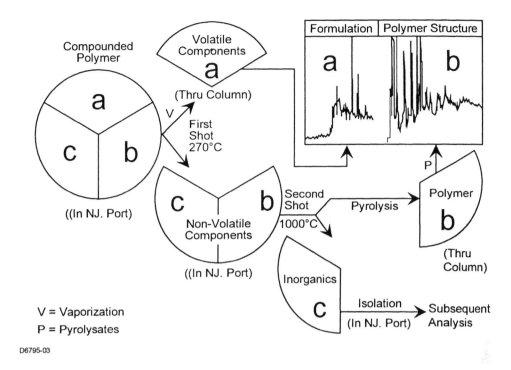

FIGURE 17. Typical analysis of pyrolysis gas chromatography analysis.

lower more complicated pattern represents the "B" portion. Chromatograms 17A and 17B are two different neoprenes made from the same raw material, Paracril C, depicted in 17C. Except for the variation of intensity, the pyrochromatograms of the polymers are very similar, as one would expect.

V. STATISTICAL PROCESS CONTROL TECHNIQUES APPLIED TO CHEMICAL CHARACTERIZATION TECHNIQUES

One of the problems with CC techniques concerns their use in a quality control setting. Normal specifications for materials are based on physical-type properties such as tensile strength, compression, and the like. Specification limits can fairly easily be established for these types of tests. However, the question is how one can establish limits relating to characterization techniques. Several approaches have been used. One method uses the information as a fingerprint for comparison only. This allows considerable variation for human judgment. Another method would be to use an extensive design of experiments to correlate CC testing with formulation and its effect on physical and mechanical properties. Although this method may be necessary for some materials, it may not be cost effective for many materials. Another approach is to utilize statistical process control techniques. With this method, baseline data are established, for example, over ten batches of material. Liquid chromatograph analysis may be run, T_g by DSC may be determined, or other CC techniques used. The data are determined over enough batches to establish baseline. Once this historical database is established, statistical rules may be applied on subsequent batches to assure no shift in distribution. For example, Figure 18 gives historical T_g data on neoprene rubber. The chart would indicate an out-of-control process occurring with sample number 17. Up to that point the distribution was fairly uniform and stable. With sample number 19 the indication is that something happened to the process that caused a drastic shift in distribution.

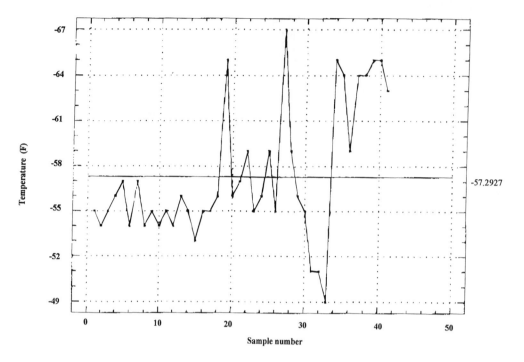

FIGURE 18.   Chemical characterization of neoprene rubber using Tg historical record.

In this particular case history, there was an actual change that occurred at that point in the formulation. The statistical rules that were used in this case were Western Electric rules.[54] This same type of analysis has been used for other types of characterization techniques with similar success. Statistical process control techniques have given us a tool to assure consistency of product from batch to batch.

VI. CONCLUSION

CC techniques are beginning to be applied in many different circumstances. Although there are still some problems with use and application, as development continues these are slowly being resolved.

A recent application is to utilize CC techniques as a screening function for quality assurance. Most materials received in an aerospace company are purchased to some specification, requiring that there be no change of formulation from that which was originally qualified. Usually "change in formulation" is not well defined, although it could be in terms of CC tests. More and more of the specifications are implementing CC testing. Present usage describes the materials to be unacceptable if there is an obvious change in formulation based on CC testing. However, used in conjunction with statistical rules there may be an "out of control" material where a change in formulation cannot be proved. The material is however outside of the baseline data distribution. In this case, other specification tests would be run to determine acceptability.

What has been described in this chapter is based on actual experience and case histories. The pitfalls described are also based on experience. It is hoped that this summary will enable one to be able to make use of these techniques in a profitable manner and be able to steer through some of the problems that are normally encountered.

REFERENCES

1. **Van Krevelen, D. W. and Hoftyzer, P. J.**, *Properties of Polymers, Correlations with Chemical Structure,* Elsevier, New York, 1972.
2. **Garn, P. D.**, *Thermoanalytical Methods of Investigation,* Academic Press, New York, 1965.
3. **Wendlandt, W. W.**, *Thermal Methods of Analysis,* John Wiley & Sons, New York, 1974.
4. **Read, B. E. and Dean, G. D.**, *The Determination of Dynamic Properties of Polymers and Composites,* Adam Hilger, Bristol, 1978.
5. **Ferry, J. D.**, *Viscoelastic Properties of Polymers,* John Wiley & Sons, New York, 1980.
6. **Lockett, F. J.**, *Nonlinear Viscoelastic Solids,* Academic Press, New York, 1972.
7. **Eisele, U.**, *Introduction to Polymer Physics,* Springer-Verlag, New York, 1990.
8. **Nielsen, L. E.**, *Mechanical Properties of Polymers and Composites,* Vol. 1 and 2, Marcel Dekker, New York, 1974.
9. **Craver, C. D.**, Ed., *Polymer Characterization,* (Advances in Chemistry Series 203), American Chemical Society, Washington, D.C., 1983.
10. **Colthup, N. B., Daly, L. H., and Wiberley, S. E.**, *Introduction to Infrared and Raman Spectroscopy,* Academic Press, New York, 1975.
11. Society for Paint Technology, Infrared Spectroscopy, Its Use in the Coatings Industry, Federation of Societies for Paint Technology, Philadelphia, 1969.
12. **Alpert, N. L., Keiser, W. E., and Szymanski, H. A.**, *IR Theory and Practice of Infrared Spectroscopy,* Plenum Press, New York, 1970.
13. **Bellamy, L. J.**, *The Infra-Red Spectra of Complex Molecules,* John Wiley & Sons, New York, 1966.
14. **Griffiths, P. R.**, *Chemical Infrared Fourier Transform Spectroscopy,* John Wiley & Sons, New York, 1975.
15. **McLafferty, F. W.**, *Interpretation of Mass Spectra,* W. A. Benjamin, Reading, MA, 1973.
16. **Hamming, M. C. and Foster, N. G.**, *Interpretation of Mass Spectra of Organic Compounds,* Academic Press, New York, 1972.
17. **McFadden, W. H.**, *Techniques of Combined Gas Chromatography/Mass Spectrometry, Applications in Organic Analysis,* John Wiley & Sons, New York, 1973.
18. **Snyder, L. R. and Kirkland, J. J.**, *Introduction to Modern Liquid Chromatography,* 2nd ed., John Wiley & Sons, New York, 1979.
19. **Yau, W. W., Kirkland, J. J., and Bly, D. D.**, *Modern Size-Exclusion Liquid Chromatography,* John Wiley & Sons, New York, 1979.
20. **Snyder, L. R.**, *Principles of Adsorption Chromatography,* Marcel Dekker, New York, 1968.
21. **Bolliger, H. R., Brenner, M., Ganshirt, H., Mangold, H. K., Seiler, H., Stahl, E., and Waldi, D.**, *Thin-Layer Chromatography, A Laboratory Handbook,* Springer-Verlag, Berlin, 1965.
22. **Kline, G. M.**, Ed., *Analytical Chemistry of Polymers,* Vol. 12, parts 1, 2, and 3, Wiley-Interscience, New York, 1962.
23. **Carroll, B.**, Ed., *Physical Methods in Macromolecular Chemistry,* Vol. 2, Marcel Dekker, New York, 1972.
24. **Campbell, D. and White, J. R.**, *Polymer Characterization, Physical Techniques,* Chapman and Hall, New York, 1989.
25. **Rosen, S. L.**, *Fundamental Principles of Polymeric Materials,* John Wiley & Sons, New York, 1982.
26. **Yau, W. W.**, *New Polymer Characterization Capabilities Using Size Exclusion Chromatography with On-Line Molecular Weight-Specific Detectors,* (Chemtracts — Macromolecular Chemistry), Vol. 1, No. 1, Jan/Feb, 1990.
27. **Hunt, B. J. and Holding, S.**, Eds., *Size Exclusion Chromatography,* Blackie & Son, Glasgow, in press.
28. **Gunderson, J. J. and Giddings, J. C.**, Comparison of polymer resolution in thermal field-flow fractionation and size exclusion chromatography, *Anal. Chim. Acta,* 189, 1, 1986.
29. Symposium on Size Exclusion Chromatography Field Flow Fractionation and Related Chromatographic Methods for Polymer Analysis, New York, Aug. 24 to 28, 1991, Division of Polymeric Materials: Science and Engineering, American Chemical Society, Washington, D.C.
30. **Yost, R. W., Ettre, L. S., and Conlon, R. D.**, *Practical Liquid Chromatography, An Introduction,* Perkin-Elmer, Norwalk, CT, 1980.
31. **Johnson, E. and Stevenson, B.**, *Basic Liquid Chromatography,* Varian Associated, San Fernando, CA, 1978.
32. **Cazes, J. and Fallick, G.**, Application of liquid chromatography to the solution of polymer problems, *Polymer News,* 3, 295, 1977.
33. **Lodding, W.**, Ed., *Gas Effluent Analysis,* Marcel Dekker, New York, 1967.
34. **Crippen, R. C**, *Identification of Organic Compounds with the Aid of Gas Chromatography,* McGraw-Hill, New York, 1973.

35. **Stevens, M. P.,** *Characterization and Analysis of Polymers by Gas Chromatography,* Marcel Dekker, New York, 1969.
36. **Huggins, M. L.,** *Physical Chemistry of High Polymers,* John Wiley & Sons, New York, 1958.
37. **Meares, P.,** *Polymers Structure and Bulk Properties,* Van Nostrand Reinhold, New York, 1965.
38. **Eisele, U.,** *Introduction to Polymer Physics,* Springer-Verlag, New York, 1990.
39. **Ravve, A.,** *Organic Chemistry of Macromolecules, An Introductory Textbook,* Marcel Dekker, New York, 1967.
40. **Billmeyer, F. W., Jr.,** *Textbook of Polymer Science,* Wiley-Interscience, New York, 1971.
41. **Williams, D. J.,** *Polymer Science and Engineering,* Prentice-Hall, Englewood Cliffs, NJ, 1971.
42. **Damusis, A.,** Ed., *Sealants,* Van Nostrand Reinhold, New York, 1967.
43. **Morton, M.,** Ed., *Introduction to Rubber Technology,* Van Nostrand Reinhold, New York, 1964.
44. **Naunton, W. J. S.,** Ed., *The Applied Science of Rubber,* Edward Arnold, London, 1961.
45. **ASTM,** *Standard Practice for Rubber and Rubber Latices — Nomenclature,* ASTM D 1418-90, American Society for Testing and Materials, Philadelphia, 1990.
46. **ASTM,** *Standard Test Methods for Rubber Products — Chemical Analysis,* ASTM D 297-90, American Society for Testing and Materials, Philadelphia, 1990.
47. **ASTM,** *Standard Practice for Rubber — Identification by Pyrolysis-Gas Chromatography,* ASTM D 3452-78, American Society for Testing and Materials, Philadelphia, 1978.
48. **ASTM,** *Standard Practice for Rubber — Chromatographic Analysis of Antidegradants (Stabilizers, Antioxidants, and Antiozontants),* ASTM D 3156-90, American Society for Testing and Materials, Philadelphia, 1990.
49. **ASTM,** *Standard Test Methods for Rubber — Identification by Infrared Spectrophotometry,* ASTM D 3677-90, American Society for Testing and Materials, Philadelphia, 1990.
50. **ASTM,** *Standard Terminology Relating to Rubber,* ASTM D 1566-91, American Society for Testing and Materials, Philadelphia, 1991.
51. **ASTM,** *Standard Test Methods for Rubber — Evaluation of NR (Natural Rubber),* ASTM D 3184-89, American Society for Testing and Materials, Philadelphia, 1989.
52. **ASTM,** *Standard Terminology Relating to Rubber and Rubber Latices — Abbreviations for Chemicals Used in Compounding,* ASTM D 3853-89, American Society for Testing and Materials, Philadelphia, 1989.
53. **ASTM,** *Standard Practice for Rubber Chemicals — Determination of Ultraviolet Absorption Characteristics,* ASTM D 2703-88, American Society for Testing and Materials, Philadelphia, 1988.
54. **Western Electric,** *Statistical Quality Control Handbook,* AT&T, Delmar Printing, Charlotte, NC, 1985.

Chapter 3

THE USE OF THERMAL ANALYSIS IN POLYMER CHARACTERIZATION

Michael P. Sepe

TABLE OF CONTENTS

0-8493-4401-8/93/$0.00 + $.50
© 1993 by CRC Press, Inc.

I. INTRODUCTION

The task of encapsulating in a single chapter a topic that can and has filled entire books, while at the same time contributing significantly to the impressive body of work already in existence, is an imposing one. In reviewing the literature, the definitive work on this subject is certainly *Thermal Characterization of Polymeric Materials,* edited by Edith Turi. The book focuses exclusively on the treatment of polymers and covers instrumentation, theory of operation, and analysis of all major classes of polymeric materials and additives. Since the book's publication a decade ago, the sophistication of computer-controlled instruments and advanced software has greatly increased the power and speed of thermal analysis. However, the work of the 1960s and 1970s is represented so well by contributors such as Bair, Prime, and Jaffe that the book necessarily provides a foundation for what follows here.

This chapter will essentially be an update on the collaborative efforts coordinated by Turi. It will assume a basic knowledge of polymer structure and classification. Further, it will assume a familiarity with the established techniques of thermal analysis, the instrumentation, and principles of operation. Brief discussions of new techniques will be included, but for detailed discussions of established methods the reader is referred to Chapter 1 of Turi's book. The chapter will be organized by analysis technique, with multiple techniques being highlighted when appropriate.

Finally, the emphasis will be on practical application involving materials and problems of commercial as well as scientific interest. While an attempt will be made to focus on elastomeric materials, useful techniques using work on rigid polymers will be discussed where they illustrate methods that may be of use in the field of elastomers.

II. STRUCTURE OF POLYMERIC MATERIALS

Wunderlich[1] has classified all materials into a three-class system based on distinguishing features in microscopic structure. Within this system, the materials generally referred to as polymers are categorized as flexible macromolecules. Much of the behavior exhibited by flexible macromolecules arises from a unique relationship between the bulk properties of the compound and the size of the basic molecular unit.

Most classical training in chemistry involves the characterization of materials that fall into the other two classes in Wunderlich's system. Rigid macromolecules such as those found in salts, metals, and ceramics and small molecules such as those that comprise low molecular weight organic compounds have a basic molecular unit that is small. This gives rise to a high degree of structural regularity. We tend to accept the concept that in a sheet of copper or a beaker of water the arrangement of copper atoms or water molecules will be approximately uniform throughout the sample. This uniformity leads to consistent and well-defined properties such as density, melting point, molecular weight, and crystal lattice parameters.

With such a background of observation, it is understandable that the chemist or engineer, confronted with the behavior of polymers, experiences some bewilderment. In producing molecules that exceed a certain minimum size, we introduce a variety of intermolecular reactions that make property characterization far more difficult. Instead of molecular weight we must speak in terms of molecular weight distributions. Crystalline and amorphous regions coexist within the same matrix and vary in local concentration as a function of processing and environmental treatment. Also, we encounter the phenomenon of viscoelasticity; polymers, whether apparently solid or fluid, are an ever-changing balance between elastic solid and viscous fluid. This makes time as important as temperature in evaluating the performance of these materials.

Nearly a century passed between the time when useful polymers were first synthesized and the period in which Staudinger presented his landmark work on the significance of molecular size to the properties of polymers. Within the same decade that Staudinger received the Nobel Prize and Flory was developing his theories, the techniques that we associate with modern thermal analysis were in their infancy. Thus, the array of methods that collectively represent thermal analysis has grown up with the commercial polymer industry. It is reasonable to say that thermal analysis and polymer science have contributed to each other's development, and thermal analysis techniques have proven uniquely suited to the task of unveiling the rich tapestry of behavior displayed by polymers.

In the 50 years since polymer science began its rapid development of useful, new materials, the focus has progressed from the roots of new polymer discoveries to the process of combining existing polymers, fillers, and additives to produce compounds that fulfill distinct property profiles. In addition, new catalyst technologies have created the potential to control polymerization of existing materials in ways that may revolutionize their end use performance. Through all these stages of evolution, thermal analysis has proven essential to the characterization process. The range of techniques presented in this chapter will illustrate the power and variety offered by thermal analysis in providing an understanding of polymer structure and properties.

III. DIFFERENTIAL SCANNING CALORIMETRY (DSC)

Differential scanning calorimetry (DSC) and differential thermal analysis (DTA) are related techniques that are capable of measuring the same thermal events with different measurement methods. DSC monitors the difference in heat flow between a sample and a reference as the material is heated or cooled while DTA measures a difference in temperature. Both techniques are capable of measuring any event in a polymer that produces a change in enthalpy. Generally, these changes are measured as a function of temperature or time. The instrumentation used in these techniques is well covered by Wendlandt and Gallagher.[2] Since the recent literature is dominated by work using DSC, we will focus most of our discussion on this technique. Figure 1 shows a generalized scheme of the physical and chemical changes in a polymer that can be characterized by enthalpic changes.

A. CRYSTALLINITY

At its most basic level, DSC can determine the presence and concentration of a crystalline phase in a polymer as well as the melting point of the crystals. Dole[3] described the measurement of degree of crystallinity in a thermoplastic polymer by dividing the measured heat of fusion by the heat of fusion for a 100% crystalline analog. This is a straightforward

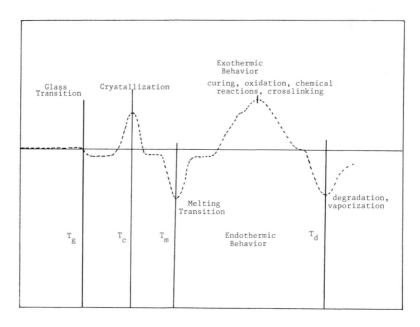

FIGURE 1.   Schematic DSC curve. (Rabek, 1983, Wiley-Interscience, New York.)

procedure for polymers that exhibit a single crystal structure. The heat of fusion is taken as the area under the curve associated with the melting event as shown in Figure 1.

It is important to distinguish between the melting point and the heat of fusion in assessing the relative crystallinity of semicrystalline polymers. It is a common perception that polymers such as nylon and poly(butylene terephthalate) are more crystalline than high-density polyethylene because of their higher melting points and room temperature rigidity. However, the heat of fusion shown in Figure 2 for several unfilled, semicrystalline polymers shows that the degree of crystallinity is highest for polyethylene and acetal homopolymer, the two materials with the lowest melting points. The higher secondary forces arising from the presence of polar groups in the polyamide and polyester backbones give rise to the higher melting points in these polymers. However, it is these same polar groups that restrict the large-scale segmental motions needed for extended crystal growth and prevent these higher melting polymers from achieving high levels of crystallinity.

Within a polymer family, however, where the groups contributing to the intermolecular attractions remain chemically the same, crystallinity and melting point can be related to one another. Figure 3 compares the heat of fusion for low- and high-density polyethylenes, polypropylene homopolymer and copolymer, and acetal homopolymer and copolymer. The reduced structural regularity of the copolymers, or the more extensively branched polyethylene, inhibits the ability of these materials to crystallize. This is evident in the reduced heats of fusion upon melting for these materials as well as the reduced melting points. This phenomenon will be reviewed later as an excellent method for evaluating the composition of thermoplastic elastomers.

The heats of fusion and recrystallization have become important considerations for processors as computer modeling of mixing, melting, and cooling of polymers has grown in popularity. Early algorithms in moldflow and moldcool software did not include a calculation for the energy associated with the phase change. This led to large errors in calculating the flow dynamics and cooling rates of semicrystalline polymers.

Recrystallization can be readily studied as a function of either temperature or time. Simple scans in dynamic heating and cooling identify the energies of bond breaking and

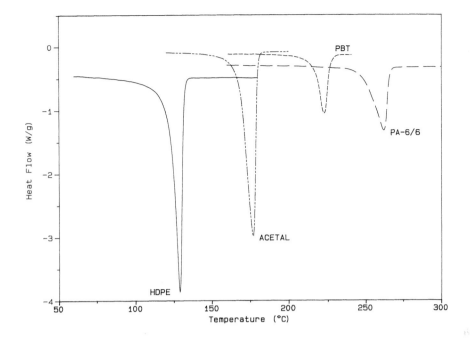

FIGURE 2. Melting endotherms of unfilled semicrystalline polymers. (From Sepe, M. P., *SPE ANTEC,* 37, 469, 1991.)

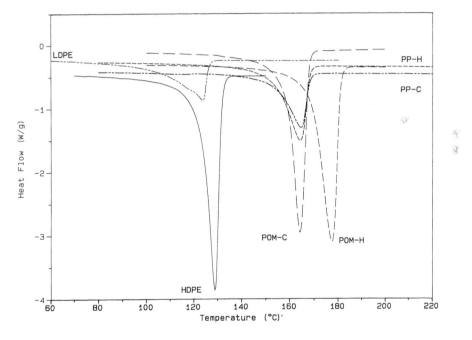

FIGURE 3. Melting endotherms as a function of structure differences. (Sepe, unpublished, 1991.)

bond formation associated with the phase change. Figure 4 shows the melting and recrystallization of an acetal copolymer at equivalent heating and cooling rates.

Unlike metals and simple organic compounds, the melting and recrystallization events in polymers are dependent on a number of factors. Cooling rates or isothermal cooling times,

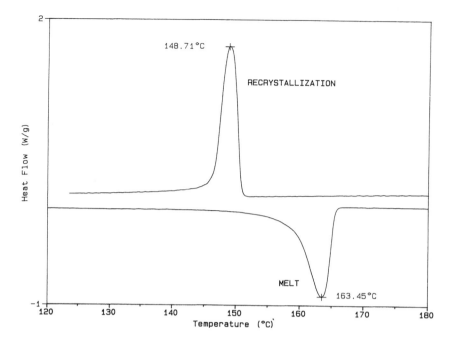

FIGURE 4.   Melting and recrystallization of acetal copolymer; heating and cooling rate of 2.5°C/min. (Sepe, unpublished, 1992.)

applied strain during cooling, molecular weight and molecular weight distribution, and structural regularity can all influence the manner in which a polymer melts and recrystallizes. Wunderlich et al.[4] have demonstrated the effect of different recrystallization times at constant temperature upon the subsequent melting behavior of polyethylene. Figure 5 illustrates an alternate approach. In this experiment, using an acetal copolymer, the crystallization time is held constant and the exposure temperature is varied. Figure 6 shows the use of different cooling rates on the crystallization of poly(*p*-phenylene sulfide) (PPS). All three of these techniques permit us to study the kinetics of crystallization, a subject that will be discussed in more detail below.

Hamada et al.[5] have examined the melting of single polyethylene crystals as a function of molecular weight. This and other studies show that crystallization occurs more slowly in higher molecular weight systems and results in less perfect crystals. This manifests upon reheating as a broader melting endotherm or multiple melting endotherms with reorganization exotherms in between.

Polymers such as nylon 6/6 and PBT polyester form two or more distinct crystal types at slow cooling rates. This tendency is suppressed at the rapid cooling rates typical in the molding process. Figure 7 shows this behavior for a PBT polyester reheated after one sample was quenched rapidly and the other was cooled slowly. Ludwig and Eyerer[6] have studied this behavior quantitatively and have augmented it with direct examinations of the microstructure. They also briefly discuss the effects of mold temperature on the dynamic mechanical properties, an area that will receive thorough treatment later in Section VI.

Some crystallizable polymers with a high degree of aromatic character in the backbone exhibit a strong recrystallization exotherm immediately above the glass transition upon reheating. The energy associated with this exotherm is strongly influenced by the cooling rate from the melt. With sufficient quenching the degree of crystallinity in polymers such as PPS and poly(ethylene terephthalate) (PET) may be reduced so that they behave as amorphous materials. This suppressed crystallization may be recovered by annealing in the solid state above the recrystallization temperature. Figure 8 shows the recrystallization

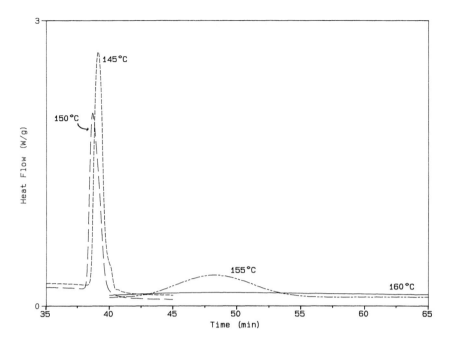

FIGURE 5. Isothermal recrystallization of acetal copolymer at temperatures indicated. (Sepe, unpublished, 1991.)

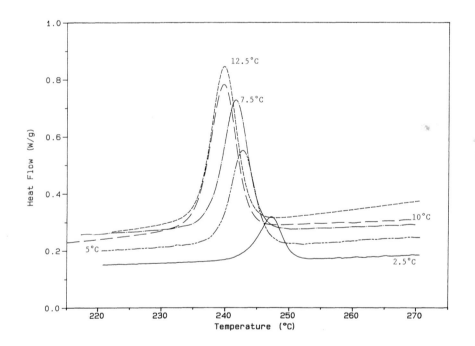

FIGURE 6. Recrystallization of PPS at different cooling rates. (Sepe, unpublished, 1988.)

exotherms for PPS samples processed at different mold temperatures by injection molding. As the mold temperature increases from below the glass transition temperature to above the recrystallization temperature, the amount of uncrystallized material decreases.

It has been shown that polymers of this type that are fully crystallized from the melt differ at the microstructure level from those that are quenched and then crystallized from

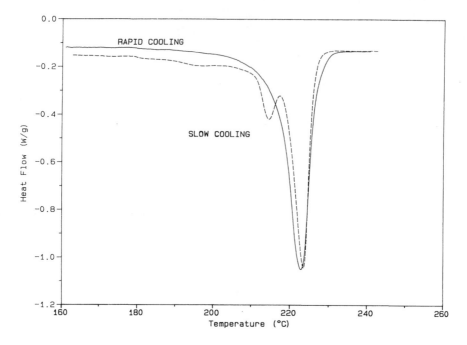

FIGURE 7. Melting endotherms of PBT polyester cooled from the melt at different rates. (Sepe, unpublished, 1989.)

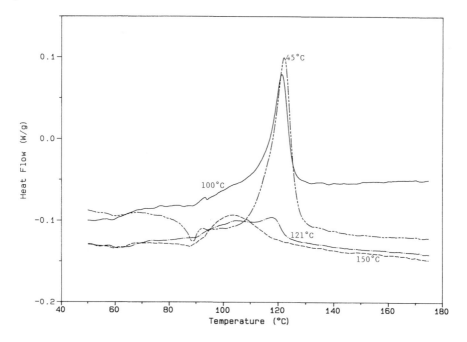

FIGURE 8. Solid-state recrystallization of PPS molded at the indicated mold temperatures. (From Sepe, M., *SPE ANTEC*, 35, 1035, 1989.)

the solid state by annealing between the recrystallization temperature, T_c, and the melting point, T_m. Cebe[7] and Seferis et al.[8] have reported on melting endotherms in PPS and poly(etheretherketone) (PEEK) that occur in the region of the annealing temperature. Figure 9 shows this behavior for PPS. It is reasonable to interpret this lower temperature endotherm

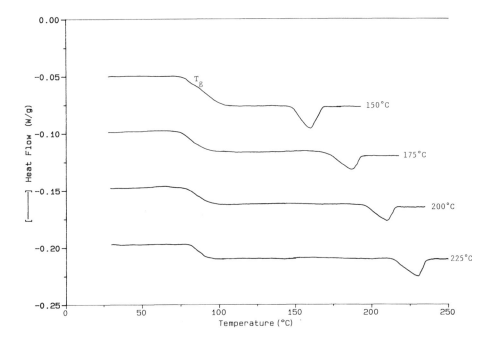

FIGURE 9. Low-melting endotherms in PPS heated to the indicated temperatures, cooled, then reheated. (Sepe, unpublished, 1992; after Cebe, P., *SPE ANTEC*, 35, 1413, 1989.)

as the melting of imperfect crystals that condense from the mobile glass during annealing. This author has observed similar behavior in PBT polyester.

Polymers such as polyethylene have subambient glass transition temperatures, T_g, and may therefore be capable of continued crystallization during storage at room temperature or below. Soni[9] has used DSC to measure the specific heat of polyethylene samples that were rapidly quenched and then annealed at $-20°C$. The results show the formation of secondary crystals that melt near the annealing temperature with a very small endotherm. At longer times the endotherms become more energetic and the melting point rises, the result of crystal thickening. This may afford a method of studying the thermal history of a sample. For a given storage temperature it may be possible to develop a plot of time vs. T_m or time vs. heat of fusion, H_f, of the secondary crystals. This plot could then be used as a calibration curve for relating secondary crystal characteristics to aging time.

It is generally accepted that secondary crystallization can occur only above T_g and below T_m. However, Bair et al.[10] have reported on the development of low-melting crystals in PBT annealed at 29°C, some 20°C below T_g. This work also examines changes in the glass transition. As-molded samples exhibit a smooth change in the specific heat, C_p, near 50°C which is associated with T_g. Aged samples show an absorption of thermal energy which increases with longer time. In addition, the T_g rises 7°C after 1000 h. The authors related this thermal behavior to a loss in ductility evidenced by a reduction in tensile elongation at yield.

The landmark work of Struik[11] on the physical aging of amorphous polymers suggests that the enthalpy changes in the glass transition of the aged PBT relate to the physical aging of the amorphous glass, which is associated with a decrease in the free volume. The formation of low-melting crystals noted by Bair may be due to the secondary crystallization of material made available because of the free volume reduction.

Since many high-performance semicrystalline polymers are supplied with reinforcing fibers and other fillers, the effects of these agents on the crystallization of the matrix polymer have been an area of interest. DSC has been the primary technique used. Ma and Yur[12] have discussed the behavior of PEEK with carbon fibers. Desio and Rebenfeld[13] have evaluated PPS crystallization with carbon, glass, and aramid fibers. In both studies isothermal crystallization was conducted at various temperatures and the data were fitted to the Avrami equation:[14]

$$1 - C = \exp(-Kt)^n \tag{1}$$

where C is the percentage of crystallization that has occurred up to time t, and K and n are the Avrami rate constant and exponent, respectively. A log-log plot of $-\ln(1-C)$ vs. time will give n and K as the slope and intercept of the line. Both studies demonstrate the considerable effect that different fibers and fiber treatments have on the crystallization rates and final crystallinity of the base polymers. In addition, microscopy studies reveal different types of crystals forming at the fiber/polymer interface.[13] Further work has been carried out on PPS by Desio and Rebenfeld[15] and on PET polyester by Reinsch and Rebenfeld.[16]

Sumita et al.[17] have demonstrated the dependency of the heat of fusion, H_f, on draw ratio of drawn polyethylene samples. H_f increased with increasing draw ratio which indicates the occurrence of strain-induced crystallization. Singh and Kamal[18] have performed extensive studies by DSC of the melting behavior in unreinforced and glass reinforced polypropylene. In this work samples were injection molded and specimens were microtomed from different layers in the moldings. DSC showed clear differences in crystal formation that were dependent on both the cooling rate and the strains induced by flow and orientation of the polymer.

Polymers that crystallize rapidly from the melt will not typically exhibit a recrystallization exotherm upon reheating. However, it has been shown above that polymers that crystallize slowly will exhibit such an exotherm, and the energy associated with this recrystallization is dependent upon the rate of crystallization from the melt. Nucleating agents are commonly added to some polymers to increase the rate of crystallization. The relative effectiveness of different nucleating agents in a given polymer can therefore be determined by measuring the recrystallization exotherm of samples produced at various mold temperatures. This author has performed evaluations on commercial PET polyesters formulated with equivalent filler amounts and types. Figure 10 shows the recrystallization exotherms for two compounds utilizing different nucleating agents. These results indicate that polymer A crystallizes faster than polymer B and will achieve a stable structure more rapidly or over the same time period at a lower mold temperature. It will be shown later that TMA and DMA measurements confirm these conclusions.

Khanna et al.[19] have studied the effects of processing history and technique on the crystallization of nylon 6. They find that extruded pellets crystallize faster than pellets from a strand of as-polymerized material. This is attributed to an orientation-induced memory effect which can survive melt processing in polymers with high intermolecular attractions such as the hydrogen bonding found in polyamides. Samples that were compression molded from the two types of pellets reflected the same disparity in crystallization from the melt as that observed for the pellets. Injection-molded samples still reveal a more rapid crystallization for the extruded pellet moldings, but the differences were reduced substantially. This would indicate that the higher shear stresses associated with injection molding disrupted the memory effects produced during pellet production. Morphology studies reveal a smaller spherulitic structure for injection-molded parts produced from extruded pellets; however, physical properties were not distinguishable from moldings made with as-polymerized pellets.

Sepe has investigated cosmetic defects in moldings made from thermoplastic polyurethane elastomers containing a polyester hard segment. DSC scans were run on regions of the moldings exhibiting a cosmetic irregularity and areas of the same parts that were free of the defects. Initial heating reveals a premature melting endotherm in the defective part

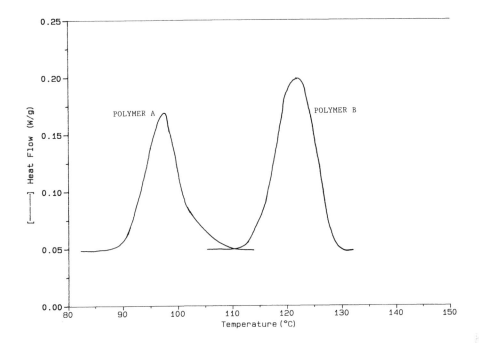

FIGURE 10. Solid-state recrystallization of cold-molded PET polyesters utilizing different nucleating agents. (Sepe, unpublished, 1990.)

that does not appear in the defect-free region. Reheating the samples after slow cooling from the melt reveals the disappearance of the premature melt. Figure 11 compares the behavior of the two samples on initial heating and Figure 12 shows the results of reheating. The premature exotherm can be duplicated in the DSC by heating a sample to a temperature below the melting point of the hard segment, maintaining this elevated temperature for a brief period, then cooling to room temperature. Reheating the sample will produce a low-melting endotherm at a temperature very close to the isothermal conditioning temperature. The presence of excess moisture inhibits the formation of these low-melting crystals and also reduces the incidence of defective moldings.

The presence of low-melting crystals is attributed to the formation of an allophonate structure. This structure is a crosslink that forms by the reaction of excess isocyanate with the urethane groups in the polymer backbone.[20] Since these crosslinks melt at approximately the temperature at which they were formed, the process engineer can examine the processing equipment for cold spots of a specific temperature region. The improvement in cosmetics with rising moisture content is explained by the hydrolysis of the isocyanate, making it unavailable for the crosslinking reaction.

Richeson and Spruiell[21] have used DSC to evaluate the effects of hard block content in a segmented copolyester ether upon the crystallization rate and T_g and T_m. It was found that as the hard segment, poly(tetramethylene terephthalate) (PTMT), increased with respect to the soft segment, poly(tetramethylene ether glycol) terephthalate (PTMEG-T), the crystallization rate increased and lower degrees of supercooling were required for crystallization to occur. Both T_g and T_m, and therefore the crystallization temperature, drop as the hard segment content is decreased. The authors note that annealing produces apparent phase separation.

Castles et al.[22] used DSC to examine the crystallization behavior of copolyester-ether block copolymers as a function of different crystallization temperatures. Multiple melting endotherms were altered as a function of the crystallization temperature.

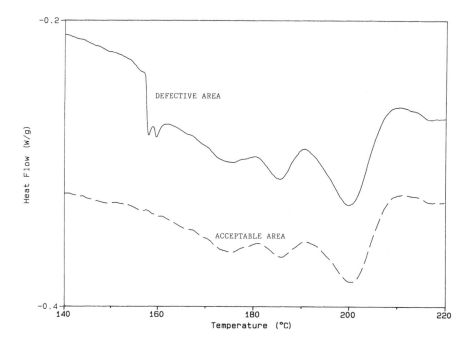

FIGURE 11.   First heating of molded thermoplastic polyurethane comparing cosmetically defective and acceptable regions. (Sepe, unpublished, 1987.)

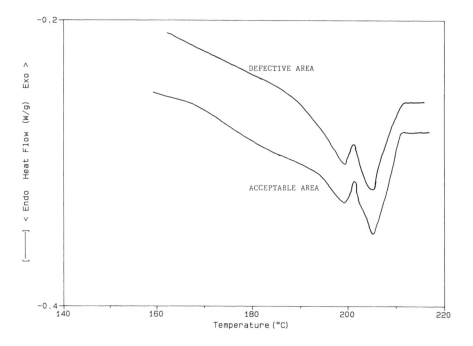

FIGURE 12.   Second heating of molded thermoplastic polyurethane showing erasure of low-melting endotherm in defective part. (Sepe, unpublished, 1987.)

Annealing studies have been conducted on a melt-processable polyimide by Hou and Bai.[23] This work is an excellent representation of the range of behavior displayed by a crystallizable polymer as a function of different annealing temperatures and times. Avrami parameters were determined.

B. THE GLASS TRANSITION

Some polymers do not crystallize under normal conditions, and even semicrystalline polymers will contain a significant amount of amorphous character. Since the amorphous phase possesses no well-defined structural order, no melting point is observed. The amorphous phase does undergo an important change in structure known as the glass transition. It is detectable by DSC as a step change in the heat capacity of the polymer.

In amorphous polymers the glass transition signifies an outright softening of the polymer matrix and an essential upper temperature limit for useful solid-state properties. In semicrystalline and crosslinked polymers it represents the onset of significant changes in a wide variety of thermal and mechanical properties. The exact measurement of these property changes can be conducted by thermomechanical analysis (TMA), dynamic mechanical analysis (DMA), and dielectric analysis (DEA), and will be detailed below. The enthalpy changes detectable by DSC can be useful in evaluating aging, degradation, crosslinking, and blend composition.

Discussions of the correct procedures for accurate determinations of the glass transition temperature, T_g, are given by Flynn,[24] Wunderlich,[25] and Shalaby.[26] Shalaby also discusses the dependency of T_g on the number average molecular weight, M_n, illustrating the combined effects of increasing M_n and changes in heating rate on measurements of T_g for polystyrene. His review, with supporting data from Blanchard et al.,[27] shows that heating rates from 5 to 80°C have only minor effects on T_g. M_n has a large influence until a critical M_n value is achieved. Beyond this point, continued increases of even an order of magnitude only raise T_g by 4°C, making small changes in molecular weight difficult to detect by T_g. In fact, T_g is often considered to be essentially constant as a function of molecular weight for a given polymer. This author has measured large changes in the T_g of polycarbonate. Figure 13 shows a plot of T_g for four different molecular weights of polycarbonate. Molecular weight was not measured directly but was determined in a relative fashion through melt viscosity determinations. The lowest T_g was measured for a color concentrate which produced brittle moldings when blended with a virgin resin at five parts per hundred of resin (phr).

The relationship of T_g to molecular weight is expressed mathematically by Fox and Flory:[28]

$$T_g = T_g(\infty) - K_g M^{-1} \tag{2}$$

where $T_g(\infty)$ is the limiting T_g at high molecular weight and K_g is a constant. Boyer[29] and Kumler et al.[30] identified three distinct straight line regions where the slope of the line relating T_g to M_n changed markedly in polystyrene. Cowie[31] proposed that a large number of polymers exhibit this three-region behavior and suggested empirical equations for defining boundaries between the regions.

A number of interesting effects of structure and composition upon the value for T_g have been summarized from various sources by Shalaby.[26] These include molecular weight distribution, crystallinity, orientation in fibers, stereoregularity, and differences between the cooling and heating rates.

The accurate determination of T_g must take the thermal history of the sample into account. Tests are ideally performed on material that has been cooled and reheated at the same rate. It must also be remembered that the glass transition is a relaxational process that occurs in

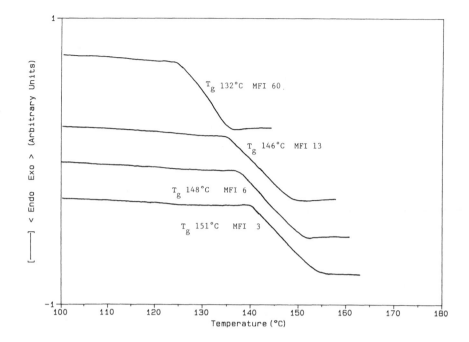

FIGURE 13. T_g for polycarbonates of different molecular weights. (Sepe, unpublished, 1988.)

the disordered regions of the polymer matrix. For materials which are predominantly amorphous, the change in heat capacity is large and easily observed. As crystallinity increases, the glass transition becomes difficult to detect. This can be seen in Figure 8 where the PPS sample with the lowest crystallinity exhibit an easily measured T_g near 90°C. The more completely crystallized specimens display no detectable activity in this temperature region. Prior knowledge of the temperature of an anticipated transition is often necessary for proper identification. Finally, sample sizes in a DSC cell are comparatively small, typically between five and fifteen milligrams. It can be risky to infer too much about the bulk properties of a fabricated part from the thermal behavior of such a small sample.

Even with these limitations, DSC continues to be a popular technique for characterization. A wide variety of sample forms such as powders and liquids can be handled easily. Also, while TMA and DMA can evaluate the physical results of changes in polymer structure, DSC is often the only way to quantify these structural and compositional changes.

C. CHARACTERIZATION OF ALLOYS AND COPOLYMERS

Measurements of the T_g and melting and recrystallization phenomena to characterize copolymers and blends has received a thorough treatment by Shalaby and Bair.[32] Some of the more significant results will be included here.

The change in heat capacity associated with the glass transition was used to make quantitative determinations of polybutadiene content in ABS[33] and ABS content as an impact modifier in PVC.[34] By comparing the heat capacity change, ΔC_p, of the pure rubber with that of the rubber modified SAN, a ratio of the two values gives a percent rubber content which agrees well with determinations using chemical techniques. The same method is used in impact-modified PVC by using the ΔC_p of the SAN and PVC glass transitions as factors in the calculations.

The above blends are treated as immiscible or semimiscible based on the retention of distinct glass transitions for each polymer. Completely miscible blends are often identified by the formation of a single T_g which is blend ratio dependent. An excellent commercial example of this is Noryl, a family of blends of polystyrene (PS) and poly(phenylene oxide) (PPO). PS has a T_g of approximately 100°C while the T_g of pure PPO is 210°C. Blends of PPO/PS exhibit one T_g which falls on a curve between the two extremes. DSC measurements of T_g for various blend ratios were performed by Prest and Porter[35] and are presented in graph form by Shalaby and Bair.[32]

Several equations have been developed to express mathematically the relationship between the T_g of a miscible blend and the ratio of its components. The simplest of these is the Fox equation:[36]

$$1/T_{g1,2} = W_1/T_{g1} + W_2/T_{g2} \tag{3}$$

Other relationships have been proposed by Gordon and Taylor[37] and rearranged by Wood[38] into what is known as the G-T-W equation:

$$T_{g12} = [k(T_{g2} - T_g)W_2/(1 - W_2)] + T_{g1} \tag{4}$$

In the above equations W_1 and W_2 are the weight fractions of the two components in the blend. In the G-T-W equation, k is a constant provided to give the best fit for the empirical data.

Couchman[39] has developed the equation:

$$\ln T_{g12} = \frac{x_1 \Delta C_{p_1} \ln T_{g1} + x_2 \Delta C_{p_2} \ln T_{g2}}{x_1 \Delta C_{p_1} + x_2 \Delta C_{p_2}} \tag{5}$$

where x_1 and x_2 are the mass fractions of the two components and the ΔC_p terms are the heat capacity changes for the two neat polymers.

Ellis has used DSC to characterize blends of amorphous and semicrystalline polyamides, using the occurrence of two glass transitions to indicate immiscibility and the existence of one T_g to signify a miscible blend. The single T_g of the miscible blends follows Couchman's equation. Ellis[40] shows that aliphatic nylons that achieve miscibility possess a combination of mobility offered by the volume fraction of methylene units and the contribution of strong interactions due to hydrogen bonding from the amide groups. Thus, nylon 4 and 4/6 are immiscible with amorphous polyamides due to a low methylene fraction while nylon 11 and 12 are immiscible with amorphous polyamides due to a low amide group concentration.

Young et al.[41] used DSC to characterize blends of polypropylene (PP) and PET and study the effects of an acrylic acid graft to the PP in reducing interfacial tension and producing a more compatible blend. Thermograms show that the acrylic acid-grafted PP had the effect of broadening the temperature range of recrystallization in the PET phase and raising the recrystallization temperature of the PP phase. The PET recrystallization became so broad that it was virtually undetectable by DSC and only a reheating of the material to produce a melting endotherm confirmed the presence of a crystalline PET phase.

A number of characterizations of liquid crystal polymer (LCP) blends have been conducted in an effort to develop polymer matrices that include a fibrillar structure. Kohli et al.[42] have worked with LCP/polycarbonate (PC), Sukhadia et al.[43] with LCP/PET, and Subramanian and Isayev[44] with LCP/PEEK. The thermograms verify morphology studies showing that each phase has little effect on the other. It was noted in the LCP/PET blends that the recrystallization temperature, T_c, of quenched films upon reheating was 20°C lower than in neat PET and may indicate a nucleating effect. In addition, the difference between the heat of fusion and the heat of cold crystallization ($H_f - H_{cc}$) for the PET increases with higher LCP content in the blend, indicating that the presence of the LCP promotes a higher degree of crystallinity in the PET.[43]

Sullivan and Weiss[45] have used DSC to characterize the compatibility of polyamide with sulfonated polystyrene. The polystyrene was prepared both with and without metallic ions. The thermograms show that polyamide (PA)/polystyrene (PS) blends displayed their typical glass transition temperatures, indicating an immiscible blend. When zinc or sodium ionomers of lightly sulfonated PS were blended with polyamide in a 75/25 mixture of PA/SPS, miscibility showed a dependency upon the degree of sulfonation. Two T_g's were still detectable with a 2.29 mol % sulfonic acid while one T_g was observed in a blend using 5.49 mol % sulfonic acid. An increase in the PA T_g was observed and is attributed to physical crosslinking between the amide groups and the sulfonate groups.

The authors point to the stoichiometry of the SO_3M and NH_2 groups to attribute miscibility to mechanisms other than that of an acid/base reaction. A reversal of the SPS/PA ratio produces a compound with a single T_g that follows the Fox equation. The miscibility is attributed to hydrogen bonding and this was verified with FTIR. DMA scans performed as part of the study help point out the need for more precise techniques to probe polymer structure. An examination of the tan δ peaks for the 75/25 blend of PA/SPS show a broadening of the relaxation but not a true single phase T_g. The blend of 75/25 SPS/PA shows a true single phase by both DSC and DMA.

Thermoplastic elastomers are receiving increased attention in industrial markets due to a combination of rubber-like properties in the solid-state and melt processability. Many of these materials are block copolymers composed of a rigid "hard" segment with a high T_g and a rubbery "soft" segment with a low T_g. A good example of this type of material is Hytrel, a family of compounds where the rigid segment is based on poly(butylene terephthalate) (PBT or PTMT) and the rubbery segment is poly(tetramethylene oxide) (PTMO). The apparent crosslinked behavior of the solid-state matrix is the result of crystalline domains of PBT which crystallize from the melt to form a system of physical crosslinks for the soft segments as well as any uncrystallized hard segment material.

DSC characterizations were performed by Lilaonitkul et al.[46] on block copolymers of varying hard block/soft block composition. DMA measurements were also performed. The DSC scans show that at all ratios the compounds exhibit a single T_g, indicating miscibility between the uncrystallized PBT and the PTMO. T_g and T_m both rise as a function of increasing hard block content. The amorphous phase appears to be dominated by the soft block since even at 84% hard block content the T_g is only $-9°C$. The T_m and percent crystallinity of the materials approach those of pure PBT as the hard block concentration increases.

The G-T-W equation was used to relate the copolymer T_g to the concentration and T_g's of the two constituents. Defining W_2 in Equation 4 as the weight fraction of PBT gave a poor fit of the data to the equation. However, treating W_2 as the weight fraction of amorphous PBT gave a good fit of the data with $k = 0.5$.

Morphology studies of these materials that go beyond the scope of this chapter have shown that these systems are, in fact, partially heterogeneous and capable of forming different spherulite structures. Interestingly, the DMA measurements made on these materials also revealed one glass transition with the peak temperature of the loss modulus giving excellent agreement with the DSC determinations.

Recently, Gallagher et al.[47] have investigated the possibility of achieving miscibility and cocrystallization in blends of PBT and PTMT-PTMO block copolymers. The identical nature of the crystal unit cell structure and interplanar spacings in PBT and the polyester-ether (PEE) block copolymers suggests that the PEE hard segment and the PBT could cocrystallize to produce crosslinks that would influence mechanical behavior. Here again DSC techniques were augmented, in this case by dielectric spectroscopy, because of the difficulty involved in making precise measurements of T_g in the semicrystalline structure by DSC alone. Dielectric measurements were made on the T_g of the PBT as well as the β relaxations associated with motion in the amorphous phase of the PEE. DSC was relied upon for evaluations of the crystal populations by melting point.

Dual β transitions were observed for blends using copolymers of 40 to 51 wt % hard block while single relaxations were exhibited in blends employing over 80% hard block. Blends using 58 to 75% hard block PEE displayed two β transitions but a downward shift in the high temperature relaxation. From this the authors conclude a progression from immiscible to miscible blends as a function of increasing hard block content.

DSC measurements show that, for blends demonstrating miscibility, quenched samples produced a single T_m and a uniform variation in T_m and percent crystallinity as a function of blend ratio. However, samples cooled slowly produced two melting points characteristic of the pure polymers. This leads to the conclusion that cocrystallization in these systems is kinetically controlled.

Chandler and Collins[48] observed that copolymers of butadiene and acrylonitrile having between 10 to 36% acrylonitrile display two T_g's detectable by DTA while those containing less than 10% or more than 36% acrylonitrile (AN) had a single T_g. A discussion of these findings is offered by Shalaby and Bair.[32] In the blends exhibiting two glass transitions, the transitions are far apart in temperature when AN content is low but converge to nearly the same temperature as AN content rises.

Shalaby and Bair conclude that the two T_g's, which are shifted away from those of the pure polymers, arise from the existence of two block types, one rich in AN and the other rich in polybutadiene (PB). If this is the case, it is apparent that the PB-rich blocks are more affected by increasing AN content since the T_g of this fraction rises rapidly as AN content increases while the T_g of the AN-rich blocks changes very little.

Machado and French[49] have studied a unique blend of polyacetal and polyvinyl phenol (PVP) using DSC and DMA to demonstrate miscibility. They explain this miscibility in terms of a strong local interaction of functional groups engaged in hydrogen bonding between phenol and ether moieties. In this study, direct calorimetry of two low molecular weight analogs, dimethoxymethane and ethyl phenol, was performed. This technique, outlined by Paul and Barlow,[50] is beyond the scope of this chapter. Briefly, the principles involved in analog calorimetry state that the free energy of mixing in polymer blends is dominated by enthalpic effects. Measuring the heat of mixing of model compounds provides thermodynamic quantities which should contribute to the behavior of polymer mixtures containing the same functional groups. In this case, analog calorimetry gave a strongly exothermic mixing arising from hydrogen bonding and yielded a negative interaction parameter which is predictive of miscibility in a polyacetal (PAc)/PVP blend.

Using the melting point depression as a means of assessing interaction strength and miscibility gave Flory interaction parameters and Van Laar interaction parameters that were in good agreement with those of the low molecular weight analogs. This indicated the existence of a miscible blend between PVP and the amorphous regions of PAc. Melting point depression and degree of crystallinity were used to demonstrate this. DMA was used to demonstrate the existence of a single, composition-dependent T_g. DSC lacked the resolution to confirm this behavior. Substantial effects on crystallinity are evident at 40% PVP content on both the melting and recrystallization temperature. At 70% PVP no crystallization from the melt is measurable at DSC.

Some blends provide such strong interactions between constituents that the T_g does not follow a composition-dependent pattern described by one of the equations presented above. Kwei[51] describes an example of this phenomenon where the T_g for blends of poly(4-vinyl-pyridine)2 and poly(4-hydroxystyrene)5 is much higher than the predicted weight average

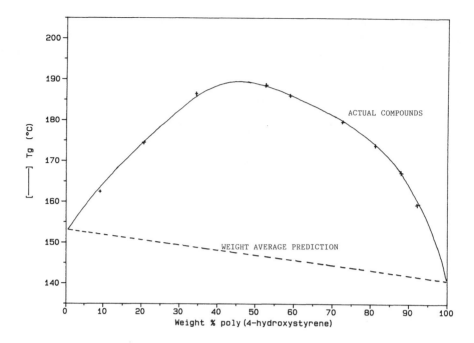

FIGURE 14. T_g's of actual blends of poly-(4-vinylpyridine)2 and poly(4-hydroxystyrene)5 and predicted weight average T_g's as a function of composition. (From Kwei, T. K., Proc. Compatibilization Polymer Blends Symp., 1990.)

value. This is attributed to hydrogen bonding between the two polymers. Figure 14 shows the plot of blend T_g as a function of weight percent of hydroxystyrene.

Studies similar to those outlined above for PTMT-PTMO block copolymers have been conducted by Cooper et al.[52] and by Hesketh and Cooper[53] for a class of polyurethane elastomers. These materials are based on a soft segment of poly(ethylene ether carbonate) polyol and hard segments based on the condensation product of diphenylmethane diisocyanate (MDI) and 1,4-butanediol. These materials possess significant potential for hydrogen bonding at the carbonate carbonyl and ether oxygen sites. This may contribute to phase mixing in a class of materials where phase mixing and demixing have a substantial effect on structural properties.

Joseph and Harris[54] describe DSC studies on materials of this type where only the hard segment composition is varied. Once, again, dynamic mechanical analysis augments the DSC work. A number of observations are made indicating the occurrence of phase mixing as a function of certain thermal treatments. Comparing the initial heating of a compound with reheating after rapid quenching shows that the quenched material has a higher T_g and a lower T_m. The T_g of the quenched material is also associated with a larger ΔC_p and is immediately followed by a crystallization exotherm. All of this is displayed in Figure 15.

The T_g's of initial heating and reheating after quenching are plotted vs. hard segment concentration to show that first-scan T_g's are nearly constant while second-scan T_g's are composition dependent, again typical of phase mixing. Plotting these second-scan T_g's vs. composition and comparing them to a plot for a completely phase mixed system by the Fox equation shows good agreement, particularly for compounds of low hard block content.

In addition, the heat of fusion of first scans was highly influenced by hard block concentration, varying from 5 J/g for a 35% hard segment material to 38 J/g for 65% hard segment. However, second-scan values varied from only 5 to 15 J/g across the same com-

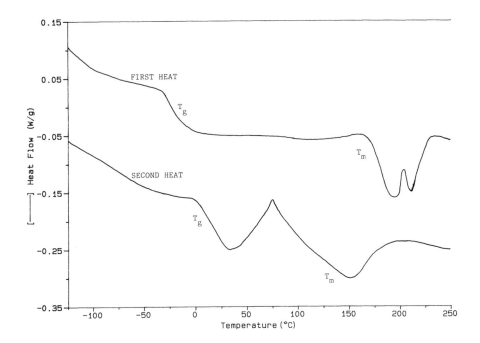

FIGURE 15. First and second heating scans of a thermoplastic urethane. (From Joseph, M. D. and Harris, R. F., *J. Appl. Polym. Sci.*, 41, 509, 1990.)

position range. The combined raising of T_g and melting point depression is attributed to amorphous hard block dissolving in the soft segment. Heating a quenched sample to a temperature above the T_{cc} exotherm but below the T_m followed by cooling and reheating revealed a reduced T_g indicating that hard segment left the soft block without actual melting. This is consistent with a model of mixing-demixing set forth by Wilkes et al.[55]

Annealing of quenched samples at various temperatures between T_{cc} and T_m cause the crystallization exotherm to vanish and a two-phase melt is established; the thermograms resemble the first heating of as-received material. Higher annealing temperatures result in a higher melting point for the low-melting crystals. At sufficiently high annealing temperatures all the crystals melt at the higher temperature while the soft block T_g shows an increase in both temperature and ΔC_p. The authors offer an explanation of this behavior in terms of mixing and demixing as a function of thermal history.

Since thermoplastic elastomers of this type have sub-ambient glass transitions, they reside in the crystallization window between T_g and T_m when stored at room temperature. While the above study shows that high temperature annealing can activate solid-state crystallization, it has also been observed that maximum physical properties are not achieved in molded thermoplastic polyurethane parts until 3 to 4 weeks of room temperature storage after molding.[56] Since the polymer cools rapidly from the melt during the injection molding process, it can be considered to be quenched. Phase demixing can occur slowly in the material because of the mobility in the polymer at room temperature, some 60°C above T_g. As an alternative to waiting for this period of time, annealing routines of 16 to 24 h at 115 to 120°C are recommended. Flexural fatigue properties are found to improve by the greatest degree, a result that would be expected from an increase in the crystalline domains.

If the crystalline structure becomes too dominant in an elastomeric system, ultimate elongation may be lost and the polymer may become brittle. This is a common occurrence after exposure to elevated temperatures for extended periods of time. Studies of Hytrel, a

PTMT-PTMO block copolymer, show that the unstabilized polymer loses a significant amount of elongation after as little as 336 h at 150°C in air.[57] It is likely that the mechanism for embrittlement is a combination of cold crystallization and oxidation. Heat stabilizers can be added to the resin system to extend the useful life of the polymer at elevated temperatures. While stabilized moldings will retain ductility, the additive produces chromophores as it is activated, causing heat-aged samples to turn a dark brown while unstabilized product remains near the as-molded color.

This author has observed that long-term storage at room temperature under UV light from fluorescent lighting fixtures will also produce chromophores in moldings produced from stabilized resin. Unstabilized moldings display little or no color shift under UV exposure. Figure 16A and B show DSC scans of unstabilized and stabilized product exposed for 8 to 10 h/d over a 2-year period. Figure 16a shows the melting endotherms for the crystalline hard block and 16b highlights the smaller-scale heat capacity changes associated with the glass transition. Several differences are apparent.

First, the heat-stabilized product exhibits the melting of some imperfect crystals near 155°C just prior to the major melting. This is typical of freshly molded products. The unstabilized product shows no preliminary melting. In addition, the stabilized molding has a heat of fusion of the major melting event of 27.45 J/g which is split between two crystal populations in a ratio of about 90/10. Scans of freshly molded product give a range of 26.8 to 28.1 J/g with a similar ratio between the two segments of the endotherm. The unstabilized molding exhibits a heat of fusion of 33.4 J/g with virtually all of the change occurring in the larger, lower temperature portion of the melt. No change in the actual melting point is observed.

At the low temperature end of the thermogram, the T_g of the two systems is identical but the ΔC_p is 50% larger for the stabilized product. A small endotherm at -5°C also has a higher heat of fusion in the stabilized molding. While no studies of morphology have been conducted, it is apparent that some phase reorganization is occurring in the unstabilized moldings while the stabilized product appears essentially unchanged from the day when it was molded. DMA evaluations at room temperature and above detect no indication of embrittlement. This would lead to the conclusion that reorganization to higher crystallinity without the occurrence of oxidation is not sufficient to produce the embrittlement observed with high-temperature aging.

Maurer[58] has described in great detail the use of DSC and DTA techniques in analyzing the composition and structure of elastomer blends using glass transition and melting point data. He reviews the benefits and pitfalls of employing equations relating T_g to composition for these systems. Problems such as dominance by one of the components or nearly equivalent T_g's in two polymers in the blend can cause experimental problems with data interpretation.

Because of these difficulties, Maurer cites work performed by Sircar and Lamond,[59,60] who evaluated degradation as measured by DSC to identify components in a blend. It was found that exothermic degradation, as in SBR 1500, endothermic degradation, as in EPDM, and mixtures of the two mechanisms in blends can often be more distinctive than T_g or T_m data. This will be discussed in more detail in Section IV.

Many of the newer thermoplastic elastomers are immiscible blends that derive many of their useful properties from the incompatibility of a particulate thermoset fraction dispersed in a thermoplastic matrix. These systems are referred to as dynamically vulcanized and the fine particle size of the vulcanized component is the key to useful properties and processability. The development of these materials has been described by Coran and Patel.[61]

When we examine the DSC thermograms of materials in these families we observe a behavior quite different from that displayed by the PTMT-PTMO block copolymers. Instead of exhibiting a T_g or T_m that varies with the percentage of soft and hard segment, we observe

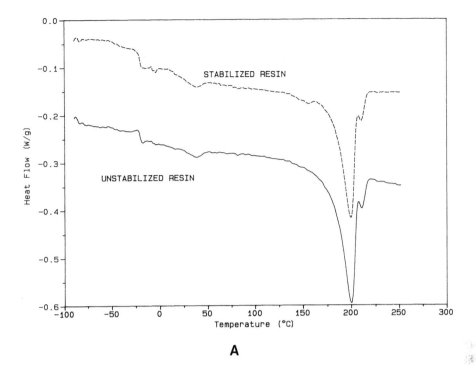

A

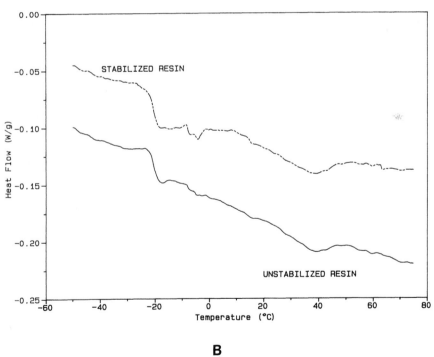

B

FIGURE 16. (A) Hytrel 5556 with and without heat stabilizer stored at room temperature and UV aged. (Sepe, unpublished, 1992.) (B) Lower temperature portion of A.

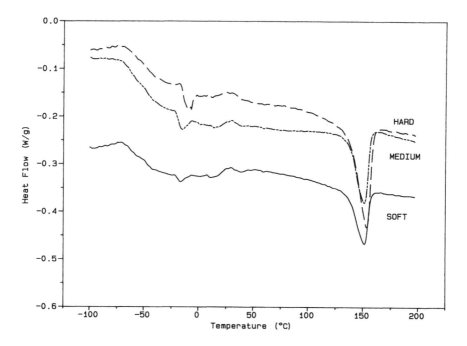

FIGURE 17. T_g and T_m behavior of Santoprene as a function of hard block/soft block content displaying little shift in transition temperatures. (Sepe, unpublished, 1991.)

transitions of constant temperature as a function of changing composition. The heat of fusion of the crystalline thermoplastic phase does change with concentration and this provides at least a relative indication of product composition. As expected, the higher hardness elastomers display a higher heat of fusion and properties such as strength, modulus, compression set, and fluid resistance will follow a regular pattern governed by the concentration of this crystalline phase. Figure 17 shows thermograms for three grades of a dynamically vulcanized compound based on polypropylene. There is a notable depression in the melting point from that of neat polypropylene, but it does not change significantly as the thermoplastic content increases. Figure 18 shows scans of an elastomeric nylon based on nylon 6/6. Here the melting point of the nylon is not shifted from that of the neat resin and again only the heat of fusion changes as the composition is altered.

D. AGING AND DEGRADATION

DSC techniques have proven useful in examining the degradation process in polymers in real time as well as in studying changes that may have occurred to products in the field due to environmental influences such as heat or weathering. Bair et al.[62] have noted that weathering and heat aging of ABS both result in a deterioration of the unsaturated polybutadiene component. This degradation, which is oxidative in nature, results in a glass transition that broadens, increases in temperature, and decreases in ΔC_p. The ΔC_p can be correlated with the polybutadiene content to quantify the change in rubber content, a factor of great significance to the impact properties of ABS products.

Many commercial polymer compounds rely on polybutadiene or another elastomeric component for a balance of impact properties and stiffness, particularly at cold temperatures. This method of evaluating degradation in the elastomer phase is applicable to a wide range of materials including high-impact polystyrene (HIPS) and blends of HIPS and PPO. In addition, many new alloys use block copolymers in the combined role of impact modifier

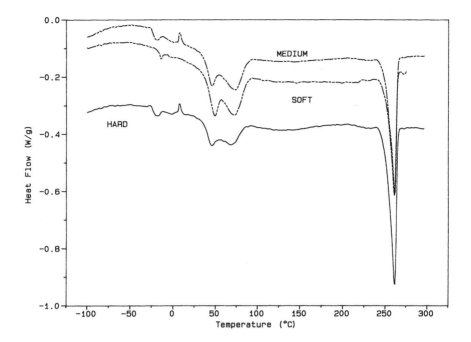

FIGURE 18. T_g and T_m of elastomeric nylons of different hard block/soft block content. (Sepe, unpublished, 1991.)

and compatibilizer, as in blends of PBT or PET with polycarbonate (PC). Changes in T_g can be useful in evaluating the response of these compounds to high-temperature aging. This is important for assessing long-term serviceability since significant aging can occur in the block copolymer at a temperature well below the short-term use temperature of the base polymers.

In semicrystalline polymers, where accurate measurement of changes in T_g can prove difficult, crystal melting can display irregularities that indicate degradation. Braunstein and Herrera[63] have related the presence of minor melting peaks below the major melt temperature to oxidative degradation in polypropylene. They demonstrate the creation of these lower melting crystals as a function of exposure time in an oxidative atmosphere at 290°C. The kinetics of the process were not examined.

Mason and Domingue[64] have used DSC as part of a simultaneous thermal analysis which acts as a combined DSC and thermogravimetric analyzer (TGA). DSC output was used to evaluate and fingerprint typical responses of degraded thermoplastic polyurethane (TPU) to controlled oxidation in the thermal analyzer. Samples of polymer which had been degraded during processing by either shear, heat, or moisture were heated through the melting point and then held at 250°C for 10 min. The same routine was performed on virgin resin. Both the melting point and heat flow responses were distinctive for each type of degradation in the specific TPU studied. DMA results and molecular weight distribution determinations by gel permeation chromatography (GPC) were also used to demonstrate the different effects that each type of degradation had on molecular structure.

This author has observed that oxidative degradation in polyacetals is detectable as a change in the heat of fusion or the heat of crystallization. In Figure 4 the H_f and H_c are essentially equivalent; no oxidation has occurred since the initial heating was performed in nitrogen to an upper temperature of 220°C. However, if the sample is heated in air to 300°C and then cooled through the recrystallization temperature, the H_c is dramatically reduced.

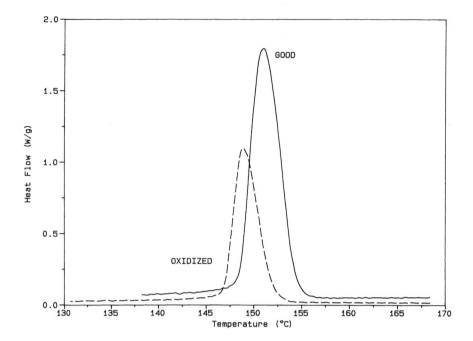

FIGURE 19. Recrystallization exotherms from the melt for acetal copolymer after normal melting and intentional oxidative degradation at cooling rate of 2.5°C/min. (Sepe, unpublished, 1992.)

Figure 19 compares the recrystallization of a degraded sample with that of a properly handled specimen. While the peak temperature decreases only 4°C for the degraded sample, the magnitude of the exotherm is nearly halved, indicating extensive destruction of the crystalline phase. Upon reheating, the same disparity appears in the melting endotherm, showing that the change in structure is permanent and not the result of the cooling rate during crystallization.

In Section A it was shown that secondary crystallization can be an indication of aging time and temperature since the melting point of secondary crystals depends upon the time and temperature at which the crystals form. Fann et al.[65] have studied this effect in telecommunication cables made from various types of polyethylene (PE) as a function of temperature and exposure to different waterproofing jellies. Work has been conducted more recently on PBT polyesters.[66]

The first study utilized four PE resins of widely different densities and therefore different crystallinities. Unaged samples of all four resins displayed a single melting endotherm. The low-density polyethylene (LDPE) and medium-density polyethylene (MDPE) exhibited a secondary endotherm within hours of initial aging at 60 and 70°C. Changes in the linear low-density PE (LLDPE) are detectable in the thermograms but are so subtle that they could not be quantified. High-density PE (HDPE) showed no tendency for the formation of secondary crystals.

The lower density materials, possessing a lower degree of crystallinity, have more uncrystallized material available for secondary crystallization. In addition, a lower crystal order implies greater freedom of movement necessary for the reorganization of polymer chains. That both the melting point and the heat of fusion of the primary melting endotherm did not change is evidence that no significant amount of material from the amorphous regions was being absorbed into the primary crystal structure. As expected, the secondary crystal T_m increased with time indicating increasing crystal thickness. Although this was a thermally activated event, the added presence of the waterproofing jellies had a measurable effect on the rate at which the T_m of the secondary crystals increased. Some jellies produced little additional effect while others such as petrolatum jelly, which has good solvent effects in PE, produced a higher secondary crystal T_m in a shorter time. The solvent effects produce increased mobility in the amorphous regions of the polymer matrix, enabling secondary crystallization to occur more rapidly. However, if solvation becomes excessive it will inhibit crystal growth by producing too much segmental motion to permit stable rearrangement. Thus, there is not necessarily a proportional relationship between absorption of the jelly and secondary crystal growth.

In PBT the secondary crystallization is measured as an enthalpy change in the major melting endotherm as well as an increase in T_m. Again the crystallization rate is temperature dependent, occurring much more rapidly at 100°C than at 70°C. The combined effects of elevated temperature and the various jellies are much greater than those of temperature alone. Time dependency is also clearly indicated.

DSC has also proven useful in evaluating the effects of solvents on amorphous thermoplastics. It has been observed that not only do solvents have the expected diluent effects measurable by shifts to lower T_g, but they can also induce crystallization. The effects of methylene chloride have been reported by Hsieh and Schneider for polyetherimide (PEI)[67] and by Hsieh for polyethersulfone (PES).[68] In the latter evaluation, Hsieh reports that the DSC thermogram of PES polymer treated with solvent shows a reduced T_g and significant baseline noise near the normal T_g. This noise is associated with the outgassing of solvent which has become strongly bonded with the polymer. The presence of a crystalline phase has not been measured by DSC but has been inferred from Raman spectra of cloudy solutions of PES in methylene chloride.

Aging of amorphous polymers below but near T_g can also be detected by DSC as changes in T_g. Measurements of T_g are most accurate when the heating rate of the experiment is equal to the cooling rate at which the sample was prepared.[24,25] If a sample is cooled at a rate which is slow compared to the rate of reheating, a higher T_g coupled with an endothermic recovery will be observed. This occurs because of the time-dependent, relaxational nature of the glass transition. The frozen-in structure of the glass has a relaxation time that is comparable to the rate of formation during cooling. When a heating rate is employed which is much faster than the cooling rate, the molecules in the glass must take on energy rapidly to keep up with the time scale of the heating experiment.

The response of the glass transition to aging is analogous on a longer time scale to the hysteresis effects observed when employing differential cooling and heating rates. This is due to the fact that amorphous polymers are not at thermodynamic equilibrium below T_g. Continued relaxation of the frozen molecular structure occurs below T_g on a time scale that can be very long. This progress toward equilibrium results in changes in free volume which are an essential feature of physical aging.[11] Physical aging is a phenomenon distinct from the oxidative or photo/chemicodegradative processes discussed above. Reductions in free volume due to physical aging result in changes in properties such as reduced impact strength and improved creep resistance at low strains; thus, evaluating physical aging has value for assessing the long-term performance of polymers.

The slow relaxations that occur in a glassy polymer at temperatures well below T_g result in a matrix characterized by long relaxation times. When these molecules are heated at normal heating rates for DSC experiments, the ΔC_p will exhibit the overshoot and hysteresis as discussed above. The degree to which this behavior differs from the true T_g of the polymer depends upon the aging time and temperature. Controlled studies at known aging conditions can be used to establish benchmarks for examining specimens of unknown thermal history. Wunderlich[25] cites an example of this technique for polystyrene annealed at 70°C for various times. An example of apparent physical aging in the amorphous glass of a semicrystalline PBT has already been discussed in Section A.

Virtually all polymers are subject to oxidative degradation. As with other types of physical and chemical changes, oxidation can occur at temperatures which might normally be considered well within the useful range. Thus, characterization of this process by thermal analysis has received a great deal of attention. While TGA is often the technique of choice for studying degradation processes, it is useful to correlate simple weight loss data with the calorimetric information.

Bair[69] describes a method for determining the oxidation induction time (OIT) by both DSC and DTA for polyethylene wire and cable insulation. The examination of field samples displaying obvious degradation can reveal differences in resistance to oxidation due to lower antioxidant content. Ezrin and Seymour reported on the DSC evaluation of dark spots in insulation made from crosslinked PE (XLPE), comparing these spots to unaffected regions. It was concluded that the dark spots resulted from oxidation brought on by transition metal salts. These salts were in turn created by the reaction of catalyst residue with hydrochloric acid from a polyvinyl chloride (PVC) jacketing. The darkened regions revealed reduced crystallinity from aging. The T_m was reduced by 4°C compared to unaged polymer and the H_f was also lower. In addition, the onset of the oxidation exotherm was 182°C for the darkened areas vs. 194°C for the unaffected regions.[70]

Effects of transition metals such as copper on the oxidative stability of PE can be evaluated by using copper sample pans instead of the more commonly used aluminum ones, or by placing a piece of copper mesh in with the PE sample before crimping the sample pan. Effects of pigments and other formulation items such as different antioxidants can also be evaluated using the OIT method. Performing the test at different isothermal conditions permits an evaluation of the kinetics of the oxidation process.

The use of a pressurized DSC cell can enhance measurements of oxidative degradation, particularly when volatility of the sample is a problem. Antioxidants have two important roles in polymer stabilization. The first is to protect the polymer during melt processing. The second is to prevent solid-state degradation that may occur during the service life of the fabricated product. For evaluating the melt-state oxidative stability, typical scans above the melting point to an oxidation onset as discussed in the OIT method are useful. However, for solid-state evaluation, extrapolating from the melt-state data can cause problems since the mobility of additives such as antioxidants is severely limited. The volatility of these additives becomes a problem when attempting to correlate thermal analysis test data to the behavior of molded articles. The moldings are relatively thick and the migration of volatile additives may be severely hindered in comparison to the small, thin DSC samples.

Pressurizing the DSC cell under air or oxygen has two advantages. It suppresses the volatilization of the additives by raising the boiling point and the more concentrated atmosphere permits lower test temperatures and/or shorter test times while still producing measurable changes in the sample. Thomas[71] illustrates the improved resolution of the onset of oxidation in polypropylene in a pressurized DSC cell compared to a cell run at ambient pressure. Figure 20 shows the results. Thomas also uses TGA data to demonstrate the problems that can arise from performing OIT measurements at temperatures above which an ingredient in the formulation is lost.

E. CURING AND CROSSLINKING

Exotherms can also be observed for curing and crosslinking. This is primarily useful in the study of thermosetting compounds which must crosslink during processing in order to achieve a property profile needed in end use applications. The process of curing defines the properties of the material. In addition, the progress of the material from uncured compound

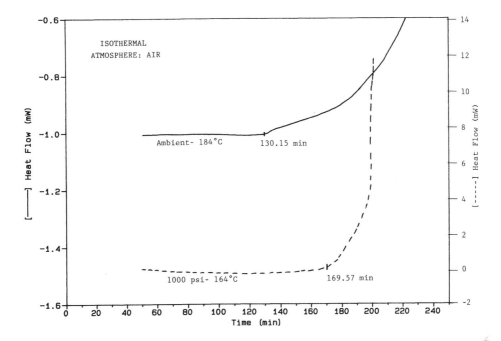

FIGURE 20. Oxidative stability determinations using ambient and elevated pressure DSC for polypropylene. (From Thomas, L, C., *Am. Lab.*, January, 1987. With permission.)

through gelation to crosslinking and vitrification governs the flow of the material, determines cycle times required for economical production, and defines conditions that can degrade the properties of the crosslinked structure. Reaction rate determinations, studies of kinetics of cure, and the measurement of extent of cure are therefore key aspects of understanding and controlling the thermoset production process.

The subject of determining the cure kinetics experimentally and a mathematical treatment of the various equations that describe the chemical reactions relating to curing are discussed in great detail by Prime.[72] This reference provides a rich source of experimental procedures, data interpretation, and an erudite discussion on the relative merits of different isothermal and dynamic heating methods in evaluating cure kinetics. It also covers the determination and significance of the rate constants and activation energies that are derived from a treatment of cure kinetics.

Determining the kinetics of the curing process allows the analyst to develop a series of time/temperature/degree of conversion profiles for a resin system. These are useful in characterizing the relative importance of different process changes to the final state of the resin in the molded part. Figure 21 shows a time vs. conversion plot at different temperatures for a phenolic material while Figure 22 shows the same data for an unsaturated polyester. The slopes of the individual lines are a measure of the effect of longer in-mold times at a given mold temperature while the spacing between the lines provides a measure of the effect of increasing temperature for a set time. These two figures show that the polyester is a faster curing system and is more responsive to changes in cure temperature than the phenolic. This is consistent with actual production experience. These profiles were developed using dynamic heating scans and method B with analysis software patterned after ASTM Method E698-79.[72]

Since physical properties are related to degree of cure, the energy release associated with the cure exotherm is an important value determined by DSC. For many resins a ratio

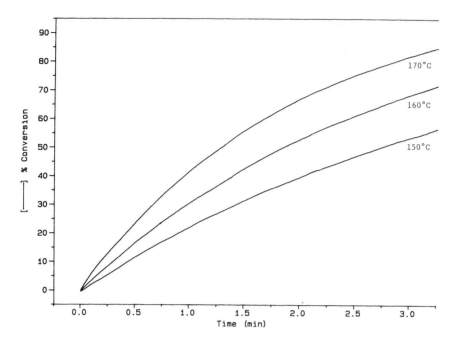

FIGURE 21. Conversion (cure) vs. time plot for a phenolic molding compound at isothermal conditions indicated. (Sepe, unpublished, 1988.)

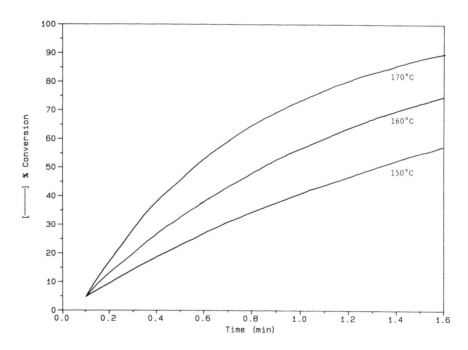

FIGURE 22. Conversion vs. time plot for unsaturated polyester compound at isothermal conditions indicated. (Sepe, unpublished, 1990.)

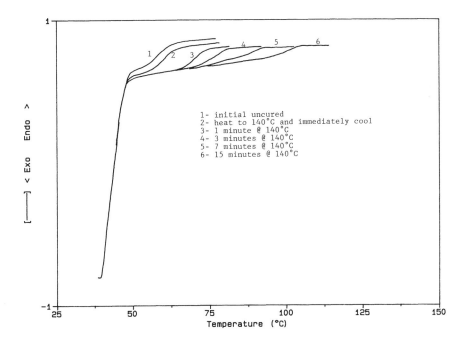

FIGURE 23. Glass transition temperatures for epoxy/glass laminates as a function of curing time at 140°C. (From Prime, R. B., *Thermal Characterization of Polymeric Materials,* Academic Press, 1981. With permission.)

of ΔH of residual cure in the molded part to that of the uncured resin is taken as the degree of conversion needed to achieve complete cure. Morrison and Waitkus[73] have proposed that changes in the structure of phenolics can occur which are often associated with post-curing even after calorimetry determinations reveal no residual exotherm. This relates to a morphology which may be unique to phenolics. At one level, initial curing produces growth sites which become the nuclei of nodules made up of highly entangled structures. Crosslinking involves the creation of bonds between the nodules. The progress of this stage of the reaction can be monitored by calorimetry and in this region the strength of the material and the transition temperatures both increase, although not monotonically.

Further attempts to increase the degree of cure will raise the transition temperature, the so-called glass transition; however, strength will actually decline. This occurs because further bridging between nodules becomes difficult due to decreased mobility in the cured matrix. At this point material may be added to the individual nodules at the expense of the internodular bridges; the transition temperature will continue to rise but strength will decrease. Changes in this stage are often referred to as post-curing; however, they are not measurable as changes in the cure exotherm and do not enhance the properties of the material.

For many other thermosetting polymers such as epoxies, T_g determinations by either DSC, DMA, or TMA are useful indicators of the state of the material. Of these techniques, only DSC provides quantitative data on the degree of conversion. Gray[74] has shown the progressive increase in the T_g of an epoxy-glass laminate, illustrated in Figure 23. Figure 24, from Reference 75, shows scans of an epoxy-anhydride cured to apparent completion at the temperatures indicated. This last illustration shows the gradual disappearance of the residual exotherm of cure as temperature increases. The slight decline in T_g at the highest cure temperature is indicative of degradation. This may occur over longer exposure times at temperatures well below the short-term upper temperature service limit for the material as determined by standard deflection tests.

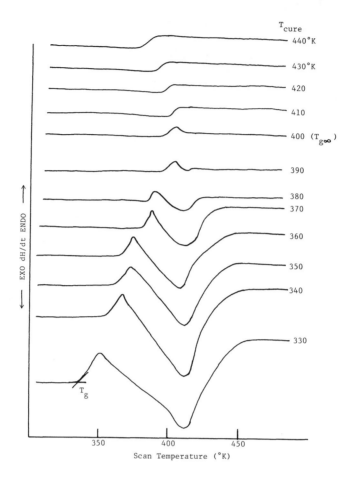

FIGURE 24. DSC scans at 8°C/min of epoxy-amine samples previously cured to apparent completion at temperatures indicated. (Fava, 1968.)

Prime also discusses the measurement by DSC of several key characteristics of curable systems. These are essentially extent of conversion determinations at various points in the cure cycle. A brief review of these will be included here. Table 1 contains a glossary of terms associated with the curing parameters.

T_{c_0} is the temperature below which no significant reaction can occur regardless of the time scale. It has practical significance as an upper limit storage temperature. Sourour and Kamal[76] demonstrated that a semilogarithmic plot of $1-\alpha_{T_c}$ vs. isothermal cure temperature can be extrapolated to T_{c_0}. $1-\alpha_{T_c}$ is the fraction of unreacted polymer due to quenching of the reaction by vitrification:

$$1 - \alpha_{T_c} = (\Delta H_{RXN} - \Delta H_{T_c})/\Delta H_{RXN} \qquad (6)$$

where ΔH_{RXN} is the total heat liberated when an uncured material is taken to complete cure and ΔH_{T_c} is the heat liberated at T_c before vitrification occurs. A standard measurement of the T_g of the unreacted fraction, T_{g_0}, should yield a value close to T_{c_0}.

T_∞ is the minimum curing temperature at which total conversion takes place. For some systems this may represent nearly 100% conversion while for others it signifies an ultimate conversion, α_{ult} which may be as low as 80%. Sourour and Kamal[76] have plotted the total

TABLE 1
Glossary of Curing Terms and Characteristic Curing Parameters

α	Extent of conversion (e.g., of epoxide groups), fraction reacted, degree of cure
α_{gel}	At the gel point
α_{ult}	Maximum achievable extent of conversion ≤ 1
t_{gel}	Time to gelation (gel time)
t_{vit}	Time to vitrification
T_c	Cure temperature
T_{c0}	T_c below which no significant reaction of the uncured resin mixture occurs, $\cong T_{g0}$ (cf. storage temperature for uncured resin mixture)
$T_{c,gel}$	T_c at which vitrification and gelation occur simultaneously
$T_{c\infty}$	Minimum T_c at which ultimate conversion occurs
T_g	Glass transition temperature
T_{g0}	T_g for thermoset with degree of conversion $\alpha = 0$
$T_{g,gel}$	T_g for thermoset with degree of conversion α_{gel}; $T_{g,gel} \geq T_{c,gel}$
$T_{g,ult}$	T_g for thermoset with degree of conversion α_{ult}
$T_{g\infty}$	T_g for thermoset with degree of conversion $\alpha = 1$
B (stoichiometric parameter)	Ratio of equivalents of comonomer 1 (e.g., amine hydrogens) to comonomer 2 (e.g., epoxide)

From Prime, R. B., *Thermal Characterization of Polymeric Materials*, Turi, E. A., Ed., Academic Press, New York, 1981. With permission.

heat of reaction, ΔH_{T_c}, at various isothermal cure temperatures, T_c, for two epoxy systems of different stoichiometries. Extrapolation of ΔH_{T_c} to ΔH_{RXN} yields a value for $T_{c\infty}$. $T_{g\infty}$ is the glass transition temperature at full conversion. If this conversion is less than 100%, the term $T_{g,ult}$ is used and $T_{g\infty}$ is achieved by extrapolation.

To measure $T_{g\infty}$ or $T_{g,ult}$ it is necessary to establish ultimate conversion by curing at a temperature above T_{c0}. Absence of a residual exotherm is one indicator of ultimate conversion; however, care must be taken to avoid curing temperatures high enough to produce thermal or oxidative degradation. Measurements should be performed on samples heated at the same rate as the cooling rate and no aging should have occurred.

For an accurate measurement of α_{ult} it is necessary to determine ΔH_{RXN}. This can be achieved by a spectroscopic measurement of reactive group concentration as has been performed by Westwood[77] in measuring residual methylol in phenolic resole.

Gelation determinations are important since beyond the gel point the thermosetting system will not flow and processing becomes difficult. Weak weld lines, trapped gases, and cosmetic defects from prematurely hardened resin in the shot can all result. DSC can measure degree of conversion but supplemental techniques are required to evaluate the physical changes that occur at gelation. Schneider et al.[78] report on experiments that combine DSC and torsional braid analysis (TBA) to establish the degree of conversion at the gel point. More work on gelation and cure determinations by DMA will be reviewed in Section VI. The time to gelation, t_{gel}, has been shown by this author to be a key factor in accounting for cosmetic defects. Two supposedly equivalent phenolic materials were utilized in long production runs; one resin produced a markedly higher percentage of rejected parts due to hard spots, localized areas on the part surface that reached gelation before the cavity had filled completely. Figure 25 shows a conversion vs. time plot for the two materials based on an evaluation of cure kinetics. This shows a faster cure and therefore a shorter t_{gel} for the more troublesome material.

Acitelli et al.[79] coupled DSC techniques with IR and dc conductivity measurements during isothermal cure of epoxy-amine. The use of conductivity measurements to examine physical changes in the crosslinking system was a precursor to the modern commercial

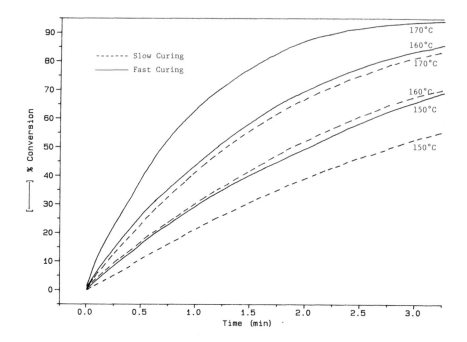

FIGURE 25. Conversion vs. time plots for phenolic compounds with different curing rates at isothermal conditions indicated, derived from kinetics. (Sepe, unpublished, 1988.)

dielectric analysis (DEA) instruments which will be discussed in Section VII. The onset of gelation produces a sudden decrease in conductivity due to the reduced ability of ionic species to migrate in the rapidly forming network. Table 2 summarizes the data obtained from both DSC and dc conductivity. The α_{gel} values agree closely with the Flory equation[80] relating α_{gel} to stoichiometry for these specific systems:

$$\alpha_{gel} = \sqrt{B/3} \tag{7}$$

Once α_{gel} is known, ΔH_{gel} can be readily determined from

$$\alpha_T = \Delta H_T / \Delta H_{RXN} \tag{8}$$

Sourour and Kamal,[76] using calculated α_{gel} values, have employed isothermal method 1 for an epoxy-amine system of varying stoichiometries. Their measurements of t_{gel} were in excellent agreement with rheologically measured gel times by cone and plate viscometry and TBA.[81]

Using the isothermal method at several temperatures, Sourour and Kamal were also able to determine $T_{c,gel}$, the cure temperature at which the total heat of reaction equals ΔH_{gel}. The reaction stops at this degree of conversion due to vitrification. $T_{g,gel}$ can be measured accurately for a material advanced to α_{gel} provided that an appropriate thermal history is applied to the sample.[24,25] Sourour and Kamal have shown that $T_{g,gel}$ was 10 to 13°C above $T_{c,gel}$ for the specific epoxy-amine system evaluated in their study.

Rheological determination of t_{gel} and identification of the gel point will be discussed in Section VI. It will be mentioned at this point that rheology can be used in conjunction with DSC to achieve accurate measurements of $T_{g,gel}$. The progress to gelation can be measured in the rheometer in real time. When the sample achieves the gel point, the test is stopped,

TABLE 2

Measurement of Degree of Conversion at the Gel Point (α_{gel}) by Combined DSC and dc Conductivity, DGEBA-m PDA (B = 1.3)

T_c (°C)	Gel time (min)[a]	α_{gel} (%)[b]
23	1860	70
56	155	68
75	70	72

[a] From dc conductivity-time plots, measured activation energy used to convert conductivity gel times (52.5 and 72.5°C) to DSC gel times at 56 and 75°C.

[b] From gel times plus measured α vs. time by DSC method 2.

From Prime, R. B., *Thermal Characterization of Polymeric Materials*, Turi, E. A., Ed., Academic Press, New York, 1981. With permission.

the sample is cooled at a specific rate, and it is then transferred to the DSC. The sample is then heated at the same rate as the cooling rate employed in the rheometer.

The ability to evaluate the curing process as described above allows the analyst to investigate the effects of formulation on the curing parameters. Catalysts, fillers, and other additives can all have an impact on the heat of reaction, activation energy, and ultimate degree of cure. The principles of these studies and specific examples are reviewed in detail by Prime.[72] Some more recent work is summarized here.

Gonzalez-Romero and Casillas[82] have used dynamic heating methods to evaluate the effect of catalyst concentration and filler content on the cure kinetics of commercial unsaturated polyesters. They determined values for the preexponential factor, A_o, and the activation energy, E, as a function of catalyst concentration using a multiple regression technique. They also demonstrated that the addition of kaolinite clay as well as fiberglass had no influence on the reaction kinetics. However, these results cannot be generalized to other formulations since interactions between fillers and other additives can produce unexpected effects. The catalyst system in the above study was bis(4-*t*-butyl cyclohexyl) peroxydicarbonate. Prime includes a review of work by Willard[83,84] on diallyl phthalate (DAP) compounds using various fillers and dicumyl peroxide as the initiator. In this instance the kaolin clay eliminated the curing process. It is proposed that the acidic clay decomposed the initiator. Figure 26 shows the heats of reaction for the various fillers. Similar but less dramatic effects from the kaolin filler were noted for *t*-butyl perbenzoate.

Since many conventional rubber compounds are formed by curing or vulcanization, the potential exists for utilizing the above techniques for the examination of thermoset elastomers. Evaluation of curing in these materials tends to focus on the effects of formulation rather than on the reaction kinetics. This is probably due to the wide variety of formulas employed by the rubber industry and the resulting complexity of interactions that can arise as the presence and concentration of different ingredients are varied. In addition, the presence of sulfur, which can result in a variety of sulfide structures,[85] complicates the kinetics analysis of the hard rubber curing reaction.

Maurer[86] stresses the need for appropriate background information on formulation details of a compound to maximize the usefulness of thermal studies. With this background, a great

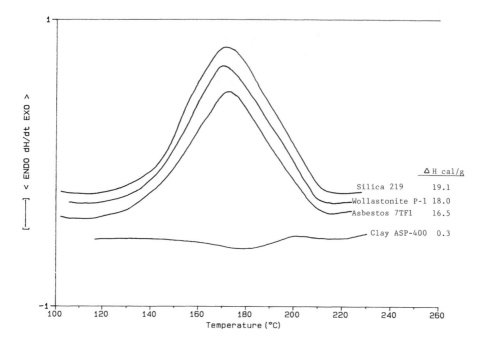

FIGURE 26. Effects of fillers (50 wt %) on curing of a DAP prepolymer catalyzed by 3 phr dicumyl peroxide. Acidic clay surface postulated to have decomposed the initiator. (Willard, Marcel Dekker, Inc., 1974. With permission.)

deal of information regarding the effects of formulation changes on the heat of vulcanization, H_v, can be derived from DSC studies. DSC techniques, in particular the use of a pressurized cell to suppress volatilization during cure and the selection of noncorroding and nonreacting thermocouple materials, overcome many of the problems encountered with early DTA studies.[87] Comparisons of uncured to cured material to examine changes in T_g, residual heat of reaction, consumption of curing agents, and the melting and degradation of the polymeric fraction can all be accomplished with pressure DSC. Physical measurements by TMA and DMA can augment this work.

Maurer also reviews work by Sircar and Lamond,[88] which illustrates the use of DSC in a nitrogen atmosphere to perform compositional analysis of ternary blends of styrene butadiene rubber (SBR), ethylene-propylene-diene rubber (EPDM), and chloro-isobutene-isoprene rubber (CIIR). It was shown that each component exhibited a characteristic exotherm or endotherm associated with degradation. The intensity of these enthalpic changes varied as a function of concentration for each ingredient.

Sircar and Lamond also demonstrated that DSC thermograms of a given blend, in this instance natural rubber and chloroprene (NR/CR), can be affected by the chemical method of curing.[89] Sulfur-cured CR in white sidewall blends displays a sharp exotherm attributed to the dehydrochlorination followed by crosslinking. The peak temperature of the exotherm shifted to higher temperatures as the CR content diminished. The peroxide-cured CR did not exhibit this behavior. Instead, only a single broad exotherm was observed which appeared to be independent of concentration in the blend.

The success of DSC techniques in the evaluation of thermoset elastomers is inextricably associated with thermogravimetric methods. TGA can often quantify that which DSC only identifies, and TGA is often needed to confirm events detected by calorimetry. We will therefore begin our discussion of thermogravimetry with a treatment of thermosetting elastomers.

IV. THERMOGRAVIMETRIC ANALYSIS

Taken together, DSC and TGA represent the most commonly used techniques in thermal analysis. Principally, TGA examines the process of weight changes as a function of time, temperature, and other environmental conditions that may be created within the apparatus. TG and DTG are often treated as separate techniques; however, here the examination of the derivative of the weight change will be regarded as an integral part of the TG analysis.

In its simplest form, TG measurements are made by heating a sample at a constant rate in a prescribed atmosphere. As different components in the sample reach the temperature of volatilization or degradation, a step change in the weight will occur. In most cases this is a weight loss, although combinatorial effects can produce weight increases in some circumstances. This weight change is highlighted by the derivative curve which begins near zero at the onset of the event, reaches a peak, then returns to zero at completion. The derivative curve can be especially useful for pinpointing the onset and endpoint of decomposition for minor components where the weight loss range covers a broad temperature range. It also has utility in multipolymer systems where the degradation temperature regions overlap and a visual examination of the weight loss curve proves difficult. While the overlap prevents a quantitative resolution of the compound into its polymeric constituents, relative values can be assigned to the weight losses bounded by the derivative curve minima. These can be used to monitor the consistency of a material on a quality control basis.

Further separation can be achieved by isothermal analysis at a temperature above the degradation onset of one ingredient but below that of the other. This can prove to be time consuming but will yield greater precision in isolating and quantifying components.

A. COMPOSITIONAL ANALYSIS

Rubber compounds offer fertile ground for compositional analysis since compounding is such a key aspect of the field performance of a finished product. More than in most polymers, the low temperature weight loss fraction constitutes a large part of the compound and is crucial to the properties of the material. This low temperature weight loss fraction is typically due to the presence of an oil extender, and a standard dynamic heating scan produces a substantial overlap between the weight loss due to oil volatilization and the weight loss from polymer degradation. In these cases, separation can be facilitated by drawing a vacuum on the instrument and then heating the sample to a temperature below the degradation onset of the polymer. Isothermal treatment for a certain time period will hasten volatilization of the oil extender without producing any loss in polymer.[90] Figures 27A and B compare the resolution of this ingredient using dynamic heating at atmospheric pressure and isothermal treatment under vacuum.

Maurer[86] presents alternatives to this method. The first involves extraction prior to TGA; the second entails establishing a correction curve based on a reference temperature for a given polymer compound; a third is a graphical resolution technique that compares unextended polymer with an oil-extended compound. While all of these methods have proven successful with given materials, they require supplemental techniques or multiple runs to verify accuracy. These are time-consuming steps that are undesirable in the context of quality control on a production compounding line. The vacuum TGA method offers the fastest resolution along with utility for a wide range of compounds.

The polymer degradation part of the thermogram provides valuable information about the relative thermal and oxidative stability of the polymer, rates of decomposition, and blend composition. For quantitative analysis of blends, background information about compound ingredients is helpful and often necessary. Maurer[86] reviews examples using TG and DTG curves to characterize thermosetting elastomers. With blends of vulcanizates that have

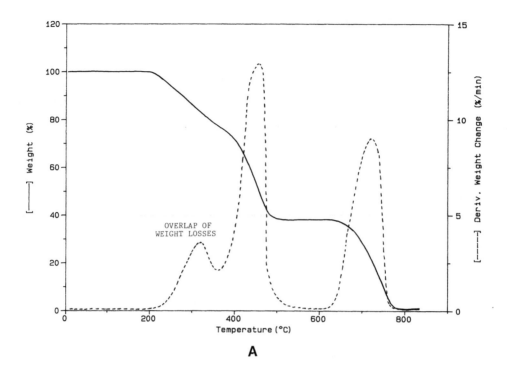

FIGURE 27. (A) TG and DTG curves of a compounded rubber showing overlap of weight losses for oil extender and polymer. (Groves and Thomas, 1988.) (B) TG and DTG curves for rubber sample analyzed under vacuum TGA with isothermal hold to separate weight losses. (Groves, I. and Thomas, L., *Res. Dev.*, February, 1988. With permission.)

distinctly different decomposition temperature ranges, such as NR and EPDM, overlaps in the DTG curves will be minimal and these provide a tool for quantitative analysis of composition. Qualitative analysis is possible with blends of vulcanizates that decompose at nearly the same temperature. Brazier and Nickel[91] present a systematic approach to analyzing polymer blend composition using TGA. They offer a method for examining materials where little or no prior knowledge is obtainable. They identify several interesting effects of degradation involving secondary decomposition in carbon black-filled isoprene. They also demonstrate the use of increased heating rates to enhance peak separation of two DTG peaks in butadiene degradation.

Sircar and Lamond[88] demonstrated good correlations between enthalpy changes due to degradation and inflections or peaks in the DTG curves of SBR/EPDM/CIIR white sidewall compounds. In addition, they evaluated quaternary white sidewall compounds of NR/SBR/EPDM/CIIR. In these instances, TGA together with DSC provided enough information to identify the presence of all four compounds. Sircar and Lamond[92] also discuss coupling T_g measurements by DSC with examination of the DTG curves to draw conclusions about components present in a blend. They showed that below a 50% butadiene (BR) content in SBR/BR blends there was a sharp dependency of T_g upon BR content, T_g declining rapidly as BR content increased.

Typically, dynamic heating is performed up to approximately 550°C in nitrogen. At this point only polymer char, carbon compounds, and inorganic fillers remain. By switching the gas stream in the TGA to oxygen or air at this point, residual organic materials oxidize to provide the final weight loss. This is an important aspect of identification for many polymers and will be discussed further below. For thermoset elastomers the primary significance is

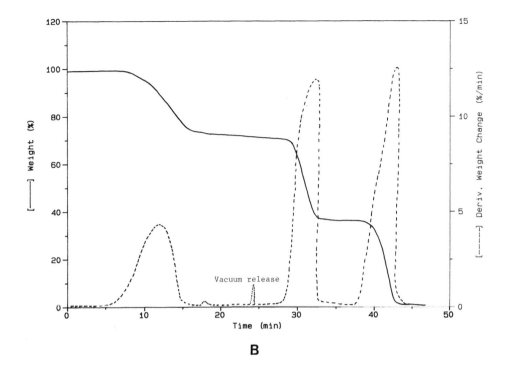

FIGURE 27 (continued).

the determination of carbon black (CB) content. To accurately determine the CB content it is necessary to distinguish the CB from polymeric char. Nitrile-butadiene rubber (NBR), for example, because of its considerable nitrogen content, forms a considerable char which will add to the CB determination. Maurer reviews a method developed by Swarin and Wims,[93] which involves stopping the heating process at 550°C, cooling to 300°C, and then introducing air while heating at 10°C/min. A small weight loss associated with char is distinguishable with only a small overlap to the main weight loss due to CB. This technique only works for CBs with oxidation temperatures high enough to be differentiated from the char.

Pautrat et al.[94] solved this problem by establishing a correlation curve between the nitrogen content of the polymer and the char that was formed. This bypasses the need for a detailed separation step. Sircar and Lamond[95] used a similar technique in which NBR composition was determined from a calibration curve correlating T_g to acrylonitrile (AN) content.

Nitrogen is only one element that can result in char formation. The other chemical group most commonly found in polymers which contributes to char formation is the backbone aromatic ring. Aromatic rings that make up a sufficient portion of an additive such as a flame retardant will also contribute to char formation; however, aromatic rings as pendant groups such as in polystyrene or SBR will not.

In polymers with no ring structure in the backbone, virtually no char is formed. Examples are polyacetal (PAc), polyethylene (PE), and polypropylene (PP). Polyamides produce a small amount of char due to the nitrogen in the amide groups. We would expect the char amounts to be higher for lower number nylons and this pattern is, in fact, observed. Partially aromatic backbones such as PBT and PET polyester produce more char than nylon, and polycarbonate produces even more due to its higher aromatic content. At this point, the higher degree of aromatic character results in a higher degree of thermal stability, observed as a higher degradation onset point. PES, PEI, and PPS represent materials at the upper end

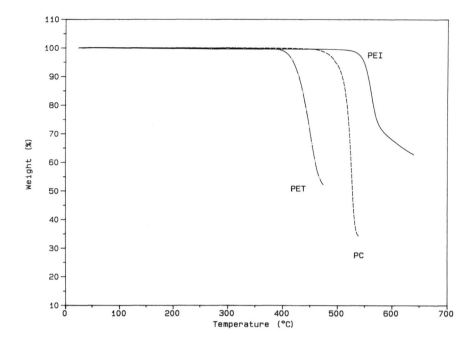

FIGURE 28.   Weight loss vs. temperature by TGA at heating rate of 20°C/min demonstrating relative thermal stability as a function of backbone aromatic character. (Sepe, unpublished, 1987.)

of this spectrum. Figure 28 shows the onset of polymer weight loss for PET, PC, and PEI. This illustrates the greater thermal stability of the aromatic polymers, which is borne out in actual processing. PET degrades at processing temperatures above 310°C, while polycarbonate can tolerate levels near 350°C and PEI can be heated above 425°C. Figure 29 shows the carbon (char) weight losses for these polymers after normalizing for filler content. Table 3 lists several polymers, their repeating unit structures, and typical char amounts formed for unfilled, natural color grades. It is interesting to note that materials which are high char formers tend to exhibit inherent flame-retardant (FR) properties.

The values of oxidized char provide a valuable tool for evaluating the composition of blends. For example, ABS/PC blends are available in a variety of blend ratios. These ratios govern physical and thermal properties, but the polymer weight loss exhibits a strong overlap between the ABS and the PC, making quantitative determinations impossible. However, pure ABS has a char content of 2 to 3% while PC measures 24 to 25%. Assuming the absence of additives such as flame retardants that will affect this value, the char content of the blend will fall on an essentially linear calibration curve between these two extremes.[96] Blends that rely on a third polymeric component acting as an impact modifier/compatibilizer will make this type of determination more difficult unless the chemical structure and percent contribution of this third component are known. However, even a relative value will be useful in establishing batch-to-batch uniformity for a given compound. PBT/PC and PPO/PA blends fall in this category.

Char formation can also be related to FR additives. These packages are designed to promote a different degradation mechanism in a polymer which would otherwise ignite. Many of these additives are halogenated biphenyls. They typically reduce the thermal stability of the compound since the orchestrated degradation of the additive package must occur before ignition of the base polymer. The aromatic character of the additive will result in substantial char formation.

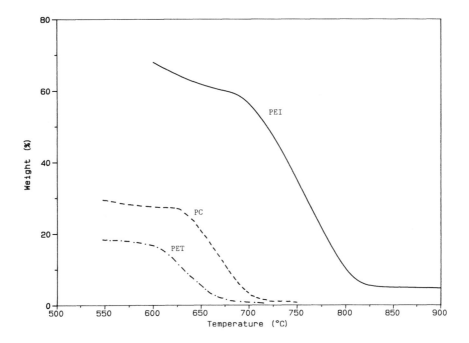

FIGURE 29. Carbon in air weight losses from scans in Figure 28. (Sepe, unpublished, 1987.)

TABLE 3
Char Formation as a Function of Polymer Backbone Structure

Material	Repeating backbone unit	Char (%)
Polyethylene	$[-CH_2-]$	<1
Polypropylene	$[-CH_2-CH]$ $\qquad\quad CH_3$	<1
Polyacetal	$[CH_2-O]$	<1
Polyamide-6	$[N-C-(CH_2)_5-]$ $\ \ H\ \ O$	3.5
PBT Polyester	$[-C-O-\bigcirc-O-(CH_2)_4-O-]$ $\quad O$	8
PET Polyester	$[-C-O-\bigcirc-O-(CH_2)_2-O-]$ $\quad O$	15
Polycarbonate	$[-O-\bigcirc-\overset{CH_3}{\underset{CH_3}{C}}-\bigcirc-O-C-O-]$	25
Polyphenylene Sulfide	$[-\bigcirc-S]$	65

Note: All ring structures are aromatic.

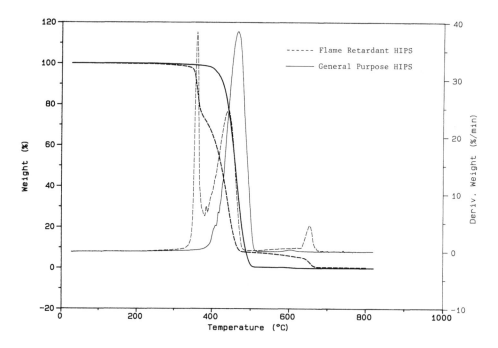

FIGURE 30. TG and DTG curves for general purpose high-impact polystyrene and flame-retardant analogue. (Sepe, unpublished 1988.)

Figure 30 illustrates the TG and DTG curves for a high-impact polystyrene (HIPS) and an FR HIPS. The early decomposition and the higher amount of char are both caused by the presence of a highly brominated diphenyl ether. Figure 31 shows a similar comparison for PC and FR PC. It is evident that less additive is used in the PC since the additional amount of char is less than that observed for the HIPS. It is also apparent that, while the FR additive degrades at a lower temperature than the base resin, the PC additive package is much more stable than that used for the HIPS. This is necessary due to higher processing temperatures required for the PC.

Inorganic or partially organic components left after the char weight loss are referred to as residue or ash. If some prior knowledge of composition exists, the ash can be identified as a filler such as clay or glass fiber. Lot-to-lot consistency of filler content can be measured on this basis. Valuable distinctions can be made between colorant systems that use dyes and organic pigments and those that use inorganic pigments. These determinations are especially useful in characterizing color concentrates, but can also be used on fully compounded materials. Figure 32 shows the TG scan after polymer decomposition of two ABS materials. While both materials were the same color, one displayed stability problems under ultraviolet light and also caused severe problems with a flake-type residue in the mold as well as in the heating cylinder. The good material has a higher level of inorganic residue indicating the use of more stable inorganic pigments. The higher char content of the bad material must be attributed to ingredients other than base polymer since it is uncharacteristically high for ABS. In this instance it was due to a less-stable pigment system. Figure 33 shows the TG traces for two red color concentrates designed for polyamides, one containing cadmium and the other free of cadmium. The cadmium-free system displays a weight loss in air that is significant while the system containing cadmium actually exhibits an initial weight gain, probably due to oxide formation when air is introduced to the instrument.

Fillers which are partially oxidized at high temperatures can be identified and assayed using the stoichiometry of the decomposition products. Calcium carbonate and talc are

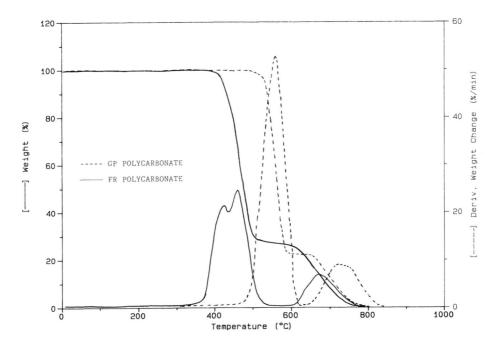

FIGURE 31. TG and DTG curves for GP and FR polycarbonates. (Sepe, unpublished, 1989.)

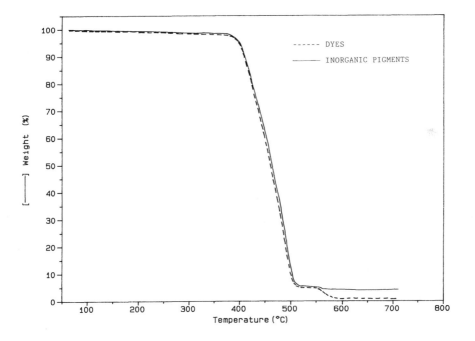

FIGURE 32. TGA scans comparing ABS compounds of the same color using different pigment systems. (Sepe, unpublished, 1989.)

minerals commonly used as fillers in polypropylene. While casual examination of a molded article is not sufficient to distinguish between the two fillers, they result in distinctly different property profiles. Figure 34 shows the char decomposition of two commercial PP materials with a nominal 20 wt % content of talc and calcium carbonate. The talc is a silicate and is

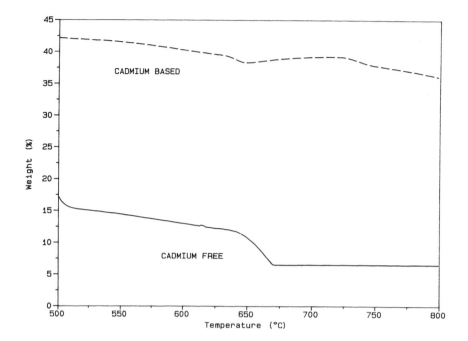

FIGURE 33.   Degradation in air of red pigment systems with and without cadmium. (Sepe, unpublished, 1991.)

stable in air at very high temperatures while the calcium carbonate decomposes readily by the reaction

$$CaCO_3 \rightarrow CaO + CO_2$$

This not only differentiates the fillers, it also permits the calculation of the amount of calcium carbonate in an uncomplicated manner since the base polymer does not contribute a significant amount of char.

The same degree of certainty can be applied to the determination of CB content in a black PE concentrate formulated for outdoor weathering applications. Since PE forms no char, the entire weight loss in air can be attributed to CB. Figure 35 shows a TGA of a 40 wt % CB concentrate. Similar work has been performed by Brennan.[97]

Because of the small sample sizes employed in most TGA instruments (50 to 200 mg), there is some debate regarding the representative nature of a sample of this size in characterizing a large lot of material. Depending upon mixing and dispersion during compounding, it is possible that samples may not be statistically representative. However, historically these measurements have agreed with muffle furnace ashing of much larger samples (15 to 30 g) to within 1%.[98] Cassel has applied the ash content measurements to the problem of uneven distribution of fillers in molded parts.[99] Here the technique should be even more effective than in the raw pellets because the molding process applies another heat and shear history to the material.

Maurer[86] discusses in some detail the analysis of the CB phase in thermoset elastomers. He shows that different CBs exhibit distinct oxidation characteristics based on differences in surface area properties. He also discusses the identification of mixed blacks. Also covered are the effects of accelerators and other agents that can influence the oxidation characteristics of CBs and cause discrepancies in the data between CBs in a vulcanizate and the same CBs analyzed in the raw state.

Even when a great deal of information is available about the details of a formulation, overlaps of DTG peaks can cause insurmountable difficulties for the analyst in quantifying

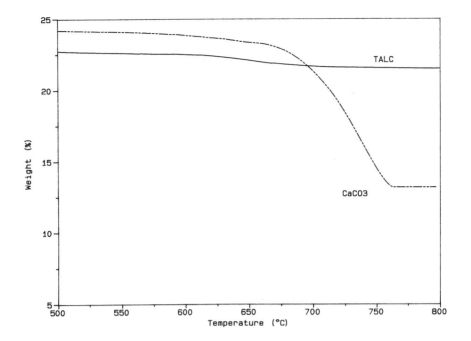

FIGURE 34. Carbon weight loss in air for CaCO₃ and talc in a polypropylene copolymer. (Sepe, unpublished, 1989.)

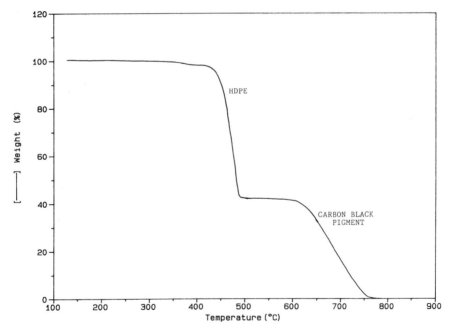

FIGURE 35. Carbon black weight loss in HDPE concentrate. (Sepe, unpublished, 1987.)

the concentration of different constituents. Figure 36 shows the TG and DTG curves for a general purpose ABS. While it is well known that there are three distinct components present in the material, the DTG curve gives only a broad decomposition range with a shoulder. Figure 37 shows DTG plots for a standard nylon 6/6 and an impact modified analog. While the secondary peak can be qualitatively attributed to the impact modifier, the overlap prevents

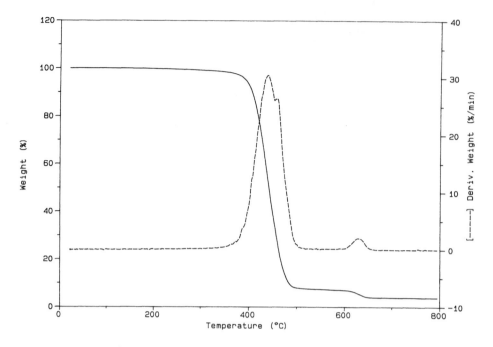

FIGURE 36. TG and DTG curves for GP ABS. (Sepe, unpublished, 1987.)

a quantitative evaluation. Figure 38 shows a commercial PC/PBT alloy with impact modifier. In this case three stages of decomposition can be delineated from the DTG curve, but again quantitative separation is not possible. Some commercial materials such as copolymers of PE and PP contain base polymers with such similar decomposition ranges that only a slight asymmetry in a single derivative peak gives any indication that the product is not a homopolymer.

While these problems are resolvable with techniques such as infrared spectroscopy, the desirability of improved resolution with TGA has given rise to various attempts at modifying the standard test methods. Rouquerol[100] reported on the use of a vacuum similar to the approach outlined above for the separation of heavy oils from thermoset elastomers. Heating rate control was then based on the sensing of pressure increases associated with the evolved gases of decomposition. Paulik and Paulik[101] devised a system that switched from dynamic heating to isothermal treatment when the weight loss per unit time reached a specified level. Sorensen[102] used a stepwise isothermal control to improve resolution of the weight loss event. Criado,[103] in an adaptation of the Paulik's method, used a heating rate adjusted to achieve a constant rate of weight loss. A less sophisticated approach is to simply slow the heating rate to 1 to 2°C/min from the normal 10 to 20°C/min.

All of these methods had the objective of slowing or stopping the heating process during a significant weight loss event in order to prevent the mixing of decomposition phases. Improved resolution was reported in all cases, but test times became extended beyond that which is useful for functions like quality control.

Within the last year a coupling of improved computer control and instrument design has given rise to a commercial instrument that incorporates into algorithms some of the principles used by the earlier experimenters. A smoothly changing dynamic heating rate that varies as a function of weight loss, a heating rate tuned to a preselected constant reaction rate similar to Criado's, or a stepwise isothermal routine keyed by the onset of a specific weight change

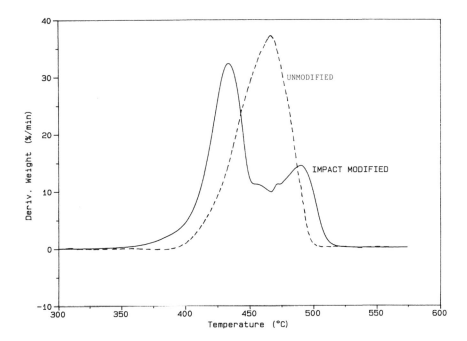

FIGURE 37. TG and DTG curves for an impact-modified and an unmodified nylon 6/6. (Sepe, unpublished, 1988.)

as in Sorensen's method can be selected by the operator as part of a programmable method. Key aspects of the hardware are as follows

1. A single thermocouple located close to the sample to facilitate rapid temperature response. Most older systems use a dual thermocouple design, one for monitoring sample temperature and one for furnace control.
2. A low mass furnace-necessary for rapid response.
3. Horizontal purge gas flow to insure good interaction between sample and purge gas and to remove decomposition products that could cause recombination and broadening of weight loss ranges. Inert gases such as helium, which has a very high thermal conductivity, further enhance the technique. However, nitrogen or argon are typically suitable.

Early work with this technique is presented by Lundgren and Michgehl.[104] The resolution of minor components in poly(methyl methacrylate) (PMMA), the improved resolution of vinyl acetate in ethylene-vinyl acetate (EVA) copolymers, and the separation of ABS decomposition into three distinct weight loss steps are all reported. Improved separation of PE and PP is also demonstrated. It is highly probable that many of the resolution problems discussed above will yield to this new capability.

Additional benefits of this methodology may be realized in the so-called hyphenated techniques of TGA-FTIR and TGA-MS. In the past 5 years commercially viable interfaces between TGA instrumentation and either infrared or mass spectroscopy have become available. Their use has been reported in the literature.[105] Prior to commercialization of such hybrid instruments, other scientists had constructed interfaces between TGA and MS instrumentation for the purpose of performing evolved gas analysis (EGA). Later, when FTIR

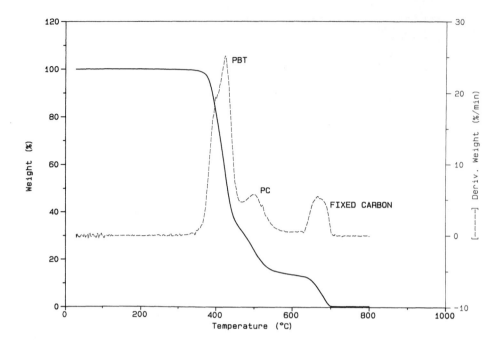

FIGURE 38. TG and DTG curves for an impact-modified PC/PBT alloy. (Sepe, unpublished, 1988.)

techniques made such rapid monitoring practical, Khorami et al.[106] described the use of a TGA-FTIR system for the same purpose. Joining these techniques provides valuable information regarding the chemical composition of decomposition products and assists in the identification of the degraded component. Lewis et al.[107] have used both a batch set up and an on-line TGA-FTIR technique to study the by-products of cure and degradation in high-temperature polyimides.

The benefits of a high-resolution TGA instrument should prove especially significant in these areas since unidentified shoulders in the DTG curves are frequently mentioned in the literature as an obstacle to accurate characterization.

B. ADDITIVES

The determination of oxidation induction time (OIT) by DSC has been discussed above. Bair[108] has demonstrated the use of TGA to achieve the same evaluations. Bair demonstrates that TGA has the advantage of not requiring a minimum temperature for measurement of an exotherm as in DSC. In TGA methods the onset of oxidation is indicated by a weight gain which reaches a maximum before volatile constituents are evolved and weight loss begins. OIT is measured as the time to initiation of weight gain at a given temperature. The relationship between induction time and antioxidant concentration in PE samples is shown. Constructing a calibration curve of induction time vs. antioxidant content permits the evaluation of unknown samples based on OIT. Sensitivity can be increased by reducing the temperatures of the test in order to spread out the induction times. Bair has used this technique to analyze field samples for remaining antioxidant and has shown very good agreement with determinations by UV absorption spectra.[109]

The effect of a FR additive on the dynamic TG scan of a polymer has been discussed above. Isothermal measurements of the volatilization of an FR-modified resin can be useful in determining the effect of the additive on the upper limit temperature for melt processing. It may also serve as a measure of the effectiveness of the FR system since in many cases

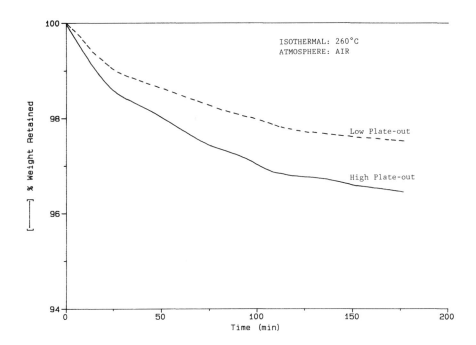

FIGURE 39. Weight loss vs. time for two ABS compounds at 260°C. (From Bozzelli, J., Dow Plastics Injection Molding Tech. Bull. Magnum ABS Resins, 12, 1987.)

the self-extinguishing capabilities of the compound depend upon an additive that vaporizes to scavenge free radicals. Bair[110] has performed quantitative analysis of the FR percentage in a compound based on ABS. Comparisons of isothermal weight loss for FR-ABS and general purpose ABS provided results that correlated well with measurements made on the effects on T_g.

Processors are familiar with a phenomenon known as plate out. Under conditions of elevated pressure and temperature, low molecular weight additives such as plasticizers and lubricants as well as the decomposition products of FR systems can volatilize and be deposited on the molding surface or in the surrounding vents. This can cause cosmetic defects such as reduced gloss on a highly polished surface and can lead to structural problems associated with trapped gases at the polymer flow front. Figure 39 shows work performed by Bozzelli[111] relating isothermal weight losses in ABS resins to the required frequency of cleaning and polishing of injection molds. Higher weight losses in the TGA studies relate to larger amounts of plate out and the need for more mold maintenance during production. Sepe has used the same technique to identify the cause of poor heat welding bonds in assemblies of glass-reinforced PPO. Figure 40 shows isothermal weight losses for a poorly bonded and a well-bonded sample. The high concentration of volatiles relating to poor joint strength was associated with the location of ejector pins which often trap gases in the narrow clearances between the pin and the mating hole. TGA scans at various depths showed that the unusual concentration of volatiles was a surface effect, extending only 0.015 in. into the part. The problem was solved by molding raised bosses on each ejector pin and then sanding them flush with the rest of the part surface before welding.

C. EXTENT OF CURE

Cassel[112] has discussed the use of residual weight loss in cured thermoset systems to determine degree of conversion. This is an alternative to residual enthalpy measurements

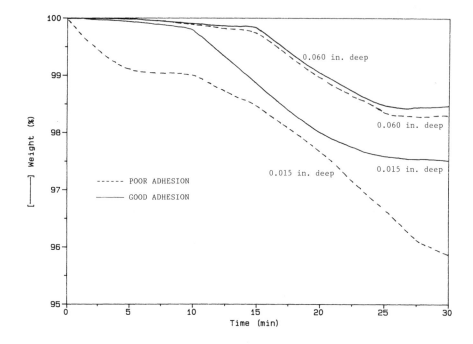

FIGURE 40. Weight loss vs. time for samples of 30 wt % glass-reinforced PPO taken from molded parts exhibiting good and poor adhesion during heat welding, scanned at 320°C. (Sepe, unpublished, 1987.)

by DSC and is feasible when the curing chemistry involves the loss of condensation by-products. It is important that absorbed water and residual solvent be removed before testing since they will interfere with the accuracy of these measurements. Using properly dried samples and following the DTG curve will optimize the results. Cassel demonstrates this technique for calculating the degree of cure in a phenolic binding resin. Figure 41 shows the weight loss and derivative curves for the uncured resin. These provide the basis for degree of cure determinations. The degree of cure or conversion is taken as a simple ratio of the weight loss in the cured product to that of the uncured product. This quotient is then deducted from the ultimate degree of conversion.

D. THERMAL STABILITY

Inevitably, any discussion of TGA methods involves determinations of thermal and oxidative stability. The first of these is typically determined by dynamic heating in an inert atmosphere while the latter uses oxygen or air as the purge gas. Thermal stability comparisons are derived from comparisons such as those shown in Figure 28. Higher onset of degradation temperatures are associated with greater thermal stability. The same technique is often used for additives. Beekman et al.[113] have compared alumina trihydrate with magnesium hydroxide by TGA to conclude that magnesium hydroxide, when used as a flame retardant in a polymer, should permit higher processing temperatures because of a higher onset temperature for degradation. These comparisons can also be made using isothermal techniques. Thomas[71] has compared the volatility of antioxidants in this manner.

While valuable information can be derived from weight loss vs. temperature comparisons, several experimental conditions must be observed to make the comparisons valid. First, the atmosphere used for all tests must be equivalent. The presence of oxygen will have a dramatic effect on the mechanism of degradation in polymers, and the flow rate of

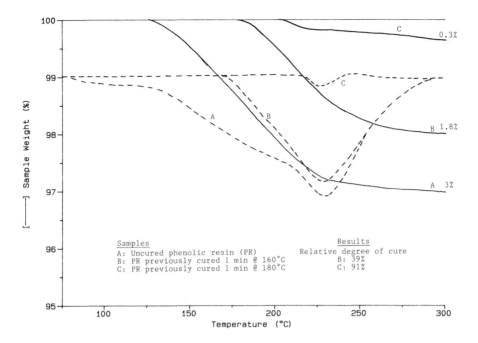

FIGURE 41. Residual weight loss to determine degree of cure in a phenolic bonding resin by TG/DTG. (From Prime, R. B., *Thermal Characterization of Polymeric Materials,* Academic Press, 1981. With permission.)

the purge gas can influence diffusion which can be a rate-controlling step. Flow rate can also have an effect on the likelihood of recombination phenomena.

The heating rate must be kept constant for comparative tests. Since the activation energy of polymer decomposition varies for different compounds, a single heating rate may be inadequate as a basis for comparison. This will be discussed further in Section E. Sample sizes must be as close to equivalent as is practical and sample shape must be maintained within a narrow range. If surface to volume ratios vary, the effects on weight loss rate can be significant even for equivalent sample weights. If multiple furnaces are used to reduce turnaround time, all furnaces must be temperature calibrated since even slight variations in furnace mass can shift the weight loss events to higher or lower temperatures. Figure 42 shows two polycarbonate samples of the same weight and surface-to-volume ratio run at a heating rate of 20°C/min in two different furnaces. A shift of 16°C is observed between the two weight loss events.

E. KINETICS AND LIFETIME PREDICTIONS

All of the above considerations become increasingly crucial when the TGA method is expanded to the task of lifetime stability predictions for polymer systems. Many papers have been written on the use of isothermal and dynamic heating methods to accelerate degradation and extrapolate the results to milder conditions for the purpose of predicting service life.

Most of the carefully performed work in this area, however, inevitably leads to the conclusion that accurate correlations between actual service performance at a given temperature and predicted behavior based on kinetics of decomposition are difficult to achieve. This author has investigated the kinetics of decomposition of an arc-quenching material composed of acetal copolymer and uncured melamine. Under high electrical loads the acetal is designed to degrade to form formaldehyde gas. The uncured melamine then reacts with the formaldehyde to form a new species *in situ*. This in turn acts to quench the electrical

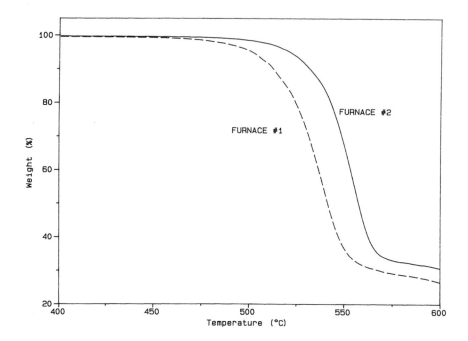

FIGURE 42. TG curves from the same instrument using two different furnaces calibrated to the same standard. Sample is polycarbonate, heating rate 20°C/min. (Sepe, unpublished, 1988.)

arc. Figure 43 shows the decomposition of this material in air. Unlike an unmodified acetal, which decomposes completely in one step with no resulting char, the arc-quenching material decomposes only partially. The weight loss essentially stops as the melamine-formaldehyde reaction occurs, and the resulting compound then degrades slowly over a broad temperature range. This reaction is crucial to the proper functioning of the product in service and it must not occur during melt processing. The kinetics of acetal decomposition are therefore of considerable interest.

Samples were prepared with weights of 8.01 to 8.15 mg and pressed into thin squares of uniform area and thickness. Both nitrogen and air were utilized; the results of the studies in air are presented here. Heating rates of 0.25 to 10.00°C/min were used and extrapolations were not made below the melting point of the acetal for reasons which will be discussed below. Figure 44 shows the conversion plots for six heating rates and illustrates the shift in temperature for several weight loss values as a function of the heating rate. Figure 45 shows a semilogarithmic plot of heating rate vs. reciprocal temperature. The slope of the lines represents the activation energy of decomposition for weight losses from 3 to 15%.

In spite of the good agreement of the data points with the plots of log heating rate vs. 1/T and the consistency of the activation energy determinations across the range of weight losses, Table 4 shows that lifetime predictions for a weight loss of 10% do not compare well with actual results of isothermal tests. The software used to analyze the kinetics is based on the Flynn and Wall method.[114] Disappointing results in terms of quantitative accuracy were also reported by Day et al.[115] for evaluations of PEEK and by Hunter and Sheppard[116] in studying fiber-reinforced composites.

Flynn[117] has discussed in some detail the problems that arise from an indiscriminate use of kinetics to predict long-term behavior in polymers. He cites several examples of oxidative degradation where substantial errors occur due to changes in the degradation mechanism. These changes are the result of crossing a threshold temperature such as T_g or some other temperature where increased mobility of the reacting species takes place. Attempts to ex-

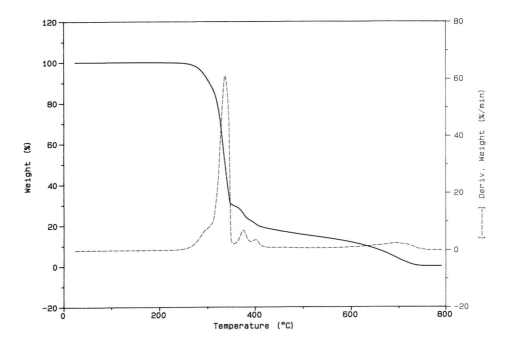

FIGURE 43. Decomposition of an arc quenching acetal-melamine compound in air. (Sepe, unpublished, 1990.)

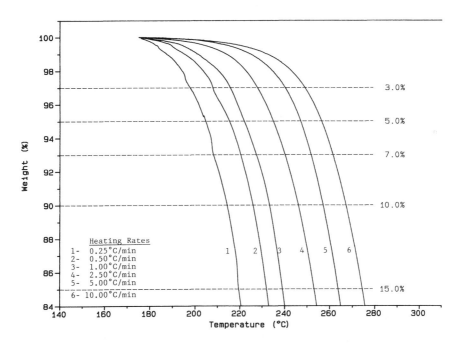

FIGURE 44. Weight losses for acetal-melamine at indicated heating rates. (Sepe, unpublished, 1990.)

trapolate from an experimental point above such a threshold temperature to a region below it result in data that are not useful.

In addition, Flynn and Dickens[118] have shown that even small errors in the determination of the activation energy can result in very large uncertainties in the lifetime prediction if the temperature extrapolation region is large. Flynn[117] reviews a number of other concerns

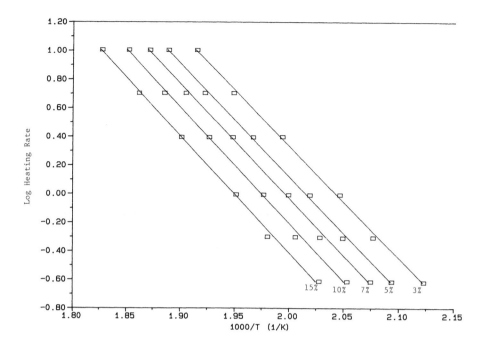

FIGURE 45. Log heating rate vs. 1/T at different conversion levels for acetal-melamine. (Sepe, unpublished, 1990.)

TABLE 4
Lifetime Predictions from TGA
Kinetics vs. Actual Test Results for
an Acetal-Melamine Compound
(Based on a 10% Weight Loss)

Temperature (°C)	Predicted time (min)	Actual test (min)
225	16.68	33.2
235	7.26	18.1
245	3.26	9.8
255	1.53	5.7
265	0.72	2.8

such as instrument precision and accuracy, an appropriate emphasis on intensifying factors that are relevant to the long-term application being modeled, (i.e., temperature, mechanical stress, atmospheric conditions, chemical agents, etc.), a knowledge of the degradation chemistry, and the establishment of a correlation between weight loss and property deterioration. Thermal history has also been shown to be a complicating factor.[119]

The effect of transition temperatures such as T_m and T_g on transport and diffusion phenomena will also introduce serious errors for extrapolations that cross these temperature barriers. It is for this reason that extrapolations for the acetal-melamine compound discussed above were not carried below the T_m of the acetal copolymer. Also, specific properties will decay at different rates for a given polymer. Polyamides aged at 130°C, for example, will exhibit a rapid decay in dielectric properties, a slower deterioration in impact resistance, and virtually no change in rigidity. The weight loss associated with long-term exposure in a 130°C environment will have a different degree of significance for the various properties.

Proper interpretation of predictive methods must rely on actual field experience relating to the important aspects of performance for a specific application.

In general, the researchers working with this technique conclude that useful relative comparisons of long-term thermal stability can be made, particularly for a given compound prepared by different methods. A thorough knowledge of transition temperatures and degradation chemistry by supplemental techniques such as DSC, FTIR, and rheometry are crucial to the prudent use of kinetics. Accuracy may be optimized by using isothermal scans at several temperatures or by employing isoconversion plots at very slow heating rates. A single degradation mechanism should be treated. Experimental precision in instrumentation and sample preparation are, of course, very important.

Even with all these considerations, accurate quantitative predictions are hampered by multiple influences in the actual application environment and by differences in the surface-to-volume ratios between TGA samples and actual products.

V. THERMOMECHANICAL ANALYSIS

Thermomechanical analysis (TMA) is a method for evaluating dimensional changes as a function of temperature, time, and environment. While DSC and TGA are concerned with the energetics of physical and chemical changes, TMA measures the dimensional effects associated with these changes. The samples are typically pieces of fabricated items rather than raw material. The sample is placed on a quartz stage which is surrounded by a furnace and a Dewar bottle for heating and cooling. A quartz expansion or penetration probe is brought to rest on the sample. As the sample expands or contracts, changes in the position of the probe are accurately monitored and translated as the dimensional change of the sample. Different loads may be applied to the sample.

With an expansion probe and running in a dynamic heating mode, TMA can be used to measure dimensional change vs. temperature. This permits a calculation of the coefficient of linear thermal expansion (CLTE), an important property for identifying thermal transitions and in determining the suitability of a material for applications that may experience large swings in temperature.

This is particularly important when semicrystalline polymers experience service conditions both below and above T_g. Figure 46 shows a TMA scan of dimensional change vs. T for PEEK, a material frequently used above its T_g. The T_g is identified as the temperature at which a sudden change in the CLTE occurs. The CLTE can be two to five times greater above T_g than below. Property data for these materials often provides only the lower value and this can lead to serious errors where inadequate clearances in an assembly may cause an unacceptable stack-up of tolerances at elevated temperatures. See Reference 72 for a discussion on sample preparation requirements for TMA. As with DSC, accurate determinations of T_g are performed on samples where the previous thermal history has been erased and the sample is cooled and reheated at the same rate.

Using a penetration probe and a sufficient load, the T_g is identified by a change from a positive to a negative slope in the dimensional change vs. T plot as the softened product allows the probe to become embedded in the sample. As with DSC studies, the heating rate will affect the measured T_g. Schwartz[120] studied the effect of both heating and cooling rates on the determination of T_g for a carboxy-terminated polybutadiene. Extrapolation of both the heating and cooling rates to zero yielded the same T_g.

TMA can offer a more sensitive means of determining the T_g in highly crystalline polymers where small enthalpy changes make such measurements by DSC difficult. Figures 47 to 49 illustrate the complementary nature of DSC and TMA for these types of materials. Figure 47 shows the heat flow vs. T plot prior to melting for a polypropylene copolymer.

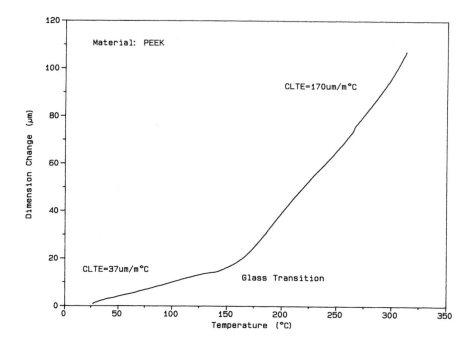

FIGURE 46. TMA scan of 30 wt % glass-reinforced PEEK showing T_g and change in CLTE. (Sepe, unpublished, 1989.)

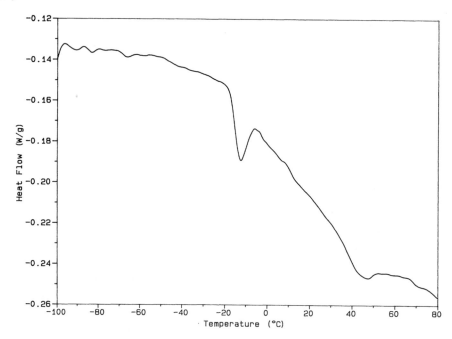

FIGURE 47. Low temperature DSC scan of polypropylene. (Sepe, M., unpublished, 1990.)

Two transitions are evident, one at $-15°C$ and another near 40°C. By TMA the second transition appears clearly as the inflection point in the slope of the dimensional change. However, the subambient transition is not readily detectable by TMA. Only by magnifying the derivative curve in this temperature region is it possible to detect a temperature range where the CTE increases locally. The midpoint of this region agrees well with the DSC

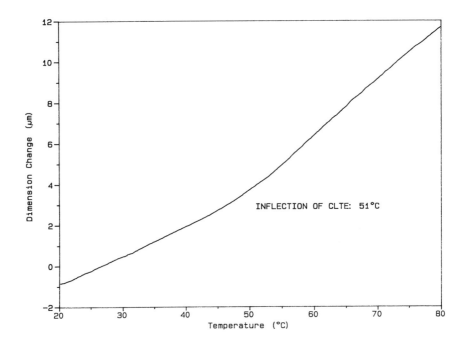

FIGURE 48. TMA showing change in CLTE for PP associated with DSC transition near 48°C. (Sepe, M., unpublished, 1990.)

transition. Figure 48 shows the major TMA transition while Figure 49 illustrates the magnified subambient region.

Many blends consisting of a semicrystalline and an amorphous component have dual T_g's which can be difficult to detect by DSC. Figure 50 shows a DSC thermogram of a PPO/PA alloy prior to the melting endotherm of the nylon. Even with advanced knowledge of where to look, the glass transitions of the polyamide and PPO are difficult to detect. However, a plot of CLTE vs. temperature by TMA clearly shows two maxima associated with the T_g's of the two components. This is shown in Figure 51.

TMA can be used to determine the T_g of phenolic materials, and the effects of different postcure routines on the T_g and the difference in CLTE above and below T_g can be readily observed. Figure 52 plots the dimensional change vs. T for phenolic moldings as produced and subjected to 24-h postcure routines at various temperatures. While the CLTE below T_g is nearly the same for all samples, the behavior above T_g changes as a function of postcure temperature. The CLTE above T_g decreases as the material becomes more fully cured. The T_g also becomes higher and eventually is indistinguishable as the CLTE becomes essentially constant from room temperature to 300°C. The material postcured at 150°C appears to depart from the pattern, displaying a very high CLTE above T_g. This is explained by an examination of DSC scans of the as-molded parts. The postcure exotherm does not begin until 168°C; consequently, the postbake routine at 150°C did not advance the crosslink structure. This is further confirmed by comparing the postcure exotherm of as-molded parts to those postbaked at 150°C. The energy of this exotherm is not reduced by the postbake cycle.

Cassel[99] demonstrated the use of a loaded penetration probe to illustrate the difference in the softening temperature and degree of penetration between two polyester gel coat samples exhibiting different degrees of crosslinking. The more highly crosslinked sample displayed a lower magnitude of penetration occurring at a higher temperature. More recently, Bair and Pryde[121] used the same technique to study crosslinking in benzocyclobutene films as a function of curing time.

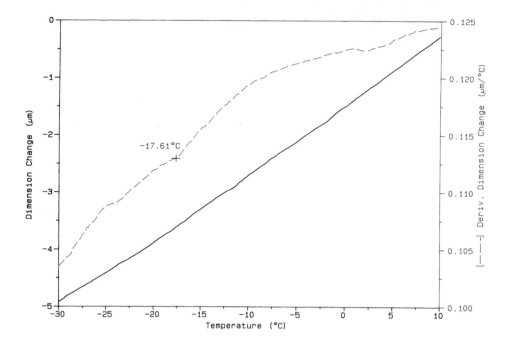

FIGURE 49. TMA scan showing dimensional change and rate of dimensional change associated with subambient transition found by DSC in Figure 47. CLTE increases by only 10% making detection difficult. (Sepe, unpublished, 1992.)

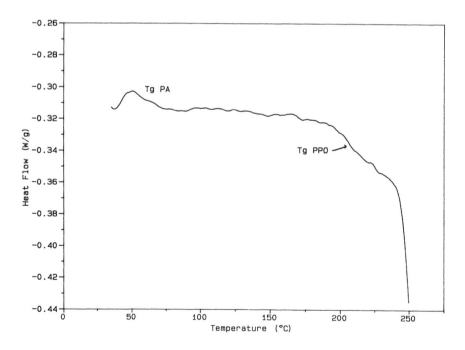

FIGURE 50. Identification of T_g's by DSC for a PPO/PA blend. (Sepe, unpublished, 1991.)

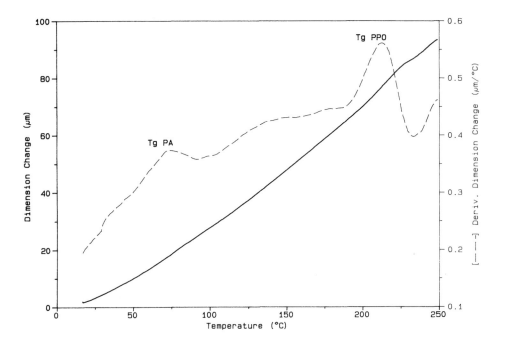

FIGURE 51. TMA using derivative curve to enhance T_g determination in PPO/PA blend. (Sepe, unpublished, 1991.)

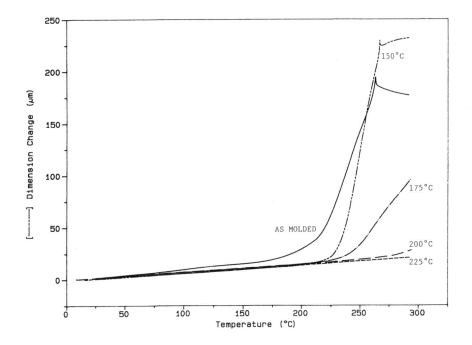

FIGURE 52. TMA of phenolic samples as molded and postcured for 24 h at temperatures indicated. (Sepe, unpublished, 1992.)

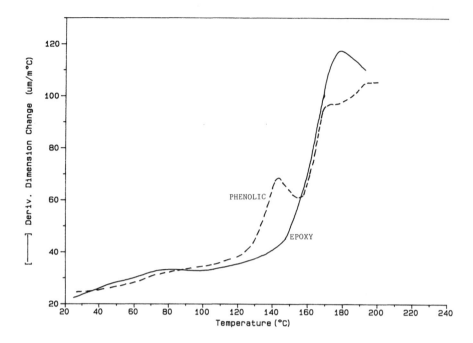

FIGURE 53. TMA showing CLTE for layers from an epoxy/phenolic laminate that failed due to delamination at elevated temperature. Contraction in phenolic is due to postcure during TMA scan. (Sepe, unpublished, 1989.)

It has been demonstrated that an expansion probe with a sufficiently high load (more than 10 g on a 3-mm diameter probe) can be used to detect undercuring in thermosets[122] and undercrystallization in semicrystalline thermoplastics.[123] Figure 53 shows a TMA thermogram of an epoxy/phenolic laminate that failed upon heating by delamination. An analysis of the two layers showed that the phenolic was undercured. While CLTE vs. T plots show the two materials to have similar responses to heating below 120°C, contraction of the phenolic due to postcure at higher temperatures is evident and was responsible for the separation of the layers.

Studies of PPS and PET, two slow crystallizing thermoplastics, show that improperly crystallized moldings exhibit contractions associated with recrystallization upon reheating. Figure 54 shows dimensional change vs. T plots for PPS samples molded at different mold temperatures. The degree of contraction agrees qualitatively with residual exotherms of recrystallization measured by DSC. The temperatures for the DSC exotherms and the TMA contractions are in good agreement. It is uncertain whether the contraction is due to excessive creep from the softening of the amorphous phase or whether it is due to a contraction of the polymer matrix as a more highly ordered crystal structure is achieved. This technique can be used to establish a minimum mold temperature at which the crystallinity of the molded part will be sufficient to prevent dimensional instability at elevated temperatures. It provides greater precision than the measurement of recrystallization exotherms by DSC since it uses a larger sample and examines dimensional stability directly. Henderson et al.[124] have reported on the same behavior in quenched and conditioned samples of PEEK.

Annealing of semicrystalline materials is often performed at a temperature between T_g and T_m to insure complete crystallization of material that has been quenched from the melt during processing. Sepe[123] has evaluated the effects of various annealing temperatures on the thermal expansion of quenched PPS moldings. The study shows that, once the annealing

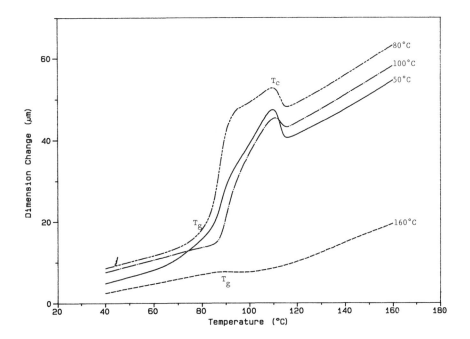

FIGURE 54. TMA showing T_g and T_{cc} behavior for PPS samples produced at the indicated mold temperatures. (Sepe, unpublished, 1989.)

temperature exceeds the recrystallization temperature, the contractions typical of quenched material vanish. The same behavior for PEEK is reported by Henderson et al.[124]

A close examination of the plots of CLTE vs. T for PPS reveals that, while annealing of quenched samples may produce the desired recrystallization, the types of crystals that form during annealing are different from those that form when the material is cooled slowly from the melt. This was discussed in Section III.A for both PPS and PEEK. For annealed samples, the CLTE curve displays a local maximum in the vicinity of the annealing temperature.[123] Figure 55 illustrates this behavior.

TMA has also proven useful for identifying the effects of orientation in both filled and unfilled polymers. High shear stresses tend to cause oriented flow in polymers as well as fibrous reinforcements. At a practical level this results in anisotropic physical properties and dimensional stability. Figure 56 shows TMA scans performed on different segments of a single part molded from 40 wt % glass fiber-reinforced PPS. The x-direction indicates measurement in the direction of polymer flow, the y-direction measures properties transverse to flow, while the z-axis measures in a direction perpendicular to the plane of resin flow. Across the entire temperature region from room temperature to near the T_m, the highly oriented material in the direction of flow only expands about one third as much as in the other two directions. Most of this difference occurs above T_g, where the less-oriented material exhibits a CLTE that is about twice as large as that measured below T_g. In a restrained state, parts exhibiting this type of orientation can warp at elevated temperatures due to differential expansion.

It is interesting to note that the highly oriented material displays a small contraction at T_g instead of the customary increase in the CLTE. This is due to the partial relaxation of the highly aligned polymer. The detection of sudden contractions or expansions in the vicinity of a transition temperature can be a useful tool for evaluating molded-in stresses that may cause a reduction in impact properties. Figure 57 shows a plot of dimensional change vs.

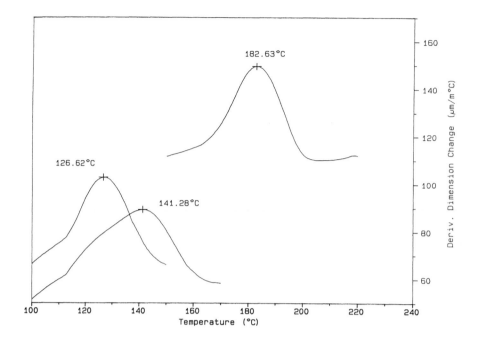

FIGURE 55. CLTE for PPS samples annealed at 120, 140, and 180°C showing local maximums near the annealing temperature. (From Sepe, M., *SPE ANTEC*, 37, 469, 1991.)

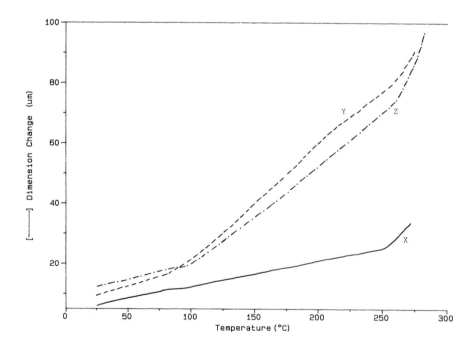

FIGURE 56. CLTE of 40 wt % glass-reinforced PPS as a function of directional flow during injection molding. (From Sepe, M., unpublished, 1989.)

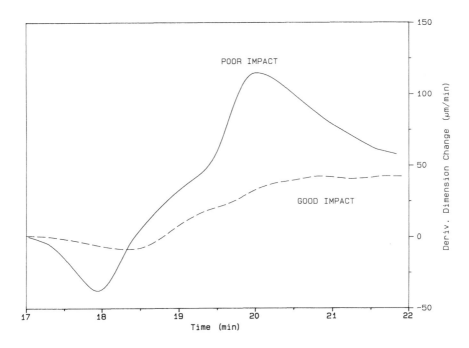

FIGURE 57. Expansion rate at isothermal conditions at T_g for molded SAN samples. (Sepe, unpublished, 1992.)

time as SAN samples exhibiting different levels of birefringence are held isothermally at the glass transition. The sample displaying a high level of birefringence shows a rapid expansion in a very short time after achieving T_g, while the other sample, exhibiting less birefringence, reveals little dimensional change. After slowly cooling both samples from T_g and then reheating, no significant expansion was observed in either specimen. This indicates that the initial heating routine in the TMA had annealed out the stresses, particularly in the one highly stressed sample. The moldings from which the two samples were taken displayed great differences in impact resistance. The highly stressed sample had fractured easily. In amorphous polymers, impact resistance can be reduced dramatically by overpacking the cavity during molding. This forces too much polymer into a given volume, creating a molecular overcrowding. Since amorphous materials undergo a very small volumetric change as they cool, the polymer has little opportunity to relax. This strained state may be corrected by annealing near T_g; the large expansion observed by TMA is caused by the relaxation of the polymer into the larger volume required for optimal spacing between polymer chains. Since dimensional changes near T_g are a qualitative indicator of molded-in stress levels, this method can also be used to evaluate the effectiveness of different annealing routines. Improper annealing will fail to relieve the stresses in the part and may actually worsen the situation. Not shown here is a scan of an SAN sample which had been annealed but still failed the impact test. An examination of the TMA scan revealed the same rapid expansion near T_g that characterized the original, unannealed samples. The sample had been "annealed" at too low of a temperature and had experienced thermal aging without achieving stress relaxation.

TMA is also a valuable technique for measuring dimensional changes in response to fluid environments. This has been performed for a variety of fluid/polymer combinations such as chlorobenzene/polyethylene[125] and water/nylon 6.[126] Figure 58 shows plots of dimensional change vs. time for samples of nylon 6 immersed in water at three different temperatures. In addition, Kettle[127] used dilatometry to determine the T_g of nylon six samples

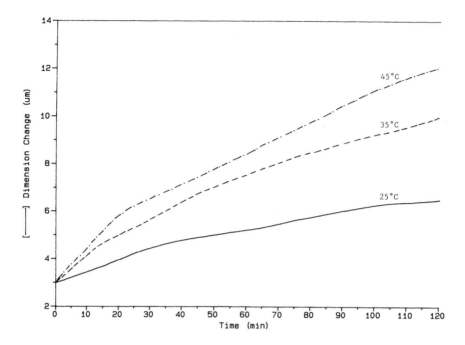

FIGURE 58.   Expansion of 30 wt % glass-reinforced nylon 6 immersed in water at temperatures indicated. (Sepe, unpublished, 1988.)

at various levels of water absorption. This technique can be applied to the characterization of elastomers in various fluids and oils at different temperatures. Volume swell can be accurately determined and the effects of changing hard block/soft block content on fluid resistance can be studied.

Prime[72] and Maurer[86] both describe the use of TMA to determine crosslink density by solvent swelling for both rigid and elastomeric thermosets, although Prime recommends TGA methods to eliminate inaccuracies due to problems with isotropy and homogeneity.[128] Prime demonstrates a correlation between TMA swell characteristics and Young's modulus which allows for very precise determinations of degree of cure. Barrall and Flandera[129] used penetration TMA to calculate the elastic modulus of crosslinked, two-component methyl silicone rubber by an equation from Gent:[130]

$$E_m = (F/p^{1/2})(9/16r^{1/2}) \tag{9}$$

where E_m is the elastic modulus, F is the load, p is the penetration, and r is the probe radius.

Many of the techniques discussed above for measuring CLTE and T_g in rigid systems are also applicable to elastomeric compounds. Use of an unloaded expansion probe will give the T_g as the onset of an increase in CLTE as already shown for rigid materials. A loaded expansion probe will produce an indentation at T_g which Brazier and Nickel[131] have related to differences in hardness. Figure 59 shows that both the extent of indentation and the temperature range over which it occurs are dependent on the plasticizer type used in the nitrile vulcanizate studied. In the same study, the authors demonstrated a linear correlation between the T_g as determined by indentation or tension TMA and the Gehmann rigidity modulus. Exact agreement in actual temperatures was not observed due to differences in methods.

Fogiel et al.[132] used TMA to define the factors that govern the shrinkage of fluoroelastomers during processing. This technique, if workable, has the obvious value of aiding in

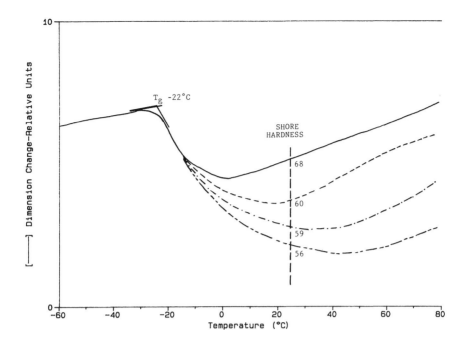

FIGURE 59. Penetration TMA beyond the glass transition for rubber compounds of different hardness. (From Maurer, J. J., *Thermal Characterization of Polymeric Materials,* Academic Press, 1981. With permission.)

the prediction of dimensional differences between the mold cavity and the fabricated article. Maurer[86] discusses this work in some detail. Good agreement was established between the experimental points and the derived equation relating filler content and cure temperature to shrinkage. It was necessary, however, to calibrate the instrument using vulcanized elastomer standards of known CTE since metal standards gave TMA displacements that were too small over the relevant temperature range of 25 to 200°C.

TMA can be used over a limited load range to evaluate compressive creep. This is done simply by placing various static loads on samples and monitoring indentation over time at different temperatures. Effects of differences in composition can be observed and useful temperature ranges for materials can be defined. Figure 60 shows dimensional change vs. time plots for two different hardnesses of a dynamically vulcanized thermoplastic olefin using two different loads. No quantitative calculations of compressive modulus have been performed using these data.

Using a weighted penetration probe, Fair et al.[133] have demonstrated the use of TMA to determine the thickness of individual film layers in a multilayer system. In this study, DSC techniques were used to identify the contractions as melting events. Comparison of the DSC results with a library of known melting endotherms helps to identify the specific materials in the film. The magnitude of the negative dimensional changes associated with each melt represents the thickness of each layer. Substrates that do not have a transition, such as metal foils, can be calculated by difference. The peaks of the melting endotherms agree with the onset of rapid dimensional change to within 2°C.

The capabilities of the TMA instrument can be extended by the addition of supplemental equipment which has become commercially available in recent years. Most of the literature on the use of these techniques, however, comes from pioneers who worked with custom-designed accessories some 20 to 50 years ago. The first of these is a parallel plate rheometry attachment. Prime[72] discusses the development of this technique, the apparatus configuration, and some of the theoretical considerations for determining viscosity as a function of

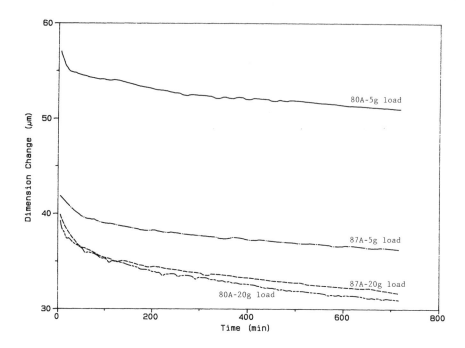

FIGURE 60. Compressive creep over time as a function of material hardness and applied load. Probe diameter was 3 mm, isothermal scan at 100°C. (Sepe, unpublished, 1992.)

temperature, time, or shear rate. Primary references are Cessna and Jabloner[134] and Blaine and Lofthouse,[135] who used the method to study thermoset cure. More recently, Hassel[136] has described the use of an updated instrument with a parallel plate rheometry probe to show differences in PE samples as a function of aging. Only raw data are presented; no viscosity determinations are made in the examples cited. Further discussion of this technique will appear in Section VI.

A second significant adaptation of the TMA instrument is an extension of the tension mode of operation for the specific purpose of studying fibers and films. As with the PPR technique, older instruments were fitted with custom-built attachments for the purpose of solving the special measurement problems presented by such samples. Prime et al.[137] describe some of these devices and discuss experimental considerations for performing accurate tests. Jaffe devotes an entire chapter to the thermal analysis of fibers and gives a thorough discussion of the TMA instrumentation and its use in measuring CTE in fibers as a function of processing history and microstructure. He outlines three essential characteristics of fiber behavior as a function of temperature: shrinkage, shrinkage rate, and shrinkage force. Figures 61 and 62 give a general scheme for the behavior of these characteristics.[138]

Newly introduced instruments permit the application of force to the sample by use of an electromechanical coil. This eliminates the need for mechanical weights and allows the experimenter to use force as a test variable, making it simple to conduct constant stress or constant strain scans. This is especially useful for characterizing fibers. Hassel[136] provides several examples of this application of the TMA instrument. Particularly useful is a constant strain test where the shrinkage force, defined as the force required to maintain the applied strain, is monitored as a function of temperature. This is used in one example to examine the processing history of two polyolefin fibers as shown in Figure 63. Interpretation of similar curves is discussed by Jaffe.[138] Hassel also discusses the generation of stress/strain measurements on fibers and the advantages of using the lower mass and inertia of the TMA

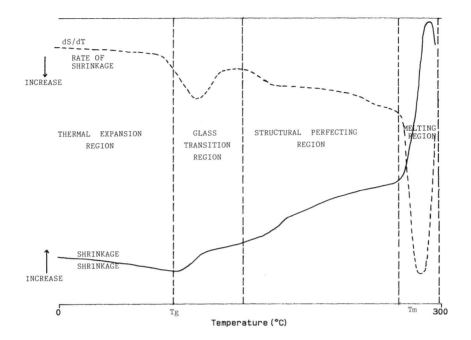

FIGURE 61. Schematic TMA scan of a fiber. (From Jaffe, M., *Thermal Characterization of Polymeric Materials,* Academic Press, 1981. With permission.)

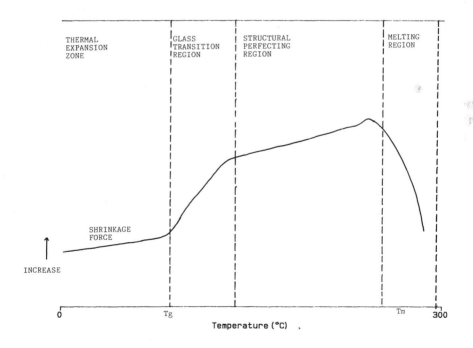

FIGURE 62. Schematic TMA scan for fibers showing shrinkage force as a function of temperature. (From Jaffe, M., *Thermal Characterization of Polymeric Materials,* Academic Press, 1981. With permission.)

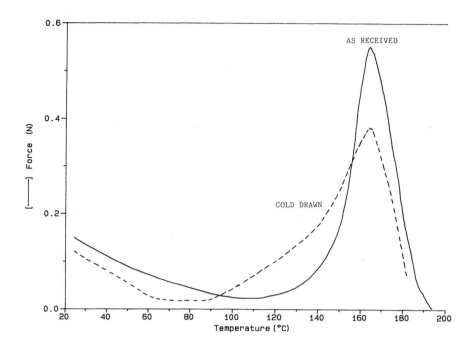

FIGURE 63.   Thermal stress analysis of polyolefin fiber by TMA. (From Hassel, R., *Am. Lab.*, January 30, 1991. With permission.)

clamping system as opposed to conventional physical testing machines for achieving accurate results. Figure 64 shows a TMA-generated stress/strain plot for a polyamide filament.

VI. DYNAMIC MECHANICAL ANALYSIS

Dynamic mechanical analysis (DMA) and its precursor, torsional braid analysis (TBA), have emerged as the most versatile and powerful thermal analysis tools in use today. For that reason, a significant amount of time will be devoted to discussing principles and applications. Since most of the instruments on the market today are DMA apparatus, this review will deal primarily with this technique.

At a commercial level, DMA instrumentation is not a standard part of a thermal analysis system. Only a few suppliers of thermal analysis equipment offer a DMA while most manufacturers of DMAs do not build DSC, TGA, and TMA equipment. This is largely due to the fact that DMA emerged initially as a rheological tool and is used to measure aspects of polymer behavior which cannot be monitored with the previously discussed techniques.

Nevertheless, DMA does identify many of the aspects of polymer behavior that are measured by DSC and TMA, and adds a wealth of information regarding polymer structure and properties as well. This makes it an indispensable tool to the thermal analyst.

A. PRINCIPLES OF OPERATION

DMA measures the viscoelastic properties of materials. This is a singularly important aspect of polymer characterization since all polymeric materials simultaneously exhibit elastic and viscous behavior. A complete description of viscoelasticity is provided by Ferry.[139]

Most classical materials exhibit one of two types of behavior in response to an applied stress, elastic or viscous. Elastic responses are typically observed in solids. When a stress

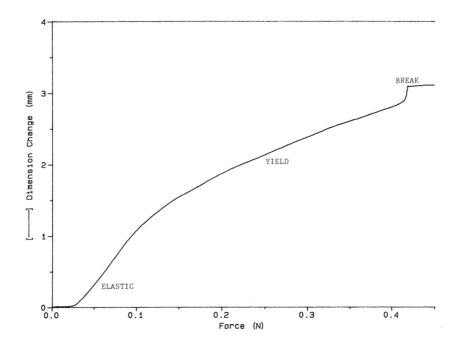

FIGURE 64.   Stress/strain plot of a polyamide filament by TMA. (From Hassel, R., *Am. Lab.*, January 30, 1991. With permission.)

is applied to an elastic system it deforms proportionally by a quantity termed the strain. The modulus is a ratio of applied stress to resultant strain:

$$\tau = G \cdot \gamma \tag{10}$$

where τ is stress in shear, γ is the strain, and G is the shear modulus. The response of an elastic system to applied stress is instantaneous and completely recoverable. We say that the system stores the energy and can return it completely upon relaxation of the stress. The above equation is known as Hooke's Law and the familiar model for materials governed by this law is that of a spring. Figure 65 illustrates this behavior.

Viscous behavior is a characteristic of fluids, materials where the bond energies necessary for large-scale translational order have been overcome. In these systems an applied stress results in a strain which increases proportionally with time until the stress is removed. The strain is not recoverable; the deformation is completely retained. We say that the system has lost energy. The familiar model of the dashpot is used as an analogy. Newton's Law governs this behavior, shown graphically in Figure 66:

$$\tau = \eta \cdot \dot{\gamma} \tag{11}$$

where τ is stress, $\dot{\gamma}$ is strain rate, and η is viscosity.

Because of molecular size and conformational variety found in polymers, these materials display a combination of viscous and elastic responses. The balance between these two aspects changes for a given material across a wide range of temperatures. In the solid state this balance is reflected in terms of load-bearing properties, time-dependent behavior such as creep and stress-strain relaxation, and impact properties. In the fluid state, viscoelasticity provides information on molecular weight, molecular weight distribution, thermal stability,

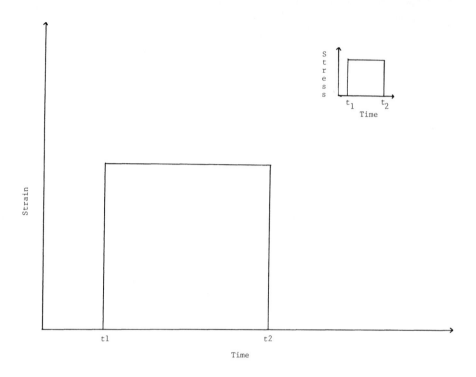

FIGURE 65. Elastic response to step change in stress. (From Rheometrics, Inc., February 22, 1990.)

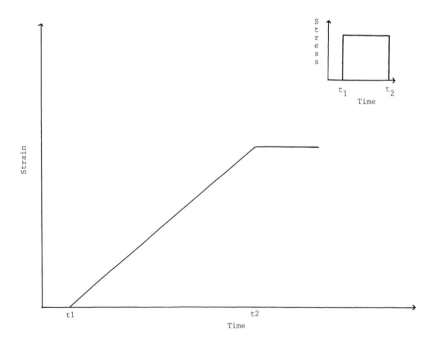

FIGURE 66. Viscous response to step change in stress. (From Rheometrics, Inc., February 22, 1990.)

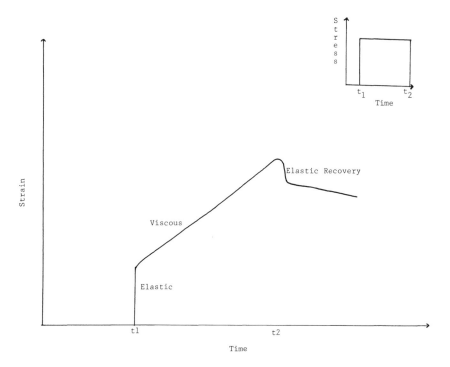

FIGURE 67.   Viscoelastic response to step change in stress. (From Rheometrics, Inc., February 22, 1990.)

and crosslinking. The equation relating stress to strain in viscoelastic systems introduces the aspect of time dependency which is so crucial to an accurate characterization of polymeric systems:

$$\tau = G(t) \cdot \gamma \tag{12}$$

where $G(t)$ is the stress relaxation modulus. The material initially responds elastically, then as a viscous fluid. When stress is removed, the elastic portion recovers over an extended period of time. Figure 67 illustrates this behavior.

Determining the proportion of the elastic and viscous components in a polymer, and the factors that cause that balance to change, is crucial to understanding the performance of a material and can yield valuable information regarding structural changes. DMA accomplishes this resolution.

The DMA instrument applies a sinusoidal deformation to a sample. For a perfectly elastic material the applied stress and the resulting strain will be in phase as shown in Figure 68. For an ideal fluid the stress will lead the strain by 90° ($\pi/2$ radians) as shown in Figure 69. A viscoelastic material will give a hybrid of these two responses. The stress and strain will be out of phase by some quantity known as the phase angle delta (δ). A small phase angle indicates high elasticity while a large phase angle indicates highly viscous properties. The complex response of the material is resolved into the elastic or storage modulus (G') and the viscous or loss modulus (G''). If measurements are made in the flexural mode, E' and E'' are used. Table 5 provides a glossary of key viscoelastic terms.

There are many different sample geometries possible within the framework of this method and various instruments operate on slightly different principles. We are more concerned here with the information that can be derived from these tests. For an excellent discussion

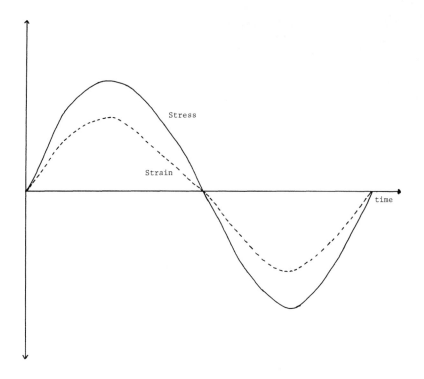

FIGURE 68.   Dynamic mechanical analysis using sinusoidal oscillation for an elastic material; stress and strain in phase. (From Rheometrics, Inc., February 22, 1990.)

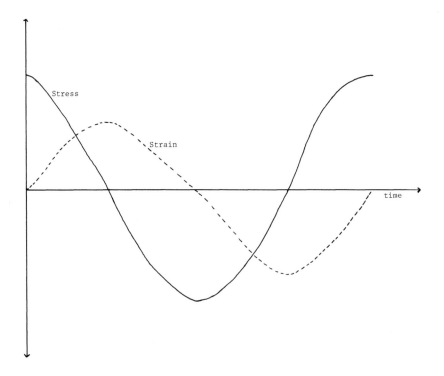

FIGURE 69.   Dynamic mechanical analysis for a viscous fluid; stress and strain out of phase 90°. (From Rheometrics, Inc., February 22, 1990.)

TABLE 5
Glossary of Viscoelastic Terms

Complex modulus	G^* or $E^* \times \tau^*/\gamma$
Elastic modulus	G' or $E' = \tau'/\gamma = (\tau^*/\gamma)\cos\delta$
Viscous modulus	G'' or $E'' = \tau''/\gamma = (\tau^*/\gamma)\sin\delta$
Complex viscosity	$n^* = G^*/w$
Loss tangent	$\tan\delta = G''/G'$

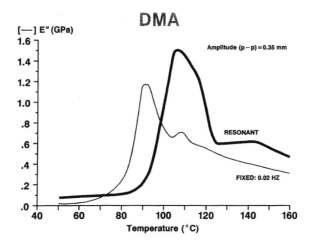

FIGURE 70. Loss modulus for undercrystallized PPS scanned at resonant frequency and a fixed frequency of 0.02 Hz. (Sepe, 1989.)

of different instruments and the resolution of apparent discrepancies in experimental results between these instruments see Reference 140.

One distinction that will be made is that of resonant frequency measurements vs. non-resonance or fixed frequency modes of oscillation. Many early instruments operated solely in a resonant frequency mode. The frequency of oscillation in this mode is dependent upon the rigidity of the sample. As a sample is heated and the modulus changes, the change is measured in terms of a variation in the frequency, which is then converted to storage modulus. In rigid systems the resonant frequency will typically fall in the 15- to 30-Hz range. While this technique can be useful for making rapid and approximate determinations of transition temperatures, the high frequencies make resolution of transitions due to different structures difficult as shown in Figure 70. Here a low fixed frequency clearly resolves a glass transition into multiple responses while the resonant frequency scan does not. In addition, relaxational processes in polymers, such as the glass transition, are frequency (inverse of time) dependent. Thus, scanning a material at several frequencies across a temperature region can yield valuable data about the activation energy of a transition. For this reason, most of the data presented here will consist of fixed frequency measurements.

DMA instruments can also be operated in the creep mode and the stress-strain relaxation mode. Short-term responses of the change in strain at a constant load (creep) or the decay in stress required to maintain initial strain (stress-strain relaxation) can be measured at different temperatures. These can be used within limits to develop extrapolations of behavior over time periods or frequency ranges which are not examined directly. This technique, known as time-temperature superposition, will receive considerable attention below.

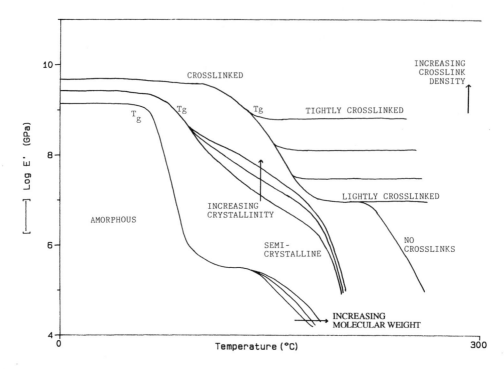

FIGURE 71.   Storage modulus behavior as a function of temperature for various structural changes in polymers. (From Soni, P., SPE Seminars, June 17, 1991.)

B. TYPES OF TESTS AND DATA PRESENTATION

The information from DMA tests can be configured in a variety of ways depending upon the design of the test and the data of interest. In the solid state, the most common experiment is a temperature scan. A constant heating rate is selected along with a frequency and amplitude of oscillation which are maintained throughout the experiment. Most commonly the resulting output is a plot of the storage and loss moduli and tan δ vs. temperature. Figure 71 shows a generalized plot of this type of experiment using only the storage modulus output. It illustrates the effects of molecular weight, crosslinking, and crystallinity as well as identifying the glass transition and the onset of melting. If we add the loss modulus and tan δ curves we can see that, as rapid changes in storage properties occur, the loss properties rise to local maxima which are associated with the onset of short-range molecular motions as the sample passes through the glass transition. Figure 72 shows this behavior for polycarbonate, an amorphous thermoplastic, and Figure 73 shows the same data for PBT, a semicrystalline material.

Assigning an exact value to T_g from these data is a subject of some disagreement. If the primary concern is the practical effect of the transition on the load-bearing characteristics of the polymer, the onset of a sharp reduction in storage modulus may be used. This is straightforward when a plateau in the storage modulus before T_g is well defined. However, it can be difficult for materials where the region of storage modulus decline is broad such as in the unsaturated polyester shown in Figure 74. Alternatively, the peak of either the loss modulus or tan δ curve may be used, the former giving the best general agreement with DSC and TMA determinations. Morrison[141] has suggested the use of the onset in the increase of the tan δ since this represents the beginning of the change in the viscoelastic balance within the polymer system. For the purpose of consistency, the peak temperature of the loss modulus curve will be used to identify the transition temperature in this discussion.

If multiple frequencies are used during a temperature scan, the temperature is increased in regular increments and then equilibrated before making multiple measurements. Typical

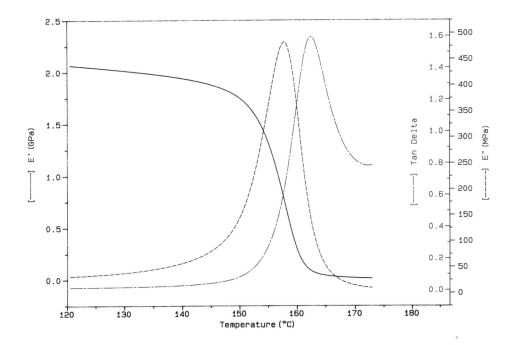

FIGURE 72. Storage and loss properties for polycarbonate.

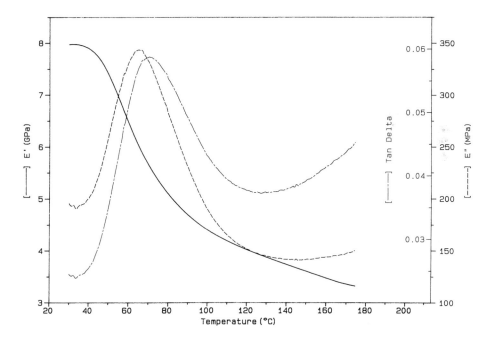

FIGURE 73. Storage and loss properties for PBT polyester.

output from this type of test appears in Figure 75 and these data can be used to develop master curves of a variety of viscoelastic properties vs. frequency or time. The degree to which a transition is frequency dependent is an indication of the activation energy associated with that transition. Similarly, short-term creep or stress relaxation scans made at various

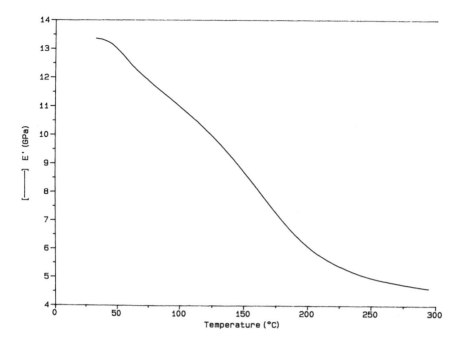

FIGURE 74. Storage modulus for a cured unsaturated polyester. (Sepe, M., unpublished, 1990.)

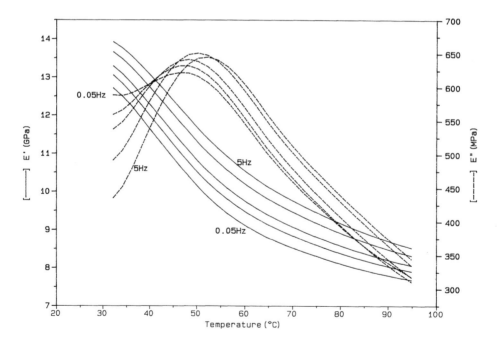

FIGURE 75. Multiple frequency scan of the glass transition region for a 30 wt % carbon fiber-reinforced nylon 6/6. (Sepe, unpublished, 1991.)

temperatures can be displayed as a series of modulus vs. time plots. These can be used in raw form to examine the interplay of time and temperature for a material or they can be superposed upon a reference curve in a manner similar to fixed frequency data.

Melt-state testing provides opportunities for more varied test configurations. Frequency sweeps can be conducted at a given temperature to provide a plot of viscosity vs. frequency

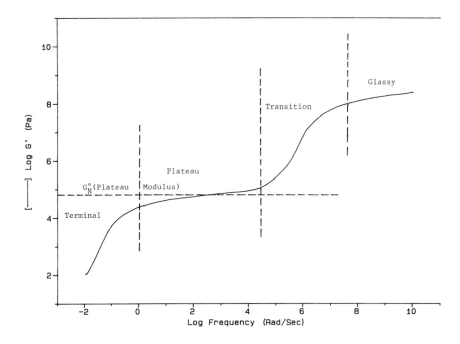

FIGURE 76. G′ mastercurve for a typical linear amorphous polymer. (Tuminello, W. H., *SPE ANTEC*, 33, 990, 1987.)

which is similar to plots of viscosity vs. shear rate provided by capillary rheometry. In addition, DMA gives the response of G′ and G″ as a function of frequency. Figure 76 shows a generalized plot of G′ vs. frequency and illustrates the different regions of behavior that can be observed. Tuminello[142] gives an excellent description of the different regions and the information that can be derived about polymer structure from tests of this type. Figure 77 shows a narrower range of frequencies and includes G″ data.

Strain sweeps are useful in determining the limits of linear viscoelasticity. Figure 78 displays G′, G″, and n* as a function of varying strain of poly(dimethylsiloxane). Time sweeps at a constant temperature can be useful in identifying crosslinking or breakdown of a structure. Figure 79 shows G′ vs. time for two ABS materials held at a constant temperature in the melt state. The increase in modulus of the bad sample indicates crosslinking of the unsaturated sites in the butadiene which will cause undesirable viscosity changes during processing and a loss of impact resistance in the molded product. It will also be likely to result in unwanted alterations in product color. Tests of this type can also be performed on solid-state samples but the time scale needed for observable change may be prohibitively long.

C. EFFECTS OF PROCESSING

Structural changes that occur as the result of changes in fabrication parameters are readily examined by studying viscoelastic transitions. While the glass transition is the most significant of these, many polymers exhibit secondary transitions associated with specific types of molecular motion. One of the best known cases is that of polyamides which exhibit two subambient transitions at approximately −130 and −60°C. These are generally believed to result from the rotation of methylene groups and carbonyl groups, respectively. ABS exhibits a subambient transition near −90°C associated with the polybutadiene phase. This provides a relative measurement of the rubber content.

For most elastomeric products T_g is subambient and no transitions are observed above room temperature. For these types of materials the T_g is closely related to the brittleness

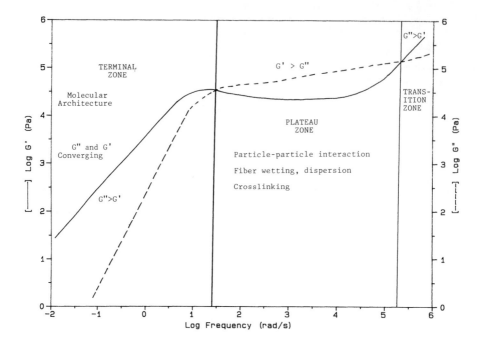

FIGURE 77. Dynamic moduli of a polymer melt vs. frequency at constant temperature. (From Rheometrics, Inc., February 22, 1990.)

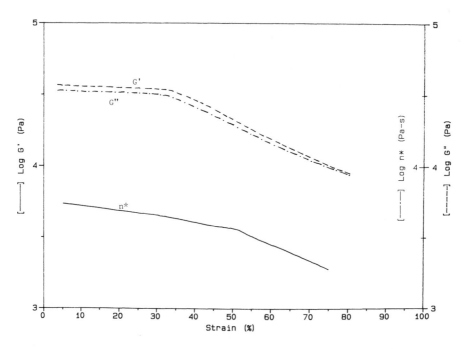

FIGURE 78. Strain sweep of poly(dimethyl siloxane) showing dynamic moduli and complex viscosity. (From Rheometrics, Inc., February 22, 1990.)

temperature and can be seen to vary with the ratio of hard and soft segment material. Figure 80 shows the loss modulus, E'' vs. temperature, for two hardnesses of Hytrel.

Transitions that occur above T_g may also provide information on structural differences. Acetal copolymers and homopolymers have T_g's below room temperature, but due to their

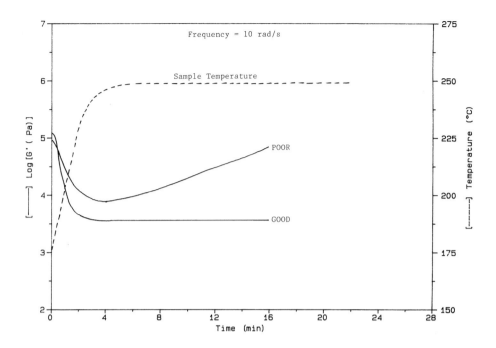

FIGURE 79. Effect of degradation on elasticity of injection-molded ABS parts. (From Rheometrics, Inc., February 22, 1990.)

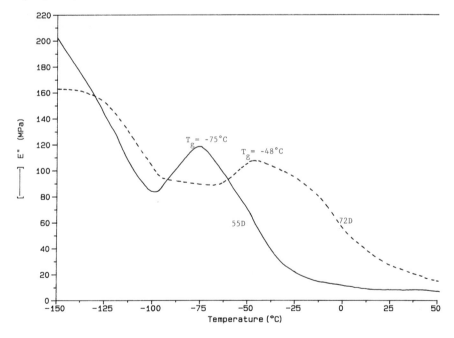

FIGURE 80. Loss modulus for two hardnesses of Hytrel showing the glass transition. (Sepe, unpublished, 1988.)

high degree of crystallinity they exhibit semirigid properties at and above room temperature. Above room temperature these materials exhibit a broad transition that peaks at 75 to 85°C for copolymer and 95 to 105°C for homopolymer. Figure 81 shows E'' vs. T for unfilled copolymer moldings produced at four different mold temperatures while Figure 82 shows both E' and E'' vs. T for the same variations in a homopolymer. It is well known that higher

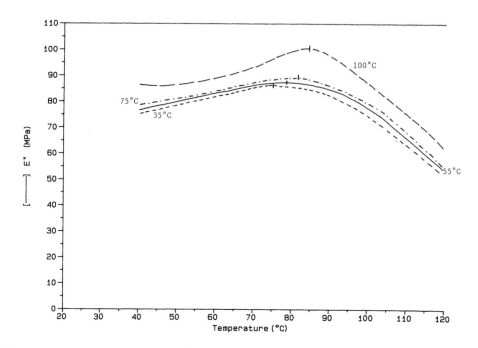

FIGURE 81. E″ vs. T for acetal copolymer samples produced at the mold temperatures indicated. Peak temperature shifts to higher values with increasing crystallinity. (Sepe, unpublished, 1991.)

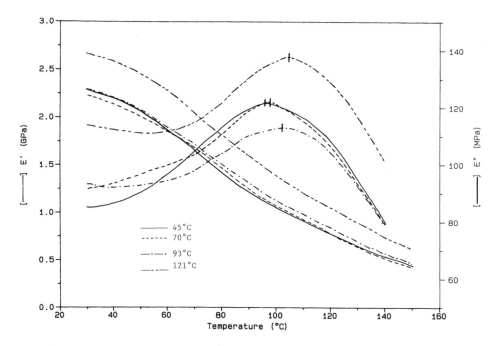

FIGURE 82. E′ and E″ vs. T for acetal homopolymer samples molded with different mold temperatures. (Sepe, unpublished, 1989.)

mold temperatures result in higher degrees of crystallinity for these materials; however, these differences can be difficult to measure by DSC. The peak temperature of E″ shows a measurable shift to higher temperatures as crystallinity increases and E′ also increases over the entire temperature range. It is significant that the greatest improvement in rigidity as

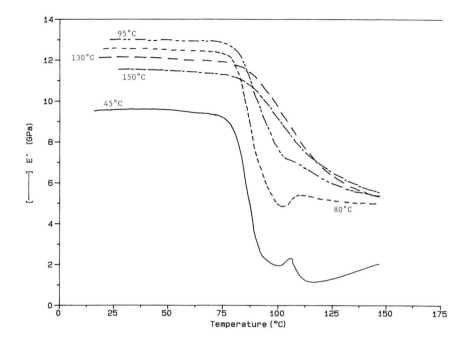

FIGURE 83. E′ vs. T for PPS samples produced at the mold temperatures indicated. (From Sepe, M., *SPE ANTEC*, 37, 469, 1991.)

well as the best molded surface finish are achieved at mold temperatures above the E″ peak temperature.

Materials such as PPS, PET, and PEEK have received a great deal of attention because their impressive properties are highly dependent upon processing. Adequate crystallization requires mold temperatures above a characteristic recrystallization temperature as discussed in Sections III and V. Most of the studies performed on these materials have been conducted by DSC. DMA, while it cannot quantify the enthalpy changes useful in developing models for crystallization kinetics, provides valuable information on changes in properties as a function of varying crystallinities. Figure 83 shows E′ vs. T for 3.18-mm thick tensile bars molded from 40 wt % glass fiber-reinforced PPS at different mold temperatures. The curves show a progression that relates to the recrystallization exotherms observed by DSC as well as the dimensional changes measured by TMA.

At mold temperatures below the onset of DSC recrystallization the resulting structure is highly amorphous. This is clearly indicated by a steep drop in E′ followed by a recovery that signals the onset of solid-state recrystallization. Mold temperatures within the range of the DSC exotherm produce increased levels of crystallinity. The room temperature modulus increases, the slope of decline in E′ associated with the glass transition becomes less severe, the plateau modulus rises, and the modulus recoveries vanish. At temperatures above the recrystallization endpoint, the highest levels of crystallinity result in the highest plateau moduli, even though room temperature rigidity is reduced slightly. Figure 84 plots tan δ vs. T in the glass transition regions and shows the shift in properties as crystallinity increases. Undercrystallized samples exhibit high peak values at reduced temperatures followed by a secondary peak. As the mold temperature rises one peak emerges at a higher temperature and the height is progressively reduced as crystallinity increases. The loss properties of a material have been related to impact resistance. In this particular case, there is an observable relationship between the peak height of the E″ and tan δ curves and the impact strength of the PPS.[144]

Maffezolli et al.[145] have used both DSC and DMA data to perform an Avrami analysis

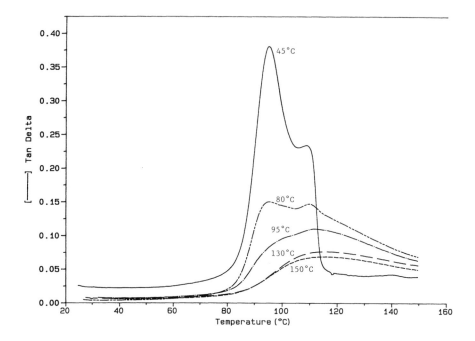

FIGURE 84. Tan δ curves for the glass transition for the various PPS samples from Figure 83. (From Sepe, M., *SPE ANTEC,* 37, 469, 1991.)

on melt- and cold-crystallized carbon fiber-reinforced PPS composites. They demonstrate good agreement between the two techniques. The authors utilize isothermal crystallization measurements by DSC and DMA to develop Avrami plots of degree of crystallinity vs. time. They use storage modulus and storage compliance to calculate the volume fraction of crystallinity by

$$X_{vcr} = X_{vc}/X_{vc,max} = M = \frac{(E' - E'_0)}{(E'_{max} - E'_0)} \tag{13}$$

where E'_0 is the modulus of the fully amorphous material at the crystallization temperature, E'_{max} is the maximum modulus reached in a given isothermal experiment, and X_{vcr} is the relative volume fraction of crystallinity as related to the maximum crystallinity, $X_{vc,max}$, developed in the same experiment. M is the normalized modulus.

Storage compliance was used in the following equation:

$$X_{vcr} = (S' - S'_0)/(S'_{max} - S'_0) \tag{14}$$

where S' is the measured storage compliance, S'_0 is the storage compliance of the fully amorphous material, and S'_{max} is the maximum storage compliance measured during the isothermal experiment.

The mathematical relationships are based on the work of Takayanagi et al.[146] The relationship between the storage modulus and the degree of crystallinity is based on a connection in parallel between the viscoelastic elements representing the behavior of the

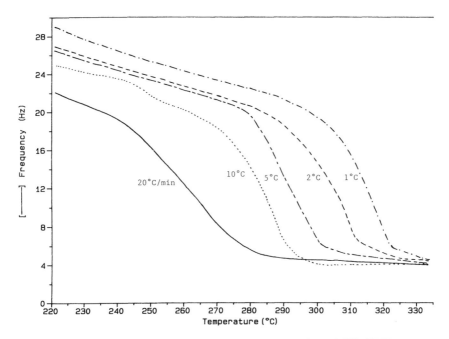

FIGURE 85. Effect of cooling rate on the crystallization of PEEK. (Sichina and Gill, 1986.)

crystalline and amorphous phases, while the relationship to storage compliance is based on a series connection between the two phases. The equations are of the form:

$$E^* = (1 - X_{vc})E_a^* + X_{vc}E_c^* \tag{15}$$

and

$$S^* = (1 - X_{vc})S_a^* + X_{vc}S_c^* \tag{16}$$

where E^* and S^* are the complex modulus and compliance and the subscripts a and c refer to the amorphous and crystalline phases, respectively.

While the actual calculations for time to maximum crystallinity agree well with theoretical predictions, the times calculated by DMA are longer for a given crystallization temperature than those determined by DSC. It has been suggested that the equations using modulus are not equivalent to the classical equations using heat flow. It is also possible that the difference in times are related to the sample geometry and the resulting heat transfer mechanisms. DSC samples are smaller and thinner and are in intimate contact with the instrument thermocouple. DMA determinations may prove to be more representative of processing in actual parts. Godovsky and Slonimsky[147] have noted similar discrepancies between studies of this type using DSC and dilatometry. They have proposed that DSC is more sensitive to the development of two-dimensional lamellae while dilatometry may be influenced by the formation of three dimensional spherulites.

Sichina and Gill[148] have illustrated the use of DMA to monitor the increase in modulus of a PEEK/carbon fiber prepreg as it is cooled from the melt at different cooling rates. The faster cooling rates shift the crystallization process to lower temperatures. Figure 85 shows this.

This author has examined the effects of mold temperature and packing pressure on the crystallinity and moisture absorption of a glass and aramid fiber-reinforced nylon 6/6. While

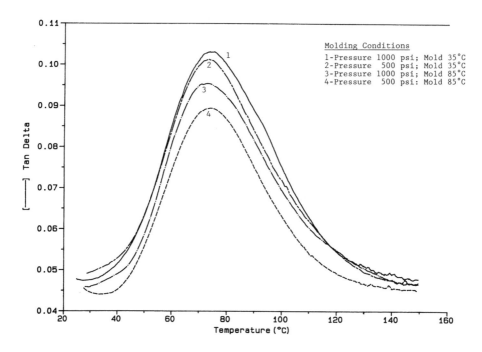

FIGURE 86. Influence of processing conditions on crystallinity of glass/aramid reinforced nylon 6/6 as measured by tan δ peaks associated with T_g. (Sepe, unpublished, 1990.)

elevated mold temperatures can aid crystallization by providing opportunities for the molecular mobility required, excessive packing pressures can work to restrict this motion. Crystals formed under high pressure during injection molding have a tendency to remelt and reform during the reduced pressures of the packing phase of the cycle. However, if the pressure is kept at elevated levels, until the polymer cools to below the remelt point, this second stage of crystallization will not occur. Patel and Nunn[149] have noted the effects of reduced packing pressures on the increase in crystallinity of PPS.

Figure 86 plots the tan δ curves vs. temperature for samples of a 30 wt % glass and 10 wt % aramid fiber-reinforced nylon 6/6 molded at four conditions which made use of varying mold temperatures and packing pressures. Both parameters had a significant effect on the material properties. The lower tan δ peaks indicate higher levels of crystallinity, the highest level being achieved by a combination of high mold temperature and low packing pressure. It is interesting to note that the effect of packing pressure in reducing crystallinity was more noticeable at the higher mold temperature. Moisture absorption and dimensional stability studies verified the differences in structure. Samples displaying the lowest crystallinity absorbed moisture to a level of 0.99% by weight while highest crystallinity samples only absorbed 0.90% after conditioning. Even more striking was the difference in dimensional change. On a 153-mm length the properly crystallized product only swelled 0.1 mm while the more amorphous moldings grew 0.3 mm. The greater effect of moisture absorption in the less-crystalline moldings is explained by the fact that moisture only effects the amorphous regions of the nylon polymer.

This is clearly shown by another study that examines the effects of different levels of moisture absorption on the dynamic mechanical properties of unfilled nylon 6. It is well known that as nylon moldings absorb water their properties change; they become less rigid and more resistant to impact. However, the room temperature modulus of nylon 6 is due to a combination of the amorphous glass, still below T_g, and the crystalline matrix. It has

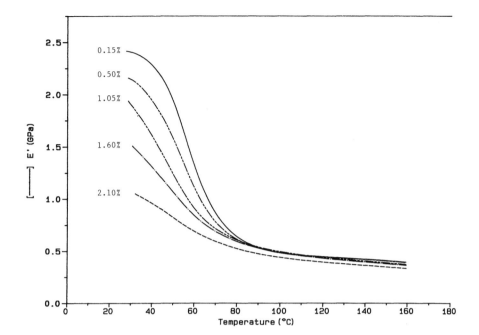

FIGURE 87. Effect of moisture absorption of E′ of unfilled nylon 6. (Sepe, unpublished, 1992.)

already been shown[127] that the T_g of nylons declines to subambient temperatures as water absorption increases. We would expect a reduction in room temperature modulus as a function of increasing moisture content. Figure 87 shows that this does occur. E′ at room temperature is reduced to 40% of the dry-as-molded value at a moisture content of 2%. However, these curves also show that the portion of E′ attributable to the crystalline phase is not changed by the absorbed moisture. Moisture content determinations conducted on samples removed from the DMA after heating above T_g showed that the samples retained 80% or more of their equilibrium moisture content. This eliminates the possibility that the constant plateau modulus above T_g was due to a drying-out process in the samples.

Melt temperature and direction of flow can be shown to have a large influence on viscoelastic properties. In the case of flow direction, the orientation of polymer chains and fibrous reinforcements will have a dramatic effect on the storage modulus. Employing temperatures just above the melting point can contribute to this orientation by increasing the extensional shear on the material as it flows through the restriction of the gate. Figure 88 shows E′ vs. T for an unfilled PP copolymer with a melting point of 165°C. Molding samples with a melt temperature of 170°C produces a modulus nearly 30% higher at room temperature than in samples molded at 205°C. The mold temperature was 30°C. Figure 89 shows E′ vs. T for samples molded at the same melt temperatures but with a mold temperature of 70°C. The higher mold temperature reduces the modulus of the cold melt samples but has virtually no effect on the parts made at the higher melt temperature. It is probable that the higher mold temperature permitted some relaxation of the highly oriented cold melt.

When fibrous reinforcements are added to polymers, the effects are even more dramatic since these fibers never melt during processing and are therefore highly aligned in the flow direction. Figure 90 plots E′ vs. T for a thermoplastic polyurethane elastomer reinforced with 30 wt % glass fiber. Samples were cut from 100 mm × 100 mm plaques in both the direction of flow and transverse to flow. The modulus of the transverse sample at room temperature is equivalent to that of the oriented sample near the short-term heat deflection

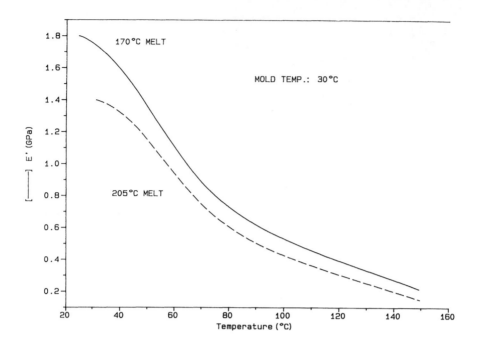

FIGURE 88.   Effect of melt temperature on E′ of a polypropylene copolymer. (Sepe, unpublished, 1992.)

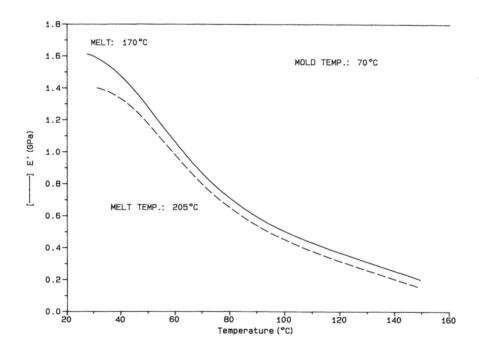

FIGURE 89.   Effect of an elevated mold temperature on the material evaluated in Figure 88. (Sepe, unpublished, 1992.)

temperature. Figure 91 illustrates another example using a liquid crystal polymer (LCP) with 30 wt % glass fiber. Samples were prepared in the same manner as for the filled polyurethane. Here the modulus of the transverse direction sample at room temperature is equal to the modulus of the oriented sample at 205°C.

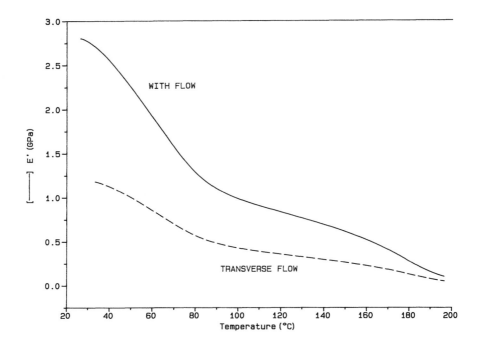

FIGURE 90. Effect of polymer and fiber orientation on the storage modulus of a 30 wt % glass-reinforced polyurethane. (Sepe, M., unpublished, 1990.)

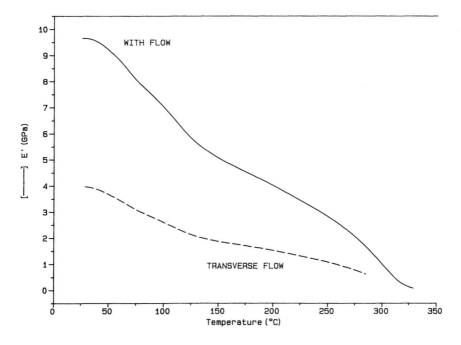

FIGURE 91. Effect of orientation of E' vs. T behavior of a 30 wt % glass-reinforced liquid crystal polymer. (Sepe, M., unpublished, 1990.)

Another possible effect of processing is that of frozen-in stresses. Detection of this phenomenon by TMA was discussed in Section V. DMA proves to be a valuable technique for examining this aspect of polymer behavior. Zafran and Beatty[150] have investigated the effects of tensile deformation on the dynamic mechanical properties of crosslinked

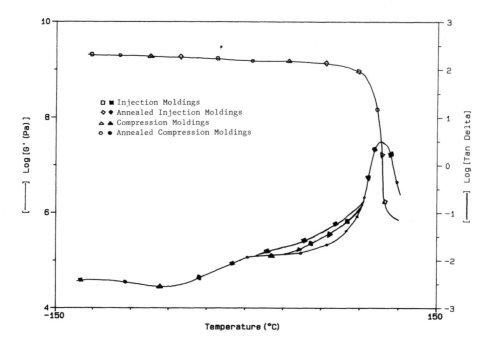

FIGURE 92. G′ and tan δ vs. T for polystyrene samples produced with different methods as indicated. Tan δ curves exhibit dispersion prior to T_g. (From Rohn, C. and Herh, P., *SPE ANTEC*, 34, 1135, 1988.)

polyurethane. When samples were stretched in tension above T_g and then quenched below T_g, the sub-T_g modulus increased with increasing tensile strain and the T_g decreased slightly. Above T_g the modulus was virtually unaffected. The reduction in the T_g was also detected by a shift of the tan δ peak to lower temperatures. The area under the tan δ curve was also reduced.

Rohn and Herh[151] have demonstrated the use of DMA to evaluate metastable structures arising in both amorphous and semicrystalline polymers during the molding process. In amorphous polymers such as polystyrene and ABS, the magnitude of tan δ in the temperature region just below T_g is greater for highly oriented samples. Injection-molded samples exhibit the highest stress levels while annealed compression moldings show the lowest levels. Both unannealed injection moldings and unannealed compression moldings show intermediate behavior. Figure 92 shows that, while loss properties are very sensitive to these stress levels, storage properties seem unaffected.

In semicrystalline materials the strains relate to differences in crystal structure brought about by shear and thermal gradients. Consequently, the differences in viscoelasticity are most evident above T_g where the crystalline phase is the dominant structural component. Figure 93 shows the effects of different stress levels on the storage modulus and tan δ for polypropylene. The modulus above T_g increases as molded-in strains and imperfect crystals become minimized by annealing and/or compression molding. The tan δ values above T_g are highest for injection-molded, unannealed samples and lowest for annealed, compression molded specimens.

D. EFFECTS OF FILLERS

The effects of fillers on polymeric systems can be examined by both solid- and melt-state DMA. In solid-state DMA it is customary to examine the effects of the filler on the loss modulus curve. In the melt state the frequency dependency of G′ and G″ in the terminal zone are evaluated.

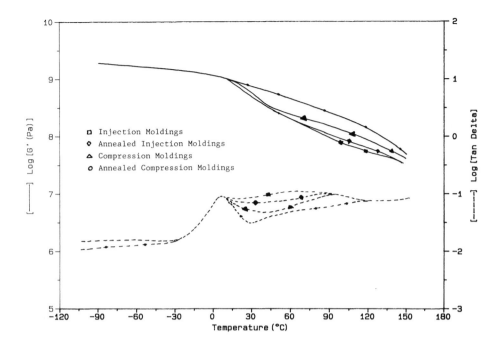

FIGURE 93. G′ and tan δ curves for polypropylene samples prepared as indicated. (From Rohn, C. and Herh, P., *SPE ANTEC,* 34, 1135, 1988.)

Salovey and Gandhi[152] have examined the effects of various CBs on the rheology of polystyrene and polybutyl methacrylate. The CBs were varied in concentration and in surface area. Their work reveals a general trend of decreasing frequency dependency in both G′ and G″ as a function of increasing CB concentration.

Trevino[153] has studied the solid-state behavior of unsaturated polyesters with varying percentages of mica. Temperature sweeps at constant frequency show several effects of increasing filler content. E′ and T_g both increase while the height of E″ and tan δ peaks was reduced. Multiple frequencies were used to establish a relaxation map of the various compounds and provide activation energy (E_a) measurements for the glass transition relaxation. The shift in the peak temperature of tan δ as a function of frequency was used for these determinations. Filler content was found to have little effect on E_a.

This general effect of fillers has been documented by Nielsen.[154] Other examples include the addition of talc to polypropylene[155] and glass fibers to PBT. Figure 94 shows tan δ vs. T plots for two glass-reinforced PBT polyesters. One contains a short glass fiber and the other employs a long glass fiber with a special wetting agent to improve polymer-fiber adhesion. The peak temperature of the long glass material is shifted higher and the peak height is reduced. Both materials are filled at 30 wt %.

Some fillers are incorporated to achieve specific end use properties; however, they do not necessarily enhance strength and stiffness. Apparent increases in short-term rigidity may decay under long-term loads. Sepe[156] has used master curves of apparent modulus vs. time constructed from short-term creep tests to show that an FR PBT polyester with a higher initial modulus creeps more rapidly than the non-FR analog. In the same study a glass fiber-reinforced PET was compared to a material filled with a combination of glass fiber and particulate mineral. While both materials displayed the same short-term modulus, the material containing mineral showed higher strain over time. Figure 95 compares the PBT compounds and Figure 96 shows the curves for the PET materials.

In the solid state, fiber-matrix adhesion has been studied by measuring G″ and tan δ values for fiber/polymer composites. Fibers are considered to be essentially elastic while

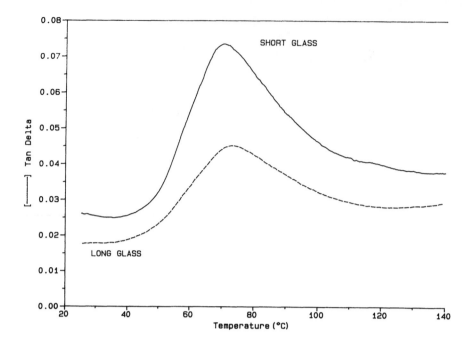

FIGURE 94.   Tan δ associated with T_g for PBT polyesters reinforced with different types of glass and coupling agents. (Sepe, unpublished, 1989.)

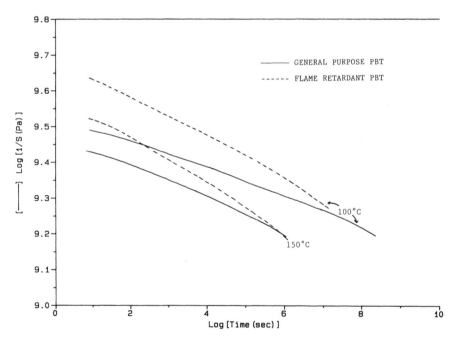

FIGURE 95.   Mastercurves of apparent modulus vs. time for GP and FR PBT with 30 wt % glass fiber. (From Sepe, M., *Plastics Design Forum*, 16(6), 45, 1991.)

the polymer and the fiber/polymer interface are viscoelastic. Therefore, energy dissipation will occur in the polymer matrix and at the interface and a stronger interface will be characterized by less energy dissipation. Chua[157] has used a resonant frequency instrument to study glass-reinforced unsaturated polyesters. Tan δ peak heights at T_g were correlated

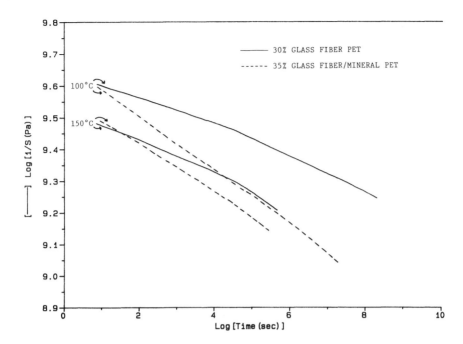

FIGURE 96. Mastercurves of apparent modulus vs. time for PET polyesters utilizing different fillers. (From Sepe, M., *Plastics Design Forum,* 16(6), 45, 1991.)

to interfacial shear strength in the longitudinal direction and to transverse flexural strength in the transverse direction. Different silane treatments on the glass were evaluated as well as coupling agent concentrations and the effects of fiber diameter. Within a given directionality a reasonably good inverse relationship was found between interfacial strength and tan δ peak heights. It is anticipated that the agreement may prove to be better for tests conducted at fixed frequencies.

Zorowski and Murayama[158] developed a method for determining the quality of interfacial adhesion in reinforced rubber using tan δ measurements in the relationship

$$\tan \delta_{adh} = \tan \delta_c - \tan \delta_{exp} \tag{17}$$

where $\tan \delta_{adh}$ is the energy dissipation due to poor adhesion, $\tan \delta_c$ is the calculated energy dissipation for the compound, and $\tan \delta_{exp}$ is the experimentally determined energy loss for the composite. Tan δ_c is expressed by

$$\tan \delta_c = \frac{\tan \delta_f \cdot E_f \cdot V_f + \tan \delta_m \cdot E_m \cdot V_m}{E_f \cdot V_f + E_m \cdot V_m} \tag{18}$$

where E is the modulus, V is the volume fraction, and the subscripts f and m refer to the fiber and the matrix, respectively. Tan δ_{adh} was found to correlate inversely to interfacial strength.

However, this relationship was not found to yield useful correlations in the evaluation of the glass fiber-reinforced polyester in the study by Chua. In addition, it was not possible to employ a simple rule of mixtures to calculate tan δ for the composite as a function of the volume fraction of matrix and fiber.

Edie et al.[159] have investigated the relationship of dynamic loss properties on the interfacial strength of epoxy/carbon fiber composites prepared with different types of carbon fiber. This study employed fixed frequency measurements at multiple frequencies. The

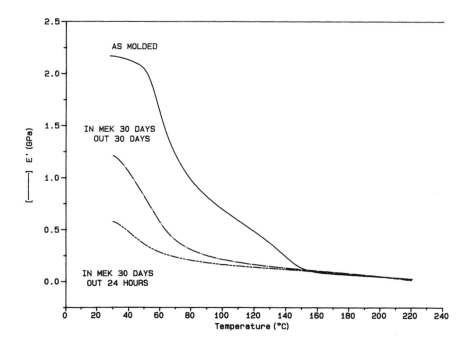

FIGURE 97.   Effects of MEK absorption on E' of a PC/PBT alloy. (Sepe, unpublished, 1989.)

inverse relationship between tan δ and interfacial strength was observed. The frequency dependency of storage and loss properties as a function of volume fraction of fiber was also examined.

It is important in studies of this type that other factors affecting viscoelasticity such as formulation, sample preparation, and storage conditions prior to testing be carefully controlled. Even factors such as room temperature aging and moisture absorption can interfere with measurements, especially of loss properties.

Chartoff[160] has demonstrated the effect of platelet graphite on the damping efficiency of rubber compounds. Compounds of chlorobutyl 1066 were prepared with platelet graphite in varying concentrations and particle size distributions. Various surface treatments to promote adhesion were also evaluated. DMA was used to characterize changes in E' and tan δ as a function of composition changes. The broadening of the tan δ peak without a reduction in peak height indicates an increased damping efficiency over a broader range of frequencies. This can be demonstrated by making multiple frequency measurements at temperatures spanning the limits of the tan δ peak and then creating master curves using time-temperature superposition. By selecting different reference temperatures across the transition region, it can be shown that the range of frequencies where useful damping occurs increases. Chartoff also notes that particle-polymer slippage is a key aspect of damping effectiveness since silane-treated filler gives a compound with a narrower tan δ peak than that obtained with an untreated filler.

Solvent effects can also be examined by DMA. Figure 97 shows E' vs. T for a polycarbonate/PBT alloy before and after immersion in methyl ethyl ketone (MEK). It is apparent that the material is severely plasticized by the solvent. Drying the sample in air for a period of time recovers a portion of the lost properties. However, a comparison of the loss moduli for each sample shows that the polycarbonate fraction of the blend is completely dissolved by the solvent. Figure 98 illustrates this.

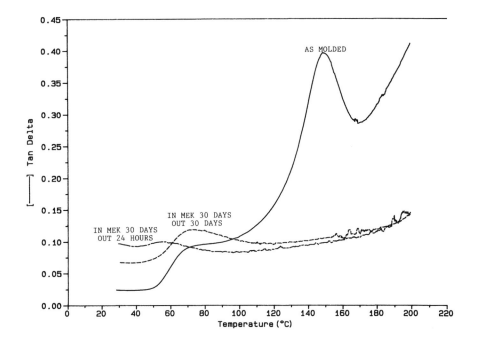

FIGURE 98.   Effects of MEK absorption on tan δ for PC/PBT alloy. The peak associated with T_g of PC phase does not reappear when sample dries out, indicating permanent damage to this phase of the blend. (Sepe, unpublished, 1989.)

E.   MULTIPLE FREQUENCY CHARACTERIZATIONS

References to multiple frequency tests above point out some of the uses for these kinds of evaluation. Viscoelastic responses are time dependent and therefore frequency dependent, since time and frequency are related by the equation

$$t = \frac{1}{2\pi f} \tag{19}$$

The measurement of the activation energy of a transition can be accomplished by making multiple frequency measurements of the damping peak across the temperature range of the transition. Typically, higher frequencies result in a higher transition temperature and a lower peak height. Figure 99 illustrates a multiple frequency sweep of the glass transition for a glass fiber-reinforced nylon 6, plotting E'' vs. T. Figure 100 shows the same evaluation for a glass fiber-reinforced PP. It is evident that the PP is much more sensitive to frequency changes. The slope of the line connecting the transition peaks is much steeper for the nylon material and thus the E_a for the nylon glass transition is greater.

In the melt state, higher frequencies (shorter times) result in an increase in elasticity and a decrease in viscosity. Frequency-dependent behavior in the melt state can be used to study aspects of polymer structure such as molecular weight and molecular weight distribution, chain branching, and crosslinking. Figure 101 plots G', G'', and n^* vs. frequency for two HDPE resins of different molecular weight distributions (MWD). The broader MWD material displays a higher elasticity, a lower frequency dependency, and a lower frequency of crossover between G' and G''. The higher elasticity and lower degree of frequency dependency in the broad MWD material is attributable to the high molecular weight tail.

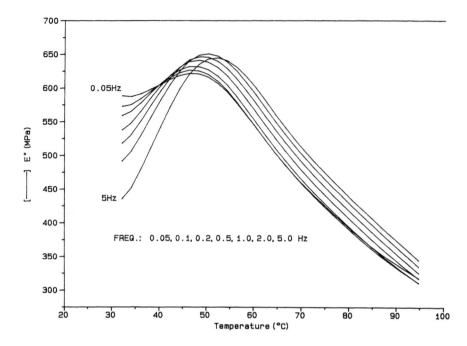

FIGURE 99. Loss modulus scanned at multiple frequencies for the glass transition region of a 50 wt % long glass fiber-reinforced nylon 6. (Sepe, unpublished, 1990.)

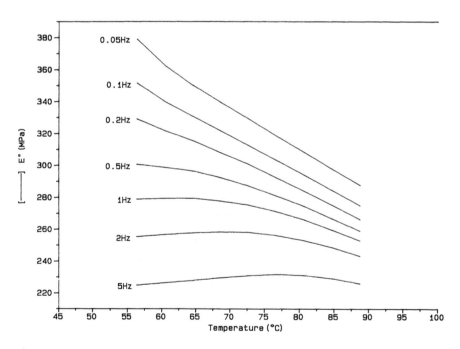

FIGURE 100. Loss modulus scan at multiple frequencies for a 40 wt % long glass fiber-reinforced polypropylene showing a large dispersion of the curves. (Sepe, unpublished, 1990.)

Note that the broad MWD material also shear thins more effectively; viscosity declines more rapidly with rising frequency (shear rate) for the broad MWD material.

Similar behavior is observed for long-chain branching as compared to a linear material. Figure 102 shows this. This distinction can appear in processing as a difference in controlling

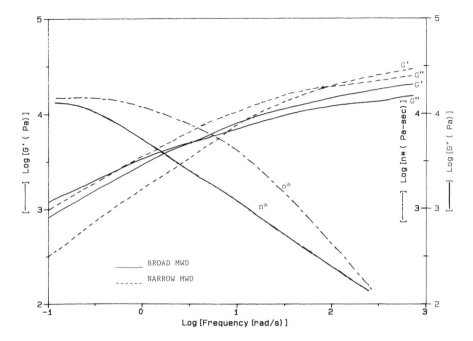

FIGURE 101. Effects of molecular weight distribution on viscoelastic properties for polyethylene. (From Product Performance Through Rheology, *course notes,* Rheometrics, Inc., Rosemont, IL, February 22, 1990.)

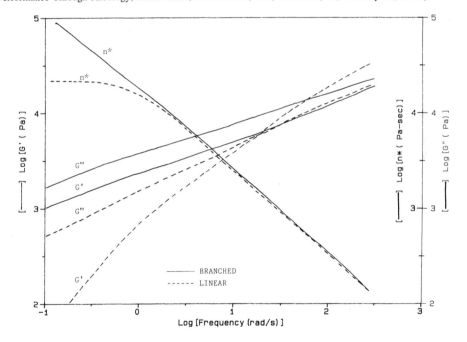

FIGURE 102. Effects of long-chain branching in ethylene-propylene copolymer on viscoelastic properties. (From Product Performance Through Rheology, *course notes,* Rheometrics, Inc., Rosemont, IL, February 22, 1990.)

flash. An excellent example is PPS. Highly branched PPS compounds tend to flash less because, as the shear rate is reduced as the injection process switches over from rapid filling to packing, viscosity builds more rapidly in the branched material than in the linear material. The polymer is more likely to reach a no-flow state before it can fill the vent regions around the perimeter of the mold cavity.

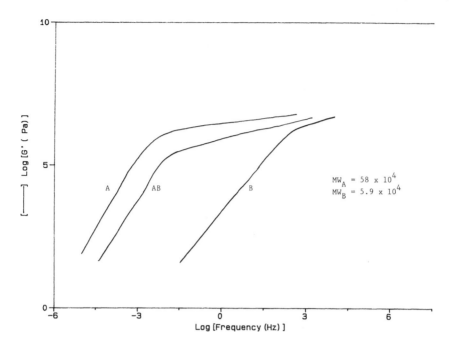

FIGURE 103.   Effects of molecular weight on G' vs. frequency. (From Product Performance Through Rheology, *course notes,* Rheometrics, Inc., Rosemont, IL, February 22, 1990.)

Increases in molecular weight shift the frequency range of the plateau modulus for G' to a lower initial frequency as shown in Figure 103. This is associated with the longer relaxation times of the more highly entangled long chains. Once the terminal zone is reached, the slope of the log-log plot of G' vs. frequency is essentially independent of molecular weight assuming the same polydispersity. Tuminello has demonstrated this.[142] Tuminello[161] has also discussed a method of quantitatively determining MWD from the G' master curve and has related it successfully to results from size exclusion chromatography. Studies were performed on narrow and broad MWD resins as well as bimodal blends. Letton and Tuminello[162] have illustrated the usefulness of this method for polycarbonate, a condensation polymer that exhibits higher polydispersity than the polystyrenes that were originally evaluated.

The development of master curves in the solid state using multiple-frequency measurements is discussed by Starkweather[163] for the β transition in PMMA and by Sichina[164] for the glass transition of polycarbonate. Sichina[165] has constructed master curves of E' vs. frequency for PEEK composites cooled at different rates. At high frequencies the modulus values for all the samples are nearly the same, but at lower frequencies the moduli curves diverge, declining more rapidly for quenched samples which are of lower crystallinity. If the modulus data are plotted vs. time instead of frequency, the graph essentially appears as the mirror image of the E' vs. frequency plot. The modulus curves will begin as equivalent at short times and diverge at longer times, the quenched samples displaying a more rapid decline than the properly crystallized specimens. Modulus vs. time plots generated from multiple-frequency measurements over a range of temperatures will provide the same information as master curves developed from superpositioning of short-term creep tests made at different temperatures. Sichina[165] has demonstrated this for epoxy/fiberglass laminates. Figure 104 shows the master curves developed by both methods. While multiple-frequency data can be used to provide information on damping (loss properties), creep measurements only supply data about the storage modulus.[166]

The relationship between modulus vs. frequency and modulus vs. time master curves is illustrated in Figures 105 and 106. Figure 105 shows a master curve of E' vs. frequency

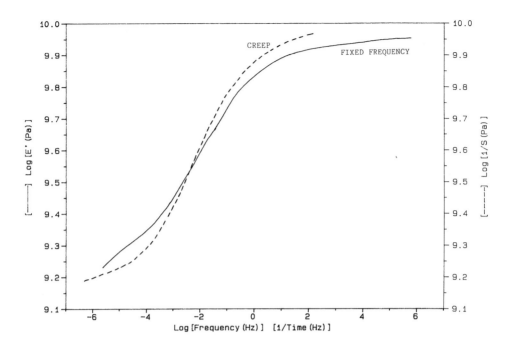

FIGURE 104. Master curves of E′ vs. frequency generated from creep data and fixed frequency data. (From Sichina, W. J., internal communication, 1988.)

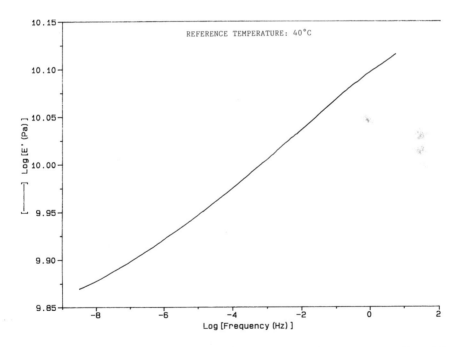

FIGURE 105. E′ vs. frequency master curve from multiple frequency data for 30 wt % carbon fiber-reinforced nylon 6/6. Shift factors fitted to a WLF model. (Sepe, unpublished, 1991.)

based on a multiple-temperature sweep at seven frequencies spanning 2 decades. Figure 106 shows E′ vs. time using the same data. The material is a 30 wt % carbon fiber-reinforced nylon 6/6. Since the temperature region involved the glass transition, the shift factors used in superposing the curves onto the reference curve were fitted to the WLF equation:

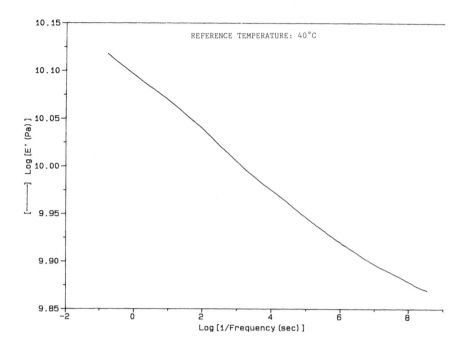

FIGURE 106.   E' vs. time master curve derived from the same raw data used in Figure 105 after inversion from frequency to time. Plot is a mirror image of Figure 105. (Sepe, unpublished, 1991.)

$$\log (a) \;=\; -\; \frac{C_1(T - T_0)}{C_2 + T - T_0} \tag{20}$$

where a is the shift factor, T is the shifted temperature, T_0 is the reference temperature, and C_1 and C_2 are the WLF coefficients. Figures 107 and 108 show the plot of log (a) vs. temperature. The WLF coefficients for the master curves agree very well.

The same reciprocal relationship between master curves based on frequency and time is also observed for damping properties. Figures 109 and 110 show plots of E″ vs. frequency and E″ vs. time, respectively. For a reference temperature of 40°C the peak value for E″ is achieved at a frequency of 0.08675 Hz. With time as the x-axis the same peak point for E″ is achieved in 12.38 s. This is the relaxation time for the glass transition when the material is at 40°C.[167] Since frequency and time are reciprocals, the two determinations should multiply to unity. The actual product of the two values is 1.073. This illustrates the utility of calculating time-dependent behavior from frequency-dependent data and vice versa. Figure 111 plots the frequency at the peak of the E″ master curve as a function of the reference temperature. The shifting of the peak value to higher frequencies as temperature increases is equivalent to the shortening of relaxation times.

The use of master curves provides an excellent means of demonstrating the interplay between time and temperature in affecting the load-bearing properties of polymers. Any temperature step in a multiple temperature scan may be used as a reference temperature. Thus, a series of master curves at different reference temperatures may be plotted vs. time to illustrate regions of structural stability and of rapid change. Figure 112 shows a plot of apparent modulus vs. time for an epoxy system with a T_g of 185°C. This transition temperature was determined from the peak temperature of E″ during a dynamic heating scan. These curves show that the relaxation of the polymer matrix goes to shorter times as the reference temperature increases toward T_g. More importantly, these relaxations occur at times sufficiently short to be of concern in structural applications, despite the fact that the short time

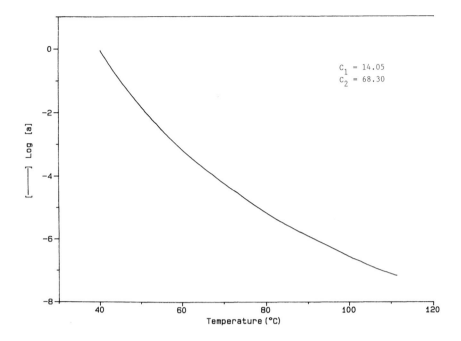

FIGURE 107.   Plot of shift factors vs. temperature showing the WLF coefficients for the data treatment in Figure 105. (Sepe, unpublished, 1991.)

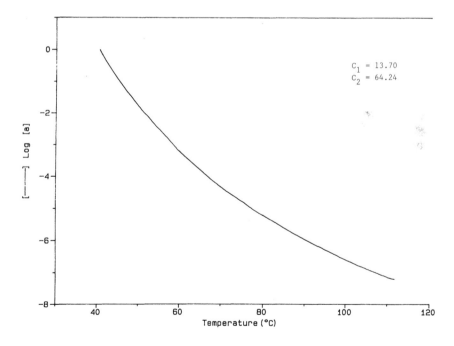

FIGURE 108.   Plot of shift factors for master curve in Figure 106. (Sepe, unpublished, 1991.)

T_g has not been achieved. The declines in apparent modulus are an indication of the molecular relaxations associated with the glass transition. This illustrates the importance of considering the time scale of measurements made on viscoelastic systems as well as the temperature scale. Sepe[156] has shown that this type of evaluation can provide useful engineering data on

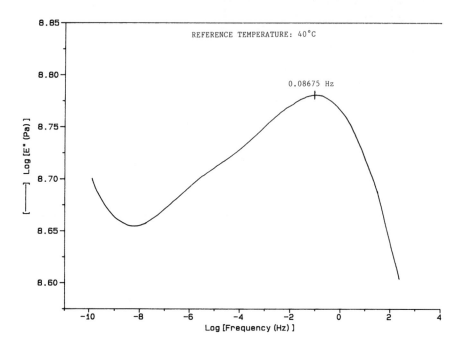

FIGURE 109.   E″ master curve vs. frequency from the same scan used in Figures 105 to 108.

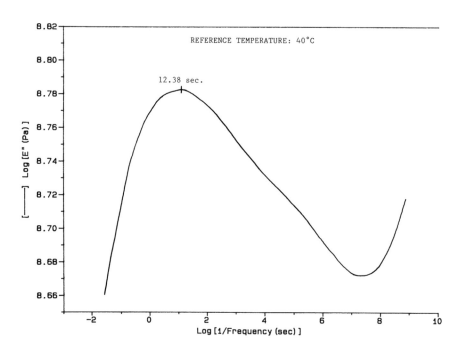

FIGURE 110.   E″ master curve with the horizontal axis inverted from frequency to time.

the time-dependent behavior of polymers under load. These data may often contradict decisions regarding appropriate material selection based on short time measurements.

While useful comparisons may be made between polymers by using master curves, the technique involves extrapolation and a number of precautions must be taken in developing the data. While Sepe[156] has demonstrated good short-term agreement between master curves

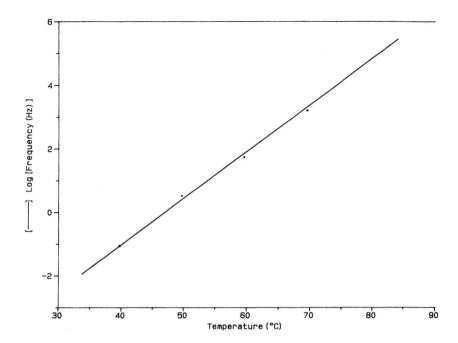

FIGURE 111. Peak frequency of E″ master curves as a function of reference temperature for material used in Figures 105 to 110.

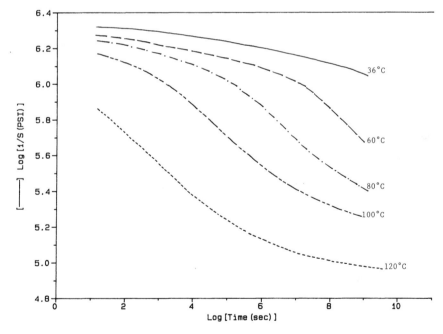

FIGURE 112. Apparent modulus master curves at various temperatures for a cured epoxy. (From Sepe, M., *Plastics Design Forum,* 16(6), 45, 1991.)

and actual creep behavior for a few materials, a number of pitfalls can arise that can result in very poor agreement.

First, it is important that the temperature region being scanned does not give rise to changes in a polymer which occur at elevated temperatures but which will not take place at

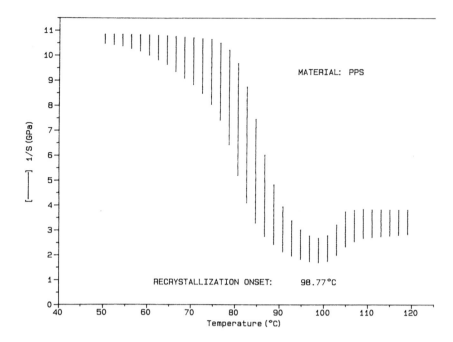

FIGURE 113.   Apparent modulus changes for 30-min creep tests at a range of temperatures for undercrystallized PPS. Modulus recovery indicates recrystallization. (Sepe, M., *SPE ANTEC,* 38, 1152, 1992.)

the reference temperature. Processes such as residual cure or crystallization, stress relief, oxidative degradation, or melting of imperfect crystals can either augment or compete with the relaxation processes associated with creep and stress relaxation. This will make the process of superposing data from different temperatures difficult. Figure 113 shows a plot of apparent modulus for various temperatures. These data are the result of short-term creep tests performed on an undercrystallized PPS sample. When the sample reaches 100°C, residual crystallization occurs, resulting in a modulus recovery similar to that observed by dynamic heating in Figure 83. If we attempt to use data from beyond the recrystallization temperature in developing a master curve for a lower reference temperature, the shift factors will become smaller with increasing temperature. Figure 114 shows the plot of log (a) vs. temperature. Instead of the higher temperature data providing an extension of the master curve to longer times, the modulus recovery results in shifts to shorter times. The shift factors cannot be fitted to a model such as the WLF equation or an Arrhenius fit. The only means available to circumvent this problem is to vertically shift data from this region to a lower modulus and empirically fit the points to the appropriate model. This method is rather arbitrary and the accuracy of the results may therefore be doubtful. More importantly, the recrystallization will not occur over any time frame at temperatures below T_g where the reference temperatures are being selected.

Rohn[168] has identified internal stresses in amorphous polymers as a hindrance to accurate use of time-temperature superpositioning for loss modulus data near the glass transition. In addition, using creep data from short-term tests near T_g leads to an overprediction of creep using the master curve. This occurs because the initial displacements selected for the lower temperatures at which the test is started can result in yielding of the polymer at higher temperatures. Figure 115 shows a master curve for apparent modulus vs. time at a reference temperature of 84°C along with the results of an actual test using the same initial displacement for polycarbonate. The divergence between the predicted and actual behavior results from the use of data in the high temperature region in building the master curve. In this region, nonlinear viscoelastic behavior occurs and the data are not useful. This situation can be

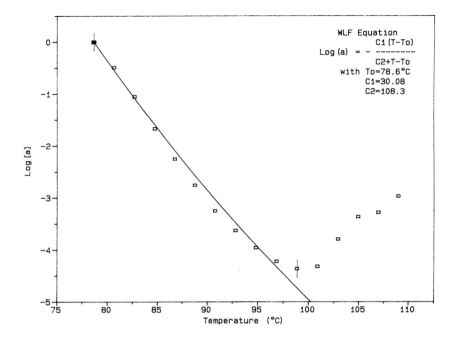

FIGURE 114. Shift factors for an apparent modulus master curve showing the divergence of the shifts from the WLF model after recrystallization begins. (Sepe, unpublished, 1992.)

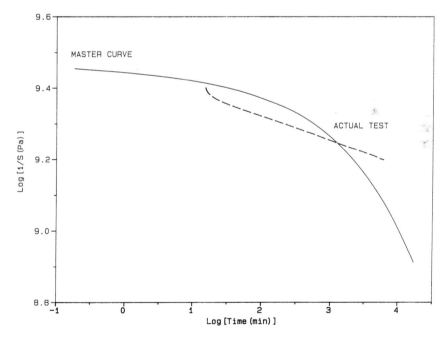

FIGURE 115. Apparent modulus vs. time plots for polycarbonate at 84°C comparing actual test results with predicted behavior from a master curve. (Sepe, unpublished, 1991.)

improved by adjusting the applied strain to lower magnitudes as the temperature is increased. Some commercial instruments now employ such an "autostrain" option.

Finally, Tryson[169] has discussed the effects of physical aging on the creep performance of amorphous polymers. The examination of physical aging has been discussed in Section

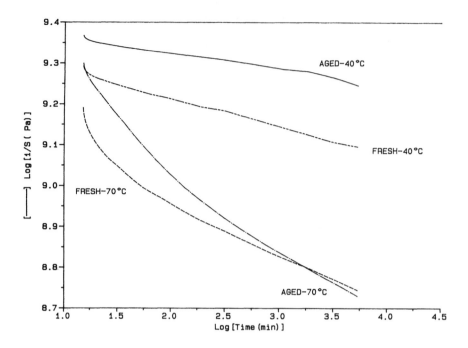

FIGURE 116. Apparent modulus vs. time plots for actual 100-h tests conducted on fresh and aged moldings of PPO. (Sepe, unpublished, 1992.)

III.D. Since the aging process affects the relaxation time of the frozen in structure in the glass, it will have an effect on creep performance. Tryson points out that frequently master curves developed from short-term creep tests tend to underpredict creep performance at short times and overpredict creep at longer times because the aging history of the sample is not accounted for. The difference in creep performance between aged and quenched samples of two amorphous materials, a polycarbonate and a PPE/HIPS alloy, shows a distinct difference in creep performance. Long-term creep data for aged and de-aged samples show that aged samples initially exhibit less creep than samples with an erased aging history. After a period of time the curves converge. This is attributed to the erasure of aging at the test temperature.

This de-aging time will be dependent upon the time the sample was aged before creep testing was initiated as well as the actual test temperature during the creep study. Higher temperatures will result in shorter de-aging times. Since the technique of time-temperature superpositioning involves making short-term creep measurements at a series of temperatures, this will have an effect on the shape of the individual segments and consequently on the master curve. At low temperatures the time of the creep experiment will be much shorter than the time required to erase aging; at higher temperatures the time scale of the creep experiments will approach or even exceed the de-aging time. This is what causes the master curve to underpredict creep at the short time scales and then overpredict it at the longer time scales.

At the low stresses generated by a dynamic mechanical analyzer, the de-aging process can be readily observed. Figure 116 plots the apparent modulus vs. time for 100-h creep tests performed at 40 and 70°C on samples aged for 36 months and for specimens that were molded 24 h prior to testing. The initial displacements employed in these tests were very small, resulting in strains of less than 0.1%. At 40°C the aged sample maintains a higher modulus for the duration of the test, indicating that the effects of 3 years of aging were not reversed in 100 h at 40°C. However, at 70°C the convergence of the apparent modulus curves occurs very quickly and is essentially complete within the first 10 h of the test.

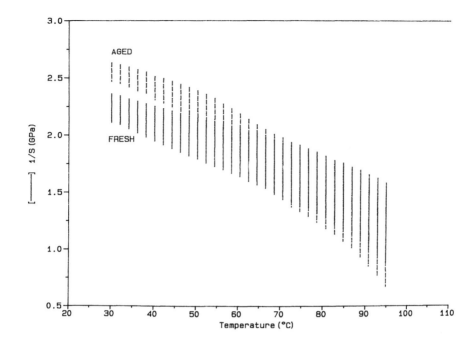

FIGURE 117. Changes in apparent modulus for 30-min tests conducted at various temperatures on fresh and aged PPO. (Sepe, unpublished, 1992.)

Short-term creep measurements made at progressively higher temperatures indicate the same pattern. Figure 117 shows the modulus declines for each test segment using aged and fresh samples. Once again, the higher initial modulus and lower creep of the aged sample at low temperatures are erased as the test temperature rises. In these tests, each segment temperature had a duration of 1 h; 30 min of this involved actual load application. The rest of the time was devoted to a relaxation cycle and temperature equilibration. Even at this short time scale, the two samples become essentially equivalent at 80°C.

It is interesting to note that master curves constructed from the data in Figure 117 for reference temperatures of 40 and 70°C reveal good agreement with the actual 100-h tests shown in Figure 116 with the exception of the aged sample at 70°C. In this case the typical underpredict-overpredict condition reported by Tryson is noted. It would be expected that longer test times or the use of higher stresses would reveal a similar condition for the aged samples tested at 40°C. In this study the combination of the short time scale and the low strains employed is not sufficient to produce de-aging and thus the correlations are good. The results of this comparison are shown in Figure 118 for the freshly molded sample and in Figure 119 for the aged samples.

It should also be noted that, because DMA instruments develop very low strains on solid samples, it is difficult to extrapolate results of creep tests into nonlinear regions of behavior where phenomena such as creep rupture can occur. For this, conventional physical testing equipment capable of exerting higher stresses must be used.

In spite of the limitations of extrapolated modulus or compliance curves, and the tendency for instrument manufacturers to overlook the experimental and theoretical pitfalls, a great deal of useful information can be derived from such studies. Figure 120 plots apparent modulus vs. temperature for 30-min test segments comparing PPS sample which were molded to produce large differences in crystallinity. The undercrystallized product shows drastic losses in modulus at relatively low temperatures compared to the more highly crystallized molding. Generating master curves from these data leads to a prediction that undercrystallized

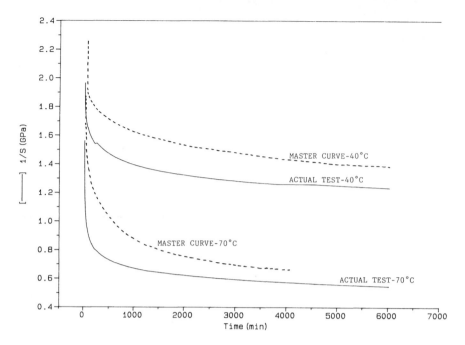

FIGURE 118. Linear plots of apparent modulus vs. time for freshly molded PPO showing good relative agreement between master curve projections and actual test results. (Sepe, unpublished, 1992.)

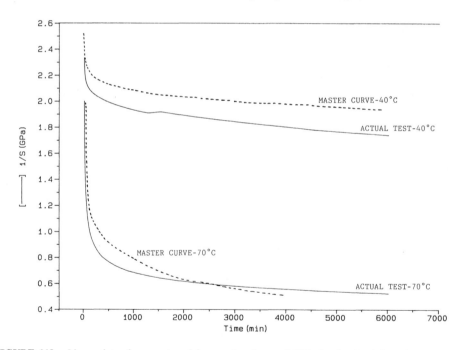

FIGURE 119. Linear plots of apparent modulus vs. time for aged PPO showing breakdown in agreement at 70°C due to de-aging during short-term multiple temperature scans. (From Tryson, G., *SPE ANTEC*, 37, 805, 1991.)

samples placed under load at temperatures below the glass transition will exhibit very large deformations over time. Figure 121 compares master curves for the two types of moldings at reference temperatures of 60 and 80°C. While quantitatively the agreement between the master curves and the results of actual 100-h creep tests is only fair, Figure 122 shows the

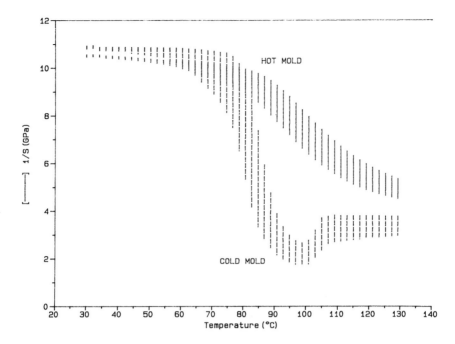

FIGURE 120. Short-term apparent modulus changes at various temperatures for PPS of low and high crystallinity influenced by mold temperature. (From Sepe, M., *SPE ANTEC*, 38, 1152, 1992.)

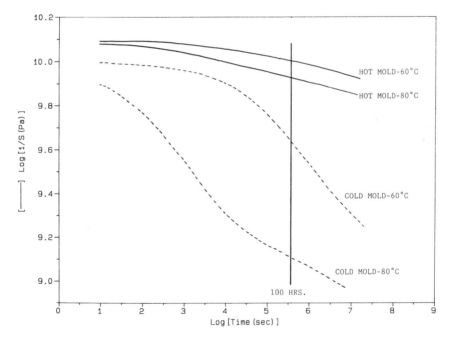

FIGURE 121. Master curves for PPS constructed from data in Figure 120. Vertical line marks the 100-h point to which actual confirmation tests were taken. (From Sepe, M., *SPE ANTEC*, 38, 1152, 1992.)

actual apparent modulus vs. time plots to provide good qualitative verification of the performance disparity predicted by the master curves.[170]

In another example of the usefulness of this method, a new method of impact modification in epoxy resins was investigated to determine the effect of the modifier on the creep resistance

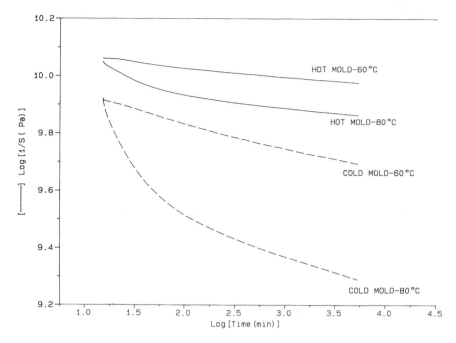

FIGURE 122. Actual apparent modulus vs. time plots for 100-h tests conducted at 60 and 80°C. (From Sepe, M., *SPE ANTEC*, 38, 1152, 1992.)

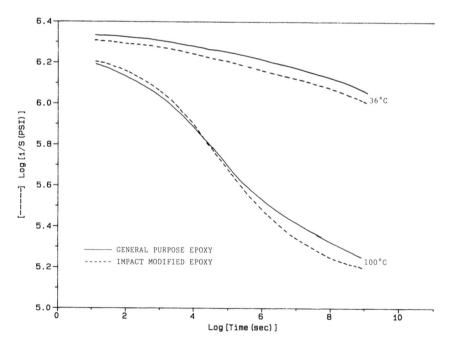

FIGURE 123. Master curves of apparent modulus vs. time for a general purpose epoxy and an impact modified epoxy. (Sepe, unpublished, 1990.)

of the compound. Typical routes to impact modification involve the incorporation of elastomers which reduce modulus and creep resistance. Figure 123 shows a comparison of master curves for the new formulation and an unmodified epoxy generated from short time creep tests. They show no detrimental effects upon load-bearing properties. Actual field studies verified these conclusions.

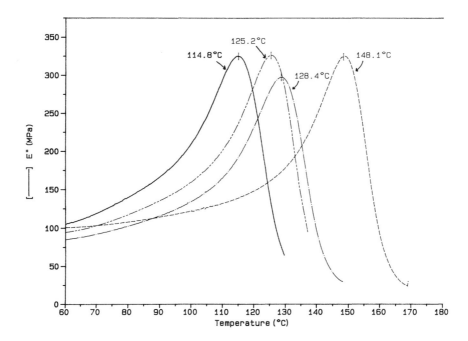

FIGURE 124. E″ vs. T curves for blends of PPO/HIPS showing a single T_g which increases with higher PPO content. (Sepe, unpublished, 1990.)

F. CHARACTERIZATION OF BLENDS

Because of the excellent resolution of the glass transition provided by DMA through the measurement of loss properties, this analytical technique offers an excellent means for examining polymer blends. Many of the rules for determining miscibility that were discussed in Section III.C are also applicable for DMA studies. However, with DMA, transitions in semimiscible blends that may be difficult to resolve by DSC are readily determined. As with DSC results, miscible blends such as PPO/HIPS will give a single T_g that will shift to higher temperatures as the PPO phase increases in amount. Figure 124 shows the behavior of E″ vs. temperature for four different compounds that vary in ratio of PPO to HIPS. Semimiscible blends such as PC/PBT will exhibit two T_g's, but one or both of the loss modulus peak temperatures will be shifted from that of the pure polymers. Figure 125 shows E″ peaks for the neat PBT and PC and a commercial blend, illustrating the shift in T_g for both polymers. Figure 126 shows the DSC plots for all three materials without the PBT melting endotherm. The improved resolution of the transitions by DMA is readily apparent.

Figure 127 shows plots of E′ and E″ for a blend of PPO and nylon 6/6 along with a neat nylon 6/6. Both materials contain 10 wt % glass fiber and were characterized in the dry-as-molded state. In this instance the T_g's are not shifted from their typical temperatures, indicating an immiscible blend. The higher modulus above the nylon T_g along with the broad plateau shows that the PPO acts essentially as a reinforcement in the nylon up to the PPO T_g. By contrast, a comparison of the E′ vs. T curves for PBT and PC/PBT reveals that the partial miscibility of the amorphous PC reduces the plateau modulus of the semicrystalline PBT as shown in Figure 128. Impact resistance evaluations of these two blends show that the semimiscible PC/PBT is tougher than the PPO/PA immiscible blend. This trade-off between retained load-bearing properties and ductility is a common phenomenon in blends of crystalline and amorphous polymers.

Chiang and Hwung[171] have discussed the viscoelastic properties of PC/ABS blends as a function of the changing ratio of the two polymers. They relate the dynamic mechanical scans to physical tests and morphology investigations that indicate a variation in interfacial

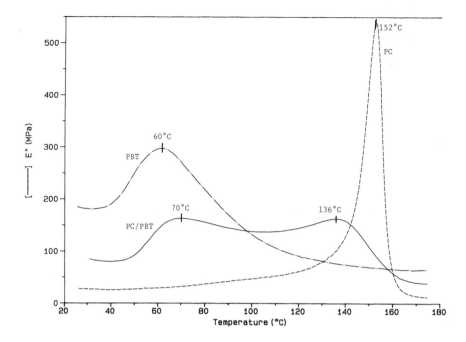

FIGURE 125. E″ vs. T for neat PBT and polycarbonate and a PC/PBT alloy. Partial miscibility is indicated by the shift of peak temperatures closer together. (Sepe, unpublished, 1989.)

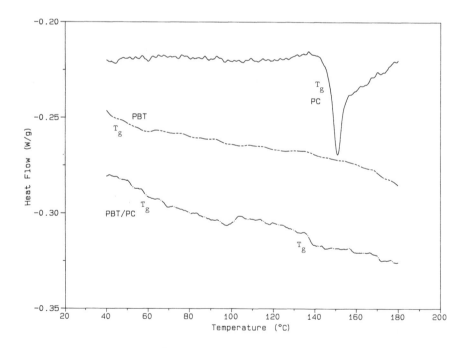

FIGURE 126. DSC scans of PC, PBT, and PC/PBT blend. (Sepe, unpublished, 1987.)

adhesion as a function of blend ratio. An overlap is observed in the glass transition of the two polymers; however, a bimodal β transition is also noted. Since the α transition in ABS is attributable to SAN and the β transition is due to polybutadiene, this indicates miscibility between PC and SAN but immiscibility between PC and PB. The study illustrates good

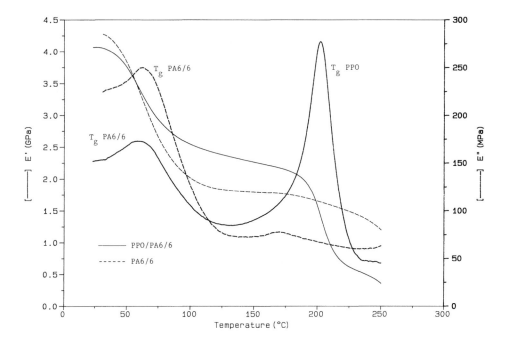

FIGURE 127. E′ and E″ curves for nylon 6/6 and an alloy of nylon 6/6 with PPO. Lack of shift in nylon T_g indicates an immiscible blend. (From Sepe, M., *Plastics Design Forum,* 16(6), 45, 1991.)

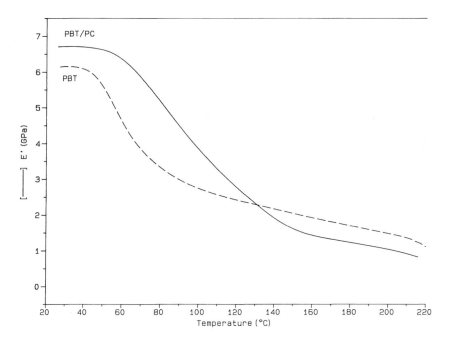

FIGURE 128. E′ curves for glass-reinforced PBT and PC/PBT showing reduced plateau modulus of the blend. (Sepe, unpublished, 1990.)

correlations between changes in DMA transitions, impact strength, and grain structure in blends using different proportions of ABS and polycarbonate.

The peak heights of E″ and tan δ are uncharacteristically low compared to measurements made by this author and Sichina,[172] particularly for the polycarbonate. This is probably due

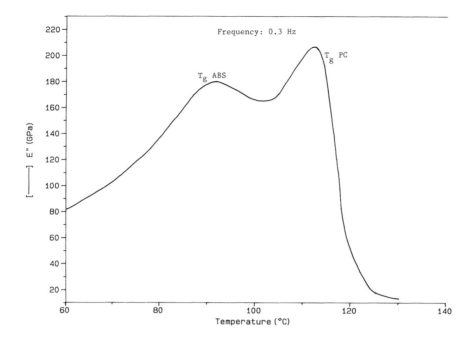

FIGURE 129. E″ curve in the glass transition region for a PC/ABS blend. The low scan frequency separates the two transitions. (Sepe, unpublished, 1987.)

to the difference in frequency employed. Chiang and Hwung used a fixed frequency of 110 Hz while the scans performed by Sepe and by Sichina independently were 2 to 3 decades lower. It is likely that even more information could be gathered about the blends at these lower frequencies. For example, Figure 129 shows a plot of E″ vs. T for a PC/ABS blend scanned at 0.3 Hz. The apparent overlap of T_g's at high frequencies resolves into two distinct transitions at low frequencies. This underscores the importance of frequency sweeps in performing complete characterizations of polymer blends.

Han et al.[173] have discussed the effects of blend ratio and block copolymer molecular weight on the dynamic mechanical properties of mixtures of poly(α-methyl styrene) and different grades of Kraton, a family of styrene block copolymers. For Kraton 1102/Pα-MS the plateau modulus of the blends increases with rising molecular weight of the Kraton. Increasing the concentration of Pα-MS has a similar effect. Figure 130A shows a plot of G′ vs. T for 70/30 blends of Kraton 1102/Pα-MS using different molecular weight Kraton. Figure 130B shows the behavior of another 70/30 blend utilizing Kraton 1107 to replace Kraton 1102. Here the plateau modulus of the blend reaches a maximum with a block copolymer molecular weight of 1.9 K. With increases beyond this point the plateau modulus declines. This is attributed to reduced association between the phases and is a compound-specific response relating to the restricted movement of the higher molecular weight blocks. Note that the moduli using higher molecular weight Kraton are more stable and extend to higher temperatures. This study provides an excellent example of the usefulness of DMA in detecting structural changes related to polymer-polymer interactions in blends.

Lin and Manson[174] have discussed the effects of changing hard block/soft block composition on the damping properties of energy-absorbing polyurethanes. In a thorough discussion of a number of viscoelastic parameters they illustrate the ability of DMA to evaluate compounds for different vibration-damping applications. Use is made of commonly observed attributes such as E″ and tan δ peak heights and temperature as a function of frequency. In addition, the authors make use of less-utilized factors such as the α-transition slope, the

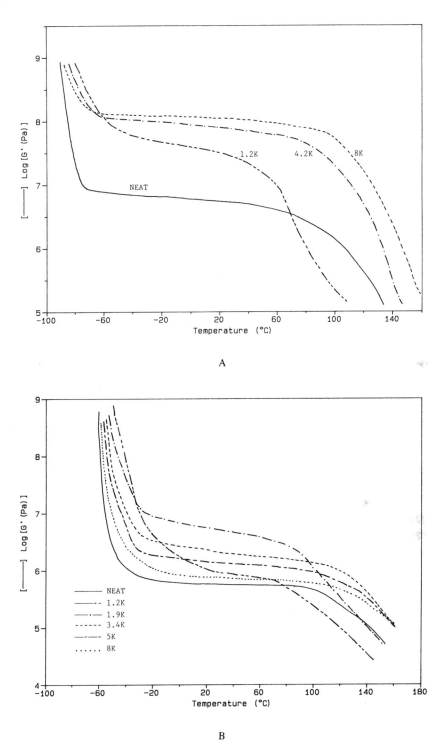

A

B

FIGURE 130. (A) G′ curves for blends of Kraton 1102 and poly(α-methyl styrene) at a 70/30 ratio as a function of the molecular weight of the block copolymer. (Han et al., 1989.) (B) G′ curves for 70/30 blends of Kraton 1107/P-αMS as a function of block copolymer molecular weight. (Han et al., 1989.) Reprinted in Proceedings of Compatibilization of Polymer Blends Symposium, May 22–24, 1990.

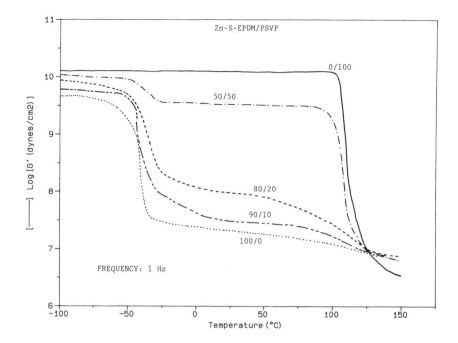

FIGURE 131. Storage modulus behavior for different blend ratios of Zn-S-EPDM/PSVP. (From Dudevan, I., Peiffer, D. G., Agarwal, P. K., and Lundberg, R. D., *SPE ANTEC,* 33, 1310, 1987.)

area under the tan δ peak, and complex plane (E′ vs. E″) plots. This last relationship is useful in determining the range of relaxation times exhibited by a compound. It also provides an alternate method for calculating the activation energy of a viscoelastic transition. The authors also point out the distinction between E″ and tan δ in making practical determinations concerning the type of vibration damping desired.

The melt-state rheological properties of PPO/PS blends and mixtures of PPO/PS with different molecular weight styrene-ethylene-butene-styrene (S-EB-S) block copolymers have been investigated by Yoshimura and Richards[175] using a dynamic mechanical analyzer with a parallel plate configuration. The frequency-dependent behavior of the shear storage modulus and complex viscosity were characterized between 220 and 280°C and superpositioned to a reference temperature of 280°C. The study illustrates the effects of the molecular weight of the block copolymers as well as their concentration on the rheological properties of the resulting compounds. An excellent discussion of the relationship between domain structure and the frequency-dependent behavior of G′ is given. The formation of a low-frequency plateau in the G′ vs. frequency plot is attributed to the existence of a network structure resulting from interparticle interactions between dispersed phase domains.

Dudevan et al.[176] have used DMA to characterize blends of zinc-neutralized sulfonated EPDM (Zn-S-EPDM) with polystyrene (PS) and poly(styrene-co-4-vinylpyridine) (PSVP). Plots of G′ vs. T in Figure 131 reveal an increasing plateau modulus as the ratio of PSVP increases in the blends. These changes are small until the blend ratio changes from 80/20 to 50/50 Zn-S-EPDM/PSVP. A similar abrupt change is observed in the G″ vs. T plots shown in Figure 132. Here the high-temperature damping transition associated with the PSVP is suppressed in the 90/10 and 80/20 blends but is readily detectable as a strong transition in the 50/50 blend. At low levels of PSVP, the plateau modulus remains low and the transitions are not distinct due to partial miscibility; the zinc cations strongly associate with the pendant amine in the pyridine ring. At higher PSVP levels the emergence of a high

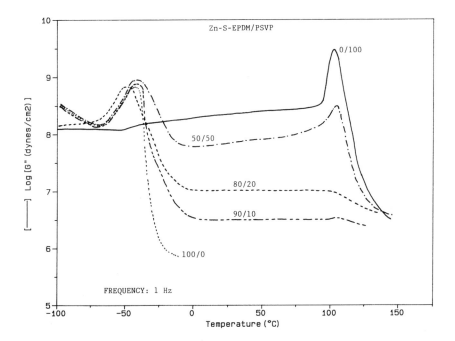

FIGURE 132. Loss modulus curves for different blend ratios of Zn-S-EPDM/PSVP. (From Dudevan, I., Peiffer, D. G., Agarwal, P. K., and Lundberg, R. D., *SPE ANTEC,* 33, 1310, 1987.)

plateau modulus and a high-temperature damping transition indicate immiscibility. Investigations of the morphology show a clear change in microstructure between the 80/20 blend, where no distinct phases are observed, and the 50/50 blend where phase separation is evident.

The dynamic mechanical properties of reactively compounded blends of unsaturated polyesters with elastomers has been discussed by Sakai et al.[177] Both nitrile rubber (NBR) and methacrylate butadiene styrene (MBS) were reactively processed in a twin screw extruder with unsaturated polyester along with the appropriate monomers and initiators for viscosity control and proper curing. Material from the extruder exit port was fed directly to a compression molding press where material was cured to final form.

E′, E″, and tan δ properties were measured as a function of temperature. The results for all three compounds are shown in Figures 133 to 135. The unmodified polyester displays a very weak transition in E″ and tan δ above 150°C. The log scale of the y axis makes detection of this event difficult. The MBS-modified material shows a large transition at 70°C while the NBR-modified resin produces two transitions, one at 70°C and one at 0°C. Subambient relaxations are often associated with improvements in impact strength and in this study the correlation was very good. Figure 136 shows the result of impact tests as a function of elastomer concentration. This illustrates that large amounts of MBS are required to produce limited gains in impact resistance while much smaller amounts of NBR yield much better results. Since elastomer content has an obvious negative effect on rigidity and creep resistance at elevated temperatures, these results are an instrumental part of the development of new compounds.

The integrity of a polymer depends upon proper processing techniques as well as the effects of environmental exposure. The effects of mold temperature and injection pressure have been reviewed above as they relate to the formation of crystalline structures. Another key aspect is the temperature of the molten polymer during processing. The integrity of a fabricated part is typically assessed by measuring key physical properties such as impact

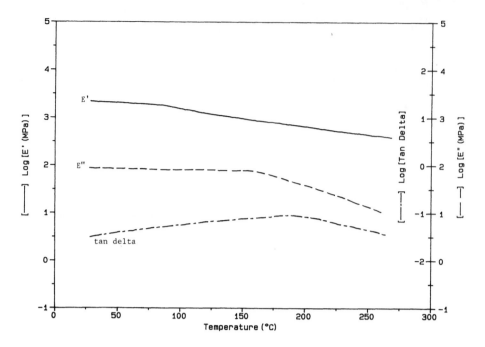

FIGURE 133.   Viscoelastic properties of unsaturated polyester base resin. (From Sakai, J., Nakamura, K., and Inoue, S., *SPE ANTEC,* 36, 1912, 1990.)

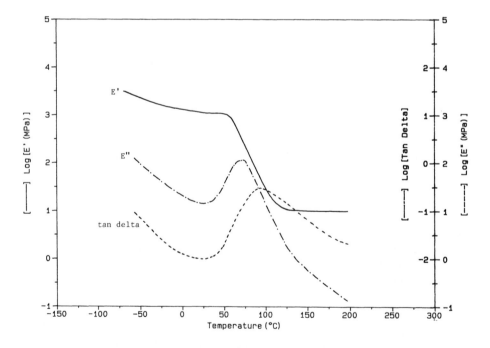

FIGURE 134.   Viscoelastic properties of MBS-modified polyester. (From Sakai, J., Nakamura, K., and Inoue, S., *SPE ANTEC,* 36, 1912, 1990.)

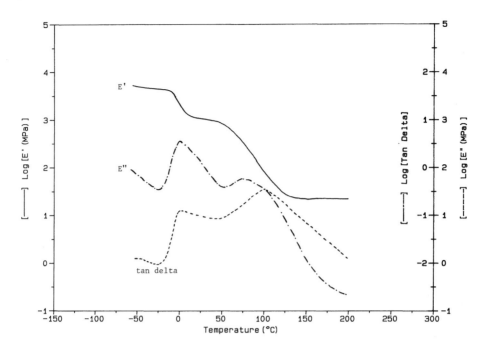

FIGURE 135. Viscoelastic properties of NBR-modified polyester. (From Sakai, J., Nakamura, K., and Inoue, S., *SPE ANTEC,* 36, 1912, 1990.)

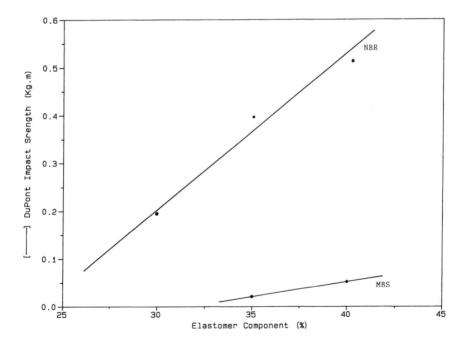

FIGURE 136. Effect of elastomer concentration on impact strength. (From Sakai, J., Nakamura, K., and Inoue, S., *SPE ANTEC,* 36, 1912, 1990.)

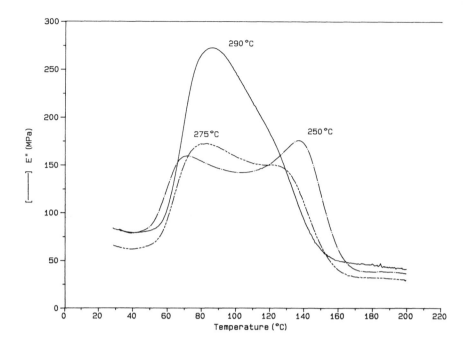

FIGURE 137. Effect of melt temperature on the loss properties of PC/PBT alloy. (Sepe, unpublished, 1988.)

strength or flexural modulus. DMA measurements can often be correlated with these property determinations to identify the nature of changes in polymer structure that can cause property deterioration.

Figure 137 shows E″ vs. T curves for a PC/PBT blend that has been processed at different melt temperatures. The properly processed material displays the two transitions typical of the blend. As melt temperatures rise to destructive levels the damping peak associated with polycarbonate gradually vanishes while the height of the PBT peak increases and becomes shifted to a higher temperature. This is an indication of transesterification initiated by the thermal degradation of the PBT fraction. Reductions in melt viscosity also occur.

The effects of long-term heat aging on the structure of a blend can also be readily observed. Figure 138 shows tan δ vs. temperature curves for a PC/PBT blend as molded and after different heat aging times at 150°C. This reveals that in a short period of time the peak temperatures associated with T_g migrate toward the glass transitions of the base polymers. This movement is especially notable for the PC phase. After additional time the height of the tan δ peak for the PC fraction is reduced. Studies on the heat aging of impact-modified PBT by Bier and Rempel[178] have shown that impact resistance is reduced as a function of heat aging time and temperature. This reduction is primarily due to the degradation of the impact modifier. Since the impact modifier in a blend has the additional role of acting as a compatibilizer, it is reasonable to hypothesize that the chemical reactions that reduce the toughening influence of the rubber modifier also affect the compatibilizing function. The movement of the PC T_g appears to be an indication of microphase separation. It occurs more slowly as the heat aging temperature is reduced.

The reduction in the tan δ peak height for the PC phase also indicates a reduction in ductility. Heat aging studies of neat polycarbonate show that significant embrittlement occurs at temperatures as low as 115°C.[179] The loss of impact strength in this blend after long-term exposure to temperatures above 140°C is well documented. The changes in the loss modulus show that this property decline is caused by a combination of microphase separation and

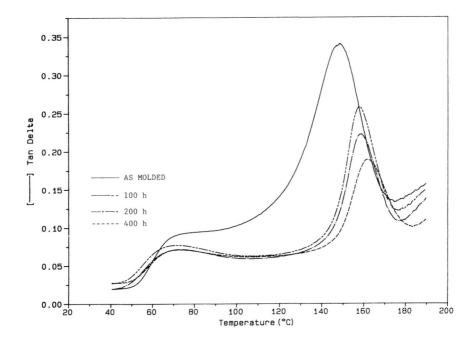

FIGURE 138. Effect of heat aging on the phase miscibility in PC/PBT alloy. (Sepe, unpublished, 1988.)

degradation of the PC phase. No quantitative correlations between impact strength reductions and viscoelastic behavior have been determined thus far for this particular family of compounds. The effects of aging in other polymers is discussed below.

G. AGING

Aging of polymers at temperatures well below the short-term deflection temperature produces a variety of structural changes that are detectable by DMA. Figure 139 plots E′ vs. T for PPS samples both as molded and after 500 h of heat aging at various temperatures. The increase in the plateau modulus is due to an increase in crystallinity at the lower aging temperatures. At the 250°C aging temperature this increase is augmented by crosslinking. Figure 140 shows the tan δ vs. temperature plots for the same samples. The reduction in peak height indicates a progressive loss in ductility. The coincident increase in the peak temperature after aging at 250°C provides further evidence of crosslinking; increases in crystallinity do not typically raise the T_g.

Figure 141 illustrates the behavior of E′ vs. temperature for two samples of PBT polyester. One sample was tested as molded while the other was heat aged for 8 h at 160°C. Once again an increase in the plateau modulus is observed due to increased crystallinity. A slight step change in the modulus of the aged sample occurs in the vicinity of the aging temperature. This is due to the melting of the imperfect crystals that were formed during the short time aging. This is another manifestation of the same behavior that produced low melting crystals detectable by DSC in materials such as PPS and HDPE.

These changes can occur after very short aging times. Figure 142 shows E′ and tan δ vs. T plots for two samples of 30 wt % glass-reinforced PEEK. One sample was scanned as molded while the other was first subjected to a 3-h heat cycle from room temperature to 310°C and then back to room temperature. The heat-cycled sample shows the familiar effects of increased crystallinity, higher modulus, and reduced damping indicating a stiffer but less ductile material.

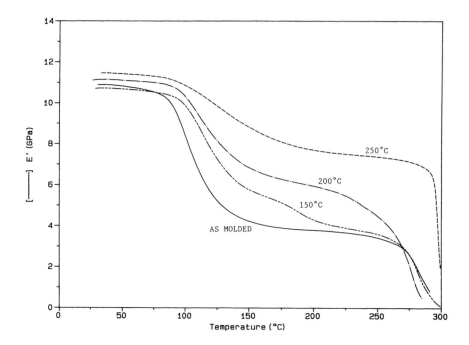

FIGURE 139.   Effect of heat aging on E' of PPS. (From Sepe, M., *SPE RETEC,* Rosemont, IL, March, 1990.)

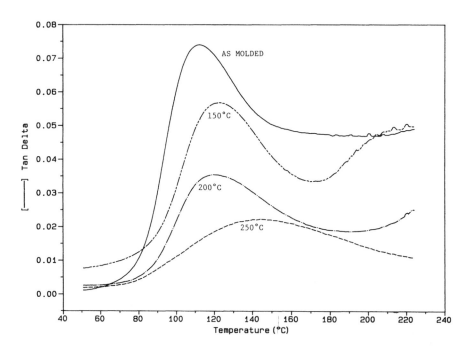

FIGURE 140.   Effect of heat aging on the tan δ peak-associated with T_g. (From Sepe, M., *SPE RETEC,* Rosemont, IL, March, 1990.)

Aging of thermosets can result in a change in the crosslink density of the material. This is readily detectable by examining the behavior of E' and tan δ as a function of temperature. Figure 143 plots E' vs. T for four phenolic samples. The samples were as molded, postcured for 24 h at 150°C, heat cycled in a manner similar to the PEEK samples above, and heat

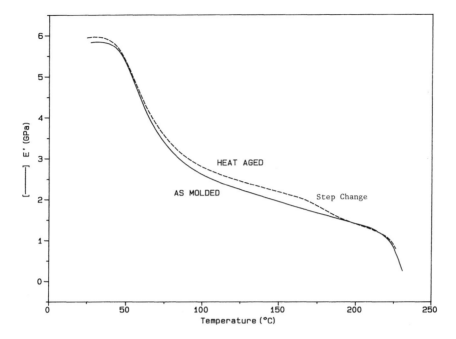

FIGURE 141. Effect of 8 h of heat aging at 160°C on the storage modulus of PBT polyester. (Sepe, unpublished, 1991.)

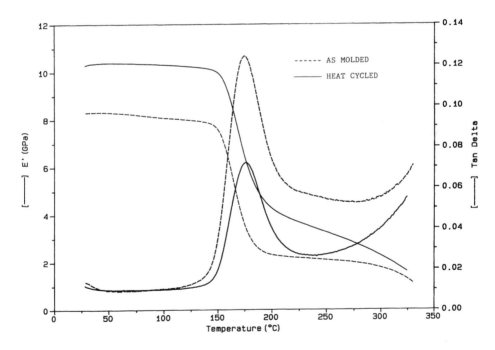

FIGURE 142. Effect of short-term elevated temperature exposure on the viscoelastic properties of 30 wt % glass-reinforced PEEK. (From Sepe, M., *Plastics Design Forum,* 16(6), 45, 1991.)

aged for 96 h near the short time deflection temperature. The as-molded sample exhibits an evident rapid decline in modulus followed by a modulus recovery typical of residual cross-linking. Postcured and heat-cycled samples produce a more stable material with no evidence of residual crosslinking. The heat-aged sample displays a severe decline in properties at an

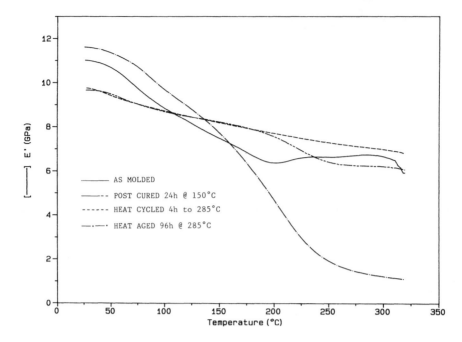

FIGURE 143. Effect of various heat exposure levels on the storage modulus of a high heat phenolic. (From Sepe, M., *Plastics Design Forum,* 16(6), 45, 1991.)

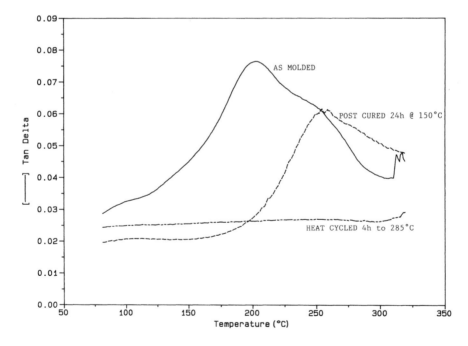

FIGURE 144. Effect of various heat exposure levels on the tan δ peaks associated with T_g for a high heat phenolic. (From Sepe, M., *Plastics Design Forum,* 16(6), 45, 1991.)

unexpectedly low temperature. This response is likely a combination of the effects of oxidation and a disruption of the crosslinks between domains discussed by Morrison and Waitkus.[73] Figure 144 plots tan δ vs. T for the samples as molded, postcured, and heat cycled. This illustrates the increase in the peak temperature and the reduction in the peak height as the material becomes more fully crosslinked. The heat-cycled sample exhibits no damping. Similar behavior has been reported by Gill as presented by Prime.[72]

Suttner[180] has studied the differences in G' and tan δ between injection-molded and annealed samples of PS and PE as a function of aging time and temperature. For both materials the viscoelastic properties were different in the injection-molded samples, G' being lower, and tan δ being higher for as-molded specimens. However, with aging time the properties of the as-molded samples approached those of the annealed samples. In the amorphous PS this is attributed to a reduction in free volume while in the semicrystalline PE it is caused by increasing crystallinity during aging. Frequency sweeps of PS samples aged at different times and temperatures reveal that increased aging time and decreased aging temperature both shift the modulus data (G') to lower frequencies. Master curves of G' and tan δ vs. frequency are essentially the same regardless of whether they utilize time-dependent data at constant temperature or temperature-dependent data taken at constant time.

Sichina[181] has shown that short-term aging of polycarbonate film results in a storage modulus that is higher and less frequency dependent than that of unaged samples. This is consistent with the data of Suttner[180] and of Tryson.[169] If the modulus vs. frequency plot is inverted to a modulus vs. time plot it can be seen that the aged samples have greater creep resistance associated with the longer relaxation times of the glass. A temperature sweep of aged and unaged polycarbonate shows that the subambient transition in the polymer broadens into two merged peaks after 3 h of aging at 130°C.

H. CURING AND CROSSLINKING

DMA provides a powerful tool for characterizing the curing and crosslinking processes in thermosetting materials. It has been mentioned above that the molded thermoset can be examined for evidence of completion of cure. Both the storage and loss properties are useful in this evaluation. An incompletely cured material will exhibit a recovery in the storage modulus as residual cure takes place during heating in the DMA. A more completely cured sample will not display such a recovery and the loss peak associated with T_g will be shifted to a higher temperature and a lower magnitude. It should be noted that the ambient storage modulus is not necessarily increased by postcuring; the benefits are seen at elevated temperatures. This is shown clearly in Figure 143. This effect is analogous to the behavior of thermoplastics such as PPS as they proceed to higher degrees of crystallinity.

Landi and Merserau[182] have used DMA scans of molded phenolic specimens in conjunction with DSC techniques to characterize the kinetics of T_g development during curing and postcuring. They verify the behavior outlined above and study in greater detail the dependence of dynamic mechanical properties on different mold temperatures and cycle times. By defining the T_g as the point at which tan δ achieves a value of 0.03, the authors develop a series of plots of the onset of T_g vs. time for several mold temperatures. They utilize these data to construct master curves for a given reference temperature. This is designed to provide a prediction of the resultant T_g for a given time-temperature profile inside the mold.

This author has observed that the development of T_g is also wall thickness dependent. DMA scans of E' vs. T were conducted on sections of molded phenolic parts produced at varying mold temperatures and cycle times. Figure 145 shows the results for parts molded at three different cycle times in wall sections 2.5 mm thick while Figure 146 shows the same data for wall sections of 6.0 mm. The degree of modulus recovery becomes smaller at a faster rate for the thin sections, indicating that prolonged exposure to the high temperature of the mold cavity results in a greater change in the degree of cure for the thinner walls. Figure 147 plots the modulus recovery vs. cycle time for both wall section thicknesses to illustrate the response more clearly.

This behavior is sensible in terms of the thermodynamics of heat exchange between the mold and the molded part. Thinner wall sections are more rapidly influenced by conditions

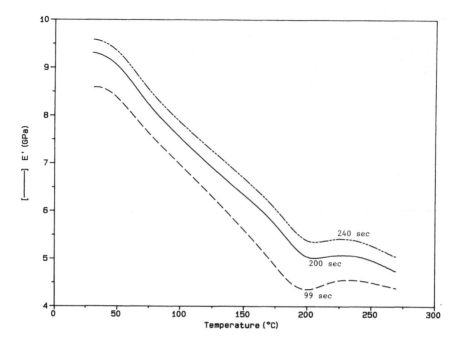

FIGURE 145. Effect of cycle time on the modulus recovery of molded phenolic articles with a wall thickness of 2.5 mm. Modulus recovery indicates postcuring during heating in the DMA. (Sepe, unpublished, 1990.)

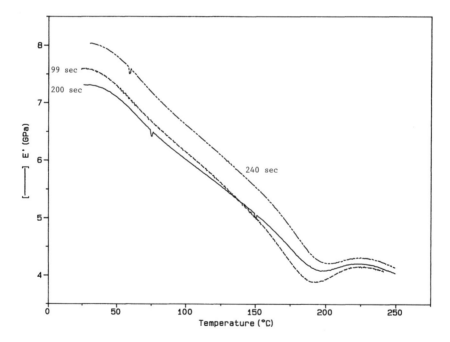

FIGURE 146. Effect of cycle time on the modulus recovery of molded phenolic articles with a wall thickness of 6.0 mm. (Sepe, unpublished, 1990.)

in the mold because a greater percentage of the total wall section comprises the skin layers that are in close proximity to the mold steel. In a thick section a large portion of the mass is insulated from the mold steel by the surface layers of the part. Because polymers have poor thermal conductivity, attempts to influence the curing process at the center of a thick

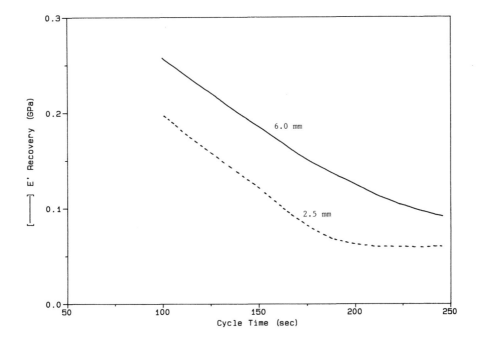

FIGURE 147. Degree of recovery of E′ as a function of cycle time for both wall thicknesses. Mold temperature was 165°C. (Sepe, unpublished, 1990.)

section by changing the temperature of the mold steel are not very successful. Analogous behavior is observed in the crystallization of semicrystalline thermoplastics. Thin-walled parts will respond dramatically to mold temperature in terms of degree of crystallinity while thick parts are less effected.[183]

Hawkins[184] has reported on structural changes in PPS when it is exposed to temperatures above T_m in air. Thermally and oxidatively induced crosslinking is observed. Ma et al.[185] have used dynamic mechanical measurements to illustrate the effects of crosslinking. Measurements of dynamic viscosity as a function of time at various isothermal conditions were employed. Viscosity rises over time, increasing more rapidly and reaching a higher plateau value as the temperature is increased. Figure 148 shows these results. The authors also use DSC and X-ray diffraction techniques to illustrate a relationship between crosslinking and the reduced ability of the material to undergo subsequent crystallization.

Sichina and Gill[186] have demonstrated the use of DMA to evaluate the curing process in epoxy/carbon fiber prepreges. Plots of G′ and G″ vs. time at various isothermal conditions provide an evaluation of the crossover point as a function of time and temperature. This crossover point is considered by many researchers to signify the gel point.[187,188] While the exact equivalency of the gel point with the modulus crossover point is still a matter of some debate,[189] it offers a useful reference point for identifying key structural changes during the curing process. An Arrhenius plot of the natural log of crossover time vs. reciprocal temperature can be constructed from multiple isothermal experiments. The slope of the resulting line is E_a/R, where E_a is the activation energy associated with cure. This method of determining E_a agrees well with the results based on DSC kinetics for this resin system.

Sichina and Gill[186] have used this method to evaluate the effects of aging during storage on the gelation time of the prepreg. Isothermal scans of a fresh and an aged sample reveal a much shorter gel time for the aged sample. In addition, the complex viscosity of the sample can be plotted vs. time. These show a decrease to minimum viscosity followed by an increase due to either curing or vitrification.

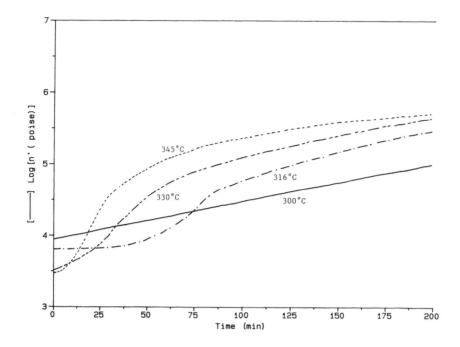

FIGURE 148. Dynamic viscosity vs. time at various isothermal conditions. All scans performed at 1 Hz. (Ma et al., 1986.)

At temperatures too low to initiate cure the viscosity rises very slowly with time and then levels off. A slight increase may continue in some resin systems due to solvent loss. Grentzer et al.[190] have noted this behavior in sheet molding compounds based on unsaturated polyester. Figures 149A to D plot G', G", and complex viscosity vs. time at various isothermal conditions for an unsaturated polyester that has undergone 30 d of room temperature aging. In this system it is interesting to note that at low temperatures the storage modulus starts out higher than the loss modulus and a crossover point is never reached. Once the temperature rises high enough to soften the material, G" starts above G' and a crossover point is observed which shifts to shorter times with increasing temperature.

Grentzer et al.[190] have shown that unsaturated polyesters stored for long times at room temperature display a considerable increase in complex viscosity while refrigerated samples are more stable. A fresh sample would then be expected to have a room temperature tan δ greater than unity. Repeating the isothermal tests shown in Figures 149A to D shows this to be the case; G" begins at a higher value and crossover points are observed for all temperatures. At the lower temperatures the gelation is not a prelude to crosslinking since no rapid increase in viscosity is seen. With the polyesters, the induction time before cure is actually shorter and initiates at lower temperatures for the fresh material. This can be seen by comparing the gel points in Figure 149C with those in Figure 151A, and by examining the G', G" and n* vs. time plots for material scanned at 80°C in Figures 149A and B for aged material and Figures 150A and B for fresh material. It is clear that some degree of curing has begun in the fresh samples at 80°C while the aged sample exhibits no steep increase in viscosity at the same temperature. This is in contrast to the conclusions of Sichina and Gill.[186]

This may be explained by the difference in chemistries between the epoxy/carbon fiber systems studied by Sichina and Gill and the polyesters treated here. It is evident that the aged material had vitrified without curing. This vitrified matrix had to undergo softening before curing could take place. The longer times in the aged material can be attributed to

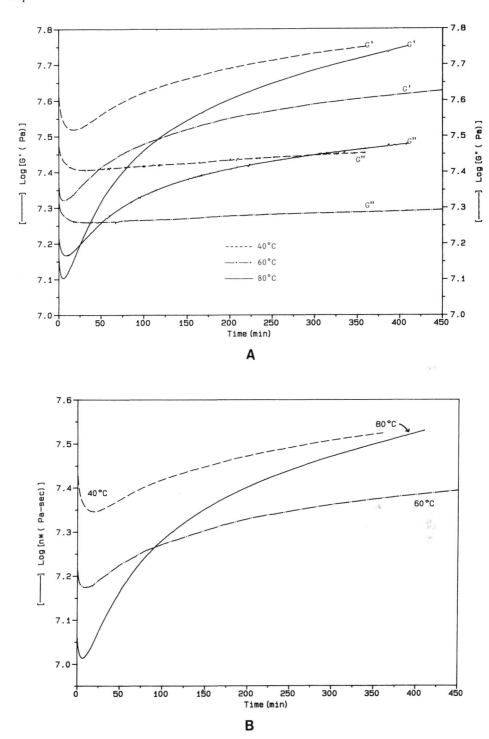

FIGURE 149. (A) G′ and G″ for isothermal scans at the indicated temperatures for an uncured polyester room temperature aged for 30 d prior to scanning. (Sepe, unpublished, 1991.) (B) Complex viscosity for the same material and test conditions in A. (Sepe, unpublished, 1991.) (C) G′ and G″ for isothermal scans of aged, uncured polyester at temperatures sufficiently high to induce cure. Crossover points are shown. (Sepe, unpublished, 1991.) (D) Complex viscosity for material and conditions in C with a plot for 80°C showing no cure for contrast. (Sepe, unpublished, 1991.)

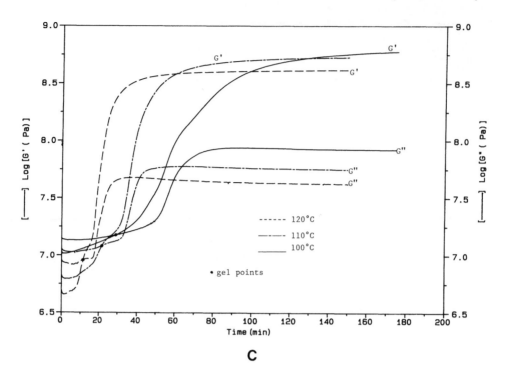

C

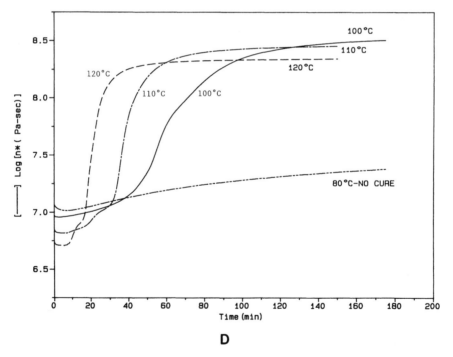

D

FIGURE 149 (continued)

the interval required for the polymer to absorb the thermal energy needed to remobilize the matrix. It is also likely that volatilization of styrene, which serves as crosslinking agent in polyesters, had occurred. This is indicated by the fact that the plateau viscosities for the cured product from fresh samples are generally higher than those attained by the aged samples at the highest curing temperature studied.

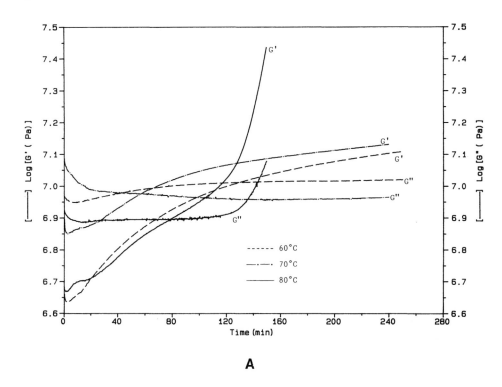

A

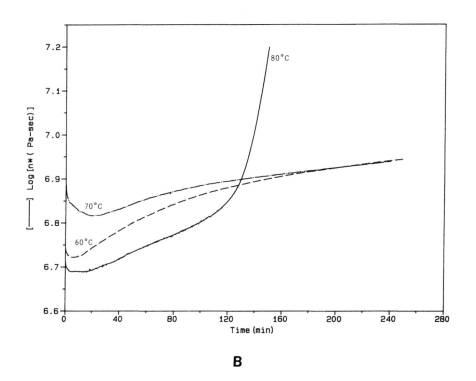

B

FIGURE 150. (A) G′ and G″ for isothermal scans using a polyester resin stored at −15°C prior to scanning. (Sepe, unpublished, 1992.) (B) Complex viscosity for material and conditions in A. (Sepe, unpublished, 1992.)

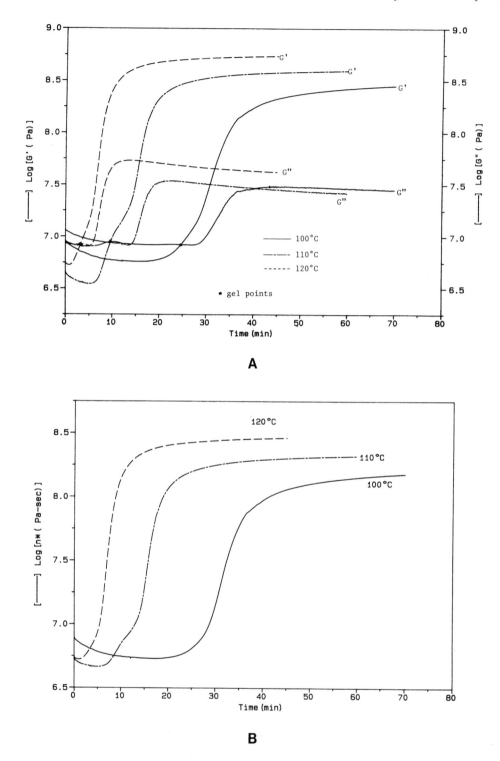

FIGURE 151. (A) G′ and G″ for scans of fresh, uncured polyester at temperatures sufficiently high to induce cure. Note that crossover times are shorter than for aged material. (Sepe, unpublished, 1992.) (B) Complex viscosity for material and scan conditions in A. (Sepe, unpublished, 1992.)

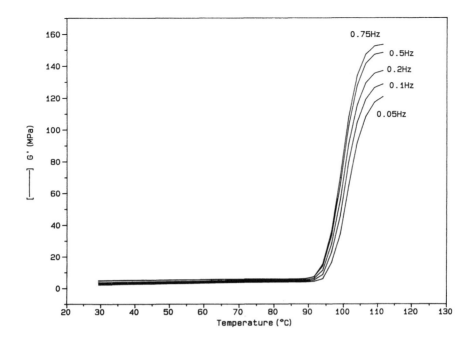

FIGURE 152. A multiple frequency scan of uncured polyester at 2°C increments. Modulus becomes frequency dependent as the system undergoes cure. (Sepe, unpublished, 1990.)

The cure onset temperature can be identified by raising the temperature of a sample in small increments and measuring the storage modulus as a function of temperature. Figure 152 shows the results of a test run at multiple frequencies with measurements taken at increments of 2°C. The modulus shows little dependency upon frequency until the initiation of cure. Although the software used in this analysis does not include complex viscosity as a menu item, it can be calculated from

$$n^* = \frac{(G'^2 + G''^2)^{1/2}}{w} \tag{21}$$

where w is the angular frequency, related to the DMA frequency by $w = 2\pi f$. These calculations can in turn be used to construct viscosity vs. frequency plots for the cured compound which are analogous to viscosity vs. shear rate plots. Note that the onset of cure is not frequency dependent.

Dynamic heating of a prepreg can also be employed to characterize the progression to minimum viscosity, gelation, and cure. With dynamic heating, however, the thermal lag inherent in the method will shift the temperature data to higher values. Figure 153 shows G' vs. T determined by the dynamic method and by the stairstep method and reveals a 40°C shift in the onset of cure temperature. Nevertheless, dynamic heating data can provide useful information on lot-to-lot consistency since it employs a single test that may take as little as 1 h to perform. Figure 154 plots G' vs. T for two different unsaturated polyesters that use different initiator chemistries. Despite the relative nature of the data, the shift in cure onset temperature permitted the processor to make a mold temperature adjustment which corresponded to the DMA determination to within 1°C.

Quan et al.[191] have discussed the use of DMA to study the effects of hydride:vinyl stoichiometry on the cure and postcure behavior of poly(dimethyl siloxane) gels. An increase

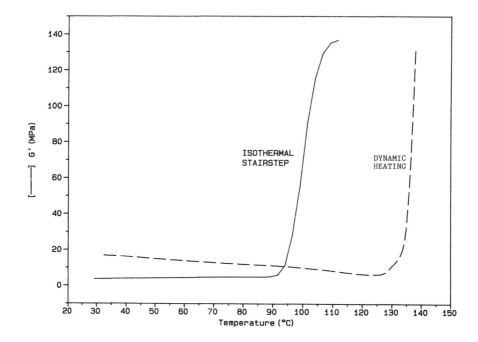

FIGURE 153. Comparison of onset of cure determination by isothermal stairstep and dynamic heating. (Sepe, unpublished, 1990.)

in G′ with time is used to indicate the occurrence of cure. Phillips et al.[192] have studied the effects of varying crosslink densities in polyethylene on the α, β and γ transitions in the dynamic mechanical spectra of the polymer. They have correlated observed changes, most notably in the α and β relaxations, to structure and property alterations in the material.

Driscoll[193] has demonstrated the effects of different postcure times on the development of T_g in polyimide laminates. As expected, the G′ plateau modulus extends to higher temperatures at longer postcure times while the G″ peak temperature moves to correspondingly higher temperatures. Driscoll has also discussed the use of DMA to evaluate the effects of hardener chemistry on the temperature behavior of cured epoxy resins. Storage modulus declines and tan δ peaks associated with the glass transition will be dramatically altered by the type of hardener employed.

The examination of G′ or complex viscosity as a function of time can also be extended to the study of rubber vulcanization in the same manner as is used for the rigid thermosets discussed above. Adhesive pot life can also be evaluated by monitoring complex viscosity vs. time for a variety of proposed storage conditions. A premature increase in complex viscosity would obviously be detrimental to process application as well as final performance.[194]

Finally, Prime and Adibi[195] have discussed the effect of substrate selection and experimental techniques in the evaluation of coatings. The authors outline experimental considerations and illustrate the usefulness of thin metallic supports as opposed to the more commonly used glass fiber ribbons. While this study involved the evaluation of cured coatings, it is also possible to evaluate the curing process by scanning the coated substrate while inducing curing in the DMA instrument.

The DMA used in the above study was a resonant frequency device. Consequently, frequency changed as a function of temperature, sample thickness, and arm spacing. In-

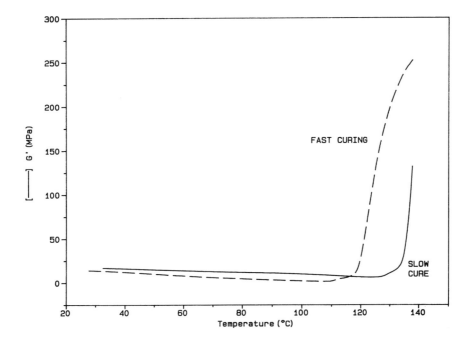

FIGURE 154. Comparison of cure onset for thermoset polyesters employing different initiators. (Sepe, unpublished, 1992.)

vestigations in a fixed-frequency mode will likely simplify the experiment and make quantitative determinations easier.

VII. OTHER TECHNIQUES

In recent years, techniques have been developed that extend the capabilities of the core techniques that are generally thought of as thermal analysis. Many of these methods were foreshadowed by customized adaptations made to existing instruments by pioneers in the field. Some of these have been commercialized while others are still in the development stages. This last section provides a brief review of these methods and instruments.

A. DIELECTRIC ANALYSIS

Dielectric analysis (DEA) is a technique closely related to DMA. In materials where the movement of dipoles or other ionic species can be electrically stimulated, DEA provides a means of examining changes in ionic conductivity. This can be used to draw conclusions about structural changes in a material. The data plots look similar to those from DMA tests and comparisons of the two methods will be discussed here.

While DMA subjects samples to an oscillatory mechanical stress, DEA utilizes a periodic electrical field. The response of the sample provides information on the capacitive and conductive properties of materials and quantifies them as a function of time, temperature, and frequency.

Four major properties are measured by DEA. Permittivity, e', also referred to as the dielectric constant, is proportional to the capacitance — the ability of a material to store electric charge. It is a measure of dipole alignment. The loss factor, e'', is proportional to conductance. It is the energy required to align dipoles and move ions. Since ionic movement is not significant in a polymer in the glassy state, e'' essentially represents only the energy required to align dipoles below and through the glass transition. Above T_g, e'' is a factor in calculating the ionic conductivity from the equation

$$e'' = \sigma/we_0 + e''_d \tag{22}$$

When e''_d, the energy required to align dipoles, becomes small, ionic conductivity is expressed as

$$\sigma = e''we_0 \tag{23}$$

where σ is the ionic conductivity, w is the angular frequency ($w = 2\pi f$), and e_0 is the absolute permittivity of free space, a constant. As with the loss modulus in DMA, e'' exhibits a peak as the polymer passes through the glass transition. Ionic conductivity can be correlated with viscosity and is useful for following the process of softening, gelation, cure, and vitrification in thermosetting systems. The fourth quantity, tan δ (e''/e'), is the dissipation factor. It is measured as the phase angle between the input voltage and the output current and is analogous to the mechanical tan δ.

DEA, like DMA, can be considered a rheological tool for examining molecular relaxations and viscosity changes. Advances in sensor technology have made the technique available through commercial instrumentation in recent years. Product literature from companies such as TA Instruments (formerly DuPont Instruments) and Polymer Laboratories provides information on instrument features, sample configurations, and analysis software.

The history of DEA began over 30 years ago when, in 1960, Warfield and Petrie[196] noted a correlation between the degree of cure and log resistivity in thermoset epoxies. In 1971, Acitelli et al.[79] performed measurements of dc conductivity as a function of time at various temperatures to determine the gel times of epoxy-amine systems. These studies document a rapid decrease in conductivity as the gel point is reached and curing proceeds. More recently, Ramanathan and Harper[197] used aluminum sample dishes and copper lids as electrodes in examining the heat release of oven-cured epoxy resins. Conductance, capacitance, and dissipation factor were reported as a function of time.

Bidstrup et al.[198] used more sophisticated instrumentation to study the development of epoxy/amine systems during cure. In this study, measurements of dielectric properties were performed at multiple frequencies. The ionic conductivity was calculated by Equation 23 and related to T_g and degree of conversion data determined by DSC. Different isothermal cure routines were used.

The effects of cure temperature and time on the ionic conductivity and T_g were demonstrated and a WLF relationship was shown relating ionic conductivity to T_g:

$$\log a_T = \log \frac{\sigma(T)}{\sigma(T_g)} = \frac{C_1(T - T_g)}{C_2 + (T - T_g)} \tag{24}$$

C_1 and C_2 are material-dependent coefficients. C_1 was found to be independent of molecular weight while C_2 and log $\sigma(T_g)$ were shown to exhibit a simple linear dependence upon T_g.

Assuming this linear dependency, Sheppard[199,200] has shown that the ionic conductivity can be modeled with a pseudo-WLF equation:

$$\log a_T = \log \sigma(T) - [C_5 + C_6 T_g] = \frac{C_1(T - T_g)}{C_3 + C_4 T_g + (T - T_g)} \tag{25}$$

where $C_5 + C_6T_g$ replaces log $\sigma(T_g)$ and $C_3 + C_4T_g$ replaces C_2 in Equation 24. Values for each coefficient were calculated and fits of log σ vs. T_g and log a_T vs. $T - T_g$ were applied to the data derived from isothermal cure studies at temperatures ranging from 135 to 188°C. The agreement of the raw data with the model is reasonably good.

While these models offered reasonable approximations, an attempt to fit T_g to the degree of conversion from DSC measurements resulted in significant error. A relationship was proposed between T_g and degree of conversion, α, by using a DiBenedetto equation:[201]

$$\frac{T_g - T_{g°}}{T_{g°}} = \frac{(E_x/E_m - F_x/F_m)\alpha}{1 - (1 - F_x/F_m)\alpha} \tag{26}$$

where E_x/E_m is the ratio of lattice energies for the crosslinked and uncrosslinked polymer, F_x/F_m is the corresponding ratio of segmented mobilities, and $T_{g°}$ is the glass transition temperature of the segmented materials. Because of the data scatter around the plot derived from the equation, the exact accuracy of the relationship between log σ and α derived by combining Equation 26 with the WLF model was in doubt.

Day et al.[202] examined the WLF model using a higher temperature cure. They developed different WLF constants. However, an attempt to fit the actual data of T_g vs. cure time to the model yielded poor agreement regardless of which coefficients were used. As an alternative, the relationship between log σ and T_g was measured for isothermal cure. A crossplot of log resistivity vs. T_g, shown in Figure 155, verifies previous work showing a nearly linear relationship between the two properties. The authors propose a linear model for calculating T_g using the equation

$$T_g = \frac{\log \sigma - \log \sigma_0}{\log \sigma_{ult} - \log \sigma_0} \times (T_{g,ult} - T_{g°}) + T_{g°} \tag{27}$$

where log σ_{ult} and $T_{g,ult}$ are the log conductivity and T_g of the fully cured system and log σ_0 and $T_{g°}$ are the same properties of the uncured system. This model, when combined with the temperature dependency of log σ_0 and log σ_{ult}, yields a plot of T_g vs. time that is in very good agreement with the actual DSC data. This work laid the groundwork for using microdielectrometry as a cure-monitoring technique.

It is important to note that this early work was performed on epoxies. These materials do not change in ionic concentration during cure and the curing process is driven by a single, dominating reaction mechanism. Conductivity is also not influenced by the loss of volatiles. DEA proves useful when these material conditions are met. Consequently, the technique has proven valuable for unsaturated polyesters, acrylics, and some thermosetting polyurethanes, but not for phenolics and polyimides.

Like DMA, DEA can be performed at multiple frequencies to yield information about relaxations in polymer structures. In addition, the range of frequencies available in DEA is broader and extends to much higher frequencies. Most commercial DEA instruments will cover 8 decades from 10^{-3} to 10^5 Hz while DMA devices typically cover from 10^{-3} to 10^2 Hz. Several researchers have investigated the relationship between DMA and DEA results through simultaneous measurements of both sets of properties.[203-205] These studies show useful correlations between mechanical and dielectric responses to resin softening, cure, and vitrification. However, because of the molecular scale at which different transitions occur, each technique is particularly sensitive to specific aspects of the curing process. Wetton et al.,[206] for example, illustrate that the DMA tan δ exhibits a strong peak as a liquid resin undergoes gelation, the point of initiation of crosslink formation. No corresponding dielectric

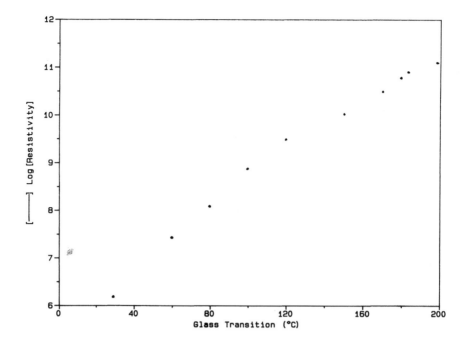

FIGURE 155.   Cross plot of T_g and log resistivity measured by dielectric analysis during isothermal cure of an epoxy-amine system. (From Day, D., Wall, A., and Sheppard, D., *SPE ANTEC,* 34, 964, 1988.)

response is evident because crosslinking occurs in a long-range elastic network while dielectric relaxation and conductivity are related to local, small-scale events. However, because of this sensitivity to local changes, dielectric measurements are more sensitive to changes in the resin during the later stages of cure. At this point the long-range effects on the network have become negligible but small-scale motions of ionic species are still possible. Both methods display the frequency dependency of the newly formed glass and dielectric measurements produce a greater dispersion of the curves to longer times. Both methods also show a strong response to the vitrification process which signals completion of cure.

Gotro and Yandrasits[203] have correlated viscosity measurements by parallel plate rheometry with dielectric loss factor (e''). This work shows a relationship between the minimum viscosity and a maximum in e'', although they do not strictly coincide in time. Both values are sensitive to the heating rate of the experiment at a given frequency, peak heights changing by over an order of magnitude as the heating rate increases from 2.5 to 16°C/min. Of interest is the fact that e'' will remain high for systems where $T_{g,ult}$ is below the cure temperature since ionic conductivity will be considerable in a material with large amounts of segmental mobility. For polymers with a $T_{g,ult}$ that exceeds the cure temperature, e'' will drop sharply and then exhibit a frequency-dependent shoulder indicative of vitrification. Figures 156 and 157 show examples of each type of behavior. This illustrates the ability of dielectric measurements to characterize the small-scale molecular motions associated with the movement of ions.

DEA is not limited to the characterization of curing systems. It can be used to examine the transitions of chemically mature systems such as thermoplastics and cured thermosets. Electrical properties can be evaluated over a wide range of temperatures and frequencies, expanding upon the usual data which are limited to room temperature characterizations at one or two frequencies. Figures 158 and 159 show e' and e'' plots vs. temperature at multiple

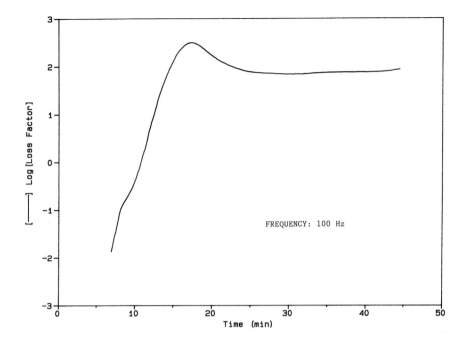

FIGURE 156.   Loss factor, e″, vs. time for FR-4 epoxy heated at 8°C/min. (From Gotro, J. and Yandrasits, M., *SPE ANTEC*, 33, 1039, 1987.)

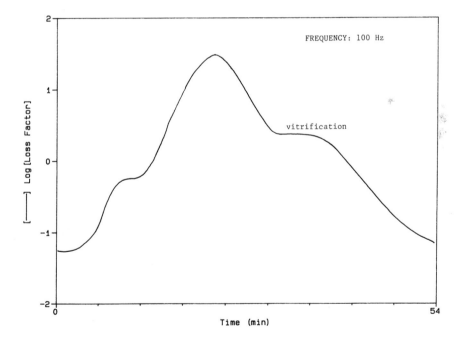

FIGURE 157.   Loss factor vs. time for bis-maleimide triazine/epoxy heated at 8°C/min. (From Gotro, J. and Yandrasits, M., *SPE ANTEC*, 33, 1039, 1987.)

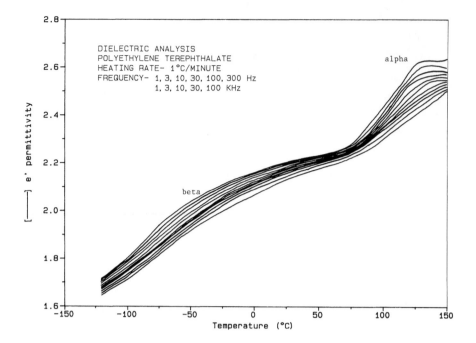

FIGURE 158. Multiple frequency scan of permittivity across a range of temperatures for PET polyester film showing the α and β transitions. (TA DuPont Instruments, 2970 DEA Technical Brochure, Wilmington, DE, 1990.)

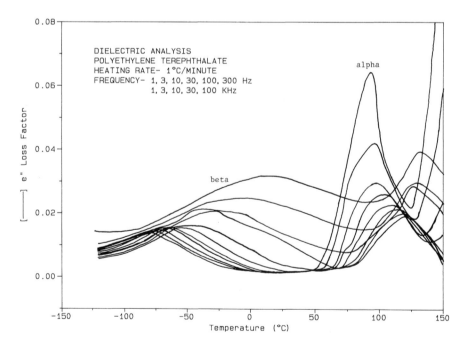

FIGURE 159. Multiple frequency scan of loss factor for PET. (TA DuPont Instruments, 2970 DEA Technical Brochure, Wilmington, DE, 1990.)

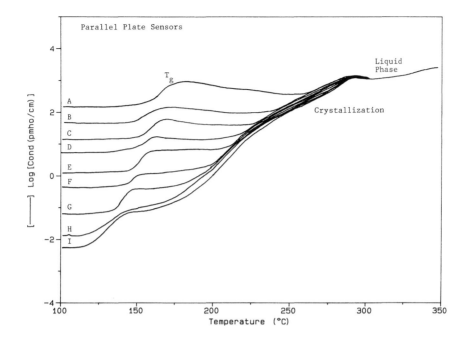

FIGURE 160. Ionic conductivity for PEEK film cooled from 380°C at 1°C/min using multiple frequencies. A — 3, B — 10, C — 30, D — 100, E — 300, F — 1000, G — 3000, H — 10,000 and I — 30,000 Hz. (From Sichina, W. J., DuPont Instruments TA Hotline, Wilmington, DE, Spring, 1989.)

frequencies for amorphous PET. The loss properties are particularly useful in examining the transitions due to segmental motion. This plot bears a strong resemblance to a plot of E'' vs. T by DMA.

The ability to scan a material at multiple frequencies allows the calculation of the activation energy for molecular motions associated with relaxations in a polymer. The β transition for PMMA has been evaluated and the E_a calculated by DEA agrees well with DMA determinations.[207]

Sichina has examined the cooling process from liquid to solid in PEEK using DEA as shown in Figure 160. This plot of conductivity vs. T shows the scans at different frequencies to be indistinguishable in the molten polymer. This is due to the fact that ionic motion dominates conductivity in the molten state and contributions from dipolar molecular motions, which are frequency dependent, are comparatively small. As crystallization begins the ionic mobility is reduced and the curves begin to exhibit frequency-dependent behavior. This response becomes increasingly pronounced as the polymer cools and the viscosity of the amorphous phase increases. At 160°C the amorphous phase vitrifies as the material cools through the glass transition. This event is frequency dependent and the dispersion of the curves becomes even greater as dipole motions dominate.[208]

Figure 161 shows a plot of log e′ vs. T during the same cooling experiment. The polymer displays a large increase in permittivity as crystallization begins at 310°C. The curves reach a frequency-dependent maximum near 270°C and then decline, indicating that the crystallization event is essentially complete. The curves converge as the polymer cools toward T_g, and below T_g e′ exhibits little dependency on frequency.

Polymer quenched from the melt and then reheated will display a strong peak at T_g. Figure 162 shows the ability of the DEA scan to detect T_g, cold crystallization, and imperfect crystal melting upon reheating of a rapidly quenched semicrystalline material. This study

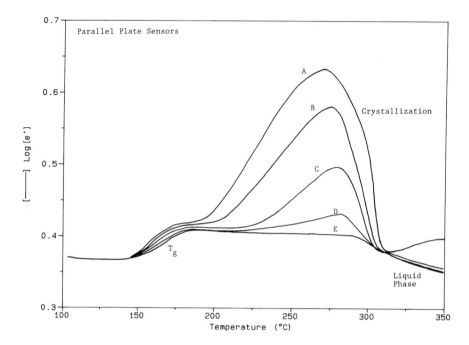

FIGURE 161. Log e′ vs. T for PEEK film cooled from the melt at 1°C/min at multiple frquencies. A — 300, B — 1000, C — 3000, D — 10,000, and E — 30,000 Hz. (From Sichina, W. J., DuPont Instruments TA Hotline, Wilmington, DE, Spring, 1989.)

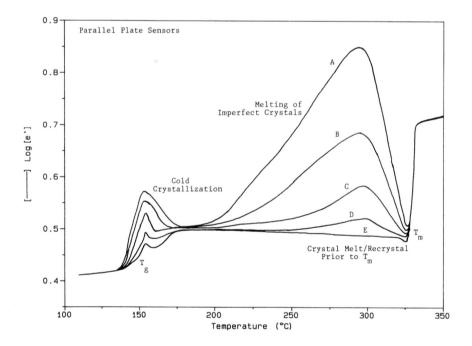

FIGURE 162. Log e′ vs. T for PEEK film heated from room temperature through T_m at 1°C/min scanning at multiple frequencies. A — 300, B — 1000, C — 3000, D — 10,000, and E — 30,000 Hz. (From Sichina, W. J., DuPont Instruments TA Hotline, Wilmington, DE, Spring, 1989.)

illustrates the superior sensitivity that DEA demonstrates over DMA in detecting structural changes in the crystalline phase below T_m. DEA also permits the examination of the liquid and the solid state in a single experiment.[208]

Day et al.[209] have reported on the use of DEA to monitor moisture content changes in epoxy. Changes in permittivity with time were used as indicators for changes in the moisture content of epoxy films. It was found that epoxies with a higher T_g, indicative of a higher crosslink density, absorbed more water. This was in agreement with earlier experiments by Enns and Gillham,[210] who employed weight gain methods to study the same phenomenon. The ability to quantitatively relate changes in dielectric properties to moisture content in polymers has given rise to a benchtop technique for moisture measurement that agrees well with the respected Karl-Fischer titration method. It has also been developed into an on-line monitoring technique suitable for many polymers. Since hydrolytic degradation of some polymers during melt processing is a major contributor to poor end-use performance of fabricated articles, this on-line system has great importance for the processing sector.

Another adaptation of the DEA technique to on-line monitoring is that of in-mold cure monitoring. By placing dielectric sensors similar to those used in benchtop instruments directly into the production mold, data can be generated in real time concerning changes in viscosity during the flow and cure of the resin. Day et al.[211] have reported on the use of this technique for unsaturated polyester/styrene systems known commercially as sheet molding compounds (SMC), and for reaction injection-molded (RIM) polyurethane thermosets.

In a production setting, the primary property of interest is the ion viscosity which is derived from the loss factor. Use of the derivative of ion viscosity with respect to time highlights the changes in ion viscosity and assists in identifying a plateau that can be considered a practical benchmark for completion of cure. DEA is unique among the thermal analysis techniques in its adaptability to on-line measurement and control. Qualifying compound behavior as a function of processing conditions, evaluating material consistency on a lot-to-lot basis, and even controlling the molding process based on a predetermined endpoint are practical possibilities using DEA in a real-time mode. Commercial systems are currently being evaluated in production environments for epoxies, unsaturated polyesters, and RIM systems based on polyurethane and polyurea.

Finally, in the area of elastomers, Johnson et al.[212] have studied block copolymers of various composition using dielectric spectroscopy. The loss factor, e'', was used to assess the structure of microphase-separated and thermodynamically homogeneous diblock copolymers of 1,4-polybutadiene-1,2-polybutadiene. Microphase-separated compounds exhibited four distinct transitions consisting of a glass transition and a secondary transition for each component. Homogeneous compounds show various degrees of transition overlap depending upon the amount of local phase mixing of the diblocks. This phase mixing is governed by the interaction parameter χN.

This study shows that, even in thermodynamically homogeneous blends with χN far from the microphase separation transition (MST), dielectric methods can identify local concentrations of a particular component. In this case the β transition for the 1,4 diblock appeared at the same temperature and peak intensity regardless of the degree of mixing. In addition, the β transition for the 1,2 diblock, which is barely detectable in the homopolymer by dielectric methods, becomes easily observable in the blends. This indicates that incorporation of 1,4 segments facilitates local motion along the chain in the 1,2 segments. As expected, lower frequencies produce the best resolution of the multiple transitions.

B. THERMALLY STIMULATED CURRENT/RELAXATION MAP ANALYSIS

The technique of thermal-stimulated current (TSC) was pioneered in the mid-1960s to study point defects in alkyl halides. It was found to be useful in studying polymers in the early 1970s and today a commercial instrument for conducting TSC analysis is available from Solomat. The company has cataloged over 100 references for TSC/RMA techniques. These have been published or released as informational newsletters between 1973 and today. The vast majority of this work deals with polymeric materials and has been performed over the last decade. A few of these applications will be discussed in order to illustrate the utility of the technique.

In its simplest form the output from a TSC scan is very similar to a plot of tan δ vs. temperature derived from a DMA or DEA test. The technique consists of orienting polar constituents within a polymer by applying a static high-voltage electrical field, E, at a given temperature, T_p. When the polarization, P, has reached its equilibrium value the sample is quenched to freeze in the configuration. The current is then turned off.

Subsequently, the sample is heated at a controlled rate and at a temperature associated with a specific relaxation the sample will depolarize. Instrumentation measures the resulting depolarizing current, which is an indication of dipolar conductivity, and the rate of depolarization is related to the relaxation times of internal motions by

$$t = P/\sigma E \qquad (28)$$

Plots of current density, J, or dynamic conductivity, σ, vs. temperature show the relaxations as peaks. It has been shown that, through correct selection of the polarization temperature, TSC can elucidate aspects of microstructure that DSC, and sometimes DMA, are incapable of discerning.[213] This work, performed by Cebeillac et al., was conducted on latex copolymers of styrene and n-butyl acrylate which were intentionally prepared to yield variations in microstructure. The limitations of DSC have already been discussed; therefore, the inability of DSC scans to delineate subtle differences in microstructure is expected. The TSC scans successfully identified multiple transitions in the block copolymers while the DMA plot of E″ vs. T revealed little detail.

Before discussing the second aspect of this technique, relaxation map analysis (RMA), it is necessary to point out that the majority of the references available concerning TSC/RMA are authored, at least in part, by people associated with the development of the technique or the marketing of the instrumentation. Much of the discussion regarding TSC/RMA is designed to contrast its superior resolution capabilities with those of alternative method such as DSC, DMA, and DEA. Unfortunately, the DMA and DEA experiments used by the authors to compare the techniques do not represent the best that DMA and DEA can offer. While a recent study by Ibar et al.[214] acknowledges the improved resolution available by DMA and DEA through the use of lower frequencies, no examples of actual work done in the low-frequency range is offered for comparison. While TSC/RMA offers an excellent set of capabilities for polymer study, the researcher uninitiated in alternate methods such as DMA and DEA are cautioned concerning the accuracy with which these competing methods are represented in the TSC/RMA literature.

A substantial increase in the precision of the TSC technique was achieved when Lacabanne[215] developed the method of windowing polarization. This involves polarizing the sample at T_p for a time t_p which is adjusted to allow for the orientation of only a fraction of the dipoles. The sample is then quenched to a temperature T_d which is 5 to 10°C below T_p. The polarizing voltage is then cut off. The sample is held at T_d for a time sufficient to allow for the depolarizing of oriented dipoles which are still capable of motions of relaxation. The sample is then quenched to a temperature far below T_d and reheated at a constant rate. This method yields a depolarization spectrum consisting of a single relaxation time which is strictly a function of temperature. When this process is repeated for multiple polarization temperatures it becomes possible to isolate elementary relaxation modes that can be difficult to separate with other techniques. Figure 163 shows the difference between a global relaxation

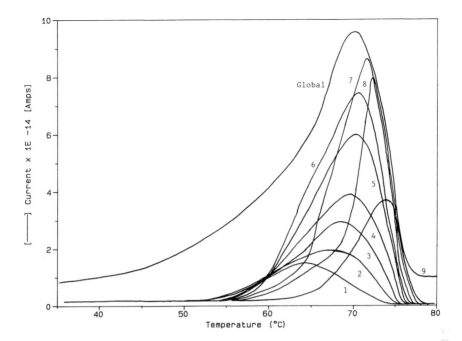

FIGURE 163. Comparison of global transition for Vectra 900 as measured by TSC and the individual peaks detected by thermal windowing. (From Ibar, J. P., Saffell, J. R., Matthiesen, A., and McIntyre, R., *SPE ANTEC,* 37, 1655, 1991.)

for Vectra 900 by standard TSC methods and the individual components of that relaxation identified by TSC/RMA. Temperatures from cryogenic to above the T_m of the polymer can be used. No substrate support is required as is for some DMA techniques.

The theory and practice of RMA is described in brochures for the commercial instrument and by Ibar et al.[216] It is summarized briefly here. When polarization is due to a distribution of relaxation times, the windowing polarization technique is used to resolve the spectra and produce a relaxation map. The elementary retardation time described by a Kelvin-Voigt model is given by

$$t_i(T) = P(T)/J(T) \qquad J(T) = \dot{P}(T) \tag{29}$$

where $t_i(T)$ is the retardation time at a given temperature in the scan, P is the area under the depolarization peak at temperature T, and J is the depolarization current density value at that temperature. Each resolved spectrum yields a temperature-dependent retardation time, t_i, which follows either an Arrhenius equation

$$t_i(T) = t_{0i} \exp(\Delta H_i/kT) \tag{30}$$

where t_{0i} is the pre-exponential factor, ΔH_i is the activation energy, and k is the Boltzmann constant, or a Vogel equation relating retardation time to free volume

$$t_i(T) = t_{0i} \exp[a_v(T - T_\infty)]^{-1} \tag{31}$$

where t_{0i} is the pre-exponential factor, a_v is the average thermal expansion coefficient for the free volume, and T_∞ is the critical temperature at which the retardation time becomes infinite.

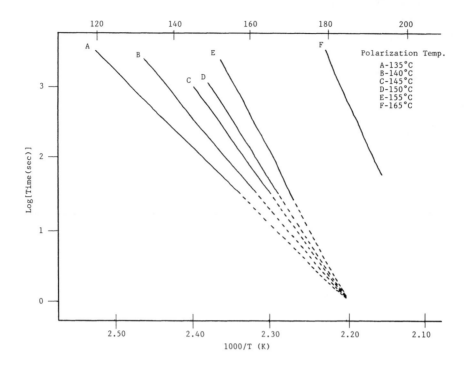

FIGURE 164. Relaxation map of nylon 14. (Solomat, 1988.)

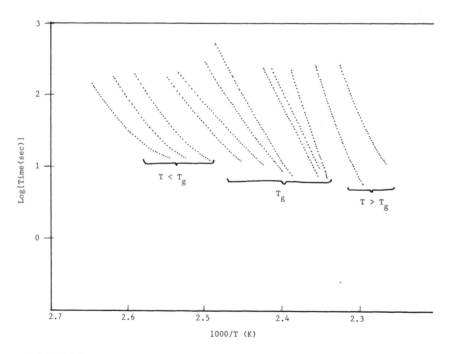

FIGURE 165. Relaxation map of polycarbonate annealed for 5 min at 235°C. (Solomat, 1988.)

Figure 164 shows a typical relaxation map for nylon 14, a material governed by an Arrhenius model. The elementary enthalpies (ΔH_i) are represented by the slopes of the lines that form the plot of log t vs. reciprocal temperature, while the pre-exponential factor is the intercept of these lines. Figure 165 shows a relaxation map for polycarbonate, a material

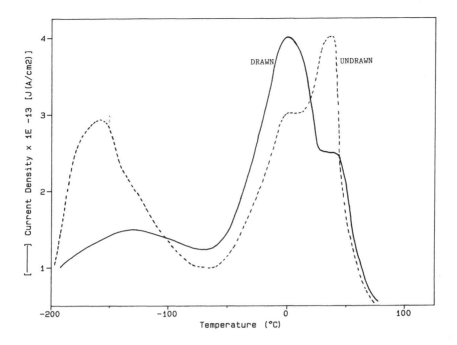

FIGURE 166. TSC spectrum of drawn and undrawn polypropylene. (Solomat, 1988.)

that exhibits a series of curved plots, particularly at temperatures below T_g. For this nonlinear behavior the WLF free volume parameters of coefficient of free volume expansion and temperature of frozen mobility are determined in addition to the pre-exponential factor. Variations that occur in these factors with respect to parameters such as molecular weight, chemical structure, orientation, and thermodynamic history have been studied for a variety of amorphous polymers by Bernes et al.[217]

The TSC technique can operate at low frequencies, typically 10^{-2} to 10^{-4} Hz. This, combined with the windowing polarization method, allows for the detection of events such as the glass transition in terms of a series of relaxations rather than one global event. This can shed new light on structure property relationships. For example, some amorphous, transparent polymers such as PS and PMMS are hard and brittle while a material like polycarbonate (PC) exhibits excellent toughness. RMA scans of PS reveal that the major motions responsible for internal flow below T_g break out into elementary mechanisms that follow an Arrhenius equation. This signifies the dominance of an activated process effect. However, just below T_g, polycarbonate relaxation processes follow a WLF-type equation which indicates the predominance of a free volume effect in the elementary mechanisms. This latter type of behavior is distinctive to tough amorphous polymers.

In semicrystalline polymers it is possible to distinguish between amorphous regions and those macromolecules that are trapped in the interlamellar regions. Figure 166 shows a TSC scan of polypropylene in the drawn and undrawn states. In both cases the glass transition is resolved into two relaxations, a lower-temperature one attributed to regions free from constraint and a higher-temperature one corresponding to amorphous regions where chain ends may be included in the crystallites. The change in intensity of the peaks as the sample is drawn indicates a change in the relative concentration of the amorphous forms. Presumably the thermal and mechanical history associated with drawing orients the polymer chains. This gives rise to a more highly ordered crystal structure and amorphous material trapped between crystallites will be expelled into the interspherulite region. This capability may be useful in answering fundamental questions regarding mechanisms of solid-state crystallization in polymers such as PPS and PEEK. In particular, the effects of fibrous reinforcements on the crystallization behavior of these polymers may be a rewarding area of investigation where up until now DSC, and to a lesser extent, DMA techniques have been used.

Ibar et al.[218] have proposed the characterization of polymers using a quantity derived from the relaxation map analysis. This parameter, given the term degree of disorder (DOD) number, is essentially a characterization of the polymer glass. It relates to changes that can occur in the amorphous glass as a result of processing, curing, thermal degradation, aging, etc. The authors describe the process for deriving the DOD number from the relaxation times and the Gibbs free-energy equation. Compensation phenomena, the convergence of Arrhenius plots of relaxation time vs. reciprocal temperature to a single point, are also related to T_g.

This work illustrates several practical uses of the DOD number. Differences in the miscibility of two polymers used in creating block copolymers are demonstrated. DOD numbers for the homopolymers can be compared with each phase of the block copolymer. Higher DOD numbers indicate better phase interpenetration. In the study of a polyether block amide (PEBA), the DOD is shown to be dependent upon the molecular weight of the polyamide.

The effects of stresses caused by different methods of cooling for polystyrene moldings are measurable by the DOD number. Rapidly cooled specimens exhibit a low DOD which indicates a highly ordered glassy state. This will translate to properties such as an elevated T_g, greater stiffness, and reduced impact strength. DOD numbers have been shown to increase upon annealing. It is likely that physical aging in amorphous polymers would be measurable by DOD determinations. The effects of both orientation and pressure are demonstrated for polystyrene.

Finally, DOD numbers are presented as an indication of the degree of cure. Here the correlation is unexpectedly poor due to an anomalous data point for which no explanation is given. More study is needed before this method can be used to quantitatively establish the degree of cure. The authors do not indicate the alternate measurement method that was used to establish degree of cure in this study.

C. FREE VOLUME MICROPROBE

In recent years a new instrument has been developed for making measurements of free volume in polymers. The free-volume microprobe (FVM) has been reported by Mayo et al.[219] The microprobe uses positrons emitted during the radioactive decay of ^{22}Na. A 1.28-MeV γ ray accompanies the "birth" of the positron and this is detected by a scintillator mounted in front of a photomultiplier tube. The positron loses energy through collisions until it interacts with an electron to form the short-lived pseudo-atom, positronium. Mutual annihilation of the positron/electron pair results in a characteristic 0.511-MeV γ ray, which is detected by a second photomultiplier. The interval between the birth event and the death event is recorded as the lifetime of the positron. Positronium exists in two forms, para and ortho, with the ortho form having a longer lifetime. The measurement principle relates the decay time of ortho-positronium to the polymer free volume; a longer lifetime indicates a larger free volume.

The method has been applied to various polymer blends to examine miscibility as a function of free volume. However, the work presented is preliminary and the number of compositions examined for a given blend were limited. In addition, this study proposes a relationship between reduced free volume and increased miscibility. However, insufficient data exist to verify this correlation. It also seems to run counter to the findings of TSC/RMA techniques which relate higher DOD numbers and therefore higher free volumes, with increased miscibility.

The FVM may be well suited to the measurement of free volume as a tool for examining phenomena such as thermal and physical aging; however, much work remains to be done

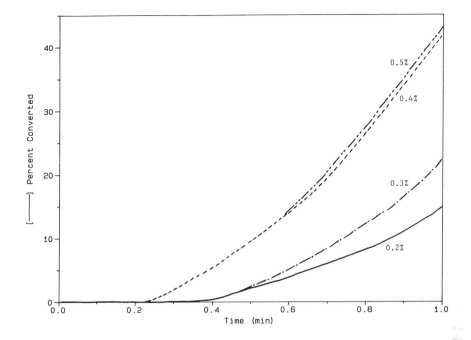

FIGURE 167. Photospeed as a function of initiator concentration. (DuPont Instruments, 1989.)

in correlating experimental results of FVM work with established techniques such as DMA, DEA, and TSC. This author has found no references to the technique or its development into a commercial analytical instrument since 1988.

D. DIFFERENTIAL PHOTOCALORIMETRY

Differential photocalorimetry (DPC) is an adaptation of DSC methods which was foreshadowed by work performed in the late 1970s by Wight and Hicks,[220] Evans et al.,[221] and Tryson and Schultz.[222] In this work, a standard DSC was modified to accept a UV light source. Evans et al. studied the photoinitiated cure of unsaturated polyester/styrene systems at various isothermal conditions. Wight and Hicks examined the effects of atmosphere and composition on the UV curing of acrylated urethane and Tryson and Schultz determined the kinetics of photoconversion for multifunctional acrylates.

A commercial instrument and support software for standard analysis and kinetics studies was introduced in 1987 by TA Instruments (formerly DuPont Instruments). Thomas[71] has described the use of the instrument in differentiating UV curing profiles as a function of photopolymer composition. Photopolymers can also be scanned by traditional DSC techniques after different irradiation times in order to examine changes in T_g and residual heat of reaction due to curing. These techniques can assist in optimizing photopolymer composition, initiator concentration, and irradiation times. Product literature also covers the effects of wavelength and atmosphere on the rate of cure. Figure 167 shows the effect of initiator concentration on the rate of conversion, Figure 168 shows the influence of radiation wavelength on the rate of cure, and Figure 169 illustrates DSC scans of a photopolymer after different UV exposure routines in the DPC instrument.

E. THERMAL CONDUCTIVITY ANALYSIS

Thermal conductivity and thermal diffusivity are essential properties of polymers in both the solid and the molten state. In the solid state these properties have an effect on the build-up of hot spots in an assembly. They can also assist in the determination of safe operating

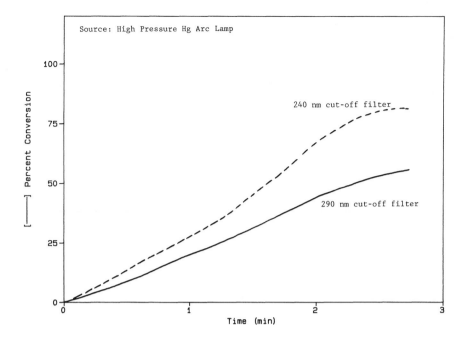

FIGURE 168. Effect of UV wavelength exposure on cure. (DuPont Instruments, 1989.)

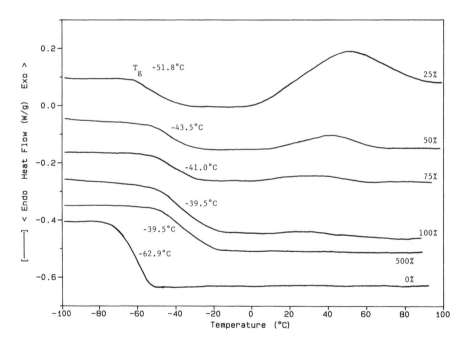

FIGURE 169. DSC scans showing the effects of UV exposure on T_g and degree of residual cure. Note that no residual cure is detectable by DSC until cure has been initiated by UV exposure. (DuPont Instruments, 1989.)

temperatures for a fabricated item. In the molten state, these properties have become important in a more accurately modeling the flow and cooling of polymers in simulated mold-filling and mold-cooling computer programs. Experimentally, the apparatus for measuring these attributes in polymers has been cumbersome to operate and has required test times that were long enough to produce degradation in many materials.

Lobo and Cohen[223] have described an alternative to the standard "guarded-hot-plate" device that is specified by ASTM F-433. Called a line-source method, it was developed for polymer melts where the problems of handling and polymer degradation were the greatest. The device and its calibration are described. Measurements were performed on a variety of molten thermoplastics and experimental results agreed with literature values to within 10%.

Coumou and Henry[224] have described a commercial instrument based on the guarded-hot-plate apparatus. Marketed by Holometrix, Inc., and called a guarded heat flow meter, it is adapted from technology originally developed at General Electric. It can scan both solids and melts over a temperature range from 30 to 300°C. Up to 20 temperature points can be selected and the instrument runs unattended. Different cells have been designed to handle thin films and melts. Data output is in the form of thermal conductivity vs. temperature.

F. MELT ELASTICITY INDEX

Finally, Maxwell[225] has described an instrument that measures the elasticity of melts by a simple method of shear deformation followed by recovery after the removal of the applied stress. The apparatus, since commercialized by Custom Scientific Instruments, provides a set of standard methods for rapidly heating a polymer to an equilibrium state and applying a strain large enough to reach the steady-state region of the polymer melt. Recovery time can be varied and the recovered strain can be monitored as a function of recovery time, melt temperature, and aging time. The effect of these test variables has been described for various polyethylenes.[226]

While not as sophisticated as DMA in measuring elasticity, this technique offers a more economical means of determining elasticity in polymer melts. This behavior has a practical bearing on the behavior of these materials as they are processed through an extruder or an injection molding machine. In addition, strain recovery as a function of time can be correlated with details of structure such as molecular weight, MWD, and long-chain branching. This instrument has primarily been offered as a quality control tool for augmenting typical melt flow information about polymer batches. Consequently, data have largely been reported in terms of a melt elasticity index, which is given in terms of arbitrary units over a standard time interval. However, the real power of this instrument lies in the information it can provide as these standard parameters are varied. Much of this work remains to be done.

VIII. CONCLUSIONS

Thermal analysis remains a dynamic field of analytical study particularly well suited to the study of polymer properties and structure. New techniques have improved capabilities in probing the microstructure and providing analysts with an increased understanding of the behavior of polymeric materials. The enhancements of computerized experiment control and data analysis have increased efficiency and precision, and the hardware itself is growing in sensitivity.

More importantly, these advances can help to make the knowledge gained from these methods intelligible to the engineers, designers, and processors who must create useful products from these polymers. The great challenge of the future will be to translate the rich understanding we have gained regarding polymers into better material selection, compounding, and fabricating techniques while continuing to satisfy the needs of the research scientist in his quest for greater understanding of fundamental principles.

ACKNOWLEDGMENTS

I would like to thank my assistant, Mary Hesprich, for her help in performing the many tests required to build this body of information over the years. I also wish to thank the rest of my technical staff who carried on so capably while I disappeared regularly to work on this project. I want to thank my editor, Dr. Nicholas Cheremisinoff, for his patience and for giving me the opportunity to work on this project. It has been a marvelous learning experience. Thanks also go to Marvin Grzys of TA Instruments who sold me my first thermal analysis system and started on the whole passionate process. However, my very special appreciation is reserved for my wife, Audrey, and my daughter, Kristina, who have endured far too many weekends and evenings without their father and husband.

APPENDIX OF ABBREVIATIONS

ABS	Acrylonitrile-butadiene-styrene
AN	Acrylonitrile
BR	Butadiene rubber
CB	Carbon black
CIIR	Chloro-isobutene-isoprene rubber
CLTE	Coefficient of linear thermal expansion
C_P	Heat capacity
CR	Chloroprene rubber
DAP	Diallyl phthalate
DEA	Dielectric analysis
DMA	Dynamic mechanical analysis
DOD	Degree of disorder
DPC	Differential photocalorimetry
DSC	Differential scanning calorimetry
DTA	Differential thermal analysis
DTG	Derivative thermogravimetry
EGA	Evolved gas analysis
EPDM	Ethylene-propylene-diene rubber
FTIR	Fourier transform infrared
FVM	Free volume microprobe
H_c	Heat of crystallization
H_{cc}	Heat of cold crystallization
HDPE	High-density polyethylene
H_f	Heat of fusion
HIPS	High-impact polystyrene
H_v	Heat of vaporization
LCP	Liquid crystal polymer
LDPE	Low-density polyethylene
LLDPE	Linear low-density polyethylene
MBS	Methacrylate butadiene styrene
MDI	Diphenylmethane diisocyanate
MDPE	Medium-density polyethylene

MEK	Methyl ethyl ketone
M_n	Number average molecular weight
MS	Mass spectroscopy
MW	Molecular weight
MWD	Molecular weight distribution
NBR	Nitrile-butadiene rubber
NR	Natural rubber
OIT	Oxidation induction time
PA	Polyamide
PAc	Polyacetal
Pα-MS	Poly(α methyl styrene)
PB	Polybutadiene
PBT	Poly(butylene terephthalate)
PC	Polycarbonate
PE	Polyethylene
PEE	Polyester-ether
PEEK	Poly(ether ether ketone)
PET	Poly(ethylene terephthalate)
POM	Poly(oxymethylene) (same as polyacetal)
PP	Polypropylene
PPE	Poly(phenylene ether)
PPO	Poly(phenylene oxide)
PPR	Parallel plate rheometry
PPS	Poly(phenylene sulfide)
PS	Polystyrene
PSVP	Poly(styrene-co-4-vinylpyridine)
PTMEG-T	Poly(tetramethylene ether glycol) terephthalate
PTMO	Poly(tetramethylene oxide)
PTMT	Poly(tetramethylene terephthalate) (same as PBT)
PVP	Polyvinyl phenol
RIM	Reaction injection molding
RMA	Relaxation map analysis
SBR	Styrene-butadiene rubber
S-EB-S	Styrene-ethylene-butene-styrene
SMC	Sheet-molding compound
SPS	Sulfonated polystyrene
TBA	Torsional braid analysis
T_c	Crystallization temperature
TCA	Thermal conductivity analysis
T_{cc}	Cold crystallization temperature
T_g	Glass transition temperature
TG	Thermogravimetry
TGA	Thermogravimetric analysis
T_m	Melting point
TMA	Thermomechanical analysis
TPU	Thermoplastic polyurethane
TSC	Thermal-stimulated current
XLPE	Crosslinked polyethylene

REFERENCES

1. **Wunderlich, B.,** in *Thermal Characterization of Polymeric Materials,* Turi, E. A., Ed., Academic Press, New York, 1981, chap. 2.
2. **Wendlandt, W. and Gallagher, P.,** in *Thermal Characterization of Polymeric Materials,* Turi, E. A., Ed., Academic Press, New York, 1981, chap. 1.
3. **Dole, M.,** *J. Polymer Sci.,* C18, 57, 1967.
4. **Wunderlich, B., Melillo, L., Cormier, C. M., Davidson, J., and Snyder, G.,** *J. Macromol. Sci.,* B1, 485, 1967.
5. **Hamada, F., Wunderlich, B., Sumida, T., Hayashi, S., and Nakajima, A.,** *J. Phys. Chem.,* 72, 178, 1968.
6. **Ludwig, H. and Eyerer, P.,** *SPE ANTEC,* 32, 665, 1986.
7. **Cebe, P.,** *SPE ANTEC,* 35, 1413, 1989.
8. **Seferis, J. C., Ahlstrom, C., and Dillman, S. H.,** *SPE ANTEC,* 33, 1467, 1987.
9. **Soni, P.,** Course Material on Crystallization and Mechanical Behavior SPE Seminars, Chicago, IL, June 17, 1991.
10. **Bair, H. E., Bebbington, G. H., and Kelleher, P. G.,** *J. Polym. Sci.,* 14, 2113, 1976.
11. **Struik, L. C. E.,** *Physical Aging in Amorphous Polymers and other Materials,* Elsevier, Amsterdam, 1978.
12. **Ma, C.-C. M. and Yur, S.-W.,** *SPE ANTEC,* 35, 1422, 1989.
13. **Desio, G. P. and Rebenfeld, L.,** *J. Appl. Polym. Sci.,* 39, 825, 1990.
14. **Avrami, M.,** *J. Chem. Phys.,* 7, 1103, 1939.
15. **Desio, G. P. and Rebenfeld, L.,** *SPE ANTEC,* 37, 2088, 1991.
16. **Reinsch, V. E. and Rebenfeld, L.,** *SPE ANTEC,* 37, 2075, 1991.
17. **Sumita, M., Miyasaka, K., and Ishikawa, K.,** *J. Polym. Sci.,* 11, 1079, 1977.
18. **Singh, P. and Kamal, M. R,** *SPE ANTEC,* 35, 235, 1989.
19. **Khanna, Y. P., Kumar, R., and Reimschuessel, A. C.,** *SPE ANTEC,* 35, 681, 1989.
20. Personal communication with Nelson Hess, B. F. Goodrich.
21. **Richeson, G. L. and Spruiell, J. E.,** *SPE ANTEC,* 33, 482, 1987.
22. **Castles, J. L., Cooper, S. L., and McKenna, J. M.,** *SPE ANTEC,* 32, 769, 1986.
23. **Hou, J. H. and Bai, J. M.,** *SPE ANTEC,* 33, 946, 1987.
24. **Flynn, J. H.,** *Thermochim. Acta,* 8, 69, 1974.
25. **Wunderlich, B.,** in *Thermal Characterization of Polymeric Materials,* Turi, E. A., Ed., Academic Press, New York, 1981, chap. 2.
26. **Shalaby, S. W.,** in *Thermal Characterization of Polymeric Materials,* Turi, E. A., Ed., Academic Press, New York, 1981, chap. 3.
27. **Blanchard, L. P., Hesse, J., and Malhotra, S. L.,** *Can. J. Chem.,* 52, 3170, 1974.
28. **Fox, T. G. and Flory, P. J.,** *J. Appl. Phys.,* 21, 581, 1950.
29. **Boyer, R. F.,** *Macromolecules,* 7, 142, 1974.
30. **Kumler, P. L., Keinath, S. E., and Boyer, R. F.,** *J. Macromol. Sci.,* B13, 631, 1977.
31. **Cowie, J. M. G.,** *Eur. Polym. J.,* 11, 297, 1975.
32. **Shalaby, S. W. and Bair, H. E.,** in *Thermal Characterization of Polymeric Materials,* Turi, E. A., Ed., Academic Press, New York, 1981, chap. 4.
33. **Bair, H. E.,** *Polym. Eng. Sci.,* 14, 202, 1974.
34. **Bair, H. E.,** *Anal. Calorim.,* 2, 51, 1970.
35. **Prest, W. M. and Porter, R. S.,** *J. Polym. Sci.,* 10, 1639, 1972.
36. **Fox, T. G.,** *Bull. Am. Phys. Soc.,* 1(2), 123, 1956.
37. **Gordon, M. and Taylor, J. S.,** *J. Appl. Chem.,* 2, 493, 1952.
38. **Wood, L. A.,** *Polym. Sci.,* 28, 319, 1958.
39. **Couchman, P. R.,** *Macromolecules,* 11, 1156, 1978.
40. **Ellis, T. S.,** Proceedings of High Performance Polymer Alloys Symposium, Ann Arbor, MI, May 23–25, 1989.
41. **Young, M. W., Xanthos, M., and Biesenberger, J. A.,** *SPE ANTEC,* 35, 1835, 1989.
42. **Kohli, A., Chung, N., and Weiss, R. A.,** *SPE ANTEC,* 35, 1843, 1989.
43. **Sukhadia, A. M., Done, D., and Baird, D. G.,** *SPE ANTEC,* 35, 1847, 1989.
44. **Subramanian, P. R. and Isayev, A. I.,** *SPE ANTEC,* 36, 489, 1990.
45. **Sullivan, M. J. and Weiss, R. A.,** *SPE ANTEC,* 37, 964, 1991.
46. **Lilaonitkul, A., West, J. C., and Cooper, S. L.,** *J. Macromol. Sci.,* B12, 563, 1976.
47. **Gallagher, K. P., Zhang, X., Runt, J. P., and Huynh-Ba, G.,** *SPE ANTEC,* 37, 1041, 1991.
48. **Chandler, L. A. and Collins, T. E.,** *J. Appl. Polym. Sci.,* 13, 1585, 1969.
49. **Machado, J. M. and French, R. N.,** *SPE ANTEC,* 37, 1586, 1991.

50. **Paul, D. R. and Barlow, J. W.,** *Polymer Compatibility and Incompatibility: Principles and Practices,* Solc, K., Ed., Michigan Molecular Institute Press, 1980.
51. **Kwei, T. K.,** Proceedings of Compatibilization of Polymer Blends Symposium, Dearborn, MI, May 22–24, 1990.
52. **Cooper, S. L., West, J. C., and Seymour, R. W.,** *Encycl. Polym. Sci. Technol. Suppl.,* 1, 521, 1976.
53. **Hesketh, T. R. and Cooper, S. L.,** *Org. Coat. Plast. Prepr.,* 37(2), 509, 1977.
54. **Joseph, M. D. and Harris, R. F.,** *J. Appl. Polym. Sci.,* 41, 509, 1990.
55. **Wilkes, G. L., Bagrodia, S., Humphries, W., and Wildnauer, R.,** *J. Polym. Sci.,* 13, 321, 1975.
56. Pellethane Design and Processing Guide, Dow Chemical, Midland, MI, 1981.
57. DuPont Bulletin HYT-111, Wilmington, DE, Table IV.
58. **Maurer, J. J.,** in *Thermal Characterization of Polymeric Materials,* Turi, E. A., Ed., Academic Press, New York, 1981, chap. 6.
59. **Sircar, A. K. and Lamond, T. G.,** *Rubber Chem. Technol.,* 45, 329, 1972.
60. **Sircar, A. K. and Lamond, T. G.,** *J. Appl. Polym. Sci.,* 17, 2569, 1973.
61. **Coran, A. Y. and Patel, R.,** *Rubber Chem. Technol.,* 53, 141, 1980.
62. **Bair, H. E., Boyle, D. J., and Kelleher, P. G.,** *Polym. Eng. Sci.,* 20, 995, 1980.
63. **Braunstein, C. H. and Herrera, E. C.,** *SPE ANTEC,* 36, 829, 1990.
64. **Mason, J. W. and Domingue, L. A.,** *SPE ANTEC,* 35, 1096, 1989.
65. **Fann, D.-M., Lin, H.-F., and Shih, H.-Y.,** *SPE ANTEC,* 35, 1178, 1989.
66. **Lin, H. F., Fann, D. M., Lin, Y. C., Hsiao, J. M., and Shih, H. Y.,** *SPE ANTEC,* 37, 1590, 1991.
67. **Hsieh, A. J. and Schneider, N. S.,** *ACS Polymer Preprints,* 31, 259, 1990.
68. **Hsieh, A. J.,** *SPE ANTEC,* 37, 2365, 1991.
69. **Bair, H. E.,** in *Thermal Characterization of Polymeric Materials,* Turi, E. A., Ed., Academic Press, New York, 1981, chap. 9.
70. **Ezrin, M. and Seymour, D.,** *SPE ANTEC,* 34, 889, 1988.
71. **Thomas, L. C.,** *Am. Lab.,* January, 1987.
72. **Prime, R. B.,** in *Thermal Characterization of Polymeric Materials,* Turi, E. A., Ed., Academic Press, New York, 1981, chap. 5.
73. **Morrison, T. E. and Waitkus, P.,** SPI Phenolic Molding Division Technical Conference, Cincinnati, OH, 46–51, 1987.
74. **Gray, A. P.,** Perkin-Elmer TAAS-2, Norwalk, CT, 1972.
75. **Fava, R. A.,** *Polymer,* 9, 137, 1968.
76. **Sourour, S. and Kamal, M. R.,** *Thermochim. Acta,* 14, 41, 1976.
77. **Westwood, A. R.,** *Thermal Analysis,* Vol. 3, Buzas, I., Ed., Heyden, London, 1975, 337.
78. **Schneider, N. S., Sprouse, J. F., Hagmauer, G. L., and Gillham, J. K.,** *Polym. Eng. Sci.,* 19, 304, 1979.
79. **Acitelli, M. A., Prime, R. B., and Sacher, E.,** *Polymer,* 12, 333, 1971.
80. **Flory, P. J.,** *Principles of Polymer Chemistry,* Cornell University Press, Ithaca, NY, 1953.
81. **Babayevsky, P. G. and Gillham, J. K.,** *J. Appl. Polym. Sci.,* 17, 2067, 1973.
82. **Gonzalez-Romero, V. M. and Casillas, N.,** *SPE ANTEC,* 33, 1119, 1987.
83. **Willard, P. E.,** *J. Macromol. Sci.,* A8(1), 33, 1974.
84. **Willard, P. E.,** in *Polymer Characterization by Thermal Methods of Analysis,* Chiu, J., Ed., Marcel Dekker, New York, 1974, 33.
85. **Bateman, L.,** Ed., *The Chemistry and Physics of Rubber-Like Substances,* John Wiley & Sons, New York, 1963.
86. **Maurer, J. J.,** in *Thermal Characterization of Polymeric Materials,* Turi, E. A., Ed., Academic Press, New York, 1981, chap. 6.
87. **Brazier, D. W. and Nickel, G. H.,** *Rubber Chem. Technol.,* 48, 26, 1975.
88. **Sircar, A. K. and Lamond, T. G.,** *Rubber Chem. Technol.,* 48, 631, 1975.
89. **Sircar, A. K. and Lamond, T. G.,** *Rubber Chem. Technol.,* 48, 640, 1975.
90. **Groves, I. and Thomas, L.,** *Res. Dev.,* February, 1988.
91. **Brazier, D. W. and Nickel, G. H.,** *Rubber Chem. Technol.,* 48, 661, 1975.
92. **Sircar, A. K. and Lamond, T. G.,** *Rubber Chem. Technol.,* 48, 301, 1975.
93. **Swarin, S. J. and Wims, A. M.,** *Rubber Chem. Technol.,* 47, 1193, 1974.
94. **Pautrat, R., Metrivier, B., and Morteau, J.,** *Rubber Chem. Technol.,* 49, 1060, 1976.
95. **Sircar, A. K. and Lamond, T. G.,** *Rubber Chem. Technol.,* 51, 647, 1978.
96. **Sepe, M.,** unpublished data.
97. **Brennan, W. P.,** *Thermochim. Acta,* 18, 101, 1977.
98. **Sepe, M.,** unpublished historical data.
99. **Cassel, B.,** Perkin-Elmer TAAS-19, Norwalk, CT, 1977.
100. **Rouquerol, J.,** *Bull. Soc. Chim.,* 31, 1964.

101. **Paulik, F. and Paulik, J.,** *Anal. Chim. Acta,* 56, 328, 1971.

102. **Sorensen, S.,** *J. Therm. Anal.,* 13, 429, 1978.

103. **Criado, J. M.,** *Thermochim. Acta,* 28, 307, 1979.

104. **Lundgren, C. J. and Michgehl, C.,** TA Instruments Hotline, Wilmington, DE, Vol. 3, 1991.

105. **Compton, D.,** *SPE ANTEC,* 34, 1260, 1988.

106. **Khorami, J., Chauvette, G., Lemieux, A., Menard, H., and Jolicoeur, C.,** *Thermochim. Acta,* 103, 221, 1986.

107. **Lewis, M. L., Nam, J.-D., Seferis, J. C., and Prime, R. B.,** *SPE ANTEC,* 35, 1092, 1989.

108. **Bair, H. E.,** *Polym. Eng. Sci.,* 13, 435, 1973.

109. **Bair, H. E.,** in *Thermal Characterization of Polymeric Materials,* Turi, E. A., Ed., Academic Press, New York, 1981, chap. 9.

110. **Bair, H. E.,** *Anal. Calorim.,* 3, 797, 1974.

111. **Bozzelli, J.,** Dow Plastics Injection Molding Technical Bulletin for Magnum ABS Resins, Midland, MI, 12, 1987.

112. **Cassel, B.,** Perkin-Elmer MA-29, Norwalk, CT, 1976.

113. **Beekman, G. F., Petrie, S., and Keating, L.,** *SPE ANTEC,* 32, 1167, 1986.

114. **Flynn, J. H.,** *Polymer Lett.,* 4, 323, 1966.

115. **Day, M., Cooney, J., and Wiles, D.,** *SPE ANTEC,* 34, 933, 1988.

116. **Hunter, A. and Sheppard, C.,** *SPE ANTEC,* 34, 937. 1988.

117. **Flynn, J. H.,** *SPE ANTEC,* 34, 930, 1988.

118. **Flynn, J. H. and Dickens, B.,** in *Durability of Macromolecular Materials,* Eby, R. K., Ed., (ACS Symp. Ser. No. 95), American Chemical Society, Washington, DC, 1979, 97.

119. **Flynn, J. H.,** Analysis of the Kinetics of Thermogravimetry: Overcoming Complications of Thermal History, National Bureau of Standards, Washington, DC.

120. **Schwartz, A.,** *J. Therm. Anal.,* 13, 489, 1978.

121. **Bair, H. E. and Pryde, C. A.,** *SPE ANTEC,* 37, 1550, 1991.

122. **Sepe, M.,** unpublished data.

123. **Sepe, M.,** *SPE ANTEC,* 35, 1035, 1989.

124. **Henderson, J. B., Stegmayer, A., and Emmerich, W.-D.,** *SPE ANTEC,* 37, 2362, 1991.

125. **Thomas, R. W. and Cadwallader, M. W.,** TA Instruments Hotline, Wilmington, DE, June, 1990.

126. **Sepe, M.,** unpublished data.

127. **Kettle, G. J.,** *Polymer,* 18, 742, 1977.

128. **Prime, R. B.,** *Thermochim. Acta,* 26, 165, 1978.

129. **Barrall, E. M., III and Flandera, M. A.,** *J. Elastomers Plast.,* 6, 16, 1974.

130. **Gent, A. N.,** *Trans. Inst. Rubber Ind.,* 34, 46, 1958.

131. **Brazier, D. W. and Nickel, G. H.,** *Thermochim. Acta,* 26, 399, 1978.

132. **Fogiel, A. W., Frensdorff, H. K., and MacLachlan, J. D.,** *Rubber Chem. Technol.,* 49, 34, 1976.

133. **Fair, P. G., Chaney, R. D., and Dallas, G.,** TA Instruments Hotline, Wilmington, DE, November, 1990.

134. **Cessna, L. C., Jr. and Jabloner, H.,** *J. Elastomers Plast.,* 6, 103, 1974.

135. **Blaine, R. L. and Lofthouse, M. G.,** DuPont Appl. Brief TA-65, Wilmington, DE, 1974.

136. **Hassel, R.,** *Am. Lab.,* January 30, 1991.

137. **Prime, R. B., Barrall, E. M., II, Logan, J. A., and Duke, P. J.,** *AIP Conf. Proc.,* 17, 72, 1974.

138. **Jaffe, M.,** in *Thermal Characterization of Polymeric Materials,* Turi, E. A., Ed., Academic Press, New York, 1981.

139. **Ferry, J. D.,** *Viscoelastic Properties of Polymers,* John Wiley & Sons, New York, 1980.

140. **Pournoor, K. and Seferis, J. C.,** *SPE ANTEC,* 35, 1103, 1989.

141. **Morrison, T. E.,** personal communication.

142. **Tuminello, W. H.,** *SPE ANTEC,* 33, 990, 1987.

143. Delrin Processing Guide, DuPont, Wilmington, DE, 29.

144. **Sepe, M.,** *SPE ANTEC,* 35, 1035, 1989.

145. **Maffezzoli, A., Kenny, J. M., and Nicolais, L.,** *SPE ANTEC,* 37, 2079, 1991.

146. **Takayanagi, M., Imada, K., and Kajiyama, T.,** *J. Polym. Sci. C,* 15, 263, 1966.

147. **Godovsky, Yu. K. and Slonimsky, G. L.,** *J. Polym. Sci.,* 12, 1053, 1974.

148. **Sichina, W. J. and Gill, P. S.,** *SPE ANTEC,* 32, 441, 1986.

149. **Patel, U. G. and Nunn, R. E.,** *SPE ANTEC,* 36, 292, 1990.

150. **Zafran, J. and Beatty, C. L.,** *SPE ANTEC,* 32, 669, 1986.

151. **Rohn, C. and Herh, P.,** *SPE ANTEC,* 34, 1135, 1988.

152. **Salovey, R. and Gandhi, K.,** *SPE ANTEC,* 34, 1083, 1988.

153. **Trevino, D.,** *SPE ANTEC,* 34, 1906, 1988.

154. **Nielsen, L. E.,** Damping and Mechanical Properties of Filled Polymeric Systems, Proc. of the Workshop on Acoustic Attenuation Materials Systems, Rep. No. NMAB, National Materials Advisory Board, National Academy of Sciences, Washington, DC, 1978, 63.

155. Product Performance Through Rheology, *course notes, Rheometrics, Inc.,* Rosemont, IL, February 22, 1990.
156. **Sepe, M.,** *Plastics Design Forum,* 16(6), 45, 1991.
157. **Chua, P. S.,** SPI Compos. Inst. 42nd Annual Conf. Proc., Session 21A, 1987.
158. **Zorowski, C. F. and Murayama, T.,** Proc. 1st Int. Conf. on Mechanical Behavior of Materials, Vol. 5, Society of Material Scientists, Kyoto, Japan, 1972, 28.
159. **Edie, D. D., Kennedy, J. M., Cano, R. J., and Ross, R. A.,** *SPE ANTEC,* 37, 2248, 1991.
160. **Chartoff, R.,** *SPE ANTEC,* 34, 1143, 1988.
161. **Tuminello, W. H.,** *Polym. Eng. Sci.,* 26(19), 1339, 1986.
162. **Letton, A. and Tuminello, W. H.,** *SPE ANTEC,* 33, 997, 1987.
163. **Starkweather, H. W.,** Proc. of 15th NATAS Conf., 1986, 40.
164. **Sichina, W. J.,** internal communication.
165. **Sichina, W. J.,** internal communication.
166. **Terosky, J. C. and Sichina, W. J.,** internal communication.
167. **Sichina, W. J.,** *SPE ANTEC,* 34, 1139, 1988.
168. **Rohn, C. L.,** *SPE ANTEC,* 35, 870, 1989.
169. **Tryson, G.,** *SPE ANTEC,* 37, 805, 1991.
170. **Sepe, M.,** *SPE ANTEC,* 38, 1152, 1992.
171. **Chiang, W.-Y. and Hwung, D.-S.,** *SPE ANTEC,* 32, 492, 1986.
172. **Sichina, W. J.,** internal communication.
173. **Han, C. D., Kim, J., Kim, J. K., and Chu, S. G.,** *Macromolecules,* 22(8), 1989.
174. **Lin, J.-S. G. and Manson, J. A.,** *SPE ANTEC,* 32, 518, 1986.
175. **Yoshimura, D. K. and Richards, W. D.,** *SPE ANTEC,* 32, 688, 1986.
176. **Dudevan, I., Peiffer, D. G., Agarwal, P. K., and Lundberg, R. D.,** *SPE ANTEC,* 33, 1310, 1987.
177. **Sakai, J., Nakamura, K., and Inoue, S.,** *SPE ANTEC,* 36, 1912, 1990.
178. **Bier, P. and Rempel, D.,** *SPE ANTEC,* 34, 1485, 1988.
179. Properties of Calibre, Dow Plastics, Midland, MI, 1985.
180. **Suttner, L.,** *SPE ANTEC,* 36, 833, 1990.
181. **Sichina, W. J.,** *Am. Lab.,* January, 44, 1992.
182. **Landi, V. R. and Merserau, J. M.,** *SPE ANTEC,* 32, 1369, 1986.
183. Delrin Processing Guide, DuPont, Wilmington, DE.
184. **Hawkins, R. J.,** *Macromolecules,* 9, 191, 1976.
185. **Ma, C.-C., Hsia, H.-C., Shieh, B.-Y., Liu, W.-L., Hu, J.-T., and Liu, R.-S.,** *SPE ANTEC,* 32, 539, 1986.
186. **Sichina, W. J. and Gill, P. S.,** *SPE ANTEC,* 33, 959, 1987.
187. **Tung, C. M. and Dynes, P. J.,** *Composite Materials: Quality Assurance and Processing,* ASTM STP 797, Browning, C. E., Ed., American Society for Testing and Materials, Philadelphia, 1983, 38.
188. **Hinrichs, R. J.,** *Composite Materials: Quality Assurance and Processing,* ASTM STP 797, Browning, C. E., Ed., American Society for Testing and Materials, Philadelphia, 1983, 29.
189. **Winter, H. H.,** *SPE ANTEC,* 33, 1106, 1987.
190. **Grentzer, T., Sauerbrunn, S., and Gill, P.,** *SPE ANTEC,* 34, 976, 1988.
191. **Quan, X., Pratt, M., Adams, J., Mujsce, A., and Reents, W.,** *SPE ANTEC,* 34, 1079, 1988.
192. **Phillips, P. J., Lambert, W. S., and Thomas, H. D.,** *SPE ANTEC,* 37, 1575, 1991.
193. **Driscoll, S. B.,** Using ASTM D-4065-82 for predicting processability and properties, in *High Modulus Composites in Ground Transportation and High Volume Applications,* ASTM STP 873, Wilson, D. W., Ed., American Society for Testing and Materials, Philadelphia, 1985, 144.
194. Rheometrics Course Notes on Product Performance Through Rheology, Rosemont, IL, February 22, 1990.
195. **Prime, R. B. and Adibi, N.,** *SPE ANTEC,* 32, 451, 1986.
196. **Warfield, R. W. and Petrie, M. C.,** *Polymer,* 1, 178, 1960.
197. **Ramanathan, R. and Harper, D. O.,** *SPE ANTEC,* 32, 1376, 1986.
198. **Bidstrup, S., Sheppard, N. F., Jr., and Senturia, S. D.,** *SPE ANTEC,* 33, 987, 1987.
199. **Sheppard, N. F., Jr. and Senturia, S. D.,** Proc. of IEEE Conf. on Electrical Insulation and Dielectric Phenomena, Claymont, DE, Nov. 2–6, 1986.
200. **Sheppard, N. F., Jr.,** Ph.D. Thesis, Massachusetts Institute of Technology, Cambridge, 1986.
201. **DiBenedetto, A. T.,** *J. Macromol. Sci.,* C3, 69, 1969.
202. **Day, D., Wall, A., and Sheppard, D.,** *SPE ANTEC,* 34, 964, 1988.
203. **Gotro, J. and Yandrasits, M.,** *SPE ANTEC,* 33, 1039, 1987.
204. **Lane, J. W. and Khattak, R. K.,** *SPE ANTEC,* 33, 982, 1987.
205. **Sichina, W. J. and Marcozzi, C. L.,** *SPE ANTEC,* 36, 908, 1990.
206. **Wetton, R., Marsh, R., Foster, G., and Connolly, M.,** *SPE ANTEC,* 37, 1170, 1991.
207. DuPont Instruments 2970 DEA Technical Brochure, Wilmington, DE.
208. **Sichina, W. J.,** DuPont Instruments TA Hotline, Wilmington, DE, Spring, 1989.

209. **Day, D. R., Sheppard, D. D., and Craven, K. J.,** *SPE ANTEC,* 36, 1045, 1990.
210. **Enns, J. B. and Gillham, J. K.,** *J. Appl. Polym. Sci.,* 28, 2831, 1983.
211. **Day, D. R., Lee, H. L., Sheppard, D. D., and Sheppard, N. F.,** *SPE ANTEC,* 37, 946, 1991.
212. **Johnson, G., Anderson, E., Quan, X., and Bates, F.,** *SPE ANTEC,* 34, 1170, 1988.
213. **Cebeillac, P., Fauran-Clavel, M.-J., Lacabanne, C., and Ibar, J. P.,** *SPE ANTEC,* 35, 1855, 1989.
214. **Ibar, J. P., Saffell, J. R., Matthiesen, A., and McIntyre, R.,** *SPE ANTEC,* 37, 1655, 1991.
215. **Lacabanne, C.,** Ph.D. Thesis, University of Toulouse, France, 1974.
216. **Ibar, J., Barnes, P., Denning, P., Thomas, T., Saffel, J., Jones, P., Bernes, A., Rodrigues, M., and Lacabanne, C.,** *SPE ANTEC,* 34, 1049, 1988.
217. **Bernes, A., Boyer, R. F., Chatain, D., Lacabanne, C., and Ibar, J. P.,** in *Order in the Amorphous State of Polymers,* Keinath, S. E., Ed., Plenum Press, New York, 1987, 305.
218. **Ibar, J. P., Denning, P., and DeGoys, C.,** *SPE ANTEC,* 36, 866, 1990.
219. **Mayo, B., Mangaraj, D., Pfau, J., and Macarus, J.,** *SPE ANTEC,* 34, 1111, 1988.
220. **Wight, F. R. and Hicks, G. W.,** *Polym. Eng. Sci.,* 18, 378, 1978.
221. **Evans, A. J., Armstrong, C., and Tolman, R. J.,** *J. Oil Colour Chem. Assoc.,* 61, 251, 1978.
222. **Tryson, G. R. and Schultz, A. R.,** *J. Polym. Sci.,* 17, 2059, 1979.
223. **Lobo, H. and Cohen, C.,** *SPE ANTEC,* 34, 609, 1988.
224. **Coumou, K. G. and Henry, K.,** *SPE ANTEC,* 36, 1012, 1990.
225. **Maxwell, B.,** *Plastics Eng.,* 43(9), 41, 1987.
226. **Maxwell, B.,** *SPE ANTEC,* 35, 1659, 1989.

Chapter 4

TENSILE YIELD IN POLYMERS

Bruce Hartmann and David Sudduth

TABLE OF CONTENTS

0-8493-4401-8/93/$0.00 + $.50
© 1993 by CRC Press, Inc.

259

I. INTRODUCTION

Because polymers are sometimes used as fillers in rubber products, some understanding of the yield properties is needed. This presentation will take a phenomenological look at these properties, that is, what are the observed properties? Some possible theoretical interpretations are presented in outline form only. The presentation will cover crystalline and amorphous polymers giving available experimental data for the temperature, strain rate, and pressure dependence of common polymers. Some comparisons of the yield properties of polymers with those of metals and ceramics are mentioned as a way of putting the polymer results in perspective.

Yield is defined as the onset of plastic deformation in a material under an applied load. This is an important process because it determines the practical limit of use more so than does ultimate rupture. Also, yield is the first step in polymer processing by forming, rolling, or drawing. While the idea of yield is a simple concept, its experimental determination is not easy. Definitions of terms are particularly important since not everyone agrees even on the operational definition of the word yield.

Two classes of polymers are considered here: crystalline and amorphous. The term crystalline refers to semicrystalline, bulk polymer specimens that are also called polycrystalline. Bulk polymers are, as a rule, only partly crystalline and are partly amorphous. The degree of crystallinity of a polymer (weight percent of the total polymer that is in the crystalline state) is typically 40%. While polymer crystals have different physical properties in different crystal directions, in a bulk specimen the individual microscopic crystals are oriented at random so that the macroscopic properties are the same in every direction. The physical properties are then isotropic. It will be shown that yield properties depend on the degree of crystallinity so the designation of a polymer by a single term, such as polyethylene, does not fully specify the material. There are significant variations in yield properties among different polyethylenes.

The term amorphous refers to polymers with no long-range order. Amorphous polymers can be either in the glassy or rubbery state, though only glassy polymers exhibit yield. Glassy amorphous polymers are unique because their molecules are not in the lowest energy equilibrium configuration. Depending on the process by which they are made, e.g., the rate of cooling from the molten state, the yield properties will vary somewhat. In addition, their properties can change with time, by a process known as physical aging, as the polymer slowly approaches equilibrium.

Yield behavior in polymers depends on the test conditions used. Yield varies not only with the test temperature but also with the speed at which the test is made and the magnitude of the test pressure. In some cases, even the gaseous environment (air, nitrogen, oxygen, helium, etc.) can have an effect. Thus, yield depends on both material properties and test conditions.

This presentation is organized into three major sections: basic concepts, experimental results, and theoretical interpretations. Basic concepts are important for the understanding of this topic and include definitions of terms used later. Experimental results will be given for various common crystalline and amorphous polymers as functions of temperature, strain rate, and pressure. Polyethylene will be used as the typical example. Finally, theoretical interpretations of the experimental results will be considered when the theories make predictions of practical use in extrapolating and interpolating experimental data and predicting behavior under conditions that have not been measured.

II. BASIC CONCEPTS

The basic concepts necessary to make proper engineering use of tensile yield measurements will be presented in this section. Not only will definitions of terms be given, but also

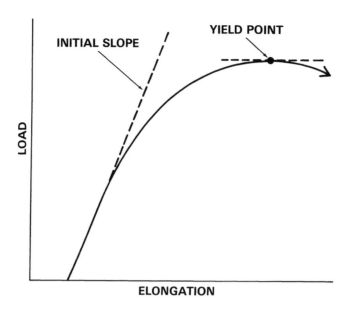

FIGURE 1. Schematic load-elongation curve for a material with a yield point.

limitations of their application and comparisons with related terms, to give an understanding why the definition is formulated the way it is.

A. YIELD

As stated earlier, yield is defined conceptually as the onset of plastic deformation in a material under an applied load. A plastic deformation is one that remains after the load is removed and is also called a permanent or nonrecoverable deformation. By contrast, at small-enough loads, deformation is elastic and is recovered after the load is removed (i.e., the specimen returns to its original length). Yield thus represents the transition from elastic to plastic deformation. As an example, consider a typical material under an applied tensile load. The length of the specimen will increase, as measured by the elongation or increase in length of the specimen, when the load is applied. A schematic drawing of a typical tensile load-elongation curve is shown in Figure 1. As the elongation increases, the load at first increases linearly but then increases more slowly and eventually passes through a maximum, where the elongation increases without any increase in load. The peak in the load-elongation curve is the point at which plastic flow becomes dominant and is commonly defined as the yield point.

Not all load-elongation curves look like Figure 1, which is typical of a material that clearly exhibits plastic deformation before rupture. Other materials fracture before reaching a maximum, as shown in Figure 2. Even though no maximum is reached, there may be plastic deformation and it is usually assumed that yield occurs at some arbitrarily chosen value of strain. Strain is defined as elongation divided by original length. For polymers, the strain at which plastic deformation begins is often taken to be 2%,[1] while in metals a value of 0.2% is common,[2] and in ceramics 0.05% is used.[3] One then draws a line parallel to the initial linear portion of the load-elongation curve but offset by the appropriate strain. The point where this line intersects the load-elongation curve is called the offset yield point, as illustrated in Figure 2. Use of an offset yield point is common in metals and ceramics where plastic deformation follows immediately after linear elastic deformation. Polymers, however, are more complex and generally exhibit considerable nonlinear elasticity before plastic deformation. Nonlinear elasticity also produces a deviation from the initial linear

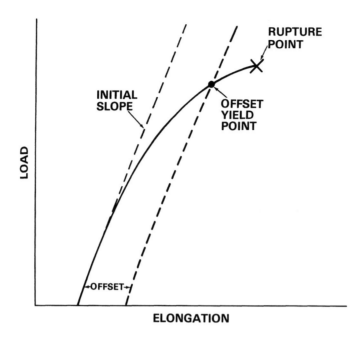

FIGURE 2.   Schematic load-elongation curve for a material with an offset yield point.

behavior but is fully recoverable. Since it is not obvious whether the deviation from linearity is caused by plasticity or nonlinearity, yield in polymers is preferably defined as the maximum in the load-elongation curve. Offset yield point is used only if there is no maximum.

In some cases, the load increases to fracture with no plastic deformation. The material is then said to be brittle in distinction to the above behavior where plastic deformation occurs and the material is said to be ductile. A given material will exhibit ductile or brittle behavior depending on the test conditions. A material that is brittle at low temperature will become ductile at high temperature. The brittle-to-ductile transition temperature, T_{bd}, is then an important material characteristic. For polyethylene, T_{bd} is about 125 K,[4] depending somewhat on the specific polyethylene and the test conditions. A brittle-to-ductile transition is also observed in metals and ceramics. For transition metal carbides (TiC, ZrC, VC, NbC, and TaC), T_{bd} is about 1500 K while the melting temperatures are about 4500 K so T_{bd} is about one third of the melting temperature, T_m.[5] In comparing these results with those for polymers, since the melting temperature of polyethylene is 413 K,[6] its T_{bd} is also about one third of its melting temperature. These observations suggest that comparisons between different materials should be made at equivalent homologous temperatures, T/T_m. When this is done, polymers and ceramics are both brittle below about $T/T_m = 0.3$ and ductile above this temperature. They differ only in the magnitude of their melting temperatures.

The brittle-ductile transition temperature can be interpreted as a competition between the shear banding mechanism of ductile flow and the crazing mechanism of brittle fracture. The two processes are independent and have different temperature, pressure, and strain rate dependence.[7,8] High temperature and low strain rate promote ductile yielding while low temperature and high strain rate favor brittle fracture. Pressure dependence is more complicated. Polyethylene, polypropylene, and Teflon all become more brittle under an applied hydrostatic pressure while polycarbonate and polyoxymethylene become more ductile.[9] Matsushige[7] finds that T_{bd} in polystyrene and polymethylmethacrylate can be induced by the superposition of hydrostatic pressure. Sauer[10] states that Teflon is brittle at all pressures.

TABLE 1
Typical Density and Transition Temperatures

Crystalline Polymers

	Density (g/cm³)	T_{bd} (K)	T_m (K)	Ref.
Polyethylene	0.945	125	413	4
Polypropylene	0.909	<295	437	12
Poly(4-methyl pentene-1)	0.835	273	513	13
Nylon 6,6	1.147	<295	538	14
Poly(vinylidene fluoride)	1.772	<295	441	15
Poly(chlorotrifluoroethylene)	2.139	<295	488	15
Polyoxymethylene	1.425	295	448	16
Polytetrafluoroethylene	2.151	>295	615	10, 17

Amorphous Polymers

	Density (g/cm³)	T_{bd} (K)	T_g (K)	Ref.
Polycarbonate	1.19	273	415	18, 19
Polyvinylchloride	1.40	<295	353	18
Polyethyleneterephthalate	1.33	<295	345	11, 20
Polystyrene	1.05	363	379	7, 20
Polymethylmethacrylate	1.19	323	376	7, 18

A further complication is that polymethylmethacrylate, for example, is brittle in tension but ductile in compression.[11]

As the test temperature of the polymer increases, the magnitude of the yield stress decreases. As will be shown in more detail later, the yield stress approaches zero as the temperature approaches the melting temperature of a crystalline polymer or the glass transition temperature of an amorphous polymer. Yield is therefore bounded by the brittle-ductile transition at low temperature and at high temperature by either the melting point (in crystalline polymers) or the glass transition temperature (in amorphous polymers). Table 1 is a collection of typical transition temperatures for common polymers, taken from various sources.[4,7,11-20] These values are representative only since they depend not only on the particular test conditions but also on the specific polymer used. In polycarbonate, for example, T_{bd} varies from -40 to $40°C$ as a function of molecular weight.[19] However, one can see what to expect in general. Typical density values are also listed to further characterize the polymers.

B. STRESS AND STRAIN

The magnitude of a load-elongation curve depends on the dimensions of the specimen being tested and is therefore not a basic material property. Rather than load, the properly normalized variable is stress, the load per unit cross-sectional area of the specimen. Stress, σ, then has units of pressure and is expressed in Pascal (1 MPa = 1 MN/m^2 = 10^7 dyne/cm^2 = 145 psi). Common engineering practice is to calculate σ on the basis of the initial (unloaded) cross-sectional area, A_0, so the engineering (or nominal) stress is given by

$$\sigma(\text{engineering}) = W/A_0 \tag{1}$$

where W is the applied load. In a tensile test, the cross-sectional area of the specimen decreases as the specimen elongates, and the true stress is given by

$$\sigma(\text{true}) = W/A \qquad (2)$$

where A is the instantaneous cross-sectional area at any given elongation. Since the relative change in lateral dimension divided by relative change in length is called Poisson's ratio, μ, σ (true) can be calculated using the formula

$$\sigma(\text{true}) = \sigma(\text{engineering})/(1 - \mu)^2 \qquad (3)$$

Accurate values of $\sigma(\text{true})$ require a knowledge of μ which is usually not available, and it is often assumed that the volume of the specimen is constant so $\mu = 0.5$. If, in addition, the strain is small, the ratio of true stress to engineering stress is $1 + \epsilon$. Thus, the true stress is somewhat higher than the engineering stress. As an example, for polyethylene at the yield point, $\sigma(\text{true}) = 33$ MPa while $\sigma(\text{engineering}) = 29$ MPa.[6]

Rather than elongation, the properly normalized variable is strain, the elongation divided by the initial length. Strain is therefore dimensionless. Engineering strain is calculated on the basis of the initial length, L_0,

$$\epsilon(\text{engineering}) = (L - L_0)/L_0 \qquad (4)$$

where L is the instantaneous length. True strain is defined as

$$\epsilon(\text{true}) = \int dL/L = \ln(L/L_0) \qquad (5)$$

It is almost universal to use engineering stress and strain when discussing polymers. Therefore in the following discussion, σ and ϵ will mean engineering values unless otherwise noted. Since a load-elongation curve can be converted to a stress-strain curve by a simple change of scale without change of shape, Figures 1 and 2 can equally well be considered stress-strain curves because no scales are given. The peak in Figure 1 can then be called the yield point and is specified by a yield stress, σ_y, and a yield strain, ϵ_y. For polyethylene, $\sigma_y = 29$ MPa at room temperature. Other polymers have yield stresses as high as 90 MPa[16] while typical metals vary from 30 to 600 MPa[21] and ceramics are usually in the range from 100 to 800 MPa.[5] Thus, there is an overlap in the range of σ_y values for the different types of materials. The yield strain of polyethylene at room temperature is 0.1 and other polymers vary from 0.05 to 0.2, while for ceramics, at about 1500 K, the yield strain is in the range from 0.001 to 0.03.[5]

An alternate definition of yield in terms of true stress and true strain is that the yield point occurs when the slope of the true stress-strain curve equals the true stress. This point can be located using a geometric construction known as Considère's method. A tangent is drawn to the curve from a true strain of -1. The point of tangency is the true yield point. Details of the method are given by Ward.[1] While this procedure is conceptually attractive, it is more difficult to carry out and the results are usually not qualitatively different from the simpler engineering approach.[12] For these reasons, the Considère method is not common and will not be used here.

C. PRE-YIELD BEHAVIOR

The initial, pre-yield, stress-strain behavior of a material is of interest here because it is intimately related to the yield behavior. The initial linear portion of the stress-strain curve is a measure of linear elasticity, where stress and strain are linearly proportional, by Hooke's law, with a constant of proportionality known as the modulus of elasticity. In the case of tensile loading, the modulus of elasticity is the tensile modulus or Young's modulus, Y (the symbol E is also often used),

$$\sigma = Y \epsilon \qquad (6)$$

The stress or strain at which deviation from linearity begins is called the proportional limit. This term is not well defined since it depends on the sensitivity of the test and the degree of deviation considered significant. As mentioned before, deviations from linearity occur as a result of nonlinear elasticity. In the nonlinear region, the modulus is given at any point of the stress-strain curve by

$$Y = d\sigma/d\epsilon \qquad (7)$$

and varies with stress (or strain). Y is of interest for several reasons. It is a basic characteristic of the material and can be calculated from the stress-strain curve used to determine the yield point without further testing. It will be shown that most variables that affect σ_y also affect Y so results for modulus are directly applicable to yield. As an example, if it is found that the Young's modulus of a crystalline polymer varies with the density of the specimen, one would expect the yield stress to be similarly affected. For polyethylene at room temperature, Y = 1.4 GPa and varies from 1 to 3 GPa for other polymers. In ceramics, Y is commonly in the range from 3 to 30 GPa,[3] though diamond is higher. For metals, the range is typically from 30 to 300 GPa,[22] though lead is lower.

Another material characteristic that can be obtained from the stress-strain curve is the yield energy, defined as the work (per unit volume) required to produce yield in the polymer. Yield energy, E_y, is given by the area under the stress-strain curve from zero strain to ϵ_y,

$$E_y = \int_0^{\epsilon_y} \sigma \, d\epsilon \qquad (8)$$

Yield energy has units of joules per cubic centimeter (1 J/cm^3 = 0.239 cal/cm^3). In metals this quantity is called resilience.[15] For polyethylene at room temperature, E_y = 2.5 J/cm^3.

The final important aspect of preyield behavior is time dependence. Polymers are viscoelastic materials so their response to mechanical loading is time dependent. While some materials display instantaneous elastic behavior followed by yield or fracture, polymers typically exhibit delayed elasticity or time dependence. For this reason, the value of the yield stress depends on the rate at which the load is applied to the polymer, as measured by the strain rate, $\dot{\epsilon} = d\epsilon/dt$. The units of $\dot{\epsilon}$ are then reciprocal time, 1/min. Since significant variations in yield stress occur only over several decades of reciprocal time, yield stress is usually plotted as a function of log $\dot{\epsilon}$.

D. POST-YIELD BEHAVIOR

Post-yield behavior will not be considered here in any detail, but some qualitative observations are relevant. Referring to Figure 1, the stress at high strain is starting to decrease. Beyond this point a dramatic and sudden narrowing of the specimen cross-section, called necking, may occur as the stress increases. The neck forms at the weakest point along the narrow section of the test specimen and then propagates, at constant stress, along the length of the specimen, a process known as cold drawing. Once the neck has traveled the entire length of the narrow section, the stress may start to increase again, a phenomenon known as strain hardening. Rupture may occur at a stress higher than σ_y. Since the strength of a polymer is defined as the highest stress it can support, σ_y may or may not be strength of the polymer depending on the post-yield behavior.

The ultimate properties, i.e., those at rupture, are used in defining the toughness of a polymer. This is the total work (per unit volume) required to produce rupture. Toughness is given by the area under the stress-strain curve from zero strain up to rupture,

$$\text{toughness} = \int_0^{\epsilon_r} \sigma \, d\epsilon \tag{9}$$

where ϵ_r is the value of the strain at rupture. Toughness is then similar to yield energy.

The post-yield behavior of some polymers is marked by crazing, the formation of multiple microscopic cracks throughout the specimen. When this occurs in an initially transparent polymer, strong light scattering is observed, a phenomenon called stress whitening. Occasionally, this behavior can even be observed below the yield point.[13]

E. TYPES OF LOADING

The most common type of loading used to test polymers is uniaxial tension, but other types of leading are also used and can add insight to the tensile results. The simplest variation of the tensile test is the uniaxial compression test (not to be confused with hydrostatic compression, in which the load is applied from all sides). It is found that compressive stresses are higher than tensile stresses for a given strain. In particular, at the yield point of polyethylene, the compressive yield stress is about 5% higher than the tensile yield stress,[23,24] while for polycarbonate the difference is about 14%.[24,25] This is a greater effect than seen with metals.[24] The other common type of loading is one that produces yield in shear. The test is usually described as a torsional test. For polyethylene, the shear yield stress, τ_y, is 13 MPa[25] compared with the tensile yield stress of 29 MPa.[6] For polycarbonate at room temperature, $\sigma_y = 83$ MPa while $\tau_y = 48$ MPa.[10] From basic mechanics it is known that a shear deformation occurs as a shape change at constant volume (at least for small strains). In contrast, under hydrostatic pressure a material undergoes a volume change at constant shape. Since a tensile deformation involves both a shape change and a volume change, a tensile stress can be viewed as a combination of a shear component and a hydrostatic component. For this reason, knowledge of shear yield and the effect of pressure on yield are important for an understanding of tensile yield.

In some cases, yield measurements have been made under a superimposed hydrostatic pressure.[1,7,9-11] The practical interest in these measurements is partly because polymers are sometimes processed under pressure and partly because tensile tests involve a pressure component. The tensile yield stress and initial modulus increase under pressure. Some polymers that were brittle can become ductile under pressure and ductile ones can become brittle.

F. EXPERIMENTAL PROCEDURES

The most common experimental procedure used for tensile testing is that described in American Society for Testing and Materials (ASTM) test method ASTM D638 Tensile Properties of Plastics. The specimen has a dumbbell or dogbone shape as shown in Figure 3 for specimens of sufficient thickness (type I). There are specificiations for smaller specimens, but these are not preferred. The ends of the specimen are wide to provide easy gripping, while the central section is narrower so elongation will occur preferentially in that region. A strain gauge extensometer is attached to the specimen in the narrow region, typically with initial length (or gauge length) of 2.54 cm. The ends of the specimen are gripped by

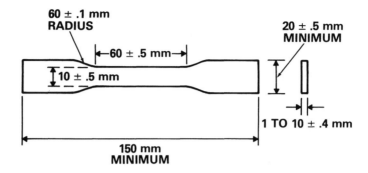

FIGURE 3.   Tension test specimen, ASTM D638, type I.

heavy, self-aligning metal grips and the load is applied by a Universal Test Machine. (A commonly used machine is manufactured by the Instron Corp. and such machines are sometimes simply called an Instron.) The top grip is stationary while the bottom grip, attached to a cross-head, moves down. Cross-head speeds available on the typical hydraulic screw-driven machine run from 0.05 to 50 cm/min.

The output of the strain gauge extensometer is fed to the x-axis of an x-y recorder while the y-axis is the applied load from the load cell. Thus, the output of the measurement is a continuous load-elongation curve at a given cross-head speed and temperature. When measurements as a function of temperature are desired, an environmental chamber is used. Pressure measurements are more involved and details of the equipment are given in the literature.[9] After the test, the load-elongation curve is converted to a stress-strain curve. Note that the cross-head speed is used only to specify the rate at which the test is run, not to determine the strain. Calculating strain from the cross-head speed rather than from an extensometer is not accurate. The test is run until, by eye, the stress goes through a maximum and then is manually stopped. Thus, the test always goes a little past the yield point, making it difficult to make accurate measurements of permanent deformation at yield.

Because of the variations in polymer specimens and test results, a minimum of five replicates should be run under each set of test conditions and the mean and standard deviations reported.

Shapes other than Figure 3 can be used, some of which, such as a hollow tube, are described in ASTM D638 and others, such as a cylinder with threaded ends and a narrower central diameter, are described in the literature.

Compressive yield testing uses a flat-ended cylinder for the test specimen, as described in ASTM D695 Compressive Properties of Rigid Plastics. Recall that the sides of the specimen must be unconstrained as compression refers to uniaxial compression. Torsional testing, to determine shear yielding, is described in the literature.[27]

III. EXPERIMENTAL RESULTS

Experimental yield properties for various common crystalline and amorphous polymers at room temperature and pressure and a strain rate of 2 min^{-1} are given in Table 2, taken from various sources.[6,9,12-15,18,24,28-32] The polymers are all commercial products. The values in Table 2 are in good agreement with other published values when the variability of the results with material properties and test conditions are taken into account. For example, the yield stress in a series of polyethylenes varied from 6 to 30 MPa depending on their density and thermal history.[33] For the same series of polymers, Young's modulus varied from 0.2 to 1.3 GPa and was well correlated with the yield stress: the higher the modulus, the higher

TABLE 2
Typical Tensile Properties at Room Temperature

Polymer	Y (GPa)	σ_y (MPa)	ϵ_y	E_y (J/cm³)	Ref.
Crystalline Polymers					
Polyethylene	2.5	29	0.10	1.4	6
Polypropylene	3.0	38	0.11	1.6	12
Poly(4-methyl pentene-1)	1.0	28	0.047	1.6	13
Nylon 6,6	12.5	76	0.18	3.0	14
Poly(vinylidene fluoride)	4.1	50	0.11	1.4	15
Poly(chlorotrifluoroethylene)	1.9	46	0.069	1.5	15
Polyoxymethylene	3.1	73	0.16	—	28
Polytetrafluoroethylene	—	9	0.04	0.4	9
Amorphous Polymers					
Polycarbonate	2.3	69	0.08	3.9	18, 24
Polyvinylchloride	—	58	0.1	2.8	18, 29
Polyethyleneterephthalate	2.4	52	0.1	—	30, 31
Polystyrene	2.8	Brittle			32
Polymethylmethacrylate	2.8	Brittle			32

the yield stress. This is an example of the observation that any variable that affects modulus will also affect yield. This would include density, degree of crystallinity, and thermal history (by changing the morphology).

To examine these polymers in more detail, their behavior as functions of test conditions are considered below. The results will be presented in order of the test variable considered: temperature, strain rate, pressure, and environment.

A. TEMPERATURE DEPENDENCE

As a function of test temperature, stress-strain curves for polyethylene[6] change progressively as shown in Figure 4. As the temperature increases, the yield stress, Young's modulus, and yield energy all decrease, while the yield strain increases. This qualitative behavior is observed over the entire temperature range from the brittle-ductile transition temperature to the melting temperature. Tabular results for polyethylene are given in Table 3.

A plot of yield stress as a function of temperature for polyethylene is shown in Figure 5. Over most of the temperature range shown, the decrease is linear, and σ_y extrapolates to zero near the melting temperature. Some polymers exhibit more curvature at higher temperature but are still roughly linear near room temperature. The rate of change of yield stress with temperature near room temperature for various polymers is listed in Table 4, taken from various sources.[6,11-15,18,25,28,34] For comparison, the temperature dependence of the shear yield stress of polyethylene is -0.13 MPa/K,[27] about one third of the tensile value. The tensile values are not much different from those for amorphous polymers such as polycarbonate.[25]

A plot of yield energy vs. temperature for polyethylene is shown in Figure 6. A linear decrease with temperature is observed, extrapolating to zero near the melting temperature. Slopes of yield energy vs. temperature near room temperature for various polymers are given in Table 4. Recalling that yield energy is the area under the stress-strain curve, the decreases

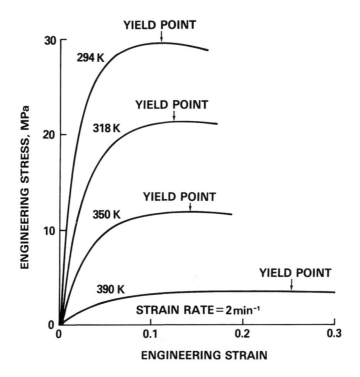

FIGURE 4. Stress-strain curves for polyethylene.

TABLE 3
Tensile Properties for Polyethylene

Temperature (K)	Yield stress (MPa)	Yield strain	Young's modulus (GPa)	Yield energy (J/cm³)
294	29.3	0.102	1.39	2.47
302	27.6	0.101	1.18	2.31
310	24.7	0.113	1.07	2.29
318	21.2	0.132	0.70	2.31
326	19.0	0.135	0.57	2.09
334	16.3	0.145	0.45	1.91
342	13.7	0.146	0.39	1.61
350	11.7	0.150	0.30	1.42
358	9.6	0.158	0.24	1.22
366	7.7	0.165	0.17	1.02
374	6.1	0.176	0.13	0.87
390	3.7	0.260	0.06	0.63

Note: strain rate $= 2$ min^{-1}.

From Hartmann, B., Lee, G. F., and Cole, R. F., Jr., *Polym. Eng. Sci.,* 26, 554, 1986. With permission.

of E_y as the temperature is raised shows that the decrease in yield stress dominates the increase in yield strain.

These results for polyethylene are qualitatively similar to most other polymers, but nylon 6,6 [poly(hexamethylene adipamide)], shows unusual behavior.[14] The yield stress vs.

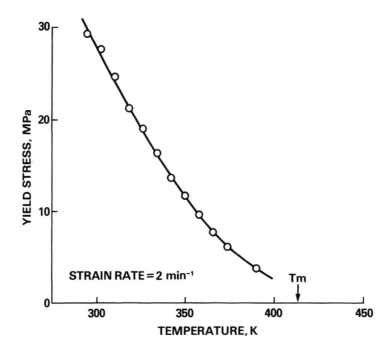

FIGURE 5.   Yield stress vs. temperature for polyethylene.

TABLE 4
Typical Temperature and Strain Rate Dependence

Polymer	$-d\sigma_y/dT$ (MPa/K)	$-dE_y/dT$ (J/cm³K)[a]	$d\sigma_y/\log \epsilon$ (MPa/dec)[b]	Ref.
Crystalline Polymers				
Polyethylene	0.30	0.021	4.1	6
Polypropylene	0.37	0.025	4.3	12
Poly(4-methyl pentene-1)	0.29	0.005	3.6	13
Nylon 6,6	0.54	0.016	2.5	14
Poly(vinylidene fluoride)	0.54	0.031	4.5	15
Poly(chlorotrifluoroethylene)	0.42	—	3.9	15
Polyoxymethylene	0.47	—	—	28
Amorphous Polymers				
Polycarbonate	0.32	0.033	2.9	18, 25
Polyvinylchloride	0.59	0.052	4.9	18, 34
Polyethyleneterephthalate	0.60	0.045	6.8	11, 18

[a]   At a strain rate of 2 min⁻¹.
[b]   At room temperature.

temperature, Figure 7, shows the typical decrease with temperature, but the yield strain vs. temperature, Figure 8, has a change in slope at 433 K. The yield energy vs. temperature, Figure 9, also shows a change at 433 K, contrary to the results for other polymers. Since yield energy is the area under the stress-strain curve, the change in slope of the yield energy

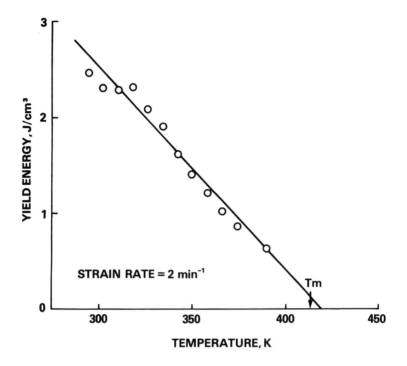

FIGURE 6.   Yield energy vs. temperature for polyethylene.

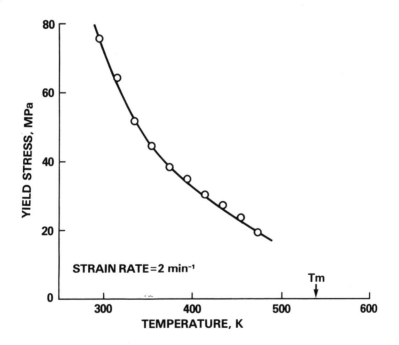

FIGURE 7.   Yield stress vs. temperature for nylon 6,6.

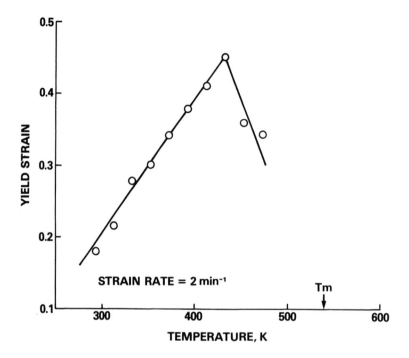

FIGURE 8.   Yield strain vs. temperature for nylon 6,6.

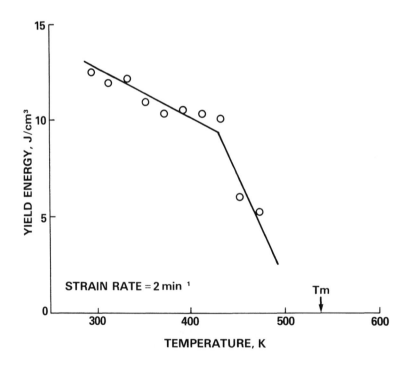

FIGURE 9.   Yield energy vs. temperature for nylon 6,6.

TABLE 5
Strain Rate Dependence of Tensile Properties for Polyethylene

Strain rate (min^{-1})	Yield stress (MPa)	Yield strain	Young's modulus (GPa)	Yield energy (J/cm^3)
		Temperature = 294 K		
0.02	20.5	0.125	0.88	2.12
0.08	23.6	0.112	0.93	2.14
0.2	25.0	0.105	1.09	2.18
0.8	28.3	0.098	1.30	2.20
2.0	29.3	0.102	1.39	2.47
8.0	30.8	0.105	1.51	3.28
		Temperature = 318 K		
0.02	13.4	0.137	0.38	1.48
0.08	15.2	0.133	0.50	1.62
0.2	17.0	0.131	0.61	1.82
0.8	20.3	0.126	0.72	2.14
2.0	21.2	0.132	0.70	2.29
8.0	24.0	0.122	0.77	2.39
		Temperature = 342 K		
0.02	7.9	0.142	0.19	0.89
0.08	9.2	0.142	0.24	1.05
0.2	10.1	0.143	0.26	1.17
0.8	11.7	0.149	0.31	1.43
2.0	13.7	0.146	0.39	1.61
8.0	15.7	0.176	0.40	2.27

From Hartmann, B., Lee, G. F., and Cole, R. F., Jr., *Polym. Eng. Sci.*, 26, 554, 1986. With permission.

is a direct result of the yield strain behavior. In particular, the yield stress shows no special behavior at 433 K. It was determined that the change at 433 K is caused by a reversible crystal-crystal transition from a triclinic lattice at room temperature to a pseudo-hexagonal lattice at 433 K. The transition, called the Brill transition, occurs because the **b** lattice constant increases with temperature until the **b** constant equals the **a** constant. In some other polyamides, the **b** constant also increases with temperature, but melting occurs before **b** equals **a**.

B. STRAIN RATE DEPENDENCE

As a function of strain rate, yield stress, Young's modulus, and yield energy all increase while yield strain is almost unchanged. This behavior is observed over the entire range of strain rates available. Results for polyethylene are given in Table 5.

A plot of the yield stress as a function of strain rate for polyethylene is shown in Figure 10. Within experimental uncertainty, yield stress is a linear function of log strain rate, though curvature in such plots is sometimes reported[11] over wide ranges of strain rate. All polymers have qualitatively similar behavior and the rate of change of yield stress with respect to log strain rate is listed in Table 4. This behavior is observed for both crystalline and amorphous polymers. The differences is quantitative only; the values of the slope are lower for crystalline

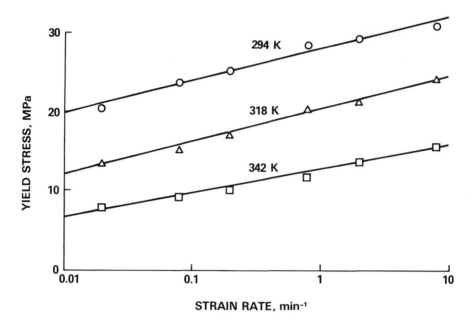

FIGURE 10.   Yield stress vs. strain rate for polyethylene.

polymers.[35] An even greater slope is observed for the strain rate dependence of yield stress in metals[2,21] and ceramics.[5]

Note that the qualitative strain rate dependence of yield stress, Young's modulus, and yield energy are all opposite to the temperature dependence of these quantities. This can be explained by the viscoelastic nature of polymers and suggests the possibility that time-temperature superposition might be applicable to yield[36] and that yield may be related to creep and other viscoelastic properties. Keep in mind, however, that yield is a nonlinear, large strain property while time-temperature superposition is generally considered in the linear, small strain region.

C. PRESSURE DEPENDENCE

As pointed out earlier, the pressure dependence of tensile yield is an important topic because tensile stress has a pressure component and understanding pressure effects is helpful in interpreting tensile yield results. In addition, various processing and forming operations are done under pressure and properties can change significantly under these conditions. For example, some materials that are brittle under atmospheric pressure become ductile under elevated pressure.

As hydrostatic pressure increases, the tensile yield stress and Young's modulus both increase. Typical results for polyethylene yield stress, at a strain rate of 0.2 min^{-1}, are shown in Figure 11.[37] A linear pressure dependence is observed. Similarly, the Young's modulus increases with pressure as shown in Figure 12.[37] A summary of pressure-dependence results for various crystalline polymers is given in Table 6. The values were taken from a variety of sources,[9,10,16,24,37,38] tested at differing strain rates, on materials of different density and thermal history. Some of the yield results are for offset yield and some are not. For these reasons, the zero pressure values are not the same as those in Table 2.

Polymers have a high pressure dependence compared with metals because polymers yield at a stress which is very high in relation to their modulus. The ratio of yield stress to bulk modulus is 0.001 to 0.01 for polycrystalline metals while it is 0.02 or higher for

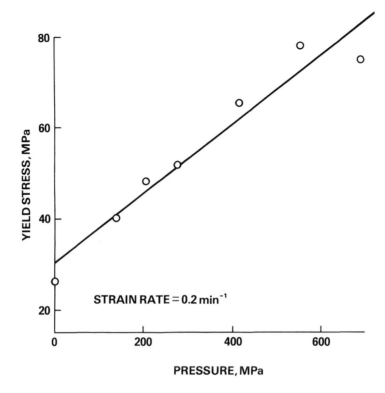

FIGURE 11.   Yield stress vs. pressure for polyethylene.

polymers.[32] Yield in metals is often said to be independent of hydrostatic pressure, but recent measurements[24] have shown that this is not the case. The effect is relatively small only because the pressures used are small compared to the modulus.

D. ENVIRONMENT

Gaseous environments can affect the tensile behavior of polymers, usually by causing them to craze at low temperature.[39] Helium is inert, but other gases may produce crazing depending on the test temperature and the partial pressure of the gas. This effect has been observed with nitrogen, argon, oxygen, carbon dioxide, and water vapor. Since gaseous environments can have an effect on tensile properties, it is possible that a liquid environment could also. For this reason, one must be careful in making pressure measurements since the pressure transmitting fluid could interact with the specimen.

IV. THEORETICAL INTERPRETATION

Polymer yield is often interpreted using phenomenological theories, frequently adapted from metals theory, while other theories are very specific to polymers. A satisfactory theory must explain the difference between yield in tension and in compression, the existence of shear yield, the temperature, strain rate, and pressure dependence and their interrelations. This is an area of active development and only an outline will be given here.

A. RATE THEORY

The most widely used theory for yield is based on the activated state rate theory of Eyring for viscosity.[1] This theory is used for crystalline and amorphous polymers as well

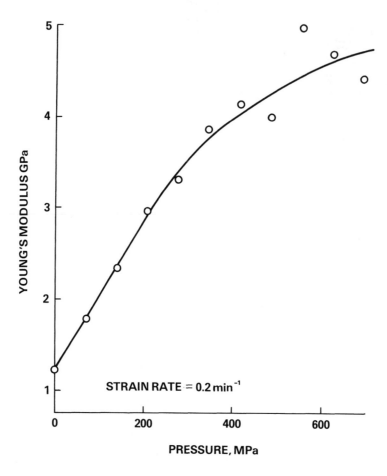

FIGURE 12. Young's modulus vs. pressure for polyethylene.

TABLE 6
Typical Pressure Dependence Parameters

Polymer	σ_y (MPa)	$d\sigma_y/dP$	Y (GPa)	dY/dP	Ref.
Crystalline Polymers					
Polyethylene	26.1	0.094	1.24	8.7	37
Polypropylene	28.6	0.205	1.52	19.0	37
Nylon 6,6	31.7	0.12	0.50	1.4	38
Poly(chlorotrifluoroethylene)	37.9	0.23	1.16	3.3	10
Polyoxymethylene	86.2	0.26	2.69	7.3	16
Polytetrafluoroethylene	9.0	0.08	0.41	3.3	9
Amorphous Polymers					
Polycarbonate	64	0.13	2.35	3	24

Note: Evaluated at room temperature.

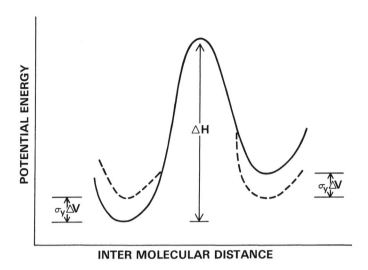

FIGURE 13.   Eyring model for yield.

as metals[39] and ceramics.[5] Rate theory assumes that, without any applied stress, the polymer segments are in potential wells formed by inter- and intramolecular steric hindrances and are separated from one position of stable equilibrium to another by an activation energy (strictly speaking, an enthalpy), ΔH, as shown in Figure 13. Under an applied stress, σ, one equilibrium potential is raised by an amount $\sigma \Delta V$ and the other lowered by a like amount, as shown in Figure 13, where ΔV is called the activation volume. Thus, the stress biases the activation energy, allowing segmental motion in the direction of the stress. The yield stress, σ_y, is the value at which the internal viscosity falls to a value such that the applied strain rate is the same as the plastic strain rate predicted by the Eyring equation,

$$\dot{\epsilon} = \dot{\epsilon}_0 \exp[(\Delta H - \sigma_y \Delta V)/RT] \tag{10}$$

where $\dot{\epsilon}_0$ is a constant (that includes the concentration of segments, entropy effects, etc.). Solving for σ_y,

$$\sigma_y = \Delta H/\Delta V + (RT/\Delta V) \ln(\dot{\epsilon}/\dot{\epsilon}_0) \tag{11}$$

so σ_y is a linear function of log strain rate, at constant temperature. Behavior of this type is often observed and this fact is usually taken as the justification for the use of Eyring Theory. Differentiating Equation 11 gives

$$(\partial \sigma_y / \partial \ln \dot{\epsilon})_T = RT/\Delta V \tag{12}$$

and activation volume can be determined from a plot of σ_y vs. log strain rate. Values of ΔV at room temperature for various crystalline polymers are given in Table 7, taken from various sources.[6,12-15] These values are a factor of two or three times smaller than for amorphous polymers,[1] which vary from 5 to 20 nm^3. For comparison, the volumes of one polymer segment, v, are also listed, where

$$v = M/\rho N_A \tag{13}$$

TABLE 7
Typical Activation Parameters

Polymer	ΔV^a (nm³)	v (nm³)	$\Delta V/v$ (kJ/mol) min⁻¹	ΔH (kJ/mol) min⁻¹	ϵ_0 (kJ/mol) min⁻¹	Ref.
Polyethylene	2.3	0.048	48	84	4×10^{12}	6
Polypropylene	2.1	0.077	27	71	5×10^{4}	18
Poly(4-methyl pentene-1)	2.6	0.17	15	79	5×10^{6}	10
Nylon 6,6	3.8	0.33	11	—	—	19
Poly(vinylidene fluoride)	2.1	0.060	35	87	1×10^{5}	20
Poly(chlorotrifluoroethylene)	2.4	0.090	27	163	5×10^{17}	21
Polyoxymethylene						

[a] At room temperature.

and M is the molecular weight, ρ is the density, and N_A is Avogadro's number. The ratio of these two numbers is also given in Table 7 and shows that yield involves the cooperative movement of many chain segments (10 to 50).

From a series of plots of σ_y vs. log strain rate at different temperatures, as shown in Figure 10, the value of ΔV at different temperatures can be obtained. ΔV increases as the temperature increases. In polyethylene, for example, ΔV is 2.3 nm³ at 294 K, 2.4 nm³ at 318 K, and 3.6 nm³ at 342 K. Thus, the correlated length of chain movement increases as the temperature increases. Similar behavior is observed with glassy polymers but not with metals.[40]

The interpretation of yield as resulting from large-scale motion is borne out by the size of the activation energy, which is intermediate between that for the glass transition and for a secondary transition.[41] Values are listed in Table 7. These values were calculated from Equation 11 using the temperature-dependent ΔV and assuming that ΔH and $\dot{\epsilon}_0$ are constant. For each pair of temperatures, values of ΔH and $\dot{\epsilon}_0$ were determined. The activation energy from pairs of temperatures tends to decrease as the temperature increases, even becoming negative at high-enough temperature, as was found[14] for nylon 6,6, indicating a limitation of this approach. Average values of activation energy are given in Table 7.

In some cases where a plot of σ_y/T vs. log strain rate over a wide range of values is not linear but shows some curvatuve, the above theory has been extended to include two Eyring processes acting in parallel. The yield stress then consists of two additive terms of the form of Equation 11. The processes have been interpreted as molecular transitions in the polymer.[25,27] Thus, there is a relation between yield and other measurements, such as dynamic mechanical and dielectric, that are used to study transitions.

The above discussion has been for tensile yield at atmospheric pressure. The theory has been extended to include the effect of hydrostatic pressure in the following manner:[1]

$$\dot{\epsilon} = \dot{\epsilon}_0 \frac{-(\Delta H - \tau_y \Delta V + P\Delta\Omega)}{RT} \tag{14}$$

where τ_y is the shear yield stress, P is hydrostatic pressure, ΔV is the shear activation volume, and $\Delta\Omega$ is the pressure (or bulk) activation volume. This analysis fits the data reasonably well.

B. INTERNAL ENERGY THEORY

Assume that a certain amount of energy is required to thermally induce large-scale cooperative segmental motion over inter- and intramolecular restrictions. Yield energy should then be related to the internal energy change in the polymer. This theory was originally developed for glassy polymers[18,42,43] and later extended to crystalline polymers.[13] The basis of the theory is that yield energy is related to internal energy, U, by the relation

$$E_y = \frac{b}{b'} \int_T^{T_g} \left(\frac{\partial U}{\partial T}\right)_P dT \tag{15}$$

where b is the fraction of thermal energy available to overcome the activation energy barrier for flow and b' is the fraction of mechanical energy available to overcome the same barrier. The result can be approximated by

$$E_y = \frac{b}{b'} \bar{\rho} \, \bar{C}_P (T_g - T) \tag{16}$$

where $\bar{\rho}$ and $\bar{C}_P$ are the average values of density and heat capacity. For crystalline polymers, Equation 16 becomes[18]

$$E_y = \frac{b}{b'} \bar{\rho} \, \bar{C}_P (T_m - T) \tag{17}$$

As shown earlier, experimentally E_y is generally found to be a linear function of temperature, extrapolating to zero near the melting temperature. For polyethylene, $b/b' = 0.012$, while for polypropylene $b/b' = 0.011$, and for poly(4-methyl pentene-1) $b/b' = 0.003$. For nylon 6,6 $b/b' = 0.007$ up to 433 K and $b/b' = 0.055$ above 433 K. For PCTFE, $b/b' = 0.007$ and for PVF2, $b/b' = 0.015$.

Because directed mechanical energy is more efficient than random thermal motion in causing flow, the value of b/b' is less than 1. For glassy polymers, b/b' varies between 0.02 and 0.03,[18] which is larger than for crystalline polymers, showing that for crystalline polymers thermal energy is even less effective compared to mechanical energy in surmounting the barrier. An exception is nylon 6,6 above 433 K. In this case, b/b' is greater than for amorphous polymers and shows a significant increase in the relative effectiveness of thermal energy to produce flow.

C. VOLUME-DEPENDENT THEORIES

It has been suggested[44] that yield stress depends on polymer volume and not on temperature or pressure except as they affect the volume. This idea was based on two observations: Young's modulus is a function of volume and yield stress is a function of Young's modulus. It was shown experimentally that the Young's modulus of polymers is a function of volume, and similar results were presented for the closely related case of bulk modulus.[45] The volume dependence of the bulk modulus of crystalline polymers also has a theoretical basis and has been derived directly from the intermolecular potential.[46] Thus, Young's modulus depends on temperature and pressure only because the volume depends on these variables.

The relation between yield stress and Young's modulus was mentioned earlier with regard to density and thermal treatment differences between different specimens of the same material.[33] Also, for a given specimen, both yield stress and Young's modulus decrease as temperature is raised and both increase as pressure is raised. These observations suggest that yield stress and Young's modulus are related. Since Young's modulus is a function of volume, it follows that yield stress is a function of volume. The temperature and pressure dependence of the yield stress is then a result of the temperature and pressure dependence

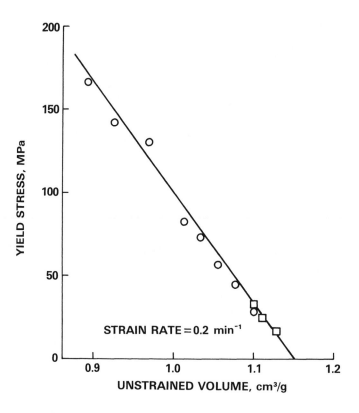

FIGURE 14.   Yield stress vs. unstrained volume for polypropylene.

of the volume. The temperature and pressure derivatives of the yield stress are then not independent but are related by the expression[12]

$$\left(\frac{\partial \sigma_y}{\partial P}\right)_T = \left(\frac{\beta}{\alpha}\right) \left(\frac{\partial \sigma_y}{\partial T}\right)_P \tag{18}$$

where α is the thermal expansion coefficient and β is the isothermal compressibility (reciprocal of the isothermal bulk modulus). The validity of Equation 18 has been tested for polypropylene since all of the input data was not available for the same density polyethylene. From Table 4, the temperature derivative of the yield stress is -0.37 MPa/K. The compressibility[47] is 2.87×10^{-4} MPa^{-1} and the thermal expansion coefficient[48] is 5×10^{-4} K^{-1} so from Equation 18 the predicted pressure derivative of the yield stress is 0.21, in good agreement with the value of 0.20 found experimentally.[37] Since yield stress is a function of volume, it should be plotted vs. volume and not temperature or pressure. Using the above values for α and β to calculate volume, the temperature dependence[12] and the pressure dependence[37] of yield stress are well correlated, as shown in Figure 14. The practical significance of this result is that reliable estimates of the pressure dependence of yield stress can be made without having to make these difficult measurements. While the assumption that yield stress is a function of volume works well for polypropylene and also is reported to be applicable to amorphous polymers such as polymethylmethacrylate and polyvinylchloride,[44] results for polycarbonate do not show correlation between temperature and pressure effects.[31,49] At any specified volume change, changes in temperature affect the yield stress in polycarbonate more than do changes in pressure. It is still possible to get correlation, but a shifting process must be used. It has been suggested[50] that better agreement for polycarbonate would result if the plot were made against the change in fractional free volume rather than against the relative change in total volume. The compression of free volume under pressure is a relatively small proportion of the total compression, while the contraction of free volume with decreasing temperature is a relatively large proportion of the total contraction. The free volume changes can be estimated by taking the temperature and pressure coefficients as the differences between the macroscopic coefficients above and below T_g. Phenomena involving molecular mobility are related to free volume, not to total volume. If yield falls in this category, free volume would be the preferred variable.

It was shown above that yield stress and Young's modulus are related, but the form of the relation was not specified. The simplest assumption is that they are directly proportional and that their ratio is a constant. It has been found[30] that the ratio of yield stress to Young's modulus is about the same for many amorphous polymers, 0.025 ± 0.010. The values in Table 2 also fall in this range. Thus, as a rough estimate, yield stress is about 1/40 of Young's modulus. This conclusion was reached through an empirical comparison of various polymers at room temperature. A theoretical justification for this result can be made directly from the intermolecular potential.[51] This theory also correctly predicts strain softening below the yield point, specifically that the stress-strain plot has a maximum. The nonlinear stress strain behavior may be explained by the nonlinear force fields between molecules. This implies that yield is not simply an accelerated form of the creep at very low strains, as sometimes suggested.

It follows from the relation between yield stress and Young's modulus that the yield stress for a given polymer at various temperatures, strain rates, and pressures is a result of the temperature, strain rate, and pressure dependence of Young's modulus. This simple assumption is only partly true. The temperature dependence of the ratio of yield stress to Young's modulus for polyethylene is shown in Figure 15, using data from Table 3. The ratio increases with temperature, but changes less than either of the individual variables. The strain rate dependence of the ratio (taking data from Table 6) shows considerable scatter but there is a small decrease with increasing strain rate. Similarly, the pressure dependence of the ratio[37] also shows a small decrease with increasing pressure. In summary, some but not all of the temperature, strain rate, and pressure dependence of yield stress can be explained by variations in Young's modulus, and this relationship should be incorporated into the theory of yielding. Similar conclusions have been reached concerning shear yield, where the governing modulus is the shear modulus.[52,53]

D. ROBERTSON THEORY

Robertson[54] pointed out that free-volume approaches to yield are principally concerned with intermolecular forces and neglect the equally important intramolecular forces. He assumes that, in an amorphous polymer, free rotation can take place about single bonds in the backbone. It is further assumed there are only two rotational conformations, the trans low-energy state and the cis high-energy state, called the flexed state. Stress causes the fraction of flexed bonds to increase from that existing in the glass and leads to yield.

For some orientations of the bonds with respect to the applied stress, the application of a stress will increase the fraction of flexed bonds while for other orientations the fraction will decrease. In the first case, this corresponds to an increase in temperature and in the second case to a decrease. The crux of the theory is the assumption that the rate at which the polymer approaches equilibrium is dependent on temperature, following the WLF equation, so that the rate for the elements that flex under stress is much faster than for the others. This approach was later modified to include the effect of pressure.[11]

This theory has the advantage of providing an explanation for the temperature, pressure, and strain rate dependence of yield in amorphous polymers. While the theory has not been

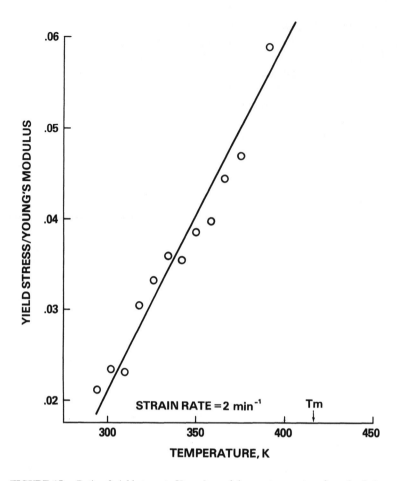

FIGURE 15.   Ratio of yield stress to Young's modulus vs. temperature for polyethylene.

applied to crystalline polymers as yet, one would assume that a similar approach would be applicable to that case as well.

E. MATSUOKA THEORY

A different fundamental approach to yield is through the distribution of relaxation times associated with the process. Matsuoka[55] represented the nonlinear viscoelastic properties of polymers by a distribution of relaxation times described by the Williams-Watts stretched exponential function. This approach leads to a simple equation for yield. Based on the form of this equation, Matsuoka derives scaling rules to calculate strain rate dependence from the properties at a reference strain rate and to calculate temperature dependence from the properties at a reference temperature.

Scaling rules for crystalline polymers differ in only one respect from those for glassy polymers. As noted above experimentally, yield strain vs. temperature has the opposite sign for crystalline and amorphous polymers and this fact is accounted for in Matsuoka theory.

Fitting parameters for this approach for several crystalline and amorphous polymers are available,[55] as well as general rules for fitting data for new polymers.

V. SUMMARY

The tensile yield properties of polymers depend on polymer characteristics such as density, degree of crystallinity, and thermal history as well as test conditions such as temperature, strain rate, and pressure. The behavior of crystalline polymers does not differ qualitatively from that of amorphous polymers (except yield strain vs. temperature) or even from metals and ceramics when temperature and pressure are scaled to melting temperature and modulus, respectively.

Because of the general nature of yield, one would expect that the theoretical explanation would not require a specific detailed mechanism applicable to polymers only. In fact, many of the polymer theories are adapted from theories for metals. While existing theories are more or less successful in predicting some aspects of yield, none of them is applicable over all conditions.

REFERENCES

GENERAL REFERENCE

This chapter is expanded and updated from B. Hartmann, in *Handbook of Polymer Science and Technology,* Vol. 2, Cheremisinoff, N. P., Ed., Marcel Dekker, New York, 1989, ch. 3. That chapter is a general reference for this chapter.

SPECIFIC REFERENCES

1. **Ward, I. M.,** *Mechanical Properties of Solid Polymers,* 2nd ed., John Wiley & Sons, New York, 1983.
2. **LeMay, I.,** *Principles of Mechanical Metallurgy,* Elsevier, New York, 1981.
3. **Kingery, W. D.,** *Introduction to Ceramics,* John Wiley & Sons, New York, 1960.
4. **Brown, N. and Ward, I. M.,** *J. Mater. Sci.,* 18, 1405, 1983.
5. **Tressler, R. E. and Bradt, R. C., Eds.,** *Deformation of Ceramic Materials,* Vol. 2 (Materials Science Research, Vol. 18), Plenum Press, New York, 1984.
6. **Hartmann, B., Lee, G. F., and Cole, R. F., Jr.,** *Polym. Eng. Sci.,* 26, 554, 1986.
7. **Matsushige, K., Radcliffe, S. V., and Baer, E.,** *J. Appl. Polym. Sci.,* 20, 1853, 1976.
8. **Jang, B. Z., Uhlman, D. R., and VanderSande, J. B.,** *J. Appl. Polym. Sci.,* 29, 3409, 1984.
9. **Sauer, J. A., Mears, D. R., and Pae, K. D.,** *Eur. Polym. J.,* 6, 1015, 1970.
10. **Sauer, J. A.,** *Polym. Eng. Sci.,* 17, 150, 1977.
11. **Duckett, R. A., Rabinowitz, S., and Ward, I. M.,** *J. Mater. Sci.,* 5, 909, 1970.
12. **Hartmann, B., Lee, G., and Wong, W.,** *Polym. Eng. Sci.,* 27, 823, 1987.
13. **Hartmann, B. and Cole, R. F., Jr.,** *Polym. Eng. Sci.,* 23, 13, 1983.
14. **Hartmann, B. and Cole, R. F., Jr.,** *Polym. Eng. Sci.,* 25, 65, 1985.
15. **Hartmann, B. and Lee, G. F.,** *Polym. Eng. Sci.,* 31, 231, 1991.
16. **Sardar, D., Radcliffe, S. V., and Baer, E.,** *Polym. Eng. Sci.,* 8, 290, 1968.
17. **Zoller, P.,** *J. Polym. Sci.,* 22, 633, 1978.
18. **Macosko, C. W. and Brand, G. J.,** *Polym. Eng. Sci.,* 12, 444, 1972.
19. **Chang, F.-C., Wu, J.-S., and Chu, L.-H.,** *J. Appl. Polym. Sci.,* 44, 491, 1992.
20. **Van Krevelen, D. W.,** *Properties of Polymers,* 3rd ed., Elsevier, Amsterdam, 1990.
21. **Dieter, G. E., Jr.,** *Mechanical Metallurgy,* McGraw-Hill, New York, 1961.
22. **Brostow, W.,** *Science of Materials,* Krieger, Malabar, FL, 1985.
23. **Ibrahim, N., Shinozaki, D. M., and Sargent, C. M.,** *Mater. Sci. Eng.,* 30, 175, 1977.

24. **Spitzig, W. A. and Richmond, O.,** *Polym. Eng. Sci.,* 19, 1129, 1979.
25. **Bauwens-Crowet, C., Bauwens, J-C., and Homes, G.,** *J. Mater. Sci.,* 7, 176, 1972.
26. **Rabinowitz, S., Ward, I. M., and Parry, J. S. C.,** *J. Mater. Sci.,* 5, 29, 1970.
27. **Truss, R. W., Clarke, P. L., Duckett, R. A., and Ward, I. M.,** *J. Polym. Sci. Phys. Ed.,* 22, 191, 1984.
28. **Hartmann, B. and Sudduth, D.,** unpublished data.
29. **Vincent, P. I.,** *Polymer,* 1, 7, 1959.
30. **Brown, N.,** *Mater. Sci. Eng.,* 8, 69, 1971.
31. **Christiansen, A. W., Baer, E., and Radcliffe, S. V.,** *Phil. Mag.,* 24, 451, 1971.
32. **Whitney, W. and Andrews, R. D.,** *J. Polym. Sci. C,* 16, 2981, 1967.
33. **Phillips, P. J. and Patel, J.,** *Polym. Eng. Sci.,* 18, 943, 1978.
34. **Bauwens-Crowet, C., Bauwens, J. C., and Homès, G.,** *J. Polym. Sci. A-2,* 7, 735, 1969.
35. **Lazurkin, J. S.,** *J. Polym. Sci.,* 30, 595, 1958.
36. **Wissbrun, K. F .,** *J. Polym. Sci.,* 23, 216, 1983.
37. **Mears, D. R., Pae, K. D., and Sauer, J. A.,** *J. Appl. Phys.,* 40, 4229, 1969.
38. **Pugh, H. Ll. D., Chandler, E. F., Holliday, L., and Mann, J.,** *Polym. Eng. Sci.,* 11, 463, 1971.
39. **Imai, Y. and Brown, N.,** *Polymer,* 18, 298, 1977.
40. **Holt, D. L.,** *J. Appl. Polym. Sci.,* 12, 1653, 1968.
41. **Boyer, R. F.,** *Rub. Chem. Tech.,* 36, 1303, 1963.
42. **Starita, J. M. and Keaton, M.,** *SPE ANTEC,* 29, 67, 1971.
43. **Beatty, C. L. and Weaver, J. L.,** *Polym. Eng. Sci.,* 18, 1109, 1978.
44. **Ainbinder, S. B.,** *Mekh. Polimerov,* 3, 449, 1969.
45. **Ainbinder, S. B.,** *Mekh. Polimerov,* 4, 986, 1968.
46. **Broadhurst, M. G. and Mopsik, F. I.,** *J. Chem. Phys.,* 52, 3634, 1970.
47. **Warfield, R. W.,** *J. Appl. Chem.,* 17, 263, 1967.
48. **Foster, G. N., III, Waldman, N., and Griskey, R. G.,** *Polym. Eng. Sci.,* 6, 131, 1966.
49. **Radcliffe, S. V.,** in *Deformation and Fracture of High Polymers,* Kausch, H. H., Hassell, J. A., and Jaffee, R. I., Eds., Plenum Press, New York, 1973, 191.
50. **Ferry, J. D.,** in *Deformation and Fracture of High Polymers,* Kausch, H. H., Hassell, J. A., and Jaffee, R. I., Eds., Plenum Press, New York, 1973, 233.
51. **Struik, L. C. E.,** *J. Non-Cryst. Solids,* 131–133, 395, 1991.
52. **Argon, A. S.,** *Phil. Mag.,* 28, 839, 1973.
53. **Bowden, P. B. and Raha, S.,** *Phil. Mag.,* 29, 149, 1974.
54. **Robertson, R. E.,** *J. Chem. Phys.,* 44, 3950, 1966.
55. **Matsuoka, S.,** in *Failure of Plastics,* Brostow, W. and Corneliussen, R. D., Eds., Hanser, New York, 1986, 24.

Chapter 5

MECHANOCHEMISTRY OF POLYMERS DEFORMATION AND FRACTURE PROCESSES

Cleopatra Vasiliu Oprea and Marcel Popa

TABLE OF CONTENTS

I. INTRODUCTION

Mechanochemistry, the transitional science developed at the boundary between mechanics and chemistry, has as the object of study the totality of transformations activated within the structure of a body which has absorbed elastic energy, transformations which are based on the reciprocal conversion of mechanical and chemical energies. One has to underline the interaction of mechanochemistry with other domains such as physics, biology, biochemistry, and molecular biology, bionics, physicochemistry, and polymer technology, rheology, etc.

The diversity of ways in which mechanical energy can be applied, the variety of concrete working conditions (involving interactions with environment or other energy forms), as well as the types of materials investigated determine a multitude of effects which raise the question of whether they belong in their totality to polymer mechanochemistry. According to the current stage of evolution of this science, one may assert that from its domain only the processes of material destruction under the influence of mechanical forces, when no chemical bond splits, or those due to nonmechanical forces (thermal, light energy, etc.) will have to be excluded. Any other transformation taking place under the action of mechanical energy, even in the conditions of interactions with other factors, is of a mechanochemical nature.

The mechanochemical process represents a sequence of physical and chemical phenomena found in the mechanochemical interaction in changing the structure, properties, and function of the material sublayer, in which these develop. In the case of polymers, this process influences all the levels of structural organization and is the more complex according to the intricacy of the structure and its relation with the environment.

The chemical transformations undergone by polymer materials are influenced by the field of forces acting on them; in their turn, the chemical reactions initiated cause a spatial display of the structural elements, beginning with the atomomolecular level and continuing with the supermolecular and morphological ones.

Consequently, the mechanochemical process cannot be reduced only to its mechanochemical feature, defining the conversion of mechanical energy into chemical energy, but it also includes, necessarily and inseparably, the chemomechanical one, consisting of the release of mechanical energy as a result of chemical reactions.

The place of these structural-energetic conversions is given by the structure of these bodies, specifically of the polymers with all organizing levels that entail a wide diversity of motion forms, from the simplest one — the physical motion — to the most complex one, an attribute of the higher-organized matter-biological motion.

If the mechanochemical feature approaching a wide range of aspects (referring to the polymers fracture as well as to the chemical changes induced within these by the mechanical effort) is well represented in numerous studies,[1-12] the chemomechanical one was proved by Kuhn[13] and Katschalsky[14] in their inaugural experiments carried out on a series of artificial systems and which were afterward taken over by other scientists.[15-26] The importance of these experiments derives, on the one hand, from their technical potential, demonstrating the feasibility of producing mechanical work by the conversion of chemical energy and, on the other hand, from the fact that it models the importance of biological phenomena, as well as the molecular contraction, whose mechanochemical mechanism has been proved.[27-34]

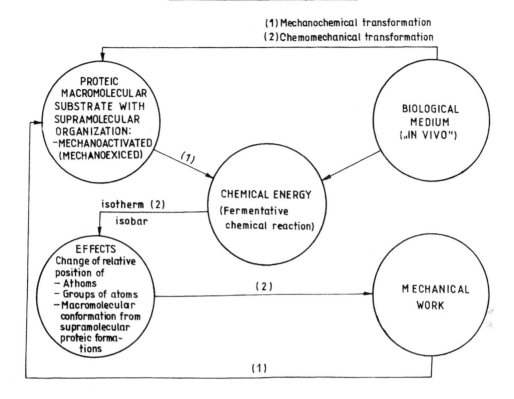

FIGURE 1. Schematic representation of the biomechanochemical process.

The most complex process generating mechanical work is the biochemomechanical one, which has a cycling character. In the case of molecular contraction, the biochemomechanical process develops at the level of supermolecular organized proteic macromolecules, mechanoexcited by a mechanical work produced during a previous cycle. In this state they become the place of some chemical reactions which develop isobarically and isothermally in their interaction with the biological environment, achieving thus the mechanochemical side of the cycle. The immediate consequence is the change of relative position of atoms, groups of atoms, and macromolecule conformation. These modifications at the atomomolecular level are translated in compression and expansion effects at supermolecular formation levels, resulting in mechanical work and the achievement of chemomechanical feature of the cycle.[34]

Schematically, the biomechanochemical process is represented in Figure 1.

II. POLYMER DEFORMATION AND FRACTURE — MECHANOCHEMICAL PROCESSES

The behavior of polymers as finite objects at various mechanical stresses is of great theoretical and practical interest, due to the widespread use of these materials in all technical branches. This promotes a thorough knowledge of the way in which the macromolecular compounds respond to mechanical stresses in the conditions of their service life, and of the factors determining this behavior, either with the aim of improving some properties or for a judicious exploitation.

Special attention is being paid, in this context, to deformation processes, especially polymer fracture under load, resulting in the loss of functions (technical, biological) of the objects manufactured from them.

Fracture, according to Andrews,[35] means the creation of new surfaces within a solid body. Apparently simple, this definition is in fact comprehensive including not only the fracture produced under the action of some exterior mechanical forces, but also the one caused by thermal fatigue, corrosion (which involves the energy released by chemical reactions), etc.

Contrary to strength, fracture is the negative side of a single process, namely that of body destruction under the action of a mechanical force, reflecting, as a matter of fact, the ultimate deformation which the body can undergo.

For a long time the deformation, fracture, and fatigue processes have been treated mostly on the basis of the physical nature of fault development.

For the past 3 decades, however, a great number of articles and monographs have drawn attention to the chemical changes induced within polymers during their mechanical stressing. It has been proved that deformation and fracture develop due to a specific mechanism of concentrating mechanic energy (a consequence of nonuniform distribution of internal stresses) on individual chemical bonds in a certain macromolecule fragment, situated, generally, in the area of a structural defect; the critical stresses thus developed, comparable in value with the bond energy, bring about their homolytical splitting and formation of macroradicals that cause a series of reactions responsible for all subsequent transformations of the polymer.

Before dealing with the important aspects concerning macromolecular compound deformation and fracture, one has to emphasize the fact that organic micromolecular materials manifest a different behavior, as compared to the polymer, i.e., generally they do not display chemical modifications.

The groups of molecules with reduced molecular weight respond to stress application by splitting the intermolecular (physical) bonds that macroscopically may be translated by the modification of the body form and microscopically by molecule displacement. This happens not only in a liquid state (for instance, under the action of shear force) but also in a solid state. The destruction of the crystals of amorphous micromolecular compounds does not lead, generally, to the formation of free radicals, which indicate the fracture of chemical bonds. These can be identified only in polymers, when mechanically stressed.

Polymer deformation and fracture are generally studied at three different structural levels, namely

- The macroscopic level; the polymer is regarded as a continuous material and for facilitating the mathematical analysis of this extremely complex level, the atomic and structural reality is avoided.
- The microscopic (supermolecular) level; in this case, the laws of statistical mechanics (used at the macroscopic level) as well as the chemical effects are, but marginally, taken into consideration.
- The atomomolecular level in which case the chemical bond-splitting processes are investigated without taking into account the superstructure effects (molecular geometry, details concenring surfaces dislocation, etc.) while the macroscopic ones are out of the question.

Although the division of these three aspects of the fracture process has existed and proceeded in the past, currently we are witnessing their sensible convergence, which has led to the process of decoding and understanding.

A. DEFORMATION AND FRACTURE PROCESSES INTERDEPENDENCE

Fracture means the degradation of a polymer material under the action of mechanical energy, and it is produced when a force acts on it. However, when a force is applied on a body, a deformation will occur, so fracture is always preceded by a deformation. Consequently, the study of polymer fracture has to be associated with that of body deformation as a previous stage.

As a principle, the interdependence of these two processes may be easily argued if it can be proved that they have a common basis. To this purpose let us analyze first the *deformation process* emphasizing that the attainment of these critical conditions necessary for chemical bond fracture depends both on the amount of energy stored in a single molecule and on the time that the molecule remains in the mechanoexcited state. As it has been assessed in numerous quantitative measurements, only a part of the elastic energy absorbed by the polymer produces the splitting of chemical bonds, the greatest part of it being dissipated in other forms in nonchemical relaxation processes. They include the sliding of macromolecular chains in relation to each other (enthalpy relaxation) or conformational changes (entropy relaxation). Chemical bond splittings should, therefore, be rather uncommon events as compared to other energy-dissipating processes.

In kinetic terms, the nonchemical relaxation processes are competing with chemical bond splitting. The more rapidly they occur, the more important becomes nonchemical relaxation time as a controlling factor of bond splitting. If we take the chemical bond fracture time as a constant, then the increase of nonchemical relaxation time will lead, naturally, to a greater number of split bonds and vice versa.

This fact takes place in solid polymers (especially the amorphous ones) under the temperature of vitreous or semicrystalline transition. Due to the strong interactions between their macromolecules, the nonchemical relaxation time has greater value as the energy for activating the conformational modifications or the reciprocal displacement of chains is very great. The macromolecules do not have the necessary time for "rearrangement" under stress, so that the polymer deformation will take place predominantly by splitting the chemical bonds. Similar situations occur even within polymers in molten state when viscosity is high, either due to the increased molecular weight or to the strong intermolecular interactions produced by a specific chemical structure. It is obvious, therefore, that at the basis of polymer deformation in solid state and partially of the viscous flow lies the mechanochemical splitting of covalent bonds in the basic chain of the polymer.

A simple experiment has proved the mechanochemical essence of this process. It has been noticed that UV ray irradiation of a polymer subjected to uniaxial static stress produced an increase of creep velocity. On the other hand, it is well known that this type of irradiating polymer structure has as a direct consequence the oxidation of valence bonds. In conclusion, the acceleration of stationary creep can only be the consequence of superposing two phenomena — deformation and photolysis — that split homolytically the valence bonds and form free macroradicals.

Analyzing the *polymer fracture* has shown that the process is triggered macroscopically by the existence of some cracks which, in turn, are generated by microcracks, i.e., discontinuities of the material acting as stress concentrating factors.[36] It has been considered for a long time that the microcracks preexist in the material, being generated by the polymer-processing conditions. Quite recently, however, Jurkov et al.[37-39] have advanced an interesting hypothesis concerning the microcracks formation (at least, partially) as a consequence of some chemical bond homolytical splitting processes. Investigating polyethylene and polypropilene by X-ray diffraction, the author has found a close correlation between the concentration of free radicals (detected by electron spin resonance), microcracks, and the final functional groups within the polymer (determined by IR spectroscopy). Thus, the number of final groups is several orders of magnitude higher than that of free radicals, the latter being practically equal to that of microcracks. Hence, the author concludes that a pair of macroradicals generates a microcrack (according to a mechanism proposed in Section B). Therefore, the macroscopic fracture process due to some magistral crack propagation, either by crack coalescence or by the development of one of them up to a critical magnitude, is based on the mechanochemical act generating the microcrack.

It results that both deformation and fracture have the same physical basis, being in a close interdependence.

The relation between these two processes becomes evident if their kinetic aspect is taken into account, in which the strength criterion is the durability of the loaded body while the deformation criterion is the velocity of deformation build-up.

Thus, the velocity of deformation build-up, defined as the reverse of relaxation time has the form (see Reference 40):

$$\dot{\epsilon} = \frac{1}{\tau_r} = \dot{\epsilon}_0 \exp\{-(Q_0 - \alpha * \sigma)/kt\} \qquad (1)$$

and by logarithmation it enables the obtainment of the dependence of the applied stress, σ, corresponding to nonelastic deformation occurrence and temperature (T) and stress rate:

$$\sigma_D = \frac{Q_0}{\alpha} - \frac{kT}{\alpha} \ln \frac{\dot{\epsilon}}{\dot{\epsilon}_0} \qquad (2)$$

where Q_0 — activation energy of stress relaxation process; τ, α — polymer constants (α — activating volume); and k — Boltzman's constant.

A similar relation correlates polymer durability under load (τ_r) with temperature and applied stress (Jurkov's equation).[41] The relation is based on the idea that fracture is a process controlled by the kinetics of split chemical bonds by thermal fluctuations:

$$\tau_r = \tau_0 \exp\{-(U_0 - Y * \sigma)/kT\} \qquad (3)$$

where U_0 — energy for activating chemical bond splitting; τ, γ — polymer constant (γ — structural parameter); k — Boltzman's constant.

By logarithmating the relation, one obtains the estimation of the polymer fracture stress, σ_r, function of temperature, and durability:

$$\sigma_r = \frac{U_0}{\gamma} - \frac{kT}{\gamma} \ln \frac{\tau_r}{\tau_0} \qquad (4)$$

Therefore, the stress producing nonelastic deformation, σ_D, and the one producing the chemical bond splitting have identical expressions, which suggests the correlation of these two processes.

Equations 3 and 4 have proved their validity in narrow temperature and stress rate ranges at high parameter values, but on the curves $\sigma_D = f(T)$ and $\sigma_r = f(T)$ there appear inflexions (Figures 2 and 3) demonstrating that some terms intervening in the equation (Q_0, α, and U_0, γ, respectively) are not constants as initially supposed.

It has been found that the position of these inflexions does not depend on the polymer stressing but on its chemical nature and physical state. Thus, by correlating the results obtained by mechanical loading with these obtained by RMN spectroscopy, it is obvious that the leaps appearing on the two stress-variation curves as a consequence of constant

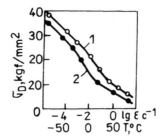

FIGURE 2. Temperature dependence of σ_D for poly(methyl methacrylate) at different deformation rates: 1 — $5 * 10^{-1}$ cm/s; 2 — $5 * 10^{-4}$ cm/s. (Based on Reference 41.)

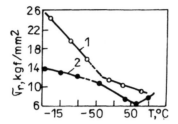

FIGURE 3. Temperature dependence of σ_r. (Based on Reference 41.)

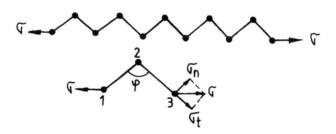

FIGURE 4. Linear macromolecule submitted to a tensile stress.

modifications from Equations 3 and 4 are produced at the same temperatures at which rotatory oscillating movements of some groups of individual molecules or chain segments on the macromolecular chains are made evident. The result is a modification of the volume of deformation activating (α) and most likely of the structural parameter (γ) as well as of the values U_0 and Q_0. It may be thus considered that deformation and fracture processes are connected with important structural modifications of the polymer (Figure 2).

The remarking of structure depends on the deformation regime applied (stress or stressing rate) being characterized by various values of activating energy or of the magnitude of kinetic elements taking part in the elementary act of deformation or fracture. When the polymer structure does not change during deformation, up to fracture, Equations 3 and 4 remain constant. However, when the macromolecules undergo conformational modifications or chemical bond scissions, equation parameters are modified, leading to the appearance of inflexions on the two stress-variation curves.

From the creep polymers data at different temperatures, it was determined that two various components of the stress are responsible for deformation and fracture. In order to illustrate this, let us consider a stretched macromolecule subjected to a tensile stress, σ (Figure 4).

The level of structural unity can be divided into two components: the normal one will act in altering the valence angle ψ; the other will act in the same direction with the chemical bond lengthening it. Converting into polar coordinates, the normal component becomes the deviating part of the stress tensor, tangent stress, respectively (t_m), while the second component becomes the spherical component of the stress tensor (σ_1). It may be easily noticed that tangent stresses are responsible for the polymer deformation, by increasing the valance angle.

Q_0, expressing deformation activating energy, constitutes the sum of all interactions preventing the elementary development of this process (inter- and intramolecular bonds), so

$$Q_0 = \Sigma\, q_i \qquad\qquad (5)$$

thus

$$\dot{\epsilon} = \dot{\epsilon}_0 \exp\{-(\Sigma\, q_i - \alpha * t_m)/kT\} \qquad\qquad (6)$$

With high strains, irrespective of the fact that fracture is brittle or ductile, it is the spheric component of the stress tensor that is responsible for the polymer durability.

$$\tau_r = \tau_0 \exp\{U_0 - \gamma * \sigma_1/kT\} \qquad\qquad (7)$$

Let us suppose that between deformation and fracture there is no connection — both being various, independent physical processes — taking place simultaneously in the stressed body. Under stress σ the process becomes dominant that develops faster. Consequently, the way in which the stressed polymer will evolve is determined by the rate ratio of the two processes.

Fracture rate may be deduced by asserting that it is the opposite of durability.

$$\bar{v} = \frac{A}{\tau_0} \exp\{-(U_0 - \gamma * \sigma_1)/kT\} \qquad\qquad (8)$$

The ratio of these two rates will be required on the one hand by Σq_i and U_0, as well as by the tangent and normal tension ratio ($\beta = t_m/\sigma_1$). With linear, rigid polymers, for low stress values the deformation appears mostly by the displacement of macromolecules, or of their associations, which becomes possible by diminishing the interaction force among the (Σq_i). If, however, the sum of these interactions is higher than U_0 (chemical bond energy), the deformation will not be able to occur without chemical bond splitting, having very low value at the moment of fracture.

The normal and tangent component ratio of the stress, β, offers information on polymer deformability under various mechanical regimes. Thus, in stretching, $\beta = 0.5$ because under these conditions σ_1 will be, obviously, higher than t_m. In torsion, due to the equivalence of the two stresses, $\beta = 1$. However, when the sample is in a compression state, when σ_1 is practically nil, $\beta \to \infty$; this fact suggests a great deforming capacity, without any fracture.

Fracture is an irreversible process being based on the scission of chemical bonds; by tensile stressing, the polymer resistance might be expected to decrease. Experimentally, a reverse phenomenon can be noticed. In this way, by stressing the polymers at higher temperatures, due to the decrease of Σq_i, the deformation rate increases very much, so the polymer deformation acquires a high value while the local tensions expressed by the γ parameter decrease. This is an effect of the conformational changes of macromolecules that have the tendency to uncoil themselves and orient in the sense of stress action making chemical bonds stressing from the cross-sectional unit become more uniform. The immediate effect is that of increasing the strength expressed as durability.

It results that deformation and fracture condition each other, the inflexion points on the curves $\sigma_r(T)$ and $\dot{\epsilon}(t)$ having to coincide, which was experimentally verified by Joseph.[42]

Often, between the two processes there is a close direct relation. Thus, the deformation of linear polymers, rigid at low tensions and low temperatures, is directly dependent on fracture, its low values being attributed to elastic deformation in the proximity of the microcrack opening. The explanation lies in the fact that deformation activating energy is in this case very high (Σq_i) exceeding by far the energy of the chemical bond (U_0); therefore, its fracture becomes more probable. This hypothesis was checked on numerous polymers, among which were polystyrene and poly(methyl methacrylate), at $-196°C$.[43]

The same phenomenon happens with linear, highly oriented polymers subjected to tensile stress. As in this case the majority of macromolecular chains are already perfectly tensioned, there being no other possibilities of conformational modification; Σq_i becomes extremely high and, as a consequence, the deformation undergone by these polymers is due to chemical bond fracture.

Referring to the highly reticulated polymers loading, it must be stated that their chemical bond splitting under the action of the stress tangent tensor cannot be avoided. In this case, the mechanochemical act will occur because there exist few possibilities of conformational changes, due to the high frequency of cross bonds between the macromolecular chains.

It results, therefore, that both processes of thermofluctuational nature, characterized by practically equal activating energies, are in a close dependence, verified at present (see References 44 and 45).

B. MECHANOCHEMICAL MECHANISM OF POLYMER DEFORMATION AND FRACTURE

As thorough research has concluded, the mechanochemical process, consisting of the valence bond splitting from the macromolecular chains under the mechanical energy action absorbed by the body, lies at the basis of the polymer deformation and fracture processes. However, from the moment of stress applying up to the final macroscopic fracture of the polymer, several stages occur, as a consequence of some molecular processes that chain sequentially, as shown by Andrews and Reed.[46]

The mechanochemical polystadial mechanism of a concatenated character comprises the following stages:

- Elastic energy absorption and macromolecular chain deformation (called also mechanoactivation or mechanoexcitation).
- Splitting of covalent bonds excites and overloads under thermal fluctuations with microcracks initiating.
- Nascent microcracks with magistral cracks formation; due to their propagation, the macroscopic fracture will take place, being accompanied by the supermolecular structure reorganization and by some new surface formation.

In a restricted sense, by mechanoactivation one understands the acceleration of increase of chemical process effectiveness under the action of mechanical energy with the deformed states playing an important role in these transformations. It is evident that by stressing the chemical activity of some polymers increases and, at the same time, their reaction with the environment is facilitated. For instance, natural rubber monoaxially stressed is more easily oxidated and its crosslinked forms undergo a faster destruction under the combined action of mechanochemical energy and ozone.

The cause lies in the concentration of mechanic energy on certain chemical bonds of some fragments of molecules crossing areas of structural defects of the polymer, which are veritable destruction centers.

Therefore, the presence of structural defects in polymers is responsible for the nonuniform dispersion on internal stresses which brings new macromolecules to a mechanoactivated state.

The nature of these defects which determine the response at the molecular level of a polymer subjected to mechanical loading depends on its morphology and physical state. Let us suppose, first of all, a linear amorphous polymer below the glass transition temperature. When a load is applied, the statistically coiled macromolcules tend to uncoil and orient in the direction of its action. Due to the strong macromolecule interactions as well as the steric hindrance, the stress magnitude must be high in order to reach this effect (different from the high elastic polymers which require lower magnitudes of the stress).

Due to polymer polydispersion, however, the shorter macromolecular chains will reach the tensioned conformation faster than those with a high polymerization degree; for this reason, they will take over a more important fraction of the stress, passing into a supertensioned state. If the polymer is in a high elastic state, the regions in which the orientation of some macromolecules segments will take place have the tendency to crystallize, the passage from one state to another being achieved through the still-unoriented states (amorphous domains) which become thus overstressed.

The number of mechanoexcited bonds within a polymer increases with the coiling degree of macromolecules, since the "nodes" between these are more frequent and act in the same manner with the physical interactions.

With previously oriented polymers the same effect will be produced by mechanical loading. Due to their polydispersed character, the stress will be taken over mainly by the already tensioned chains passing into a mechanoactivated state. The physical bonds will play the role of redistributing the stress on individual chemical bonds belonging to these chains (see Reference 47).

The crystalline polymers, according to Peterlin's model,[48,49] display in their structure crystallites interconnected by "linking" molecules of various sizes. When the effort is applied, the shorter ones or the ones already tensioned will take higher local stress which will bring them to an overstressed state, preliminary to fracture. Therefore, the chemical bonds of the chain crossing these amorphous domains become mechanoexcited, concentrating the polymer destruction centers, while the chain fragments passing through the crystalline zones being uniformly stressed do not reach the overstressed state.

With crosslinked polymers the localization of mechanoactivated bonds will become obvious being required by the frequency of transverse bridges, acting similarly to the physical intermolecular forces, but much higher.

The shortest chain fragments between crosslinks will concentrate on the mechanoexcited bonds, as they attain more quickly the tensioned conformational state.

The research of mechanoactivation stage by some chemical bonds in a stressed polymer has been demonstrated by spectral methods (IR with Fourier transformation).[50-53] Thus, in the case of poly(ethylene terephthalate) films, displacements and splittings of absorption bands were evidenced. For a deformation of 20% from the fracture one, the IR bands attributed to "gauche" conformations of $->C-O-$ bonds, neighbors with $->C-C<-$ bonds, were affected.

Once the mechanoactivating stage is reached a question arizes regarding the direction in which the overstressed chemical bonds will evolve, if the mechanical loading continues. The mechanical energy concentrated on some chemical bonds results in reducing the energy and activating their splitting, which takes place according to Jurkov's theory by thermal

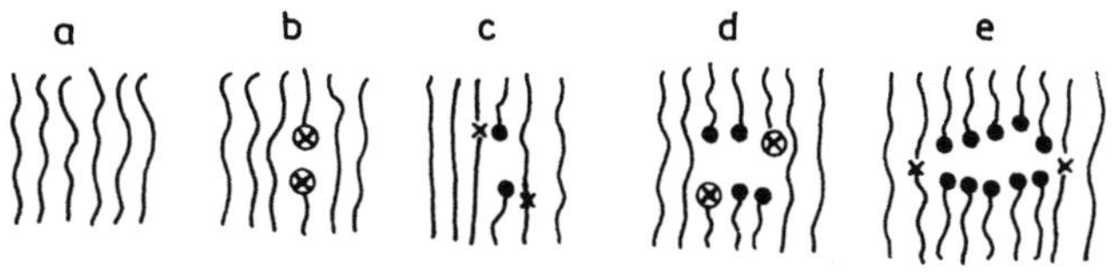

FIGURE 5. Jurkov's model for microcracks generation. (Based on Reference 54.)

fluctuations.[37,38] The feasibility of bond splitting is determined by the $\exp[-(U_0 - \gamma^*\sigma)/kT]$ factor from the durability equation discussed above.

The homolytical splitting of the bonds leads to the formation of primary macroradicals (or mechanoradicals), extremely reactive, which immediately react, according to a specific chained mechanism, with the neighboring macromolecules or with substances found in the environment generating secondary radicals.

The great number of split bonds existent at a certain moment within a microvolume generates the occurrence of microcracks confirmed by X-ray diffraction measurements.

Another interested mechanism for microcracks occurrence in amorphous solid polymers subjected to high stresses was proposed by Jurkov,[54] which was taken over by other investigators.[5,55] According to Jurkov, the higher the parameter of Equation 7, the higher the overload on the bonds to be split and the lower the durability, τ_r. On the contrary, if the external load is uniformly distributed, γ decreases and the real strength of the material increases.

When a chemical bond splits, a pair of primary mechanoradicals results (Figure 5).

The most probable reaction of macroradicals, according to experimental data, is the pulling out of a hydrogen atom belonging to a neighbor macromolecule, a process which takes place at very high rate even at low temperatures (130 K). The forming of a new, secondary macroradical, with the free electron on the chain, weakens the neighboring interatomic bonds considerably. Thus, the activation energy for splitting the bond in position α — as related to the unpaired electron — is one third lower than that for splitting any other bonds; as a consequence, their splitting takes place easily, other species of macroradicals being formed. The newly formed macroradicals will initiate, according to a chained mechanism, the splitting of other neighboring macromolecules; in according with existent data, a pair of mechanoradicals may cause the splitting of about 1000 neighboring chemical bonds, in keeping with the following mechanism, briefly presented below:

$$R_1 - R_2 \xrightarrow{\text{mechanical energy}} R_1^{\cdot} + R_2^{\cdot}$$

$$R_1^{\cdot} + R_2^{\cdot} + 2(R-CH_2-CH_2-R') \longrightarrow R_1H + R_2H + 2(R-\overset{\cdot}{C}H-CH_2-R')$$

$$R-\overset{\cdot}{C}H-CH_2-R' \longrightarrow R-CH{=}CH_2 + \overset{\cdot}{R}'$$

$$\overset{\cdot}{R}' + R_1 - R_2 \longrightarrow R'H + R_1 - R_2^{\cdot}$$

As a consequence, the reduced number of primary mechanoradicals formed under the action of mechanical energy by thermal fluctuations contributes to the splitting of a great number of molecules. This mechanochemical reaction represents the fundamental act of polymer fracture micromechanics, and the making up of active radical species has been proved by numerous researchers, using various direct or indirect techniques, the most used being the RES method.[5,56-61]

The accumulation of a great number of split chemical bonds within a microvolume generates a microcrack, according to the above-mentioned mechanism. The microcracks are the first microscopic act of fracture, being finalized by the appearance of new surfaces. On these surfaces new final groups are localized; as a result of macroradical stabilization, their concentration is nearly three orders of magnitude higher than that of microcracks (detected by X-rays), or of macroradicals (detected by RES). These results allowed Jurkov to prove that microcracks appear due to chemical bond splitting, not by loosening of physical, intramolecular bonds. Besides, the dependence of microcrack number on stress and temperature is of a thermofluctuational character, demonstrating that their formation is inevitably connected with the mechanochemical act.

Microcracks are, however, of too small dimensions to cause the final fracture of the sample, as stated by Peterlin.[62] They have to grow or merge until a critical crack is formed, according to Griffith's criterion. Therefore, the magistral crack is either the result of microcrack development (nucleation) or of their coalescence.

Within a fibrillated polymer, for instance, microcracks can grow either radially, by the breaking of adjacent microfibrils, or axially, by the separation of adjoining microfibrils. In the former case all macromolecules binding the neighboring crystallites around the growing microcrack are broken step by step, leading finally to the fracture of the microfibril containing these formations. In the latter case, however, only the macromolecules binding the microfibrils are fractured, that is, those which bind the microfibrils on the opposite sides of the growing microcracks. Such growing is facilitated especially on the boundary area of fibrils, where more of their ends are concentrated. The coalescence of the microcracks appeared in the microfibrillary structure defects; on the boundary area of the same microfibril, it tends to separate it from the rest of material out of a long, axial magistral crack — the phenomenon being known under the name of fibrillation, an unwanted feature for materials used as fibers.

It must be noted, however, that, despite the way in which the magistral crack is formed, its occurrence involves mechanochemical acts.

Magistral crack appearance means another stage in the fracture process. For its finalization it is necessary that the magistral crack propagate through the sample. This propagation involves both the creation of new surfaces (which, in turn, appear by chemical bond splitting) and energy dispersion by unelastic losses (which does not involve chemical bond splitting).

The method of X-ray diffraction at small angles as well as perfected microfilming techniques enabled the observation of the magistral crack formation and its propagation process. The stress concentration in the top of magistral crack, as well as the splitting of these chemical bonds, which after macroradicals are stabilized provide new functional groups, was investigated by IR spectroscopy. Thus, Korsukov[63] experimenting on polyethylene and polypropylene films to which magistral cuts, simulating the magistral crack, were performed and which were subjected to tensile stresses, succeeds in determining the magnitude of real stresses and of fractured bond concentration in the top of the cut. It was found that, at a load of $\sigma = 12$ Kgf/mm^2 near the crack top, there appear local stresses of 100 Kgf/mm^2 leading to the overloading of individual macromolecules (in concentrations of approximately 0.01 to 0.1%) up to stress values of 1000 Kgf/mm^2. These molecules split, as a consequence, even at room temperature under thermal fluctuations, after relatively short durations (minutes, sometimes hours). The distribution of split bond concentration, depending on the distance from the crack top corresponds to stress distribution (Figures 6 and 7).

The rate at which the crack propagates through the sample is not constant because overcoming of obstacles (macromolecules, chain strings, inclusions, etc.) takes different times. For the same reason the crack front has an extremely irregular configuration.

Over the entire crack perimeter, for low stress values, a state of equilibrium between the elastic mechanical forces and those of chemical bonds is attained, crack propagation

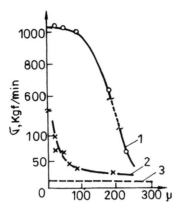

FIGURE 6. Stress distribution near the top of the magistral crack. 1 — the maximum stress on isolated macromolecules; 2 — in the crack top; 3 — the medium stress. (Based on Reference 63.)

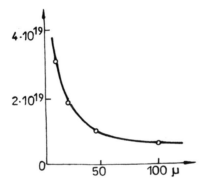

FIGURE 7. The distribution of the split bonds concentration near the top of the magistral crack. (Based on Reference 63.)

being a consequence of the disturbance of this equilibrium. If, at a certain moment, the elastic forces have taken hold of an equilibrium position as part of a freshly completed destruction act, in order to annihilate the subsequent obstacles, it will take time, during which the redistribution of stresses will occur. It thus produced an important accumulation of elastic energy along the entire crack front while at the crack-polymer limit there appears an overloading which materializes in a force impulse. The elastic stress field and the impulse energy enable the displacement of the crack front at a certain rate, determined by the total energy reserve existent at the moment of crack "jump" initiation.

In this way, in the highly mechanically loaded areas of a polymer, under some fast or slow stresses and due to the stresses developing mostly in the magistral crack top flow phenomena occur accompanied by destruction processes consisting of chemical bond splitting. These processes do not proceed simultaneously within the whole loaded domain, but propagate very fast by successive acts.

When the crack front is displaced, its propagation becomes more accelerated, because the cross section of the sample decreases continuously and the local stresses increase considerably.

Through systematic experiments with magistral crack development, it has been proved that the rate of this process (v) is related to temperature (T) and the applied stress (σ), through the relation[64-67]

$$\nu = c * \exp \frac{U_0 - \alpha * \sigma}{kT} \qquad (9)$$

where U_0 — activation energy of magistral crack growth; α — parameter of top crack stresses; k — Boltzman's constant; and c = constant, having magnitudes of the order 10^{10} cm s^{-1}.

The value of activation energy in magistral crack development, U_0, was proved to coincide with the chemical bond splitting energy in the investigated polymers, thus demonstrating the thermofluctuational character of the mechanism of the process (see Reference 68).

By using spectral methods, the chemical bond splitting from the magistral crack top was directly specified by Korsukov and Vettegren,[67] the obtained results explaining the role of the real stresses in splitting interatomic bonds. The concentration in new, final functional groups, determined on the magistral crack surface, attains the magnitude of approximately $0.5 \cdot 10^{22}$ cm^{-3}, equal in magnitude order with the number of chemical bonds within 1 cm^3. Consequently, the microcrack surface consists practically, as a whole, in chain ends.

It may be estimated on the basis of the data stored so far that macroscopic fracture can be divided into a crack nucleation regime and a propagation one. The way in which sample fracture parameters such as durability under load and fracture strength depend on the processes involved in these two areas is a function of the physical properties of the material, sample geometry, and load conditions (rate, temperature, environmental factors). Although chemical bond splitting is inevitable under both regimes, its role and the interaction with other physical processes are significantly different from the two regimes.

It may be concluded that the fracture processes, from the moment of its initiation up to the macroscopic fracture of the material, is a consequence of chemical bond splitting and to a less extent of some physical processes involving the participation of intermolecular bonds. Thus, polymer fracture as well as their deformation is of a mechanochemical nature.

III. MECHANOCHEMISTRY OF THE DEFORMATION AND FRACTURE PROCESSES OF THE POLYMERS IN DIFFERENT PHYSICAL STATES

The scission of the chemical bonds from the basic chain of the mechanically stressed polymers, an elementary act which lies at the basis of the deformation and fracture processes, initiating and accompanying them during their entire evolution, is influenced by a series of factors which belong both to their structure at all their organizing levels and to the parameters of the mechanical regime or medium factors.

The chemical structure constitutes a less important factor because of the close values of the activating splitting energy for all the types of covalent bonds between the organogenic elements which enter the structure of each polymer. Numerous experiments on various polymer classes have led to the conclusion that one can obtain high and close concentrations of free radicals in the compounds with very different chemical structure under similar stress conditions. This factor determines only the connecting type where the mechanochemical act is being produced and determines the nature of the formed macroradicals, but not the intensity of the macromolecules scission. The physical state of the polymer (correlated with the conditioning temperature) and the parameters of the mechanical regime (stress intensity, deformation, stress rate, etc.) are decisive factors in the evolution of the deformation and fracture processes.

For a better understanding of the importance of physical and phase state of the polymers in the deformation and fracture processes, we have to assume that the breaking of the polymer chain is a competitive response to its deformation; if the deformation can be more easily accommodated to the molecular flow or to the crystallographic processes, the breaking of the chemical bonds will be less evident and vice versa.

A great importance is given in this context to the nature and strength of the forces which manifest themselves between the macromolecules, namely,

1. The linear chains amorphous polymers, below the vitreous transition temperature, will suffer the deformation and fracture by macromolecules scission, but without excluding the shape changes or their relative displacement.
2. The linear amorphous polymers in a high elastic and flowing state will accomplish the deformation by conformational changes and by sliding of the molecules, the mechanochemical act manifesting itself, especially on the final stages of the process preceding the macroscopic fracture.
3. The reticulated and semicrystalline polymers will strain and fracture exclusively by the splitting of the chemical bonds from the basic chain.

A. MECHANOCHEMICAL PROCESSES IN THE DEFORMATION AND FRACTURE OF THE SOLID-STATE POLYMERS

Solid-state polymers represent complex systems formed by the aggregation of the macromolecular chains by means of the low-intensity temporary physical-type or high-intensity permanent chemical-type bonds. The great variety of their mechanical properties is the result of the difference existing between the various forms of aggregation (superstructure) and the different molecular structures (the chemical structure of the chains, their mobility, and the reticulation density).

That the macromolecules are long, anisotropic, and flexible gives to these bodies a series of characteristic properties such as the anisotropy of the macroscopic properties, the microscopic nonuniformity, the nonlinearity, and the strong dependence on time of the response to the mechanical stresses.

The deformation and fracture of the solid bodies will constitute complex processes, influenced both by some structural features (i.e., interatomic bonds, surface energy, amorphous-crystalline structure) and by some processes of mechanical nature (i.e., the sliding and the twisting of the crystallites, the phase transformations induced by the stress, the molecular relaxation, etc.). All these parameters depend on the method of obtaining polymers (the "history" of the sample) and on the real conditions under which the stress takes place. For instance, by changing the duration of the force action or the temperature, the same polymer can present a variety of responses, ranging from that of the brittle glass (the vitreous state) to that of a very viscous liquid (the fluid viscous state). Even for the same physical state, the polymer response under different time and temperature conditions indicates that some deformation and fracture processes (varying from brittle to ductile fracture) are dominant.

No matter what the real stress conditions are, the strong interactions of the physical nature, the cross bridges from the reticulated structures, or the crystalline domains existing in the solid-state polymers have the role of redistributing the stress on the chemical bonds from the basic chain by concentrating the stress on the structural fault zones, thus determining the splitting of some bonds, an elementary act which generates and controls the deformation and the fracture processes.

1. Characteristics of the Deformation and Fracture Processes of the Amorphous Polymers under T_g

Linear amorphous polymers consist of ball-shaped macromolecules among which physical forces or hydrogen bonds are manifested; their intensity depends on their chemical structure. In this case the relative movement of the macromolecules, even of the chain segments, is strongly restricted compared to the high elastic state or especially to the flowing one. Consequently, the polymer deformation will be the result of splitting the chemical bonds and, to a lesser extent, of the physical ones, which allows the accomplishment of the shape changes and the flow. Being tightly coiled, the macromolecules can be "knotted", such knots acting as real crosslinkings; this fact makes the splitting of the chemical bonds unavoidable during the mechanical stress. However, the concentration of the free radicals detected by RES spectroscopy is more reduced for the same stress conditions as compared to the nonreticulated amorphous or semicrystalline polymers.

The studies dealing with the fracture of the amorphous polymers under T_g are not so numerous, being performed especially on polystyrene and poly(methyl methacrylate);[67-71] however, they point out that concentrations of the primary macroradicals of the 10^{14} spins/cm^3 order do appear. The inevitability of splitting the chemical bonds of the linear amorphous polymers under T_g is demonstrated, as well, by the following assumptions (see Reference 72).

The fracture of the mechanically stressed solid-state micromolecular compound (ethane at very low temperature) is exclusively done by the scission of the physical bonds. The fracture activating energy estimated empirically is of about 1 kcal/mol. Considering a poly(methylenic) chain (having a structure similar to that of ethane) made up of n structural units, one can assume that the energy necessary for the activation of its displacement from the polymer matrix is n*1kcal/mol. At the same time, knowing that the energy of the bond $->C-C<-$ is 83 kcal/mol, it follows that for a value of the polymerization degree n = 83, the sum of the interactions energies between the macromolecules exceeds the value of the bond energy $->C-C<-$.

Accordingly, the fracture of such a solid-state macromolecular compound will not be achieved by the displacement of the macromolecules but by the splitting of the chemical bonds. Thus, it can be understood why the splitting of the chemical bond has not been noticed in paraphine where the number of the C atoms in a molecule does not exceed 80.

The above-given explanation can be extended to all linear amorphous polymers where the presence of the substituents or of the double bonds on the main chain make the adjacent bonds $->C-C<-$ much more unstable.

The experimental data demonstrate that the linear amorphous polymers in a vitreous state have a brittle breaking, preceded by the microcracking phenomenon,[73] which is characterized by the accomplishment of highly oriented polymers fibrils in the cracks which will cause the macroscopic fracture. The macroradicals appear as a result of the splitting of these fibrils; however, the ESR signal is evident only in the fibrils in which the flow process cannot occur (all the macromolecules are oriented in the direction of the stress action), the only possible deformation mechanism being the splitting of the chemical bonds. It is also possible that the splitting of the molecules could be involved in the nucleation phenomenon of the preexisting cracks, but less probable in the vitreous state thermoplastics.

The vitreous linear amorphous polymers subjected to a mechanical stress undergo at first an elastic deformation produced by nonchemical processes (the shape change, the relative displacement of the macromolecules), after which the splitting of the chemical bonds starts; this act is responsible for the appearance of the plastic deformation and macroscopic fracture. That is why at the brittle fracture of the isotropic polymers we do not always obtain detectable ESR signals, a fact possible only after the induction of the material plastic flow.

a. Static Tensile Stress

A polymer deformation under a constant stress (creep) can have the following consequences:

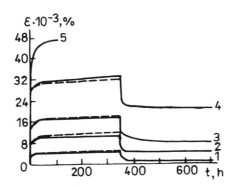

FIGURE 8. Experimental (full line) and theoretical (dotted line) creep curves for PVC. 1 — $\sigma = 105$ daN/cm²; 2 — 193 daN/cm²; 3 — 249 daN/cm²; 4 — 364 daN/cm²; 5 — 400 daN/cm².

- The sample fractures almost immediately after the application of the stress
- The sample fractures after a certain period of time
- The sample supports the load of an undefined time

From a practical point of view we are interested in the third type of polymer behavior.

Although the creep has been studied for a great variety of polymers to obtain its characteristic equations, the studies dealing with the mechanochemical aspects of this process are less numerous, especially for the linear amorphous polymers under T_g. Among these might note the studies concerned with the tensile stress under creeping conditions of the poly(vinyl chloride) (PVC), of the PVC modified with aromatic diamines, or of PVC subjected to an accelerated aging by thermal UV treatments.[76-79]

The study of the PVC creeping was aimed at obtaining the creep equations, to make evident the character of the fracture as well as of the change of some physicomechanical properties induced by stress, respectively. The mechanical stress has been accomplished on an original testing stand (see Reference 76) at stress values ranging between 100 to 400 daN/cm², at the environment temperature and humidity.

The experimental rheological curves have the typical shape of the creeping curves, emphasizing the viscoelastic behavior of the material (Figure 8), i.e., elasticity, creep, and reverse creep.

Consequently, the suggested rheologic model is a variant of the Burgers model. The elasticity of the material due to the change in the atom position and valence angle in macromolecules (Figure 9) is expressed by the E_1 spring. The viscoelasticity expressed by the retarded elasticity is conveyed by the Voight generalized model where the springs E_{2k} express the modification of the macromolecule deformation while the piston cylinders λ_{2k} express the reaction of the environment.

The element λ_3 characterizes the motion of the macromolecule segments provoked by the shape changes but especially by the destruction of some chemical bonds, which leads to the appearance of a nonreversible viscous deformation.

The processing of the obtained experimental results allowed the computation of the coefficients and exponents from the rheologic equation corresponding to the proposed model; thus, for the studied case the equation has the form:

$$\epsilon(\%) = 10^{-4}[0.5 \, \sigma^{1.73} + 10(1 - e^{-5*10^{-6}t}) + \sigma^{1.63}] \tag{10}$$

A good correlation between the theoretical equation (dotted line) and the experimental creeping curve has been demonstrated (Figure 8), thereby confirming that the proposed rheological model is the proper one.

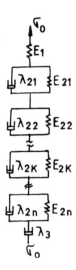

FIGURE 9. The rheological model for PVC creep.

Similarly, we have proceeded to establish the creep equations characteristic for the polymer modified with benzidine or subjected to accelerated aging under the action of high temperature and UV rays. These equations have the following form which also demonstrate a good correlation with the experimental creep curves, specifically for

1.   The benzidine-modified PVC

$$
\epsilon(\%) = 10^{-4}\left[\frac{1}{2}\right]\lg(1 + \sigma) + \sigma\left(1 - e^{-t/3*10^{-6}}\right)\left(0.5 + \frac{1}{t * 10^{-6} + 1}\right)
$$
$$
+ 0.01256\,\sigma^{1.65}\Bigg] \tag{11}
$$

2.   The heat subjected PVC for 150 h at 70°C

$$
\epsilon(\%) = 5 * 10^{-4}[\sigma^{0.31} + 2\sigma^{-0.058}\mathrm{th}(2 * 10^{-5}t) + 2(0.3 + 0.8 * 10^{-7}t)] \tag{12}
$$

3.   The PVC subjected to temperature (above-mentioned conditions) combined with the UV-ray treatment

$$
\epsilon(\%) = 5 * 10^{-4}[\sigma^{0.29} + 2\sigma^{-0.087}\mathrm{th}(2 * 10^{-5}t) + 2(0.3 + 0.8 * 10^{-7}t)] \tag{13}
$$

4.   The benzidine-modified PVC subjected to the combined action of heat and UV rays

$$
\epsilon(\%) = 5 * 18^{-4}[\sigma^{0.25} + 2\sigma^{-0.186}\mathrm{th}(2 * 10^{-5}t) + 2(0.3 + 0.8 * 10^{-7}t)] \tag{14}
$$

The PVC modification by two-roll processing in the presence of benzidine results in the structure strengthening manifested at the level of the physicochemical properties, namely, the reduction of the values of the instantaneous elastic deformation and of the remaining

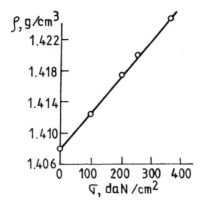

FIGURE 10. Picnometric density variation of PVC with static tensile stress.

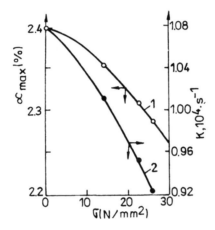

FIGURE 11. Maximum swelling degree (1) and the swelling constant rate (2) variation for PVC modified with benzidine, as a function of static tensile stress.

one, respectively, for the same stress value. The process of polymer aging acts similarly, being provoked by heating and especially by the simultaneous action of heat and UV rays. Both the witness polymer and the one modified chemically or by physical treatments suffer during creeping, important changes which affect its structure at all its organizing levels.

A first effect of the stress action is the partial shape change of the macromolecules in their elongation and orientation which lead to the increase of the polymer packing degree and to the adequate modification of some physical properties. Thus, the material density is increased with the stress intensification at the same stress duration (Figure 10).

Accordingly, the behavior in specific swelling agents (acetone, toluene) is affected too. The increase of the packing degree of the stressed macromolecules leads to the intensification of the interactions between them. That is why the swelling agent penetrates with more difficulty in polymer while the maximum swelling degree and the swelling constant rate, respectively, are reduced with the applied stress. This phenomenon has been registered for all studied PVC types, being exemplified for the benzidine-modified PVC in Figure 11.

The mechanical energy absorbed by the polymer, concentrated in the fault zones of the structure, is consumed almost to the extent of deformation of both the chemical bonds and the valence angles, so that the polymer passes into a mechanoexcited state. During the stress relaxation, a part of the accumulated elastic energy is used up for the homolytical splitting

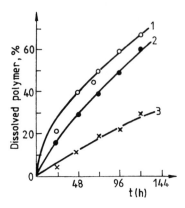

FIGURE 12.   Time variation of the dissolved PVC amount. 1 — witness sample; 2 — benzidine modified PVC statically stressed ($\sigma = 253$ daN/cm^2); 3 — benzidine modified PVC, witness sample.

of the chemical bonds and for the formation of free radicals which, by stabilization, lead to the accumulation of some reduced moelcular weight fractions; therefore, the polymer solubility in specific solvents must increase, an effect rendered evident in Figure 12.

The lower solubility of the benzidine-modified PVC compared to that of the unmodified one is the result of some crosslinking reactions produced during synthesis.[80-83] However, this solubility has superior values in the mechanically stressed polymer due to the reduction of the molecular weight provoked by the mechanochemical degradation reaction induced by the stress (curve 2 compared with curve 3).

The prolonged action of some stresses, even lower than the fracture one, causes the polymer fracture. To provide information on the fracture type of this rigid material, some samples were subjected to a stress $\sigma = 400$ daN/cm^2. The macroscopic fracture took place after 70 to 80 h and the fracture surfaces were analyzed by means of electron microscopy (Figure 13, a to c).

Their marked relief leads to the conclusion that the material response to the mechanical stress is of the brittle type. A careful analysis shows, as well, the stages of the mechano-chemical process up to the sample fracture.

The splitting of the chemical bonds takes place in microvolumes which become destructive centers. Under the combined action of stress, thermal fluctuations, and mechanoradicals generated on the deformation incipient stages, the great number of bonds split in microvolumes leads to the apperance of a discontinuity in the material (i.e., the microcrack, obvious if we analyze Figure 13c). The microcracks continually increase, either by macromolecules splitting or by coalescence, up to some critical dimensions. In this stage, the stress values are below those of the breaking one, so that the relaxation times are comparable with the stress ones. Consequently, the macromolecules can be rearranged under stress so that the relief of the fracture surfaces is less marked. The appearance of the main crack is the first stage of the macroscopic fracture. Because the stress is lower than the one necessary for the macroscopic fracture, the main crack propagates slowly through the sample, making new surfaces with a relatively coarse aspect. The stored elastic energy is partially dissipated in the material by nonelastic deformations which do not imply the splitting of the chemical bonds. When the sample section attains a critical value and the stress is comparable with the fracture one, the propagation of the main crack is accelerated, thus becoming "catastrophic". In this stage the chemical bonds split almost exclusively because of the mechanical stress, the participation of the thermal fluctuations being strongly diminished. Although the propagation of the crack is rapid, it is controlled by the nonelastic properties of the polymer.

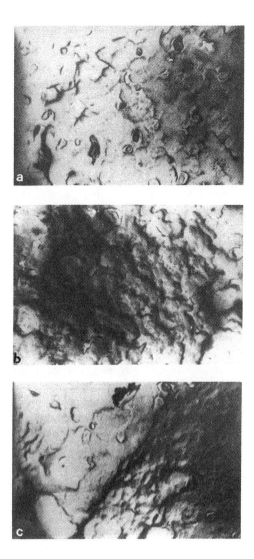

FIGURE 13. Micrographs of fracture surfaces of statically stressed PVC (σ = 400 daN/cm^2; 75 h; magnification × 9900).

As the relaxation times are now much higher than those of the stress, the overstressed macromolecules cannot rearrange themselves under stress; therefore, the relief of the breaking surface becomes very coarse (Figure 13b), a characteristic for the rigid polymers which do not present energy losses due to their plastic and viscoelastic properties. The transition from the regime characterized by the slow propagation of the crack to that with a rapid propagation becomes evident from the analysis of the microphotographs. A clear delimitation of two zones which constitute the front of the main crack in the moment when its "catastrophic" propagation starts is clearly shown in Figure 13c.

These results prove that the deformation and the fracture of PVC under prolonged static stress at temperatures lower than the vitreous transition are based on the mechanochemical act and pass compulsorily through all the stages of the polystadial chained mechanism, specific to the mechanochemical processes.

b. Dynamic Tensile Stress

The dynamic stressing of the polymer is often encountered in engineering practice when polymer-made objects are employed; therefore, the study of their behavior under these conditions is justified. In this case an important parameter is the stress rate which determines to a high extent the ''response'' of the polymers.

The fracture character is, in fact, the result of the competition between the relaxation process of the stressed macromolecules and the fracture process of the chemical bonds. The rapid stressing determines a superior relaxation time as compared to the stress action time, so that the dissipation of the elastic energy accumulated in material would take place predominantly by means of chemical processes (the splitting of covalent bonds) and to a lesser degree by nonchemical ones (shape changes and displacement of macromolecules, one against the others); consequently, the fracture will be of the ''brittle'' type, taking place at small deformations.

The slow stress will favor the relaxation processes, so that the ultimate obtained deformation will be much higher and the fracture will acquire a ductile character, determined by the possibility of the stressed structure reorientation.

PVC, a preeminently rigid polymer, is used very much in engineering. The study of its behavior at tensile stress with small deformation rate has led to a variety of information on the deformation and fracture mechanism. The studies have been done on 50-μm average thickness polymer films stressed by means of a device which allows their immersion in various liquid mediums.

By starting from the hypothesis of mechanoradical formation, even at deformations lower than the macroscopical fracture, we have proceeded to stimulate their reaction with some radicalic acceptors (DPPH, aromatic diamines) or with styrene in methanolic solutions. Thus; by straining the films for a 2-h period at a deformation that increases slowly and uniformly in time up to the vaue of 20% out of the fracture one, in the presence of DPPH, the appearance of a violet tinge on the sample active sector has been discovered, a tinge that is stable in time as well as at repeated extractions in methanol and which indicates the acceptor chemical bonding. This conclusion has been confirmed with the UV spectra of the sample by the existence of a wide absorption band from 350 nm which is present both on DPPH and on the analyzed sample.

By knowing the radical acceptor character of aromatic diamines,[82] we have produced the mechanical stress of some films at a duration ranging between 1 and 6 h to attain the maximum deformation in o-phenylene diamine (o-PhDA) and p-phenylene diamine (p-PhDA) methanolic solutions with concentrations ranging between 0.1 to 1%. The recording of the UV spectra of these samples revealed the existence of some absorption bands specific to the micromolecular compound (Figure 14).

The absorption bands ranging between 310 and 250 nm (curve 3) characteristic for p-PhDA, and 300 and 250 nm (curve 5) specific to o-PhDA, respectively, prove the chemical bonding of those two aromatic diamines to the chain fragments with a radical character, generated by the stress action.

The quantity of reacted diamine depends on the concentration of the solution in which the stressing takes place, a conclusion drawn from the analysis of Figure 15 in which the absorption read in the UV spectrum was considered as an evaluation criterion of the reaction efficiency of the micromolecular compound with the polymer. Thus, it results indirectly that the p-PhDA is more reactive as compared with the o-isomer, being found in a greater quantity in the stressed films. Its 0.5% concentration is practically enough to consume all the mechanoradicals formed as a result of the mechanical stress, the absorbance remaining, as a matter of fact, constant above this concentration.

The stress duration is an important parameter of the mechanical regime. One can establish that the efficiency of the p-PhDA reaction with the polymer decreases with the duration

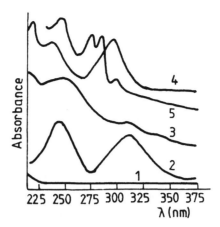

FIGURE 14. UV spectra of some PVC films stressed in methanolic solutions of aromatic diamines. 1 — witness; 2 — p-PhDA; 3 — sample stressed in methanolic solution of p-PhDA (0.75%); 4 — o-PhDA; 5 — sample stressed in a methanolic solution of o-PhDA (0.75%).

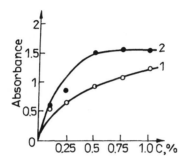

FIGURE 15. Influence of diamine concentration on the reaction with PVC macroradicals. 1 — o-PhDA; 2 — p-PhDA.

increase, a 20% deformation out of the fracture one being attained on this time interval (Figure 16).

This effect is normal if the relaxation phenomena taking place in the stressed polymer are considered. The increase of the stress duration to attain the same deformation means, in fact, the reduction of the deformation rate. At 1-h duration, the deformation rate is high; the relaxation times of the stress are high, as well as at 20°C, so that the polymer is "brittle" to the stress action, breaking at a smaller elongation. Because there are many macroradicals, the diamine consumption is also high. With the increase of the stress duration, the deformation rate is reduced. The relaxation times of the stress become comparable to the straining ones, there appearing high plastic deformations as a result of the convenient arrangement of the stressed macromolecules. The number of formed macroradicals is reduced; thus, the quantity of chemically bonded diamine is diminished. The mechanoradical·formation under the studied conditions has been proved, as well, by initiating some block or graft copoolymerization reactions in the presence of a vinylic monomer, i.e., styrene. The appearance of some new absorption bands at 775, 1480, and 1600 cm^{-1} in the IR spectra of the stressed films in methanolic styrene solution has demonstrated the chemical bonding of the graft of sequence-shaped monomer to the PVC macroradical fragments. The conclusion is also proved by the appearance of the 295-nm band in the UV spectra (Figure 17).

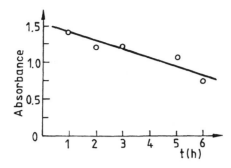

FIGURE 16. Variation of chemically bound p-PhDA amount with stressing duration (1% solution concentration).

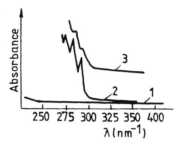

FIGURE 17. UV spectra of PVC films. 1 — witness sample; 2 — styrene; 3 — sample stressed in styrene methanolic solution (15% concentration).

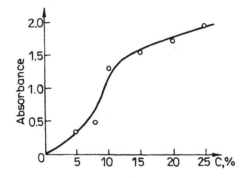

FIGURE 18. Variation of the amount of chemically bound styrene with monomer concentration.

The quantity of chemically bonded monomer, indirectly reflected by the value of the absorbance (from the UV spectrum), is correlated with the concentration of the styrene methanolic solution in which the stressing takes place (Figure 18). This quantity is continuously increased as a result of the increase of the grafted ramifications length under the conditions in which the macroradical concentration is kept practically constant.

A supplementary argument concerning the styrene grafting on the surface of the stressed films is the microphotographs obtained on the electron microscope (Figure 19, a to c).

Sample a presents a relatively smooth surface with perpendicularly formed graves on the stress direction being distinguished. The film stressed in the presence of 15% styrene concentration in a methanolic solution has a marked relief surface on which the grafting centers of the monomer are visible (b). They are evident, as well, in the sample stressed in

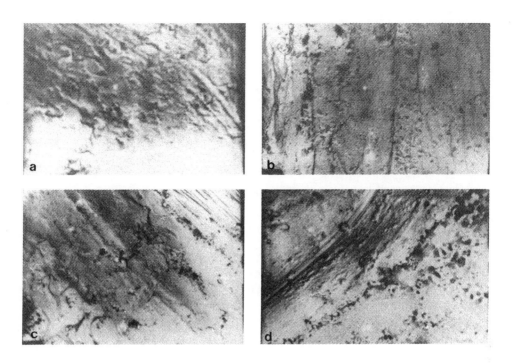

FIGURE 19. Microphotographs of a — PVC stressed in the absence of styrene; b — PVC stressed in the presence of styrene (15% concentration); c — PVC stressed in the presence of styrene (20% concentration, magnification *4400); d — PVC stressed in the presence of styrene (*12800).

a 20% concentration solution (c), being much more numerous in the vicinity of the domain in which the film fracture begins, the domain characterized by numerous graves (d).

The results confirm that, during the PVC dynamic stressing, even at values of the deformation lower than those of the fracture, macroradicals generated by mechanochemical processes are produced. On the basis of these results, the following molecular mechanism can be stated:

$$\cdots -CH_2-CH-CH_2-C\ H- \cdots \xrightarrow[\text{energy}]{\text{mechanical}} \cdots -CH_2-\overset{\bullet}{C}H + \overset{\bullet}{C}H_2-CH- \cdots$$
$$\underset{Cl}{|} \qquad \underset{Cl}{|} \qquad\qquad\qquad \underset{Cl}{|} \qquad \underset{Cl}{|}$$

The radical fragments can be stabilized by radical acceptors (DPPH or aromatic diamines):

$$\cdots -CH_2-\overset{\bullet}{C}H + H_2N-Ar-NH_2 \longrightarrow \cdots -CH_2-CH_2 + H\overset{\bullet}{N}-Ar-NH_2$$
$$\underset{Cl}{|} \qquad\qquad\qquad\qquad\qquad \underset{Cl}{|}$$

$$\cdots -CH-\overset{\bullet}{C}H_2 + H\overset{\bullet}{N}-Ar-NH_2 \longrightarrow \cdots -CH-CH_2-HN-Ar-NH_2$$
$$\underset{Cl}{|} \qquad\qquad\qquad\qquad\qquad \underset{Cl}{|}$$

These fragments initiate grafting and block-copolymerization reactions in the presence of vinylic monomers (styrene):

$$\cdots -CH_2-\overset{\cdot}{C}H + nCH_2=CH \longrightarrow \cdots -CH_2-CH-(-CH_2-CH-)_{n-1}-CH_2-\overset{\cdot}{C}H$$

with Cl, C6H5, Cl, C6H5 substituents

blockcopolymer

$$\cdots -CH_2-\overset{\cdot}{C}H + \cdots -CH_2-CH-CH_2-CH- \cdots \longrightarrow \cdots -CH_2-CH_2 +$$

with Cl, Cl, Cl, Cl substituents

$$\cdots -CH_2-\overset{\cdot}{C}-CH_2-CH- \cdots$$

with Cl, Cl substituents

$$\cdots -CH_2-\overset{\cdot}{C}-CH_2-CH- \cdots + mCH_2=CH \longrightarrow \cdots -CH_2-C-CH_2-CH- \cdots$$

with Cl, Cl, C6H5 substituents; product with Cl substituents and

$$(CH_2-CH-)_{m-1}-CH_2-\overset{\cdot}{C}H$$

with C6H5, C6H5 substituents

graftcopolymer

A continuous interdependence exists between the deformation and fracture processes in the mechanically stressed polymer, the main factor being the splitting of the chemical bonds.

c. Cyclical Stress

Under operating conditions, the polymers can be subjected either to the action of constant stresses (i.e., creeping), to the continuous increasing in time stresses, or to some periodical, cyclical ones. The process developed in the second case is termed fatigue.

The fracture under fatigue conditions is produced not only by the periodical action of some high value stresses, but also by the application of some prolonged stresses and by less-intense stresses. The dependence of the durability on stress and temperature, dependence expressed by the kinetic theory of polymer fracture, proves the nature of the thermofluctuational processes of chemical bond splitting; it is not different from the elementary acts of fracture under the continuous static or monotonously increasing stresses.[84] The equation which describes the durability has the form:

$$\int_0^{t_r} \frac{dt}{\tau[\sigma(t)]} = 1 \qquad (15)$$

where the denominator is replaced by the expression known from the Jurkov equation, the τ_0, U_0, and γ parameters being determined from creep experiments. However, the difference consists in the fact that, after every loading-unloading cycle, a quantity of elastic energy is dissipated in the polymer; this energy is primarily of the thermal type and produces local overheatings, leading to the structure modification in the main crack propagation zone. Consequently, the structural parameter is changed. These effects are taken into account by introducing in Equation 15 both the real temperature T at the top of the main crack and the actual value for γ.

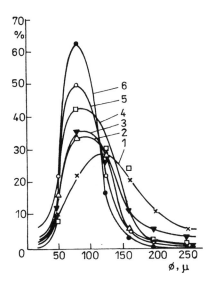

FIGURE 20.   Influence of ultrasonic treatment duration on the granulometric distribution of PVC. 1 — witness; 2 — PVC ultrasonically treated for 30 h; 3 — for 40 h; 4 — for 60 h; 5 — for 90 h; 6 — for 120 h.

The appearance of the microcracks suggests an induction period of the fatigue fracture which, however, can be absent in the preexisting crack polymers. Whether this period exists or not, the process develops continuously by incremental increases as shown by Andrews[35] and not by the catastrophic propagation of the main crack as it happens for other types of stresses. Thus, the fatigue implies the noncatastrophic fracture propagation through the material, the main crack increasing with each cycle but only to a small extent, insufficient to produce the macroscopic failure of the polymer. The mechanism of the gradual incremental propagation of the main crack in cyclically stressed polymers has been checked out for a wide range of macromolecular compounds (see Reference 85). On studying, for instance, the PVC behavior on fatigue, regular grooves are evident on the surface of the samples, the distance between them being correlated with the application rate of the stress.[86,87] The number of the grooves represents about 2% of the cycles numbers, which lead to the conclusion that the fatigue fracture of this material represents a discontinuous process, the main crack propagating once for every two cycles. Similar results have been obtained by Constable et al.[86] in the case of poly(methyl methacrylate).

A particular case of mechanical fatigue stress is the ultrasonic treatment when reduced intensity but very high frequency stresses act on the polymer. Some data concerning the molecular mechanism and the structural-morphological changes caused by the ultrasonic treatment are revealed by the study of poly(vinyl chloride).[88,89] Ultrasonic treatment has been accomplished at PVC water slurry, the frequency being 40 kHz and the duration ranging between 10 and 120 h. The PVC used was obtained by the suspension process, having a molecular weight of about 55,000 kDa and the granulometric distribution ranging between 40 and 250 μm.

Macroscopically, the ultrasonic action on polymers is materialized in its intense dispersing and in the increase of weight of the 80-μm diameter fraction. This effect is, obviously, the result of the formation of new surfaces by the cleavage of the monolith structure of the grains due to the accumulation of mechanical stresses on the faults by the mechanism of crack appearance and propagation. The longer the duration of the ultrasonic treatment, the more intense the dispersion will be (Figure 20), as manifested by the prevailing accumulation of the 80-μm diameter fraction in the material.

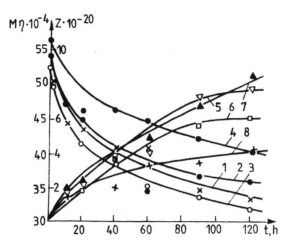

FIGURE 21.   Variation of the molecular weight ($\overline{M}_n$) and of the split bonds number (Z) as a function of ultrasonic treatment duration of PVC in aqueous suspension. 1 — $\overline{M}_n$ in unfractionated polymer; 2 — $\overline{M}_n$ in the 80-μ fraction; 3 — $\overline{M}_n$ in the ψ 200 fraction; 4 — $\overline{M}_n$ in the ψ 40-μ fraction; 5 — Z in the unfractionated polymer; 6 — Z in the ψ 80-μ fraction; 7 — Z in the ψ 200-μ fraction; 8 — Z in the ψ 40-μ fraction.

The mechanochemical act of the homolytical splitting of chemical bonds with the formation of macroradicals is the basis of the mechanodispersion process; this act is demonstrated both by the continuous diminishing of the molecular weight and by the increase of the number of split chemical bonds during the ultrasonic treatment (Figure 21).

The shape of the variation curve of the molecular weight is typical for the mechanodestructive processes, tending to a limiting value which, however, has not been attained on the experimental interval. It is also evident that a maximum intensity of the destruction is registered at the 200-μm diameter fraction, upon which the mechanical energy is especially concentrated.

The mechanodispersion process in the given conditions has, at the same time, led to the appearance of new functional groups by the reciprocal stabilization of the active destructing fragments or by their reaction with radical acceptors. The IR spectra have proved that, as a result of the ultrasonic treatment, the concentration of the end nonsaturations increases and numerous oxidated functions of the hydroxyl, carboxyl, or peroxidic types appear. All these bring about the increase of the newly formed surface reactivity, an increase rendered evident by the growth of the plasticizer sorption processes or by the acceleration of the reactions in the heterogeneous system. Therefore, the nonultrasonically treated 200-μm diameter fraction has a soaking period ranging between 92 and 105°C; after the ultrasonic treatment, even after the first 10 h, this fraction practically absorbs instantaneously the plasticizer, but the swelling and the dissolution are produced at the 25 to 70°C interval.

In light of these results, mechanodispersion has to be considered not only as a process of forming new surfaces, but also as a process which ensures the increase of the system free energy entails the displacement of the chemical and phase equilibriums. The existence in the solid body of an excess of free energy leads to the appearance of an unstable thermodynamic state.

We can consider that the ultrasonic treatment produces in every polymer grain a structural reaction made up of two interconditioning components, namely, mechanodegradation (or mechanocracking) in microdomains which constitute the active germs of mechanodispersion. In every moment of the process an equilibrium is established between the number of split molecules defining the decrease of the molecular weight and the corresponding specific surface, which is an evaluation criterion of mechanodispersion. The interdependence of these

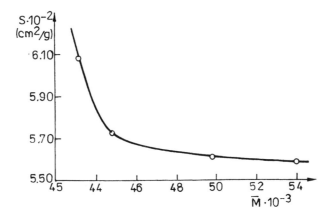

FIGURE 22. Specific surface variation (S) with molecular weight for different durations of ultrasonic treatment (0, 60, 90, 120 h).

effects brought about by the mechanical energy is clearly made evident for the 80-μm diameter fraction (Figure 22).

Each polymer grain is an imperfect conglomerate of small solid bodies kept together by adhesive forces (as a result from suspension polymerization technology), containing numerous faults such as pores, air bubbles, embeddings, and local breakings of the cohesive and adhesive bonds. These faults are, as a rule, statistically distributed inside the granulitic formation; the higher the dimensions of the grains, the more numerous the faults. Critical stresses appear because, under ultrasonic treatment, the mechanical energy is concentrated primarily on these very faults. They will strain the adhesive-type intergranulitic bonds, determining the grain dispersion up to certain dimensions but will cause, as well, the splitting of some chemical bonds from the polymer mass. Therefore, two types of bonds splitting — of adhesive and cohesive nature — occur. The place and the nature of the breaking will be decided by the position and the nature of the fault point from the domain in which the mechanical energy was located.

Taking into account the PVC polydispersed granulometric nature and the higher number of faults existing in the higher dimensioned grains, one could expect that both the mechanodestruction and the mechanodispersion should take place predominantly in these grains. The conclusion is supported by the distribution curves (Figure 20) which indicate the polymer uniformization by its concentration in fractions characterized by an optimum diameter for a given mechanical regime (80-μm diameter for the discussed case).

Figure 23 clearly illustrated the fact that mechanodestruction is rigorously correlated with the grains dimensions, i.e., the highest destruction rate (expressed by the value of the k constant) is obtained at the fraction with 200-μm diameter, even when the ultrasonic treatment is accomplished in the presence of mersolate (curve 2).

An important role in accomplishing the mechanodispersion is played by the liquid, as a function of its nature; it can loosen some physical interactions, especially under the conditions of penetrating in microcracks or gaps where it acts just like a "wedge", thus favoring both the mechanodestruction and mechanodispersion.

We have to emphasize that both processes are characterized by a limit determined by the polymer chemical nature and by the mechanical regime (nature of medium, working parameters, temperature). The diminishing of the polymer grains up to an equilibrium value takes place at the beginning, after which the dispersion stops but the molecular phenomenon still goes on in the stressed material, the result being a decrease in the molecular weight.

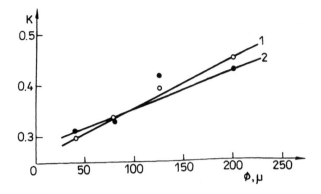

FIGURE 23.   Variation of mechanodestruction constant rate with PVC diameter particles. 1 — PVC ultrasonated in aqueous suspension with 0.1% mersolate; 2 — PVC ultrasonated in aqueous suspension.

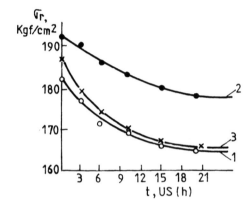

FIGURE 24.   Tensile strength variation with ultrasonic treatment duration. 1 — unfractionated PVC; 2 — ψ 45μ; 3 — ψ 100 μ.

The molecular weight reduction, which is a first consequence of the mechanochemical act, is macroscopically reflected in the change of some physicomechanical properties, among them being the tensile strength σ_R (see Reference 89). Accordingly, the σ_R variation with the parameters of the ultrasonic treatment process, especially with the time, can constitute an estimation criterion for the intensity of the mechanochemical processes which develop in the mechanically stressed material.

The σ_R evolution along with ultrasonic treatment duration both for the unreacted PVC and for the 45- and 100-μm diameter fractions, respectively, is shown in Figure 24. The shape of the curves is identical and typical for the mechanodestruction processes when the discussed property tends asymptotically to a limiting value.

From the granulometric distribution viewpoint, superior mechanical properties are to be noted for the more homogeneous fractions, as compared to the unfractioned polymer.

Having been given the curves's shape shown in Figure 24, we proceeded to the mathematical modeling of the PVC ultrasonic treatment fatigue process, the following equation being suggested:[90]

$$\sigma_R(t) = \frac{\sigma_R(cr)}{1 + \left[\dfrac{\sigma_R(cr)}{\sigma_R(in)} - 1\right] e^{-\sigma_R(cr)*K*t}} \tag{16}$$

where $\sigma_R(t)$ — σ_R value at the t time; $\sigma_R(cr)$ — the limiting value to which σ_R tends; $\sigma_R(in)$ — σ_R value corresponding to the nonultrasonically treated polymer; t — ultrasonic treatment time (h); and K — coefficient whose value is determined from experimental data by means of the least squares method:

$$ K = \frac{\nu}{\nu_R(cr)} * \frac{\Sigma \, y_i N_i}{\Sigma \, N_i^2} \tag{17} $$

where ν — frequency of the ultrasonic field (Hz) and N_i — number of stress cycles at the t time;

$$ y_i = \ln \frac{\dfrac{\sigma_R(cr)}{\sigma_R(in)} - 1}{\dfrac{\sigma_R(cr)}{\sigma_R(t)} - 1} \tag{18} $$

The theoretically obtained curves are practically identical to the experimental ones, thus demonstrating that the proposed mathematical model has been properly chosen.

The above-presented results has led to conclude that the fatigue process of the amorphous polymers under T_g has a mechanochemical basis, evolving practically according to the same laws as the fracture process of the polymers in a constant or variable monotonously increasing field of forces. From a macroscopic viewpoint, this fatigue process is manifested by forming new surfaces, rich in energy and with increased reactivity.

d. Vibratory Milling Processing

The intensive mechanical stress of rigid polymers, especially by vibratory milling, determines morphological changes materialized in the considerable dispersion of the material. The process consists in the reduction of the grain's geometrical dimensions and the formation of new surfaces, a phenomenon accompanied by the appearance of free macroradicals due to the splitting of the chemical bonds from the basic chain.

In the amorphous polymers case under T_g, the dispersion takes place up to very small geometrical limits of the grains (microns size), being accomplished by the molecular weight decrease. The process is determined by the polymer chemical nature, the mechanical processing regime, the reaction medium nature, etc. Generally speaking, the milling under these conditions goes up to 1- to 3-μm dimensions of the particles, after which the dispersion degree does not practically change, but instead, the reduction of the molecular weight keeps on going, as a result of the mechanochemical destruction process.

During vibratory milling, the polymer stores up the elastic energy received as shock and friction forces. The physicochemical changes produced under the action of the shock waves determined by the impact with the milling bodies as well as with the walls of the equipment take place extremely quickly (in less than 10^{-6} s). It is considered that, at the impressions which are developed in the front of the shock wave, the polymer is changed into an activated complex similar to the metallic state, characteristic for high degrees of compression. A relaxation effect takes place after the passing of the shock wave and a specific state of the matter, corresponding to the dilatation degree attained at the moment the activated complex decomposition is established.

The origin and nature of the extremely complicated physicochemical phenomena produced during the mechanodispersion of the rigid amorphous polymers are explained to some extent by the "magma-plasma" model,[91] Figure 25.

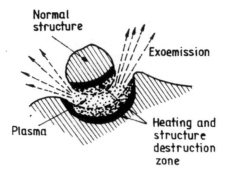

FIGURE 25. The "magma-plasma" model. (Based on Reference 91.)

According to this model, the energy generated by impact or friction, determines, due to the reduced thermal conductivity of the polymers, not only strong local overheatings but also the polymer passing into a new state made up of ions, radicals, electrons, i.e., a plasma state.

The anisotropy of the physicomechanical properties constitutes an important factor which controls the mechanochemical destruction. Thus, the polymers characterized by isotropic properties are milled, forming particles with a reduced degree of asymmetry, so that a determined direction of the destruction process could not be noticed. This fact evidently comes out from the increase of the polymer anisotropy, implicitly characterized, as well by the increase of the asymmetry degree of the obtained particles.

The nature of the liquid medium and their interaction with the polymer will greatly determine the evolution of the mechanodispersion process. In the first stages of the process, the liquid penetrates into the faulty zones of the polymer structure (preexisting cracks, interfibrilar spaces, nascent microcracks, etc.) giving rise to pressures which lead to the intensification of the material destruction. The nature of the gaseous atmosphere is very important, too. The higher the geometrical dimensions of the gas molecules are, the greater the "wedge" effect exerted by them on the top of the main crack will be, thus determining the increase of the mechanodispersion and mechanodegradation efficiency. An "active" gaseous atmosphere will determine, as well, reactions with the radical centers appeared on the material fracture surfaces, forming some new functional groups.

The most important phenomenon, a result of the mechanodispersion process, is the increase of the specific surface by forming new surfaces. The mechanism according to which this effect is accomplished covers all the stages of the fracture mechanism being based on the mechanochemical act. There is a relation between the attained degree of milling (reflected by the specific surface) and the molecular weight ($\overline{M}$), a relation proved to be true for many polymers, the amorphous ones included.

$$\overline{M} = f(S) \tag{19}$$

Thus, at the vibratory milling of polystyrene and poly(methyl methacrylate) in a gaseous medium, the molecular weight decrease is produced only at the same time with specific surface increase and with the grain dimensions decrease, respectively, while the established function is determined by the working temperature, the frequency of the mechanical stress, and the nature of the medium (Figures 26 and 27) (see Reference 92).

A similar effect appears of the PVC vibratory milling either in a dry medium or in the presence of methanol.[93,94] On analyzing the results shown in Figures 28 and 29 we have learned that the molecular weight is reduced (according to an exponential law characteristic

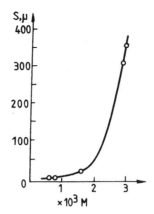

FIGURE 26. Variation of particles dimensions with molecular weight for polystyrene dispersed by vibratory milling.

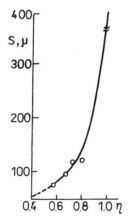

FIGURE 27. Variation of particle dimension with η for poly (methyl methacrylate) dispersed by vibratory milling.

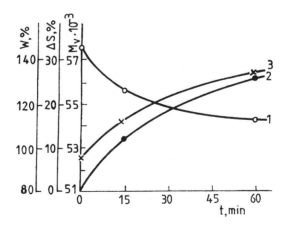

FIGURE 28. Influence of vibratory milling duration (in dry medium) on some characteristics of PVC-S. 1 — $\overline{M}_\eta$; 2 — specific surface (ΔS, %); 3 — plasticizer sorption (%).

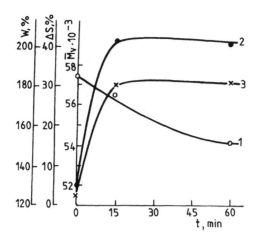

FIGURE 29. Influence of vibratory milling duration (in methanol) on some characteristics of PVC-S. 1 — $\overline{M}_\eta$; 2 — specific surface (ΔS, %); 3 — plasticizer sorption (%).

to the mechanochemical processes) at the same time with the increase of the vibratory milling durati/ong (curve 1), while the specific surface registers a relative continuous increase (curve 2).

One can note a higher efficiency of the mechanodispersion process in the presence of methanol (Figure 29), a consequence of this polar liquid corrosive effect, which penetrates in microcracks and in the main crack accelerating the nucleation process and the propagation of the latter one. Accordingly, the molecular weight of the polymer milled under these conditions registers a more pronounced decrease as compared to that in the dry medium. The increase of the specific surface as well as the appearance and multiplication of the functional groups from the newly formed surfaces increase the polymer reactivity and improve some physicochemical properties. Thus, the plasticizer sorption increases during the dry milling from 95% (the initial polymer) to 135% after 60 min processing (curve 3, Figure 28). This coefficient has a much higher value in methanol, reaching the 220% value only after 15 min of milling. All these changes are based on the mechanochemical act starting when obtaining a critical amount of elastic energy stored up in the polymer during milling, thus, macroradicals are generated; those present during the entire mechanical stress can be pointed out by reactions with specific radical acceptors or aromatic diamines.

Therefore, by the PVC vibratory milling in the presence of benzidine and p-phenylene diamine (p-PhDA), products made up of two fractions (one soluble in cyclohexanone and quantitatively dominant — over 80% — the other one insoluble) are obtained. The chemical bonding of the micromolecular compound to the PVC destruction fragments has been proved both by spectral methods and by reactions specific to the aromatic aminic group. In the soluble fraction IR spectra, there appear new absorption bands, such as those from 1490 and 1600 cm^{-1} typical of the bond $->$C–N$<$, and from 1700 cm^{-1} attributed to the –C=N– bond. A marked maximum value at 3400 cm^{-1} characteristic for the primary imidic and aminic groups is to be noted, as well.

Considering that the aromatic diamine works as a radical acceptor through one of the functional groups, the other one remaining free on the destruction fragments end, we have proceeded to accomplish the diazotization reaction and subsequent coupling with phenolic components. In this way we have obtained products with a color that is specific to the newly formed azoic groups, the color intensity being a function of the coupled diamine amount, an indirect reflection of the mechanical regime parameter's influence on the reaction.

TABLE 1
The Influence of the Milling Duration on
the Chemically Bonded Diamine Quantity
(10 g of PVC + 2 g of Aromatic Diamine

	Duration of the mechanical activation (h)		
	24	72	96
	Chemically bonded diamine amount (%)		
Benzidine	1.60	4.80	5.46
p-Phenylene diamine	1.58	3.39	5.40

TABLE 2
The Influence of the Diamine Quantity on
the Reaction Efficiency with the
Mechanoradicals Formed by PVC
Vibratory Milling (10 g, 24 h)

	Quantity of diamine from the initial mixture (g)		
	1	2	3
	Chemically bonded diamine amount (%)		
Benzidine	1.60	4.07	4.60
p-Phenylene diamine	1.58	2.78	3.30

The analysis of the reacted diamine amount has therefore pointed out its dependence on the milling duration, on the diamine quantity from the PVC mixture subjected to the reaction, and on the micromolecular compound chemical nature.

The influence of the milling duration is reflected by the results given in Table 1. One can note the continuous increase of the reacted diamine quantity with this parameter of the mechanical regime, a consequence of the free radical continuous generation at the newly formed surfaces.

The diamine amount from the milling-subjected mixture (10 g of PVC) influences, as well, the reaction efficiency (Table 2) in what it is increasing.

From the above-given data we can conclude that the PVC reaction with aromatic diamine is conditioned by the chemical nature of the micromolecular compound. The most active seems to be benzidine, whose content in synthesis products is constantly higher, providing, as well, deeper colors of the diazotizing and coupling products. The mechanism of the reaction between the PVC radical fragments and the aromatic diamine is the one previously presented (Section b), a mechanism which explains both the reduction of the polymer molecular weight and the presence of the free aminic groups.

It must be mentioned that new products synthesized by PVC vibratory milling in the benzidine and p-PhDA presence are characterized by their particular thermostability pointed out both by the losses in weight with the temperature, losses which are inferior to the PVC witness, and by the calcination residue, which is superior (at 1200°C) demonstrating the formation of some structures with a remarkable specific thermal stability at temperature.

The molecular weight decrease and the possibility of obtaining some reactions of the PVC with the aromatic diamine-type radical acceptors are conclusive arguments for a radical destructive mechanism of the polymer under vibratory milling conditions. Consequently, mechanodispersion is a mechanochemical-type process, developing itself according to the general laws of polymer fracture under stress.

2. Characteristics of the Deformation and Fracture Processes of the Semicrystalline Polymers

The majority of the studies dealing with the deformation and fracture processes of solid-state polymers has been accomplished on semicrystalline macromolecular compounds, especially on fibers subjected to high stresses, mainly tensile ones.[95-102] High concentrations, i.e., up to 10^{17} to 10^{18} spins/g of free radicals even at small deformations (8%) are obtained in the case of such polymers. The explanation consists of the presence of numerous and strong interactions between the macromolecules which aggregate in crystalline groups, interactions which act in the same direction with the crosslinks; thus, the chain's relative displacement is prevented, the mechanical energy being concentrated on the chemical bonds.

Which are the fault zones at a supermolecular and morphological level upon which the stress is being concentrated? The data obtained up to the present for the semicrystalline oriented polymers of the textile fibers types, for instance, have pointed out the existence of two heterogeneity levels in their supermolecular organization, namely, micro- and macrofibril (for example, see Reference 103). Microfibrils are supermolecular elements with a diameter of several hundred angstroms and a length microns; in turn, they group together (hundreds, even thousands, of them) in great individual units called macrofibrils.[104] The X-rays diffraction at small angles reveals regions of maximum order (crystallites) and irregular ones (amorphous) along the microfibril axis. The structure of those two domains is the major problem; among the numerous models that have been proposed, the most adequate is the one suggested by Hoseman and Bonart. According to this model, based on the discovery of polymer unique crystals, the oriented microfibrillar crystallites contain folded macromolecules, among them existing a reduced number of connecting macromolecules. The microfibril ends are localized especially at the boundary between them, a fact which represents a reduction or even an interruption of the axial bonds between microfibrils, bonds ensured by the connecting macromolecules which pass through the amorphous zone. These bonds can be intrafibrillar, when they connect the crystalline blocks of the same microfibril (being responsible for the fiber's axial modulus of elasticity) or interfibrillar, when they connect the blocks of some different microfibrils (Figure 30).

The amorphous regions in the oriented semicrystalline polymers have small dimensions,[105] their length ranging between 15 and 17 Å under the conditions in which the crystalline regions occupy 70 to 80% from the microfibrile volume.

The strong interchain bonds from the crystalline domains will guide the stress upon the amorphous domains, and upon the connecting macromolecules which interconnect the crystallites, respectively. At the uniaxial stress, the crystallites pass into the mechanoexcited state, being subjected to the mechanocracking process finalized by the macroradical formation, evidenced by the ESR method.

It has been experimentally found out that the number of macroradicals formed by uniaxial extension in the nylon 6 fibers is smaller than that of the connecting macromolecules which cross the macroscopic crack, a fact which enabled Peterlin to suggest a facture model of the fibrillar semicrystalline polymers, schematically represented in Figure 31;[106] this model has been based on the following considerations:

1.   The free radical concentration does not depend on the stress but on the deformation.

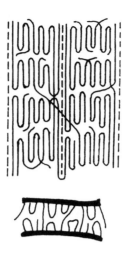

FIGURE 30. Microfibrillar structure schematic representation of oriented semicrystalline polymers. (Based on Reference 104.)

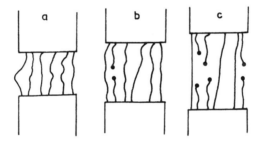

FIGURE 31. Peterlin's schematic model for deformation and fracture of oriented semicrystalline polymers. (Based on Reference 106.)

2. The macroradical concentration is not modified after the stress is removed.
3. Not even a pair of mechanoradicals is formed during a new stress until the deformation accomplished at the previous load has not been attained.

The interfibrillar connecting macromolecules which cross the fault zones in an unstrained sample are not all elongated; their length between two crystalline blocks varies significantly according to the distribution function presented in Figure 32.[107,108] The crystalline blocks are gradually moved at the stress application so that the shortest macromolecules are elongated up to their maximum possible length (Figure 31a), then they broke forming a macroradical pair (b). By the crystallite's continuous displacement, more and more macromolecules attain the extended conformation, being overloaded and thus generating new mechanoradical pairs (c); consequently, their concentration increases at the same time with the deformation.

The more coiled is the conformation of the connecting macromolecules, the higher are the values of the deformation at which the fracture occurs. This hypothesis is proved by Figure 33 which shows the results obtained for two nylon 6 fibers, one of them being previously heated at 199°C. The applied thermal treatment led to the relaxation of the initial oriented macromolecules which, thus, became the shape of a statistically coiled ball. To obtain the same concentration of the split chemical bonds (here expressed by the fracture probability), deformations higher than those needed for the thermally untreated sample should be imposed.

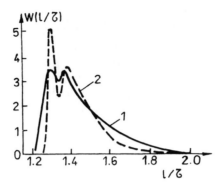

FIGURE 32. Distribution function of macromolecular chains length in the disordered domains of nylon 6. 1 — $\overline{M}_w$ = 3 * 10⁴; λ = 5.2Å; 2 — $\overline{M}_w$ = 5.8 * 10⁴; λ = 5.2Å (λ — the distance between crystalline blocks). (Based on Reference 107.)

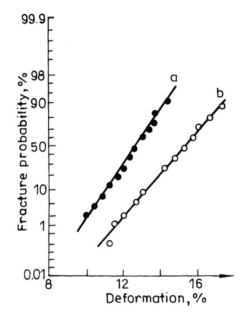

FIGURE 33. Fracture probability as a function of nylon 6 tensile deformation. a — witness; b — treated at 199°C. (Based on Reference 108.)

The formed macroradicals accumulated in microvolumes generate subsequently the microcrack, i.e., germ of macroscopic fracture, according to the previously discussed mechanism.

The model suggested by Peterlin for the deformation and fracture of the semicrystalline polymers which assumes that the crystallites are not affected in the process of mechanical stress is supported by experimental data. Thus, ESR signal characteristic for the formed macroradicals presents anisotropy, while the mechanoradicals can initiate graft and block copolymerization reactions of some specially introduced monomers which penetrate only in the amorphous zones. Finally, the results obtained by Sohma et al. are very suggestive.[110] It has been learned that, by straining a polyethylene sample previously treated with fuming nitric acid, a detectable ESR signal cannot be obtained, while the untreated sample easily produces the ESR spectrum. It is evident that this result is a consequence of the fact that

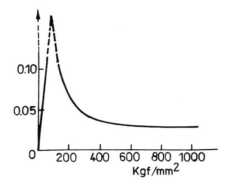

FIGURE 34. Stress distribution on chemical bonds from isotactic polypropilene. (Based on Reference 111.)

the nitric acid destroyed both the amorphous domains and the connecting macromolecules, which interconnects the crystallites.

The structural models accepted nowadays for the semicrystalline polymers imply however the existence of some faults even in the semicrystallites constitution. Thus, it is explainable why the mechanical stressing of such polymers under certain conditions can lead to the strong reduction of crystallinity. In this case, too, the mechanism of the deformation and fracture processes passes through the already known stages, having a chained polystadial feature. At its basis lies the mechanochemical act of splitting the covalent bonds from the crystallite faults points. In vibratory ball mills, for instance, when important shock and friction forces act on the polymer, the mechanodispersion is accompanied both by the reduction of the molecular weight and of the crystallinity, and by the increase of the specific surface.

a. Static Tensile Stress

Owing to its outstanding practical importance, the time dependence of polymer capacity of being subjected to constantly applied mechanical stresses has been thoroughly studied. Only a few studies approach the creeping subject with semicrystalline polymers used as yarns and fibers and even fewer refer to the elucidation of the molecular aspects implied into this process.

Among these, the ones referring to the behavior of polypropylene, poly(ethylene terephthalate), polyamide 6, and copolymer based on acrylonitrile-vinyl acetate-α-methyl styrene[111-125] are worth mentioning.

Thus, using the IR spectroscopy for the study of polypropylene monofilament creep, important data concerning the effort distribution on chemical bonds have been obtained. This fact allowed the determination of the fraction of the chemical bonds passing into a mechanoactivated state, previous to fracture, at a certain stress value (Figure 34).

The monofilament creeping study based on poly(acrylonitrile-co-vinyl acetate-co-α-methyl styrene) permitted the attainment of the experimental rheological curves, as well as the study of the monofilament behavior at different stages of their obtaining technology: after spinning, wet stretching; avivage, dry stretching.[115]

These ones are shown in Figure 35 for monofilaments with 6-denier fineness, stressed at a constant stress in their domain of viscoelastic behavior on the 0- to 60-min interval.

It may be remarked that the typical shape of creep curves for all the analyzed samples but also differences in some of their characteristics are evident. Thus, the instantaneous deformation as well as the remanent one decrease in value from the fiber which is immediately samples after spinning to the stretched and dried one, pointing out the structure rigidizing tendency by the applied mechanical treatment.

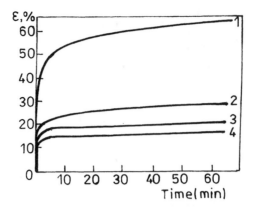

FIGURE 35. Experimental creep curves for monofilaments of poly (acrylonitrile-co-vinyl acetate-co-α-methyl styrene) (6 den fineness) at different stages of their obtaining technology. 1 — after spinning; 2 — after wet stretching; 3 — after avivage; 4 — after stretching and drying.

TABLE 3
Rheological Parameters Calculated from the Monofilament Creeping Experimental Curves (6 den)

	Rheological parameters			
Processing phase	Relaxation time Θ_r (s)	High elasticity modulus, G_1 (gf/mm²)	Viscoelasticity modulus, G_2 (gf/mm²)	Viscosity, $\eta * 10^{-10}$ (gf * s/mm)
Spinning	242	1.714	0.75	2.64
Wet stretching	273	2.000	2.83	3.96
Avivage	296	2.107	4.09	4.16
Stretching drying	460	2.727	7.50	5.94

Note: Based on poly(acrylonitrile-co-vinyl acetate-co-α-methyl styrene).

The same effect is reflected by increasing tendency of some rheological parameters, which are calculated on the basis of creeping equation having the form:

$$\epsilon = A + B * t - C * D^{-t} \qquad (20)$$

their values being presented in Table 3.

The increase of the crystallinity index of the orientation factor as well as of the dissolving heat along with stretching — a well-known effect — suggests at the same time the structure stiffening by intermoelcular interaction intensification, due to the increase of macromolecules packing degree.

The existence of a deformation plastic component, even in the case of the dry-stretching monofilament (curve 4, Figure 35) proves also that macromolecule fracture takes place in the polymer during stressing. These bond splittings, nevertheless, are not so frequent as to be pointed out by the molecular weight decrease.

For establishing whether under the tensile creeping conditions the studied copolymer undergoes mechanochemical reactions, it has been proceeded to the tensioning of some monofilaments immersed in a radical acceptor solution. Thus, it has been found out that after a 60-min stressing under a constant load and in the presence of DPPH, the monofilaments become colored on the area immersed in solution, the color being stable to repeated extractions with methanol.

Similarly, the monofilament stressed in methanol solution of aromatic diamines acquires a shade specific to the micromolecular compound and which lasts even after the copolymer dissolving and reprecipitation; this fact becomes outstanding proof that the diamine has been chemically bonded to the copolymer chains.

An argument in this sense might also be considered the possibility of diazotizing and coupling of the free aminic groups from the chemically bounded aromatic diamine with phenolic compounds (resorcinol, naphthol) and which causes the appearance of colors specific to the formed azo dyes.

Filaments reacted with benzidine could also be directly dyed in weak shades with acid dyes. Although it was not possible to accomplish a quantitative correlation between the used radical acceptor amount and the value of the deformation obtained at different creeping times, an experimental finding is worth mentioning. The color of the monofilaments stressed at short durations (10 to 20 min) is very low and, practically, it intensifies along with the stressing duration. After about 50 min of creeping its intensity remains invariable. By correlating this result with the creep curve shape, it becomes obvious that mechanodegradation intensity depends to a lesser degree on the stress value and is more evidently dependent on the deformation value which after 30 min of stressing remains practically steady. Similar results have been also obtained for some other types of fibers, as, for instance, the ones based on poly[p-2(hydroxyetoxy)benzene carboxylic acid] (see References 116 and 117) or nylon 6 (see Reference 118).

It may be concluded that, during the static stress of a semicrystalline polymer of poly(acrylonitrile-co-vinyl acetate-co-α-methyl styrene) type under monofilaments form, there appear mechanochemical reactions which finish with free radical formation even at deformations that are lower in comparison with the macroscopic fracture one. This monofilament deformation mechanism is thus of a mechanochemical type, behaving this way on the whole stress period up to their fracture.

b. Dynamic Tensile Stress

The elaboration and development of the molecular theories concerning polymer fracture under load have been achieved on the basis of some experimental data obtained by using some various techniques and methods for pointing out the chemical bond scission, especially in spinable semicrystalline polymers stressed in alternative force fields. A direct way of pointing out the chemical splitting induced by the load is ESR spectoscopy. Thus, the experiments accomplished by Jurkov proved that, at the dynamic stress of nylon 6 fibers, the free radical concentration grows faster in the case of a constant deformation rate than with a constant stress.[126] Therefore, the stress increase associated with a constant rate accelerates the chemical bond splitting. At the same polymer as well as at poly(ethylene terephthalate) or poly(hexamethylene sebachamide), the ESR signal is visible at about 0.4 σ_R, and is generally characteristic to the secondary radicals formed by a chain transfer.[126,127] This radical concentration, especially in nylon 6, diminishes severely in the presence of picric acid (radical acceptor), as shown by Butt et al.[128] and DeVries and Roylance.[129]

Using the ESR method, the influence of temperature on the free radical concentration could be established, this effect being, as a matter of fact, suggested by Jurkov's equation. By carrying on such experiments on polyamide fibers,[130] Johnson finds the relation between free radical concentration increase, with the temperature inversion. It must be nevertheless pointed out that temperature does not influence only the free radical generation but also the apparent distribution of chain lengths which are to split. It seems that a higher temperature stress allows higher breaking deformations and thus a wide distribution of chain lengths which can be split.

TABLE 4
The Free Radical Relative Concentration and the Content of Grafted Vinyl Chloride as a Result of Mechanical Stressing of Monofilaments Based on Poly(Acrylonitrile-co-Vinyl Acetate-co-α-Methyl Styrene) in Different Conditions

Mechanical parameters			Free radical	Grafted vinyl
Deformation (% from the fracture one)	Deformation rate (mm/min)	DPPH concentration (relative units, %)	concentration (relative units, %)	chloride content (%)
50	5.0	72.0	18.0	2.07
75	5.0	22.0	78.0	2.85
100	5.0	21.0	79.0	2.92
100	2.4	48.5	51.5	2.36
100	8.1	12.0	88.0	2.87
100	13.5	8.5	91.5	3.02

The fracture process is also influenced by the rate with which the stress or deformation is applied. A higher concentration of radical active centers at the given deformation is accomplished concurrently with the deformation rate increase, in the case of nylon 6 fibers (see Reference 131).

Recent experiments by ESR have been carried by Popa[115] on poly(acrylonitrile-co-vinyl acetate-co-α-methyl styrene) monofilaments which were dynamically stressed. Singlet-type spectra have been obtained, which are like those recorded at thermal degradation of the same polymer.

It can be certainly asserted that the macroradicals which are mechanically generated in these monofilaments exhibit the unpaired electron located on a tertiary C atom; this fact mainly suggests the splitting of that chemical bond in which the quaternary carbon atom from α-methyl styrene structural unit is implied.

An indirect technique has been used for accomplishing a quantitative correlation of the generated macroradical concentration function of the mechanical regime parameters. Ternary copolymer monofilaments have been soaked with DPPH and then subdued to stressing. The analyses were based on the ESR dosing of the concentration of DPPH stable radical concentration that were not consumed in the reaction with the newly formed macroradicals in the stressed sample; on the basis of these results, the relative concentration of mechanically generated macroradicals has been calculated.

Some results concerning mechanical stress effect are pointed out in Table 4. It must be pointed out that the macroradicals are produced in high concentrations even at deformations much inferior to the fracture one, in accordance with Peterlin mechanism concerning the fracture of fibrillar semicrystalline polymers.[106]

The intercrystalline domains are passed by linking macromolecules with more or less stretched shapes. When the monofilament is subdued to deformation, the linking macromolecules are gradually brought to an extended state, being in turn submitted to homolytic splitting. Therefore, the fracture process is influenced, in a higher degree, by the deformation value and, to a lesser extent, by the stress value.

The stressing rate increase also has as an effect on the appearance of great macroradical concentration as a consequence of the fact that the stress action duration becomes comparable with the relaxation time. The overstressed macromolecules lose their capacity to rearrange under load, mainly in the case of the copolymer under study, which is preeminently rigid and, consequently, undergoes the homolytic splitting.

It must be pointed out that macroradicals appear at low deformations within the whole volume of the sample, but they are placed in its amorphous domains, as Becht and Fischer have demonstrated[127] using an indirect analyzing technique. It has been thus found out that

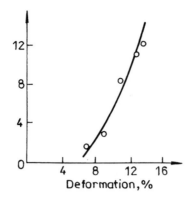

FIGURE 36. Molecular weight variation (relative units) with deformation for nylon 6 fibers. (Based on Reference 135.)

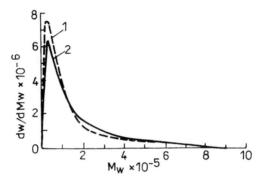

FIGURE 37. Molecular weight distribution before (2) and after (1) deformation and fracture of nylon 6 fibers. (Based on Reference 133.)

by stressing some polyamide 6 monofilaments immersed in vinyl monomers that polymerize easily (acrylic acid) the micromolecular compound penetrates the fiber and, diffusing obviously in the amorphous regions with a lowered packing degree, it polymerizes in contact with the macroradicals.

Oprea and Popa, making use of the same method, have succeeded in accomplishing the vinyl monomer polymerization initiated by mechanoradicals by various rate stressing or at various deformations of some ternary copolymer monofilament bundles based on acrylonitrile and in the vinyl chloride presence. Although reduced, the amount of grafted or block copolymerized monomer is correlated with the mechanical parameters and is in a perfect accordance with ESR spectroscopy (Table 4).

As some research has demonstrated,[132-135] the great number of chemical bonds from the main chain, split under stress action up to the 10^{18} scissions/g polymer, suggests that, as a consequence of the mechanical stress, important molecular weight changes take place. Thus, Becht and Fischer discovered considerable changes consisting of the molecular weight reduction with up to 25% of its initial value. The deformation increase reduces in a higher degree this characteristic of nylon 6 as compared with the stress increase, as shown by Becht and Fischer[135] (Figure 36). Molecular weights distribution, before and after fracture at room temperature, for the same polymer also illustrates the average value reduction of molecular weight, corresponding to a number of splittings of 3×10^{18} scissions/g, in a good accordance with ESR spectroscopy data and with Mehta's viscometric measurements;[133] at the same time, a diminishing of the polydispersion range can be observed (Figure 37).

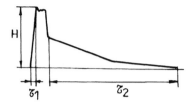

FIGURE 38. IR signal obtained by polymers tensile stressing. (Based on Reference 141.)

Stoeckel's study[136] about semicrystalline polymers in fibers forming [nylon 6,nylon 6,6,poly(ethylene terephthalate), polypropylene] correlates the split chemical bond concentration with molecular weight changes, which permits at the same time some results to be obtained that are in accordance with ESR or IR spectral analyses. The chemical bonds splitting in dynamically stressed semicrystalline polymers may be pointed out by IR spectral analysis, which permits both the cognition of the way in which the stress is distributed on the chemical bonds and the pointing out of the final functional groups which appear as a result of the macroradical stabilization.[136-138] By using the same method, Lisnevski[139] obtained information concerning the accumulation of the $>C=C<$ double bonds, or hydroxyl and carboxyl groups in polypropylene fibers. The same method is also used by Gafurov,[140] who studied the changes produced in poly(ϵ-caprolactame) by dynamic stress under the conditions of temperature increase considering as a criterion for macromolecules breaking the final carboxylic group accumulation.

Some other indirect methods for pointing out the macromolecule splitting under load are based on the study of the associated phenomena: thermic, acoustic, electronic, or photonic emissions.

The systematic study of thermal effect on a variety of semicrystalline polymers [poly(ethylene terephthalate),nylon 6, polypropylene] subdues to stretching in dynamic conditions has been accomplished by Tomasevski and Egorov[141,142] on the principle of recording the emitted IR radiations. The shape of the obtained IR signal form is about the same for all the studied polymers (Figure 38). The time τ_1 is correlated with the appearance of the exothermal effect in the polymer "incandescent" zone, close to the top of the magistral crack, ultimately indicating its increasing rate (about 400 m/s). τ_2 represents the necessary time for cooling this zone after the fracture surface formation. The amount of heat generated in the sample in the case of poly(ethylene terephthalate), for instance, is situated in the interval $(2 \text{ to } 5) \times 10^{-3}$ cal/mm^2.

Acoustic and electronic emissions, very useful for studying the rigid amorphous polymers fracture, are less applicable on semicrystalline polymers (see Reference 143). The obtained results demonstrate that the acoustic emission is a consequence of microcrackings and magistral crack formation. The phenomenon is due to the difference in energy between the mechanoactivated state and the equilibrium state attained by the macromolecules after the mechanochemical act takes place; the energy in this case dissipates in the polymer under the form of acoustic energy (eventually thermal or luminous energy).[144,145] Therefore, it results that this phenomenon may become very useful for establishing the stress and tensile strain which can be supported by a polymer without suffering irreversible changes of the structure that could modify its properties. Therefore, premises for polymer characterization are created not so much by stress and strain (used frequently at present to determine the mechanical properties of a macromolecular compound), but by the stress and strain corresponding to the crack formation; therefore, valuable information can be obtained regarding the more judicious and longer-lasting use of the polymers.

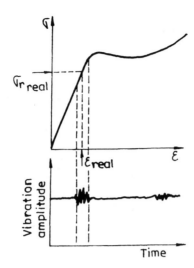

FIGURE 39. Acoustic diagram correlation with the rheological curve of a polymer.

Starting from these premises and from the assertion that during dynamical tensile stress of monofilaments based on poly(acrylonitrile-co-vinyl acetate-co-α-methyl styrene) there are emitted sharp sounds which can be easily perceived even by the ear, a device able to detect the acoustic emission has been created by Vasiliu Oprea et al.[145] The sonic sounder, the essential piece of the device, takes over the vibrations produced by the acoustic emission in the monofilament bundle and transform them in electrical pulse which is simultaneously recorded with the stress-strain curve (marked by the dynamometer). In this way, the representation from Figure 39 is obtained.

In the main, the point on the rheological curve corresponding to the maximal number of impulses in acoustic diagram identifies the values of the stress and deformation at which the microcracks are generated in polymer. These values are comprised in the interval 80 to 90% out of those corresponding to the stress and macroscopic fracture deformation.

The luminescence at polymer deformation and fracture during the tensile stress, compression, mechanical dispersion, fatigue, etc. has been discovered and thoroughly studied by Butiaghin.[147,148] The mechanism of its appearance has been less investigated, but at present it is an accepted hypothesis that chemiluminescence plays an important role in its appearance, owing to the free radical reactions.

The phenomenon has been recently pointed out under the conditions of shearing stresses of ternary copolymer based on acrylonitrile, vinyl acetate, α-methyl styrene films,[115] it being recorded on a photographic film (Figure 40). Luminous punctiform pulses of higher intensity and frequency can be very clearly observed toward the polymer film extremities, where the stress rate as well as the shearing stress value are maximal.[148,149]

Even if this phenomenon has not been well explained so far, its effect can be used for determining the stress areas (photoelasticity) from the polymeric materials, with obvious practical applications.

The gathered experimental results concerning the behavior of the monofilaments of poly(acrylonitrile-co-vinyl acetate-co-α-methyl styrene) in various mechanical stress conditions have become valuable arguments for the radical feature of this process and permitted the formulation of a mechanism of their deformation and fracture.

The supermolecular structure of this ternary copolymer can be well represented by an alternative of Hoseman-Bonnart model (Figure 41). The macromolecules go successively through the ordered and amorphous domains, adopting folded conformations in the

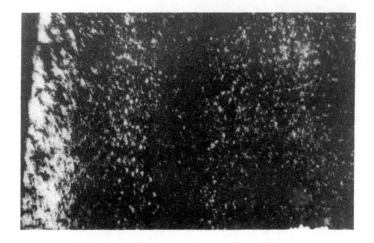

FIGURE 40. Chemiluminescence produced by shear stressing of a poly(acrylonitrile-co-vinyl acetate-co-α-methyl styrene) film.

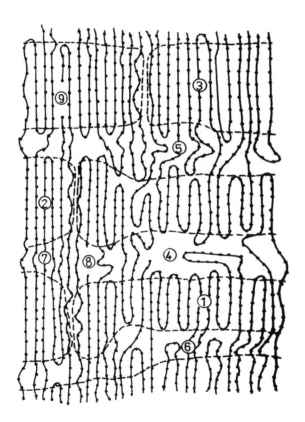

FIGURE 41. Supermolecular structure of a monofilament based on poly(acrylonitrile-co-vinyl acetate-co-α-methyl styrene). 1 — monocrystals; 2 — monofibrils constituted during the wet-drawn process; 3 — monofibril bundless formed by hot-drawn process; 4 — holes; 5 — long folded chains; 6 — short folded chains; 7 — link macromolecules; 8 — chain ends in the amorphous domains; 9 — chain ends in the crystalline domains.

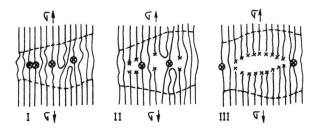

FIGURE 42. Schematic representation of a microvolume from amorphous domain passed by a link macromolecule (⊗ ~ — mechanoactivated macromolecule; ● ~ — macroradicals).

crystallites and statistically coiled shapes, respectively, in the zones having a reduced arrangements degree (amorphous). The crystalline phase is made up of monocrystals (1), monofibrils constituted during the wet-drawn process (2), and fibrils formed at hot-drawn process (3). These maximum-ordered domains alternate with the amorphous ones, made up of holes (4), long (5) or short (6) folded chains, more or less coiled macromolecules (7), and chain ends (8).

In some cases the order of crystalline domains can be easily disturbed by chain head existence (9). In mechanical stress conditions, the stress is concentrated on the structural faults which, at the overmolecular level, are represented by the amorphous zones; these amorphous domains will be the first in taking over this load. The overall deformation will be the consequence of the accumulation of all deformations in these zones which are gone through by macromolecules interconnecting the crystalline domains. The macromolecule's reciprocal shifting will be impeded because they are fixed in the structure matrix due to the crystalline zones or to some strong physical interactions (hydrogen bonds, van der Waals forces, etc.). Along with the deformation increase, the macromolecules will uncoil themselves, acquiring an elongated conformation, while the already stretched ones will store up a quantity of elastic energy passing in a mechanoactivated state. The phenomenon is produced at low deformations of the sample, below 50% of the fracture one.

Due to the nonhomogenity of the structure, certain macromolecular chains become oriented before some others, so that the number of the completely stretched ones depends on the deformation to which the monofilament is subjected.

Let us consider a microvolume from the amorphous domain of the polymer, passed through by link macromolecules (Figure 42). At the stress applied, the mechanical energy is first concentrated on the chains presenting an already perfectly stretched conformation, a fact which brings them in a mechanoactivated state (Figure 42I). Under the combined action of the continuous increasing stress, as well as of the thermal fluctuations, the homolytic splitting will be produced, according to the fracture kinetic theory, when the energy stored on some chemical bonds will reach values commensurable with the activating energy of their splitting.

At the chemical structure level, the stress concentration factor is made up by the most ramified carbon atom (the quaternary one). Consequently, the neighboring bonds or those from the structural unit of α-methyl styrene will be split, generating new primary macroradicals (Figure 43). In the microvolume taken into consideration, more mechanochemical acts are simultaneously produced, supplying the same number of macroradicals pairs (Figure 42II). As a result of some macromolecule splitting, the stress is unequally redistributed on the ones left undamaged in the considered volume, a function of their stretching degree. If the stress or deformation is still constant, the rate of the chemical bond splitting is produced up to its suppression. The process advances only if the stress or deformation increases for bringing new macromolecules in a completely stretched state.

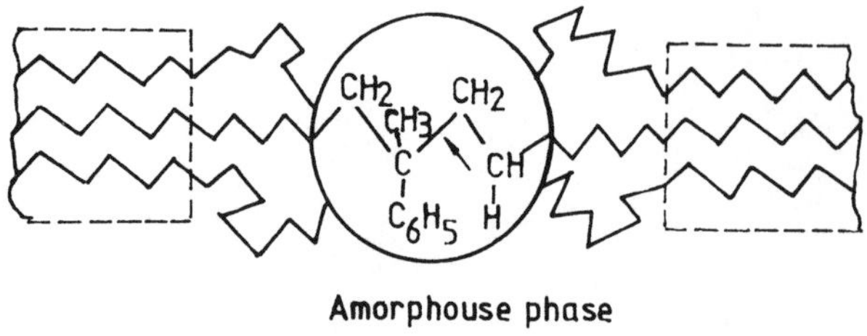

Amorphouse phase

FIGURE 43. A fragment of a strained link-macromolecule (from amorphous domain) which splits the more labile chemical bond.

The macromolecule fractures have the tendency to accumulate in a limited volume, because the stress is distributed in a restricted zone of the sample. To these primary splittings generated by stress, there are added some other chemical bonds scissions induced by the primary mechanoradicals which initiate a chained reaction. They immediately attack the neighboring macromolecules by a chain transfer, leading to the formation of a stable macromolecule as well as of a secondary radical with its electron situated on the chain.

$$\cdots -CH_2-\overset{\overset{\textstyle CH_3}{|}}{\underset{\underset{\textstyle C_6H_5}{|}}{\dot{C}}} + \cdots -CH_2-\overset{\overset{\textstyle CH_3}{|}}{\underset{\underset{\textstyle R}{|}}{CH}}-CH_2-\overset{\overset{\textstyle CH_3}{|}}{\underset{\underset{\textstyle C_6H_5}{|}}{C}}- \cdots \longrightarrow$$

$$\longrightarrow \cdots -CH_2-\overset{\overset{\textstyle CH_3}{|}}{\underset{\underset{\textstyle C_6H_5}{|}}{CH}} + \cdots -CH_2-\overset{\overset{\textstyle CH_3}{|}}{\underset{\underset{\textstyle R}{|}}{\dot{C}}}-CH_2-\overset{\overset{\textstyle CH_3}{|}}{\underset{\underset{\textstyle C_6H_5}{|}}{C}}- \cdots$$

where $R = -CN$ or $-OCOCH_3$.

The snatch of a hydrogen atom from the chain reduces the activation energy of splitting neighboring bond, which becomes, according to the literature, about 1/3 from the one characteristic to the usual bond $->C–C<-$. Consequently, the macromolecular chains split easily near the central radical, leading to the formation of a stable final group and of a new end radical, which is still active.

$$\cdots -CH_2-\overset{\overset{\textstyle CH_3}{|}}{\underset{\underset{\textstyle R}{|}}{\dot{C}}}-CH_2-\overset{\overset{\textstyle CH_3}{|}}{\underset{\underset{\textstyle C_6H_5}{|}}{C}}- \cdots \longrightarrow \cdots -CH=CH + \dot{C}H_2-\overset{\overset{\textstyle CH_3}{|}}{\underset{\underset{\textstyle C_6H_5}{|}}{C}}- \cdots$$

The new formed radical generates another cycle of reactions, the process continuing this way as long as on the strained macromolecule a tertiary radical will exist and it will stop if this one is stabilized by a chain transfer or if it is produced by the stress relaxation.

Being accumulated in a limited volume, the split chemical bonds will have as a consequence a microcrack formation (Figure 42III). Concurrently with the stress intensification, a greater and greater number of macromolecules will reach the limit conformation. The local stress reaches faster and faster its critical value because the load is distributed on a more and more reduced number of chains. A great number of "nascent" microcracks appear in

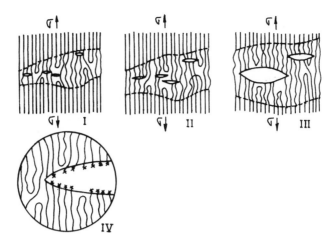

FIGURE 44. A microdomain from amorphous phase of the ternary copolymer in which the microcrack nucleation (I), microcrack growth (II), microcrack coalescence and magistral crack formation (III), and magistral crack propagation by link-macromolecules splitting are produced.

the polymer, great quantities of energy being released at their opening (acoustic, luminous, etc.) (Figure 44I). These microcracks, to which there are added both the holes from the material mass and the domains in which end chains extremities exist, have nevertheless, too reduced dimensions for producing the monofilament macroscopic fracture. The microcracks increase (Figure 44II) or coalesce until at least one of them reaches critical dimensions (Thomas-Rivlin criterion) generating thus the magistral crack (Figure 44III).

The formation or nucleation of the magistral crack is only the first step which causes the macroscopic fracture beginning. In order that this macroscopic crack should take place, the magistral crack must propagate through material, a fact which implies the creation of new areas (by molecular breakings), the energy dissipation by nonelastic losses (which do not produce chemical bonds breakings) or by some other forms of energy.

Magistral crack propagation is accomplished at a very high rate (''catastrophal'') according to a front passing through the minimal resistance fields (evidently, the amorphous ones). Along all its trajectory, in the magistral crack top continuous splittings of macromolecules are produced, being pointed out by reactions with radical acceptors or with vinylic monomers.

$$\ldots -CH_2-\underset{\underset{C_6H_5}{|}}{\overset{\overset{CH_3}{|}}{C}}\cdot \; + \; nCH_2=\underset{\underset{Cl}{|}}{CH} \; \text{----} \rightarrow \; \ldots -CH_2-\underset{\underset{C_6H_5}{|}}{\overset{\overset{CH_3}{|}}{C}}-(-CH_2-\underset{\underset{Cl}{|}}{CH}-)_{n-1}-CH_2-\underset{\underset{Cl}{|}}{\overset{\cdot}{C}H}$$

blockcopolymer

$$\ldots -CH_2-\underset{\underset{\cdot}{|}}{\overset{\overset{R}{|}}{C}}-CH_2-\ldots \; + nCH_2=\underset{\underset{Cl}{|}}{CH} \; \text{----} \rightarrow \; \ldots-CH_2-\overset{\overset{R}{|}}{\underset{\underset{(CH_2-CH)_{n-1}-CH_2-\overset{\cdot}{C}H}{|}}{C}}-CH_2-\ldots$$

graftcopolymer

The defects from the crystalline network (among these the most frequently met being the chain ends), can be stress concentration factors at the level of the supermolecular structure. Their sliding, under load, in crystalline zone diminishes the number of the macromolecules able to support the mechanical stress, so that some chemical bonds in which the quaternary C atom is implied should split homolytically. From this moment the fracture develops similarly to the mechanism from the amorphous phase, having as a result the crystal destruction. The crystalline index diminishes (an effect observed after the operation of fiber crimping) while on the monofilament surface artefacts, pointed out by the electron microscopic analysis, appear.

It may be asserted, therefore, that the deformation and fracture process of the crystalline polymers, particularly of monofilaments based on acrylonitrile-vinyl acetate-α-methyl styrene, develops by radical mechanism and is based on the homolytical splitting of the chemical bonds from the polymer main chain.

c. Cyclical Stress

Most of the papers dealing with the study of the semicrystalline polymer fatigue are concentrated on the supermolecular and macroscopic levels of the structure. The molecular mechanism of this process with special reference to the poly(ε-caprolactam) (see References 150 to 154) has been approached.

Thus, the study of the mechanical properties of poly(ethylene terephthalate) and nylon 6,6 fibers offered the possibility of establishing a distinct mechanism of fatigue fracture, as shown by Bunsell and Hearle.[155] The crack is initiated close to the fiber surface and is propagated along it at a small angle, reported to its axis, as if a wedge should act; the cross section reduces gradually and the fiber fractures finally by a simple creep process. Some other recent studies on polyamide and polyester fibers have pointed out the importance of the minimal stress they are subjected to. It has been thus proved that under this stress no typical fatigue process is noticed and that its value depends on the type as well on the chemical structure of the fiber (see Reference 157). The fatigue process which leads to specific morphologies of the fracture surface has been explained by the fiber supermolecular structure. In the polyester fibers, the macrofibrils have a cylindrical shape, with the diameter between 200 and 500 nm, in which the constituent microfibrils have a 5- to 20-nm diameter. The IR spectral analysis as well as the X-rays diffraction one have pointed out the microfibril

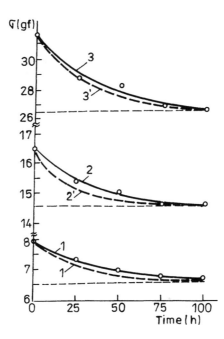

FIGURE 45. Influence of ultrasonic treatment duration on the tensile strength of poly(acrylonitrile-co-vinyl acetate-co-α-methyl styrene)-based monofilaments, with different finesses. 1 — 2.5 den; 2 — 6 den; 3 — 15 den (with dotted lines are represented the theoretical curves).

crystallinity decreasing under fatigue conditions, concurrently with the creation of an amorphous zone extended in the fiber and proved by electron microscopy. The fracture fatigue is propagated along this band by chemical bond splitting, creating holes on its trajectory. It seems that the amorphous band is formed by the crystalline material destruction and the amorphous zones coalescence observed close to the crack top. The stress in a fatigue regime (accelerated) has been achieved by the ultrasounding of the monofilaments of poly(acrylonitrile-co-vinyl acetate-co-α-methyl styrene) in aqueous suspension, in the interval of 10 to 100 h.

The tensile strength (σ_r) has been considered as an estimation criterion of the ultrasounding consequences. The results obtained for three types of monofilaments having a different thickness are shown in Figure 45. In all the studied cases a diminishing with time of σ_r which has the tendency of reaching a limit value after 100 h of ultrasonic treatment may be noted. The σ_r variation curve is specific to the mechanodegradation process, the inferior limit of the pursued property being dependent on the polymer chemical structure.

The fatigue process could be mathematically modeled according to Urzimtsev's equation,[90] used also with good results in the PVC ultrasonic treatment. The obtained theoretical curves are in good accordance with the experimental ones, as it can be seen in Figure 45. The lowering of the tensile strength must be attributed to the modification of some of the polymer structural characteristics, both at the molecular level and at the morphological one.[115]

For establishing if this fatigue regime affects the macromolecules, it has been proceeded to the stimulation of the polymer reaction with aromatic diamines in an ultrasonic field. The obtained samples in the presence of benzidine exhibit a pink-red color which is stable in the diamine solvents (methanol) and has an intensity which varies in accordance with the ultrasonic treatment duration. The diamine chemical bounded has also been tested by the possibility of diazotizing and coupling of the free aminic groups, when colors, specific to the formed azo dyes, have been obtained. Therefore, it can be appreciated that, under the conditions of fatigue caused by ultrasonic treatment, the mechanochemical act lies at the basis of the mechanical strength diminishing of the stress fibers. The process evolves according to the already classical mechanism of the deformation and fracture of the mechanically stressed polymers and finishes with the changes of their structural and morphological characteristics as well as with their physicomechanical properties, as a reflection of the former ones.

d. Vibratory Milling Processing

The mechanical stress of the macromolecular compounds including also the action of the shock and friction forces — both of them being developed by vibratory milling — materialize first with the appearance of the mechanoexcited states. The molecule's reactivity in this state is superior to that of some corresponding unstressed molecules, favoring thus the development of some reactions which, under other conditions, would be very difficult to occur.

Along all its duration and regardless of the way in which the solid body is stressed, the fracture process is accompanied by the mechanochemical act that generates the macroradicals. It finishes with the formation of new and rich energy surfaces which are continuously renewed and constitute the basis of numerous chemical reactions such as graft and blockcopolymerization, polycondensation, complexing, homo- and copolymerization, etc.

A characteristic behavior implying the deep modification of the properties has been found out in the case of highly oriented semicrystalline polymers [polyacrylonitrile, polyamides, poly(ethylene terephthalate)] processed by vibratory milling.

The first noticed effect is, of course, the mechanodispersion which consists of the change of the form as well as of the geometric dimensions of the material particles. The character of the microstructure modification depends on the conditions under which the mechanical processing is developed. Thus, the polyamides modify distinctly depending on the implied temperature range. From the very first moments of the process, the dispersion without cooling points out both the fiber's deformation and their accumulation while the specific surface diminishes. For instance, by cooling the milling bodies as well as the polymer (at $-60°C$), a marked stiffening of the polymer is produced, determining the increase of the mechanodispersion efficiency, materialized in a sensitive increase of the specific surface, and in a reduction of the dimensions particles under 8 μm.

The polyacrylonitrile, owing to its peculiar structure consisting in a strong dipolar momentum determined by the nitrile groups' presence, exhibits an advanced stiffness even without cooling and leading to a corresponding advanced mechanodispersion.

The milling without cooling of the poly(ethylene terephthalate) passes through a maximum efficiency due to the fiber's shortening in the dispersion incipient stages; subsequently, particle agglomeration is produced owing to the self-adhesion phenomenon.

The main consequence of the mechanical stress by vibratory milling of the semicrystalline polymers is based, therefore, on the mechanodispersion; but, at the same time, the mechanodestruction is produced concurrently with this effect. The latter one is pointed out by the molecular weight reduction as well as by the appearance and increase of the concentration in the newly formed final groups that are formed by means of the macroradical stabilization. In this case, similar to the amorphous polymers may be also noticed a close dependence between the geometric dimensions of the processed powders and the diminishing of the molecular weight; this dependence was checked, for instance, with the polyamide vibratory milling (see Reference 159).

The vibratory milling of the cellulose, polyacrylonitrile, polyesters, and polyamides materializes in the molecular weight continuous reduction, which in time tends to a limit value, characteristic for the applied mechanical regime, simultaneously with the increase of the number of the carboxylic, aminic, or hydroxy groups or of terminal unsaturations. These changes at a molecular level are also reflected by the physicochemical properties modification: solubility, swelling capacity, crystallinity, etc. Apparently, the crystallinity should not undergo any changes because the mechanical energy is concentrated especially on the amorphous domains, which constitute defect zones of semicrystalline polymer structure. Under the conditions of the very intensive shock and friction mechanical stresses which appear at the vibratory milling, the stress supported even by the crystalline domains is very high. The stress rate being also high, the macromolecular crystals takes over in a rigid manner the mechanical energy because it has no time for redistributing it in the amorphous zones. The latter ones, in turn, take over a high stress. The shock energy, concentrated on the crystal defect zones, causes its destruction which is macroscopically displayed by the crystallinity index diminishing.

The formation of the macroradicals which accompanies permanently the mechanodispersion and mechanodestruction processes of semicrystalline polymers can be pointed out by specific reactions or even used for obtaining some modified polymers having new chemical and physicomechanical properties. The basis of a new type of synthesis, i.e., of a mechanochemical type, has been established including polycondensation, complexing, graft and block copolymerization in polymer-monomer and polymer-polymer systems.

The idea of the mechanochemical polycondensation achievement has been suggested by a phenomenon observed during the polyamide mechanochemical destruction, when it has been learned that, besides the homolytical splitting of basic chains, there are also produced condensation reactions; the latter ones imply the interaction of the polymer carboxylic and aminic groups concurrently with their severe reduction in the medium.[158-160] Of course, such a reaction in a solid-liquid or even a solid-solid heterogeneous system at the ambient temperature is accomplished only if the reaction partners are in a high excited state, owing in this case to the mechanical stress. Therefore, the mechanoactivation, as a stage preceding both the molecular splitting and the macroscopic fracture, represents a way for stimulating reactions that are almost or even impossible in normal conditions.

The accomplishment of the condensations between various functional groups belonging to the destruction fragments has also suggested the possibility of carrying on this reaction among those of a polymer and a judicious selected micromolecular compound. By the convenient selection of the reaction partners, it became possible for the condensation process to attain a dominant feature leading to the acquiring of some products with remarkable properties.

Mechanochemical polycondensation has been achieved by working with semicrystalline macromolecular compounds from polyamides, polyesters, and polysaccharides class and using aliphatic or aromatic diamines as well as dicarboxylic acid chlorides as micromolecular components. The reaction develops not only in the preexistent functional groups, but also in those which are continuously formed as a result of macroradical stabilization, generated by mechanodegradation.[159]

The systematic study of this type of synthesis led to the conclusion that mechanochemical polycondensation is able to develop owing to the macroradicals themselves, due to the capacity of some difunctional compounds to act as radical acceptors (aromatic diamines, especially). Valuable arguments for such a mechanism may be considered the results obtained with ethylene diamine, by vibratory milling. Thus, it may be learned that the nitrogen percentage in the reaction products — an indirect measure of the chemical-bonded diamine amount — is generally inferior to the product obtained in the diamine absence.

Higher values of the mentioned percentage are recorded when the acceptor itself contains nitrogen in its structure.[161]

Thus, in this way, the achievement of another type of mechanochemical polycondensation became possible, namely, the one between carbochained semicrystalline polymers — which neither contain nor generate condensable functional groups — and aliphatic or aromatic diamines. The mechanism of such a type of reaction may be formulated as in Figure 46.

The reaction is influenced by the mechanic regime parameters (filling ratio, duration, temperature) as well as by the chemical nature of reaction partners. The acquired products present a limited solubility, a high thermostability, and contain free aminic groups on the ends of the destruction fragments reacted with the micromolecular compound. By the aminic groups diazotizing and coupling various range of colors and shades (an indirect proof of diamine bonding to the polymer) have been achieved. Frequently, the products of mechanochemical polycondensation of poly(ethylene terephthalate) and poly(ϵ-caprolactam) with aromatic diamines present semiconductor characteristics which amplify as the vibratory milling conditions become more energic.

The great number of the nitrogen and often oxygen atoms from mechanical polycondensation products, with electronic availabilities, favors complexing reactions of some metals. It has been found that the reaction efficiency depends both on the nitrogen percentage from the ligand and on the nature of metallic ions. Therefore, Mn^{2+} ion has a higher capacity to complex at a ligand obtained by mechanochemical polycondensation reaction of poly(ethylene terephthalate) with ethylene diamine, in comparison with the Fe^{3+} ion.

The PET polychelate structure is of the following type:

/162-164/

In the case of poly(ϵ-caprolactam) the complexing could also be achieved in the absence of the ethylene diamine, as a condensation agent, by the mechanical activation of the polypeptidic groups from the polymer structure, as stated by Simionescu et al.[165-167]

Owing to the presence of metallic atoms, the synthesized polychelates present semiconductor and paramagnetic properties as well as a high thermostability.

Some macromolecular complexes able to insolubilize the enzymes by coupling reactions have been obtained by the vibratory milling of the microcrystalline cellulose, in the presence of some transitional metal salts (Ni, Fe, Mn, Co, Cu, Ti, V). Co complexes proved to be very efficient in invertase coupling, in which case an immobilized enzyme with an activity close to the free one has been acquired, but which offers the possibility of being reused in more hydrolytic cycles (see Reference 168).

The inevitable mechanoradical formation during the polymer vibratory milling, the semicrystalline ones, inclusively, led to the possibility of some block and graftcopolymer synthesis, working in a polymer-polymer or monomer-polymer system. The basic reactions implied in the process or monomer-polymer system. The basic reactions implied in the process are the following ones:

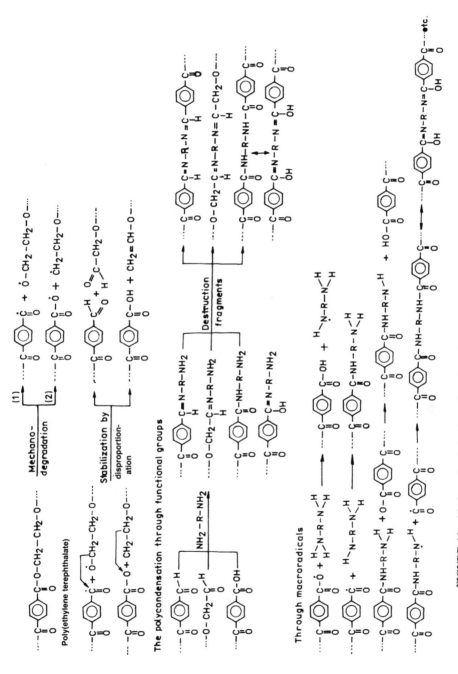

FIGURE 46. Mechanochemical polycondensation scheme for poly(ethylene terephthalate) with aromatic diamines.

Polymer chain splitting $\quad\quad$ $P_{m-n} \xrightarrow[\text{energy}]{\text{mechnical}} P_m^{\cdot} + P_n^{\cdot}$

$$R_{s-t} \xrightarrow{\text{energy}} R_s^{\cdot} + R_t^{\cdot}$$

Polymerization initiation $\quad\quad$ $P_m + M \longrightarrow P_m M^{\cdot}$

Cross recombination $\quad\quad$ $P_n^{\cdot} + R_s^{\cdot} \longrightarrow P_n - R_s$

$$P_m^{\cdot} + R_t^{\cdot} \longrightarrow P_m - R_t$$

Chain transfer $\quad\quad$ $P_n^{\cdot} + R-R-R \longrightarrow P_n + R-\overset{\cdot}{R}-R$

Recombination $\quad\quad$ $R-\overset{\cdot}{R}-R + P_m^{\cdot}(R_s^{\cdot}) \longrightarrow R-R-R$
$$|$$
$$P_m(R_s)$$

Disproportionation $\quad\quad$ $P_n^{\cdot} + R_s^{\cdot} \longrightarrow P_n + R_s$

In this way the attainment of some graft and block copolymers of polyacrylonitrile with butadiene and vinyl chloride,[169] of poly(ϵ-caprolactam) with acrylic acid, acrylonitrile, styrene,[169,170] of poly(ethylene terephthalate) with acrylonitrile,[171] vinyl acetate,[172] and vinyl chloride[173] has been accomplished.

Other studies have had in view the synthesis of the graft and block copolymers in the polymer-polymer system. The reaction products represent usual mixtures of block and graft copolymers, the synthesis being practically impossible to be directed in the sense of a single type of polymer acquirement. However, depending on the chemical nature of the reaction partners, one of the products can be predominantly obtained. Thus, if the mechanical degradation rate of the two polymers is approximately the same, the probabilities of two types of macroradical appearance in the system are equal. They will be "cross" stabilized and, in this way, block copolymers are predominantly obtained. If one of the polymers presents a higher degradation rate, its macroradicals will predominantly appear in the system. The macroradical stabilization will be mainly made by chain transfer, with the participation of the undamaged polymer macromolecules, on whose basic chains radically active centers appear. In the next stage the "cross" recombination of end macroradicals takes place with the middle ones resulting, thus, the graft copolymer. In this way, by vibratory milling there have been obtained block and graft copolymers of the cellulose and of the cellulose triacetate with vinyl polymers.[174]

The results presented here prove therefore that, at the basis of the mechanodispersion and mechanodestruction processes by vibratory milling of the semicrystalline polymers, lies the mechanochemical act; it has, as a consequence, the continuous formation in the material of highly activated new surfaces. These are able to initiate chemical reactions that end with the formation of polymers exhibiting new properties.

e. The Increase of the Mechanical Strength of the Semicrystalline Polymers

Once the molecular mechanism of deformation and fracture of polymers have been cleared up, there arose the problem of improving mechanical strength by intervening on a certain stage of the process. Extremely interesting results have been obtained in the PET case, demonstrating that the modern theory of the macromolecular compound fracture permits the establishment of some concrete ways for improving the polymer strength. Starting from the homolytical splitting of some chemical bonds from the main chain, the idea was advanced that inhibiting the free radical reactions made up with some stabilizers could prevent the material from further destruction and improve its mechanical strength (see Reference 175).

By determining the interaction intermolecular energy in the amorphous zones of PET,

$$E_{ii} = U_0 - E_d \tag{21}$$

and introducing this expression in the durability equation, we obtain

$$\sigma\left(1 + \frac{\epsilon}{100}\right) = \left(E_d + E_{ii} - RT \ln \frac{\tau}{\tau_0}\right)\gamma^{-1} \tag{22}$$

From the above-mentioned equation we can determine the highest value of the PET strength. The maximum E_d corresponds to the thermal degradation process, this being of 235 kJ/mol.[176] The interaction energy among the macromolecules within the amorphous domains cannot exceed the one existing in the crystalline domains which has the value of 72 kJ/mol. The coefficient γ, calculated by the commonly known methods, is 0.12 m³/kmol.

The analysis of the experimental data regarding the stress-strain properties of the samples studied at 293 K shows that the unstabilized oriented PET strength is 1.6 times lower than its highest value. The polymer strength is increased by introducing 4-[4'-phenyl azophenil amino)5-metoxy]1,2 benzoquinone (AQ), while the oriented PET fibers which were intermolecularly stabilized with melamine (MA) copolyesterified with ethylene glycol have the tendency to stabilize at the strength maximum value. This stabilization accomplished by the two additives is a consequence of their influence on E_d and E_{ii}. From this viewpoint, the lower strength of the oriented stabilized PET as compared to the highest value is the result of the lower values of E_d and E_{ii} (191 and 59 kJ/mol) in comparison with the maximum values of 238 and 72 kJ/mol.

Thus, the stabilizer's introduction increases E_d (to 210 and 230, respectively, for AQ and MA); this parameter is consequently determined by the chemical nature and the stabilizer concentration. The E_d value increase is determined by the efficiency of the macroradical inhibiting reactions in an oxygen atmosphere, occurring in a field of mechanical forces.

B. MECHANOCHEMICAL PROCESSES AT THE DEFORMATION AND FRACTURE OF POLYMERS IN HIGH-ELASTIC STATE

The high-elastic state is exclusively specific to the polymers, being a consequence of the linear macromolecule flexibility.

Taking into account the macromolecule's great dimensions as well as their capacity of aggregation in various supramolecular forms, the polymer viscosity becomes so high that the flowing process does not occur even after long durations of stress application. Consequently, the tension developed with material has not the possibility of relaxing and causes the deformation of the flexible high macromolecular chain, in the sense of their uncoiling and orientation according to the mechanical stress direction, and which appears as a reversible high-elastic deformation. Therefore, this property appears only when the macromolecule deformation rate is much higher than the displacement one, so that the irreversible deformation can be neglected. That is why even if we often speak about an equilibrium high-elastic deformation of a linear polymer; in reality this equilibrium is only partial. Referring to the high-elastic polymers we must admit that low-intensity relaxation phenomena — more pronounced in the transition domains — may be always noticed, especially at deformations close to the fracture one.

By virtue of the polymer polydispersion character, the more reduced length macromolecules will reach the strained conformation faster, thus supporting a higher fraction of the local stress, a fact which entials their preferential splitting. However, at the same time, due to the chain's statistically coiled ball appearance, there exist numerous "knots" between them which act as real cross-linkings. When stress is applied, the knotted macromolecules will undergo strong stresses in their contact points, a fact which entials some chemical bond homolytical splittings. Also, at very high deformations, when the mechanical vitrifying phenomenon or even the crystallization process take place within the elastomer, the mechanochemical processes become very frequent due to the strong orientation of the macromolecules and to the intensification of the interaction between them, having as a result the macroradical formation.

Therefore, the mechanochemical phenomena are also present in the case of the high-elastic polymer mechanical stressing even if they are more reduced in intensity in comparison with vitreous state polymers.

1. Characteristics of the Deformation and Fracture of the High-Elastic Polymers

The high elasticity is conditioned both by the thermodynamic and kinetic flexibility. That is why a high elastic macromolecular compound is that which exhibits such a chain flexibility as to provide a maximum deformation as well as a corresponding and sufficient developing rate to avoid mechanical losses which could lead to the material mechanical destruction.

For example, by freezing an elastomer, its macromolecules are fixed into certain conformations and an outer force action produces, due to the much increased viscosity, such a slow modification that practically the high elasticity does not occur any longer. Reed et al.,[178-180] studying a variety of elastomers beginning with the natural rubber up to the silicone rubber, learned that the prestressed samples that were broken at low temperatures emit stronger ESR signals than the ones broken at the same temperatures but without being submitted to a previous load; these conclusions were also confirmed by Brown.[181] Taking into account these results, the authors consider that cooling and preloading produce in the polymer a greater number of slightly oriented regions through which no microcrack occurs as a result of the broken chemical bonds. However, as the deformation increases microcracks occur in the polymer unoriented regions (analogous to the semicrystalline polymers), their increase leading to the free radical formation, detectable by the ESR method. The model proposed by the authors suggests therefore that the elastomer ductility at low temperatures can be explained only by its conversion into a semicrystalline-oriented polymer, as a result of the prestressing and of the thermal treatment.

The macromolecule fixation in networks by means of cross-linkings so that the polymer high-elastic properties should be preserved permits the increase of free radical concentration under stress. Thus, if for a 250% strain a slight ESR signal ($\sim 10^{14}$ spins/g) is observed with an uncross-linked elastomer submitted to tensile stress at low temperatures, the slight cross-linking has as a consequence its strong intensification.[178-182] For a 200% deformation of the elastomer, the concentration of the free radicals increases with two orders of magnitude; the lower the temperature at which the load is applied, the higher the increase (Figure 47).

The obtained results suggest that the cross-linkings play a leading part in preventing the total flow of the macromolecules one to another. This situation is arrived at as soon as there exist two cross-linkings in one macromolecule, which are enough to develop a continuous lattice.

A similar dependence of the free radical concentration on the network density was obtained with the cis-polyisoprene stressing (see Reference 183). The appearance of some great concentrations of radicals at lower temperatures (123 K) and network densities of

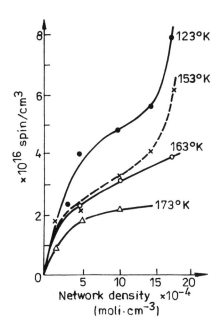

FIGURE 47. Effect of temperature and network density (moles network chains per cubic centimeter) on the radical generation intensity (tensile strain, $\epsilon = 200\%$).

10^{-3} mol/cm³ reveals the fact that the macromolecules split even before they are completely strained. The appearance of some gas bubbles producing a foaming phenomenon at the elastomer surface may also be noticed under the same temperature conditions. The gas chromatography proved that H_2 is that gas.

The molecular mechanism of the crack increase was also studied for the same elastomer by IR spectroscopy.[181] Thus, it was noted that, in the neighboring zones of the crack practiced in a rubber sample, as a result of the mechanical load there are accumulated important concentrations of functional groups which were either initially present in the polymer or completely newly formed ones (unsaturated bonds in conjugated system and aldehydic groups, respectively), an obvious consequence of the mechanochemical processes.

Very important in developing mechanodestructive processes with elastomers are proved to be the stress applying rate as well as the mechanical conditions. Thus, a rapidly applied stress (shock) at temperatures a little above T_g determines the polymer fracture in a similar way to vitreous, amorphous ones.

The application of some intense shearing stresses, even at temperatures much higher than T_g (e.g., under the conditions of two-roll mixing or mastication processing), has also as a result the mechanodestruction intensification, which consists in the appearance of a high concentration of macroradicals and finally in the decrease of the elastomer molecular weight. The newly formed radical species can recombine with radical acceptors of aromatic diamines type or enable the initiation of some graft and block copolymerization reactions with vinylic monomers or with other macromolecular compounds that are present in the reaction medium.

2. Static Tensile Stress

The elastomer creep has been widely studied because these polymers do not show a deformation remnant component; when the force action stops, the deformation is totally recovered. The discovery and wide utilization of the thermoplastic elastomers such as

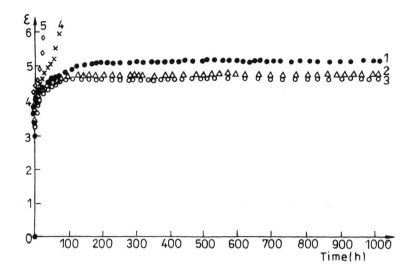

FIGURE 48. Creep curves of the thermoplastic elastomer in different gaseous media. 1 — air; 2 — nitrogen; 3 — vacuum; 4 — vinyl chloride; 5 — nitrogen oxide.

TABLE 5
The Deformation Values Calculated from the Creep Curves of
Estane for Different Gaseous Atmospheres (σ = 42.20 MPa)

Deformation type	Medium				
	Air	Vacuum	Nitrogen	NO	Vinyl chloride
Instantaneous deformation (mm/mm)	3.25	3.10	3.00	3.50	3.45
Delayed elastic and flowing deformation (mm/mm)	1.84	1.60	1.55	2.50	2.15
Total strain	5.09	4.70	4.55	5.95	5.65

polyurethanes in recent years determined the serious investigation of their behavior in different mechanical conditions. Such a linear polyurethanic elastomer is Estane 5707, a urethanic polyester synthesized from 4,4′-diphenyl methane, diisocyanate, poly(tetramethylene adipate) determined into two hydroxylic groups ($\overline{M}_w$ = 2000) and buthane diol as a chain extender. Blocks of polyester and urethanic segments alternate therefore on its chain. The creep of this polymer was studied on films achieved by solution deposit, at definite values of the stress in inert or active gaseous media, sometimes in solution of radical acceptors or vinylic monomers, by Vasiliu Oprea and Constantinescu.[185]

The creep curves for different gaseous media at a stress σ = 42.2 MPa are shown in Figure 48. The values of the three components of the deformation calculated from these curves (Table 5) show considerable differences.

It may be noted that in active media (air, NO, vinyl chloride) all the deformation components, the overall one, too, have higher values than in inert media. The static tensile stress in the presence of NO and of the monomer finalized with the rapid fracture of the samples (after 22 and 65 h, respectively). This effect suggests the radical mechanism of the elastomer stressing process. Both gaseous substances can react with the macroradicals made up by the chemical bond scission of the polymer, thus preventing their rapid stabilization through specific reactions (recombination, disproportionation).

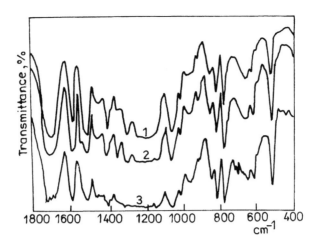

FIGURE 49. IR spectra of the thermoplastic elastomer after creep. 1 — witness; 2 — stressed in the presence of nitrogen oxide; 3 — stressed in the presence of vinyl chloride.

The strain value relative increase in active media is mainly due to the increase of the viscoelastic and flow component. Knowing that the flow strain is the main consequence of the molecular fracture, it is obvious that the mechanochemical processes intensify in the presence of the active gaseous substances. Thus, the experimental results obtained in the study of the urethanic polyester creep are good arguments for a radical mechanism of the deformation and fracture processes.

The different behavior of the statically stressed polymers in the presence of NO and vinyl chloride compared to the sample stressed in an inert medium is therefore the consequence of some of its reactions with micromolecular compounds, activated by load. The polymer IR spectra submitted to creep point out some considerable changes (Figure 49). Thus, the spectrum characteristic to the sample stressed in the presence of NO (Figure 49) shows an intensification of the absorption band from $1360 \, cm^{-1}$ specific to the R-NO group. At the same time, the bands from 1600 and $620 \, cm^{-1}$ are intensified, being attributed to the R-ONO group, which is made up through the nitrogen oxide reaction with the macroradicals obtained as a result of the esteric group splitting from the polymeric chain.

The presence of the band from $700 \, cm^{-1}$ (curve 3) also shows the linkage of the vinyl chloride to the same macromolecules. Its missing from the unstressed polymer spectrum is an argument, too, for the vinyl monomer reaction with the macroradicals.

Conclusions on the deformation and fracture mechanism of the polyurethanic elastomer have also been obtained by studying the creep in a liquid medium, in the presence of some radicalic acceptors, and vinylic and dienic monomers. Thus, it has been observed that, by stressing the polyurethane film in absolute ethylic alcohol, it does not fracture macroscopically and the deformation becomes practically constant after about 20 h (Figure 50, curve 1). On the other hand, in alcoholic solutions of DPPH, acrylonitrile, and isoprene, the macroscopic fracture takes place after 8, 48, and 140 h from the beginning of the stress. These outcomes also prove the radical character of the mechanism of the polymer deformation and fracture processes and the capacity of the newly formed macroradicals to initiate graft and block copolymerization reactions of the vinylic and dienic monomers which are present in the medium.

The DPPH reaction with the stressed film can be also seen when visually examining the sample; there appears a violet color on its active sector — a characteristic of the radical acceptor — which is resistant to washing and repeated extractions in methanol. The

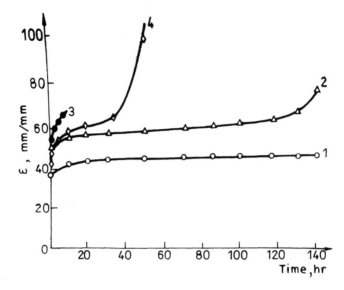

FIGURE 50. Creep curves of the thermoplastic elastomer in different liquid media. 1 — ethanol; 2 — isoprene; 3 — DPPH (methanolic solution, 0.5% concentration); 4 — acrylonitrile.

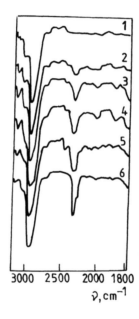

FIGURE 51. IR spectra of the thermoplastic elastomer films after creep in the presence of acrylonitrile (different postreaction durations). 1 — witness; 2 — stressed for 24 h; 3 — stressed for 48 h; 4 — stressed for 72 h; 5 — stressed for 96 h; 6 — stressed for 120 h.

appearance of an absorption band at 350 nm in the UV spectrum of the same film proves also its reaction with the micromolecular compound. The acrylonitrile chemical bonding to the stressed polyurethane has been proved by the IR spectroscopy (Figure 51). The spectra exhibit a new absorption band at 2350 cm^{-1} attributed to the >C=N− group. Studying the influence of the duration, the stressed film was kept in acrylonitrile after the macroscopic fracture; we can notice the amount of chemically linked monomer increase, with the post-

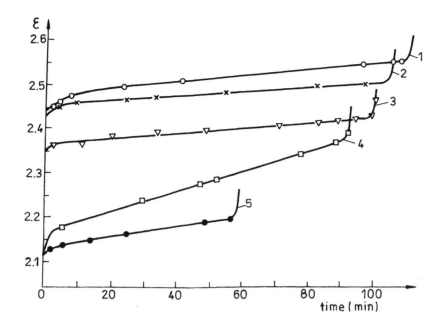

FIGURE 52. $\epsilon - t$ curves for polyurethane samples characterized by different cross-linking degrees. 1 — M_c = 7700; 2 — 7100; 3 — 6600; 4 — 6000; 5 — 5500.

reaction time pointed out by the intensification of the band from 2350 cm^{-1}. The increase of the acrylonitrile content in the stressed film, in time, is the consequence of the grafted chain length increase, when the concentration of the macroradicals initiating the reaction is kept practically constant.

The behavior of the polyurethane elastomer with different cross-linking degrees at the static tensile stress is also described by typical creep curves (see References 186 and 187). The value of the overall deformation and of its components depends on the cross-linking degree, indirectly expressed by M_c constant. This represents the average molecular weight of the chain segment between two knots of the polymer network and it is expressed by the relation (see Reference 189):

$$E = \frac{3R\rho T}{M} \tag{23}$$

where E — dynamic elasticity modulus; R — gases universal constant; ρ — polymer density; and T — temperature.

For the tested samples, M_c varies from 5500 to 7700.

Interesting conclusions can be obtained if we consider — to make the comparison easier — the curves t belonging to five elastomers with different cross-linking degrees within the interval of 0 to 120 h (Figure 52). The sample with the lowest cross-linking degree has, as it was expected, the highest deformation value, the deformation decreasing as M_c is reduced (signifying the cross-linking degree increase). Although strong and numerous, the interactions between the linear macromolecules, determined by the presence of the –CO–NH– groups in the chain, allow nevertheless the achievement of the structural changes under load and the manifestation of a high elasticity, respectively. However, the introduction of some chemical cross bridges reduces the possibility of the linear segment structural change, stiffening the structure and contributing thus to the deformation decrease.

TABLE 6
The Characteristic Deformations at Simultaneous
Creep and Aging of a Cross-Linked Polyurethane
(M_c = 5500, σ = 9.0 MPa)

The creep conditions [deformation, ϵ (mm/mm)]	Witness	With accelerating aging
Instantaneous elastic deformation, ϵ_0 (mm/mm)	0.880	0.885
Delayed elastic and flowing deformation, ϵ_c (mm/mm)	0.255	1.435
Maximum deformation, ϵ_{max}	1.135	2.320

On the other hand, it may be noted that the cross-linking degree increase has as a consequence the decrease of the polymer durability, expressed by the time between the stress application and the macroscopic fracture (Figure 52). We can find the explanation in the structural changes caused by the cross-bridge introduction of the allophanate type, which remove the macromolecules one from the other, reducing the hydrogen bond frequency. During the mechanical loading they will take over the stress by splitting themselves or determining the neighboring chemical bond splitting from the main chain. As a result, the resistance to creeping decreases as the cross-linking degree increases.

The loading of some samples submitted to an accelerated aging determined by the UV irradiations and the simultaneous heating at 60°C leads to greater deformations compared to the witness sample, for the same duration of the mechanical loading. Thus, a pronounced increase of the deformation irreversible component may be observed (i.e., the flowing component), an evident consequence of the intensification of the chemical bond splitting from the main chain, as well as of the cross-links, as stated by Vasiliu Oprea and Constantinescu[188] (Table 6).

Also noted is the intensification of the microcrack formation process which can be detected by the visual observation of the stressed samples. The acceleration of the elastomer deformation and fracture process during the simultaneous thermal and UV stressing is a good argument for the common mechanism of the mechanochemical, photochemical, and thermal destruction. At its basis lies the macroradical formation by the homolytic splitting of the chemical bonds from the polymer main chain.

3. Cyclical Stress

During exploitation the elastomers, cyclic mechanical stresses are met very frequently. The importance of the dynamic fatigue study is given by the fact that, under the action of some fluctuating in time stresses, having a certain frequency, the polymers are destroyed at values of the stress which are lower than the ones that the polymers support during loading with a monotonous variation of the stress.

Under the conditions of the fatigue stressing, the polyurethanic elastomers are broken after a number of cycles N_F which depend on the applied stress value and the polymer cross-linking degree. It can be seen that the fatigue strength decreases as the cross-linking degree increases (the decrease of M_c, respectively), an effect which has the same explanation as in the case of static fatigue (Section 3) (Table 7).

The durability decreases as the stress amplitude increases for a constant load frequency. The same effect is registered by the fatigue stressing of the samples, concurrently with their submission to an accelerated aging process by using the UV radiation and heating at 60°C. However, the concomitantly aged samples exhibit a much diminished fatigue strength due

TABLE 7

The Influence of the Cyclically Applied Stress and Cross-Linking Degree (M_c) on the Fatigue Strength (N_F) for Witness Sample and Samples Submitted to Accelerated Aging

	MPa							
	0.810		**0.850**		**1.000**		**1.110**	
M_c,g/mol	**1**	**2**	**1**	**2**	**1**	**2**	**1**	**2**
7700	30850	21750	20500	14500	8500	5640	6350	2500
7100	23250	18900	16850	13250	7900	5050	5500	1800
6600	23200	17750	14300	9800	5800	4430	5000	1100
6000	21400	13500	9700	6400	5000	3220	4250	750
5500	20100	12100	9500	5950	3500	2950	2400	690

From Baramboim, N. K., *Mechanochemistry of Polymers,* MacLaren, London, 1964. With permission.

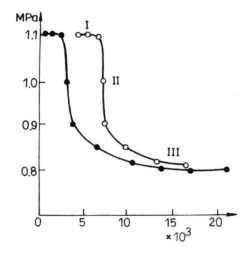

FIGURE 53. Wöhler curves for the thermoplastic elastomer (M_c = 6000 g/mol; ν = 4 Hz). 1 — witness; 2 — after accelerated aging.

to the acceleration of the valence bond splitting, exerted by the UV radiations as well as by the temperature. This fact becomes more obvious by comparing Wohler curves for one of the samples (Figure 53). There are distinguished three regions on these curves. In the first region (I) the fracture is caused by the crack increase in the microcracked material, followed by the crack catastrophic propagation throughout the sample. The number of cycles until fracture occurs is much influenced by the stress amplitude, probably due to the great dependence on the number of microcracks at this level of the stress.

The second region (II) is characterized by the fact that the magistral crack catastrophic propagation is preceded by a nucleation period of the micorcracks. The N_F dependence on stress is reduced in this case.

The third region (III) represents the limit strength of the material, the N_F dependence on stress being very reduced. The crack initiation is produced after a long induction period.

The durability of the cyclically stressed polyurethane (N_F) depends also on the stress frequency (ν_a) and on the deformation imposed to the sample at one cycle (ϵ_a), respectively. The regression equation that describes this dependence has the form

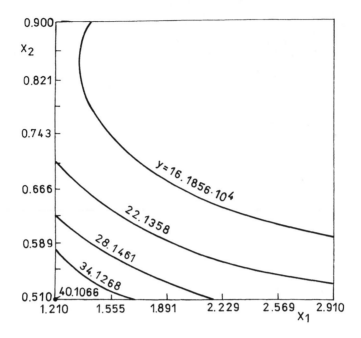

FIGURE 54. Constant level curves in the experimental plane $\nu_a - \epsilon_a$.

$$N_F = 13.41 - 3.62\,\nu_a - 5.63\,\epsilon_a + 1.55\,\nu_a^2 + 1.58\,\nu_a\epsilon_a + 3.67\,\epsilon_a^2 \qquad (24)$$

The constant level curves in the experimental plane $\nu_a - \epsilon_a$ point out the fact that the polymer durability decreases as both the ν_a and ϵ_a increase (Figure 54).

The temperature, a component of the polymer aging, does not influence so much the durability. However, it is worth mentioning that as the temperature increases the elastomer durability improves (within the interval 11 to 55°C). This can be explained on the basis of polymer thermal plastifying which implies the intensification of the macromolecules slipping and, therefore, a delay of their splitting.

Thus, when the elastomers are submitted to dynamic fatigue, mechanochemical reactions occur and determine the strength decrease and finally their macroscopic fracture. The same conclusion can be clearly derived from the fact that the UV radiations are able to homolytically split the covalent bonds from the polymer main chain, that fact demonstrating that the mechanical stress as well as UV radiation have the same cumulating effects and determine the acceleration of the material macroscopic fracture. The interatomic bonds splitting either from the main chain or from the cross-linkings represents the decisive stage of the mechanochemical process of the elastomer deformation and fracture.

4. Vibratory Milling Processing

The mechanochemical reactions taking place during the vibratory milling are activated at low values of the elastic energy, but it is concentrated in the unit of volume of the mechanically stressed polymer, reaching commensurable values as compared to those belonging to the energy of the covalent chemical bonds. The most commonly used method for pointing out the effects of this type of processing is to determine the molecular weight or the concentration of the active particles, formed as a result of the mechanical destruction by using the ESR spectroscopy or chemical methods.

The vibratory milling processing of the polyurethanic elastomer of Estane type is accompanied by a continuous decrease of the viscosimetric molecular weight and tending to

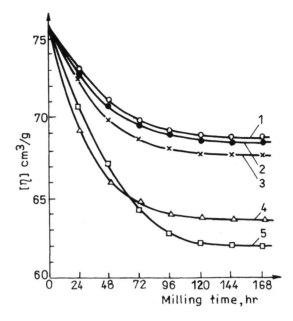

FIGURE 55. Intrinsic viscosity variation of the thermoplastic elastomer with the milling duration in different media. 1 — vacuum; 2 — alcohol; 3 — nitrogen; 4 — benzidine; 5 — air.

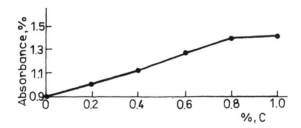

FIGURE 56. Variation of the reacted benzidine amount as a function of diamine concentration in methanolic solution (absorbance corresponding to $\lambda = 280$ nm from the UV spectrum). (Based on Reference 190.)

reach a constant value after about 120 h (Figure 55), regardless of the nature of the gaseous atmosphere or of the substances from the reaction medium. A strong decrease of the intrinsic viscosity (a molecular weight measure) is produced in the air compared to the inert atmosphere or vacuum, as a result of the accelerating action performed by oxygen on the radical processes that take place (curve 5 compared to curves 1 and 2).

The polyurethane vibratory milling in liquid medium (ethylic alcohol) leads to a less-advanced destruction of the polyurethane comparable to the registered in vacuum or inert medium (curve 2). The explanation lies in the fact that the introduced liquids absorb partially the mechanical energy and prevent the occurrence of the maximum destruction effects.

However, working in benzidine alcoholic solutions (a very well-known radicalic acceptor) the intrinsic viscosity reduction is again pronounced, comparable to that registered when the vibratory milling is performed in air (curve 4). UV spectra of the milling products in the benzidine presence show an intensification of the absorbance at $\lambda = 280$ nm, specific to the aromatic nuclei, with the diamine concentration increase. Representing graphically the absorbance corresponding to this wavelength function of the diamine concentration, the linkage of some higher quantities of benzidine may be noted in the polymer, concurrently with this parameter increase (Figure 56).

At a concentration of 0.8%, a maximum of the reacted diamine quantity is reached, which means that the benzidine from the solution is sufficient to react with all the macroradicals produced as a result of the mechanodestruction.

The formation of macroradicals during the vibratory milling of the polyurethanic elastomers was also assayed by the possibility of initiating some graft and block copolymerization reactions with vinylic monomers (vinyl chloride, acrylonitrile) and dienic monomers (isoprene). The obtained products have specific colors (gray in acrylonitrile, orange in isoprene) and, depending on the vibratory milling duration, a variable content of these monomers. Generally, in the first 48 h, the chlorine content (due to acrylonitrile), increases owing to the graft of greater and greater quantities of monomers, then a decrease in time is recorded due to the mechanodegradation which reduces the grafts or segments length from the synthesized blockcopolymers.

The results are good arguments for the radicalic mechanism of the elastomers mechanodestruction during the processing by vibratory milling, according to the following reactions:

$$\cdots -OOC-NH-C_6H_4-CH_2-C_6H_4-NH-COO-CH_2-CH_2-OOC-(CH_2)_4-COO- \cdots \xrightarrow{\text{mechanical energy}}$$

$$\cdots -OOC-NH-C_6H_4-CH_2-C_6H_4-NH-COO^{\cdot} + {}^{\cdot}CH_2-CH_2-OOC-(CH_2)_4-COO- \cdots$$

These radicals are stabilized by means of the disproportionation reactions which explains the decrease of the polymer molecular weight by vibratory milling:

$$\cdots -C_6H_4-NH-COO^{\cdot} + {}^{\cdot}CH_2-CH_2-OOC-(CH_2)_4-COO- \cdots \longrightarrow$$

$$\cdots -C_6H_4-NH-COOH + CH_2=CH-OOC-(CH_2)_4-COO- \cdots$$

In the presence of radical acceptors of the aromatic diamine type, ''mechanochemical polycondensation'' reactions are produced, which finalize with the formation of the following structure:

$$\cdots -C_6H_4-NH-COO-NH-Ar-NH-CH_2-CH_2-OOC-(CH_2)_4- \cdots$$

The macroradicals also bring about the vinylic monomers polymerization which are present in the medium, leading to the formation of the graft and blockcopolymers.

$$\cdots -C_6H_4-NH-COO^{\cdot} + nCH_2=CH \rightarrow \cdots -C_6H_4-NH-COO-(CH_2-CH)_{n-1}-CH_2-\overset{\cdot}{C}H$$
$$\qquad\qquad\qquad\qquad\quad | \qquad\qquad\qquad\qquad\qquad\qquad\qquad\qquad\quad | \qquad\qquad\quad |$$
$$\qquad\qquad\qquad\qquad\quad Cl \qquad\qquad\qquad\qquad\qquad\qquad\qquad\qquad\quad Cl \qquad\qquad Cl$$

block copolymer

$$\cdots -C_6H_4-NH-COO^{\cdot} + \cdots -C_6H_4-NH-COO-CH_2-CH_2-OOC- \cdots \longrightarrow$$

$$\cdots -C_6H_4-NH-COOH + \cdots -C_6H_4-NH-COO-\overset{\cdot}{C}H-CH_2-OOC- \cdots$$

$$\cdots -C_6H_4-NH-COO-\overset{\cdot}{C}H-CH_2-OOC- \cdots + mCH_2=CH \rightarrow$$
$$\qquad\qquad\qquad\qquad\qquad\qquad\qquad\qquad\qquad\qquad\qquad\qquad\qquad | $$
$$\qquad\qquad\qquad\qquad\qquad\qquad\qquad\qquad\qquad\qquad\qquad\qquad\qquad Cl$$

$$\longrightarrow \cdots -C_6H_4-NH-COO-CH-CH_2-OOC- \cdots$$

$$\underset{\displaystyle \overset{|}{\underset{Cl}{(CH_2-CH)_{m-1}-CH_2-\overset{\cdot}{C}H}}}{\underset{\displaystyle \underset{Cl}{|}}{}} \quad \text{graftcopolymer}$$

5. Chemical Stress Relaxation

When a polymer is submitted to the action of some outer forces, this is taken out from its equilibrium state while its inner energy increases. The possibility of reaching a new state of equilibrium, able to adapt the polymer to the new conditions, becomes valid when some transformations at the molecular level are produced, and which mainly consist in the macromolecule conformation change. The process, on the whole, is called relaxation and has as a consequence the stress continuous decrease, which tends asymptotically toward a limiting value. With the high elastic polymers, due to their macromolecule flexibility, a fast uncoiling and orientation of these take place in the direction of the stress action so that the relaxation phenomenon appears less intensely while the load maintains in the sample at a value close to the initial one until the action of the stress stops, and when it decreases immediately to zero. The introduction of the low-density cross-linkings, as it usually happens during the vulcanizing process, does not modify the elastic properties so that the polymer behavior under stress should be similar to the linear elastomer. However, surprisingly, keeping a sample slightly cross-linked at a constant deformation, a rapid stress relaxation or even its decrease to zero has been noticed (see Reference 190). The phenomenon was attributed to the splitting of the chemical bonds from the rubber tridimensional network and their remaking into other position after the displacement of some of its fragments, following the chemical relaxation.

Some studies have had in view the establishing of some correlations between the number of splittings from the main chain per unit of volume $[q_m(t)]$ and the stress relaxation, expressed by the subtraction of the ratio $\sigma(t)/\sigma_0$ — with $\sigma(t)$ and σ_0 represent the stress at the moment t and the initial one.[191-193] Thus, it was established that the stress chemical relaxation is well approximated by a Maxwellian subtraction given by the equation:

$$q_m(t) = -N_0 \ln \sigma(t)/\sigma_0 \tag{25}$$

$$\sigma(t)/\sigma_0 = g^{-k_i t} \tag{26}$$

There, N_0 is the initial number of the network chains from a unit of volume (cm³).

This theoretical conclusion was very well checked up on all the hydrocarbon rubbers and especially on the natural one (Figure 57, curve 1). It was experimentally observed that the process of chemical relaxation depends also on temperature as it can be seen from Figure 58. Thus, it could be calculated the activation energy of chemical bonds splitting of $->C-C<-$ type, the value of $30+2$ kcal/mol being thus stabilized.

The experiments carried out in the presence of oxygen at different temperatures underline a special behavior which does not follow a Maxwellian law. In this case, the chemical stress relaxation develops according to a law which can be expressed mathematically by the relation:

$$\ln \sigma(t)/\sigma_0 = -k_1 t - b * \exp(ct) \tag{27}$$

where b and c are the polymer constants.

The registered effect is a consequence of the acceleration of the chemical bond splitting due to the oxygen presence, according to the below reaction scheme:

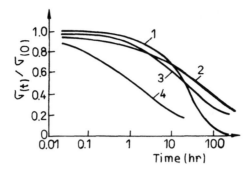

FIGURE 57. Stress-chemical relaxation for some vulcanized elastomers (100°C). 1 — natural rubber; 2 — butadiene-styrene rubber; 3 — butilic rubber; 4 — neoprene rubber. (Based on Reference 190.)

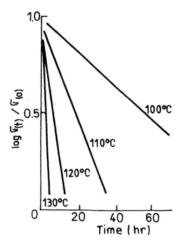

FIGURE 58. Temperature effect on the stress chemical relaxation for natural rubber vulcanizates. (Based on Reference 190.)

$$R\text{–}R \longrightarrow 2R$$

$$R^{\cdot} + O_2 \longrightarrow R\text{–}OO^{\cdot}$$

$$R\text{–}OO^{\cdot} + RH \longrightarrow R\text{–}OOH + R^{\cdot}$$

$$R\text{–}OOH \longrightarrow R\text{–}O^{\cdot} + {}^{\cdot}OH$$

New changes of the relaxation curves appear when the elastomer mechanooxidative degradation is controlled by the diffusion process of the O_2 in the polymer.

Other studies started from the assumption that the bonds splitting from the cross-linkings is responsible for the chemical relaxation. Considering C_0 the number of these bonds, $q_c(t)$ the number of splittings at time (t), and n_0 the initial density of the cross-linkings, the stress relaxation will be expressed in this case by the following equation:

$$\sigma(t)/\sigma_0 = 1 - 2q_c(t)/n_0 \tag{28}$$

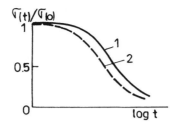

FIGURE 59. Theoretic (1) and experimental (2) curves for a vulcanized natural rubber. (Based on Reference 195.)

If the network is ideal (the length of the fragments between two cross-linkings is the same) i.e., $2c_0 = U_0$, a new equation is obtained which shows that the stress chemical relaxation follows after a typical Maxwellian curve.

The most general case of stress relaxation in a cross-linked polymer consists of the splitting with the same probability of both the main chain and of the cross-linkings (see References 194 to 196). The equation correlating the total number of splittings, $Q(t)$, produced in the main chain and in the reticulations, too, represents the sum of the splittings of the two individual types of network fragments found by the aid of Equations 25 and 28, having the form:

$$Q(t) = -n_0 \ln \sigma(t)/\sigma_0 + n_0/2(1 - e^{-kt}) \tag{29}$$

The explanation of the ratio $\sigma(t)/\sigma_0$, which expresses the stress decrease in time, leads to a complicated equation but which characterizes very well the process as a whole. Thus, it may be found, from Figure 59, a good agreement between the theoretical curve and the experimental one. Although the experimental curve is placed under the theoretical one, the main reason for this discrepancy is given by the changes taking place in the strength of the main chain, close to cross-linking.

The stress relaxation in the vulcanized elastomers is therefore of a mechanochemical nature and it is based on chemical bond splitting. An argument in this sense is the experiment achieved with a polysulfonic rubber which undergoes a faster relaxation process in the presence of n-butylethylsulphhydrate than in the air, due to the chemical reaction which takes place:

$$R–S–S–R + Bu–SH \longrightarrow R–SH + Bu–S–S–R$$

Such reactions are also possible between the polymer macromolecules which are referred to as interchange reactions.[197,198] They become important if the elastomer is kept up at a certain deformation for a long period of time, being of the type:

$$
\begin{array}{ccc}
\text{R–S–S–R} & & \text{R–S} \quad \text{S–R} \\
+ & \longrightarrow & | \; + \; | \\
\text{R'–S–S–R'} & & \text{R'–S} \quad \text{S–R'}
\end{array}
$$

$$
\begin{array}{ccc}
\text{R–S–H} & & \text{R–S} \quad \text{S–H} \\
+ & \longrightarrow & | \; + \; | \\
\text{R'–S–S–R'} & & \text{R'–S} \quad \text{S–R'}
\end{array}
$$

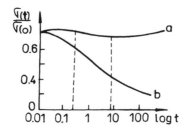

FIGURE 60. Stress chemical relaxation in continuous (b) or intermittent (a) determinations. (Based on Reference 199.)

$$
\begin{array}{ccc}
\text{R–S–Na} & & \text{R–S} \quad\ \ \text{Na} \\
+ & \longrightarrow & |\ +\ | \\
\text{R'–S–S–R'} & & \text{R'–S} \quad \text{S–R}
\end{array}
$$

where R and R′ represent the macromolecular chains.

Such reactions appear not only in the polysulfuric elastomers but also in those based on silicone. They imply both covalent bonds and salt-like bond splittings, the interchanges having to be regraded in this case as two-staged processes where the bond splitting and remaking are accomplished at an equal rate.

During the stress relaxation at different temperatures, some vulcanized rubbers become gradually softer while others exhibit a continuous stiffness. The vulcanized rubber shows both stages; that means that, first, it softens and then it stiffens. Thus, it has been inferred from this aspect that, during stressing, in this elastomer there are produced simultaneous splittings and reticulations whose rates are not essentially affected by the sample stress or deformation (Figure 60). The reactions mentioned above change the curves of the stress chemical relaxation.

It results therefore that the number of the splittings and subsequent reticulations, respectively, can be established by measuring the stress decrease in time (see References 199 and 200).

In the case of natural rubber, the hydrogen release by experiments of stress chemical relaxation at low temperatures is the consequence of such reticulation reactions, for which the following mechanism has been suggested:

Obviously, at the basis of this reaction lies also the macroradicals generating mechano-chemical act which in their tendency to become stable lead to the polymer structural organization.

A process of chemical relaxation takes also place in the linear polymers submitted to some, not too high, deformations which should not cause the fracture. Although the studies on the topic are not too numerous (see References 201 and 202), they demonstrate nevertheless, at present, that the chemorheological treatment of the linear polymers is possible by considering both the physical relaxation and the chemical one which takes place in the nonelastic flow of the uncross-linked polymers.

C. MECHANOCHEMICAL PROCESSES AT POLYMER FRACTURE IN FLOWING STATE

During processing in fluid-viscous state, the polymers are submitted to some intensive shearing stresses which have as an effect both important changes in their viscous-elastic properties and the reduction of their molecular weight due to the chemical bonds splitting from the main chain.

The study of the mechanochemical aspects implied in the polymer melts processing meets some difficulties due to the overlapping of the thermal and oxidative degradation effects; it is difficult to separate only one component of this process of thermo-mechano-oxidative degradation.

The continuous stressing with shearing stresses of the polymer melts has as a result the appearance of some discontinuities in the material which give birth to new surfaces, according to the general laws of the fracture mechanism. Even if the relaxation times are low, suggesting the easy and rapid adaptation of the macromolecules to the stress, the high rate stressing leads to the appearance of some tensions which exceed the material resistance, determining its fracture.

When a polymer melt is forced to pass through a capillary, the phenomenon referred to as the "melt fracture" occurs, it being observed only to the non-Newtonian fluids, never at the micromolecular compounds. The phenomenon takes place at the melts exit from the capillary and was given the following explanation. During the flow through the capillary, the macromolecule orientation takes place in the direction of the force action. According to Boltzman theory, any molecular orientation process corresponds to a lower entropic state than the one characteristic to the completed disordered state. Thus, the polymer entropy at the exit of the capillary is inferior to the initial one. On the other hand, the Brownian movement tends to disorder the system and, if the flow is too slow, this process may become dominant, preventing the orientation. The orientation effect is very pronounced for the rapid flow of the very great macromolecules. What, then happens to the flow if the Brownian movement fails to maintain the macromolecules in a disordered state and the stress field continues to increase? There exists only the following two alternatives:

1. The force field may modify itself so that to reduce or to remove the macromolecular orientation.
2. The great macromolecules may break into small fragments which are more mobile and, therefore, more susceptible to perform Brownian movements of disorder.

The latter alternative is more probable, suggesting the mechanochemical aspects implied in the "melts fracture".[203-205]

Important contributions to the explanation of the polymer fracture mechanism under the conditions of melts flow are given by Porter's studies.[206-208] The importance of the molecular knots participation on the fracture phenomenon is taken into account here. Under the shearing conditions the mechanical stress may be placed either in the middle of the macromolecule at the intersection point of two chains or in the middle of two adjacent coilings. Either the former or the latter of the possibilities are preferred, depending on the duration of the stressing process, working temperature, the probability of chain relative slippage, and the chemical bond strength from the main chain. Although Bueche theory postulates the preferential macromolecule splitting in the middle, the experimental data obtained by Porter do not entirely confirm this mechanism; there can be seen greater destruction toward the chain extremity when examining the distribution curves of the molecular weight.

Intense fracture processes develop not only when the melts are forced through capillaries but also when they are processed by extrusion, mastication, two-roll mixing, calendering, and injection. The formation of the strongly activated new surfaces under these conditions leads to some reactions with micromolecular compounds or with other polymers which are, at the same time, arguments for a radical mechanism of the process.

1. Polymers Processing by Melt Spinning

Melt spinning is a typical case of forced flow through capillaries of a fluid-viscous polymer when intense shear stresses act on it, they being determined by the melt rate gradient from the wall toward the center of the capillary. Although the mechanochemical changes in the polymer melts are known, few studies dealt with the laws governing the process (see References 209 and 210). Of these, the results obtained by the melt spinning of poly(ethylene terephthalate) are valuable due to the industrial importance of this polymer.

The changes undergone by the PET in the melt are determined both by the mechanical shear and thermal degradations. The results of these changes materialize in the reduction of the molecular weight, in the increase of both carboxylic groups and the splitted bond concentration.

In order to develop a mathematical model of the mechanochemical degradation in the melt, the following hypotheses were started from[211]

- The mechanical degradation rate is proportional to the shearing stress.
- Polyester melt is a pseudoplastic fluid.

In order to describe the process, the following equation was suggested:

$$D_m = C_m * \tau_w \tag{30}$$

where D_m = mechanical degradation rate; C_m — mechanical degradation coefficient; and τ_w — wall shearing stress.

Within the range of τ_w small values ($\tau_w = 9.65 * 10^{-5}$ dyn/cm^2), the melt viscosity (η), wall shearing rate ($\dot{\gamma}_w$), and inherent viscosity number (IV) are correlated in the following way:

$$\eta = \frac{\tau_w}{\dot{\gamma}_w} = 1.55428 * 10^4 \exp\left(-11.975 + \frac{6802.1}{T}\right) (IV)^{5.145} \tag{31}$$

For high values of the shearing stress, the following equation can be used:

$$\eta = \frac{\tau_w}{\dot{\gamma}_w} = 4.8569 * 10^4 \exp\left(-11.975 + \frac{6802.1}{T}\right)^{0.65} (IV)^{3.638} \tag{32}$$

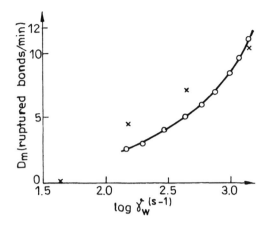

FIGURE 61. Number of split bonds as a function of shear rate (270°C; IV = 0.948; ○ — theoretical values; + — experimental values).

Some experimental results were obtained working with a polymer having the IV = 0.870 (0.948, respectively) submitted to shear stresses in an Instron capillary rheometer, at rates of 45 to 1500 s^{-1}, in the temperature range of 270 to 310 cm^{-1}; these results allowed the calculation of the C_m coefficient. The value obtained (C_m = 1.69 $*$ 10^{11} split bonds/min $*$ dyn/cm^2) was used later to determine the number of split bonds at different shear stresses, calculated by using Equations 30, 31, and 32. It can be observed (Figure 61) that the number of scissions determined by the mechanical stress, calculated on the basis of the mathematical model, agrees satisfactorily with the values which have been obtained experimentally.

Thus, it can be estimated, on the basis of the suggested mathematical model, the effect of the mechanical component of the PET melt degradation during spinning if the experimental conditions are properly chosen.

The size of the shearing stress acting on the molten polymer also determines different values of the shearing rate as well as of the stationary time in the extruder which then in turn will influence the intensity of the polymer degradation.

The decrease of its molecular weight is influenced, at the same time, by temperature, extrusion rate, and the nature of the gaseous atmosphere in which the process occurs. The influence of the temperature on IV at different stationary times is shown in Figure 62; a continuous decrease in time of this property may be observed, this decrease being higher at temperature increases. Thus, at temperature higher than 290°C, the macromolecule significant splitting occurs pointed out by the IV pronounced decrease.

Starting from Jellinek's equation[212] for statistic degradation of a polymer and applying it to the PET, the following relation is arrived at:

$$\frac{1}{(IV)_n^{1.47}} = \frac{1}{(IV)_0^{1.47}} + 3.92 * 10^4 k_{IV} t \tag{33}$$

where k_{IV} is the IV reduction rate constant.

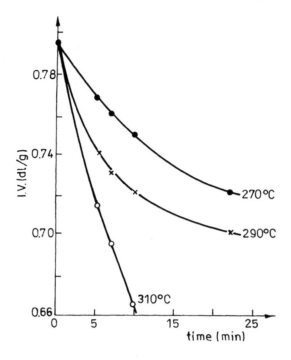

FIGURE 62. Influence of the stationary time in the extruder on IV (at different melt temperatures).

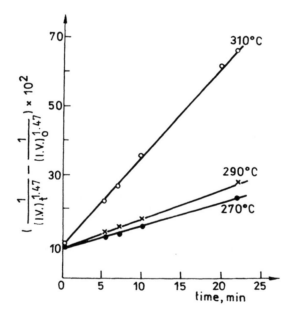

FIGURE 63. Plotting of Equation 33 for different temperatures.

The values of the k_{IV} constant for different temperatures were determined from experimental data using Equation 32; Figure 63 gives the linear plottings that were obtained. It may be seen that the degradation is more pronounced at temperatures higher than 290°C and the degradation rate constant changes as function of temperature according to an Arrhenius law. Showing this dependence, the value of the activation energy of PET degradation was obtained (E_a = 84 kJ/mol) when the melt is forced through the capillaries.

Consequently, the equation describing the reduction of PET inherent viscosity in time has the following form:

$$\frac{1}{(IV)_t^{1.47}} = \frac{1}{(IV)_o^{1.47}} + 3.29 * 10^4\left(e^{2.782} - \frac{84}{RT}\right)t \tag{34}$$

The equation allows the calculation of IV for the extruded PET as a function of the stationary time inside the extruder at different temperatures; in its turn, this includes both the preheating and the proper extrusion time.[213]

Compared to the thermal degradation, the PET degradation under the combined action of the temperature as well as mechanical stresses needs a lower activation energy (84 kJ/mol as compared to 119 kJ/mol in stationary conditions). During all the experiments the shearing stress was maintained at high values (0.15 to 0.5 MPa) so that the mechanical splitting of the chemical bonds may develop predominantly. It was expected that the shearing mechanodegradation should intensify as the shearing stress increases and the stationary time inside the extruder decreases. As it has resulted from Figure 62, the bond splitting was more intense at great stationary times. This fact as well as the temperature influence on the polymer degradation suggests that temperature has the main contribution to the reduction of PET inherent viscosity during melt spinning.

Although the mechanical stress is not the main cause, it has nevertheless an important role in reducing sensibly (with over 25%) the value of the destruction activation energy. The two destructive processes complete each other, their contribution, and occurring, according to the same mechanism of free radicals, formation by the homolytical splitting of the macromolecular chains.

2. Polymers Processing by Two-Roll Mixing

Polymer mechanical and thermal stressing when processed by mastication, two-roll mixing, extrusion, and injection has as a first result the appearance of the mechanoactivated state followed by the covalent bonds splitting from the main chain. Thus, great concentrations of active species of a radical nature are formed which are placed on the new surfaces that are continuously discovered, under the action of the shearing stresses. The processed polymer becomes extremely reactive, being able to chemically interact with other micro- or macromolecular components, which are present in the medium and have a radical acceptor feature.

The possibility of activating some polymer reactions with certain substances, under the conditions of their melt shearing, is a good argument for the radical mechanism of the process. At the same time, modified polymers with properties superior to the initial ones can be achieved by a judicious selection of the substances able to react with the free radicals obtained by mechanocracking. This way of polymer chemical change has more advantages, among these, the most important one consisting of the fact that it can be practically carried out on devices commonly used in their processing.

An important polymer which was widely studied from the point of view of the mechanochemical processes taking place during the melt shearing stress was poly(vinyl chloride) (PVC) under the two rolling conditions. Kargin and Sokolova[214,215] developed the theory of "chemical flow" which, in the authors opinion, consists in the continuous destruction and recombination of the chemical bonds. This concept was used to explain the PVC behavior during injection. At high speeds of the screw, the mechanical properties deteriorate, as stated by Savelev et al.;[216,217] as the processing temperature increases, the destruction curve of the molecular weights narrows in the zone of the high values of the molecular weight and the branches increase as well. At a sufficient high temperature, the cross-linking reactions become predominant; consequently, the PVC mechanical properties go through a maximum

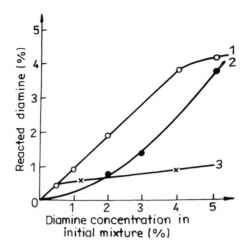

FIGURE 64. Reacted diamine amount variation with diamine content in the initial mixture. 1 — benzidine; 2 — p-PhDA; 3 — m-PhDA.

function of the processing temperature. The "chemical flow" process is neglectable for the PVC having a reduced molecular weight and it is completely absent in the case of low viscosity melts. However, surprisingly, an improvement of the properties can be obtained by PVC recycling and processing under the optimum known conditions, as a result of a better distribution of the used ingredients and of the homogeneity increase.

The chemical flow implies, in fact, the splitting of the covalent bonds from the main chain and their remake in another zone of the melt where, after some time, the fragment of the broken macromolecule is arrived at; naturally, free macroradicals will be generated during this process.

Their presence was pointed out by stimulating some PVC reactions in two-rolling processing conditions, with aromatic diamines which are well-known radical acceptors. It was proved, by means of spectral and chemical methods, that reaction products containing chemically linked aromatic diamine are obtained by the simultaneous two-rolling of the PVC with benzidine m- or p-phenylene diamine. Thus, the IR spectra reveal absorption bands characteristic for the aromatic nucleus which are absent from the spectrum of the witness polymer (see Reference 218). It was also possible to establish the quantity of reacted diamine for different values for the process parameters by the photocolorimetering of the solutions in cyclohexanone, of the reaction products. It was thus found out that the percentage of the bonded diamine in the reaction product is related to the diamine concentration in the mixture subjected to processing, which varies as Figure 64 shows (see Reference 219).

Benzidine proves to be most reactive because it is consumed almost entirely, regardless of its initial concentration in the polymer (Figure 64, curve 1). An indirect method which allowed the qualitative pointing out of the chemically bonded diamine was also given by the possibility of diazotizing the reactive products and their subsequent coupling with resorcin or β-naphthol.

The reaction took place obviously with the free aminic groups enabling the achievement of some different color ranges whose intensity amplifies as the processing time as well as the diamine initial concentration in the mixture increase. It must also be pointed out that diamine-modified products have the best coloring capacity while the p-phenylene diamine-modified ones have less-intensive coloring effects. The explanation of this phenomenon is given by the establishment of the way in which the polymer molecular weight varies in function of the processing duration (Figure 65). Thus, it can be noticed that with the samples containing m-phenylene diamine (curve 1) it is recorded a continuous decrease of the vis-

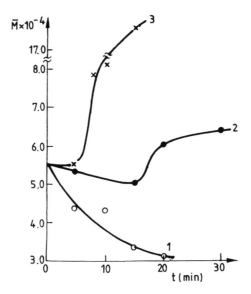

FIGURE 65. Molecular weight variation with two-roll mixing duration. 1 — PVC + m-PhDA; 2 — PVC + benzidine; 3 — PVC + p-PhDA.

cosimetric molecular weight according to a law typic for mechanodestruction, a limit value being obtained at durations exceeding 45 min. The diamine acts, in this case, as a radical acceptor through only one aminic group, stabilizing macroradicals formed by mechanodestruction at reduced values of the macromolecular compound. The second functional group remains free and it can be diazotized and coupled. The increase in time of the reacted m-phenylene diamine quantity leads therefore to the color intensification of the diazotizing and coupling products. With benzidine products, a minimum is acquired and afterward $\overline{M}$ begins to increase as the two-rolling time increases. It thus results that the latter aminic group becomes also activated at durations exceeding 15 min and thus it can act as a radical acceptor bonding other fragments obtained as a consequence of the destruction. Therefore, the color intensity of the diazotizing and coupling products decreases in time.

The p-phenylene diamine samples show, nevertheless, a continuous increase of the polymer molecular weight (curve 3). The well-known reactivity of this micromolecular compound determines its reaction, through both functional groups, with macromolecule fragments resulted from mechanodestruction having as a final effect the M increase.

The micromolecular compound chemical bonding to PVC gives birth to deep changes of the polymer structure which is mirrored in the pronounced modification of some properties. Thus, the fracture strength registers a considerable increase which depends to a large extent on the diamine quantity in the reaction initial mixture (Figure 66).

The best results are obtained in the presence of the benzidine. Consequently, a new material is obtained having a fracture strength 140% higher than the one of the initial polymer. It must be also noted that, at a benzidine concentration higher than 3% in the mixture with PVC, the quantity of thermal stabilizer (Pb salts) could be gradually diminished up to 1% as compared to the polymer. The obtained results certify the function of thermal stabilizer of the aromatic diamine, proving once again the radical character of the process (see Reference 221).

The efficiency of the PVC reaction with benzidine, reflected in the modification of the mechanical properties, depends not only on the micromolecular compound concentration (C_d), but also on the thermal stabilizer concentration (C_s), on the two-rolling time (t) as well

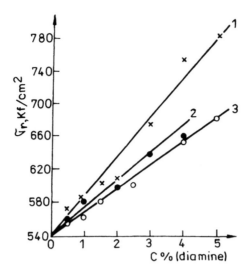

FIGURE 66. PVC tensile strength variation with diamine content. 1 — PVC + benzidine; 2 — PVC + p-PhDA; 3 — PVC + m-PhDA.

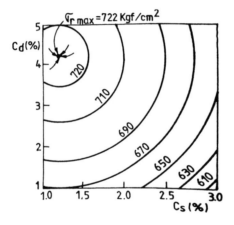

FIGURE 67. Influence of thermal stabilizer concentration (c_s) and aromatic diamine concentration (c_d), respectively, on the tensile strength of PVC.

as on the temperature (T). The function showing the dependence of the fracture strength (σ_r) on these parameters has the form:

$$\sigma_r = 703 - 16 * C_s + 12.2 * C_d - 5 * C_s^2 - 5.3 * C_d^2 + 7.6 * t * T$$
$$+ 9.4 * t * C_s + 11.4 * t * C_d - 9.3 * T * C_s - 11.8 * T * C_d \quad (35)$$

Figure 67 shows the variation of the followed mechanical property according to the concentration of the diamine (C_d) as well as of the thermal stabilizer one (C_s) which allows to establish the optimum values of these parameter functions of the maximum value of σ_r: $C_d = 4.2\%$, $C_s = 1.2\%$.

The PVC processing under these conditions led to a material having a fracture strength of 722 kgf/cm² (see Reference 220).

The possibility of PVC to develop a reaction with some micromolecular compounds, based on the free radical formation, suggested the achievement of some graft and block copolymerization reactions. As the purpose was to improve the PVC shock strength, the simultaneous processing by two-roll mixing with different elastomers was accomplished.

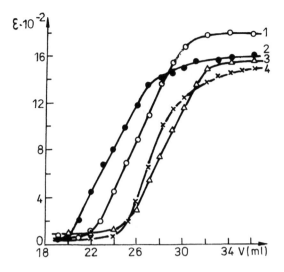

FIGURE 68. Turbidimetric titration curves of some products obtained by PVC and polyurethanes simultaneous two-roll mixing. 1 — witness; 2 — PVC + polyurethane (with ethylene glycole); 3 — PVC + polyurethane (with triethylene glycole); 4 — PVC + polyurethane with glycerol.

Using polyurethanes based on dibenzyl diisocyanate which were distinguished by the alcoholic component (ethylene glycol, triethyl glycol, glycerol) high shock-resistant products were obtained.

The proof of block and graft copolymer formation on the basis of the two macromolecular compounds was given by the IR spectra and by the turbidimetric titration curves.[222] It can be thus observed in the IR spectra, the presence of the 1520 and of 1580 cm^{-1} bands, attributed to the −CO−NH− groups, as well as the band from 1690 cm^{-1} corresponding to the −CO−O− group, as well as of the band from 3300 cm^{-1} characteristic to amidic NH. The turbidimetric curves shown in Figure 68 prove that the obtained products are not physical mixtures of the two polymers, because they have only one level corresponding to a unique compound which cannot be but the result of their reaction.

The presence of elastomer sequences on the chain of the resulted products (or as branches) leads to the increase of the macromolecule flexibility which at a macroscopic level is expressed by the shock-resistance increase. It can be observed from Figure 69 that the shock strength (a_k) increases continuously as the polyurethane concentration increases (curve 1) but being also affected by the other parameters of the process: temperature and duration. Thus, it may be seen that, as temperature increases, a_k increases too for the polyurethanic elastomer based on ethylene glycol.

However, the mechanical properties vary differently for the polyurethane based on glycerol (curve 3) when a maximum is registered at T = 170°C. The shock strength decrease beyond this temperature is the consequence of some cross-linking reactions of the destruction fragments by means of the lateral hydroxylic groups, when, as a result, the structure stiffens. The same effect becomes evident when the rolling time increases (curve 2). The shock strength up to 20 min of processing is due to the presence of a higher and higher quantity of elastic component in the reaction product structure, even if slight cross-linkings are also produced. However, at longer reaction times the cross-linking process intensifies determining the structure stiffening and the a_k increase.

The dependence of the PVC mechanical properties on the two-rolling process parameters was systematically aimed at working in the presence of a polyurethane based on dibenzyl diisocyanate, ethylene glycol, and adipic acid (see Reference 223). There have been obtained

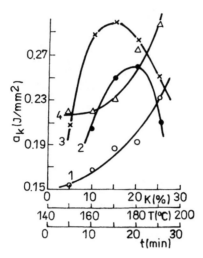

FIGURE 69. Shock strength variation with polyurethane concentration. 1 — PVC + polyurethane (with glycerine); with temperature (3 — PVC + polyurethane with glycerol) and 4 — PVC + polyurethane (with ethylene glycole); with duration (PVC + polyurethane with glycerol).

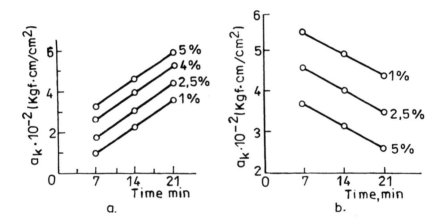

FIGURE 70. Constant level curves in the experimental plane of temperature-polyurethane concentration (a_k).

regression equations with particular forms for the tensile (σ_r) as well as shock resistances (a_k).

$$\sigma_r = 635.88 + 14.83 * c + 8.64 * c * T \tag{36}$$

$$a_k = 0.03575 + 0.0028 * T - 0.0052 * c * T - 0.0046 * t * T \tag{37}$$

where c — elastomer concentration (%); t — two-rolling time; and T — temperature.

Figure 70 points out the correlated infuence of the temperature and processing time on the shock strength. At low temperatures (150°), as the rolling time increases, a_k increases continuously, because the cross-linking reactions do not take place, and the elastomer quantity in the obtained product increases.

However, at high temperatures the effect is opposite: as the polyurethane concentration and processing time increase, the shock resistance decreases. The explanation lies in the fact that the cross-linking reactions intensify, being performed on the basis of the elastomer.

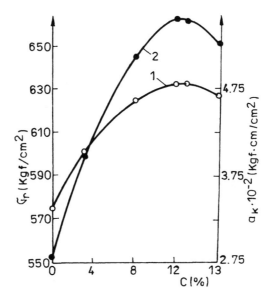

FIGURE 71. PVC tensile strength (1) and shock strength (2) variation with nitrile rubber concentration (T = 170°; t = 14 min).

However, the tensile strength increases both with the polyurethane concentration in the initial blend and with the temperature, it being favored by the cross-linking reactions which are produced. The optimum conditions of the two macromolecular compound simultaneous rolling can be established on the basis of these results, according to the goal aimed at. High tensile strength but low shock-resistant products can be obtained by introducing polyurethane concentrations up to 5% at short rolling times (7 min) and high temperatures (190°C). A superior shock strength correlated with a tensile strength close to that of the PVC is obtained at short rolling times (7 min), high temperatures (190°C), but smaller polyurethane quantities (1 to 2%) in order to prevent the cross-linking process.

High shock and tensile strength products can be obtained when working with high polyurethane concentrations at long durations and low temperatures (150°C).

Similar results were obtained by Vasiliu Oprea and Popa[224] when PVC two-rolling processing with nitrile rubber is performed. In this case, too, the determination of the regression equations expressing the dependence of the graft and block copolymer mechanical properties on the process parameters permitted process optimization. Watching the variation of these properties in Figure 71 (for T = 170°C and t = 14 min), according to the elastomer concentration, it follows that the best results are obtained when the nitrile rubber has a proportion of 12% in the initial mixture.

It follows that the processing of the polymer melts when there are manifested some intense shearing stresses, combined with the temperature action, inevitably accompanied by fracture phenomena which cause the genration of new surfaces with high concentration of free radicals. The judicious selection of some micro- or macromolecular reaction partners allows the exploitation of the inevitable effects of the thermomechanical destruction for the purpose of improving the physicomechanical properties of some macromolecular compounds. The process is advantageous because these changes can be carried out on traditional processing devices without an essential modification of the respective technologies. Thus, the mechanochemistry of the polymer melt fracture provides solutions for getting some new products with superior exploiting properties.

IV. MECHANOCHEMICAL PROCESSES AT INORGANIC SURFACES MECHANICALLY ACTIVATED

The mechanical stressing of micro- and macromolecular solids has as a consequence their potential energy increase by the elastic energy concentration, transmitted in different forms on structural defects. In these zones the inner stresses often reach critical values, diminishing thermal activating energy of chemical bond splitting or causing electron emissions from the atoms peripheral layers. The accumulation in microvolumes of a great number of fracture elementary acts by a chained process ends with the magistral crack formation and the appearance of new active surfaces. The presence of some micromolecular substances whose molecules can act as a "wedge" at the top crack determines the crack-pronounced propagation by promoting a "corrosive splitting" process. Such substances may be the micromolecular compounds themselves, whose transformation are watched (the monomers, for instance). The acceleration of the crack propagation may be often brought about by the gaseous substances which are present in the reaction medium, the effect being intensified by their molecular dimension increase.

It may be appreciated that along its duration, regardless of the way of stressing the inorganic network, the fracture process is accompanied by mechanochemical acts, consisting of the interatomic bond splitting. It finishes with the formation of new rich-in-energy surfaces, which become permanently renewed, and constitute the support of numerous chemical reaction developments. The possibility of accomplishing such reactions that cannot be developed in other conditions proves, at the same time, the presence of the molecular mechanism of the inorganic networks fracture and allows the appreciation of the active centers nature which are formed by their mechanical stressing.

A. MECHANOCHEMICAL GRAFTING ON INORGANIC SURFACES

The mechanical dispersion by vibratory milling of some crystalline inorganic combinations is accompanied by the process of some new surface formation with a high chemical reactivity, due to the occurrence of some intense electronic flows. In the presence of vinyl monomer, the polymerization reaction can be initiated which finishes with the polymer formation, grafted to the inorganic support.

Such syntheses have been first accomplished by Plate and Kargin[225-227] and developed later by Baramboim.[228-230] The experiences have been developed in the presence of some extremely varied inorganic combinations, including metal powders (Mg, Fe, Ca, Al, Cr, Bi, Sn),[226,231] nonmetals (graphite, black carbon), metallic and nonmetallic oxides (SiO_2, TiO_2, ZnO, Al_2O_3, BaO, Fe_2O_3),[226,227,232,233] and salts (NaCl, NaF, KCl, KBr, KI, LiF, CaF_2, HgCl, HgI, ZnS, BaS, $SnSO_4$, $CaSO_4$, $CaCO_3$),[234-236] on which numerous vinyl monomers have been grafted (methyl methacrylate, styrene, acrylonitrile, acrylic acid, isoprene, maleic anhydride, α-methyl styrene, acrylamide, etc.). It has been found out that the polymerization initiation takes place at the inorganic surface, a polymer grafted on this surface being obtained; afterward, as a consequence of the destructive effects developed during the mechanodispersion, the graft homolytical splitting as well as the soluble polymer accumulation are produced. To prove that vinyl monomer polymerization is really initiated in the presence of inorganic substances remains the fact that the reaction does not occur in their absence. At the same time, when heating the monomers at 50 to 60°C (attained in the milling vessels, even when these are cooled), it is not possible to obtain polymers. A large range of monomers has been polymerized on newly formed surfaces by the mechanical desintegration of an alumino-silicate (SiO_2 — 91.4%; Al_2O_3 — 4.1%, H_2O). The polymerization reaction evolution has been watched by determining the variation of the monomer vapor pressure; it decreases in time and proves both the organic substance chemosorption and its chemical reactions (vinyl monomers) (see Reference 237). The styrene polymerization on before-hand

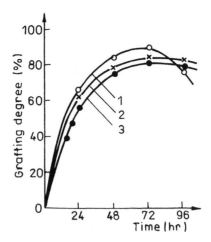

FIGURE 72. Grafting degree variation with vibratory milling duration. 1 — PAN grafted on caolin; 2 — on mica; 3 — on volcanic tuff.

milled NaCl crystals has also been achieved. The electron microscopy analyses have pointed out that the polymer chains are initiated and grow only in the crystalline parts of the inorganic underlayer, forming globular formations. The graft size depends on the energy of the used crystalline network and of the ionic bond strength, respectively, which determine the duration necessary for the accomplishment of a certain grafting degree.

The caolin, volcanic tuff, and mica are also efficient grafting supports for some great quantities of acrylonitrile under vibratory milling conditions. The grafting degree changes together with the milling duration (Figure 72) reaching a maximum value after about 72 h as stated by Vasiliu Oprea and Weiner.[238] The grafted polymers have been used as filling materials at PVC processing, contributing to the essential increase of its mechanical properties, i.e., tensile and shock strength.

The polymerization mechanism in the presence of the inorganic substances is a complex one, depending on the nature of the used substance. For substances with $->C-C<-$ bonds (black carbon, graphite) or $-Si-O-$ bonds (silicates) the mechanism is a radical one; it is proved by the oxygen inhibitory action and has been formulated by Baramboim.[228]

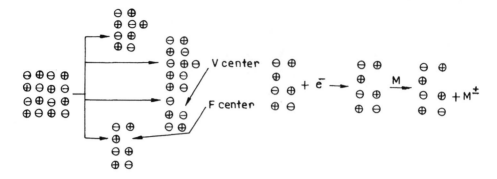

FIGURE 73. Polymerization initiation scheme by crystalline inorganic substances mechanodispersing. (Based on Reference 227.)

TABLE 8
The Influence of the Inorganic Substance Chemical Nature on the Styrene Polymerization Efficiency

Inorganic substance	Dissociation energy (kcal/mol)	Bond ionization in crystal (%)	Electronic emission intensity (imp/s)	Copolymerization yield (%)
BaO	143 ± 3	88.0	25	0.013
BaS	100 ± 5	69.0	95	3.90
HgCl$_2$	23 ± 2	60.0	30	0.06
HgI$_2$	8.2 ∓ 0.4	31.0	40	0.80
ZnO	65 ± 10	62.5	25	0.90
ZnS	48 ± 3	46.0	35	1.10

The polymerization in the presence of desintegrated salts (ionic compounds) implies a more complex mechanism in which a very important part is played by the reaction medium.

The active centers which initiate the reaction appear to be due to an electron beam emitted from the peripheral layers of the metallic atoms, as shown by Plate and Kargin.[227]

$$e^{(-)} + CH_2{=}CH \longrightarrow {}^{(-)}CH_2{-}\dot{C}H$$
$$\qquad\qquad\; | \qquad\qquad\qquad |$$
$$\qquad\qquad R \qquad\qquad\qquad R$$

The chain propagation is made through radical extremity, while the anionic one deactivates itself due to the nucleophile additives from the system (water traces). Baramboim reaches the same conclusion and advances an initiation scheme based upon the formation of an "F" center appeared by ZnS crystal destruction (Figure 73).

Investigating the styrene polymerization on the surface of some alkaline halides (NaF, LiF), Ianova points out intense electron beams as well as electrical charges of the newly formed surfaces.[235] Two types of products are obtained under these conditions: a polymer grafted on crystalline surfaces, by an ionic mechanism, influenced by water traces, and a homopolymer resulted from a radical mechanism initiated by electron beams.

The more reduced the dissociation energy of the ionic bond, the more intense the electronic emission and, therefore, the capacity of initiating the polymerization (Table 8) (see Reference 234).

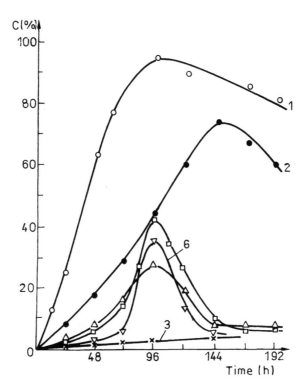

FIGURE 74. Time variation of conversion for some monomers polymerization. 1 — acrylonitrile; 2 — styrene; 3 — vinyl acetate; 4 — acrylamide; 5 — methyl methacrylate; 6 — ε-caprolactam (T = 18 + 2°C).

B. MECHANOCHEMICAL POLYMERIZATION

During the vibratory milling process the mechanical energy is equally concentrated both on the monomer introduced in the reaction vessel and on its walls, as well as on the milling bodies. In the former case, the consequence is the mechanical energy accumulation at a molecular level where it produces the deformation of both the chemical bonds and of the valence angles as well as bringing the substance into an extremely reactive mechanoexcitated state. In the latter case, there appear electronic beams as well as metal particles in a colloidal state representing fine divided and strongly activated surfaces, which are also responsible for the initiation of some specific chemical reactions.

Starting from these premises, they accomplished the mechanochemical polymerization by vibratory milling of a large range of vinyl, acryl, dienic, or cyclic monomers (Figure 74). It must be noticed that in this type of synthesis the liquid monomers generally present a higher reactivity as compared to the one belonging to the solid monomers. The explanation consists in the fact that the monomers diffuse in the microcracks generated by the shock and friction forces in the walls of the vessel and of the milling bodies as well; they act as a ''wedge'', favoring the detachment of some extremely fine metal particles and even their colloidal dissolution and leading to the faster and easier initiation of the polymerization.

The shape of the curves is generally the same, presenting a maximum of conversion at different durations, a function of the monomer chemical nature. The descendent part of the kinetic curve is the consequence of the fact that after a certain duration, when an enough quantity of polymer has been accumulated into the reaction medium, the mechanodegradation reaction prevails; this one finishes with the polymer molecular weight decrease, even with the formation of the oligomers which finally are not found any longer in the milling product.

The spectral analyses of such polymerization products have pointed out that not only the classical vinyl monomers but also the cyclical substances, the aliphatic nitriles, or even the aromatic compounds can participate in the reaction (see References 250, 251, 253, and 256).

The polymer appears on the milling bodies or on the vessel walls under the form of a dried powder (when the conversion is total) or under the form of a wetted powder with monomer (at low values of conversion). Two fractions separate from the reaction product: one soluble in specific solvents and another insoluble one. Therefore, it becomes evident that the polymerization initiation takes place at the surface of the metal colloidal particles detached from the equipment, on which the polymer is grafted. When the length of these grafts reached a certain value, their splitting is produced concurrently with the release of the soluble fraction in the reaction medium. The insoluble fraction represents therefore a grafted polymer on the metallic surfaces of the colloidal particles.

For clearing up the nature of the active centers which initiate the polymerization determining upon the reaction mechanism, it has proceeded to syntheses in the presence of radical acceptors or in an inert gaseous atmosphere (nitrogen), in an active one, respectively, (oxygen, air) and in some cases in the presence of liquids with different values of the dielectric constants.

It has been thus established that the gaseous atmosphere infuences the reaction, because the vinyl monomers polymerize harder in the presence of the oxygen from the air, a fact which suggests a radical mechanism. The polymerization of some aliphatic nitriles is influenced to a much lesser degree by the presence of air.

The polar liquids diminish the conversion in the polymer while the liquids with low dielectric constant favor the reaction. It follows that, in the function of the chemical nature of the liquid which is present in the reaction medium, the mechanism can be predominantly radical or ionic, depending on whether the polymerization takes place at a vinylic bond or at the nitrilic group. In this latter case, a cationic mechanism through a solvated electron may be considered taken into account with the highest probability.

For the vinyl monomer polymerization, a radical mechanism can be consequently taken into account, when the leading part belongs to the electron detached from the peripheral electronic layer of the metallic atom (Figure 75). A radicalic anion is therefore made up which, at its anionic end, is stabilized by the metallic ion. The propagation develops through the radical end, and the interruption is possible by disproportionation and recombination. Another way may also be considered: the transfer of a new electron at the metallic cation. On this basis the increase of the Fe content chemically bonded in the polymer with the milling duration can be explained (Figure 76).

For the aliphatic nitrile polymerization, an anionic mechanism — as Figure 77 shows — must be taken into consideration.

The presence of metal in the mechanochemical polymerization product conveys properties which differ from the ones belonging to the analogs synthesized by classical methods. Among these, there must be distinguished their thermal stability, semiconductor properties, as well as their paramagnetism. At the same time, their polychelate characteristics permit their utilization as macroinitiators in the polymerization of some vinyl monomers.

The possibility of obtaining polymers under the vibratory milling conditions of some vinyl and unconventional monomers in the presence or the absence of the inorganic substances by grafting on their surfaces or at those of the equipment metal becomes a serious argument in favor of the molecular mechanism of the solid fracture process. Only in this way may there appear active centers able to initiate reactions which are difficult to be obtained in other conditions.

Consequently, it results that both the micro- and the macromolecular compounds are based on a common deformation and fracture mechanism which implies in its first part the

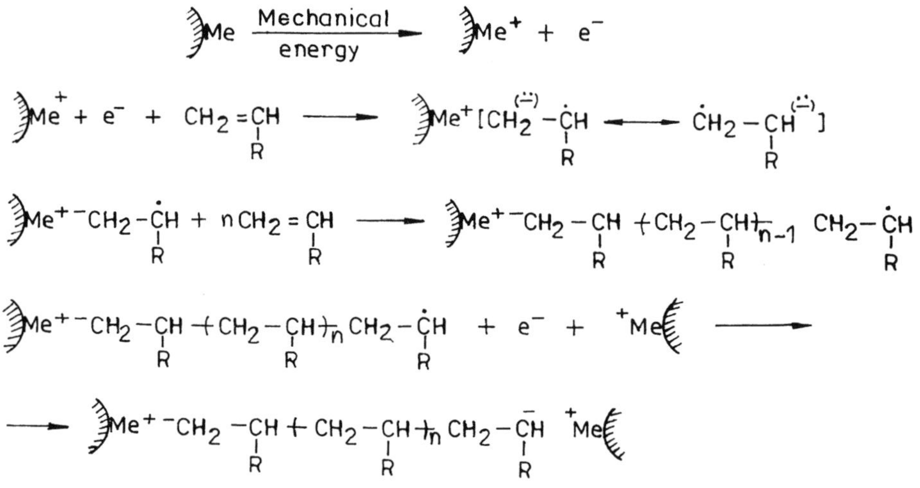

$$\rangle\!\!\!\rangle Me \xrightarrow[\text{energy}]{\text{Mechanical}} \rangle\!\!\!\rangle Me^+ + e^-$$

$$\rangle\!\!\!\rangle \overset{+}{Me} + e^- + CH_2 = \underset{R}{CH} \longrightarrow \rangle\!\!\!\rangle Me^+ [\underset{}{CH_2^{(\cdot\cdot)}} - \underset{R}{\overset{\cdot}{CH}} \longleftrightarrow \overset{\cdot}{CH_2} - \underset{R}{CH^{(\cdot\cdot)}}]$$

$$\rangle\!\!\!\rangle Me^{+\,-}CH_2 - \underset{R}{\overset{\cdot}{CH}} + nCH_2 = \underset{R}{CH} \longrightarrow \rangle\!\!\!\rangle Me^{+\,-}CH_2 - \underset{R}{CH} \left(CH_2 - \underset{R}{CH} \right)_{n-1} CH_2 - \underset{R}{\overset{\cdot}{CH}}$$

$$\rangle\!\!\!\rangle Me^{+\,-}CH_2 - \underset{R}{CH} \left(CH_2 - \underset{R}{CH} \right)_n CH_2 - \underset{R}{\overset{\cdot}{CH}} + e^- + {}^+Me \langle\!\langle \longrightarrow$$

$$\longrightarrow \rangle\!\!\!\rangle Me^{+\,-}CH_2 - \underset{R}{CH} \left(CH_2 - \underset{R}{CH} \right)_n CH_2 - \underset{R}{\overset{-}{CH}} \; {}^+Me \langle\!\langle$$

FIGURE 75. Mechanochemical polymerization mechanism of vinyl monomers.

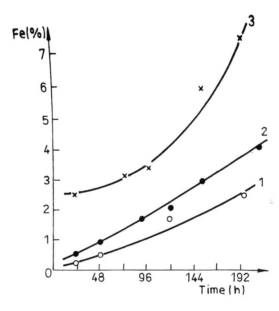

FIGURE 76. Fe content variation with vibratory milling duration. 1 — polyacrylonitrile (soluble fraction); 2 — unseparated; 3 — insoluble fraction.

formation of microcracks in microvolumes in which there were concentrated numerous chemical bonds split by the mechanical stress; this process is followed afterward by the microcrack nucleation or coalescence in a magistral crack which propagates rapidly through the material and forms new and very reactive surfaces.

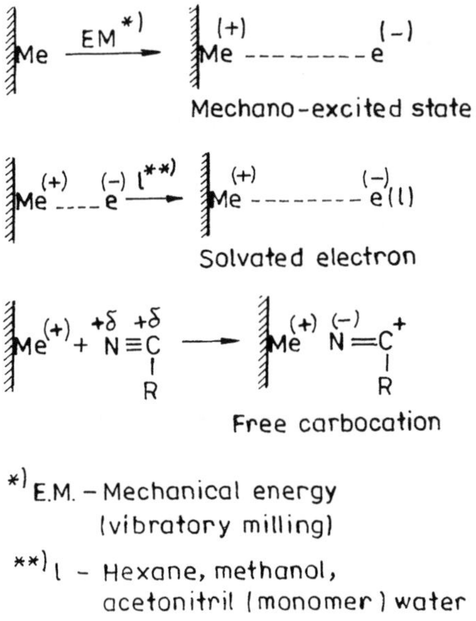

FIGURE 77.   Mechanochemical polymerization mechanism of aliphatic nitrils.

REFERENCES

1. **Simionescu, Cr. and Vasiliu Oprea, Cl.**, *Mecanochimia Compusilor Macromoleculari*, Bucharest, 1967.
2. **Baramboim, N. K.**, *Mechanochemistry of Polymers*, MacLaren, London, 1964.
3. **Watson, W. F.**, Mechanochemical reactions, in *Chemical Reactions of Polymers*, Fetters, E. M., Ed., John Wiley & Sons, New York, 1964.
4. **Allport, D. C.**, Mechanochemical syntheses, in *Block Copolymers*, Allport, D. C. and Iones, W. H., Eds., Applied Science, London, 1973.
5. **Sohma, J. P. and Sakaguchi, M.**, ESR studies on polymer radicals produced by mechanical destruction and their reactivity, *Adv. Polym. Sci.*, 20, 111, 1976.
6. **Casale, A. and Porter, R. S.**, *Polymer stress reactions*, Vol. 1 and 2, Academic Press, New York, 1978.
7. **Kausch, H. H.**, *Polymers, Properties and Applications*, Vol. 2, Springer-Verlag, Berlin, 1978.
8. **Andrews, E. H. and Reed, P. E.**, Molecular fracture in polymers, *Adv. Polym. Sci.*, 27, 1, 1978.
9. **Williams, J. C.**, Applications of linear fracture mechanics, *Adv. Polym. Sci.*, 27, 67, 1978.
10. **Bucknall, C. B.**, Fracture and failure of multiphase polymers, *Adv. Polym. Sci.*, 27, 121, 1978.
11. **Murakami, K.**, Mechanical degradation, in *Aspects of Degradation and Stabilization of Polymers*, Jellinek, H. H. G., Ed., Elsevier, Amsterdam, 1978.
12. **Schnabel, W.**, *Polymer Degradation. Principles and Practical Applications*, Hanser International, Vienna, 1981.
13. **Kuhn, W.**, *Experentia*, 5, 319, 1949.
14. **Katschalsky, A.**, *Experentia*, 5, 319, 1949.
15. **Zwick, M.**, *J. Polym. Sci.*, 16, 221, 1955.
16. **Hargitay, B., Katschalsky, A., and Eisenberg, H.**, *Nature*, 765, 1955.
17. **Hoffman, H.**, *Biochim. Biophys. Acta*, 27, 247, 1958.
18. **Idfson, S., Michaeli, I., and Zwick, M.**, Elementary mechanochemical processes, in *Contractile Polymers*, Pergamon Press, London, 1960.
19. **Rand, A. and Walters, D. H.**, *Contractile Polymers*, Warnek, H., Ed., Pergamon Press, Oxford, 1960.
20. **Katschalsky, A.**, Equilibrium mechanochemistry of collagen fibers, in *Structure and Function of Connective and Skeletal Tissue*, Butterworths, London, 1965.
21. **Robinson, D. R. and Jencks, W. P.**, *J. Am. Chem. Soc.*, 87, 2470, 1965.
22. **Steinberg, I. Z., Oplatka, A., and Katschalsky, A.**, *Nature*, 210, 568, 1966.
23. **Yonath, J. and Oplatka, A.**, *Biopolymers*, 6, 1129, 1968.

24. **Oplatka, A. and Yonath, J.**, *Biopolymers,* 6, 1147, 1968.
25. **Reich, S., Katschalsky, A., and Oplatka, A.**, *Biopolymers,* 6, 1159, 1968.
26. **Sherebin, M. H. and Oplatka, A.**, *Biopolymers,* 6, 1169, 1968.
27. **Flory, P. J.**, *J. Cell. Physiol.,* 49 (suppl. 1), 175, 1957.
28. **Vorobiev, V. I.**, *Biochimia,* 22, 3, 597, 1957.
29. **Kuhn, W., Ramel, M., Walters, D., Ebnek, G., and Kuhn, H.**, *Adv. Polym. Sci.,* 1, 340, 1960.
30. **Katschalsky, A. and Kedem, O.**, *Biophysics,* 2, 53, 1962.
31. **Frenkel, I. Ia., Kuhareva, L. V., Ginsburg, B. M., Kasparsan, K. A., and Vorobiev, V. I.**, *Biofizika,* 10, 5, 735, 1965.
32. **Blumenthal, K. and Katschalsky, A.**, *Biochim. Biophys. Acta,* 173, 51, 1969.
33. **Oster, G., Perelson, A., and Katschalsky, A.**, *Q. Rev. Biophys.,* 6, 1, 1973.
34. **Volkenstein, M. V.**, *Obsceaia Biofizika,* Nauka, Moscow, 1978.
35. **Andrews, E. H.**, *Fracture in Polymers,* Olivier and Boyd, London, 1968.
36. **Griffith, A. A.**, *Trans. R. Soc. London, Ser. A,* 221, 163, 1920.
37. **Jurkov, S. N., Zakrevskii, V. A., Korsukov, V. E., and Kuksenko, V. S.**, *Fiz. Tverdogo tela,* 13, 2004, 1971; *Soviet Phys. solid state,* 13, 1680, 1972.
38. **Jurkov, S. N. and Kursukov, V. E.**, *Fiz. Tverdogo tela,* 15, 2071, 1973; *Soviet Phys. solid state.,* 15, 1379, 1974.
39. **Zakrevskii, V. A. and Korsukov, V. E.**, *Vysokomol. Soedin. B,* 13, 105, 1971; *Vysokomol. Soedin. A,* 14, 955, 1972.
40. **Stepanov, V. A.**, *Mehanika Polimerov,* 1, 95, 1975.
41. **Jurkov, S. N. and Narsulaev, B. N.**, *J. Tech. Phys. USSR,* 23, 1677, 1953.
42. **Joseph, S. H.**, *J. Polym. Sci., Polym. Phys. Ed.,* 16, 1071, 1978.
43. **Jitkov, V. V. and Pesceanskaia, N. N.**, *Mehanika Polimerov,* 1, 65, 1972.
44. **Gul, V. E.**, *Mehanika Polimerov,* 3, 205, 1971.
45. **Karabelnikov, Yu. G.**, *Mehanika Polimerov,* 4, 412, 1971.
46. **Andrews, E. H. and Reed, P. E.**, *Advances in Polymer Science,* Springer-Verlag, Berlin, 1978, 27.
47. **Jurkov, S. N. and Korsukov, V. E.**, *J. Polym. Sci., Polym. Phys. Ed.,* 12, 385, 1974.
48. **Peterlin, A.**, *J. Polym. Sci., C,* 32, 1970.
49. **Peterlin, A.**, *ESR Application to Polymer Research,* Kink, P. O., Ranby, B., and Remstrom, V., Eds., Almquist, Wiksell, Stokholm, 1973.
50. **Siesler, H. W.**, *Makromol. Chem.,* 180, 1979, 2261.
51. **Vettegren, V. I. and Novak, I. I.**, *Fiz. Tverdogo Tela,* 15, 1973, 1417.
52. **Garton, A., Carlsson, D. I., Holmes, L. L., and Wiles, D. M.**, *J. Appl. Polym. Sci.,* 25, 1505, 1980.
53. **Holland-Mortiz, K. and Siesler, H. W.**, *Polym. Bull.,* 4, 165, 1981.
54. **Jurkov, S. N., Zakrevskii, V. A., Korsukov, V. A., and Kuksenko, V. S.**, *J. Polym. Sci. A,* 2, 10, 1972.
55. **Jansson, J. F., and Terselius, B.**, *J. Appl. Polym. Sci., Appl. Polym. Symp.,* 35, 1979.
56. **Jourkov, S. N. and Tomasevski, E. E.**, Proc. Conf. Physical Basis of Yield and Fracture, Oxford, 1966.
57. **DeVries, K. L., Roylance, D. K., and Williams, M. L.**, *J. Polym. Sci., A,* 8, 26, 1970.
58. **Verma, G. S. P. and Peterlin, A.**, *Koll. Polym.,* 236, 11, 1970.
59. **Becht, J., and Fischer, H.**, *Koll. Polym.,* 240, 775, 1970.
60. **Kausch, H. H.**, *Makromol. Chem.,* 5, 97, 1970.
61. **Andrews, E. H. and Reed, P. E.**, in *Deformation and Fracture in High Polymers,* Ed. Plenum Press, New York, 1973.
62. **Peterlin, A.**, *Polym. Eng. Sci.,* 2, 19, 1979.
63. **Korsukov, V. E. and Vettegren, V. I.**, *Mehanika Polimerov,* 4, 350, 1972.
64. **Jurkov, S. N. and Tomasevski, E. E.**, *J. Tehn. Fiz.,* 27, 1248, 1957.
65. **Leksovski, A. M. and Regel, V. R.**, *Fizika Tverdogo Tela,* 12, 11, 1970.
66. **Regel, V. R. and Leksovski, A. M.**, *Fizika Tverdogo Tela,* 12, 253, 1970.
67. **Korsukov, V. E., and Vettegren, V. I.**, *Mehanika Polimerov,* 4, 621, 1972.
68. **Toggenburger, R. and Newman, S.**, *J. Polym. Sci., B,* 2, 7, 1964.
69. **Jurkov, S. N. and Tomasevski, E. E.**, *Physical Basis of Yields and Fracture,* Institute of Physics, London, 1966.
70. **Jurkov, S. N., Zakrevskii, V. A., Korsukov, V. E., and Kuksenko,, V. S.**, *J. Polym. Sci., A,* 10, 1509, 1972.
71. **Grayson, N. A., Clarence, J. W., Levy, R. L., and Miller, D. B.**, *J. Polym. Sci., Polym. Phys. Ed.,* 14, 1601, 1976.
72. **Sawashima, T., Shimada, S., Kashiwabara, N., and Sohma, J.**, *Polym. J.,* 5, 135, 1973.
73. **Andrews, E. H.**, *The Physics of Glassy Polymers,* Haward, R. N., Ed., Applied Science, London, 1973.
74. **Andrews, E. H. and Reed, P. E.**, *Advances in Polymer Science,* Vol. 27, Springer-Verlag, Berlin, 1978.

75. **Vasiliu Oprea, Cl., Popa, M., and Birsanescu, P.,** Section IV, "Structure and properties", IUPAC MACRO '83, preprints, p. 553.
76. **Vasiliu Oprea, Cl., Popa, M., and Birsanescu, P.,** *Mehanika Kompoz. Materialov,* 3, 401, 1985.
77. **Vasiliu Oprea, Cl., Popa, M., Ioanid, A., and Birsanescu, P.,** *Colloid Polym. Sci.,* 263, 738, 1985.
78. **Vasiliu Oprea, Cl., Popa, M., Constantinescu, Al., and Ioanid, A.,** *IPI Bull.,* 31, 59, 1985.
79. **Vasiliu Oprea, Cl., Popa, M., and Birsanescu, P.,** *Plaste Kautschuk,* 35, 282, 1988.
80. **Vasiliu Oprea, Cl. and Popa, M.,** *Colloid Polym. Sci.,* 263, 25, 1985.
81. **Vasiliu Oprea, Cl. and Popa, M.,** *Colloid Polym. Sci.,* 258, 371, 1980.
82. **Vasiliu Oprea, Cl. and Popa, M.,** *Rev. Roum. Chim.,* 26, 291, 1981.
83. **Vasiliu Oprea, Cl. and Popa, M.,** *Mater. Plast.,* 20, 185, 1983.
84. **Regel, V. R. and Leksovski, A. M.,** *Mehanika Polimerov,* 1, 70, 1969.
85. **Gurevici, S.,** *Teor. Prikl. Mekh.,* 9, 117, 1978.
86. **Constable, J., Williams, J. G., and Burns, D. J.,** *J. Mech. Eng. Sci.,* 12, 1, 1970.
87. **Mills, N. J. and Walker, N.,** Polymer, 17, 335, 1976.
88. **Vasiliu Oprea, Cl., Negulianu, Cl., and Weiner, F.,** *Rev. Roumaine Chim.,* 29, 683, 1984.
89. **Vasiliu Oprea, Cl. and Popa, M.,** *IPI Bull.,* 27(31), 115, 1981.
90. **Urzimtev, Iu. S.,** *Mehanika Polimerov,* 3, 467, 1967.
91. **Vasiliu Oprea, Cl. and Negulianu, Cl.,** *IPI Bull.,* 25(29), 115, 1979.
92. **Baramboim, N. K. and Gorodiev, V. N.,** *Vysokomol. Soedin.,* 2, 197, 1960.
93. **Vasiliu Oprea, Cl., Negulianu, Cl., Popa, M., and Simionescu, Cr.,** *Plaste Kautschuk,* 9, 604, 1977.
94. **Simionescu, Cr., Vasiliu Oprea, Cl., Negulianu, Cl., and Popa, M.,** *Plaste Kautschuk,* 10, 689, 1977.
95. **Peterlin, A.,** *Polym. Eng. Sci.,* 2, 19, 1979.
96. **Kausch, H. H., and Becht, J.,** *Kolloid Polym.,* 250, 169, 1972.
97. **Crist, B. and Peterlin, A.,** *Makromol. Chem.,* 171, 453, 1973.
98. **Ballenger, T. F., Chen, J., Crowder, J. P., Hagler, G. E., Bogne, D. C., and White, J. L.,** *Trans. Soc. Rheol.,* 15, 195, 1971.
99. **Gafurov, J. G.,** *Dokl. Akad. Nauk. Uzb. SSSR,* 8, 30, 1978.
100. **Regel, V. R., Leksovski, A. M., and Pozdniskov, O. F.,** *Mehanika Kompoz. Mater.,* 2, 211, 1979.
101. **Bartenev, G. M. and Tulinov, B. M.,** *Mehanika Polimerov,* 1, 83, 1977.
102. **Bartenev, G. M. and Kartasov, E. M.,** *Plaste Kautschuk,* 28, 241, 1981.
103. **Marihin, V. A.,** *Makromol. Chem.,* 7(suppl.), 147, 1984.
104. **Peterlin, A.,** *J. Makromol. Sci.,* 8, 277, 1973.
105. **Marihin, V. A.,** *Acta Polym.,* 30, 8, 507, 1979.
106. **Peterlin, A.,** *J. Makromol. Sci.,* 19, 40, 1980.
107. **Egorov, E. A. and Zizenkov, V. V.,** *J. Polym. Sci., Polym. Phys. Ed.,* 20, 1089, 1982.
108. **Zizenkov, V. V. and Egorov, E. A.,** *J. Polym. Sci., Polym. Phys. Ed.,* 21, 1264, 1983.
109. **Lloyd, B. A., DeVries, K. L., and Williams, M. L.,** *J. Polym. Sci., A,* 2, 1415, 1972.
110. **Kawashima, T., Shimada, S., Kashiwabara, H., and Sohma, J.,** *Polym. J.,* 5, 135, 1973.
111. **Tudose, R. Z., Ciovica, S., and Condeanu, M.,** *IPI Bull.,* 25(29), 21, 1979.
112. **Vasiliu Oprea, Cl. and Popa, M.,** *Rev. Roumaine Chim.,* 34, 365, 1989.
113. **Yaunas, I. V.,** *J. Polym. Sci.,* 9, 163, 1974.
114. **Gaube, E., Diedrich, G., and Muller, W.,** *Kunststoffe,* 66, 2, 1976.
115. **Popa, M.,** Doctoral thesis, Polytechnic Institute of Jassy, Jassy, Romania, 1989.
116. **Nagamura, T. and Takayanagi, Y.,** *J. Polym. Sci.,* 12, 609, 1974.
117. **Nagamura, T. and DeVries, K.L.,** *Polym. Eng. Sci.,* 19, 2, 1979.
118. **Bressler, S. and Osminskaia, A.,** *Kolloid Z.,* 20, 403, 1958.
119. **Pakhomova, P. M.,** *Vysokomol. Soedin., A,* 24, 1072, 1982.
120. **Studentov, V. N. and Mitiasov, V. N.,** *Khim. Prom.,* 1, 25, 1980.
121. **Fancioni, B. M. and DeVries, K. L.,** *Polymer,* 23, 1027, 1982.
122. **Bartenev, G. M. and Savin, E. S.,** *Vysokomol. Soedin., A,* 22, 1420, 1980.
123. **Hlouskova, Z., Placek, J., and Szoks, F.,** *Eur. Polym. J.,* 5, 387, 1986.
124. **Sakaguki, M. and Kawashibara, H.,** *J. Polym. Sci.,* 18, 563, 1980.
125. **Bartenev, G. M.,** *Vysokomol. Soedin., B,* 24, 625, 1982.
126. **Jurkov, S. N., Savostin, A. I., and Tomasevski, E. A.,** *Dokl. Akad. Nauk. SSSR,* 151, 1, 1964.
127. **Becht, J. and Fischer, H.,** *Kolloid Polym.,* 229, 167, 1969.
128. **Butt, A. D., Mony, W. B., and Chingas, G. C.,** *Polym. Eng. Sci.,* 19, 2, 1979.
129. **DeVries, K. L. and Roylance, D. K.,** *Prog. Solid State Chem.,* 8, 283, 1973.
130. **Johnson, V. and Klinkenberg, D.,** *Kolloid Polym.,* 251, 843, 1974.
131. **Lloyd, B. A., DeVries, K. L., and Williams, M. L.,** *Rheol. Acta,* 13, 352, 1974.
132. **DeVries, K. L.,** *J. Appl. Polym. Sci. Appl. Polym. Symp.,* 35, 1979.
133. **Mehta, R. E.,** *J. Macromol. Sci., B,* 8, 961, 1973.

134. **Birch, W. M. and Williams, J. G.,** *Int. J. Fract.,* 14, 169, 1978.
135. **Becht, J. and Fischer, H.,** *Kolloid Polym.,* 240, 775, 1970.
136. **Stoeckel, T. M., Blassius, J., and Crist, B.,** *J. Polym. Sci., Polym. Phys. Ed.,* 16, 3, 1978.
137. **Wool, R. P.,** *Polym. Eng. Sci.,* 20, 805, 1980.
138. **Dupuis, J., Geguela, R., Somcket, B., Legrand, P., and Reetsch, F.,** *Polym. Bull.,* 18, 323, 1987.
139. **Lisnevski, V. A. and Jmindo, A. V.,** *Vysokomol. Soedin., A,* 25, 4, 702, 1983.
140. **Gafurov, V.,** *Mehanika Polimerov,* 649, 1971.
141. **Tomasevski, E. E.,** *Fiz. Tverdogo Tela,* 12, 3202, 1970.
142. **Egorov, E. A. and Jijenkov, V. V.,** *Fiz. Tverdogo Tela,* 17, 1, 1975.
143. **Grabec, J. and Peterlin, A.,** *J. Polym. Sci.,* 14, 651, 1976.
144. **Peterlin, A.,** *Polymer Preprints,* 18, 2, 1977.
145. **Vasiliu Oprea, Cl., Popa, M., and Ioanid, E. G.,** Romanian Patent, 99613/27.10, 1989.
146. **Butiaghin, P. I., Erofeev, V. S., Misaelian, I. N., Patrikeev, G. A., Streletki, A. N., and Suliak, A. D.,** *Vysokomol. Soedin., A,* 12, 290, 1970.
147. **Streletki, A. N. and Butiaghin, P. I.,** *Vysokomol. Soedin., A,* 20, 1893, 1973.
148. **Teodorescu, H. N., Cristea, D., Sofron, E., and Popa, M.,** *Proc. 8th Int. Conf. on Experimental Stress Analysis,* Wieringa, H., Ed., Martinus Nijhoff, Amsterdam, 1986, 605.
149. **Teodorescu, H. N., Popa, M., Sofron, E., and Simionescu, Cr.,** Proceedings of Experimentelle Mechqnik in Forschung und Praxis, Ausburg, May, 7–8, 1987.
150. **Sauer, J. A. and Richardson, G. C.,** *Int. J. Fract.,* 16, 6, 1980.
151. **Friedrich, K.,** *Polym. Compos.,* 5, 65, 1982.
152. **Lyons, W. G.,** *Text. Res. J.,* 40, 60, 1970.
153. **Radon, J. C.,** *Int. J. Fract.,* 16, 533, 1980.
154. **Takemori, M. T. and Matsuoto, D. S.,** *J. Polym. Sci., Polym. Phys. Ed.,* 20, 2027, 1980.
155. **Bunsell, D. R., and Hearle, J. W. S.,** *J. Appl. Polym. Sci.,* 18, 267, 1974.
156. **Oudet, Ch. and Bunsell, D. R.,** *J. Mater. Sci.,* 3, 295, 1984.
157. **Oudet, Ch. and Bunsell, D. R.,** *J. Appl. Polym. Sci.,* 29, 436, 1984.
158. **Simionescu, Cr. and Vasiliu Oprea, Cl.,** in *Polymerization Kinetics and Technology,* Plastzer, N., Ed., (Advances in Chemistry Series, Vol. 128), 1973, 68.
159. **Vasiliu Oprea, Cl.,** Doctoral thesis, Technische Hochschule fur Chemie, Leuna, Merseburg, RDG, 1965.
160. **Simionescu, Cr., Vasiliu Oprea, Cl., and Negulianu, Cl.,** *J. Polym. Sci., Polym. Symp.,* 64, 149, 1978.
161. **Vasiliu Oprea, Cl. and Popa, M.,** *Polym. Plast. Technol. Eng.,* 28, 1025, 1989.
162. **Simionescu, Cr., Vasiliu Oprea, Cl., and Negulianu, Cl.,** *Makromol. Chem.,* 148, 155, 1971.
163. **Simionescu, Cr., Vasiliu Oprea, Cl., and Negulianu, Cl.,** *Makromol. Chem.,* 163, 75, 1973.
164. **Negulianu, Cl., Vasiliu Oprea, Cl., and Simionescu, Cr.,** *Makromol. Chem.,* 175(2), 371, 1974.
165. **Simionescu, Cr., Vasiliu Oprea, Cl., and Negulianu, Cl.,** *Angew. Makromol. Chem.,* 44, 17, 1975.
166. **Simionescu, Cr., Vasiliu Oprea, Cl., and Negulianu, Cl.,** *Makromol. Chem.,* 181, 1579, 1980.
167. **Simionescu, Cr., Vasiliu Oprea, Cl., and Negulianu, Cl.,** *Angew. Makromol. Chem.,* 115, 1, 1983.
168. **Vasiliu Oprea, Cl. and Popa, M.,** *Cell. Chem. Technol.,* in press.
169. **Becht, J. and Fischer, H.,** *Kolloid Polym.,* 240, 775, 1970.
170. **Stoeckel, T. M., Blassius, I., and Crist, B.,** *J. Polym. Sci., Polym. Phys. Ed.,* 16, 3, 1978.
171. **Joseph, S. H.,** *J. Polym. Sci., Polym. Phys. Ed.,* 16, 1071, 1978.
172. **Stepanov, V. A.,** *Mehanika Polimerov,* 1, 95, 1975.
173. **Hakeem, E. and Kulver, H. A.,** *J. Appl. Polym. Sci.,* 22, 2689, 1978.
174. **Deters, W. and De Huang, H. C.,** *Faserforsch. Textiltech.,* 14, 58, 1963.
175. **Prokopchuk, N. R.,** *Dokl. Akad. Nauk. SSSR,* 26, 1020, 1982.
176. **Prokopchuk, N. R., Matusevici, I. I., and Krul, L. P.,** *J. Polym. Sci.,* 25, 503, 1987.
177. **Matusevici, I. I. and Krul, L. P.,** *Termochim. Acta,* 97, 351, 1986.
178. **Reed, P. E.,** *J. Polym. Sci.,* 12, 2, 1971.
179. **Mead, W. T. and Reed, P. E.,** *Polym. Eng. Sci.,* 14, 22, 1974.
180. **Andrews, E. H. and Reed, P. E.,** *Polym. Eng. Sci.,* 14, 22, 1974.
181. **Brown, R. D., DeVries, K. L., and Williams, M. L.,** *Polymer Network: Structural and Mechanical Properties,* Plenum Press, New York, 1971.
182. **Mead, W. T.,** Molecular Fracture, in Mechanically Deformed Polymers, Ph.D. thesis, University of London, 1975.
183. **Natarajan, R. and Reed, P. E.,** *J. Polym. Sci., A,* 2, 585, 1972.
184. **Drida, V. V. and Nadumi, V. P.,** *Kauchuk Rezina,* 1, 37, 1975.
185. **Vasiliu Oprea, Cl. and Constantinescu, A. C.,** *Polym. Plast. Technol. Eng.,* 27, 173, 1988.
186. **Vasiliu Oprea, Cl. and Constantinescu, A. C.,** unpublished data.
187. **Vasiliu Oprea, Cl., Constantinescu, A. C., and Birsanescu, P.,** *Colloid Polym. Sci.,* 264, 590, 1986.

188. **Vasiliu Oprea, Cl. and Constantinescu, A. C.,** *Mat. Plast.,* 20, 2, 1986.
189. **Perepechko, I. I.,** *An Introduction to Polymer Physics,* Mir. Moskow, 1981, 230.
190. **Tobolsky, A. V., Prettyman, I. B., and Dillon, J. H.,** *J. Appl. Phys.,* 15, 309, 1944.
191. **Bueche, F.,** *J. Chem. Phys.,* 21, 114, 1953.
192. **Berry, J. P. and Warson, W. F.,** *J. Polym. Sci.,* 18, 201, 1955.
193. **Yu, H.,** *Polym. Lett.,* 2, 631, 1964.
194. **Murakami, K. and Tamura, S.,** *J. Polym. Sci., B,* 11, 529, 1973.
195. **Tamura, S. and Murakami, K.,** *Polymer,* 14, 569, 1973.
196. **Murakami, K. and Tamura, S.,** *J. Polym. Sci.,* 13, 317, 1973.
197. **Osthof, R. O., Bueche, A. M., and Grubb, W. I.,** *J. Am. Chem. Soc.,* 76, 4659, 1954.
198. **Tobolsky, A. V.,** *Polym. Lett.,* 2, 823, 1964.
199. **Murakami, K.,** *J. Soc. Mater. Sci. Jpn.,* 15, 312, 1966.
200. **Nagamura, S. and Yakahashi, Y.,** *Kobunshi Kagaku,* 7, 705, 1970.
201. **Murakami, K., Tamura, S., and Nakanishi, H.,** *J. Polym. Sci.,* 11, 313, 1973.
202. **Murakami, K. and Nakanishi, H.,** *J. Polym. Sci., Polym. Chem.,* 14, 489, 1976.
203. **Mointire, I. V.,** *J. Appl. Polym. Sci.,* 16, 2901, 1972.
204. **Han, C. D.,** *J. Appl. Polym. Sci.,* 17, 1403, 1973.
205. **Balmer, R. T.,** *J. Appl. Polym. Sci.,* 18, 3127, 1974.
206. **Goetze, K. P. and Porter, R. S.,** *J. Polym. Sci., C,* 35, 189, 1971.
207. **Whitlock, L. R. and Porter, R. S.,** *J. Polym. Sci., A,* 2, 877, 1972.
208. **Whitlock, L. R. and Porter, R. S.,** *J. Appl. Polym. Sci.,* 17, 276, 1973.
209. **Abbas, K. B.,** *Polymer,* 22, 836, 1981.
210. **Abbas, K. B. and Porter, R. S.,** *J. Appl. Polym. Sci.,* 20, 1289, 1976.
211. **Vasiliu Oprea, Cl. and Savin, A.,** IUPAC MACRO '83, section III, Polymer Processing, Bucharest, 120.
212. **Jellinek, H. H.,** Aspects of Degradation and Stabilization of Polymers, Elsevier, New York, 1978.
213. **Vasiliu Oprea, Cl. and Savin, A.,** unpublished data.
214. **Kargin, V. A. and Sogolova, T. I.,** *Dokl. Akad. Nauk. SSSR,* 108, 662, 1956.
215. **Kargin, V. A. and Sogolova, T. I.,** *Zh. Fiz. Khim.,* 31, 1328, 1957.
216. **Savelev, A. P., Shilov, G. I., Malyshev, L. N., and Braginskii, V. A.,** *Plast. Massy,* 12, 25, 1971.
217. **Savelev, A. P., Malyshev, L. N., Braginskii, V. A., and Minsker, K. S.,** *Plast. Massy,* 6, 56, 1973.
218. **Vasiliu Oprea, Cl. and Popa, M.,** *Rev. Roum. Chim.,* 26, 291, 1981.
219. **Vasiliu Oprea, Cl. and Popa, M.,** *Mater. Plast.,* 20, 185, 1983.
220. **Vasiliu Oprea, Cl., Petrovan, S., and Popa, M.,** *Mehanika Kompoz. Mater.,* 3, 977, 1979.
221. **Vasiliu Oprea, Cl. and Popa, M.,** *Colloid Polym. Sci.,* 258, 371, 1980.
222. **Vasiliu Oprea, Cl. and Popa, M.,** *Acta Polymer.,* 36, 675, 1985.
223. **Vasiliu Oprea, Cl. and Popa, M.,** *Mehanika Kompoz. Mater.,* 4, 679, 1982.
224. **Vasiliu Oprea, Cl. and Popa, M.,** *Colloid Polym. Sci.,* 260, 570, 1982.
225. **Kargin, A. and Plate, N. A.,** *Vysokomol. Soedin.,* 1, 330, 1959.
226. **Plate, N. A. and Kargin, V. A.,** *J. Polym. Sci., C,* 4, 1027, 1963.
227. **Plate, N. A. and Kargin, V. A.,** *Vysokomol. Soedin., A,* 14, 440, 1972.
228. **Baramboim, N. K. and Protasov, W. G.,** *Die Technik,* 2, 73, 1975.
229. **Baramboim, N. K. and Antonova, L. A.,** *Vysokomol. Soedin., A,* 18, 675, 1976.
230. **Baramboim, N. K. and Antonova, L. A.,** *Kolloid. Zh.,* 38, 961, 1976.
231. **Kuznetov, V. A., Ianova, L. P., and Tolstaia, S. N.,** *Kolloid. Zh.,* 40, 590, 1978.
232. **Ianova, L. P., Kuznetov, V. A., and Taubman, A. B.,** *Kolloid. Zh.,* 37, 614, 1975.
233. **Takasui, I. and Tazuke, S.,** *Chem. Lett.,* 5, 589, 1981.
234. **Antonova, L. A., Hrustaleev, Yu. A., Baramboim, N. K., and Krotova, N. A.,** *Kolloid Zh.,* 38, 535, 1976.
235. **Ianova, L. P. and Tolstaia, S. N.,** *Plaste Kautschuk,* 4, 198, 1979.
236. **Moustafa, A. B.,** *Angew. Makromol. Chem.,* 39, 1, 1974.
237. **Momose, L., Yamada, K., and Nagayama, K.,** *J. Polym. Sci.,* 12, 635, 1974.
238. **Vasiliu Oprea, Cl. and Weiner, F.,** *J. Appl. Polym. Sci.,* 31, 951, 1986.
239. **Vasiliu Oprea, Cl., Avram, I., and Avram, R.,** *Angew. Makromol. Chem.,* 68, 72, 1978.
240. **Vasiliu Oprea, Cl. and Popa, M.,** *Angew. Makromol. Chem.,* 90, 13, 1980.
241. **Vasiliu Oprea, Cl. and Popa, M.,** *Angew. Makromol. Chem.,* 92, 73, 1980.
242. **Vasiliu Oprea, Cl., Negulianu, Cl., Popa, M., and Weiner, F.,** *IPI Bull.,* 1–2, 87, 1980.
243. **Simionescu, Cr., Vasiliu Oprea, Cl., and Negulianu, Cl.,** *Eur. Polym. J.,* 15, 1037, 1979.
244. **Vasiliu Oprea, Cl. and Weiner, F.,** *Angew. Makromol. Chem.,* 106, 207, 1982.
245. **Simionescu, Cr., Vasiliu Oprea, Cl., and Nicoleanu, J.,** *Eur. Polym. J.,* 19, 6525, 1983.
246. **Vasiliu Oprea, Cl. and Popa, M.,** *Angew. Makromol. Chem.,* 116, 125, 1983.

247. **Vasiliu Oprea, Cl. and Popa, M.,** *Acta Polym.,* 34, 612, 1983.

248. **Vasiliu Oprea, Cl. and Weiner, F.,** *Angew. Makromol. Chem.,* 126, 89, 1984.

249. **Vasiliu Oprea, Cl. and Popa, M.,** *Polym. Bull.,* 11, 269, 1984.

250. **Vasiliu Oprea, Cl. and Popa, M.,** *Angew. Makromol. Chem.,* 127, 49, 1984.

251. **Vasiliu Oprea, Cl., Popa, M., and Hurduc, N.,** *Polym. J.,* 16, 191, 1984.

252. **Vasiliu Oprea, Cl. and Popa, M.,** *Acta Polym.,* 35, 261, 1984.

253. **Vasiliu Oprea, Cl., Popa, M., and Cascaval, C. N.,** *Br. Polym. J.,* 16, 123, 1984.

254. **Simionescu, Cr. and Vasiliu Oprea, Cl.,** *J. Polym. Sci. Polym. Chem. Ed.,* 23, 501, 1985.

255. **Vasiliu Oprea, Cl. and Popa, M.,** *Acta Polymer.,* 37, 177, 1986.

256. **Vasiliu Oprea, Cl. and Weiner, F.,** *Acta Polym.,* 37, 19, 1986.

257. **Vasiliu Oprea, Cl. and Weiner, F.,** *Acta Polym.,* 38, 429, 1987.

258. **Vasiliu Oprea, Cl., Popa, M., and Dumitriu, S.,** *Angew. Makromol. Chem.,* 157, 93, 1988.

259. **Vasiliu Oprea, Cl. and Popa, M.,** *Polymer,* 26, 452, 1985.

260. **Vasiliu Oprea, Cl. and Constantinescu, Al.,** *Polymer-Plast. Technol., Eng.,* 28, 9, 1989.

Chapter 6

MORPHOLOGY AND MECHANICAL PROPERTIES OF NATURAL RUBBER/POLYOLEFIN BLENDS

Joachim Sunder

TABLE OF CONTENTS

0-8493-4401-8/93/$0.00 + $.50
© 1993 by CRC Press, Inc.

I. INTRODUCTION

There are many properties of plastics and rubber which are being changed or even improved by a multiphase structure of the polymeric material. Over the past years, this finding, which was determined by empirical methods, has been leading to intense scientific research on a large variety of multiphase systems.

EPDM/PP is one of the most usual combinations of a rubber and a thermoplastic, but polyolefin/natural rubber combinations are available commercially, too.

To begin, a general view on the developments of natural rubber/polyolefin blends is given. In the following, we shall introduce some investigations to the reader. They were performed on a blend that consists of linear low-density polyethylene (LLDPE) and a natural rubber with stabilized viscosity (CV50). In the course of the investigations, mixing ratios, portions of fillers, and cross-linking agents were analyzed during processing in a discontinuous internal mixer. Moreover, we will present some extensions added to the emulsion theory, which is to help assess a processing window. These evaluations are transferred to the compounding of polypropylene/natural rubber blends.

II. PREVIOUS STUDIES ON NATURAL RUBBER/POLYOLEFIN BLENDS

As early as the 1970s several studies[1,2] investigated natural rubber/polyolefin blends. These thermoplastic mixtures with natural rubber (TPNR) pertain to the group of thermoplastic elastomers (TPE). They are obtained in a non-crosslinked form, or with dynamic vulcanization, i.e., crosslinking of the rubber phase at the end of the mixing cycle. For an evaluation and comparison with other TPEs, see Table 1.[3,4]

Up to now, there is a large share of investigations available on natural rubber/polypropylene blends. For the purpose of improving mechanical properties and stabilizing the microscopic structure otherwise changed by processing, peroxide or else the TMTD sulfur donor, stearic acid, and zinc oxide can be employed. Using peroxide, antioxidants serve as stabilizers aimed at cushioning the peroxide radicals dissolved in the thermoplastic phase following dynamic vulcanization. Peroxide leads to destruction of polypropylene.[1,3,5] The following mixing cycle presented in Table 2 may be used for compounding of carbon-black-loaded NR/PP blends in an internal mixer with a tangential rotor system.

In this case, output temperature must be above polypropylene melt temperature. The investigations performed in Reference 1 found this figure to be between 180 and 200°C. The mixtures produced in this way can either be drawn into stripes on the rolling mill or processed on an extruder directly related to the internal mixer. The extruder, in a continuous way, can produce the stripes, which can then be chopped.[3,5] For processing on an injection molding machine (REP-B43), tempering was adjusted as follows (Table 3).

As injection velocities were at their maximums and parts were 2 mm thick, processing periods — disregarding discharge — were down to 40 s, which is much below times required for pure rubber mixtures. Since these mixtures can be reused, trials of multiple processing were also performed. Having processed the material ten times by injection molding, the loss in the flexural modulus at room inside temperature was 8 to 12%, for the tensile strength 7 to 10%, and for the fracture strain it was 4 to 14%. It showed that the compounds could be reused at least two or three times, with no significant deterioration; innovative developments[6] can even be processed up to five times. Investigations concerning phase modifications for NR/PP blends by means of m-phenylenbismaleimide (HVA-2) revealed another possibility to improve mechanical properties, apart from the possibility of cross-linkage.[7-9] In this application, HVA-2 does not only modify polypropylene, the phase of natural even cross-links.

TABLE 1
Comparison between Individual Types of TPE

TPE type	Shore hardness	Density (g cm^{-3})	Tensile strength (MPa)	Elongation at break (%)	Related average cost to NR blends
NR blends	55A–65D	0.9–1.0	6–20	200–500	1
EPDM blends	55A–65D	0.9–1.0	5–20	200–500	1.6
SBS, SEBS	40A–95A	0.9–1.1	10–25	600–1200	1.7
PU elastomer	70A–75D	1.1–1.3	20–50	200–700	2.9
Polyester	35D–72D	1.2–1.3	25–40	350–450	3.2
Polyamide	60A–70D	1.0–1.15	14–51	150–680	3.9

TABLE 2
Mixing Times for Compounding in the Internal Mixer

0 min	Adding carbon black, polypropylene, NR
3–4 min	Adding dicumyl peroxide
6–7 min	Adding antioxidants
6, 5–7, 5 min	Discharge

TABLE 3
Temperature Adjustment for Injection Molding

Feed zone	170–190°C
Cylinder temperature/middle	180–200°C
Cylinder temperature/front	190–220°C
Nozzle temperature	190–220°C
Mold temperature	30–100°C subject to flow marks occurring

Reference 10 also added 10% polyolefin modified by maleic anhydride. Another polymer such as chlorinated polyethylene (CPE), ethylene propylene dien rubber (EPDM), epoxy natural rubber (ENR), and sulfonated EPDM (S-EPDM) were added in order to form NR polyolefin intercopolymers; 20 and 30% modificators were introduced. Nevertheless, NR/PP mechanical properties were not improved, since polypropylene crystallite production was made impossible here.

NP/PE blends are being compounded while temperatures are lower. If these mixtures are submitted to cross-linkage with peroxides, the polyethylene also cross-links. Temperatures are lower here, and it is therefore recommended to employ sulfur donor systems for they entail no covulcanization. These mixtures can be processed with parameters similar to those of NR/PP blends. The mechanical characteristics are only slightly below those of polypropylene blends; temperature resistance, in turn, is only at 90°C. Jentzsch et al.[11-13] were the first to point to the high level of interdependence between structure and process parameters, which is the object of investigation in this thesis. If LDPE is employed, the service temperature will decrease, ultimate strength will only be about 7 MPa, while breaking strain will be up 400%.[13]

NR/polystyrene (PS) blends have been playing only a minor role, yet. In this respect, Tinker[13] describes the graft reaction, while azodicarboxylated functionalized polystyrene is used in an internal mixer. With mixing times up to 10 min, torsional speeds of 150 min^{-1} and 90°C temperature, graft proportions up to 76% are being achieved. The material appears to be more or less transparent, subject to graft efficiency. With room temperatures, ultimate strength reaches values of above 20 MPa, with elongation however low. As a possible alternative of polystyrene high-impact modification, natural rubber[14] may just as well be used. The blends' viscosities rise as the rubber portion is increased; nevertheless, they always remain below those of pure polystyrene.

Current studies almost disregard the process parameters' effects on structure and mechanical characteristics. This relation will be subject to the following presentation.

III. STUDIES ON NATURAL RUBBER/POLYETHYLENE BLENDS

A. ANALYSIS OF MORPHOLOGIES AND MECHANICAL PROPERTIES OF THE UNVULCANIZED BLENDS

Unvulcanized blends of LLDPE and natural rubber CV50 were processed on the internal laboratory mixer GK-1,5 E of Werner & Pfleiderer and the effect of machine adjustment parameters on their morphologies and mechanical properties[15] examined. Three mixtures with different PE ratios (30, 50, and 70%) were examined, varying the parameters of torque, cooling fluid temperatures of chamber and rotors, ram pressure, and cycle time.

In the following, we shall use a 30% NR/70% PE mixture and a variation in rotor speeds to present an example of how essential is the effect exerted by the processing parameters.

1. Variation of Rotor Speed

For the purpose of promoting the melting of PE, some preliminary trials served as a basis for temperature adjustment to 130°C in the chamber heating cycle and 150°C in the rotor heating cycle.

Figure 1 shows the morphologies during several rotor speeds. The upper half of the figure shows the morphology referring to 60 min^{-1} rotor speed, while the lower half is related to 100 min^{-1}. The disperse phase (NR, in this case) is by one order of magnitude higher. Morphologies are subject to influence by the shear stress and thus viscosity ratios. Unless they keep to a certain interrelation, the value of the dispersed phase will fail to reach the optimum. In the case under consideration, higher temperatures resulting from higher rotor speeds as well as increased shear and elongational speeds lead to severe changes in viscosity ratios, since the pseudoplastic behavior of natural rubber is much more pronounced than that of PE. This may be the reason for morphologies differing at such an extreme rate. Figure 2 shows the upper half the amount of particles around 400-μm^2 size, used as a measure of changes in morphology, while rotor speeds are altered. With 30 and 50% PE contents, optimum adjustment is at 40 min^{-1}, whereas with 70% PE the highest number of particles and finest morphology entailed is at 60 min^{-1}.

These differences are reflected in the mechanical properties, too. Figure 2 presents in the lower half an example of the elongations at break, while mixing ratios and rotor speeds are changed. Except for the case of rotor speeds being 100 min^{-1}, the differences in morphologies correspond to the mechanical properties (here: elongation at break). In the example presented in Figure 1, the differences in elongations at break are between 690% with good dispersion and 75% with a poor one. Even if morphological differences are slight, still, there will be differences of 150% resulting, in terms of elongation at break (rotor speed 40 min^{-1} and 60 min^{-1}, PE content 70%).

2. Variation in Cycle Time

Apart from changes in rotor speeds, variation in cycle times must be considered as another important parameter influencing the morphology and mechanical properties. In this respect, Figure 3 shows the diameters of particles, while cycle times are varied.

Neither in this field can a uniform evaluation of operational points be performed for all kinds of mixtures. If the PE content is 30%, it will be recommended to have short mixing

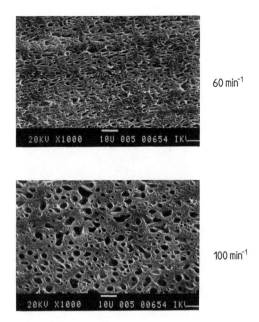

FIGURE 1. Difference in morphology caused by variation of rotor speed.

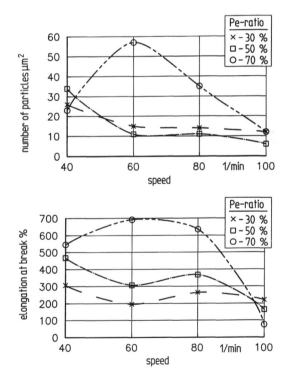

FIGURE 2. Variation of speed PE/NR blend unvulcanized.

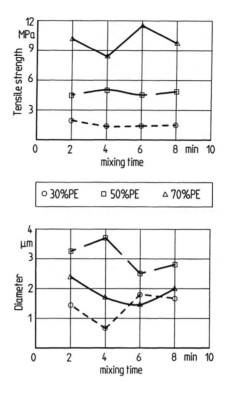

FIGURE 3. Variation of the mixing time on unvulcanized blends.

times of between 2 and 4 min. For 50% PE contents, the optimum mixing time increases to 4 min, and for 70% PE contents it comes up to even 6 min. The latter case is the one to reveal the highest level of submittance to the mixing time. Provided mixing times are short, PE contents of 50% lead to larger equivalent diameter in the network structure, but, at the same time, step up the mechanical properties of the blend.

3. Variation in Temperature of the Tempering Medium

The temperature of the tempering medium is one of the parameters interesting and important for variations in operational points.

In a first step during the examinations, the flow temperature in the chamber cycle was reduced by 30°C. This measure succeeded in refining microstructures. However, only with 30 and 50% PE, this incurred better mechanical properties. As PE contents were high, properties deteriorated slightly, despite the fact that structures were refined. In addition, instead of the chamber, the rotor reduced its temperature by 30°C. Structures and mechanical properties behaved in a uniform way refining or improving, respectively, which can be seen in Figure 4. Therefore, it can be said that the tempering conditions are particularly important when it comes to producing polymer blends.

4. Variations in ram Pressure

The pressure of the ram in the internal mixer was also reduced. It revealed, however, that lower ram forces deteriorate morphologies, homogeneities, and mechanical properties. This results from part of the material being pressed into the ram shaft remove from the actual zone of shear and mixing, if ram forces are low. This parameter variation can only be carried out with maximum ram pressure and was therefore turned down.

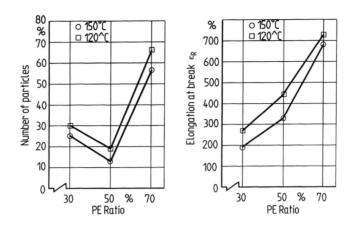

FIGURE 4. Variation of the rotor temperature PE/NR blend (unvulcanized).

The shore hardness is being effected by the parameter variations mentioned above, i.e., most of all in case of low PE portions. It corresponds to the mechanical characteristics; as a result the following ranges are obtained:

PE portion (%)	Shore hardness (A)
30	35–55
50	80–88
70	90–96

The trials described above can serve to indicate a processing window for compounding on the laboratory machine.

Rotor speed	40–60 min^{-1}
Mixing time	3–5 min
Temperature of tempering medium in the chamber	110–120°C
Temperature of the tempering medium of the rotors	130–140°C
Mass temperature	160–175°C

B. ANALYSIS OF THE MECHANICAL PROPERTIES OF FILLED AND VULCANIZED BLENDS

Since the blends' properties of use and application were to be further analyzed, an optimum operational point was taken as a basis to examine the effects of filler materials and cure, in the case of 50 and 70% PE contents. As a possibility of stabilizing morphologies for further processing, dynamic vulcanization of the rubber phase during mixing must be considered. The melting point of PE crystallities is higher than 130°C; therefore, tetramethylthiuramdisufid (TMTD) with zinc stearate or zinc oxide with stearin, respectively, were used for cure. As fillers, one light and one black one were used: 35 phr carbon black N 332 and 21 phr silica VN3.

Figure 5 shows that the power curve is well suited to clearly present the course of the process. Power consumption during production of a blend filled with carbon black is used as an example here. Having added rubber, fillers and PE, the filler is incorporated, in a first step, which is followed by a brief melting phase of the PE. After the dispersion phase, i.e., after mixing has finished, the linking agent is being added. The blend is being discharged after cure is concluded and the power curve has reached a maximum.

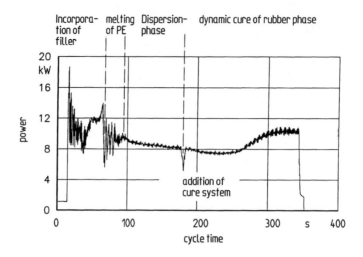

FIGURE 5.   Course of process according to the power curve.

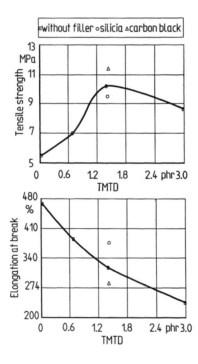

FIGURE 6.   Variation of the mechanical properties on a 50% CV50/50% PE for different amounts of cure system.

Figure 6 shows how tensile strengths and elongations at break change, in case PE contents are 50% and portions of linking agents are different, incurring different degrees of vulcanization. The tensile strength has its maximum point at 1.4 phr TMTD, whereas elongation at break decreases in an exponential way, with the blend of 3 phr thus being only 50% of the unvulcanized. For 1.4 phr TMTD, carbon black was found to entail slight reinforcement of the material, whereas the light filler improved elongation at break, but deteriorated tensile strength. The influence of the filler was not examined with high TMTD contents, since, in this case, it would only have been possible to process the materials if large quantities of oil had been added. This has a steep impact on morphologies. For the same reason, neither mixtures with 30% PE contents were examined.

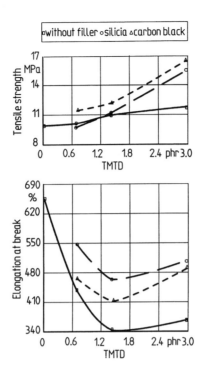

FIGURE 7. Variation of the mechanical properties on a 30% CV50/70% PE for different amounts of cure system.

Figure 7 shows how the mechanical properties change in the case of 70% PE content. Unfilled blends reveal severe reductions in elongations at break, whereas tensile strengths increase only slightly. Adding a filler, elongations at break as well as, in most of the cases, tensile strengths are stepped up, if compared to unfilled blends. In terms of tensile strength, improvement reaches up to 50%. In Reference 16, Hofmann describes the possibility of adding ethylene/vinyl acetate (EVA) and thus improve the ozone resistance of natural rubber. EVA then also acts as an additional plasticizer, which is very well compatible. There were mixtures examined with different EVA contents. For financial reasons, involving competitiveness with EPDM/PP blends, only EVA contents up to 7.5 phr were investigated. Ozone resistance of pure NR/PE blends is very good. No cracks were produced and merely the surface of the treated specimens turned slightly dull. In fact, neither EVA adding could improve this significantly. There were specimens of commercial EPDM/PP examined for reasons of comparison. Their surfaces revealed only very minor changes. The specimens' hardness was not effected by the EVA. Merely the tensile strength could be improved, while elongation at break remained unchanged. At the same time, the optimum EVA content rises as the PE content is stepped up (Figure 8).

Ozone resistance was found to be only insignificantly lower than for EPDM/PP blends, while the service temperature range is lower by 10 to 15°C with mechanical characteristics having slightly increased.

To give thermoplastic vulcanizates of good mechanical properties, a few characteristics of the pure rubber and plastic components can be named. These characteristics are surface energies, crystallinity of the hard phase material, and the critical chain length of the rubber molecules for entanglement.[17] Physical properties not only affect material properties but also the processing conditions during compounding.

The trials described above can then serve as a basis to make up a processing window for manufacturing the blends on a laboratory machine. The processing window had before

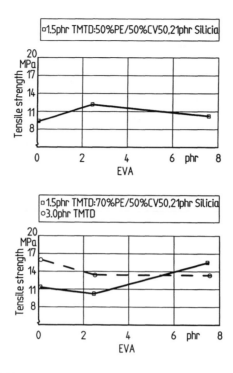

FIGURE 8.   Influence of EVA parts on tensile strength.

been defined by merely empirical methods and must be determined for each individual machine. The following considerations will therefore deal with the question of whether morphological calculations can predetermine a processing window.

IV. MORPHOLOGICAL CALCULATION USING A PROCESS MODEL OF AN INTERNAL MIXER AND THE EMULSION THEORY

A. PROCESS MODEL OF THE INTERNAL MIXER

IKV scientists have developed a model to calculate the dissipation in the internal mixer.[18] On the basis of an integral consideration of the mixing chamber divided into volume elements, the model calculates the average velocities of shear and elongation. Geometrical changes in the area of the rotor intermeshing region can serve to determine the elongations of materials. This makes it possible to find out the shear velocity and elongational deformation submitted to the actual geometry and rotor speed.

B. CONCEPT FOR CONSIDERATION OF THE INFLUENCE EXERTED BY STRAIN DEFORMATION

There are several models available referring to the thermodynamic-rheological effects during compounding.[19-22]

Generally speaking, these approaches start from the microstructure having the shape of small drops or layers. For drop structures, they deliver a description of the way the size of the disperse phase is related to the ratio between surface stress and shear viscosities as well as to the principal normal stress difference. These model calculations found particles to be up to one order of magnitude larger than had been found in microtome analysis. Moreover, the changes subject to the process parameters were reflected only incorrectly. An attempt

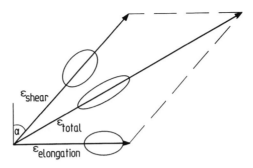

FIGURE 9. Vector addition of droplet deformation.

was therefore made to take the effect of strain deformation into account. The considerations are based on the assumption that a drop deformed it breaks into pieces. A severely deformed drop, however, breaks into very small pieces; therefore, superposition of the deformation from shear and elongation appears to be indispensable (Figure 9). Following Taylor,[23] the angle of the drop deformed by shear is 45° related to the direction of flow. Deformation from elongation as a result of geometrical changes takes place in the 0° angle. Overall deformation

$$\epsilon \quad = \quad \sqrt{\epsilon_{shear}^2 + \epsilon_{shear}\,\epsilon_{elong}\,\sqrt{2} + \epsilon_{elong}^2}$$

$$\epsilon_{elong} \quad = \quad \left(\frac{1 + D_{elong}}{1 - D_{elong}}\right)^{2/3} - 1$$

$$\epsilon_{shear} \quad = \quad \left(\frac{1 + D_{shear}}{1 - D_{shear}}\right)^{2/3} - 1$$

$D_{(elong,shear)}$ = deformation as a result of elongation taken from the process model, deformation as a result of shear

η_R = viscosity ratio

C. CALCULATION OF THE DROP SIZE

Table 4 can help explain the general strategy for the determination of drop sizes.

A first step determines the rheological data, i.e., the viscosity function and the first normal stress difference. As was described above, the particle deformation is calculated on the basis of superposition of the shear and elongational fraction. The process model for the calculation of dissipation in the internal mixer serves to determine the shear rate of the continuous phase; in order to find out the shear rate of the disperse phase, the Elmendorp equation[20] is employed.

$$\gamma_d = \frac{1 - D}{1 + D}\,\gamma_c\,\sqrt{\eta_r\,(\eta_r + 2)}$$

where γ_d shear rate of the disperse phases (s^{-1}) and η_r viscosity ratio.

The viscosities are determined for the individual shear or elongational rates in the blend. Since the shear rate in the disperse phase is submitted to the deformation of the small drops, a numerical iterative method is required for the determination of all values.

TABLE 4
Flow Chart of Droplet Calculation

1. Rheological data
   Viscosity function
   Function of principal normal stress difference
2. Calculation of deformation of the droplets
   Deformation and elongation caused by shear stress with Taylor's theory
   Elongation in an internal mixer with the process model by J. Sunder
   Total elongation and deformation
3. Calculation of shear rate
   Shear rate of continuous phase with the process model by J. Sunder
   Shear rate of dispersed phase with estimation by J. J. Elmendorp
4. Calculation of "elastic" interfacial tension
   Interfacial tension between dispersed and continuous phase with a linear estimation
   Addition of the principal normal stress differences to the interfacial tension by suggestion of Vanoene
5. Calculation of droplet size
   Estimation of A. P. Plochocki

The interfacial tensions are then being determined, on the one hand, for the individual temperatures on the basis of a linear shifting approach. On the other hand, Vanoene[22] adds the interfacial stress and the principal normal stress difference, thus suggesting an additional approach that aims at consideration of elastic effects. Actual determination of the drop diameter is then performed by means of an equation by Plochocki:[19,24]

$$d = \frac{40 \; 10^3 (\sigma_{11} - \sigma_{22})_d \; \sigma_{el} \; (\eta_r + 1)^2}{(19 \; \eta_r + 16)^2 \; \eta_d^2 \; \gamma_c^2 \; w}$$

where $\sigma_{11} - \sigma_{22}$ = principal normal stress difference, σ_{el} = interfacial stress (Vanoene) mN/m, η_r = viscosity ratio, η_d = viscosity of the dispersed phase, and γ_c = shear rate of the continuous phase.

V. COMPARISON BETWEEN SIZES OF BLEND PARTICLES MEASURED AND CALCULATED

A. TRIALS ON A LABORATORY SCALE

An operational field with rotational speeds of 50, 60, and 70 min^{-1} was run on a laboratory machine, compounding blends of 70% PE/30% NR and 1.5 phr TMTD. This was performed to examine the approaches as to calculate the particle size. The mechanical properties in Figure 10 show that tensile strength and elongation at break decrease as rotor speeds are reduced, with their maxima at 3.5 min. In the earlier stages, the material is distributed less homogeneously and, as the rotor speed is further stepped up, the mechanical properties deteriorate. The particle size, as is shown in the upper diagram of Figure 11, takes an analog course. As the rotor speed is increased, the particle sizes are reduced, which means that they are not in their optimum, in this stage. It is not so clearly submitted to the mixing time. Higher material load is reflected in the curve of 70 min^{-1} rotor speed, considering the increasing particle size, since, as early as in this stage, a segregation process is on its outset. In the area of values, calculated using the temperatures achieved each time after mixing, correspondence and approximation to the measured values are satisfactory. However, this deviates by up to one order of magnitude with strain deformation being left apart in the considerations, and it completely fails to describe the course. Because of the curing of the rubber phase a decrease of the particles takes place, which is not considered by the calculation. With 30 min^{-1}, the particles are largest, while with 70 min^{-1} the rise

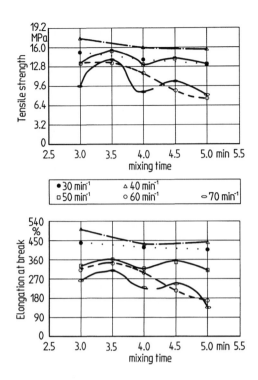

FIGURE 10. Variation of the mechanical properties vs. mixing time (laboratory mixer) of 70% PE/30% NR/1.5 phr TMTD.

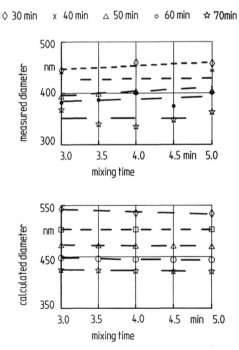

FIGURE 11. Measured and calculated droplet size at the laboratory mixer.

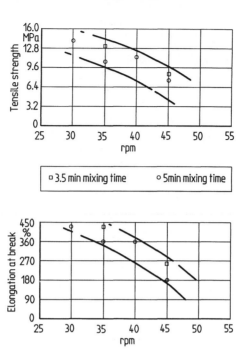

FIGURE 12.   Variation of the mechanical properties vs. rotor speed (45L mixer) — 70% PE/30% NR/1.5 phr TMTD.

in particle sizes points to high material load and initial segregation. In addition, the high rotor speed at 5-min cycle time must not be adjusted for mere energetic reasons, since the particle size remains the same as at 3-min cycle time.

Alterations in the rotor speed of the laboratory machine lead to only relatively minor changes in the operational point. We will therefore introduce some trials performed on a 451 production machine.

B. TRIALS ON THE PRODUCTION MACHINE

On the production machine, an operational field was run with thermal boundary conditions being constant (70°C chamber temperature and 90°C rotor temperature). The mechanical properties (Figure 12) deteriorate as the rotor speed is stepped up, while down to 40 min^{-1} characteristic values are the same as were found in the trials on the laboratory machine.

Morphologies are not as homogeneous as were determined at the laboratory machines, and measured particle sizes are between 500 and 800 nm. The calculated values are in the same range. As early as with 45 min^{-1} and 5-min mixing time, morphologies are inhomogeneous and reveal also large particles. Up to now, it is impossible to evaluate microtome analyses statistically. Therefore, no presentation is available here.

C. DETERMINATION OF A PROCESSING WINDOW

By the help of the calculated diameters a process area can be defined, which leads to an optimized point of operation. Therefore, the dimensionless Weber (We) number is plotted vs. the viscosity ratio (Figure 13). The Weber number is defined as follows:

$$We = \frac{\eta \gamma d}{2\sigma}$$

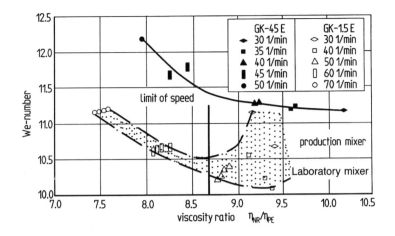

FIGURE 13. We number vs. viscosity ratio at a production and laboratory mixer.

where η = viscosity of the continuous phase, γ = shear rate of the continuous phase, d = calculated diameter, and σ = interfacial tension.

If the We number is at the minimum, the ratio of viscous and interfacial forces reaches an equilibrium, which leads to an optimal operation point. At this point the material load is relatively small to get a good dispersion.

In the case of the laboratory machine the values of the We number, which were calculated with the temperatures measured at the end of the cycles, are the smallest for 40 min^{-1}. For the production machine the values step up with increasing rotor speed. The course of the We-number correlates reversed to the mechanical properties (Figures 10 and 12). At the minimum We number the maximum values for mechanical properties can be seen. For that reason the limit of rotor speed can be determined at the minimum of the We number for the two machines.

VI. COMPOUNDING OF NR/PP BLENDS

For the purpose of checking the approach aimed at determining the optimum operational points, the blends were produced also on a laboratory and a production kneader with the ratios being 40/60 or 60/40, respectively. The blends were compounded without a filler and vulcanized dynamically with dicumylperoxide. The peroxide dissolved in the thermoplastic phase was cushioned by adding antioxidants, just as has been done in Reference 1. The power curve for the production kneader presented in Figure 14 serves to describe the cycle. Following the addition of PP and NR, in a first step the thermoplastics were mixed and melted at a high rotational speed (60 min^{-1}), in order to reduce cycle times. Thermoplastic mixing and melting is finished at performance maximum. Afterward, PP is dispersed. For this purpose, the operational point is reduced down to different values. These phase times were kept to a constant 2^{1}/$_{2}$ min. As adequate thermal boundary conditions are selected, dispersion is effected with temperature almost constant and homogeneity satisfactory. Water temperature was between 80 and 120°C, subject to the torsional speed. Then the peroxide is added. Cross-linking is concluded, when the performance has reached a constant level. In order to introduce the antioxidant, the torsional speed is further reduced to 20 min^{-1}. This keeps the output temperature below 180°C and no damage is caused to the rubber.

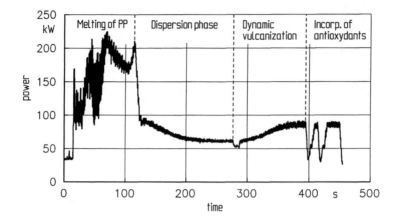

FIGURE 14. Course of the process at the production mixer NR/PP blend.

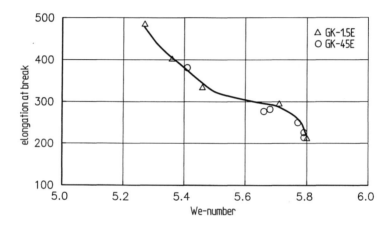

FIGURE 15. Correlation of We number and elongation at break NN/PP (40/60).

VII. REVIEWING THE APPROACH TO DETERMINE OPTIMUM OPERATIONAL POINTS

For the tests performed on the laboratory and production equipment, the We number was determined. For this aim, the temperature at the end of the dispersion phase and the shear speeds taken from the torsional speeds were used. Plotting it above the breaking strain for the combination of 40% NR and 60% PP (Figure 15) shows that minimum We number yields best mechanical properties. The breaking strain was between 10 and 12 MPa, in all of the trials. No torsional speed limit was found, since minimum torsional speeds yield best results. Table 5 shows that, for the combination of 40% NR and 60% PP, there is a correlation revealing in the We number and directly in the calculated diameter, which describes the course of the measured values and has minor deviations only in terms of absolute height. As the We number decreases, mechanical properties improve. Provided the We number remains constant, a small diameter also entails improved mechanical characteristics (Table 5).

Even for other polyolefin/rubber mixtures, a correlation between diameter and mechanical properties was found.[25] Here, as well, mechanical properties deteriorate as particles grow. In the area of optimum particle size, a high range of dependence reveals corresponding

TABLE 5
Correlation between Morphology, We number, and Mechanical Characteristics

Machine	Speed (rpm)	Tensile strength (MPa)	Elongation at break (%)	We number	Calculated diameter (nm)
GK-1.5 E	30	8.1	380	15.09	960
GK-1.5 E	60	7	360	15.44	820
GK-45 E	45	6.5	330	18.96	1200
GK-45 E	35	5.1	200	18.98	1300

to changes in terms of percent, which shows in the example above. However, no calculations were performed for this case.

VIII. OUTLOOK

In conclusion, it can be said that consideration of the elongational deformation in the drop deformation which was calculated using the emulsion theory allows for assessment of a processing window. Minimum We number corresponds to maximum material properties. By a diagram the We number, at different rotor speeds, is plotted vs. the viscosity ratio at a limit of rotor speed that can be obtained. This is possible thanks to the fact that the values from calculation correspond to those measured on laboratory as well as production machines. When it comes to taking account of strain deformation, it is presently assumed to be identical for flow and drops. This assessment certainly calls for complementary model considerations to improve the description.

ACKNOWLEDGMENTS

The author would like to thank the companies Uniroyal Englebert Reifen GmbH, Dow Chemical, BASF, and Werner & Pfleiderer GmbH for providing us with material testing, material, and machinery. The grant from the "Deutsche Forschungsgemeinschaft" (DFG) is gratefully acknowledged.

REFERENCES

1. **Campbell, D. S., Elliot, D. J., and Wheelans, M. A.,** *Thermoplastische Naturkautschukmischungen Naturkautschuktechnologie,* 9, 21, 1978.
2. **Mullins, L.,** Thermoplastic Natural Rubber Scandinavian Rubber Conference, Copenhagen, April 2–3, 1979.
3. **Elliott, D. J.,** *Developments in Rubber Technology,* Natural Rubber Systems, 1982, 203.
4. **Elliott, D. J.,** Comercial Prospects of Thermoplastic Natural Rubber, International Conference Developments in the Plastics and Rubber Product Industries, Kuala-Lumpur, July 5–16, 1987.
5. **Elliott, D. J.,** Entwicklungen mit Thermoplastischen, Naturkautschuk-Verschnitten, *Naturkautschuktechnologie,* 12, 59, 1981.
6. **Tinker, A. J., Icenogle, R. D., and Whittle, I.,** Thermoplastic NR opens TPEs to new uses, *Rubber Plastics News,* October 17, 1989.
7. **Tinker, A. J.,** Factors influencing the impact properties of polypropylene/natural rubber blends, International Conference Plastics and Rubber Institute, London, July 2–4, 1985.
8. **Tinker, A. J.,** Preparation of polypropylene/natural rubber blends having high impact strength at low temperatures, *Polym. Commun.,* 25, 325, 1984.

9. **Mathew, N. M. and Tinker, A. J.,** Impact-resistant polypropylene/natural rubber blends, *J. Nat. Rubber Res., 1,* 240, 1986.

10. **Choudhury, N. R. and Bhowmick, A. K.,** Compatibilization of natural rubber-polyolefin thermoplastic elastomeric blends by phase modification, *J. Appl. Polym. Sci.,* 38, 1091, 1989.

11. **Jentzsch, J., Nguyen, Q., and Krause, K.-H.,** Mechanische Eigenschaften und Alterungsverfahren (Mischungen aus Polyethylen und Naturkautschuk), *Plaste Kautschuk,* 37, 224, 1990.

12. **Qin, C., Yin, J., and Huang, B.,** Mechanical properties, structure, and morphology of natural-rubber/low-density-polyethylene blends prepared by different processing methods, *Rubber Chem. Technol., 63,* 77, 1989.

13. **Campbell, D. S., Mente, P. G., and Tinker, A. J.,** Natural rubber analogues of styrene-diene thermoplastic rubbers, *Kautschuk + Gummi Kunststoffe,* 34, 636, 1981.

14. **Kurian, P. and Mathew, N. M.,** Polystyrene-natural rubber blends — mechanical and rheological properties, *Kautschuk + Gummi Kunststoffe,* 43, 1098, 1990.

15. **Ossendorf, P.,** Verarbeitungsparameter, Morphologie und mechanische Eigenschaften bei der Aufbereitung von unvernetzten Blends im Innenmischer, unpublished diploma thesis, IKV, 1990.

16. **Hofmann, W.,** *Rubber Technology Handbook,* Hanser, Munich, 1989.

17. **Coran, A. Y., Patel, R. P., and Williams, D.,** Rubber-thermoplastic composites. V. Selecting polymers for thermoplastic vulcanizates, *Rubber Chem. Technol.,* 55, 117, 1982.

18. **Sunder, J.,** Prozessmodelle am Innenmischer liefern Möglichkeiten zur verbesserten Prozessführung, *Kautschuk + Gummi Kunststoffe,* 42, 587, 1990.

19. **Plochocki, A. P.,** Effect of rheological parameters on phase morphology of polyblends: a route to computer aided design in development of industrial polyblends, *Polym. Eng. Sci.,* 26, 82, 1986.

20. **Elmendorp, J. J.,** A study on polymer microrheology, *Polym. Eng. Sci.,* 26, 418, 1986.

21. **Plochocki, A. P., Dey, S. K., and Kiani, A.,** Computer simulation of melt viscoelasticity functions for binary polyblends, *Int. Polym. Processing,* 4, 119, 1989.

22. **Vanoene, H.,** Modes of dispersion of viscoelastic fluids in flow, *J. Colloid Interfac. Sci.,* 40, 448, 1972.

23. **Taylor, G. I.,** The viscosity of a fluid containing small drops of another fluid, *Proc. R. Soc., A, 132,* 41, 1932.

24. **Plochocki, A. P., Dagli, S. S., and Mack, H. H.,** *Morphologie von Polymergemischen und Kunststoffe,* 78, 254, 1988.

25. **Kresge, E. N.,** Polyolefin thermoplastic elastomer blends, *Rubber Chem. Technol.,* 64, 469, 1990.

Chapter 7

PROPERTY MODELING OF VISCOELASTIC ELASTOMERS

H. So and Ue-De Chen

TABLE OF CONTENTS

0-8493-4401-8/93/$0.00 + $.50

I. INTRODUCTION

According to results obtained by experiments, it is found that many mechanical properties or even mechanical behaviors of pure or particulate-filled elastomers can be predicted by using mathematical methods. It will save much time for engineers or designers to employ mathematical models instead of performing experiments to obtain design data, when they are using elastomers as engineering materials.

The properties of viscoelastic elastomers depend upon many factors such as the type and volume content of fillers, the relative humidity and temperature of the environment which the elastomers are exposed to, the loading conditions acting on the elastomers, etc.; such factors may markedly affect the properties and mechanical behavior of elastomers. In practice, the use of solid particles in elastomers is not uncommon, as they can improve the mechanical, physical, or chemical properties of the elastomers. On the other hand, the solid particles can cause the composite to be susceptible to moisture and temperature. The former enhances the dewetting effect on the composite, while the latter decreases the maximum elongation of the composite. Such effects are detrimental to the composite.

The addition of solid fillers to elastomers also changes the viscoelastic behaviors of the composites when they are subjected to external loads.

Therefore, the effects of the influent factors mentioned above on the mechanical properties of elastomers are so complicated that they should be studied with tedious experimental works. Fortunately, it has been proved with intensive experiments that the effects of the influent factors on the mechanical properties of some specific particulate-filled elastomers can be predicted by mathematical models. However, it should be pointed out that in this article the basic elastomer used in experiments to evaluate the models was prepared from hydroxyl-terminated polybutadiene (HTPB) cured with 4,4-diphenyl methane diisocyanate (MDI), which we call polybutadiene rubber. The particulate-filled elastomers were obtained by adding glass beads or salt particles in quantities necessary to obtain different volume loadings into the HTPB-MDI mixture. More detailed information on the preparation of the elastomeric specimens is given by Chen and So.[1]

II. STRENGTH MODELING

The strength of elastomers is affected by many factors as stated previously. To avoid ambiguous or complex results obtained under the influence of more than one factor, it is wise to study the effects of an individual influent factor on the strength of elastomers. The following results are obtained under such an idea.

A. EFFECTS OF FILLER CONTENT

The type and volume loading of fillers used in an elastomer can affect the stress-strain behaviors as well as the tensile strength of the composites. According to the rubber elasticity the relation between stress, σ, and stretch, λ, can be written as

$$\sigma = G(\lambda - 1/\lambda^2) \tag{1}$$

where $\lambda = 1 + \epsilon$, ϵ is the engineering strain; G is the shear modulus which depends on the filler content.

Equation 1 is almost correct for static tests for particulate-filled elastomers in small strains or for pure elastomers. As the strain rate or filler content increases, Equation 1 tends to deviate the experimental results. Typical results are shown in Figure 1. The solid lines in Figure 1 are obtained from tensile tests in different strain rates; the symbols are given

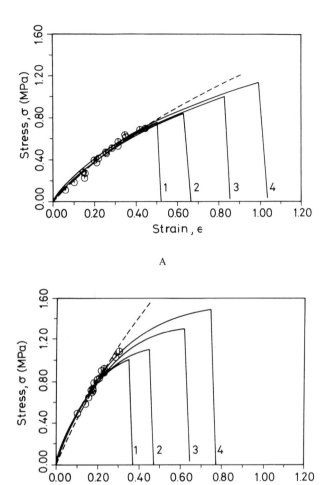

FIGURE 1. Typical stress-strain curves of polybutadiene rubber for different strain rates: $\dot{\epsilon}$ = (1) 0.0025, (2) 0.01, (3) 0.1, (4) 0.5 (1/s). The dashed lines are obtained with Equations 1 and 14. The symbols are the data from creep tests for (A) pure elastomer and (B) 30% glass-filled rubber.

from creep tests in the primary creep stage; the broken lines are the regressive results with Equation 1. The modulus G for particulate-filled elastomers obeys the modified Kerner's equation given by[2,3]

$$\frac{G}{G_p} = \frac{1 + ABV_f}{1 - B\phi V_f} \tag{2}$$

where G and G_p are moduli of particulate-filled and pure elastomers, respectively; $A = 1.5$ is a material parameter for Poisson's ratio $v = 0.5$; $B = 1$ for rigid fillers; and ϕ is given by

$$\phi = 1 + \left(\frac{1 - V_m}{V_m^2}\right) V_f \tag{3}$$

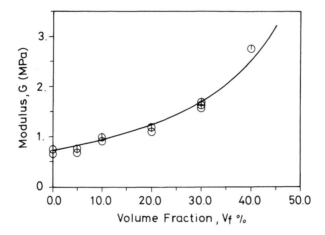

FIGURE 2.   Variation of initial modulus with volume fraction of solid filler. Symbols denote experimental results; the solid line is from Equation 2.

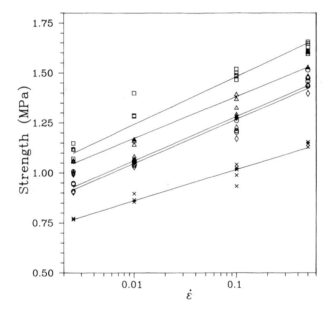

FIGURE 3.   Variation of tensile strength with logarithmic strain rate, ln $\dot{\epsilon}$ for polybutadiene rubber filled with glass beads in different volume fraction, V_f = (□) 40%, (△) 30%, (○) 20%, (◇) 10%, (×) 0%.

where V_f is the volume fraction of solid fillers, V_m is the maximum value of V_f and is set to 0.65 for polybutadiene rubber. Figure 2 indicates the agreement of Equation 2 with experiments.

It can be shown that the effect of strain rate on the tensile strength of particulate-filled elastomers satisfies a logarithmic function of strain rate $\dot{\epsilon}$,

$$\sigma_u = A \ln \dot{\epsilon} \tag{4}$$

where A is a material parameter. Typical results for the plot of tensile strength against logarithmic strain rate are shown in Figure 3 for glass-filled elastomers with different filler contents.

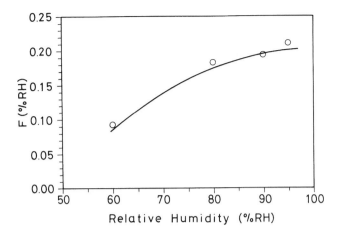

FIGURE 4. The relation between the function F and relative humidity; the solid line is found by a method of curve-fitting while the symbols are the experimental data.

B. EFFECTS OF MOISTURE
1. Semi-Dewetting Case

If the solid particles added to the polybutadiene rubber do not react chemically with the matrix phase, the binding forces between the solid fillers and the matrix are dominated by hydrogen bonds and the van der Waals force. Such binding conditions are easily destroyed by water molecules when the composite is exposed to a humid environment. This is known as a dewetting effect. Subjected to such an effect, the strength of the composite will decrease corresponding to the amount of water molecules penetrating into it. However, the quantity of water molecules $C(t)$ penetrating into the composite depends upon time t and can be calculated by the Fick's diffusion law,[4-6]

$$C(t) = C_o[1 - \sum K_i \exp(-t/\tau)] \tag{5}$$

where C_o is the concentration of water molecules in the environment which the composite exposes to; τ is the relaxation time depending on the coefficient of diffusion, K_i. It is sensible to assume that the reduction in strength of the composite caused by the diffusion of water molecules obeys a similar function as Equation 5; then we arrive at

$$\sigma_u = \sigma_{uo}\{1 - F(\% RH)[1 - \exp(-t/\tau)]\} \tag{6}$$

where σ_{uo} is the tensile strength of the composite in a dry environment, $F(\%RH)$ is a function of the concentration of water molecules equivalent to the relative humidity of the atmosphere. When $\tau = 3.69$, the function $F(\%RH)$ exhibits strongly nonlinear characteristics and can be obtained by a method of curve fitting according to the experimental data as shown in Figure 4. Figure 5 shows some typical stress-strain curves for the 30% glass-filled polybutadiene rubber exposed to humid environments for 7 d at 25°C. Figure 6 indicates the variation of strength with time of exposure to the atmosphere of different relative humidity at 25°C. The solid lines are the results given by Equation 6, while the symbols are from experiments. It is found that, when the relative humidity is over 80%, the amount of reduction in strength will approach the same limiting value. However, this value may depend upon the filler content of the composite.

It must be pointed out that the strength of pure elastomers is not affected by water molecules even though they are exposed to the atmosphere of 100% relative humidity. This

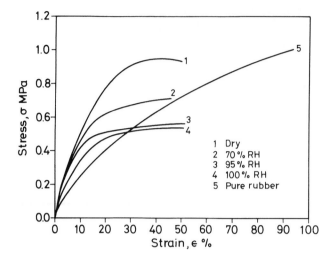

FIGURE 5.   Stress-strain curves for the solid-filled rubbers exposed to humid environments for 7 d at 25°C for $V_f = 30\%$, $D = 180$ μm.

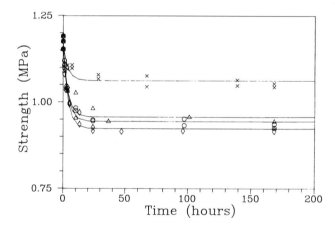

FIGURE 6.   Variation of tensile strength of glass-filled rubber with time of exposure to atmospheres of different relative humidity of ($\times$) 60%, ($\triangle$) 80%, ($\circ$) 90%, ($\diamond$) 95% at 25°C for $V_f = 30\%$, $\dot{\epsilon} = 0.1$ s^{-1}.

can be proved by the experimental results shown in Figure 7. If the solid filler can dissolve in water, the percentage of relative humidity will become unimportant and the final strength will reach the same value. Equation 6 is still correct except that the function $F(\%RH)$ converges to a single function satisfying all the relative humidity over 60%. Such a condition is indicated in Figure 8.

2. Fully Dewetting Case

As mentioned previously, the strength of the particulate-filled polybutadiene rubber in the same filler loading will reduce to the same limiting value when it is exposed to the atmosphere of higher than 80% relative humidity. This means that the binding condition between the solid fillers and the matrix phase is completely destroyed. Such a condition is known as "no adhesion". Under the no adhesion condition, the tensile strength of the composite will depend on the net cross-sectional area of the gum elastomer matrix; the area occupied by the filler may be considered as holes in the matrix. Therefore, the tensile

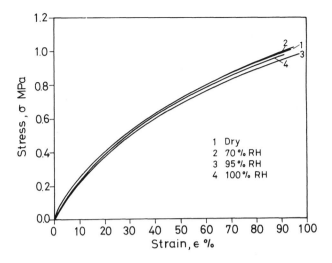

FIGURE 7. Stress-strain curves for the pure elastomer exposed to humid atmospheres for 7 d at 25°C.

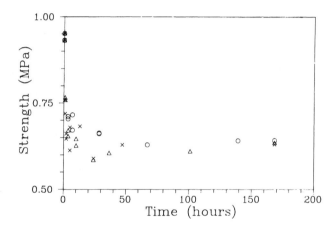

FIGURE 8. Variation of tensile strength of salt-filled rubber with time of exposure to atmosphere of different relative humidity of (○) 60%, (△) 80%, (×) 90% at 25°C for $V_f = 30\%, \dot{\epsilon} = 0.1 \text{ s}^{-1}$.

strength of the composite in the no adhesion condition only relates to the net cross-sectional area of the gum elastomer matrix and can be approximated with a simple relation given by Nicolais and Narkis[7] based on a model of one single particle placed in a unit volume of the composite material. They proposed that

$$\sigma_u = \sigma_{up}(1 - 1.21V_f^{2/3}) \tag{7}$$

where σ_{up} is the tensile strength of the gum elastomer.

Another approximated relation was proposed by Piggott and Leidner;[8] they assumed that the area fraction of the solid filler in any cross-section is equal to the volume fraction of the solid filler in the composite. They concluded that the decrease in the tensile strength of the composite was of the form

$$\sigma_u = \sigma_{up}(a - bV_f) \tag{8}$$

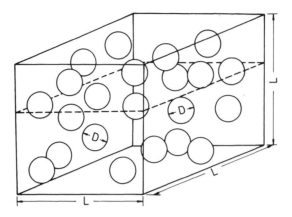

FIGURE 9. Schematic diagram for a particulate-filled cube with edge length L and particle diameter D.

where the constant a and b depend on the shape and adhesive characteristics between particles and the matrix phase as well as on the rigidity and strength of the matrix phase.

The net cross-sectional area of the matrix is obtained by subtracting the area of a diametral plane of the filler from the unit area of the composite in the unit volume model.[7] In such a case, the net cross-sectional area of the matrix phase results in a minimum value. Hence, the predicted tensile strength of the composite will be underestimated. Equation 8 implies that the solid particles are regularly distributed in the matrix, which will overestimate the net cross-sectional area of the matrix and so will the tensile strength of the composite as well.

In fact, the solid particles in a particulate-filled elastomer are randomly distributed. The maximum area fraction of the particles in some specific cross section must be greater than their volume fraction in the composite, but less than that given by the unit volume model, because it is difficult to find a cross-section passing through all the diametral planes of the particles. Chen and So[1] proposed a statistical model for predicting the strength of solid-filled elastomers.

In this model, it is assumed that there is no interaction between the matrix phase and the filled particles. Under such a condition, it is assumed that fracture will occur at the cross-section at which the area fraction of the matrix is a minimum. If the effect of stress redistribution during loading is neglected, the reduction of strength will be in proportion to the area fraction of the filler along that plane. It should be noted that this plane is not likely to pass through all the equators of the particles lying on that plane, which are assumed to be spherical. The procedure for calculating the maximum area fraction of the filler is described as follows.

Consider a cube of the composite of dimension L taken from a tensile specimen, in which are distributed N particles of diameter D, as shown in Figure 9. The volume fraction of the filler, V_f, is given by $\pi ND^3/6L^3$. The coordinates (x_i, y_i, z_i) of the ith particle likely to be found in the particular cube chosen are given by random values generated by a personal computer with the condition that

$$D/2 < x_i, y_i, z_i < L - D/2 \qquad i = 1, 2, ..., N \qquad (9)$$

and

$$[(x_i - x_j)^2 + (y_i - y_j)^2 + (z_i - z_j)^2]^{1/2} > D \qquad j = 1, 2, ..., i - 1 \qquad (10)$$

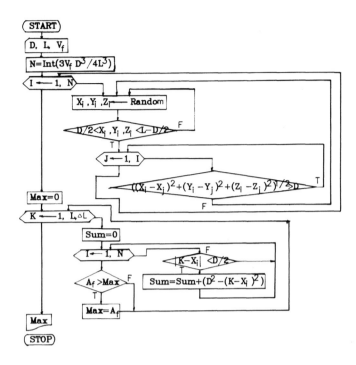

FIGURE 10. Computer flow chart for statistical model.

Equation 9 ensures that all N particles are located within the cube, while Equation 10 ensures separation between the particles. This cube is now divided on each edge into 1000 parts. The probable area fraction of the filler on any one of the 1000 cross-sectional area is then determined. The cross-section that accommodates the largest area fraction A_f of the particles is set aside for later calculation. This calculation was repeated more than 150 times. The average value of these maxima was then found. This is considered to be the expected value of the maximum area fraction for random distribution of the particles. The flow chart for the statistical calculation is shown in Figure 10.

We employed particles of 50, 100, 150, and 200 units in diameter compared with 1000 units of the cube on each edge, respectively, and volume fraction of filler of 5, 10, 15, and 20%. It was found that the area fraction in the cross-sections obeyed a normal distribution centered about A_f. Figure 11 shows the area fraction of the matrix as a function of the volume loading. It will be observed that all area fractions lie between the values of $1 - 1.21\ V_f^{2/3}$ and $1 - V_f$, and that the particle size has a significant effect on the area fraction. Furthermore, the particles are never uniformly distributed.

In comparing the experimental results with those of the statistical calculations, it should be noted that the ratio of the particle diameter D to the dimension of the composite cube L becomes the ratio of D to the thickness of the specimen, 2 mm. In our case, for the particle sizes chosen, this is 0.09 and 0.042, respectively. Figure 12 plots the strength ratio, that is, the ratio of the specimens exposed to atmosphere of 100% relative humidity to those stored in dry air, against the volume fraction of the filler, both from statistical computations and from experiments. It can be seen that good agreement is obtained. The results according to Nicolais' and Piggott's models are also plotted in the same figure. The simplified Piggott's model becomes the upper bound while Nicolais' model is the lower bound of the real condition.

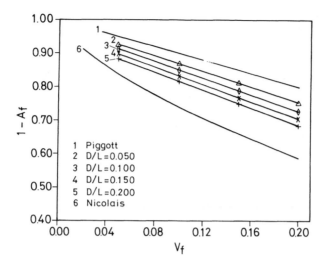

FIGURE 11.   Variation of area fraction of the matrix phase, $1 - A_f$, with volume fraction of the solid filler, V_f.

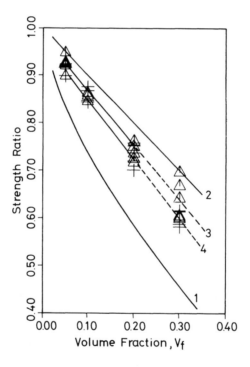

FIGURE 12.   Variation of strength ratio (strength of wetted filled-rubber/strength of dry filled-rubber) with volume fraction of the filler, V_f. (1) Nicolais, (2) simplified Piggott, (3) statistical model $D/L = 0.042$, (4) statistical model $D/L = 0.09$, ($\triangle$) experiment $D/L = 0.042$, (+) experiment $D/L = 0.09$.

C. EFFECTS OF TEMPERATURE

The significant effect of high temperature on many elastomers is to increase their tensile strengths resulting from the process of cross linking, which can carry on at high temperature. From many tensile tests following a treatment of temperature aging on polybutadiene rubber, it is found that, when the aging temperature is higher than 60°C, the effect of temperature is pronounced. Figure 13 shows some typical results. Much more results are presented by So and Chen.[9]

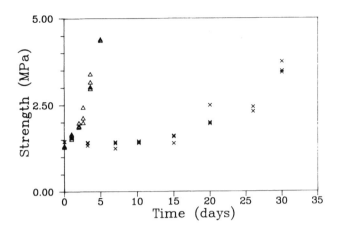

FIGURE 13. Variation of strength with time of aging for 30% glass-filled rubber at different temperature: ($\triangle$) 120°C, ($\times$) 90°C; $\dot{\epsilon} = 0.1$ s^{-1}.

From the results indicated in Figure 13, it is found that the effects of temperature on the tensile strength of elastomers must be dependent on the total time that the elastomers last at that temperature. It can thus be concluded that the combined effects of time and temperature on the tensile strength of particulate-filled elastomers obey the time-temperature superposition principle, such that

$$\sigma_u(T,t) = \sigma_u(T_o, \zeta) \tag{11}$$

where $\sigma_u(T,t)$ denotes that the tensile strength is a combined function of temperature T and the time of aging t, while T_o is a reference temperature; ζ is the shift time and is given as

$$\zeta = t/a(T) \tag{12}$$

where $a(T)$ is the shift factor which denotes the amount of horizontal shift of the time scale due to any temperature T differing from the reference temperature T_o. If the temperature varies from time to time, the shift time can be calculated by integrating with respect to time as given below.

$$\zeta = \int \frac{dt}{a(T)} \tag{13}$$

If a reference temperature is assumed to be 90°C, a master curve for tensile strength against shift time ζ is plotted in Figure 14, in which the shift factor $a(T)$ is given by an Arrhenius plot shown in Figure 15 according to the modified WLF equation with an energy term.[10]

Furthermore, the expression for the master curve in Figure 14 can be obtained easily by a method of curve fitting.

III. VISCOELASTIC MODELING

A. CREEP BEHAVIORS

Solid fillers added to elastomers always enhance the nonlinear viscoelastic characteristics of the composites; thus, the Boltzmann superposition principle cannot be applied. Some investigators[11-18] proposed some useful models to describe the nonlinear relationship between

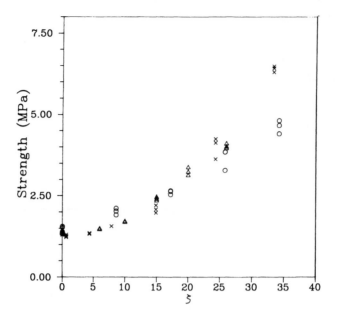

FIGURE 14. Variation of strength with shift time, ζ for 40% glass-filled rubber at different temperature of aging: ($\times$) 60°C, ($\triangle$) 90°C, ($\circ$) 120°C.

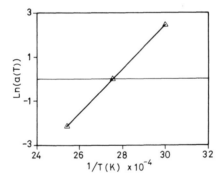

FIGURE 15. Variation of logarithmic shift factor $\ln[a(T)]$ with reciprocal temperature $1/T$. The reference temperature is 90°C.

stress and strain for polymeric materials. Among these models, the nonlinear mechanical model proposed by So and Chen[18] is recommended for its simplicity.

According to the theory of rubber elasticity, the relationship between stress and stretch is nonlinear and can be written in the form as follows:

$$\sigma = G_i(\lambda - 1/\lambda^2) \tag{14}$$

where G_i is the initial modulus of elasticity. If a nonlinear stretch parameter Λ is defined in the form

$$\Lambda = \lambda - 1/\lambda^2 \tag{15}$$

then a linear relation between stress and the stretch parameter will have the form as follows:

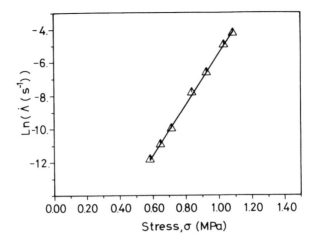

FIGURE 16. Variation of the time rate of nonlinear stretch parameter $\dot{\Lambda}$ with applied stress σ in secondary creep stage.

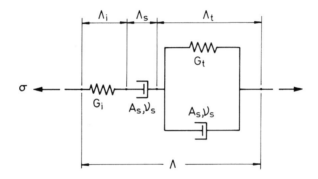

FIGURE 17. The schematic diagram of the nonlinear mechanical model.

$$\sigma = G_i \Lambda \tag{16}$$

We assumed that the time rate of change of nonlinear stretch parameter, $\dot{\Lambda}$ satisfies the Eyring equation for activated non-Newtonian viscous materials. In fact, this assumption can be approved by experiments as shown in Figure 16. Then, we obtain

$$\dot{\Lambda} = A \sinh \left(\frac{\nu \sigma}{RT} \right) \tag{17}$$

where A is associated with the activation energy, ν is the activation volume, and R is the gas constant. According to such characteristics of the stretch parameter Λ, we can introduce a mechanical model with nonlinear elements as shown in Figure 17, and we will obtain the following relations between stress and the stretch parameter or the time rate of stretch parameter,

$$\sigma = \text{constant} \tag{18}$$

$$\Lambda = \Lambda_i + \Lambda_s + \Lambda_t \tag{19}$$

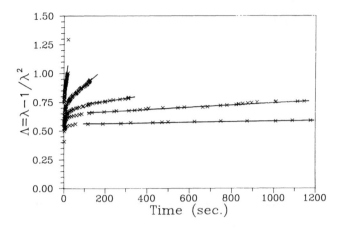

FIGURE 18.   Variation of nonlinear stretch parameter Λ with time for $V_f = 30\%$.

$$\Lambda_i = \sigma/G_i \tag{20}$$

$$\dot{\Lambda}_s = A_s \sinh(v_s\sigma) \tag{21}$$

$$\dot{\Lambda}_t = A_t \sinh[v_t(\sigma - G_t\Lambda_t)] \tag{22}$$

and find an equation describing the creep behavior as follows

$$\Lambda(t) = \frac{\sigma}{G_i} + A_s \sinh(v_s\sigma)t + \frac{\sigma}{G_t} \times$$

$$\left\{ 1 - \frac{2}{v_t\sigma} \tanh^{-1}\left[\tanh\left(\frac{v_t\sigma}{2}\right) \exp\left(-A_t v_t G_t t\right)\right]\right\} \tag{23}$$

There are six material parameters, namely, G_i, A_s, v_s, G_t, A_t, and v_t, in the nonlinear model as well as in the creep equation, Equation 23. In general, some experimental data from two creep tests subjected to two different stresses are required. According to these data the six parameters are determined by using the method of least squares.

For instance, the initial modulus G_i can be obtained from the initial value on a $\Lambda - t$ curve as shown in Figure 18 in which the stress is constant for each curve. The data at initial time in Figure 18 can be transformed to the stress-strain curves as shown in Figure 1. The parameters A_s and v_s are the main factors governing the steady-state creep. They can be obtained from the linear parts in a $\dot{\Lambda} - t$ diagram as shown in Figure 18. On the linear part of a $\Lambda - t$ curve, the rate of stretch $\dot{\Lambda}$ is a constant; hence, the two parameters, A_s and v_s, can be determined by two $\Lambda - t$ curves. The last three parameters, G_t, A_t and v_t, are determined by making use of Equation 23 with the data on the two $\Lambda - t$ curves used before, and by using the method of least squares.

It should be pointed out that the six material parameters are not independent of one another. On the contrary, they depend on one another. Furthermore, they depend on the stretch as well as the rate of stretch. Therefore, the model expressed by Equations 18 to 23 and Figure 17 can be known as a nonlinear mechanical model.

To show how well the present nonlinear mechanical model is in agreement with the experiments, intensive creep tests on pure and particulate-filled polybutadiene rubbers mixed with different volume loading of glass beads have been conducted. Some typical results are shown in Figure 19. As mentioned before, the mechanical behaviors of particulate-filled and unfilled polybutadiene rubbers are affected by the effects of temperature aging markedly. On this behalf, some specimens were subjected to temperature aging followed by creep tests. The results are indicated in Figure 20. The solid lines in Figures 19 and 20 are found by the model with the values of the six parameters listed in Table 1, while the symbols are obtained from experiments. It is clear that the present model agrees with experiments extremely well.

Furthermore, from these results it is found that the effects of viscous flow on the creep stretch increase with increasing filler content of the composite. This viscous flow is caused by the dewetting effect between solid particles and the elastomer matrix. For low filler content, say, 10%, the creep stretch is hindered by the three-dimensional cross-linking network of the molecular chains. Hence, the effect of viscous flow on the creep stretch becomes less dominant in the secondary creep stage. Such a phenomenon can be applied to gum elastomers, the structure of which is dominated by three-dimensional network chains; this implies that viscous flow will not occur in the secondary creep stage, so that the strain in creep tests will diminish as time increases. Therefore, the dashpot connected in series with the spring in the model (Figure 17) can be dropped. The model becomes an equivalent standard solid with three nonlinear elements. The six material parameters reduce to four, namely, G_i, G_t, A_t, and v_t. The creep behaviors for pure elastomers are given by

$$\Lambda(t) = \frac{\sigma}{G_i} + \frac{\sigma}{G_t} \left\{ 1 - \frac{2}{v_t \sigma} \tanh^{-1} \left[\tanh\left(\frac{v_t \sigma}{2}\right) \exp(-A_t v_t G_t t) \right] \right\} \qquad (24)$$

This model also agrees very well with experiments. The typical results are shown in Figure 21.

When the particulate-filled polybutadiene rubbers are exposed to the atmosphere of 100% relative humidity for 15 d or more, the composites will reach a fully dewetting condition. Under such a condition, the viscous flow in the secondary creep stage will not occur when the composites are subjected to a creep test. This means that the three-element model (Equation 24) can describe the creep behaviors of the composites under fully dewetting condition. Figure 22 shows some typical results. The solid lines are obtained by the model (Equation 24) with the values of the four material parameters listed in Table 2, while the symbols are obtained from experiments.

B. TIME TO FAILURE

The nonlinear mechanical model proposed above cannot predict the rupture condition in creep processes. It is found that the theory of cumulative damage presented by Saylak and Beckwith[19] can be employed, which is given by

$$\ln(\sigma t_f) = A - v\sigma \qquad (25)$$

where t_f is the total time to failure in a creep test when the specimen is subjected to a constant stress, σ; A is the material parameter; v is the activation volume of the polymer. In general, A and v depend upon the filler content of the composite. Typical results are shown in Figure 23. It is also found that Equation 25 can be applied to the composite under the influence of moisture or after a treatment of high temperature aging.[9]

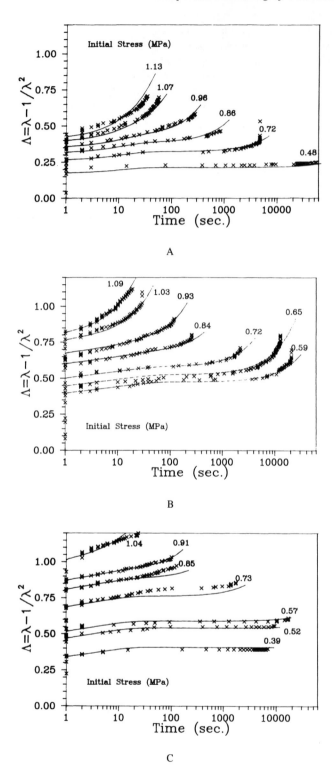

FIGURE 19. Comparison of present model with experiments for different volume fraction of filler. Symbols: experiments, solid lines; model, numbers; initial stresses for (A) $V_f = 40\%$, (B) $V_f = 30\%$, (C) $V_f = 20\%$.

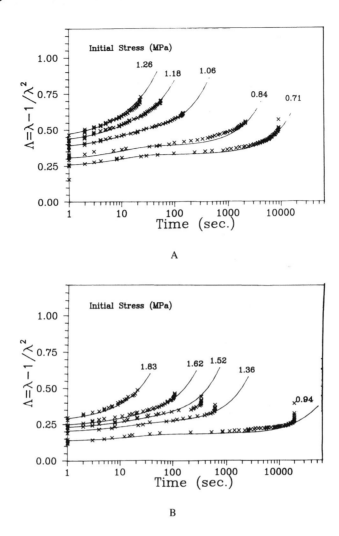

FIGURE 20. Comparison of present model with creep tests with aged effect at 90°C for (A) 10 and (B) 20 d of aging.

TABLE 1
The Aggressive Values for the Six Material Parameters

Days[a]	V_f (%)	G_i (MPa)	A_s (s⁻¹)	v_s (MPa⁻¹)	G_t (MPa)	A_t (s⁻¹)	v_t (MPa⁻¹)
0	40	2.75	5.28×10^{-10}	15.2	12.0	6.05×10^{-3}	1.16
0	30	1.70	1.37×10^{-9}	15.5	4.49	8.88×10^{-4}	10.8
0	20	1.28	6.89×10^{-11}	18.8	3.82	3.63×10^{-3}	9.62
10	40	2.81	1.55×10^{-8}	10.9	9.71	3.97×10^{-3}	1.98
20	40	6.81	2.38×10^{-9}	8.50	20.5	6.44×10^{-4}	2.38

[a] Duration of temperature aging before creep test.

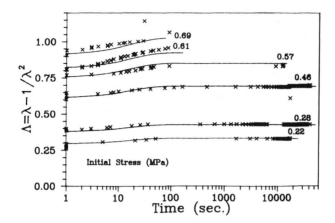

FIGURE 21. Comparison of present model with experiments for gum polybutadiene rubber. Symbols: experiments, solid lines; model, numbers; initial stresses.

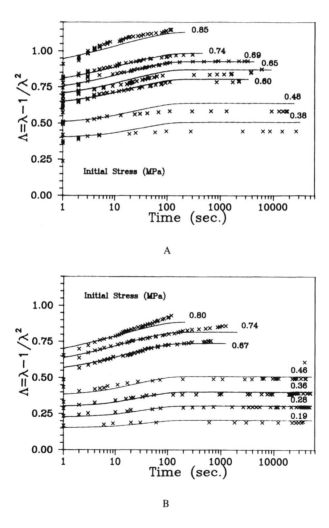

A

B

FIGURE 22. Comparison of present model with experiments for glass-filled polybutadiene rubbers exposed to the atmosphere of 100% relative humidity for 15 d. Symbols: experiments, solid lines; model, numbers; initial stresses for (A) $V_f = 20\%$ and (B) $V_f = 30\%$.

TABLE 2
The Aggressive Values for the Four
Material Parameters

V_f (%)	G_i (MPa)	G_t (MPa)	A_t (s^{-1})	v_t (MPa^{-1})
0	0.76	5.94	2.32×10^{-2}	0.39
20	0.95	3.54	9.24×10^{-4}	6.54
30	1.22	3.51	1.06×10^{-3}	6.13

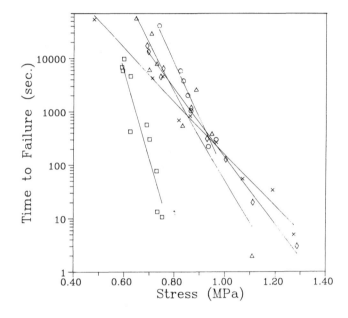

FIGURE 23. Variation of time to failure with applied stress under creep tests. The solid lines are from Equation 25 while the symbols are from experiments for (□) pure elastomer, (○), $V_f = 10\%$, (△) $V_f = 20\%$, (◇) $V_f = 30\%$, (×) $V_f = 40\%$ glass-filled rubbers.

C. RELAXATION BEHAVIORS

If the nonlinear mechanical model (Figure 17) is employed to describe the relaxation behavior of particulate-filled polybutadiene rubbers, it will have the form

$$\Lambda = \Lambda_i + \Lambda_s + \Lambda_t = \text{constant} \tag{26}$$

$$\Lambda_i = \sigma/G_i \tag{27}$$

$$\Lambda_s = \int A_s \sinh(v_s \sigma) dt \tag{28}$$

$$\Lambda_t = \int A_t \sinh[v_t(\sigma - G_t \Lambda_t)] dt \tag{29}$$

where the six material parameters G_i, A_s, v_s, G_t, A_t, and v_t are the same as those determined by creep tests, e.g., Tables 1 and 2. Typical results are indicated in Figure 24. The model coincides with the experiments very well for both pure (Figure 24A) and particulate-filled (Figure 24B) polybutadiene rubbers.

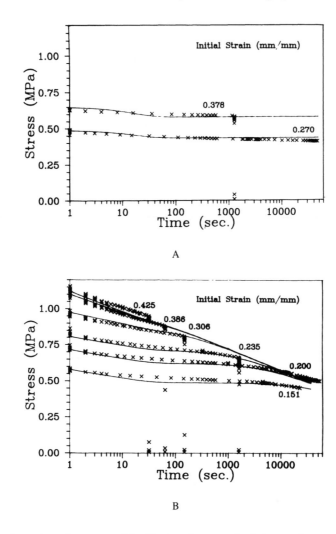

FIGURE 24. Comparison of present model with relaxation tests. Symbols: experiments, solid lines; model, numbers; initial strains for (A) unfilled polybutadiene rubber and (B) filled rubber, $V_f = 30\%$.

IV. CONCLUDING REMARKS

The effects of filler content, moisture, temperature aging, etc. on polybutadiene rubbers are studied. Some mathematical methods describing the relationships between some properties or behaviors and the influent factors are summarized in this article. They are found to be in good agreement with experiments. The properties which can be predicted by mathematical models with confidence are summarized as follows.

1. The modulus of elasticity for particulate-filled elastomers can be predicted with Equations 2 and 3.
2. The relation between tensile strength and strain rate is given by Equation 4.
3. The reduction in tensile strength of particulate-filled elastomers exposed to humid atmosphere can be modeled by Equation 6.
4. The tensile strength of particulate-filled elastomers in no adhesion condition can be predicted by a statistical model.

5. The tensile strength of elastomers subjected to the combined effects of temperature and time of aging can be determined by a modified WLF equation.

6. The nonlinear viscoelastic behaviors of pure and particulate-filled elastomers can be predicted by the nonlinear mechanical models. Equations 23 and 24 are applied to creep behaviors, while Equations 26 to 29 are employed in the prediction of relaxation behaviors of pure and particulate-filled elastomers.

7. The time to failure in a creep test can be predicted with Equation 25.

NOMENCLATURE

A	Activation energy or a constant
A_f	Area fraction of filled particles on a cross-section of a composite
$A_{s,t}$	Material parameters
$a(T)$	Shift factor
D	Diameter of the filler
G_i	Initial modulus or a material parameter
G_t	Material parameter
L	Edge length of a cube
T	Temperature
t	Time
t_f	Time to failure
V_f	Volume fraction of filler
x, y, z	Coordinates of a particle in the model cube
ϵ	Engineering strain
$\dot{\epsilon}$	Time rate of strain
ζ	Shift time
Λ	Nonlinear stretch parameter
$\dot{\Lambda}$	Time rate of Λ
λ	Stretch $(= 1 + \epsilon)$
ν	Activation volume
$\nu_{s,t}$	Material parameters
σ	Engineering stress
σ_u	Tensile strength

REFERENCES

1. **Chen, U. D. and So, H.,** *Polym. Eng. Sci.,* 29, 1614, 1989.
2. **Lewis, T. B. and Nielsen, L. E.,** *J. Appl. Polym. Sci.,* 14, 1449, 1970.
3. **Nielsen, L. E.,** *J. Appl. Phys.,* 41, 4626, 1970.
4. **Frisch, H. L.,** *Polym. Eng. Sci.,* 20, 2, 1980.
5. **Schneider, N. S., Illinger, J. L., and Cleaves, M. A.,** *Polym. Eng. Sci.,* 26, 1547, 1986.
6. **Schneider, N. S., Dusablon, L. V., Spano, L. A., and Hopeenberg, H. B.,** *J. Appl. Polym. Sci.,* 12, 527, 1968.
7. **Nicolais, L. and Narkis, M.,** *Polym. Eng. Sci.,* 11, 194, 1971.
8. **Piggott, M. R. and Leidner, J.,** *J. Appl. Polym. Sci.,* 18, 1619, 1974.
9. **So, H. and Chen, U. D.,** *J. Polym. Eng.,* 11, 245, 1992.
10. **Williams, M. L.,** *J. Phys. Chem.,* 59, 95, 1955.
11. **Green, A. E. and Rivlin, R. S.,** *Arch. Rational Mech. Anal.,* 1, 1, 1957.
12. **Green, A. E. and Rivlin, R. S.,** *Arch. Rational Mech. Anal.,* 4, 387, 1960.
13. **Green, A. E., Rivlin, R. S., and Spencer, A. J. M.,** *Arch. Rational Mech. Anal.,* 3, 82, 1959.

14. **Onaran, K. and Findley, W. N.,** *Trans. Soc. Rheol.,* 9, 299, 1965.

15. **Schapery, R. A.,** *Polym. Eng. Sci.,* 9, 195, 1969.

16. **Brueller, O. S.,** *Polym. Eng. Sci.,* 27, 144, 1987.

17. **Christensen, R. M.,** *Trans. ASME, J. Appl. Mech.,* 47, 762, 1980.

18. **So, H. and Chen, U. D.,** *Polym. Eng. Sci.,* 31, 410, 1991.

19. **Saylak, D. and Beckwith, S. W.,** Research Report of Air Force Rocket Propulsion Laboratory, UTECSI 69-070, 1969.

Chapter 8

INTERPENETRATING POLYMER NETWORKS

Daniel Klempner and Daniel Sophiea

TABLE OF CONTENTS

0-8493-4401-8/93/$0.00 + $.50

I. INTRODUCTION

Interpenetrating polymer networks (IPNs) are a broad class of polymer alloys classically defined as materials containing two or more intimately mixed network polymers with at least one being polymerized or crosslinked in the presence of the other.[1]

IPNs differ from other multicomponent polymer alloys because they possess, to some extent, topological catenanes,[2] Figure 1. These topological catenanes, or physical crosslinks, influence the polymer morphology most often by limiting the growth of large-phase domains, which normally occurs with increasing degree of polymerization. In cases of extreme immiscibility between constituent monomers, complete phase separation that may occur without crosslinking can be circumvented. In this way, mixing on the monomer scale followed by rapid crosslinking, can be thought of as kinetic control. However, when mixing components with higher compatibility the formation of IPNs may actually result in a decreased level of phase mixing.

Copolymers, grafted polymers and A-B type network polymers contain intermolecular covalent bonds which increases the level of mixing between the dissimilar components but essentially makes them part of the same molecule, Figure 2. Mechanical blends, on the other hand, do not contain these types of bonds and do not possess catenane structures. The degree of molecular mixing in polymer blends is, therefore, limited by thermodynamics to those polymer pairs which exhibit an overall negative free energy of mixing.

IPNs were traditionally classified by their synthesis route. In sequential synthesis, monomer I is network polymerized first, then swollen in monomer II along with the cross-linking agent and catalyst, and then network II is formed. In sequential IPNs, the first network formed regulates the formation of network II, influencing the continuity and domain size. In simultaneous synthesis, the different monomers are intimately mixed with their respective cross-linking agents and catalysts and both are polymerized simultaneously by way of non-interfering mechanisms, such as free-radical initiated chain and acid-catalyzed step-growth mechanisms shown in Figure 3. Another often used method, *in situ* polymerization, refers to sequential polymerization of the respective monomers, but where network I is polymerized in the presence of monomer II. The interpretation of IPNs today has broadened the definition to include semi- or pseudo-IPNs, in which only one polymer component is a network; thermoplastic IPNs, in which specific physical interactions replace physical crosslinks; gradient-IPNs, in which interpenetration within the host network varies with depth on a macroscopic scale; and cross-intercross IPNs in which a few (3 to 10%) intentional grafts between the dissimilar polymers are formed, Table 1. If we omit thermoplastic types, then IPNs are characterized by (1) the presence of catenane-like structures which limit the phase domain size, (2) phase separation processes which occur simultaneously with polymerization, and (3) the final product is a thermoset, which resists flow and dissolution.[3]

Standardized nomenclature of IPNs has yet to become widely used in polymer science in general. Currently, most authors use common names based on the monomer structure. That is, polymerized vinyl chloride monomer is referred to most often as poly(vinyl chloride) or PVC. Sperling[4] provided a system encompassing the types of polymers combined, the principle modes of combining them, and the time sequence of reactions. The system was documented as "Source-Based Nomenclature for Polymer Blends, IPNs and Related Materials" by the Division of Polymer Chemistry Nomenclature Committee in 1984.[5] However, presumably due to its complexity, it has yet to be utilized widely. More recently, the IUPAC Commission on Macromolecular Nomenclature has submitted proposed prefix and connective identifiers applicable to IPNs and related materials, Table 2.[6] Thus, as an example, a polyurethane(PU)/poly(vinyl chloride) semi-IPN would be identified as *net*-polyurethane-*s-inter*-poly(vinyl chloride).

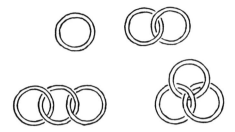

FIGURE 1. Idealized view of topological catenanes.

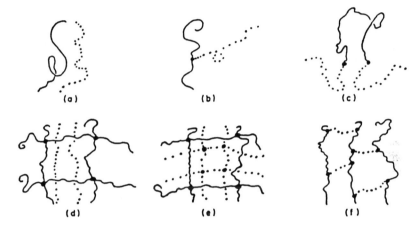

FIGURE 2. Simple two-polymer combinations illustrate (a) polymer blend, (b) graft copolymer, (c) block copolymer, (d) semi-IPN, (e) full-IPN, and (f) A-B type crosslinked polymer.

Commercialization of IPNs is fast becoming reality, and is often limited by the ability to process them cost effectively. As yet, no general aspect covering IPN processing has been mentioned. Processing, in general, is similar to other cross-linked systems with respect to final products, such as coatings. However, it is important to realize that sequential IPNs pose additional requirements. Sequential IPN synthesis is slower than simultaneous formation due to the time it takes for swelling equilibrium to occur. Also, sequential synthesis usually requires an additional curing cycle, adding additional expense. There are currently about 30 commercial materials specifically designated as IPNs, although countless others exist, especially in the rubber industry. The purpose of this paper is to give an introduction of IPN technology and exemplify some of the latest studies covering structure property relationships, miscibility and phase behavior, and IPN synthesis. Interested readers are directed to earlier review papers, including IPN synthesis prior to 1984,[7] sequential IPN morphology,[3] IPNs as composite matrices,[8] and others.[9-16]

II. MISCIBILITY AND PHASE BEHAVIOR

Immiscibility in polymeric systems arises from dissimilar molecular structures or lack of positive interactions (enthalpy contribution) and their inherently high molecular weight (entropy contribution). In mechanical blending of high polymers, the contribution of entropy of mixing to the overall free energy is low. Small molecules occupy only a single lattice site, Figure 4, and are free to move to any unoccupied lattice site (top). Polymers, however, occupy numerous contiguous lattice sites (bottom) and random movement is limited. That

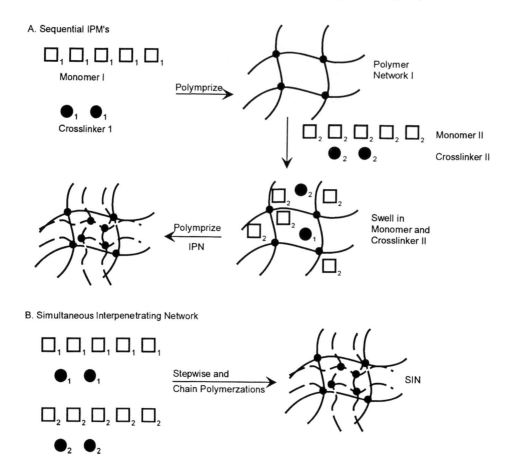

FIGURE 3. Idealized two-dimensional view of sequential IPNs (A) and (B) simultaneous IPNs.

TABLE 1
Classification of IPNs

Category	Definition
Full IPN	Material containing two or more network polymers in which there are no chemical crosslinks between the dissimilar polymers
Semi-I IPN	A sequential IPN in which polymer I (first formed) is a network and polymer II is linear
Semi-II IPN	A sequential IPN in which polymer I is linear and polymer II is a network
Pseudo-IPN	Any simultaneous IPN in which one polymer component is linear and the other is a network
Latex IPN	Sometimes called interpenetrating elastomeric networks, IEN; IPNs prepared in latex form
Gradient	Materials which possess a macroscopic concentration gradient due to the degree of diffusion of polymer II into the host matrix
Cross-intercross	IPNs possessing intentional grafts between the dissimilar polymers
Thermoplastic	Also called interpenetrating polymer blends, IPB; true thermoplastics containing physical crosslinks that prevent flow at normal operating temperatures
Sequential	Sequential polymerization of monomers, swelling network I in monomer II, then network polymerizing
Simultaneous	Simultaneous polymerization of monomers I and II by way of noninterfering modes
In situ	Sequentially prepared IPNs in which polymerization of polymer I is in solution or mixture of monomer II

TABLE 2
Proposed Prefix and Suffix
Nomenclature[6]

Identifier	Description
cyclo-	Cyclic
star-	Starlike
net-	Network
m-net-	Micronetwork
-blend-	Polymer blend
-inter-	IPN
-s-inter-	Semi-IPN
-compl-	Interpolymer complex

From L. H. Sperling, *Polym. Mat. Sci. Eng.*, 65, 80 (1991). With permission.

is, a single polymer repeat unit cannot move without coordinated movement of surrounding mers. Entropy is also reduced because the polymer chains cannot cross and their individual mer units cannot exchange lattice sites. In IPN formation, mixing is affected with low molecular weight species, then crosslinking is carried out rapidly. This type of kinetic control was initially believed to circumvent thermodynamic equilibrium and result in homogeneity, thus, leading to the term "interpenetrating polymer networks". Today, it is known that most IPNs are heterogeneous, possessing phase sizes in the range of hundreds of Angstroms and larger.

In any multicomponent polymer system, the end-use properties are dependent on the equilibrium phase state as well as chemical composition. This is true, especially when mixing thermodynamically immiscible polymers such as used in many IPNs. The degree of phase separation is characterized by the size and distribution of the domains which develop simultaneously with polymerization. Microphase separation in IPNs has been studied extensively using small angle X-ray scattering (SAXS),[17-20] small angle neutron scattering (SANS),[21-22] scanning electron microscopy (SEM),[23-24] and transmission electron microscopy (TEM).[25-26] Lipatov suggests, in an extensive review of the literature,[20] that IPNs may exist in a state of "forced" miscibility as well as forced microphase separation that may be characterized quantitatively with only the degree of segregation. This parameter can be found using electron microscopy data or viscoelastic properties.[27] In an idealized diagram of loss tangent vs. temperature, Figure 5, the degrees of component segregation are shown to vary from fully separated (top) in which each phase is characterized by its own maximum loss tangent (glass transition) to compatibility on the molecular scale (bottom) and the formation of a single phase. Further, the degree of segregation is quantified between $\alpha = 1$ (complete separation) and $\alpha = 0$ (complete mixing). It is suggested that the absolute value of tan δ maximum is more sensitive to structural change than a shift in the value with respect to temperature. An empirical derivation of the degree of segregation is given by:

$$a = [h_1 + h_2 - (l_1 h_1 + l_2 h_2 + l_m h_m)/L]/(h_1^o + h_2^o) \tag{1}$$

where h_1 and h_2 are the values of mechanical loss tan δ for each component at different degrees of segregation, h_1^o and h_2^o are the values of the pure components (or at complete phase separation), l_m are the shifts of the maxima with respect to temperature, and L is the difference between the maximums.

The effect of polydimethylsiloxane (PDMS) cross-link density and volume fraction of polydeuterostyrene (PDS) on PDMS/PDS sequential IPN morphology was studied[28] using

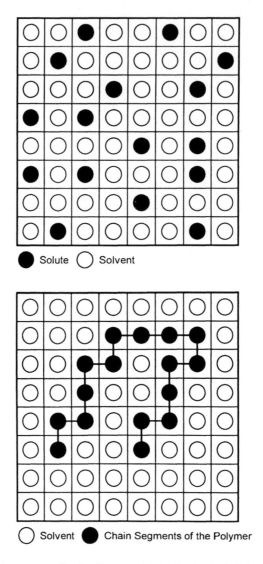

FIGURE 4. Lattice site theory — small molecules occupying single sites (top). Polymer occupying contiguous lattice sites (bottom).

SANS, low angle light scattering, membrane osmometry, and gel permeation chromatography. The domain sizes were characterized and theories for predicting domain sizes were compared. The IPNs were prepared by swelling network PDMS in a deuterostyrene solution containing 0.5% w/v azobisisobutyronitirle (AIBN) and 0.5% w/v divinylbenzene (DVB) in darkness to prevent photoinitiated polymerization from occurring before the desired swelling equilibrium was reached. PDS regions were 20 nm in size at low PDS content to over 50 nm at PDS levels greater than 50% v/v. The variations in domain size as a function of PDS concentration were shown to correlate with predicted values using Yeo's theory[25] while Donatelli's theory[26] failed to predict the quest polymer dimensions. No correlation was found between cross-link density of PDMS and the domain size of PDS.

The kinetics of phase separation in PU/polystyrene (PS) *in situ* semi-I IPNs were investigated by light transmission loss (turbidity) and Fourier transform infrared spectroscopy (FTIR).[29] The degree of styrene polymerization with respect to the degree of PU conversion

tan δ

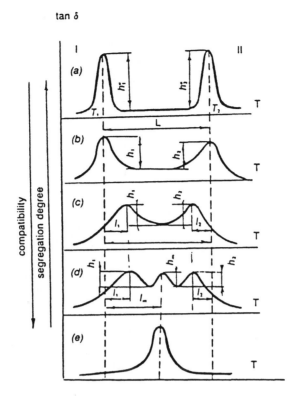

FIGURE 5. Temperature dependence of mechanical loss tan at various degrees of component segregation.

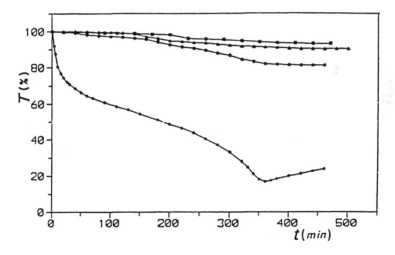

FIGURE 6. Light transmission curves for semi-I PU/PS IPNs (30/70) at 70°C with different t_{Tam}: 6 h. (■); 2 h (▲); 15 min. (●); and 5 min. (*).

was investigated. Polyurethane was based on aliphatic polyisocyanate and poly(oxypropylene) glycol of 4000 MW. All monomers and their respective catalysts were mixed together at room temperature. The degree of PU polymerization was varied by varying the time before the mixture was heated to 70°C, causing AIBN initiation of the styrene monomer. The time at which urethane was allowed to remain at room temperature (t_{Tam}) was varied from 5 to 120 min, Figure 6. The degree of polystyrene (PS) conversion was shown to be high and

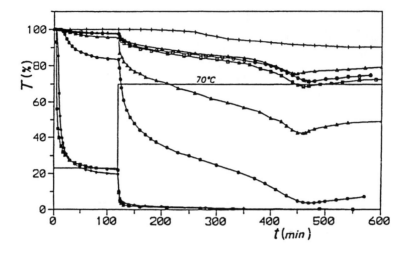

FIGURE 7. Light transmission curves for semi-I PU/PS IPNs (30/70) with 0.5% PS of differing Mn: 2000 ($\triangle$); 4900 ($\circ$); 22,000 ($\square$); 27,500 ($\blacktriangle$); 40,300 ($\bullet$); 52,000 (*); 10,500 ($\blacksquare$), and without PS ($+$).

independent of PU conversion, although the rate was decreased for $t_{\text{Tam}} = 120$ min. The corresponding light transmission behavior, Figure 7, suggests that phase separation occurred only when PU was not allowed to fully cure. The PU gel time was approximately 20 min. This suggests that the network formation of PU inhibits phase separation by not allowing the system to equilibrate. Also demonstrated was the effect of PS molecular weight. As PS molecular weight increased, phase separation occurred earlier in the reaction and with higher magnitude.

Stein et al.[30] recently reviewed phase separation processes in PS/polyvinyl methyl ether (PVME) semi-IPNs. The effect of molecular weight on the reduction of entropy of mixing and the resultant decrease in phase mixing was discussed. Thus, the lower critical solution temperature (LCST) of a blend such as PS/PVME should decrease as the molecular weight of either component is increased. Also, network formation reduces the entropy of mixing and phase mixing should thus decrease, as proven by Binder both theoretically[31] and experimentally.[32-33] However, experimental data do show that introducing crosslinks can impede phase separation, and is the primary reason for utilizing IPN techniques to control morphology. In general, this suggests that morphology and phase behavior is as much a function of kinetics as thermodynamics. The authors also assert that so-called molecular composites comprised of crystalline structures dispersed in flexible chain matrices cannot be prepared from solution. Upon evaporation of the solvent, equilibrium phase separation occurs in most cases and dispersion is not affected on the molecular scale.

III. BIOCOMPATIBLE IPNs

A series of semi-IPNs and semi-IPNs containing some intentional grafts were prepared for use as biocompatible and biodegradable materials. When poly(ϵ-caprolactone) glycol (PCL) was incorporated into poly(2-hydroxy ethyl methacrylate) hydrogel (PHEMA) the mechanical properties were enhanced[34-35] and natural tissue growth in implanted specimens occurred. The PHEMA hydrogel is biocompatible and readily accepted in the body while the PCL segments are biodegradable, leaving voids in the implant through which natural tissue can grow and intertwine.[36]

Grafting PCL to the PHEMA network was done by first reacting PCL (2000 mol wt) with itaconic anhydride according to Figure 8. To form either grafted or nongrafted semi-

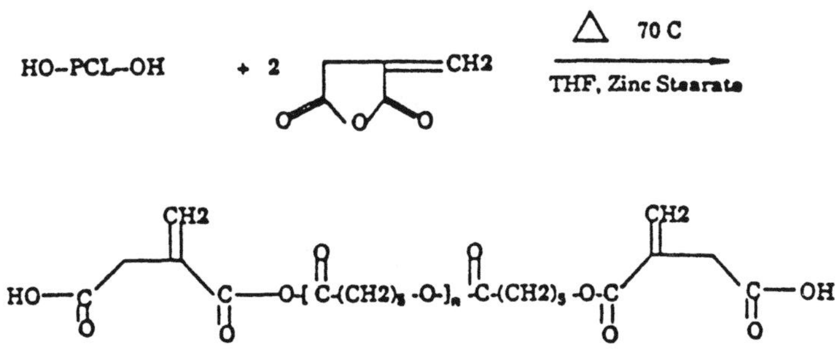

FIGURE 8. End-capping PCL with itaconic anhydride.

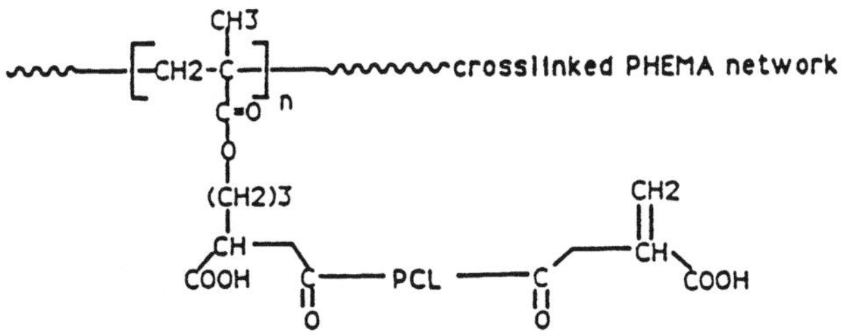

FIGURE 9. PCL grafted as a pendant group of PHEMA.

IPNs (Figure 9), the vinyl end-capped PCL or PCL diol, was dissolved in HEMA monomer at 70°C with addition of 1% ethylene dimethacrylate (EDMA) cross-linking agent and 1% benzoyl peroxide initiator. The solutions were cured 1 h at 90°C between Mylar®-coated glass plates. Pure PHEMA was clear both in solution and in the cured state, whereas PHEMA/PCL diol was turbid. The semi-IPNs containing a few grafts were semi-turbid, suggesting higher compatibility than nongrafted semi-IPNs. No data on the biodegradation efficiency of the PCL segments in either the grafted or nongrafted forms was given.

IPN synthesis was used as an alternative route to the preparation of polymer-supported reagents.[37] Polymer-supported reagents are highly selective to targeted ions and are used to simplify organic/biomolecular reactions and separations. Alexandratos and Grady[37-39] have applied IPN technology to form dual mechanism bifunctional polymers (DMBP) that possess two different types of ligands which enhance the degree of ionic and molecular selectivity. Two classes of DMBPs can be formed. Class I contains ion exchange/redox resins and class II contains ion exchange/coordination resins. The DMBPs were prepared from lightly cross-linked (0.6% DVB) PS beads. The beads were exposed to toluene solution of basic monomer, such as 4-vinylpyridine or N-vinylimidazole, and an acidic monomer such as ethyl acrylate or methyl methacrylate as well as a free-radical initiator. Class II DMBPs were produced when the former was fully polymerized then extracted and hydrolyzed to generate the acid sites.

Initial studies utilized IPNs that were synthesized with N-vinylimidazole (VIm) and adsorbed onto a PS bead matrix. After extraction, conditioning with NaOH/H_2O/HCl/H_2O resulted in the imidazole ligands being complexed with HCl. Table 3 gives the percent metal complexed by the ion exchange/coordination resins for IPNs prepared in a 1:1 molar ratio with MMA, acrylamide (AM), and styrene (Sty) along with VIm and MMA-supported agents as controls. The ligands' ability to complex metal ions was shown to be a function of its microenvironment. Further studies of the ligands' ion recognition with respect to the performance of the different acid/base monomers employed, their respective ratios, and on the degree of crosslinking is underway.

TABLE 3
Percent Metal Ion Complexed by Ion
Exchange/Coordination IPNs

	VIm	VIm/MMA	VIm/AM	VIm/acid	VIm/Sty	MMA
%Cu	93.3	95.5	96.3	73.8	52.9	79.9
%Co	27.9	62.1	55.0	87.9	7.9	79.5

IV. GRADIENT IPNs

IPNs that possess varying compositions as a function of position within the material are known as gradient IPNs. These materials are valued for their gradient properties and found their first application as an integral polymeric-constrained, layer damping system. Gradient IPNs are prepared sequentially by swelling a host network polymer in monomer to a controlled degree. Diffusion of monomer into the host matrix is terminated prior to equilibrium and then polymerization of monomer II is carried out. The desired properties of these IPNs are, therefore, dependent on the usual parameters such as composition, monomer type, and degree of phase mixing as well as the macroscopic degree of diffusion of monomer II into the host matrix. Alkoveli et al.[40] first studied two series of gradient IPNs: PS with a gradient of polyacrylonitrile and poly(methyl methacrylate) with a gradient of poly(methyl acrylate). The rate of acrylonitrile diffusion into the host polymer was measured and compared to theory. The diffusion rate was found to be dependent on monomer concentration in solvent, monomer type, and compatibility with the host matrix and the cross-linking agent, among others. Recent applications include a patented molecular composite prepared from a soft viscoelastic core constrained with a rigid epoxy used to dampen vibrations.[41]

Dror et al.[42] prepared a series of gradient IPNs based on a polyether urethane — urea as one phase with acrylamide, 2-oxyethyl methacrylate, or N-vinylpyrrolidone for use in biomedical applications as blood compatible hydrogels.

Lipatov et al.[43] prepared thermally stable polyamide fibers with improved tensile properties via gradient interpenetration using a phenol-formaldehyde oligomer. Two levels of heterogeneity were reported to exist: the first was attributed to the composition gradient (macroheterogeneity) and the second to microphase separation (microheterogeneity), common to all IPNs. Again, the overall performance was dependent on the diffusion gradient and the distribution of phases.

Lipatov and Karabanova[44] recently studied optical gradient elements of gradens based on IPNs. Three steps are necessary to prepare gradens. First, a prepolymer matrix is formed from a bifunctional monomer exhibiting a high refractive index and possessing a cylindrical rod form. The prepolymer mixture, (80% monomer) is then immersed with another monomer of lower refractive index, wherein diffusion occurs. Diffusion is interrupted when a parabolic profile of the composition has been formed within the prepolymer rod mixture, then polymerization is carried out. The resultant materials are transparent when based on miscible polymer components and act as light-focusing elements. A comparison was made of gradient IPNs to classical IPNs using PU as the host network and butyl methacrylate-triethyleneglycol dimethacrylate copolymer as either the guest network or gradient network. The ordinary IPNs were shown to be heterogeneous, Figure 10, while cross-sectional areas taken of the gradient IPNs were shown to be homogeneous in the range of 10 to 60% copolymer, Figure 11.

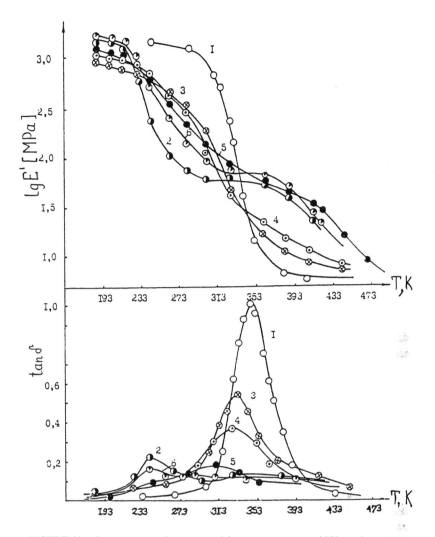

FIGURE 10. Loss tangent and storage modulus vs. temperature of PU/copolymer IPNs.

V. SYNTHESIS AND STRUCTURE

Polyorganophosphazine polymers are characterized by their alternating P and N atoms, forming flexible chains. Visscher et al.[45] chose to prepare IPNs using polyorganophosphazines as one component due to the broad range of available pendant-side groups. Polyorganophosphazines can be prepared via ring opening polymerization as shown in Figure 12. Six components were chosen to form the second matrix via sequential synthesis, Figure 13. The IPNs were characterized by FTIR, differential scanning calorimetry (DSC), electron microscopy, and two NMR isotope techniques. Polymerization was monitored by following the loss of protons over time, while DSC was used to evaluate the degree of phase mixing. The materials ranged from soft and pliable to hard and brittle depending on the constituent monomers and their composition. Applications of these new materials are being investigated.

The phase behavior of IPNs as a function of grafting for *in situ* polymerized semi-II IPNs was investigated. Methyl methacrylate macromonomer was reacted with various reactive endgroups, Table 4, then grafted to DVB-crosslinked PS *in situ*.[46-47] The macromonomer reactive endgroups were chosen because they exhibited different grafting efficiencies

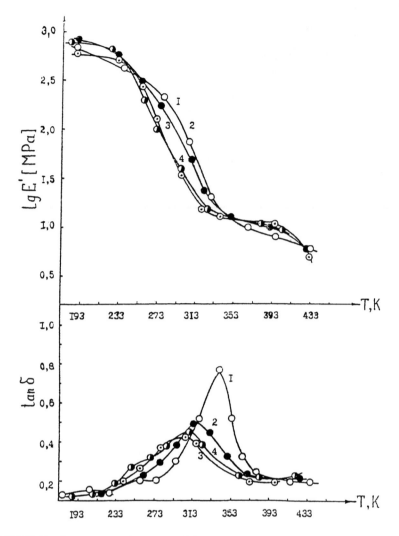

FIGURE 11.   Loss tangent and storage modulus vs. temperature of PU/copolymer gradient IPNs.

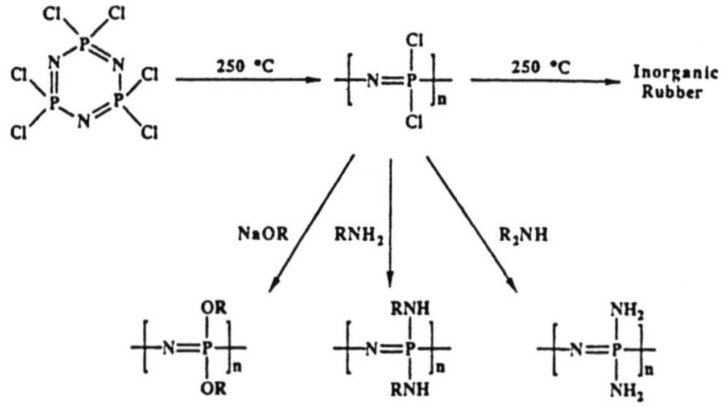

FIGURE 12.   Ring opening polymerization to prepare polyorganophosphazines.

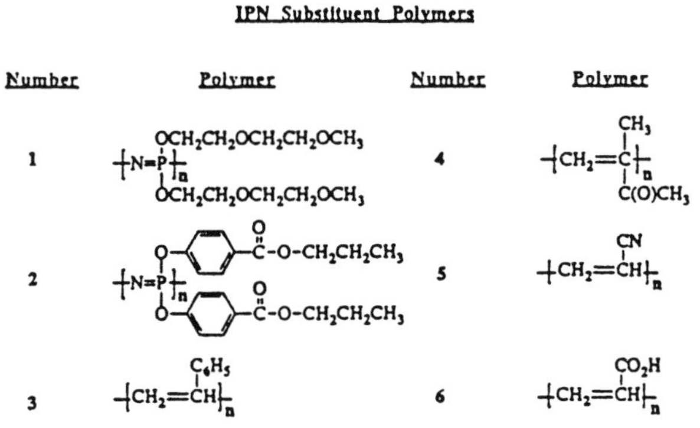

IPN Substituent Polymers

Number	Polymer	Number	Polymer

FIGURE 13. Polyorganophosphazine IPN substituent polymers.

TABLE 4
Macromonomer Characterization

Endgroup	MW	Mn	% Functionalized
Alkacrylate	7200	4200	>90
Methacrylate	4600	3800	96
Acrylate	7100	4600	95
α-Methyl styrene	6400	4300	>90
None	7900	3600	0

and, thereby, were used to control the amount of grafting. All samples were bulk polymerized 18.5 h at 60°C followed by 3 h at 110°C using 0.2 wt% AIBN as initiator and 1 wt% DVB as a cross-linking agent. The degree of grafting was characterized by GPC analysis of the toluene extract. No grafting attributable to chain transfer or other unwanted mechanisms was observed, suggesting that grafting occurred via the reactive endgroups only. Thermomechanical analysis (TMA) and SAXS were used to characterize the extracted specimens. Figure 14 shows the scattering intensity vs. scattering vector obtained from SAXS, whose shapes and relative intensity reveal information on the phase mixing. In general, it was shown that increasing the number of grafts resulted in increased phase mixing and additional control of morphology. Table 5 gives the relative scattering intensity for each of the reactive endgroups, the percent monomer reacted, and the optical clarity of the resultant IPNs.

IPNs containing PDMS as one component have been studied relatively few times.[48-53] He et al. have used the IPN approach to take advantage of PDMS properties such as low surface energy,[54] high gas permeability,[55] and biocompatibility[56] while overcoming poor mechanical properties.

In situ polymerization requires that all components be mixed simultaneously, but network formation occurs sequentially. A mixture of α,ω-dihydroxy polydimethylsilane, tetraethylorthosilicate, stannous octoate, methyl methacrylate, trimethylolpropane trimethacrylate, 4-*tert*-butylpyrocatechol (TBPC), and AIBN was poured into a mold and held at room temperature, allowing a PDMS network to form. Methyl methacrylate monomer was then polymerized by heat-initiated dissociation of AIBN. It was found that the PDMS catalytic of stannous octoate was inhibited by the formation of radicals, Figure 15, and incomplete gelation occurred. Addition of 1 wt% TBPC, a radical inhibitor commonly used as a monomer stabilizer, resulted in increased reaction rates and higher degrees of polymerization, Figure 16. The IPNs possessed improved mechanical properties over PDMS while retaining important surface characteristics. The light transmission properties, Table 6, demonstrate, in

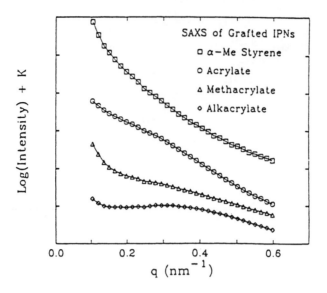

FIGURE 14.   SAXS log(scattering intensity) vs. vector q for PMMA/PS IPNs.

TABLE 5
SAXS and Extraction Results of Grafted
PS/PMMA IPNs

Endgroup	SAXS	% MM reacted	Optical
Alkacrylate	4	90	Clear
Methacrylate	3	72	Clear
Acrylate	2	82	Clear
α-Methyl styrene	1	36	Cloudy
None	∞	0	Opaque

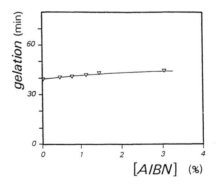

FIGURE 15.   Gelation of PDMS as a function of AIBN concentration.

this case, that the *in situ* prepared IPNs possess a higher degree of phase mixing than grafted IPNs of similar composition.

VI. ENERGY ABSORBING IPNs

The ability of materials to dampen vibrations is characterized by the storage modulus (E′) and the loss modulus (E″), which are the quantities of energy stored through elasticity

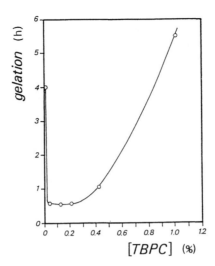

FIGURE 16. Gelation of PDMS as a function of TBPC concentration.

TABLE 6
Percentage of Light Transmission of
Full and Grafted IPNs

PDMS/PMMA	Mn = 4200		Mn = 860	
	Full	Graft	Full	Graft
5/95	0	0.1	0.5	88.1
10/90	0	23.7	0.6	85.1
20/80	0.4	56.0	41.7	85.7
30/70	0.9	45.0	49.5	84.6
40/60	0.4	35.0	41.0	79.5
50/50	0.4	25.0	39.0	81.0
60/40	0.4	41.0	28.0	83.0
70/30	0.4	46.0	18.0	91.5

and energy lost through conversion to heat, respectively. The conversion of vibrational energy to heat energy is greatest in the polymers' viscoelastic region, Figure 17, which is the transition between the hard, glassy state to the soft, rubbery state. The loss tangent (tan δ) is another useful term used to quantify dampening performance. The expressions are as follows:

$$E^* = E' + iE'' \tag{2}$$

$$\tan \delta = E''/E' \tag{3}$$

$$H = \pi E'' \epsilon_o^2 \tag{4}$$

where H is actually the quantity of heat gained by the material for each cycle and ϵ_o^2 represents the magnitude of deformation.

IPNs can be particularly efficient energy absorbers due to the ability to control their morphology. Broad and high glass transition regions, easily prepared using IPN technology, relate directly to damping over broad temperature and frequency ranges.

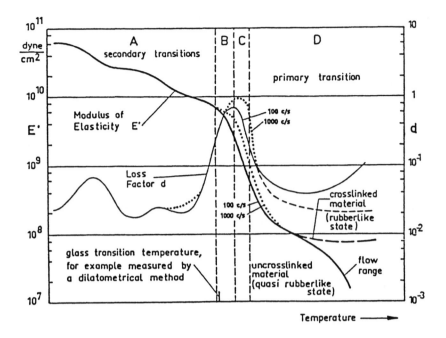

FIGURE 17. Modulus of elasticity and loss factor of viscoelastic material as a function of temperature.

Huelck et al.[58] first reported latex IPNs based on polyethylacrylate and polymethyl-methacrylate that possessed tan δ values that stayed high over a broad temperature range. Further work led to quantifying dampening performance by expressing it as the area under the tan δ vs. temperature curve.[59] Recently, the performance has been better characterized by the area under the linear loss modulus vs. temperature, LA, which is given by:

$$LA = E'' \, dT = (E'_G - E'R)R/(E_a)_{avg} \, [\pi/2] \, T_g^2 \tag{5}$$

where E'_G and E'_R are the storage moduli in the glassy state and rubbery state, respectively, T_G and T_R are the glassy and rubbery temperatures just below and above the T_g, $(E_a)_{avg}$ is the average activation energy of the relaxation process, and R is the gas constant. Also, by assuming that the total loss area is a combination of the weight additive contribution of the individual structural repeat units, Figure 18, the loss area can be given by:

$$LA = \sum (LA)_i \, M_i/M = \sum G_i/M \tag{6}$$

where M_i is the molecular weight of the ith group in the mer, M is the molecular weight of the whole mer, G_i is the molar loss constant for the ith group, $(LA)_i$ is the loss area contributed by the ith group, and n represents the number of units in the mer. The molar loss constants and contribution to the loss area have been reported for a significant number of structural groups.[60]

A series of pseudo-IPNs based on network PU and linear PVC were prepared and the morphology and phase behavior was characterized by dynamic mechanical spectroscopy (DMS).[61] The PU was based on a mixture of two polyols: PCL and polyoxypropylene diol (PPG). PCL is known to be fully miscible in all proportions with PVC and when it is included into polyurethanes as the soft segment, the PU/PVC alloys, thus, formed are homogeneous. PPG diol is immiscible with PVC and results in phase separation between PU and PVC. To obtain pseudo-IPNs with microheterogeneous morphologies, the polyurethanes were based

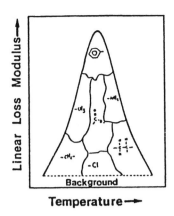

FIGURE 18. The idea of group contribution analysis for dynamic mechanical spectroscopy. Storage modulus vs. temperature.

on a mixture of these two polyols. The microheterogeneous morphology resulted in a broadened T_g and frequency dampening range, Figure 19. The materials were annealed up to 200°C for various lengths of time to induce phase separation and determine the phase behavior. IPNs based on 100% PCL exhibited a nearly linear composition dependence on T_g and did not undergo phase separation; therefore, no LCST could be found. Intermediate compositions based on 80% PCL and 20% PPG exhibited broader transitions that did phase separate upon annealing. This is shown for a single composition in Figure 20. These results suggest a co-continuous microheterogeneous morphology in which the intermediate compositions lay near the phase boundaries, ideal for dampening applications. The phase diagram, derived from the annealing studies, Figure 21, shows the LCST to exist at 120°C.

IPN semiflexible foams based on cross-intercross PU and polyvinylester hybrid were prepared and evaluated to determine the viscoelastic response (damping curves) and its relationship to low frequency sound absorption properties. The effect of cell structure, cell orientation, density, and foam thickness were demonstrated as well.[62] The foams were prepared via the one-shot, free-rise method at room temperature using a high speed, constant torque mixer. Three isocyanate/polyol combinations were used to prepare PU foams as controls and later to be used as one component in the IPNs: toluene diisocyanate (TDI)/Pluracol® 726; TDI/Pluracol® 1003; and 4,4′-diphenylmethane diisocyanate (MDI)/Pluracol® 1003. Pluracol® 726 is a high molecular weight polyether type and Pluracol® 1003 is a high molecular weight grafted polyether type used to introduce stiffness. The measured sound absorption coefficients (SAC) show, Figure 22, that a PU based on Pluracol® 1003 exhibits better low frequency (100 to 500 Hz) sound absorption properties although DMS data of the same samples suggest very little differences in the morphology, Figure 23. This behavior suggests that other parameters besides morphology may have a great effect on the sound absorption properties. This is in contrast to dampening, which is directly related to the viscoelastic nature of the material. This behavior has been exhibited in foams such as polyimides that exhibit very high SAC and no T_g in the pertinent temperature range.[63] The SAC of all the IPNs was similar, and is shown only for those based on TDI and Pluracol® 726. In general, the IPNs exhibited maximum SAC at PU/PVE = 90/10, with pure PU and PU/PVE = 80/20 exhibiting nearly equal values, Figure 24. This behavior may be due to phase separation. In the IPN composition PU/PVE = 90/10 a microheterogeneous morphology probably exists, although there is too little PVE component to result in formation of a separate phase. At a composition of PU/PVE = 80/20, phase separation does result, Figure 25, and the values are similar to pure PU.

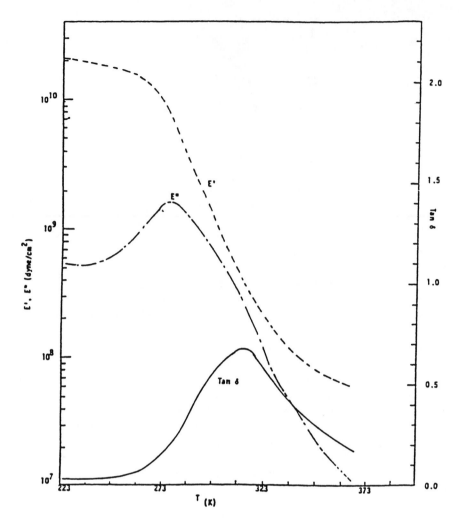

FIGURE 19. Dynamic mechanical spectrum of a pseudo-IPN composed of 50% PU (80% PCL/20% PPG) and 50% PVC.

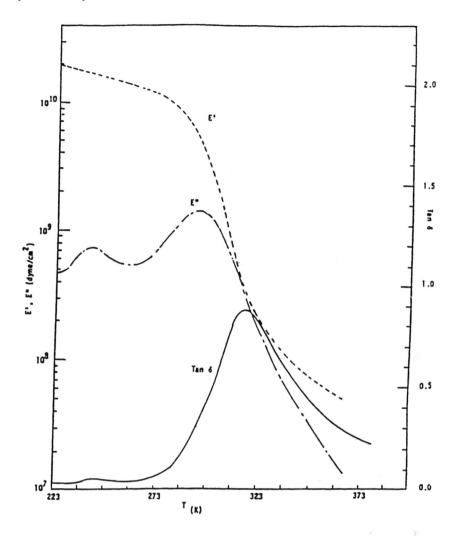

FIGURE 20. Dynamic mechanical spectrum of pseudo-IPN composed of 50% PU (80% PCL/20% PPG) and 50% PVC, annealed 15 min at 22°C.

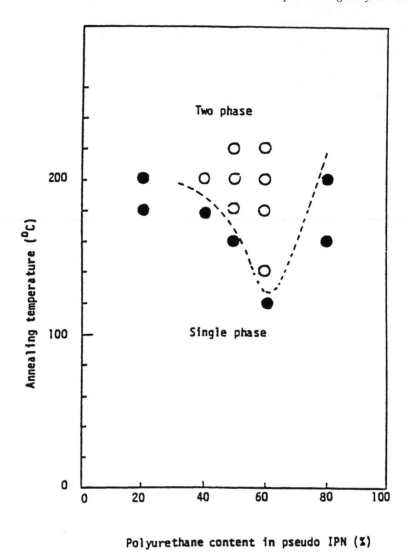

FIGURE 21.   Phase diagram of pseudo-IPNs based on PU (80% PCL/20% PPG) and PVC.

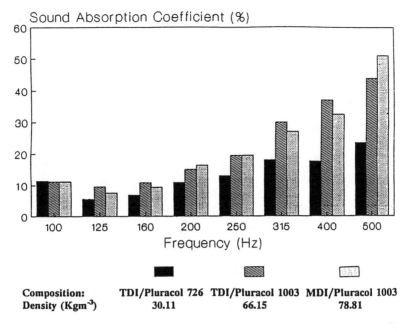

FIGURE 22. Effect of chemical structure on SACS of PU foams.

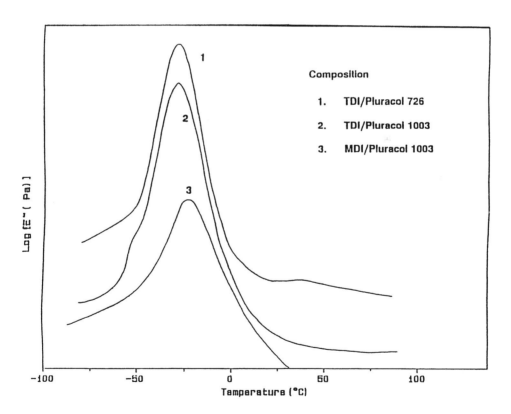

FIGURE 23. Loss modulus curves of PU forms.

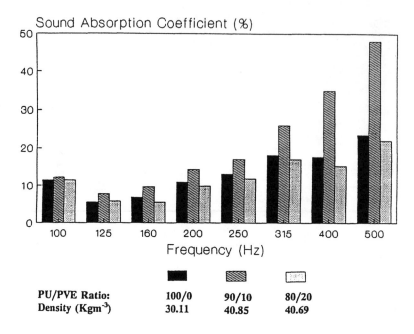

FIGURE 24. SACS of TDI/Pluracol® 726-based PU and PU/PVE IPN foams.

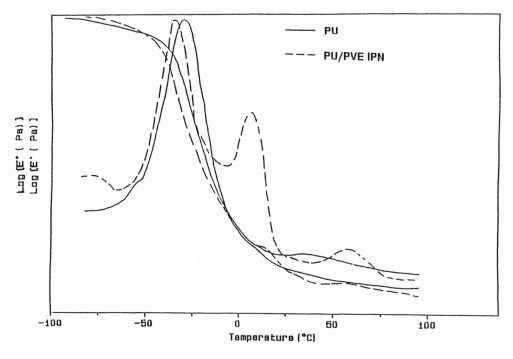

FIGURE 25. Comparison of loss modulus and storage modulus of TDI/Pluracol® 726 based PU and PU/PVE IPNs foams.

REFERENCES

1. **Sperling, L. H.,** *Interpenetrating Polymer Networks and Related Materials,* Plenum, New York, 1981.
2. **Frisch, H. L. and Wasserman, E.,** *J. Am. Chem. Soc.,* 83, 3789 (1961).
3. **An, J. H. and Sperling, L. H.,** *ACS Symp. Ser.,* 367, 269 (1988).
4. **Sperling, L. H., Ferguson, K. B., Manson, J. A., Corwin, E. M., and Siegfried, D. L.** *Macromolecules,* 9, 743 (1976).
5. **Sperling, L. H.,** *Source-Based Nomenclature for Polymer Blends, IPNs and Related Materials,* Division of Polymer Chemistry Nomenclature Committee Document, 1984.
6. **Sperling, L. H.,** *Polym. Mat. Sci. Eng.,* 65, 80 (1991).
7. **Klempner, D. and Berkowski, L.,** in *Encyclopedia of Polymer Science,* Vol. 8, 1984.
8. **Klempner, D. and Sophiea, D.,** in *Encyclopdedia of Composites,* Vol. 3, 1990.
9. **Xiao, H. X. and Frisch, K. C.,** *J. Coat. Technol.,* 61, 51 (1989).
10. **Sperling, L. H., Fay, J. J., Murphy, C. J., and Thomas, D. A.,** *Makromol. Chem., Macromol. Microsymp.,* 38, 99 (1990).
11. **Sperling, L. H.,** *Contemp. Top. Polym. Sci.,* 6, 665 (1989).
12. **Frisch, H. L.,** *Mater. Res. Soc. Symp. Proc.,* 1, 231 (1990).
13. **Tabka, M. T., Widmaeir, J. M., and Meyer, G. C.,** *Makromol. Chem., Macromol. Symp.,* 30, 31 (1989).
14. **Sperling, L. H.,** *Polym. Mat. Sci. Eng.,* 59, 150 (1988).
15. **Frisch, H. L.,** *New Polym. Mat., Proc. Int. Symp.,* 88 (1987).
16. **Chang, M. C. O., Thomas, D. A., and Sperling, L. H.,** *J. Polym. Sci., Part B: Polym. Phys.,* 26(8), 1627 (1988).
17. **Shilov, V. V., Karabanova, L. V., and Lipatov, Y. S.,** *Vysokomolek. Soed.,* A20, 643 (1978).
18. **Lipatov, Y. S., Shilov, V. V., Goza, Y., and Kruglyak, N.,** *X-Ray Methods of Polymer Systems,* Kiev, Nauk: Dumka, 1972.
19. **Lipatov, Y. S., Shilov, V., and Bogdonovich, V.,** *J. Polym. Sci. Polym. Phys.,* 25, 43 (1987).
20. **Lipatov, Y. S.,** in *Advances in Interpenetrating Polymer Networks,* Vol. I, D. Klempner and K. C. Frisch, Eds., Technomic, Lancaster, PA, 1991.
21. **An, J. H., Fernandez, A. M., and Sperling, L. H.,** *Macromolecules,* 20, 191 (1987).
22. **McGary, B. and Richards, R. W.,** *Polymer,* 27, 1315 (1986).
23. **Widmaier, J. M. and Sperling, L. H.,** *Macromolecules,* 15, 625 (1982).
24. **Lipatov, Y. S.,** *Pure Appl. Chem.,* 57(11), 1691 (1985).
25. **Yeo, J. K., Sperling, L. H., and Thomas, D. A.,** *Polym. Eng. Sci.,* 22(3), 190 (1982).
26. **Donatelli, A. A., Sperling, L. H. and Thomas, D. A.,** *J. Appl. Polym. Sci.,* 21, 1189 (1977).
27. **Lipatov, Y. S.,** *Pure Appl. Chem.,* 57, 1691 (1985).
28. **McGary, B.,** in *Advances in Interpenetrating Polymer Networks,* Vol. I, D. Klempner and K. C. Frisch, Eds., Technomic, Lancaster, PA, 1989.
29. **Jin, S. R. and Meyer, G. C.,** *Polymer,* 27, 592 (1986).
30. **Stein, R. S., Hanyu, A., Sethmudhaven, M., and Gaudiana, R.,** *Polym. Mat. Sci. Eng.,* 65, 46 (1991).
31. **Binder, K.,** *J. Chem. Phys.,* 79, 6387 (1983).
32. **Briber, R. M. and Bauer, B. J.,** *Macromolecules,* 21, 3296 (1988).
33. **Bauer, B. J., Briber, R. M., and Han, C. C.,** *Macromolecules,* 22, 940 (1989).
34. **Davis, P. A., Nicolais, L., and Haung, S. J.,** *J. Bio. Comp. Poly.,* 3 (1988).
35. **Jarrett, P., Benedict, C. V., Bell, J. P., Cameron, J. A., and Haung, S. J.,** *Polymers as Biomaterials,* S. W. Shalaby and A. S. Hoffman Eds., Plenum, New York, 1981, 181.
36. **Eschbach, F. O. and Haung, S. J.,** *Polym. Mat. Sci. Eng.,* 65, 9 (1991).
37. **Alexandratos, S. D. and Strand, M. A.,** *Macromolecules,* 19, 273 (1986).
38. **Alexandratos, S. D.,** *Sep. Purif. Methods,* 17, 67 (1988).
39. **Alexandratos, S. D. and Grady, C. E.,** *Polym. Mat. Sci. Eng.,* 65, 7 (1991).
40. **Alkoveli, G., Biliyar, K., and Shen, M.,** *J. Appl. Polym. Sci.,* 20, 2419 (1976).
41. **Sperling, L. H. and Thomas, D. A.,** U.S. Patent No. 3,833,404, 1974.
42. **Dror, M., Elsabee, M. Z., and Berry, G. C.,** *Biomat. Med. Dev. Artif. Org.,* 7 (1979).
43. **Lipatov, Y. S., Sergeeva, L. M., Novikova, O. A., and Glukhovskaja, I. I.,** *Khim. Volokna,* 4, 14 (1983).
44. **Lipatov, Y. S. and Karabanova, L. V.,** in *Advances in Interpenetrating Polymer Networks,* D. Klempner and K. C. Frisch, Eds., Technomic, Lancaster, PA, 1992.
45. **Visscher, K. B., Manners, I., and Allcock, H. R.,** *Macromolecules,* 22, 4885 (1990).
46. **Bauer, B. J. and Briber, R. M.,** *Polym. Prepr.,* 31, 578 (1990).
47. **Bauer, B. J. and Briber, R. M.,** *Polym. Mat. Sci. Eng.,* 65, 5 (1991).
48. **Sperling, L. H. and Sarge, H. D.,** *J. Appl. Polym. Sci.,* 16, 3041 (1972).

49. **Ebdon, J. R., Hourston, D. J., and Klein, P. G.,** *Polymer,* 25, 1633 (1984).

50. **McGary, B. and Richard, R. W.,** *Polymer,* 27, 1315 (1986).

51. **Klein, P. G., Ebdon, J. R., and Hourston, D. J.,** *Polymer,* 29, 1079 (1988).

52. **Frisch, H. L., Gebreyes, K., and Frisch, K. C.,** *J. Polym. Sci., Part A: Polym. Chem.,* 26, 2589 (1988).

53. **Xiao, H., Ping, Z. H., Xie, J. W., and Yu, T. Y.,** *J. Polym. Sci., Part A: Polym. Chem.,* 28, 585 (1990).

54. **He, X. W., Widmaier, J. M., Herz, J. E., and Meyer, G. C.,** in *Advances in Interpenetrating Polymer Networks,* Vol. IV, D. Klempner and K. C. Frisch, Eds., Technomic, Lancaster, PA, 1992.

55. **Gaines,G. L.,** *Macromolecules,* 14, 208 (1981).

56. **LeDuc, L., Blanchard, L. P., and Malhotra, S. L.,** *J. Macromol. Sci. Chem.,* A14, 389 (1980).

57. **Okkema, A. Z., Fabrizius, D. J., Grasel, T. G., Cooper, S. L., and Zdrahala, R. J.,** *Biomaterials,* 10, 23 (1989).

58. **Huelck, V., Thomas, D. A., and Sperling, L. H.,** *Macromolecules,* 5, 348 (1972).

59. **Keskkula, H., Turnley, S. G., and Boyer, R. F.,** *J. Appl. Polym. Sci.,* 15, 351 (1971).

60. **Fay, J. J., Murphy, C. J., Thomas, D. A., and Sperling, L. H.,** in *Sound and Vibration Damping With Polymers;* R. D. Corsaro and L. H. Sperling, Eds., American Chemical Society, Washington, D.C., 1990.

61. **Omoto, M., Kusters, J., Sophiea, D., Klempner, D., and Frisch, K. C.,** *Polym. Mat. Sci. Eng.,* 56, 459 (1988).

62. **Klempner, D., Sophiea, D., Suthar, B., Frisch, K. C., and Sendijarevic, V.,** *Polym. Mat. Sci. Eng.,* 65, 82 (1991).

63. **Lee, D. K., Chen, L., Sendijarevic, V., Sendijarevic, A., Frisch, K. C., and Klempner, D.,** *J. Cell. Plast.,* 27, 135 (1991).

Chapter 9

CATALYSTS FOR THE COPOLYMERIZATION OF ETHYLENE-PROPYLENE ELASTOMERS

Kenya Makino

TABLE OF CONTENTS

0-8493-4401-8/93/$0.00 + $.50

445

I. INTRODUCTION

Ethylene/propylene copolymers and ethylene/propylene/nonconjugated diene terpolymers are industrially important synthetic elastomers. The outstanding property of the copolymer and the terpolymer are their good weather-resistance compared with polybutadiene (BR), polyisoprene (IR), and styrene/butadiene copolymer (SBR), since they have no double bonds in the backbone of the polymer chains and, thus, are less sensitive to oxygen and ozone. Other excellent properties of these rubbers are high resistance to acid and alkali, electrical properties, and high and low temperature performance. The copolymer and the terpolymer are used in the automotive industry for hoses, gaskets, wipers, bumpers, belts, etc. Furthermore, it is used for cable insulation and for roofing. Production of these olefinic copolymer and terpolymer rubbers have been increased with increasing car production. In 1989, about 620,000 mt of the copolymer and terpolymer rubbers were produced all over the world.[1] The elastomer is produced by vanadium catalysts such as soluble $VOCl_3$ or VCl_4 and $AlEt_{1.5}Cl_{1.5}$ or $AlEt_2Cl$ as a co-catalyst. It is well known that the catalyst system was first reported by Natta et al. and provided a highly random copolymer of ethylene and propylene.[2] A number of papers have been reported regarding copolymerization of ethylene with propylene by using a hydrocarbon soluble vanadium catalyst (e.g., VCl_4, $VOCl_3$, $V(acac)_3$, $VO(OR)_3$, $VOCl(OR)_2$, $VOCl_2(OR)$, etc.) together with an organoaluminium compound.[3-8] Although the catalysts provide a highly random copolymer of ethylene with propylene, their catalytic activities markedly decrease with temperature. This phenomenon is drastic, particularly at higher temperatures.[9]

With respect to titanium catalysts for polymerization of ethylene or propylene, Ziegler synthesized the first high-density polyethylene and Natta prepared isotactic polypropylene by means of coordination catalysts about 30 years ago. The preparation methods of catalysts have been studied extensively.[10-13] These $TiCl_3$ catalysts have very high activity for homopolymerization of ethylene and propylene,[14,15] whereas, they exhibit low activity for random copolymerization of ethylene with propylene when compared to vanadium catalysts.[16]

As mentioned above, copolymerization and terpolymerization by using hydrocarbon soluble vanadium catalysts have been reviewed in detail.[17-19] Recently, some papers have been written about new titanium catalysts, zirconium compound/methylaluminoxane catalysts, and chromium catalysts which give random copolymer of ethylene with propylene.

In this chapter, I would like to review for new catalysts of the copolymerization of ethylene/propylene and the terpolymerization of ethylene/propylene/nonconjugated diene, which have been reported during the last 10 years.

II. VANADIUM CATALYST

Perchloroacetate and perchlorocrotonate improve catalytic activity for copolymerization of ethylene with propylene.[20,21] Milani et al.[21] investigated the preparation of various types of vanadium complexes with β-diketones, β,δ-triketones, ketophenols, and copolymerization of ethylene with propylene by the vanadium catalyst and activating agents such as butylperchlorocrotonate or ethyltrichloroacetate. The copolymerization was carried out in a liquid monomer system and in a heptane solution system. Attempted correlations of catalytic activity with the coordination geometry, coordination number of vanadium, and steric and electric effect of the ligands did not give regular trends. Neither improvement on the features of the copolymer produced nor on the catalyst activity was found over the commonly known vanadium catalysts.

Milani et al.[22] reported in another paper on the effect of tributyltin hydride on the various vanadium-based homogeneous catalytic systems with activating agents for the ethylene/ propylene/diene terpolymerization as shown in Table 1.[21] The activating agents used were

TABLE 1
Results of Ethylene/Propylene/ENB Terpolymerization[a]

Run no.	Vanadium compound	Tin compound	$[Al(C_2H_5)_3]$ (mmol/l)	$[Al(C_2H_5)_2Cl]$ (mmol/l)	Productivity[b] (wt%)	Propylene content (wt %)	ENB content (wt %)
1	$V(Acac)_3$	—	—	4.71	23.3	58	1.5
2	$V(Acac)_3$	—	2.36	2.36	14.0	45	2.5
3	$VO(OC_2H_5)_3$	—	2.36	2.36	14.0	49	2.1
4	$V(Acac)_3$	$(C_4H_9)_3SnH^c$	—	4.71	28.3	59	1.2
5	$V(Acac)_3$	$(C_4H_9)_3SnCl^d$	—	4.71	23.6	58	1.4
6	$V(Acac)_3$	$(D_4H_9)_3SnH$	2.36	2.36	24.0	41	1.4
7	$V(Acac)_3$	$(C_4H_9)_3SnCl$	2.36	2.36	23.7	57	2.0
8	$VO(OC_2H_5)_3$	$(C_4H_9)_3SnH$	2.36	2.36	22.7	46	1.6

[a] Conditions: temperature, 20°C; total pressure, 1 atm: (vanadium compound), 0.0589 mmol/l; concentration of chlorinated activator (butyl perchlorocrotonate), 7.07 (mEq · Cl)/l; ENB, 5-ethylidene-2-norbornene.
[b] Numerical values in gram of polymer per gram of vanadium per hour.
[c] 1.77 mmol/l.
[d] 2.36 mmol/l.

TABLE 2
Results of Polymerization of Olefins with $V(mmh)_3$ and Alkylaluminium Compounds[a]

Run no.	Co-catalyst	Monomer conc. (mol/dm³) (C_2H_4)	(C_3H_6)	Time (min)	Polymer yield (kg/ g-V)	Mn^b ($\times 10^{-4}$)	MW/Mn	Propylene content[c] (mol %)
1	$Al(C_2H_5)_2Cl$	0	8.3	180	0.62	2.98	1.22	100
2	$Al(C_2H_5)_2Cl$	0.3	8.3	40	6.5	102	1.22	40
3	$Al_2(C_2H_5)_3Cl_3$	0	8.3	180	2.7	9.51	1.55	100
4	$Al_2(C_2H_5)_3Cl_3$	0.3	8.3	10	4.5	38.2	1.62	43

[a] Polymerization conditions: amount of vanadium compound = 0.01 mmol, amount of alkylaluminium compound = 5.0 mmol, volume of toluene solution = 100 cm³, and temperature = −60°C.
[b] Number-average molecular weight equivalent of polypropylene.
[c] Determined from ¹³C-NMR.

butyl perchlorocrotonate, ethyl trichloroacetate, trichloromethylbenzene, and 1,1,2,2-tetrachloro-1,2-diphenylethane. Productivity improvements were always observed. The results are interpreted on the basis of the enhanced oxidation of the vanadium(II) species by radical intermediates resulting from the reaction between tin (IV) hydrides and the chlorinated organic activator. They said that extremely short-lived radical species involved in the hydrostannolysis reaction of the chlorinated activator, probably plays a central role as demonstrated by the enhancement of the reaction rate constant when the oxidation of V(II) to V(III) by aralkyl halides is carried out in the presence of tributyltin hydride.

Doi et al.[23] found that the soluble catalyst consisting of *trans*(2-methyl-1,3-butanedionato) vanadium, $V(mmh)_3$, and $Al(C_2H_5)_2Cl$ in toluene shows a high activity for the living polymerization of propylene and gives almost monodisperse polypropylene at temperatures below −40°C. In the catalytic system, all of the vanadium species function as active centers for the propagation of living polypropylene.[24] The same vanadium-based catalyst was used for the synthesis of a living copolymer of ethylene with propylene.[24] The copolymer obtained with $Al(C_2H_5)_2Cl$ shows a very high molecular weight (Mn = 10⁶) and narrow molecular weight distribution (MW/Mn = 1.22), indicating that a living copolymerization of ethylene with propylene take place at −60°C as shown in Table 2. In the presence of $Al_2(C_2H_5)_3Cl_3$,

TABLE 3
Constants for Ethylene and Propylene
Copolymerization with Heterogeneous
Ziegler Catalysts

Catalyst	r_1 (ethylene)	r_2 (propylene)	$r_1 \cdot r_2$
$TiCl_4$	4.2	0.14	0.59
α-$TiCl_3 \cdot 0.3AlCl_3$	4.9	0.25	1.23
$TiCl_4$/MgO (0.27 Ti)	7.8	0.13	1.01
$TiCl_4$/Al_2O_3,SiO_2(3.71 Ti)	18.5	0.24	4.44

[a]   Co-catalyst, $Al(C_2H_5)_3$; solvent, heptane, 0.1 l; temperature, 70°C.

however, the produced polypropylene and copolymer show relatively broad molecular weight distributions (MW/Mn = 1.5 to 1.6). It is most likely that a slow chain terminating reaction takes place in the presence of $Al_2(C_2H_5)_3Cl_3$.

III. TITANIUM CATALYST

A. SOLID OR SUPPORTED TITANIUM CATALYST

Baulin et al.[25,26] reported that the infuence of the chemical nature and structural properties of various supports on the efficiency of Ziegler supported catalytic systems for ethylene polymerization and copolymerization with propylene was studied. The fixation of the titanium-containing component of a $TiCl_4$-$Al(C_2H_5)_3$ catalytic system on MgO supports increase the reactivity of its active species in the elementary propagation step of polyethylene macrochains, in contrast to alumosilicate-supported and unsupported systems. The results show that Mg containing supports are not only inert ''supports'' used for the increase in the area of distribution of the titanium component, but are active promoters of Ziegler catalytic systems. Table 3 shows the copolymerization constants by the Fineman-Ross graphical method. The propagating species on the surface of aluminosilicate are more selective on ethylene than those fixed on MgO. At adequate copolymerization conditions, the MgO supported catalyst permits the production of copolymer more enriched propylene than that supported by alumosilicate.

Baulin et al.[27] also investigated the influence of temperature on the kinetic aspects of the copolymerization of ethylene with propylene on the system $TiCl_4$/MgO-$Al(C_2H_5)_3$. A rise in temperature in the reaction zone leads to a marked reduction in the copolymer yield. If the temperature rises from 30 to 70°C, the copolymer yield is reduced practically by one half over the entire range of variation in the comonomer ratio (propylene content 5 to 95 mol%). Table 4 shows the copolymerization constant measured at different temperatures.

Soga et al.[28] presented a paper showing that copolymerization of ethylene with propylene was conducted over the thermally reduced γ-Al_2O_3 supported $TiCl_4$ catalyst, both in the absence or presence of $Al(C_2H_5)_3$ or $Al(C_2H_5)_2Cl$. The structure of the polymer drastically changed from a random copolymer to a polyethylene with an increase in the concentration of $Al(C_2H_5)_3$.

$$\text{>Al–O–TiCl}_3 \xrightarrow{\text{800 °C, 4h}} \text{>Al–O–TiCl}_2$$

[1]   [2]

TABLE 4

Constants for Ethylene and Propylene
Copolymerization on Catalyst System
TiCl$_4$/MgO-Al(C$_2$H$_5$)$_3$ vs. Temperature[a]

Temp. (°C)	r$_1$ (ethylene)	r$_2$ (propylene)	r$_1$ · r$_2$
30	4.7	0.28	1.32
50	6.3	0.21	1.30
70	7.8	0.13	1.01
70	7.6	0.12	0.91

[a] Solvent — hexane, 100 ml; total concentration of comonomers in reaction zone, 0.4 mol/l; propylene content 10 mol%; TiCl$_4$/MgO = 0.3–1.0 g/l, Al(C$_2$H$_5$)$_3$ = 0.2–0.3 g/l, molar ratio of Al/Ti = 60/1 to 140/1.

TABLE 5

Results of Copolymerization of Ethylene with Propylene[a]

Run no.	Ethylene (l)	Propylene (l)	Ti-supported (matom/g-cat.)	Polymer yield (g/matom-Ti · h)	Propylene content (mol%)
1	5.0	0	0.36	9.2	—
2	3.5	3.5	0.36	9.6	15
3	2.0	5.0	0.36	5.7	24
4	1.0	6.0	0.36	3.7	38
5[b]	1.0	6.0	0.36	18	34
6	0.5	6.5	0.36	1.4	51
7	0	7.0	0.36	0.74	—

[a] Copolymerization was conducted at 65°C for 15 min over the thermally reduced (evacuated at 800°C for 4 h(γ-Al$_2$O$_3$ supported TiCl$_4$ catalyst by using 5 ml of hexane in 100 ml stainless steel reactor.
[b] 0.4 mmol of Al(C$_2$H$_5$)$_2$Cl was added.

Structure [1] was completely reduced to Structure [2] when the catalyst was treated with Al(C$_2$H$_5$)$_2$Cl (10 mmol/l, heptane) at 75°C. The reduction of [2] did not proceed at all under the condition. The copolymerization was conducted at various molar ratios of ethylene/ propylene by the thermally reduced γ-Al$_2$O$_3$ supported TiCl$_4$ catalyst in the absence of a co-catalyst. The results are listed in Table 5. On the other hand, when the catalyst was treated with Al(C$_2$H$_5$)$_3$ (10 mmol/l, heptane) at 75°C, [1] was rapidly reduced to give the mixture of [2] and further reduced titanium species (Ti^{2+} and Ti^0). While increasing the concentration of Al(C$_2$H$_5$)$_3$, the activity for propylene polymerization markedly decreased, whereas, that for ethylene polymerization slightly increased. These results strongly suggest that only the Ti^{3+} species is active for propylene polymerization, while further reduced titanium species (probably Ti^{2+}) as well as the Ti^{3+} species are active for ethylene poly-merization. From these results, they concluded that it is most important for the production of a random copolymer of ethylene and propylene to prevent the catalyst from over-reduction.

Soga et al.[29,30] also reported the copolymerization by using an SiO$_2$ supported MgCl$_2$/ TiCl$_3$ catalyst. The catalyst was prepared by treating the mixture of TiCl$_3$ · 3C$_5$H$_5$N · MgCl$_2$ · n(THF), and SiO$_2$ with Al(C$_2$H$_5$)$_2$Cl in heptane solvent. From the electron spin resonance (ESR) spectrum of the catalytic system, the titaniums supported on SiO$_2$ were assigned as

TABLE 6
Reactivity Ratios r_1, r_2, and the Product $r_1 \cdot r_2$ in the
Copolymerization of Ethylene and Propylene with
Titanium-Based Catalysts

Catalyst system	Polymer temp. (°C)	r_1	r_2	$r_1 \cdot r_2$
$TiCl_4/MgCl_2/SiO_2-Al(C_2H_5)_3$	40	7.5	0.26	1.9
$TiCl_3-Al(C_2H_5)Cl_2$	60	6.8	0.5	3.5
$TiCl_3-Al(C_2H_5)_3$	40	7.3	0.76	5.5

$Ti^{3+}Cl_3$ and $Ti^{3+}(C_2H_5)Cl_2$. The $TiCl_3$ supported was highly dispersed on SiO_2 and the Ti^{3+} species in the catalyst were very stable even in the presence of $Al(C_2H_5)_3$. Copolymerization of ethylene with propylene was carried out at 65°C for 2 h by using the catalyst combined with $Al(C_2H_5)_3$. The number of the average molecular weight of the copolymer obtained at propylene/ethylene = 1 was about 29,000 with a Q value of 6.8. The copolymers obtained were found to be random or moderately alternating from IR and ^{13}C-NMR spectra. It should be noted that the content of propylene in the copolymer obtained with the present catalyst is much higher when compared with those produced with the conventional Ziegler-Natta catalysts. The conventional Ziegler-Natta catalysts seem to contain an appreciable amount of the Ti^{2+} species in addition to the Ti^{3+} species. The marked difference in the copolymer composition may result from such a difference in the catalytic composition. Monomer reactivity ratios of r_1 and r_2 were measured from the ^{13}C-NMR spectra of the copolymer,[30] the results are shown in Table 6. The r_1 values hardly vary among the titanium catalysts, whereas, the r_2 value of SiO_2 supported $MgCl_2/TiCl_3$ catalyst is apparently smaller than the r_2 values of the $TiCl_3$ based catalysts. The value of the reactivity ratio product $r_1 \cdot r_2 = 1.9$ of the SiO_2 supported $MgCl_2/TiCl_3$ catalyst decreases toward unity as expected for an ideal random copolymerization.

Kashiwa et al.[31] reported copolymerization of ethylene with α-olefins (i.e., propylene, butene-1, 4-methylpentene-1) by a $MgCl_2$ supported $TiCl_4$ catalyst in combination with $Al(C_2H_5)_3$ at temperatures as high as 170°C. The copolymerization system was homogeneous. This catalytic system showed very high activity. Copolymers obtained have a density of 0.91 to 0.94 g/ml and have an α-olefin content of 2 to 7 mol%. In the other paper,[32] they reported ethylene/propylene/5-ethylidene-2-norbornene terpolymerization with a highly active titanium catalyst system ($MgCl_2/TiCl_4/C_6H_5COOC_2H_5-Al(i-Bu)_3/di-iso$-amyl ether) as shown in Table 7. The obtained terpolymer could be vulcanized with sulfur, but the increase of torque value through the vulcanization measured by a curelastometer for the terpolymer obtained with this catalyst was considerably lower than that for the terpolymer obtained with the conventional $VOCl_3$ catalyst system (Table 8). They thought that lower torque value would be attributed to the heterogeneous diene distribution in the obtained terpolymer, particularly lower ethylidene norbornene (ENB) content in the high molecular weight fraction.

Kakugo et al.[33] presented ethylene/propylene copolymer prepared with various heterogeneous titanium-based Zieger-Natta catalysts, i.e., δ-$TiCl_3$, β-$TiCl_3$, and $MgCl_2$ supported titanium catalysts. They fractionated the copolymers by successive solvent extraction. Wide composition distributions were observed for all samples. Composition distributions of some samples were investigated precisely by temperature-programmed elution column fractionation. The fractionation data showed that these copolymers are a mixture of polyethylene and of copolymers with different structures, i.e., random and block copolymers. In every sample, abundant random copolymers were found, the propylene sequences being present only as

TABLE 7

Ethylene/Propylene/ENB Terpolymerization with MgCl$_2$/TiCl$_4$/C$_6$H$_5$COOC$_2$H$_5$-Al(i-Bu)$_3$/(di-iso-Amyl Ether) and VOCl$_3$-Al(C$_2$H$_5$)$_{1.5}$Cl$_{1.5}$ Catalyst System

| Run no. | Conditions[a] | | | | Terpolymers | | | |
	Ti or V (mmol/l)	Ethylene/propylene in feed (mol/mol)	ENB in feed (ml/l)	Activity[b]	Ethylene content in terpolymer	ML[c]	IV[d]	ENB conversion (%)
T-1	0.03	40/60	15	1.5	61.3	27	11.1	17.7
T-2	0.03	50/50	30	1.1	72.3	43	16.7	16.3
V-1	0.5	25/75	2	0.086	63.3	62	8.7	98.3
V-2	0.5	40/60	2	0.133	59.9	32	5.6	98.6

[a] Terpolymerization conditions: T-1 and T-2 — molar ratio Al/Ti, 50; temperature, 60°C; solvent, toluene. V-1 and V-2 — Al, 5 mmol/l; V, 0.5 mmol/l, ENB, 2 ml/l; temperature, 20°C; solvent, hexane.

[b] Terpolymer yield in kg per mmol of Ti or V.

[c] ML — ML$_{(1+4, 100°C)}$.

[d] IV-iodine value.

TABLE 8
Cure Data for the Obtained Terpolymer

Terpolymer	ML^a	IV^b	T_{15} (min)	T_{90} (min)	F_{min} (kg · cm)	F_{max} (kg · cm)	ΔF (kg · cm)	Cure rate (kg · cm/min)
T-1	27	11.1	7.25	11.5	0.95	13.9	13.0	2.45
T-2	43	16.7	6.00	8.25	0.81	16.3	15.5	5.43
V-1	63	8.7	4.25	11.0	1.76	29	27	3.22
V-2	32	5.6	6.00	15.5	0.88	23	22	1.82

[a] $ML - ML_{(1+4,100°C)}$.
[b] IV — iodine value.

TABLE 9
Results of Ethylene/Propylene Copolymerization[a]

Catalytic system	Activity (mol/g-Ti · h)	Propylene in copolymer (mol%)
$TiCl_4-Mg(C_6H_{13})_2$	16.8	36
$TiCl_4-Mg(C_2H_5)(C_4H_9)$	6.9	25
$TiCl_4-Mg(C_4H_9)_2 · 2Al(C_2H_5)_3$	1.3	13
$TiCl_4-Al(C_2H_5)_3$	3.3	10

[a] Copolymerization was carried out at 40°C for 1 h at an initial
    total pressure of 400 torr (ethylene = 152 torr, propylene =
    248 torr) using 0.01 mmol of $TiCl_4$ and 0.15 mmol of co-catalyst
    in 10 ml of heptane.

pentad (*mmmm*). These data suggest that a random copolymer is formed on an isospecific
site.

Corbelli et al.[34] reported ethylene/propylene copolymers with highly active catalysts.
They said that the copolymers were prepared at an industrial plant. Other members of Dutral
S.p.A. reported the characterization of the copolymers.[35-37]

B. SOLUBLE TITANIUM CATALYST

Soga et al.[38] reported the copolymerization of ethylene with propylene with the titanium
catalyst system in which $TiCl_4$ and di-*n*-hexylmagnesium were contacted *in situ*. The catalytic
system was found to be highly active in giving a transparent copolymer of an alternating-
like distribution ($r_1 · r_2 = 0.39$). Copolymerization results were shown in Table 9. Both the
composition and the microstructure of the copolymers depend strongly upon the co-catalyst
used. $Mg(C_6H_{13})_2$ gave an alternating-like copolymer in which the inversion of propylene
units was negligible. In contrast, $Al(C_2H_5)_3$ gave a block-type copolymer containing long
methylene chains. Such a marked difference in the copolymer structure may result from the
difference in the reducing power of the co-catalysts, i.e., MgR_2 gives the Ti(III) species,
while $Al(C_2H_5)_3$ gives a considerable amount of the Ti(II) species, together with the Ti(III)
catalysts.

Kashiwa et al.[39] reported that copolymerization of ethylene with propylene was carried
out by using two-catalyst systems. One (Cat-a) was a solid $MgCl_2$ containing Ti treated with
ethyl benzoate together with $Al(C_2H_5)_3$, and the other catalyst (Cat-c) was based on a
homogeneous mixture of $MgCl_2$ dissolved in 2-ethylhexanol/decane and $TiCl_4$, which was
treated with $Al(C_2H_5)_2Cl$. Both catalyst systems exhibited very high activity in comparison
with the conventional catalysts of $TiCl_3-Al(C_2H_5)_2Cl$(Cat-b) and $VOCl_3-Al(C_2H_5)_2Cl$

TABLE 10
Reactivity Ratio for Various Catalyst Systems[a]

Catalyst system	r_1	r_2	$r_1 \cdot r_2$
Cat-a MgCl$_2$/TiCl$_4$/EB-Al(C$_2$H$_5$)$_3$/EB	5.5	0.36	2.0
Cat-b TiCl$_3$-Al(C$_2$H$_5$)Cl	11.6	0.35	4.1
Cat-c MgCl$_2$/TiCl$_4$/EHA-Al(C$_2$H$_5$)$_2$Cl	6.0	0.02	0.13
Cat-d VOCl$_3$-Al(C$_2$H$_5$)$_2$Cl	21.1[b]	0.018[b]	0.22[b]

[a] Copolymerization conditions: Al, 2.4 mmol/l; Ti, 0.04 mmol/l, temperature, 90°C; time, 30 min; ethylene/propylene feed ratio (in solution), varied from 4/96 to 58/42 mol/mol.

[b] From the data of Cozewith, C. and Ver Strate, G., *Macromolecules*, 10, 773, 1977.

(Cat-d). These two new catalyst systems are different from each other in microstructure of the product copolymer, i.e., the latter catalyst (Cat-c) brings about more random distribution of the monomeric units and also brings about the decreased regiospecificity concerning the arrangement of the propylene unit. The value of the reactivity ratios (r_1 and r_2) for both catalyst systems were calculated by the Fineman-Ross method. It is of interest to note in Table 10 that the values of $r_1 \cdot r_2$ for the solid Ti-based catalyst systems (Cat-a and -b) are higher than 1.0, whereas, those for the catalyst systems derived from the homogeneous titanium and vanadium catalyst components (Cat-c and -d) are lower than 1.0. Thus, the distribution of the ethylene and propylene monomeric units is more random with Cat-c and -d in comparison with the cases of Cat-a and -b.

Makino et al.[40-45] recently reported some papers for the copolymerization of ethylene/propylene and terpolymerization of ethylene/propylene/5-ethylidene-2-norbornene by using a homogeneous titanium catalyst together with AlR$_3$ as a co-catalyst.

The solubilization of TiCl$_3$ catalysts in organic solvents was tried to enhance the activity for random copolymerization of ethylene with propylene.[40] β-TiCl$_3$ · 1/3AlCl$_3$ was solubilized in halogenated hydrocarbon solvents by using the appropriate ether as a donor. The results were shown in Table 11. α-TiCl$_3$ free from AlCl$_3$ was found to be soluble in the presence of the VCl$_4$ or VOCl$_3$ ligand and a suitable ether.

Soluble TiCl$_3$ complex catalysts were prepared by reduction of TiCl$_4$ with hydrogen in the presence of Pd-carbon as a reducing catalyst and ether as a donor in a halogenated hydrocarbon solvent and copolymerization of ethylene with propylene was studied by soluble TiCl$_3$ catalysts prepared by reduction of TiCl$_4$ with hydrogen.[41,42] The extent of reduction of TiCl$_4$ markedly affected the catalytic activity for the copolymerization as shown in Table 12. Metal halides (MX$_n$) such as VCl$_4$ and VOCl$_3$ enhanced the catalytic activity of soluble TiCl$_3$ catalyst for the copolymerization.

Reductions of TiCl$_4$ by organoaluminium compounds were carried out in hydrocarbon and halogenated hydrocarbon solvents by using various ethers as donors.[43] Under appropriate conditions, TiCl$_4$ was quantitatively reduced by trialkylaluminium to give a homogeneous black solution of TiCl$_3$ · 1/3AlCl$_3$ · ether complex. Catalyst systems of the soluble TiCl$_3$ and Al(*i*-Bu)$_3$ (co-catalyst) showed high catalytic activities for random copolymerization of ethylene with propylene as shown in Table 13.

Preparations of souble TiCl$_3$ catalysts by reduction of TiCl$_4$ with some types of Gringnard reagents were carried out in halogenated hydrocarbon solvents by using appropriate ethers as donors.[44] The soluble TiCl$_3$ · MgX$_2$ · ether complex catalysts and triisobutylaluminium as co-catalyst showed high activities for the copolymerization of ethylene with propylene as shown in Table 14. It was found that the soluble TiCl$_3$ · MgX$_2$ · ether complex catalysts enhance the activities for the copolymerizations in the same manner as solid titanium catalysts

TABLE 11
Solubilization of β-TiCl₃ · 1/3AlCl₃ · Donor Catalysts and their Use for Copolymerization of Ethylene with Propylene

Run no.	Catalyst preparation[a]			Copolymerization[b]		
	Solvent	Donor	Homogeneity of β-TiCl₃ solution[c]	Polymer yield (g)	Propylene content (mol%)	Crystallinity[d] (%)
1	DCE	DEE	C. Sol.	2.1	24	1.1
2		DBE		3.3	34	0.7
3		DHE		3.3	32	0.7
4		DOE		3.6	33	0.7
5		DDE		3.4	27	0.8
6	CBz	DBE		3.6	31	1.0
7		DOE		3.3	31	1.1
8	Hexane	DOE	A. Sol.	2.0	23	1.8
9		DBE	Insol.	1.6	21	3.5

[a] Solvent, 50 ml; donor, 15 mmol; β-TiCl₃, 10 mmol; temperature, 25°C; time, 1 h.
[b] Solvent, hexane, 200 ml; soluble TiCl₃ catalyst (as Ti), 0.2 mmol; co-catalyst, Al(i-Bu)₃, 2 mmol; monomer (ethylene/propylene) gas feed, 1.0/1.5 (l/min); temperature, 30°C; time, 10 min.
[c] C. Sol. — completely soluble; A. Sol. — almost soluble; Insol. — insoluble.
[d] Crystallinity at 26.5 mol% of propylene content.

TABLE 12
Effect of the Extent of Reduction of TiCl₄ with Hydrogen and their Use for the Copolymerization

Run no.	Catalyst preparation[a]			Copolymerization[b]		
	Mole ratio (Pd/TiCl₄ × 10³)	Time (h)	Yield of TiCl₃ (%)	Polymer yield (g)	Propylene units (mol%)	Crystallinity (%)
1	—	—	0	0.5	9	4.2
2	0.23	2	48	1.6	16	1.5
3	0.23	4	77	2.2	20	1.0
4	0.94	2	95	2.7	22	0.8
5	2.35	2	95	2.7	22	0.7
6	4.70	2	95	2.8	25	0.7
7	9.40	2	96	2.6	26	0.8

[a] Amount of TiCl₄, 10 mmol; of DBE, 20 mmol; volume of DCE, 50 ml; flow rate of hydrogen, 0.2 l/min; temperature, 30°C; time, 2 h.
[b] Volume of hexane, 200 ml; amount of catalyst (as Ti), 0.2 mmol; amount of Al(i-Bu)₃, 2 mmol; flow rate of monomer gas (ethylene/propylene), 1.0/1.5 l/min; temperature, 30°C; time, 10 min.

supported on $MgCl_2$, which show high activities for homopolymerizations of olefin monomers.

$TiCl_4$ and $MgCl_2$ were dissolved in hydrocarbon or halogenated hydrocarbon solvents by using phosphate or phosphonate donors.[45] Mg(OH)Cl recated with $TiCl_4$ with HCl gas evolution above 100°C in the presence of phosphate or phosphonate donor, and a homogeneous solution of [$TiCl_3$-O-MgCl] complex was obtained.[45] The soluble catalysts and triisobutylaluminium as co-catalyst showed high activities for the copolymerization of ethylene with propylene as shown in Tables 15 and 16. The copolymers obtained were rubbery and possessed very low crystallinities.

Resulting copolymers mentioned above[40-45] were superior elastomers with low crystallinity and had outstanding high tensile strength and elongation at break in comparison with

TABLE 13
Influence of Organoaluminium Compounds on the Reduction of TiCl₄ and the Copolymerization by Resulting Catalyst Systems

Run no.	Al compound	Catalyst preparation[a]		Copolymerizatioin[b]		
		Al/Ti (mol/mol)	Yield of Ti³⁺(%)	Polymer yield (g)	Propylene units (mol%)	Crystallinity[c] (%)
1	Al(Et)₃	0.67	95	4.5	29	1.3
2	Al(*i*-Bu)₃	0.67	95	4.3	29	0.7
3	Al(*n*-Hex)₃	0.67	96	4.4	31	0.7
4	Al(*n*-Oct)₃	0.67	95	4.4	33	0.7
5	Al(*n*-Dod)₃	0.67	95	4.8	34	0.6
6	Al(Et)₂Cl	1.0	94	3.2	30	1.6
7	Al(Et)Cl₂	1.0	~0	0.8	9	3.9
8	None	—	—	0.5	9	4.2

[a] TiCl₄, 10 mmol; solvent, DCE (50 ml); donor, DOE (15 mmol); temperature, 3–5°C; time, 30 min.

[b] Solvent, hexane (200 ml); soluble TiCl₃, 0.2 mmol (as Ti); co-catalyst, Al(*i*-Bu)₃ (2 mmol); monomer gas feed (ethylene/propylene) = 1.0/1.5 l/min; temperature, 30°C; time, 20 min.

[c] Crystallinity at 26.5 mol% of propylene.

TABLE 14
Influence of Grignard Reagent on the Reduction of TiCl₄ and their Use for Copolymerization of Ethylene with Propylene

Run no.	RMgX	Catalyst preparation[a]		Copolymerization[b]		
		RMgX/TiCl₄ (mol/mol)	Yield of Ti³⁺ (%)	Polymer yield (g)	Propylene units (mol%)	Crystallinity[c] (%)
1	(*n*-Bu)MgCl	1.0	96	5.0	30	1.2
2	(*n*-Bu)MgBr	1.0	95	7.5	33	1.0
3	(*n*-Bu)MgI	0.5	47	3.9	22	1.5
4		0.75	73	6.2	30	1.0
5		1.0	95	8.2	34	0.5
6		1.25	91	7.8	34	1.0
7	EtMgI	1.0	97	6.1	20	1.5
8	(*n*-Oct)MgI	1.0	97	8.7	35	0.4

[a] Preparation conditions: solvent, DCE, 50 ml; TiCl₄, 10 mmol; DBE, 30 mmol; temperature, 0°C; time, 30 min.

[b] Copolymerization conditions: solvent, hexane, 200 ml; soluble TiCl₃ (as Ti), 0.2 mmol; Al(*i*-Bu)₃ as co-catalyst, 0.6 mmol; monomer gas feed (ethylene/propylene) = 1.5/1.0 l/min; temperature, 30°C; time, 10 min.

[c] Crystallinity at 26.5 mol% of propylene.

copolymers prepared by a conventional VOCl₃/AlEt₁.₅Cl₁.₅ catalyst system. Characterization data of copolymers suggest that the excellent tensile properties are due to a microblock type of sequences of ethylene and propylene in a polymer chain.

IV. CHROMIUM CATALYST

Soga et al.[46-48] investigated homo- and copolymerization of ethylene and propylene using a homogeneous and heterogeneous chromium catalysts. Cr(C₁₇H₃₅COO)₃/Al(C₂H₅)₂Cl gave a homogeneous catalyst system in toluene.[46] The catalytic system was found to be active only for ethylene polymerization as shown in Table 17. However, the catalytic system showed

TABLE 15

Results of Copolymerization of Ethylene with Propylene by the Soluble (TiCl$_4$ · MgCl$_2$) · Phosphine Compound/Al(*i*-Bu)$_3$ Catalyst System

| Run no. | Catalyst Preparation[a] | | Copolymerization[b] | | |
	Phosphine compound	Reduction temp. (°C)	Copolymer yield (g)	Propylene content (mol%)	Crystallinity[c] (%)
1	O=P(–O–Bu)$_3$	-10	8.0	29	0.3
2	O=P(–O–Oc)$_3$	-10	9.2	32	0.2
3	O=P(–O–Bu)$_2$(Bu)	-10	11.0	31	0.2
4	O=P(–O–Eh)$_2$(Eh)	10	11.2	29	0.2
5		0	11.7	32	0.1
6		-10	12.6	34	0.1
7		-30	13.2	37	0.1
8	None	-10	0.5	9	3.8

[a] Solubilization: MgCl$_2$, 1 g (10.5 mmol); hexane, 30 ml; phosphine compound, 42 mmol; temperature, 25°C; agitation time, 20 min; TiCl$_4$, 10.5 mmol; agitation time, 1 h.
Reduction conditions: hexane, 50 ml; soluble Ti catalyst, 0.2 mmol; Al(*i*-Bu)$_3$, 2 mmol; monomer gas (ethylene/propylene) = 0.1/0.4 l/min; aging temperature, $-10°C$; time, 1 min.

[b] Hexane, 200 ml; monomer gas (ethylene/propylene) = 1.0/2.0 l/min; temperature, 35°C, time, 10 min.

[c] Crystallinity at 26.5 mol% of propylene.

TABLE 16

Results of the Copolymerization by the Soluble (TiCl$_3$-O-MgCl) · Phosphine Compound/Al(*i*-Bu)$_3$

| Run no. | Catalyst Preparation[a] | | Copolymerization[b] | | | |
	Phosphine compound	Solvent	Copolymer yield (g)	Propylene content (mol%)	Termonomer content (I$_2$ no.)	Crystallinity (%)
9	O=P(–O–Eh)$_2$(Eh)	None	8.3	35	11	0.1
10		CBz	8.4	36	10	0.1
11	O=P(–O–Bu)$_3$		6.7	29	13	0.1
12	O=P(–O–Bu)$_2$(Bu)	Isooctane	7.9	31	13	~0
13			12.5	35	—	~0
14	O=P(–O–Eh)$_2$(Eh)		9.1	37	15	~0
15			12.9	39	—	0.1

[a] Mg(OH)Cl, 1 g (13 mmol); phosphine compound, 39 mmol; TiCl$_4$, 13 mmol; temperature, 130°C; time, 2 h. Reduction conditions are the same as Table 15.

[b] Conditions are the same as Table 15, except for addition of termonomer (hexane solution of 5-ethylidene-2-norbornene (4.8 vol%), pumping speed, 2 ml/min for 10 min).

some activity for ethylene/propylene copolymerization in toluene at 30°C, giving a random copolymer with a narrow molecular weight distribution.

Soga et al.[47,48] also investigated that homo- and copolymerization of ethylene and propylene was carried out using a heterogeneous Cr(CH$_3$COO)$_3$/Al(C$_2$H$_5$)$_2$Cl catalytic system in toluene at 40°C. The activity of copolymerization was not affected by varying the molar ratio of Al/Cr in the range of 5 to 10. The mol% of propylene in the monomer mixture affects the kinetics of copolymerization. The average rate decreases with increasing mol% of propylene. Random copolymers were obtained despite the catalytic system being heterogeneous.

TABLE 17

Results of Copolymerization of Ethylene and Propylene by the Soluble Chromium Catalyst[a]

Run no.	Ethylene (l)	Propylene (l)	Copolymer Yield (g polymer/g-Cr · h)	Propylene content (mol%)	MW	Mn	Q
1	2	0	504	—	—	—	—
2	2	4	177	10.5	30 000	14 600	2.0
3	2	6	137	11.0	22 500	12 000	1.9
4	2	8	65.4	18.3	16 500	9 200	1.8

[a] Copolymerization was conducted at 30°C in a 200 ml stainless steel reactor for 15 min with $Cr(C_{17}H_{35}COO)_3$ (100 mg) and $Al(C_2H_5)_2Cl$ (11.1 mmol, Al/Cr = 100) in toluene (20 ml).

V. ZIRCONIUM CATALYST

Since the biscyclopentadienyl zirconium compound and methylaluminoxane catalyst system for olefin polymerization was discovered by Kaminsky, there have been numerous reports on the homopolymerization of olefin.[49-52] Kaminsky et al.[53] investigated that ethylene/propylene/5-ethylidene-2-norbornene terpolymer could be prepared by means of a soluble Ziegler catalyst formed from biscycropentadienyl zirconium dimethyl and methylaluminoxane. The overall activities lie between 100 and 1000 kg EPDM/mol-Zr · h · bar. After an induction period (0.5 to 5 h) the polymerization rates increased and then leveled to a value which was constant for several days. From copolymerization kinetics reactivity ratios, r_{12} = 31.5, r_{21} = 5×10^{-3}, and r_{13} = 3.1 could be derived, and by ^{13}C-NMR spectroscopy $r_{12} \cdot r_{21}$ = 0.3 was found (1:ethylene, 2:propylene, and 3:5-ethylidene-2-norbornene). The regiospecificity of the catalyst toward propylene leads exclusively to the formation of head-to-tail enchainments. The diene polymerizes via vinyl polymerization of the cyclic double bond, and the tendency to branching is low. Molecular weights were estimated between 40,000 and 160,000. The average molecular weight distribution of 1.7 is remarkably narrow. Glass transition temperatures of -60 to -50°C could be observed.

The effects of the chiralities, steric requirements, and basicities of ligands attached to soluble Zr metallocene catalysts on propylene and ethylene homo- and copolymerization was reviewed by Ewen.[54] The alkyl substituted Cp ligands are more sterically hindered than Cp_2ZrCl_2 at the monomer coordination site. The Si-bridged derivative is strained, with a reduced Cp-Zr-Cp angle (10 degrees) and has an enlarged R-Zr-L angle. This change in geometry creates a monomer coordination site which is less sterically hindered by both the Cp ligands and the chain ends. Cp* (methyl substituted chiral cyclopentadiene) has the opposite effect. The ligand steric effects are clearly more pronounced for the coordination of bulkier monomers. Reactivity ratios of metallocenes in ethylene/propylene copolymerization is shown in Table 18. The large variation in the four zirconocene r_1 values as a function of the Cp ligand steric requirements is consistent with the steric effects on coordination being more severe for propylene than ethylene. The moderately alternating copolymer structure obtained with Cp_2Ti (IV) and the more random copolymer structure for Cp_2Zr (IV) are consistent with the former catalyst having, predictably, greater nonbonded interactions between the CH_3 groups of coordinated C_3H_6 and propylene chain-end units. The decrease in the L-Zr-R angle in $Cp*_2Zr$ (IV) relative to Cp_2Zr (IV) results in the former complex producing a more alternating copolymer, as expected. The opposite effect with Me_2SiCp_2Zr (IV) may have been masked by neglecting the regioirregularities with this relatively unhindered species.

TABLE 18
Reactivity Ratios of Metallocenes in Ethylene/Propylene
Copolymerizations

Run no.	Catalyst	r_1	r_2	$r_1 \cdot r_2$	Method
1	$Cp_2Ti = CH_2 \cdot Al(CH_3)_2Cl$	24	0.0085	0.20	NMR
2	$Cp_2Ti(Ph)_2$	19.5	0.015	0.29	IR
4	Cp_2ZrCl	48	0.015	0.72	NMR
5	$Me_2SiCp_2ZrCl_2$[a]	24	0.029	0.70	IR
6	Cp_2ZrCl_2	250	0.002	0.50	IR
7	$(MeCp)_2ZrCr_2$[a]	60	—	—	IR

[a]   Copolymer probably contains significant levels of head-head and tail-tail propylene regioirregularities.

VI. CONCLUSIONS

There are a number of papers prepared regarding the random copolymer of ethylene
with propylene. Hydrocarbon soluble vanadium catalysts are well known to give random
copolymer. However, the catalytic activity is not so high, especially at higher than room
temperature. Solid or supported titanium catalysts give extremely high activity for homopolymerization of ethylene or propylene. The copolymers obtained by the titanium catalysts
have high crystallinity.

Recently, the catalysts for copolymerization of ethylene with propylene have been improved by using some types of soluble catalysts. That is, the copolymerization with some
kind of soluble catalysts such as titanium, chromium, and zirconium complex catalyst systems
were reported. These catalyst systems gave high catalytic activity for the copolymerization
and the copolymer obtained has random sequences of ethylene and propylene. I hope that
the studying of the catalysts in this field will become more developed in the future.

REFERENCES

1. *Eur. Rubber J.*, 6, 20 (1989).
2. **Natta, G., Mazzani, A., Valvassori, A., Sartori, G., and Barbagallo, A.,** *J. Polym. Sci.*, 51, 441,429 (1961); Italian Patent 554 803, 1955.
3. **Carrick, W. L., Karol, F. J., Karapinka, G. L., and Smith, J. J.,** *J. Am. Chem. Soc.*, 82, 1502 (1960).
4. **Junghanns, E., Gumboldt, A., and Bier, G.,** *Makromol. Chem.*, 58, 18 (1962).
5. **Ichikawa, M.,** *J. Chem. Soc. Jpn. Ind. Chem. Sect.*, 68, 535 (1965).
6. **Cozewith, C. and Ver Strate, G.,** *Macromolecules*, 4, 482 (1971).
7. **Bier, G.,** *Angew. Chem.*, 73, 186 (1961).
8. **Natta, G., Crespi, G., Valvassori, A., Sartori, G., Cameli, N., and Turba, V.,** *Rubber Plast Age*, 46, 683 (1965).
9. **Oblój, J.,** *Polym. Sci. USSR*, 7, 1040 (1965).
10. **Natta, G., and Danusso, F.,** in *Stereoregular Polymers and Stereospecific Polymerizations*, Pergamon Press, New York, 1967.
11. **Sivaram, S.,** *Ind. Eng. Chem. Prod. Rev.*, 16, 121 (1977).
12. **Anderson, J. R. and Boudart, M.,** in *Catalysis Science and Technology*, Vol. 6, Springer-Verlag, Hei-delberg, 1984, 66.
13. **Boor, J. Jr.,** in *Ziegler-Natta Catalysts and Polymerization*, Academic Press, New York, 1979, 563.
14. **Dumas, C. and Hsu, C. C.,** *Rev. Macromol. Chem.*, 24, 355 (1984).
15. **Galli, P., Barbe, P. C., and Noristi, L.,** *Angew. Makromol. Chem.*, 120, 73 (1984).

16. **Bier, V. G., Gumboldt, A., and Schleitzer, G.,** *Makromol. Chem.,* 58, 43 (1962).
17. **Crespi, G., Valvassori, A., and Flisi, U.,** in *The Stereo Rubbers,* W. M. Saltman, Ed., J. Wiley & Sons, New York, 1977, 365.
18. **Cesca, S.,** *J. Polym. Sci., Macromol. Rev.,* 10, 1 (1975).
19. **Baldwin, F. P. and VerStrate, G.,** *Rubber Chem. Technol.,* 45, 709 (1977).
20. **Emde, H.,** *Angew. Makromol. Chem.,* 60/61, 1 (1977); **D. L. Christman,** *J. Polym. Sci.,* 10, 471 (1972).
21. **Milani, F., Casellato, U., Vigato, P. A., Vidali, M., Fenton, D. E., and Lealgonzalez, M. S.,** *Inorg. Chimi. Acta,* 103, 15 (1985).
22. **Giannetti, E., Mazzocchi, R., Albizzati, E., Fiorani, T., and Milani, F.,** *Makromol. Chem.,* 185, 2133 (1984).
23. **Doi, Y., Koyama, T., and Soga, K.,** *Makromol. Chem.,* 185, 1827 (1984).
24. **Doi, Y., Tokuhiro, N., Suzuki, S., and Soga, K.,** *Makromol. Chem., Rapid Commun.,* 8, 285 (1987).
25. **Ivanchev, S. S., Baulin, A. A., and Rodionov, A. G.,** *J. Polym. Sci., Polym. Chem. Ed.,* 18, 2045 (1980).
26. **Baulin, A. A., Ivanchev, S. S., Rodionov, A. G., Kreitser, T. V., and Gol'denberg, A. L.,** *Polym. Sci. U.S.S.R.,* 22, 1630 (1980).
27. **Baulin, A. A., Rodionov, A. G., Ivanchev, S. S., Gol'denberg, A. L., and Asinovskaya, P. S.,** *Int. Polym. Sci. Technol.,* 8, T/70 (1981).
28. **Soga, K., Sano, T., and Ohnishi, R.,** *Polym. Bull.,* 4, 157 (1981).
29. **Soga, K., Ohnishi, R., and Sano, T.,** *Polym. Bull.,* 7, 547 (1982).
30. **Doi, Y., Ohnishi, R., and Soga, K.,** *Makromol. Chem., Rapid Commun.,* 4, 169 (1983).
31. **Kashiwa, N., Tsutsui, T., and Toyota, A.,** *Polym. Bull.,* 12, 111 (1984).
32. **Kashiwa, N., Kajiura, H., and Minami, S.,** *Polym. Bull.,* 12, 363 (1984).
33. **Kakugo, M., Naito, Y., Mizunuma, K., and Miyatake, T.,** *Makromol. Chem.* 190, 849 (1989).
34. **Corbelli, L., Milani, F., and Zucchini, V.,** in *Ethylene-Propylene Elastomers Produced with Highly Active Catalysts,* Int. Rubber Conf., Vol. 3A, Moscow, September 4–8, 1984.
35. **Abis, L., Bacchilega, G., and Milani, F.,** *Makromol. Chem.,* 187 (1986).
36. **Avella, M., Martuscelli, E., Volpe, G. D., Segre, A., Rossi, E., and Simonazzi, T.,** *Makromol. Chem.,* 187, 1927 (1986).
37. **Maglio, G., Milani, F., Musto, P., and Riva, F.,** *Makromol. Chem., Rapid Commun.,* 8, 589 (1987).
38. **Soga, K., Ohtake, M., Ohnishi, R., and Doi, Y.,** *Polym. Commun.,* 25, 171 (1984).
39. **Kashiwa, N., Mizuno, A., and Minami, S.,** *Polym. Bull.,* 12, 105 (1984).
39b. **Cozewith, C. and VerStrate, G.,** *Macromolecules,* 10, 773, 1977.
40. **Makino, K., Tsuda, K., and Takaki, M.,** *Rubber Chem. Technol.,* 64, 1 (1991).
41. **Makino, K., Tsuda, K., and Takaki, M.,** *Makromol. Chem. Rapid Commun.,* 11, 223 (1990).
42. **Makino, K., Tsuda, K., and Takaki, M.,** *Makromol. Chem.,* 193, 341, 1992.
43. **Makino, K., Tsuda, K., and Takaki, M.,** *Polymer,* 33, 2416, 1992.
44. **Makino, K., Tsuda, K., and Takaki, M.,** *Polym. Bull.,* 26, 371, 1991.
45. **Makino, K., Tsuda, K., and Takaki, M.,** *Polym. Bull.,* 27, 41, 1991.
46. **Soga, K., Chen, S. I., Shiono, T., and Doi, Y.,** *Polymer,* 26, 1888 (1985).
47. **Gan, S. N., Chen, S. I., Ohnishi, R., and Soga, K.,** *Makromol. Chem., Rapid Commun.,* 5, 535 (1984).
48. **Gan, S. N., Chen, S. I., Ohnishi, R., and Soga, K.,** *Polymer,* 28, 1391 (1987).
49. **Kaminsky, W.,** Int. Symp. on Recent Advances in Polyolefins, Chicago, September 1985.
50. **Kaminsky, W.,** Int. Symp. on Future Aspects on Olefin Polymerization, Tokyo, July 1985.
51. **Kaminsky, W.,** 189, ACS Meeting, Miami, April, 1985.
52. **Ewen, J. A.,** *J. Am. Chem. Soc.,* 106, 6355 (1984).
53. **Kaminsky, W. and Miri, M.,** *J. Polym. Sci., Chem. Ed.,* 23, 2151 (1985).
54. **Ewen, J. A.,** Int. Symp. on Future Aspects on Olefin Polymerization, Tokyo, July 1985.

Chapter 10

THERMAL AND OXIDATIVE STABILITY OF ETHYLENE-PROPYLENE RUBBERS

George Mareş and Petra Budrugeac

TABLE OF CONTENTS

SYMBOLS

A — pre-exponential factor for the nonisothermal degradation and the thermal-accelerated aging dependent on pressure

a', b — constants of the material

E — activation energy

α — chemical reaction coefficient

K — chemical reaction rate

Q — heat

p — pressure

ϵ — elongation at break

t — time

T — temperature

ϵ_o — elongation at break for $t = 0$

n — form factor

A' — a constant of the material

R — gas constant

I. INTRODUCTION

The thermal and oxidative stability of the ethylene-propylene rubbers requires both theoretical and experimental studies, starting with the basic units of the copolymeric chain extended to the copolymer as a unit, or to the compound as a finite form in order to cover the field of application. The thermal and oxidative stability, understood in a broader sense, for the ethylene-propylene rubber corresponds to the degradation of the physical and chemical properties adequate to its application under the influence of heat and oxygen in the environment in which the material exists or under the influence of the oxygen pre-existing in the material. The heat can directly proceed from the environment (through convection or infrared radiation) or as a secondary effect of some irradiation phenomena. Under the conditions of the present technological development, one should know the thermal and oxidative stability of the polymeric materials and, implicitly, their lifetime, correspond to some final values of the properties defining the user function.[1] The ethylene-propylene rubber accelerated thermal aging in air (at atmospheric pressure or under pressure) and in oxygen. In order to emphasize the degradation of its mechanical (elongation at break and residual deformation under constant deflection) and electrical properties (dielectric rigidity, volume resistivity, etc.), a series of conditions preliminary to the experiment proper is required in order to ensure an interpretation of the results in accordance with reality. These preliminary conditions are imposed by the conclusions that resulted from the application of the thermal analyses methods thermogravimetric curve, derivative thermogravimetry, DTA, and differential scanning calorimetry (TG, DTG, DTA, and DSC) that are characteristic of the field of the thermal stability[2-6] of the degradation nonisothermal kinetic parameter values,[7-11] as well as of the changes which the polymeric materials undergo as a result of the accelerated thermal agings.[4,12,13] The following study deals with a model of the thermal and oxidative degradation. The property to be analyzed is the elongation at break, or in the case of the ethylene-propylene rubber, an EPDM elastomer. This elastomer was obtained by CATC-Piteşti in Romania, and besides the basic copolymer, consists of a hydrated aluminum oxide of the hydrol type, plastifiers, vulcanizing agents, a coupling agent (vinyl silane), and antioxidants.

II. NONISOTHERMAL DEGRADATION

The thermal analyses (TG, DTG, DTA, or DSC) point out the thermal effects and the weight variations that can occur under the conditions of the thermal polymeric material (within an inert atmosphere — N_2, Ar) or thermo-oxidative (in air or oxygen) degradation. The thermal properties of the polymeric materials can be better assessed if the DTA, DSC, TG, and DTG methods are simultaneously applied as is customary, for example, with the Q-1500D Derivatograf of the Paulik-Paulik-Erdey type manufactured by the MOM-Budapest, which we also used when we studied the thermal and thermo-oxidative degradation of the ethylene-propylene rubber. Further in this chapter the results obtained in this field are presented.

The ethylene-propylene rubber derivatogram (Figure 1) obtained within an inert atmosphere (Ar), points to the fact that at temperatures lower than 325°C, the material under study undergoes only one process of endothermic degradation that occurs with the formation of volatile products. The maximum rate of the degradation process corresponds to the DTG minimum temperature (290°C) and is practically equal to the DTA minimum temperature.

In the case of the thermal degradation process, the nonisothermal kinetic parameters (activation energy E, reaction apparent order m, and pre-exponential factor A) were assessed by means of the following methods: Coats-Redfern;[14] Coats-Redfern modified by Urbanovici and Segal;[15] and Flynn-Wall for constant heating rate.[16] These methods led to concordant

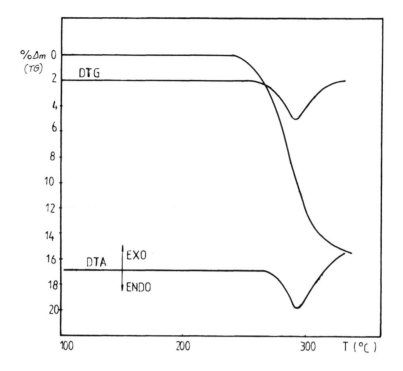

FIGURE 1. Derivatogram of ethylene-propylene rubber Ar thermal degradation. Heating rate: a = 2.5 K/min.

TABLE 1
Values of Nonisothermal Kinetic Parameters for the Thermal Degradation (in Ar) of Ethylene-Propylene Rubber

Method	E (kcal/mol)	m	A (s⁻¹)	r
Coats-Redfern	48.2	1.6	2.4905×10^{16}	-0.9943
Modified Coats-Redfern	48.05	1.5	2.2486×10^{16}	-0.9940
Flynn-Wall (a = ct.)	47.9	1.6	1.9716×10^{16}	-0.9948

E — activation energy; m = apparent reaction order; A — pre-exponential factor; r — correlation coefficient of the linear regression.

results (see Table 1), according to which the thermal degradation process of the studied rubber has a relatively high activation energy (about 48 kcal/mol).

There were five heating rates used for the air degradation of the rubber under study, and the resultant thermograms emphasize the stages of the thermal and oxidative degradation. Figure 2 shows the thermogram in the case of the air degradation for a heating rate of 2.5 K/min; for the other heating rates similar thermograms are obtained.

The TG shows that the material undergoes two processes that lead to the sample weight reduction (the processes denoted by I and III). The DTG curve shows that to process I there corresponds a rapid reduction without any return to the zero line. For the same process, the DTA curve points out a weak exothermic effect. The results show that process I is complex and could be assigned to the oxidation transformation of the ingredients (antioxidants, plastifiers, vulcanizing agents, etc.) into volatile products. The complexity of process I also results from the assessment of the activation energy by means of the Flynn-Wall method,[7]

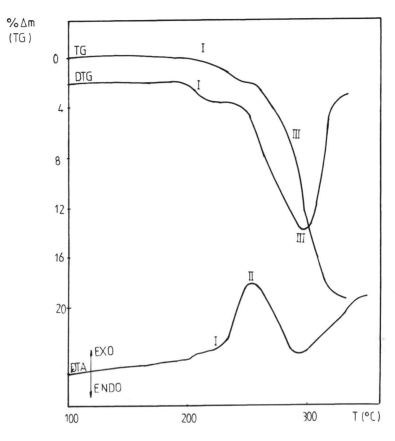

FIGURE 2. Derivatogram of ethylene-propylene rubber air thermo-oxidative degradation. Heating rate: a = 2.5 K/min.

which stresses that this energy varies the function of the degree of conversion between 63 and 55 kcal/mol. Between processes I and III there occurs an exotermic process II which is not accompanied by any variation of the sample weight. Because peak II of the DTA curve shown in Figure 2 was not noticed on the material degradation within inert atmosphere (see Figure 1), the results are that process II consists of the interaction of the oxygen with the polymer that leads to the solid product formation. The activation energy of this process assessed by means of the Kissinger method,[17] on the basis of the thermograms traced at five heating rates, is 23 kcal/mol. This value is characteristic of the polymer degradation reactions initiated by the oxygen.[18] Similar processes of thermo-oxidation were also emphasized in the case of a series of elastomers,[19] in the compound ethylene vinyl acetate (EVA),[1] in polyethylene,[20] and in polypropylene.[20] At the same time, one noticed a similar oxidation peak with the DTA curves[21] obtained for vulcanized and unvulcanized EPDM. The determinations carried out by Sircar and Lamond[21] have pointed to the fact that substantial differences do not exist between the DTA curves obtained for the EPDM elastomers that are produced by various manufacturers. Because of this, the effect of various thermomonomers is not detectable by the differential thermal analysis. In the case of process III (see Figure 2) which is accompanied by weight loss, the nonisothermal kinetic parameters were estimated by using the Coats-Redfern method.[14] The results are found in Table 2.

One can notice that the values of the activation energy and of the pre-exponential factor depend on the heating rate. As shown in Figure 3, these values are correlated through the relation:

TABLE 2
The Values of the Nonisothermal Parameters
of Process III of the Thermo-Oxidation of the
Ethylene-Propylene Rubber

a (K/min)	E (kcal/mol)	IgA	m	r
0.695	50.6	17.1988	1.5	−0.99396
1.31	47.8	16.0761	1.5	−0.99916
2.79	49.9	16.9022	1.5	−0.99723
2.91	43.5	14.1274	1.5	−0.99908
5.805	36.3	11.4931	1.5	−0.99886
11.99	33.3	10.3095	1.5	−0.99923

a — heating rate; E — activation energy; A — pre-exponential factor; m — apparent reaction order; r — correlation coefficient of linear regression.

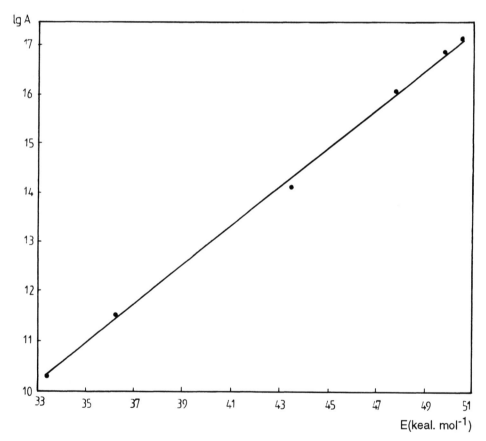

FIGURE 3.   The plotted IgA vs. E values for process III of the thermo-oxidative degradation of ethylene-propylene rubber.

$$IgA = a'E + b \qquad (1)$$

where a $= 0.398$ mol/kcal, b $= 2.9902$, and r $= -0.999$

Relation 1 corresponds to the so-called "compensation effect" which was examined for different heterogeneous solid-gas relations[22-25] and for the thermal aging of some polymeric materials.[26,27] As far as the compensation effect is concerned, Garn[22] noticed that the common element of the processes described by it consists in an unaltered main reaction in which a parameter considered as a second factor changes the variation of the reaction rate with temperature. As for the process III of ethylene-propylene rubber nonisothermal degradation, the main reaction seems to be the decomposition of the hydroperoxides formed by process II and their interaction with oxygen, thus, generating the formation of volatile products. Obviously, the secondary factor which determines the nonisothermal kinetic parameter change is the heating rate of the analyzed sample. The above-presented results stress the fact that the ethylene-propylene rubber thermo-oxidative degradation occurs according to the general mechanism of the polyolefin thermo-oxidation.[28] At the beginning of the process, macromolecular radicals R˙ occur either by the homolytic break of the more reactive C–H bond:

$$R{-}H \rightarrow R^{\cdot} + H^{\cdot} \atop \text{(polymer)} \qquad (2)$$

or by the interaction of the polymer with the activated oxygen by heating:

$$R{-}H + O_2 \rightarrow R^{\cdot} + HO_2^{\cdot} \qquad (3)$$

The radicals R˙ react with the oxygen in order to form peroxidic radicals and hydroperoxides:

$$R^{\cdot} + O_2 \rightarrow R{-}O{-}O^{\cdot} \qquad (4)$$

$$R{-}O{-}O^{\cdot} + RH \rightarrow R{-}O{-}OH + R^{\cdot} \qquad (5)$$

As seen from the therograms we obtained, the reactions that lead to the solid product formation have an exotermic effect and have a total activation energy of about 23 kcal/mol. By heating, the products that resulted from Reactions 4 and 5 are decomposed and/or react with the oxygen forming volatile substances. The total nonisothermal kinetic parameters of these processes (E and A) depend on the heating rate.

III. AIR AND PRESSURIZED OXYGEN-ACCELERATED THERMAL AGING

A. THEORETICAL PREMISES

The thermal analysis of the ethylene-propylene rubber performed in the air and inert atmosphere (Ar) emphasizes the dependence of the tested material degradation process on the experimental atmosphere. Under these conditions, the rate of the chemical reaction depends on both heat and the partial pressure of the oxygen in the enclosure atmosphere. Experimentally, one can notice a factorization of the chemical reaction rate under the form:

$$K(Q + p) = K(Q) \cdot K(p) \qquad (6)$$

where $K(p)$ is the rate of the chemical reaction due to oxygen partial pressure, $K(Q)$ is the rate of the chemical reaction due to heat, and $K(Q+p)$ is the chemical reaction rate under the simultaneous influence of heat and oxygen pressure. The previous relation is especially

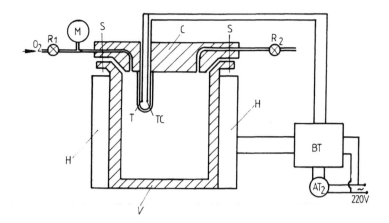

FIGURE 4. Block diagram for pressurized oxygen-accelerated thermal aging. V, stainless steel cell; C, cell cover; S, clamping screws; H, heating oven; T, thermometer; TC, Fe-constantan thermocouple; BT, thermal adjusting unit block: R_1 and R_2, metal taps; M, manometer.

suitable for the elastomer mechanical properties. The thermal and oxidative degradation of the elastomer mechanical properties of the Arrhenius type:

$$\frac{d\,P_m}{dt} = -K(Q + p)P_m^\alpha \tag{7}$$

where P_m is the mechanical property and α is the chemical reaction coefficient which is equal to the unity in the case of ethylene-propylene rubber. The previous reasons are taken into account for temperatures superior to the vitreous transition temperature.

B. EXPERIMENTAL DETERMINATION

The necessity of the experimental determination of some superior acceleration factors in the case of ethylene-propylene rubber thermo-oxidative aging imposed, besides the accelerated thermal aging within the drying chambers with air-forced circulation, the aging within special enclosures with pressurized oxygen atmosphere and temperature-controlled variations. The air-accelerated aging was performed within WSU-200 drying chambers with air-forced circulation and temperature stabilization accuracy within the limits of $\pm 2°C$. The thermal aging within an oxygen-controlled atmosphere was performed within the installation built according to ASTM D 572-73.[29] The cell (Figure 4) is provided with a pressure gauge (M), a mercury thermometer (T), a heater (H), and a Fe-constantan thermocouple (TC) for adjustment of thermal control. Oxygen is provided directly from a pressurized oxygen cylinder (purity of 99%) and the constant temperature is preserved by means of a heat-regulating system (BT) controlled by the Fe-constantan thermocouple. Temperature control was $\pm 1°C$ up to 150°C.

The samples were perpendicularly disposed in the cell to a maximum loading of 10% of the volume of the cell. Samples of ethylene-propylene rubber, having a thickness of 1 mm, were made in accordance with ASTM D-412. Elongation at the break of the samples was determined by means of a universal testing machine. Monsanto-T-10/E, which has an accuracy of $\pm 1\%$ and is provided with an extensometer and an automatic recording system.

C. AIR-ACCELERATED THERMAL AGING

The ethylene-propylene rubber air-accelerated thermal aging was performed at the following temperatures: 100, 115, 125, 135, and 140°C. The curves of the relative breaking

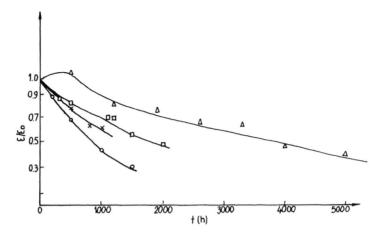

FIGURE 5. Elongation at break as a function of duration of aging in open air at various temperatures; △, 100°C, □, 115°C; x, 125°C; ○, 135°C; ✦, 140°C.

elongation variation with the aging time at the previously mentioned temperatures are given in Figure 5. The analysis of experimental data led to a degradation equation under the form:

$$\frac{d\epsilon}{dt} = -K(Q) \cdot K(p) \, \epsilon^{\alpha} \tag{8}$$

The result of the experimental value statistical processing led to a pseudodynamic equation of the relative breaking elongation dependence on temperature (T), pressure (p), and time (t), in accordance with Equations 7 and 8 under the form:

$$\epsilon(T,p,t) = \epsilon_o \cdot \exp[-K(T) \, K(p) \cdot t] \tag{9}$$

where $\epsilon(T,p,t)$ represents the relative elongation on break at the moment t. The same statistical analysis explicitly emphasizes the dependence of the chemical reaction rate on temperature and the dependence on pressure being included in the pre-exponential factor:

$$K(T) \, K(p) = A(p) \exp(-E/RT) \tag{10}$$

The following numerical values result in the medium thermo-oxidative activation energy E = 12.6 ± 1.6 kcal/mol and the pre-exponential factor A = 4461 h^{-1}. The chemical reaction rate is determined with a degree of reliability of 95%, a fact that shows that within the experimental error limit it can be accepted as constant (the activation energy fluctuations for each isotherm belong to the interval 12.6 ± 1.6 kcal/mol) and correspond to the process of oxygen diffusion in the polymeric material.

Equations 9 and 10 present the complete form of the empirical equation of the relative breaking elongation dependence on the thermal aging temperature, as well as the aging period of time given by the following equation:

$$\epsilon(T,p,t) = \epsilon_o \exp[-A\exp(-E/RT)] \tag{11}$$

D. PRESSURIZED OXYGEN-ACCELERATED THERMAL AGING

In order to emphasize the dependence of the chemical reaction rate on the enclosure oxygen pressure, the accelerated aging process was carried out under the following

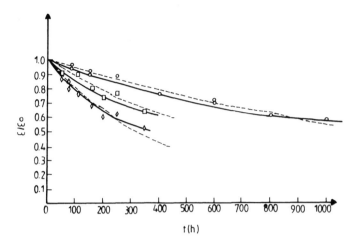

FIGURE 6. Elongation at break as a function of duration of aging in oxygen at 5 bar pressure at various temperatures: ○, 95°C; □, 110°C; ◇, 125°C; —-, experimental curve: - - - -, theoretical curve.

conditions: at the temperature of 125°C and oxygen pressures of 0.2, 5, and 10 bar; at 110°C and 5 bar; and at 95°C and 5 bar.

In order to ensure these conditions, one proceeded to a preheating of the enclosure samples in the air at a temperature 10 to 15°C lower than that corresponding to the thermal aging, to an oxygen rapid washing of the enclosure, and then to the introduction of the oxygen corresponding to the pre-established pressure. During the oxygen washing, and especially at pressures higher than atmospheric (generally higher than 2 bar), there occurs a strong exothermic reaction that results in the previously mentioned temperature rise, the temperature being automatically controlled at the specified value. The increase of the oxidation rate that occurs at higher pressures of oxygen is due to the initiation of an oxygen-direct attack against the antioxidant. The oxidation of the elastomers with elemental oxygen is called autooxidation because it is an autocatalytic process in which the hydroperoxides formed, as the primary product of the reaction decompose, to produce free radicals that initiate the free radical chain mechanism.[30]

After the free radicals formation according to Equations 2–5 and the chain reaction propagation, within the interruption stage the resultant radicals are consumed and small quantities of carbonyl compounds are obtained. The final study of the process is characterized by an accelerated increase of the oxygen uptaking rate, the temperature is increased and oxygen-based groups are rapidly obtained. The thermo-oxidation rate goes up with the oxygen pressure and temperature. The experimental results obtained in the case of oxygen-accelerated thermal aging are presented in Figure 6 by means of block curves. In order to perform the analysis of the experimental data, some simplifying hypotheses on the diffusion phenomenon and on the general thermo-oxidation process formal kinetics are also necessary.

1. The amount of oxygen diffused within the specimens during the process is negligible compared to the amount of oxygen in the enclosure (the volume of the samples is, at the most, 10% of the enclosure volume).
2. The diffusion is one-dimensional and bilateral.
3. The thermo-oxidation kinetics depend only on the concentration of the oxygen diffused within the specimen and on the temperature.

By using these simplifying hypotheses and Equation 6, an explicit dependence of the chemical reaction rate $K(Q+p) = K(T,p)$ on temperature and pressure under the form:

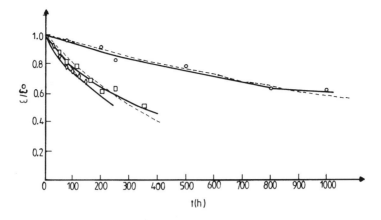

FIGURE 7. Elongation at break as a function of duration of aging in oxygen at 125°C at various pressures: O, 0.2 bar; □, 5 bar; ◇, 10 bar; —-, experimental curve; - - - -, theoretical curve.

$$K(T,p) = A' p^n \exp(-E/RT) \qquad (12)$$

is determined by means of the computer-aided processing of the experimental data, where A' and n are constants of material which, for the analyzed ethylene-propylene rubber, have the following numerical values: $A' = 9353$, $n = 0.46$.

By replacing Equation 10 by 12, the connection between A and A' is determined:

$$A = A' p^n \qquad (13)$$

Equation 11 will have the following expression:

$$\epsilon(T,p,t) = \epsilon_o \exp[-A' p^n \exp(-E/RT)] \qquad (14)$$

The concordance of this equation with the experimental results is given in Figures 6 and 7 where, by means of broken lines, the theoretical aging curves are traced.

The form factor n depends, generally, on the rubber vulcanizing degree, the environment humidity on the plastifier nature, and quantity, etc., but within the limits of some values that do not substantially modify the form of the chemical reaction rate dependence curve on pressure. Figure 8 presents a series of isotherms of Equation 12 that stress a saturation of the chemical reaction rate, with the pressure much higher at relatively low temperatures, practically in the field of the ethylene-propylene rubber application and lower at high temperatures within the range of the oxygen-accelerated aging temperatures, which, at the experimentally used pressures is not detected any longer. Figure 8 points out that up to a temperature of about 363 K, the saturation occurs around the pressure of 2 bar. As to the material analyzed for temperatures up to 90°C, the effect of an oxygen pressure of 10 bar is analogous to the effect induced by an oxygen pressure of 2 bar. The activation energy of the whole degradation process $E = 12.6$ kcal/mol is associated with the phenomenon of the oxygen diffusion within the ethylene-propylene rubber specimens. The thermo-oxidative degradation kinetic equation (Equation 12), experimentally examined for a large pressure range (0.2 to 10 bar), emphasizes the importance of the oxygen pressure as an acceleration factor of the thermo-oxidative degradation and of the oxygen diffusion as a factor necessary for the thermo-oxidative stability determination.

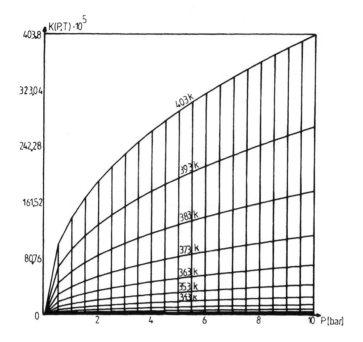

FIGURE 8.   Change in reaction rate as a function of pressure and temperature (isotherms).

IV. CONCLUSION

The study of thermal and oxidative stability of the ethylene-propylene rubber shows the dependence of the tested material degradation process on the experimental atmosphere and temperature. The derivatographic analyses showed that at progressive heating of ethylene-propylene rubber in air, the following processes occur successively:

1. Plastifier and other ingredient loss
2. Ethylene-propylene rubber interaction with oxygen which generates the forming of solid products (probably hydroperoxides)
3. Decomposition of these products or their interaction with oxygen which generates the forming of volatile products

Also demonstrated was the effect of oxygen pressure on the accelerated thermo-oxidative degradation of ethylene-propylene rubber. The comparison between the results obtained by thermal analysis with those obtained at the thermal aging study led to the conclusion that the rate determinant process of mechanic thermo-oxidative degradation is the oxygen diffusion in polymer.

REFERENCES

1. IEEE 101/1972-Guide for the Statistical Analysis of Thermal Life Test Data.
2. **Vogl, O., Sung, I., Miller, H. C., and Williams, H. W.,** *J. Macromol. Sci.,* A-2, 175 (1968).
3. **Korshak, V. V., Vinogradova, S. V., Danilov, G. V., Beribze, L. A., and Salazkin, S. N.,** *Visokomol. Svedin.,* 12, 129 (1970).
4. **Berlin, A. A., Gherasimov, S. G., and Ivanov, A. A.,** *J. Polym. Sci.,* C-41, 183 (1973).

5. **Gademer, T. J.,** *J. Macromol. Sci.,* A-8, 105 (1974).

6. **Ciutacu, S., Fătu, D., and Segal, E.,** *Thermochim. Acta,* 131, 279 (1988).

7. **Flynn, J. H. and Wall, L. A.,** *J. Res. Natl. Bur. Stand. — Phys. Chem.,* 70A, 487 (1966).

8. **Flynn, J. H.,** *Thermal Methods in Polymer Analysis,* S. W. Shalaby, Ed., The Franklin Institute Press, Philadelphia, 1977, 163.

9. **Flynn, J. H.,** *Polym. Eng. Sci.,* 20, 675 (1980).

10. **Fătu, D., Geambaşu, G., Segal, E., Budrugeac, P., and Ciutacu, S.,** *Thermochim. Acta,* 149, 181 (1989).

11. **Segal, E., Budrugeac, P., and Ciutacu, S.,** *Thermochim. Acta,* 164, 161 (1990).

12. **Karamazsin, E., Satre, P., Borhoumi, R., Liptan, G., and Liegethy, L.,** *J. Therm. Anal.,* 29, 259 (1984).

13. **Segal, E., Budrugeac, P., Ciutacu, S., and Mareş, G.,** *Thermochim. Acta,* 164, 171 (1990).

14. **Coats, A. W. and Redfern, J. P.,** *Nature,* 26, 68 (1964).

15. **Ubranovici, E. and Segal, E.,** *Thermochim. Acta,* 80, 379 (1984).

16. **Flynn, J. H. and Wall, L. A.,** *Polym. Lett.,* 4, 323 (1966).

17. **Kissinger, H.,** *Anal. Chem.,* 29, 1702 (1957).

18. **Flynn, J. H. and Florin, R. E.,** *Degradation and Pyrolysis Mechanism,* S. A. Lieberman and E. J. Levy, Eds., Marcel Dekker, New York, 1985, 149.

19. **Slusarski, L.,** *J. Therm. Anal.,* 29, 905 (1984).

20. **Iring, M., Lazlo, Z. H., Kelen, T., and Tudos, F.,** *Proc. 4th Int. Conf. Therm. Anal.,* E. Buzagh, Ed., Acad. Kiado, Budapest, 1975, 127.

21. **Sircar, A. K. and Lamond, T. G.,** *Rubber Chem. Technol.,* 45, 329 (1972).

22. **Garn, P. D.,** *J. Therm. Anal.,* 7, 475 (1975).

23. **Galwey, A. K.,** *Adv. Catal.,* 26, 24 (1977).

24. **Segal, E. and Fătu, D.,** *Introduction to Nonisothermal Kinetics* (in Romanian), Publishing House of the Academy of R. S. Romania, Bucharest, 1983, chap. 8.

25. **Agrawal, R. K.,** *J. Therm. Anal.,* 31, 73 (1986).

26. **David, P. K.,** *IEEE Trans. Electr. Insul.,* EI 22, 229 (1986).

27. **Montanari, G. C.,** *IEEE Trans. Electr. Insul.,* EI 23, 1057 (1988).

28. **Kemiya, Y. and Niki, E.,** in *Aspects of Degradation of Polymers,* A. H. Jellinek, Ed., Elsevier, Amsterdam, 1978, 79.

29. **Ciutacu, S., Budrugeac, P., Mareş, G., and Boconios, I.,** *Polym. Degrad. Stability,* 29, 321 (1990).

30. **Sheton, J. R.,** *Rubber Chem. Technol.,* 45(2), 359 (1972).

Chapter 11

THE NATURE OF SULFUR VULCANIZATION

Michael R. Krejsa and Jack L. Koenig

TABLE OF CONTENTS

I. INTRODUCTION

The use of elastomers as a material for finished products dates back to the 1830s. Unvulcanized natural rubber was originally used, but this suffered from problems of softening during the summer, hardening during the winter, and noxious odors after use during the summer.[1] Prompted by this, Charles Goodyear in the U.S. and Thomas Hancock in England mixed sulfur with natural rubber and then heated the sample.[2] This improved the strength of the rubber and limited the hardening and softening of the product with changes in temperatures. Sulfur alone was used as the vulcanizing agent up to the discovery of organic accelerators in the early part of the 20th century. It was quickly realized that the use of accelerators gave improved properties and significantly reduced the required cure times.[3] The first accelerators were amine-based compounds, with other classes of accelerators following quickly.

The commercially used accelerators can generally be categorized as belonging to one of three main categories: amine-based, alkyl thiocarbamates, and benzothiazole-based (MBT and derivatives). Examples of compounds in each class are shown in Table 1.[4] Thiocarbamate accelerators are well known for their superior accelerating capabilities, but suffer from very short (if any) induction periods, which limits processing safety. Another way of stating this is that thiocarbamate-based systems are sensitive to premature vulcanization. On the other hand, sulfenamides, amide derivatives of mercaptobenzothiazole (MBT) enjoy good processing safety due to their lengthy induction period, but do not have the accelerating power of thiocarbamates. Several newer accelerators consist of mixed thiocarbamates and benzothiazole moieties, and, thus, incorporate the higher accelerating power of the thiocarbamates with the lengthy induction period of the sulfenamides. Several of these accelerators are shown in Table 2.[5]

Another change that is occurring in vulcanization formulations is the use of binary accelerator systems [5-10] Binary accelerator formulations involved the use of two different accelerators in the system, often leading to improved properties. In many cases, a synergistic behavior is observed, leading to better properties in the combined system then would be expected by a rule of mixtures for the individual accelerator systems. This is illustrated in Figure 1 for an OTOS/OBTS system (structures in Table 2).[6]

Other compounds commonly used in vulcanization, in addition to sulfur and accelerators, are zinc oxide and saturated fatty acids such as stearic or lauric acid.[1,3] These materials are termed activators (as opposed to accelerators). Zinc oxide serves as an activator, and fatty acids are used to solubilize the zinc into the system. Rubber formulations can also include fillers such as fumed silica and carbon black, and compounds such as stabilizers and antioxidants. Further complicating the situation is the engineering practice of blending various elastomers to obtain the desired properties.

Cure of vulcanizates is commonly measured using a Monsanto cure rheometer,[4] which plots the change in torque as a function of time as is shown in Figure 2. From the percentage rise of the graph, the percent cure can be calculated. This figure also illustrates the three main regions of rubber vulcanization. The first regime is the induction period, or scorch delay, during which accelerator complex formation occurs. The second time period is the cure period, in which the network or sulfurization structures are formed. The network structures can include crosslinks, cyclics, main chain modification, isomerization, etc., which will be discussed in detail in the following sections. The third regime is the overcure, or reversion regime. During this time period network maturation occurs. The three different curves in this region represent continued crosslinking (top curve), which is also referred to as "creeping modulus"; no change, which is the middle curve; and reversion (bottom curve), which is a degradation of the network.

TABLE 1
Examples of Common Accelerators Used for Accelerated Sulfur Vulcanization

Accelerators for Sulfur Vulcanization

Compound	Abbreviation	Structure
Benzothiazoles		
2-mercaptobenzothiazole	MBT	
2,2′-dithiobisbenzothiazole	MBTS	
Benzothiazolesulfenamides		
N-cyclohexylbenzothiazole-2-sulfenamide	CBS	
N-*t*-butylbenzothiazole-2-sulfenamide	TBBS	
2-morpholinothiobenzothiazole	MBS	
N-dicyclohexylbenzothiazole-2-sulfenamide	DCBS	
Dithiocarbamates		
tetramethylthiuram monosulfide	TMTM	
tetramethylthiuram disulfide	TMTD	
zinc diethyldithiocarbamate	ZDEC	
Amines		
diphenylguanidine	DPG	
di-*o*-tolylguanidine	DOTG	

From Alliger, G. and Sjothun, I. J., *Vulcanization of Elastomers,* Von Nostrand Reinhold, New York, 1964, chap. 1. With permission.

Sulfur vulcanization can be divided into two main categories: unaccelerated and accelerated sulfur vulcanization. Unaccelerated formulations typically consist of sulfur, zinc oxide, and a fatty acid such as stearic acid, while accelerated formulations include an accelerator in the system. A subcategory of accelerated sulfur vulcanization is sulfur-free systems, also referred to as sulfur-donor systems. In these systems the sulfur needed for network formation is supplied by the accelerator, which functions as both an accelerator and sulfur-donor. It

TABLE 2
Examples of Common Binary and Mixed Accelerator Systems

	Chemical name	Abbreviation	Structure
1.	Cyclohexylthiodibutylamine	CDBA	
2.	Cyclohexyldithiobenzothiazole[6]	CDB	
3.	Cyclohexylthiodibenzopyrrole	CDBP	
4.	Cyclohexylthiodiethylamine	CDEA	
5.	Cyclohexylthiodiisopropylamine	CDIPA	
6.	Cyclohexylthiodiphenylamine	CDPA	
7.	Cyclohexylthiomorpholine[7]	CM	
8.	Cyclohexylthiopiperidine[7]	CP	
9.	Cyclohexylthiopyrrolidine[7]	CP_y	
10.	N-cyclopentamethylene-2-benzothiazole sulfenamide[8]	CPBS	
11.	Bis(Cyclopentamethylene)thiuram disulfide[9]	CPTD	
12.	N-Cyclopentamethylene thiocarbamyl-2-benzothiazyl disulfide	CTBD	
13.	N-Cyclopentamethylene-thiocarbamyl-N'-Cyclopentamethylene sulfenamide[10]	CTCS	
14.	N-Cyclohexylthiophthalimide	CTP	
15.	Dibenzopyrrole	DBP	

TABLE 2 (continued)
Examples of Common Binary and Mixed Accelerator Systems

	Chemical name	Abbreviation	Structure
16.	Dicyclohexyldithiopiperazine	DCDP$_z$	
17.	Di-β-napthyl-*p*-phenylene-diamine	DNPD	
18.	2-Mercaptobenzothiazole	MBT	
19.	Dibenzothiazyldisulfide	MBTS	
20.	4-Morpholinyl-2-benzothiazole disulfide[11,12]	MDB	
21.	*N*-Oxydiethylene-2-benzothiazole sulfen-amide	OBTS	
22.	*N*-Oxydiethylenethiocarbamyl-2-benzo-thiazyl disulfide	OTBD	
23.	*N*-Oxydiethylenethiocarbamyl cyclo-hexyl disulfide[4]	OTCD	
24.	*Bis*(oxydiethylene) thiuram disulfide[13]	OTD	
25.	*N*-Oxydiethylenethiocarbamyl-*N'*-oxydi-ethylene sulfenamide	OTOS	
26.	*N*-phenyl-*N'*-isopropyl-*p*-phenylenedi-amine	PIPPD	

From Das, P. K., Datta, R. N., and Basu, P. K., *Rubber Chem. Technol.*, 61, 760 (1988). With permission.

should be noted that while unaccelerated sulfur systems are no longer of commercial significance, they are of interest as a starting point to understanding accelerated sulfur vulcanization systems.

Work by Bevilacqua, Scheele, Craig, Dodagkin, and others[11-20] has suggested a free radical mechanism is operative, while work by Bateman, Ross, and others,[11,21-25] has led to the proposal of a polar mechanism of vulcanization. Work by others including Shelton and McDonel,[26] Wolfe,[27,28] Coleman et al.,[29-33] Duchacek,[34-43] Manik, Banerjee, and co-workers,[44-53] and others has led to the proposal that both free radical and polar mechanisms are operative depending on the formulation and curing conditions.

A major source of the uncertainty has arisen from the intractable nature of the cured vulcanizate. Cured elastomers are insoluble; this eliminates most analytical techniques for examining polymeric structures. To avoid these problems, several methods have been used, including model compound work,[11,21-25,27,28,54-59] electron spin resonance (ESR),[20,32] chemical probes,[60-65] radical scavengers,[24,26] solid-state C^{13}-NMR,[66-75] and analysis of the extra-network materials.[76-78] The chemical probe work has allowed characterization and

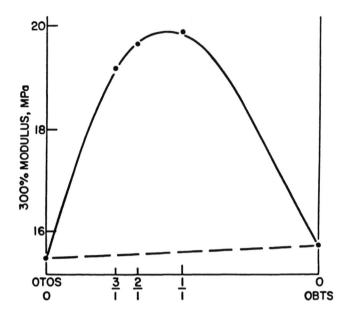

FIGURE 1.   Effect of OTOS/OBTS ratio on modulus of natural rubber,[16] illustrating the synergistic behavior of binary accelerator systems.

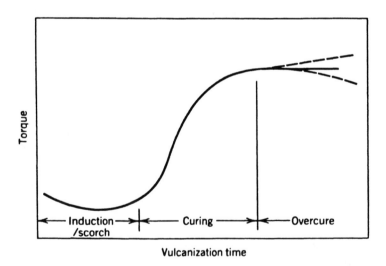

FIGURE 2.   Typical cure rheometer trace for rubber vulcanization illustrating the induction period (I), cure period (II), and overcure or reversion period. (III). (From Vulcanization, *Encyclopedia of Polymer Science and Engineering,* 2nd ed., Vol. 17, Mark, Bikales, Overberger, and Menges, Eds., Wiley, New York, 1989. With permission.)

quantification of the number of mono-, di-, and polysulfidic crosslinks. The model compound work has been useful in providing information for mechanistic studies by allowing comparison of products predicted by a mechanism to the products obtained from model compound work. Electron spin resonance and radical scavenger work has been used to examine whether radical processes are active during vulcanization with various formulations and elastomers. Solid-state NMR work has been used to help elucidate the products of vulcanization. Analysis of the extra-network material has provided insights into the nature of the vulcanization chemistry by analysis of content of various intermediates as cure progresses.

A major contributing factor to the disagreement over mechanisms is the possible reactions that sulfur and accelerators can undergo. Sulfur occurs naturally as an eight-membered ring; this ring is capable of both homolytic cleavage to form radicals or heterolytic cleavage to form ions.[79] Also, the precise interaction of the accelerator and sulfur in the vulcanization process has not been definitively elucidated. It is known that accelerator complexes are formed, but the actual sulfurating species has not been determined. This question and the possible reactions of sulfur and accelerators has made it impractical to eliminate either the radical or polar mechanisms.

II. UNACCELERATED SULFUR VULCANIZATION

The reaction mechanisms of unaccelerated sulfur vulcanization has been a subject of research interest for several decades. Despite numerous works designed to elucidate the reaction mechanisms, the actual mechanism has not been conclusively proved. Proposed mechanisms include radical[11-17,27] and polar,[11,21-25,27] schemes.

Model compound work has been useful in providing information on the nature of the cross-link structures. Model compound work for unaccelerated sulfur vulcanization has included the use of mono-[11,21,24,25,27] and diolefins.[11,21,23-25] The polyisoprene model compound work has indicated that the structures in elemental sulfur vulcanization are alkyl-alkenyl and alkyl-alkyl in nature.[11,21-25] These studies observed migration of the olefinic double bonds and formation of conjugated trienes and internally cyclized structures. The polybutadiene model compound work found significant saturation of the double bond and formation of vicinal sulfur structures.[27] Upon addition of zinc oxide to both systems, the amount of alkenyl substitution increased. Addition of zinc oxide dramatically changes the stereochemistry of the vicinal structures. Wolfe[27] attributed this to a change in reaction mechanism, with pure sulfur cure as a radical mechanism and zinc oxide-sulfur cure as a polar process.

Solid-state C^{13}-NMR has been used to study chemical structure of vulcanizates of sulfur cure.[66-68] The NMR studies indicate that isomerization is the dominant reaction occurring during vulcanization. NMR also found evidence for allylic-substituted alkenyl structures, saturation, and internal cyclization. These studies were performed both on *cis*-polybutadiene[66] and natural rubber.[67,68] These results agree well with chemical structures suggested by the model compound work.

Electron spin resonance has also been used to study the reaction mechanism. The early ESR work was done by Blokh,[20] but this was later found to suffer from problems with impurities when used to study accelerated-sulfur cure.[11] For unaccelerated-sulfur cure, Blokh found no evidence for the presence of radicals during vulcanization. This also agrees with the work of Shelton and McDonel[26] in which they observed the effect of radical traps on various accelerator systems. Their work is summarized in Table 3.

Free-radical sulfurization is believed to be initiated by homolytic scission of the sulfur ring leading to formation of radicals as illustrated in Reaction Scheme I in Figure 3.[13,14] Proton abstraction by this radical and sulfur addition to the resulting rubber radical gives rise to the propagating sulfurization species. This mechanistic scheme is able to account for alkenyl-alkyl products. Alkyl-alkyl products and vicinal crosslinks are formed if the cross-linked rubber radical added sulfur instead of abstracting a proton. Isomerization and double-bond migration is accounted for by Reaction Scheme 2 in Figure 4.[13,14] The rubber radical exists in two allylic positions. Sulfur can add to the radical at either position; addition of sulfur at the quaternary carbon results in double-bond migration. Isomerization occurs when the radical returns to the methylene carbon; the double bond is then capable of reforming in either configuration.

TABLE 3
Mechanism of Vulcanization per Work
of SHELTON and MCDONEL Using
Radical Scavengers

System	Mechanism
Sulfur	Polar
Sulfur-Guanadine	Polar
Thiuram Disulfide-sulfur	Predominantly polar
Dithiocarbamate-sulfur	Predominantly Polar
Thiazole-sulfur	Predominantly polar
Thiazole-disulfide sulfur	Polar-radical
Sulfenamide-sulfur	Polar-radical
Thiuram disulfide	Unsure

From Shelton, J. R. and McDonel, E. T., *Rubber Chem.
Technol.*, 33, 342 (1960). With permission.

The proposed polar mechanism for unaccelerated sulfur vulcanization is shown in Reaction Schemes 3 and 4 in Figures 5 and 6.[11] Figure 6 is a generalized version of Figure 5 allowing for either proton or hydride transfer, while Reaction Scheme 3 illustrates the reaction for proton transfer. The key step is the formation of the three-membered, sulfur-carbon charged ring. This reaction mechanism is also able to predict the formation of alkenyl-alkyl structures. Alkyl-alkyl structures result from the saturated persulfenyl rubber ion reacting with a rubber molecule, followed by proton transfer. Isomerization occurs through the nonsulfurated saturated ion. If this ion undergoes proton transfer then the double bond reforms; this can occur in either configuration. Formation of noncross-linked saturated structures and conjugated trienes are shown in Reaction Scheme 5 in Figure 7.[11] These structures are formed from rubber moieties that undergo proton abstraction and then either proton transfer or hydride abstraction. The formation of cyclic structures is shown in Figure 8[11] for Reaction Scheme 6. The sulfur crosslink can cleave at an S-S bond, which is a relatively labile bond. The sulfur atom then becomes bonded to a carbon to form the cyclic structure.

Related to the formation of cyclic structures is the reduction of rank of crosslinks with increasing cure time. There are several proposed reaction schemes; the mechanism that is commonly favored is an exchange reaction between the crosslink and the sulfurating intermediates. This reaction terminates upon formation of monosulfidic structures.

Based on available evidence, unaccelerated vulcanization is believed to occur by a polar mechanism. A radical mechanism should give much higher levels of migration of the double bond; this arises from the stability of the radical at the quaternary carbon. Electron spin resonance studies and the radical scavenger work also point toward polar reactions. Due to the ability of sulfur to undergo radical reactions, it is not possible to completely eliminate the possibility of radical reactions during unaccelerated sulfur vulcanization.

III. ACCELERATED SULFUR VULCANIZATION

The study of accelerated sulfur vulcanization suffers from the same problems as unaccelerated sulfur vulcanization, including the ability of sulfur to undergo both radical and ionic reactions and the intractable nature of cured vulcanizates. Further complicating the situation is the necessity of understanding how the accelerators and activators interact, and how these interactions affect the vulcanization mechanism. Proposed mechanisms have ranged from radical[11,14-19] to polar.[11,21] Several researchers have concluded that both radical

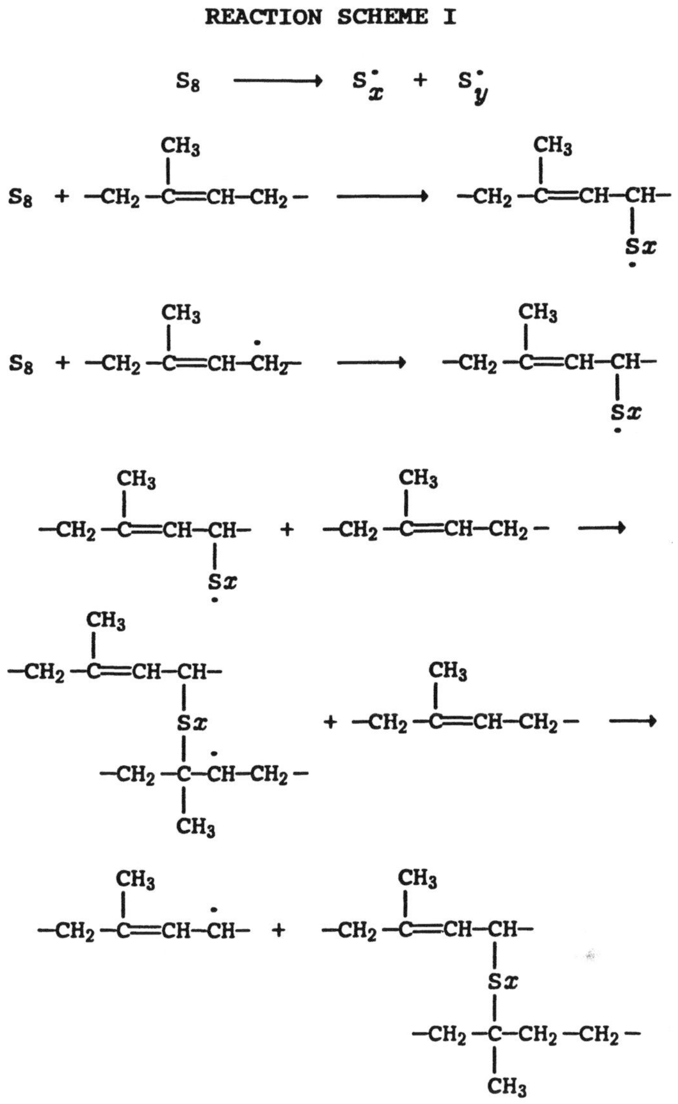

REACTION SCHEME I

$$S_8 \longrightarrow S_x^{\cdot} + S_y^{\cdot}$$

$$S_8 + -CH_2-\underset{\underset{CH_3}{|}}{C}=CH-CH_2- \longrightarrow -CH_2-\underset{\underset{CH_3}{|}}{C}=CH-\underset{\underset{\underset{\cdot}{S_x}}{|}}{CH}-$$

$$S_8 + -CH_2-\underset{\underset{CH_3}{|}}{C}=CH-\overset{\cdot}{CH_2}- \longrightarrow -CH_2-\underset{\underset{CH_3}{|}}{C}=CH-\underset{\underset{\underset{\cdot}{S_x}}{|}}{CH}-$$

$$-CH_2-\underset{\underset{CH_3}{|}}{C}=CH-\underset{\underset{\underset{\cdot}{S_x}}{|}}{CH}- \ + \ -CH_2-\underset{\underset{CH_3}{|}}{C}=CH-CH_2- \longrightarrow$$

$$-CH_2-\underset{\underset{CH_3}{|}}{C}=CH-\underset{\underset{\underset{-CH_2-\overset{\cdot}{\underset{\underset{CH_3}{|}}{C}}-CH-CH_2-}{|}}{\underset{S_x}{|}}}{CH}- \quad + \ -CH_2-\underset{\underset{CH_3}{|}}{C}=CH-CH_2- \longrightarrow$$

$$-CH_2-\underset{\underset{CH_3}{|}}{C}=CH-\overset{\cdot}{CH}- \ + \ -CH_2-\underset{\underset{CH_3}{|}}{C}=CH-\underset{\underset{\underset{-CH_2-\underset{\underset{CH_3}{|}}{C}-CH_2-CH_2-}{|}}{S_x}}{CH}-$$

FIGURE 3. Proposed radical mechanism for unaccelerated sulfur vulcanization. (From R. F. Naylor, *J. Polym. Sci.*, 1, 305 (1946) and G. F. Bloomfield, *J. Polym. Sci.*, 1, 312 (1946). Both with permission.)

and polar mechanisms are operative[26,28-53] and that the precise nature is dependent on the formulation.

A typical accelerated sulfur system contains sulfur, accelerator, zinc oxide, and a saturated fatty acid such as stearic or lauric acid.[1,3-4] Other additives can include stabilizers, antioxidants, and fillers such as silica or carbon black. There are also sulfur-free systems; the most widely used sulfurless system is zinc oxide and tetramethyl thiuram disulfide. Additionally, other accelerators can be added to thiocarbamate systems to increase the induction time. The most widely used compound for this application is MBT.[34,41,80,81]

Accelerated sulfur vulcanization has been found to consume the accelerator in the system at a rate far greater than the rate of crosslinking.[76,77] This has led to the proposal that accelerated vulcanization proceeds through an intermediate.[11,21,76,77,82,83] Subsequent research has provided strong evidence for the existence of this species.[29-53,55-58] Coran, in his work,[82]

REACTION SCHEME II

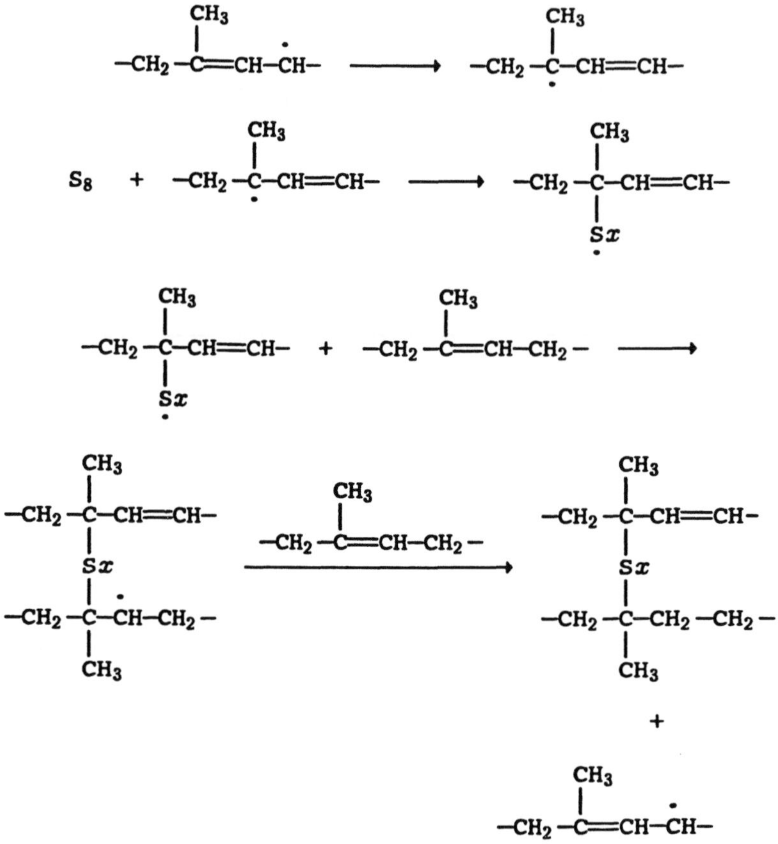

FIGURE 4. Proposed radical mechanism for unaccelerated sulfur vulcanization illustrating the migration of the carbon-carbon double bond. (From R. F. Naylor, *J. Polym. Sci.*, 1, 305 (1946). Both with permission.)

isolated a compound he identified as a complex consisting of the zinc salt of the accelerator stabilized by interactions with stearic acid. Coran[83] modeled the mechanistic scheme for formation of this complex and subsequent crosslinking/sulfurization as:

$$A \xrightarrow{k1} B \xrightarrow{k2} B^* \xrightarrow{k3} \alpha Vu$$

$$A + B^* \xrightarrow{k4} \beta B$$

where A = accelerator, B = intermediate, B* = active intermediate (sulfurating agent), Vu = crosslink, and α, β = constants to adjust stoichiometry.

Differential scanning calorimetry (DSC) has been used to study this kinetic scheme;[84] a mechanism was proposed based on the results which agreed with the results of Coran.

The proposed mechanism for formation of the complex is shown in Reaction Scheme 7 in Figure 9.[16,85-88] Zinc is inserted into the accelerator, and this zinc salt then reacts with sulfur to form a persulfenyl zinc complex. Stearic acid is believed to solubilize the zinc[1,3,11,82] allowing diffusion into the elastomer. Coran[82] found that using zinc stearate instead of zinc oxide and stearic acid produced similar vulcanizates. The stearic acids also serves to stabilize

REACTION SCHEME III

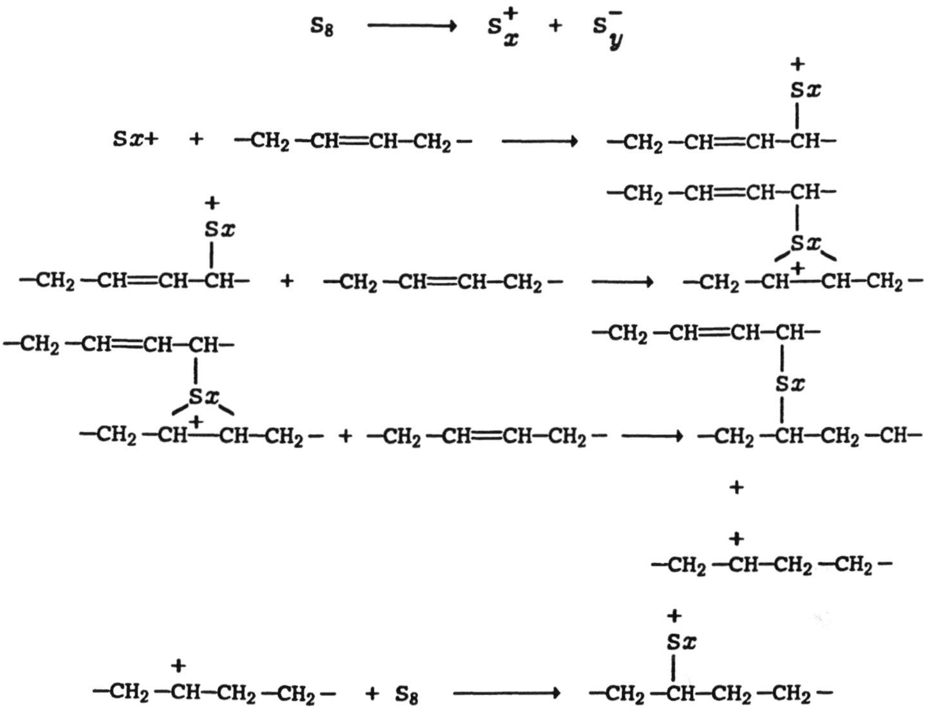

$$S_8 \longrightarrow S_x^+ + S_y^-$$

$$Sx^+ \ + \ -CH_2-CH=CH-CH_2- \ \longrightarrow \ \overset{\overset{\textstyle +}{\underset{\textstyle |}{Sx}}}{-CH_2-CH=CH-CH-}$$

$$\overset{\overset{\textstyle +}{\underset{\textstyle |}{Sx}}}{-CH_2-CH=CH-CH-} \ + \ -CH_2-CH=CH-CH_2- \ \longrightarrow \ \begin{array}{c} -CH_2-CH=CH-CH- \\ | \\ Sx \\ -CH_2-CH\overset{+}{-}CH-CH_2- \end{array}$$

$$\overset{Sx}{-CH_2-CH\overset{+}{-}CH-CH_2-} \ + \ -CH_2-CH=CH-CH_2- \ \longrightarrow \ \begin{array}{c} -CH_2-CH=CH-CH- \\ | \\ Sx \\ -CH_2-CH-CH_2-CH- \\ + \\ \overset{+}{-CH_2-CH-CH_2-CH_2-} \end{array}$$

$$\overset{+}{-CH_2-CH-CH_2-CH_2-} \ + \ S_8 \ \longrightarrow \ \overset{\overset{\textstyle +}{\underset{\textstyle |}{Sx}}}{-CH_2-CH-CH_2-CH_2-}$$

FIGURE 5. Proposed polar mechanism for unaccelerated sulfur vulcanization. (From L. Bateman, C. G. Moore, M. Porter, and B. Saville, *The Chemistry and Physics of Rubber-like Substances,* L. Bateman, Ed., Wiley, New York, 1963, chap. 15. With permission.)

the complex. Duchacek[36] noted that increasing the zinc oxide concentration can increase the rate and extent of crosslinking up to a certain zinc oxide concentration; this concentration is believed to be the minimum level of zinc oxide needed to completely convert the accelerator to the zinc-accelerator-sulfur complex. Duchacek also noted that the optimum zinc oxide content to minimize reversion was slightly greater; the rationale for this was not deduced. Coran[82] also noted that the induction time had a dependence on the zinc oxide concentration in excess of that required for formation of the accelerator-zinc complex. This suggests that the zinc exerts an influence beyond inclusion in the accelerator complexes.

Amides also are often used in vulcanization formulations; their purpose is to solubilize the zinc and to stabilize the complex.[89-92] It was found that the addition of amines to a rubber formulation decreased the induction time,[89-92] but for thiuram accelerators it decreased the maximum extent of crosslinking.[90,91] This was attributed to the reaction of the amine with the accelerator. For sulfenamide accelerators the addition of amides was found to decrease the induction time but had no affect on the rate or extent of crosslinking.[89]

Coleman et al. performed thermal decomposition studies[32] of TMTD and ZnDMDC in the presence and absence of ZnO and sulfur using Raman spectroscopy and ESR. Their ESR work indicated the presence of thiuram and persulfenyl thiuram radicals, while the Raman studies identified the zinc complex of mono- and dithiocarbamates, but was unable to find evidence for the zinc-persulfenyl-thiocarbamate complex. It was also found that the persulfurated complexes existed primarily at short cure times; at longer cure times the TMTD and ZnDMDC are degraded to thiourea and CS_2. The ESR work was careful to avoid the Cu impurities that had plagued past studies.[11,20]

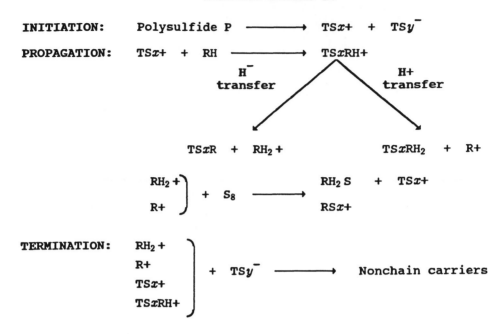

FIGURE 6.   Generalized polar reaction mechanism for unaccelerated sulfur vulcanization. T is an alkyl or alkenyl group and RH is a rubber molecule. (From L. Bateman, C. G. Moore, M. Porter, and B. Saville, *The Chemistry and Physics of Rubber-like Substance*, L. Bateman, Ed., Wiley, New York, 1963, chap. 15. With permission.)

The proposed mechanism of formation of the persulfenyl-zinc complex in a sulfurless system is shown in Figure 10,[11] Reaction Scheme 8 by a polar mechanism. Zinc oxide cleaves the thiuram moiety resulting in an anion. This ion then attacks another accelerator molecule resulting in a polysulfidic accelerator and reformation of an ion. Zinc insertion into the accelerator gives the intermediate. Compounds other than TMTD have been studied for use as sulfur-donor accelerators; this includes benzothiazyldithimorpholide[93] shown below:

Model compound work has found that for accelerated sulfur vulcanization the resulting structures are primarily alkenyl-alkenyl, with limited migration of the double bond.[11,12,28,54,55,57,58] Models used have included cyclohexene and *cis*-3-hexene for *cis*-polybutadiene[10,11,28,55,58] and 2-methyl-2-pentene and 2,6-dimethyl-octa-2,6-diene[10,11,54,55,57] for natural rubber and *cis*-polyisoprene. The model compound work did find evidence for saturated structures and internal cyclization, but these reactions are minor reactions compared to the alkeynl-alkenyl structures. This work indicates that with the addition of accelerators and activators, the reaction mechanisms undergo a change from substitutive-additive reactions to substitutive-substitutive reactions.

Model compound work on polybutadiene models led to higher levels of vicinal crosslinks and saturation then did natural rubber models.[28,55,58] Gregg and Lattimer indicated that this may be due to the reactive behavior of the polybutadiene models.[54] Their work found that cyclohexene and 1,5-cyclooctadiene did not react fully. Their work also found that a 16-

REACTION SCHEME V

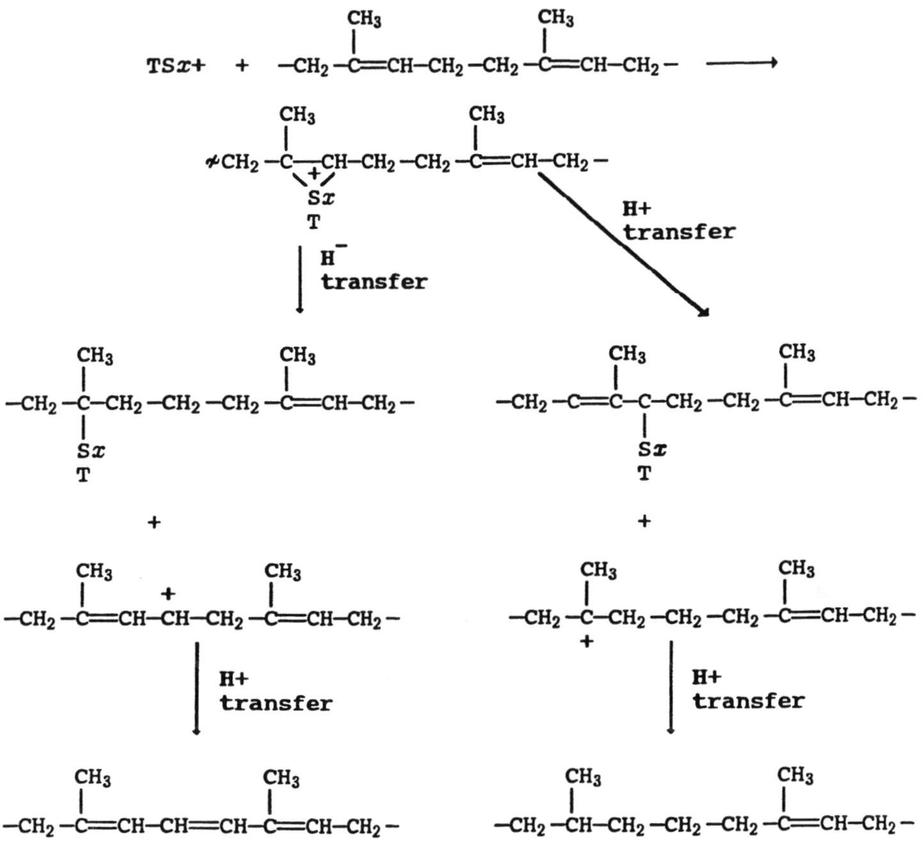

FIGURE 7. Proposed polar mechanism for the formation of saturated and conjugated triene structures in un-accelerated sulfur vulcanization. (From L. Bateman, C. G. Moore, M. Porter, and B. Saville, *The Chemistry and Physics of Rubber-like Substance*, L. Bateman, Ed., Wiley, New York, 1963, chap. 15. With permission.)

membered ring with four double bonds was a better model for polybutadiene and resulted in products similar to those from natural rubber model compounds.

Solid-state NMR studies of accelerated sulfur vulcanization have been performed on several elastomers including natural rubber/*cis*-polyisoprene[72-75] polybutadiene,[69,71] and butyl rubber.[62] These studies found similar results to those of model compound studies, but with some differences. The model compound studies found higher levels of saturation and double-bond migration then were observed by NMR studies. The sulfurization structures in the NMR studies consist almost exclusively of substitutive-substitutive structures (alkenyl-alkenyl).

Research by Manik, Banerjee, and co-workers[44-53] led to the conclusions that both polar and radical mechanisms are operative during vulcanization as is shown in Table 4. Their work indicated that several common accelerators including MBT and TMTD were capable of undergoing both polar and radical reactions. This agrees with the results of Shelton and McDonel[26] shown in Table 3 using radical scavengers, which indicated that both radical and polar reactions can occur.

The proposed radical mechanism of accelerated sulfur vulcanization is shown in Figure 11,[15] Reaction Scheme 9. The intermediate cleaves to form persulfenyl radicals, which then abstract protons. The rubber radical reacts with another intermediate to form a rubber-bound

REACTION SCHEME VI

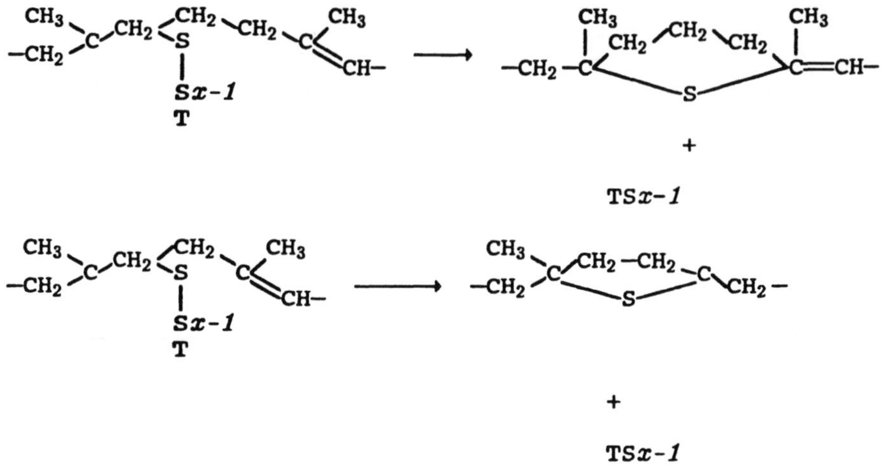

FIGURE 8.   Proposed polar mechanism for the formation of cyclic structures in unaccelerated sulfur vulcanization. (From L. Bateman, C. G. Moore, M. Porter, and B. Saville, *The Chemistry and Physics of Rubber-like Substance,* L. Bateman, Ed., Wiley, New York, 1963, chap. 15. With permission.)

REACTION SCHEME VII

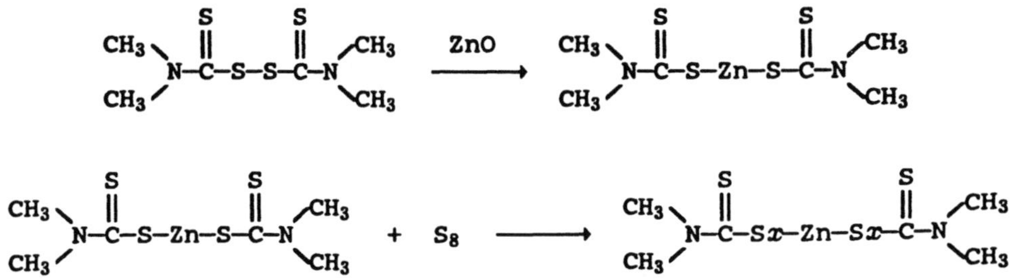

FIGURE 9.   Proposed mechanism for the formation of the accelerator complex in accelerated sulfur vulcanization.[16,85-88].

intermediate. Two rubber-bound intermediates then form the actual crosslink. Maturing of the network occurs through sulfur exchange reactions. Isomerizaton, which is widely observed in vulcanization,[11,69] occurs through R· similar to the mechanism for unaccelerated sulfur vulcanization.

The proposed polar mechanism is shown in Figure 12,[11] Reaction Scheme 10. The key step is a concerted reaction of a "ring" structure leading to the formation of the rubber-bound intermediate. Reduction of sulfur rank with this mechanism is shown in Reaction Scheme 11 in Figure 13[11] by attack of the intermediate on the sulfur crosslink. Isomerization occurs by loss of the rubber-bound intermediate; while as an ion or during a concerted displacement the double bond does not have full olefinic character and is able to isomerize.

It should be noted that there are varying opinions on the concentration of the rubber bound intermediate at any given time in the reaction process. Solid-state studies have been useful in this respect,[70,71] as have been model compound studies.[94] These studies suggest that in isoprene or isoprene-based systems such as butyl rubber there is a very low

REACTION SCHEME VIII

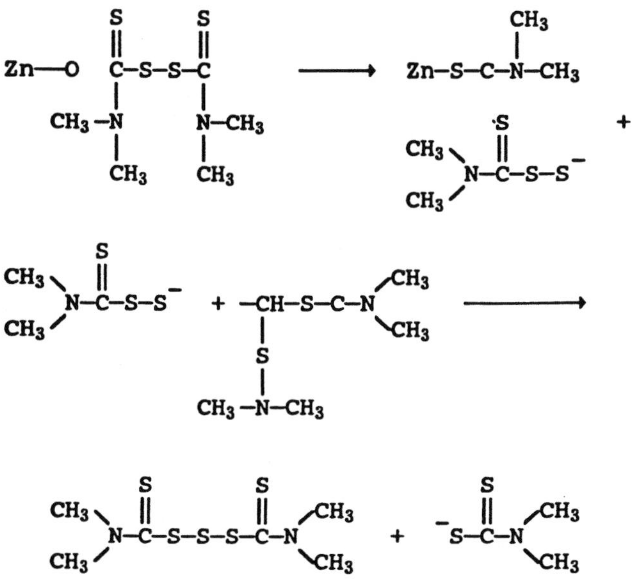

Zn—O C-S-S-C → Zn-S-C-N-CH₃ +

CH₃—N N-CH₃
CH₃ CH₃

CH₃ N-C-S-S⁻ + -CH-S-C-N CH₃ →
CH₃ CH₃
S
CH₃-N-CH₃

CH₃ N-C-S-S-S-C-N CH₃ + ⁻S-C-N CH₃
CH₃ CH₃ CH₃

FIGURE 10. Proposed polar mechanism for sulfurless vulcanization. (From L. Bateman, C. G. Moore, M. Porter, and B. Saville, *The Chemistry and Physics of Rubber-like Substance*, L. Bateman, Ed., Wiley, New York, 1963, chap. 15. With permission.)

TABLE 4
Reaction Mechanisms for Vulcanization[a]

Mechanism	System
Radical	NR/TMTD
	NR/TMTD/sulfur
	NR/CBS/sulfur
	NR/MBT/sulfur
	SBR/MBT/sulfur
Mixed radical/polar	NR/MBT/ZnO/stearic acid
	NR/CBS/sulfur/ZnO/stearic acid
	SBR/CBS/sulfur
	SBR/CBS/sulfur/ZnO/stearic acid
	SBR/sulfur/DPG/ZnO/stearic acid
	NR/MBT/TMTD
	NR/MBT/DPG
	NR/DPG/sulfur/ZnO/stearic acid
Polar	NR/TMTD/ZnO
	NR/TMTD/sulfur/ZnO/stearic acid
	NR/MBT/sulfur/ZnO/stearic acid
	NR/MBT/TMTD/sulfur/ZnO/stearic acid
	NR/DPG/sulfur
	NR/MBT/DPG/sulfur//ZnO/stearic acid
	SBR/MBT/sulfur/ZnO/stearic acid
	SBR/sulfur/DPG

[a] From References 44–53. With permission.

REACTION SCHEME IX

$$\text{XSS}x\text{ZnS}x\text{SX} \longrightarrow \text{XSS}x^{\bullet} + \text{XS}x\text{Zn}^{\bullet}$$

$$\text{XSS}x^{\bullet} + \text{RH} \longrightarrow \text{XSS}x\text{H} + \text{R}^{\bullet}$$

$$\text{R}^{\bullet} + \text{XSS}x\text{ZnS}x\text{SX} \longrightarrow \text{RS}y\text{X} + \text{XS}x\text{-}y$$

$$\text{R}^{\bullet} + \text{XS}x^{\bullet} \longrightarrow \text{RS}x\text{X}$$

$$\text{RS}x\text{X} + \text{RS}y\text{X} \xrightarrow{\text{XSS}x\text{ZnS}x\text{SX}} \text{RS}x\text{R} + \text{XS}y\text{X}$$

FIGURE 11. Proposed radical mechanism of accelerated sulfur vulcanization. X is an accelerator residue and RH is a rubber molecule.

REACTION SCHEME X

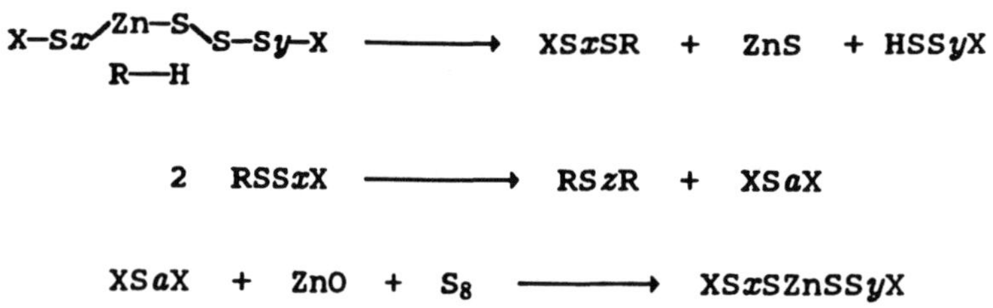

$$\text{X-S}x{\nearrow}^{\text{Zn-S}}{\searrow}\text{S-S}y\text{-X} \longrightarrow \text{XS}x\text{SR} + \text{ZnS} + \text{HSS}y\text{X}$$
$$\text{R-H}$$

$$2\ \text{RSS}x\text{X} \longrightarrow \text{RS}z\text{R} + \text{XS}a\text{X}$$

$$\text{XS}a\text{X} + \text{ZnO} + \text{S}_8 \longrightarrow \text{XS}x\text{SZnSS}y\text{X}$$

FIGURE 12. Proposed polar mechanism for accelerated sulfur vulcanization. (From L. Bateman, C. G. Moore, M. Porter, and B. Saville, *The Chemistry and Physics of Rubber-like Substance,* L. Bateman, Ed., Wiley, New York, 1963, chap. 15. With permission.)

REACTION SCHEME XI

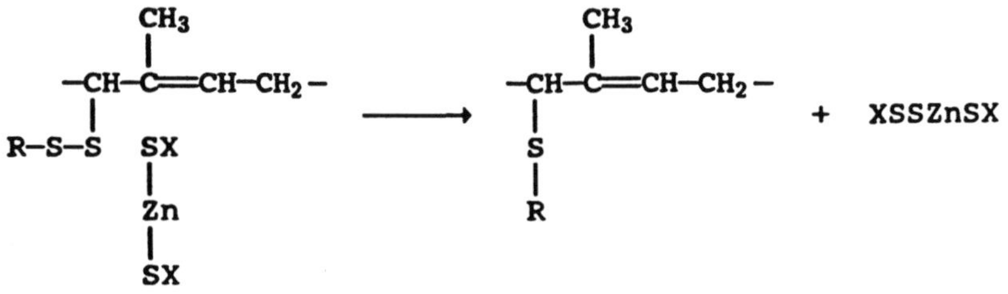

FIGURE 13. Proposed polar mechanism for reduction of sulfur rank during accelerated sulfur vulcanization. (From L. Bateman, C. G. Moore, M. Porter, and B. Saville, *The Chemistry and Physics of Rubber-like Substance*, L. Bateman, Ed., Wiley, New York, 1963, chap. 15. With permission.)

concentration of the rubber-bound intermediates, while for butadiene the concentration of rubber-bound intermediates is quite high.

The exact nature of accelerated sulfur vulcanization is still a topic of much debate. Recent advances in instrumental techniques have been able to provide more information about the resulting structures, but the actual chemistry involved is still not sufficiently clear.

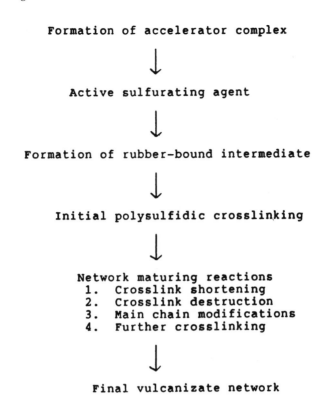

FIGURE 14. Summary of overall course of accelerated sulfur vulcanization.

It is probable that both free radical and polar mechanism are operative, and that the exact nature of the vulcanization process will vary between different curing systems.

An overall summary of the present view of accelerated sulfur vulcanization is shown in Figure 14. The accelerator complex forms, then cleaves to yield the active sulfurating agent. This results in the formation of a rubber-bound intermediate. The rubber-bound intermediate then forms polysulfidic crosslinks, which are reduced in rank or destroyed upon network maturing. As the cure continues, some chain scission will occur with further crosslinking. The end result is the final vulcanizate network.

REFERENCES

1. **Alliger, G. and Sjothun, I. J.,** *Vulcanization of Elastomers,* Von Nostrand Reinhold, New York, 1964, Chap. 1.
2. *A Centennial Volume of the Writings of Charles Goodyear and Thomas Hancock,* American Chemical Society, 1939.
3. **Hoffman, W.,** *Vulcanization and Vulcanizing Agents,* MacLean, London, 1967.
4. Vulcanization, *Encyclopedia of Polymer Science and Engineering,* 2nd ed., Vol. 17., Mark, Bikales, Overberger, and Menges, Eds., Wiley, New York, 1989.
5. **Das, P. K., Datta, R. N., and Basu, P. K.,** *Rubber Chem. Technol.,* 61, 760, 1988.
6. **Layer, R. W.,** *Rubber Chem. Technol.,* 62, 124, 1989.
7. **Adhikari, B., Pal, D., Basu, D. K., and Chaudhuri, A. K.,** *Rubber Chem. Technol.,* 56, 328, 1983.
8. **Pal, D., Adhikari, B., Basu, D. K., and Chaudhuri, A. K.,** *Rubber Chem. Technol.,* 56, 827, 1983.
9. **Mathur, R. P., Mitra, A., Ghoshal, P. K., and Das, C. K.,** *Kautsch. Gummi Kunstst.,* 36, 1067, 1983.

10. **Datta, R. N., Das, P. K., and Basu, D. K.,** *Kautsch. Gummi Kunstst.,* 39, 1090, 1986.
11. **Bateman, L., Moore, C. G., Porter, M., and Saville, B.,** *The Chemistry and Physics of Rubber-like Substances,* L. Bateman, Ed., Wiley, New York, 1963, chap. 15.
12. **Farmer, E. H. and Shipley, F. W.,** *J. Polym. Sci.,* 1, 293, 1946.
13. **Naylor, R. F.,** *J. Polym. Sci.,* 1, 305, 1946.
14. **Bloomfield, G. F.,** *J. Polym. Sci.,* 1, 312, 1946.
15. **Scheele, W., Lorenz, O., and Dummer, W.,** *Rubber Chem. Technol.,* 29, 29, 1956.
16. **Craig, D.,** *Rubber Chem. Technol.,* 30, 1291, 1957.
17. **Dogadkin, B. A.,** *Rubber Chem. Technol.,* 32, 174, 1959.
18. **Beniska, J. and Dogadkin, B.,** *Rubber Chem. Technol.,* 32, 774, 1957.
19. **Bevilacqua, E. M.,** *Rubber Chem. Technol.,* 32, 721, 1959.
20. **Blokh, G. A.,** *Rubber Chem. Technol.,* 33, 1005, 1960.
21. **Bateman, L., Glazebrook, R. W., and Moore, C. G.,** *Rubber Chem. Technol.,* 35, 633, 1962.
22. **Bateman, L., Glazebrook, R. W., Moore, C. G., Porter, M., Ross, G. W., and Saville, R. W.,** *J. Chem. Soc.,* 2838, 1958.
23. **Bateman, L., Glazebrook, R. W., and Moore, C. G.,** *J. Chem. Soc.,* 2846, 1958.
24. **Ross, G. W.,** *J. Chem. Soc.,* 2856, 1958.
25. **Bateman, L., Moore, C. G., and Porter, M.,** *J. Chem. Soc.,* 2866, 1958.
26. **Shelton, J. R. and McDonel, E. T.,** *Rubber Chem. Technol.,* 33, 342, 1960.
27. **Wolfe, J. R., Pugh, T. L., and Killian, A. S.,** *Rubber Chem. Technol.,* 41, 1329, 1968.
28. **Wolfe, J. R.,** *Rubber Chem. Technol.,* 41, 1339, 1968.
29. **Coleman, M. M., Shelton, J. R., and Koenig, J. L.,** *Rubber Chem. Technol.,* 44, 71, 1971.
30. **Coleman, M. M., Shelton, J. R., and Koenig, J. L.,** *Rubber Chem. Technol.,* 45, 173, 1972.
31. **Coleman, M. M., Shelton, J. R., and Koenig, J. L.,** *Rubber Chem. Technol.,* 46, 938, 1973.
32. **Coleman, M. M., Shelton, J. R., and Koenig, J. L.,** *Rubber Chem. Technol.,* 46, 957, 1973.
33. **Kapur, R. S., Koenig, J. L., and Shelton, J. R.,** *Rubber Chem. Technol.,* 47, 911, 1974.
34. **Duchacek, V. and Frenkel, R.,** *J. Appl. Polym. Sci.,* 23, 2065, 1979.
35. **Duchacek, V.,** *J. Appl. Polym. Sci.,* 22, 227, 1978.
36. **Duchacek, V.,** *J. Appl. Polym. Sci.,* 20, 71, 1976.
37. **Duchacek, V.,** *J. Appl. Polym. Sci.,* 19, 1617, 1975.
38. **Duchacek, V.,** *J. Appl. Polym. Sci.,* 19, 645, 1975.
39. **Duchacek, V. and Brajko, V.,** *J. Appl. Polym. Sci.,* 18, 2797, 1974.
40. **Duchacek, V.,** *J. Appl. Polym. Sci.,* 18, 125, 1974.
41. **Duchacek, V.,** *J. Appl. Polym. Sci.,* 16, 3245, 1972.
42. **Duchacek, V.,** *J. Appl. Polym. Sci.,* 15, 2079, 1971.
43. **Duchacek,** *Angew. Makromol. Chem.,* 23, 21, 1972.
44. **Manik, S. P. and Banerjee, S.,** *Rubber Chem. Technol.,* 42, 744, 1969.
45. **Manik, S. P. and Banerjee, S.,** *Rubber Chem. Technol.,* 43, 1249, 1970.
46. **Manik, S. P. and Banerjee, S.,** *Rubber Chem. Technol.,* 43, 1311, 1970.
47. **Das, C. K. and Banerjee, S.,** *Rubber Chem. Technol.,* 47, 266, 1974.
48. **Bhoumick, S. and Banerjee, S.,** *Rubber Chem. Technol.,* 47, 251, 1974.
49. **Ghosh, A. K., Das, C. K., and Banerjee, S.,** *J. Polym. Sci., Poly. Chem. Ed.,* 15, 2773, 1977.
50. **Das, C. K. and Banerjee, S.,** *J. Polym. Sci., Poly. Chem. Ed.,* 16, 2971, 1978.
51. **Manik, S. P. and Banerjee, S.,** *J. Appl. Polym. Sci.,* 15, 1341, 1971.
52. **Bandyopadhyay, P. K. and Banerjee, S.,** *Angew. Makromol. Chem.,* 64, 59, 1977.
53. **Bandyopadhyay, P. K. and Banerjee, S.,** *J. Appl. Polym. Sci.,* 23, 185, 1979.
54. **Gregg, E. C. and Lattimer, R. P.,** *Rubber Chem. Technol.,* 57, 1056, 1984.
55. **Skinner, T. D.,** *Rubber Chem. Technol.,* 45, 182, 1972.
56. **Morrison, N. J.,** *Rubber Chem. Technol.,* 57, 97, 1984.
57. **Morrison, N. J.,** *Rubber Chem. Technol.,* 57, 86, 1984.
58. **Gregg, E. C. and Katrenick, S. E.,** *Rubber Chem. Technol.,* 43, 549, 1970.
59. **Lautenschlaeger, F. K.,** *Rubber Chem. Technol.,* 52, 213, 1979.
60. **Selker, M. L. and Kemp, A. R.,** *Ind. Eng. Chem.,* 36, 16, 1944.
61. **Moore, C. G., Mullins, L., and Swift, P.,** *J. Appl. Poly. Sci.,* 5(15), 293, 1961.
62. **Moore, C. G. and Trego, B. R.,** *J. Appl. Poly. Sci.,* 5(15), 299, 1961.
63. **Moore, C. G. and Watson, A. A.,** *J. Appl. Sci.,* 8, 581, 1964.
64. **Moore, C. G. and Trego, B. R.,** *J. Appl. Poly. Sci.,* 8, 1957, 1964.
65. **Cunneen, J. I. and Russell, R. M.,** *Rubber Chem. Technol.,* 43, 1215, 1970.
66. **Clough, R. S. and Koenig, J. L.,** *Rubber Chem. Technol.,* 62, 908, 1989.
67. **Andreis, M., Liu, J., and Koenig, J. L.,** *J. Polym. Sci., Part B: Poly. Phys.,* 27, 1389, 1989.
68. **Zaper, A. M. and Koenig, J. L.,** *Rubber Chem. Technol.,* 60, 252, 1987.
69. **Zaper, A. M. and Koenig, J. L.,** *Rubber Chem. Technol.,* 60, 278, 1987.

70. **Krejsa, M. R. and Koenig, J. L.**, *Rubber Chem. Technol.*, 64, 40, 1991.
71. **Smith, S. R. and Koenig, J. L.**, *Rubber Chem. Technol.*, 65, 176, 1992.
72. **Hirst, R. C.**, Paper No. 69 Rubber Division Meeting, Detroit, MI, October 1991.
73. **Gronski, W., Hasenhindl, H., Freund, B., and Wolff, S.**, *Kautsch. Gummi Kunstst.*, 44, 119, 1991.
74. **Gronski, W., Hoffman, U., Simon, G., Wutzler, A,., and Straube, E.**, *Rubber Chem. Technol.*, 65, 63, 1992.
75. **Krejsa, M. R. and Koenig, J. L.**, *Rubber Chem. Technol.*, 65, 427, 1992.
76. **Campbell, R. H. and Wise, R. W.**, *Rubber Chem. Technol.*, 37, 635, 1964.
77. **Campbell, R. H. and Wise, R. W.**, *Rubber Chem. Technol.*, 37, 650, 1964.
78. **Sullivan, A. B., Hann, C. J., and Kuhls, G. H.**, *Rubber Chem. Technol.*, in press.
79. **Prior, W.**, *Mechanisms of Sulfur Reactions*, McGraw-Hill, New York, 1962.
80. **Das, C. K. and Millns, W.**, *Rubber India*, 30, 13, 1978.
81. **Simanenkova, L. B., Tarkhov, G. V., Pipiraite, P. P., and Shurkus, A. S.**, *Kauch. Rezina*, 42, 13, 1983.
82. **Coran, A. Y.**, *Rubber Chem. Technol.*, 37, 679, 1964.
83. **Coran, A. Y.**, *Rubber Chem. Technol.*, 37, 689, 1964.
84. **Huson, M. G., McGill, W. J., and Wiggett, R. D.**, *J. Polym. Sci., Poly. Chem. Ed.*, 23, 2833, 1985.
85. **Higgins, G. M. C. and Saville, B.**, *J. Chem. Soc.*, 2812, 1963.
86. **Krebs, H.**, *Rubber Chem. Technol.*, 30, 962, 1957.
87. **Milligan, B.**, *Rubber Chem. Technol.*, 39, 1115, 1966.
88. **Milligan, B.**, *J. Chem. Soc.*, 34, 34, 1966.
89. **Banerjee, B. and Chakravarty, S. N.**, *J. Indian Chem. Soc.*, 59, 403, 1982.
90. **Banerjee, B., Chakravarty, S. N., and Biswas, A. B.**, *J. Appl. Polym. Sci.*, 25, 1263, 1980.
91. **Banerjee, B., Chakravarty, S. N., Biswas, A. B., and Kamath, B. V.**, *J. Appl. Polym. Sci.*, 24, 683, 1979.
92. **Fegade, N. B., Deshpande, N. M., Millns, W.**, *Rubber World*, 191(1), 32, 1984.
93. **Mitra, A., Das, C. K., and Millns, W.**, *Kautsch. Bummi Kunstst.*, 36, 103, 1983.
94. **Morrison, N. J. and Porter, M.**, *Rubber Chem. Technol.*, 57, 63, 1984.

Chapter 12

PEROXIDE CROSSLINKING OF EPDM RUBBERS

Wim C. Endstra and Carel T. J. Wreesmann

TABLE OF CONTENTS

I. INTRODUCTION

The discovery of peroxide crosslinking of elastomers by peroxides dates from almost eight decades ago when experiments were described concerning the reaction of polymers with organic peroxides. In 1915, Ostromyslenski published that natural rubber could be transferred into a crosslinked state after treatment with dibenzoyl peroxide.[1] From a technical point of view, however, in those days hardly any interest existed for this possibility of crosslinking. This lack of technical interest was mainly caused by the fact that only a limited number of peroxide types were available and additionally, those types being available at that period, had a number of obvious drawbacks.

It lasted until the development and technical breakthrough of fully saturated ethylene-propylene copolymers in the early 1970s,[2] before real technical interest grew for more suitable cross-linking peroxide types. Based on this interest, a number of new peroxide types and formulations were developed that overcame the drawbacks in thermal stability, crosslinking efficiency, and handling and safety aspects of the few existing peroxide types. Particularly the rapid growth of ethylene-propylene terpolymers has given rise to a steadily increasing demand for cross-linking peroxides over the last two decades.

This report surveys various typical aspects of crosslinking of EPDM rubber compounds by organic peroxides. Special attention will be given to compounding considerations, as far as these deviate from general compounding principles applicable for conventional sulfur cure systems. Further processing characteristics of peroxide-containing EPDM compounds will be discussed, and finally, some pilot compound recipes for different application areas will be presented.

II. CROSSLINKING PEROXIDES

From a multitude of different peroxide compounds, it appeared, from theoretical considerations as well as from practical tests, that only a limited number are suitable for cross-linking purposes. The most suitable types are those which form one of the following radicals during their homolytic decomposition:

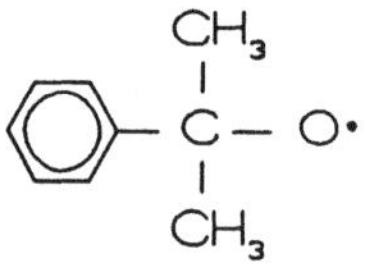

CUMYLOXY RADICAL

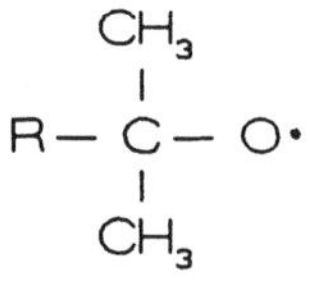

ALKYLOXY RADICAL

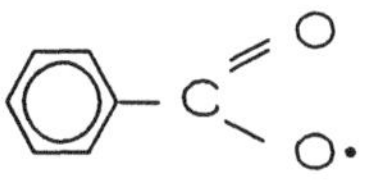

ACYLOXY RADICAL

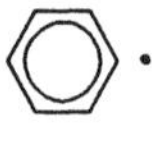

PHENYL RADICAL

$H_3C \cdot$ METHYL RADICAL

TABLE 1
Some Characteristics of Main Peroxides Suitable for Crosslinking
of EPDM Compounds

Class	Example	Ref.	Temp. (°C) $t^{1}/_{2} = 6$ min.	Typical XL temp. (°C)	Dosing range (phr)
Diakyl peroxides	2,5-Bis(*tert*-butylperoxy)-2,5-dimethyl-3-hexyne	1	173	190	6.4–10.7
	2,5-bis(*tert*-butylperoxy)-2,5-dimethylhexane	2	159	180	6.5–10.9
Alkyl-aralkyl peroxides	Bis(*tert*-butylperoxyisopropyl)benzene	3	160	180	3.8–6.3
	tert-butyl cumyl peroxide	4	160	180	4.7–7.8
Diaralkyl peroxide	Dicumyl peroxide	5	155	170	6.1–10.1
Peroxy ketales	Butyl 4,4-bis(*tert*-butylperoxy)valerate	6	143	160	7.5–12.5
	1,1-bis(*tert*-butylperoxy)-3,3,5-trimethyl-cyclohexane	7	129	150	6.8–11.3
Peroxy ester	*tert*-Butyl peroxybenzoate	8	146	140	4.4–7.3

Note: Typical crosslink temperature, as indicated in this table, of a peroxide is the temperature at which 90% of the total possible crosslinks in a low-extended EPDM compound are formed within 10 to 15 minutes (determined with a square die oscillating disc rheometer).

Parts of peroxide to be added per 100 parts of polymer, based on 40% active peroxide formulation.

These radicals are such aggressive species, that they can abstract hydrogen atoms from the polymer chain, whereby stable peroxide decomposition products[3] and polymer radicals are formed. Finally, these polymer radicals can combine, producing a stable C-C crosslink.

Of the three aforementioned successive reaction steps (formation of peroxide radicals, H abstraction, and crosslinking, respectively) the first one is rate determining. The overall cross-linking time required is, therefore, directly dependent on the decomposition rate of the peroxide. The efficiency of the total crosslinking reaction depends mainly on the type of peroxide and polymer radicals formed during the process. The relation between peroxide structure and crosslinking efficiency has been described earlier.[4]

For a first approach these statements are valid. On the other hand, however, in addition to the obvious effect of peroxides on rate and efficiency of the crosslinking reaction, polymer characteristics and other compounding ingredients can also show essential influence on the crosslinking performance, see Sections IV and V.

The main peroxide classes applied for crosslinking EPDM-based compounds are di-alkyl peroxides, alkyl-aralkyl peroxides, diaralkyl peroxides, peroxy ketales, and peroxy esters (see Table 1). Compounds of the di-alkyl peroxide class belong to the thermally most stable ones, whereas, the peroxy ester example mentioned, possesses only quite a limited thermal stability — in a rubber compound — which greatly restricts the practical suitability of this material for crosslinking purposes. In fact, the specific product listed, *tert*-butylperoxy benzoate and a halogen-containing derivative, are claimed to be effective for room temperature curing of EPDM compounds,[5] however, practical use of this phenomenon is very limited.

Thermal stability of peroxides is usually expressed by their so-called half-life (t1/2). This is the time required to decompose half of a given amount of peroxide at a specific temperature. Half-life values, indicated in Table 1, are determined by differential thermal analysis of a solution of the peroxide concerned in monochlorobenzene; these figures have to be used, however, with care, as they do not always directly correspond with the thermal stability in a polymer compound. As a rough estimate, a vulcanization time for a peroxide-containing EPDM rubber compound can be applied that corresponds with approximately seven times the half-life value at a specific temperature. This will result in a situation whereby, in a practical sense, all peroxide is decomposed.

The amounts of peroxide to be added to an EPDM rubber compound, in order to obtain an adequate state of cure, usually range from 9 to 15 mmol/100 g polymer. Features of main crosslinking peroxide types (reference numbers, see Table 1):

1. Possesses extremely high thermal stability; compounds containing this peroxide can be processed safely, without scorch phenomena up to 150°C.
2. Characterized by high thermal stability; FDA-approved for rubber articles intended for repeated use in contact with food; in corresponding BGA recommendation this product is included for EPM-based articles.
3. Shows good thermal stability; because both peroxide groups present are active in crosslinking reactions, the product is a very efficient curing agent; high addition levels can result in blooming phenomena on the surface of the final article, caused by the relatively small solubility of one of the decomposition products, *bis*(hydroxy isopropyl) benzene, in EPDM rubber compounds.[6]
4. Thermal stability comparable with products 2 and 3; technical pure form sometimes preferred because of its liquid physical form; a specific decomposition product, ace-tophenone, imparts a strong odor to the final cured product.
5. Attractive combination of good scorch safety and a relatively fast cure; smell characteristics of cured product comparable with 4.
6. Fast curing can be achieved at 160°C; cross-linking efficiency is less than for 3, 4, and 5, for vulcanization of EPM-based compounds, extra addition of a coagent is beneficial.
7. Fast curing can be achieved at 150°C; efficiency and application in EPM compounds, see 6.
8. Remarkably, the decomposition rate of this product under model conditions is not in line with the cure rate observed under practical conditions in a polymer compound; practical use of this peroxide type is limited.

III. CROSSLINKING SYSTEMS

Usually peroxide-based crosslinking systems are simple. From a chemical point of view the addition of zinc oxide and stearic acid to the recipe is superfluous, in contrast to sulfur-based cure systems, so that in many cases crosslinking can be performed by addition of a singular additive — a specific crosslinking peroxide formulation. In a number of cases, however, the combination of crosslinking peroxide with coagent can be beneficial (see Section V.E). In some other specific cases it appeared to be helpful to apply combinations of different types of crosslinking peroxides. Examples of these kinds of combinations are:

1. The combination of two peroxide compounds that possess a different thermal stability and a different crosslinking efficiency, thus, achieve a fast onset of cure, impairing dimensional stability, and finally, after crosslinking is completed, good physical properties. In particular in salt bath cures the combination of peroxide formulations based on 3 and 7 are successfully applied, see Table 15.
2. As indicated before, the application of peroxide 3 is sometimes affected with blooming phenomena. Although a complete replacement by another peroxide has been proposed,[6] a partial replacement of 3 by a peroxide type that does not suffer from blooming decomposition products and that possesses identical thermal stability (no change in cure characteristics) is beneficial. In particular, 2 is the preferred type of peroxide to be used as a partial replacement of 3.

TABLE 2
Dissociation Energy of Various
C-H Bonds

C-H bond	Dissociation energy (kJ/mol)
H-isopropyl	398
H-*tert*-butyl	385
H-allyl	362

From *Handbook of Chemistry and Physics,*
67th ed., CRC Press, Boca Raton, FL, 1987,
p. F-178.

IV. POLYMER CHARACTERISTICS

An extensive number of structural polymer parameters can be listed that have paramount influence on processing behavior and physical properties of the cured rubber compound.[7-11] As mentioned above, peroxide crosslinking proceeds via abstraction of hydrogen atoms, followed by a rapid recombination of the residual elastomer radicals. Most of the differences in peroxide-cure behavior observed for the many different EPM and EPDM types can be related to the dissociation energy of the various C-H bonds and/or to the relative stability of the intermediary elastomer radicals. The three most important parameters in this aspect are highlighted below.

A. DIENE TYPE

The number of types of third monomer used for commercial EPDM grades is limited: usually dicyclopentadiene (DCPD), ethylidene norbornene (ENB), or 1,4-hexadiene (HX) is applied. Although in the case of sulfur-based systems the influence of the diene type on the curing characteristics is pronounced, only minor differences between the diene types are usually observed when crosslinking peroxides are used.

The dissociation energy of an allylic hydrogen atom (i.e., adjacent to a double bond) is significantly lower than in the case of an aliphatic one (i.e., not adjacent to a double bond) as indicated in Table 2.[12] Therefore, it is believed that in the case of EPDM the allylic H atoms are abstracted preferentially.

Differences in curing characteristics can be expected to arise from differences in the nature and the number of these allylic H atoms present in the termonomers. Dicyclopentadiene contains only three, HX five, and ENB up to six allylic H atoms per termonomer moiety (see Figure 1). This seems to be in line with results obtained from a comparative study between a DCPD-based and an ENB-based EPDM type (see Table 3).

As indicated in Table 3, the ENB-based EPDM compounds gave rise to somewhat higher delta torque values and correspondingly increased modulus values and improved compression set figures in comparison with the corresponding DCPD-based EPDM compounds. These differences, however, cannot be attributed to the termonomer type only; additionally, other polymer parameters play a role: DCPD-containing types are usually only branched, ENB-containing polymers can be branched or linear depending on the polymerization conditions, whereas, the HX-containing polymers are usually of a linear type.[13] Following the same line of reasoning as indicated above it is obvious that the cross-link density is not only dependent on the type of diene present, but also is more or less directly proportional to the diene content for any given peroxide-based recipe.

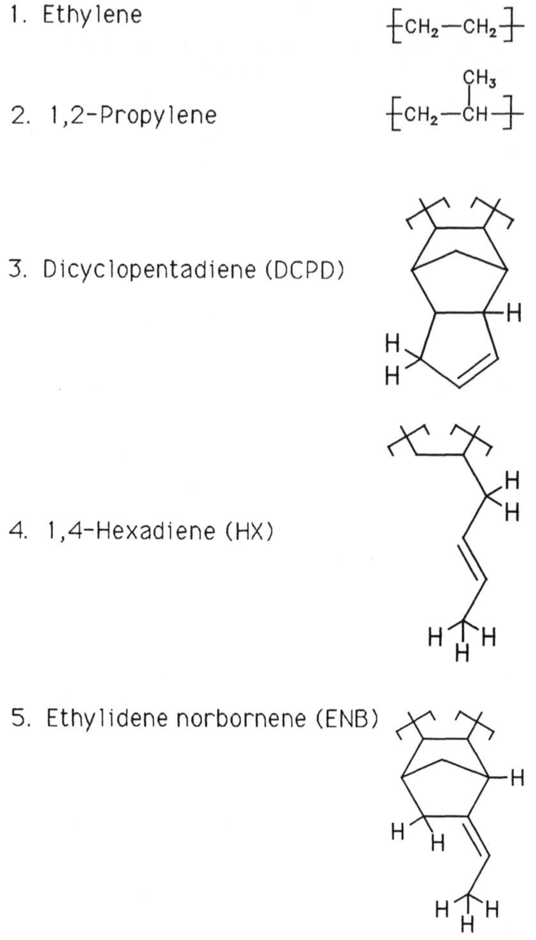

1. Ethylene

$$\left[CH_2 - CH_2 \right]$$

2. 1,2-Propylene

$$\left[CH_2 - \underset{\underset{CH_3}{|}}{CH} \right]$$

3. Dicyclopentadiene (DCPD)

4. 1,4-Hexadiene (HX)

5. Ethylidene norbornene (ENB)

FIGURE 1. Building blocks of EPDM elastomers. The allylic H atoms in the termonomer units 3, 4, and 5 are indicated with –H.

TABLE 3
Comparison of Rheological Data and Physical Properties
of a DCPD-Type and an ENB-Type EPDM Rubber

Type of EPDM		DCPD (Keltan 520®)	ENB (Keltan 512®)
Scorch time t_2	(min)	1.9	1.8
Cure time t_{90}	(min)	6.2	5.1
Torque increase	(Nm)	6.6	7.3
Hardness	(°ShA)	59	60
Tensile strength	(MPA)	15.4	15.0
Modulus 200%	(MPa)	5.5	7.0
Elongation at break	(%)	330	280
Compression set 24 h/100°C	(%)	13	19

Compound composition: EPDM (types as indicated): 100; carbon black N762: 50; Sunpar 150®: 10; Trigonox 29–40®: 7.6. Rheology: ODR-type, oscillation 3 degrees, 1.7 Hz, 160°C. Mold cure conditions: 10 min, 160°C.

B. ETHYLENE-PROPYLENE RATIO

The ethylene-propylene ratio and especially, the alternating character are important factors influencing the results of peroxide crosslinking, as in particular consecutive propylene units are prone to give rise to scission reactions in the presence of reactive peroxides. This scission reaction can be counteracted by the use of effective coagents, see Section V.E.

Although scission reactions can be expected in all polymers where adjacent propylene units are present, the effect is more pronounced for ethylene-propylene copolymers than for the terpolymer types.

C. POLYMER MOLECULAR WEIGHT

This parameter, in combination with molecular weight distribution (MWD), mainly affects rubber processing behavior. Nevertheless, an influence on the efficiency of peroxide crosslinking can be expected too. Crosslinking of high molecular weight material will be more efficient with the identical amount of peroxide than a low molecular weight material. The same holds for a narrow MWD in comparison with a broad MWD polymer. The former type can be crosslinked more efficiently, due to the absence of a low molecular weight polymer.

V. INFLUENCE OF OTHER COMPOUNDING INGREDIENTS ON PEROXIDE CURE

A. GENERAL

As the radicals formed during decomposition of the crosslinking peroxide are very reactive species, it is essential that care is taken that mainly those reactions occur that lead to a crosslink, and that other unwanted side reactions are prevented as much as possible. Compounding ingredients, known to react with peroxide radicals and in this way inhibiting the formation of polymer-polymer crosslinks, are antidegradants, certain types of fillers and extender oils. On the other hand, compounding ingredients are also known that will favorably affect the degree of crosslinking of a peroxide-cured compound, i.e., coagents. The use of these compounding ingredients, in particular with respect to their effect on the peroxide cure, will be commented on below.

B. ANTIDEGRADANTS

The use of these materials in combination with a peroxide-cure system needs special attention, since the main function of antidegradants is to inactivate radicals. Consequently, reactions between antidegradants and peroxide and/or polymer radicals can be expected, resulting in a reduced crosslink density and inactivation of the antidegradant.

EPDM polymers, well known for their good heat resistance and excellent ozone and weathering resistance, can be additionally stabilized by relatively small additions of anti-degradants. From practical experience it appeared that a number of types are well suitable for peroxide-cured compounds. The most common ones and their main functions are listed in Table 4.

In combination with peroxide-cure systems other antidegradant compounds, such as thiophosphates,[16] paraphenylene diamines,[17] and modified triazines,[18] are also reported to be suitable.

A good balance between the dosage level of both antidegradant and cross-linking peroxide should be sought to guarantee a good degree of crosslinking and an optimal stability. These somewhat contradictory interests are demonstrated with an EPDM-based test recipe (see Table 5 and Figure 2). From the rheological and physical data indicated, it appears that the compound without any stabilizer is well crosslinked, even heat resistance, determined after

TABLE 4
Antidegradants Suitable for Peroxide-Cured Rubber Compounds

Chemical	Trade name	Function
Polymerized 1,2-dihydro-2,2,4-trimethyl quinoline (TMQ)	Agerite Resin D® Flectol H® Permanax TQ®	Heat stabilizer
4,4′-Thiobis(6-*tert*butyl-m-creosol)	Santonox R®	Heat stabilizer
Tetrakis[methane(3,5-di*tert*-butyl-4-hydroxy hydrocinnamate)]methane	Irganox 1010®	Heat stabilizer[14]
2-Mercapto benzimidazole	Vulkanox MB®	Metal desactivator
Zn-salt of methylmercapto benzimidazole	Vulkanox ZMB-2® Vanox ZMTT®	Heat stabilizer[15]
Synergistic blend	Permanax CNS®	Copper desactivator
N,*N*-bis[3(3′,5′-di*tert*-butyl-4′hydroxy phenyl)propionyl] hydrazine	Irganox MD-1024®	Metal desactivator

TABLE 5
Effect of Antidegradant Dosage on Physical Properties of a Peroxide Crosslinked EPDM Compound

Permanax TQ® (phr)		0	0.5	1.0	2.0
Scorch time, t_2	(min)	2.9	2.9	3.0	3.2
Cure time, t_{90}	(min)	11.3	12.3	12.5	12.9
Torque increase	(Nm)	7.5	6.9	6.4	5.5
Hardness	(°ShA)	58	57	57	57
Tensile strength	(MPa)	14.3	14.5	15.3	16.7
Modulus 200%	(MPa)	6.4	5.6	4.7	3.8
Elongation at break	(%)	310	340	370	450
Compression set, 24 h/120°C	(%)	10	10	13	16

Compound composition: Keltan 520®: 100; carbon black N762: 50; Sunpar 150®: 10; Permanax TQ®: as indicated; Perkadox 14-40®: 4.2. Rheology: ODR-type, oscillation 3 degrees, 1.7 Hz, 180°C. Mold cure conditions: 15 min, 180°C.

7 d aging at 120°C, is quite acceptable. However, increasing the aging test temperature up to 150°C, shows a detrimental effect on retained tensile strength and elongation at break, this effect can only be compensated for by adding an effective heat stabilizer, such as Permanax TQ.® Figure 2 clearly indicates that optimum heat resistance for this specific test compound can be achieved at a stabilizer dosage of approximately 1 phr. Although Table 5 shows that the negative effect of an increasing amount of stabilizer on the physical properties is well measurable, the data observed for a compound containing 1 phr Permanax TQ® are still acceptable.

C. FILLERS

Because EPDM rubber is a noncrystallizing elastomer, the use of reinforcing fillers is a necessity to achieve acceptable mechanical properties. Types usually applied in rubber industry such as carbon black, fumed and precipitated silica, silicates, and clays are added to EPDM compounds also. Additionally, semi- or nonreinforcing fillers such as talc and whiting are practically applied. A number of these fillers can show strong interference reactions with a peroxide-cure system. These interference reactions can be based either on the high surface area of the filler, or on the acidity of the material.

In the former case, it is expected that the crosslinking peroxide will be adsorbed on the filler surface during compounding and storage of the uncured stock. As a consequence, the

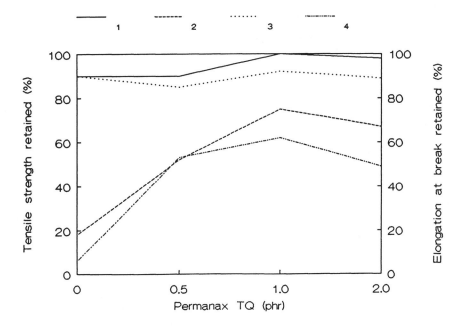

FIGURE 2. Effect of antidegradant dosage on physical properties of a peroxide crosslinked EPDM compound, after aging in air, 7 d at 120 and 150°C, respectively. For compound composition, see Table 5. Indicated are the retained values (in %) for tensile strength and elongation at break. 1 — Tensile strength retained after aging at 120°C; 2 — Tensile strength retained after aging at 150°C; 3 — Elongation at break retained after aging at 120°C; 4 — Elongation at break retained after aging at 150°C.

formation of peroxide radicals during the curing step is less homogeneous throughout the compound, and the crosslinking efficiency will be reduced. In the latter case, the peroxide can partly decompose by heterolytic reaction, whereby no peroxide radicals are formed, see Figure 3.

As an example, dicumyl peroxide **1** is protonated under strongly acidic conditions to yield oxomium-ion **2**, which on its turn decomposes into one molecule of cumyl hydroperoxide **3** and cation **4**. The latter cation may either react with water to give phenylisopropanol or be converted into α-methylstyrene. Cumyl hydroperoxide eventually decomposes into phenol and acetone. In this way, depending on the type and amount of filler, the peroxide cure may be inhibited almost completely.

It is, therefore, recommended whenever this kind of fillers are used, to add pH-neutral or mildly basic ingredients, that will be adsorbed preferentially to the fillers surface. In this way, the unwanted interference reaction of filler and crosslinking peroxide can be prevented. Examples of such additives with proven practical suitability in this respect are guanidines, such as DPG and DOTG; ethylene glycols, such as DEG and PEG; triethanol amine, and quinolines, e.g., TMQ. These additives must always be added to the rubber mix prior to the crosslinking peroxide. In the case of certain mineral fillers, e.g., clay types, filler-polymer bonding can be promoted by a silane treatment, e.g., with tris(methoxy-ethoxy)vinyl silane, which renders the surface of filler particles completely apolar. Consequently, the interaction with the polymer matrix is enhanced, while the adsorption of crosslinking peroxide is greatly reduced. This favorable effect in peroxide-cured compounds is illustrated in Table 6.

The interference of different types of carbon black with different classes of peroxide is demonstrated in a basic test recipe, see Table 7. The results of the test series containing a peroxy ketale (Trigonox 29-40®) illustrate the differences in the reinforcing effect that could be expected between the different types of carbon black: the smaller particle size material,

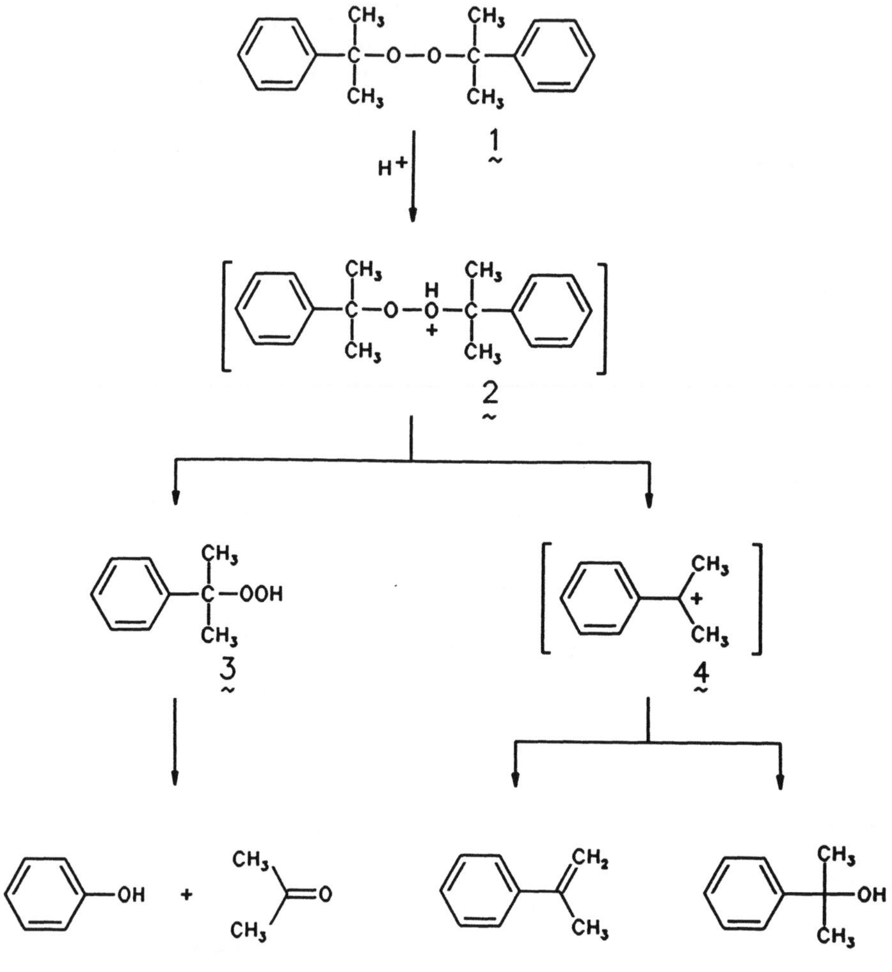

FIGURE 3. Heterolytic decomposition of dicumyl peroxide.

TABLE 6
Effect of Surface Treatment of a Mineral Filler on Physical
Properties of a Peroxide-Cured EPDM Test Compound

Type of clay pH (10% suspension)		Calcined clay 5.0	Silane treated calcined clay 7.0
Compound properties			
Hardness	(°ShA)	62	62
Tensile strength	(MPa)	6.7	9.2
Modulus 100%	(MPa)	1.8	3.3
Elongation at break	(%)	670	240
Compression set, 24 h/100°C	(%)	22	11

Compound composition: Keltan 520®: 100; zinc oxide: 5; lead oxide: 4; clay (types as indicated): 120; Sunpar 150®: 25; Permanax TQ®: 1; Perkadox 14–40®: 4.2. Mold cure conditions: 15 min, 180°C.

TABLE 7

Influence of Different Types of Carbon Black on Physical Properties of an EPDM Test Compound, Vulcanized by a Peroxy Ketale and a Peroxy Ester, Respectively

			Cross-linking system	
		Carbon black type	Trigonox 29–40® (7.6 phr) (1)	Trigonox C-50® (5.8 phr) Perkalink 300® (1.0 phr) (2)
Hardness	(°ShA)	N326	60	54
		N330	62	55
		N774	59	56
Tensile strength	(MPa)	N326	18.1	1.6
		N330	19.1	1.6
		N774	15.1	10.3
Elongation at break	(%)	N326	370	305
		N330	315	300
		N774	320	350
Compression set 24 h/100°C	(%)	N326	24	69
		N330	17	62
		N774	18	35

Compound composition: Keltan 520®: 100; carbon black (types as indicated): 50; Sunpar 150®: 10; crosslinking system: as indicated. Mold cure conditions: (1) 15 min, 150°C; (2) 15 min, 140°C.

such as N326 and N330, shows a greater reinforcement than the coarser material, N774. Interference of these carbon black types with this type of peroxide can be neglected. Effects of other types of carbon black on peroxide-cured EPDM are reported earlier.[19] On the other hand, the results of the test series containing a peroxy ester (Trigonox C-50®) show a substantial influence of carbon black type on physical properties after cure. Particular smaller particle size materials show a detrimental effect on final crosslinking density (even with the extra addition of a coagent). Only by applying the coarse type of carbon black can a certain state of cure can be achieved. This effect on cure is caused by an adsorption of peroxide to the carbon black surface, where a heterolytic decomposition can be expected; in particular, peroxy ester compounds are more prone to acid catalyzed decompositions than peroxy ketales or alkyl- or aralkyl peroxides.

D. EXTENDER OILS

Extender oils for application in peroxide-cured rubber formulations have to be selected carefully. Synthetic ester plasticizers, such as phthalates, sebacates, and adipates can be well applied in combination with crosslinking peroxides without affecting the cure reaction. Also, some alkylated benzene derivatives are known for their very small consumption of free radicals, which is, of course, desirable. Many types of mineral oil, however, can react with free radicals, due to the presence of double bonds, tertiary C atoms, and heterocyclic aromatics containing nitrogen, sulfur, or oxygen.[20,21]

With respect to possible cure interference, the paraffinic mineral oil types are preferred over the naphthenic types. These types usually undergo an extra treatment to guarantee optimum results in peroxide-cured compounds. This kind of refined mineral oil is offered by a number of suppliers. Aromatic mineral oils are usually not applied in peroxide-cured EPDM rubber compounds.

Free-radical consumption of an extender oil is not the singular criteria to select a suitable type; compatibility with the polymer and volatility, during the curing process and during

TABLE 8
Extender Oils for Peroxide-Cured EPDM Compounds

Type	EPDM compatibility	Free-radical consumption	Volatility
Ester based plasticizers	Small	Small	Low/moderate
Alkylated benzenes	Good	Small	High
Mineral oils:			
Paraffinic, standard	Good	Considerable	Low/moderate
Paraffinic, refined	Good	Moderate	Low
Naphthenic, standard	Very good	Considerable	Low/moderate
Naphthenic, refined	Very good	Moderate	Low
Aromatic	Moderate	Very high	Low/moderate

end-use, also play an important role. In practice, particularly low-volatile oil types are preferred, even if free-radical consumption becomes measurable; where necessary this effect can be counteracted by increasing the dosage of crosslinking peroxide, or by extra addition of a coagent. Table 8 summarizes the most important characteristics of different types of extender oils.

E. COAGENTS

Coagents are reactive compounds that can improve the efficiency of peroxide crosslinking. Most of the practically applied coagents belong to the classes of methacrylates or allyl-containing derivatives, but polymeric materials with a high vinyl content are also known to act in a similar way.[22] Further, sulfur or sulfur donors are also applicable. A comprehensive survey of other compounds investigated for this application field has been published.[23]

The improvement of the efficiency of peroxide crosslinking can be explained by the favorable influence that coagents exhibit on suppressing unwanted side reactions of polymer radicals.[24] The most common side reactions to be expected are disproportionation and scission of the polymer radical, see Figure 4. In case of a disproportionation reaction, one H atom is transferred from one polymer radical to another instead of establishing a crosslink between the two polymer chains. This results in the formation of one saturated and one unsaturated polymer molecule. Although the latter species may act as a weak coagent, the overall result is the loss of one peroxide moiety.

A scission reaction requires the presence of two adjacent propylene moieties. The C-C bond between the secondary and tertiary C atoms next to the tertiary C atom carrying the radical is broken up. The final result is the formation of two smaller chains, one with a terminal double bond and one with a radical C atom. This also eventually results in a less efficient crosslinking reaction.

Depending on their chemical class, coagents react in peroxide-crosslinked unsaturated elastomers either by addition and hydrogen abstractions (e.g., methacrylate compounds), or by an addition reaction only (e.g., allyl-containing compounds and sulfur).[25] A detailed insight of how coagents enhance the crosslink density during a peroxide cure is still lacking. Kinetic studies of the homopolymerization of triallylcyanurate (TAC) and other di- and triallyl derivatives in the presence of a radical source have revealed that in TAC residues two out of every three allyl moieties are involved in an "alternating intra-intermolecular polymerization or cyclopolymerization reaction".[26] Whether this mechanism also applies when di- and triallyl derivatives are involved in a peroxide cure is still an open question. The effect of coagents on compound properties can be considerable. The magnitude is not only dependent on the type of coagent used, but also on the type of polymer and the type of crosslinking peroxide.[27]

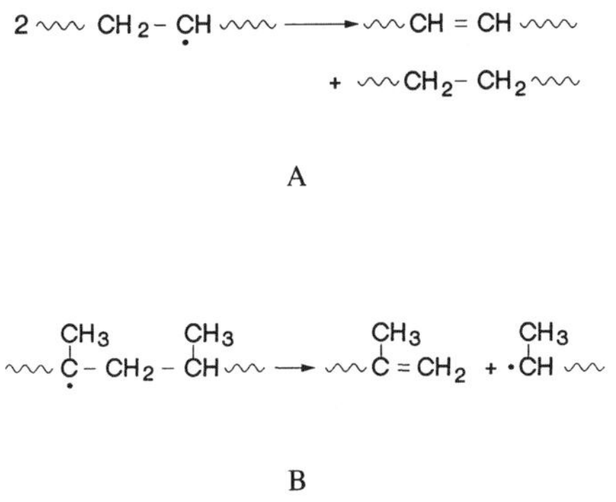

$$2 \sim\!\sim\!\sim CH_2 - \overset{\bullet}{C}H \sim\!\sim\!\sim \longrightarrow \sim\!\sim\!\sim CH = CH \sim\!\sim\!\sim$$
$$+ \sim\!\sim\!\sim CH_2 - CH_2 \sim\!\sim\!\sim$$

A

$$\sim\!\sim\!\sim \overset{\overset{\displaystyle CH_3}{|}}{\underset{\bullet}{C}} - CH_2 - \overset{\overset{\displaystyle CH_3}{|}}{CH} \sim\!\sim\!\sim \longrightarrow \sim\!\sim\!\sim \overset{\overset{\displaystyle CH_3}{|}}{C} = CH_2 + \cdot \overset{\overset{\displaystyle CH_3}{|}}{CH} \sim\!\sim\!\sim$$

B

FIGURE 4. Disproportionation (A) and scission (B) reactions of polymer radicals.

As far as the effect of different types of coagents on processing is concerned, it appeared that usually both allyl-containing derivatives, methacrylates, and polymers with a high vinyl content (liquid 1,2-poly-butadiene resins) will reduce the viscosity of a rubber mix. They act, more or less, as a plasticizer and facilitate processing of the rubber compound in this way. In general, methacrylate compositions can be expected to exhibit an enhanced tendency to scorch, whereas, allyl-containing compounds behave more or less indifferently.[25] Concerning the time necessary for optimum cure, similar considerations hold.

With respect to the physical properties of the rubber compound after cure, marked influence of all coagents on cross-linking density can be expected. This is usually reflected in terms of increased hardness and modulus values, whereas, simultaneously elongation at break is reduced. Compression set figures are usually improved. The combination of good processability before and high hardness after crosslinking can be achieved by combining an elevated level of methacrylate coagents and a peroxide-based cure system.[28] This combination can be favorably used, for example, in roll covering applications.

In most cases, tensile strength of peroxide-cured compounds is barely influenced by addition of coagents. The greatest effects can be expected in the case where less efficient peroxides are used. In particular, if saturated polymers, which are prone to scission reactions, e.g., EPM, are crosslinked by this type of peroxide, a remarkable positive influence of coagent addition on all physical properties, including tensile strength, can be observed (see Table 10 and Figures 5 and 6).

The effect of the addition of a small amount of sulfur as a coagent to a peroxide cross-linking system is, in most respects, contrary. Usually, hardness remains unaffected, whereas, the modulus values decrease and tensile strength and elongation at break figures increase. Compression set and also high temperature resistance, however, are influenced in an unfavorable sense. These phenomena can be explained from the fact that with the use of sulfur, in combination with a cross-linking peroxide, some sulfidic crosslinks are formed. Although these linkages can be of disulfidic nature,[24,29] it is known that even these links are characterized by a much smaller bond dissociation energy than the carbon-carbon bonds usually obtained during peroxide crosslinking.[30]

Although the application of sulfur as a coagent is cost-price attractive, the practical application today is only limited. In addition to the above-mentioned negative influence on heat resistance and compression set properties, the use is often connected with an unacceptable smell of the end product, due to formation of mercaptanes. Usually, for peroxide crosslinking the following coagents find practical interest: allyl-containing compounds, such as diallyl terephthalate, triallyl cyanurate, triallyl isocyanurate, methacrylate compounds,

TABLE 9
Summary of the Effects of the Addition of Various Coagents on Rheological and Physical Properties of Peroxide-Cured Compounds

Property	Type of coagent		
	Allyl-containing derivatives	Methacrylates	Sulfur
Rheology			
Minimum viscosity	Reduced	Reduced	No effect
Scorch time	Small/no effect[a]	Reduced	No effect
Cure time	Small/no effect[a]	Reduced	No effect
Degree of X-linking	Increased	Increased	Increased
Physical properties			
Hardness	Increased	Increased	Small/no effect
Tensile strength	Small effect[b]	Small effect[b]	Increased
Modulus	Increased	Increased	Increased
Elongation at break	Reduced	Reduced	Increased
Tear strength	Small effect	Small effect	Increased
Compression set	Reduced	Reduced	Increased
Heat resistance	Improved	Improved	Reduced

[a] Dependent on type of base polymer a reducing effect can also be observed.
[b] For saturated elastomers crosslinked by less efficient peroxides, a remarkable increase can be expected (see Table 10).

TABLE 10
Effect of Co-agent/Peroxide Combinations on Properties of Various EPM Compounds

Trigonox 29-40®	Perkadox 14-40®	Perkalink 300®
7.6	—	—
7.6	—	0.8
—	4.2	—
—	4.2	0.8

EPM (Type 1, 2, or 3, respectively): 100; carbon black N762: 50; Sunpar 150®: 10; Permanax TQ®: 1; crosslinking system: as indicated. EPM 1: C_2 content approximately 40%; EPM 2: C_2 content approximately 50%; and EPM 3: C_2 content approximately 60%.

such as ethyleneglycol dimethacrylate, butylene glycol dimethacrylate, and trimethylolpropane trimethacrylate, and various reactive compounds, such as N,N'-m-phenylene dimaleimide, and 1.2-*cis*-polybutadiene. The influence of different classes of coagents on various properties of peroxide-cured compounds are summarized in Table 9.

Some of the aforementioned effects are demonstrated in an EPM-based test compound in which both the type of polymer and the type of crosslinking peroxide are subject to change. Compared are combinations of an allyl-containing coagent and peroxides varying in cross-linking efficiency: Trigonox 29®[7] (Table 1) being far less efficient than Perkadox 14®[3] (Table 1). Both peroxides are used as a single curative and in combination with Perkalink 300® (TAC) as base polymer EPM types are selected differing in C_2/C_3 ratio (see Table 10). Physical properties of the compounds like tensile strength and compression set after vulcanization are indicated in Figures 5 and 6, respectively.

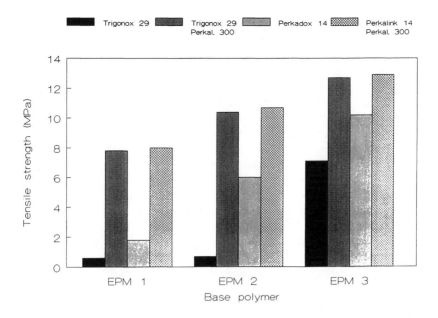

FIGURE 5.   Effect of coagent/peroxide combinations on tensile strength of various EPM compounds.

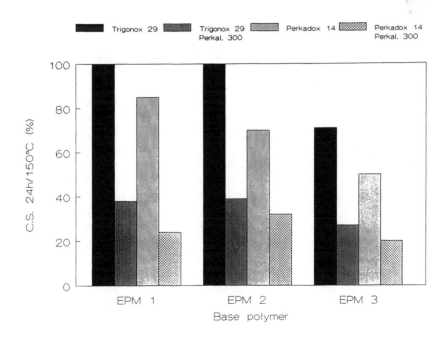

FIGURE 6.   Effect of coagent/peroxide combinations on compression set of various EPM compounds.

It appears from the results that considerable differences exist between various EPM types. In the absence of a coagent, types 1 and 2 hardly show any crosslinking with a less efficient peroxide such as Trigonox 29,® and only a limited degree of crosslinking with the efficient peroxide type, Perkadox 14.® EPM type 3, on the other hand, shows a remarkable state of cure even without the addition of a coagent.

Although other polymer characteristics (molecular structure, C_2/C_3 distribution, branching, etc.) can also have an effect, a great part of the difference observed between EPM types 1, 2, and 3 can be attributed to the ethylene-propylene ratio, the smaller this ratio, EPM type 1, the more scission reactions can be expected. In this case, the effect of coagent addition is extremely beneficial, which is demonstrated both by tensile strength and compression set figures. Furthermore, the improved heat resistance of the compounds crosslinked by coagent/peroxide combinations is obvious. This improvement is even remarkable for the compound based on EPM type 3.

F. SCORCH-RETARDING SYSTEMS

In comparison with cure systems based on conventional sulfur and accelerator combinations, peroxide-based cure systems are less versatile with respect to designing the cure curve. This is caused by the fact that decomposition of crosslinking peroxides into radical fragments does not show an induction period. It appeared, however, to be possible to influence the subsequent reactions of this peroxide radical, e.g., by the addition of radical scavengers. In the past, numerous suggestions in this direction have been made, among them the following additions were recommended:

1.   2,6-Bis(*tert*-butyl)-*p*-cresol (BHT) for PE-based compounds[31]
2.   N-nitroso diphenylamine in combination with m-phenylene bismaleimide[32]
3.   Diethylene thiourea, active in clay-containing compounds only[33]
4.   Lauryl methacrylate[34]
5.   Hydroperoxide[35]

In addition to toxicological considerations which render some of these recommendations obsolete, the main difficulty is to obtain a sufficient retardation effect without impairing crosslink density and final physical properties. To date, almost invariably the combination of a suitable radical scavenger in combination with a coagent is used to impart scorch retardation. Ready-made mixtures containing a crosslink peroxide, a coagent, and a radical scavenger,[36,37] or merely a coagent and a radical scavenger,[38] have been introduced in order to facilitate compounding of "scorch-retarded" or "scorch-safe" recipes. The compounder may, therefore, either apply these premixed systems, which always contain a fixed ratio of ingredients, or add the various ingredients up to the appropriate levels required for the fine tuning of specific processing conditions and/or eventual physical properties. An example of a typical approach to improve the scorch resistance of a peroxide-containing recipe is illustrated in Table 11 and Figure 7. In an EPDM compound scorch retardation is effectuated by the combined addition of antioxidant BHT and Perkalink 300®. First the dosage of radical scavenger is determined empirically; the decrease of delta torque should be between 25 and 50%. In most cases a dosage between 0.1 and 1.0 phr is sufficient. In this example, 0.4 phr of BHT is used. Then coagent is added until the original level of crosslink density is reached again. In most cases a dosage between 1.0 and 2.0 phr is needed. In this case, 1.0 phr of Perkalink 300® is used.

Since scorch behavior is especially important at processing temperature, in this case 120°C, the scorch time at this temperature, is a key indicator. The data of Table 11 show that the scorch time at processing temperature has been increased by a factor of two, while the cure time has only been slightly enhanced and the physical properties have remained virtually unaffected. In Figure 7 the values of the torque increase at 150°C and scorch time at 120°C have been represented graphically. After addition of 0.4 phr BHT, the torque increase can only be kept on its original level with an additional dosage of 1.0 phr Perkalink 300® (see lower graph). Addition of 0.4 phr BHT does have a strong impact on the scorch time at 120°C, but this scorch time, however, is not influenced by the additional dosage of 1.0 phr Perkalink 300.

TABLE 11
Effect of a Scorch-Retarding System in a Peroxide-Cured
EPDM Test Compound

Additives: antioxidant BHT			—	0.4	0.4
Perkalink 300®			—	—	1
Scorch time, t_2	(min)	120°C	7.6	16.8	16.3
Cure time, t_{90}	(min)	150°C	8.1	9.9	9.1
Torque increase	(Nm)	150°C	5.3	3.6	5.2
Hardness	(IRHD)		59		58
Tensile strength	(MPa)		17.2		16.0
Modulus 100%	(MPa)		1.8		1.8
Elongation at break	(%)		330		320
Compression set, 72 h/100°C	(%)		18		16

Compound composition: Keltan 520®: 100; carbon black N762: 50; Sunpar 150®: 10; Trigonox 29–40®: 7.5; scorch-retarding system: as indicated. Rheology: ODR-type, oscillation 3 degrees, 1.7 Hz, temperature: as indicated. Mold cure conditions: 15 min, 150°C

VI. MIXING

Mixing of EPDM rubber compounds can be carried out by the usual techniques as described in numerous recommendations of the polymer suppliers. Predominantly internal mixers are applied, but mills can be used also. Compounds containing peroxide-cure systems are usually compounded in a two-step procedure, whereby the curatives are incorporated during a mill operation. Specific, thermally stable peroxide compounds, however, are well suitable to be incorporated into the final rubber mix at the end of a one-step mixing procedure, provided that the temperature of the mix, as present in the internal mixer, is controllable and below certain limits. The maximum allowable mix temperature is dependent on the type of peroxide applied, but it is obvious that clean running of the mixer is also of importance (no material hold up for several mixing cycles). For such a one-step mixing procedure, mainly the following peroxide types can be applied:

Peroxide type	Max. allowable mix temp. (°C)
Bis(*tert*-butylperoxyisopropyl)benzene	130
Dicumyl peroxide	120

For application in rubber compounds explicitly phlegmatized (= diluted) peroxide formulations are mainly used, instead of peroxides in a technically pure form. This is based on improved handling of the phlegmatized types, but also on safety considerations. As phlegmatizing agents, inert mineral fillers are usually applied. Peroxides predispersed in appropriate polymers, so-called masterbatches, possess an additional advantage over peroxide formulations on mineral carriers; the time required to incorporate the peroxide into the final rubber mix is considerably reduced. This holds for both mill mixing as well as for internal mixing procedures.

VII. VULCANIZATION

A typical limitation of the use of organic peroxides for crosslinking purposes is the interference reaction of oxygen with peroxide and polymer radicals. In the presence of oxygen, polymer radicals are converted into hydroperoxy radicals (see Figure 8, step 1) at

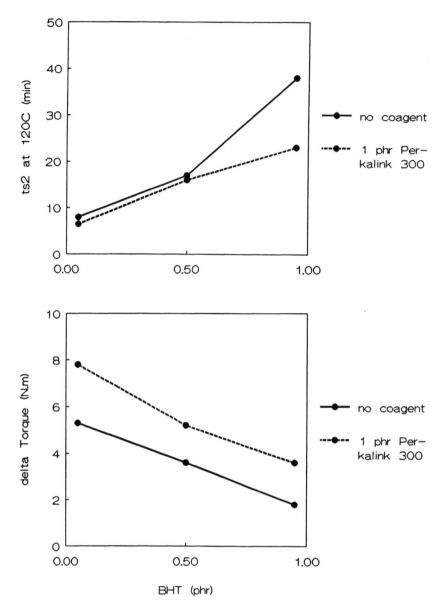

FIGURE 7.   Effect of radical scavenger and coagent on scorch time and crosslinking degree of a peroxide-cured EPDM test compound.

such an extremely high rate[39] that recombination of two elastomer radicals to form a crosslink is almost completely inhibited. Subsequently, the hydroperoxy radicals may abstract another H atom to form a hydroperoxide (step 2), which thermally decomposes into an oxy radical and a hydroxy radical (step 3). The latter radicals, on their turn, may abstract another H atom as well yield an alcohol (among others, step 4) and water (step 5), respectively.[40]

Since the H atoms are mostly abstracted from other elastomer molecules and the alcohols formed are prone to chain scission, it is clear that oxygen has a deteriorating effect on peroxide crosslinking. In this way, crosslinking of the surface of an article can be inhibited strongly. In the past several attempts have been made to overcome this inhibiting effect, e.g., by surface coating with a solution of ammonium bicarbonate, with organic metal

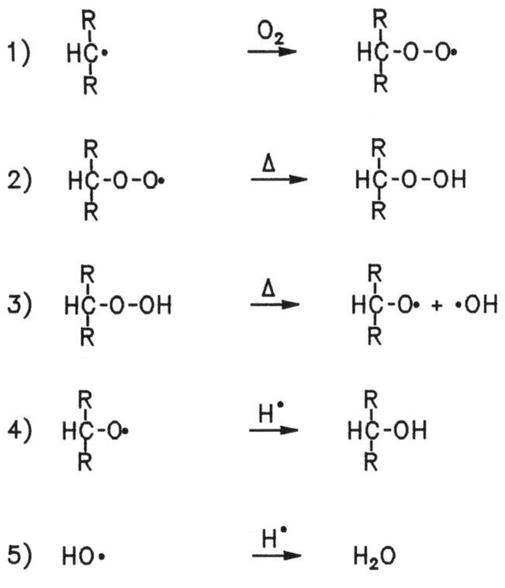

1) $HC \cdot$ (R, R) $\xrightarrow{O_2}$ $HC-O-O \cdot$ (R, R)

2) $HC-O-O \cdot$ (R, R) $\xrightarrow{\Delta}$ $HC-O-OH$ (R, R)

3) $HC-O-OH$ (R, R) $\xrightarrow{\Delta}$ $HC-O \cdot + \cdot OH$ (R, R)

4) $HC-O \cdot$ (R, R) $\xrightarrow{H \cdot}$ $HC-OH$ (R, R)

5) $HO \cdot$ $\xrightarrow{H \cdot}$ H_2O

FIGURE 8. Reaction of polymer radicals with oxygen.

compounds,[41] with boric acid,[42] or lately, with the combination of a microwave/hot air system with electron beam cure.[43] As long as direct contact of the rubber compound with oxygen is excluded, however, all current vulcanization techniques can be applied for peroxide-containing EPDM compounds. Widely applied are mold cure techniques, such as compression, transfer, and injection molding, autoclave cure in steam or nitrogen, and so-called salt bath cure, whereby the shaped compound is crosslinked continuously by transport through a mixture of molten anorganic salts with usual bath temperatures of 200°C and up. Furthermore, fluid bed lines operating with inert gas, are sometimes also recommended.

Vulcanization time and temperature conditions for optimum cure are directly dependent on the thermal stability of the specific peroxide type applied (see Table 1). However, as the C-C links formed during crosslinking are very stable, the efficiency of crosslinking will not be adversely affected by elevated vulcanization temperatures, contrary to many sulfur-based cure systems. This allows a relatively high cure temperature to be applied, consequently, connected with a reduced cure time.

VIII. APPLICATIONS

Excellent ozone and weathering resistance, good heat and chemical resistance, good low temperature flexibility and outstanding electrical properties, make EPDM rubber predestined for a great number of specific applications. To unfold most of these polymer intrinsic properties to a maximum, peroxide cure is preferred. For many years peroxide-cured, EPDM-based compounds have been applied, e.g., for window seals, automotive hoses, steam hoses, conveyor belts, roof sheeting, tank lining, roll coverings, moldings, and last but not least, for electrical insulation and jacketing compounds. Some examples are mentioned below.

A. RADIATOR HOSE

In addition to increasing demands for temperature resistance of automotive radiator hoses, up to 150°C as service temperature being quite usual, the extractables from the innertube compound in a water-glycol mixture should also be minimal. Excellent heat resistance has been reported for peroxide-cured EPDM rubber compounds for this application.[44,45] Recently, a comprehensive study has been published[46] in which an optimization of a number of compounding ingredients were included, such as polymer type, stabilizer package, and cross-linking system. The recommended compound, see Table 12, was investigated for hose production under practical conditions.

TABLE 12
Pilot Recipe for Automotive Radiator Hose
(with courtesy of Exxon Chemicals)

Physical properties determined after steam cure:
20 min, 180°C

Hardness	(°ShA)	60
Tensile strength	(MPa)	13.3
Modulus 100%	(MPa)	3.0
Elongation at break	(%)	340
Change after aging	(%)	
168 h/140°C		−4
168 h/150°C		−12
168 h/160°C		−31
Extractables, 168 h/100°C, water-glycol		<0.05
(g/100 g rubber)		

Compound composition: Vistalon 7500®: 100; carbon black N550: 100; Flexon 815®: 45; antioxidant TMQ: 1; antioxidant ZMMBI: 3; coagent EDMA: 0.5; Perkadox 14–40®: 7.

TABLE 13
Pilot Recipe for Injection Molded Article

BHT (phr)			—	0.4
Scorch time, t_2	(min)	120°C	5.8	9.1
		150°C	2.1	2.5
Cure time, t_{90}	(min)	150°C	8.4	9.6
Torque increase	(Nm)	150°C	4.4	4.0
Mold temperature		(°C)	180	200
Cure time		(min)	3.0	0.5
Hardness		(°ShA)	67	65
Tensile strength		(MPa)	12.1	9.0
Elongation at break		(%)	250	230
Compression set, 24 h/100°C		(%)	26	22

Compound composition: Keltan 578®: 100; carbon black N770: 70; carbon black N550: 70; Sunpar 150®: 70; stearic acid: 0.5; Perkalink 400®: 1; BHT: as indicated; Trigonox 29–40®: 11.4. Rheology: ODR-type, oscillation 3 degrees, 1.7 Hz, temperature 120 and 150°C, respectively. Injection molding conditions: equipment: Arburg Allrounder type 221E/150R, screw injection; temperatures (°C): barrel : 90; die : 100; mold: see above. Injection pressure: 1500 bar. Injection time: up to 6 s.

B. INJECTION MOLDING COMPOUND

Injection molding technique allows short production cycle times by increasing mold temperature. Applying fast reacting, cross-linking peroxide types, short cycle times can be realized. The maximum allowable cure temperature for a compound is determined by its scorch resistance and the processing time necessary to fill a specific mold. The compound example shown in Table 13, could be injection molded up to a mold temperature of 180°C; at higher mold temperatures scorch problems were observed, and the mold could not be filled completely. By addition of a radical scavenger, BHT, the mold temperature could be increased up to 200°C without scorch problems. At this temperature, the vulcanization time could, of course, be reduced considerably.

TABLE 14
Influence of Addition of EPDM to an NR Compound

Polymer blend:		100	90	80	50
RSS 1 Keltan 520®		—	10	20	50
Perkadox 14–40®		1.0	1.3	1.6	2.6
Hardness	(°ShA)	45	48	48	55
Tensile strength	(MPa)	13.5	13.6	12.6	13.6
Modulus 200%	(MPa)	2.3	2.7	2.9	3.8
Elongation at break	(%)	470	470	450	430
Compression set, 24 h/100°C	(%)	18	18	18	17
Outdoor testing; crack formation					
First cracks	(d)	2	13	>80	>80
Type of cracks		Very many small cracks	Many great cracks	No cracks	No cracks

Compound composition: polymer blend NR/EPDM: 100; stearic acid: 0.5; carbon black N762: 50; Sunpar 150®: 10; Perkadox 14–40®: as indicated. Mold cure conditions: 15 min 180°C.

C. COVULCANIZATION OF HIGHLY UNSATURATED ELASTOMER WITH EPDM

Improvement of weathering resistance of highly unsaturated rubbers, such as natural rubber, can be achieved by blending with EPDM. By applying conventional sulfur, accelerator combinations, the EPDM portion in such blends will hardly be vulcanized, due to the great difference in reactivity between both polymers and due to the differences in solubility of the accelerators in both polymers. Applying peroxides as a crosslinking agent will result in a homogeneous crosslinked blend, whereby the cross-linked EPDM rubber acts as a nonleachable stabilizer. The example shown in Table 14 demonstrates this effect. The data indicate that already a replacement of 20 parts NR by EPDM, results in a considerable improvement of weathering resistance.

D. SALT BATH CURED SEALING COMPOUND

Window sealing rubber compounds are preferentially based on EPDM compounds. To achieve excellent set properties, the application of a peroxide-cure system is obvious, particularly if a molten salt bath curing line is used. To prevent deformation of the often highly complicated soft profile in the first part of the salt bath, a fast onset of cure is required (fast decomposing peroxide), whereas, the physical properties after cure must also be of a good level (efficiently curing peroxide). In the example shown in Table 15, two crosslinking systems are compared: an addition of a singular peroxide and an addition of a peroxide combination possessing a considerably reduced scorch time. The faster onset of cure for the compound containing the peroxide combination is not only indicated by scorch and cure time data, but is also reflected by the physical properties after step-wise salt bath cure.

E. ELECTRICAL INSULATION

The polymer characteristics of EPDM rubber makes this material very suitable for electrical applications. Additionally, the electrical properties of a peroxide-cured compound are considered to be much superior to those obtained with sulfur-based vulcanization systems.[47] For insulation purposes mineral filled compounds are usually applied. An example of a pilot compound composition for a medium voltage power cable insulation is shown in Table 16.

IX. OUTLOOK

Crosslinking of EPDM-based compounds by organic peroxides is a growing technique. On the one hand, the technical requirements for specific rubber end products are steadily

TABLE 15
Cross-linking System for Salt Bath Cured Sealing Compound

Cross-linking system							
Perkadox 14–40®		6.3			3.8		
Trigonox 29–40®		—			4.5		
Mooney viscosimeter							
t_5-Scorch time	(min)	32.7			6.3		
Rheometer							
Scorch time, t_2	(min)	0.8			0.3		
Cure time, t_{90}	(min)	5.4			3.3		
Torque increase	(Nm)	0.9			0.9		
Cure conditions	(min/°C)						
Compression molding		15/180			15/180		
LCM			2/220	3/220		2/220	3/220
Hardness	(°IRHD)	55	50	51	55	51	51
Modulus 100%	(MPa)	1.4	1.2	1.2	1.6	1.3	1.3
Modulus 300%	(MPa)	6.1	5.1	5.2	7.1	5.6	5.3
Tensile strength	(MPa)	11.8	11.7	12.2	12.7	12.0	11.8
Elongation at break	(%)	460	500	510	430	480	490
Compression, 24 h/120°C	(%)	15	28	11	16	14	17

Compound composition: Keltan 778®: 100; zinc oxide: 5; stearic acid: 0.5; carbon black N762: 100; Omya BSH®: 40; Sunpar 2280®: 70; Dispergator DS®: 7; Caloxol W3®: 8; Perkalink 400®: 1; crosslinking peroxides: as indicated. Rheology: Mooney viscosimeter: 121°C; MDR-type: oscillation 0.5 degrees; 1.7 Hz, temperature: 180°C. Cure conditions: as indicated.

TABLE 16
Pilot Recipe Medium Voltage Insulation Compound
(with courtesy of Du Pont Comp.)

Physical properties after cure are indicated as such and the change after aging 7 d in air, 135°C

Scorch time, t_2	(min)	1.3	
Cure time, t_{90}	(min)	8.5	
Torque increase	(Nm)	5.6	
		Test sheet	**Test wire**
Hardness	(°ShA)	62	n.d.
Tensile strength	(MPa)	7.4	6.8
change	(%)	+11	+7
Modulus 100%	(MPa)	2.5	2.0
Elongation at break	(%)	305	380
change	(%)	−27	−17
Insulation resistance	(MΩ·km)	n.d.	6800

Compound composition: Nordel 2522®: 50; Nordel 1040®: 50; Polestar 200R®: 100; Sunpar 2280®: 15; zinc stearate: 1; antioxidant TMQ: 1; Ucarsil RC-1®: 2; Perkadox 14–40MB®: 8. Rheology: ODR-type, oscillation 3 degree 1.7 Hz, temperature 180°C. Cure conditions: test sheets: 15 minutes, 180°C and test wires: 2 min steam 16 bar; 1.0 mm insulation on No. 14 AWG solid aluminum.

n.d. = no data

increasing (e.g., improved high temperature resistance and reduction of extractables). On the other hand, product improvements (e.g., easy dispersable peroxide formulations in masterbatch form) or system improvements (application of peroxide-coagent combinations, scorch-retarding systems) are offered today to the rubber processing industry. Future developments to reduce some of the aforementioned unwanted interactions of other compounding ingredients with cross-linking peroxides will certainly further stimulate the wider application of peroxide-based cure systems.

REFERENCES

1. **Ostromyslenski, I. I.,** *J. Russ. Phys. Chem. Soc.,* 47, 1467, 1915; *India Rubber J.,* 52, 470, 1916.
2. **Natta, G., Crespi, G., DiGiulio, E., Ballini, G., and Bruzzone, M.,** *Rubber Plastics Age,* 42, 53, 1961.
3. **Berg, A. L.,** *Plastics Additives,* 3rd ed., Gächter, R. and Müller, H., Eds., Hanser Publishers, Munich, 1990, 844.
4. **Endstra, W. C.,** *Proc. 8th SRC,* Copenhagen, 1985, 775.
5. NL 7600409, 1976.
6. **Endstra, W. C.,** *Kautsch. Gummi. Kunstst.,* 42, 414, 1989.
7. **Fujioka, H.,** *Int. Pol. Sci. Technol.,* 5, T/65 1978.
8. **Allen, R. D.,** *Gummi. Asbest. Kunststoffe,* 36, 534, 1983 3rd ed.,
9. **Easterbrook, E. K. and Allen, R. D.,** *Rubber Technology,* 3rd ed., Morton, M., Eds., Van Nostrand Reinhold, New York 1987, p. 263.
10. **Keller, R. C.,** *Rubber Chem. Technol.,* 61, 238, 1988.
11. **Stella, G. and Cheremisinoff, N. P.,** Proc. PRI Conf., Leuven, Belgium, paper XVII, 1991.
12. *Handbook of Chemistry and Physics,* 67th ed., CRC Press, Boca Raton, FL, 1987, p. F-178.
13. **Easterbrook, E. K. and Allen, R. D.,** *Rubber Technology,* 3rd ed., Morton, M., Ed., Van Nostrand Reinhold, New York, 1987, p. 264.
14. **Raue, D. P.,** *Elastomerics,* 115, 17, 1983.
15. **Spenadel, L.,** *Rubber Chem. Technol.,* 56, 113, 1983.
16. **Tueseev, A. P., Bukalov, V. P., Potashova, G. N., and Vald'man, A. I.,** *Int. Pol. Sci. Technol.,* 7, T/27, 1980.
17. **Lévy, M.,** *Kautsch. Gummi. Kunstst.,* 42, 129, 1989.
18. **Nakamura, Y., Mori, K., Tamura, K., and Saitoh, Y.,** *Int. Pol. Sci. Technol.,* 12, T/56 1985.
19. **Amberg, L. O.,** *Vulcanization of Elastomers,* Alliger, G., and Sjothun, I. J., Eds., Reinhold Publishing, New York, 1984, p. 318.
20. NN, Rubber Process and Extender Oils, Technical Bulletin 93; Sun Oil Company Philadelphia, PA.
21. **Hamilton, G. G.,** Process Oils for Peroxide Cured Rubbers, poster Rubbercon, Harrogate, England, June 1981.
22. **Drake, R. E.,** *Elastomerics* 114, 28, 1982.
23. **Hofmann, W.,** *Kautsch. Gummi. Kunstst.,* 40, 308, 1987.
24. **Loan, L. D.,** *Rubber Chem. Technol.,* 40, 149, 1967.
25. **Perkins, G. T.,** Peroxide curing of Nordel. ND-310.2, p. 12. Du Pont Company Wilmington, DE.
26. **Odian, G. G.,** *Principles of Polymerization,* 2nd ed., John Wiley & Sons, New York, 1981, p. 488.
27. **Endstra, W. C.,** *Kautsch. Gummi. Kunstst.,* 43, 790, 1990.
28. **Blow, C. M. and Hepburn, C.,** *Rubber Technology and Manufacture,* Butterworths, London, 1987, p. 383.
29. **Fujio, R., Kitayama, M., Kataoka, N., and Anzai, S.,** *Rubber Chem. Technol.,* 52, 74, 1979.
30. **Hofmann, W.,** *Vulkanisation und Vulkanisationshilfsmittel.* Bayer AG, Leverkusen, Germany, 1965, p. 31.
31. **Benning, C. J.,** *SPE J.,* 21, 1083, 1965.
32. **Chow, Y. W. and Knight, G. T.,** *Plastics Rubber Process.,* 2, 115, 1977.
33. **Purakel, T. L., and Purper, R. L.,** *Elastomerics,* 109, 19, 1977.
34. **Schober, D. L.,** *Wire J.,* 11, 84, 1978.
35. US 4015058, 1977; US 4025706, 1977.

36. **Kmiec, C. J. and Kamath, V. R.,** *Rubber World,* 194, 33, 1986.
37. EP 346863, 1989.
38. US 4857571, 1988.
39. **Landolt-Bönstein,** *Numerical Data and Functional Relationships in Science and Technology. New Series: Group II: Atomic and Molecular Physics: Vol 13: Radical Reaction Rates in Liquids,* Hellwege K-H. and Madelung, O., Eds., Springer-Verlag, Berlin, 1984.
40. **Gugumus, F.,** *Plastics Additives,* 3rd ed., Gächter, R., and Müller, H., Eds., Hanser Publishers, Munich, 1990, p. 1.
41. DE 2936906, 1979.
42. **Groepper, J.,** *Kautsch. Gummi Kunstst.,* 36, 466, 1983.
43. **Aoshima, M., Jinno, T., and Sassa, T.,** *Cell. Polym.,* 10, 359, 1991.
44. **Dunn, J. R., Keller, D., and Patterson, J.,** *Gummi Asbest Kunstst.,* 42, 516, 1989.
45. **Keuper, D.,** *Kunststoffberater,* 35, 26, 1990.
46. **Keller, R. C. and Mills, T. A.,** *Kautsch. Gummi Kunstst.,* 44, 1032, 1991.
47. **Perkins, G. T.,** Peroxide curing of Nordel, ND-310-2, DuPont Company, Wilmington, DE, p. 21.

Chapter 13

PROPERTIES AND DEGRADATION OF NITRILE RUBBER

Susmita Bhattacharjee, Anil K. Bhowmick, and Bhola Nath Avasthi

TABLE OF CONTENTS

0-8493-4401-8/93/$0.00 + $.50

519

I. INTRODUCTION

Nitrile rubber (NBR) is a copolymer of butadiene and acrylonitrile. The butadiene unit imparts elasticity, cross-linking site, low temperature flexibility, and all the rubbery properties like other diene rubbers. It is the acrylonitrile unit which makes NBR a special elastomer. The acrylonitrile unit offers oil and fuel resistance and high strength to the material. The structural formula of NBR is as follows:

$$\left[(-CH_2-CH=CH-CH_2-)_n- \quad -(-CH_2-CH-)_m \atop \qquad\qquad\qquad\qquad\quad CN \right]$$

Nitrile rubber was first commercialized by I.G. Farbenindustrie, Germany, in 1937 under the trade name of Buna N. In commercially available nitrile rubbers, the acrylonitrile content (ACN) ranges from 18 to 51%. The grades of nitrile rubber are differentiated by the nitrile content, which mainly determines the ultimate properties of the elastomer. Due to the tougher requirements in industrial and automotive applications, there has been a continuous demand to improve the heat, ozone, and weather resistance of nitrile rubber. Hence, many attempts have been made to modify the nitrile rubber by chemical and physical methods to improve its properties and degradation behavior. Chemical modifications include structural changes of the polymer chain by chemical reaction, whereas, physical modification involves the mechanical blending of NBR with other polymers or chemical ingredients to achieve the desired set of properties. Carboxylated and hydrogenated nitrile rubbers (XNBR and HNBR) are the two most important chemically modified NBRs. Physically modified NBRs are mainly the blends with different rubbers and plastics. Many compounding recipes are also being developed for superior aging resistance of NBR. Literature on NBR technology is available from the manufacturers. The focus in this chapter is on the more recent advances with reference to aging properties and degradation behavior. Also, a systematic analysis on the chronological development is presented.

II. PREPARATION OF NITRILE RUBBER

A. NITRILE RUBBER

Nitrile rubber is manufactured by emulsion copolymerization of butadiene and acrylonitrile.[1,2] The monomers are emulsified in water and a catalyst is added to generate free radicals and initiate polymerization. Generally, anionic emulsifiers such as alkali salts of "fatty" carboxylic acids, sulfates, or sulfonates are used in nitrile rubber polymerization. They can be washed out during coagulation and dewatering. The catalyst is a water-soluble peroxide activator[1] such as potassium, ammonium salt of persulfuric acid, organic peroxides, hydroperoxides, and peracids. The polymer chain is initiated when a free radical formed by decomposition of a catalyst reacts with a monomer molecule. The initiated chains quickly leave the water phase and enter the micelles and the polymer chain grows. The catalyst is inactivated by a short stop and the residual monomers are removed from the emulsion. Usually, the short stops are water-soluble reducing agents such as sodium hydrogen sulfite, sodium dithionite, hydroxylamine, hydrazine and its salts, or sodium dimethyldithiocarbamate. A stabilizer or antioxidant is added to the latex for protection during drying and storage. *Bis-p*-cresol and alkylated phenols are the most effective stabilizers. The polymer is coagulated with an aqueous solution of inorganic salt. The rubber crumb is washed with water several times and dried by hot air.

B. CARBOXYLATED NITRILE RUBBER

Carboxylated nitrile rubber is synthesized by copolymerizing butadiene, acrylonitrile, and an acidic organic monomer in an emulsion in the presence of a stabilizer.[2,3] The acrylonitrile and acid groups are distributed randomly in a polymer molecule. Generally, the active acidic group is present at a concentration level of less than 10%. The polymerization techniques adopted are similar to that of NBR.

C. HYDROGENATED NITRILE RUBBER

Hydrogenated nitrile rubber is obtained by selective reduction of olefinic unsaturation in a butadiene unit. The hydrogenation reaction proceeds in the presence of a catalyst precursor, which offers activity toward a carbon-carbon double bond in a polymer containing other reducible functional groups such as nitrile, carboxyl, etc. Both heterogeneous and homogeneous catalyst systems can be used.[4] Heterogeneous catalysts include metallic palladium,[5,6] ruthenium,[7] rhodium[8] or nickel[9] for hydrogenation. Homogeneous catalysis involves the activation of molecular hydrogen by transition metal complexes in solution and subsequent hydrogen transfer to the unsaturated substrate. Homogeneous catalysts offer better selectivity as compared to heterogeneous systems. Tris(triphenylphosphine)rhodium chloride,[10,11] palladium acetate,[12] cyclopalladate complex of 2-benzoyl pyridine,[13] and other complexes based on ruthenium,[14] nickel,[15] and titanium[16] are used as homogeneous catalysts for NBR hydrogenation.

The polymer is dissolved in a suitable organic solvent such as chlorobenzene, acetone, or 2-butanone. A requisite amount of catalyst is added to the polymer solution and the mixture is autoclaved at high temperatures under hydrogen pressure. The reaction can be controlled up to the desired level of hydrogenation by proper monitoring of reaction pressure, temperature, time, and catalyst concentration. After the reaction is stopped and the mixture cooled down, the polymer is coagulated by using methanol, washed repeatedly with methanol, and dried under a vacuum.

III. PROPERTIES OF NITRILE RUBBER

Nitrile rubber exhibits good processing characteristics and can be vulcanized by a variety of curing systems.[1,2,17-19] It shows poor to moderate tack. The processing properties of NBR

TABLE 1
Typical Formulation of a Nitrile Rubber Vulcanizate and the Variation of Physical Properties with ACN

Nitrile rubber		100					
Carbon black		40					
Stearic acid		1					
Zinc oxide		3					
Sulfur (2% MgCO$_3$ coated)		1.5					
TBBS		0.7					
Cure time		40 min					
Temperature		150°C					
Acrylonitrile content (%)	29	24	31	34	40	45	50.5
Mooney viscosity (ML 1 + 4 at 100°C)	65	55	53	50	50	50	75
Ultimate elongation, (%)	460	440	500	520	460	600	420
Tensile strength, (MPa)	16.6	20.0	24.0	27.0	26.0	23.5	24.5
Modulus 100%, (MPa)	7.9	8.8	9.8	10.5	12.1	8.2	15.7

From Bayer-Polysar Technical Centre N.V., Antwerp, Belgium. With permission.

are mainly affected by plasticity and gel content. Gel content generally ranges from 0 to 80% in NBR. Permanent gel or cross-linking affects the tensile strength, elongation, and abrasion resistance in cured NBR compounds. Microgel, however, results in increased green strength with a sacrifice of surface tack and mold flow characteristics. Mooney viscosity of the rubber, generally, has a great influence on the processing and the properties of the finished product. With the increase of Mooney viscosity of the compound, the time of sheet formation, and filler incorporation, injection and extrusion increases. Mixing temperature, mill shrinkage, green strength, and modulus also increases with Mooney viscosity. The properties which decreases are the porosity and compression set. If the branching increases in the polymer without any change in Mooney viscosity, there is an increase in processing time, improvement in compound green strength, and reduction in shrinkage and compression set. The ACN has the greatest effect on the rubber properties. As the amount of bound acrylonitrile increases in the polymer, density, processibility, cure rate, thermoplasticity, and stiffness increase.

A. MECHANICAL PROPERTIES

Nitrile rubber is not a strain crystallizing rubber. Hence, its vulcanizate without reinforcing filler has poor tensile strength and low tear resistance when compared with natural rubber.[1,2,17] Tensile strength of NBR increases with an increase in ACN and also depends upon the kind of reinforcing filler. The modulus and hardness of NBR vulcanizates can be varied over a wide range with proper selection of compounding ingredients. Usually hardness increases with ACN. Table 1 gives a typical compounding recipe for NBR with varied ACNs and the vulcanizates properties.

The permanent set characteristics of NBR is very similar to natural rubber. The factors affecting compression set are the nitrile content, the vulcanization system, and the nature of fillers. Generally, an increase in the nitrile content of the rubber has an unfavorable effect on the set properties. However, NBR with high viscosity shows an improvement in compression set. Peroxide-cured NBR vulcanizates have a low compression set.

The elasticity of NBRs is less than the corresponding compounds of natural or styrene-butadiene rubber. The rebound resilience depends upon the ACN of the polymer. With the decrease in nitrile content the resilience improves. The addition of plasticizers can markedly improve the elastic properties.

Nitrile rubber compounds absorb more energy than the corresponding natural rubber compounds under dynamic conditions. However, under prolonged dynamic stressing NBR vulcanizates generate heat rapidly. Since NBR is more heat resistant than natural rubber, slightly higher heat generation can be tolerated in some applications.

The abrasion resistance improves with increasing ACN of the polymer. Nitrile rubber reinforced with fillers shows improved abrasion resistance. The abrasion loss of NBR is almost similar to natural rubber under high filler loading. The abrasion properties of NBR are also better than those of styrene-butadiene rubber (SBR). The above discussion is limited to room temperature properties. This may change at other temperatures, as viscoelastic properties of rubber are a function of temperature.

B. SWELLING PROPERTIES

The principal characteristics of NBR vulcanizates are their resistance to swelling in contact with organic liquids and retention of properties over extended periods at moderate to elevated temperatures.[1,2,17,20] The solubility parameters of NBR vary from 9.0 to 10.5 as the bound acrylonitrile increases from 20 to 45%. If the difference in the solubility parameter of a fluid and NBR is greater than 1.5 points, the swelling of the elastomer is less than 25%, when it is immersed in the fluid.[2] Thus, NBRs are resistant to lubricating oils (δ = 7.5 to 8.0), whereas SBR (8.4) and butyl (7.8) are not. All three elastomers are resistant to methanol (14.5). Oil resistance is the most important property of NBR. Swelling resistance varies directly with the concentration of acrylonitrile in the polymer, volume loading of insoluble fillers, and the cross-link density of the elastomer. There is a significant reduction in volume swell when a higher molecular weight polymer, or a more reinforcing filler is employed at the same volume loading, particularly for the polymer containing low ACN. However, the swelling resistance of NBR is restricted to nonpolar or slightly polar substances, such as mineral oils, liquid fuels with low aromatic content, and greases.[21-23] Highly polar substances such as ester, ketones, and aromatic hydrocarbons cause severe swelling of NBR. The fuel resistance of NBR depends upon aromatic content of the fuel, such as gasoline. The swelling resistance of NBR reduces drastically when the aromatic liquid fuels are modified with alcohols. Oxidized gasoline resistance can be improved by using bound antioxidants and a sulfur-cure system.[24,25] By changing the compounding recipe, NBR vulcanizates with improved resistance to service fluid can be prepared.[26] The oil resistance of NBR depends upon the viscosity, paraffinic, naphthenic, and aromatic content of the oil. However, the loss of physical properties and volume swell decreases with increase in ACN in the elastomer.

C. LOW TEMPERATURE FLEXIBILITY

Nitrile rubber exhibits moderate to good low temperature flexibility.[1,2] The glass transition temperature increases with the ACN in the polymer. The flexibility of NBR vulcanizates at low temperature is high at lower ACN content. To improve low temperature flexibility, nitrile elastomers are often blended with NBR of low ACN content, natural rubber, styrene-butadiene, or polybutadiene rubbers. The brittle temperature of such blends is determined by their ACN. The Gehman T_{10} temperature is increased when a homogeneous NBR is replaced by a blend of polymers having the same average ACN content. If the polymers in the blend have very different stiffening temperatures, the Gehman T_{10} temperature of the blend will be similar to that of the polymer with the higher stiffening temperature.

D. GAS PERMEABILITY

Nitrile elastomers show very good resistance to permeation by gases.[1,2] Gas permeability of NBR is markedly less than natural rubber (NR) or SBR and is almost equal to that of butyl rubber vulcanizates in many instances. The gas permeation increases with increased temperature and deceases with increasing bound ACN. Nitrile rubber vulcanizates also offer similar resistance to permeation of gasoline vapors and freon.[23]

E. ELECTRICAL PROPERTIES

Nitrile rubber is a semiconductor due to the polar effects of its nitrile group. Hence, it cannot be used for electrical insulation. Further, its specific resistance can be significantly brought down by the addition of conductive carbon black. For oil resistant electrical insulations, the specific resistance can be increased by the addition of phenolic resins. The addition of an antistatic plasticizer can also enhance electrical conductivity in nonblack NBR compounds.[1]

The heat conductivity and coefficient of expansion of NBR vulcanizates are almost similar to those of other vulcanized polymers. Their magnitudes depend upon the bound acrylonitrile and compounding ingredients in vulcanizates.[1]

F. RESISTANCE TO AGING, HEAT, AND OZONE

Nitrile rubber has long been recognized as having better resistance to aging than NR and polybutadiene rubbers.[2,3,27,28] At elevated temperatures, the polymer chains possess a higher thermal energy and are in a mobile condition. The cross-linked network responds rapidly to imposed stress as it goes to a new equilibrium position. Consequently, there is deterioration of physical properties at elevated temperatures. The heat resistance of NBR is directly related to the increase in ACN of the elastomer. However, the presence of the double bond in the polymer backbone makes it more susceptible to heat, ozone, and light. Several strategies have been adopted recently to improve the aging resistance of NBR. The air and fluid aging can be improved by the proper choice of compounding and cure system.[29-31] The use of cadmium and magnesium oxides in place of zinc oxide as activator results in stabilization toward oxidation. However, cadmium-containing materials are toxic.[32] Polymers designed with noninterfering stabilizers, when vulcanized with peroxide, show very good aging resistance. The addition of white fillers also imparts better resistance to aging through catalytic destruction of chain-initiating hydroperoxide.[33] Volatilization or fluid extraction of antioxidants are major problems during NBR aging. To avoid this, polymers containing an antioxidant moiety bound to the chain during the polymerization have been developed.[34] These polymers show excellent resistance to air aging when compounded with heat-resistant cure systems in the presence of white fillers. The thermo-oxidation resistance can be enhanced significantly by the synergistic mixture of primary antioxidants (phenolic and amine types) and secondary antioxidants [β-(n-alkylthio)propionates of polyethylene glycols].[35] Grafting of high molecular weight antioxidants to the polymer backbone also eliminates the extraction and improves the aging resistance.[36] Antioxidants which copolymerize with butadiene and acrylonitrile, such as *N*-(4-anilinophenyl)methacrylic amide, can also be used.[37] Compounds based on symmetrical dithiophenols or phenol- or amine-substituted sulfur compounds may be attached to the polymer chain during or after polymerization to enhance aging resistance.[38,39]

The weather and ozone resistance of NBR is the same as that of other diene rubbers. The oxidation process is enhanced in the presence of sunlight. The addition of paraffinic hydrocarbon and antiozonants during compounding does not improve ozone resistance significantly, as is observed in the case of NR and SBR. Recently, various techniques have been developed to bring about a considerable improvement in ozone resistance, (Sections IV.A and IV.L.7).[40,41] However, NBR vulcanizates show better durability than other diene rubbers during aging in natural climatic conditions.[42]

Like other diene polymers, the properties of NBRs are slowly deteriorated by the action of nuclear radiations; the hardness increases slowly while the tensile strength and elongation decrease. Nitrile rubber vulcanizates containing medium nitrile proportions and a small amount of fillers, offer considerable resistance to degradation by radiations. However, the total process depends upon the nature of compounding ingredients, fillers, curing system, and ACN content.

IV. MODIFICATION OF PROPERTIES OF NITRILE RUBBER

A. HYDROGENATION OF NITRILE RUBBER

The principal driving force behind the development of hydrogenated NBR is the increasing performance demanded of elastomers by the automative and oil drilling industries. The conventional oil resistant rubbers such as NBR, chloroprene rubber, and chlorinated polyethylene are reaching their performance limits. Fluoroelastomers, which are the possible substituents in these applications, are highly expensive and possess processing difficulties. Hence, hydrogenated NBR has been developed to bridge the price-performance gap between general purpose oil resistance rubbers and fluoroelastomers. The double bonds present in the diene part of NBR makes it less resistant to heat and ozone. The selective hydrogenation of olefinic unsaturation in NBR upgrades the product performance in many of its traditional applications. The development HNBR is a great success in improving almost all the major drawbacks of nitrile rubber as well as in widening its range of applications.[40-48]

The properties of HNBR are mainly determined by the degree of saturation and the ACN. The degree of saturation is responsible for the heat resistance. It also influences the rheological behavior and processibility of the polymer. The residual unsaturation present in the elastomer gives a favorable balance between efficient crosslinking by the sulfur/accelerator system and minimized loss in heat, ozone, and chemical resistance. The level of oil and fuel resistance is determined by the bound acrylonitrile. The rebound resilience decreases with increase in ACN, as observed in NBR. However, hydrogenated nitrile rubber is significantly better in the range between 30 to 40% ACN.

For raw polymers, the tensile strength is much higher in HNBR than NBR (Figure 1). With the degree of hydrogenation the stress-strain properties improve. Hydrogenated nitrile rubber vulcanizates have much higher tensile strength (Figure 2) and wear resistance than NBR and other heat- and oil-resistant elastomers. Hydrogenated nitrile rubber is able to maintain a high level of physical properties, even under a combination of many adverse conditions. In comparison with other heat-resistant elastomers, such as FPM and acrylic rubber (ACM), vulcanizates based on HNBR have considerably better mechanical properties at elevated temperatures. This strength retention is essential for sealing where extremely high pressures at elevated temperature are involved. Table 2 gives a comparison of physical properties of HNBR and NBR at a particular composition.

Low temperature flexibility of HNBR is better than NBR. With the increase in degree of hydrogenation, the glass transition temperature of the elastomer decreases.[11-13] Tg can be further lowered without noticeable loss in heat resistance by adding a plasticizer.

Peroxide-cured vulcanizates of HNBR show good results in a high temperature compression set. Partially hydrogenated grades of NBR offer favorable compression set values at low temperatures. Hydrogenated nitrile rubber vulcanizates almost completely retain their hardness over a wide range of elevated temperatures. Hydrogenated nitrile rubber is an excellent abrasion-resistant elastomer[49] (Figure 2). Like fluoroelastomers, the abrasion resistance of HNBR is hardly affected by high environmental temperature.

Rubber-to-metal bond strength of HNBR vulcanizates exceeds its inherent tear strength in the presence of a suitable rubber-to-metal bonding agent. Bond strength remains almost the same after hot air aging of a metal-rubber composite. Suitably formulated HNBR offers good adhesion to different reinforcing fabrics and cords.

Hydrogenated nitrile rubber is comparable with other elastomers in dynamic mechanical properties due to its excellent mechanical strength, relatively high elasticity retention of

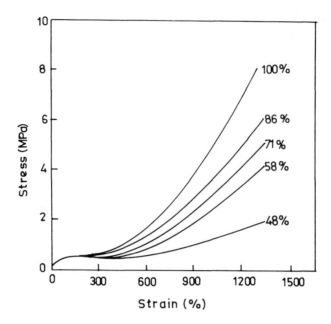

FIGURE 1. Stress-strain curves of HNBR containing 40% ACN with different degrees of hydrogenation. (From *Indust. Eng. Chem. Res.*, 30, 1086 (1991). With permission.)

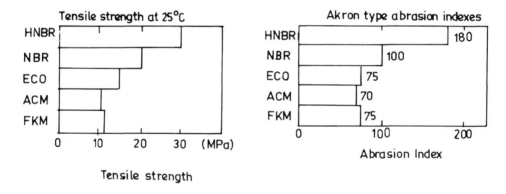

FIGURE 2. Tensile strength and abrasion resistance of HNBR in comparison with other rubbers. (From Nippon Zeon Co., Ltd., Japan. With permission.)

physical properties over a wide range of temperature, as well as very good retention of dynamic behavior and flex cracking resistance over long periods at elevated temperatures, due to its excellent heat aging and ozone resistance.[50] Hydrogenated nitrile rubber is slightly superior to NBR in gas and water vapor permeation.

1. Resistance to Aging

Hydrogenated nitrile rubber sets itself apart from other high performance elastomers in maintaining properties under adverse conditions. For example, HNBR offers resistance against a combination of the following aggressive media, which can singly cause rapid failure of other elastomers.

1.   Hot air (after 10,000 h/130°C retained elongation of 100%)
2.   Ozone (more than 500 pphm at 40°C and 50% relative humidity)
3.   Steam (up to 180°C)

TABLE 2

Comparison of Technical Properties of Nitrile and Hydrogenated Nitrile Rubbers

	Hydrogenated nitrile rubber (TORNAC A 3855®)	Nitrile rubber (KRYNAC 3850®)
Polymer	100	100
Stearic acid	1	1
Zinc oxide	5	5
MgO (Maglite D®)	10	10
Substituted diphenyl amine (Naugard 445®)	2	2
Vanox ZMTI® (Zinc 2-marcaptotolyl imidazole)	2	2
N550 black	50	50
Curative — Varox DBPH-50®	10	3.5
Coagent(1,2-polybutadiene 65% on calcium silicate) — RICON 153-D®	6.5	2.6
Acrylonitrile content (%)	38	38
Compound viscosity, ML 1 + 4 (100°C)	118	86
Scorch time (min at 125°C)	28	28
Cur time (min)	14	14
Hardness, ShA	76	76
Modulus at 100% elongation (MPa)	7.7	9.3
Tensile strength (MPa)	25.7	21.9
Ultimate elongation (%)	250	180
Tear strength, Die C(kN/m)	35.3	16.7
Compression set (%)		
70 h at 150°C	24	20
168 h at 150°C	25	27
Abrasion resistance		
NBS %	344	118
DIN (mm³ loss)	122	260
Low temperature properties (°C)		
Brittle point	-51	-22
Gehman low temp. stiffening $-T_2$	-15	-12
$-T_{10}$	-24	-21
Low temp. refraction (°C) $-TR_{10}$	-22	-21
$-TR_{70}$	-6	-11
Ozone Resistance (50 pphm at 40°C, 20% strain)		
Time 1st crack (h)	>168	<24
Aged in Air—168 h at 150°C-Change		
Hardness, ShA	$+6$	$+17$
Tensile (%)	Nil	Brittle
Ultimate elongation (%)	-32	Brittle
Volume Change (%) after 168 h at 150°C		
ASTM Oil #1	-2	-2
ASTM Oil #1 + 1% NACE B[a]	$+1$	$+2$
ASTM Oil #3	$+18$	$+11$
SAE 90 gear oil	$+5$	$+1$
Mineral hydraulic fluid	$+4$	$+3$
Automatic transmission fluid	$+5$	$+4$

[a] NACE B is an amine-based corrosion inhibitor

From Bayer-Polysar Technical Centre N.V., Antwerp, Belgium. With permission.

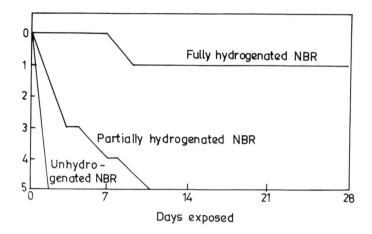

FIGURE 3.   Ozone resistance of NBR and HNBR. (From Carl Hanser Verlag, Munich Germany. With permission.)

4.   Oils (especially containing additives and H_2S)
5.   Chemicals (especially amines, H_2S, alkaline, and oxidative chemicals)
6.   High energy radiation

The most outstanding advantage of HNBR is its ozone resistance. Unlike NBR, completely hydrogenated NBR is unaffected by ozone, even under extreme exposure conditions. Figure 3 represents the ozone resistance of nitrile and partially and completely hydrogenated NBR. Hydrogenated nitrile rubber remains unaffected when exposed to pressurized steam containing traces of alkali. It may be noted that the above condition causes degradation of fluoroelastomers.[46-48] Similarly, hot water has no significant effect on HNBR.

Hydrogenated nitrile rubber inherits its excellent oil and swelling resistance from the parent NBR. It surpasses NBR in maintaining its oil resistance at much higher temperatures (Figure 4) and also shows higher resistance to oil contaminants such as amines, hydrogen sulfide, and sour crude oils which are poisonous to NBR and many other heat- and oil-resistant elastomers. In comparison to NBR, HNBR with the same ACN, offers better resistance to fuels, especially in maintaining the properties for a prolonged period.[51,52] Service temperature and life of HNBR in comparison with NBR, CR, and ACM is presented in Figure 5. Hydrogenated nitrile rubber offers good resistance to antifreeze. It is particularly resistant to alkaline substances such as ammonia and caustic soda solution and is sufficiently resistant to acids for many applications. It is not recommended for contact with aromatic and organic polar substances, such as toluene, MEK, etc. Table 3 gives the resistance of HNBR to a wide variety of chemicals. Oil filed applications involve often exposure to oils with acids and amines as well as hydrogen sulfide and carbon dioxide. This kind of "sour crude oil" resistance of HNBR is considerably better than FPM, NBR, and XNBR (Figure 6).[53] In oil filed applications, resistance to explosive decompression is often an essential factor. Hydrogenated nitrile rubber shows better resistance than FPM, NBR, epichlohydrine rubber (ECO) and EPDM at high temperature and pressure.[46-48] A relative performance of hydrogenated nitrile and other heat- and oil- resistant rubbers is given in Table 4. HNBR also shows superiority in extrusion resistance. Hydrogenated nitrile rubber offers excellent resistance against high energy radiation. Another special feature of HNBR is retention of abrasion resistance after exposure to high energy radiation. Figure 7 gives the overall performance profile of HNBR, NBR, ECO, NBR-PVC, and FPM.

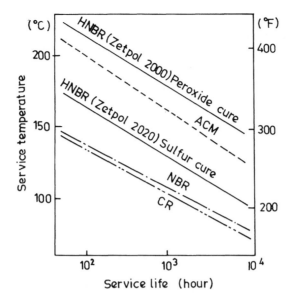

FIGURE 4. Service temperature and life of HNBR as compared to other rubbers. (From Nippon Zeon Co. Ltd., Japan. With permission.)

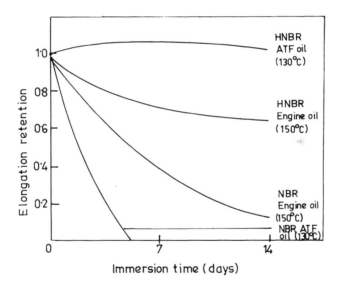

FIGURE 5. Elongation retention of HNBR and NBR after immersion in engine and ATF oil. (From Bayer A-G, Germany. With permission.)

B. CARBOXYLATED NITRILE RUBBER

A modification of the NBR molecule by incorporation of carboxyl moiety in the polymer backbone results in XNBR with the usual oil resistance characteristics of NBR and much improved abrasion resistance.[2,3,17,20] The terpolymer of butadiene, acrylonitrile, and a diene monomer with a carboxylic acid group possesses an additional cross-linking site due to the reactive carboxylic side group.[54] The carboxylic group can react with multifunctional reagents in a heteropolar reaction to crosslink the polymer chain. Ionic crosslinking takes place through

TABLE 3
Chemical Resistance of Hydrogenated Nitrile Rubber

		HNBR	NBR	EPDM	ECO	CR	CSM	Q	FKM	ACM
Organic acid	Steam (150°C/302°F)	B	U	A	U	U	U	U	C	U
	Acetic acid (30%)	B	B	A	B	A	A	A	B	U
	Hydrochloric acid (25%)	A	B	A	U	B	A	B	A	U
Acid	Phosphoric acid (20%)	A	B	A	—	B	A	B	A	—
	Nitric acid (25%)	B	U	B	U	A	A	B	C	U
Alkali	Sodium hydroxide (30%)	A	B	A	B	U	A	B	B	—
	Aqueous ammonia (28%)	A	A	A	B	A	A	A	B	U
Salt	Sodium chloride (30%)	A	A	A	A	A	A	A	A	—
	Sodium carbonate (10%)	A	A	A	—	A	A	A	B	—
Oxidizing agent	Hydrogen peroxide (3%)	B	C	B	—	C	A	A	A	—
	Sodium Hypochlorite (5%)	B	U	B	—	U	B	B	A	U
Aliphatic hydrocarbon	Iso octane	A	A	U	A	B	B	U	A	A
Aromatic hydrocarbon	Toluene	C	C	U	U	U	U	C	A	U
Chlorinated hydrocarbon	Trichloroethylene	C	C	U	U	U	U	U	A	—
Alcohol	Methyl alcohol	A	A	A	B	A	A	A	C	U
	Ethyl alcohol	A	A	A	A	A	A	A	A	U
Ether	Ethyl ether	C	C	C	B	U	U	U	U	U
Ester	Ethyl acetate	U	U	B	U	C	C	U	C	—
Ketone	Methyl ethyl ketone	U	U	A	U	U	U	U	U	U
Aldehyde	Furfural	B	C	A	U	B	B	A	U	U
Amine	Triethanol amine	A	C	A	—	A	A	U	U	U
	Carbon disulfide	C	C	U	U	U	U	—	A	—

Rating system employed: A — recommended — little or minor effect; B — minor to moderate effect; C — moderate to severe effect; and U — not recommended.

From Nippon Zeon Co. Ltd., Japan. With permission.

carboxylic groups in the presence of zinc oxide. Hence, during conventional vulcanization using a sulfur accelerator and zinc oxide, along with sulfur crosslinks, zinc salt bridges are formed. Carboxylated nitrile rubber has a tendency to scorch in the presence of zinc oxide. This problem may be avoided by using coated particles of ZnO with zinc sulfide or zinc phosphate or by the use of zinc peroxide masterbatches in place of zinc oxide.[55] The good abrasion resistance of XNBR is due to the ionic bonds formed between metal oxides and carboxyl groups.[56] Compared to NBR, XNBR vulcanizates are a harder and tougher material with high modulus, tensile strength, and tear strength. However, they are less flexible at low temperatures and less resilient than noncarboxylated grades. Unvulcanized XNBR sticks to a mill or mold surface during processing. Carboxylated nitrile rubber has a relatively slower cure rate due to restriction of freedom of motion by steric hindrance and molecular interaction. The peroxide-cured XNBR has a low compression set. The extraordinary cohesive strength of XNBR is useful in adhesives. Carboxylated nitrile rubber has a greater affinity for nonblack fillers because of its hydrophilic character. Nonblack fillers such as

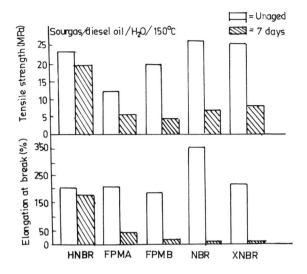

FIGURE 6. Comparison of HNBR with FPM, NBR, and XNBR in "sour crude oil". (From Bayer A-G, Germany. With permission.)

TABLE 4
Comparison of Hydrogenated Nitrile Rubber with Other Heat- and Oil-Resistant Elastomers

Rubber type	Oil resistance	Heat resistance	Mechanical properties	Resistance to aggressive media	Low temperature flexibility
HNBR	2	2–3	1	1–2	2
EPM	1	1	3	1–2	4
Q	3	1	4	3	1
ACM	2	2	4	2	4
EAM	3	2	4	2	3
CR	3	4	1–2	3	1–2
NBR	2	4	1–2	4	2

Note: This property depends very much on the medium concerned.

Legend: 1 = excellent; 2 = very good; 3 = good; and 4 = satisfactory.

From Bayer A-G, Germany. With permission.

silica and clay give better reinforcement in XNBR than in NBR. Carboxylated nitrile rubber vulcanizates have good resistance to hydrogen sulfide, in particular the modulus retention, at high levels of carboxylic acid.[22] The deterioration of XNBR, in comparison with other rubbers, by sour environment is shown in Figure 8. Carboxylated nitrile rubber crosslinked through carboxylic groups shows better solvent swell resistance and abrasion resistance than conventional NBR vulcanizates. Some of the outstanding features of XNBR are its water resistance, low metal corrosion, oil and fuel resistance, and good retention of properties at relatively high temperatures. Hydrogenation of XNBR further improves its heat and chemical resistance.[13]

C. POWDERED NITRILE RUBBER

Modification of plastics, such as polyvinyl chloride (PVC), acrylonitrile butadiene styrene (ABS), and phenolic resins by NBRs results in highly improved performances of the

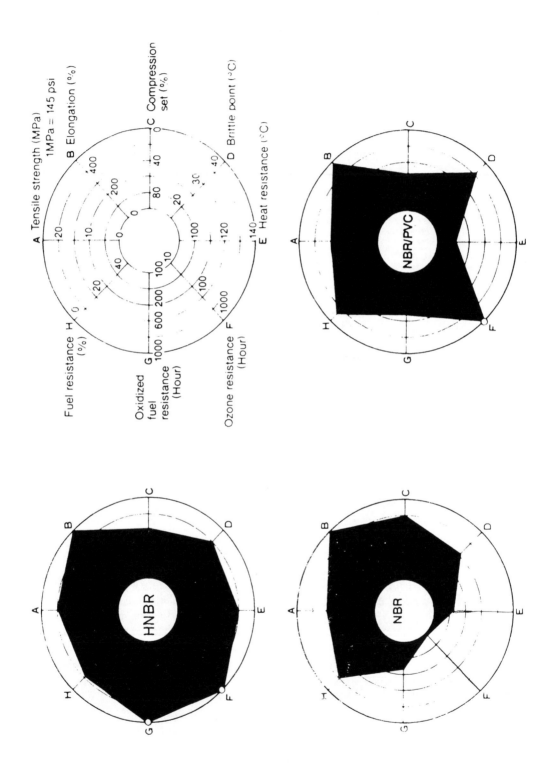

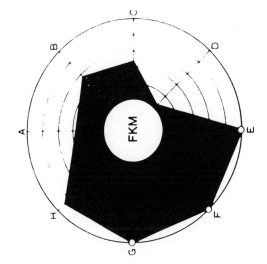

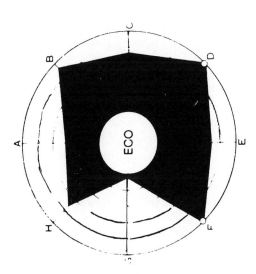

FIGURE 7. Performance profile of HNBR, NBR, NBR-PVC, ECO, and FKM. (From Nippon Zeon Co., Ltd. Japan. With permission.)

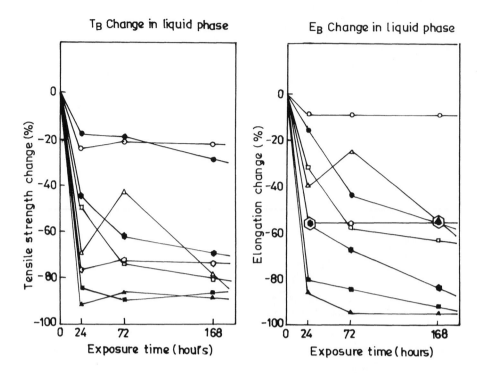

FIGURE 8. Deterioration of tensile strength and elongation at break by sour environment. Test conditions: temperature 150°C; pressure 6.9 MPa; gas phase—hydrogen sulfide 5%; CO_2 20%; methane 75%; Liquid phase — diesel oil 95%; water 4%; and NACE amine B—1%. ○, HNBR (higher degree of hydrogenation); ● HNBR (lower degree of hydrogenation); ◇, EPDM; △, TFE/P; □, FKM-GF; ◆, FKM-E; ■, NBR: and ▲, XNBR. (From Nippon Zeon Co. Ltd., Japan. With permission.)

end products. The particulate form of NBR allows fast and easy incorporation and a homogeneous distribution in plastic materials. The combination of easy processing characteristics and improved physical properties of final products has established powdered NBRs in a wide variety of NBR-modified thermoplastic and thermoset products.[2,3,20] Flexible PVC plasticized with conventional, liquid plasticizer may loose plasticizer and become rigid and brittle when subjected to heat or in contact with oils, greases, and solvents or in contact with rigid, nonplasticized materials. Powdered NBR acts as permanent plasticizers in these blends. Their polymeric structure prevents migration and volatilization and they are not extracted, even under aggressive conditions. Hence, they ensure retention of flexibility and long service, in addition to imparting oil and fuel resistance to the finished products.[57,58]

Powdered NBR contain antioxidants and stabilizer systems, which increase the resistance to environmental aging in plastic materials.[2] Modification with powdered NBRs improves the abrasion resistance of certain flexible thermoplastic materials. The poor compression set properties of thermoplastic materials can be improved by blending them with powdered NBR which, being an elastomer, improves recovery after deformation. Modification with powdered NBRs improves the processibility, the mold stability, and the dimensional stability of finished articles in contact with chemicals. The technical advantages of powdered NBRs over NBR are their excellent storage stability, low heat building during mixing, low dump temperature, low risk of search, less polymer degradation during mixing, short mixing time, etc.

D. LIQUID NITRILE RUBBER

Low molecular weight NBR in liquid form is produced as a modified polymer with carboxyl, hydroxyl, mercaptan, or halogen groups along the chains or at the chain ends.

These groups serve as the active site or chain extension and crosslinking.[3,20] The structure of a represented polymer may be shown as

$$\text{HOOC–[(CH}_2\text{–CH=CH–CH}_2\text{)–} \quad \text{(CH}_2\text{–CH)–]–COOH}$$
$$|$$
$$\text{CN}$$

The liquid NBR has been developed as nonextractable and nonvolatile plasticizers and binders in polymeric materials. They are also added in NBR compounds with high viscosity to ease the processing. Carboxyl terminated liquid NBRs are mainly used in elastomer-modified epoxy resins for applications such as structural adhesives, glass-fiber-reinforced composites, coatings, and electrical encapsulations. The investigation on the potential application of these materials and various ways of crosslinking are still in progress. One of the curative systems for carboxyl-terminated polymers is liquid epoxy resin (diglycidylethers of 4,4-isopropylidine diphenol) in presence of amine catalysts.[59] The vulcanizates, thus, obtained have good mechanical properties and resistance to aging. The carbon black filled vulcanizates show greatly enhanced strength. The blend of carboxyl-terminated NBR with epoxy resin is castable.[60,61] The rubber, coreacted and reinforced with resin, possesses a tensile strength equivalent to conventional NBR vulcanizates. The vulcanizate shows high modulus, lower elongation, good retention of properties after air aging, oil and water swell comparable to solid NBR, and poor compression set. Optimum properties are obtained when the distance between two cross-link sties is greater than the chain lengths of the prepolymers. Hence, chain extension is necessary.

However, the manufacture of rubber goods based on NBR telechelics has failed to establish itself because the available materials are not exactly difunctional. The reinforcing fillers, which are added to enhance mechanical properties, make it difficult to process and the prepolymers are quite expensive. So far, liquid NBR has found application as a toughening agent for thermoset resins. The highly reactive mercaptan-terminated liquid NBR is used as a flexibilizer. The liquid XNBR increases the rupture energy of epoxy resins and vinyl NBR increases that of the unsaturated polyester resins.

The liquid NBR has moderate heat stability. Hence, their olefinic unsaturation is selectively hydrogenated to improve heat resistance.[62,63] Liquid XNBR on hydrogenation exhibits improved thermal stability and low temperature properties. The glass transition temperature decreases significantly with hydrogenation of double bonds.

E. CROSS-LINKED NITRILE RUBBER

During the extrusion of NBR, the problems of die swell or shrinkage are very common. The use of cross-linked NBR alone or in blends with conventional NBR improves the quality of the finished goods.[2] The advantages of cross-linked NBRs over NBR are easy extrusion and calendering, limited die swell and shrinkage and improved surface characteristics and dimensional stability. Cross-linked NBR is produced by using a termonomer as a cross-linking agent during polymerization or by stopping the polymerization reaction at very high conversions. Compared to NBR, cross-linked NBR shows a slight drop in mechanical properties. however, for the majority of applications, the slight drop is not a detrimental factor.

F. NITRILE-BLACK MASTERBATCH

Another development in the NBR technology is the introduction of NBR carbon black masterbatches. Masterbatches are also available in crumb form. Carbon black is incorporated during the polymerization stage and any proportion of polymer and filler can be maintained. Crumb masterbatches have all the advantages claimed by nonpigmented NBR powders for reducing compounding time and cost.[2,3,20] These masterbatches show higher elongation and lower modulus than the comparable compounds based on bale rubber. This suggests adsorption of some of the curatives by the masterbatch. The energy of mixing is much lower with masterbatches than that with corresponding bale form. Since NBR is used in a wide variety of recipes, it is difficult to obtain masterbatches of all those varieties commercially. Although they offer many advantages over bale rubber, they have failed to become widely established.

G. PREPLASTICIZED NITRILE RUBBER

Plasticizer-modified NBR grades are also available on a commercial scale. The addition of ester and ether plasticizers to NBR during polymerization reduces the long compounding time of baled rubbers.[3,20] This is possible only when high proportions of plasticizers are incorporated. Nitrile rubbers can also be extended with oil during the coagulation stage. The advantages of oil-extended NBR over regular nitrile rubbers in soft compounds are excellent filler dispersion, easier mixing, reduced mixing time, better physical properties and elasticity, reduced weighing errors, and no loss of plasticizer during mixing. They are useful in the production of soft vulcanizates.

H. THERMOPLASTIC NITRILE RUBBER

The development of thermoplastic NBR is still in progress. Several methods are being suggested to produce thermally or mechanically reversible crosslinks in NBR which can be a route to thermoplasticity. One of the methods involves the introduction of unstable or labile cross-linking sites by reaction between organic polyhalogen compounds and tertiary amines.[3,20] Tertiary amines are added during the copolymerization of butadiene and acrylonitrile as a co-monomer of acrylates or methacrylates of the formula

$$
\begin{array}{cc}
R & C \\
| & \| \\
H_2\,C{=}C{-}C{-}O{-}\!-(R')N(CH_3)_2
\end{array}
$$

where R = H or CH_3, R' = aliphatic hydrocarbon and secondary or tertiary amine-substituted aliphatic hydrocarbon with two to four carbon atoms.

The polymers have increased green strength and the scraps formed from the molding process can be reused. Unlike regular thermoplastics, the unstable cross-linking sites result in poor processibility in thermoplastic NBR.

I. ALTERNATING NITRILE RUBBER

Polymerization of acrylonitrile and butadiene in the presence of a suitable catalyst gives an alternating equimolar butadiene — acrylonitrile copolymer.[3,20] The catalyst consists of alkyl aluminium halides and transition metal compounds such as VCl_3 and $TiCl_4$.[64,65] The catalyst is complexed with acrylonitrile before adding butadiene to achieve, almost completely, an alternating copolymer. The butadiene is bonded in a *trans*-1,4- configuration in the copolymer chain.

The alternating NBR vulcanizates show a higher tensile strength and elongation than a random copolymer with an equal amount of acrylonitrile.[66] This is attributed to the regular structure of the polymer which offers chain flexibility and strain-induced crystallization. However, the vulcanizates have lower hardness and modulus than the random copolymer. The high temperature resilience and low temperature retraction of the alternating polymer are almost the same as that of the random copolymer. In spite of its advantages over the corresponding random copolymer, the technology of an alternating butadiene acrylonitirle copolymer is yet to be commercially exploited. The reason is that the alternating copolymer must be of high ACN which is expensive and not as widely used as the polymers with lower ACN.

J. ACRYLONITRILE/ISOPRENE RUBBER

The copolymer of acrylonitrile and isoprene can be considered as a structurally modified NBR.[20] Both alternating and random copolymers have been prepared. Compared to NBR, acrylonitrile/isoprene rubber (NIR) is more easily depolymerized, easier to extrude, and has higher tack. Strain crystallization and high tensile strength are observed in NIR vulcanizates, since it contains a high proportion of triads.[67] The elasticity and low temperature flexibility are considerably inferior to that of NBR. Acrylonitrile/isoprene rubber has been commercially available for the past few years, but it has not yet acquired much technological importance.

K. OTHER MODIFICATIONS OF NITRILE RUBBER

Modified NBR containing epoxy groups show good interaction with the silanol group of silica fillers.[20,68] The product has yet to be commercialized.

A reactive aldehyde group may be incorporated in the double-bond site of the polymer backbone by hydroformylation reaction.[69] The resultant hydroformylated nitrile rubber (HFNBR), containing an extremely reactive aldehyde group in the copolymer chain, can be converted to hydroxyl- or carboxyl-containing NBR having a saturated backbone. They can be used as binders and coupled with other rubbers due to the presence of polar aldehyde groups. Various other properties of the HFNBR and their commercial viability are yet to be explored.

L. BLENDING WITH OTHER POLYMERS

Modification of NBR by blending with other rubbers or thermoplastics or thermosetting resins is performed to reduce cost, improve processibility, and achieve desired physical or chemical properties of the end products. The properties of the NBR blends depend on the effectiveness of the physical blending process and compatibility of the polymers. Some common rubber — rubber and rubber — plastic blends of NBR are discussed in the following sections.

1. NBR-NR and NBR-SBR Blends

NR is blended with NBR for improvement in tack properties. NBR-SBR blends are used to compensate the volume decrease in oil seal application.[20,70]

2. NBR-CR Blends

Nitrile rubber is blended with neoprene to improve ozone and fatigue resistance without much sacrifice of oil resistance properties.[20,70]

3. NBR-EPDM Blends

NBR-EPDM blends have been developed with the idea of combining the ozone resistance of EPDM with the oil resistance of NBR.[3,20] Due to the difference of polarity of the two materials, the blend forms a heterogeneous phase. As a result, NBR-EPDM blends exhibits poor physical properties which arise from improper partitioning of curatives between the rubber phases.[71] Many methods have been developed to overcome the problem of an unequal rate of crosslinking and accelerator solubilities in the blend. For example, addition of chemicals with a homogenizing effect improves the properties of the blend. Dithiocarbamates with long chain alkyls can be used as a curative which is compatible with both rubbers. Tetraethyl disulfide when used as an accelerator gives vulcanizates with tensile strength varying linearly with blend composition. Lead diethyldithiocarbamate can be used for co-vulcanization, since it is insoluble in the polymer and, thus, remains evenly distributed. The red lead-zinc oxide combination also imparts good physical properties to the NBR-EPDM blend. Although peroxide crosslinking gives good ozone resistance, overall properties are not improved, as NBR is overcured and EPDM is undercured in the system. Yet another method is the preparation of EPDM masterbatch with carbon black, plasticizers, and activators, prior to addition of NBR, but at lower EPDM content it is difficult to prepare the masterbatch.

The advantages of the swelling resistance of NBR and ozone resistance of EPDM can be achieved in the blend only when the heterogeneity of the system is reduced. For good ozone resistance EPDM must be dispersed in the medium finely and for fair retention of oil resistance, the ratio of EPDM must be less. in NBR-EPDM blends, the mixing procedures, compounding systems, and curing methods determine the ultimate properties. Vulcanizable HNBR-EPDM blends offer better strength and ozone resistance than NBR-EPDM composition.[72]

4. NBR-BR Blends

Blending of polybutadiene rubber (BR) with NBR and XNBR improves the performance in cold bending tests and reduces the brittleness temperature without changing the glass transition temperature.[20,70,54] Addition of BR improves the flow behavior in injection molding and reduces extrusion energy. However, due to the difference in polarity, the blend is heterogeneous and adverse effects on the physical properties of the vulcanizates cannot be avoided.

5. NBR-EVA Blends

When 30% ethylene vinyl acetate (EVA) rubber (vinyl acetate content > 40%) is added to NBR, the vulcanizates offer oil and ozone resistance.[20,64] These blends are more resistant to aging than NBR-PVC blends. Thiuram disulfide or peroxide can be used for crosslinking. The physical properties are inferior to that of NBR vulcanizates. The small addition of NBR to EVA compounds improves the mold release and the peelability.

6. NBR-CIIR Blends

NBR-CIIR blends are heterogeneous and impart good ozone resistance with deterioration of physical properties.[4,70]

7. NBR-PVC Blends

Blending of PVC with an NBR matrix has been known since 1936. NBR-PVC blends have attracted extensive commercial importance because they offer a combination of ozone and chemical resistance of PVC with the crosslinkability elastic properties and oil resistance of NBR.[2,3,20,73,74] The properties of blend vary with the blend ratio. With up to 50% PVC content, the properties of NBR predominate and are processed like rubbers. When PVC content is more than 50%, the properties of thermoplastics are predominant. Nitrile rubber is not crosslinked in the blends with high PVC content and it serves as a polymeric plasticizer or impact modifier.[57,58] The NBR-PVC system is a homogeneous one as long as the ACN of NBR is about 34 to 40%. Below 20% ACN, the system fails to form a homogeneous phase. The compatibility of the blend results in an interesting balance of properties.

The principal technical advantage of NBR-PVC blends is the significant improvement of ozone and weather resistance of NBR without sacrificing other properties of NBR. Optimum ozone resistance is obtained when PVC content is more than 30% in the blend. The improvement in ozone resistance is attributed to the blocking of double bonds of NBR by the PVC segments. In an NBR-PVC blend, NBR contributes to low temperature flexibility and resistance to crack growth on flexing. The acrylonitrile level chosen represents a compromise between low temperature flexibility and oil resistance and the PVC contributes to abrasion, ozone, and weather resistance.[75] Polyvinyl chloride acts as a reinforcing material in NBR-PVC blends and increases tensile strength and tear resistance. Table 5 shows the effect of increasing PVC in an NBR-PVC blend. The additional characteristics of NBR-PVC compounds are fast smooth calendering and extrusion, long storage stability, antistatic

TABLE 5
Effect of PVC Level in NBR-PVC Blends

Test recipe		phr		
NBR or NBR/PVC		100		
Sulfur		0.25 Stearic acid		
1 N550 black		40 Zinc oxide		
5 TMTM		0.5		
KRYNAC 34-50®	100	85	70	55
PVC resin	—	15	30	45
Compound Properties				
Mooney viscosity, ML 1 + 4 (100°C)	67	61	58	62
Garvey Die extrusion				
#1 Royle, 104°C, 70 rpm,				
Rate (cm/min)	80	94	121	135
Die swell (%)	137	106	60	44
Vulcanizate Properties				
Cured at 166°C (min)	8	10	10	12
Hardness, ShA	63	69	73	82
Modulus at 100% elongation (MPa)	2	3.1	4.6	7.5
Modulus at 300% elongation (MPa)	10.1	10.5	12.0	13.5
Tensile strength (MPa)	22.3	20.2	17.3	15.8
Ultimate elongation (%)	600	600	530	460
Tear resistance, Die C (kN/m)	51	55	62	70
Compression set, 22 h at 100°C (%)	37	53	48	64
NBS abrasion index (%)	157	287	441	352
Low temperature stiffening — unaged — (°C)				
Gehman T10®	−22	−20	−18	−9
Aged in ASTM Oil #2, 168 h at 100°C — change				
Volume (%)	+9	+6	+3	NIL
Aged in Fuel 8, 168 h at 23°C — change				
Volume (%)	+30	+24	+19	+16
Dynamic ozone resistance 50 pphm, 20% extension, 40°C				
Rating after 100 h	3	4	2	0

Rating: 0—No cracks visible under $10\times$ magnification; 1—little cracks visible under $20\times$ magnification; 2—few little cracks visible; 3—visible cracks of 1 mm width; 4—visible cracks of 2 mm width; and 5—large number of visible cracks.

From Bayer-Polysar Technical Centre N.V., Antwerp, Belgium. With permission.

and flame-resistant properties by proper selection of compounding ingredients, and moisture resistance. Tensile and tear strength and abrasion resistance increases when NBR is replaced by XNBR in the blend.[54] Recently, an HNBR-PVC blend with better oil, cold, heat, and ozone resistance has been developed.[76,77] Figure 9 illustrates better performance of HNBR-PVC over NBR-PVC blends.

8. NBR-PVA Blends

The objective of blending polyvinyl acetate (PVA) with NBR is to obtain better low temperature flexibility, compression set behavior, and aging resistance than NBR-PVC blends.[20] Although the solubility parameters are similar for both polymers, the blend has

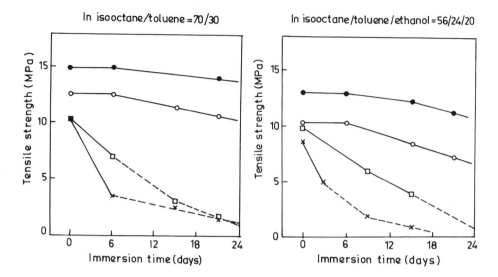

FIGURE 9.   Deterioration of tensile strength in model fuel containing 3% lauryl peroxide at 40°C. ●, HNBR-PVC (70/30): ○, HNBR; □, NBR-PVC (70/30), X, NBR; and ——, elongation at crack initiation. (From Nippon Zeon Co., Ltd., Japan. With permission.)

been found to be heterogeneous. Thus, all the mechanical properties and ozone resistance, except the resistance to aging, are inferior to NBR-PVC blends.

9. NBR-Nylon Blends

NBR-nylon compositions have been prepared by melt mixing of the polymer and subsequent dynamic vulcanization. They are elastomeric materials, which can be fabricated as thermoplastic vulcanizates and exhibit good strength and excellent hot oil resistance.[73,74] The characteristics of the blend depend upon the melting point and polarity of the nylon resin, types of NBR (nitrile content, viscosity, susceptibility to self-curing, etc.) and curing system. Curatives used during dynamic vulcanization of the blend are accelerated sulfur, activated bismaleimide or peroxide. Plasticizers and fillers can be added to the composition to improve fabricability and hardness, respectively. Both plasticizer and filler can be used together to retain thermoplasticity and extensibility of the material. An increase in the amount of rubber reduces stiffness and strength, but improves resistance to permanent set. However, excess of rubber results in poor fabricability. Blends of XNBR with nylon shows excellent low temperature impact resistance.[54] Thermoplastic HNBR-nylon blends exhibit good physicomechanical properties, toughness, and resistance to weathering.[78,79]

10. NBR-Cellulose Ester Blend

The formation of hydrogen bridge bonds restricts the formation of a homogeneous phase in NBR-cellulose ester acetate, propionate, or acetate butyrate blends.[20] The vulcanizates exhibit high hardness with good resilience, but the compounds are very viscous and difficult to process. Moreover, the presence of two separated phases decreases the ozone resistance of the blend.

11. NBR-Phenolic Resin Blend

Nitrile rubber can be crosslinked with phenolic resins in the presence of activators such as organic acids or halogen-containing compounds.[20] They can also be added to NBR for reinforcement as fillers. For sufficient reinforcement, hexamethylene tetramine is added to the mixture. The resultant blend offers high tensile strength and hardness. When NBR is added in small amounts in phenolic resins, it raises the impact resistance of the resins.

TABLE 6
Properties of Thermoplastic Elastomer Vulcanizates
of NBR with Different Plastics (60:40 Blend Ratio)

Plastics	Tensile strength (MPa)	Elongation at break (%)	Tension set (%)
PP	17.0	201	31
PE	17.6	190	—
PS	7.7	20	—
ABS	13.6	164	—
SAN	25.8	196	55
PMMA	10.8	56	—
PTMT	19.3	350	25
PA	21.5	320	44
PC	18.2	130	—

From Carl Hanser Verlag, Munich, Germany. With permission.

12. NBR-Thermoplastic Vulcanizates

Like NBR-nylon blends, a large number of NBR-plastic compositions have been developed.[73,74] They are prepared by dynamic vulcanization during melt mixing of the polymer. Curative systems used for vulcanization are dimethylolphenylic, bismaleimide, bismaleimide-peroxide-MBTS, bismaleimide-peroxide, organic peroxide, organic peroxide-coagent, accelerated sulfur, or sulfur donor. The properties of a few such compositions are given in Table 6. In the case of NBR-polyolefin blends, where the plastic phase is not compatible with the elastomeric phase, technological compatibilization is done by grafting. To prepare a compatible nitrile rubber-polypropylene (NBR-PP) blend, polypropylene is modified by amine-terminated liquid NBR to obtain a compatible NBR-PP composition. When dimethylol phenolic compounds are used as curatives, they act as compatibilizers in NBR-PP mixture. NBR-PP blends are gaining importance because of their excellent hot oil resistance. They exhibit slightly poor low temperature properties, which however, can be improved with a minimal sacrifice of hot oil resistance by further blending with EPDM-PP thermoplastic elastomer. The hardness of NBR-PP ranges from shore A of 80 to shore D of 50. Unlike covalently crosslinked NBR, its thermoplastic vulcanizate offers good swelling resistance to automotive oil and fuel as well as excellent ozone resistance. The resistance to compression set is better even at 100°C. Therefore, to save cost and ease the processing, conventionally crosslinked NBR may be replaced by NBR-PP thermoplastic vulcanizates in many applications. Blends of XNBR with various polyolefins possesses high modulus and tensile strength.[54] Recently, thermoplastic elastomeric polyolefin-NBR-HNBR blends with good mechanical properties have been prepared.[80] HNBR-Cross-linkable polyethylene (XLPE) blends serve as peelable outer semiconductive layers on XLPE insulation layers in electric cables.[81]

V. DEGRADATION

Degradation involves the chemical modification of the polymer by its environment, which is detrimental to the performance of the material. The polymer loses its desired properties by mechanical action, heat, radiation, or the chemical action of oxygen, ozone and atmospheric pollutants, water, acids, or bases. Polymer degradation is associated with chain cleavage, reduction in molecular weight, or crosslinking. A few degradative processes have now been established for the degradation of NBR.[82,83] It is now possible to anticipate the reaction steps which are important for the deterioration of the copolymer. The types of degradation described in the following sections are thermal, photooxidative, ozone, radiation, and aggressive media.

A. THERMAL DEGRADATION OF RAW NITRILE AND HYDROGENATED NITRILE RUBBERS

Nitrile rubbers are susceptible to thermo-oxidative degradation due to the presence of carbon-carbon double bonds in the diene unit. In the last few years, as the environmental conditions in many industrial applications are becoming more and more severe, NBRs have reached their performance limits. To improve their thermal stability, HNBRs have been developed. In this section, the degradation behavior of raw NBR, in comparison with HNBR, is discussed. The advantages offered by HNBR toward thermo-oxidative degradation is highlighted.

1. Low Temperature Degradation

During low temperature aging (75 to 150°C) of NBR[84] in the presence of air, the molecular weight of the polymer decreases with aging time. On the other hand, for HNBR, molecular weight initially increases with aging time and on prolonged aging, the same decreases. When the NBR and HNBR samples are aged at elevated temperatures, there is a change on the polymer surface toward greater wettability. This can be observed by measuring the contact angle of the aged samples. The contact angle decreases with aging time and temperature. The enhanced wettability of the polymer surface is attributed to the oxidation of surface and incorporation of carboxyl, peroxide, and other oxygen-containing groups. However, HNBR samples show more resistance to surface changes with aging time and temperature than NBR.

Nitrile rubber forms gel on aging, but there is no gel formation in HNBR at high temperature even after prolonged aging. The reason for gelation in NBR is due to the crosslinking of butadiene units through the olefinic bonds. Since HNBR has an almost saturated backbone, gel formation is not favored.

On degradation, there is an increase in the carbonyl ($>$C=O) functionality in both NBR and HNBR which is marked by the increase in the absorbance at around 1735 cm^{-1} in infrared spectra. The concentration of the carboxyl group in the aged samples can be quantified by infrared spectroscopic analysis. The increase in carbonyl functionality can be measured by taking the ratio of absorbances at 1735 cm^{-1} (due to C=O) to 1446 cm^{-1} (due to CH$_2$ deformation). The rate of formation of the carbonyl group increases initially and then remains almost constant. The higher the temperature of aging, the higher the amount of carbonyl group formed. The rate of increase in carbonyl group is higher in NBR than HNBR (Figure 10). The concentration of nitrile group reduces rapidly at elevated temperatures in HNBR when compared to NBR. However, infrared spectroscopic studies cannot give detailed information regarding the degradation, as it determines the bulk properties while oxidation during aging mostly takes place at the surface.

Electron spectroscopy for chemical analysis is employed to study the mechanism of degradation, particularly at the polymer surface.[85] The binding energy for N1s electron of –C≡N in HNBR (400.8 eV) shifts to higher order (406.0 eV) on oxidation (Figure 11). This indicates that, during oxidation the –C≡N group is affected and may be converted to –C–NH by the following mechanism.

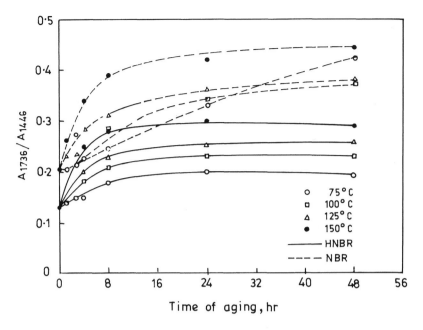

FIGURE 10. Change of absorbance ratio (A_{1736}/A_{1446}) with aging time of NBR and HNBR. (From *Polym. Degrad. Stabil.*, 31, 71, (1991). With permission.)

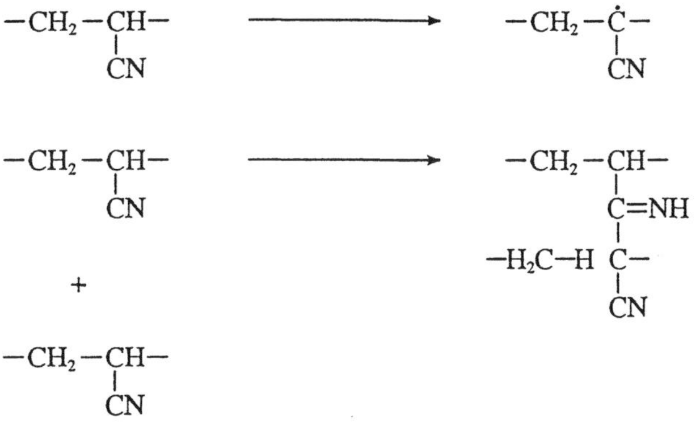

Scheme 1

The molecular weight increases on aging if the degradation proceeds through the above mechanism. The degradation of NBR occurs, however, through the attack of carbon-carbon double bonds.

2. High Temperature Degradation

Hydrogenated nitrile rubber is much more stable than NBR at elevated temperatures. For example, the degradation (in nitrogen atmosphere) starts at 357°C for NBR containing 40% ACN, whereas for a completely HNBR with similar ACN content, the onset of degradation is at 419°C.[84] The maximum degradation in HNBR takes place at 495°C. On the other hand, the corresponding NBR shows two maximum at 433 and 476°C (Figure 12).

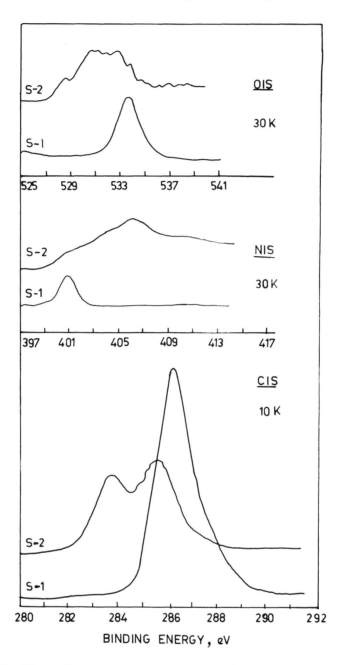

FIGURE 11. C1s, N1s, and O1s spectra of unaged and aged HNBR. (From *Polym. Degrad. Stabil.*, 31, 71, (1991). With permission.)

The intermediate products formed during oxidation and their stability are different in NBR and HNBR. The order of degradation reaction is unity in HNBR, whereas the order is 2 in NBR. The activation energies of degradation are 350 and 175 kJ/mol for HNBR and NBR, respectively. This indicates the better thermal stability of HNBR over NBR.

In NBR, there is electron withdrawal from α-methylene carbon by nitrile groups as well as by carbon-carbon double bonds during oxidation. When these double bonds are reduced, the attack on α-methylene groups by oxygen is also reduced. Hence, the thermo-oxidative stability is significantly improved by hydrogenation. The mechanism of degradation of HNBR

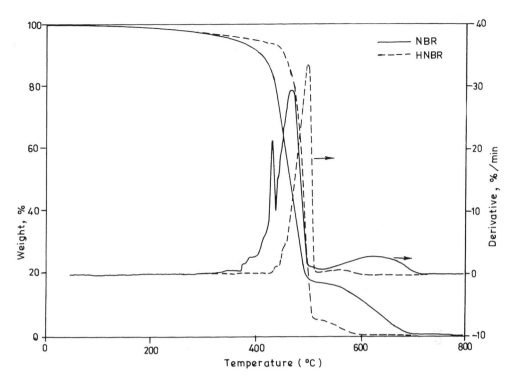

FIGURE 12. TG and DTG curves of NBR and HNBR (in nitrogen atmosphere). (From *Polym. Degrad. Stabil.*, 31, 71, (1991). With permission.)

is shown in Scheme 1 (under section V.A.1) However, in the case of NBR, the prominent attack of oxygen is on the double bonds with the generation of a hydroperoxide radical.[86,87]

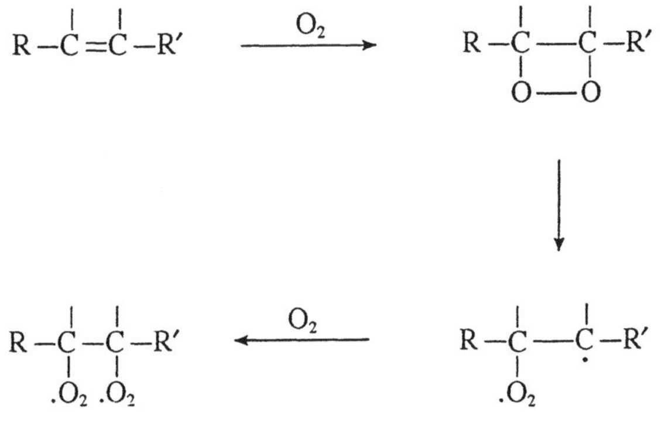

$$R-\overset{|}{C}=\overset{|}{C}-R' \quad \xrightarrow{\quad O_2 \quad} \quad R-\overset{|}{\underset{O-O}{C}}-\overset{|}{C}-R'$$

Scheme 2

$$R-\overset{|}{\underset{.O_2 \, .O_2}{C}}-\overset{|}{C}-R' \quad \xleftarrow{\quad O_2 \quad} \quad R-\overset{|}{C}-\overset{|}{\underset{.O_2}{\overset{\cdot}{C}}}-R'$$

The presence of residual double bond in HNBR may also initiate the free radical reaction.

3. Thermal Degradation of Liquid Carboxyl-Terminated Nitrile Rubbers

Carboxyl-terminated liquid NBRs exhibit excellent stability up to a temperature of 50°C. However, prolonged exposure of these reactive liquid polymers to a high temperature environment during storage sometimes causes changes in their physical properties, disappearance of terminal carboxyl functional groups, and an increase of polymer viscosity.

A detailed study on thermal aging of these liquid polymers[88] shows that at temperature beyond 50°C, the viscosity of the polymers increases and the terminal carboxylic groups start disappearing. The rates of viscosity increase and the disappearance of carboxyl groups are faster at higher temperatures. Polymers containing higher amounts of acrylonitrile undergo the above deterioration at a faster rate. The major reason for rapid viscosity increase and disappearance of carboxy groups is the formation of crosslinks between terminal carboxylic acid groups and nitrile groups through imide linkage.

$$\sim\sim COOH \quad + \quad \begin{array}{c} \sim\sim\sim \\ | \\ CN \end{array} \quad \longrightarrow \quad \begin{array}{c} \sim\sim C=O \\ | \\ NH \\ | \\ \sim\sim C=O \\ | \\ \sim\sim\sim \end{array} \qquad \text{Scheme 3}$$

Crosslinking among the unsaturation in butadiene segments also contributes to the slow and steady viscosity increase. However, at elevated temperatures, acid anhydride formation or decarboxylation may also take place in the polymer. Okamoto[88] has reported that the above two processes are not observed during thermal aging of liquid polymers at higher temperatures. Thermal stability of liquid rubbers can be improved to a considerable extent by hydrogenation of carbon-carbon double bonds.[63]

4. Thermal Degradation of Nitrile Rubber Vulcanizates

Thermal degradation of NBR vulcanizates presents some complication due to the nitrile functionality in the polymer. The degradation of the vulcanizates can be thoroughly investigated by total thermal analysis involving thermogravimetry (TG), derivative thermogravimetry (DTG), differential scanning calorimetry (DSC), and glass transition temperature (Tg).[89]

When NBR vulcanizates is thermally degraded in a nitrogen atmosphere, an exothermic reaction takes place due to cyclization of butadiene units present in the elastomer. In addition to this, there is cyclization of the nitrile component of the rubber. During the degradation process, various components of the vulcanizates separate out in a series. Oil, plasticizer, sulfur, accelerator, antioxidants, stearic acid, etc. are the first to volatilize followed by polymer and carbon black at a higher temperature. The residue consists of mostly mineral filler and zinc oxide. When NBR vulcanizate is degraded at high temperature, there is formation of additional carbon residue due to the acrylonitrile component. The higher the nitrile content, the higher the amount of carbon formed as the residue (Figure 13). Moreover, carbon skeleton in NBR decomposes faster in air than the added carbon black. Hence, the weight loss step due to nitrile carbon can be isolated from the DTG curves.

Siracar and Lamond[89] have introduced a correction factor to determine the composition of NBR vulcanizates of different nitrile contents during high temperature degradation. Table 7 demonstrates how the composition of NBR vulcanizates could be derived from the thermal analysis.

B. PHOTODEGRADATION

Nitrile rubber is very susceptible to photo-oxidation. The butadiene unit of the copolymer is photo-oxidized by molecular oxygen or single oxygen when subjected to UV irradiation. The mechanism of oxidation by molecular oxygen follows a free radical pathway and that by singlet oxygen follows the ene-type oxidation mechanism.[82,83] The acrylonitrile unit is much more stable than the butadiene unit toward photo-oxidation at room temperature. However, at elevated temperatures the rate of oxidation increases.[90]

During photoinduced initiation, the polymer alkyl radicals (P) are formed which further participate in a chain auto-oxidative reaction in a photochemical process. The formation of free radicals can be explained by electronic-vibrational coupling which causes the excitation of specific molecular vibration and leads to bond breakage. In butadiene and acrylonitrile

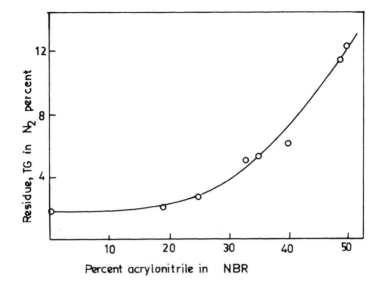

FIGURE 13. Variation of residual weight in nitrogen with ACN of NBR. (From *Rubber Chem. Technol.*, 51, 647, (1978). With permission.)

units, the hydrogen in methylene groups (CH_2) are involved in high frequency molecular vibrations. The sigma-electrons of carbon-hydrogen bonds in the methylene groups (CH_2) cannot be excited at wavelengths exceeding 200 nm. The photochemical hydrogen abstraction takes place by preferential electronic energy transfer to stretching vibrations involving hydrogen.

In the propagation step, the polymer alkyl radicals react with molecular oxygen to form the hydroperoxide groups. The overall reaction can be represented by the following steps:

$$PH + h\nu \longrightarrow P\cdot + H\cdot$$

$$P\cdot + O_2 \longrightarrow POO\cdot$$

$$POO\cdot + PH \longrightarrow POOH$$

The formation of hydroperoxide group scan be measured by infrared spectra by calculating the ratio of absorbance at 3450 cm^{-1} due to hydroperoxy or hydroxyl groups to that at 2840 cm^{-1} due to the CH of CH_2 groups. The rate of formation of hydroperoxide groups decreases when the copolymer contains higher amounts of acrylonitrile.

The hydroperoxide groups further decompose by the unimolecular or bimolecular reaction process

$$POOH \longrightarrow PO\cdot + \cdot OH$$

$$2POOH \longrightarrow PO\cdot + POO\cdot + H_2$$

Carbonyl groups are formed by the decomposition of hydroperoxide groups. Associated hydroperoxides are primarily formed in the allylic position of the butadiene unit. Unsaturated hydroperoxides then decompose into secondary allylic alcohols and α-β-unsaturated ketones.[91,92] The presence of acrylonitrile units favor the formation of hydrogen bonding between hydroperoxides and alcohols.

TABLE 7
Composition of Vulcanizates from TG and DTG Curves

Polymer	% acrylonitrile (Kjeldahl)	Recipe weight (phr)	Differential	Oil and other volatiles (phr)	Differential	Carbon black (phr)	Differential	ZnO ash, (phr)	Differential
True value		189.5		24.5		60		5–6	
Hycar 1024	18.5	197	+7.5	25.8	+1.3	64.2	+4.2	6.9	+0.9
Nysyn 25-8	25.0	192.2	+2.7	22.1	−2.4	64.0	+4.0	6.0	0.0
		191.0	+1.5	21.5	−3.0	64.7	+4.7	4.9	−1.1
Nysyn 30-5	28.8	194.0	+5.4	23.5	−1.0	65.2	+5.2	6.2	+0.2
Hycar 1042	32.2	189.3	−0.2	21.6	−2.9	60.6	+0.6	6.8	+0.8
		190.1	−0.6	21.1	−3.4	62.1	+2.1	6.7	+0.7
Nysyn 35-3	34.1	190.0	+0.5	21.9	−2.6	61.0	+1.0	6.9	+0.9
		187.4	−2.1	21.0	−3.5	62.1	+2.1	6.1	+0.1
Nysyn 40-5	38.5	188.4	−1.1	20.0	−4.5	63.8	+3.8	5.4	−0.6
Hycar 1000X	47.0	192.7	+3.2	20.1	−3.4	65.6	+5.6	8.1	+2.1
Nysyn 50-8	47.5	190.1	+0.6	20.5	−4.0	63.4	+3.4	5.8	−0.2
Mean		191.2		21.7		63.3		6.3	
Standard deviation		2.8		1.6		1.7		0.9	
Coefficient of variation		1%		7%		3%		15%	

From *Rubber Chemistry and Technology*, 51, 647 (1978). With permission.

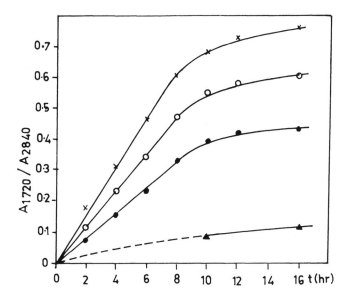

FIGURE 14. Kinetics of formation of carboxyl groups in ▲, pure polyacrylonitrile; ○, NBR (21.7% ACN); ●, NBR (41.6% ACN); and X, polybutadiene during UV irradiation in air. (From *Polym. Degrad. Stabil.*, 5, 173, (1983). With permission.)

$$-CH_2-OOH \longrightarrow -CH_2-O \cdot + \cdot OH \longrightarrow \overset{\overset{\displaystyle O}{\displaystyle \|}}{\underset{\displaystyle H}{-C}} + H_2O$$

Carbonyl groups are also formed by the β-scission process of polymer alkoxy radicals (PO·)

$$\overset{\overset{\displaystyle O\cdot}{\displaystyle |}}{-CH_2-CH-CH_2-} \longrightarrow \overset{\overset{\displaystyle O}{\displaystyle \|}}{\underset{\displaystyle H}{-CH_2-C}} + \cdot CH_2-$$

This can be detected by the infrared spectroscopy from the $>$C=O absorption band at 1720 cm^{-1}. The formation of carbonyl functionality can be quantified by taking the ratio of absorption at 1720 cm^{-1} (due to $>$C=O) to that at 2840 cm^{-1} (due to C–H of CH$_2$). The rate of formation of carbonyl group reduces when the acrylonitrile structures are in the copolymer (Figure 14). The nitrile group of the polymer remains unaffected during photo-oxidation reaction. Formation of carbonyl groups by decomposition of hydroperoxide is favored at higher temperatures. Carbonyl groups can also be formed by metathesis reactions of PO or polymer alkylperoxy (POO·) radicals.

The formation of hydroperoxide or hydroxyl and carboxyl groups can be understood by UV/visible spectroscopy or electron spectroscopy for chemical analysis. ESCA experiments of photo-oxidized NBR samples show small changes in the overall core-binding energies for C1s (285 eV), N1s (399.3 eV) and O1s (534.5 eV) (Figure 15). Shifts in the binding energy of C1s indicate the formation of $-\overset{|}{\underset{|}{C}}$–O and $>$C=O groups.

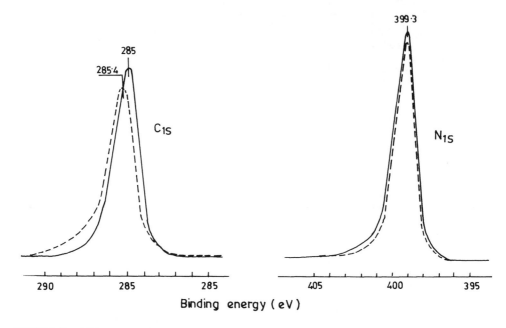

FIGURE 15. ESCA, C1s, and N1s level spectra of NBR: undegraded sample and ——, sample after 16 h UV irradiation in air. (From *Polym Degrad. Stabil.*, 5, 173, (1983). With permission.)

The general feature of the photo-oxidation of NBR shows that the degradation occurs mainly in butadiene units rather than in acrylonitrile units. The level of photo-oxidation is lowered by increasing amounts of acrylonitrile in NBR samples. The nature of degradation resembles that of polybutadiene and polyacrylonitrile. The rate of photo-oxidation of polyacrylonitrile is negligible, when compared to that of polybutadiene, under similar experimental condition. The main degradation process during photo-oxidation of NBR is associated with the susceptibility of olefinic unsaturation in the diene units of the copolymer. The major degradation products are the formation of carboxyl and hydroperoxy functionalities.

C. OZONE DEGRADATION

It has been mentioned earlier that NBR has very low stability toward atmospheric and ozone aging. In fact, NBR has a lower stability toward ozone than styrene-butadiene rubbers and particularly in *cis*-1,4-polybutadiene. The mechanism of zone degradation of NBRs in solution[93] is discussed in the following sections.

Ozonolysis of NBR decreases the molecular mass of the polymer which can be detected viscometrically. When the NBR is subjected to ozone atmosphere, significant amounts of low molecular mass fragments are formed. The number of chain scissions per molecule of reacted ozone can also be calculated from the relative viscosity (η_{rel}).

$$\eta_{rel}^{t} = f(G)$$

G = amount of reacted ozone and η_{rel}^{t} = relative viscosity of the polymer after ozonization for time t.

The dimension of the macromolecules plays a very important role in the degradation process, when the reaction is performed in solution. The larger the dimension of macromolecules, the greater the proportion of carbon-carbon double bond exposed to ozone. The structure of the rubber also has a pronounced influence upon the intensity of the degradation.

Interaction of ozone with 1,2-double bonds does not affect the main chain of the macro-molecules and, thus, cannot lead to a decrease in the molecular mass. In the case of NBRs containing higher amounts of acrylonitrile, the probability of formation of low molecular mass fragments is less. Hence, the extent of chain scission during ozone degradation is low when the 1,2- or ACN is more in the copolymer.

The rate of addition of ozone to carbon-carbon double bonds is four to eight times higher than that with $-$CH$-$CH$-$ or CH$_2$$-CH_2$$-$ bonds. The reaction of ozone with $-$C$\equiv$N group is very slow and hence, it is almost impossible to obtain products of ozone conversion of nitrile groups. During the ozonolysis of NBR, it is only the carbon-carbon double bond which takes part in the reaction. The possible interaction of ozone with $-$C$=$C$-$ are summarized in Scheme 1. The main route of the reaction is the formation of ozonide (Reactions 1, 2, 3, and 4). Reactions 5 and 10 even lead to an increase in molecular mass and their occurrence is accompanied by gel formation. The biomolecular reactions, in which the macromolecular fragments of 1,2- double bonds take part, result in an increase in the molecular mass. Cross-linking reactions are also favorable during ozonization.

The major products of ozonization of NBR are the formation of aldehyde and ozonide groups. These functional, groups can be detected by proton nuclear magnetic resonance (NMR) and infrared spectroscopies. ^{1}H-NMR spectrum of NBR reacted with O$_3$ shows signals at 9.75 ppm due to aldehyde protons and at 4.97 to 5.20 ppm due to methine protons of ozonide groups. The infrared spectrum of the same NBR sample shows two strong bands at 1731 and 1110 cm^{-1}, which are assigned to the vibration of aldehyde C$=$O bond and ozonide C$-$O bonds, respectively. There is no interaction of the zwitterion A (Scheme 4) with the C$\equiv$N bond during ozonization of NBR. The principal mechanistic step for the degradation of NBR is the deactivation of zwitterion A.

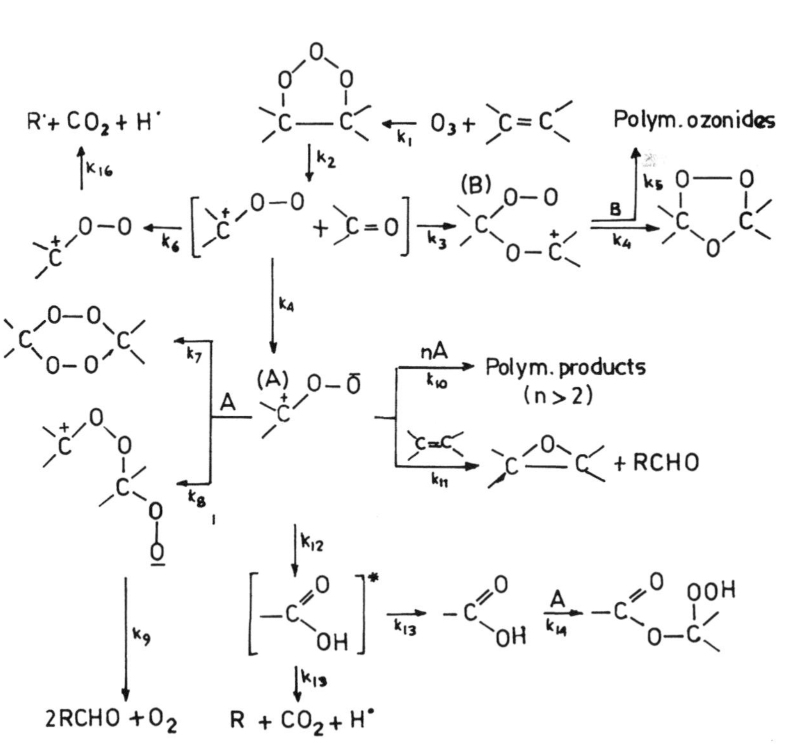

Scheme 4. From *Polym. Degrad. Stabil.,* 19, 293, (1987). With permission.

When NBR and *cis*-1,4-polybutadiene are reacted with ozone under similar experimental conditions, the amount of ozonide and aldehyde groups formed is higher in the former than the latter. The reason for the above peculiarity is due to the structural difference between both polymers. The nitrile group does not have any influence on the reaction of O_3 with $-C=C-$ bond. The only difference between the structures of *cis*-1,4-polybutadiene and NBRs is the predominance of *trans*-1,4- configuration of the C=C bond in the latter. Ozone shows a greater reactivity toward *trans*-olefins than *cis*-olefins. This explains the poor stability of NBR toward ozone in comparison with *cis*-1,4-polybutadiene.

However, with the recent development of NBR technology, ozone resistance of NBRs has been improved (Sections IV.A and IV.L.7). The major breakthrough in this area is the introduction of HNBR. Since, it contains almost no saturated carbon skeleton, no favorable reactive site is available for ozone degradation (Figure 3).

D. RADIATION-INDUCED DEGRADATION

The effect of ionizing radiations on polymers is apparent from the deterioration of nuclear power station components, sterilization of medical supplies, controlled crosslinking, and radiation-initiated graft polymerization. Also, NBRs are directly being subjected to γ-rays, X-rays, and electron beams in many industrial applications. The interaction of high energy radiation with the rubber leads to physicochemical processes which generate free radicals to initiate bond scission and crosslinking.

When sulfur-vulcanized NBR is subjected to γ-irradiation in the presence of air at room temperature, tensile properties, hardness, and swelling behavior are deteriorated.[94,95] The modulus and hardness increases and the elongation at break decreases. The degradation mainly takes place at the *cis*-1,4- double bonds of the polymer. The degradation of the polymer is heterogeneous through the sample thickness. The radiation-induced degradation in the polymer results from two different processes: (1) the formation of free radicals which react with oxygen and leads to heterogeneous oxidation on the polymer surface due to diffusion effects and (2) ozone generation by the action of ionizing radiation on the air (atmosphere) surrounding the sample. Ozone attack on the polymer surface has a major role in radiation-induced oxidative degradation because the simultaneous action of ozone and ionizing radiations are strongly synergistic in their effects on the polymeric material.

When HNBR is subjected to irradiation, oxidative degradation takes place.[96] The degradation products indicate the formation of carboxyl, hydroperoxide, and hydroxyl groups. Hydroperoxides are formed from both tertiary H–C–CN and secondary H–C–H groups of the polymer.

E. DEGRADATION IN AGGRESSIVE MEDIA

Nitrile rubber finds application where oil and fuel resistance is required over a broad range of temperatures. The environments in which NBR components are being used nowadays, have become more aggressive. However, this challenge has been met by compound development and the appropriate choice of NBR grades.

Nitrile rubber vulcanizates on aging in fuel lose elasticity and strength.[97] The aging proceeds with degradation and crosslinking. Their degree depends on aging time, fuel, and type of vulcanizing systems. The degradation in fuel is attributed to leaching of antioxidants from the surface layer of the polymer. However, the rate of degradation can be decreased somewhat by the addition of chemically bonded antioxidants.[98-100]

The presence of hydrogen sulfide in the surrounding environments severely affects the properties of NBR vulcanizates. When NBR vulcanizates are exposed to H_2S for a longer period, crosslinking and hardening take place.[101] Hydrogen sulfide introduces crosslinking and increases the hardness of the NBR vulcanizates. Nitrile rubber with low ACN offers more resistance to H_2S. The increasing ACN results in a greater modulus increase and elongation loss on hydrogen sulfide exposure. Hydrogen sulfide reacts with both double bonds and cyano groups of the polymer. Reaction of hydrogen sulfide with nitrile groups gives rise to thioamides, which accelerate the crosslinking reaction. Carboxylated nitrile rubber vulcanizates with high carboxyl content offers better resistance to H_2S, particularly in retention of properties.[22] Hydrogenated nitrile rubber offers considerably better resistance than both NBR and XNBR (Figure 8).

Gasoline is peroxidized due to the combined action of air moisture, heat, and copper ions in certain fuel injection systems. Peroxidized gasoline attacks fuel hoses, which result in hardening of NBR lines.[102,103] Resistance can be improved by increasing ACN of the polymer or by using HNBR. Increasing aromatic content of the fuel does not markedly influence the physical properties of the NBR products, but it enhances the swelling and permeability of the material.[22] However, this can be avoided by using HNBR in place of NBR. Similarly, increasing alcohol content of the gasoline affects the swelling and permeability of the material. This can be overcome, to some extent, by the use of XNBR, NBR-PVC, XNBR-PVC, and significantly, by the use of HNBR.

GLOSSARY

ABS	Acrylonitrile-butadiene-styrene resin
ACM	Acrylic rubber
BR	Polybutadiene rubber
CIIR	Polychloroprene rubber
CSM	Chlorosulfonated polyethylene
EAM	Ethylene-ethylacetate copolymer (Vamac)
ECO	Epichlohydrine rubber
EVA	Ethylene vinyl acetate
EPDM	Ethylene-propylene-diene termonomer
EPM	Ethylene-propylene monomer
FMQ	Methyl silicone rubber with fluoro groups
FKM, FPM	Fluoro, fluoroalkyl, or fluoroalkoxy group containing rubber
HNBR	Hydrogenated nitrile rubber
HFNBR	Hydroformylated nitrile rubber
IR	Polyisoprene rubber
NBR	Nitrile rubber
NIR	Acrylonitrile-isoprene rubber
NR	Natural rubber
PE	Polyethylene
PP	Polypropylene
PS	Polystyrene
PMMA	Polymethyl methacrylate
PTMT	Polytetramethylene terephthalate
PA	Polycarbonete
PVC	Polyvinyl chloride
Q	Silicone rubber
SAN	Styrene-acrylonitrile resin
SBR	Styrene-butadiene-rubber
TFE	Tetrafluoroethylene
XLPE	Cross-linkable polyethylene

REFERENCES

1. **Hofmann, W.,** *Rubber Chem. Technol.,* 36(5), 1, 1963; *Krynac Formulation Guide,* Polysar Technical Centre, Antwerp, Belgium, 1989.
3. **Dunn, J. R., Coulthard, D. C., and Pfisterer, H. A.,** *Rubber Chem. Technol.,* 51, 389, 1978.
4. **Schulz, D. N.,** in *Encyclopedia of Polymer Science and Engineering, Vol.,* 7 Mark, H. F., Bikales, N., Overberger, C. C., and Menges, G., Eds., John Wiley and Sons, New York, 1986, 807.
5. **Kubo, Y.,** *Sekiyu Gakkaishi (Jpn),* 33, 189, 1990.
6. **Kubo, Y.,** *Shokubai (Jpn),* 32, 218, 1990.
7. **Ishihara, N. and Kuramoto, M.,** (Idemitsu Kosan Co Ltd.) European Patent Appl. EP 322, 731, 1989.
8. **Fukawa, I., Morita, H., and Oshima, S.,** (Asahi Chemical Industry Co. Ltd., Japan) French Demande 2, 468, 618, 1981.
9. **Burfield, D. R.,** Int. Conf. Rubber Rubber-Like Materials, Jamshedpur, India, November 6–8, 1986.
10. **Mohammadi, N. A. and Rempel, G. L.,** *Macromolecules,* 20, 2362, 1987.
11. **Bhattacharjee, S., Bhowmick, A. K., and Avasthi, B. N.,** *Ind. Eng. Chem. Res.,* 30, 1086, 1991.
12. **Bhattacharjee, S., Bhowmick, A. K., and Avasthi, B. N.,** *J. Polym. Sci. Part A, Polym. Chem.,* 30, 1992.
13. **Bhattacharjee, S., Bhowmick, A. K., and Avasthi, B. N.,** *J. Appl. Polym. Sci.,* 41, 1357, 1990.
14. **Buding, H., Fiedler, P., Koenigshofen, H., and Thormer, J.,** (Bayer A-G, Germany) German Offen. DE 3, 433, 392, 1986.
15. **Molova, J., Svoboda, P., Hetflejs, J., and Sufeak, M.,** Czech, CS 227, 105, 1985.
16. **Kishimoto, Y. and Masubuchi, T.,** (Asahi Chemical Industry Co. Ltd., Japan) Jpn Kokai Tokyo Koho. JP 61, 28, 507, 1986.
17. **Hofmann, W.,** *Rubber Technology Handbook,* Hanser Publishers, New York, 1989, 157.
18. **Morrell, J. P.,** in *Vanderbilt Handbook,* Babbit, R. O., Ed., Vanderbilt, 1978, 169.
19. **Seil, D. A. and Wolf, F. R.** in *Rubber Technology, 3rd Ed.,* Morton, M., Ed., Von Nostrand Reinhold, New York, 1987, 322.
20. **Bertram, H. H.,** in *Developments in Rubber Technology,* 2nd ed., Whelan, A. and Lee, K. S., Eds., Elsevier Applied Science Publishers, Barking, 1981, 51.
21. **Dunn, J. R. and Vara, R. G.,** Presented at the 128th Rubber Division Technical Meeting, Cleveland, OH, October 1–4, 1985.
22. **Dunn, J. R.,** Polysar Technical Literature, Polynote No. N.2, Polysar Technical Centre, Antwerp, Belgium 1985.
23. **Pfisterer, H. A. and Dunn, J. R.,** *Rubber Chem. Technol.,* 53, 357, 1980.
24. **Killgoar, P. C., Jr. and Lemieux, M. A.,** *Rubber Chem. Technol.,* 56, 853, 1983.
25. **Hofmann, W.,** *Kautsch Gummi Kunstst.,* 34, 1017, 1981.
26. **Thompson, G. and Steve, S.,** *Rubber World,* 200, 28, 1989.
27. **Dunn, J. R.,** *Plast. Rubber Process. Appl.,* 2, 161, 1982.
28. **Dunn, J. R.,** in *Developments in Polymer Stabilization-4,* Scott, G., Ed., Applied Science Publishers, London, 1981, 223.
29. **Paulin, D. A.,** *Rubber Age,* 101, 69, 1969.
30. **Hofmann, W.,** *Plast. Rubber Process. Appl.,* 5, 209, 1985.
31. **Dunn, J. R.,** *Rubber World,* 190, 16, 1984.
32. **Walter, G.,** *Rubber Chem. Technol.,* 49, 775, 1976.
33. **Dunn, J. R.,** *Rubber Chem. Technol.,* 47, 960, 1974.
34. **Horvath, J. W., Grimm, D. C., and Stevick, J. A.,** *J. Elastomers Plast.,* 7, 337, 1975.
35. **Dean, P. R., II and Kuczkowski, J. A.,** *Rubber Chem. Technol.,* 59, 842, 1986.
36. **Scott, G.,** *Macromol. Chem.,* 8, 319, 1973.
37. **Kline, R. H. and Miller, J. P.,** *Rubber Chem. Technol.,* 46, 96, 1973.
38. **Weinstein, A. H.,** *Rubber Chem. Technol.,* 50, 641, 1977.
39. **Weinstein, A. H.,** *Rubber Chem. Technol.,* 50, 650, 1977.
40. **Dunn, J. R.,** Polysar Technical Literature, Polynote No. N.3, Polysar Technical Centre, Antwerp, Belgium, 1983.
41. **Dunn, J. R.,** Polysar Technical Litratrue, Polynote No. N.6, Polysar Technical Centre, Antwerp, Belgium, 1987.
42. **Shmekov, A. G., Kazymova, V. K., and Ivanov, A. N.,** *Kauch Rezina (Russia),* 12, 10, 1989.
43. **Hashimoto, K. and Todani, Y.,** in *Handbook of Elastomers — New Developments and Technology,* Bhowmick, A. K. and Stephens, H. L., Eds., Marcel Dekker, New York, 1988, 741.
44. **Thormer, J., Mirza, and Buding, H.,** presented at the PRI Rubber Conference 1984, Birmingham, Michigan, March 12–15, 1984.
45. **Rawlinson, A. and Djuricis, N.,** Polysar Technical Literature (Polysar Technical Centre/Jugoauto), Sarajevo November 8–10, 1988.

46. Technical Literature on Zetpol, hydrogenated nitrile rubber, Nippon Zeon Company Limited, Tokyo, Japan, 1989.
47. Technical Literature on Therban, hydrogenated nitrile rubber, Bayer A-G, Germany, 1990.
48. Technical Literature on Tornac, hydrogenated nitrile rubber, Polysar Technical Centre, Antwerp, Belgium, 1991.
49. **Thavamani, P. and Bhowmick, K.,** *Rubber Chem. Technol.,* 65, 1992.
50. **Thavamani, P. and Bhowmick, K.,** *J. Mater. Sci.,* 26, 1992.
51. **Wiseman, W. A. and Ridland, J. J.,** *Kunstst, Rubber,* 43, 12, 1990.
52. **Hashimoto, K., Watanabe, N., and Yoshioka, A.,** *Rubber World,* 190, 32, 1984.
53. **Thormer, J., Marwede, G., and Budding, H.,** *Kautsch. Gummi Kunstst.,* 36, 269, 1983.
54. **Dunn, J. R.,** in *Handbook of Elastomers — New Developments and Technology,* Bhowmick, A. K. and Stephens, H. L., Eds., Marcel Dekker, New York 1988, 503.
55. **Hallenbeck, V. L.,** *Rubber Chem. Technol.,* 46, 78, 1973.
56. **Bryant, C. L.,** *J. Inst. Rubber Ind.,* 4, 202, 1970.
57. **Woods, M. E., Morsek, R. J., and Whittington, W. H.,** *Rubber World,* 167, 42, 1973.
58. **DeMarco, R. D., Woods, M. E., and Arnold, L. F.,** *Rubber Chem. Technol.,* 45, 1111, 1972.
59. **Darke, R. S. and McCarthy, W. J.,** *Rubber World,* 159, 51, 1968.
60. **Kalfogolou, N. K. and Williams, H. L.,** *J. Appl. Polym. Sci.,* 17, 1377, 1973.
61. **Siebert, A. R.,** *J. Elastomers Plastics,* 8, 177, 1976.
62. **Hashimoto, K., Oyama, M., Nakagawa, T., Murakata, K., Saya, T., and Aimura, Y.,** Paper presented at 136th Meeting of the Rubber Division, A.C.S., Detroit, MI, October 17–20, 1989.
63. **Bhattacharjee, S., Bhowmick, A. K., and Avasthi, B. N.,** *J. Polym. Sci., Part A, Polym. Chem.,* 30, 1992.
64. **Furukawa, J. and Iseda, Y.,** *J. Polym. Sci., Polym. Lett.,* 7, 47, 1969.
65. **Furukawa, J. and Iseda, Y.,** *J. Polym. Sci., Polym. Lett.,* 7, 561, 1969.
66. **Furukawa, J., Nishioka, A., and Kotani, T.,** *J. Polym. Sci., Polym. Phy.,* 8, 24, 1970.
67. **Gatti, G. and Carbonaro, A.,** *Makromol. Chem.,* 175, 1627, 1974.
68. **Wolff, S.,** *Rubber Chem. Technol.,* 50, 447, 1977.
69. **Bhattacharjee, S., Bhowmick, A. K., and Avasthi, B. N.,** *Die Ang. Makromol. Chem.,* 1992.
70. **Stollfuss, B.,** Technical Information Bulletin, Perbunan N, No. 2.7.1, Bayer A-G, Germany, September 1977.
71. **Woods, M. E. and Davidson, J. A.,** *Rubber Chem. Technol.,* 49, 112, 1976.
72. **Saito, Y., Fujino, A., and Ikeda, A.,** (Nippon Zeon Co. Ltd.) German Offen. DE 3, 918, 929, 1989.
73. **Coran, A. Y.,** in *Thermoplastic Elastomers, A Comprehensive Review,* Legge, N. R., Holden, G., and Schroeder, H. E., Eds., Hanser Publishers, New York, 1987, 133.
74. **Coran, A. Y. and Patel, A.,** in *Handbook of Elastomers — New Developments and Technology,* Bhowmick, A. K., and Stephens, H. L., Eds., Marcel Dekker, New York, 1988, 249.
75. **Schwarz, H. F. and Edwards, W. S.,** *Appl. Polym. Symp.,* 25, 243, 1974.
76. **Watanabe, N.,** in *Thermoplastic Elastomers from Rubber — Plastic Blends* De, S. K. and Bhowmick, A. K., Eds., Ellis Horwood, New York, 1990, 198.
77. **Imaeda, T.,** (Tokai Rubber Industries Ltd.), Japan Kokai Tokyo Koho JP 02, 228, 341, 1990.
78. **Eichenauer, H., Budding, H., Marten, J., and Ott, K. H.,** (Bayer A-G), German Offen. DE 3, 827, 529, 1990.
79. **Plaumann, H. P. P.,** (Polysar Ltd.), European Patent Appl. EP 364, 859, 1990.
80. **Imai, T., Nagano, M., and Motai, M.,** (Japan Synthetic Rubber Co.), Japan Kokai Tokyo Koho JP 02, 228, 343, 1990.
81. **Ishikawa, T., Takahashi, S., Hatada, M., Nagai, K., and Igarashi, M.,** (Fujikura Ltd.), Japan Kokai Tokyo Koho JP 02, 160, 850, 1990.
82. **Dunn, J. R.,** *Rubber Chem. Technol.,* 51, 686, 1978.
83. **Carlsson, D. J. and Wiles, D. M.,** in *Encyclopedia of Polymer Science and Engineering, Vol. 4* Mark, H. F., Bikales, N. M., Overberger, C. G., and Menges, G., Eds., John Wiley and Sons, New York, 1986, 603.
84. **Bhattacharjee, S., Bhowmick, A. K., and Avasthi, B. N.,** *Polym. Degrad. Stab.,* 31, 71, 1991.
85. **Clark, D. T., Dilks, A., and Thomas, H. R.,** in *Developments in Polymer Degradation I,* Grassie, N., Ed., Applied Science Publishers, London, 1977, 87.
86. **Moiseev, D. I., Potseluev, V. M., Yu. K., Gusev, and Sozov, O. V.** *Kauch Rezina (Russia),* 6, 11, 1986.
87. **Moiseev, D. I., Yu. K., Gusev, Sigov, O. V., and Dorofeeva, T. P.,** *Prom-st. Sint. Kauch. (Russia),* 11, 9, 1983.
88. **Okamoto, Y.,** *Polym. Eng. Sci.,* 23, 222, 1983.
89. **Sircar, A. K. and Lamond, T. G.,** *Rubber Chem. Technol.,* 51, 647, 1978.
90. **Skowronski, T. A., Rabek, J. F., and Randy, B.,** *Polym. Degrad. Stab.,* 5, 173, 1983.
91. **Adams, C., Lacoste, J., and Lamaire, J.,** *Polym. Degrad. Stab.,* 27, 85, 1990.

92. **Lemaire, J., Adam, C., and Lacoste, J.,** *Polym. Sci., Symp. Proc. Polym. 1991, 1,* Sivaram, S., Ed., Tata McGraw-Hill, New Delhi, 1991, 383.
93. **Anachkov, M. P., Rakovsky, S. K., Stefanova, R. V., and Shopov, D. M.,** *Polym. Degrad. Stab.,* 19, 293, 1987.
94. **Clough, R. L. and Gillen, K. T.,** *J. Poly. Chem., Part A, Polym. Chem.,* 27, 2313, 1989.
95. **Kohjiya, S., Matsumura, Y., Yamashita, S., Matsuyama, T., and Yamaoka, H.,** *Nippon Gomu Kyokaishi (Japan),* 59, 472, 1986.
96. **Carlsson, D. J., Chemla, S., and Wiles, D. M.,** *Makromol. Chem., Makromol. Symp.* 27, 139, 1989.
97. **Seregunova, L. I., Andreeva, A. I., and Donstov, A. A.,** *Kauch Rezina (Russia),* 5, 18, 1988.
98. **Bayers, J. T., Hewitt, N. L., and Tultz, J. P.,** *Gummi Fasern. Kunstst.,* 42, 436, 1989.
99. **Seregunova, L. I., Andreeva, A. I., and Donstov, A. A.,** *Kauch Rezina (Russia),* 7, 16, 1988.
100. **Dunn, J. R.,** *Rubber India,* 35, 9, 1983.
101. **Pfisterer, H. A., Dunn, J. R., and Vukov, R.,** *Rubber Chem. Technol.,* 56, 418, 1983.
102. **Trexler, H. E.,** *Rubber Chem. Technol.,* 54, 155, 1981.
103. **Nakagama, T., Oyama, M., Yagishita, S., and Todani, Y.,** *Kautsch. Gummi Kunstst.,* 42, 395, 1989.

Chapter 14

THERMOPLASTIC ELASTOMERS: A RISING STAR

Marc T. Payne and Charles P. Rader

TABLE OF CONTENTS

I. INTRODUCTION

Thermoplastic elastomers (TPEs) continue to grow in commercial importance. These materials combine the functional properties of comparable thermoset elastomers with the fabrication advantages of thermoplastics. As a class, TPEs comprise several types of materials such as elastomeric alloys (EAs), styrenic block copolymers, copolyesters, and thermoplastic polyurethanes. This chapter will describe each type of commercially important TPE, the characteristics of these materials, typical properties, and examples of applications.

Thermoplastic elastomers are defined by ASTM D 1566 as "a family of rubber-like materials that, unlike conventional vulcanized rubber, can be processed and recycled like thermoplastic materials". A rubber is defined[1] as "a material that is capable of recovering from large deformations quickly and forcibly,..." and "retracts within 1 min to less than 1.5 times its original length after being stretched at room temperature (18 to 29°C) to twice its length and held for 1 min before release".

The term thermoplastic elastomer has developed into the accepted generic name for materials as defined in the previous paragraph. This term is generally used as a noun, to distinguish a TPE from a conventional thermoset rubber that has not been vulcanized. Archaic terms for TPEs include "elastoplastic",[2] thermoplastic rubber, thermoplastic vulcanizate,[3] and impact modified plastic. In addition, there has been a growth of subcategories of TPEs to distinguish between the different types of materials that generally meet the TPE definition. Several examples are "thermoplastic rubber blends", "elastomeric alloys", and "block copolymers".[4] These different terms are used to indicate different morphological structures and different elastomeric performance. These various materials will be described in detail in this chapter.

There have been numerous books,[5-7] book chapters,[8,9] technical papers, and review articles,[4,10-15] patents, trade literature, and symposia[16-24] on TPEs, most of which were published or held within the last 10 to 15 years. Several new organizations with TPEs as a focus have been formed, such as the Thermoplastic Elastomers Special Interest Group (SIG) of the Society of Plastics Engineers and a similar Topical Group within the Rubber Division of the American Chemical Society. As evidenced by this growth in information, the technology of TPEs has grown steadily and, sometimes, by quantum jumps. There is now a wide selection of different TPEs for the materials technologist to consider, depending on specific application needs.

II. RUBBER, PLASTICS, AND INNOVATION

An analogy showing TPEs as a man with one foot in the rubber industry and the other in the plastics industry has been used to describe the developmental history and current thinking of TPEs (Figure 1).[25] As described earlier, TPEs are either block copolymers or combinations of a rubber-dispersed phase and a plastic continuous matrix. The attribute contributed by the rubbery phase—such as butadiene or ethylenebutylene in an S-E-S or S-EB-S styrenic block copolymer, or the completely vulcanized EPDM rubber particles in a polypropylene (PP)/EPDM EA—is classical elastomeric performance.

The elastic properties of a rubber result from long, flexible molecules that are coiled in a random manner. When the molecules are stretched, they uncoil and have a more specific geometry than the coiled molecules. The uncoiled molecules have lower entropy because of the more restricted geometry and, since the natural tendency is an increase in entropy, the entropic driving force is for the molecules to retract, giving elasticity.[26] The "soft" butadiene or ethylenebutylene segments in the styrenic block copolymer are coiled segments held together by the polystyrene "hard blocks". When acted upon by an external stress, the initial deformation is the uncoiling of the soft rubbery segments, since the energy for

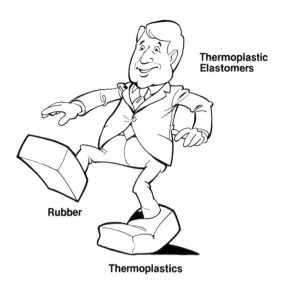

**Thermoplastic
Elastomers**

Rubber

Thermoplastics

FIGURE 1. Thermoplastic elastomers have one foot in the plastics industry and the other in the rubber industry. Their growth rate is, and will continue to be, much greater than that of either of these two mature industries.

displacement is less for the soft segments than for the hard. This behavior gives elastic performance up to the point where the strain is so great that permanent deformation occurs due to loss of bonding between the hard and soft segments, or the hard segment is deformed beyond its elastic limit, or the temperature is raised to the point where the hard segment softens or even melts. Therefore, under definable conditions of temperature and stress, TPEs behave with classic elastomeric characteristics, just like thermoset rubbers. With this type of performance, TPEs can and are used in many typical thermoset rubber applications, and thus, have one foot in the rubber industry.

The primary morphological difference between TPEs and the thermoset rubbers is the presence of soft rubbery domains bonded to hard plastic domains (with a distinct melting point above which the TPE is molten and suitable for fabrication). The hard plastic domains of a TPE can be formed, destroyed, and reformed repeatedly through the simple process of adding or removing heat energy. Their formation is thus reversible. This *capability* for the formation of these hard domain "crosslinks" is essentially irreversible. Melting the hard domains by conventional plastic fabrication processes, such as injection molding, blow molding, extrusion, thermoforming, etc., is why TPEs also have their other foot in the plastics industry. While most TPEs do not have typical thermoplastic physical properties, such as high load deflection, they are made into finished articles by typical thermoplastic processing equipment and techniques. It is the nature of TPEs that has allowed thermoplastic processors to expand into the manufacture of rubber parts for merchant production and captive use.

The plastics industry, since John Wesley Hyatt, and the rubber industry, since Charles Goodyear, have both grown into major industries with worldwide sales in the tens of billions of dollars. Although both industries are based on polymer science and technology, there has been little interaction between them. Plastic and rubber materials are processed quite differently. Outside of polyvinyl chloride, few plastics are extensively modified by compounding before fabrication into end-use articles. On the other hand, the rubber polymer is simply the base for a rubber compound developed for specific performance characteristics. The generation of useful rubber articles has created a whole technology based on the compounding and production of specific rubber compounds with a desired set of properties. In addition, conventional thermoset rubbers are vulcanized by chemically crosslinking reactive sites in

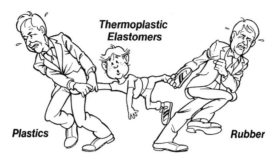

FIGURE 2.   Both the rubber and the plastics industries now claim TPEs. Success has a thousand fathers; failure is an orphan.

the base polymer; this operation requires specialized processing equipment for the preparation of rubber parts. Thermoplastic processing is a simpler fabrication process because no chemical modification of the material is required to form the final article. These differences in the rubber and plastics industries have set them apart from each other, to the point where until recently, few companies have done both rubber compounding and/or part manufacture and thermoplastic processing into end-use articles.

It is now possible to manufacture rubber parts by using TPEs with plastics processing equipment. Since TPEs behave as rubber up to the temperature-dependent limit of permanent deformation, rubber parts end users have adopted them as rubber. However, since TPEs are processed (fabricated) into rubber parts on conventional thermoplastic processing equipment, the plastics industry has also claimed them (Figure 2). The advent of TPEs has resulted in "rubber-only" companies investing in thermoplastics processing equipment and in plastics processors fabricating rubber parts, thus, spanning the gap between these two major industries.

III. POSITION OF TPEs IN THE RUBBER INDUSTRY

While some TPEs, such as EAs, were developed as offshoots of research to produce castable pneumatic tires, current TPEs are not candidates for use in the manufacture of these tires. TPEs are capable of replacing a major portion of the approximately 7 million metric tons of rubber used annually in the nontire market.[27] As of 1990, TPEs had a volume of approximately 8 to 9% of the nontire market, or 650,000 metric tons.[28] It is forecasted that TPEs will grow to 15% or more of the nontire rubber market by the year 2000. Figure 3 shows the growth history to 1990 and the growth projection beyond based on three different rates of growth. Present expectation is that the TPE industry as a whole will grow by 8 to 9% per year through the 1990s.[28] On the basis of past growth of TPEs from 1970 through 1989, which was 8 to 9% on a compounded basis, the future forecast looks quite realistic, particularly when much of the higher growth has occurred in recent years.

The birth of TPEs is generally regarded to be the invention and commercialization of thermoplastic polyurethanes by B.F. Goodrich in 1959.[29] Following this development were the introductions of styrenic block copolymers by Shell Chemical Company[30] in the 1960s, copolyesters by E.I. duPont Company,[30] thermoplastic elastomeric olefins (TEOs) by Uniroyal in the 1970s, EAs by Monsanto Chemical Company[31] in 1981, and block copolymers of polyamides by Atochem in 1982.

The target area for growth of TPEs is primarily in thermoset rubber replacement. A second area is new elastomeric applications where thermoset rubber would rationally be considered. The third area is in soft thermoplastic replacement where, greater flexibility is needed or the existing material (such as PVC in Europe, particularly Germany) is to be

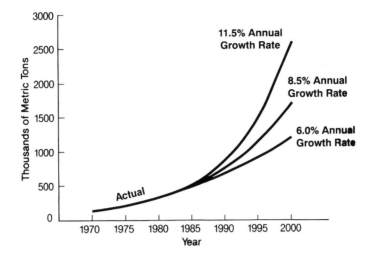

FIGURE 3. Past and future growth curves for TPEs. Actual growth rate for 1970 to 1988, projected for 1988 to 2000.

TABLE 1
Rational Substitution of Thermoset Rubbers by TPEs

Thermoset rubbers	TEOs	Styrenics	EAs	Polyurethanes	Copolyesters	Polyamides
Styrene-butadiene rubber (SBR)	X	X	X			
Natural rubber (NR)	X	X	X			
Butyl rubber (IIR)			X			
EPDM		X	X			
Neoprene (CR)			X	X	X	
Chlorosulfonated polyethylene (CSM)			X	X	X	
Nitrile rubber (NBR)			X	X	X	
Epichlorohydrin (ECO)			X	X	X	X
Acrylate (ABR)			X	X	X	X
Fluoroelastomers						X

replaced for toxicological and ecological reasons. Table 1 shows a grid of opportunities for substitution of thermoset rubbers by TPEs. The various classes of TPEs offer a wide range of performance that will be detailed later in this chapter. This range of performance allows the replacement of a variety of thermoset rubbers ranging from fluoroelastomers to styrene-butadiene rubber (SBR). Elastomer alloys generally offer the widest range of rubber substitution.

In many applications TPEs are able to replace thermoset rubbers because they offer equivalent or better performance, generally with a significant cost saving due to part fabrication cost differences. Understanding the specific requirements of an application is very important in selecting the right material and optimum part design for the specific application needs. In many cases, even greater cost savings and improved performance can be obtained by redesigning a thermoset rubber part to exploit the properties of the chosen TPE.

The design flexibility of TPEs has resulted in rubber applications previously nonexistent because of the design and fabrication limitations of thermoset rubbers. The primary design advantages come from the tighter dimensional tolerance of a TPE part and the capability of heat welding the TPE to a rigid compatible material, employing an automatic process such

Fabrication Steps

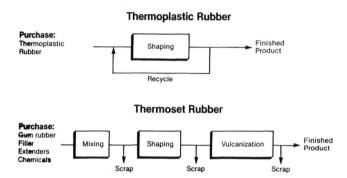

FIGURE 4.   Thermoplastic processing is much simpler and more efficient than thermoset rubber processing.

as insert injection molding, dual injection molding, dual blow molding, and multiextrusion. Thermoplastic elastomers enable a synergistic combination of state-of-the-art materials technology with sound design engineering.

IV. TPEs VS. THERMOSET RUBBERS

Thermoplastic elastomers have both advantages and disadvantages in comparison to conventional thermoset rubbers.

A. ADVANTAGES

1. There is little to no compounding or mixing required for TPEs. Unlike thermoset rubbers that require mixing of curatives, stabilizers, process aids, and other additives, most TPEs are fully compounded and ready for fabrication into parts. Part-to-part consistency is improved because there is no compound difference that can result from variations in weighing and metering of ingredients.
2. The processing operation is simpler for TPEs and has fewer steps, as shown in Figure 4. Since there is no mixing operation or vulcanization step with TPEs, the cost of a part can be lower and the reproducibility of the fabrication is greater.
3. Thermoplastic processing typically has shorter cycle times due to the types of processes involved (i.e., injection molding of a thermoplastic vs. compression molding of a thermoset rubber) and no need for vulcanization. Generally, rubber process cycle times are measured in minutes, whereas TPE cycle times are in seconds. The production rate for TPE parts is much greater from a given piece of equipment.
4. While scrap occurs in both operations, TPE process scrap can be reground and reprocessed with no significant loss in performance. Thermoset rubber process scrap has limited reuse capability. Since more unusable scrap is generated in thermoset rubber processing, the production cost is made greater by the loss of material and disposal cost of the scrap.
5. The thermoplastic process uses less total energy due to more efficient processing and shorter cycle times.
6. Thermoplastic elastomer processes are suitable for high speed automation and amenable to robotic assistance, relative to thermoset rubber processing that is often very labor intensive.
7. Some thermoplastic processes—such as blow molding, heat welding, and thermoforming—are not possible with thermoset rubbers. These processes can offer very

economical and unique part fabrication that is not possible with a conventional thermoset rubber.

8. Since there are fewer steps in fabrication, part quality is improved and cost of quality control is reduced.

9. Thermoplastic elastomers are generally lower in density than the thermoset rubber compounds they replace; this results in a greater number of parts per kilogram of material, thereby reducing final part cost still further.

B. DISADVANTAGES

1. Thermoplastic elastomers are relatively new as materials for widespread thermoset rubber replacement, and thus, unfamiliar to the typical thermoset rubber processor. Also, the past lack of communication between the thermoset rubber and thermoplastic industries has been a barrier to thermoset rubber processors embracing this new technology.

2. Thermoplastic processing equipment is different from thermoset rubber processing equipment. Therefore, capital investment is required for the thermoset rubber processor to produce TPE parts.

3. For replacement of existing rubber parts, there has to be sufficient volume to recover the investment in a new thermoplastic mold or extrusion die. Thermoplastic molds typically cost more than a compression mold for thermoset rubber.

4. While the hardness range and number of TPE types are constantly increasing, TPEs are still fewer in number than the essentially limitless variation of rubber compounds. There are still a limited number of TPEs below 65 Shore A hardness. However, EAs and styrenic block copolymers have made significant recent advances with materials of 30 to 50 Shore A hardness.

5. By the very nature of the material, TPEs melt or soften at a specific temperature, above which they do not function as rubbers. A relatively brief exposure at or above the melt or softening temperature of a TPE will result in permanent deformation. Brief exposure of thermoset rubber to high temperature would probably not result in significant deformation.

6. Many TPEs require drying of the material (by desiccant dryers) prior to processing into fabricated articles.

The suitability of a TPE in any given rubber application must be determined by direct comparison of these advantages and disadvantages. There are specific applications that are still more suitable for thermoset rubbers than the TPEs available today. However, the many thousands of applications that have been successfully commercialized with TPEs are a testimony to the value of these materials in the rubber products market.

Many of the successful TPE applications of today resulted from the end-user of a rubber part being convinced of the improved performance, greater design options, and lower cost obtainable with TPEs vs. thermoset rubbers. Since the end-user generally does not have a vested or, perhaps better, an invested interest in selecting thermoset rubber over TPEs, he/she is better able to make an objective selection of the best material for his/her application. Because of the success of TPEs and end-users specifying them, there is a growing number of traditional thermoset rubber compound and part suppliers making the investment in thermoplastic processing equipment, and realizing greater profitability from their TPE operation than from supplying thermoset rubber parts. Table 2 gives an example of relative cost and performance from a comparison of an EPDM thermoset rubber automotive overslam bumper and the same part made from an EA TPE.

TABLE 2
Cost/Performance Comparison of Fabricating
Automobile Overslam Bumper from EPDM/PP EA TPE
vs. EPDM Thermoset

Material	EPDM/PP Elastomeric alloy 64 Shore A	EPDM thermoset
Material cost, $/lb	1.40	0.70
Part weight, g	13.9	18.0
Material cost/part, cents	4.3	2.3
Process type	Injection molding	Compression molding
Annual number of parts	350,000	350,000
Cost per molded part, cents	13	25

Features/advantages/benefits of TPE in this part
- Weight reduction, 20%
- Cost per part reduction, 40%
- High productivity (cycle time, 30 s)
- Improved dimensional control (2 to 3 times tighter)
- No demolding distortion at deep undercuts
- No flash to trim
- Improved material uniformity
- Color capability

V. GENERIC CLASSES OF TPEs

Table 3 shows a generic classification of TPEs based on morphology and chemistry. The commercially important TPEs fall into three primary categories: (1) block copolymers; (2) rubber/thermoplastic blends (where the rubber phase is not cured or crosslinked); and (3) EAs. The specific types of block copolymers are styrenic-rubbers, copolyesters, polyurethanes, and polyamides. Rubber/plastic blends are of the type EPDM/polyolefin and nitrile rubber/polyvinyl chloride. EAs are a growing class of TPEs arising from the dynamic vulcanization of an elastomer (EPDM, nitrile, natural, and butyl rubber), in the presence of a thermoplastic (PP).

As with the various thermoset rubber polymers available, the various TPEs on the market cover a wide range of product performance and cost. Styrenics and TEOs offer low to mid-range performance in a price range of 2.40 to $4.40/kg. Performance and price increase through the EAs to the thermoplastic urethanes, copolyester, and polyamides, which range in price from 3.00 to $30.00/kg. Table 4 gives a listing of representative TPEs in the various classes. A relative performance spectrum of both TPEs and thermoset rubbers is shown in Figure 5. A given TPE is a rational replacement candidate for the thermoset rubbers at the same position in the charts. Thus, the EAs have found much use as replacements for EPDM and polychloroprene rubbers.

A. BLOCK COPOLYMERS
1. Styrenics
This class of TPEs consists of block copolymers of styrene and a diene. The block copolymer structure can be indicated as:

$$(S-E)_n X$$

TABLE 3
Generic Classification of Thermoplastic Elastomers

Block Copolymers

Styrene-dienes (styrenics)
Copolyesters
Polyurethanes
Polyamides

Elastomeric Polyolefin (TEOs)

EPDM/polyolefin
Nitrile/polyvinyl chloride

Elastomeric Alloys (EAs) in Thermoplastic Matrix

EPDM/polypropylene
Nitrile rubber/polypropylene
Natural rubber/polypropylene
Butyl rubber/polypropylene

where X is a junction point (usually difunctional, n = 2), S denotes a hard polystyrene chain segment with typical molecular weights of 5000 to 8000, and E indicates a soft polydiene chain segment with molecular weights in the 800 to 10,000 range.

The polydienes commonly available in an S-E-S TPE are butadiene (B), isoprene (I), and ethylenebutylene (EB). The chemical structures of these three types of styrenics are shown in Figure 6. The relative ratio of the polystyrene (PS) blocks and those of the diene rubber segments determines the morphology of the material and affects the overall performance characteristics of the S-E-S material. Generally, the relative ratio is from 15/85 to 50/50 S/E. Figure 7 illustrates a simple artistic depiction of the change in morphology from a low level of PS (S) content, where the TPE is soft and rubbery with relatively low tensile strength, through the phase inversion where PS is the dominant phase and a glossy, impact-modified plastic results.[32-35] Up to the 50/50 S/E level indicated earlier, the S-E-S will have rubber-like characteristics, up to the melting point of PS (approximately 110°C).

It is important that the structure is S-E-S, and not E-S-E, for useful performance as an elastomer. As one would expect, the elastomeric component gives the material the characteristics of a rubber, but only when it is "tied" on both ends to a stiff, nonflexible material like PS. The PS block serves essentially the same function as a chemical crosslink (or carbon black) in a conventional thermoset rubber. The bonding between the PS segment and the elastomeric segment is through covalent bonding. The two-phase structure occurs because of the chemical incompatibility of the two segments. Since the commercially useful styrenics contain a predominance of the elastomeric segments, the continuous phase is the elastomer, with the PS serving as hard connectors of the flexible rubber segments. With the PS segments limiting mobility, the materials can be stretched or compressed to the elastic limit of the elastomeric phase and return to their original state when the stress is removed.

The specific soft block chosen — whether one with unsaturation, such as butadiene, or a hydrogenated material, such as ethylenebutylene — also gives the characteristic performance to the S-E-S. Unsaturated polymers give poorer resistance to oxidative and thermal stability, due to the presence of olefinic unsaturation in the polymer backbone. The specific

TABLE 4
Representative Commercial Thermoplastic Elastomers

Generic class	Supplier	Trade Name[a]
Stryene-diene	Shell Chemical	Kraton
	Concept Polymers	C-Flex
	Dow/Exxon	Dexco
	Enichem	Europrene
	Fina	Finaprene
	J-Von	J-Plast
	Nippon Zeon	Quintac
EPDM/polyolefin blend	Advanced Elastomer Systems, L.P.	Vistaflex
	DNS Plastics	Renflex
	Ferro Corp.	Ferroflex
	Himont	Dutral
		Dutralene
	J-Von	Hercuprene
	Mitsui	Milastomer
	Nippon Petrochemicals	Softlex
	Schulman	Polytrope
	Teknor Apex	Telcar
Nitrile/polyvinyl chloride	Alpha (Dexter)	Vynite
	Nippon Zeon	Elastar
	Schulman/Mitsubishi	Sunprene
Elastomeric alloy	Advanced Elastomer Systems, L.P.	Santoprene
		Geolast
		Vyram
		Dytron
		Trefsin
	DuPont	Alcryn
Copolyester	Akzo	Arnitel
	DuPont	Hytrel
	Eastman	Ecdel
	General Electric	Lomod
	Hoeschst Celanese	Riteflex
Polyurethane	BASF	Elastollan
	Dow Chemical	Pellethane
	B.F. Goodrich	Estane
	Imperial Chemical Ind.	Permuthane
	Mitsui Nisso	Hiprene
	Mobay (Bayer)	Texin
	Schulman	Polypur
	Thiokol	Plastothane
Polyamide	Atochem	Pebax
	Emser Ind.	Grilamid
	Huels	Vestamid

[a] Many of the above tradenames are registered trademarks of the respective suppliers.

thermal and oxidative stability is, of course, dependent on the amount of unsaturation along the backbone of the polymer chain.

Styrenics usually have high viscosity and are highly non-Newtonian in flow properties, due to the two-phase structure. Therefore, increasing shear stress is more effective in increasing flow than temperature alone. There is some difference in rheological behavior

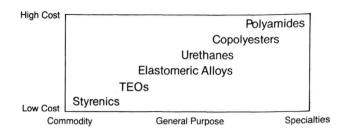

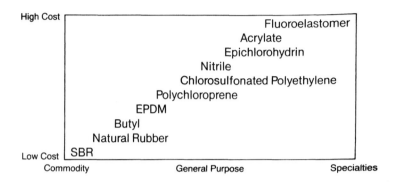

FIGURE 5. Relative cost and performance of generic classes of thermoplastic elastomers and thermoset rubbers.

B: $\left(CH_2CH\right)_a\left(CH_2CH = CHCH_2\right)_b\left(CHCH_2\right)_c$

I: $\left(CH_2CH\right)_a\left(CH_2C = CHCH_2\right)_b\left(CH_2CH\right)_c$
         CH_3

EB: $\left(CH_2CH\right)_a\left(CH_2CH_2CH_2CH_2CH_2CH\right)_b\left(CH_2CH\right)_c$
                CH_3CH_2

FIGURE 6. Structures of common styrenic block copolymer TPEs; a and c $= 50$ to 80, b $= 20$ to 100.

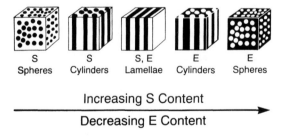

S   S   S, E   E   E
Spheres Cylinders Lamellae Cylinders Spheres

Increasing S Content

Decreasing E Content

FIGURE 7. Change in morphology of S-E-S block copolymer TPE with S/E ratio.

between the linear (n $= 2$) and star-block (n>2) copolymers that are more highly branched than typical S-E-S materials. A more highly branched material exhibits an even greater dependency of viscosity on shear; so at equivalent shear rates typical of thermoplastic processing equipment (i.e., 10^2 to 10^4 s^{-1}), the star-block styrenics have lower melt viscosity than dibranched styrenics.

TABLE 5
Properties of Different Classes of Thermoplastic Elastomers

Property	Styrenics	Copolyester	Polyurethane	Polyamide	TEO	EA
Specific gravity[a]	0.90–1.1	1.1–1.3	1.1–1.3	1.0–1.2	0.89–1.0	0.94–1.0
Shore hardness	30A–75D	35D–72D	60A–70D	75A–65D	60A–75D	55A–50D
Low temperature limit, °C	−70	−65	−50	−40	−60	−60
High temperature limit, °C (continuous)	100	125	120	170	120	135
Compression set resistance, at 100°C	P	F	F/G	F/G	P	G/E
Resistance to aqueous fluids	G/E	P/G	F/G	F/G	G/E	G/E
Resistance to hydrocarbon fluids	P	G/E	F/E	G/E	P	F/E

[a] Does not include grades containing a special flame-retardant package, which generally raises the specific gravity 20 to 30 %.

P = poor, F = fair, G = good, and E = excellent.

Typical property ranges for styrenic TPEs are shown in Table 5.[36] The hardness ranges from 30 Shore A to over 50 Shore D. True rubber-like properties are obtained at hardnesses below 85 to 90 A. The low temperature limit of −70°C results from the glass transition temperature of the elastomeric phase. Below this temperature the TPE is very hard and rigid. As mentioned earlier, the upper temperature limit is dictated by the melt temperature of the hard, immobile PS phase. Above this temperature, the PS segments no longer behave as "crosslink" sites. The melting of the plastic phase is what makes these materials thermoplastic, and its temperature primarily determines the theoretical upper temperature limit for retention of rubber properties. This ability to form, destroy, and reform the aggregates of hard segments gives the flexibility of thermoplastic processing.

Fluid resistance of S-E-S materials is good against polar fluids, such as water and aqueous solutions, but poor to oils, fuels, and organic solvents, due to swelling of the rubber phase and possible dissolution of the plastic phase. The S-E-S materials that contain butadiene or isoprene are very susceptible to degradation from oxygen, ozone, and UV radiation because of the unsaturation of the elastomeric segment. The ethylenebutylene (S-EB-S) materials give better resistance because the elastomer is fully saturated.

2. Copolyesters

This class of TPEs consists commonly of block copolymers with alternating segments of a hard crystalline component such as poly (1,4-butanediol terephthalate) and a soft segment that is amorphous. The soft segment is usually a long chain poly (alkylene ether terephthalate), such as a polytetramethylene ether or polypropylene glycol terephthalate. Figure 8 pictorially shows the structures of a typical copolyester where the hard segment contains 16 to 40 mers and the soft segment 16 to 40 mers. As indicated, the structure is A-B-A-B.[37]

Copolyesters function as TPEs in a manner very similar to the styrenic block copolymers. That is, the hard segment serves as the crosslink at temperatures below its melting point and the soft segment gives the material rubbery characteristics above its glass transition temperature. Due to the chemical nature of these materials, the range of hardness and flexibility available is limited. As shown in Table 5, the softest copolyester TPE is 35 Shore D (90 Shore A), which is typically harder than most thermoset rubbers.

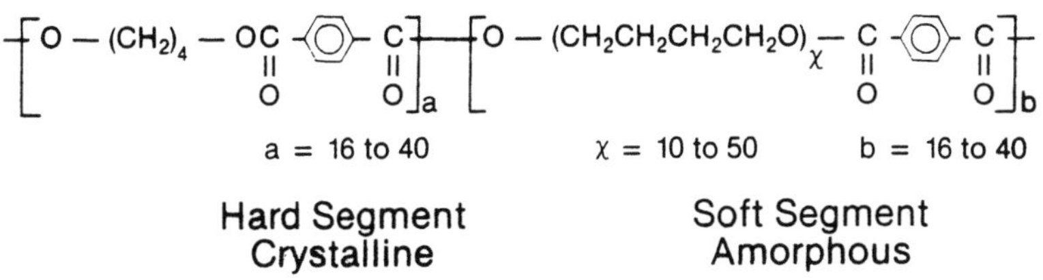

$$\left[O-(CH_2)_4-OC-\langle\bigcirc\rangle-C \right]_a \left[O-(CH_2CH_2CH_2CH_2O)_\chi-C-\langle\bigcirc\rangle-C \right]_b$$

a = 16 to 40	χ = 10 to 50	b = 16 to 40

Hard Segment   **Soft Segment**
Crystalline   **Amorphous**

FIGURE 8. Chemical composition and morphology of copolyester block copolymer TPEs.

Copolyesters can replace thermoset rubbers in many applications even though the per se material properties may be significantly different. This can be done by recognizing the material differences and designing the part appropriately Copolyesters exhibit elastic performance only within a range of 5 to 25% elongation, which is much lower than most thermoset rubber compounds and most other TPEs. However, within this elastic range these materials have very good creep resistance, resilience, and fatigue resistance. This combination of properties makes copolyesters suitable for demanding dynamic applications, such as automotive CVJ boots, keeping the elastic limit in mind when designing the part. Copolyesters also have very good load bearing characteristics and impact resistance.

The useful temperature range of copolyesters is from -55 to $140°C$. The temperature range is limited on the upper end due to oxidative degradation of the polymer backbone, and on the lower end by the glass transition temperature of the soft segments. Copolyesters have very good resistance to organic fluids, with the exception of halogenated solvents. They are susceptible to degradation from acids or bases. Copolyesters are hygroscopic and must be dried before processing to prevent degradation during injection molding, extrusion, etc. The specific gravity of copolyesters is higher than that of styrenics, and ranges from 1.1 to 1.3.

3. Polyurethanes

Thermoplastic polyurethanes (TPUs), the oldest of the TPEs, are block copolymers with urethane backbone linkages. They are composed of hard and soft segments, as are the copolyesters and styrenics. The hard segments are polymer blocks from the reaction of a diisocyanate and a short chain diol. As with copolyesters, the hard segment is crystalline and the soft segment amorphous. The soft segments are either polyether or polyester oligomeric diol blocks with molecular weights of 800 to 3500. The basic chemical structure is shown in Figure 9. Thermoplastic polyurethanes have an alternating -A-B-A-B- structure like the copolyesters. The hard segment is composed of polar materials that can form carbonyl to amino hydrogen bonds to give higher modulus and tensile strength and greater crystallinity, which strongly influences the morphology of the system.

Similar to copolyesters and styrenics, the soft blocks give the characteristic elasticity and low temperature behavior. When the soft block is a polyester, tear strength is enhanced, as is abrasion, toughness, and resistance to polar solvents. Polyether soft blocks have better resilience, low temperature flexibility, hydrolytic stability, microbial resistance, and lower specific gravity. The hard blocks contribute stiffness, hardness, and upper-use temperature. The tensile properties of softer TPUs are quite rubber-like, especially at higher temperatures (Figure 10). Thermoplastic polyurethanes have excellent abrasion resistance and are the standard by which the abrasion of other TPEs is compared. They also offer a low coefficient of friction. This combination of properties makes them the material of choice in certain caster wheel uses, shoe heels, and a variety of applications where a nonwear, low-friction surface is desirable. Compression set resistance is good up to $100°C$, as shown in Table 5. Since there is virtually no unsaturation in TPUs, they have good resistance to ozone and

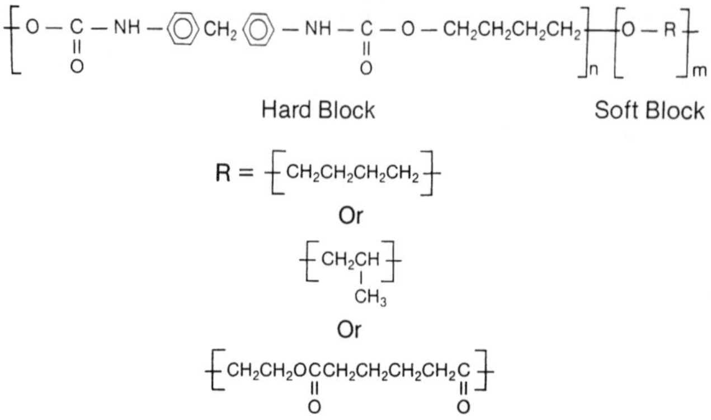

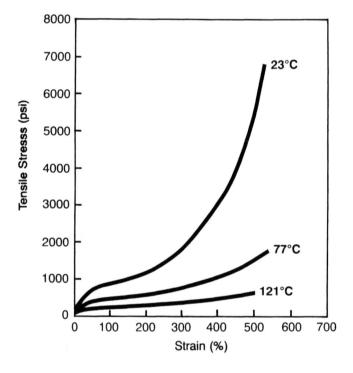

FIGURE 9. Chemistry and morphology of polurethane TPE.

FIGURE 10. Tensile stress-strain curves of 86 Shore A TPU at different temperatures.

UV degradation. Gas permeability resistance is also quite good. Thermoplastic polyurethanes generally are the TPEs (with the exception of butyl rubber-based EAs) closest to thermoset butyl rubber in low permeability to gases.

The upper-use temperature of TPUs is 120°C, as defined by the hard segment melting point, and a lower temperature of −40°C, as defined by the soft segment glass transition temperature. Thermal degradation occurs above 200°C. Thermoplastic polyurethanes have good resistance to nonpolar fluids, as well as water and aqueous solutions with a pH of 5 to 9. They are hydrolytically unstable to acid and basic solutions. Because of their inherent hemocompatibility and low toxicity, TPUs are used in biomedical applications. The available Shore hardness range of TPUs is from 60 A to 70 D, with a range in specific gravity of 1.1 to 1.3.

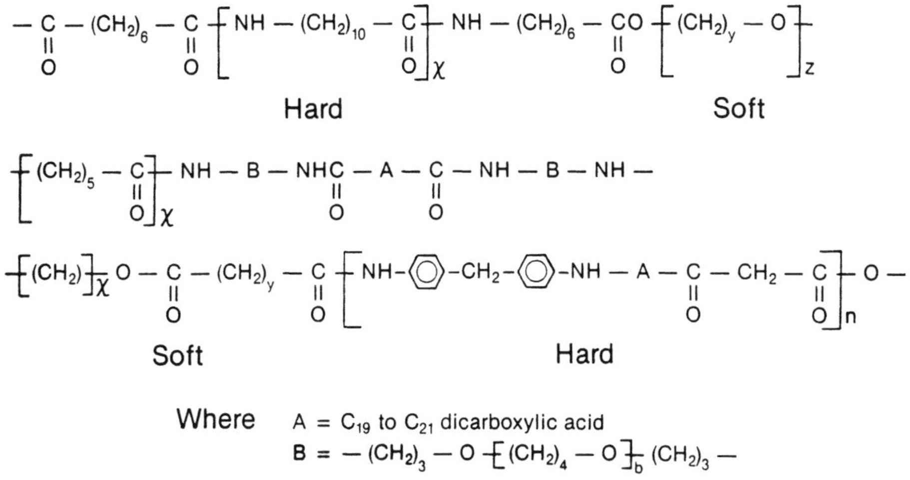

$$- C - (CH_2)_6 - C \left[NH - (CH_2)_{10} - C \right] NH - (CH_2)_6 - CO \left[(CH_2)_y - O \right]_z$$

Hard   Soft

$$\left[(CH_2)_5 - C \right]_x NH - B - NHC - A - C - NH - B - NH -$$

$$\left[(CH_2) \right]_x O - C - (CH_2)_y - C \left[NH-\bigcirc-CH_2-\bigcirc-NH - A - C - CH_2 - C \right]_n O -$$

Soft   Hard

Where   A = C_{19} to C_{21} dicarboxylic acid

B = $- (CH_2)_3 - O \left[(CH_2)_4 - O \right]_b (CH_2)_3 -$

FIGURE 11.   Basic structure and morphology of polyamide TPEs.

4. Polyamides

The basic chemical description of a thermoplastic polyamide (PEBA) is a block copolymer of hard and soft segments connected by amide linkages. The hard segments are polyamides and the soft segments are either polyether, polyester, or polyetherester. The basic structures of the three types are shown is Figure 11. The polyamide blocks determine the melting point and maximum service temperature of the PEBA, whereas the soft segments give characteristic fluid resistance and low temperature flexibility. The specific type of soft segment (i.e., polyether or polyester) also influences the performance of the PEBA. Polyester segments give better oxidation stability and fluid resistance, and polyether segments give better hydrolytic stability and functional properties at low temperatures.

The PEBAs are relatively high cost TPEs that give high strength and flexibility (Table 5). They have very good high temperature stability, with an upper temperature limit of 170°C. They are susceptible to hydrolysis in humid air, particularly polyester-based PEBAs. These materials have very good tear strength, as well as good fatigue and abrasion resistance. Thermoplastic polyamides are commercially available with a range in hardness of 75 A to 65 D, with those materials below 90 A giving elastomeric performance. The low-temperature flexibility down to -40°C and below makes these materials suitable for automotive applications. The PEBAs have very good resistance to hydrocarbon fluids such as oils and greases, moderate resistance to organic polar fluids, and good resistance to aqueous fluids, except very acidic or basic solutions at elevated temperatures.

B. ELASTOMER/THERMOPLASTIC BLENDS (TEOs)

1. EPDM—Polypropylene

The most common example of a commercially available rubber/plastic blend is that of PP and unvulcanized ethylene-propylene diene monomer rubber (EPDM). Blends of polyethylene (PE) with EPDM and with ethylene-propylene rubber (EPR) are also commonly found. Less common are blends of either PE or PP with natural rubber, styrene-butadiene rubber or other rubber types. These systems are most often simple physical blends of unvulcanized rubber with the thermoplastic. The rubber component contains the usual ingredients found in thermoset rubber compounds, such as extender oil, carbon black (though usually in much smaller amounts), fillers, and stabilizers. There are also systems with sufficient peroxide, crosslinking resin or sulfur to result in partial vulcanization of the rubber

TABLE 6
Properties of TEO Thermoplastic Elastomers of Different Hardness

Material Properties

Physical Properties			
Hardness, 5 Shore	70 A	84 A	35 D
Specific gravity	0.87	1.01	1.00
Mechanical Properties			
Tensile strength, MPa	8.3	7.0	8.0
Ultimate elongation, %	810	670	570
100% Modulus, MPa	2.4	4.1	6.9
Tension set, %	20	33	56
Tear strength, die C, kN/m	44	49	61
Compression set, 22 h/100°C, %	26.1	32.6	56.2
Brittle point, °C	−60	−56	−60
Thermal Properties			
Hot air aging, 168 h/150°C			
Ultimate tensile ret., %	105	109	104
Ultimate elongation ret., %	112	94	121
100% Modulus ret., %	115	118	121
Weight gain, %	−4.4	−6.9	−2.4
Hardness change, points	+3	0	+5
Fluid Resistance			
ASTM No. 3 oil, 70 h/125°C			
Hardness change			to 60 A
Tensile ret., %	Sample disintegrated	Sample disintegrated	79
Elongation ret., %			35
100% Modulus ret., %			75
Weight gain, %			+124

phase. Most often, the plastic phase is a homopolymer, isotactic PP, although copolymers of propylene are used as well.

In those blends where the rubber is completely unvulcanized, there is but little true rubber performance. As shown in Table 6, simple blends have poor set resistance and poor hydrocarbon fluid resistance. Blends are most commonly used to replace plastics when improved impact resistance is required, lower modulus is desired, or more rubber-like service is needed. For rubber-like performance, these blends (TEOs) are limited to ambient temperature and softer grades (i.e., 80 Shore A and softer). With unvulcanized blends, there is a limit to the amount of EPDM that can be used with the retention of a suitable morphology and fabricability. Abdou-Sabet and Patel[38] have described the changes in morphology as a function of rubber content. To maintain a dispersed rubber phase, no more than 20 to 30% EPDM can be used.

A blend of EPDM and PP is thermoplastic, by the nature of the PP phase. To maintain good fabricability and a balance of rubber-like performance at room temperature, the morphology of these blends has to be one of a dispersed EPDM rubber in a continuous matrix of PP. A pictorial description is shown in Figure 12.[39] For compatible systems such as EPDM and PP, a dispersed-phase morphology can be formed simply by mechanically mixing the two materials, as long as the EPDM phase is less than approximately 30 wt%, as noted earlier. Vulcanization of the rubber phase does result in a more stable dispersed-phase morphology, so partially vulcanized blends are capable of containing higher concentrations of EPDM that result in a more rubber-like performance.[38]

Characteristic properties of partially vulcanized systems are shown in Table 7. Blends of EPDM and PP, PE, or copolymers of the two thermoplastics behave elastomerically when

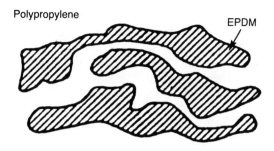

Polypropylene

EPDM

FIGURE 12. Morphology of EPDM/PP TEO. The discontinuous phase (elastomer) has to be dispersed as small particles so the interfacial surface area is sufficient for the attractive forces between the two phases to be adequately strong.

TABLE 7
Properties of Partially Vulcanized TEOs

Material Properties

Physical Properties						
Hardness, 5 s Shore	53 A	65 A	75 A	85 A	45 D	55 D
Specific gravity	1.0	1.0	0.99	0.97	0.95	0.94
Mechanical Properties						
Tensile strength, MPa	3.9	4.1	6.2	8.6	16.1	27.6
Ultimate elongation, %	450	490	600	670	740	760
100% Modulus, MPa	1.8	2.3	3.4	5.8	11.7	13.2
Compression set, 22 h/70°C	30.0	36.2	35.0	35.7	43.3	45.7
Tear strength, kN/m	23	32	39	53	83	123
Thermal Properties						
Brittle point, °C	<-60	<-60	<-60	<-60	<-60	<-50
Fluid Resistance						
ASTM No. 3 oil, 22 h/125°C						
Weight Gain, %	181	192	155	113	151	139
Hot air aging, 168 h/150°C						
Ultimate tensile ret., %	81	86	96	110	92	91
Ultimate elongation ret., %	84	73	75	73	41	72
100% Modulus ret., %	96	100	110	121	120	122
Weight change, %	-7.1	-5.8	-6.0	-5.6	-3.5	-3.9
Hardness change, points	$+2$	$+4$	$+6$	$+6$	$+5$	$+7$

the EPDM is well dispersed as small (approximately 10 μm or less) particles. Unlike block copolymers, there is no actual chemical bond between the rubber phase and the plastic except for weak van der Waals attraction. The primary attraction is mutual solubility. When a compressive or tensile stress is applied to this system, EPDM functions as the spring (as in the Maxwell model), with the plastic phase being the rigid nodes. For unvulcanized TEOs, the rubber phase does not have a very extensive elastic range. Since the rubber content is low as well, properties such as compression set are fair-to-good at ambient temperature and poor at elevated temperatures. Partial vulcanization of the rubber phase results in a more stable morphology, as noted earlier, and extends the elastic range of the rubber.

TEOs are relatively low cost TPEs that provide a range of performance with fair-to-good mechanical, electrical, and environmental properties. More rubber-like performance is obtained with hardnesses less than 80 to 85 Shore A and with partial vulcanization of the rubber. Rubber performance is best at room temperature, particularly regarding stress-strain, tear strength, compression set, and hydrocarbon fluid resistance. Elevated temperature performance is limited to 100 to 125°C due to oxidation in hot air. The upper-use temperature limit is also dictated by the selection of the plastic phase. A PP-based material will have a higher temperature capability than a PE-based TEO. The practical upper temperature limit is 80°C where strength retention is required. The lower service temperature is dependent on the rubber phase (approximately $-60°C$ for EPDM blends) and the amount of rubber in the blend.

For EPDM and polyolefin plastics like PP or PE, there is very good resistance to ozone and air oxidation up to the upper temperature range, because of the very low level of unsaturation in both phases. These materials have excellent resistance to aqueous solutions and polar organic solvents, but poor fluid resistance to hydrocarbons and halocarbons.

2. NBR — Polyvinyl Chloride

Thermoplastic polyurethanes were the first true TPEs, as indicated earlier; however, blends of nitrile rubber (NBR) with polyvinyl chloride (PVC) were also developed early as materials that give some characteristic properties of TPEs. In 1940, a patent was granted to B.F. Goodrich Company on NBR/PVC blends, and marketing of them was begun in 1947.[40] Materials of this type have been used commercially since that time for applications where improved elastomeric performance of flexible PVC is needed. Blends of this type are predominantly composed of particles of gelled NBR admixed with a PVC compound. The NBR particles are generally rather large, on the order of 50 μm in diameter. Because of the polar nature of both PVC and NBR, there is a compatibility between the two materials, so good fabricability and physical properties are obtained.

The PVC/NBR blends have better hydrocarbon fluid resistance than PP/EPDM blends. While the inclusion of NBR into a blend with flexible PVC improves the elastomeric performance of PVC, these materials do not have the elastomeric performance of most TPEs. Applications tend to be directed at a level of performance just above flexible PVC in terms of compression set resistance, resiliency, or impact resistance.

C. ELASTOMERIC ALLOYS
1. Distinction between Alloys and Blends

Elastomeric alloys are a class of TPEs generated from a chemical combination of two or more polymers to give an alloy having better elastomeric properties than those of the corresponding blend.[41] The blends discussed in the previous section are either noncrosslinked or slightly crosslinked rubber systems admixed with a compatible thermoplastic phase. Elastomeric alloys are highly vulcanized rubber systems with the vulcanization having been done dynamically in the melted plastic phase.

Through dynamic vulcanization of an elastomer in the presence of a compatible thermoplastic or an incompatible thermoplastic with a compatibilizer, there is a bonding that occurs between the rubber and the plastic with the formation of a stable morphology.[42]

There is also an EA consisting of a single phase embracing a synergistic combination of plasticized chlorinated polyolefin, ethylene vinyl acetate, and acrylic ester. The ethylene vinyl acetate is crosslinked *in situ*.[43] The generic name given to this single-phase material is melt-processable rubber (MPR).

The synergistic result of the alloying of a two-phase system vs. an uncrosslinked blend is seen in Table 8. This example compares the differences between a blend of EPDM and PP and a dynamically vulcanized EA made from the same rubber and plastic composition. The crosslinking and alloying process increases the tensile strength by over three times that of the uncrosslinked system. Similar improvements in performance are seen in compression set, tension set, resistance to oil swell, and retention of properties at elevated temperature. The degree of crosslinking is important to the physical properties and fabricability of an

TABLE 8
Comparison of EA and TEO with the Same
EPDM Rubber/PP Composition[a]

Property	TEO	EA
Hardness, Shore A	81	84
Ultimate tensile strength, MPa	4.0	13.1
Ultimate elongation, %	630	430
100% Modulus, MPa	2.8	5.0
Compression set, %	78	31
Tension set, %	52	14
Swell in ASTM No. 3 oil, %	162	52

[a] Parts by weight: EPDM rubber, 91.2; PP, 54.4; extender oil, 36.4; carbon black, 36.4.

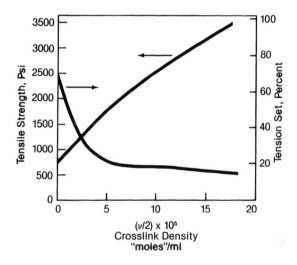

FIGURE 13. Effect of dynamic vulcanization on properties of EA.

EA. Figure 13 shows how tensile strength increases progressively with crosslink density, and tension set improves as well.

Another feature of EAs vs. TEOs is the stability of the dispersed (rubber) phase morphology. Uncrosslinked or partially crosslinked TEOs have either a dispersed-rubber phase morphology or, in some instance a cocontinuous system.[38] The uncrosslinked rubber particles of a blend can coalesce after removal of the temperature and shear of dynamic mixing. In an EA, the rubber particles have been crosslinked while dispersed in the less viscous molten thermoplastic and cannot reaggregate. Thus, they are trapped in this dispersed state. The surface area generated by their fine dispersion enables them to interact intimately and synergistically with the thermoplastic, to give a highly elastomeric composition.

The process of dynamic vulcanization, when done under prescribed conditions and with the optimum crosslinking system produces a dispersed phase of particulate rubber in a continuous phase of plastic.[44,45] This morphology has been disputed but, Abdou-Sabet and Patel[38] have shown conclusively that this is the structure of dynamically vulcanized EAs. A graphical depiction of this morphology is shown in Figure 14. The smaller the average rubber particle size, the better the stress/strain properties of the material (Figure 15). The

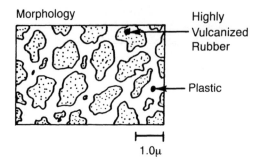

FIGURE 14. Morphology of EA TPE: highly vulcanized rubber particles in a thermoplastic matrix.

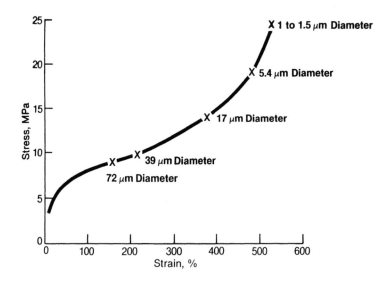

FIGURE 15. Effect of rubber particle size on EA tensile strength.

mechanics of such as system — where there are vulcanized rubber particles finely dispersed, with a high degree of packing in a thermoplastic matrix — gives the true thermoset rubber performance of EAs, which no other TPE can match.

The bulk characteristic properties of an EA are essentially those of the thermoset rubber used as the elastomeric phase, within the elastic and thermal limits of the thermoplastic phase. The elastic limit of the thermoplastic phase only indirectly limits the elasticity of the EA because of the included rubber-phase morphology and the low thickness of thermoplastic between the crosslinked rubber particles. When a tensile or compressive stress is applied to the EA system, the stress is transferred from the stronger plastic phase to the weaker rubber particulate, resulting in strain of the rubber particle in the direction of tensile stress or orthogonal to compressive stress. The spatial disturbance of the thermoplastic phase is minimal due to the small thickness of the thermoplastic matrix around the dispersed rubber particles. This mechanism is shown in Figure 16. Because there is sufficient bonding of the rubber particulate to the thermoplastic matrix, there should be little or no permanent deformation due to loss or displacement at the rubber/plastic interface. This mechanism is also responsible for the outstanding flex fatigue resistance of EAs relative to thermoset rubbers, as seen in Figure 17. The improvements of EAs over TEOs due to dynamic vulcanization are:

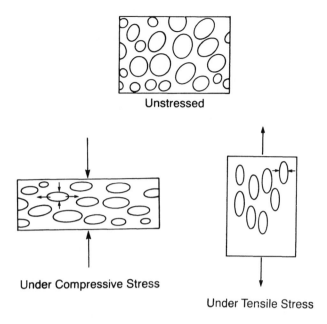

FIGURE 16. Changes in EA morphology on application of tensile and compressive stress.

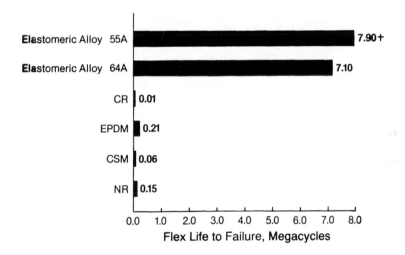

FIGURE 17. Fatigue resistance of EPDM/PP EA relative to that of thermoset vulcanizates. Monsanto fatigue-to-fail test, 100% elongation.

- Reduced permanent set
- Improved ultimate mechanical properties
- Improved fatigue resistance
- Greater resistance to attack by fluids
- Improved high temperature utility
- Greater stability of phase morphology in the melt
- Greater melt strength
- More reliable thermoplastic fabricability.

TABLE 9
Properties of Different Elastomeric Alloy TPEs

Property	EPDM/PP (64 A)	EPDM/PP (40 D)	NBR/PP (70 A)	NBR/PP (40 D)	NR/PP (70 A)	NR/PP (40 D)	IIR/PP (70 A)
Hardness,[a] Shore	64 A	40 D	70 A	40 D	70 A	40 D	70 A
Specific gravity[b]	0.97	0.95	1.00	0.97	1.04	1.01	0.98
Tensile strength,[c] MPa	6.9	19.0	6.2	16.6	7.6	14.3	6.2
Ultimate elongation,[c] %	400	600	265	420	380	540	410
Stress at 100% elongation,[c] MPa	2.3	8.6	3.3	9.0	3.7	8.2	2.8
Tear strength,[d] kN/m	25	65	32	76	29	73	20
Brittle point,[e] °C	< −60	−57	−40	−28	−50	−40	−56
Tension set,[c] %	10	48	10	37	16	39	15
Compression set,[f] %, 22 h/ 100°C	34	65	28	46	32	60	38
Weight change,[g] %, ASTM No. 3 oil, 166 h/100°C	74	30	0	5	97	26	75[h]

[a] ASTM D 2240, 5 s delay.
[b] ASTM D 297.
[c] ASTM D 412.
[d] ASTM D 624.
[e] ASTM D 746.
[f] ASTM D 395B, 25% compression.
[g] ASTM D 471, after 70 h/100°C.

EAs have very good mechanical properties as shown in Table 9, outstanding flex fatigue (better than specially compounded thermoset rubbers), excellent retention of properties at elevated temperatures during short (Figure 18) and extended (Figure 19) periods of time, better fluid resistance in many systems than the comparable thermoset rubber, and low temperature properties dependent on the glass transition temperature of the elastomer.

2. EPDM/Polypropylene

The earliest commercially available EA was a dynamically vulcanized, completely cross-linked system of EPDM and PP.[46] This material has the morphology depicted in Figure 14, with the PP phase continuous and the highly crosslinked EPDM phase discontinuous, even for the EAs with a high EPDM/PP ratio. As this ratio increases, the layer of PP between the small EPDM particles (about 1 μm in diameter) becomes progressively thinner. Consequently, the EA/TPE will become more rubber-like in properties, and less like a thermoplastic. Thus, it will have a lower hardness and modulus, lower set (both compression and tensile), and a more linear tensile stress-strain curve.

Figure 20 shows the tensile stress-strain curves for four different EPDM/PP EAs with progressively increasing hardness. The softest EA has a very rubber-like curve, approaching linearity and passing through the origin. With increasing hardness the curves have a progressively more distinct knee, which actually evolves into a slight yield point for the 50 Shore D hardness EA.

As typical EAs, the EPDM/PP EAs are distinctly anisotropic upon injection molding. Thus, an injection-molded EA will have significantly greater ultimate elongation and ultimate tensile strength in the direction perpendicualr (strong direction) to flow of molten TPE into the injection mold, and significantly greater modulus (at lower elongation) in the direction parallel (weak direction) to the flow. Figure 21 depicts this anisotropy on a typical injection-molded EA plaque. Table 10 shows stress-strain data for tensile pulls of two different EPDM/PP EAs in the strong and weak directions. The magnitude of this anisotropy will be deter-

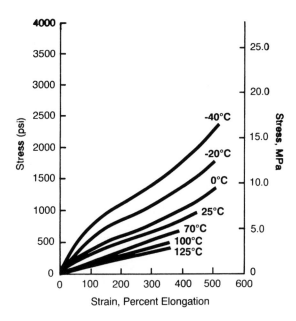

FIGURE 18.   Tensile stress-strain curves for 64 Shore A EA at different temperatures.

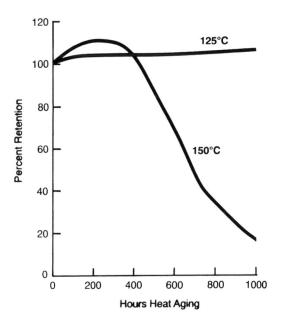

FIGURE 19.   Ultimate elongation heat aging data for 64 Shore A EPDM/PP EA.

mined by the mold dimensions (length, width, thickness), its gating, its venting, and the specific molding conditions (injection pressure, melt temperature, molding cycle) used.

Typical properties of two commercial EPDM/PP EAs are given in Table 9. These interesting materials are now available in hardnesses ranging from 45 Shore A up to 50 Shore D, with specific gravity between 0.95 and 1.00, and tensile strength from 3 to 28 MPa. The tensile strength of EAs, and TPEs in general, is significantly below that of a

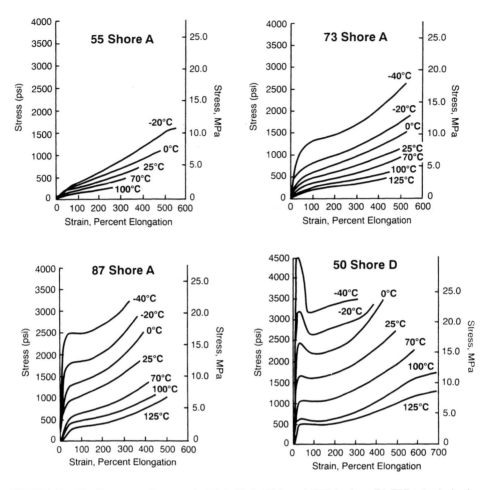

FIGURE 20.   Tensile stress-strain curves for 55 A, 73 A, 87 A, and 50 D hardness EA TPEs. As the hardness increases and the temperature decreases, the EA becomes less rubber-like and more like a thermoplastic.

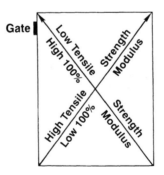

FIGURE 21.   Anistropy of injection-molded TPE plaques.

thermoset rubber of the same hardness. This is of little practical consequence since very few rubber articles are used at an elongation anywhere close to the ultimate tensile limit. In fact, most are used in compression, shear, torsion, or some combination thereof.

The useful temperatures of EPDM/PP EAs range from −60 up to 135°C. These materials have resistance to attack by a broad spectrum of fluids. In water, aqueous solutions, and

TABLE 10
Variation of Injection-Molded EA Tensile
Properties with Direction[a]

Property	EA Hardness, Shore A		
	64	73	87
Ultimate elongation, %			
Strong direction	502	490	550
Weak direction	317	260	390
Ultimate tensile strength, MPa			
Strong direction	7.06	8.28	14.83
Weak direction	4.76	5.86	11.31
Stress at 100% elongation, MPa			
Strong direction	2.31	3.24	6.5
Weak direction	2.83	4.07	7.66

[a] Injection-molded plaques (11.3 cm × 8.06 cm × 0.297 cm).
Strong and weak directions are as indicated in Figure 21.

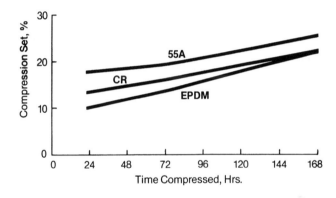

FIGURE 22.   Compression set of 55 Shore A EA TPE and compounded neoprene and EPDM thermoset rubbers at 100°C.

other polar fluids, their resistance is excellent. In hydrocarbon fluids — fuels, oils, solvents — it is fair (i.e., good retention of properties, but significant swelling), and in heavily halogenated organic fluids it is generally poor.

ASTM specification D 5046[47], "Standard Specification for Fully Crosslinked Elastomeric Alloys", covers the EPDM/PP EAs. This specification distinguishes these materials from the TEO blends by means of their lower compression set, lower oil swell, and greater retention of modulus at elevated temperature. The crosslinking of the elastomer phase and its synergistic interaction with the continuous thermoplastic phase gives rise to these superior properties of EPDM/PP EAs over those of the corresponding TEOs.

EPDM/PP EAs offer highly unusual combinations of very good compression set and compression stress relaxation for sealing applications, coupled with outstanding fatigue resistance.[46,48,49] The compression set (and sealing capability) of EPDM/PP EAs is comparable to that of specially compounded thermoset rubber stocks. Figure 22 compares the compression set of a 55 Shore A hardness EPDM/PP EA to that of polychloroprene and EPDM thermoset rubbers (with which EPDM/PP EAs compete directly in the rubber products marketplace). In thermoset rubbers, low compression set and good fatigue resistance tend to be mutually exclusive. Thus, a thermoset stock compounded for good compression set will generally have poor fatigue resistance, and vice versa. The EPDM/PP EAs have been found to have good compression set and outstanding fatigue resistance.

3. Nitrile Rubber/Polypropylene

The alloying NBR with PP is somewhat more difficult than that of EPDM, due to the much greater incompatibility of the two polymers. This derives from their marked difference in solubility parameter, chemical composition, and polarity.[50] Thus, the generation of suitable EAs from these two polymers has required compatibilization techniques for rendering them less incompatible, but not thermodynamically compatible. Coran and Patel pioneered this technology by resorting to compatibilizing block copolymers, chemical modification of the two polymers, graft polymerization, and combinations thereof.[51,52]

Evolving from this basic compatibilization technology was a commercial NBR/PP elastomeric alloy aimed at complementing the properties of the established EPDM/PP EA.[53] This NBR/PP EA was engineered for applications in which low swell was required in oils, fuels, and other hydrocarbon media. Table 11 compares the swell of NBR/PP to that of an EPDM/PP of the same hardness, as well as their respective retention of hardness and tensile stress-strain properties. The swell of the NBR/PP EA is much less than that of the EPDM/ PP one; whereas, the latter retains its properties well. Thus, the degree of permissible swell will be highly significant in the material selection process.

Commercial NBR/PP EAs range in hardness from 70 Shore A to 50 Shore D. Table 9 gives some of the basic properties of two of these EAs. These properties parallel those of EPDM/PPs of comparable hardness, with the exception of (1) their lower oil swell and (2) their slightly higher brittle point, the latter determining the lower temperature limit of their elastomeric properties. The brittle point is determined by the glass transition temperature of the rubber phase of the EA, and it is well known that this temperature for EPDM is 20 to 35°C below that of NBR.[54]

The upper temperature service limit for NBR/PP is 125°C, approximately 10 degrees below that of EPDM/PP. This can be understood readily by considering that this limit is determined inversely by the ease with which the elastomer backbone can be attacked by oxygen. The backbone unsaturation in the butadiene units of NBR renders it more susceptible to oxidative attack than does the chemically saturated backbone of EPDM.

4. Natural Rubber/Polypropylene

Pioneering research of the 1970s and early 1980s gave rise to the first commercial natural rubber (NR)/PP EA in the late 1980s.[55,56] Though preceded by other NR/PP compositions,[57] the compositions arising from the work of Coran and Patel[55] were the first true EAs to become commercial. The preceding compositions had a degree of crosslinking well below that of the Coran Patel ones and may be classed as TEO blends.

Like the EPDM/PP EAs, the NR/PP ones are prepared by dynamic vulcanization, yet their morphology is somewhat different.[9] The NR/PP EAs with a low NR/PP ratio have a morphology with a continuous PP phase and a discontinuous (particulate) crosslinked NR phase, similar to the EPDM/PP EAs. As the NR/PP ratio is increased, however, the morphology shifts progressively toward two interpenetrating cocontinuous phases (Figure 23). For the EA to be thermoplastic, it is necessary that the PP phase be continuous. However, the aggregation of the crosslinked NR particles to form an apparently continuous phase cannot be very strong and breaks up at or near the melting point of the continuous PP phase. It is tempting to speculate that in the low ratio NR/PP EAs, the crosslinked NR particles have little or no PP between them and may actually touch, with the attraction between them being weak van der Waals forces. In contrast, the low ratio EPDM/PP EAs will likely have a layer of PP between the crosslinked EPDM particles.

TABLE 11

Resistance of EPDM/PP and NBR/PP Elastomeric Alloys to Different Hydrocarbon Fluids[a]

Fluid/Exposure Conditions[b]	Hardness change Shore A units		% Retention Ultimate Tensile		% Retention Ultimate Elongation		% Retention 100% Modulus		% Weight change	
	EPDM/PP	NBR/PP	EPDM/PP	NBR/PP	EPDM/PP	NBR/PP	EPDM/PP	NBR/PP	EPDM/PP	NBR/PP
Cyclohexane, 168 h/23°C	−9	−7	98	81	102	91	87	80	51.7	2.1
Turpentine 168 h/23°C	−14	−4	70	70	77	83	82	79	50.5	10.6
ASTM No. 1 oil, 168 h/100°C	−9	0	82	103	87	103	94	109	17.3	−10.8
ASTM No. 2 oil, 168 h/100°C	−12	2	79	105	81	97	96	104	33.4	−2.1
ASTM No. 3 oil, 168 h/100°C	−15	−4	71	91	70	89	93	94	48.4	2.1
ASTM Ref. fuel A (isooctane), 168 h/23°C	−11	−3	76	85	83	94	84	86	18.9	−7.0
ASTM Ref. fuel C (isooctane/toluene, 50/50), 168 h/23°C	−14	3	76	61	75	66	79	79	18.3	21.2
Diesel fuel, 168 h/23°C	−11	−11	67	72	74	80	84	84	37.3	13.2

a EPDM/PP and NBR/PP both 80 Shore A hardness.
b ASTM D 471, total immersion.

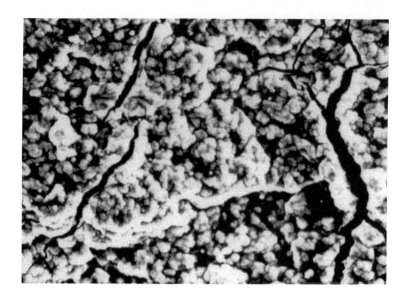

FIGURE 23. Photomicrograph of NR/PP EA.

TABLE 12
Comparison of Ozone Resistance of NR/PP
and EPDM/PP Elastomeric Alloys[a]

	Time to initial crack	Shell crack rating[b]
Natural rubber/polypropene	No cracking	10
EPDM/polypropylene	No cracking	10

[a]   ASTM D 1149 — bent loop, 100 parts per hundred million ozone conc., 72 h exposure, 38°C. Under these test conditions, diene rubbers — NR, SBR, neoprene — fail catastrophically due to massive crack formation.

[b]   0 to 10 scale — 0 denotes catastrophic failure due to crack formation and 10 denotes no perceptible cracks in rubber.

Table 9 gives typical properties of a soft (70 Shore A) and a hard (40 Shore D) NR/PP EA. The tensile stress-strain and set properties are comparable to those of the EPDM/PP and NBR/PP alloys. As expected, the resistance to oil and other hydrocarbon media is inferior; however, it is only moderately so relative to the EPDM/PP. The hydrocarbon resistance of the NR/PP EA has been found clearly superior to EPDM/PP TEO blends, a result of the crosslinking of the NR phase.

The useful temperature range of the NR/PP EAs extends from a low of approximately −50°C to a high of approximately 100°C, the lower limit being determined by the brittle point and the upper by hot air aging. In a variety of fluids, the NR/PP has been found only slightly inferior to the EPDM/PP. It is surprising that the NR/PP EA has good resistance to ozone attack (Table 12); whereas, thermoset NR is so susceptible to it. The explanation lies in the molded NR/PP articles having a thin skin of PP (a saturated polymer) which protects the olefinic double bonds of the NR backbone from ozone attack.

The NR/PP EA is slightly lower in elevated temperature performance and resistance to environmental and fluid attack than either the EPDM/PP or the NBR/PP. Yet, there is a

TABLE 13
Relative Vapor Permeability of Butyl/PP TPE
and Thermoset Butyl Rubber

		Relative permeability	
Permeant gas	ASTM test	Butyl thermoset[a]	Butyl/PP TPE
Air	D 1434	1.00	1.0
Oxygen	D 3985	1.00	1.2
Water vapor	F 1249	1.00	1.1

[a] Arbitrarily set at 1.00 to serve as comparison base.

distinct niche for NR/PP in the elastomeric materials spectrum due the low cost and abundant supply of natural rubber. Thus, EPDM/PP and NBR/PP fill performance niches of 135 and 125°C, respectively, and NR/PP fills a niche at 100°C. This gives the materials technologists a range in the desired cost and performance as they seek to fill a need for an elastomeric material by compromising the two basic parameters of cost and performance.

5. Butyl Rubber/Polypropylene

Appearing commercially in the early 1990s was an EA based on the dynamic vulcanization of a butyl rubber (IIR)/PP system as previously reported in the scientific literature.[58-60] This IIR/PP EA has a morphology (fine crosslinked rubber particles in continuous PP matrix) very similar to the EPDM/PP EA.[61] Its properties are also quite similar as shown by the data in Table 9. The useful temperature range of this IIR/PP EA is very comparable to that of the EPDM/PP one, ranging from −55 to 135°C. The fluid resistances of the two EAs are also quite close, as one would suspect from the chemical similarity of butyl and EPDM rubber.

The principal advantage of the IIR/PP EA is its low permeability to gases, such as oxygen, nitrogen, water vapor, etc. Whereas the EPDM/PP EA has a permeability an order of magnitude greater than butyl/halobutyl thermoset rubbers, the IIR/PP EA has a permeability that is quite comparable (Table 13) to thermoset butyl rubber. This low permeability will likely enable the IIR/PP EA to find a variety of applications (e.g., bladders, air containment vessels, stoppers, etc.) where low permeability is essential.

VI. PROCESSING AND FABRICATION

In fabricating useful articles from the different commercial TPEs, it is of basic importance that only thermoplastic and not thermoset rubber processing methods be considered. Thus, the fabrication of a styrenic, polyurethane or EA TPE should be viewed as essentially fabricating a rigid thermoplastic such as PP, PS, or nylon. In some cases it is possible to use thermoset rubber processing equipment (e.g., extrusion) with TPEs. Though possible, it is seldom recommended due to the narrow processing window that renders adequate process control very difficult, if not impossible. The use of thermoplastic equipment and techniques will give a much wider processing window that is far more practical to control.

A. EFFICIENCY OF THERMOPLASTIC PROCESSING

The speed and efficiency of thermoplastic processing has, for three decades, provided the principal driving force behind the rapid growth of TPEs, with the bottom line result being lower cost per fabricated part (as illustrated in Table 2). The functional basis for this economy has been discussed in a previous section. The skill and expertise of the rubber compounders of today have left very few performance gaps in the spectrum of thermoset rubbers; however, the inefficiency and variability of thermoset rubber processing leave much room for improvement in the cost per fabricated rubber part.

In most cases, the material cost for a TPE (weight basis) will exceed that of thermoset rubber. Counterbalancing this disadvantage is an immense cost saving (0.60 to $2.00/kg of fabricated rubber parts) from the greater economy of thermoplastic processing. Often the use of a TPE can justify a redesign of the desired part to optimize the performance of new material, with a concomitant reduction of needed material for proper part performance. For more than a century, rubber companies have tended to solve their problems through compositional changes (or chemistry); whereas, plastics companies have solved their problems primarily through engineering and design. The use of a TPE will enable rubber part manufacturers to use both of these powerful problem solving techniques.

B. ADDITIVES

Contrary to thermoset rubber and soft PVC practice, the compounding of TPEs is generally not done and the use of additives to them is discouraged. This approach is different from the normal experience of fabricators of thermoset rubber and plasticized PVC. Most commercial TPEs have been formulated for an optimum balance of properties and the addition of more than a small percentage of a different material will most likely detract from one or more of the desirable properties of the TPE. Additives to a commercial TPE should only be used where there is a compelling need for them and their effect on the performance properties clearly defined.

The principal exception to this "no additives" practice is the use of colorants. In numerous cases a colored TPE part is desired. Most commercial TPEs can be given a desired color — red, blue, green, yellow, white, etc. — by blending to homogeneity with the desired inorganic pigment, preferably as a concentrate in a thermoplastic carrier compatible with the TPE. Thus, PP is a suitable carrier for TEOs and EAs based on this thermoplastic.

C. REGRIND AND RECYCLE

The recent accelerating need for reducing the amount of solid waste in our society has further enhanced the status of TPEs relative to thermoset rubbers.[62] To properly address this enhancement, it is necessary to precisely define "regrind" and "recycle". Regrind is the reprocessing of in-process scrap from a fabrication operation and recycle is the reprocessing and subsequent reuse of a fabricated article at the end of its useful lifetime. The thermoplastic nature of TPEs makes them amenable to both regrinding and recycling. With thermoset rubbers, neither regrind nor recycle is possible without severe chemical modification of the material.

The compatibility of TPEs with the current need for protecting the environment by decreasing our amount of solid waste has recently been articulated.[63] It is well established that the regrinding of a TPE generates essentially no significant change in the material.[46] On the other hand, the recycling of used TPE parts will only be practical if the material has not deteriorated (from oxidation, chemical attack, solvent attack, etc.) significantly. In the U.S.A., a massive effort is now underway[64] to recycle automotive parts. It is probable that other industries will be impacted by a similar effort.

D. MOISTURE AND DRYING

Most TPEs, even those of hydrocarbon chemistry, can easily pick up sufficient moisture (Figure 24) to give processing problems in both extrusion and molding operations. This moisture pickup can give poor rubber parts with unacceptable dimensions, porosity, and surface roughness. In high humidity conditions, injection molding of an undried TPE can be dangerous, due to the buildup of high pressure steam, particularly in a hot runner molding system.

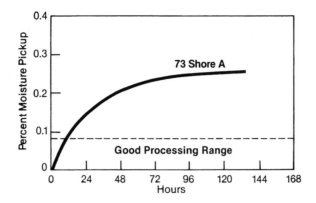

FIGURE 24. Moisture pickup curve for EA.

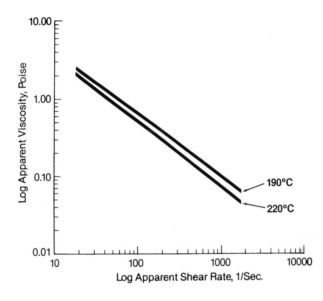

FIGURE 25. Elastomeric alloy viscosity dependence on shear rate and temperature, 55 Shore A.

Most TPE suppliers recommend drying for 2 to 6 h at 70 to 100°C immediately before processing. The moisture removal should use only dried, filtered air. The drying of regrind should be 1 to 3 h longer than virgin materials. In addition, it is important to dry color concentrates that are added to the TPE.

E. RHEOLOGY

Molten TPE have a rheology (flow behavior) which is unusual, complex, and highly sensitive to shear (extremely non-Newtonian). Figure 25 gives the variation of the viscosity of two EA TPEs over shear rates of three orders of magnitude. At high shear rates (as found in injection molding) the materials flow well, but at low shear rates they flow poorly. This high viscosity at low shear rates enables (1) the rapid demolding of injection molded parts (where the interior of the molded part is still molten), (2) the stability of a parison in blow molding, and (3) the very good dimensional stability of extruded profiles.

In contrast to its shear sensitivity, the flow of a TPE is much less sensitive to changes in barrel or melt temperature brought about by conductive rather than shear heating. Figure 26 depicts the viscosity variation of two EA TPEs with temperature. TPE molding operations

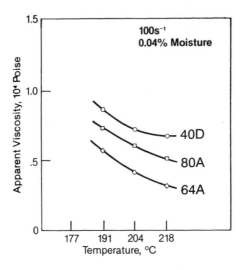

FIGURE 26. Viscosity variation of different hardness EA with temperature.

FIGURE 27. Injection-molding of TPE parts.

should exploit these relative variations of viscosity with shear rate and melt temperature, by increasing shear (higher pressure) rather than relying on conductive heat for best flow. This is different from the usual practice for thermoplastics, such as PP or nylon, where flow is impacted more by conductive heat than higher injection speeds and pressures.

F. INJECTION MOLDING

The most important single method for fabricating TPE parts is injection molding (Figure 27), which exploits the various processing advantages of TPEs over thermoset rubbers.[65]

The lower cycle times, reuse of scrap, and virtual absence of compounding make this method a preferred one for fabricating TPE parts, ranging in size from a 0.10 g medical part to a 3000 g automotive fire wall sound-deadening mat. Many volume injection molders of TPEs now use hot runner molding, a rapidly growing method which generates essentially no in-process scrap.[66] Both dual and coinjection molding have been found suitable for TPEs and take advantage of the capability of TPEs to easily bond (by heat welding) to compatible thermoplastics.[67]

The equipment and techniques commonly used for polyolefin injection molding should be suitable for most TPEs. The clamping pressure of the molding equipment should be in the 40 to 70 MPa range over the total projected shot area, with a barrel capacity of six shots or less. Nozzles, sprues, runners, and gates should be moderate in size. Minimum shrinkage on cooling and strong weld lines result from good mold packing and use of balanced flow into the cavity so pressure drop and flow distance are equalized throughout the part during molding. Short runners will minimize the pressure drop, with a full round configuration (minimum surface/volume ratio) preferred. Removal of entrapped air from the mold is facilitated by vents at least 0.03 mm deep, located as far as possible from the gate and where a weld line occurs. Full peripheral venting is recommended for the EA TPEs. At the shear rates normally used in injection molding (>500 s^{-1}), most TPEs flow well. The mold should be filled as rapidly as possible (usually 4 s or less), followed by a short holding time (1 to 10 s) and subsequent cooling (8 to 60 s, depending on part size). The necessary cooling is for the part to develop a sufficiently thick skin for extraction from the mold without distortion, with the interior still molten. Mold releases are generally not needed and are discouraged to eliminate buildup on the mold.

Melt temperatures should be 20 to 50°C above the TPE melting point. With proper mold packing and melt temperature, the mold shrinkage should be in the 1.5 to 2.5% range, depending on cross-sectional thickness and flow length. Control of mold shrinkage, generally, can be sufficient for dimensional tolerances two to three times closer than with thermoset rubber molding.

The various suppliers of TPEs provide detailed injection molding and mold design recommendations for the materials they market and the reader should contact the supplier for the latest information.

G. EXTRUSION

Sheeting, tubing, and complex profiles can readily be fabricated by simple extrusion of TPEs, and coextrusion with a compatible material if the two materials have similar melting points.[65] Cross-head extrusion is widely used to apply TPEs over a reinforced hose or electrical wire and cable.

Only a thermoplastic extruder is recommended for TPE extrusion. The extruder should have a length to diameter (L/D) ratio of the barrel and screw of at least 20/1 and preferably 24/1 to 32/1. Conductive heating capability should be 170 to 250°C. Polyolefin type screws with a 2:1 to 4:1 compression ratio and no screw cooling are commonly used with TPEs, though other designs — pin mixing, Maddock mixing, flighted barrier, and other — have been successfully used. To generate even flow and pressure and to insure clean extrudate, screen packs of 20, 40, and 60 mesh screens are commonly used.

Dimensional control of the TPE extrudate requires control of die swell, which generally, is less than that of rigid thermoplastics and carbon black-loaded thermoset rubbers. This swell (Figure 28) increases with increasing extrusion shear rate and hardness of the TPE, and decreases with increasing temperature.

The TPE extrusion, unlike thermoset rubber extrusion, does not require a final vulcanization step. Thus, only rapid cooling of extrudate to room temperature is needed, obviating the need for capital intensive vulcanization equipment.

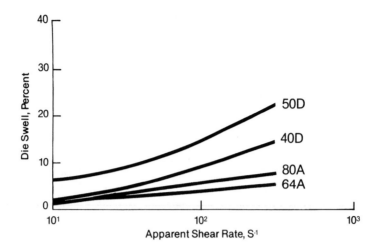

FIGURE 28.   Die swell at 204°C for different hardness EA TPEs.

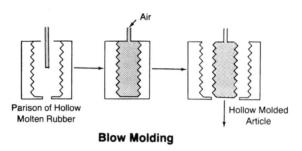

FIGURE 29.   Schematic description of blow molding of TPE.

The preferred method for fabricating TPE sheeting is extrusion. Calendering can be used for generating TPE sheets, but it is a more difficult process due to the lower shear rates in calendering and the viscosity shear dependence of TPEs. The TPE sheet can be extruded in uniform thicknesses from 4 mm down to 0.2 mm, and widths of up to 3 m. They can also be extruded into blown film with a thickness of 0.05 mm for harder grades and 0.2 mm for the softer grades.

H. COMPRESSION MOLDING

Like calendering, compression molding is possible with TPEs under some conditions, but is limited because of the low shear rates involved, and is not economical compared to injection molding. Compression molding is used primarily in the preparation of standard laboratory test specimens. Compression molding from preformed slabs is possible at melt temperatures of 175 to 200°C and pressures of 1.5 to 4.0 MPa. For total mold filling, a 10% excess of material should be used, generating flash along the part periphery.

I. BLOW MOLDING

Hollow rubber articles — bottles, boots, bellows, etc. — can readily be blow molded from TPEs, with very significant cost savings compared to injection molding similar parts from thermoset rubber. The thermoset rubber part must be injection-molded over a collapsible core that has to be removed by hand. Both extrusion and injection-blow molding are feasible.[68,69] Figure 29 schematically depicts extrusion blow molding. The parison (a hollow

TABLE 14
Cost Comparison for Blow Molding TPE Automotive
Rack and Pinion Boots vs. Injection Molding from
Thermoset Rubber

Material	Neoprene	Copolyester	Elastomeric Alloy
Part weight, g	125	65	54
$/kg	2.64	5.83	3.19
$/part	0.330	0.380	0.172
$ scrap or regrind/part	0.050	0.015	0.015
Total material cost, $	0.380	0.395	0.187
Molding Cost			
Injection	0.420	—	—
Blow	—	0.100	0.100
Total manufacturing cost, $	0.420	0.100	0.100
Grand total cost/part, $	0.800	0.495	0.287

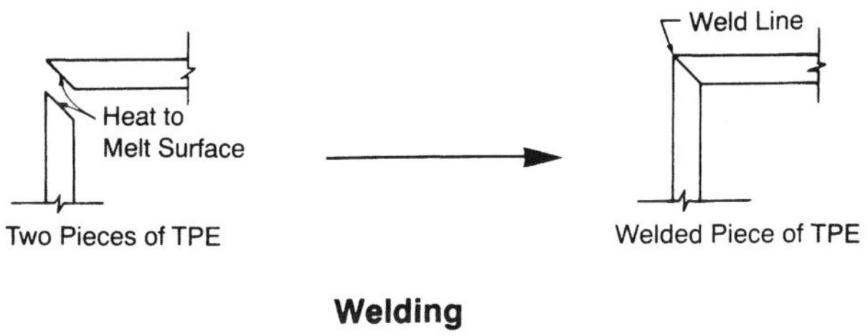

Welding

FIGURE 30. Heat welding to a compatible thermoplastic material. The surfaces are heated to fusion and then rapidly joined together.

tube of molten TPE) is extruded vertically downward into the open mold. Blowing takes place as the mold is clamped around the parison, expanding the molten TPE against the water-cooled mold. When the part has cooled sufficiently (after 10 to 240 s) to have structural integrity, the mold opens and the part is removed.

The blow ratio (mold diameter to parison diameter) should be as small as practical. Melt temperatures are essentially those of TPE extrusion and injection molding. The high TPE viscosity at low shear rates (Figure 25) enables the molten parison to retain its shape and integrity prior to blowing and the exploitation of this efficient fabrication method that is not practical for thermoset rubbers. Table 14 gives a cost analysis for the blow molding of convoluted rack and pinion automotive boots from two different TPEs vs. the injection molding of the thermoset rubber previously used.

J. OTHER PROCESSING METHODS

The thermoplastic nature of TPEs enables them to be heat welded to themselves or to another compatible material (Figure 30). Heat welding is both rapid (4 to 8 s) and simple. The surfaces to be bonded must be heated above their melt point(s), rapidly joined, and cooled until solidification takes place. A variety of methods — hot air, radiation, vibration, etc. — may be used to melt the surfaces. Properly done, the weld strength will be 70 to 80% of the tensile strength of the material.

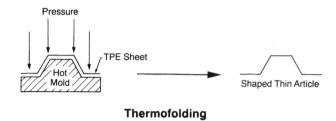

Thermofolding

FIGURE 31.   Thermoforming of shaped rubber article from extruded TPE sheet.

A principal virtue of heat welding is that it avoids the need for an adhesive system for bonding the TPE. Adhesive systems and adhesion techniques have been developed for bonding TPEs to dissimilar materials, though the process is more complex than heat welding.[70] Proper surface cleaning is critical for a good adhesive bond.

Thermoforming (Figure 31) can be readily utilized to fabricate shaped articles from sheets of extruded TPE. The sheet is heated on a shaping mold to a temperature 10 to 40°C above its softening point. Either external or vacuum pressure is used to conform the molten sheet to the mold.[71] Upon removal it has the desired shape. With decreasing TPE hardness, thermoforming becomes progressively more difficult, and is not practical for the softest TPES.

Extrusion foaming is a TPE processing area that has seen significant progress in the recent past.[72] Processes have been developed for extruding foams with densities 10 to 90% that of the solid TPEs. The blowing agent can be chemical (azodicarbamide), physical (chlorofluorocarbon), or a combination of the two. A recent advance in TPE extrusion foaming uses water as a mechanical foaming agent, eliminating the need for any type of chlorofluorocarbon.[73]

VII. INNOVATING WITH TPEs

During the past half century, one of the greatest changes to impact the nontire segment of the rubber industry has been TPEs.[74] Clearly, TPEs are proving to be a change of the highest magnitude. Innovation can properly be defined as the commercialization of invention or of new technology. The invention of the block copolymer and EA TPEs has been followed by commercialization efforts epitomizing the entrepreneurial development of the TPE business.

The first step in the commercialization process was the selection of the prime area on which to concentrate the marketing effort.[46] For the EA TPEs this was nontire thermoset rubber. For more than a century, the rubber industry has tended to solve technical problems through compounding or compositional change; whereas, the plastics industry has tended to use more engineering and design. The commercialization of TPEs has used a combination of both of these different approaches. Quite early, rubber processors were advised to use thermoplastic equipment and techniques in the fabrication of TPE articles. In most cases they possessed neither the equipment nor the techniques. As mentioned previously, the use of thermoplastic techniques gave a far more reliable fabrication process. It also provided an opportunity for the thermoplastic processor to enter the rubber products market.

A basic part of the TPE commercialization effort has been "end-use selling", marketing the material to the end-user of rubber parts as well as the rubber fabricator.[75] This has entailed the effectuation of a close working relationship between the TPE supplier and (1) the technologist selecting the material, (2) the engineer specifying the part, (3) the fabricator molding or extruding the rubber part, (4) the mold or die designer, (5) the mold or die maker, and (6) production people in the fabricator's plant. Involved is a high level of technical assistance from the TPE supplier — from the initial material selection to commercial plant processing.

The innovation process with a TPE contains, either explicitly or implicitly, the following sequential steps:

1. Preliminary assessment to delineate suitability of a TPE
2. Selection of candidate TPEs
3. Testing and selection of the prime candidate TPE
4. Generation of prototype parts from prime candidate
5. Evaluation of prototype parts
6. Design of part to optimize performance of TPE chosen
7. Design of mold or extrusion die
8. Production start-up
9. Commercial production

It is quite natural that the automotive industry, with its hundreds of rubber parts per vehicle, would be the first to be penetrated by TPEs.[76] Parts fabricated from TPEs are found throughout today's new vehicles — bumpers, fascias, hose, tubing, gaskets, under-the-hood parts, weather stripping, etc. Close on the heels of automotive was the penetration of the mechanical rubber goods (MRG) market by TPEs.[77] An MRG is a rubber part used in the assembly of a nonautomotive useful device such as a household appliance, power tool, telephone, toy, or office machine.

Building construction has also provided a ready market for TPEs, with window glazing, plumbing gaskets, and expansion joints leading the penetration. The plumbing applications, as well as food-contact and medical uses, exploit the greater "cleanliness" of TPEs over conventional sulfur-vulcanized thermoset rubber.[78] For these uses, the latter contains chemical residues, not found in most TPEs, which are capable of leaching out and giving significant problems with toxicity and safety.

The many thousands of uses of a broad variety of TPEs throughout the industrialized world offer evidence of their suitability for applications in the nontire industrial rubber market. Many of these uses resulted from following the nine steps listed previously. There is much reason to expect that the existing uses of TPEs will be joined by thousands more — not to mention many new TPEs — before the end of the 20th century. The star of TPEs has already risen and will continue to rise even higher, at least for the next decade, and likely well beyond.

ACKNOWLEDGMENTS

The authors wish to thank the many contributors of the information contained within this chapter, Jennifer L. Digiantonio for her capable assistance in preparation of the manuscript, and Advanced Elastomer Systems, L.P. for permission to publish.

REFERENCES

1. **ASTM D 1566, American Society for Testing and Materials,** *Annual Book of ASTM Standards,* Vol. 9.01, 1992.
2. **Walker, B. M., Ed.,** *Handbook of Thermoplastic Elastomers,* Van Nostrand Reinhold, New York, 1979.
3. **Coran, A. Y. and Patel, R. P.,** *Rubber Chem. Technol.,* 53, 140, 1980.
4. **Rader, C. P. and Stemper, J.,** *Progress in Rubber and Plastics Technology,* RAPRA Technology, Vol. 6, No. 1, 50, 1990.

5. **Whelan, A. and Lee, K. L., Eds.,** *Developments in Rubber Technology — 3. Thermoplastic Rubbers,* Applied Science Publishers, London, 1982.
6. **Legge, N. R., Holden, G., and Schroeder, H. E., Eds.,** *Thermoplastic Elastomers — A Comprehensive Review,* Hanser Publications, Munich, 1987.
7. **Walker, B. M. and Rader, C. P., Eds.,** *Handbook of Thermoplastic Elastomers, Second Edition,* Van Nostrand Reinhold, New York, 1988.
8. **Arnold, R. L. and Rader, C. P.,** in *Handbook of Plastics, Elastomers and Composites,* C. Harper, Ed., McGraw-Hill, New York, 1992, ch. 7.
9. **Rader, C. P. and Abdou-Sabet, S.,** in *Thermoplastic Elastomers from Rubber Plastic Blends,* S. K. De and A. K. Bhowmick, Eds., Ellis Horwood, Chichester, England, 1990, ch. 6.
10. **West, J. C. and Cooper, S. L.,** *Nippon Gomu Kyokaiski,* 45, 1984.
11. **Rader, C. P.,** Thermoplastic elastomers, *RAPRA Review Reports,* R. Meredith Ed., Vol. 1, No. 3, Pergamon Press, London, 1987.
12. **Thorn, A. D.,** Thermoplastic elastomers: a review of current information, Rubber and Plastics Research Association of Great Britain, Shawbury, U.K., 1980.
13. **Holden, G.,** in *Rubber Technology, Third Edition,* M. Morton, Ed., Van Nostrand Reinhold, New York, 1987, ch. 16.
14. Laboratoire de Recherche et de Controle du Caoutchouc, *Etude Bibliographique des Caoutchouc Thermoplastiques, Rapport Technique,* 122, Moutroque, France, 1980.
15. **Nicaud, J.,** *Rev. Inst. Fr. Pet.,* 44(2), 245, 1989.
16. **Legge, N. R., Davison, S., De Le Mare, H. E., Holden, G., and Martin, M. K.,** in *Block Polymers and Related Materials,* R. W. Tess and G. G. Poehlein, Eds., *ACS Symp. Ser.* 285, American Chemical Society, Washington, D.C. 1985.
17. Thermoplastic Elastomers Symp., 121st Nat. Meeting Rubber Division, American Chemical Society, Philadelphia, 1982.
18. Thermoplastic Elastomers Symp., 127th Nat. Meeting Rubber Division, American Chemical Society, Los Angeles, 1985.
19. Thermoplastic Elastomers Symp., 134th Nat. Meeting Rubber Division, American Chemical Society, Cincinnati, 1988.
20. Thermoplastic Elastomers and Engineering Properties and Structure Symp., ANTEC 1988, 46th Ann. Tech. Conf., Society of Plastics Engineers, Atlanta, 1988.
21. Thermoplastic Elastomers — Threat or Opportunity?, Symp. Organized by *Eur. Rubber J.* and RAPRA Technology Ltd., London, 1988.
22. Thermoplastic Elastomers — II. Processing for Performance Symposium, Organized by *Eur. Rubber J.* and RAPRA Technology Ltd., London, April 1989.
23. TPE '88, First Int. Conf. Thermoplastic Elastomer Markets and Technol. Conducted by Schotland Business Research, Inc. Orlando, FL March 9–11, 1988.
24. Thermoplastic Elastomers Symp. 138th Nat. Meeting of the Rubber Division, American Chemical Society, Washington, D.C., 1990.
25. **Rader, C. P.,** Thermoplastic Elastomers — Their Future is Bright, keynote address, Int. Rubber Conf., Sydney, Australia, October, 1988.
26. **Allinger, N. L., et al.,** *Organic Chemistry,* Worth Publishers, 1971.
27. **International Rubber Study Group,** *Rubber Stat. Bull.* 46, No. 5, 1992.
28. **Rader, C. P.,** Thermoplastic Elastomers: A Major Innovation in Materials, A. D. Little, *Spectrum Ser.* 1991.
29. **Schollenberger, C. S., Scott, H. S., and Moore, G. R.,** *Rubber World,* 137, 549, 1958.
30. **Legge, N. R.,** *Chemtech* , 13, 630, 1983; Brown, M. and Witsieppe, W. K., *Rubber Age,* 104, 35, 1972.
31. **Coran, A. Y. and Patel, R. P.,** Presented at the 116th Meeting of the Rubber Division, American Chemical Society, Cleveland, OH, October 23–26, 1979.
32. **Canevarolo, S. V. and Birley, A. W.,** *Br. Polym. J.,* 19, 43, 1987.
33. **Gohil, R. M.,** *Coll. Polym. Sci.,* 264, 847, 1986.
34. **Pakula, T., et al.,** *Macromolecules,* 18, 1294, 1985.
35. **Mitani, T., et al.,** *Kobunski Ronlsun,* 40, 653, 1983.
36. **Holden, G.,** in *Handbook of Thermoplastic Elastomers, Second Edition,* B. M. Walker and C. P. Rader, Eds., Van Nostrand Reinhold, New York, 1988, chap. 2.
37. **Sheridan, T. W.,** in *Handbook of Thermoplastic Elastomers, Second Edition,* B. M. Walker and C. P. Rader, Eds., Van Nostrand Reinhold, New York, 1988, chap. 6.
38. **Abdou-Sabet S. and Patel, R.,** Presented at 138th Meeting of the Rubber Division, American Chemical Society, Washington, D.C., October 9–12, 1990; *Rubber Chem. Technol.,* 64, 769, 1991.
39. **Umeda, I. and Makino, K.,** *Jpn Plast. Age,* July-August, 1985, p. 27, 1985.
40. **Bregar, R.,** *Plast. News,* May 21, 8, 1990.
41. **Coran, A. Y. and Patel, R. P.,** *Rubber Chem. Technol.,* 54, 892, 1981.

42. **Coran, A. Y., Patel, R. P., and Williams, D.,** *Rubber Chem. Technol.,* 55, 116, 1982.
43. **Wallace, J. G. and Paonessa, L.,** *J. Elastomers Plast.,* 137, 1989.
44. **Coran, A. Y., Das, B., and Patel, R. P.,** U.S. Patent 4,130,535, 1978.
45. **Abdou-Sabet, S. and Fath, M. A.,** U.S. Patent 4,311,628, 1982.
46. **O'Connor, G. E. and Fath, M. A.,** *Rubber World,* 25; December 1981; *Rubber World,* 26, January, 1982.
47. **ASTM D 5046, American Society for Testing and Materials,** *Annual Book of ASTM Standards,* Vol. 8.03, 1992.
48. **Burton, T., Delanaye, J. L., and Rader, C. P.,** *Rubber Plast. News,* 16, December 26, 1988.
49. **Rader, C. P. and Kear, K. E.,** *Rubber Plast. News,* May 6, 1986.
50. **Barton, A. F. M.,** *Handbook of Solubility Parameters and Other Cohesion Parameters,* CRC Press, Boca Raton, FL, 1983, 1.
51. **Coran, A. Y. and Patel, R. P.,** *Rubber Chem. Technol.,* 56, 1045, 1983.
52. **Coran, A. Y., Patel, R. P., and Williams-Headd, D.,** *Rubber Chem. Technol.,* 58, 1014, 1985.
53. **Abdou-Sabet, S., Wang, Y. L., and Chu, E. F.,** *Rubber Plastics News,* 20, November 4, 1985.
54. **Morton, M.,** *Rubber Technology, Third Edition,* Van Nostrand Reinhold, New York, 1987, chap. 9 and 11.
55. **Coran, A. Y.,** in *Thermoplastic Elastomers — A Comprehensive Review,* N. R. Legge, G. Holden, and H. E. Schroeder, Eds., Hanser Publications, Munich, 1987, chap. 7.
56. **O'Connor, G. E.,** TPE 1988, First Int. Conf. Thermoplastic Elastomer Markets-and Technol., Schotland Business Research, Orlando, FL, March 9–11, 1988.
57. *Rubber World,* 191, 49, 1985.
58. **Payne, M. T. and Wang, Y. L.,** Presented at 136th Nat. Meeting of Rubber Division, American Chemical Society, Detroit, MI, October 17–20, 1989.
59. **Coran, A. Y. and Patel, R. P.,** U.S. Patent 4,130,534, 1978.
60. **Coran, A. Y. and Patel, R. P.,** *Rubber Chem. Technol.,* 55, 116, 1982.
61. **Ouhadi, T. and Wang, D. S. T.,** Presented at Fifth Int. Schotland Conf. on Thermoplastic Elastomer Markets and Products, Luxemborg, October 22–23, 1991.
62. **Gonzalez, E. A., Purgly, E. P., and Rader, C. P.,** Presented at 140th Meeting of Rubber Division, American Chemical Society, Detroit, MI, October 8–11, 1991.
63. **Purgly, E. P., Gonzalez, E. A., and Rader, C. P.,** Presented at ANTEC 1992, Society of Plastics Engineers, Detroit, MI, May 5, 1992.
64. **SAE J 1344,** *Marking of Plastic Parts,* Society of Automotive Engineers, Warrendale, PA, 1992.
65. **Rader, C. P. and Richwine, J. R.,** *Rubber Plast. News,* February 11, 1985.
66. **Rosato, D. V. and Rosato, D. V.,** *Injection Molding Handbook,* Van Nostrand Reinhold, New York, 1986.
67. **Digiantonio, R. J. and Lawrence, G. K.,** Presented at ANTEC 1992, Society of Plastics Engineers, Detroit, MI, May 4, 1992.
68. **Raia, T. M., Jr., Richwine, J. R., and Rader, C. P.,** Presented at 129th Nat. Meeting of Rubber Division, American Chemical Society, New York, April 8–11, 1986.
69. **D'Auteuil, J. G., Peterson, D. E., and Rader, C. P.,** *J. Elastomers Plast.,* 265, 1989.
70. **Rader, C. P.,** in *Handbook of Thermoplastic Elastomers, Second Edition,* B. M. Walker and C. P. Rader, Eds., Van Nostrand Reinhold, New York, 1988, p. 130.
71. **Van Issum, E., Peterson, D. E., and Weider, B. K.,** ANTEC 1986, Society of Plastics Engineers, Boston, April 28, 1986.
72. **Peterson, D. E. and Agrawal, P. D.,** *Cell. Polym.,* 7, 475, 1988.
73. **Dumbauld, G. L.,** U.S. P. 5,070,111, 1991.
74. **School, R.,** Presented at 134th Meeting of Rubber Division, American Chemical Society, Cincinnati, OH, October 18–21, 1988.
75. **O'Connor, G.E.,** Presented at 128th Meeting of Rubber Division, American Chemical Society, Cleveland, OH, October 1–4, 1985.
76. **Killgoar, P. E.,** in *Handbook of Thermoplastic Elastomers, Second Edition,* B. M. Walker and C. P. Rader, Eds., Van Nostrand Reinhold, New York, 1988, chap. 10.
77. **Muhs, J. H.,** in *Handbook of Thermoplastic Elastomers, Second Edition,* B. M. Walker and C. P. Rader, Eds., Van Nostrand Reinhold, New York, 1988, chap. 12.
78. **Williams, J.,** in *Handbook of Thermoplastic Elastomers, Second Edition,* B. M. Walker and C. P. Rader, Eds., Van Nostrand Reinhold, New York, 1988, chap. 14.

Chapter 15

Markets for Thermoplastic Elastomers

Rudy J. School

TABLE OF CONTENTS

0-8493-4401-8/93/$0.00 + $.50

I. INTRODUCTION

The worldwide consumption of thermoplastic elastomers (TPEs) will approach 1 million metric tons—2.24 billion pounds—by 1995, making their usage a $2.6 billion market.

The materials, now in their fourth decade of usage, continue to show rapid growth rates of between 7 and 8% per year compared to thermoset rubbers (virtually no growth) and many thermoplastic polymers (4% per year through 1995). The size of the 1995 market will be 67% greater than the 1991 global market, which was estimated by numerous experts to be between 600,000 and 650,000 mt. Table 1 lists market sizes in pounds for the U.S., Europe, and Japan for 1985, 1990, and 1995.

Of that total, end-use applications in North America will account for 963 million pounds (430,000 mt) by 1995 compared with 618 million pounds (276,000 mt) in 1991.[1] This represents 43% of global usage, a greater amount than in 1988, when 37% of global usage, a greater amount than in 1988, when 37% of global TPE usage was centered in North America.[2]

Since their introduction in the early 1960s, the growth of TPE usage has centered on transportation goods ranging from footwear to components on automobiles. Automotive parts and replacements for mechanical rubber goods remain the major end-use markets for TPEs in the 1990s. Automotive applications will represent a market valued at $255.5 million in 1994 compared with $173.5 million in 1989.[3]

While many studies predict growth rates of 8 to 12% annually for many types of TPEs, the explosive growth does not mirror the past growth of the material, especially in the early years of development. The TPE industry grew slowly following the invention in Europe of the first thermoplastic urethane, or TPU, by Bayer A.G. in the 1930s. The first U.S. based production of TPU occurred in 1959. The next TPE was marketed in the 1960s—Shell Chemical Company's Kraton® line of styrenic block copolymer (SBS) TPEs.[1]

Long periods of time passed between major commercial introductions of these various types of materials, which have the properties of thermoset elastomers and the processing characteristics and economies of thermoplastics. Olefinic TPEs came on the scene in the early 1970s, but the next major surge in technology occurred in 1981 when Monsanto Company introduced Santoprene.®

That product, which is olefinic in nature, is made by a dynamic vulcanization process in which EPDM is cured during production of the TPE. Because of major differences in the performance of the product and properties compared with other olefinics, the industry now classes these types of materials as thermoplastic vulcanizates.

The debut of Santoprene® also started the fast growth of the TPE industry. Within a decade Santoprene® would be the basis of one TPE empire valued at $100 million and a joint venture between two polymer giants—Monsanto and Exxon Chemical Company.

Atochem Polymers, Inc. followed the Santoprene® introduction with a polyamide-based TPE in 1982, naming its product Pebax.® A few years later Du Pont marketed a halogenated polymer—a single-phase "melt processible rubber"— claiming to be the first TPE capable of being processed on either rubber or plastics machinery. Trade named Alcryn®, the product languished somewhat while its processing abilities were honed, but by the early 1990s it had found many market niches in the world of thermoplastic elastomers.

In addition, specialty TPEs have emerged, including silicone-modified styrenic block copolymer (SBC) products marketed under the C-Flex® trade name by Concept Polymers, Inc. Daikin Industries, Ltd. of Japan first announced, in 1988, the development of a thermoplastic fluoroelastomer (Dai-el®),[1] although the first reports of the product being available in North America did not appear until 1991.

Plastics processors have led the move to thermoplastic elastomers as replacements for other materials—mainly at the expense of manufacturers of rubber products.

TABLE 1
Market Sizes for Thermoplastic Elastomers

Market region	Market size in pounds (million pounds)			Market size in dollars ($ million)		
	High	Low	Average	High	Low	Average
Worldwide						
1985	1110.0	1040.0	1080.0	—	—	—
1990	1470.0	1030.0	1370.0	2000.0	1600.0	1870.0
1995	2240.0	1980.0	2060.0	—	—	2660.0
2000	—	—	3640.0	—	—	—
North America						
1985	—	—	—	—	—	—
1990	618.0	543.9	579.9	—	—	836.5
1995	963.0	791.8	891.6	1600.0	1200.0	1400.0
U.S.						
1985	392.0	325.5	358.8	—	—	—
1990	—	—	546.0	—	—	—
1995	—	—	799.0	—	—	—
Europe						
1985	515.2	515.0	515.1	—	—	—
1990	617.4	533.1	565.5	—	—	—
1995	730.4	703.4	716.9	—	—	—
Japan/Far East						
1985	201.6	196.0	198.8	—	—	—
1990	308.7	250.9	279.8	—	—	—
1995	—	—	343.7	—	—	—

From a compilation of data from numerous industry sources and market studies.

For those end-users with global manufacturing operations, TPEs offer an advantage because compounds based on thermoplastic elastomers do not have to be certified in each country as do the thermoset rubber compounds; they are available worldwide as standard grades. Processors only have to empty the pellets from the bag into the processing equipment.

In addition to not needing factory-based compounders to mix and formulate TPE compounds as you do for thermoset rubbers, the materials are capable of being reused—a very critical advantage with the move in the 1990s to take better care of the environment.

The balance of this chapter will examine the historical market development for thermoplastic elastomers up to 1983, TPE markets during the rapid growth period of 1983 to 1989, and the decade of the 1990s. It will also examine in more detail the future expectations for market penetration and growth.

This article discusses only the marketing aspects of TPEs. Legge[4] provides an excellent review of precursor research that led to the development of TPEs and divides the R&D growth into three time intervals of accomplishments.

II. THE BIRTH AND DEVELOPMENT OF THE TPE INDUSTRY (1937–1983)

Thermoplastic elastomers result from a blend of technologies. Scientists have combined the elasticity required for thermoset rubber applications with the melt flow properties and reusability of thermoplastics to create a class of materials that are a hybrid of both.

The first use of rubber dates to the early Mayan civilization in Mexico, where ruins dating to 500 B.C. indicate that the indians had a court game similar to basketball.[5] Spanish explorers also described rubbery materials in the 15th and 16th century, and in 1770 Joseph

Priestley named the material ''rubber'' because the gummy material rubbed out lead pencil marks.

It wasn't until 1892 that scientists first synthesized isoprene, or natural rubber. Germany produced methyl rubber during World War I, but it wasn't until the late 1930s that research produced the styrene rubbers (SRs) still in use today. Emulsion polymerization was developed in 1923, and the 1930s saw the invention of polychloroprene and polysulfide elastomers.

World War II factions curtailed natural rubber (NR) shipments from the Far East and Southeast Asia, forcing the Allied and Axis powers to further develop synthetics. By 1951 the U.S. had a capacity to produce 850,000 short tons of general purpose styrene-butadiene rubber (SBR) annually.

The plastics industry in the U.S. has its roots in the 1860s, when cellulose nitrate first was developed by John Wesley Hyatt. This was parallel to the development several years earlier by English inventor Alexander Parkes of dried collodion that produced a material to waterproof fabric. Collodion was a cellulose solution in an alcohol-ether mixture, and the solid residue that was left after the evaporation of the solvent produced the elastic, waterproof substance.[6] Collodion also was used by Hyatt in an 1869 patent to coat billiard balls, as an alternative to ivory.

Cellulose acetate, a thermoplastic, was introduced in 1927 as a molding compound.[6] This was followed in the 1930s and 1940s by the major developments of today's thermoplastics: polyvinyl chloride (PVC); low-density polyethylene; polystyrene (PS); and polymethyl methacrylate. The first decade after World War II saw the development of polypropylene (PP) and high-density polyethylene.

While the thermoset rubber and thermoplastics industries were developing as two distinct entities, Bayer created a gray area in 1937 with its TPU development efforts that marked the beginning of the TPE industry. The TPUs became the first TPEs used in the U.S., debuting in 1959 (Mobay Corporation's Texin®, made by Bayer) and 1960 (Estane® by B.F. Goodrich).

Competition appeared in the marketplace in 1966 when Shell Chemical Company introduced Kraton® to processors in the U.S. The material wasn't commercialized in Europe until 1972. Thermoplastic urethane soon lost market share to the material in the footwear industry, and Kraton® also made strides into the adhesive and sealant arena. These three markets today are the mainstay for SBC products.

Shell monopolized the SBC market until 1968, when Phillips Chemical Company unveiled Solprene®. Solprene® was withdrawn from the U.S. market in 1983, but production was continued in Mexico. At that time, the Phillips product reportedly controlled 40% of the U.S. SBC market.[7]

Olefinics first appeared in U.S. markets in 1972 when Uniroyal began selling TPR® for hose and some wire and cable markets. The product, which was the only olefinic available in the 1970s, changed ownership three times in the 1980s before becoming part of the Monsanto Chemical Company TPE business group (now Advanced Elastomer Systems L.P., a joint venture with Exxon Chemical Company).

The olefinics (TPO) offered a cost reduction compared with TPU and began garnering such automotive applications as sight shields and filler panels. Their impact on overall TPE usage in the U.S. is apparent: 79,400 mt in 1978 compared with 45,300 mt in 1983, of which 8.4% of the 1978 total was olefinic. Olefinics accounted for only 1.9% of the 1978 usage.

The work of R. D. Lundberg, C. M. Alsys, and B. J. Walker[8] also indicates stable demand for SBC during the 1973 to 1978 period as a percentage of the overall volume, although actual demand rose to 30,000 mt in 1978 from 16,500 in 1973.

By 1983, the global demand for TPEs had reached 421,700 mt: 148,000 tons in the U.S., 199,800 tons in Western Europe, and 73,900 in Japan.[9]

Table 2 summarizes important introductions of thermoset and thermoplastic materials, many of which eventually led to the development of TPEs.

A. HISTORIC DEVELOPMENT OF THE AUTOMOTIVE MARKET FOR TPE

The automotive industry played a major role in the rapid increase in TPE consumption in the 1960s and 1970s. The needs of the assembly lines in Detroit made the materials a natural because of processibility, aging characteristics, and paintability by painting systems then in use.

The biggest breakthrough came in 1969 when a TPU was used for the "B" nose on the General Motors Corporation Pontiac Bonneville. The component was featured in television commercial segments, making this the first TPE application widely known to the consuming population.

Gradually, the automakers shifted to other TPEs and engineering plastics for that application, and as a result TPU usage in cars had diminished in the last two decades.

Other TPE applications in the automotive sector prior to 1983 include: styrenic wire and cable applications; styrenic bumper rub strips, grommets, holders and plugs; TPU sight shields, filler panels, and bumper systems; TPU blow-molded bellows; olefinic wire harnesses, bumper covers, air dams, and air ducting; and copolyester gasoline tank caps, seat belt locking devices, and door latch covers.

B. HISTORIC DEVELOPMENT OF THE FOOTWEAR USES FOR TPE

The footwear market of the 1970s embraced TPU and styrenic TPEs, using TPU in 55% of the world's ski boot production by 1979—about 3500 mt of TPU annually.[10] The 1974 production of shoe soling in the U.K. consumed 2500 mt of TPU, which represented 5.2% of the overall soling market there. Thermoset rubber was used in 47.5% of U.K. soling applications in 1974, accounting for 37,000 mt of consumption.

However, styrenics accounted for most TPE usage in footwear in the early 1970s—83%, or 60,000 mt, in 1973. A decade later the styrenics would capture 70,000 mt of usage in footwear in the U.S. and the U.K.—a major increase considering most footwear production had moved offshore to developing nations by the early 1980s.

Lantz, Sanford, and Young[11] reported in 1976 that TPEs helped U.S.-based footwear producers move to less labor-intensive methods of production and compete more favorably with imports. Styrenics and plasticized PVC became the materials of choice in the U.S. and Europe.

Imports also forced the U.S. footwear industry to switch manufacturing to mostly unit soles in the late 1970s because of the lower costs associated with bonding to uppers, extensive labor savings, and styling flexibility. Thermoplastic elastomers captured about half of the unit sole market in 1975.[1]

Easy processing by injection molding, low compound cost, improved walking traction, temperature resistance, and design flexibility all played a role in the inclusion of TPEs into footwear.

However, long molding runs were needed to keep costs down, and thick soles required long cycle times in the molding operations. Thermoplastic elastomers also offered poor oil resistance and wear in work shoes and poor wear in athletic shoes.

C. HISTORIC DEVELOPMENT OF THE WIRE AND CABLE TPE MARKET

Plastics such as PVC, polyethylene, and PP accounted for about 70% of all polymers used in wire and cable insulation and jacketing when TPEs entered the marketplace in the mid 1970s. Thermoplastic urethane was the first entrant for jacketing, especially in geophysical cable where polychloroprene was replaced and the cable redesigned to eliminate sheathing and reinforcing braid.

TABLE 2
Timeline of Introduction of Polymers

Date	Material	Original or modern uses
500 BC	Natural rubber	Game balls for Mayan Indian
1700s	Gum natural rubber	Waterproofing for clothing
1868	Cellulose nitrate	Eyeglass frames
1909	Phenol formaldehyde	Knitting needles
1927	Polyvinyl chloride	Wall coverings
1927	Polysulfide rubber	Putty, cements, and gaskets
1930s	Polyurethane	Seating and wheels
1930	Nitrile rubber	Oil--resistant products
1931	Polychloroprene rubber	Wire and cable, hose, and belts
1936	Acrylic	Brush backs
1936	Polyvinyl acetate	Flash bulb lining
1938	Polystyrene	Housewares
1939	Melamine-formaldehyde	Tableware
1942	Unsaturated polyester	Boat hulls
1942	Low-density polyethylene	Packaging
1942	Styrene-butadiene rubber	Tires
1942	Butyl rubber	Tire inner tubes
1943	Fluoropolymers	Industrial gaskets
1944	Silicone rubber	Seals and gaskets
1947	Epoxy	Tools
1947	Polyacrylate rubber	Harsh environment application
1948	ABS	Luggage
1950s	Chlorinated PE	Mechanical rubber goods
1951	Chlorosulfonated PE	Pond linings, roofing
1954	Styrene-acrylonitrile	Housewares
1957	High-density PE	Milk jugs
1957	Polypropylene	Safety helmets
1957	Fluoroelastomers	Seals, gaskets, and hose
1957	Polycarbonate	Appliance parts
1959	PU	U.S. based automotive parts
1960	Synthetic polyisoprene	Mechanical rubber goods
1960s	Halogenated butyl rubber	Tubeless-tire innerliners
1960s	Epichlorohydrin rubber	Hose, tubing, and gaskets
1964	Polyphenylene oxide	Battery cases
1964	Polyimide	Bearings
1964	Ethylene vinyl acetate	Adhesives and coatings
1966	Styrenic TPE	Footwear and adhesives
1970	Thermoplastic polyester	Electrical/electronic parts
1970s	Polyphosphazine rubber	Military/aerospace parts
1971	Copolyester TPE	Automotive parts
1972	Olefinic TPE	Wire and cable and automotive
1975	Ethylene/acrylic rubber	Automotive parts
1978	Linear low-density PE	Extruded film
1981	Thermoplastic vulcanizates	Automotive/architectural
1988	Thermoplastic NR	Medium-performance automotive
1982	Polyetherimide	Electrical/electronic parts
1982	Polyamide TPE	Sporting goods
1984	Aromatic copolyester	Electrical/electronic parts
1985	Liquid crystal polymer	Electrical/electronic parts
1986	TP fluoroelastomer	Electrical/electronic parts

From R. T. Vanderbilt Co. Inc. and Crain Communications, Inc.

Cross-linked polyethylene, or CPE, opened the door for olefinic TPEs in power cable insulation and jacketing. Auchter estimated[12] that by 1981 between 40 and 50% of all TPU, TPO, and copolyester TPE used in the U.S., perhaps 42,500 of the total 85,000 mt consumed, was in the manufacture of wire and cable. Processors in the U.S. in 1973 used only 4500 mt of the material.

The global wire and cable industry had embraced oil-resistant TPEs by 1983, with Monsanto's Santoprene® and Uniroyal Chemical Company's TPR® meeting standards for applications where wire and cable comes into contact with hydrocarbon fuels and oils.

D. HISTORIC DEVELOPMENT OF OTHER TPE MARKETS

The 1970s saw styrenic TPEs emerge as the material of choice to replace thermoset rubber in adhesive applications, with one estimate[13] in 1974 setting adhesives usage as 12% of all TPEs consumed in the U.S. Simpson noted[14] that the polystyrene domains created outstanding resistance to creep for pressure-sensitive adhesives because they could undergo viscous flow only by detachment of the radial or linear polymer blocks and subsequent reattachment to other domains.

The early 1980s saw the styrenics grow in usage for hot-melt adhesives, many of which are used on automotive assembly lines for the attachment of various components.

Architectural applications for thermoplastic vulcanizates and melt-processible rubbers (Santoprene® and Alcryn®, respectively) began to appear by 1983, including window gaskets and glazing on commercial buildings.

Other TPEs found use as impact modifiers for plastics such as PP and polystryene. In a 1980 review,[15] Thorn listed numerous other applications for TPEs:

1. Styrenics: bath mats, phonographic turntable mats, door stops, refrigerator door seals, garden hose, and billiard table cushions
2. Olefinics: pipe gaskets, pond linings, garden hose, washing machine hose, lawn mower wheels, and gaskets
3. TPUs: components for textile machinery; general engineering components such as gears, caster tires, shock absorber pads, and mining separation screens; and animal ear tags
4. Copolyester: textile spinning wheels, flexible couplings, support rings for hydraulic cylinders, backing pad for flexible grinding discs, and shoe sole inserts

These applications set the stage for an era of rapid development of new types of TPEs and many new applications. New entrants in the early 1980s broke up monopolies such as Du Pont's Hytrel® copolyester, which had been the only copolyesters in the U.S. since 1974. Expansions of existing lines, most notably Monsanto's Santoprene® olefinic-based products, would continue in the 1980s and eventually result in major consolidations and joint venture formation.

Table 2 lists many of the early applications for thermoplastic elastomers.

III. THE MIDDLE YEARS OF TPE MARKET GROWTH (1983 TO 1989)

The relatively slow growth in TPE consumption during the 1960s and 1970s changed abruptly in 1981 with Monsanto's introduction of Santoprene®. The business development efforts for the TPE products centered around an entrepreneurial-style management that was needed to move the business from the grassroots level to being the dominant force in the TPE industry by 1990.

This success of Santoprene®, and later Monsanto's oil-resistant Geolast® TPE, resulted in many companies accelerating their efforts to enter the marketplace. By 1989, the value of the U.S. market for TPE usage was pegged at about $836 million.

O'Connor noted in 1988[16] that nontire automotive parts accounted for 80 million pounds of thermoplastic rubber in 1986, while mechanical rubber goods used 150 million pounds. Usage of TPE in mechanical goods was expected to increase 63% to 240 million pounds by 1990, while thermoset rubber usage in that area was expected to grow 20 million pounds over the period to 620 million pounds annually. Automotive parts, excluding tires, consumed roughly 400 million pounds of thermoset rubber annually in the late 1980s.

Overall, 25% of all mechanical rubber goods produced in 1986 were made from TPEs 68,000 of the 272,000 mt—and were expected to account for 35.3% of the 308,000-mt market by 1990.[17]

While automotive applications spearheaded much of the growth of TPEs between 1960 and 1980, the materials in the mid-1980s were finding many new niches in architectural, medical, appliance, and wire and cable applications.

A. AUTOMOTIVE MARKETS

Swift presented data in 1986[18] that put TPE usage at 8.6 pounds per car made in the U.S. in 1985—48.7 million pounds for the 8 million cars built that year. By 1990, each car was predicted to use 11.2 pounds of TPE. Most of the applications were exterior parts, although TPE usage in the passenger compartment would begin to grow steadily after 1989.

Usage of TPE still remained small compared with the thermoset rubber usage of about 280,000 mt in 1983 and 274,000 mt by the end of the decade.[19]

Globally, the automakers were reducing weights of cars in the 1980s by incorporating more polymers in each vehicle's makeup. Mazda, which used plastics (including TPE) in its RX7 13BSI model at a level of 7% by weight in 1985, predicted that by the end of the 1980s that level would be 10% by weight and 40% by volume.[20] Plastics content had been 5.5% by weight in 1980.

Similarly, Nissan predicted plastics use would reach as high as 25% by weight for its non-high-performance cars by the end of the 1980s vs. 7.3% in 1986.[21]

Applications using TPEs varied from country to country in the 1980s. For example, Japan-based car producers used copolyester TPE to make constant-velocity joint boots, while those in the U.S. used polychloroprene rubber and Santoprene® thermoplastic vulcanizate, in addition to copolyester TPE. Europe-based automakers mostly shunned TPEs for this application because of the higher heat environments within their cars and the higher speeds on highways in Europe.

General Motors Corporation became the first to specify TPE for CV boots in 1984 for its now-defunct Chevette model, citing rising temperatures in the engine compartment as the reason. Wright, Hamblin, and Rader[22] presented forecasts that TPEs would replace more than 5 million pounds of thermoset rubber in this application during the 1980s.

Producers in the U.S. considered TPE for glass run channels, while Japan producers specified PVC for the application.

Gabris also noted[23] in 1986 that olefinic TPEs (including the thermoplastic vulcanizates and melt-processible rubbers) would grow 17% the balance of the 1980s in automotive applications. Bumpers and sight shields would dominate in the U.S., while EPDM-modified PPs — the basic olefinic TPOs—would capture 12,000 mt of bumper business in 1986. Japanese producers used 25,000 mt of olefinics in 1985 for bumpers, window gaskets, and bellows.

The growth in olefinic usage reduced the need for TPUs in automobiles because of the improved performance of the former materials at a much lower cost. About 7000 mt of TPU was used annually in the early 1980s to produce sight shields and filler panels.

The engineered olefinics such as Santoprene® and Alcryn® began replacing thermosets such as polychloroprene and chlorosulfonated polyethylene (CSM) in hose applications. For example, Ford in 1985 replaced a nylon tube/CSM cover with a flame-resistant Santoprene® and expected to consume 1.5 million pounds annually of TPE in that application.

Numerous other automotive applications were available by 1989,[1] including:

- Polyamides—glove box door retainer strap at Ford Motor Company
- Styrenics—miscellaneous grommet insulators and shock absorber dust shield
- Thermoplastic vulcanizates—shielding applications, front-wheel-drive boots, and power steering gears
- Olefinics—sight shields and bumper covers
- Copolyesters—transmission vibration shock insulators, air conditioning vacuum tubes, seat belt mechanisms, and (in Europe) plugs for exterior use

B. MEDICAL GOODS MARKETS

Carew predicted[24] that TPE usage in medical goods would account for at least 18% of the U.S. market by 1989—a market valued at $3.1 billion. A clear copolyester TPE (Ecdel®) from Eastman Chemical Company and C-Flex®, a silicone-modified styrenic from Concept Polymer Technologies, Inc., were the 1986 leaders in product developments. Ecdel® found use in i.v. bags in 1985 as a replacement for glass bottles, while C-Flex® was finding use in tubing, bottle nipples, and other applications. C-Flex® was also targeted at the $212 million annual market for urology devices, cardiovascular devices (a $134 million market), and other areas such as radiological components, i.v. drug delivery systems, and nutritional feeding products.

While PVC was the material of choice in the 1980s for tubing, holding 80% of the market in the U.S., TPEs and silicone each owned a 10% share of the market. Although TPEs are inert to most drugs and body fluids, some reactions occur with PVC.

C. WIRE AND CABLE MARKETS

Flexible cord, automotive primary wiring, and wiring for truck/trailer interconnects constituted the major markets for wire and cable applications in the U.S. during 1985 for TPEs. Automotive primary wiring was a growing application that consumed about 1300 mt in 1989. Growth rates from 1985 through 1989 had put the total usage of elastomers and plastics in wire and cable at 1.5 billion pounds annually in the U.S.

By 1989, TPEs had replaced some of the polyethylenes, fluorplastics, and PVC in wire and cable. In 1988, School reported[2] that olefinic TPEs in 1988 held between 10 and $30 million of the annual U.S. Wire and cable industry compared with $10 million or less each for butyl, nitrile, polyisopoprene, TPU, and styrenic TPEs. The cable and jacketing share of the marketplace used 50,000 pounds of TPEs annually and 200,000 pounds of thermoset rubbers.

Europe's standard for wire and cable in 1986 did not permit the use of TPEs, although experimental work was under way in 1988 to add standards. That market was largely PVC (403,000 mt per year) and thermoset rubber (less than 59,000 mt). A fiberoptic cable brought on line in 1988 to link the U.S. with the U.K. and France used 200,000 pounds of copolyester TPE as part of the 11.2-million-pound telecommunications cable.

Japan's wire and cable industry relies on olefinics as the primary TPE in wire and cable. Gabris reported[23] the major end-use market in 1986 for wire and cable was telecommunications, where 2000 mt was used that year. Wire and cable jacketing in Nissan and Toyota cars amounted to less than 1000 mt of olefinic TPE consumption that year.

D. OTHER MARKETS

Small inroads were made into the oil-field industry in the 1983- to 1989-period. Specifically, Santoprene® was specified in some applications where temperatures did not exceed 150°C downhole. Deeper wells that were drilled in the 1980s produced much higher temperatures that required replacing EPDM and polychloroprene parts with parts made from carboxylated nitrile rubber and (XNBR) NR.

An outer jacket for reinforced stainless steel tubing used in the transport of corrosion inhibitors, control fluids, antifreeze, and other chemicals was one application developed for the thermoplastic vulcanizates. The material does not kink or crush at bends in the well, and it acted as a replacement for nylon.

The later 1980s saw most TPEs being used in sealing devices, such as styrenics for pipe joints in irrigation pipes. Growth in sealing applications continued because of reduced labor and energy costs during manufacture compared with thermoset elastomers.

In the appliance area, various vacuum cleaner parts were produced from TPE beginning about 1986, and about 600,000 parts were made annually by 1988. Refrigerator crisper seals, used to keep air from getting to vegetables and causing them to freeze, was on application.

Architectural applications for TPE grew after 1986, with the thermoplastic vulcanizates and melt-processible rubbers finding many market niches. Flexible PVC was replaced by Alcryn® in seals and weatherstripping applications, markets that combined were worth about 125 million pounds annually—TPE accounted for 3 million pounds of the total. In addition, the market in 1988 used 45 million pounds of other plastics and 75 million pounds of thermoset rubber.

IV. THE FOURTH DECADE OF CONSUMPTION

While the automotive market remains the largest single end-use for TPEs, the fastest growing sector through 1995 will be medical goods.[25]

Overall consumption of TPEs will hit 920 million pounds in North America by 1995, which represents an annual increase of 7.4% from 1991 levels. The market's value in 1995 will be $1.5 billion. Of that total, automotive applications will approach 270 million pounds—nearly 40% of the total North American TPE market—compared with 170 million in 1989.

Table 3 details information on the market in North America by TPE type, while Table 4 examines the 1990 usage of the three types of olefinic TPEs in North America. The table includes the recently introduced technology of reactor-modified TPOs, which are made via reactions of the elastomer with PP in the reactor during polymerization. Production costs for in-reactor TPOs are about half that of physical-blended TPOs.

Blum advised[26] that end-use property requirements will result in the emergence of high-performance TPEs in the 1990s that offer improvements over engineering TPEs such as polyamides and copolyesters. Such high-performance products will be based on higher performance rubber alloys.

Globally, the TPE industry will produce 1 million metric tons of products worth about $2.6 billion by 1995. In addition to the U.S. consumption, European processors will use up to 762 million pounds.[27] Japan-based processors will consume 359 million pounds of various TPEs as they play catch up to processors in the West in developing applications for TPEs.

TABLE 3
Thermoplastic Elastomer Consumption in North America (million pounds)

Category of TPE	1990	1995	Annual growth (%)
Styrenic	315	463	8.0
Olefinic	130	225	11.6
Thermoplastic vulcanizates	50	96	14.0
Polyamide	15	24	9.5
Urethane	70	94	6.0
Copolyester	38	61	10.0
TOTAL	618	963	10.0

From Phillip Townsend Associates.

TABLE 4
North American TPO Usage in 1990 (million pounds)

Application	Physical blends	Vulcanizates	Reactor-modified	Total
Auto bumpers	22	—	30	52
Other auto	24	18	—	42
Wire and cable	16	4	—	20
Hose and tube	6	8	—	14
Film	—	—	3	3
Molded goods	16	8	—	24
Other	8	10	0	18
TOTAL	92	48	33	173

From *Chem. Week.*

Medical products such as tubing, diaphragms, and drug-delivery devices will show the greatest growth, but copolyester TPE demand will leap 12% annually for industrial product applications to become a 95-million-pound product market. Other applications for copolyester include pipe and hose for oil and gas well sites, refineries, and petrochemical facilities.

Other TPEs will show slower growth demand. Annual growth for styrenics will not exceed 4.2%, while that of TPUs will grow less than 7% annually to reach between 80 and 95 million pounds in North America by 1995. Aiding that growth, however, are new blends of TPU with plastics such as ABS and polycarbonate.

Adhesives, sealants, and coatings, which primarily are styrenic TPE applications, will grow in value to be worth $317.1 million in North America in 1994. This represents a significant increase over the market's value of about $230 million in 1989. Dexco Inc. unveiled a styrenic product in 1991 based on 100% triblock technology that is designed to provide special performance properties in adhesive applications.

Producers of TPEs continue to search for new niche end-use markets for their products. This will represent fast-growing areas in the 1990s.

In comparing TPEs to other materials, Rader stated[28] that TPEs have grown at an annual rate of between 8 and 9% since 1970 while thermoset rubber and thermoplastics have grown at annual rates of 0 to 2% and 1 to 3%, respectively. The forecast calls for TPEs to continue that growth curve through the year 2000. Synthetic rubber usage in 1995 in the U.S. will be 17.34 million metric tons compared with 15.69 million in 1991.

Muhs has predicted[17] that TPEs will be used in 45.3% of all elastomer needs in the mechanical rubber goods sector by 1995—154,000 mt of the 340,900-mt market.

A. AUTOMOTIVE MARKETS IN THE 1990s

Thermoplastic elastomer usage in cars made in the U.S. will be worth $255.5 million annually to the TPE producers by 1994.

Meanwhile, Europe is bracing for the onslaught of Japanese car production on that continent that began in the early 1990s. Nissan Motors will produce 200,000 vehicles per year in the U.K. by 1994, while Toyota will produce the Carina model there at levels of 200,000 per year by 1997 or 1998.[29] Honda began limited production in 1991 and forecasts an output of 100,000 per year by 1994.

A report by *European Chemical News*[30] noted that the continued establishment of Japanese car production in Europe could boost consumption of TPEs by around 6000 mt annually by 1993, moving the automotive sector ahead of footwear as the dominant end-use for TPEs on the continent. It could account for 25 to 30% of all TPEs used in Europe.

The report indicated that the weight of TPE per car will increase to 6.5 kg by 1995 from 4 kg in 1990. About 9 million units were produced in Germany, France, and Italy in 1990, which put TPE consumption at 79.2 million pounds. By 1995, sources[31] predict the three major producing nations will build 9.4 million vehicles that will require 134.4 million pounds of TPE.

The increase in TPE consumption in automotive applications reflects an overall trend to reduce vehicle weight that began in the late 1970s. The overall use of plastics, including TPEs, in U.S. produced cars rose to 229 lb per vehicle in 1990 compared with just 168 lb per vehicle in 1977. During the same period, the use of steel and iron dropped to 2171 lb per car from 2742 lb.[32]

Eller projects[33] the total potential automotive market for TPO (physical-blended or produced-in-the-reactor olefinic TPEs) is 340,000 mt worldwide for bumper fascia alone. The weight used in U.S. made passenger cars is about 20 lb per vehicle, or 55,000 mt, while in Japan and Europe the usage of TPO for automotive applications is about 18 lb (106,000 mt annually in Europe and 79,000 in Japan).

Trucks produced in the U.S. require 24 lb of TPO for bumpers, while elsewhere in the world the figure is 22 lb. Currently, TPOs are used in 15% of the fascia market in North America but has the major share of the bumper fascia market in Europe and Japan.

Svigel[25] predicted automotive-market demand for TPEs in North America will account for 26.6% of the total 920 million pounds of TPE used in 1995 vs. 24.7% of the 643 million pounds consumed in 1990. That trend will continue, with automotive applications grabbing 28% of TPE usage by the year 2000. The market division between the U.S. and Canada will remain stable over the entire period with 91% of usage in the U.S.

Table 5 compares data for TPE usage by the automotive industry and other markets in North America through the turn of the century. In addition to automotive applications garnering more of the market share, the pounds of TPE used in North America for all applications per 1000 lb of thermoplastics used will increase to 12.2 in 1995 and 15.0 in 2000 from 10.1 in 1990.[25]

Usage of TPEs per 1000 lb of thermoset rubber used will grow to 108 lb in 1995 and 143 lb in 2000 compared with 81 lb in 1990. By the year 2000, TPE usage will be 14.3% that of thermoset rubber compared with 8.1% in 1990. The numbers when compared to plastics are similar; 15% and 10.1%, respectively for 2000 and 1990.

While TPEs are grappling for a bigger piece of the automotive pie, the general trend in materials is to use more plastics. Table 6 lists the top 10 automotive applications for plastics by 2001 in North America. Some TPE usage, such as the 52 million pounds of TPO used in bumpers, is included in the figures.

TABLE 5
Thermoplastic Elastomers by Market in North America
(million pounds)

Market segment	1980	1985	1990	1995	2000
Light vehicles/other transportation	56	98	159	245	370
Industrial products	60	101	157	240	355
Footwear	96	129	145	135	130
Medical products	5	24	60	110	185
Wire and cable	15	28	48	80	120
Other markets	20	42	74	110	160
U.S. usage	230	385	586	838	1202
Canada usage	22	37	57	82	118
Pounds of TPE/thousand pounds of plastic	6.7	8.7	10.1	12.2	15.0
N.A. Plastics demand	37,460	48,340	63,750	75,400	87,900
Pounds of TPE/thousand pounds of rubber	40	64	81	108	143
N.A. Rubber demand	6272	6601	7915	8550	9200

From Freedonia Group.

TABLE 6
Top 10 Plastic Applications by the
Year 2001 (million pounds)

Application	1991	2001
Bumper systems	241.5	299.0
Gas tanks	29.5	45.0
Door outlet	13.7	35.0
Front hoods	7.4	19.0
Fender	3.5	20.0
Intake manifolds	0.0	51.0
Rocker arm covers	6.4	26.0
Fuel pump	0.2	1.3
Nonpneumatic spare tire	0.0	26.0

From *Ward's Auto World* September, 1991.

B. MEDICAL MARKETS IN THE 1990s

Applications in the medical field represent a major growth area for TPEs in the 1990s. Low production costs and relatively inexpensive material costs have made TPEs a viable choice to replace thermoset rubber in many single-use, disposable medical products. Increasing concerns over some toxicity of NR-based articles has prompted the evaluation of TPEs as a way to prevent anaphylactic shock in some patients. Thermoplastic elastomers also avoid the dangers from plasticizer migration that are present in products made from flexible PVC.

One such product is a TPE glove that was introduced in 1991. Despite a cost that is 50 times that of NR-based gloves, the TPEs do not have chemical irritants or proteins that natural rubber does. Efforts are under way to develop TPE-based condoms, catheters, and diagnostic probe covers.

The U.S. and Europe each will show an annual growth rate of 10% in usage of TPEs in medical applications. Table 7 lists many current and potential applications that make up

TABLE 7
Medical Goods Applications for Thermoplastic Elastomers

Application	TPE type	Advantage over other materials
Wound dressing	Copolyester	Prevent bacterial entry, allow moisture transmission
Drug delivery	Copolyester	Patches tailored to proper moisture vapor transmission rates and chemical resistance suitable for particular drug being administered
Blood collection	TPV	Stoppers made from Trefsin® seal out air after removal of syringe from vial, permits vacuum in the collection vial, easier to process than butyl rubber
Dialysis tubing	TPV	Improves flex life of tubing $20\times$ vs. silicone rubber
Infant nipples	Styrenic	Improved bite and tear resistance vs. natural rubber or silicone rubber feeding nipples
Tubing	Styrenic	Improved cleanliness and dimensional stability vs. NR, easy to sterilize, chemically inert
Instrument handles	Numerous	Design flexibility compared with thermoset rubber
Needle shields	Numerous	Cost reduction, low extractables
Aerosol valve seal	TPV	Improved flexibility, no plasticizer

TPV—thermoplastic vulcanizates.

a small part of a market worth nearly $40 billion annually excluding equipment and machinery. Included in the market figures are such things as syringes/needles, disposable kits/trays, catheters, clinical lab supplies, and many other products. Urinary catheters alone, most of which are produced from NR latex or silicone rubber, account for nearly $300 million of that total.

Wound dressings and drug-delivery patches are two other potential products. The dressings prevent bacterial entry but allow simulation of the moisture transmission of a natural blister. The process promotes healing with less pain and less scar formation.[34] Such applications represent a potential use of between 5 and 10 million pounds of TPE in the U.S. Drug-delivery patches can be tailored to give the moisture vapor transmission rate and chemical resistance suitable for the particular drug being administered.

Intraveneous site dressings made from copolyester TPE reduce infection rates. Their low friction also reduces the incidence of catheter dislodging.

Disposable surgery gowns produced from copolyester TPE could represent more than a 5-million-pound market in the U.S. and a 16-million-pound worldwide market by the mid-1990s.

Syringe plunger tips, tubing for peristaltic pumps, stoppers for blood collection vials, panels for treating jaundiced babies, and tubing for kidney dialysis machines are some of the many medical applications currently used for thermoplastic vulcanizates. Some applications, such as the vial stoppers, replace butyl rubber and other elastomers.

Styrenic TPEs have found use recently in infant feeding nipples as replacements for NR nipples. The TPE provides precision moldings with good strength and durability. Its durability and bite resistance are rated better than that for NR.[34]

Kraton® styrenic also is used for medical tubing because of its cleanliness, dimensional stability, ease of sterilization, and good gas- and vapor-transmission properties. The TPE is also chemically inert and produces excellent clarity. At least one firm, Kent Latex Products, Inc., overcoats the TPE with NR latex to provide a more rubbery feel.

Some catheter and dental tubing applications have switched to TPU to eliminate plasticizer migration and extraction. The products are flexible at low temperatures and have low gas and vapor permeability.

Overall, the market segment of disposable medical products uses about 260 million pounds of polymers annually and is growing at double-digit rates. The continuing need for cleaner and safer materials will advance the use of TPEs in place of thermoset rubber in medical goods.

C. OTHER MARKETS IN THE 1990s

Adhesives, sealants, and coatings will remain the largest end-use segment for TPEs in the 1990s. The market value for these products will reach $317 million by 1994 compared with $230 million in 1989.[35] For example, Japan used 58,000 mt of hot-melt styrenics in 1990, about a 15% increase from the 1988 level of 50,250 mt, and the trend is expected to continue.[36] Smaller increases in adhesives made from polychloroprene or NR are predicted. Overall use of synthetic rubber latexes for adhesives in Japan will remain steady at about 65,000 mt per year.

The asphalt-modifier market, which primarily uses styrenic TPEs, will be worth $38.2 million in 1994 compared with $23.7 million in 1989.

General growth in the use of olefinic TPEs is pegged at 10% per year through 1995 for North America.[37] Hose and tubing, which in 1990 used 6 million pounds of physical-blend TPO and 8 million pounds of thermoplastic vulcanizates, will use 9.7 million pounds of TPO and 12.9 million pounds of thermoplastic vulcanizates annually by 1995. Nonautomotive uses of reactor-modified TPO will be limited to films, with growth in that area going to 4.8 million pounds in 1995 from 3 million in 1990.

Continued growth in TPE consumption is expected for the appliance market. Co-polyesters in the 1990s will find use in "soft-touch" pads for handles of small appliances and for hair curlers. Hoses, tubes, and seals for washing machines represent growth areas for thermoplastic vulcanizates in Europe, while components for dryers and washing machines in the U.S. will find additional uses for the TPEs. Thermoplastic vulcanizates will also find use for soft-touch handles on irons and other small appliances.

Arm pads for office chairs first were produced from TPE in 1991, and the office furniture industry is expected to find more uses for that application because of design flexibility and feel. The pads are molded with no part lines, flow lines, or depressions.

Also under development are V-belts and a manufacturing process that uses a plastics extruder to produce the materials.[38]

D. THE VIEW FROM THE SUPPLY SIDE FOR THE 1990s

Potential new suppliers of existing types of TPEs will face a difficult situation in trying to enter the marketplace in the 1990s, unless they offer unique products. Acquisitions, consolidations, and joint ventures are creating giant-sized producers equipped with massive marketing and technical departments to fight for new niches in the polymer arena.

A $100 million TPE empire was created in 1991 with the formation of Advanced Elastomer Systems, L.P., which is a joint venture between Monsanto Chemical Company and Exxon Chemical Company. Prior to forming the venture, Monsanto had purchased assets of some Bayer A.G. TPEs, most of the TPR® olefinic TPE line from B.P. Performance Polymers, and introduced numerous new products such as Vyram® (an NR-based thermoplastic vulcanizate) and Dytron®, a specialty product for wire and cable that can be cross-linked further by irradiation.

Shell Chemical Company, which currently holds about 50% of the market share for styrenic TPEs in North America, is now facing further competition from Dexco, Inc., a joint venture of Dow Chemical Company and Exxon. Further competition will appear in late 1992 when Enichem S.p.A. starts up its plant to make styrene-butadiene-styrene and styrene-isoprene-styrene SBC products in the U.S. Thus far, Shell remains the only producer of styrenics with the ethylene-butylene soft segment, although in late 1991 Enichem S.p.A. had begun producing trial quantities of the saturated material at a pilot plant in Italy. The product is scheduled for commercialization in 1992.

In addition, Shell is considering building a plant in Europe by 1994 to produce styrene-ethylene butylene-styrene (SEBS). Currently, all SEBS used there are imported from the U.S.

Like Shell and Advanced Elastomer Systems, the major segment of each TPE market in the U.S. is held by those firms that pioneered the products:

- B.F. Goodrich controls 35% of the TPU market, although BASF Corporation recently opened a TPU plant in Michigan.
- Du Pont has a 75% share of the copolyester TPE market with General Electric Plastics being the nearest competitor.
- Atochem products account for 80% of the polyamide market in the U.S., where they find a lot of use in sporting goods and automotive applications.

Producers of TPEs in search of greater market share or market entry also promise many new products as the industry progresses to the 21st century. Softer copolyesters will permit automakers to paint parts without primer through the use of one- and two-component paints as well as some water-borne systems.[39] Potential end-uses include body-side cladding and grilles on automobiles.

The replacement of NR, SBR, and EPDM thermoset elastomers is the target of Vyram®, a dynamically vulcanized TPE based on NR. It is usable in applications where rubber parts do not reach 110°C and offers good wear and compression set characteristics. Expected applications in the 1990s include caster wheels, grips and handles because of its excellent dimensional stability, and UV-resistant window seal profiles. The latter application could feature a coextrusion with rigid PP.

A vinyl-based TPE marketed by B.F. Goodrich Chemical Company combines the traditional advantages of plasticized vinyl with superior combustion properties. Introduced in 1991, potential applications by the year 2000 include wire and cable used in plenum spaces without conduits, such as voice/video/data communications cable.

In-reactor olefinic TPEs represent the biggest single potential growth area in TPE production as the industry progresses to the mid-1990s. Downstream compounding steps are eliminated by the combination of ethylene and propylene monomers at the polymerization step to reduce manufacturing costs by 50% compared with the traditional physical-blend TPOs.

Numerous producers in Europe and the U.S. had introduced in-reactor TPOs by late 1991, while many others are set for introduction before 1995.[33] In addition to finding use in interior trim applications, the ability of the in-reactor TPOs to be extruded make them a viable option to PVC in interior skin applications during the 1990s.

Touted as an alternative to in-reactor TPOs, ethylene methacrylate TPEs are under study by Chevron Chemical Company for introduction in the 1990s. The acrylate functionality contributes to better bondability than that offered by TPO with numerous polar and nonpolar polymers. Potential applications include molded seals, stretch film, small bumper pads, tubing, squeeze bottles, and bellows.

Alloys of TPU and such materials as polycarbonate and ABS resins will create new markets for TPUs through the year 2000. In addition to potential usage in bumper fascia and rocker panels, forecasts call for the alloys to be used in hard-sided luggage and furniture armrests. Other suppliers are developing medical-grade TPUs that are free of plasticizer and offer flexibility and biocompatibility. Potential applications include flexible tubing, multi-lumen catheters, connectors, and manifolds.

Japanese based technology has come to U.S. shores in the form of thermoplastic fluoroelastomers produced by Daikin. While thermoplastic, the materials can be cured through peroxide, polyol, or radiation methods. The products consist of a fluoropolymer hard segment and a fluoroelastomer soft segment and are usable for products made via extrusion, injection molding, and compression molding. The 3M Company also has under development a fluorinated TPE that has a fluoroelastomer continuous phase and a polyamide dispersed phase. It also can be crosslinked via radiation methods.

Dynamically vulcanized TPEs based on chlorinated polyethylene could appear in the marketplace by 1993. The CPE is crosslinked in a thermoplastic matrix and is expected to find use in applications that require resistance to hydrocarbons, weathering, and heat aging.

E. RECYCLING AND ENVIRONMENTAL FACTORS IN FUTURE MARKETS FOR TPEs

Protecting the environment has become the buzz phrase of the 1990s, and the process will be a driving force throughout the decade in the growth of TPEs.

Producers of the materials touted the reuse of TPE scrap, often up to 10 times, without having a major impact on end-use properties of the part being made. This reuse compares with the single use of most thermoset rubber compounds, although methods of recycling thermosets are beginning to appear: including pyrolysis to recover oils, carbon black, and metal; and grinding of rubber to micron-sized particles for reuse in asphalt and mechanical rubber goods.

However, the oldest method of rubber recycling has fallen out of favor as quality becomes the critical factor in product acceptance. Reclaimed rubber, once used as about a third of all rubber in a compound, has dropped to being just 5% of the total.[40]

Producers of TPOs are working to win market share in Japan for calendered in-reactor TPO as a replacement for dry-powder, slush-molded PVC and thermoformed PVC/ABS skins on the basis of cost savings, nonfogging (absence of volatile plasticizer), and recyclability in monomaterial sandwich constructions.[33] Legislation mandating recycling in Japan will take effect in 1994 or 1995.

Meanwhile, recycling most of all components of cars has prompted automakers in the U.S. and Germany to take a proactive stance. Germany wants its automakers to have a plan to recycle all old cars by the mid-1990s.[32] Bayerishe Moteren Werke AG has opened a plant to do just that, while Daimler Benz has joined forces with another company to develop recycling technology.

In the U.S., car makers are focusing their efforts on plastics recycling. Steel usage has dropped 22% in the last 15 years, while the use of plastics has climbed about 35%. Efforts to develop recycling are also being pushed by exploding disposal costs as landfills in the U.S. reach capacity and close.

Today only 13% of the 320 billion pounds of solid waste produced in the U.S. is recycled, three quarters of it goes to landfills. Because of this, only 6000 of the 18,500 municipal landfills that were in operation in 1979 still accept waste.[40]

To counter this problem, Business Communications Company reports[41] that the amount of plastic resins being recycled could triple by 2000 to as much as 3.6 billion pounds per year. The total includes 1.4 billion pounds of polyethylene terephthalate and 389 million pounds of PP (more than double thecurrent 155 million pounds).

As stricter controls are put on landfilling much of the 14 million metric tons of thermoset rubber produced annually, TPEs will begin to find more acceptance with rubber product manufacturers, and they will accept the need for installing plastics processing equipment as a method of remaining competitive with the plastics industry.

F. FUTURE OUTLOOK FOR TPE MARKETS

The phenomenal growth of TPE markets will continue into the 21st century because of numerous benefits to the processor, end-user, and global population.

Stringent legislation that will require processors to dramatically increase the average fuel economy of its products will enhance the usage of lightweight TPE components. High-strength composites are expected to replace steel frames, while rigid thermoplastics are replacing steel in many vertical body components. TPEs will replace heavier elastomers and metals in many interior and exterior automobile components.

Recycling of scrap will help TPEs to replace more thermoset rubber during the 1990s. Added to this are state and federal requirements for the disposal of packaging that, in some states, will ban by 1994 the landfilling of any bags that contained chemicals. Because TPEs are supplied in a precompounded state, only one bag must be disposed of instead of numerous bags for the various compounding ingredients used to make a rubber batch.

Precompounded pellets also enable users to more easily receive returnable, nestable drums that then can be shipped back to the TPE supplier for refilling. The rubber industry is pursuing a similar return system for its products but faces a more difficult task because many ingredients, such as antioxidants and accelerators, are supplied in powder form.

In addition, producers of TPEs continue to expand the performance range of their products to include higher in-service temperature ranges, oil resistance comparable to thermoset rubber, and compression set values making them more acceptable for sealing and other applications. The TPE producers, however, are consolidating market share through acquisitions, joint ventures, and other actions that will make it difficult for new players to enter the production arena in the 21st century unless they have a unique TPE product to offer.

With all things considered, the dawning of the 21st century will see usage of TPEs nearly triple that of 1990—approaching 3.7 billion pounds per year worldwide.

V. CONCLUSION

The rubber and plastics industries no longer follow never-crossing, parallel paths. The advent of TPEs in the 1950s, and the rapid growth of these materials since 1982, have created a very large gray area for processors.

Processors of traditional thermoplastics have embraced the materials to a much greater extent than processors of thermoset elastomers, because they already had the necessary processing equipment in place. The labor-intensive rubber industry has lacked the desire to make the capital investment necessary to add this equipment.

As the auto industry rewrites its parts specifications to allow more TPE usage, the rubber processor of the 21st century will be forced to buy the tools needed to produce TPE parts. The process has begun.

REFERENCES

1. **D'Amico, E. B.,** A need to stretch, *Chemical Marketing Reporter,* Vol. 240, No. 18, SR15 1991.
2. **School, R.,** Markets for thermoplastic elastomers, in *Handbook of Thermoplastic Elastomers, 2nd ed.,* C. P. Rader and B. Walker, Eds., Van Nostrand Reinhold, New York, 1988, 285.
3. *Elastomerics,* Vol. 122, No. 6, 19, 1990.
4. **Legge, N. R.,** Thermoplastic elastomers—three decades of progress, Shell Chemical Co., Houston, paper presented at a meeting of the Rubber Division, American Chemical Society, Cincinnati, October 18–21 1988.
5. **School, R.,** "Rubber, Natural and Synthetic," *Rubbicana Rubber Directory & Buyer's Guide,* 12th ed., Vol. 18, No. 13, Crain Communications, 1988.
6. **Chrisman, C. and School, R.,** Internal business proposal to form a trade newspaper for the plastics industry, Crain Communications, Detroit, p. 4, 1988.
7. *Plastics World,* 11, 1983.
8. **Lundberg, R. D., Alsys, C. M., and Walker, B. J.,** *Handbook of Thermoplastic Elastomers,* B. M. Walker, Ed., Van Nostrand Reinhold, 309, 1979.
9. **Chem Systems Inc.,** data from *Process Evaluation/Research Planning* study, 1985.

10. **Hamson, N.,** *Plastics Today* 1, 25, 1978.
11. **Lantz, W. L., Sanford, J. A., and Young, J. F.,** History and future of kraton thermoplastic rubber in shoe soling applications, Shell Chemical Co., Houston, 1976.
12. **Auchter, J. F.,** Outlook for TPEs in the 1980s, Shell Chemical Co., Houston, 1981.
13. *Modern Plastics,* 51, 1974.
14. **Simpson, B. D.,** The use of solprene elastomers in adhesives and mechanical goods, Phillips Chemical Co., paper presented at meeting of the Rubber Division, American Chemical Society, San Francisco, October 5–8, 1976.
15. **Thorn, A. D.,** Thermoplastic elastomers—a review of current information, Rubber and Plastics Research Association of Great Britain, Shawbury, England, 261, 274, 277 and 282, 1980.
16. **O'Connor, G. E.,** Business Director for Thermoplastic Elastomers, Monsanto Chemical Co., St. Louis, personal correspondence with author, 1988.
17. **Muhs, J.,** Mechanical rubber goods, in *Handbook of Thermoplastic Elastomers, 2nd ed.,* C. P. Rader, and B. Walker, Ed., Van Nostrand Reinhold, New York, 331, 1988.
18. **Swift, K. T.,** Industry study 140: high technology materials in automobiles, The Freedonia Group, Cleveland Heights, OH, 1986.
19. **Gabris, T.,** *Elastomers in the Automotive Industry,* Gabris International, Tolland, CT, 1985.
20. Personal correspondence with Mazda, July 1, 1986.
21. Personal correspondence with Nissan Motor Co. Ltd., Aug. 1, 1986.
22. **Wright, M. A., Hamblin, N. R., and Rader, C. P.,** The impact of thermoplastic elastomers in the U.S. automotive market, Monsanto Chemical Co., St. Louis, 1985.
23. T. Gabris International, Tolland, CT., personal correspondence with author, March 15, 1986.
24. **Carew, R.,** "New materials for urological and ENT implants," Concept Polymer Technologies Inc., St. Petersburg, FL, a paper presented at Business Opportunities in Biomaterials and Implants Seminar, Minneapolis, May 28, 1986.
25. **Svigel, D.,** Thermoplastic elastomers to 1995, Freedonia Group, Cleveland Heights, OH, 1991.
26. **Blum, H. R.,** Thermoplastic elastomers—a global perspective. The decade of deeper penetration, Chem Systems Japan Ltd., Tarrytown, NY, a paper presented at TPE Europe 1991 Conference, Luxembourg, October 23, 1991.
27. *Chem. Week,* February 27, 1991.
28. **Rader, C. P.,** Thermoplastic elastomers: a major innovation in materials, Advanced Elastomer Systems, L. P., St. Louis, 1991.
29. Prepared for success, *Automotive News Insight,* June 24, 1991.
30. Thermoplastic elastomers face mixed prospects, *European Chemical News,* Feb. 12, 1990.
31. *Eur. Plastics News,* 17, 4, 1990.
32. **Miller, K.,** On the road again and again and again: auto makers try to build recyclable car, *Wall Street Journal (New York),* April 30, 1991.
33. **Eller, R.,** Global automotive applications for thermoplastic elastomers, Charles River Associates, Boston, a paper presented at the TPE Europe 1991 Conference, Luxembourg, October 23, 1991.
34. Clean TPEs find medical uses, *Eur. Rubber J.,* 173, 28, 1991.
35. TPE market will hit $1.2 billion by 1994, *Elastomerics* 122, 19, 1990.
36. Adhesives industry, *Plastic Ind. News,* 36, 1990.
37. Elastomers in the '90s, *Chem. Week,* 148, 1991.
38. *Plastics News,* 3, 16, 1991.
39. Lots of new materials for automotive debut at SAE '91, *Plastics Technol.,* 28, 1991.
40. **Gonzalez, E. A., Purgly, E. P., and Rader, C. P.,** Thermoplastic elastomers—friends of the environment, Advanced Elastomer Systems L. P., Akron, a paper presented at the Rubber Division, ACS, Detroit, October 8–11, 1991.
41. Plastics recycling: boom ahead, *Chem. Week,* Aug. 14, 1991.

Chapter 16

DEGASSING OF THERMOPLASTICS DURING EXTRUSION WITH SINGLE-SCREW AND CO-ROTATING TWIN-SCREW EXTRUDERS

Gerhard Martin and Werner Schuler

TABLE OF CONTENTS

0-8493-4401-8/93/$0.00 + $.50
© 1993 by CRC Press, Inc.

I. INTRODUCTION

In view of today's tightened safety and health regulations, the increasing environmental consciousness, as well as the compulsory demand for raw material and energy saving production processes, degassing has become an important process step when producing, compounding, and processing thermoplastics. To state an example: polystyrene should not contain more than 0.03% (corresponding to 300 ppm) styrene as residual monomer. Regarding food packing material made of SAN or ABS, the required residual content of acrylonitrile is to be lower than 0.0005% (corresponding to 5 ppm) while the content of formaldehyde in polyoxymethylene (polyacetal) must be below 0.0001% (corresponding to 1 ppm).

Since the late 1970s, the degassing of polyethylene has become a requirement by the air pollution control board. Up to that date, there was basically no interest in the degassing of polyethylene because all remaining monomers could be vented in the silos by silo-degassing. In a few cases, rear venting of polyethylene extruders has been used for economic reasons. In these cases, the pressure of the low-pressure separator was about 4 bars, to eliminate the compressor stage between low-pressure separator and main compressor. These systems reduced the monomer content from about 3000 to 4000 ppm in the low-pressure separator and to 750 to 900 ppm in the rear vent zone. Based on this experience, we developed degassing extruders for polyethylene with forward venting down to 200 ppm. Today, we reach 20 ppm with these systems.

These examples are taken from a wide range of thermoplastics and their demands can only be met if a degassing process takes place during or after the production of polymer.

Degassing means the removal of low-molecular substances from polymers. Under process conditions occurring during the extrusion, these substances are converted to gas and discharged as such. The polymers are then in a comparatively high-viscous phase. Table 1 shows examples of degassing tasks to be fulfilled during the production, compounding, and processing of thermoplastics.

At atmospheric pressure, the substances to be removed usually have a boiling temperature far below the processing temperature of polymers. In any case, they have a considerable vapor pressure, rising clearly above 1 mbar at processing temperature. On the other hand, the vapor pressure of typical thermoplastics is very low and practically unmeasurable.

It seems at first to be a very easy separation task, but practice shows the opposite. In highly concentrated polymer solutions and melts, the individual coiled chain molecules diffuse themselves considerably as the chains are partially looped, whereby the mobility of the molecule is very restricted. In the case of such polymer systems, this factor leads to high viscosities and slow diffusion processes.

The requested degassing tasks led to a whole string of process-technical solutions, partially involving a large amount of machine equipment. Figure 1 shows examples of appliances and machines which are suitable for a continuous degassing of liquids and/or melts with different viscosities.

The extruders we will be dealing with in detail in the following, are particularly suitable for the degassing of high-viscous polymer solutions and/or polymer melts within a viscosity range of approximately 10^1 to 10^6 Pas. Figure 2 shows a degassing extruder with two forward degassing sections.

II. SOME BASIC STATEMENTS WHICH GENERALLY APPLY TO DEGASSING WITH EXTRUDERS

Characteristic of the degassing process is the mass transfer from the liquid, high-viscous phase into a gaseous one, Figure 3. A mass transfer of the volatile components is only

TABLE 1
Degassing Tasks During Production and Processing of Polymers

Degassing tasks

1. Concentration of polymersolutions
    HDPE   → hexan
    PSU   → chlorobenzene
    PS   → styrene
2. Final degassing of monomers
    PMMA   → methyl methacrylate
    ABS, SAN → acrylonitrile
    PS   → styrene
    LDPE   → ethylene
3. Degassing of reaction products
    PA 6.6   → water
    PET   → alcohol
4. Degassing at processing
    PA, PMMA, PC,...   → water
    Polymers feed as powder   → air

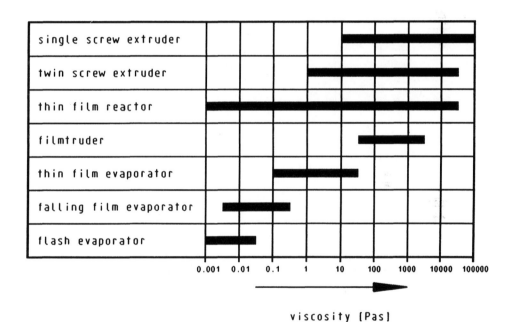

FIGURE 1.   Viscosity ranges for different degassing apparatus.

possible as long as the concentration status in the liquid phase (melt) is higher than in the gaseous one. The momentaneous mass flow (gas flow) $\dot{m}_G$ depends on the difference of concentration and on the size of the phase boundary surface and can be described in principle by the equation

$$\dot{m}_G = A \cdot \beta \cdot (C_{in} - C_{eq}) \tag{1}$$

If the overall mass transfer coefficient, the size of the phase boundary surface A, and the difference of concentration $(C_{in} - C_{eq})$ were known, the degassing capacity could be

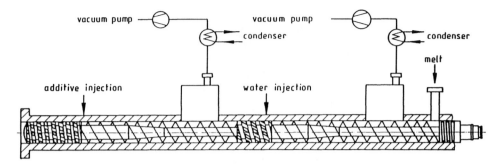

FIGURE 2. Two-stage degassing extruder with water injection.

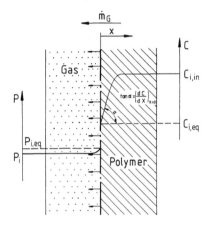

First Fick's law : $\dot{m}_G = A \cdot D\left(\dfrac{dc}{dx}\right)_{x=0}$

Second Fick's law : $\dfrac{dc}{dt} = D \cdot \dfrac{d^2c}{dx^2}$

FIGURE 3. Mass transfer.

determined immediately. Unfortunately, the overall mass transfer coefficients for thermoplastics under extrusion conditions are unknown. The mass transfer on the gaseous side of the phase boundary is very good, as compared to the mass transfer in the liquid phase, which is effected by diffusion.

In the liquid phase, the momentaneous mass flow $\dot{m}$ depends on the diffusion coefficient D and on the concentration gradient dc/dx on the boundary surface (first Fick's Law).

$$\dot{m}_G = A \cdot D \cdot \left(\frac{dc}{dx}\right)_{x=0} \tag{2}$$

From the point of view of the melt particles, degassing is an unsteady process. When degassing with extruders, the mass flow is time-dependent. Just after the formation of a new liquid layer, the concentration gradient is at its highest and becomes smaller in the course of time until a new layer is formed. A frequent formation of new layers leads, therefore, to a higher average overall mass transfer coefficient and, thus, to a better degassing.

The diffusion from an even layer, which shows a uniform concentricity at the time t = 0 and on whose surface an equilibrium concentration is set and kept constant, can be described by the second Fick's Law.

$$\frac{\delta c}{\delta t} = \frac{\delta}{\delta x}\left(D \cdot \frac{\delta c}{\delta x}\right) \tag{3}$$

In Figure 4 you can see the concentration profiles of the polymer melt, depending on residence time and thickness.

From this equation it can be seen that the change of the gradient of concentration $\delta C/\delta X$ on the phase boundary surface depends strongly on the residence time on this surface. Therefore, the surface of the polymer melt to be degassed should be frequently renewed.

This knowledge is emphasized more by the fact that the diffusion coefficient in polymer solutions or melts is very low, i. e., between 10^{-10} and 10^{-6} cm²/s. (In comparison, the diffusion coefficient in low-molecular liquids is between 10^{-5} and 10^{-4} cm²/s).

This fact can be explained by the following model. Molecules of the size of monomer or solvent molecules diffused by the polymer melt require, for every migration process of the dissolved molecule, a synchronized motion of several segments of adjacent, convoluted polymer chains. This model shows clearly that the diffusion is temperature-dependent — as already proven by tests. The diffusion speed increases as the temperature rises, since the whole system is loosening itself and becoming more permeable because of the increased thermal motion of the chain segments.

Moreover, the diffusion coefficient shows a distinct dependence on concentration. The dissolved monomer and/or solvent molecules, which are small as compared to the macromolecules of the polymers, increase the mobility of the polymer chain segments and consequently, the diffusion rate. From the above statements, one can emphasize that the degassing process is quicker at high temperatures and that the final degassing concentration of the volatile components becomes slower and more difficult.

The temperature can be raised only up to a certain limit. At too high temperatures and long residence times, decompositions may occur, i.e., monomers could be formed anew. In this respect, it cannot always be clearly distinguished whether the decomposition is caused by the temperature alone or by a high mechanical stress (shearing) of the polymer melt in the extruder. Decompositions can be determined for example in polystyrene, SAN, or PMMA. Figure 5 shows the forming of monostyrene dependent on the temperature and residence time.

With regard to the polymer temperature, it is to be noted that the evaporation of the volatile components absorbs energy from the polymer melt, i. e., the evaporation enthalpy of the solvent chlorobenzene (C_6H_5Cl) is 325 kJ/kh and the one of water (H_2O) is 2257 kJ/kg. This leads to a cooling of the melt, which is even a requirement in some cases. You can calculate the increase of temperature with the following equation:

$$\Delta\vartheta = \frac{C_{in} \cdot r}{C_{poly}} \tag{4}$$

(Theoretically, the polymer melt cools down by approximately 10 K at a water evaporation of 1% and by approx. 1.5 K at a solvent evaporation of 1%). Usually, the loss of energy incurred in this way has to be frequently compensated by external heating or mechanical energy dissipation (shearing) by means of the extruder screw; a procedure which, in the rule, is more efficient.

Equation 1 shows that the larger the difference in concentration is, the better the degassing. The equilibrium concentration of the volatile components dissolved in the plastics melt, which appears on the latter's boundary surface to the gas phase, depends on the plastics melt temperature and on the pressure in the gas phase. In case of two or more gases or vapors in the gas phase, the equilibrium concentration depends on the relevant partial pressure.

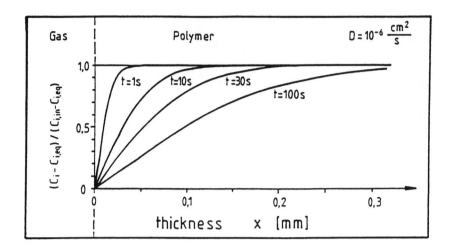

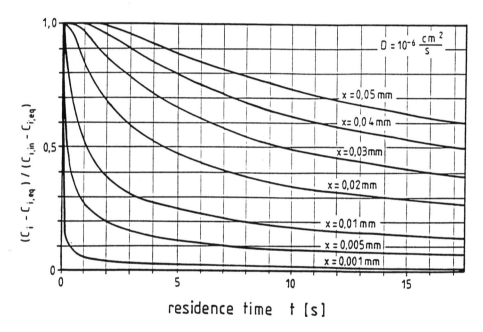

FIGURE 4. Concentration profiles.

Like low concentrations, there is a linear relation between the equilibrium concentration of the volatile components dissolved in the plastics melt and the partial pressure. Only at a high partial pressure does the curve deviate from the straight line. The linear relation can be described by the simplified Henry's Law:

$$P_i = C_i \cdot H_{(T)} \tag{5}$$

However, Henry constants which are applicable for the degassing of thermoplastics are scarce to find in the available literature. Figure 6 shows the equilibrium concentration of ethylene and vinyl acetate dissolved in PE-LD vinyl acetate copolymer (EVA). These

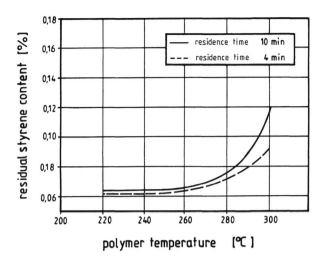

FIGURE 5. Degradation of polystyrene.

Henry's law: $\quad p_i = c_i \cdot H_i$

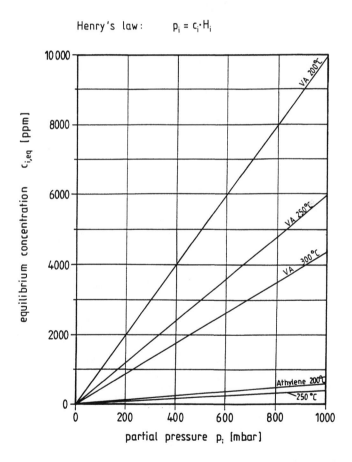

FIGURE 6. Equilibrium concentration of ethylene and vinyl acetate in LDPE.

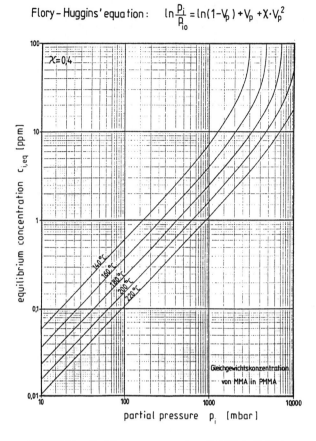

Flory–Huggins' equation: $\ln\dfrac{P_i}{P_{io}} = \ln(1-V_p) + V_p + X\cdot V_p^2$

FIGURE 7. Equilibrium concentration of MMA in PMMA.

equilibrium concentrations allow the judging as to which residual content of volatile components in polymers the degassing can be best effected. In general, one can conclude from the diagram that low partial pressures and, on the other hand, high temperatures, are favorable for the degassing.

Like high concentration the relation can be described by the Flory and Huggins equation. Figure 7 shows the Flory-Huggins diagram of PMMA-MMA solutions.

$$\ln \frac{Pi}{P_{io}} = \ln(1 - Vp) + Vp + x \cdot Vp^2 \qquad (6)$$

Low partial pressures can be obtained by a low total pressure, i.e., a good vacuum system in the degassing section of the extruders. Lowering the total pressure below 10 mbars, however, involves, in many cases, high expenses for the appropriate vacuum system itself and for the sealing of the extruder.

Lowering the partial pressure below the total pressure in the degassing section can be achieved by the addition of so-called "stripping agents", such as water or inert gases, Figure 8. These stripping agents are injected into the extruder, carefully blended with the melt, and degassed together with the volatile components to be degassed. Even small quantities of low-molecular substances lead to such a reduction of partial pressure.

Equations 5 and 6 show the relation between the partial pressure of the gas to be removed P_1, the total pressure in the degassing section P as well as the quantity portions m_1 and m_2 and the relevant molecular weights M_1 and M_2.

● Reduction of partial pressure:

$$P_i = x_i \cdot P_{vac}$$

● Decreasing of polymer temperature:

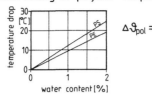

$$\Delta \vartheta_{pol} = \frac{c_w \cdot \Delta h_w}{c_{p,pol}}$$

● Increasing of degassing surface:

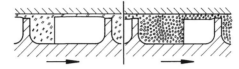

● Azetropic mixture:

Boiling temperature of the mixture styrene–water is lower than boiling temperature of the single components.

FIGURE 8. Effect of stripping agents.

$$P_i = P_{vac} \cdot \frac{m_1/M_1}{m_1/M_1 + m_2/M_2} \tag{7}$$

Figure 9 shows the reduction of the styrene's partial pressure by the addition of water. Stripping agents support the degassing in their ways as well. If the stripping agent is blended homogeneously under pressure in the melt and heated, the polymer melt in the degassing zone foams up, forms bubbles, and enlarges the inner degassing surface considerably. Simultaneously, this procedure causes the forming of thin layers between the bubbles, which favor the diffusion of the polymer melt. In some cases, the stripping agent and the volatile components to be removed from an azeotropic mixture whose boiling temperature is below the one of the individual components.

III. INFLUENCE OF THE BOUNDARY CONDITIONS ON THE RESIDUAL VOLATILE CONTENT

Corresponding to the equilibrium theory, the residual content becomes lower by lowering the partial pressure in the degassing section. In Figure 10 you can see the influence of the vacuum on the residual content. According to the first Fick's Law, a high concentration gradient on the polymer surface is necessary for a high gas flow. This can be reached by frequent surface renewal. The influence of the throughput per surface renewal relation on the residual content is shown in Figures 11 and 12.

IV. SCALE-UP FROM TEST EQUIPMENT TO PRODUCTION MACHINES

For the scale-up of the test results, the boundary conditions must be kept constant. This means the type and viscosity of the polymer, the percentage of volatiles and stripping agents,

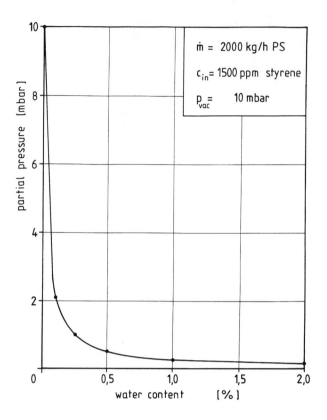

FIGURE 9.   Reduction of partial pressure due to water injection.

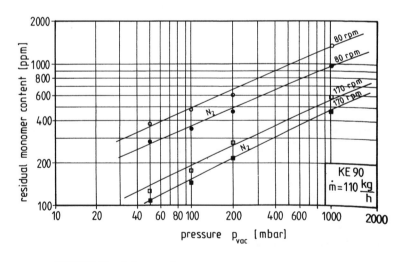

FIGURE 10.   Influence of pressure on the residual monomer content.

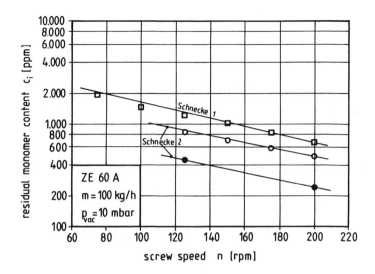

FIGURE 11. Influence of screw speed on the residual monomer content.

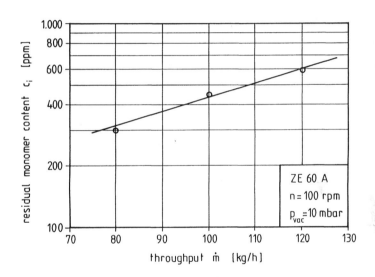

FIGURE 12. Influence of throughput on the residual monomer content.

as well as the temperature and the pressure in the degassing zone have to be kept constant. Knowing that an effective degassing is only possible via renewed surfaces (Figure 13), we have to provide equal output per renewed surface on the lab machine and on the production machine. This means that the specific surface loading

$$\pi_{\text{deg}} = \frac{\dot{m}}{\dot{A}} \qquad \left[\frac{\text{kg}}{\text{m}^2}\right] \tag{8}$$

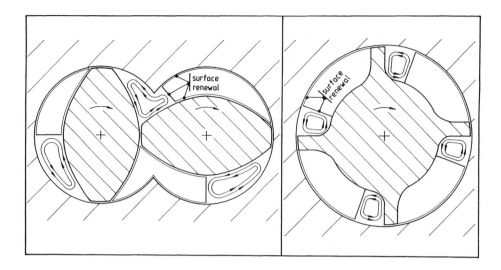

FIGURE 13.   Partially filled screw zones of single- and twin-screw extruders.

and the medium residence time

$$\bar{t} = \frac{V}{\dot{V}} \cdot f \tag{9}$$

have to be equal. The calculation of the surface renewal rate and the specific surface loading is shown on Figure 14.

Figure 15 shows the residual monomer content above the specific surface loading. The results have been obtained by tests with a twin-screw extruder of 43 mm diameter, a twin-screw extruder of 96 mm diameter, and on a single-screw extruder of 90 mm diameter. All tests result in one curve for degassing without a stripping agent and in one curve with a stripping agent. This diagram confirms the theory that the renewed surface is of essence. It is possible to scale-up from the process laboratory tests with ethylene, vinyl acetate, hexene, butene, and other monomers injected in the primary extruder. Production results confirm the theory of surface renewal.

V. PLANNING OF DEGASSING TRIALS

To understand the degassing process, extensive studies were effected. The real difficulty in simulating polyethylene degassing in process in laboratory tests is the volatility of the ethylene. To simulate the degassing process for ethylene and for copolymers, for example, vinyl acetate has to be injected and incorporated in the polyethylene. To prove a good incorporation of the ethylene, extensive mixing studies were done. The test set-up is shown in Figure 16. In a first melting and mixing extruder, the polymer is molten and the volatiles like ethylene, vinyl acetate, or higher boiling monomers are injected in the extruder and mixed into the polymer using special mixing elements. Special emphasis has to be put on good mixing, otherwise all degassing tests are unreliable. If the production extruder is fed with melt, we always need a test set-up similar to Figure 16 with a melting extruder and a degassing extruder.

In order to predict the residual content depending on the throughput of the production line, it is necessary to make a degassing diagram with the help of the values realized during the trial (Figure 15). This is done with the characteristics for the specific surface loading of the test machine (Figure 17) and the residual content values measured.

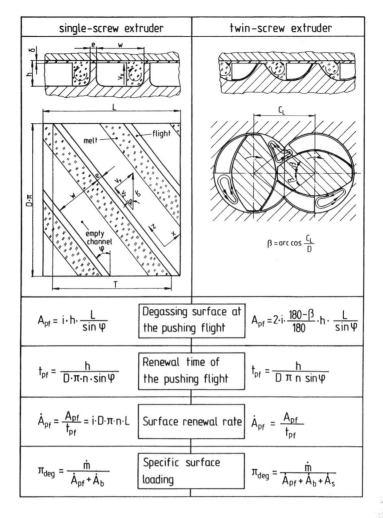

single-screw extruder		twin-screw extruder
$A_{pf} = i \cdot h \cdot \dfrac{L}{\sin \varphi}$	Degassing surface at the pushing flight	$A_{pf} = 2 \cdot i \cdot \dfrac{180-\beta}{180} \cdot h \cdot \dfrac{L}{\sin \varphi}$
$t_{pf} = \dfrac{h}{D \cdot \pi \cdot n \cdot \sin \varphi}$	Renewal time of the pushing flight	$t_{pf} = \dfrac{h}{D \, \pi \, n \, \sin \varphi}$
$\dot{A}_{pf} = \dfrac{A_{pf}}{t_{pf}} = i \cdot D \cdot \pi \cdot n \cdot L$	Surface renewal rate	$\dot{A}_{pf} = \dfrac{A_{pf}}{t_{pf}}$
$\pi_{deg} = \dfrac{\dot{m}}{\dot{A}_{pf} + \dot{A}_b}$	Specific surface loading	$\pi_{deg} = \dfrac{\dot{m}}{\dot{A}_{pf} + \dot{A}_b + \dot{A}_s}$

FIGURE 14. Calculation of the specific surface loading.

In case of a certain throughput it is possible to find out the minimum residual content. Otherwise in case of an allowed maximum residual content, the maximum throughput can be determined. For the degassing diagram, the specific surface load of the test machine has to be varied. Figure 18 shows how to vary the surface loading when the degassing conditions remain constant.

VI. PRODUCTION EXPERIENCE

Production extruders with rear and forward degassing have been built with an output range up to 28.000 kg/h for polyethylene. They are either used for degassing of ethylene only or for copolymers like vinyl acetate and acetic acid as well as higher boiling volatiles like butene, octene, and others. When degassing monomers, waxes are also removed. The vacuum system has to take this into account and to allow enough surface for condensing the stripping agent and the waxes. In some cases, gas washers are advantageous.

Figure 19 shows the monomer content before and after degassing. On an extruder with 550 mm diameter and an output of about 25.000 kg/h, we have extensive results of degassing polyethylene and copolymers. The monomer A content is reduced from 780 to 120 ppm,

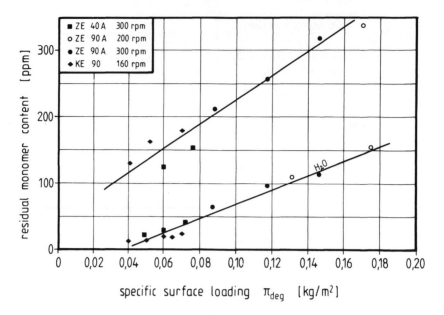

FIGURE 15. Degassing diagram.

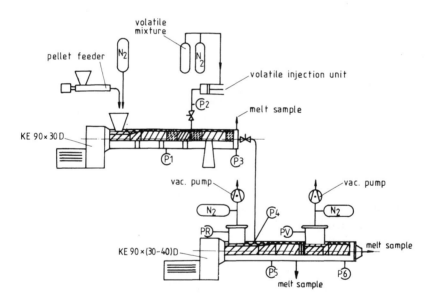

FIGURE 16. Set-up for degassing LDPE-copolymer.

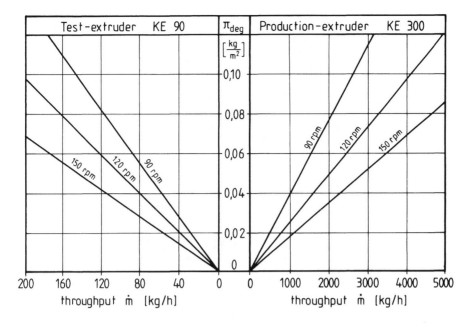

FIGURE 17. Characteristics for the specific surface loading.

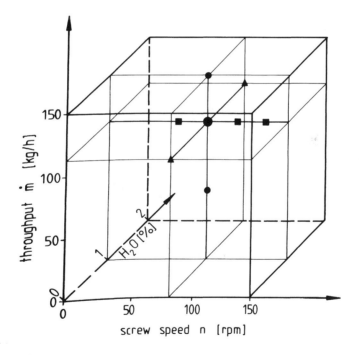

FIGURE 18. Variation of the degassing conditions.

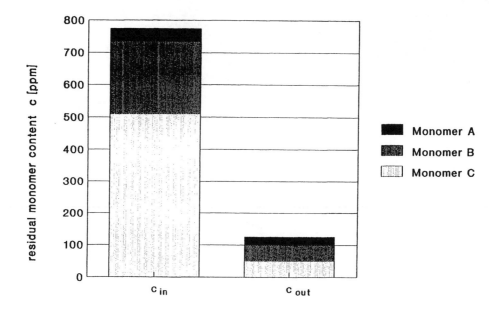

FIGURE 19.   Residual monomer content on production extruder.

TABLE 2
Comparison Between Single- and Twin-Screw
Degassing Extruder

Processing characteristics	Single-screw	Twin-screw
Melt feeding	+	+
Solid feeding, $\dot{m} < 3000$ kg/h	+	+
Solid feeding, $\dot{m} > 3000$ kg/h	−	+
Cooling capacity	+	O
High volatile concentration at extruder inlet	O	+
Self-cleaning	−	+
Vacuum in rear vent below 100 mbar	O	+
High pressure generation at a low temperature increase	+	O
Distribution of stripping agent and additives in polymer melt	+	+

+, possible without restrictions; O, possible with restrictions; and −, not possible.

the monomer B from 740 to 100 ppm and the ethylene from 510 to 50 ppm. The vacuum level was 30 mbars.

VII. MAIN USE OF SINGLE- AND TWIN-SCREW EXTRUDERS

For the majority of degassing tasks with melt feeding, it is possible to use both, the single- and twin-screw extruder. Table 2 shows a general valuation for the best extruder, single- or twin-screw, depending on the demand or boundary conditions. The fundamental observations made about the degassing led to the demands as classified in Table 3. These demands apply to both, single- and twin-screw extruders.

TABLE 3
Requirements for Degassing Extruder

Completely molten polymer
Screw channels partially filled
Large surface area
Frequent surface renewal
Ample volume of screw channel
Large vent opening
Long but controlled residence time
High vacuum, separate vacuum stages
High energy input
Addition and mixing in of flashing agents

VIII. SUMMARY

The renewal of the boundary surface proves to be the governing factor in extruder degassing. The scale-up of test results from laboratory extruders to production size extruders shows positive results. Production scale lines with 25,000 kg/h are in service.

Chapter 17

EPOXIDIZED RUBBERS

Sanjoy Roy, B. R. Gupta, and S. K. De

TABLE OF CONTENTS

0-8493-4401-8/93/$0.00 + $.50
© 1993 by CRC Press, Inc.

635

I. INTRODUCTION

Natural rubber (NR) is produced by the tree, *Hevea brasiliensis*. Chemically, it is *cis*-1,4-polyisoprene. It possesses excellent physical properties which ensure its usefulness as a general purpose rubber. However, natural NR along with other synthetic diene elastomers like styrene-butadiene rubber (SBR) and polybutadiene rubber (BR) have disadvantages, like low oil resistance, high air permeability, and low aging resistance. To overcome some of these drawbacks, chemical modification of the structure has been attempted by introducing different functional groups, like epoxy. This results in modified rubbers with reduced levels of unsaturation, showing enhanced resistance to oil and solvents, improved air permeability resistance, good bonding to metal, ceramics, and textiles, and good wet-skid resistance.

The introduction of epoxy groups along the polymer backbone is one of the most promising methods of modifying polydienes and has been known for a long time. Natural rubber was first epoxidized by Pummer and Burkhard[1] as early as in 1922. The chemistry of epoxidation of unsaturated compounds and subsequent ring-opening reactions have been studied by Rosowsky[2] and Swern.[3] Swern[4] and Witnauer[5] showed that epoxidation is a stereospecific reaction and the rate of epoxidation is governed by the substituents on the double bond. The ease and positioning of ring-opening of epoxy groups is again controlled by the neighboring groups.[6]

In earlier times, the epoxidation of NR and polydiene rubbers was carried out in solution. However, latex-stage epoxidation is economically advantageous for polymers which are available in latex form. Most of the earlier attempts to epoxidize NR resulted in nonrubbery products of complex mixture which often could not be reproduced. Gelling[7] patented a process of producing epoxidized NR (ENR) in latex-stage with the retention of rubbery properties comparable to those of NR.

II. METHODS OF EPOXIDATION OF DIENE ELASTOMERS

Epoxidation of diene elastomers can be achieved by the action of various peroxides and peracids. The different epoxidizing agents used are tabulated in Table 1. Epoxidizing agents can be broadly classified into five groups.

A. PEROXIDES AND HYDROPEROXIDES

Benzoyl peroxide has been used to epoxidize polyisoprene[8] while BR has been epoxidized with *t*-butylhydroperoxide in the presence of dioxomolybdenum *bis* (acetyl acetonate) as catalyst.[9] Although peroxides have been used for epoxidation, peracids appear to be more effective.

B. DIRECT USE OF PERACIDS

Peracids are used to epoxidize diene rubbers either in solution or as latex. Thus, BR and NR dissolved in solvents have been epoxidized using peroxybenzoic acid.[10-11]

Monoperphthalic acid was used by Roux, et al.[13] to epoxidize various diene rubbers, like *cis* and *trans* polyisoprene, SBR, BR, and polychloroprene (CR). Dreyfuss and Kennedy[14] epoxidized ethylene propylene diene rubber (EPDM) and butyl rubber (IIR) with the same reagent.

Among the various peroxycarboxylic acids used for epoxidation of diene polymers, m-chloroperbenzoic acid has been found to be the most efficient.[15] Dreyfuss and Kennedy[16] showed that this peracid reacts with the double bonds of diene polymers quantitatively.

One of the most widely used peracids is peracetic acid. Gelling and Smith[17] reported the use of this peracid to epoxidize NR in the latex-stage. Nippon Zeon Company Ltd.[18] patented the process of solution epoxidation of BR with peracetic acid at 60°C, but this

TABLE 1
Different Types of Epoxidizing Agents

Epoxidizing agent	Elastomer	Reaction	Ref.
Benzoyl peroxide	Polyisoprene	Solution	8
t-Butyl hydroperoxide + dioxomolebdenum *bis*(acetyl-acetonate)	Polybutadiene	Solution	9
Perbenzoic acid	Polybutadiene, natural rubber	Solution	10,11,12
Monoperphthalic acid	Polybutadiene, natural rubber	Solution	13,14
m-Chloroperbenzoic acid	EPDM, butyl, polybutadiene, polyisoprene	Solution	15,16
Peracetic acid	Polybutadiene, natural rubber	Solution latex	17,18,20, 21
Hydrogen peroxide + acetic acid	Polybutadiene, natural rubber	Solution	22,23,24, 25,26
Hydrogen peroxide and acetic acid + toluene-*p*-sulfonic acid	Polybutadiene, natural rubber	Solution	28
Hydrogen peroxide + formic acid	Natural rubber, polybutadiene, EPDM, SBS	Solution and latex solution	26,29–34
Hydrogen peroxide + formic acid + A second acid	SBS and SIS TPE, natural rubber	Solution Latex	35,36 37,38
N-phenylcarbamoyl azoformate	Natural rubber	Solution	40
$H_2O_2 + Ce(SO_4)_2 \cdot 4H_2O$ + polyether or polyether ester	Polybutadiene	Solution	39
H_2O_2 + ammonium-tetrakis (diperoxotungsto)-phosphate (3-) $[(C_8H_{17})_3NCH_3]_3[PO_4[W(O)(O_2)_2]_4]$	SBS (TPE), butyl EPDM, SBR	Solution	41,42

reagent possesses problems as a health hazard and loss of available oxygen.[19] Burfield et al.[20] studied the kinetics of epoxidation with peracetic acid. Gelling also reported the epoxidation process of NR latex with peracetic acid at temperatures <10°C.[21]

C. *IN SITU* GENERATED PERACIDS

This process accomplishes the charging of hydrogen preoxide and a carboxylic acid with the epoxidizing substrate in a reaction vessel. Hydrogen peroxide and the carboxylic acid reacts to generate the peracid *in situ*, which in turn reacts with the olefinic double bonds to convert them to epoxy groups. A schematic representation of the reaction is given below, (Reactions 1 and 2)

$$R-C\overset{O}{\underset{OH}{\diagup}} + H_2O_2 \longrightarrow R-C\overset{O}{\underset{O-O_{\diagdown H}}{\diagup}} + H_2O \cdots \quad [1]$$

$$R-C\overset{O}{\underset{O-O_{\diagdown H}}{\diagup}} + >C-C< \longrightarrow >C\overset{O}{\underset{}{\diagdown\diagup}}C< + R-C\overset{O}{\underset{OH}{\diagup}} \cdots \quad [2]$$

The *in situ* generated peroxy carboxilic acid has been used for the production of low-molecular weight products and is found to be a better epoxidizing agent for diene polymers, in terms of reaction rate, product purity, and yield. Mainly acetic and formic acids have been used to generate *in situ* peroxy carboxylic acids to epoxidize various diene polymers.

1. Acetic Acid/H_2O_2

Acetic acid and hydrogen peroxide react to form peracetic acid at a temperature higher than 40°C.[22] However, at temperatures below 60°C the rate of generation of peracetic acid is rather slow. Thus, to obtain a reasonable rate, the reactions are, generally conducted at higher temperatures. Nonagaki and Moroshita[23] patented the use of a mixture of acetic acid and hydrogen peroxide for epoxidation. Terry and Jacobs[24] patented the process to produce anticorrosive paints by further modification of epoxidized polydienes produced by the action of acetic acid and hydrogen peroxide at 60 to 70°C. Zuchowska[25] used the same route to epoxidize BR of various microstructures. Hayashi et al.[26] also used acetic acid/H_2O_2 to epoxidize NR and BR in solution.

In an acetic acid/H_2O_2 system, use of additional catalyst has also been reported to accelerate reaction rate and/or to perform the reaction at lower temperatures. Thus, acetic acid and 50% solution of hydrogen peroxide was used in the presence of an ion-exchange resin (Dowex 50W-X12®) to epoxidize liquid BR in benzene at 80°C.[27] Badran and Abdel-Bary[28] improved this method by the reduction of reaction temperatures to 40°C by replacing Dowex resin by toluene-*p*-sulfonic acid.

2. Formic Acid/H_2O_2

Subsequently, the *in situ* generated performic acid process was developed for epoxidation of polydiene elastomers and was found to be very effective in the absence of catalyst.[26,29-31] Gelling patented a process[7] of producing epoxidized NR with the *in situ* formed performic acid in the presence of 1 phr of 2,5,di-*tert*-phenylhydroquinone at a reaction temperature of 60°C to get a 50 mol% epoxidized NR. Ryabova et al.[32] described the epoxidation of EPDM in carbon tetrachloride solution at a temperature of 20 to 35°C with formic acid/H_2O_2 using a ratio of H_2O_2: $>C=C< = 3:1$ and HCOOH:H_2O_2 in the range of 1:1 to 2:1. The *in situ* generated performic acid route was also followed by Bradbury and Perera[33] to epoxidize NR. Later, Perera[34] reported the surface epoxidation of NR sheets and strips. The epoxidation of stripes (laces) produced a mixture, in which 10% of the polymer was 50 mol% epoxidized while 90% remained unmodified.

3. H_2O_2/Formic Acid in the Presence of a Second Acid

The *in situ* formed performic acid can be used with another low-molecular weight acid, e.g., glacial acetic acid, acetic anhydride, or propionic acid.[35,36] Thus, styrene-butadiene-styrene (SBS) triblock copolymer was epoxidized with this system by Udipi.[35] Roy et al.[37,38] used the *in situ* generated performic acid route in the presence of acetic acid to epoxidize NR latex and reported the effect of the acetic acid concentration on the overall epoxidation rate and the product distribution. Few typical feed compositions for the epoxidation of NR latex by *in situ* generated performic acid in the presence of a second acid is described in Table 2.[37] Figure 1 shows the reaction time vs. conversion plots for the above feed compositions.

D. HYDROGEN PEROXIDE IN THE PRESENCE OF AN ORGANIC OR ORGANOMETALLIC CATALYST

The yield of the product from the *in situ* formed peracid route was good when the epoxidation level was less than 50%, but above the 50% level formation of the side products becomes a concern. Moreover, trace amounts of unreacted acid present in the rubber matrix

TABLE 2
Some Typical Feed Compositions

	1	2	3
60% NR latex, ml	50.0	50.0	50.0
H_2O_2, g · mol · dm^{-3}	2.93	2.31	2.95
Formic acid, g · mol · dm^{-3}	4.40	2.31	1.48
Acetic acid, g · mol · dm^{-3}	—	2.31	2.21
Emulvin T,[a] g	1.0	1.0	1.0
Distilled water, g	100.0	100.0	100.0

[a]   Emulvin T™ is the registered TM of Bayer (India) Ltd., which is a
nonionic surfactant.

From S. Roy, B. R. Gupta, and B. R. Maiti, *J. Elast. Plast.*, 22, 280,
(1990). With permission.

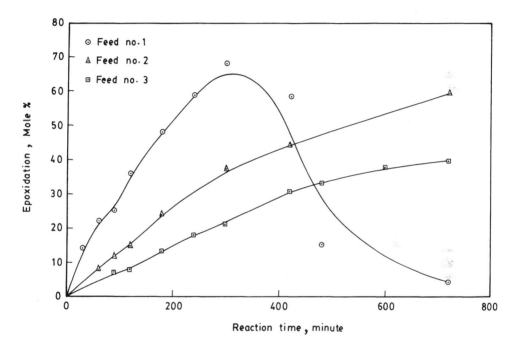

FIGURE 1.   Reaction time vs. conversion plots for different feeds as shown in Table 1.

becomes detrimental from a stability point of view. Thus, attempts have been made to
epoxidize diene rubbers with hydrogen peroxide in the presence of a basic or an organometallic catalyst. The Nippon Zeon Company, Ltd. patented,[39] in 1981, the process of
epoxidizing BR with H_2O_2 at 40 to 70°C in the presence of $Ce(SO_4)_2 \cdot 4H_2O$ and an organic
compound of the following type: glycerol, $RO(CH_2CH_2O)_nH$, $RCO_2(CH_2CH_2O)_nH$, and
$RCO_2(CH_2CH_2O)_nCOR$, where n = 1 to 200. It was claimed that this process produced a
byproduct-free epoxidized BR.

An unsuccessful attempt was made by Burfield, et al.[20] to epoxidize NR with hydrogen
peroxide in the presence of acetonitrile and sodium tungstate separately. However, NR has
been epoxidized using ethyl N-phenylcarbamoyl azoformate.[40]

Very recently Jian and Hay[41] reported successful use of a noble catalyst, viz., methyltrioctyl ammonium tetrakis (diperoxo tungsto) phosphate (3-), $[(C_8H_{17})_3NCH_3]_3$

$[PO_4[W(O)(O_2)_2]_4]$ with hydrogen peroxide to produce side-product-free epoxidized SBR, SBS, EPDM, and IIR rubbers.

In a typical catalytic epoxidation run[42] with an organometallic catalyst and hydrogen peroxide, 5.0 g of polymer was dissolved in 50 ml of 1,2-dichlorobenzene (or toluene) in a 100 ml three-necked flask fitted with a thermometer, reflux condenser, and a stirrer. The polymer solution was heated to maintain the desired temperature. The required amount of catalyst, methyltrioctyl ammonium tetrakis(diperoxotungsto)phosphate (3-) $[(C_8H_{17})_3NCH_3]^+{}_3[PO_4\{W(O)(O_2)_2\}4]^{3-}$ and 30% hydrogen peroxide were introduced with vigorous stirring. After completion of the reaction the organic layer was washed with distilled water to remove unreacted H_2O_2 followed by the polymer precipitation with methanol or hexane. The precipitated polymer was dried under vacuum at ambient temperature for 24 h.

E. REACTIVITY OF DIFFERENT DIENE POLYMERS TOWARD EPOXIDATION

The reactivity of different double bonds toward epoxidation for a particular reagent were found to be as follows.[13,25] (Reactivity decreases downward).

Cis-1,4-polyisoprene
Trans,1,4-polyisoprene
1,2 and 3,4-polyisoprene
Cis-1,4-polybutadiene
trans,1,4-polybutadiene
1,2-polybutadiene
Styrene-butadiene copolymers
Polychloroprene

The reason behind such behavior is that, any group adjacent to the double bond which enriches the –electron density of C=C, will contribute to a higher rate of epoxidation. Polychloroprene, having an electron withdrawing group, shows the least affinity toward epoxidation.

F. SECONDARY REACTIONS DURING EPOXIDATION

During the epoxidation process secondary products of different types are formed in varying degrees, depending upon reaction parameters like temperature, type of reagent, and acid concentration. The major secondary reaction products are tetrahydrofuran and molecules containing hydroxyl, esters, carbonyls, and ether groups. The schematic representation of different secondary reactions are given below. (Reactions 3 and 4).

It has been found that the higher the acid concentration and reaction temperature, the higher the extent of secondary product formation, but lowering of acid concentration and reaction temperature lowers the reaction rate. Optimum acid concentration for maximum rate and minimum secondary reaction has been determined by Roy, Gupta, and Maiti[38] in the case of systems with H_2O_2, formic acid, and acetic acid.

The undesirable secondary reaction groups are detectable by ^{1}HNMR and IR spectrometry. The characteristic IR absorption peaks of these groups and the characteristic ^{1}H-NMR resonances are tabulated in Table 3.

Often, the 3.85 ppm signal due to the proton to the -OH group is found to be missing, even when the -OH group has been detected by IR. D_2O-wash, in this case, may give more prominent indication for the presence of hydroxyl group in the ^{1}H NMR spectra. The concentration of hydroxyl group may be determined by the integral of the resonance peak area due to the evolved H_2O molecules by isotope exchange process.

$$> C \overset{O}{\underset{}{\diagup\!\!\diagdown}} C < \quad \xrightarrow{[H_2O]} \quad \underset{OH \;\; OH}{> C - C <} \quad \xrightarrow{RCO_2H} \quad \underset{OH \quad O}{> C - C <} \underset{R}{\overset{}{\underset{\underset{O}{\parallel}}{C}}} \qquad \cdots \quad [3]$$

$$\underset{X \;\; OH}{> C - C <} \quad \xrightarrow{\underset{X \;\;\; OH}{> C - C <}} \quad \underset{X \quad O}{> C - C <} \underset{X}{\overset{}{> C - C <}} \qquad \cdots \quad [4]$$

(vertical arrow labeled HX from equation [3] to equation [4])

TABLE 3
Characteristic IR and ^{1}H-NMR Peaks

Functional group	IR absorption (cm^{-1})	^{1}H-NMR signals (ppm)
Hydroxyl (–OH)	3600–3200	—
–C–H OH	—	3.85
Ester	1720–1740	—
Carbonyls	1710–1730	—
Formate	1725	9.75
Tetrahydrofuran	1065	3.65
Aliphatic ether	1115	—
Epoxy (*cis*-1,4-polyisoprene)	870, 1240	2.70

When epoxidation is carried out at comparatively higher temperatures (>25°C) using excess acid concentration, beyond certain epoxidation level, the formation of epoxy groups is overshadowed by the formation of the secondary reaction products. If the reaction is further continued, depletion of the epoxy group results.[37] This phenomenon is depicted in Figure 1. The maxima in the conversion vs. the reaction time curve depends on the initial concentration of the acid and peracid, at a constant reaction temperature.[38]

III. ANALYSIS AND CHARACTERIZATION

A. CHEMICAL ANALYSIS

Several direct titration methods are available for the determination of epoxy content. One such method involves the titration with hydrogen bromide in acetic acid.[43] Carboxylic acids, aldehydes, ethers, esters, and peroxides do not interfere with this reaction, but hydroperoxide reacts slowly. The titration of the epoxidized diene polymer is performed by dissolving it in a suitable solvent.

Another titrimetric method[44] involves the use of perchloric acid in the presence of a soluble quaternary ammonium salt.

The third method[45,46] uses dioxane to dissolve the rubber sample and then allowed to stand for 2 h in a known concentration of excess hydrochloric acid. The unreacted acid was back titrated with standardized caustic soda solution.

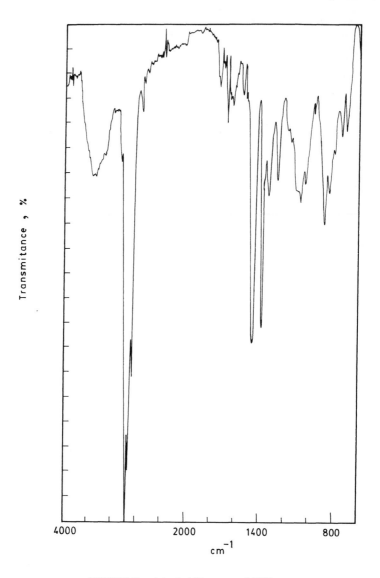

FIGURE 2.   A typical IR spectra of ENR.

Elemental analysis for C, H, and O has also been used to determine the epoxy content after the removal of residual moisture by hot pressing the sample at 110°C for 1 min prior to analysis.[47]

B. SPECTROSCOPIC ANALYSIS
1. IR Spectrophotometry

For IR analysis, different methods of sample preparation are in practice. Films may be cast from the polymer solution directly on potassium bromide or sodium chloride plates.[13] If the polymer contains high gel, a gelled piece may be pressed between two KBr or NaCl plates.[48] Alternatively, the dry ENR sample is dispersed in polyethylene matrix and a thin film prepared from this blend by hot pressing around 120°C may be used directly to analyze the sample by IR.[49] Films may also be prepared directly on a silver halide plate from the epoxidized latex and drying it under a vacuum for 2 min at 100°C[47] Figure 2 is a typical IR spectra of ENR.

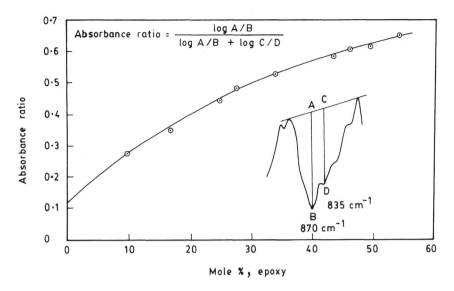

FIGURE 3. Calibration curve for determination of mol% epoxy content of ENR from IR analysis. (Reproduced from *Br. Polym. J.*, 16, 134 (1984) by permission of the publisher, Elsevier Applied Science Publishers Ltd.)

Oxirane oxygen bands appear at 800 to 950 cm^{-1}, depending upon the backbone structure of the parent diene polymer. When *cis*-1,4-polyisoprene backbone is converted to the *cis*-epoxy structure, a new band near 870 cm^{-1} appears.[13] The quantitative estimation of epoxy content may be performed by calibration against an internal standard (generally methyl deformation band at 1375 cm^{-1}) or an external standard as per ASTM method.[50] Relative percent conversion of double bonds to epoxy groups may be calculated by the Davey and Loadman method.[47] They calculated the relative absorbance ration, A$_r$, which is defined as,

$$A_r = \frac{a_{870}}{a_{836} + a_{870}} \tag{1}$$

where, a$_{870}$ and a$_{836}$ are the absorbance values at 870 cm^{-1} and 836 cm^{-1} due to *cis*-epoxy and *cis*-double bond, respectively. The calibration curve showing variations of A$_r$ with mol% epoxy content is shown in Figure 3.[47]

2. NMR Analysis

a. ^{1}H-NMR Technique

The ^{1}H-NMR method is rapid and provides useful structural information. Dubertaki and Miles[51] first developed the method for the analysis of epoxy compounds by ^{1}H-NMR technique. In diene polymers the olefinic methine proton resonance appears at 5.14 ppm, which on epoxidation, gradually disappears with the appearance of a new signal around 2.7 ppm. This was assigned by Golub[52] to the epoxy methine proton resonance. The epoxy content can be determined from the integrated area of epoxy methine proton resonance divided by the sum of the area of the epoxy and olefinic methine proton resonances,[32,53,54] by the following formula:

$$Mol \% \ epoxy = \frac{A_{2.70}}{A_{5.14} \times A_{2.70}} \times 100 \tag{2}$$

where, A$_{2.70}$ and A$_{5.14}$ represent the integrated area under the signal at 2.70 and 5.14 ppm, respectively. When the backbone of the parent diene possesses mixed microstructures, as

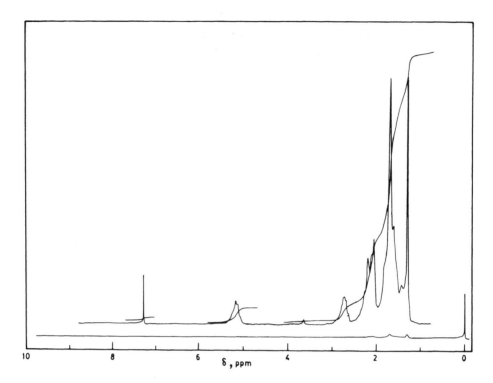

FIGURE 4.   A typical ¹H-NMR spectra of ENR.

in an SBS triblock copolymer, BR, and SBR where 1,2-, *cis*-1,4-, and *trans*-1,4- structures may coexist, Equation 2 should be modified appropriately to accommodate the contributions for all the respective olefinic groups and epoxy signals.[43] Figures 4 and 5 represent typical ¹H-NMR spectra of epoxidized NR and BR, respectively.

b. ¹³C-NMR Technique

The ¹³C-NMR resonances of epoxidized NR were first assigned by Gemmer and Golub.[15] Later the fine structure in ¹³C-NMR spectra of various epoxidized polydienes, viz., 1,4-polybutadiene[30] and 1,4-polyisoprene[32,47,54] were assigned by other workers. With a *cis*-1,4-polyisoprene backbone, it was found that the olefinic methine carbon resonance occurred at 125 ppm and for epoxy methine carbon at 64 ppm. They had similar nuclear overhauser enhancements and short spin-lattice relaxation times.[54] Thus, it is possible to determine the extent of epoxidation from the areas of the signal at 64 ppm divided by the sum of the area of the signals at 64 and 125 ppm.[54] Figure 6 is a typical ¹³C-NMR spectra of ENR.

C. DIFFERENTIAL SCANNING CALORIMETRY

The glass transition temperature, Tg, determined by differential scanning calorimetry (DSC) has been found to increase linearly with the epoxidation level.[37,47,53,55] Thus, epoxidation levels can also be determined by measuring Tg. The dried samples of ENR are encapsulated in aluminium crimples and heated to 150°C at the rate of 300 degrees/min⁻¹. It is then equilibrated at − 110°C by quenching in liquid nitrogen. The DSC scan is taken at the heating rate of 20 degrees/min⁻¹ up to 120°C in nitrogen.

D. DENSITY

One of the simplest methods of estimation of epoxy content is to measure the density which is also found to vary linearly with the epoxy content. Figure 7 shows the variation of Tg and density with epoxidation level of ENR.

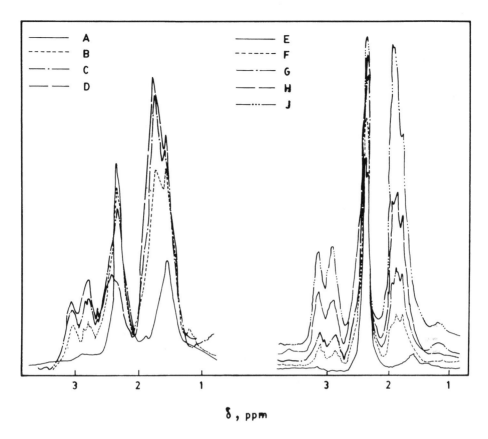

FIGURE 5. ^{1}H-NMR spectra of: (A) 1,4 BR; (B) 1,4 EBR (0.27 mol/100 g); (C) 1,4 EBR (0.35 mol/100 g); (D) 1,4 EBR (0.46 mol/100 g); (E) 1,2 BR; (F) 1,2 EBR (0.27 mol/100 g); (G) 1,2 BR; EBR (0.40 mol/100 g); (H) 1,2 EBR (0.49 mol/100 g); and (J) 1,2 EBR (0.57 mol/100 g). (Reproduced from *Polymer*, 21, 514 (1980) by permission of the publisher, Butterworth Heineman Ltd, U.K.)

E. COMPARISON OF METHODS

The various methods for determination of epoxy content have been compared by different workers.[47,53,54] The accuracy of quantitative analysis by ^{1}H-NMR depends on the level of epoxy content, since at low levels or very high levels of epoxidation, the error becomes large in the integration of the peak area of epoxy or olefinic methine protons. In the range of 20 to 70 mol% of epoxidation, the error in measurement is within 1 mol%, but secondary reaction products of epoxy groups may lead to erroneous results. Thus, it is desirable to have a knowledge of the extent of secondary reactions before the actual calculations of epoxy level. Differential scanning calorimetry and IR spectrophotometry provide very precise results but require calibration. However, at higher epoxidation levels, the hydrofuranization of epoxy groups leads to erroneous DSC results, since, hydrofuran groups also contribute to the increase in Tg. Direct titration methods give accurate results below 10 mol% of epoxidation. The reason of deviation of the results in titrimetric methods has been explained as follows. At higher epoxidation level the blocks of epoxy groups are more, as soon as hydrohalic acid is added to the epoxidized polydiene sample and hydrofuranization reaction is initiated at one epoxy group of a block, the remaining epoxy groups of the block undergo hydrofuranization instantly and, as such, very low hydrohalogenation reactions can occur.[47] However, at low epoxidation levels (<10%) due to the randomness of the reaction, no block of epoxy group exists and, thus, the titrimetric results obtained are accurate.[47] However, when properly calibrated, IR spectrophotometry provides accurate results with the

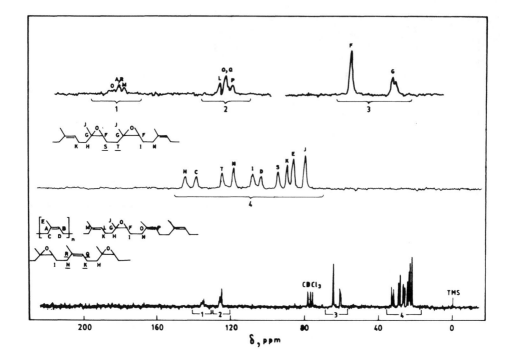

FIGURE 6. ^{13}C-NMR spectra of ENR.

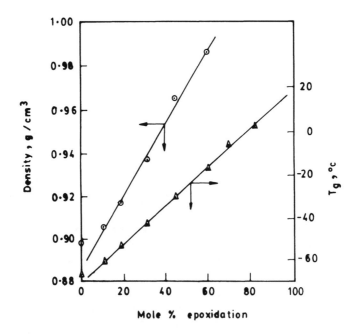

FIGURE 7.   Variation of T_g and density with epoxidation level of ENR. (Reproduced from *J. Appl. Polym. Sci.*, 29, 1661 (1984) by permission of the publisher and copyright owner, John Wiley & Sons, Inc.)

TABLE 4A
Epoxy Content Determined by Various Methods

Sample no.	Epoxide by DSC (mol%)	Mol% Epoxide by IR (mol%)	Epoxide by titration (HCl) (mol%)
1	1.0	1.5	<1.0
2	8.5	8.0	8.0
3	14.0	14.0	12.0
4	20.0	19.5	15.0
5	26.0	26.0	22.0
6	57.5	60.0	25.5

From Davey, J. E. and Loadman, M. J. R., *Br. Polym. J.*, 16, 134, 1984. With permission.

TABLE 4B
Comparison of Epoxy-Content (mol%) of ENR as
Determined by NMR and Titrimetric Method (HCl)

Sample no.	Epoxide by Titration (HCl) (mol%)	Epoxide by ^{1}H NMR (mol%)
1	9.9	10.0
2	15.5	17.0
3	20.5	25.0
4	24.5	34.5
5	27.0	44.0
6	26.0	50.0

From Davey, J. E. and Loadman, M. J. R., *Br. Polym. J.*, 16, 134, 1984. With permission.

epoxidation level of 10 to 75 mol%. Table 4A and B show that at low epoxidation levels, the titrimetric method provides results which are similar to the other methods, but at higher epoxidation levels, (>10 mol%), the titrimetric results are erroneous.

IV. STRUCTURES OF EPOXIDIZED POLYDIENES

The ^{13}C-NMR method gives more information than ^{1}H NMR regarding the microstructure of epoxidized polydienes. Hayashi et al.[26,30] studied the ^{13}C-NMR of epoxidized polybutadiene and polyisoprene, while Bradbury and Perera[33] studied the epoxidized polyisoprene alone. The methylene resonance in ^{13}C-NMR spectra has been assigned on the basis of diad and triad sequence. The integrals of these resonance peaks are used to calculate the diad and triad sequence distribution and compared with the calculated values assuming a random distribution. The random distribution of epoxy groups has been confirmed by the good agreement between these two sets of data.[26,33,47]

The DSC study performed by Burfield[20] revealed that single inflection for Tg appeared in the thermogram, which ruled out the possibility of the existence of an alternating copolymer. However, at higher epoxidation levels (>50%), the broadening of the inflection for glass transition occurs, reflecting inhomogenity of the system. At still higher (>60%) extents of reaction, two distinct Tgs became visible.[37,47] This has been attributed to the fact that at

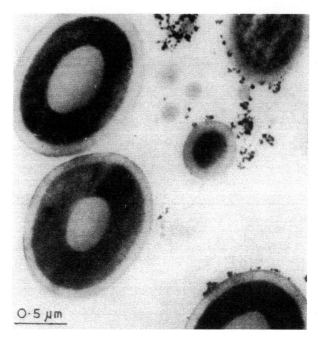

Fresh sample

FIGURE 8.   Electron micrograph of fresh latex particles.

higher extents of epoxidation, especially in the presence of excess acid in the *in situ* generated peracid route, due to the subsequent hydrofuranization process, some tetrahydrofuran blocks might have been generated, thereby inducing the incompatibility which is reflected in two distinct inflections in glass transition measurements.[47] Transmission electron microscopic (TEM) studies with fixation technique of NR latex particles during epoxidation with per-formic acid showed dark-ringed structures.[55] Figures 8 and 11 show two such TEMs. This was interpreted to indicate heterodispersity of latex so that at any instant a spectrum of particles with varying levels of epoxidation might exist, giving rise to a random distribution of epoxy groups in the overall product.

The reaction of m-chloroperbenzoic acid with double bonds of polydiene is quantitative, highly selective, and gives very pure product.[15] This fact has been used to study the crystal structure of 1,4-polybutadiene. Epoxidation was carried out on a solution of suspended polymer crystals and ^{1}H NMR data was used to evaluate the fraction of surface folds and cilia.[56-58] The work was extended to calculate the fold lengths using ^{13}C-NMR data.[59] ^{13}C-NMR data of 100% modified polymer also revealed that the addition of oxygen to the double bond may take place in two modes, and thus, gives rise to diastereomers in diad structures.[33,59]

Davies et al.[60] found that epoxidized *cis*-1,4-polyisoprene underwent strain crystallization up to a 50 mol% modification level, but above 50% though the extent of strain crystallization is considerably low, it could be detected up to as high as 95 mol% of epoxidation. X-ray studies showed[60] that the size of the crystal cell increased with the extent of epoxidation, which was attributed to the inclusion of epoxy rings in lieu of the double bonds.

Tutorskii et al.[12] studied the change in solution viscosity of epoxidized polybutadiene and polyisoprene produced by the action of perbenzoic or, monoperphthalic acid in benzene and benzene-dioxane mixture. The viscosity vs. percent epoxidation plot showed maxima at 50% using monoperphthalic acid and in the range of 30 to 40% for perbenzoic acid. The existence of maxima was attributed to two opposing phenomena: (1) at low level of

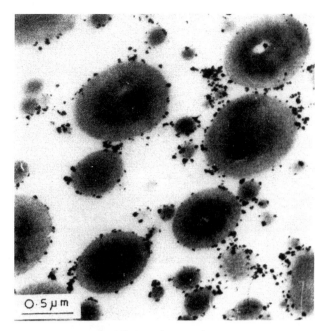

Matured sample

FIGURE 9.   Electron micrograph of matured latex particles. (Reproduced from, *Proc. Intl. Rubber Conf.*, 1985, Kuala Lumpur, by permission from the publisher, The Rubber Research Institute of Malaysia, Kuala Lumpur, Malaysia.)

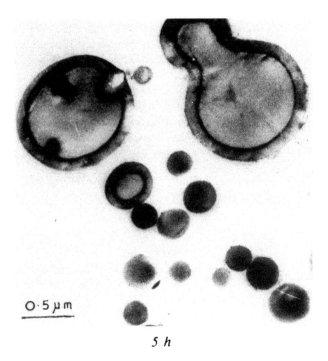

5 h

FIGURE 10.   Electron micrograph of latex particles after 5 h of reaction. (Reproduced from, *Proc. Intl. Rubber Conf.*, 1985, Kuala Lumpur, by permission from the publisher, The Rubber Research Institute of Malaysia, Kuala Lumpur, Malaysia.)

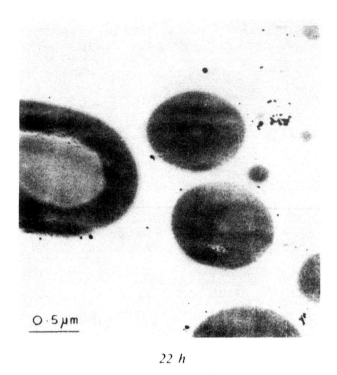

22 h

FIGURE 11. Electron micrograph of latex particles after 22 h of reaction. (Reproduced from, *Proc. Intl. Rubber Conf.*, 1985, Kuala Lumpur, by permission from the publisher, The Rubber Research Institute of Malaysia, Kuala Lumpur, Malaysia.)

epoxidation there was an increase in the coil-structure size due to the replacement of the double bonds by epoxy rings and electrostatic repulsion and (2) increased polarity of the polymer backbone at higher conversion level resulting in the mismatch of the cohesive energy density (or solubility parameter) of solvent and polymer system, causing partial collapse of the coil structure, and thus, reduction in viscosity. Addition of dioxane to benzene, however, improved the solubility parameter of the solvent system toward the epoxidized polydiene and, thus, maxima of the curve shifted toward the higher value of epoxidation. This conclusion was confirmed by the determination of solubility parameters of epoxidized polybutadiene[25] and ENR.[61,62] Viscosity studies also confirmed that there was no main chain cleavage, provided the reaction was performed under controlled conditions.[33,63]

V. KINETICS OF EPOXIDATION REACTION

Burfield et al.[20] suggested that the epoxidation reaction of NR with peracetic acid, like the epoxidation of low-molecular weight olefins, was also found to be second order. This has been confirmed by Roy et al.[37] who used performic acid. They also found that with the *in situ* generated performic acid route, if hydrogen peroxide concentration is kept higher than that of acid, a pseudo-first-order kinetics results.[37,38] Subsequently, they described a kinetic model for predicting maximum possible epoxidation level for a given feed composition at certain temperatures.[64] The relevant equations are given below:

$$\frac{(C_E)_{max}}{C_{RO}} = [\beta]^{\beta/(1-\beta)} \qquad (3)$$

$$\beta = \frac{(K_2 + K_3)\, C_{AO}}{K_1\, C_{PO}} \tag{4}$$

where K_1, K_2, and K_3 are the rate constants for the following steps,

$$\text{Rubber} + \text{Peracid} \xrightarrow{\quad K_1 \quad}$$
$$\underset{(R)}{} \quad \underset{(P)}{}$$

Epoxidized Rubber + Acid

(E)   (A)

K_2 K_3

Tetrahydrofuran ←⎯⎯ ⎯⎯→ Ring-opened products
(F)   (H)

where C_E = concentration of epoxidized rubber, C_{RO} = initial concentration of $>C=C<$ in the rubber, C_{AO} = initial acid concentration, and C_{PO} = initial peracid concentration.

VI. RAW RUBBER PROCESSING

The mill breakdown and mill shrinkage at any given temperature for epoxidized rubber has been found to be higher than the unmodified counterpart.[65] The difference in mill shrinkage behavior has been explained by the restricted chain mobility of the epoxidized polydiene compared to the unmodified one, resulting in reduced orientation along the applied stress. The lower chain mobility of the epoxidized polymer is also evident from the lower spin-lattice relaxation time of ^{13}C NMR resonances, compared with those in NR.[33]

A recent study by Kumar et al.[66] on the milling behavior of gum rubbers showed that during high temperature (100°C) milling, ENR-25 formed bands on the back roll, but at a lower temperature (60°C) it banded on the front roll. ENR-50 formed bands in both rolls and the bands were not uniformly formed. Epoxidized rubber sticks onto the roll surface more than the unmodified rubber.[65,66] ENR with 50 mol epoxy content was found to be a dry adhesive for bonding aluminum to aluminum.[67]

VII. CROSS-LINKING REACTIONS

Gelling and Morrison[68] first studied the cross-linking reactions of epoxidized NR. They found that the sulfur vulcanization of ENR is faster than that of NR. The faster rate of sulfur vulcanization was attributed to the prevention of cyclization reaction (found in NR vulcanization) due to the existence of adjoining epoxy groups.[69] The sulfenamide acceleration of ENR has been reported to be slower as compared to NR. The reason was assigned to be the reaction of sulfenamides with epoxy groups, which was confirmed by the model compound study.[68] However, Roux et al.[13] found that for a particular sulfur curing system, the cure rate slightly reduced with increasing level of epoxidation.

The pH of the water extracts has been found to have a profound effect on the scorch behavior of ENR.[65,68] The optimum scorch time is found to be at a pH of 6.5. Acid-catalyzed secondary reaction of epoxy groups occurs at lower pH and cross-linking reactions take place via secondary ring-opened products like hydroxyl groups, resulting in aliphatic ether linkage.[68] This can be prevented by the incorporation of sodium carbonate in the mix, but excess sodium carbonate again reduces the scorch safety which is overcome by the use of a scorch-delaying reagent like cyclohexyl thiophthalimide (CTP).[68]

The peroxide cross-linking reaction of epoxidized 1,2-polybutadiene with dicumyl peroxide (DCP) has been studied by Zuchowska.[70] It is found that the crosslinking efficiency of the modified elastomer is less than that of the unmodified one. However, a decrease in double bond content by epoxidation of 1,4-polybutadiene has no affect on the crosslinking efficiency. This may be attributed to the promotion of inter- and intramolecular rearrangement of the epoxy rings by polymeric radicals in the 1,4-polybutadiene chain. The rearrangement

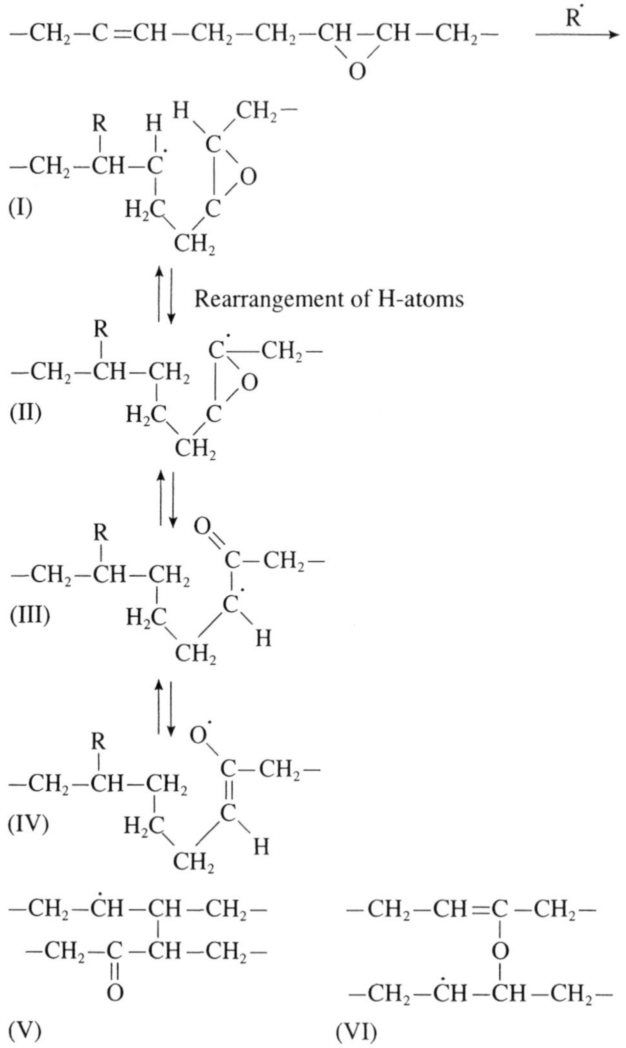

$$-CH_2-C=CH-CH_2-CH_2-CH-CH-CH_2- \quad \xrightarrow{R^\cdot}$$

with epoxide (O bridging the CH–CH).

$$\cdots \quad [5]$$
Structure (I): $-CH_2-CH-C$ with R and H substituents, bearing a ring containing CH_2, O, H_2C, C, CH_2 groups.

$$\Big\updownarrow \quad \text{Rearrangement of H-atoms}$$

$$\cdots \quad [6]$$
Structure (II): $-CH_2-CH-CH_2$ (R substituent) with ring $C-CH_2-$, O, H_2C, C, CH_2.

$$\Big\updownarrow$$

$$\cdots \quad [7]$$
Structure (III): $-CH_2-CH-CH_2$ (R substituent) with ring $O=C-CH_2-$, H_2C, C, H, CH_2.

$$\Big\updownarrow$$

$$\cdots \quad [8]$$
Structure (IV): $-CH_2-CH-CH_2$ (R substituent) with ring $O^\cdot-C-CH_2-$, H_2C, C, H, CH_2.

Structure (V):
$$-CH_2-\overset{\cdot}{C}H-CH-CH_2-$$
$$-CH_2-C-CH-CH_2-$$
$$\overset{\|}{O}$$

Structure (VI):
$$-CH_2-CH=C-CH_2-$$
$$\overset{|}{O}$$
$$-CH_2-\overset{\cdot}{C}H-CH-CH_2-$$

mechanism[70] is schematically represented by Reactions 5, 6, 7, and 8. Further rearranged structures of Reactions 7 and 8 may react with the 1,4-butadiene unit of EBR to form crosslinks as shown by Structures V and VI.

A typical compound formulation for ENR is given in Table 5 and Figure 12 shows the Monsanto rheograph of the compound.

Epoxidized polydiene elastomers can also be crosslinked by nonconventional means through the reaction of epoxy groups. Thus, dibasic acids, polyamines, and other curatives for diglycidyl-ether-based epoxy resins may be used for crosslinking epoxidized rubbers.[27,71] A detailed study on the cross-linking behavior of ENRs with saturated dicarboxylic acid has revealed that 1.5 to 3.0 phr of adipic acid without any catalyst or accelerator gives a moderate state of cure.[72] The scorch safety increases with increasing chain length of the dicarboxylic

TABLE 5
Typical Formulations for Gum Vulcanizates

	I	II	III
Polymer	100	100	100
Zinc oxide	5	5	5
Stearic acid	2	2	2
Poly-1,2-dihydro-2,2,4-tri-methyl-quinoline	2	2	2
Sulfur	1.5	0.3	2.5
N-oxydiethylenebenz-thiazole-2-sulfenamide	1.5	2.4	0.6
Tetramethylthiuram disulfide	—	1.6	—

From Baker, C. S. L., Gelling, I. R., and Newell, R., *Rubber Chem. Technol.*, 58, 67, 1985. With permission.

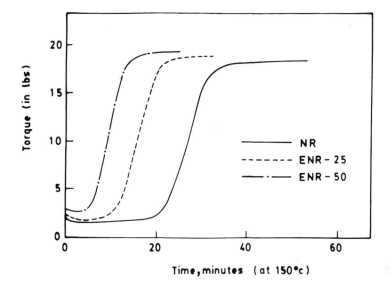

FIGURE 12.   Monsanto rheographs of NR and ENRs. (Reproduced from *Rubber Chem. Technol.*, 58, 86 (1985), by permission from the publisher, Rubber Division, A.C.S., Akron OH.)

acid. The vulcanizates show better heat aging, oil resistance, lower compression set, and air permeability as compared to the sulfur vulcanizates.

VIII. PROPERTIES OF EPOXIDIZED RUBBERS

The properties of the vulcanized product depend very much on the type of curing agent used. When epoxy groups are at the curing site, after the reaction the epoxy groups are destroyed, except when less than stoichiometric amounts of the curing agents are employed. Properties can be widely varied by varying the chain lengths of the curing agents. Thus, long chain dibasic-acid-cured products are rubbery and flexible while with the decrease in chain length of the dicarboxylic acid, the cured product progressively becomes hard and brittle.[27,36] The amine-cured epoxidized polybutadiene gives a very high flexural and tensile strength, while the diacid-cured product gives the lowest tensile strength. Maleic-anhydride-cured products give the highest heat deflection temperature. Epoxidized polybutadiene exhibits excellent electrical properties, especially the dielectric properties have been reported to remain constant over a wide range of temperatures.[27]

TABLE 6
Physical Properties of Gum Vulcanizates[a]

Property	ENR-25	ENR-50
Modulus 100%, MPa	0.69	0.74
Modulus 300%, MPa	1.50	1.56
Tensile strength, MPa	24.3	28.3
Elongation at break, %	770	770
Compression set	30	36
(24 h/70°C)	9[b]	8[b]
Hardness (IRHD)	34	36
Dunlop resilience, % (23°C)	71	45
Akron abrasion (mm³/500 rev)	160	20
Ring fatigue, kHz		
0–100 % extension	134	234
50–150 % extension	915	880
Stress relaxation, % decade	1.0	1.1

[a] Formulation I from Table 5 except where mentioned otherwise.
[b] Formulation II from Table 5.

From Baker, C. S. L., Gelling, I. R., and Newell, R., *Rubber Chem. Technol.*, 58, 67, 1985. With permission.

Epoxidized styrene-butadiene block copolymer (ESBS) based thermoplastic elastomer showed improved oil and solvent resistance without affecting the domain morphology and mechanical properties of the parent thermoplastic elastomer.[36] Improved adhesive strength, heat stability, and oil and solvent resistances of epoxidized 1,4-polybutadiene has been reported by Minoura et al.[73]

Oil resistance[74] of the sulfur-vulcanized ENR-50 has been found to be comparable with that of cured NBR-33 and the gas permeability resistance comparable to IIR.[76] Moreover, the tensile properties of cured ENR-50 have been reported to be better than NBR-33 or IIR. The tensile, hardness and compression set of conventional sulfur cured ENR are comparable to that of NR but Dunlop resilience, fatigue, and heat buildup properties are poorer, while wet-skid resistance and abrasion properties are better.[74] Tearing behavior of ENR gum vulcanizates is similar to that of NR.[75] The tearing energy is less sensitive to rate and higher than that of SBR even at higher temperatures. The wet-skid ranking of different ENR-black and ENR-silica composites are as follows:[74] ENR 25/silica > ENR 30/black > ENR 25/black = OESBR/black > ENR 20 black > 80:20 NR-ENR 50/black > NR/black.

The skid resistance[74] on smooth and rough concrete at 20°C also increases with the extent of epoxidation up to a 50 mol% level, but at higher epoxidation levels the skid resistance goes down. The skid resistance on ice (− 10°C), however, decreases progressively with the increasing epoxidation level. Table 6 shows the physical properties of a few typical ENR gum vulcanizates.[74]

The damping property[74] of ENR is found to be increasing with the level of epoxidation as can be seen from Figure 13, which depicts that the maxima of tan δ vs. temperature plots shift progressively with increasing levels of epoxidation.[74]

It has been found that ENR can be reinforced with silica fillers without any coupling agent, and gives vulcanizates with properties as good as those obtained with highly reinforcing blacks.[74,76]

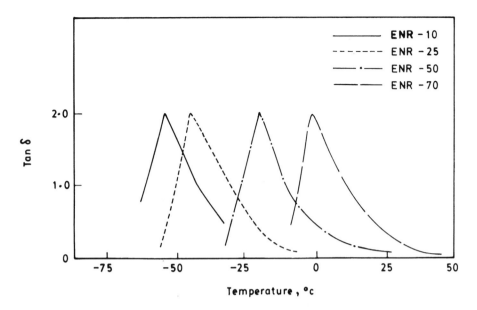

FIGURE 13. Tanδ vs. temperature plots for ENR of varying epoxy content. (Reproduced from *Rubber Chem. Technol.*, 58, 67 (1985), by permission from the publisher, Rubber Division, A.C.S., Akron, OH.)

IX. DEGRADATION AND AGING

Kumar et al.[77] have recently reported that during high temperature milling, the increase in carboxyl functionality of ENR-50 is much faster in rate and higher in extent compared to those for NR, but during low temperature (45°C) milling, the rate and extent of change of the carbonyl functionality for ENR-50 and NR are nearly similar. However, the XPS study (Figure 14) has shown that although the increase in oxygen content, at 45°C after 15 min milling of ENR, is somehow less than that of NR, but broadening in appearance of a second peak in SC bands for milled ENR reflect that some transformation among the oxygenated functionality has taken place. This phenomena may be related to the ring expansion of epoxides to five-membered tetrahydrofuran rings. This has been confirmed by IR studies.

The thermal aging behavior of raw ENRs has been investigated by Roy et al.[78] by thermal gravimetric analysis (TGA), IR, and gel-content techniques. Chaki et al.[79] studied the aging of ENR by dielectric, sol-gel, Mooney viscosity, and stress-strain analyses. The activation energy of degradation determined by the Freeman-Carroles method and integral procedural decomposition temperature (ipdt) determined,[78] for different ENR samples from the TGA traces, that with increase in epoxidation level, both idpt and activation energy go down, i.e., the apparent thermal stability decreases. The IR study has revealed that many new absorption bands appear due to the formation of secondary and tertiary alcohols, ether crosslinks, cyclic ethers, etc.[78] Gel-content determination of aged and unaged ENR samples showed that in the case of ENR, gel content increases after aging, which supports the formation of ether crosslinks.[78,79] Interestingly, it has been found[13,78] that the addition of 0.5 to 1.0 phr of 2,4-dinitrophenyl hydrazine during the epoxidation process improves the aging behavior of ENR to some extent.

The thermal stability of epoxidized polydienes, cured with some derivatives of triazines and phthalimides, has been studied by derivative thermogravimetry (DTA) and TGA

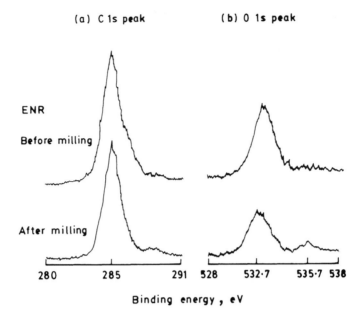

FIGURE 14.   X-ray Photoelectron Spectra (XPS) of ENR, before and after milling.

techniques.[80] It was found that the elasticity of the products decreased after aging. Epoxidized natural rubber vulcanizate with conventional sulfur curing showed poor aging properties.[17,68] These vulcanizates have been found to be hardened after aging. It is believed that sulfur acids are formed during aging from the polysulfidic linkages, which act on the epoxy groups to cause ring opening and ring expansion.[68] Ring opening leads to further increases in cross-link density, as explained earlier. UV (ultra-violet) and semi-EV-cured ENR vulcanizates show improved aging behavior due to a lower proportion of polysulfidic linkages.

X. BLENDS OF EPOXIDIZED NATURAL RUBBER WITH OTHER RUBBERS

A blend can offer a set of properties that may give it the potential of entering into application areas not possible with the single polymers comprising the blend. Blends of ENR and PVC have been reported to be miscible.[81-83] Usefulness of ENR/BR blends in tire-tread applications has been reported by Baker, Gelling, and Palmer.[84] They have reported this blend as the best practical solution for tire treads, in spite of the cost and doubtful wear resistance of ENR. Varughese, Tripathy, and De[85] showed that the ENR/BR blend is immiscible in all blend ratios. Processing and vulcanization behavior of both rubbers improves on blending.

Recently De and co-workers[86] have shown the potential of ENR as a chemically reactive polymer for self-vulcanizing rubber blends in the absence of any external curing agents. Such blends include ENR with XNBR,[86-88] ENR with CSM,[89,90] and ENR with polychloroprene.[91] These rubber-blend vulcanizates behave like the conventionally vulcanized rubber systems.[87] The blends have been characterized by Monsanto rheometry, DSC, IR spectroscopy, solvent swelling, and physical properties. It has also been reported that high temperature molding of the melt-mixed blends of ENR with PVC and with chlorinated rubber results in self-crosslinkable rubber-plastic blends.[92,93]

XI. REACTIONS OF EPOXIDIZED POLYDIENE ELASTOMERS

The epoxy groups may be utilized as a reaction site for introducing a variety of other functional groups in the polymer backbone.[94] For example, epoxy groups may be converted to esters, ammonium derivatives,[95] and halohydrines.[96,97]

As discussed earlier, acid-catalyzed ring expansion of epoxidized polydiene results in a tetrahydrofuranyl backbone,[21,98,99] which has been found to be a very effective chelating agent. However, only the parent polymers with all *cis*-1,4-olefin backbones converted to tetrahydrofuranyl groups, having a sequence length higher than six, have been reported to be effective chelating agents.[99]

Epoxidized natural rubber has been isomerized in the presence of an aluminum isopropoxide catalyst to allylic alcohol groups.[100]

XII. APPLICATIONS

The improved oil resistance of epoxidized polyalkadienes find their applications in areas such as oil hoses, seals, connectors, and LPG tubing. The increased air permeability resistance may be used for its blend with NR, SBR, or BR in tire liners and inner tubes.[101] ENR-25 may be considered for tire-tread application,[101] due to its wet grip property balanced with rolling resistance. Low resilience of epoxidized rubbers should find application in many engineering areas.[74]

Epoxidized natural rubber can be used to bond PVC to a basically noncompatible material as NR or SBR.[21,82-84] The epoxidized (10 to 50%) 1,2-polybutadienes have been used as hot-melt adhesives.[102] Epoxidized polybutadiene rubber latex has found applications in resorcinol-formaldehyde latex dipping compositions for tire cords.[103] Epoxidized polymers can be used as macromolecular initiators for cationic grafting.[14,104-106]

REFERENCES

1. **Pummer, R. and Burkhard, P. A.,** *Uber Kautschuk. Ber.,* 55, 3458, 1922.
2. **Rosowsky, A.,** in *Heterocyclic Compounds with Three and Four Membered Rings,* Interscience, New York, 1964, Part 1.
3. **Swern, D.,** in *Organic Peroxides,* Wiley Interscience, New York, 1971, Vol. 2.
4. **Swern, D., Billen, G. N., and Scalan, J. T.,** *J. Am. Chem. Soc.,* 68, 1504, 1946.
5. **Witnauer, L. P. and Swern, D.,** *J. Am. Chem. Soc.,* 72, 3364, 1950.
6. **Parker, R. E. and Isaacs, N. S.,** *Chem. Rev.,* 59, 737, 1959.
7. **Gelling, I. R.,** British Patent 2 11 36 92, 1984.
8. **Tutorskii, I. A., Khodzhaeva, I. D., and Dogadkin, B. A.,** *Vyskomol Soedin. Ser. A.,* 16, 157, 1974.
9. **Koshel, N. A., Turov, B. S., Papov, V. V., Ustavshchikov, B. F., Nikitina, T. P., and Musabekov, Y. Y.,** *Chem. Abstr.,* 86, 172762, 1977.
10. **Mairs, J. A. and Todd, J.,** *J. Chem. Soc.,* 38C, 1932.
11. **Saffer, A. and Johnson, B. L.,** *Ind. Eng. Chem.,* 40, 538, 1948.
12. **Tutorskii, I. A., Khodzhaeva, I. D., Novikov, S. V., and Dogadkin, B. A.,** *Vyskomol. Soedin. Ser. A.,* 15, 2282, 1974.
13. **Roux, C., Pautrat, R., Cheritat, R., Ledran, F., and Danjard, J. C.,** *J. Polym. Sci., Part C,* 16, 4687, 1969.
14. **Dreyfuss, P. and Kennedy, J. P.,** *J. Appl. Polym. Sci., Appl. Polym. Symp.,* 30, 165, 1977.
15. **Gemmer, R. V. and Golub, A.,** *J. Polym. Sci., Polym. Chem. Ed.,* 16, 1985, 1978.
16. **Dreyfuss, P. and Kennedy, J. P.,** *Anal. Chem.,* 47, 771, 1975.

17. **Gelling, I. R. and Smith, J. F.,** *Proc. Int. Rubber Conf.,* Milan, Italy, 140, 1979.
18. **Nippon Zeon Co. Ltd.,** Japanese Patent JPN 8040, 604 (Cl.C08 F8/08), 1980.
19. **Food Machinery and Chemical Corporation, Plastics Department,** Oxirane Resin in Premix Molding, Technical Bulletin, 5, British Patent 769127, 1957.
20. **Burfield, D. R., Lim, K. L., and Law, K. S.,** *J. Appl. Polym. Sci.,* 29, 1661, 1984.
21. **Gelling, I. R.,** *Rubber Chem. Technol.,* 58, 86, 1988.
22. **Tobalsky, A. V. and Mesrobion, R. B.,** *Organic Peroxides,* Interscience, New York, 1954, 35.
23. **Nonagaki, S. and Moroshita, H.,** Japanese Patent 74769 983, 1974.
24. **Terry, R. W. and Jacobs, A. F.,** British Patent Appl. No. 2008 125, 1979.
25. **Zuchowska, D.,** *Polymer,* 21, 514, 1980.
26. **Hayashi, O., Takahashi, T., Kurihara, H., and Uneo, H.,** *Kobunshi Ronbonshu.,* 37, 195, 1980.
27. **Greenspan, F. P.,** *Chemical Reactions of Polymers,* Fettes, E. M., Ed., Vol. 19, Interscience, New York, 1964, 152.
28. **Badran, B. M. and Abdel-Bary, E. M.,** *Chem. Ind.,* 314, 1977.
29. **Pandele, C., Andrusen, E., Corciovei, M., Nistor, D., Dan, L., and Selegean, D.,** *Chem. Abstr.,* 80, 122123q, 1974.
30. **Hayashi, O., Kimura, H., Ooi, Y., and Uneo, H.,** *Kobunshi Ronbonshu.,* 37, 327, 1980.
31. **Hayashi, O., Takahashi, T., Kurihara, H., and Uneo, H.,** *Polym. J. Jap.,* 13, 215, 1981.
32. **Ryabova, M. S., Sautin, S. N., Volin, Y. M., Lazarev, S. Y., and Shibaev, V. A.,** *Zh. Prikh. Khim.,* 56, 1116, 1983.
33. **Bradbury, J. H. and Perera, M. C. S.,** *J. Appl. Polym. Sci.,* 30, 3347, 1985.
34. **Perera, M. C. S.,** *J. Appl. Polym. Sci.,* 34, 2591, 1987.
35. **Udipi, K.,** U.S. Patent 4 131 725, 1978.
36. **Udipi, K.,** *J. Appl. Polym. Sci.,* 23, 3301, 1979.
37. **Roy, S., Gupta, B. R., and Maiti, B. R.,** *J. Elast. Plast.,* 22, 280, 1990.
38. **Roy, S., Gupta, B. R., and Maiti, B. R.,** *Ind. Eng. Chem. Res.,* 30, 1991.
39. **Nippon Zeon Co. Ltd.,** Japanese Patent, 8118 605, 1981.
40. **Baker, C. S. L. and Gelling, I. R.,** *Rubber India,* 37, 9, 1985.
41. **Jian, X. and Hay, A. S.,** *J. Polym. Sci., Polym. Lett.,* 28, 285, 1990.
42. **Jian, X. and Hay, A. S.,** *J. Polym. Sci. Part A. Polym. Chem.,* 29, 1183, 1991.
43. **Durbetaki, A. J.,** *Anal. Chem.,* 28, 2000, 1956.
44. **Jay, R. R.,** *Anal. Chem.,* 36, 667, 1964.
45. **Swern, D., Findley, T. W., Billen, G. N., and Scanlan, J. T.,** *Anal. Chem.,* 19, 414, 1947.
46. **King, G.,** *Nature,* 164, 706, 1949.
47. **Davey, J. E. and Loadman, M. J. R.,** *Br. Polym. J.,* 16, 134, 1984.
48. **Ng, S. C. and Gan, L. H.,** *Eur. Polym. J.,* 17, 1073, 1981.
49. **Roy, S. and De, P. P.,** *Polym. Testing,* 10, 1991.
50. **ASTM** *Annual Book of ASTM Standards,* ASTM, Philadelphia, Vol. 37, Test method No. D3677, 823, 1981.
51. **Dubertaki, A. J. and Miles, C. M.,** *Anal. Chem.,* 37, 1231, 1965.
52. **Golub, M. A., Hsu, M. S., and Wilson, L. A.,** *Rubber Chem. Technol.,* 48, 953, 1975.
53. **Burfield, D. R., Lim, K. L., Law, K. S., and Ng, S.,** *Polymer,* 25, 995, 1984.
54. **Dorsey, J. G., Dorsey, G. F., Rutenberg, A. C., and Green, L. A.,** *Anal. Chem.,* 49, 1144, 1977.
55. **Lau, C. M., Gomez, J. B., and Subramanium, A.,** *Proc. Int. Rubber Conf.,* Kuala Lampur, 1985, 510.
56. **Stellman, J. M. and Woodward, A. K.,** *J. Polym. Sci., Part B. Polym. Lett.,* 7, 755, 1969.
57. **Wichacheewa, P. and Woodward, A. E.,** *J. Polym. Sci. Polym. Phys. Ed.,* 16, 1849, 1978.
58. **Stellman, J. M. and Woodward, A. E.,** *J. Polym. Sci. Part A-2, Polym. Phys.,* 9, 59, 1971.
59. **Schilling, F. C., Bovey, F. A., Tseng, S., and Woodward, A. E.,** *Macromolecules,* 16, 808, 1983.
60. **Davies, C. K. L., Wolfe, S. V., Gelling, I. R., and Thomas, A. G.,** *Polymer,* 24, 107, 1983.
61. **Roy, S.,** private communication.
62. **Alex, R.,** private communication.
63. **Tutorskii, I. A., Khodzhaeva, I. D., and Dogadkin, B. A.,** *Chem. Abstr.,* 81, 106968j, 1974.
64. **Roy, S., Namboodri, C. S. S., Maiti, B. R., and Gupta, B. R.,** *Polym. Eng. Sci.,* in press.
65. **Amu, A. B., Rahman Ismail, K. A., and Sedek, B. D.,** *Proc. Int. Rubber Conf.,* Kuala Lumpur, 1985, 289.
66. **Kumar, N. R., Bhowmick, A. K., and Gupta, B. R.,** *Plast. Rubber Proc. Appl.,* 14, 119, 1990.
67. **Varughese, S., Tripathy, D. K., and De, S. K.,** *J. Adhesion Sci. Technol.,* 4, 847, 1990.
68. **Gelling, I. R. and Morrison, N. J.,** *Rubber Chem. Technol.,* 58, 243, 1985.
69. **Ross, G. W.,** *J. Chem. Soc. London,* 10, 2856, 1958.

70. **Zuchowska, D.,** *Polymer,* 22, 1073, 1981.
71. **Colchough, T.,** *Trans. Inst. Rubber Ind.,* 38, 11, 1962.
72. **Loo, C. T.,** *Proc. Int. Rubber Conf.,* Kuala Lumpur, 1985, 368.
73. **Minoura, Y., Yamashita, S., Yamaguchi, H., Khojiya, S., Yamada, K., Muko, M., and Nishimura, T.,** *Nippon Gomu Kyokaishi,* 52, 517, 1979.
74. **Baker, C. S. L., Gelling, I. R., and Newell, R.,** *Rubber Chem. Technol.,* 58, 67, 1985.
75. **Samsuri, A. B., Thomas, A. G., Gelling, I. R., and Southern, E.,** *Proc. Int. Rubber Conf.,* Kuala Lumpur, 1985, 386.
76. **Davies, K. and Rowley, R. J.,** *Mater. Appl.,* 3, 23, 1978.
77. **Kumar, N. R., Roy, S., Bhowmick, A. K., and Gupta, B. R.,** *J. Appl. Polym. Sci.,* in press.
78. **Roy, S. and Chaki, T. K.,** private communication.
79. **Chaki, T. K., Bhattacharya, A. K., and Bhowmick, A. K.,** *Kautschuk Gummi Kunst.,* 42, 408, 1990.
80. **El. Fayoumi, A. A. Z.,** *J. Therm. Anal.,* 23, 135, 1982.
81. **Margaritis, A. G. and Kalfoglou, N. K.,** *Polymer,* 28, 497, 1987.
82. **Varughese, K. T., Nando, G. B., De, P. P., and De, S. K.,** *J. Mater. Sci.,* 23, 3894, 1988.
83. **Varughese, K. T. and De, P. P.,** *J. Appl. Polym. Sci.,* 37, 2537, 1989.
84. **Baker, C. S. L., Gelling, I. R., and Palmer, J.,** *Proc. Int. Rubb. Conf.,* Kuala Lumpur, 1985.
85. **Varughese, S., Tripathy, D. K., and De, S. K.,** *Kautschuk Gummi Kunst.,* 43, 871, 1990.
86. **Alex, R., De, P. P., Mathew, N. M., and De, S. K.,** *Plast. Rubber Process. Appl.,* 14, 223, 1990.
87. **Alex, R., De, P. P., and De, S. K.,** *Polymer,* in press.
88. **Alex, R., De, P. P., and De, S. K.,** *Polymer,* 32, 2546, 1991.
89. **Mukhopadhyay, S. and De, S. K.,** *Polymer,* 32, 1223, 1991.
90. **Mukhopadhyay, S. and De, S. K.,** *J. Appl. Polym. Sci.,* 42, 2773, 1991.
91. **Alex, R., De, P. P., and De, S. K.,** *Kauschuk Gummi Kunst.,* 44, 333, 1991.
92. **Ramesh, P. and De, S. K.,** *J. Mater. Sci.,* 26, 2846, 1991.
93. **Ramesh, P.,** private communication.
94. **Brosse, J. C., Soutif, J. C., and Pinazzi, C.,** *Macromol. Chem.,* 180, 2109, 1979.
95. **Lye, P. H. and Toh, H. K.,** *J. Polym. Sci., Polym. Lett.,* 22, 327, 1984.
96. **Blausi, J. and Grimaud, E.,** French Patent, 1 503 552, 1967.
97. **Mayhan, K. G., Janssen, R. A., and Drake, R. F.,** U.S. Patent Appl., 577 300, 1984.
98. **Perera, M. C. S., Elix, J. A., and Bradbury, J. H.,** *J. Polym. Sci., Polym. Chem. Ed.,* 26, 637, 1988.
99. **Schultz, W. J. and Katritzky, A. R.,** U.S. Patent, 4 309 516, 1982.
100. **Campbell, D. S.,** *Br. Polym. J.,* 5, 55, 1973.
101. **Loh, P. C. and See, T. M. S.,** *Proc. Int. Rubber Conf.,* Kuala Lumpur, 1985, 312.
102. **Ube Industries Ltd.,** Japanese Patent 8 098 202, 1979.
103. **Gonsvskaya, T. B., Ponomarev, F. G., Puluektov, P. T., and Khavatova, L. K.,** *Chem. Abstr.,* 85, 6978, 1976.
104. **Dreyfuss, P. and Kennedy, J. P.,** *J. Polym. Sci., Polym. Symp.,* 60, 47, 1977.
105. **Dreyfuss, P. and Kennedy, J. P.,** *J. Appl. Polym. Sci., Appl. Polym. Symp.,* 30, 153, 1977.
106. **Dreyfuss, P. and Kennedy, J. P.,** *J. Appl. Polym. Sci., Appl. Polym. Symp.,* 30, 179, 1977.

Chapter 18

TOUGHENING CONCEPT IN RUBBER-MODIFIED HIGH PERFORMANCE EPOXIES

Hung-Jue Sue, Eddy I. Garcia-Meitin, and Dale M. Pickelman

TABLE OF CONTENTS

0-8493-4401-8/93/$0.00 + $.50

I. INTRODUCTION

Having versatility in its chemical forms as well as the corresponding impressive physical and mechanical properties, epoxy resins have been, and will still be, one of the mainstream materials for coatings, adhesives, electrical laminates, and structural components applications. Depending on specific needs for certain physical and mechanical properties, a combination or combinations of choices of epoxy resin chemistries and/or curing agents can usually be formulated to meet the market demands. However, when the application is structural, i.e., for load-bearing purposes, epoxy resins are unfortunately, not different from other engineering polymers; they are either brittle or notch sensitive, or both. As a result, tremendous effort has been focused on toughness improvement of epoxy resins.

The term "toughness", in a broader sense, is a measure of a material's resistance to failure. Depending on the application and preference of researchers, toughness is usually measured as either the stress or the energy required to fail a specimen under a specific loading condition. More specifically, toughness can be defined as either (1) the tensile strength, (2) the area under the tensile stress-strain curve, (3) the Izod impact strength, (4) the Charpy impact strength, (5) the plane strain critical strain energy release rate (G_{IC}), or (6) the plane strain critical stress intensity factor (K_{IC}). Although the definition of toughness varies from application to application and from researcher to researcher, it is generally agreed that when applications are safety related, the "worst case scenario" toughness definition, i.e., either K_{IC} or G_{IC}, should be considered. Since the present work focuses on toughening high performance epoxies for both automotive and aerospace applications, K_{IC} or G_{IC} (or J_{IC}, if the material is extremely ductile) will be used to determine the toughness of modified epoxy systems in the context of this chapter.

The commonly known approaches for toughening brittle epoxies, which typically have G_{IC} values of less than 200 J/m^2, include (1) chemical modification of a given rigid epoxy backbone to a more flexible backbone structure, (2) increase of epoxy monomer molecular weight, (3) lowering of the cross-link density of the cured resin via mixtures of high-monomer molecular weight epoxies or use of low functionality curing agents, and (4) incorporation of dispersed toughener phase(s) in the cured epoxy matrix. Among these approaches, toughening via dispersed toughener phase(s) has been shown to be most effective and can provide an order of magnitude toughness improvement.[1-6] However, only the highly toughenable epoxies modified with rubber tougheners are known to produce such an impressive toughening effect. These highly toughenable ductile epoxies usually exhibit rather low glass transition temperatures (Tg) and/or low cross-link densities. Consequently, they are not suitable for high performance structural applications. Thus, research efforts are mainly focused on toughness improvement of highly cross-linked high performance epoxies, where toughening effects have, thus, far been less effective.

For high performance, high Tg brittle epoxies, the toughening effect via rubber modification is usually only incremental. The cause for such a disappointing result is largely attributed to the high cross-link densities of epoxies, which greatly reduces the local molecular mobility. As a result, crack deflection, crack bifurcation, crack pinning, and crack bridging types of energy absorption mechanisms (all relatively low in energy absorption) are among the dominant toughening mechanisms. Nonetheless, the recent work conducted by Kinloch et al.[7] and Glad[8] demonstrates that the tightly cross-linked high performance epoxies can undergo strain softening and strain hardening when tested under compression. These findings indicate that the high cross-link density brittle epoxies can undergo shear yielding/banding as long as the stress state favors such mechanisms. It is, therefore, conceptually possible to toughen high cross-link density epoxies via shear yield/banding mechanisms, as long as the toughener phase can effectively alter the crack-tip stress state from one that favors brittle fracture to one that promotes shear yielding.

In order to understand how an epoxy resin can be effectively toughened, it is important to know the role(s) the toughener phase plays during the toughening process and the toughening theories behind these processes. In the last two decades, there have emerged many plausible toughening theories for both thermosetting and thermoplastic polymers. Nevertheless, there is still uncertainty regarding the role(s) that the rubber particles play in rubber-toughened polymers. These uncertainties include: the interfacial adhesion, degree of dispersion, size, and type of the rubber particles affect the toughening process; what is the sequence of toughening events, and causal relationships among the toughening mechanisms. These unsettled issues still await clarification.

This chapter focuses on providing a comprehensive up-to-date understanding in dealing with the toughening of epoxy resins via rubber modification. This includes a brief review of the recent developments in epoxy toughening, a brief overview on the use of effective experimental and numerical tools for studying the fracture behavior of epoxy systems, illustrations of important toughening mechanisms observed in rubber-toughened epoxies, and the mechanics and physics relating to the observed toughening mechanisms. The existing toughening concepts and theories in epoxy toughening are reviewed. The detailed procedure for producing these toughened epoxy systems is then discussed. It is the intent to convey the recent understanding and technology in epoxy toughening, and the hope that the readers will not be limited by the scope of this work in developing advanced toughened epoxy systems for both existing and new applications.

It should be noted that the toughening principles to be discussed in this chapter are derived from fundamental materials science and mechanics understandings. Therefore, even though the operating conditions for effective toughening may vary from polymer to polymer, the toughening principles should be universally applicable to other thermosetting and thermoplastic polymers, and even applicable to the toughening of composite and adhesive materials if care is taken.

II. TOUGHENING PRINCIPLES

In order to effectively toughen epoxies, it is important that the fundamental physics of toughening be understood, i.e., knowing under what circumstances the desirable toughening mechanism(s) can be promoted and the toughening effect be optimized. To do so, it is essential that one know all the possible operative toughening mechanisms and their relative effectiveness in toughened epoxies, as well as the toughening theories behind the mechanisms.

A. PROMOTING EFFECTIVE TOUGHENING MECHANISMS

Thus far, there have been more than 15 operative toughening mechanisms observed in rubber-modified epoxy systems.[9,10] These operative toughening mechanisms are schematically shown in Figures 1–3. Among the operative toughening mechanisms, some of the toughening mechanisms are found to be more effective than others. In order for certain desirable toughening mechanism(s) to take place, the material needs to possess appropriate physical and mechanical properties. This section focuses on describing all the important toughening mechanisms that can operate in rubber-modified epoxies, and the conditions under which these mechanisms can be activated.

1. Shear-Yielding/Banding

Shear-yielding/banding is among the most effective toughening mechanisms known in polymers. The terms yielded (plastic) zone, localized shear, and diffuse shear are sometimes used interchangeably to describe the failure events of ductile polymers. Nevertheless, more strictly speaking, they have different characteristic features. The yielded zone is usually

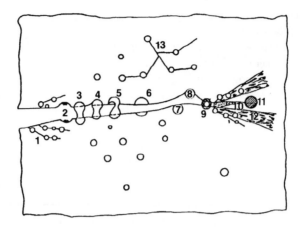

FIGURE 1.   Toughening mechanisms in rubber-modified epoxies: (1) shear-band formation near rubber particles; (2) fracture of rubber particles after cavitation; (3) stretching, (4) debonding and (5) tearing of rubber particles; (6) transparticle fracture; (7) debonding of hard particles; (8) crack deflection by hard particles; (9) voided/cavitated rubber particles; (10) crazing; (11) plastic zone at craze tip; (12) diffuse shear-yielding; (13) shear band/craze interaction. (From Garg, A. C. and Mai, Y. W., *Comp. Sci. Technol.*, 31, 179, 1988. With permission.)

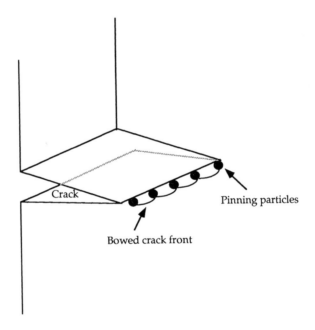

FIGURE 2.   Cracking-pinning mechanism. The crack is pinned, and therefore, bowed by the toughener particles.

used to describe the crack-tip plastic zone (Figure 1). Since the size of the crack-tip-yielded zone is often too small to be detected, it is frequently estimated using the elastic/plastic crack-tip stress field and the choice of the yielding criterion.[11] Diffuse shear and localized shear can occur anywhere in the sample, depending on the testing conditions. The physical dimensions of the diffuse shear and localized shear are different. The diffuse shear involves less localized shear straining (only a few percent shear strain more than the surrounding material), and covers a larger volume of material.[12] In contrast, the localized shear banding phenomenon is highly localized, and the shear strain in the band can reach as high as 250%.[13] The characteristics of the shear bands, either diffuse shear or localized shear, are thought

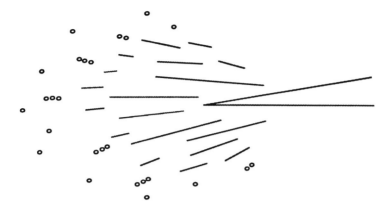

FIGURE 3. Croiding mechanism. Highly localized rubber cavitation line arrays form around the crack tip. The matrix adjacent to the rubber cavitation line arrays is plastically dilated and shear yielded.

to be mainly due to the strain-rate-sensitivity of polymers, which is material-dependent.[12] Nevertheless, the shearing characteristic is also found to be affected by the size and type of the toughener particles incorporated.[2,3,10]

Although these yielding mechanisms have different physical characteristics, from the mechanical and molecular aspects, the requirements for these yielding phenomena to occur are nevertheless the same. The details concerning the physical definition of polymer yielding have been given by Bowden.[12] Here, only the factors that affect the behavior of shear yielding are discussed.

From the mechanical aspect, for a shear yielding mechanism to operate, it is necessary that the material experiences a shear stress component exceeding the critical shear stress value for yielding. This critical shear yielding stress, depending on the shear yielding criteria, either Tresca[14] or von Mises,[15] can be defined as follows:

$$\sigma_1 - \sigma_3 = 2\tau_T \text{ (Tresca)}$$

$$(\sigma_1 - \sigma_2)^2 + (\sigma_2 - \sigma_3)^2 + (\sigma_3 - \sigma_1)^2 = 6\tau_M^2 = 9\tau_{oct}^2 = 2Y^2 \text{ (von Mises)}$$

where σ_1, σ_2, and σ_3 are the principal stresses and $\sigma_1 > \sigma_2 > \sigma_3$, τ_T and τ_M are the critical shear stress for Tresca and von Mises, respectively, Y is the uniaxial yield stress, and τ_{oct} is the critical octahedral shear stress. Both Tresca and von Mises criteria are originally proposed for metals. Tresca suggested that yielding will occur when the resolved shear stress on any plane in the material reaches a critical value. Later, von Mises mathematically improved the awkwardness of the Tresca criterion, which possesses discontinuities at the corners of the hexagonal yield surface. The von Mises criterion can be interpreted as either an octahedral shear stress (τ_{oct}) criterion or a shear strain energy density criterion (i.e., the material will yield when the elastic shear-strain energy density reaches a critical value) for elastic/plastic materials. In the case of polymers, it is found that most fall between the predictions of Tresca and von Mises. Further, the yielding behavior of polymers is also known to be pressure-dependent. Modifications to the Tresca and von Mises criteria are needed to account for the pressure-dependence of polymers in yielding.[16-19] In general, the pressure sensitivity coefficient is usually rather small (≈ 0.1 to 0.2) for most polymers, therefore, it is not critical to use either the unmodified or modified Tresca and von Mises criteria to describe the yielding behavior of polymers.

From the molecular point of view, all polymers are both viscoelastic and viscoplastic in nature. Even when the stress state favors the occurrence of a shear-yielding mechanism,

Dynamic Mechanical Behavior

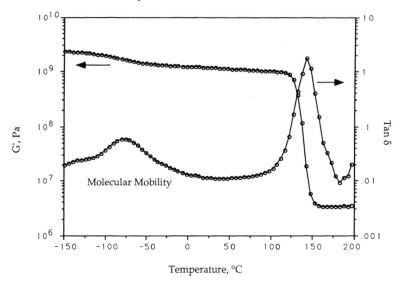

FIGURE 4.   A typical DMS spectrum showing both the dynamic shear modulus (G') and tan δ curves plotted against temperature. When the testing frequency is increased, the curves will shift toward the right. This can potentially cause the material to behave brittlely.

the rate of testing and the temperature the polymer experiences will both affect the local mobility of the macromolecules. This, in turn, will alter the value of the critical shear stress for yielding.[20] When the critical stress value for other mechanisms, such as crazing and cracking, is reached earlier than that for shear yielding, catastrophic failure may precede the preferred ductile shear failure.

The local molecular mobility of polymers, in a given temperature range, is usually detected using dynamic mechanical spectroscopy (DMS). The tan δ curve can provide insight concerning the available molecular motion at specific testing temperature ranges and rates of testing.[21,22] In general, the tan δ curve retains its shape and shifts its positions toward the higher temperature region as the loading frequency during the test is increased. Therefore, when the rate of testing is higher than a critical value, even at room temperature, the available molecular mobility for dissipating mechanical energy may disappear,[20-22] and the shear-yielding mechanism is then suppressed. Only fracture surface energy-related mechanisms can operate, which usually yield low fracture toughness. On the other hand, if the testing rate is extremely slow, the tan δ curve will shift toward the low-temperature regime. The originally unavailable molecular mobility is now present. Consequently, the brittle material can undergo larger-scale yielding. It is, therefore, useful to use the DMS technique as a screening tool to check whether or not the candidate epoxy system is toughenable under the specific testing conditions. A typical DMS spectrum is shown in Figure 4. Detailed discussion on this subject can be found elsewhere.[20-22]

2. Normal Yielding

The commonly recognized normal-yielding phenomenon is crazing, which is known to take place in most thermoplastics.[23] A fracture toughness improvement of less than a few hundred percent via the crazing mechanism is not uncommon. However, in thermosetting polymers, crazing is seldom found to exist. As a result, the physical and mechanical requirements for normal yielding in thermosetting polymers are frequently neglected.

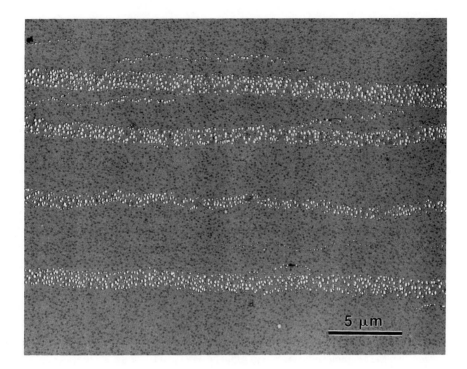

FIGURE 5. A TEM micrograph taken at the crack wake of the DN-4PB damage zone. The croids can be readily observed. The crack propagates from left to right.

Recently, a new form of normal yielding was discovered and termed "croiding" (derived from "crack" and "void").[10] The croids, which exist in the form of highly localized rubber particle cavitation line arrays, were found to take place in certain core-shell, rubber-modified epoxy systems (Figures 3 and 5). Although the physics of crazing and croiding are different, the mechanics of how these two mechanisms are triggered are similar. The well-established mechanical criteria for craze initiation, and therefore, possibly for the croiding formation, are described below.

The earliest and simplest criteria for normal yielding were proposed by Maxwell et al.[24] and Bucknall et al.[25] They attributed the formation of crazes to either a critical applied strain or a critical applied stress. However, it is known that craze formation involves volume dilatation. As a result, maximum dilatation[26,27] and maximum total strain energy criteria[28-31] have since been proposed. In most cases, these criteria can be applied for some systems, but are not generally applicable to others.

Sternstein and Ongchin[32] were the first to propose a so-called stress-bias criterion which includes the dilatational stress component. The criterion is expressed as follows:

$$\sigma_b = |\sigma_1 - \sigma_2| = A_1 + \frac{B_1}{I_1} \qquad (3)$$

where σ_b is the stress bias between the two principal stresses ($\sigma_3 = 0$, in their experiment), $I_1 = (\sigma_1 + \sigma_2 + \sigma_3)$, and A_1 and B_1 are time-temperature-dependent material parameters. The shortcomings of this criterion are that no plausible physical interpretation is given,[33,34] and that, in reality, since $\sigma_3 \neq 0$, application of this criterion becomes very difficult.

A more comprehensive critical strain criterion was suggested by Bowden and Oxborough.[34] The criterion is given following:

$$\sigma_1 - \nu\sigma_2 - \nu\sigma_3 = E\epsilon_1 = Y + \frac{X}{I_1}$$

where σ_1, σ_2, and σ_3 are the principal stresses, ϵ_1 is the critical strain, ν is Poisson's ratio, E is Young's modulus, and X and Y are time-temperature-dependent material parameters. This critical tensile strain criterion does offer a better physical interpretation for normal yielding and provides a way to accommodate the triaxial stress condition for practical uses.

3. Particle Bridging

The concept of particle bridging was first proposed by Merz et al.[35] in rubber-modified polystyrene systems. Kunz-Douglass et al.[36] later utilized the same concept to study and model the toughening effect in rubber-modified epoxy systems. The particle-bridging mechanism is a rather straightforward phenomenon (Figure 1). When the crack advances in the rubber-modified epoxy system, it has a tendency to grow preferentially in the more brittle epoxy matrix phase, and therefore, around the rubber particles, at least in the initial stage of the crack growth. As a result, when the crack begins to open up, the rubber particles span between the two separating crack planes. Since the rubber particles are extremely ductile and strain-harden rapidly, the fracture energy required to drive the crack to grow is moderately increased. This toughens the brittle epoxies.

In principal, the bridging particle needs to be able to stretch between the two crack planes. Therefore, the particle needs to be ductile. The size of the particle must be several times larger than the characteristic crack-tip radius in order to function as an effective bridge. Also the interfacial adhesion between the particle and the matrix needs to be stonger than the cohesive strength of the particle itself. If all of the above requirements are met, the particle bridging mechanism should take place in the toughened epoxy system. An illustration of the crack-bridging mechanism, under which no significant shear yielding of the matrix occurs, can be seen in Figure 6.

4. Other Toughening Mechanisms

Other useful toughening mechanisms include crack-tip blunting,[37-39] zone shielding,[40-43] transformation toughening,[44] crack bifurcation,[45,46] crack deflection,[47,48] crack pinning,[49,50] and microcracking.[51] The shear-yielding process at the crack tip will create a plastic zone, thus, blunting the crack. This will reduce the crack-tip stress intensity factor, which, in turn, will require a higher load to drive the crack to propagate. Consequently, the system is toughened. When the wake of the propagating crack is surrounded by widespread crazes, microcracks, and dilated voids, the so-called "zone shielding" mechanism may be operative. This will induce an attendant dilatation at the crack wake, which will then suppress the crack from opening up. A significant toughening effect can be obtained if the volume of the zone-shielding material is optimized.[42] For a crystalline toughener phase embedded in a brittle matrix, as the crack propagates into the matrix, the crystalline inclusion may change its crystalline form and expand its volume when activated by the high stress field around the crack tip. The dilatation of the crystalline inclusion invokes an internal stress that suppresses the opening of the crack, which, in turn, toughens the system.

Under static loading, crack bifurcation, crack deflection, and microcracking usually occur only when there are mechanical inhomogeneities in the system. When the toughener phase has elastic constants (i.e., Young's modulus and Poisson's ratio) different from those of the matrix, the stress field in front of the crack tip will be perturbed by the toughener particles. This will alter the crack to grow at an angle from the original direction, bifurcate the crack into two cracks, or simply form microcracks around the toughener particles. These phenomena will both create new fracture surfaces and reduce the crack-tip stress intensity factor, thus, toughening the material. In the case of crack pinning, the crack front is pinned

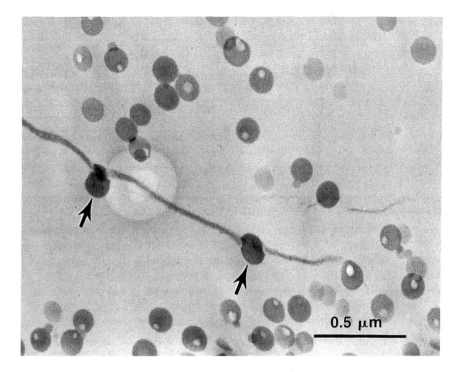

FIGURE 6. A TEM micrograph taken at the crack-tip region of a highly cross-linked CSR-modified epoxy system. In cases when the matrix shear-yielding is not significant, the crack/particle bridging mechanism can be observed (see arrows). The crack propagates from left to right.

by the toughener particles. As a result, the crack front will bow between the pinning particles. This also requires additional energy to drive the crack to grow, and thus, toughens the material.

Crack blunting, zone shielding, transformation toughening, crack bifurcation, crack deflection, microcracking, and crack-pinning mechanisms may furnish significant contributions to epoxy toughening. These mechanisms should also be seriously considered if both shear yielding and normal yielding mechanisms cannot be easily activated.

B. REVIEW OF EXISTING TOUGHENING CONCEPTS AND THEORIES

Sultan and McGarry[52] were the first to utilize the concept of rubber toughening in an epoxy matrix. In their study, they used the carboxyl terminated acrylonitrile (CTBN) liquid rubber to toughen Epon® 828 diglycidyl ether of bisphenol A (DGEBA) epoxy resin. They attributed the observed toughening effect mainly to the crazing of the epoxy matrix, based on the facts that the toughening effect is strongly dependent on the rubber particle size and the higher pressure sensitivity under biaxial tension of the large rubber-modified epoxy, compared to those of the small rubber-modified, and unmodified epoxies. As a result, rubber particle size and interfacial adhesion between the rubber particles and the epoxy matrix are thought to be critical for promoting crazing in toughened epoxies. Nevertheless, Bascom et al.[53] accredited the high toughness value of CTBN-modified epoxy to an increase in the plastic zone size. The interpretation was that the triaxial stress associated with the crack tip caused the cavitation of the rubber particles. These cavitated rubber particles induced plastic flow, which is now recognized as massive shear banding, of the matrix around the particles.

Rubber stretching and tearing, also known as the rubber-bridging mechanism, are proposed by Kunz-Douglass and co-workers[36,54] as a major toughening mechanism for rubber-modified plastics. The phenomenon does find support from some experimental observations.[3,36,54] The rubber particle bridging theory proposed by Kunz-Douglass et al. predicts that about two-fold improvement in toughening can be achieved. However, more recent experimental investigations conducted by others[1-6] indicate that the toughening effect due to rubber stretching and tearing alone cannot account for an order of magnitude increase in toughness improvement. Rather, it is the shear yielding of the matrix that should account for the impressive toughness improvement.[55]

A more recent and plausible elastomer toughening concept that uses the ground of mechanics to describe the toughening mechanisms of rubber-toughened polymers is proposed by Yee and Pearson.[1,2] In studying the toughening mechanisms of CTBN rubber-modified epoxy systems, they attributed an order of magnitude increase in toughness to the cavitation of the rubber particles, followed by large scale shear yielding of the epoxy matrix. They emphasized the importance of the sequence of toughening events.

In an attempt to further support the above toughening concept, Yee and co-workers[2,3,56-59] conducted a series of experiments and clearly demonstrated that in order for significant plastic shear banding to operate under constrained conditions in both thermoplastic and thermoset systems, cavitation of the toughener phase, via either internal rubber particle cavitation, debonding at the interface, or crazing mechanisms, is essential. In other words, there is a sequence of toughening events (i.e., cavitation occurs first, followed by shear banding) and a causal relationship (i.e., without the cavitational process, the shear banding mechanism cannot take place) involved in the toughening process.

III. EXPERIMENTAL WORK

In order to study the toughening mechanisms and the sequence of failure events in toughened epoxies, it is important that one microscopically investigate and mechanically understand exactly how the crack is evolved during the failure process. To do so, an effective mechanical testing method has to be conducted to create and preserve the crack evolution process, followed by an achievable microscopy technique to observe the failure events. Furthermore, fracture mechanics tools relating to the crack evolution events are required to gain fundamental knowledge of how epoxies can be toughened. These experimental tools are introduced and discussed below.

A. MATERIALS

Throughout this work D.E.R.® (The Dow Chemical Company) 332 (DGEBA) epoxy resin cured stoichiometrically with 4,4'-diaminodiphenylsulfone (DDS) is used to make model high cross-link density epoxy matrices.

The rubber modifiers used in this study include core-shell butadiene rubbers (CSR)[60] with various shell compositions (Table 1), dispersed acrylic rubber (DAR),[61,62] and Proteus® 5025 particle (obtained from B.F. Goodrich). A curing schedule of 180°C for 2 h and 220°C for 2 h was used for both toughened and untoughened epoxies.

B. DAMAGE CREATION

After the 1/4'' (0.635 cm) sample plaques were completely cured, they were slowly cooled to room temperature inside the oven. Single-edge-notch, three-point-bend (SEN-3PB) specimens, having dimensions of 2.5'' × 0.5'' × 0.25'' (6.35 cm × 1.27 cm × 0.635 cm), were used for K_{IC} measurements. Since the damage evolution process ahead of the crack tip is obliterated by the SEN-3PB test, information concerning the sequence of failure events and the role(s) the rubber particles play in the toughening process cannot be definitively obtained. Also, knowledge concerning the possible interactions among the operative tough-

TABLE 1
Compositions of the Core-shell Rubber Particles

Rubber Particle	Core/Shell	Particle Diameter (nm)		Shell Composition[c]				δ^d
		Experimental[b]	Theory	S	MMA	AN	GMA	
Core[a]	100/0	119	—	—	—	—	—	
CSR-A	84/16	Not measured	126	3.6	3.6	4.0	4.8	10.3
CSR-B	84/16	127	126	4.8	4.8	4.0	2.4	10.2
CSR-C	84/16	127	126	6.0	6.0	4.0	0.0	10.1
CSR-D	84/16	Not measured	126	7.0	7.0	2.0	0.0	9.7
CSR-E	84/16	Not measured	126	8.0	8.0	0.0	0.0	9.3
CSR-F	75/25	133	131	7.5	7.5	6.25	3.75	10.2
CSR-G	65/35	140	138	10.5	10.5	8.75	5.25	10.2
Neat epoxy								10.3

[a] Core composition, 7% styrene and 93% butadiene.
[b] Brice-Phoenix Universal Light Scattering Photometer.
[c] Parts of shell, S = styrene, MMA = methyl methacrylate, AN = acrylonitrile, GMA = glycidyl methacrylate.
[d] Obtained by the Small Group Method.

ening mechanism cannot be understood. Therefore, the double-notch, four-point-bend (DN-4PB) method[63,64] is utilized to generate and preserve the crack evolution process.

The DN-4PB technique, a relatively well-known technique in both ceramics and metals, is a simple and yet effective tool for obtaining information concerning the fracture behavior of polymers.[3,10,56-59,65,66] This technique was first successfully carried out by Sue and Yee[59] to study the failure mechanisms of a toughened nylon and has proven to be extremely useful for unambiguous understanding of the failure behavior of toughened polymers. The DN-4PB technique is now widely utilized to study the fracture behavior of toughened polymers under both static and impact loading conditions.[67,68]

The principle of the DN-4PB technique is rather simple. Two nearly identical cracks are generated on the same edge of a rectangular bar (Figure 7). This bar is loaded in a four-point bending geometry with the nearly identical cracks positioned on the tensile side and inside the two inner loading points (Figure 8). The portion of the bar between the two inner loading points is subjected to a constant bending moment. Thus, the two cracks experience nearly identical stresses.[63] As the load is applied, plastic zones form in front of the crack tips. Since the two cracks cannot be exactly identical, one crack will propagate unstably leaving behind the other crack with a nearly critically developed process zone at its tip. Since this crack is arrested, the events in the crack tip process zone are not obliterated by the final failure which often involves tearing of the plastic ligaments spanning the crack faces. Various sectioning and microscopy techniques, to be discussed below, can then be applied to probe the failure mechanisms which occur at the crack tip and its wake.

In this work, the DN-4PB specimens, having dimensions of 5'' × 0.5'' × 0.25'' (12.7 cm × 1.27 cm × 0.635 cm), were used to generate a subcritically propagated crack for the investigation of the crack-tip damage evolution process of rubber-toughened epoxies.

C. MICROSCOPY

In order to understand the toughening mechanism(s) and the sequence of failure events in rubber-modified epoxies, it is imperative that the damage zone of the modified systems be studied. Ease of use and straightforward sample preparation procedures make reflected optical microscopy (OM) and scanning electron microscopy (SEM) among the most utilized microscopy techniques for studying both morphology and failure mechanisms of all materials.

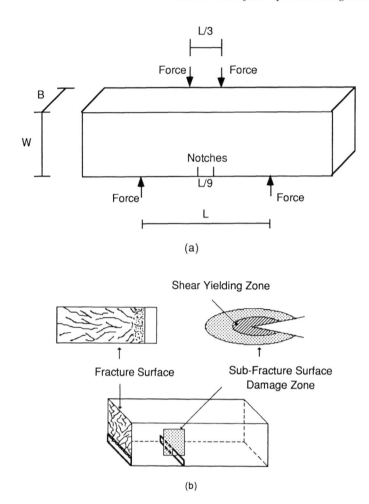

(a)

(b)

FIGURE 7.   (a) Schematic of the DN-4PB geometry. (b) Schematic of the regions from the DN-4PB specimen for SEM, OM and TEM investigations.

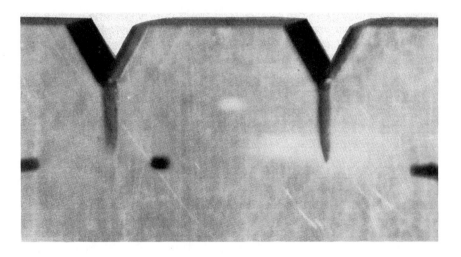

FIGURE 8.   An OM micrograph showing a pair of nearly identical cracks created inside the two inner loading points of a DN-4PB geometry.

These techniques are especially useful for determining any surface features that exhibit topographical variations due to the fracture surface failure characteristics and phase morphology of modified systems. These microscopic techniques, however, only provide information relating to the fracture surface, such as the crack path, the rubber particle size, shear lip, plastic drawing of the material, remnants of crazes (if any), crack pinning, basic longitudinal texture, and mackerel pattern, etc. When information concerning the subfracture surface zone (SFSZ) is of interest, investigation of the SFSZ of the damaged specimen using transmitted OM (TOM), and/or transmission electron microscopy (TEM) becomes necessary. Also, damage features relating to the SFSZ, which usually account for most of the fracture energy dissipation in polymers, cannot be obtained from fracture surface studies. As a result, the SFSZ investigation is essential for complete understanding of the contributing toughening mechanisms in rubber-modified epoxies.

This paper only emphasizes the sample preparation procedures for OM and TEM investigations. Theories and techniques regarding microscopy will not be covered here as they are described elsewhere.[69,70]

1. TOM Sample Preparation Procedures

Following Holik et al.,[71] Yee and Pearson[1-3] utilized the petrographic thin-sectioning technique to study the SFSZ of CTBN rubber-modified epoxies using OM. Using post mortem SEN-3PB specimens, Pearson and Yee were able to ascertain that (1) shear banding of the matrix is the major toughening mechanism accounting for most of the fracture energy absorption and (2) cavitation of the rubber particles precedes the shear-yielding of the matrix. This indicates the importance of TOM in toughening mechanism investigations.

In preparing TOM thin sections for the present work, the damaged DN-4PB sample is cut to an appropriate size ($\approx$1.27 cm $\times$ 1.27 cm) around but away from the region of interest, such as the plane-strain-core region or the plane-stress-skin region (Figure 7). The region of interest is then secured by embedding liquid mounting material, such as the D.E.R.™ 331 epoxy resin cured with diethylenetriamine (12:1 ratio by weight) at 38°C for 16 h, to avoid further damage of the sample. One side of the sample surface is ground using 1200 grit silicon carbide paper, followed by polishing with a 0.3 μm alumina powder water suspension, thus, eliminating any scratches on the sample surface that would scatter light. The polished side of the sample is then glued to a petrographic glass slide. The unwanted portion of the sample away from the polished surface is cut off, using a cut-off wheel, to a sample thickness of $\approx$150 μm. The unpolished sample surface is then polished, following the above polishing procedure. A final thin-section thickness is reached at $\approx$40 μm. To gain information on the SFSZ, the sections were made normal to the fracture surface but parallel to the cracking direction throughout this study. The thin sections were then examined using an Olympus Vanox-S® optical microscope both under bright field and crosspolarization.

2. TEM Sample Preparation Procedures

When the rubber particles are small (<1 μm) and information concerning the understanding of the role(s) the rubber particles play in the toughening process is needed, application of the TEM technique becomes essential.

In the TEM work, the region of interest (Figure 7) around the damage zone is carefully trimmed to an appropriate size (Figure 9a), i.e., an area of $\approx$5 mm $\times$ 5 mm, and embedded in D.E.R.™ 331 epoxy resin/diethylenetriamine (12:1 ratio by weight). The embedment is cured at 38°C for 16 h (Figure 9b). The cured block is then further trimmed to a size of $\approx$0.3 mm $\times$ 0.3 mm with the crack tip in the damage zone roughly at the center of the trimmed surface. An Olympus Vanox-S® optical microscope is used in the reflectance mode to observe the sample during the trimming process to assure good crack location on the thin section (Figure 9c). A glass knife is then utilized to face off the trimmed block. The faced-off block (except for the DAR-modified epoxy) is placed in a vial containing 1 g of 99.9%

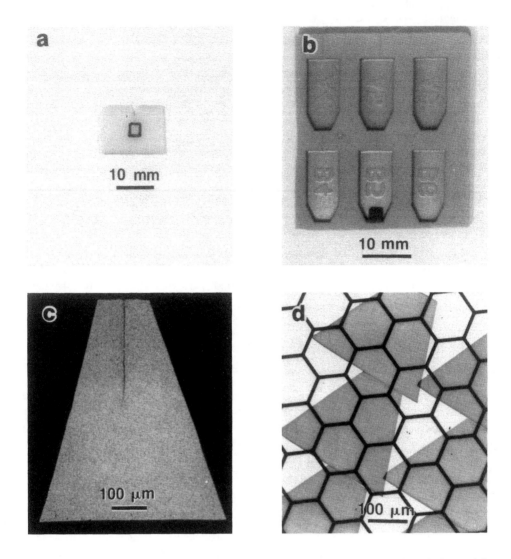

FIGURE 9.   TEM sample preparation procedures. (a) Isolation of the area of interest, (b) embedment of the sample, (c) a trimmed block showing the crack tip, and (d) thin sections on a formvar-coated copper grid for TEM investigation.

pure OsO_4 crystals and stained for 65 h. The stained block is microtomed to produce thin sections ranging from 60 to 80 nm thickness using a Reichert-Jung Ultracut E microtome with a diamond knife at ambient temperature.

Two different staining techniques are utilized for the DAR-modified epoxy systems. The first technique involves the room temperature thin-sectioning and post-staining by RuO_4. However, owing to the fact that DAR particles cannot be successfully stained and thin-sectioned using conventional approaches, a novel two-stage styrene/OsO_4 staining method is also used to study whether or not DAR particles cavitate.[66] This novel method is found to be effective in staining the otherwise unstainable DAR particles. As a result, understanding of the role of DAR particles in epoxy toughening can be pursued.

The TEM thin sections are placed on 200-mesh, formvar-coated copper grids (Figure 9d). The thin sections are examined using a JEOL 2000FX® ATEM operated at an accelerating voltage of 100 kV.

D. MECHANICS TOOLS

1. Fracture Toughness Measurements

Polymeric materials are, as pointed out earlier, viscoelastic and viscoplastic in nature. Strictly speaking, the conventional linear elastic fracture mechanics (LEFM) approach[11] cannot be applied to characterize polymer fracture behavior. Nevertheless, when the crack-tip plastic zone is small, compared to the width and thickness of the specimen, and the testing rate is significantly higher than the characteristic polymer relaxation time constant, the LEFM approach may still hold. Care must be taken to assure valid fracture toughness measurements.

In order to measure a valid K_{IC} of the rubber-modified epoxies, a mechanical testing procedure following ASTM D399-83 is conducted in this study. A Sintech screw-driven mechanical tester, with a cross-head speed of 0.02''/min (0.0508 cm/min), is used throughout the work.

To assure a valid K_{IC} value, the peak load is plotted against $B \cdot W^{1/2}/Y$ (B, thickness; W, width; and Y, correction factor) for each epoxy system. For each testing geometry and technique, the correction factor is different and can be found in published literature.[72] The correction factor used for the SEN-3PB geometry with a span to width ratio of four as in this study is given below:

$$Y = 11.6(a/W)^{1/2} - 18.4(a/W)^{3/2} + 87.2(a/W)^{5/2} - 150.4(a/W)^{7/2} + 154.8(a/W)^{9/2}$$

(where a is crack length)

The least-squares slope of the line drawn from a plot of the peak load against $B \cdot W_{1/2}/Y$ (for at least six specimens) is defined as K_C. If this line is straight and the K_C value fulfills the following conditions:[73]

$$B, (W - a), \text{ and } a \geq 2.5 \cdot (K_C/\sigma_{ys})^2$$

(where σ_{yi} is yield stress)

then, a valid K_{IC} value is attained. All of the tests in the current study fulfill the above requirements for a valid K_{IC}. The G_{IC} value is then calculated from the following equation:

$$G_{IC} = \frac{K_{IC}^2 \cdot (1 - \nu^2)}{E}$$

(where E is Young's modulus and ν is Poisson's Ratio)

The K_{IC} and G_{IC} values of all the rubber-modified epoxy systems are summarized in Tables 2 and 3. A typical plot of the peak load plotted against $B \cdot W_{1/2}/Y$ is shown in Figure 10.

2. Stress Analysis

Toughening phenomena in rubber-modified epoxies result, per se, from the differences in mechanical response of each component (i.e., the rubber inclusion or the epoxy matrix) to the external mechanical disturbance, such as tensile load or impact load. Furthermore, since all polymers are viscoelastic and viscoplastic in nature, finite time is required to build up the stress level for each component to undergo irreversible deformation. Therefore, adequate mechanics tools are needed for definitive descriptions of the role(s) the rubber particles play and the sequence of toughening events in the polymer toughening process.

The conventional LEFM and elastic plastic fracture mechanics (EPFM) approaches[11,74] directly characterize the fracture behavior of toughened polymers using basic material parameters, such as Young's modulus and Poisson's ratio. However, these parameters are sometimes inappropriate in connection with the operative micromechanisms in the damage zone.

TABLE 2
Summary of Fracture Toughness of Various Rubber-modified
Epoxy Systems

Material	K_{IC} (MPa · m$^{1/2}$)	G_{IC} (J/m^2)	Tg[a] (°C)
DGEBA Epoxy/DDS	0.83 ± 0.03	180	220
DGEBA Epoxy/DDS/10% CSR-B Rubber	1.20 ± 0.04	490	219
DGEBA Epoxy/DDS/10% DAR Rubber	1.08 ± 0.06	422	193
DGEBA Epoxy/DDS/10% Proteus Particle	0.97 ± 0.03	303	212

[a]   Second-heat mid-point Tg value is reported.

TABLE 3
Summary of K_{IC}, G_{IC}, and T_g of Designed CSR-Modified
Epoxy Systems

DGEBA Epoxy/DDS	K_{IC},[a] MPa·m$^{1/2}$	G_{IC},[b] J/m^2	Tg,[c] °C
Neat Resin	0.83 ± 0.02	180	220
Modified with 10 wt. % CSR-			
A	1.20 ± 0.07	490	222
B	1.20 ± 0.04	490	219
C	1.30 ± 0.09	580	223
D	1.07 ± 0.05	390	229
E	1.05 ± 0.03	380	224
F	1.33 ± 0.07	620	223
G	1.37 ± 0.06	640	220

[a]   A 63% confidence interval is used.
[b] $G_{IC} = K_{IC} (1-\nu 2)/E$, where E: Young's Modulus and ν: Poisson's ratio.
[c]   Second-heat mid point Tg is reported.

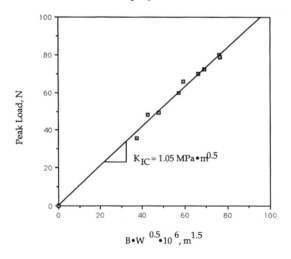

FIGURE 10.   A typical plot of the peak load vs. B·W$^{0.5}$/Y. The slope of the straight line is defined as the K_C.

Micromechanics, using both stress and energy approaches,[75,76] was developed to link the micromechanical behavior of polymers to their global mechanical behavior. Unfortunately, the lack of knowledge concerning the fundamental physics of how and why a micromechanism is triggered makes it impossible to directly utilize fracture mechanics tools to predict the toughening event(s) upon fracture. Hence, alternative approaches, i.e., classical continuum mechanics and finite element methods (FEM),[77] are implemented and found to be extremely useful as supplemental tools for qualitative or semiquantitative modeling of the micromechanical behavior of polymers.[78-80] In other words, only the well-established mechanics tools, such as the slip-line field theory,[81-84] Irwin,[85] Goodier,[86] Dewey,[87] Eshelby equations,[88] and FEM[77] are used to study both the crack-tip stress field and the stress state under which a toughening mechanism can be triggered. Applications of combination(s) of these approaches, in turn, can greatly help the qualitative design of toughened polymers.[78-80]

In studying the stress disturbance due to the mismatch of elastic constants between the matrix and the toughener phase, the well-established classical equations derived by Goodier, Dewey, Good, and Eshelby can all be used to explicitly describe the linear elastic stress and strain fields around the inclusion phase, as long as the inclusion is either spherical or ellipsoidal in shape. In order to estimate the stress and strain fields around an inclusion in front of the crack tip, superpositions of Irwin's equation, slip-line field theory, and one of the above classical equations (e.g., Dewey's equation), as a first approximation, can help describe how the toughener phase responds to the crack perturbation.[80] This approach has been shown to provide plausible interpretation of why cavitation of the rubber particles is crucial in promoting the shear yielding/banding mechanism in polymer toughening.

When the material is nonlinear in nature and/or the morphology of the toughened system is too complex, the FEM is found to be extremely powerful in elucidating the unusual postyielding behavior of polymers.[78-80] However, it should be noted that care has to be taken when nonlinear behavior of polymers is incorporated into the FEM simulation process. Experimental verification of simulation results is recommended, due to the oversimplification of the FEM analysis.

IV. TOUGHENING MECHANISMS STUDIES

In an effort to shed some light on how to effectively toughen high-performance epoxies, research on effects of types of rubber, particle size, and particle dispersion characteristics in toughening is conducted. Based on these efforts, along with the relevant work conducted by others, it is hoped that a more definitive concept for toughening can be established.

As shown in Tables 2 and 3, it is clear that the types of rubber, particle size, and particle dispersion characteristics do affect the toughening effect in brittle epoxies. In order to determine the exact role(s) the rubber particles play and investigate the operative toughening mechanisms in the toughening process, the DN-4PB damage zone of each brittle epoxy system is studied using TOM and TEM. Only on rare occasions when the fracture surface accounts for most of the toughening effect, will the SEM be utilized.

A. CORE-SHELL RUBBER MODIFICATION

In this study, 10 wt% of CSR-B particles (Table 1), having a uniform particle size of approximately 0.12 μm and a Tg of ≈ −80°C, was incorporated into DGEBA epoxy resin to study the effectiveness of small rubber particles in toughening brittle epoxies. This system exhibits random dispersion and greater toughening effect than those using DAR, Proteus® particle (Table 2), or CTBN rubber modification.[2] It is, therefore, important to discover the role of the CSR particles in the toughening process.

Since the size of the CSR-B particles is too small for TOM observation, only TEM is utilized. The TEM micrographs of the DN-4PB crack-tip damage zone, as shown in Figure 11, indicate that the small 0.12 μm CSR-B particles (see Table 1) are quite effective in promoting extended shear yielding of the brittle epoxy matrix. The matrix has undergone approximately 60% plastic deformation based on the fact that the rubber particles at the crack tip elongated by as much as 60%, compared to the undeformed spherical rubber particles (Figure 12). The major toughening mechanisms in this system appear to be cavitation of the CSR-B particles, followed by shear-yielding of the matrix. In other words, the cavitation of the rubber particles helped relieve the plane strain constraint induced by the thick specimen and the sharp crack; the octahedral shear stress component arount the crack tip is, as a result, greatly raised and causes extended yielding of the matrix. For clarity, the schematic of the toughening sequence of events is drawn in Figure 13.

Interestingly, the rubber particle at the crack tip (Figure 11b) appears to be partially broken. It is, thus, intuitive that prior to the crack advancement, the rubber particles in front of the crack tip first cavitate and induce extended shear-yielding of the matrix by as much as 60% plastic strain. After cavitation and crack advance, the elongation of the rubber particles can reach as much as several hundred percent; crack bridging due to the rubber particles is then possible. However, the rubber particles were cavitated before particle bridging took place. The rubber particles were also partially elongated prior to particle bridging. As a result, the toughening effect due to particle bridging is probably minimal, if any.

In addition to the rubber particle cavitation/matrix shear yielding and the possible rubber-bridging mechanisms, crack bifurcation, and crack deflection mechanisms are also observed (Figures 14 and 15). These mechanisms are supplemental to the toughening of the brittle epoxy matrix. Although the crack appears to follow a path through the CSR particles in front of the crack tip, the degree of crack deflection is minimal. This is because the size of the CSR particles is too small to effectively deflect the crack.

It is noted that some of the CSR particles in the wake and at the crack tip do not show cavitation. This abnormality is mainly due to the OsO_4 staining solution which tends to penetrate through the crack path faster (due to capillary action) than through the matrix. As a result, the rubber particles adjacent to the crack face are overstained, and the cavities of these rubber particles are covered by the stain. This has also been observed and verified in other systems.[66,89]

B. DISPERSED ACRYLIC RUBBER MODIFICATION

Having exactly the same curing schedule and weight percent of rubber, the DAR-modified brittle epoxy system exhibits a lower toughening effect than that of the CSR-modified epoxy system (Table 2). The DAR particle has an average size of about 0.4 μm and a Tg of about −60°C, which are different from those of the CSR particles.

When the DAR is used to toughen the brittle epoxy matrix, the response of the DAR particles to the advancing crack appears to be quite different from that of the CSR particles. Except for the particles that are in contact with the crack planes, the DAR particles stay spherical around the damaged crack tip (Figure 16). In other words, no sign of extended shear-yielding of the matrix is found. This is probably the reason that the DAR modification is less effective than the CSR modification in toughening brittle epoxies.

The major toughening mechanisms in DAR-modified epoxy are found to be crack deflection and crack bifurcation (Figures 17 and 18). Since the DAR particle size (≈0.4 μm) is bigger than that of the CSR particles (0.12 μm), the crack deflection mechanism is more effective in the DAR-modified epoxy than in the CSR-modified system. Interestingly, voids in the epoxy matrix can also help promote the crack deflection mechanism (Figure 19). Therefore, it is conceivable that well-dispersed voids with appropriate sizes (preferably

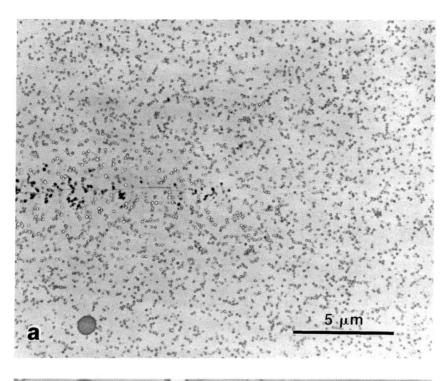

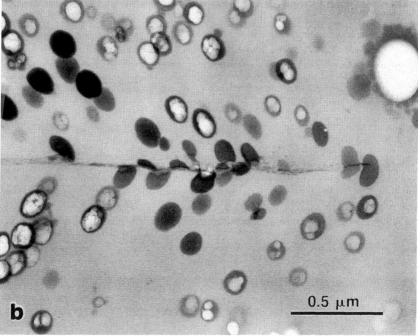

FIGURE 11. TEM micrographs taken at the crack tip of the damaged DN-4PB specimen of CSR-B-modified epoxy system at (a) low magnification and (b) at high magnification. A 60% elongation of the CSR rubber particles at the crack tip is found. This indicates that the brittle epoxy is capable of undergoing extended plastic deformation. The crack propagates from left to right.

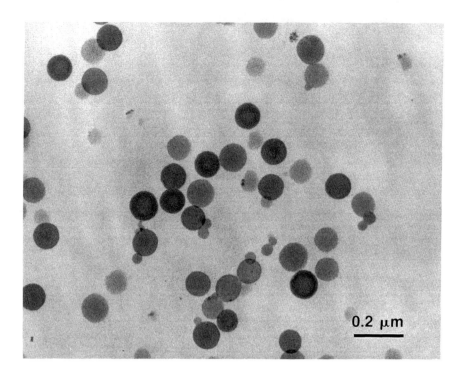

FIGURE 12.   A TEM micrograph taken at the undamaged region of the CSR-modified epoxy system. The CSR rubber particles appear to be spherical. No cavities are found inside the rubber particles.

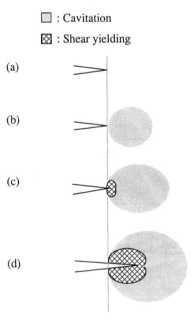

FIGURE 13.   A sketched sequence of the toughening mechanisms of the CSR-modified epoxy system: (a) the initial starter crack; (b) formation of a cavitation zone in front of the crack tip when the specimen is initially loaded; (c) formation of initial shear-yielded plastic zone around the crack tip when the hydrostatic tension is relieved due to the cavitation of the rubber particles; and (d) once the buildup of shear strain energy reaches a critical value, the material begins to undergo shear yielding, and the crack propagates, leaving a damage zone surrounding the propagating crack before the crack grows unstably. Note that the size of the plastic zone is not drawn to scale.

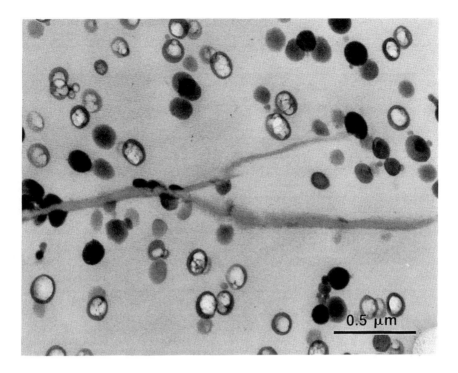

FIGURE 14. A TEM micrograph taken at the damaged crack wake of the CSR-modified epoxy system. Crack bifurcation is observed. The crack propagates from left to right.

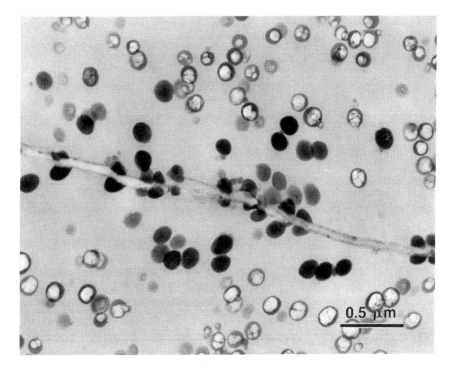

FIGURE 15. A TEM micrograph taken at the damaged crack wake of the CSR-modified epoxy system. The crack propagates through the rubber particles, instead of propagating around the rubber particles. The rubber particles appear to have deflected the crack path. The crack propagates from left to right.

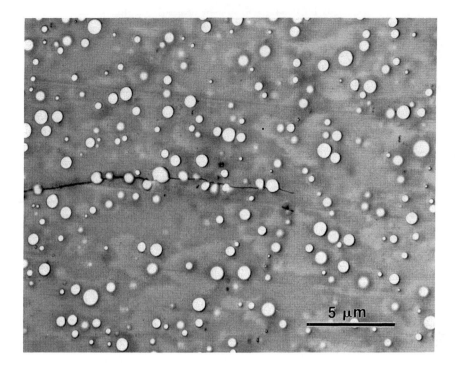

FIGURE 16. A TEM micrograph taken around the crack tip of the damaged DN-4PB specimen of the DAR-modified epoxy. Only the particles that contacted the crack faces are deformed. The adjacent rubber particles appear to be unaffected by the propagated crack. The crack propagates from left to right.

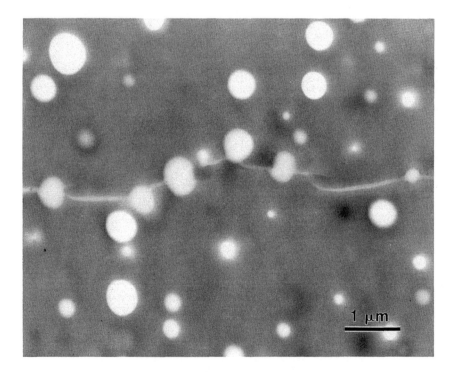

FIGURE 17. A TEM micrograph taken at the damaged crack wake of the DAR-modified epoxy. The crack is deflected by the DAR rubber in front of it. The crack propagates from left to right.

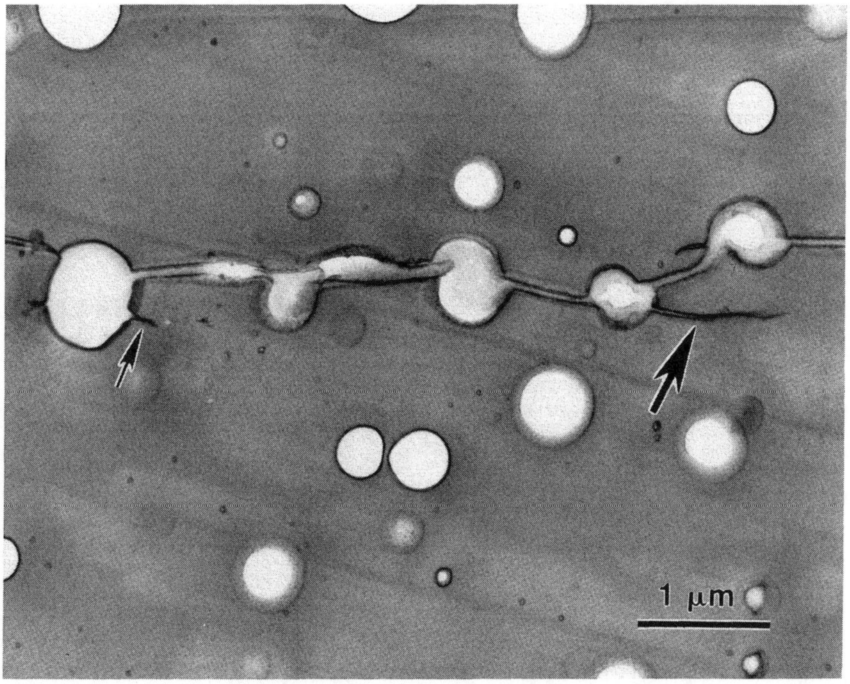

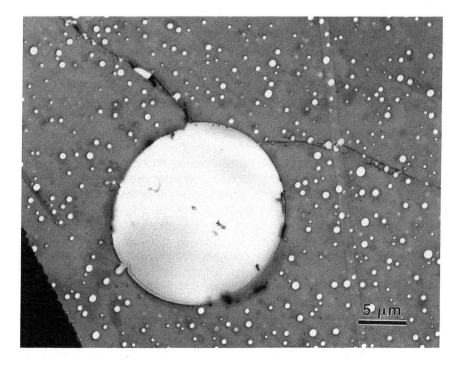

FIGURE 18. A TEM micrograph taken at a region behind the crack tip of the DAR-modified epoxy. The crack bifurcation mechanism (big arrow) can be easily observed. Microcracking (small arrows) appears to take place, as well. In addition, a possible particle/crack bridging mechanism appears to be present. The crack propagates from left to right.

FIGURE 19. A TEM micrograph taken at the damaged crack wake of the DAR-modified epoxy. The crack is deflected by the void. The crack propagates from left to right.

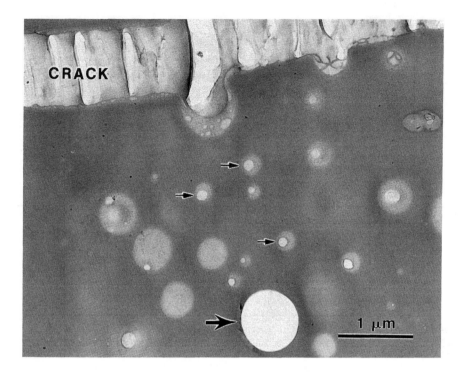

FIGURE 20.   A TEM micrograph taken at the core region of the DN-4PB sample. The thin sections are stained with styrene/OsO$_4$ and treated with periodate anion.[66] It is evident that the DAR particles do cavitate (indicated by small arrows) at the SFSZ. The big arrow indicates a hole, which is formed due to rubber pull-out during the thin-sectioning process. The crack propagates from left to right.

a few times bigger than the characteristic crack-tip radius) are as effective as the rubber particles in deflecting the crack. Crack bifurcation appears to occur in the DAR-modified epoxy (see big arrows in Figure 18), as well. Microcracking induced by the DAR particles also seems to operate in this system (small arrow in Figure 18). In general, however, the crack deflection, crack bifurcation, and microcracking types of toughening mechanisms are less effective than the shear-yielding mechanism in toughening epoxies.

Owing to the fact that the DAR particles cannot be effectively stained via any existing staining technique, a novel selective solvent sorption and staining technique[66] was developed to stain the crack-tip damage zone of the DAR-modified epoxy system. As indicated in Figure 20, the DAR particles are, in fact, cavitated around the crack tip and the crack wake. This finding implies that the cavitation of the rubber particles alone is not sufficient to cause extended shear yielding of the matrix. The stress state at which the rubber particles cavitate should also exceed a certain critical value to activate the yielding of the matrix.[90]

It is noted that even with the use of the novel selective solvent sorption and staining technique, no crack/particle bridging mechanism can be observed. Therefore, it is still not certain whether or not crack bridging takes place in this system.

C. PROTEUS® PARTICLE MODIFICATION

To find out whether or not big composite rubber particles are effective in toughening, 10 wt% of the Proteus® particle is used to modify the brittle epoxy matrix. The Proteus® particle, having an average particle size of about 20 μm, is composed of numerous small (0.2 μm) rubber particles (Figure 21). The toughening effect due to the Proteus® particle

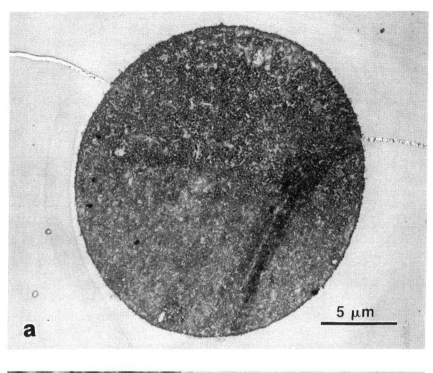

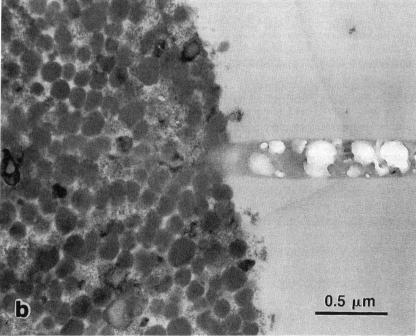

FIGURE 21. TEM micrographs, (a) at low magnification and (b) at high magnification, showing that when the crack propagates through the Proteus® particles, no signs of rubber particle cavitation or deformation are observed. The crack propagates from left to right.

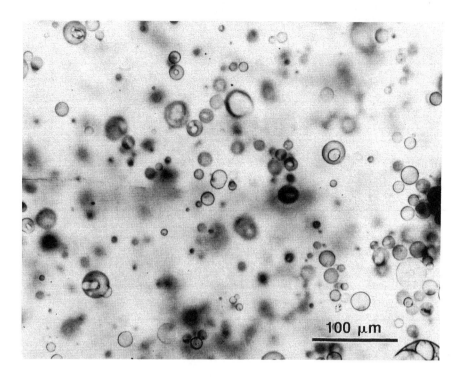

FIGURE 22.   A TOM micrograph taken under bright field at the damaged crack of the Proteus® particle-modified epoxy. The crack deflection is observed. The crack propagates from left to right.

modification is found to be comparable to that of the DAR modification. Since the Proteus® rubber particle size is about 20 μm, TOM and SEM are effective in identifying the toughening mechanisms in this system, TEM will only be used to supplement the investigation.

The Tg of the matrix inside the Proteus® particle is about 28°C, the Proteus® particle is therefore, quite rigid at room temperature (25°C). Also, as shown in Figure 21, the agglomerated small rubber particles inside the Proteus® particle appear to be inactive, i.e., no rubber cavitation and stretching occur, with the advancing crack. Therefore, the plane strain constraint cannot be relieved via the Proteus® particle modification. As expected, the major operative toughening mechanisms in Proteus® particle-modified epoxy are found to be mainly crack deflection and crack pinning (Figures 22 and 23). Since the size of the Proteus® particle (≈20 μm) is far greater than the characteristic crack-tip radius (<0.5 μm), the crack deflection angle is rather small. Therefore, the toughening effect due to crack deflection is probably not significant. On the other hand, the Proteus® particles appear to be capable of promoting extensive crack pinning, which is evidenced by the steps or the tails ahead of pinned particles (see arrows in Figure 23), on the fracture surface.

Despite exhaustive effort, the crack/particle bridging mechanism is not observed.[65] This is probably due to the fact that the Proteus® particles are brittle in nature. They are not drawable. As a result, only the crack pinning mechanism dominates the toughening.

D. VARIATION OF CORE-SHELL RUBBER INTERFACIAL CHARACTERISTICS

In an attempt to study whether or not the dispersity of rubber particles and the grafting efficiency of rubber particles to the matrix affect the toughening effect, the composition and thickness of the shell on the CSR particle are varied. A total of seven types of CSR particles are investigated and they are listed in Table 1. The fracture toughness as well as the Tg of each of the CSR-modified high performance epoxies are shown in Table 3.

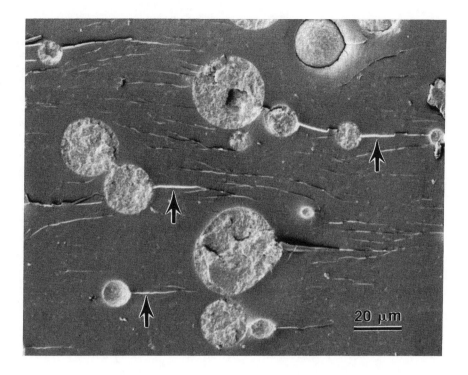

FIGURE 23. An SEM micrograph taken of the fracture surface of the Proteus® particle-modified epoxy. The crack-pinning mechanism, shown as a tail in front of the rubber particle (see arrows), is observed. The crack propagates from left to right.

1. Variation of Glycidyl Methacrylate Content in the Shell

Since glycidyl methacrylate (GMA) contains an epoxide functional group, the GMA component in the shell is believed to chemically react with, and therefore, graft to, the epoxy matrix. Consequently, the GMA concentration can affect not only the dispersion of the CSR in the matrix via solubility parameter (δ) changes (Table 1), but also the chemical grafting efficiency to the epoxy matrix. The variation of the GMA content from 30 to 0% by weight in the shell (i.e., CSR-A, CSR-B, and CSR-C) appears to alter the dispersion of the CSR particles in the epoxy matrix somewhat, i.e., from multimodal dispersion, to random dispersion, and to globally random, but locally clustered dispersion (Figures 11, 24, and 25). Nevertheless, the toughening effect observed for the three systems is practically the same (Table 3). This implies that the chemical bonding between the rubber particles and the epoxy matrix is not critical in toughening by CSR. In other words, the possible molecular intermixing and/or the physical polarity interactions at the interface between the shell of the CSR and the epoxy matrix may be sufficient to maintain the interfacial integrity. This also implies that the cavitational strength of the CSR particle is weaker than that of the resultant interfacial strength. The toughening mechanisms in these systems are quite similar. Except for the CSR-C system where an additional crack deflection mechanism is operative (due to local clustering), the major toughening mechanisms in these three systems are cavitation of the rubber particles, followed by shear yielding of the matrix.

2. Variation of Acrylonitrile Content in the Shell

The above study on CSR-A-, CSR-B-, and CSR-C-modified epoxies suggests that the GMA content does not affect the toughening effect. Consequently, the GMA is omitted hereafter.

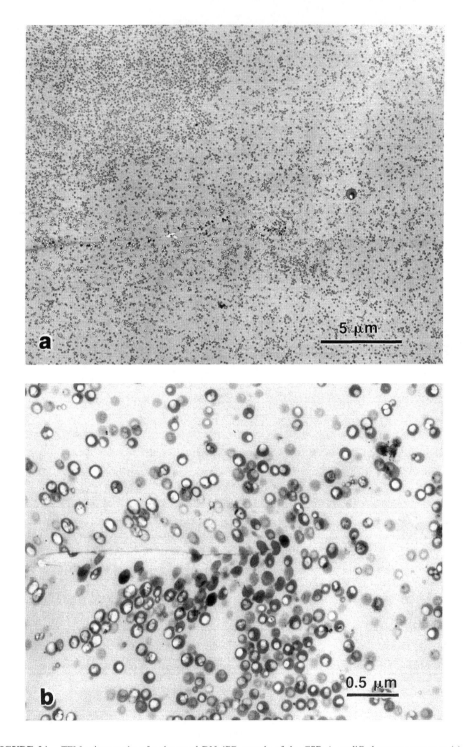

FIGURE 24. TEM micrographs of a damaged DN-4PB sample of the CSR-A-modified epoxy system at (a) a lower magnification and (b) at a higher magnification. A multimodal dispersion of the CSR particles is observed. The crack propagates from left to right.

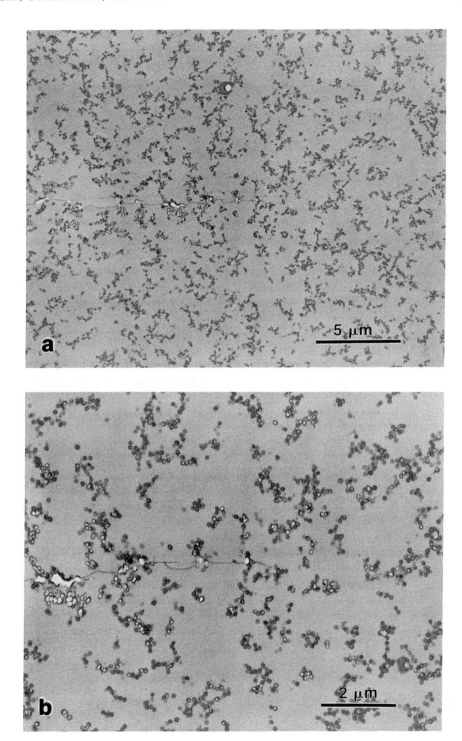

FIGURE 25. TEM micrographs of a damaged DN-4PB sample of the CSR-C-modified epoxy system taken both at (a) a low magnification and (b) a high magnification. The CSR particles are globally well dispersed, while locally clustered. The crack propagates from left to right.

When the acrylonitrile (AN) content of the shell is varied from 25 to 0% by weight (i.e, CSR-C, CSR-D, and CSR-E), the δ of the shell varies from about 10.1, to 9.7, and to 9.3. Since the δ for the epoxy matrix is ≈10.3, the dispersion of the CSR particles in the epoxy matrix will likely be changed from good dispersion to poor dispersion. Indeed, as shown in Figures 25–27, the dispersion of the CSR particles is altered from globally random but locally clustered dispersion, to more local particle clustering, and to large scale agglomeration of the CSR particles in the epoxy matrix. The toughening effect is, as anticipated, good for the CSR-C-modified system and poor for both the CSR-D- and CSR-E-modified systems.

The toughening mechanisms in these systems are quite different; rubber particle cavitation, matrix shear-yielding, and limited crack deflection are found for the CSR-C-modified system, while only the crack deflection mechanism is observed for both the CSR-D- and CSR-E-modified systems.

The investigation of the toughening mechanisms in this series suggests that, qualitatively speaking, a certain degree of rubber particle clustering is desirable. Local clustering of the CSR particles will not affect the toughening effect. However, when the clustering increases in magnitude, the local stress field overlap causes the clustered rubber particles to act like a big irregularly shaped rubber particle, which functions like the Proteus® particle (Figure 21). As a result, the individual rubber particles can no longer cavitate to relieve the plane strain constraint. This results in brittle failure of the matrix.

3. Variation of Shell Thickness

The thickness of the shell may not only help retain the shape of the CSR particle during part fabrication, but also potentially has an effect on how the copolymers arrange themselves in the shell. The possible rearrangement of the copolymers in the shell will likely affect the dispersity of the CSR particles in the epoxy matrix. This will, in turn, alter the failure process in the CSR-modified epoxy systems.

When the shell thickness is varied by changing weight percent from 15 to 25%, and to 35% in the shell (i.e., CSR-B, CSR-F, and CSR-G) while all other parameters are kept the same (Table 1), the dispersions of the CSR particles do change from random dispersion (Figure 11) to globally well dispersed, but locally clustered dispersions (Figures 28 and 29). This implies that the copolymers in the shell have somehow rearranged themselves due to shell thickness variations. Questions concerning how the shell thickness affects copolymers rearrangements still await further investigation.

The toughening effect exhibited by the three systems is quite surprising. Random dispersion of the CSR particles does not give optimal toughening (Table 3). Instead, there appears to be a synergistic toughening effect due to the nonrandom dispersion of the rubber particles. An average of ≈640 J/m^2 in G_{IC} can be obtained for the CSR-G-modified brittle epoxy system. In comparison, the G_{IC} is only ≈490 J/m^2 in the CSR-B-modified system, where the CSR particles disperse randomly. As shown in Figure 29, it is apparent that the local clustering of the CSR-G particles in the epoxy matrix not only preserves the important rubber cavitation and matrix shear-yielding mechanisms, but also triggers a vigorous crack-deflection mechanism. This additional crack-deflection mechanism is probably the main reason for such a synergistic toughening effect.

The study of the effect of rubber particle dispersion in epoxy toughening brings up a valuable concept concerning approaches for toughness optimization. That is, the dispersion of the rubber particles can influence, and even alter, the toughening mechanisms in toughened epoxies. Furthermore, based on this effort, it is clear that if combinations of certain toughening mechanisms coexist during the toughening process, then, a synergistic toughening effect may be obtained.

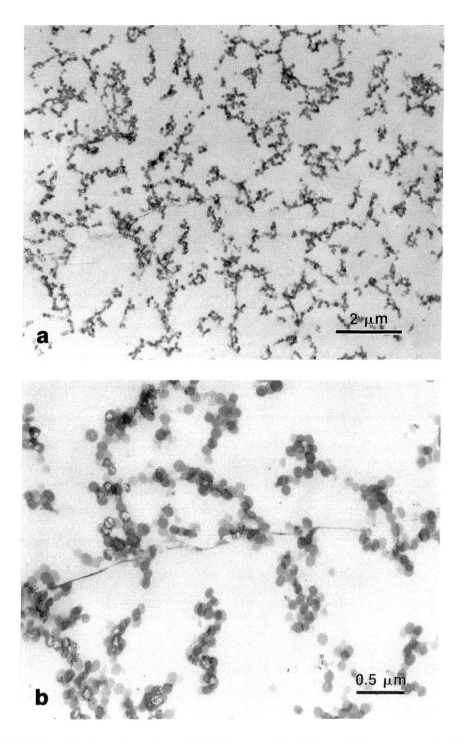

FIGURE 26. TEM micrographs of a damaged DN-4PB sample of the CSR-D-modified epoxy system taken both at (a) a low magnification and (b) a high magnification. Cavitation of CSR particles is highly suppressed by the poor dispersion of the CSR particles. The crack propagates from left to right.

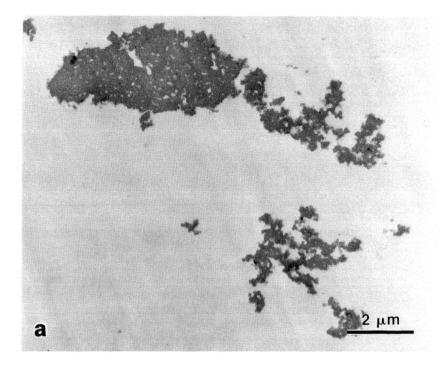

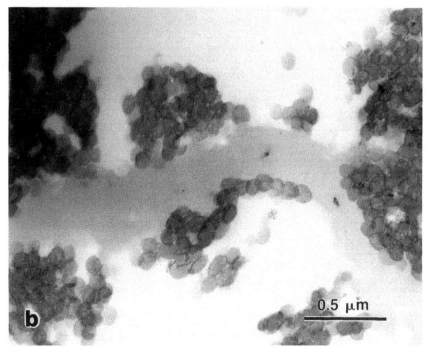

FIGURE 27. TEM micrographs of the CSR-E-modified epoxy system. (a) Without GMA and AN in the shell, the CSR particles severely agglomerate and form irregular domains. (b) At the damage zone, only crack deflection is observed and rubber particle cavitation is suppressed. The crack propagates from left to right.

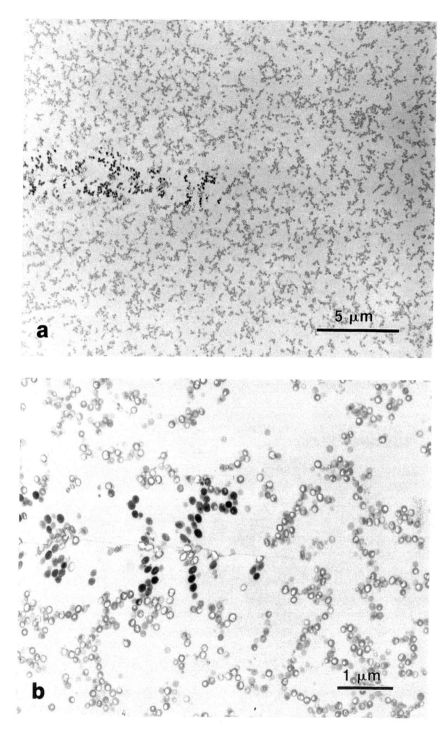

FIGURE 28. TEM micrographs of a damaged DN-4PB sample of the CSR-F-modified epoxy system taken both at (a) a low magnification and (b) a high magnification. The CSR particles are globally well dispersed. However, the particles are locally clustered. The crack propagates from left to right.

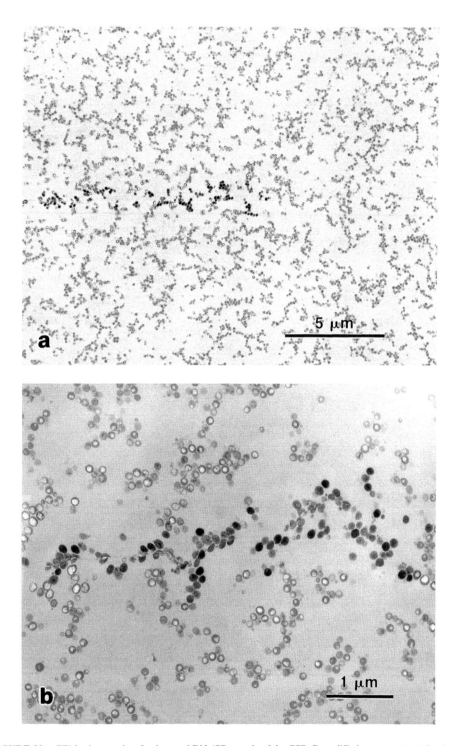

FIGURE 29. TEM micrographs of a damaged DN-4PB sample of the CSR-G-modified epoxy system taken both at (a) a low magnification and (b) a high magnification. The CSR particles are globally well dispersed. However, the particles are locally clustered. The crack propagates from left to right.

V. APPROACH FOR MAKING TOUGHER EPOXY SYSTEMS

To toughen epoxies, the traditional chemistry oriented approaches have been to chemically link the rigid epoxy molecules to more flexible chains, physically blend flexible monomers into the rigid monomers, use less reactive curing agents, or to combine the above. Indeed, these approaches do produce ductile epoxies (for engineering applications) exhibiting better molecular chain flexibility, which can improve the tensile elongation to break from several percent to over 20%.[91] However, the notch sensitivity of the epoxies made via these modifications is usually not altered much. As a result, the brittle epoxies modified through chemical means are still brittle in nature and are not suitable for structural applications.

An alternative approach, which utilizes the presence of a dispersed inhomogeneity in the epoxy matrix to promote an extended damage zone when fracture occurs, is found to be, in general, far more effective in toughening epoxies. This approach, however, does not always guarantee effective toughening of brittle epoxies. The type of rubber, particle size, and dispersion can all greatly affect toughening. The understanding of the exact role(s) the dispersed phase (usually the rubber particles) plays in the toughening process is still mostly unsettled. The present work, which is only part of a larger effort in an attempt to optimize the toughening effect in epoxy systems, focuses on using materials science and mechanics tools to help the understanding, and therefore, the improvements of the highly versatile but brittle epoxy matrices. Based on the understanding from the present experimental work, as well as the related work conducted by others, an up-to-date toughening concept is discussed below.

A. MATERIALS SELECTION

The process of selecting an appropriate epoxy matrix for structural applications usually begins with the requirements of its Tg, stiffness, and possibly, its physical and chemical properties. However, owing to its inherent brittleness and/or notch sensitivity, the epoxy matrix needs to be toughened. The physical and chemical properties of the epoxy matrix, which are material characteristics and cannot be altered, are governed by the chemical species in the molecule. The Tg and stiffness of the epoxy matrix is, however, strongly dependent on the structural arrangement of the molecules. In other words, the rigidity of the molecular chain and the cross-link density of the epoxy may dictate both the Tg and stiffness of the epoxy matrix.

The work conducted by Yee and Pearson[1-3] convincingly demonstrates that the low cross-link density epoxies are far more toughenable than the high cross-link density epoxies. Combining the experimental observations by Yee and Pearson and the understanding that Tg is governed by the rigidity of the epoxy backbone, Portelli et al.[92] are able to develop high Tg and highly toughenable fluorene epoxies. The fluorene epoxies are synthesized and cured in such a way that the local rigidity of the molecule is enhanced by the fluorene backbone structure, while allowing the molecular weights between the crosslinks to be high. With only 7.5% of CSR modification (obtained from Rohm & Haas), an improvement of over 300% in G_{IC} can be obtained without sacrificing the matrix Tg.[92] Therefore, in selecting an epoxy matrix for structural applications, the preferred choices are low cross-link density and high local rigidity epoxy molecules.

In the case of selecting the rubber tougheners, it appears that, depending on the applications, the type of rubber, size, and dispersion of the rubber particles all play important roles in toughening of brittle epoxies. The use of soluble rubbers to toughen epoxies can result in (1) lowering of the Tg of the epoxy matrix, (2) difficulty in morphology control, and (3) poor reproducibility of the product performance.[93,94] These shortcomings are highly undesirable in high performance aerospace applications. Consequently, preformed rubber particles with controllable and reproducible particle size and shape have recently been developed. These preformed rubber particles include the DAR and CSR particles investigated in this work.

The investigation of the CSR-modified epoxy systems shows that the highly cross-linked brittle epoxy system can undergo shear yielding around the propagated crack tip, even when the crack experiences the plane strain mode-I loading condition. The major toughening mechanisms for these systems are found to be the cavitation of the rubber particles, followed by the formation of a shear-yielded zone around the propagated crack. If this rather small-scale yielding can be further extended, then, potentially an order of magnitude increase in fracture toughness can be achieved. Nonetheless, not all the rubbers can produce the same result as that of the CSR particle. The CTBN rubber used by Yee and Pearson[2] and the DAR and Proteus® 5025 particle studied in this work all show negative results in inducing the extended shear-yielding mechanism. Even with the use of CSR particles, if the particle dispersion is poor (i.e., the CSR-D and CSR-E systems discussed in this study), the rubber particle cavitation and matrix shear-yielding mechanisms are suppressed. Thus, the type of rubber, rubber particle size, and particle dispersion in the matrix will induce entirely different toughening mechanisms in modified brittle epoxies.

The degree of interfacial adhesion between the toughener phase and the matrix is also critical in toughening. The strength of the adhesion necessary at the interface to both relieve the triaxial tension and trigger localized shear-banding is still not clear. The present study and previous works on polyethylene-modified polycarbonate[95] and rubber-modified nylon[96] do, however, suggest that chemical bonding at the interface is not essential to assure sufficient interfacial bonding. Molecular interlocking, physical interactions, and thermal stress effect can all either enhance or deteriorate the interfacial adhesion between the toughener phase and the matrix. In general, it is believed that an intermediate interfacial bonding, i.e., neither too strong nor too weak an interfacial strength, should be beneficial for effective toughening. This conjecture, nevertheless, cannot be verified until quantitative measurement of interfacial strength can be established.

B. PROMOTING DESIRABLE TOUGHENING MECHANISMS

The toughening principles for relatively ductile polymers have been reviewed and discussed by Yee and Pearson[1-3] and Kinloch.[4] In the case of brittle epoxies, Garg and Mai[9] have summarized the important mechanisms for toughening. Based on the experimental work conducted in the present study, an approach for toughening high performance epoxies is discussed below.

The present work shows that shear yielding can occur in brittle epoxy systems. It also shows that shear yielding and crack/particle bridging mechanisms cannot both be dominant toughening mechanisms in uniform particle size rubber-modified systems.[57-60,65] At this stage, however, it is still not clear if an order of magnitude increase in fracture toughness may be attained in brittle epoxy systems. In order to clarify this uncertainty, it is necessary that the nature of shear yielding and shear banding in brittle epoxy systems as well as the quantitative estimation of the crack-tip stress field before and after the rubber particles cavitate be understood.

To test whether or not a synergistic toughening effect due to both shear-yielding and crack/particle bridging exists, a bimodal size distribution of rubber particles has been examined by Pearson and Yee.[3] They manage to incorporate both the 1 to 2 and 10 to 20 μm CTBN rubber particles in D.E.R.™ 331 epoxy resin and cure with piperidine. They conclude that the bimodal rubber particle modification does not further improve the fracture toughness of the system, compared to that of the unimodal (1 to 2 μm particle size) distribution system.

More recently, a new toughening mechanism, being termed as the croiding mechanism,[10] is found to exhibit both high shear plasticity and high dilatational plasticity in the epoxy matrix. This mechanism, though it has only been investigated in low cross-link density epoxy systems, may be an alternative route for effective toughening of brittle epoxies. Although there is still much to be done to understand exactly how the croid is formed and the conditions to activate it, toughening epoxies by croiding is shown to be as effective as, if not better than, use of the shear-yielding mechanism. An order of magnitude improvement in toughness via the croiding mechanism has been reported.[97] It appears that the intrinsic properties of both the matrix and the rubber particles, as well as the rubber particle concentration, play significant roles in causing the formation of croids. It is also shown that the stress state under which croiding occurs must be highly triaxial.[10,97] Since the croiding mechanism can operate in a highly localized manner, it can potentially be utilized for both adhesive and composite toughening applications, where the stress is almost always highly triaxial and the matrix material is very thin.

If, somehow, the shear yielding and plastic dilatation of the matrix cannot be induced for other reasons, alternative mechanisms, such as crack/particle bridging,[35,36,54] microcracking,[51] crack deflection,[47,48] crack bifurcation,[49,50] and crack pinning,[45,46] should be considered. These mechanisms appear to occur in the rubber-modified systems quite readily. As pointed out earlier, in order for the crack-bridging mechanism to occur, the rubber particles need to be large with respect to the characteristic crack-tip radius. The interfacial adhesion between the rubber particle and the matrix needs to be stong. The rubber particle needs to be drawable, as well. Therefore, the composite rubber particles, such as the Proteus® particles and occluded rubber particles which cannot be extensively stretched, are not suitable for promoting the crack-bridging mechanism.

For the microcracking mechanism to occur, the interfacial adhesion between the matrix and the toughener phase does not need to be strong. Debonding at the interface, internal cavitation of the toughener phase, and microcracking and crazing of the inclusion phase can effectively serve the purpose of shielding the crack and impeding crack growth. The above dilatational processes can also result in the matrix microcracking, depending on the physical nature of the matrix material. Consequently, a toughened system is obtained.

For the crack deflection and crack bifurcation mechanisms to occur, the toughener phase needs to generate sufficient stress disturbance in front of the crack tip. When the crack propagates, the crack path will then be altered by the stress disturbance. This causes crack deflection and crack bifurcation mechanisms to occur. The rubber particle size also appears to play an important role in deflecting the crack. By comparing Figure 15 with Figures 17 and 19, it is evident that a larger rubber particle (or a hole) can deflect the crack more effectively than a smaller one. Further, when the smaller particles cluster together, they can be mechanically treated as a big particle (i.e., the CSR-C, CSR-F, and CSR-G systems). Consequently, the crack deflection mechanism can be enhanced. Therefore, a larger rubber particle size or a local clustering of small particles while maintaining a good global particle dispersion should be utilized when the crack-deflection mechanism is to be promoted.

For the crack-pinning mechanism to take place, the toughener phase needs to adhere to the matrix strongly. Thus, when the crack grows around the particles, the crack front is bowed and, thus, more energy is required to propagate the crack. It has been shown that the particle size and concentration are critical in toughness optimization via crack pinning.[45] It should be noted that the maximum stress concentration around a rigid inclusion is at the polar region. Thus, when the crack grows slowly, the crack will tend to grow toward the polar region of the particle. This discourages occurrence of the crack-pinning mechanism. Only when the crack grows fast, which is usually the case for a brittle epoxy matrix, will the crack grow around the rigid toughener particles and possibly cause crack pinning to occur. When the toughener particle is soft, as in the case of CTBN rubber particles, and when the matrix does not undergo shear-yielding, the crack-bridging mechanism is also likely to take place right after the pinning mechanism occurs. In cases where the toughener particles are softer than the matrix and are not stretchable, like the Proteus® particles, then the crack-bridging mechanism becomes inoperative and the crack-pinning mechanism becomes dominant.

When failure occurs in epoxy systems, there is usually more than one toughening mechanism taking place. In order to optimize the toughening effect, it is desirable to promote more than one toughening mechanism when fracture occurs. Therefore, it is important to first consider promoting the most effective toughening mechanisms, such as shear yielding and croiding, and then, without sacrificing the above, promote other effective mechanisms. This concept can be best demonstrated by the CSR-G-modified epoxy system the rubber cavitation and extended matrix shear-yielding mechanisms are preserved, while an additional crack deflection mechanism is promoted through the local clustering of the CSR-G particles.

Furthermore, since practically all the operative toughening mechanisms in rubber-modified epoxies are viscoelastic and viscoplastic in nature, they are time- and temperature-dependent. The toughening approach, for a given material, may as a result vary from one testing condition to another. Therefore, caution should be exercised in any attempt to apply fundamental knowledge learned in toughening under one testing condition to another.

Finally, the mechanics tools introduced in this chapter are quite useful in fundamental understanding of how and why certain toughening mechanisms are affected by the morphology of the toughened system. The importance of relieving the plane strain constraint in triggering matrix shear-yielding can also be rationalized via the mechanics approach. Detailed descriptions on how to effectively utilize these tools will not be covered in this chapter due to the complexity and the stand-alone nature of this subject. They can be found elsewhere.[11,80-88]

VI. CONCLUDING REMARKS

This chapter, while concentrating on the recent experimental efforts conducted by the authors in toughening brittle epoxies, focuses on using the more generic materials science and mechanics tools to study routes for toughening epoxies. This will inevitably help provide a broader utilization of the present understanding in toughening other types of polymers as well as utilization in other applications. Effects due to rubber type, particle size, interfacial adhesion, and particle dispersion in epoxy toughening are discussed. Methodology regarding how to make a tough epoxy system is also addressed. It should be cautioned that in the field of polymer toughening, many toughening concepts and theories are still tentative. Only the experimental data are reliable. Care needs to be taken when utilizing any toughening theories and models to design new products.

NOTICE

The information in this paper is presented in good faith, but no warranty is given, nor is freedom from any patent to be inferred.

ACKNOWLEDGMENTS

The authors would like to thank Professor A. F. Yee for his constant valuable discussion concerning this work. The authors also would like to thank Professor R. S. Porter and Ms. C. Stamm for permitting the use of the previously published work in *Polymer Engineering and Science*, and the ACS Book Series No. 233 (*Toughened Plastics: Science and Engi-*

neering). Special thanks are given to R. E. Jones, D. L. Barron, C. C. Garrison, N. A. Orchard, D. W. Hoffman, C. E. Allen, C. J. Bott, R. D. Peffley, L. M. Kroposki, and T. E. Fisk for their input, experimental assistance, and material supplied for this work.

REFERENCES

1. **Yee, A. F. and Pearson, R. A.,** *J. Mater. Sci.,* 21, 2462, 1986.
2. **Pearson, R. A. and Yee, A. F.,** *J. Mater. Sci.,* 24, 2571, 1989.
3. **Pearson, R. A. and Yee, A. F.,** *J. Mater. Sci.,* 26, 3828, 1991.
4. **Kinloch, A. J., Shaw, S. J., Tod, D. A., and Hunston, D. L.,** *Polymer,* 24, 1341, 1983.
5. **Low, I. M. and Mai, Y. W.,** *Comp. Sci. Technol.,* 33, 191, 1988.
6. **Levita, G.,** in *Rubber-Toughened Plastics, Adv. in Chem. Ser.,* C. K. Riew, Ed., 222, 93, 1989.
7. **Kinloch, A. J., Finch, C. A., and Hashemi, S.,** *Polym. Commun.,* 28, 322, 1987.
8. **Glad, M. D.,** Ph.D thesis, Cornell University, Ithaca, New York, 1986.
9. **Garg, A. C. and Mai, Y. W.,** *Comp. Sci. Technol.,* 31, 179, 1988.
10. **Sue, H.-J.,** *J. Mater. Sci.,* 27, 3098, 1992.
11. **Broek, D.,** *Elementary Engineering Fracture Mechanics,* Noordhoff International Publishing, The Netherlands, 1974, p. 91.
12. **Bowden, P. B.,** in *The Physics of Glassy Polymers,* R. Haward, Ed., Applied Science Publishers, London, chap. 5, 1973, p. 279.
13. **Bowden, P. B. and Raha, S.,** *Phil. Mag.,* 22, 463, 1970.
14. **Tresca, H.,** *Comptes Rendus Acad. Sci. (Paris),* 59, 754, 1864.
15. **von Mises, R.,** *Gott. Nach., Math.-Phys., Klasse,* 582, 1913.
16. **Schofield, A. N. and Wroth, C. P.,** *Critical State Soil Mechanics,* MacGraw-Hill, Maidenhead, 1968.
17. **Sternstein, S., Ongchin, L., and Silverman, A.,** *Appl. Polym. Symp.,* 7, 175, 1969.
18. **Bauwens, J. C.,** *J. Polym. Sci.,* A-2, 5, 1145, 1967.
19. **Bauwens, J. C.,** *J. Polym. Sci.,* A-2, 8, 893, 1970.
20. **Ward, I. M.,** *Mechanical Properties of Solid Polymers,* 2nd ed., John Wiley & Sons, New York, 1983.
21. **Jho, J. Y. and Yee, A. F.,** *Macromolecules,* 24, 1905, 1991.
22. **Ferry, J. D.,** *Viscoelastic Properties of Polymers,* John Wiley & Sons, New York, 1980.
23. **Bucknall, C. B.,** *Toughened Plastics,* Applied Science, London, 1977.
24. **Maxwell, B. and Rahm, L. F.,** *Ind. Eng. Chem.,* 41, 1988, 1948.
25. **Bucknall, C. B. and Smith, R. R.,** *Polymer,* 6, 437, 1965.
26. **Strella, S. J.,** *Polym. Sci.,* A-2, 4, 527, 1966.
27. **Newman, S. and Strella, S. J.,** *Appl. Polym. Sci.,* 9, 2297, 1965.
28. **Spurr, O. K., Jr. and Niegisch, W. D.,** *J. Appl. Polym. Sci.,* 6, 585, 1962.
29. **Regel, V. R.,** *J. Tech. Phys. (USSR),* 26, 359, 1956.
30. **Sato, Y.,** *High Polym. Chem. (Japan),* 23, 69, 1966.
31. **Gesner, B.,** *Encyclopedia of Polymer Science and Technology,* Vol. 10, H. Mark et al., Eds., Wiley, New York, 694, 1969.
32. **Sternstein, S. and Ongchin, L.,** *Polym. Preprint,* 10, 1117, 1969.
33. **Kambour, R. P.,** *Macromol. Rev.,* 7, 1, 1973.
34. **Bowden, P. B. and Oxborough, R. J.,** *Phil. Mag.,* 28, 547, 1973.
35. **Merz, E. H., Claver, G. C., and Baer, M.,** *J. Polym. Sci.,* 325, 22, 1956.
36. **Kunz-Douglass, S., Beaumont, P. W. R., and Ashby, M. F.,** *J. Mater. Sci.,* 16, 2657, 1981.
37. **Wlliams, J. G. and Hodgkinson, J. M.,** *Proc. R. Soc.,* A375, 231, 1981.
38. **Kinloch, A. J. and Wlliams, J. G.,** *J. Mater. Sci.,* 15, 987, 1980.
39. **Gledhill, R. A. and Kinloch, A. J.,** *Polym. Eng. Sci.,* 19, 82, 1979.
40. **Ritchie, R. O. and Yu, W.,** Short crack effects in fatigue: a consequence of crack tip shielding in *Small Fatigue Cracks,* R. O. Ritchie and J. Lankford, Eds., TMS-AIME, Warrendale, PA, 1986.
41. **Schmidt, R. A. and Lentz, T. J.,** ASTM STP, 678, 166, 1979.
42. **Evans, A. G. and Faber, K. T.,** *J. Am. Ceram. Soc.,* 67, 255, 1984.
43. **Evans, A. G. and Cannon, R. M.,** *Acta Metall.,* 34, 761, 1986.
44. **Burns, S. J. and Swain, M. V.,** *J. Am. Cer. Soc.,* 69, 226, 1986.
45. **Lange, F. F. and Radford, K. C.,** *J. Mater. Sci.,* 6, 1199, 1971.
46. **Lange, F. F.,** Fracture of brittle matrix particulate composites, in *Composite Materials, Vol. 5: Fracture and Fatigue,* L. J. Broutman, Ed., Academic Press, New York, 2, 1974.

47. **Faber, K. T. and Evans, A. G.,** *Acta Metall.,* 31, 565, 1983.
48. **Faber, K. T. and Evans, A. G.,** *Acta Metall.,* 31, 577, 1983.
49. **Clark, A. B. J. and Irwin, G. R.,** *Exp. Mech.,* 6, 321, 1966.
50. **Ramulu, M. and Kobayashi, A. S.,** *Exp. Mech.,* 23, 1, 1983.
51. **Evans, A. G. and Faber, K. T.,** *J. Am. Ceram. Soc.,* 67, 255, 1984.
52. **Sultan, J. N. and McGarry, F. J.,** *J. Polym. Sci.,* 13, 29, 1973.
53. **Bascom, W. D., Cottington, R. L., Jones, R. L., and Peyser, P. J.,** *Appl. Polym. Sci.,* 19, 2545, 1975.
54. **Kunz, S.,** The toughening of epoxy-rubber particulate composites, Ph.D. thesis, University of Cambridge, 1978.
55. **Kinloch, A. J.,** in *Toughened Plastics, Adv. in Chem. Ser.,* C. K. Riew, Ed., 67, 222, 1989.
56. **Pearson, R. A. and Yee, A. F.,** *J. Mater. Sci.,* 21, 2475, 1986.
57. **Yee, A. F., Pearson, R. A., and Sue, H.-J.,** 7th Int. Conf. Fracture, 4, 2739, 1989.
58. **Parker, D. S., Sue, H.-J., Huang, J., and Yee, A. F.,** *Polymer,* 31, 2267, 1990.
59. **Sue, H.-J. and Yee, A. F.,** *J. Mater. Sci.,* 24, 1447, 1989.
60. **Henton, D. E., Pickelman, D. M., Arends, C. B., and Meyer, V. E.,** U.S. Patent 4,778,851, 1988.
61. **Hoffman, D. K. and Arends, C. B.,** U.S. Patent 4,708,996, 1987.
62. **Hoffman, D. K. and Arends, C. B.,** U.S. Patent 4,789,712, 1988.
63. **Sue, H.-J.,** *Polym. Eng. Sci.,* 31, 270, 1991.
64. **Sue, H.-J., Pearson, R. A., Parker, D. S., Huang, J., and Yee, A. F.,** *Polym. Preprint,* 29, 147, 1988.
65. **Sue, H.-J.,** *Polym. Eng. Sci.,* 31, 275, 1991.
66. **Sue, H.-J., Garcia-Meitin, E. I., Burton, B. L., and Garrison, C. C.,** *J. Polym. Sci. Polym. Phys.,* 29, 1623, 1991.
67. **Sue, H.-J. and Yee, A. F.,** *J. Mater. Sci.,* Nov., 1992.
68. **Sue, H.-J. and Chou, J.,** in preparation.
69. **Roulin-Moloney, A. C.,** *Fractography and Fracture Mechanisms of Polymers and Composites,* Elsevier Applied Science, New York, 1989.
70. **Sawyer, L. C. and Grubb, D. T.,** *Polymer Microscopy,* Chapman & Hall, New York, 1987.
71. **Holik, A. S., Kambour, R. P., Hobbs, S. Y., and Fink, D. G.,** *Microstruct. Sci.,* 7, 357, 1979.
72. **Towers, O. L.,** *Stress Intensity Factor, Compliance, and Elastic h Factors for Six Geometries,* The Welding Institute, Cambridge, England, 1981.
73. **ASTM Standard,** E399-83.
74. **Knott, J. F.,** *Fundamentals of Fracture Mechanics,* Butterworths, London, 1976.
75. **Mura, Y.,** *Micromechanics of Defects in Solids,* Martinus Nijhoff, Boston, 1982.
76. **Evans, A. G., Ahmad, Z. B., Gilbert, D. G., and Beaumont, P. W. R.,** *Acta Metall.,* 34, 79, 1986.
77. **Zienkiewicz, O. C.,** *The Finite Element Method,* McGraw-Hill, New York, 1977.
78. **Sue, H.-J. and Yee, A. F.,** *Polymer,* 29, 1619, 1988.
79. **Sue, H.-J., Pearson, R. A., and Yee, A. F.,** *Polym. Eng. Sci.,* 31, 793, 1988.
80. **Sue, H.-J.,** Mechanical modeling and experimental observations of toughened rigid-rigid polymer alloys, Ph.D. thesis, The University of Michigan, Ann Arbor, 1988.
81. **Johnson, W. and Mellor, P. B.,** *Engineering Plasticity,* Van Nostrand Reinhold, New York, 1973.
82. **Hill, R.,** *Plasticity,* Clarendon Press, Oxford, 1950.
83. **Thomsen, E. G., Yang, C. T., and Kobayashi, S.,** *Mechanics of Plastic Deformation in Metal Processing,* Macmillan, New York, 1965.
84. **Johnson, W., Sowerby, R., and Haddow, J. B.,** *Plane-Strain Slip-Line Field: Theory and Bibliography,* American Elsevier, New York, 1970.
85. **Irwin, G. R.,** *Trans. Am. Soc. Mech. Eng., J. Appl. Mech.,* 24, 361, 1957.
86. **Goodier, J. N.,** *J. Appl. Mech.,* 1, 39, 1933.
87. **Dewey, J.,** *J. Appl. Phys.,* 18, 578, 1947.
88. **Eshelby, J. D.,** *Proc. R. Soc. (London),* A241, 376, 1957.
89. **Yang, P. C., Woo, E. P., Sue, H.-J., Bishop, M. T., and Pickelman, D. M.,** *PMSE, ACS,* 63, 315, 1990.
90. **Yee, A. F.,** Modifying matrix materials for tougher composites, *Toughened Composites,* ASTM STP 937, N. Johnston, Ed., American Society for Testing and Materials, Philadelphia, 1986, 377.
91. **Walker, L. and Bertram, J. L.,** Dow Chemical U.S.A., private communication.
92. **Portelli, G. B., Schultz, W. J., Jordan, R. C., and Hackett, S. C.,** *Polym. Comp.,* 2, 381, 1989.
93. **Tong, S., Chen, C., and Wu, P. T. K.,** in *Adv. in Chem. Ser.,* C. K. Riew, Ed., 222, 1989, 376.
94. **Verchere, D., Pascault, J. P., Sautereau, H., and Moschiar, S. M.,** *J. Appl. Polym. Sci.,* 43, 293, 1991.
95. **Sue, H.-J., Huang, J., and Yee, A. F.,** *Polymer Commun.,* 33, 4868, 1992.
96. **Wu, S.,** *Polymer,* 26, 1855, 1985.
97. **Sue, H.-J. and Garcia-Meitin, E. I.,** *J. Polym. Sci.,* Polym. Phys., Nov., 1992.

Chapter 19

CHARACTERIZATION OF PHASE BEHAVIOR IN POLYMER BLENDS

Toshiaki Ougizawa and Takashi Inoue

TABLE OF CONTENTS

0-8493-4401-8/93/$0.00 + $.50
© 1993 by CRC Press, Inc.

I. INTRODUCTION

It is well known that most pairs of high molecular weight polymers are immiscible. This is so because the combinatorial entropy of mixing of two polymers is dramatically less than that of two low molecular weight compounds and the enthalpy of mixing is often a positive quantity.[1-4] Therefore, dissimilar polymers are only miscible if there are favorable specific interactions between them, leading to a negative contribution to the Gibbs free energy of mixing. Miscible polymer-polymer mixtures tend to phase separate at elevated temperatures. This lower critical solution temperature (LCST) behavior is interpreted in terms of the equation of state or the free volume contribution.[2-4] About 30 pairs of dissimilar polymers have been found to exhibit the LCST behavior. Some miscible polymer mixtures also exhibit phase separation at low temperatures. This upper critical solution temperature (UCST) has been observed only when one or two components have relatively low molecular weight, i.e., oligomers.[5,6]

The LCST behavior provides a new prospect for the design of supermolecular structure in polymer blends. For instance, if a homogeneous mixture of dissimilar polymers is allowed to undergo a rapid temperature jump from below LCST to above LCST, spinodal decomposition takes place and a highly interconnected two-phase morphology with uniform domain size (so-called "modulated structure") develops. Of course, by quencing the phase-separated system below the glass transition temperature after an appropriate time of phase separation, one is able to fix or freeze this characteristic morphology.

Unfortunately, the formation of modulated structure by thermally induced phase separation is limited to binary polymer systems having an LCST phase diagram. If we use a common solvent for both polymers, we are able to prepare a polymer blend with modulated structure by solution casting, even for combinations of immiscible polymers.[7] This implies that the potential for the design of polymer blends with modulated structure is expanded to a much wider variety of combinations of dissimilar polymers.

For such designs of new materials by polymer blending, one needs a basic understanding of phase behavior: phase equilibrium; kinetics of phase separation; and kinetics of phase dissolution, in binary polymer-polymer systems. In this article, we indicate that the spinodal decomposition and the modulated structure are very useful for the characterization of phase behavior in polymer blends.

First, in Section II we indicate the thermodynamics of phase diagram. It is shown how the Flory-Huggins equation and the equation of state theory explain the phase behavior in polymer blends. In Section III we describe the procedure to estimate the phase diagram. We estimate the phase diagram from measurement of concentration fluctuation and characterize the phase behavior in polymer blends.

II. THERMODYNAMICS OF PHASE DIAGRAM

We start with the discussion of phase stability. The phase behavior of any mixture at constant pressure P and temperature T is directed by the Gibbs free energy of mixing. ΔG^M, which is given by:

$$\Delta G^M = \Delta H^M - T\Delta S^M \qquad (1)$$

where ΔH^M and ΔS^M are the enthalpy and the entropy of mixing, respectively. According to the second law of thermodynamics, two components will only mix if the Gibbs free energy of mixing is negative:

$$\Delta G^M < 0 \qquad (2)$$

Furthermore, the condition for phase stability in a binary mixture of composition ϕ (volume fraction) at fixed temperature and pressure is:

$$\left(\frac{\partial^2 \Delta G^M}{\partial \phi^2}\right)_{P,T} > 0 \tag{3}$$

In order to discuss the phase stability in polymer mixtures, one needs an accurate expression of ΔG^M. Some basic models to describe a mixture were proposed and used to explain the phase behavior.

A. FLORY-HUGGINS EQUATION

Flory and Huggins estimated the entropy of a polymer mixture ΔS^M by using a simple lattice model.[1] It is assumed that polymers are composed of r_i segments and each segment is placed in a lattice. N_1 chain molecules composed of component 1 having r_1 segments and N_2 chain molecules composed of component 2 having r_2 segments are placed in the lattice. In this case, ΔS^M is given by

$$\Delta S^M = -k(N_1 \ln \phi_1 + N_2 \ln \phi_2) \tag{4}$$

where k is the Boltzmann constant and ϕ_i is the volume fraction of i component, i.e.,

$$\phi_1 = N_1 r_1/(N_1 r_1 + N_2 r_2)$$

$$\phi_2 = N_2 r_2/(N_1 r_1 + N_2 r_2) = 1 - \phi_1$$

Equation 4 expresses "the combinatorial entropy of mixing", which is estimated from the increment of a number of ways in arranging. This is fundamentally similar to the expression of the entropy of mixing in an ideal solution. Thus, the difference of ΔS^M between low and high molecular weight systems does not exist qualitatively, but quantitatively. The value of ΔS^M in a high molecular weight system is small, and it can be shown that the combinatorial entropy of mixing is proportional to l/r of each component.

By using, for convenience, χ_{12} parameter for the free energy change except for the combinatorial entropy term, ΔG^M is given by

$$\frac{\Delta G^M}{RT(V/V_r)} = \frac{\phi_1}{r_1} \ln \phi_1 + \frac{\phi_2}{r_2} \ln \phi_2 + \chi_{12} \phi_1 \phi_2 \tag{5}$$

where V is the total volume and V_r the molar volume of segment. Equation 5 is the famous Flory-Huggins expression. χ_{12} is often called the Flory-Huggins interaction parameter. χ_{12} includes ΔH^M and the effect of the entropy except for the combinatorial entropy term.

We consider only the effect of exchange energy for χ_{12}.

$$\chi_{12} = \frac{z}{kT}\left[\frac{1}{2}(\epsilon_{11} + \epsilon_{22}) - \epsilon_{12}\right] \tag{6}$$

where ϵ_{ij} is the energy of contact between component i and j, and z is a coordination number. Equation 6 uses the assumption of random mixing and that probability of contact is approximated by fraction, ϕ_i and ϕ_j.

As the mixture between nonpolar molecules does not have any specific interaction like hydrogen bonding, the molecules can be expected to obey the geometric rule

$$\epsilon_{12} \sim \sqrt{\epsilon_{11} \cdot \epsilon_{22}} \qquad (7)$$

In this case, χ_{12} is given by

$$\chi_{12} = \left(\frac{V_r}{RT}\right)(\delta_1 - \delta_2)^2 \qquad (8)$$

where δ_i is the solubility parameter. The difference between the values of δ in two polymers significantly affects ΔG^M.

If both polymers have high molecular weight, the value of the combinatorial entropy term, composed of the first and second terms in the right hand of Equation 5, is a very small negative. Accordingly, ΔG^M is a positive value even if the difference between δ_1 and δ_2 is small, i.e., mutual solubility does not occur.

It is apparent from Equations 5 and 8 that almost all polymer pairs are generally immiscible.

In order to estimate the phase stability more qualitatively, one uses the value of χ_{12} at critical point. The actual critical value of χ_{12} is found through application in Equation 5 of the critical conditions, i.e.,

$$\frac{\partial^2 \Delta G^M}{\partial \phi_2^2} = \frac{\partial^3 \Delta G^M}{\partial \phi_2^3} = 0 \qquad (9)$$

If χ_{12} is independent on ϕ_2, from Equation 9 the critical condition is given by

$$\phi_{2,\text{crit}} = \frac{1}{1 + \sqrt{r_1/r_2}} \qquad (10)$$

$$\chi_{12,\text{crit}} = \frac{1}{2}\left(\frac{1}{\sqrt{r_1}} + \frac{1}{\sqrt{r_2}}\right) \qquad (11)$$

If we put actual values of r_i, T, and V_r into Equations 8 and 11, we can understand that a small value of $|\delta_1 - \delta_2|$ is necessary for miscibility in polymer mixtures. For a certain value of $|\delta_1 - \delta_2|$, we can predict how a high temperature UCST exists.

Generally, we should discuss this using Equation 6. For positive $\Delta\epsilon$ systems, it is unfavorable to mixing (repulsive system). For negative $\Delta\epsilon$ systems, it is favorable to mixing (attractive system). Since the values of $\Delta\epsilon$, z, and V_r are insensitive to temperature, the value of χ_{r2} decreases or increases monotonically with temperature for repulsive or attractive systems, respectively. Figure 1 shows the temperature dependence of χ_{12}. In a repulsive system, χ_{12} decreases with increasing temperature and intersects the value of $\chi_{12,\text{crit}}$. The UCST exists at the crossing point between χ_{12} and $\chi_{12,\text{crit}}$. If the molecular weight of both polymers is higher or the value of $\Delta\epsilon$ is larger, UCST moves to a higher temperature. In an attractive system, on the other hand, the value of χ_{12} is always negative and below the value of $\chi_{12,\text{crit}}$. Accordingly, the attractive system is miscible at all temperatures, i.e., both UCST and LCST do not appear.

The contact energy of a van Laar-type for the Flory-Huggins χ_{12} parameter is considered. From the Flory-Huggins equation, it is indicated that (1) different polymers are generally immiscible, (2) UCST appears in repulsive systems, and (3) molecular weight and $|\delta_1 - \delta_2|$ dependence of UCST can be explained. However, this theory cannot explain the existence of LCST.

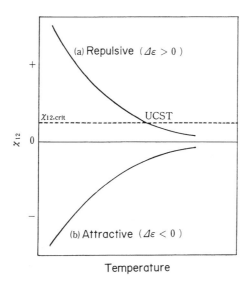

FIGURE 1. Schematic representation for temperature dependence of χ_{12} in the Flory-Huggins equation. For a repulsive system, UCST appears.

B. EQUATION OF STATE THEORIES

Simple lattice theories do not adequately describe the behavior of polymers, polymer solutions, and mixtures. These do not consider the volume change in mixing ΔV^M. Various other theories have been developed which allow for the possibility of volume changes. In order to obtain an accurate expression for a polymer mixture, one has to start from obtaining an accurate expression for polymer molecules. Accordingly, one constructs an equation of state describing the pressure-volume-temperature (PVT) behavior of polymer liquids.

In order to obtain the equation of state, one constructs the partition function from a model of a polymer liquid. There exists two basic models for a polymer liquid: cell model and hole theory. In the cell model the volume change is expressed by expansion or shrinking of lattice size. In the hole theory, the model allows the existence of vacant lattice (hole) and the system expands or shrinks by hole fraction.

Here we briefly show the most famous equation of state of Flory and co-workers, which belongs to the cell model category,[8,9] according to the derivation by Dee and Walsh.[10] The Flory model was constructed on the basis of the method of Prigogine.[11] Polymer molecules are modeled as having two sets of modes. Internal modes are those associated with internal motions of the molecules (rotation, vibration, etc.), which do not contribute to the equation of state properties of liquid. The external modes are those associated with the intermolecular interaction (translation, etc.), which will affect the PVT properties of the system.

If one accepts the idea of a clear separation between internal and external contribution for the partition function, the partition function for a system with N-interacting polymer molecules having r-mers is of the form

$$Z = Z_{int}(T) \, Z_{ext}(T,V) \, \exp(-rNE_o/kT) \tag{12}$$

where E_o is the configurational or mean potential energy per mer when all mers are at their cell centers, and $Z_{ext} = (gv_f)^{3Ncr}$ where 3c is the external degree of freedom per segment, g is the geometric factor, and v_f is the free volume term, which is given by

$$v_f = \int d\vec{a} \, \exp[-(E(\vec{a}) - E_o)/kT] \tag{13}$$

where $E(\vec{a})$ is the interaction energy of a molecule at position a within its cell and the integration is over the cell volume. By replacing the exact potential with a square well approximation,[12,13] one obtains the following expression

$$Z = Z_{int} (T) \{(v^{1/3} - v_{hc}^{1/3}) g(V)\}^{3Ncr} \exp(-rNE_o/kT) \qquad (14)$$

where v_{hc} is the hard-core cell volume and constant multiplicative factors were neglected. To evaluate the equation of state from Equation 14, one needs to know the dependence of g and E_o on V. We set g equal to a constant, arguing that over the thermodynamic range of interest to us g is a weak function of V.

E_o is written by the form

$$E_o = s\eta\psi(R)/2 \qquad (15)$$

where s is the number of contacts per mer, η is the characteristic mer-mer interaction energy, ψ is the interaction potential, and R is the intermer separation. The form of the potential used to derive the Flory equation is given by

$$\psi(R) = -2(\sigma/R)^3 \qquad (16)$$

The cell volume v is related to R via the following relationship

$$v = \gamma R^3 \qquad (17)$$

where γ is a geometrical factor which is determined by the cell geometry ($\gamma = 1$ for simple cubic geometry in Flory model). Using Equation 17, we can relate the hard-core cell volume to σ as follows:

$$v_{hc} = \sigma^3 = v^* \qquad (18)$$

This enables us to write the potentials of Equation 16 as functions of the cell volume v. Using Equations 17 and 18, Equation 16 is rewritten by

$$\psi(v) = -2v^*/v = -2/\widetilde{V} \qquad (19)$$

We can then write Z in the following form:

$$Z = Z_{int} \{g(v^{1/3} - v^{*1/3})\}^{3Ncr} \exp\{Ncr/(\widetilde{T}\widetilde{V})\} \qquad (20)$$

where $\widetilde{T} = T/T^*$ with $T^* = s\,\eta/(ck)$ and $\widetilde{V} = v/v^*$. The equation of state can then be computed by using Equation 20 and

$$p = kT(d(\ln Zn)/dV)|_T \qquad (21)$$

to obtain

$$\frac{\widetilde{P}\widetilde{V}}{\widetilde{T}} = \frac{\widetilde{V}^{1/3}}{\widetilde{V}^{1/3} - 1} - \frac{1}{\widetilde{T}\widetilde{V}} \qquad (22)$$

TABLE 1
Equations of State for Polymer Liquid

	Equations of state	Lattice size	Hole	Potential
Flory FM[8,9]	$$\frac{\widetilde{P}\widetilde{V}}{\widetilde{T}} = \frac{\widetilde{V}^{1/3}}{\widetilde{V}^{1/3}-1} - \frac{1}{\widetilde{T}\widetilde{V}}$$	Change	—	1/V
Cell model CM[10]	$$\frac{\widetilde{P}\widetilde{V}}{\widetilde{T}} = \frac{\widetilde{V}^{1/3}}{\widetilde{V}^{1/3}-2^{-1/6}} - \frac{2}{\widetilde{T}}\left(\frac{1.2045}{\widetilde{V}^2} - \frac{1.011}{\widetilde{V}^4}\right)$$	Change	—	L.J.6–12[a]
Modified CM MCM[14]	$$\frac{\widetilde{P}\widetilde{V}}{\widetilde{T}} = \frac{\widetilde{V}^{1/3}}{\widetilde{V}^{1/3}-q\cdot2^{-1/6}} - \frac{2}{\widetilde{T}}\left(\frac{1.2045}{\widetilde{V}^2} - \frac{1.011}{\widetilde{V}^4}\right)$$	Change	—	L.J.6–12
Sanchez-Lacombe SL[15]	$$\frac{\widetilde{P}\widetilde{V}}{\widetilde{T}} = -\widetilde{V}[\ln(1-1/\widetilde{V}) + \frac{1}{\widetilde{V}}(1-1/r)] - \frac{1}{\widetilde{T}\widetilde{V}}$$	Fix	Hole	1/V
Simha-Somcynsky SS[16]	$$\frac{\widetilde{P}\widetilde{V}}{\widetilde{T}} = \frac{(y\widetilde{V})^{1/3}}{(y\widetilde{V})^{1/3}-2^{-1/6}y} - \frac{2y}{\widetilde{T}(y\widetilde{V})^2}\left(1.2045 - \frac{1.011}{(y\widetilde{V})^2}\right)$$	Change	Hole	L.J.6–12
van der Waals	$$\frac{\widetilde{P}\widetilde{V}}{\widetilde{T}} = \frac{8V}{3\widetilde{V}-1} - \frac{3}{\widetilde{T}\widetilde{V}}$$			1/V

[a] Lenard-Jones 6–12 potential.

where $\widetilde{P} = P/P^*$ with $P^* = ckT^*/v^*$. We obtain the usual form of Flory equation of state by Equation 22. Equation 22 is the reduced equation of state and is available for all polymer liquids.

Table 1 shows some popular equations of state that belong to the category of the cell model and hole theory for polymer liquids and these characteristic features. For reference, the reduced equation of state of van der Waals is indicated in Table 1. These equations of state have both the free volume and potential term. It should be noted that models of this type have been most successful at describing PVT data for small molecular liquids.

We describe ΔG^M for a polymer mixture of two components based on the Flory equation of state.[17-18] When we apply the equation of state to a mixture we need some definitions and mixing rules.

Consider a polymer mixture containing N_1 molecules of the first component composed of r_1 mers and N_2 molecules of the second composed of r_2 mers ($N = N_1 + N_2$). The two components are characterized by their reduction parameters P_i^*, V_i^*, and T_i^*, where $i = 1, 2$. c_i and s_i are the number of effective external degrees of freedom per mer and the number of nearest-neighbor contacts per mer for i component, respectively. Then one defines the following quantities:

$$x_1 \equiv N_1/N = 1 - x_2 \tag{23}$$

$$r \equiv x_1 r_1 + x_2 r_2 \tag{24}$$

$$\phi_1 \equiv r_1 N_1/rN = 1 - \phi_2 \tag{25}$$

$$s \equiv \phi_1 s_1 + \phi_2 s_2 \tag{26}$$

$$c \equiv \phi_1 c_1 + \phi_2 c_2 \tag{27}$$

With these definitions, Flory assumed the following rules (the combining rule) in order to combine the expression of mixture with one of the pure components.

1.   The hard-core mer volumes are defined to be equal:

$$v_1^* = v_2^* = v^* \tag{28}$$

2.   The hard-core volume of the mixture, V^*, is equal to the sum of the hard-core volumes of the components:

$$V^* = rNv^* \tag{29}$$

3.   The number of pair interactions in the mixture is equal to the sum of the pure-component pair interactions, i.e., if N_{ij} is the number of i, j pair interactions, then

$$N_{11} + N_{12} + N_{22} = 1/2(s_1 r_1 N_1 + s_2 r_2 N_2) = srN/2 \tag{30}$$

With the above definitions and rules, the partition function for the mixture can be written in the following form:

$$Z = Z_{int} \cdot Z_{comb} \{v^{1/3} - v^{*1/3}\}^{3(N_1 C_1 r_1 + N_2 C_2 r_2)} \exp(-E_o/kT) \tag{31}$$

where Z_{int} is the contribution of the internal modes, and Z_{comb} is the same as the expression of the Flory-Huggins combinatorial entropy, which is given by

$$\ln Z_{comb} = N_1 \ln \phi_1 + N_2 \ln \phi_2 \tag{32}$$

The third factor of Equation 31 is the so-called "free volume" term where the cell potential has been approximated by a square-well potential. The function $E_o(V)$ is the interaction energy computed when all mers are at their cell centers, and for mixtures it is given by the form

$$E_o(V) = (N_{11}\eta_{11} + N_{12}\eta_{12} + N_{22}\eta_{22}) \, \psi(v) \tag{33}$$

where η_{ij} is the characteristic energy of interaction between mers of component i and j, and $\psi(v)$ is the interaction potential. For the Flory model, $\psi(v)$ is of the same form as in Equation 19

$$\psi(v) = -2/\widetilde{V} \tag{34}$$

If random mixing of the two mers is assumed, then

$$N_{11} = 1/2(s_1 r_1 N_1 \theta_1), \qquad N_{22} = 1/2(s_2 r_2 N_2 \theta_2), \qquad N_{12} = s_1 r_1 N_1 \theta_2 \tag{35}$$

where θ is the site fraction and $\theta_1 = s_1 r_1 N_1 / srN = 1 - \theta_2$. Using these definitions, we can write

$$E_o/rN = 1/2(\epsilon_1 \phi_1 + \epsilon_2 \phi_2 - s_1 \phi_1 \theta_2 \Delta\eta) \psi(\widetilde{V}) \tag{36}$$

where $\Delta\eta = \eta_{11} + \eta_{22} - 2\eta_{12}$ and $\epsilon_i = s_i \eta_{ii}$. If we define an interaction parameter X_{12} (not Flory-Huggins χ_{12} parameter) so that

$$x_{12} = s_1 \Delta\eta / v^* \tag{37}$$

then Equation 33 becomes

$$E_o/rN = 1/2(\epsilon_1 \phi_1 + \epsilon_2 \phi_2 - v^* X_{12} \phi_1 \theta_2) \psi(\widetilde{V}) = \epsilon \psi(\widetilde{V})/2 \tag{38}$$

Since P* and T* are defined as

$$P^* = \epsilon/v^* \qquad T^* = \epsilon/(ck) \tag{39}$$

Equation 38 leads to a definition of P* and T* for the mixtures of the form

$$P^* = \phi_1 P_1^* + \phi_2 P_2^* - X_{12} \phi_1 \theta_2 \tag{40}$$

$$T^* = \frac{\phi_1 P_1^* + \phi_2 P_2^* - X_{12} \phi_1 \theta_2}{\phi_1 P_1^*/T_1^* + \phi_2 P_2^*/T_2^*} \tag{41}$$

Finally, if we define $\widetilde{P} = P/P^*$ and $\widetilde{T} = T/T^*$, the partition function can be written in the form

$$Z = \alpha Z_{comb} V^{*Ncr} (\widetilde{V}^{1/3} - 1)^{3Ncr} \exp\{Ncr/(\widetilde{T}\,\widetilde{V})\} \tag{42}$$

where we have neglected the temperature dependence resulting from the motion of the mers within the cells.

Equation 42 is the same form as the partition function for one component, Equation 20, except for the Z_{comb} term. Accordingly, the reduced equation of state for the systems composed of two components is fundamentally the same form as that for one component.

By using the partition function, the Helmholz free energy for mixture is given by

$$\Delta F^M = -kT \ln\{Z/(Z_1 \cdot Z_2)\} \tag{43}$$

ΔF^M is obtained from the partition functions for mixture Equation 42 and pure components, Equation 20. For liquids having a high density, the Gibbs free energy of mixing is given by

$$\frac{\Delta G_M}{kT} = N_1 \ln\phi_1 + N_2 \ln\phi_2 + \frac{rNv^*}{kT} \left[\phi_1 P_1^* \left(\frac{1}{\widetilde{V}_1} - \frac{1}{\widetilde{V}} \right) + \phi_2 P_2^* \left(\frac{1}{\widetilde{V}_2} - \frac{1}{\widetilde{V}} \right) \right.$$

$$\left. + \phi_1 P_1^* \widetilde{T}_1 \ln \frac{\widetilde{V}_1^{1/3} - 1}{\widetilde{V}^{1/3} - 1} + \phi_2 P_2^* \widetilde{T}_2 \ln \frac{\widetilde{V}_2^{1/3} - 1}{\widetilde{V}^{1/3} - 1} + \frac{\phi_1 \theta_2 X_{12}}{\widetilde{V}} \right] \tag{44}$$

Using Equation 42, we can then compute the chemical potential of mixing using the expression

$$\Delta\mu_1 = \mu_1 - \mu_1^0 = kT\left(\frac{\partial(\ln Z)}{\partial N_1}\right)_{T,V,N_2} = \left(\frac{\partial \Delta F^M}{\partial N_1}\right)_{T,V,V_1,N_2}$$

$$+ \left(\frac{\partial \Delta F^M}{\partial \widetilde{V}}\right)_{T,V,N_1,N_2} \cdot \left(\frac{\partial \widetilde{V}}{\partial N_1}\right)_{T,V,N_2} + \left(\frac{\partial \Delta F^M}{\partial \widetilde{V}_1}\right)_{T,V,N_1,N_2} \cdot \left(\frac{\partial \widetilde{V}_1}{\partial N_1}\right)_{T,V,N_2}$$

$$= kT\{\ln\phi_1 + (1 - r_1/r_2)\phi_2\} + r_1\epsilon_1^*\left\{3\widetilde{T}_1 \ln \frac{\widetilde{V}^{1/3} - 1}{\widetilde{V}^{1/3} - 1}\right.$$

$$+ \widetilde{P}_1(\widetilde{V} - \widetilde{V}_1) + \left(\frac{1}{\widetilde{V}_1} - \frac{1}{\widetilde{V}}\right) + \frac{\phi_2\theta_2 X_{12}}{P_1^*\widetilde{V}}\right\} \tag{45}$$

Equation 45 is the form corrected by the indication of McMaster[19] and Sanchez.[20]

Using Equation 45 and a similar expression for $\Delta\mu_2$, we can compute $\Delta\mu_1$ and $\Delta\mu_2$ and locate the phase diagram for this binary system by seeking solutions to the equations

$$\Delta\mu_1^I = \Delta\mu_1^{II} \qquad \Delta\mu_2^I = \Delta\mu_2^{II} \tag{46}$$

where the superscripts I and II indicate two different phases. In practice these solutions, if they exist, are determined numerically.

The spinodals are determined by

$$\left(\frac{\partial^2 \Delta G^M}{\partial\phi_2^2}\right)_{P,T} = 0 \tag{47}$$

or by finding solutions of the equation

$$\left.\frac{d\Delta\mu_1}{-\partial\phi_2}\right|_{\phi_2^s} = \left.\frac{d\Delta\mu_2}{\partial\phi_2}\right|_{\phi_2^s} = 0 \tag{48}$$

where ϕ_2^s denotes the value of ϕ_2 at which Equation 48 is satisfied. These values are determined numerically for each equation of state.

From the Flory-Huggins equation, the chemical potential is given by[1]

$$\Delta\mu_1/(kT) = \ln\phi_1 + (1 - r_1/r_2)\phi_2 + \phi_2^2\chi_{12} \tag{49}$$

Comparing Equation 49 with Equation 45, the combinatorial entropy term is completely the same. From the equation of state theory, χ_{12} in the Flory-Huggins equation is expressed by

$$\chi_{12} = \frac{r_1\epsilon_1^*}{\phi_2^2 kT}\left\{3\widetilde{T}_1\ln\frac{\widetilde{V}^{1/3} - 1}{\widetilde{V}^{1/3} - 1} + \widetilde{P}_1(\widetilde{V} - \widetilde{V}_1) + \frac{1}{\widetilde{V}_1} - \frac{1}{\widetilde{V}} + \frac{\phi_2\theta_2}{P_1^*\widetilde{V}}X_{12}\right\} \tag{50}$$

The temperature dependence of χ_{12} from Equation 50 is very different from the simple one taking into account only exchange energy of van Laar-type, since Equation 50 includes the free volume term. However, existence of many parameters makes Equation 50 slightly complex. Thus, Flory carried out the series expansion of Equation 50 and eliminated neg-

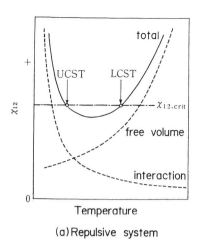
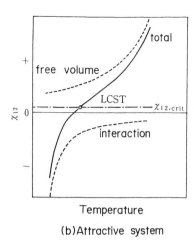

(a) Repulsive system   (b) Attractive system

FIGURE 2. Schematic representation for temperature dependence of χ_{12} in the Flory equation of state theory: (a) repulsive system; (b) attractive system. (From D. Patterson and A. Robard, *Macromolecules*, 11, 690, 1969. With permission.)

ligible higher order terms and rewrote χ_{12}.[9,21] Patterson also obtained an almost similar form for χ_{12} as[22-23]

$$\chi_{12} = \frac{P_1^* V_1^*}{RT_1^*} \left[\frac{\widetilde{V}_1^{1/3}}{\widetilde{V}^{1/3} - 1} \left(\frac{X_{12}}{P_1^*} \right) + \frac{\widetilde{V}_1^{1/3}}{2(4/3 - \widetilde{V}_1^{1/3})} \tau^2 \right] \tag{51}$$

where $\tau = 1 - T_1^*/T_2^*$, and reflects the difference in free volumes or degrees of thermal expansion of the components. In Equation 51, the first term in the bracket is called the interactional term and the second term, the free volume term. The interactional term corresponds to the exchange energy term of van Laar-type, Equation 6. It is considered that the free volume term is added to the interactional term by the equation of state theory.

Figure 2 shows the temperature dependence of χ_{12} expressed by Equation 51.[24] The free energy term increases monotonically with increasing temperature. Figure 2(a) shows the case of a repulsive system ($X_{12} > 0$). The total value of χ_{12}, including the interaction and free volume terms, becomes a U-shaped curve and intersects with $\chi_{12,crit}$ at two points; UCST in lower temperatures and LCST in higher temperatures appear. This is coexistent behavior of UCST and LCST and familiar to polymer solutions. From Equations 11 and 51, we can expect that this behavior takes place even for polymer mixtures having relative low molecular weight. In an attractive system ($X_{12} < 0$) of Figure 2(b), the total value of χ_{12} increases monotonically and the sign changes from negative to positive. Accordingly, χ_{12} and $\chi_{12,crit}$ intersects at one point and LCST appears. This is an interpretation of an LCST-type phase diagram appearing in a polystyrene (PS)/poly (vinyl methyl ether) (PVME) system. It is interpreted from Equations 11 and 51 that higher molecular weight polymer makes the LCST move to lower temperatures and the two-phase region expands. Introducing the free volume term by the equation of state theory, the behavior of LCST can be explained.

For phase behavior in polymer mixtures, the equations of state theory can explain (1) UCST in polymer mixtures having relative low molecular weight, (2) coexistent behavior of UCST and LCST if LCST is below the degradation temperature of polymers, and (3) two phases in all measurable region for repulsive systems (except for the edge of phase diagram) or LCST for attractive systems in polymer mixtures having high molecular weight; only two cases exist for high molecular weight polymer blends.

For other equations of state theories, one can also get the expression of ΔG^M and $\Delta\mu$ and calculate the phase diagram. Qualitatively, similar results for phase diagrams are obtained.

The possibility of coexistent behavior of UCST and LCST in a polymer mixture having high molecular weight may exist if the detail of some parameters is examined.[25,26] Since the information about equation of state parameters is too little, it is necessary to discuss the more in detail in the future.

C. EQUATION OF STATE PARAMETERS AND PHASE DIAGRAMS

Solving Equation 22 first for $\widetilde{T}$, differentiating the resulting expression, and finally eliminating $\widetilde{T}$, one obtains[8]

$$(\alpha T)^{-1} = \frac{1}{3(\widetilde{V}^{1/3} - 1)} - 1 + \frac{2\widetilde{P}\,\widetilde{V}^2}{\widetilde{P}\,\widetilde{V}^2 + 1} \tag{52}$$

where α is the coefficient of thermal expansion and given by

$$\alpha = \frac{1}{V}\left(\frac{\partial V}{\partial T}\right)_P$$

The values of P* in general polymers are about 500 MPa and one can approximate $\widetilde{P} \fallingdotseq 0$ at atmospheric pressure (P = 0.1 MPa). In this case, Equation 52 becomes

$$\widetilde{V}^{1/3} - 1 = \frac{\alpha T}{3(1 + \alpha T)} \tag{53}$$

One can obtain $\widetilde{V}$ from the value of α and V_{sp}* (hard-core specific volume) from V_{sp}. At atmospheric pressure, Equation 22 becomes

$$\widetilde{T} = \frac{\widetilde{V}^{1/3} - 1}{\widetilde{V}^{4/3}} \tag{54}$$

One can obtain $\widetilde{T}$ from $\widetilde{V}$ and determine T*.

Solving Equation 22 for $\widetilde{P}$ and defferentiating the resulting expression by $\widetilde{V}$, one obtains the coefficient of compressibility κ and the thermal pressure coefficient γ.

$$\kappa = -\frac{1}{V}\left(\frac{\partial V}{\partial P}\right)_T = \frac{3(\widetilde{V}^{1/3} - 1)\widetilde{V}^2}{\{1 - 3(\widetilde{V}^{1/3} - 1)\}P^*} = \frac{\alpha T\widetilde{V}^2}{P^*} \tag{55}$$

$$\gamma = -\frac{\alpha}{\kappa} = \left(\frac{\partial P}{\partial T}\right)_V = \frac{P^*}{T\widetilde{V}^2} \tag{56}$$

P* is obtained from κ or γ. Flory et al. estimated the characteristic parameters (equation of state parameters) for some liquids.[8,27] However, P*, V_{sp}*, and T* which are obtained, thus, depend on temperature and are not very accurate because of using the approximation at atmospheric pressure. Equations 52 to 56 are useful for indicating the relationship between the characteristic and thermodynamic parameters.

TABLE 2
Values of s² and Characteristic Parameters

	s^2	P*(MPa)	V_{sp}^*(cm³/g)	T*(K)
PE(0<P<200 MPa, 152<T<262°C)				
SL	16.97	445.1	1.1051	625.5
FM	9.25	516.8	1.0076	6821.6
CM	1.942	587.6	1.1265	4340.9
SS	1.49	796.5	1.1670	10207.2
MCM	0.85	563.9	1.0571	5569.3
PS(0<P<180 MPa, 121<T<320°C)				
SL	30.6	408.3	0.90675	772.0
FM	11.1	474.5	0.83153	8104.8
CM	1.64	608.2	0.92275	5144.5
SS	0.975	778.6	0.96390	12381.7
MCM	0.51	564.6	0.8706	6705.6

From G. T. Dee and D. J. Walsh, *Macromolecules*, 21, 811,
1988 and G. T. Dee and D. J. Walsh, *Macromolecules*, 21,
815, 1988. With permission.

Another method for the estimation of the characteristic parameters is fitting the actual data for the liquid state of PVT behavior to the equations of state. It is considered that this method is more accurate than the one from the approximation of atmospheric pressure and the direct method. However, the values of the characteristic parameters depend on the size of the data block, because the equations of state cannot explain the behavior of polymer liquids completely.

We measured the PVT behavior of a number of polymers and fitted the melt PVT data to the equations of state in Table 1 for each of the polymers. We compared the five equations of state with the same sets of PVT data. We performed a nonlinear least squares fit of each equation by minimizing the quantity

$$s^2 = \sum_i \{P_i(\text{data}) - P_i(\text{fit})\}^2/(N - 3)$$

where N is the number of data points, P(data) is the measured pressure at a given value of (V,T) for the system, and P(fit) is the value of the pressure predicted by the relevant equation of state. A smaller value of s^2 means better fitting of the equation of state to PVT data sets. Table 2 shows values of s^2 and the characteristic parameters for selected polymer liquids.[10,14] We see that in each polymer the pattern is the same in that the MCM or SS equation provides the best fit to the data. The Flory equation provides a moderately good fit and the SL equation provides the worst fit. The improvement in the fit between the FM and CM equations is due to the different potential type.

From Equations 44 to 48, by using these parameters and the interaction parameter for each mixture, we can calculate the phase diagrams for the polymer mixtures. Figures 3 and 4 show the examples of actual and calculated phase diagrams. The interaction parameter X_{12} can be obtained from, for example, the heats of the mixing of low molecular weight analogs. On the other hand, one can use the interaction parameter as an adjustable parameter to fit the phase diagram at some points, typically the maximum or minimum of the curve, and compare this value of X_{12} with one from the experiment. This agreement is good in mixtures of Figures 3 and 4.

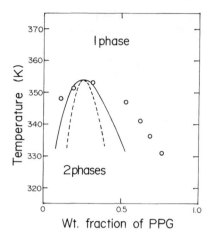

FIGURE 3.   UCST-phase diagram for methoxylated poly(ethylene glycol) (MW = 600)/poly(propylene glycol) (MW = 4000) mixture. The open circles are cloud-points. The solid and broken curves correspond to the calculated binodal and spinodal curve by the equation of state theory and value of $X_{12} = 7.4$ J/cm³, respectively.

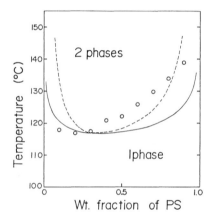

FIGURE 4.   LCST-phase diagram for PS(MW = 114,000)/PVME(MW = 99,000) mixture. The open circles are cloud-points. The solid and broken curves correspond to the calculated binodal and spinodal curve by the equation of state theory and value of $X_{12} = -1.75$ J/cm³, respectively.

We also estimate X_{12} for miscible polymer mixtures from PVT properties. The characteristic pressure and temperature in a mixture are defined in Equations 40 and 41, respectively. The characteristic parameters in mixtures are obtained from the fitting of PVT properties to the equation of state in the same way as pure components. By using values of the characteristic parameters for pure components and a mixture, we can estimate the value of X_{12},[28,29] Figure 5 shows the characteristic pressure, P*, plotted against the hard-core specific volume fraction of PS in PS/PVME[28]

$$\phi_2 = m_2 V^*_{sp2}/(m_1 V^*_{sp1} + m_2 V^*_{sp2}) \qquad (57)$$

where m_i is the mass and $V_{spi}*$ is the hard-core specific volume of component i. The fit of the equation to the data is least sensitive to P* of the parameters, and this is evidenced by the scatter in this quantity. However, the composition dependence of P* is not additive, and

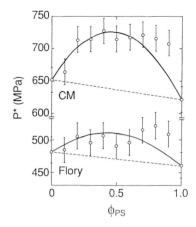

FIGURE 5. Characteristic pressure, P*, of PS/PVME blends plotted against the hard-core volume fraction of PS from two equations of state. The solid lines are the expected values for $X_{12} = -150$ J/cm³(Flory) and -300 J/cm³(CM). (From T. Ougizawa, G. T. Dee, and D. J. Walsh, *Macromolecules,* 24, 3834, 1991. With permission.)

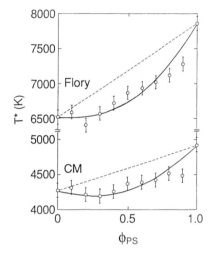

FIGURE 6. Characteristic temperature, T*, of PS/PVME blends plotted against the hard-core volume fraction of PS from the two equations of state. The solid lines are the expected values for $X_{12} = 100$ J/cm³(Flory) and 200 J/cm³(CM). (From T. Ougizawa, G. T. Dee, and D. J. Walsh, *Macromolecules,* 24, 3834, 1991. With permission.)

P* is obviously larger in the blends, indicating that X_{12} is negative. The solid curve in Figure 5 represents the value of $X_{12} = -150$J/cm³ for the Flory equation of state calculated from Equation 40.

Figure 6 shows the characteristic temperature, T*, plotted against ϕ. T* of blend is much lower than the additive. X_{12} can, therefore, be estimated by using the value of T* in the blends and the characteristic parameters of the pure components in Equation 41. The solid curve in Figure 6 is drawn for $X_{12} = +100$ J/cm³ for the Flory equation of state. This is obviously inconsistent with the results of X_{12} obtained from P*. Equation 41 indicates that a negative X_{12} leads to a larger T* which is predicted by the assumption of additivity in the external degrees of freedom, Equation 27.

This discrepancy could be explained if the assumption of Equation 27 is incorrect. The c value of the blend can then be given by

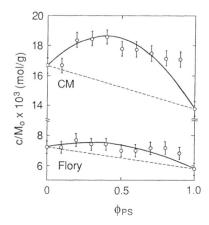

FIGURE 7. Values of the degrees of freedom divided by the segment molecular weight, c/M_o, plotted against the hard-core volume fraction of PS from the two equations of state. (From T. Ougizawa, G. T. Dee, and D. J. Walsh, *Macromolecules,* 24, 3834, 1991. With permission.)

$$c = \phi_1 c_1 + \phi_2 c_2 - \phi_1 \phi_2 c_{12} \tag{58}$$

The relationship between T^* and c is given by

$$T^* = P^* v^* / (ck) \tag{59}$$

and thus c/M_o is given by

$$c/M_o = P^* V_{sp}^* / (RT^*) \tag{60}$$

where R is the gas constant and M_o is the molecular weight per segment (not necessarily equivalent to a monomer). Figure 7 shows a plot of c/M_o against ϕ. Clearly, c/M_o calculated in this way is larger than the additive, and c_{12} is calculated to be negative.

It is, however, unusual that in spite of a negative X_{12}, the external degrees of freedom in the blend are predicted to be higher than the additive. If X_{12} is negative and specific interaction exists in the system, the external degrees of freedom would be expected to be restricted. It was indicated that this could arise from a temperature dependence of the interaction, as an interpretation.[28]

At present the equation of state theories for polymer liquids and mixtures are not yet perfect, but improvement of these theories is helpful to resolve the problems in the phase behavior of polymer blends.

III. PHASE DIAGRAM IN POLYMER BLENDS

In this section we describe the estimation of a phase diagram in polymer blends. In a phase diagram, the binodal and spinodal curves exist as shown in Figure 4. However, it is very difficult to estimate the binodal curve strictly. Thus, we may generally regard the cloud point curve as the binodal curve. On the other hand, we can estimate the spinodal curve by the thermodynamic behavior, i.e., the behavior of concentration fluctuation. We show some methods of estimation of the phase diagram, especially, the light scattering method.

A. CLOUD-POINT MEASUREMENTS (BINODAL)

All scattering experiments are based on the existence of variations in the homogeneity of the scattering medium. These inhomogeneities result in the existence of a concentration fluctuation in polymer mixtures. If the concentration fluctuation causes variation in the physical properties relevant for scattering a particular type of radiation, its radiation is scattered by a sample. These properties are the refractive index for light, the electron density for X-rays, and the scattering length density for neutrons. The measurement of the cloud point utilizes this scattering phenomenon.

If a sample is thrust from a one-phase region into a two-phase region, the phase separation takes place. As the size of particles increases and the concentration variation becomes more extreme, the sample may become turbid unless the components of the mixture have identical refractive indexes. The cloudiness is relatively easy to detect either as a reduction in transmitted light, or an increase in scattered light. The exact location of its onset usually depends on a number of factors like heating or cooling rate, molecular mobility, etc. Cloud-point curves are used as indications of the miscibility limit on a phase diagram, especially the light scattering method which is wildey used for simplicity.

For the polymer mixtures where phase separation takes place at higher temperatures, as shown in Figure 4 (LCST), we begin by preparing one-phase and transparent samples in order to measure cloud-points. It is sometimes difficult to prepare a one-phase sample by using conventional mechanical mixing. Thus, we dissolved both polymers in common solvent and prepared a one-phase sample by evaporating the solvent (casting). In this case, we have to remove the solvent from the polymer sample by drying completely. Another method is pouring the ternary polymer solution, including two polymers, into poor solvent. Polymer precipitate was dried and pressed at a higher temperature than the T_g of blends. Thus we can obtain a one phase sample.

For the mixture including a crystallizable polymer, we use very fast casting or pouring the polymer solutions into a poor solvent in order to freeze the one-phase state or prevent the liquid-solid phase separation from progressing.

The most simple measurement of cloud-point is judging if the sample on the heating stage setting at a constant temperature is transparent or opaque after constant time. In order to judge the transparency of the sample, the variation of the intensity in transmitted or scattered light is measured. In this case, the optical microscope is occasionally used in order to confirm the existence of a phase-separated structure. Similar experiments arc carricd out for blend specimens in various compositions and various temperatures; the cloud-point curve is then drawn at the boundary between transparent-opaque.

After annealing at constant temperature for a constant time, the sample is quenched below the T_g and judged as a one- or two-phase state by availing of the difference in physical properties. For example, a phase-separated sample indicates two T_gs and a one-phase sample only a single T_g, for example, by differential scanning calorimetry (DSC) measurement. This method is useful for the mixture composed of polymers having similar refractive indexes in which we can select proper measurements, but this method needs a long time for measurements.

Instead of setting at a constant temperature, one can measure the cloud-point with a constant heating rate. In this case, the change of light scattering intensity or profile is measured and the cloud point T_d is determined. For example, T_d is the onset temperature either of decreasing intensity of transmitted light, or increasing intensity of scattered light. T_d, of course, depends on heating rate. One can determine T_d by extrapolation to 0 a heating rate from measurements of some heating rate. However, one cannot accurately determine the T_d for the mixture in which the phase separation takes place very slowly.[30]

Next, for the system that has a UCST-type phase diagram as shown in Figure 3, we can use a similar method as the case of the LCST-type phase diagram, i.e., after annealing the sample on the heating stage setting at a constant temperature, we judge if the sample is one- or two-phase. However, for UCST-type phase diagrams, it takes a long time for measurement at low temperature, especially for the mixture including high T_g polymers.

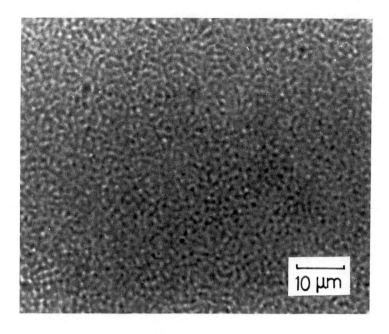

FIGURE 8.   Light micrograph of the modulated structure in solution cast film of BR/SBR-45(50/50) blend.

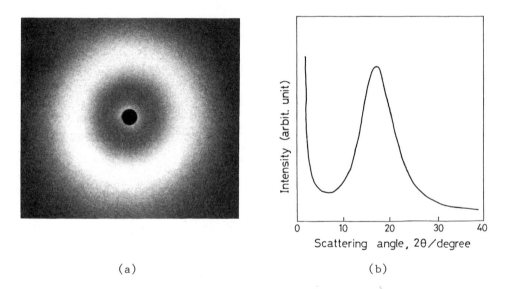

(a)   (b)

FIGURE 9.   (a) Light-scattering pattern (V_v) from the BR/SBR-45 blend film of Figure 8, and (b) its goniometer trace. (From T. Inoue, T. Ougizawa, O. Yasuda, and K. Miyasaka, *Macromolecules,* 18, 57, 1985. With permission.)

We show an example of measurement of phase diagrams in a polymer mixture. In the casting process from a ternary polymer solution in order to prepare blend samples, there is a possibility that the spinodal decomposition takes place and a regularly phase-separated structure (modulated structure) is formed as shown in Figure 8.[7] The ring pattern, like Figure 9(a), appears in the light scattering pattern from this structure, i.e., the light scattering profile from the as-cast film with modulated structure has a peak as shown in Figure 9(b).

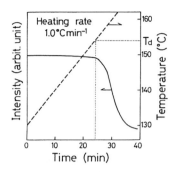

FIGURE 10. Change in the peak intensity of scattered light during heating at a constant rate of 1°C/min to obtain T_d. BR/SBR-45 = 50/50. (From T. Ougizawa, T. Inoue, and H. W. Kammer, *Macromolecules*, 18, 2089, 1985. With permission.)

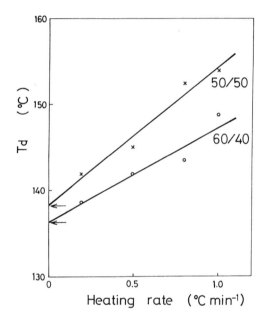

FIGURE 11. T_d vs. heating rate plots. (From T. Ougizawa, T. Inoue, and H. W. Krammer, *Macromolecules*, 18, 2089, 1985. With permission.)

Figure 10 shows the results of a heating experiment for the sample with modulated structure, which has a scattering peak. When the temperature is elevated, the light scattering profile does not change up to a certain temperature T_d. Above T_d the peak intensity of scattered light decreases with temperature, keeping the scattering peak angle constant, until the scattering peak finally disappears. T_d corresponds to the onset temperature of phase dissolution, as discussed later; T_d varied with heating rate and selected results of T_d as a function of the heating rate are shown in Figure 11. The intercept of T_d, at which the heating rate is zero, may correspond to the binodal temperature. Similar experiments were carried out for blend specimens of various compositions. The binodal points, thus, estimated are indicated by open circles in Figure 12. UCST-type phase diagram for polybutadiene (BR)/poly(styrene-co-butadiene) (SBR-45, containing 45 wt% styrene) system is shown in Figure 12.[31,32]

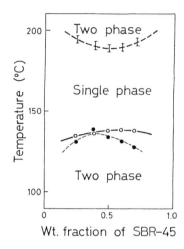

FIGURE 12. Phase diagram of BR/SBR-45. ○, binodal points estimated by extrapolation to zero heating rate, ●, spinodal points estimated by kinetic study of phase dissolution (D/T vs. T plots). The coexistent behavior of UCST and LCST appears. (From T. Inoue and T. Ougizawa, *J. Macromol. Sci.: Chem.*, A26, 147, 1989. With permission.)

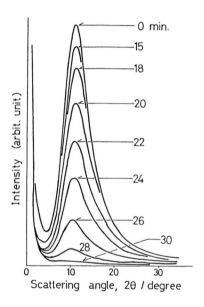

FIGURE 13. Change of light-scattering profile with annealing at 155°C. Numbers are times after temperature jump. BR/SBR-45 = 30/70. (From T. Ougizawa, T. Inoue, and H. W. Kammer, *Macromolecules*, 18, 2089, 1985. With permission.)

The UCST behavior was confirmed by isothermal experiments, as follows. The film specimen with the modulated structure was allowed to undergo a rapid temperature jump from room temperature to various higher temperatures set isothermally below and above the UCST. Below the UCST, no appreciable change in the scattering profile with the time of isothermal annealing could be detected on a time scale of 2 h. This means that no appreciable structural change took place. On the other hand, above the UCST, the scattered intensity decreased with the time of annealing at a constant peak angle. A typical example is shown in Figure 13. The intensity decay in Figure 13 corresponds to the phase dissolution of the modulated structure at a constant periodic distance, and in a one-phase sample also thrust into a constant temperature of a two-phase region, phase-separation was observed.

In Figure 12 there is a two-phase region at higher temperatures. This LCST-type phase diagram was obtained by a dissolution and temperature jump procedure: the cast film was annealed above UCST, e.g., at 150°C for ~30 min, then the annealed specimen underwent a rapid temperature jump to an isothermal setting of higher temperatures, and the structural change was observed under the microscope and by the light scattering technique. When no appreciable change took place, we judged that the system was still in the one-phase region. On the other hand, when the film became opaque and the scattering intensity increased, we judged that the system was in the two-phase region. On the basis of these observations, the LCST line was drawn somewhat arbitrarily in Figure 12. Actually, we observed development of a modulated structure during isothermal annealing in the two-phase region above the LCST. It was similar in appearance to that in Figure 8, but the periodic distance of the structure was much smaller than the original one in the cast film. This implies that the structural memory in the cast film had disappeared by annealing in the one-phase region, and a new concentration fluctuation had developed by spinodal decomposition induced thermally above the LCST. Thus, we found the coexistent behavior of UCST and LCST.

B. KINETIC STUDIES OF PHASE SEPARATION (SPINODAL)

The modulated structure indicated in the previous section appears at an initial stage of spinodal decomposition that takes place when the sample is thrust from a one-phase to a two-phase region. One can estimate the spinodal points by measurement of concentration fluctuation at the initial stage of spinodal decomposition.

The polymer mixture undergoes a rapid temperature drop from a one-phase to a two-phase region of new isothermal conditions in the phase diagram. The change of the angular distribution of scattered light intensity with time is measured during the isothermal phase separation process. Figure 14 is a typical example of the changes of scattered intensity with time at various values of q in a PS/PVME mixture.[33] The parameter q is the magnitude of the scattering vector given by $q = (4\pi/\lambda)\sin(\theta/2)$, where λ is the wavelength of light in the sample and θ is the external scattering angle. In the initial stages of phase separation, the scattered intensity increases exponentially with time. In the later stage, the intensity increase deviates from the exponential curve.

According to Cahn's linear theory of spinodal decomposition,[34] the exponential increase of the scattered intensity is described by

$$I(q,t) \propto \exp[2R(q) \cdot t] \qquad (61)$$

where I is the scattered intensity, t is the time after the initiation of spinodal decomposition, and R(q) is the growth rate of concentration fluctuation having wave number q. R(q) is given by

$$R(q) = -Mq^2 \left(\frac{\partial^2 f}{\partial c^2} + 2\kappa q^2 \right) \qquad (62)$$

where M is the mobility, κ is the gradient energy coefficient, f is the free energy of the mixture, and c is the concentration. According to Equation 62, a plot of lnI vs. time at a fixed q should yield a straight line of slope 2R(q). A linear relationship is obtained for the initial stage of phase separation, as shown in Figure 14, indicating that it can be described by the linearized spinodal decomposition theory. Linear results are also obtained for mixtures with other concentrations at various temperatures.

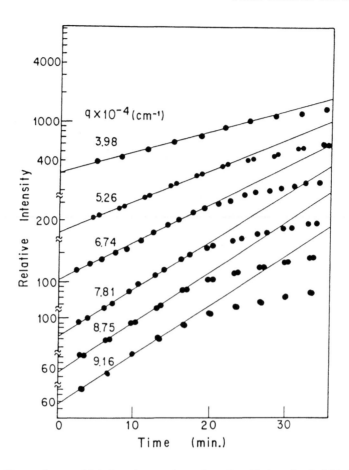

FIGURE 14. Change of scattered light intensity at various values of q with time after the initiation of the phase separation at 99.7°C for PS/PVME = 30/70, demonstrating the exponential character of I(t) at the initial stage of phase separation. (From T. Hashimoto, J. Kumaki, and H. Kawai, *Macromolecules,* 16, 641, 1983. With permission.)

In Figure 15, $R(q)/q^2$ is plotted vs. q^2. In agreement with Equation 61, the data yield a straight line, indicating again that at least the initial stage can be described within the framework of the linear theory. From the plots, one can obtain the characteristic parameters, such as q_m, q_c, and D_{app}, describing the dynamics of phase separation, where q_c is the critical (maximum) wave number of fluctuation which can grow and q_m is the most probable wave number of fluctuation having the highest rate of growth. According to Equation 62, q_c is given by the intercept on the q^2 axis, while q_m is calculated by $q_m^2 = {}^1/_2 q_c^2$. The apparent diffusion coefficient D_{app}, defined by $-M(\partial^2 f/\partial c^2)$, is given by the intercept on the vertical axis in Figure 15.

In the context of the mean-field approximation for the polymer mixtures, the parameters characterizing the initial stage of spinodal decomposition are expressed by the Flory-Huggins χ parameter and the fundamental molecular parameters;[33]

$$D_{app} = D_c (\chi - \chi_s)/\chi_s \tag{63}$$

$$q_m^2 = (18/R_o^2) (\chi - \chi_s)/\chi_s \tag{64}$$

$$R(q_m)^{1/2} = (3D_c^{1/2}/R_o) (\chi - \chi_s)/\chi_s \tag{65}$$

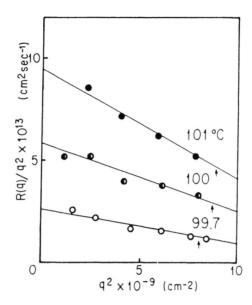

FIGURE 15. Plots of $R(q)/q^2$ vs. q^2, testing Equation 62. The arrows indicate the positions of q_m^2. (From T. Hashimoto, J. Kumaki, and H. Kawai, *Macromolecules*, 16, 641, 1983. With permission.)

where χ_s is the value of χ parameter at the spinodal point. D_c identical self-diffusivity, and R_o the unperturbed end-to-end distance of the polymer coil. Expanding χ into a series about $T = T_s$,

$$(\chi - \chi_s)/\chi_s = (\partial \ln\chi/\partial T)_{Ts} (T - T_s) + \cdots \qquad (66)$$

Since higher order terms in Equation 66 are negligible, D_{app} q_m^2, and $R(q_m)^{1/2}$ should be proportional to quench depth, $\Delta T = T - T_s$.

Figure 16 shows the temperature dependence of D_{app}. The intercept of T, at $D_{app} = 0$, gives an estimate of the spinodal temperature T_s at which $(\partial^2 f/\partial c^2) = 0$, hence D_{app} is zero. The spinodal curve, thus, estimated is indicated by a broken curve in Figure 17.[35] These seem to be reasonable, i.e., they are inside the two-phase region and slightly apart from the binodal curve.

One can also estimate the spinodal point from the plot of q^2 vs. T, according to the relationship of Equation 64.

C. KINETIC STUDIES OF PHASE DISSOLUTION (SPINODAL)

When a phase-separated sample undergoes a temperature jump from a two-phase to a one-phase region, phase dissolution takes place. However, it is difficult to measure this phenomenon for the sample having irregular morphology. On the other hand, for the sample having the modulated structure, the decay of scattered intensity takes place as shown in Figure 13. From measurement of this decay, we are able to discuss the kinetics of phase dissolution.[36] Figure 18 shows the change in peak intensity of the scattered light, I, with annealing time after the temperature jump to above the UCST. The peak intensity initially remains constant for a certain time, and then decreases exponentially with time. This seems to be a kind of induction period. The reason why an induction period exists for the phase dissolution is not obvious at present. Here, we discuss the kinetics of phase dissolution from the later stage of the intensity change, i.e., by the linear decay of $\ln(I/I_o)$ vs. time plots in Figure 18, where I_o is the peak intensity at zero annealing time. From the slope of the intensity decay, one can estimate the apparent diffusion constant $\tilde{D}$ for phase dissolution by

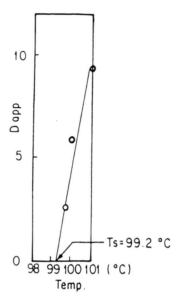

FIGURE 16. Temperature dependence of apparent diffusion constant D_{app}, from which the spinodal temperature $T_s = 99.2°C$ is deduced. (From T. Hashimoto, J. Kumaki, and H. Kawai, *Macromolecules*, 16, 641, 1983. With permission.)

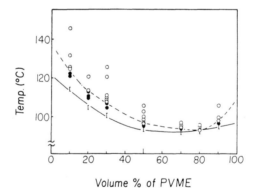

FIGURE 17. Phase diagram of PS/PVME mixture (LCST). The solid and broken curves correspond to the cloud-point and spinodal curve, respectively. The spinodal curve was determined by the kinetic method. (From J. Kumaki and T. Hashimoto, *Macromolecules*, 19, 763, 1986. With permission.)

$$\ln(I/I_o) = -2\widetilde{D}[(4\pi/\lambda) \sin(\theta_m/2)]^2 t \tag{67}$$

where t is the time. λ is the wavelength of the light in the specimen, and θ_m is the scattering peak angle. Similarly, we estimated the value of $\widetilde{D}$ at various temperatures above the UCST, as shown in Figure 19. The values of $\widetilde{D}$ are in the range of $10^{-13} \sim 10^{-12}$ cm²/s.

Phase dissolution is the reverse phenomenon of phase separation. However, the basic formulation of the dynamics may be formally common for the two phenomena. The only difference is the sign of $\widetilde{D}$. Thus, the theory of the spinodal decomposition may be applicable to the phase dissolution of the two-phase system with modulated structure. Hence, the apparent diffusion constant in dissolution is given by

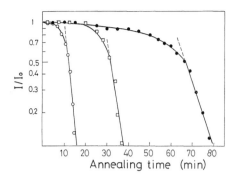

FIGURE 18. Change of scattered intensity of BR/SBR-45(50/50) system during annealing at various temperatures: ●, 135°C(1.51 μm); □, 145°C(1.65 μm); ○, 155°C(1.84 μm). Figures in parentheses are the periodic distance of modulated structure. (From T. Takagi, T. Ougizawa, and T. Inoue, *Polymer*, 28, 103, 1987. With permission.)

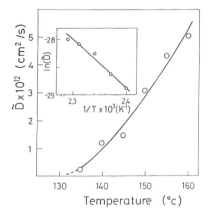

FIGURE 19. Temperature dependence of apparent diffusion constant $\widetilde{D}$ and its Arrhenius plot from the relationship of $\widetilde{D} = D_c \exp(-Q/RT)$, where Q is the apparent activation energy for diffusion. BR/SBR-45 = 50/50. (From T. Takagi, T. Ougizawa, and T. Inoue, *Polymer*, 28, 103, 1987. With permission.)

$$\widetilde{D} \propto D_c \left| \frac{\chi - \chi_s}{\chi_s} \right| \qquad (68)$$

where D_c is the self-diffusion coefficient for translational diffusion of polymers. χ is the Flory-Huggins interaction parameter, and χ_s is the value of χ at the spinodal temperature.

Employing the relationship $|(\chi - \chi_s)/\chi_s| \propto |T - T_s|$ and $D_c \propto T$ from the Stokes-Einstein equation in the region where the temperature dependence of D_c is not negligible, Equation 68 leads to

$$\widetilde{D} \propto T|T - T_s| \qquad (69)$$

Equation 69 implies that the $\widetilde{D}/T$ vs. T plot results in a straight line and that T_s can be estimated by the intercept on the T axis. Typical examples are shown in Figure 20.[32] The spinodal points, thus, estimated are indicated by the filled circles in Figure 12. They are located reasonably; they are inside the two-phase region and depart slightly from the binodal curve. In this way, we are able to estimate the spinodal curve by a kinetic study of phase dissolution.

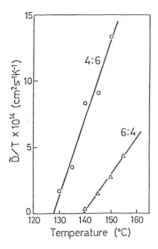

FIGURE 20. $\widetilde{D}/T$ vs. T plots, testing Equation 69. Spinodal temperature is obtained from the intercept on the T axis. (From T. Inoue and T. Ougizawa, *J. Macromol. Sci.: Chem.*, A26, 147, 1989. With permission.)

D. STATIC STUDIES IN ONE-PHASE REGION (SPINODAL)

Thus far, we have mostly discussed the light-scattering from the phase-separated systems. In this section we deal with the scattering from one-phase polymer mixtures. There is the thermal concentration fluctuation even in the one-phase region — the mixture scatters light. The scattering function in the one-phase region can be described by the random-phase approximation (RPA) results calculated by de Gennes.

De Gennes derived a scattering function S(q) for homogeneous binary polymer blends by RPA:[37]

$$S^{-1}(q) = [N_A \phi g_D(R_{gA}^2 q^2)]^{-1} + [N_B(1 - \phi)g_D(R_{gB}^2 q^2)]^{-1} - 2\chi \tag{70}$$

where N is the number of segments per polymer chain. ϕ is the volume fraction of polymer A, R_g is the radius of gyration of the chain, and g_D is the Debye function for a Gaussian chain:

$$g_D(x) = 2x^2(x - 1 + e^{-x}) \tag{71}$$

In the low q regime ($R_g \ll 1$), the Debye function is approximated by

$$g_D(x) \doteqdot 1 - x/3 = 1 - R_g^2 q^2/3 \tag{72}$$

Hence, in the case of $R_g = R_{gA} = R_{gB}$, S(q) is given by

$$S^{-1}(q) = 2(\chi_s - \chi) + 2\chi_s R_g^2 q^2/3 \tag{73}$$

where χ_s is the interaction parameter at the spinodal point. It is convenient to rewrite this equation in the standard form

$$\frac{S(q)}{S(o)} = \frac{1}{1 + \xi^2 q^2} \tag{74}$$

where ξ is the correlation length defined by

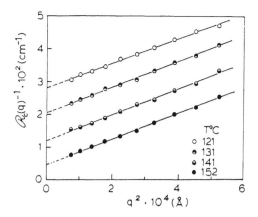

FIGURE 21. Reciprocal Rayleigh factor vs. q^2 plots (OZ plot) for deuterated PS/PVME blend at various temperatures. (From M. Shibayama, H. Yang, R. S. Stein, and C. C. Han, *Macromolecules,* 18, 2179, 1985. With permission.)

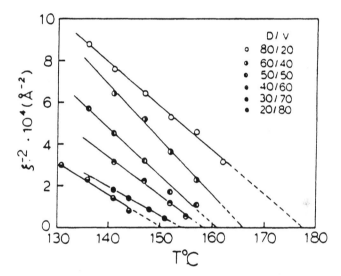

FIGURE 22. ξ^{-2} vs. temperature plots for various compositions. The intercept on the T axis gives the spinodal temperature. (From M. Shibayama, H. Yang, R. S. Stein, and C. C. Han, *Macromolecules* 18, 2179, 1985. With permission.)

$$\xi^2 = \frac{R_g^2}{3} \left| \frac{\chi_s}{\chi - \chi_s} \right| \propto |T - T_s|^{-1} \tag{75}$$

Equation 75 expresses the Ornstein-Zernike form for small q.

By employing this S(q) formulation, Shibayama et al. estimated ξ, χ, and the phase diagram in a mixture of deuterated PS and hydrogenated PVME by small angle neutron scattering (SANS).[38] Figure 21 shows plots of $l/R_c(q,T)$ vs. q^2, where $R_c(q,T)$ is the scattering Rayleigh factor. According to Equation 74, the plots yielding straight line and ξ values were obtained. The spinodal temperatures were estimated from the plots of ξ^2 vs. T as shown in Figure 22 according to Equation 75.

Bates et al. also applied the S(q) method to SANS studies for a binary mixture of protonated and deuterated 1,4-polybutadiene(PB).[39-40] They demonstrated that the mixture of protonated and deuterated PB was characterized by UCST. It was also demonstrated that a binary mixture of protonated and deuterated PS was characterized by UCST.[41] They indicated that this UCST behavior was due to a very small positive value of the Flory-Huggins χ parameter for the mixtures of protonated and deuterated polymers, i.e., this unusual UCST behavior can be explained within the framework of the Flory-Huggins equation by considering both effects of segment volume and van der Waals force as content of χ.[41]

IV. CONCLUDING REMARKS

We indicated the coexistent behavior of UCST and LCST in a mixture composed of polymers having high molecular weight. This is not specific to this particular pair. Similar phase behavior has also been found in other pairs of high molecular weight polymers such as poly(acrylonitrile-co-styrene)/poly(acrylonitrile-co-butadiene),[25] poly(methyl methacrylate)/poly(vinylidene fluoride),[42] and PS/random copolymer of carboxylated poly(2,6-dimethyl-1,4-phenylene oxide).[43] It seems to be a rather general phenomenon. However, it is hard to interpret in the framework of the current equation of state theories as shown in Section II.

Kammer et al. indicated that the coexistent behavior of UCST and LCST can be explained in terms of a refined version of the Prigogine-Flory-Patterson theory by introducing the contribution of the difference in segment size.[26]

Some modifications of the equation of state theory are under way to interpret the coexistent behavior of UCST and LCST in the blend of high molecular weight polymers.

There are very few reports discussing the relationship between phase diagram and pressure in polymer mixtures.[44-46] Equation of state theories are useful to interpret the relationship. Moreover, the information of phase behavior concerning pressure and temperature is important in order to obtain a more accurate description for polymer mixtures.

REFERENCES

1. **Flory, P. J.**, *Principles of Polymer Chemistry,* Cornell University Press, Ithaca, New York, 1953.
2. **Paul, D. R. and Newman, S., Eds.**, *Polymer Blends,* Academic Press, New York, 1978.
3. **Akiyama, S., Inoue, T., and Nishi, T.**, *Polymer Blends-Compatibility and Interface,* CMC Press, Tokyo, 1979.
4. **Olabishi, O., Robeson, L. M., and Shaw, M. T.**, *Polymer-Polymer Miscibility,* Academic Press, New York, 1979.
5. **Roe, R. J. and Zin, W. C.**, *Macromolecules,* 13, 1221, 1980.
6. **Zacharius, S. L., ten Brinke, G., MacKnight, W. J., and Karasz, F. E.**, *Macromolecules,* 16, 381, 1983.
7. **Inoue, T., Ougizawa, T., Yasuda, O., and Miyasaka, K.**, *Macromolecules,* 18, 57, 1985.
8. **Flory, P. J., Orwoll, R. A., and Vrij, A.**, *J. Am. Chem. Soc.,* 86, 3507, 1964.
9. **Eichinger, B. E. and Flory, P. J.**, *J. Trans. Faraday Soc.,* 65, 2035, 1968.
10. **Dee, G. T. and Walsh, D. J.**, *Macromolecules,* 21, 811, 1988.
11. **Prigogine, I.**, *The Molecular Theory of Solutions,* North-Holland, Amsterdam, 1957.
12. **Prigogine, I., Traffeniers, H., and Mathot, V.**, *J. Chem. Phys.,* 26, 751, 1957.
13. **Nanda, V. S. and Simha, R.**, *J. Phys. Chem.,* 60, 3158, 1964.
14. **Dee, G. T. and Walsh, D. J.**, *Macromolecules,* 21, 815, 1988.
15. **Sanchez, I. C. and Lacombe, R. H.**, *J. Phys. Chem.,* 80, 2352, 1976.
16. **Somcynsky, T. and Simha, R.**, *Macromolecules,* 2, 343, 1969.
17. **Walsh, D. J., Dee, G. T., Halary, J. L., Ubiche, J. M., Millequant, M., Lesec, J., and Monnerie, L.**, *Macromolecules,* 22, 3395, 1989.
18. **Walsh, D. J., Dee, G. T., and Ougizawa, T.**, *Makromol. Chem. Macromol. Symp.,* 38, 255, 1990.

19. **McMaster, L. P.,** *Macromolecules,* 6, 760, 1973.
20. **Sanchez, I. C.,** in *Polymer Blends,* Paul, D. R. and Newman, S., Eds., Academic Press, New York, 1978.
21. **Flory, P. J.,** *J. Am. Chem. Soc.,* 87, 1833, 1965.
22. **Patterson, D.,** *J. Polym. Soc.: Part C,* 16, 3379, 1968.
23. **Patterson, D. and Delmas, G.,** *Trans. Faraday Soc.,* 65, 708, 1969.
24. **Patterson, D. and Robard, A.,** *Macromolecules,* 11, 690, 1978.
25. **Ougizawa, T. and Inoue, T.,** *Polymer J.,* 18, 521, 1986.
26. **Kammer, H. W., Inoue, T., and Ougizawa, T.,** *Polymer,* 30, 888, 1989.
27. **Abe, A. and Flory, P. J.,** *J. Am. Chem. Soc.,* 87, 1838, 1965.
28. **Ougizawa, T., Dee, G. T., and Walsh, D. J.,** *Macromolecules,* 24, 3834, 1991.
29. **Ougizawa, T., Dee, G. T., and Walsh, D. J.,** *Polym. Prep. Jpn.,* 40, 778, 1991.
30. **Maruta, J., Ougizawa, T., and Inoue, T.,** *Polymer,* 29, 2056, 1988.
31. **Ougizawa, T., Inoue, T., and Kammer, H. W.,** *Macromolecules,* 18, 2089, 1985.
32. **Inoue, T. and Ougizawa, T.,** *J. Macromol. Sci.: Chem.,* A26, 147, 1989.
33. **Hashimoto, T., Kumaki, J., and Kawai, H.,** *Macromolecules,* 16, 641, 1983.
34. **Cahn, J. W.,** *J. Chem. Phys.,* 42, 93, 1965.
35. **Kumaki, J. and Hashimoto, T.,** *Macromolecules,* 19, 763, 1986.
36. **Takagi, T., Ougizawa, T., and Inoue, T.,** *Polymer,* 28, 103, 1987.
37. **de Gennes, P.-G.,** *Scaling Concepts in Polymer Physics,* Cornell University Press, Ithaca, New York, 1979.
38. **Shibayama, M., Yang, H., Stein, R. S., and Han, C. C.,** *Macromolecules,* 18, 2179, 1985.
39. **Bates, F. S., Wignall, G. D., and Koehler, W. C.,** *Phys. Rev. Lett.,* 55, 2425, 1985.
40. **Bates, F. S., Dierker, S. B., and Wignall, G. D.,** *Macromolecules,* 19, 1938, 1986.
41. **Bates, F. S. and Wignall, G. D.,** *Phys. Rev. Lett.,* 57, 1429, 1986.
42. **Saito, H., Fujita, Y., and Inoue, T.,** *Polymer J.,* 18, 521, 1986.
43. **Cong, G., Huang, Y., MacKnight, W. J., and Karasz, F. E.,** *Macromolecules,* 19, 2765, 1986.
44. **Suzuki, Y., Miyamoto, Y., Miyaji, H., and Asai, K.,** *J. Polym. Sci.: Polym. Lett. Ed.,* 20, 563, 1982.
45. **Rostami, S. and Walsh, D. J.,** *Macromolecules,* 18, 1228, 1985.
46. **Maeda, Y., Karasz, F. E., MacKnight, W. J., and Vuković, R.,** *J. Polym. Sci.: Polym. Phys. Ed.,* 24, 2435, 1986.

Chapter 20

COMPATIBILIZATION OF POLYMER BLENDS

Richard L. Markham

TABLE OF CONTENTS

0-8493-4401-8/93/$0.00 + $.50
731

I. INTRODUCTION

Blending two or more polymers to obtain a unique product suitable for an application has been practiced for decades. Although blending thermoplastics has been an area of high interest for the past 15 years, rubber blending has been carried out since the last half of the 19th century.

Because the thermodynamic conditions required for mixing on the molecular level are met in very few pairs of polymers, most blends require some type of action — addition of a selected component or variation in a process condition — to convert the simple mixture into a useful product. This action is "compatibilization."

Until the past five to seven years, very few results on the use of compatibilizers and compatibilization were presented in research papers. Most of the results were hidden in the obscure language of the patent literature. The recent literature, however, provides important information with which to select compatibilization approaches. This chapter provides a summary of compatibilization techniques presented in the recent literature.

II. TERMINOLOGY

For many years, the term "compatibility" was used in the literature as a synonym for "miscibility." More recently, the terms have been differentiated by most authors. In this chapter, "compatibility" will refer to the state of an immiscible blend in which it provides a combination of properties and characteristics that make it more useful than the original mixture of polymers. "Compatibilization" will refer to the approach used to reach this state. "Miscible" will refer to a blend in which the components are mixed on the molecular level.

There are many definitions of and differentiations between the terms "blend" and "alloy." Because such differentiations are rarely needed and to avoid confusion, the term "blend" will be used throughout this chapter.

III. WHY IS COMPATIBILIZATION IMPORTANT?

There are few pairs of miscible polymers. One miscible pair that yields rubbery properties upon blending is polyvinyl chloride/nitritebutadiene rubber (PVC/NBR). These polymers have specific interactions that result in a decrease in the free volume of the mixture, an important driving force for molecular mixing. Specific interactions may result from hydrogen bonding, induced dipole-induced dipole, dipole-dipole, or ionic interactions.

Miscible blends are desirable because they are more easily converted into consistent, homogeneous products, and the properties can be varied easily (within the bounds of the properties of the components) by changing the composition. In the absence of specific interactions, i.e., in immiscible mixtures, the polymeric components will remain in separate phases. In most cases, the desirable properties of each component are significantly compromised so that the blended product has little or no commercial value. The properties may be very sensitive to small changes in the processing conditions that affect the morphology of the final product.

It should be noted that immiscibility is required to meet some combinations of properties. The most important example of the value of immiscibility is the widespread use of impact-modified polymers. Considerable improvement in the impact resistance of brittle polymers is obtained by the addition of a rubbery phase. This improvement would not occur if the rubber were miscible with the matrix polymer.

Compatibilization is important in providing to immiscible blends the morphological stability, processing homogeneity, and interphase interactions required in commercial polymeric materials. In the typical immiscible mixture, the adhesion between the phases is very small. As a result, physical forces applied to the blend will not be transferred to the dispersed phase, and some properties are significantly lower than would be predicted by the weighted average of the properties of the components. Products prepared from such blends are not likely to have a desirable combination of properties. Conversion of these phase-separated blends into useful polymeric materials combining the desirable properties of each component requires compatibilization.

IV. WHAT IS COMPATIBILIZATION?

From the perspective of developing commercial polymer blends, compatibilization is the process of converting an otherwise useless polymer blend into a commercially useful product. As Utracki[1] has stated so succinctly, "The goal of compatibilization is to obtain a stable and reproducible dispersion which would lead to the desired morphology and properties."

There are two important goals in developing polymer blends: (1) a morphology that will enhance the expression of the useful properties of each component in the final product, and (2) an interaction between the phases that will stabilize the morphology and assure the transfer of stress from the matrix to the dispersed phase.

Successful compatibilization will

- Reduce interfacial energy
- Permit finer dispersion during mixing
- Provide a measure of stability against gross separation throughout the processing/conversion to the final product
- Result in improved interfacial adhesion

This may be accomplished by (1) the addition of a compatibilizer or (2) changes in the processing conditions to induce the above characteristics in the final blended product. The first approach is the most widely studied. However, more recently, the use of specific processing conditions to enhance the property profile of blends has received more interest and will be discussed below.

The number of proven approaches for the selection of the compatibilizer as well as other means of inducing compatibilization continues to increase. Compatibilization has been achieved by

- Addition of linear or star-shaped copolymers
- Addition of graft or random copolymers
- Coreaction within the blend to generate *in situ* either copolymers or interacting polymers
- Using interpenetrating polymer network (IPN) technology

- Crosslinking the blend ingredients
- Modification of homopolymers, e.g., through incorporation of acid/base groups, hydrogen bonding groups, charge-transfer complexes, ionic groups, etc.
- Addition of co-solvent
- High stress shearing[1]

V. MORPHOLOGICAL CONSIDERATIONS

A. IMPORTANCE OF MORPHOLOGY

In designing and developing polymer blends, one of the first concerns must be the desired morphology of the blend in the ultimate product. Of course, a uniform and reproducible morphology is required for commercial success. The blend will usually have a better balance of properties if the dispersed phase size is small.

Realistically, the situation is quite complex, combining the effects of the

- Rheological relationship of the components
- Compatibilizer as a stabilizer of the morphology and as an interfacial bridge to improve properties
- Processing conditions involving both the processing forces and the rheological response of the components to those forces

Understanding and predicting the resulting morphology of the blend is further complicated by the interdependencies of the effects of parameters within each of these factors. Hence, in developing compatibilized polymer blends, it is important to know the rheological properties of the components and to understand the interactions and reactions between them that may affect the compatibilization process.

The effects of the conversion process on morphology may be profound. For example, in injection molding, one must take into account the effects of (1) high shear at temperatures above the melting/glass transition temperature, (2) shear heating as the blend is injected into the mold, and (3) quenching rate influenced by the temperature of the mold.

With commercial blends, the desired morphology and interfacial effects must be obtained in the final product at many different facilities and, in cases in which the material may be recycled, after a second melting and molding. The compatibilizer can help to prevent delamination, agglomeration, "skinning", and other undesirable phase effects that may act in the conversion to the final product.

B. MELT VISCOSITIES OF COMPONENTS

One of the most important functions of the compatibilizer is to influence the dispersed phase size. An efficient compatibilizer will narrow the distribution of particle sizes as well as lower the average size of the particles. However, the ratio of the melt viscosities is also an important influence on the dispersibility of the blend components. If the viscosities of the components are significantly different at the processing temperature, it is difficult to obtain a uniform dispersion of small particles, particularly if the minor phase has the higher viscosity. The component with the higher viscosity may act as a hard sphere, and that phase cannot be broken into small dispersed particles.

Heggs et al.[2] reported the use of a viscosity-modifying additive to match more closely the viscosities at the processing conditions in a blend of polycarbonate with nylon 6. The melt viscosity of polycarbonate is about 12 times that of nylon 6. When the matrix was nylon 6, the dispersed phase size of the uncompatibilized blend was on the order of 20 to 30 μm and was nonuniform.

TABLE 1
Cooling Effects with PP/EP Blends

Blend composition			EP particle size (μm)	
iPP	EP	Comp.	Rapid cooling	Slow cooling[a]
80	20	0	5–7	large
79	19	2	5–7	10–15
76	19	5	2–4	—
72	18	10	0.3–2	—
68	17	15	0.2–1	0.2–1

[a] In slow cooling, blend was maintained in the molten,
quiescent state at 200°C for 45 min followed by cooling.

The addition of polycaprolactone, which is miscible with polycarbonate, significantly lowered the viscosity of the polycarbonate phase. Addition of 10 parts of polycaprolactone to 90 parts of polycarbonate lowered the viscosity to obtain a viscosity ratio of only 7. As a result, the average dispersed phase size decreased to less than 1 μm. In addition, the distribution of sizes was considerably narrowed. The authors note that the effect may be related to the melt elasticities of the components rather than to the melt viscosities.

C. SHEAR RATE DURING PROCESSING/CONVERSION

The type of shear and its duration have a significant effect on the morphology, and controlled variations in these factors can yield a wide variety of morphological structures. This topic is discussed further in Section XI. A useful discussion of these effects is provided by Meijer et al.[3]

D. QUENCHING RATE IN PART FORMATION

The morphology of a compatibilized blend is influenced by the quenching or cooling rate during formation of the ultimate product. Compatibilizers provide the stabilization of the morphology needed to assure that a blend will furnish the same properties within a reasonably broad range of processing conditions.

Datta and Lohse[4] reported the effects of rapid vs. slow cooling of compatibilized and uncompatibilized blends of polypropylene (PP) and ethylene-propylene (EP) copolymer. Their results are shown in Table 1.

It is apparent that in the uncompatibilized blend held at 200°C, the dispersed phase agglomerated to yield a much larger dispersed phase size. The addition of only 2% of graft compatibilizer resulted in a significantly smaller dispersed phase size with either rapid or slow cooling. With 15% compatibilizer, the blend morphology was stable even during 45 min in the molten state. Thus, variation in cooling rate has a significantly smaller effect on the morphology when the blend is efficiently compatibilized.

E. CO-CONTINUOUS PHASES

Co-continuous phases, a unique morphological structure, has high potential for providing unique combinations of properties, but this morphology is rarely observed. This morphology differs from IPN, being the product of mixing two polymers, while IPNs are prepared by simultaneous and separate polymerization of two unique monomers. The end result in both cases is that each of the two components exists as a single phase, extending throughout the structure. Hence, the bulk properties of *each* phase are expressed throughout the part. Some properties of the blend, such as glass transition temperature, chemical resistance, and strength, may be that of either or both of the components. This morphology may provide the most desirable combination of properties of the blend components for some applications.

To obtain co-continuous phases by mixing polymers, a suitable balance must be obtained among (1) blend composition, (2) relative viscosities of the components, (3) shear mixing of the blend before and during part formation, and (4) cooling rate in part formation. The first two points must be determined empirically. However, the last two, i.e., composition and component viscosity, can be used to estimate the combination of components most likely to result in co-continuous phases. If the blend fits the following simple equation (Equation 1), there is a high probability that a co-continuous phase morphology will be obtained.[5,6]

$$\frac{\eta_1}{\eta_2} \cdot \frac{\phi_2}{\phi_1} \cong 1 \tag{1}$$

where η is the viscosity of the phase at processing temperature, and ϕ is weight fraction of the phase component in the blend. However, melt elasticity rather than melt viscosity may be the property that influences morphology.

VI. TYPES OF COMPATIBILIZERS

Compatibilizers may be nonreactive, reactive, or both. The most obvious type of non-reactive compatibilizer is a copolymer of A and B for a mixture of polyA and polyB. However, other copolymers may be effective if they have specific interactions, i.e., miscibility, with one or both of the blend components.

Reactive compatibilizers are often preferred because of the greater stability of the blend morphology throughout the processing and service life of the final product. Included in this type of compatibilizer are the crosslinked or vulcanized blends.

VII. NONREACTIVE COPOLYMERS AS COMPATIBILIZERS

The use of block copolymers as compatibilizers has been a frequent topic in the literature. Paul[7] provides a review of important research results and an analysis of major points to be considered in selecting nonreactive block copolymers as compatibilizers. Xanthos[8] reviewed the use of both nonreactive and reactive compatibilizers. Included is a documented list of compatibilizers used with specific blends.

Diblock copolymers appear to have the most appropriate conformation for compatibilization. However, multiblock copolymers may be more efficient than diblock copolymers. Leibler[9] carried out thermodynamic calculations that appeared to have been confirmed by Trostyanskaya.[10] However, Leibler later made a strong argument for preference for symmetrical diblocks.[11]

There are a number of principles that can be applied in selecting the most efficient compatibilizers for specific polymer blends. The most important result is for the compatibilizer to bridge the interface providing an adhesion between the two phases. This requires a miscibility of each segment of the compatibilizer in one of the phases or, in the case of the reactive compatibilizers, reactivity with one of the phases. Figure 1 shows several possible configurations at the interface.

A. CHEMICAL STRUCTURE OF COMPATIBILIZERS

Because a molecular mixture of a segment of the compatibilizer with the base component of the phase is most desirable, the use of a block in the compatibilizer miscible in the phase should be the first option. Thus, it is imperative for the developer of polymer blends to understand the fundamentals of miscibility, even though working with immiscible blends.

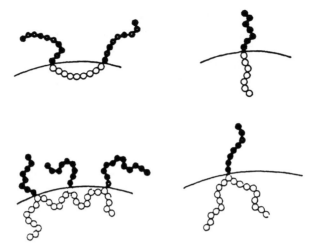

FIGURE 1.   Conformations of compatibilers at the interface.

Several good summaries of these fundamentals are available.[12,13] However, a brief discussion of some of the important points as they apply specifically to compatibilization of polymer blends is useful.

The most obvious choice to meet this miscibility requirement is to select a block co-polymer having blocks of the same polymer in the two phases. Thus, a blend of polyA with polyB may be compatibilized by a block copolymer, polyA-b-B. This approach may also use grafted copolymers.

Although this is a direct route to compatibilization, it is not always simple to carry out. Preparation of block and graft copolymers is often expensive, and many combinations of the two polymers are very difficult to prepare, such as a combination of an addition and a condensation polymer for a blend of polyethylene and nylon 6.

Seo, Jo, and Ha[14] report an innovative route to a graft copolymer to compatibilize a blend of polymethyl methacrylate (PMMA) and nylon 6. Starting with a copolymer of methyl methacrylate and methacrylic acid, the authors chlorinated the methacrylic acid groups with thionyl chloride. An oligomer of nylon 6 was prepared by hydrolysis of a commercial nylon 6. When added to the chlorinated copolymer, the amine end groups of the nylon 6 oligomer reacted with the methacrylic acid chloride in the copolymer to form a graft copolymer (see Structure 1).

Structure 1

Thermal properties indicated effective compatibilization.

If it is not feasible to use a block or graft copolymer with the same chemical structure as the blend components, the next option is to use a copolymer with segments that are miscible or reactive with the components.

A difficult-to-compatibilize mixture is polyphenylene oxide (PPO) and polyvinyl difluoride (PVDF). Teyssie et al.[15] report an interesting combination of blocks in a copolymer. The blocks are different from the base components, but each is miscible with one of the base polymers. PPO is miscible with polystyrene, and PVDF is miscible with polymethyl methacrylate. The authors demonstrated that polystyrene-b-methyl methacrylate is an effective compatibilizer for PPO-PVDF.

Silicone-modified thermoplastic elastomers (TPE) under the trade name of C-Flex are formulated from styrene-b-ethylene-butylene-b-styrene copolymers, polyolefins, and polydimethyl siloxane.[16] They combine desirable physical properties and processability with excellent biocompatibility. Addition of an organo-modified polysiloxane to the base formulation converted the original white opaque product to a transparent product.

B. "TAPERED" COPOLYMERS AS COMPATIBILIZERS

Traditionally, AB and ABA block copolymers have been the first choice for compatibilizers. It is easy to envision the bridging of the interface with this copolymer structure. However, other structures may provide advantageous positioning of the compatibilizer.

Teyssie et al.[17] demonstrated the use of "tapered" copolymers of ethylene and styrene to compatibilize polyethylene/polystyrene mixtures. The copolymers were prepared by solution polymerization of butadiene and styrene. Because of the higher reactivity of the butadiene, the first propagation incorporated mostly butadiene. As the butadiene was depleted, the styrene was added into the polymer statistically, eventually ending in a block of styrene after the butadiene was depleted. The resulting "tapered" copolymer has an essentially pure block of each monomer at each end with a section in between that has a gradually changing composition. (The butadiene polymers were hydrogenated to yield polystyrene-co-ethylene.)

The ultimate tensile strengths of polystyrene/polyethylene blends compatibilized with 9% "tapered" copolymer ($M_n = 80,000$) were better than those compatibilized with a pure diblock of styrene and ethylene having a M_n of either 80,000 or 155,000.

C. RANDOM COPOLYMERS AS COMPATIBILIZERS

Karasz, MacKnight, and co-workers have published several reports of the use of random copolymers to compatibilize blends. For example, PPO is immiscible with homopolymers of o-chlorostyrene and p-chlorostyrene. In addition, the two halogenated copolymers are immiscible.[18] The authors found that the phase diagram for blends of PPO with a statistical copolymer of the halogenated monomers has a "miscibility window." In simple terms, the interactions between the different monomers in the copolymer are so adverse that the interactions between the PPO and the copolymer are less important. As a result, the components are forced together in a molecular mixture.

D. *IN SITU* PREPARATION OF BLOCK COPOLYMERS

Specific block copolymers to compatibilize polymer blends are often costly. The preparation of the compatibilizer *in situ* is appealing as a lower cost approach, but it also offers the potential for making compatibilizers that cannot be made otherwise.

It is likely that some copolymer of the base components will be formed in almost any melt blending process due to shear-temperature effects. However, this may not yield a sufficient amount of the compatibilizer, and it may not be dependable enough for a commercial product. Hence, some effort must be made to prepare *in situ* adequate amounts of the compatibilizer reproducibly.

Scott and co-workers[19] studied the preparation of compatibilizers during the melt blending of polyethylene/polystyrene and polyethylene/polyvinyl chloride in an internal mixer. Tensile strength was improved when a radical generator was added to the blend before mixing.

Inducing an esteramide interchange by the addition of a catalyst appears to have increased the compatibility of a blend of polyester and polyamide.[20] Only one pass through the twin-screw extruder yielded enough compatibilizer to improve the properties of the final product.

On the other hand, this approach may be counterproductive. Xanthos et al.[21] studied the effects of catalysis on the ultimate properties of polypropylene/polyethylene terephthalate blends compatibilized by polypropylene-g-acrylic acid. Magnesium acetate or p-toluene sulfonic acid was used to induce ester-acid interchange. Blends with catalysts exhibited lower tensile and flexural properties than blends without catalysts.

Other interesting studies of polyester transreactions were reviewed by Porter.[22]

Some research groups have selected components to assure that a reaction takes place during the mixing process to prepare *in situ* a desirable compatibilizer. Coran et al.[23] mixed two thermoplastic vulcanizates (TPV), ethylene-propylene-diene rubber (EPDM)-polypropylene (PP) TPV and NBR-nylon TPV, yielding a product with gross incompatibility. This was remedied by the use of a reactive compatibilizer. In one case, this was effected using an added reactive component. Subsequent experiments indicated that similar results could be obtained using a modified reactive component in one of the original TPVs.

The authors state '' . . . interaction between the modified olefin polymer and the nylon results in the formation of compatibilizing segments or blocks of both nylon and polyolefin sequences. Best results are obtained when the segments are the same as those of the homopolymers whose compatibilization is intended.'' Compatibilizers included maleic anhydride-modified PP, dimethylol phenol-modified PP, PP modified by carboxymethylmaleamic acid, and a combination of the latter two modifiers. Polypropylene and EP rubber were also compared.

Coran and Patel[24] report that a mixture of NBR with polypropylene has potential for use as an oil-resistant thermoplastic elastomer. However, they are immiscible, and mixtures of the two have very poor properties. Dynamic vulcanization of such a mixture **compatibilized** with a graft copolymer formed *in situ* provided a composition with excellent hot-oil resistance and mechanical properties similar to those of EPDM/PP, which is near miscibility.

VIII. COMPATIBILIZER MOLECULAR WEIGHT

A. EFFECTS ON MISCIBILITY

The effects of the molecular weight of base polymers and of the components of the compatibilizer are complex. Simplistically, it is most desirable to match the molecular weight of the compatibilizer block with that of the corresponding base polymer. This is often not possible, and there is the complication of the polydispersity of the molecular weight of any polymer.

If the base polymers are near miscibility, a lower molecular weight block may interpenetrate the phase more easily. On the other hand, the molecular weight of the block must be adequate to entangle with the molecules of the base component, assuring adhesion between the phases. This length appears to be at least 15 to 20 mers. However, if the base polymers are not near miscibility, the interpenetration of the block will be higher if the block has a higher molecular weight than that of the polymer to be compatibilized.[7]

An important understanding in selecting the molecular weights of the blocks is the tendency of a polymer to exclude dissimilar molecules, even those with the same composition but different molecular weight. This tendency is greater at higher molecular weights of the base polymers. There is also the tendency of the blocks in the copolymer to separate into domains excluding the base component. This tendency increases as the molecular weight of

TABLE 2
Effects of Molecular Weight and Styrene Content of SEPS on Impact Strength and Melt Viscosity

Test recipe	Control	A	B	C	D
SEPS mfi	—	22	0.1	25	0
SEPS % styrene	—	13	13	65	50
Melt Index (g/10 min)	7.6	8.2	6.1	6.1	5.7
Izod impact strength (J/m) at 22°C	53	64	181	75	85
Tensile strength at yield (MPa)	22.9	20.7	21.2	23.2	21.7
Elongation at yield (%)	10	10	10	10	10
Morphology					
Part Distance (nm)	NA	7971	7526	6955	7213
No. average size (nm)	3500	2460	2512	2302	2317
Average size (nm^3)	6978	4163	3948	3500	3766

Note: NA = not applicable.

the block increases. Cocrystallization, which excludes dissimilar molecules, must also be considered.

In developing commercial polymer blends, it is not practical to predict the effects of various molecular weight combinations of blocks and base polymers. Thus, an empirical approach using various combinations of molecular weights must be employed to optimize the system.

B. EFFECTS OF BLEND VISCOSITY

Molecular weight of a polymer strongly influences its melt viscosity. In studying the relationship between impact strength and melt viscosity of the EPDM in blends with PP, Gallagher[25] found that the addition of polystyrene-b-ethylene-propylene-b-styrene (hydrogenated copolymer of styrene/isoprene/styrene) (SEPS) would enhance flow while improving impact strength. The isoprene precursor was chosen over the butadiene precursor because of an expected greater miscibility of the EP soft segment with the EPDM. Blends were filled with 20% talc to simulate commercial thermoplastic polyolefin rubber.

Table 2 shows the results of varying the molecular weight and styrene content of the SEPS compatibilizer. The blend formulation was

PP (melt index = 12)	75
EPDM	20
SEPS	5
Talc	20
Zinc stearate	1
Irganox 1076	0.5

A control was included with no SEPS.

Note that A and C use high melt-flow SEPS vs. B and D; C and D use high styrene-content SEPS vs. A and B. Neither of these two parameters has a strong influence on the tensile strength at yield.

Increasing the melt flow of the low-styrene SEPS (lower molecular weight of SEPS) increased the melt flow of the compound significantly (compare A and B). This effect was not so significant using high-styrene SEPS (C and D). Higher styrene in the SEPS also increased the melt flow of the final compound (compare A with C and B with D). The authors state that these trends are in agreement intuitively with the increased miscibility of the SEPS with EPDM.

The effects of these parameters on particle size and distribution are also interesting. As expected, the addition of SEPS significantly lowered the particle size and narrowed the distribution no matter which of the variations was used. The higher styrene resins gave smaller particle sizes than the lower styrene resins. Molecular weight did not seem to have a significant influence on particle size.

In summary, careful selection of the molecular weight and structure of the compatibilizer resulted in an improvement in the melt flow, a better dispersion of one phase in the other, and a more uniform particle size distribution.

IX. REACTIVE COPOLYMERS AS COMPATIBILIZERS

Compatibilizers that interact with one or both of the phases are desirable. Because the phases are held together by covalent bonds, the blends are less sensitive to the temperature-shear environment in eventual processes and to the service conditions of the final product. However, there are situations in which a nonreactive interaction is preferred. For example, reversible, noncovalent bonds, such as hydrogen and ionic bonds, can provide the strength and stability in the final product while being ''released'' at melt temperatures, resulting in lower melt viscosity and easier processing.

A. COPOLYMERS WITH REACTIVE GROUPS

Blends in which at least one of the base components has an available reactive group, whether in the backbone or at the chain end, are good candidates for this approach. The compatibilizer may be a block or a random copolymer. In some cases, it may be feasible to use a compatibilizer that offers a reaction with one phase but a nonreactive interaction with the other.

Some of the more commonly used polymers incorporate functional groups as part of the backbone, such as styrene-maleic anhydride or polypropylene-acrylic acid copolymers, or as adducts, such as the maleic anhydride adducts of polyolefins. These materials are good first candidates when compatibilizing a nonpolar polymer, such as polyolefin or polystyrene, with polar polymers, such as polyamide or polyester.

The preparation of adducts may be cost effective. The adduct may be prepared during the melt mixing of the blend or in a separate process.

B. GRAFTING TO OBTAIN COPOLYMERS

An important route to preparing block copolymers is to graft a block of one monomer that will compatibilize one phase onto a polymer that will compatibilize the other phase. This is most often accomplished by adding a peroxide to generate radicals on the backbone of one or both of the blend components followed by the addition of a selected monomer, which grafts to the polymer.

Another interesting approach was taken by Seo, Jo, and Ha[14] in preparing a graft of nylon 6 onto PMMA. This work was discussed earlier in Section VII.A.

C. REACTION OF FUNCTIONAL POLYMERS

The properties of blends of PP and EP copolymers are significantly improved by addition of up to 10% of a graft copolymer as compatibilizer. Lohse and co-workers[4,26] have formed graft copolymers containing dissimilar blocks by ''reaction of complementary reactive functionality.'' The compatibilizer was prepared by the reaction of an EP copolymer having amine functionality with an isostatic PP with succinic anhdyride functionality ''near the end.'' The resulting block/graft copolymers were effective in improving the impact strength of PP/EP blends by a factor of as much as four. The composition of the compatibilizer was confirmed by infrared spectroscopy, molecular weight and molecular weight distribution studies, thermal properties, morphology studies, and differential solubility in a range of solvents.

D. CHEMICAL MODIFICATION OF BLEND COMPONENTS

The guiding principle behind this approach is to add a functionality to induce specific interactions or reactivity between the components or between one of the components and the compatibilizer. The blend developer needs an understanding of suitable organic reactions and interactions that will compatibilize the mixture under consideration.

An important study of this approach involves the compatibilization of blends of polystyrene or polyethylene with polar polymers such as polyesters and polycarbonates. Pearce et al.[27] found that they could induce **miscibility** of polystyrene blends with polyesters by making the adduct of polystyrene with hexafluoroacetone. The resulting hexafluorodimethyl carbinol on the polystyrene provides a strong hydrogen bonding capability that can induce compatibility with many polar polymers (see Structure 2).

Structure 2

Similar results were obtained with functionalized polyethylene.[28]

Maleic anhydride, an unsaturated reactive monomer, is useful in rendering polymers reactive through inclusion as a comonomer or as an adduct. The maleic anhydride adduct of EPDM provides the reactivity to compatibilize the elastomer with nylon in preparing super tough commercial products.[29-31] Amine end groups on the nylon react with the anhydride to form a grafted block copolymer that compatibilizes the system. Similar compatibilizations have used acrylic acid grafts or adducts.

Ionomers offer the potential of an inorganic interaction to compatibilize appropriate polymers. Willis and Favis[32] used the ionomer Surlyn 9020 to compatibilize blends of polyolefin with nylon 6. Only 5% of the ionomer (comprising roughly 80% ethylene and 20% methacrylic acid and isobutyl acrylate; 70% of acid neutralized with Zn) was required to reduce the average dispersed phase size by a factor of as much as four.

With carefully selected polymer pairs, ionomers may also provide thermally reversible crosslinks that may result in better processing while maintaining a stronger and stiffer material.

E. CHEMICAL MODIFICATION OF SURFACES

Chemical modification of the surface of polymer particles to be added to a matrix of another polymer can induce significant interactions between the particles and the matrix. Bauman et al.[33-35] have demonstrated the effects of treating the surface of several polymer powders, including several rubbers and ground scrap tires, polyethylene, polypropylene, and polyethylene terephthalate. Significant property improvements were obtained over materials incorporating untreated fillers.

TABLE 3

Tensile Properties of Mixtures of Untreated and Surface-Treated EPDM Particles with Polyurethane

EPDM (wt%)	Untreated filler		Surface-treated filler	
	Tensile strength at break, (lb)	Elongation at break (%)	Tensile strength at break, (lb)	Elongation at break (%)
25	434	235	474	227
50	326	142	389	165

TABLE 4

Properties of Polyurethanes Unfilled and Filled with Surface-Modified Scrap Rubber

	Unfilled	15% Filler
Tensile strength (psi)	4100	3500
Elongation (%)	278	275
Tear resistance (die C)	593	522
Tear resistance (trouser)	113	104
Rebound (%)	49	48
Hardness (shore D)	50	50

The surfaces of particles were treated with a reactive gas mixture to provide sites for chemical interactions with selected matrix materials. The gas mixtures comprised fluorine, an inert gas such as nitrogen, and, in some cases, another oxidizing gas such as oxygen or sulfur dioxide.

An EPDM ground to a particle size of 150 to 850 Å (20 to 100 mesh) was treated with a mixture of fluorine (1% by volume), sulfur dioxide (40%), and nitrogen (59%) at room temperature for 30 min. Several mixtures were prepared with a polyurethane (methylene di-p-phenylene isocyanate plus 1,4-butanediol) and tested with interesting results (Table 3).

The results indicate that the interface of the mixture was strengthened by a chemical interaction between the surface of the rubber particles and the polyurethane matrix. In another example, a mixture of 15% of surface-modified scrap EPDM rubber (80 mesh) in polyurethane provided similar properties at a lower cost (Table 4).

Although the tensile strength is slightly lower than the unfilled polyurethane, the other properties are equivalent. Ordinarily, addition of an untreated ground scrap rubber would yield a composite with significantly lower properties. Thus, surface modification has provided a compatibilization.

X. CROSSLINKING TO EFFECT COMPATIBILIZATION

A. TRADITIONAL CROSSLINKING

Blending two or more rubbers has been a standard method for obtaining a new rubber with a unique property profile. The miscibility or compatibility of the components of such blends has not been a frequent topic in the recent literature. Since thermoset rubbers are crosslinked, the components are held in the morphological structure induced in the mixing and molding of the part.

It is interesting that there have been few recent reports of definitive studies on the size and shape of the domains in rubber blends. However, Inoue et al.[36] have shown that crosslinking may result in the obliteration of the distinct phase-separated structure seen commonly with thermoplastic blends.

A blend of *cis*-1,4-polybutadiene with a styrene-butadiene rubber (23.5% styrene) exhibited two T_gs. After covulcanization, a more homogeneous blend was obtained having only one T_g between the original two T_gs. There was no evidence of a two-phase morphology.

The authors suggest that this phenomenon is a broadening of the polymer-polymer interface. The effects of curing temperature, curing agents, and the periodic distance in the uncured blends on the rate of dissolution were studied through radiothermoluminescence (T_g) and light scattering. The phase diagram was estimated to have an upper critical solution temperature (UCST). Covulcanization in this study was conducted at a temperature only slightly higher than the UCST. The authors explain the observed results through the following hypothesis.

1. A small amount of the curing agent, which is a small molecule, locates at the interface where it generates block and graft copolymers at an early stage of the curing.
2. The block and graft copolymers compatibilize the blend in the interfacial region, thus thickening the interface.
3. Assuming the temperature is well above the order-to-disorder transition temperature, the block copolymer results in a strong driving force to smooth the concentration gradient at the interface.

B. DYNAMIC VULCANIZATION

In the late 1950s, Gessler[37] developed a process now called *dynamic vulcanization*, which uses crosslinking to stabilize the morphology of rubber/thermoplastic blends. Several commercial TPEs are produced by this process.

Using dynamic vulcanization to obtain TPEs is effected on a mixture of thermoplastic and rubber.[38] The blend is mixed well at a temperature above the thermoplastic's melting point. Vulcanizing agents are added, and the mixing is continued. The dispersed rubber particles are vulcanized, and the morphology of the mixture is preserved by the vulcanization. Thus, the crosslinking of the rubber compatibilizes the thermoplastic/rubber blend. The resulting compositions have

●   Reduced set
●   Improved mechanical properties
●   Improved high-temperature utility
●   Improved fatigue resistance
●   Greater resistance to attack by hot oils
●   Greater stability of melt-phase morphology during ultimate molding to a finished product
●   Greater melt strength
●   More reliable fabricability

Without the crosslinking, the mixture would not have properties of commercial value.

The number of products produced by dynamic vulcanization continues to grow. Coran[38] offers some guidance in selecting the rubber and thermoplastic components. "The best compositions are prepared when the surface energies of the rubber and plastic materials are matched, when the entanglement molecular length of the rubber is low, and when the plastic material is crystalline. It is also required that neither plastic nor rubber decompose in the presence of the other at temperatures required for melt mixing. Further, a curing system appropriate for the rubber under the conditions of melt mixing is required."

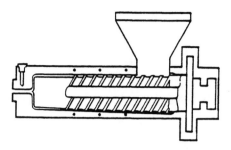

FIGURE 2.   Patfoort Plasticator, a high-speed plastifying compounder-extruder.

XI. SELECTION OF PROCESSING APPROACH

Processing conditions are very important in establishing the eventual properties of the polymer blend. Besides the temperature of the melt, the time of mixing may have a significant effect. Another important factor may be the order of addition or the manner in which the components are mixed.

A. EFFECT OF MIXING TIME

Teyssie et al.[39] investigated the effect of mixing time on compatibilization. A mixture of polystyrene and polyethylene, compatibilized by polystyrene-b-ethylene, was melt blended on a roll mill or in a corotating twin-screw extruder. Whereas the twin-screw extruder yielded a nonequilibrium fibrillar morphology, the milled blend exhibited a fine dispersion. The authors suggested that this difference reflected the longer mixing time during mill mixing which allowed a more complete diffusion of the compatibilizer to the interface.

B. EFFECT OF SHEAR

Very high shear mixing can also provide some advantage in preparing polymer blends. Utracki[40] discusses the use of the Maxwell normal stress extruder to effect high shear mixing of polymer blends. This type of extruder appears to have some potential for preparing co-continuous phases. Blends were prepared from pairs of polystyrene (PS), PMMA, and polyethylene (PE). Solvent extraction of one component left a structure which indicates that the morphology was that of two continuous phases. The blends are immiscible and were subjected to annealing. Although one would expect that this treatment would lead to further agglomeration and destruction of the co-continuous phase structure, this was not observed. The principle of the Maxwell normal stress extruder is incorporated in the Patfoort Plasti-ficator, a high-speed plastifying compounder-extruder[41,42] (Figure 2).

C. EFFECT OF ORDER OF MIXING

The order of mixing of the components can have an effect on the efficiency of the compatibilizer. In some cases, the order is obvious because a reactive compatibilizer might be added to the phase with which it will react before the addition of the second-phase polymer. In any combination, it is worthwhile to compare the results of several addition and mixing schemes to identify the optimal approach.

The interesting study by Willis and Favis[32] (discussed in Section IX.D) on ionomeric compatibilizers also investigated the effects of mode of addition of the interfacial agent. The blend was nylon 6 with polyolefin. They compared the dispersed phase size of blends prepared by two-step mixing to that obtained in a one-step mixing process. Two-step mixing involved preparation of a master batch of the ionomer in the polymer that would become the minor phase. This mixture was then mixed with the other component to obtain the compatibilized blend.

When the minor phase was nylon 6, the ionomer was mixed first with the nylon 6, and then this mixture was added to the polyolefin. The average dispersed phase size actually increased by 68% over that in the product of one-step mixing. If the minor phase was polyolefin and the inomer was added first to the polyolefin, the average dispersed phase size in the final blend was 15% smaller than that of the one-step mixed blend.

The authors attribute these results to the greater affinity between the ionomer and the nylon 6 and its better dispersion in the masterbatch. Consequently, the ionomer is less available at the interface when blended with the polyolefin. When mixed first with the polyolefin for which it has little affinity, the ionomer is poorly dispersed and moves easily to the interface where it can function to compatibilize the mixture.

Theoretically, this effect can be overcome by allowing time for the mixture to come to equilibrium. This is not practical in producing commercial blends. However, this study points out the importance of considering the effects of time in the mixing process.

D. EFFECT OF RELATIVE VISCOSITIES OF COMPONENTS

The relative viscosities of the components have a profound effect on the dispersive effects of mixing. Since the morphology of the mixture is such an important influence on the properties of the blend, consideration of approaches to match more closely melt viscosities is worthwhile. If several grades (molecular weights) are available for the components, matching the viscosities may be straightforward. However, many polymers are not offered in grades sufficiently different in molecular weight or distribution to permit this approach.

Obtaining good dispersion of the rubber phase is important in blends of PP and EPDM. Inadequate dispersion can lead to variable mold shrinkage, flow marks, and weak weld lines. Gallagher[25] reported the results of an extensive study of factors that affect the melt viscosity of the blends and, consequently, the properties of the molded product.

High density polyethylene (HDPE) was added to a blend of PP and EPDM to lower the melt viscosity of the EPDM. The grade of EPDM was chosen for its ethylene content and solubility parameter match with the HDPE. The resulting blend comprised a matrix of PP and a dispersed phase of a miscible blend of EPDM/HDPE. Table 5 compares properties of the blends with and without the HDPE.

The addition of the HDPE did lower the melt viscosity of the blend. However, the effect on impact strength was disappointing. The straight blend of PP/EPDM had a significantly higher impact strength than either of the two blends with HDPE. However, the addition of HDPE did yield a significant improvement in the tensile strength at yield.

An earlier study with similar results was reported by Lindsay et al.[43] They postulated that the improved strength resulted from a nucleating effect of the HDPE on the crystalline portions of the EPDM. The EPDM molecules would then "crosslink" the crystalline domains with results similar to that seen with dynamically vulcanized TPEs.

Gallagher[25] extended his study to the use of oil-extended EPDM. Although no increase in impact strength was noted, the melt flow increased, which should be an advantage in injection molding. In another experiment, the EPDM was modified with a lower viscosity EPDM that was oil extended. In this case, improvements in impact strength were noted while also increasing the melt viscosity of the blend. Although no information was provided of the effect of these additions on the dispersed-phase particle size, one would expect that the particles would be smaller if the melt viscosity of the dispersed phase was lowered to be a better match with that of the matrix.

Another interesting study involved the increase in the viscosity of one component. In nylon-NBR blends, the addition of a dimethylol phenolic compound as a curative can yield improved properties.[44] Although this would be the expected result of vulcanizing the rubber,

TABLE 5
Properties of PP/EPDM/HDPE Blends

Test recipe	A	E	G	H
PP	100.0	70.0	70.0	70.0
EPDM	—	30.0	15.0	15.0
HDPE, 0.3 mfi	—	—	15.0	—
HDPE, 1.7 mfi	—	—	—	15.0
Melt Index (g/10 min)	10.4	5.6	6.4	7.6
Izod impact strength (J/m) at 22°C	13	208	80	75
Tensile strength at yield (MPa)	32.1	22.0	25.9	26.8
Elongation at yield (%)	17	15	12	15

the authors found that the effect remained even when gel content of the product was in the range of 50%. They hypothesize that "a portion of the dimethylol phenolic compound is believed to react with the nylon to give chain extension or a small amount of crosslinking. This increases the viscosity of the nylon to where it is more like that of the rubber and thus, mixing, homogenization, and rubber particle size reduction are greatly enhanced." They add that crosslinking between nylon and rubber to form a graft copolymer can also enhance the homogenization.

XII. MULTICOMPONENT COMPATIBILIZERS

The search for compatibilizers generally focuses on finding an effective single component. However, in some difficult-to-compatibilize blends, multicomponent compatibilizers may prove to be effective.

Choudury and Bhowmick[45] reported the successful use of multicomponent compatibilizers in a 70/30 blend of natural rubber with polyethylene. The objective was a TPE. The effects of various combinations of three reactive polymers on tensile properties were studied: maleic anhydride-modified polyethylene (PEM), epoxidized natural rubber (ENR), and sulfonated EPDM. The combination of ENR and PEM showed the greatest improvement although a combination of S-EPDM and PEM showed similar improvement.

A patent assigned to Dexter Corporation[46] reveals a complex "compatibilizing agent" for blends of polycarbonate and polyamide, an immiscible pair. In one claim, the "compatibilizing agent" comprises a polyetherimide, an impact-modifying agent (such as maleic anhydride-grafted ethylene-propylene rubber), and an alloying agent (such as acrylonitrile-butadiene-styrene terpolymer or styrene-maleic anhydride copolymer).

XIII. CONCLUSIONS

For many years, compatibilization has been a hidden art. Very little information was available about how compatibilizers were selected for specific blends. In the past ten years, compatibilization has come into the open. Research reports by many groups worldwide are now available to the research and development teams who are developing commercial polymer blends.

Compatibilization is now a science; principles that can guide the blend developer are now easily available. This chapter has covered many of the approaches that have proven to be applicable in specific blends. Application of the principles of thermodynamics as they concern polymer mixtures in conjunction with an understanding of the reactions and interactions of organic functional groups is an important part of selecting polymer blend compatibilizers.

REFERENCES

1. **Utracki, L.,** *Polymer Alloys and Blends: Thermodynamics and Rheology,* Hanser, Munich, 1989, 124.
2. **Heggs, R. P., Macarus, J. L., Markham, R. L., and Managaraj, D.,** *Plast. Eng.,* June, 29–32, 1988.
3. **Meijer, H. E. H., Lemstra, P. J., and Elemans, P. H. M.,** *Makromol. Chem.,* 16, 113–135, 1988.
4. **Datta, S. and Lohse, D. J.,** Presentation at American Chemical Society, Atlanta, April 16, 1991.
5. **Paul, D. R. and Barlow, J. W.,** *J. Macromol. Sci. Chem., Rev. Macromol.* C18(1), 109–168, 1980.
6. **Sperling, L. H.,** in *Multicomponent Polymer Materials,* Paul, D. R. and Sperling, L. H., Eds., Advances in Chemistry Ser. 211, American Chemical Society, Washington, 1986, chap. 2.
7. **Paul, D. R.,** in *Polymer Blends,* Vol. 2, Paul, D. R. and Newman, S., Eds., Academic Press, New York, 1980, 35.
8. **Xanthos, M.,** *Polym. Eng. Sci.,* 28(21), 1392–1400, 1988.
9. **Leibler, L.,** *Makromol. Chem., Rapid Commun.,* 2, 393, 1981.
10. **Trostyanskaya, E. B., Zemskov, M. B., and Mikhasenok, O. Y.,** *Plast. Massy,* 11, 28, 1983.
11. **Leibler, L.,** *Makromol. Chem., Macromol. Symp.,* 16, 1, 1988.
12. **Olabisi, O., Robeson, L. M., and Shaw, M. T.,** *Polymer-Polymer Miscibility,* Academic Press, New York, 1979.
13. **Mangaraj, D., Pfau, J. P., and Markham, R. L.,** *Identification of Major Developments in Polymer Blends/Alloys,* Vol. 3, Battelle, Columbus, OH, 1986.
14. **Seo, S. W., Jo, W. J., and Ha, W. S.,** *J. Appl. Polym. Sci.,* 29, 567–576, 1984.
15. **Ouhadi, T., Fayt, R., Jerome, R., and Teyssie, P.,** *J. Polym. Sci., Polym. Phys. Ed.,* 24, 973, 1986.
16. **Rubber World,** August 1987, 24–29.
17. **Fayt, R., Jerome, R., and Teyssie, P.,** *J. Polym. Sci., Polym. Phys. Ed.,* 20(12), 2209–17, 1982.
18. **tenBrinke, G., Karasz, F. E., and MacKnight, W. J.,** *Macromolecules,* 16(12), 1827–1832, 1983.
19. **Haijan, M., Sadrmohaghegh, C., and Scott, G.,** *Eur. Polym. J.,* 20(2), 135–138, 1984.
20. **Pillon, L. Z. and Utracki, L. A.,** *Polym. Eng. Sci.,* 24(17), 1300–1305, 1984.
21. **Xanthos, M., Young, M. W., and Biesenberger, J. A.,** *Polym. Eng. Sci.,* 30(6), 355–365, 1990.
22. **Porter, R. S., Jonza, J. M., Kimura, M., Desper, C. R., and George, E. R.,** *Polym. Eng. Sci.,* 29(1), 55–62, 1989.
23. **Coran, A. Y., Patel, R., and Williams-Headd, D.,** *Rubber Chem. Technol.,* 58(5), 1014–1023, 1985.
24. **Coran, A. Y. and Patel, R.,** *Rubber Chem. Technol.,* 56(5), 1045–1060, 1983.
25. **Gallagher, M. T.,** *Pro. 6 Int. Conf. Thermoplastic Elastomers,* Schotland Business Research Inc., Princeton, 1992, 315.
26. **Lohse, D. J., Datta, S., and Kresge, E. N.,** *Macromolecules,* 24, 561–566, 1991.
27. **Ting, S. P., Pearce, E. M., and Kwei, T. K.,** *J. Polym. Sci., Polym. Lett. Ed.,* 18, 201–209, 1980.
28. **Schlecht, M. F., Pearce, E. M., Kwei, T. K., and Cheung, W.,** *ACS Symp. Ser.,* 364 (Chem. React. Polym.), 300–311, 1988.
29. **Hammer, C. F. and Sinclair, H. K.,** (to E. I. duPont de Nemours and Co.), U.S. Patent 3,972,961, 1976.
30. **Epstein, B. N.,** (to E. I. duPont de Nemours and Co.), U.S. Patent 4,174,358, 1979.
31. **Hammer, C. F. and Sinclair, H. K.,** (to E. I. duPont de Nemours), U.S. Patent 4,017,557, 1977.
32. **Willis, J. M. and Favis, B. D.,** *Polym. Eng. Sci.,* 28(21), 1416–142, 1988.
33. **Bauman, B. D.,** Presentation at Int. Symp. Research Developments for Improving Solid Waste Management, Cincinnati, February 7, 1991.
34. **Bauman, B. D., Burdick, P. E., and Mehta, R. K.,** U.S. Patent 4,833,205, 1989.
35. **Bauman, B. D.,** *Proc. SPI 32nd Technical/Marketing Conf. Polyurethanes,* SPI, 1989, 638.
36. **Inoue, T., Shomura, F., Ougizawa, T., and Miyasaka, K.,** Presentation at "Frontiers in Rubber Science", ACS Rubber Div. Symp., Los Angeles, 1985.
37. **Gessler, A. M. and Haslett, W. H.,** U.S. Patent 3,037,954, 1962.
38. **Coran, A.,** *Poly. Proc. Eng.,* 5(3 & 4), 317–326, 1987–88.
39. **Fayt, R., Jerome, R., and Teyssie, P.,** NRCC/IMRI Symp. "Polyblends '86", Montreal, 1986.
40. **Utracki, L.,** *Polymer Alloys and Blends: Thermodynamics and Rheology,* Hanser, Munich 1989, 18.
41. **Patfoort, G. A.,** *Plastica,* 3(22), 95, 1969.
42. *Plast. Machin. Equip.,* February, 52, 1982; June, 53, 1981.
43. **Lindsay, G. A., et al.,** *Adv. Chem. Ser.,* 176, 367, 1979.
44. **Coran, A. Y. and Patel, R.,** *Rubber Chem. Technol.,* 53(4), 781–794, 1980.
45. **Choudury, N. R. and Bhowmick, A. K.,** *J. Appl. Polym. Sci.,* 38, 1091–1109, 1989.
46. **Perron, P. J. and Bourbonais, E. A.,** (to Dexter Corp.), U.S. Patent 4,782,114, 1988.

Chapter 21

RUBBER MIXING PRINCIPLES

Nicholas P. Cheremisinoff

TABLE OF CONTENTS

0-8493-4401-8/93/$0.00 + $.50

I. INTRODUCTION TO RUBBER MIXING

Mixing practices for polymeric materials, such as elastomers, resins, and composite polymer blends, involve factors different from those involved in the mixing of Newtonian fluids. A rotating propeller immersed in a pastelike or rubberlike material affects only a small portion of the total mass, leaving the greater part of it relatively undisturbed. Semisolids, viscous liquids, and heavy pastes require mixing devices that are capable of affecting every portion of the material and achieving a homogeneous state by pushing one portion of the mass into another portion. Factors requiring special attention are the physical preparation of the materials, the behavior of the material under flow, and the effect of the order in which the ingredients are introduced into the mix. For highly viscous materials, such as elastomers, equipment is selected on the basis of its ability to shear the material at low speed or to wipe, smear, fold, stretch, or compress the mass being processed.

Of primary importance in the selection of a particular type of mixer for processing non-Newtonian materials is the behavior of the material while it is undergoing flow or shear. Many non-Newtonian materials exhibit different resistances to shear at different shear rates. Some of these materials have a greater resistance to shear, or deformability at high shear rates; others show a greater resistance at low shear rates. The behavior of non-Newtonian materials in regard to these properties determines the type of equipment used for mixing and the amount of power that must be used to achieve a desired degree of mixing.

In this Section we review the principles and practical aspects of conventional mixing practices for elastomeric materials. The discussion is largely limited to compounding and blending operations for rubbers; however, the more fundamental concepts relating energy input to mixing effectiveness are applicable to all types of polymer mixing applications. For detailed discussion of mixing practices for plactics blending and dispersing pastelike materials, the reader is referred to the text by Cheremisinoff (1987).

II. GENERAL CONCEPTS OF RUBBER MIXING

Irrespective of the individual ingredients comprising a non-Newtonian material (i.e., liquid, solid particles in suspension, or elastomers), the objective of mixing is to decrease the concentration gradient to a desired minimum. In some cases this is accompanied by a decrease in the thermal gradient in processes involving heat transfer.

The minimum gradient of concentration (or temperature) is achieved when the infinitesimal samples of the mixture randomly extracted from any location within the mixing system have the same composition. In some processes mixing is accompanied by dispersion, that is, by phenomena presenting a change in the physical characteristics of the components (e.g., intrusion of solid fillers into a resin mass or decrease in the particle size of a filler). In addition, mechanical mixing may be accompanied by physicochemical processes of plasticization.

The mixing methods and the design of mixing machines depend on the physical properties of components to be mixed. The mixing of non-Newtonian materials is basically carried out by different mechanical devices that both combine ingredients and provide intensive agitation. Common devices used are Sigma-Blade mixers, Banbury mixers, roller mills, anchor agitators, centrifugal disc mixers, and others.

Heavy pastes and materials with pronounced plastic properties require heavy machinery to achieve satisfactory mixing. Natural flow of materials and momentum effects are negligible, and because of this the mixing must be forced. Sections of the mix must be separated from the whole mass and then forced back into a different part of the mixture volume by means of shearing, kneading, or wiping actions. Highly viscous liquids and pastes can be mixed in either batch or continuous operations.

A batch mixing process as a rule is performed in a closed volume of a mixing device. The principal parameter affecting mixing efficiency in this case is the minimum mixing time necessary for providing the desired homogeneity of the mixture. The capacity of the mixing device depends on the mixing time, the times of loading ingredients, and starting, stopping, and unloading. Mixing time is established not only by the available volume of the mixing device, the design of the working elements, the rotational velocity, and the properties of the components to be mixed, but also to a large degree by the initial orientation of components and the order in which they are introduced.

A rationally designed mixer must provide an equal mixing time independent of the initial location of components. Periodic reorientation of material during mixing decreases the basic mixing time. In general, batch mixers are not very efficient at handling large capacities, mainly because of difficulties encountered when attempting to automate the operation. Over the course of a batch cycle, the properties of the mixture may change considerably. During mixing and plastification of polymer materials the power consumption as a rule decreases. This often requires the installation of a motor drive of high capacity, which must be selected in accordance with the initial characteristics of the mixture. However, there is a need to use batch mixers in many cases, especially if the required mixing times are too long.

Continuous mixing operations involve the continuous loading and unloading of components. In this case there is a drastic variation in the concentration of components at the inlet. When properly performed, mixing decreases these variations at the exit to the desired minimum. Continuous operations have the advantage of providing a stable process. Also, the power consumption is both lower and more constant than in batch operations. The parameters of the mixture as it moves from inlet to outlet vary even though the volume of the mixer remains constant. Despite the advantages, continuous mixing operations are difficult to justify in many cases because of the long mixing time requirements. Long mixing times are somewhat compensated for in continuous mixers by increasing the process intensity (i.e., operating at higher rotational velocity). This may, however, result in uneconomical increases in the dissipation of mechanical energy and thermal destruction.

III. BLADE MIXERS

Blade mixers are used to promote kneading and mixing action accompanied by the heating or cooling of different deformable or plastic solids and high consistency pastes. Their operation involves compressing the fluid mass flat, folding it over on itself, and then compressing it again. The material is usually torn apart, and high shear is produced between the moving and stationary fluid elements. The mixing is usually performed by two Σ-shaped heavy blades rotating in opposite directions at different speeds on parallel horizontal shafts.

Typical blade designs are illustrated in Figure 1. The top sketch shows a sigma blade employed for general-purpose kneading. There is great variation in commercially available blade designs, ranging from lightweight to heavyweight construction. Selection of the specific design depends on the consistency of the mix. The size of the mixer is limited primarily by considerations of power input, weight, speed of mixing, materials of construction, and methods of mix discharge. Many machines are equipped with a tilting mechanism that facilitates the discharge of the batch while the blades are turning. For special mixing requirements, the sigma blade mixer may incorporate special construction features, such as jackets for heating or cooling the troughs, dust-tight or vacuum-tight lids, and linings of special alloys for mixing corrosive or abrasive materials. Capacities of the equipment for handling heavy-consistency material range from laboratory-sized batches to several hundred gallons, and capacities for lighter consistency materials are much larger. In general, these mixers require very heavy drive mechanisms and motors.

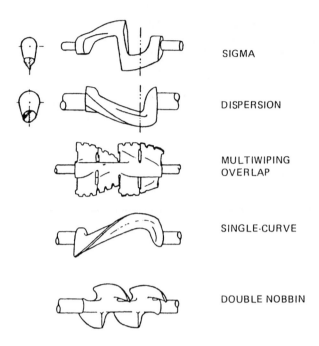

FIGURE 1. Agitator blade designs for double-arm kneaders.

The agitator blades in the sigma blade mixer may be mounted so that their paths are either tangential or overlapping. In the tangential mounting arrangement the two blades rotate side by side with their circular paths of rotation not quite touching. In the overlapping mounting, the paths of rotation overlap and consequently the blades must be designed and the speed of rotation adjusted so that the blades always clear each other. In the tangential type the blades can have any relative speed because their paths of rotation do not overlap. Many modifications may be made in the shape of blades to meet specific mixing requirements. Many manufactures provide double-arm mixers with interchangeable blades.

Examples of materials handled by double-arm kneaders of the sigma blade type are resins, putty, adhesives, baker's dough, and cellulose additives. Sigma type agitator blades are preferred for general-purpose use.

The single-curve (or spiral agitator) blade, shown by the bottom sketch in Figure 1 provides the same smooth blending and folding action as the dispersion blade design (also shown in Figure 1) but has been designed for larger overlapping mixers. The dispersion agitator blades provide excellent mixing for fiber-reinforced products. They do not have the same heavy shearing action as other types, and therefore produce a smooth homogeneous mix with no fiber breakdown. Commercial scale units provide direct drive to the front agitator shaft through a gear reducer. A master hydraulic system operates both a compression ram cover and a bowl tilting mechanism.

The double-nobbin agitator blade produces the extra shearing action required for certain materials. This agitator design resists blade distortion from stresses in the mixing load. Commercial designs are available in both overlapping and tangential action mixers. These units are most often used for heavy plastic materials.

IV. BANBURY MIXER OPERATION

There are a variety of kneading machines in which the trough or container is open so that the contents may be unloaded by tilting. Some designs have an enclosed mixing chamber,

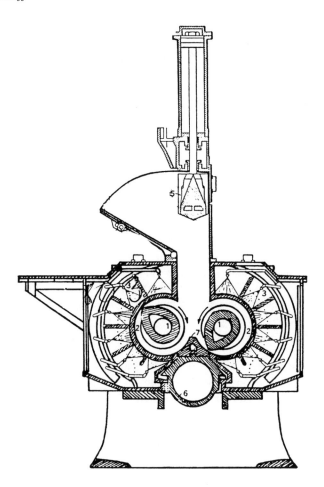

FIGURE 2. Cutaway view of Banbury mixer: (1) rotors or agitators; (2) mixing chamber; (3) cooling sprays; (4) feed hopper; (5) floating weight; (6) sliding discharge; and (7) saddle discharge opening.

however, in which case the units are known as internal mixers. This type of mixer is often used for dissolving rubber and for dispersing rubber in liquids. The Banbury mixer illustrated in Figure 2 is a typical example. It is a heavy-duty double-arm mixer in which the agitators are in the form of interrupted spirals.

The Banbury mixer resembles a double-arm kneader in that it has two adjoining cylindrical sections that meet tangentially. In the Banbury design, however, the rotating agitators are much more rugged in construction and are able to apply considerably more power per gallon of working volume. As shown in Figure 2, two rotors (1) turn slowly toward each other at slightly different speeds. Each rotor has a blade that extends along the length of the rotor, roughly in the form of a spiral. Each rotor is cored to permit cooling or heating by the passage of cooling water or an appropriate heating agent. The mixing chamber (2) may also be cooled or heated by means of sprays (3). The materials to be mixed are fed into the mixing chamber through the feed hopper (4). A floating weight (5) is operated by compressed air and rests on top of the feed, serving to confine the material to the mixing space and to exert pressure on it. The mixed batch is discharged by a power-operated sliding discharge mechanism (6). The saddle (7) between the two rotors, which is attached to the sliding mechanism, slides out from under the rotors, leaving an opening through which the mix discharges.

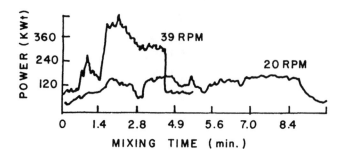

FIGURE 3. Typical power consumption cycle for batch mixing.

Because of the high degree of agitation a Banbury mixer can complete a mix in a relatively short time. It is capable of reducing to a matter of minutes mixing operations requiring hours in other types of mixers. Examples of applications include the mixing of wood flour and linseed oil in linoleum manufacture and the incorporation of fillers, softeners, and accelerators into rubber stock. The mixing of viscoelastic polymeric materials, such as rubber with different ingredients is considered a mechanical process in which, under different shear deformations, the dispersion of ingredients takes place. In addition to the mechanical processes of mixing, physicochemical processes take place between the material, fillers, and other ingredients.

The mixing process is normally carried out under laminar conditions of material flow created by the rotating parts of the machine. During this action the non-Newtonian material is subjected to different stresses and strains. Selection of the mixing regime depends on several parameters, such as the amounts of materials loaded, the order of loading times, the time of a mix, the rotational speed of the blades, and the pressure in the chamber. By varying these parameters the process may be considerably intensified. However, the basic source of intensification is an increase in blade velocity and an increase in the pressure in the mixing chamber. An increase in the rotational velocity of the blades increases the shear velocity, which leads to more frequent dividing of the mass and renewal of its surface. This promotes rapid distribution of ingredients in the mass. An increase in rotational velocity leads to temperature rises in the mixture as well as an increase in the pressure exerted against the walls of the chamber.

The largest amounts of mixing energy are consumed in applications involving viscous components and when filler materials are added in the form of powders to viscous solutions. The energy consumption in these cases is variable. It depends on various technological factors as well as the physicochemical properties of components. Figure 3 shows a typical time plot of mixer power consumption. The area limited by the power consumption curve and the x axis is equal to the energy consumption per mixing cycle. A plot like that in Figure 3 provides evaluation of the intensity of the process as well as an indication of the distribution of the ingredients in the paste. The energy consumption for a given amount of mixture of specified composition and temperature is constant and independent of the rotational velocity of the blades.

For Newtonian fluids a good approximation is to assume that the drive horsepower is proportional to the rotational velocity. For non-Newtonian materials, however, this is not a valid assumption because nonlinear relationships exist between shear stress and shear rate. Furthermore, at increased speeds the rate of share deformation also increases, which promotes temperature rises due to energy dissipation. Heat transfer promoting mixture cooling is generally poor; thus an increase in the temperature of the mixture results in lowering of the fluids' viscosity. This in turn decreases the energy consumption of mixing. It should be noted that the power of the drive and the energy consumption depend on the order of loading mixture components. For material loading in succession, a decrease in the time intervals between loadings of discrete portions demands power increases due to shortening of the mixing cycle. The maximum power of the drive occurs when all materials are loaded simultaneously. Pressure increases over the mixture influence the energy consumption when the chamber is totally filled. This is especially true when loading hard mixtures. At the same time, energy consumption increases, but not in proportion to pressure increases. Pressure rises cause temperature increases in the mixture, resulting in a delay of power consumption. During this action the specific energy consumption does not undergo significant changes but, rather, gradually decreases with mixing time.

V. RUBBER COMPOUNDING

Carbon black is widely used as a reinforcing filler; however, the unique property of improving tear and abrasion resistance in rubber vulcanizates, particularly in synthetic rubber, is not fully understood. It is likely that carbon black surface and rubber interact both chemically and by physical adsorption. The parameters important to this interaction are capacity, intensity, and geometry. The total interface between polymer and filler is expressed as square meters per cubic centimeter of vulcanizate or compound. The intensity of interaction is determined by the specific surface activity per unit area. It should be noted that adsorptive energies vary greatly in different locations on the black surface, and it is likely that this distribution is responsible for the variability in properties among different types of carbon blacks. The geometric properties are characterized by the structure, which is basically anisometric, and particle shape and porosity. Greater anisometry results in looser particle packing. The void volume is used as a measure of the packing density.

Carbon black particles form irregular structures that tend to break down during intensive mixing. It is therefore a combination of carbon black reactivity and rubber and black chemical, physical, and rheological properties, as well as the conditions of mixing, that establishes the final strength properties of the vulcanizate. As a rule of thumb, high-structure blacks impart a higher modulus (at 300% elongation) than a corresponding normal black. The high modulus is determined by both the anisometry and surface activity. The separate influence of these properties can be observed by heat treating these blacks at 2,000–3,000°C. The crystallographic structure of carbon black is altered by heat treatment, and the properties approach those of graphite. Through recrystallization, highly active sites on the carbon black surface lose their high activity. In this situation the entire surface area becomes homogeneous and adsorption energies approach their lowest state.

Microscopic examination of thin sections of vulcanizate in which carbon black or other fillers and polymer were mixed for a short time reveal only that this additive exists as coarse agglomerates in an almost pure rubber matrix, without almost any colloidal dispersion of the additive. Only after mixing has been continued at greater intensity and/or for longer time periods can one observe how these agglomerates gradually disappear, making room for an increasing amount of additive. In all cases the initial product of mixing is comprised of an agglomerate formed by the penetration of rubber in the voids between carbon black particles or other additives, such as mineral fillers (e.g., clays), under the pressure and shear that build up during Banbury mixing and on a roll mill. Once these voids are filled, the additive is incorporated but not yet dispersed. As soon as the agglomerates are formed, continuous mixing exposes them to shearing forces that break them up and eventually disperse them. In other words, both agglomerate formation and breakdown take place almost simultaneously.

The composition of the carbon black agglomerates is established by the void volume per gram or per cubic centimeter carbon black (or filler). This is the volume determined during an oil adsorption experiment. This value is determinant for the carbon black concentration in the primary agglomerates.

Carbon blacks that are compatible in size and total surface area, but vary in structure and, hence, void volume, require different mixing times to achieve the same level of dispersion. The reason for this is that since during mixing the voids between particles are filled, a greater void volume requires a longer time to be filled. Also, for most commercial blacks, increasing structure is associated with increasing surface activity. Lower adsorption activity in graphitized blacks promotes slippage of rubber molecules on the particle surface when filling interstices during black incorporation. This shortens the time needed for this process compared to the original blacks. Easy incorporation, however, does not necessarily ensure good dispersibility. After a black has been completely incorporated, additional dispersibility depends on structure and adsorption activity. As a general rule, high structure blacks tend to require more incorporation energy; however, once they are incorporated, less energy for dispersion is needed compared to that needed for lower structure blacks.

An important consideration during the mixing and dispersion of carbon black and rubber is the degree of interaction between these two species, which can lead to the formation of bound rubber. These bonds are of both a physical nature (i.e., adsorption onto highly active sites) and a chemical nature. The latter is described as a reaction of free rubber radicals with reactive carbon black surface groups, promoted by high shearing energy. A parallel relationship exists between the formation of bound rubber and shearing forces in a mixer. It is observed that bound rubber can be formed almost immediately during mixing. The maxims observed in torque during mixing typically signify that degrading forces exist. It is not clear to what extent bound rubber influences vulcanizate properties. Usually with crosslinking in bound rubber, gel tends to be low. The formation of bound rubber is accompanied by an increase in viscosity. During bound rubber formation, shearing forces at constant rotor speeds increase because of this rise in viscosity. These shearing forces affect both further dispersion of the carbon black and breakdown of the rubber. Vulcanizate properties eventually degrade after continued mixing because of a breakdown of the polymer.

The effect of fillers on cross-link density in vulcanizates can be characterized by either modulus or swelling measurements. A common problem in selecting the proper black is the difficulty of knowing whether the black interferes with the vulcanization, causing higher cross-link density. It may also lower the degree of swelling for physical reasons while the cross-link density in the rubber compound remains unchanged. A distinction can be made between adhering and nonadhering filler depending on its degree of interaction with the polymer in the vulcanizate state. Adhering fillers (most carbon blacks) tend to reduce the swelling of the vulcanizate in a manner that is proportional to the filler content. In contrast, nonadhering fillers (e.g., $CaCO_3$ or glass beads) tend to increase swell through the formation of voids surrounding the filler particles.

The term "reinforcement" usually refers to the improved tensile strength achieved by the addition of carbon black to noncrystallizing elastomers. In the case of styrene-butadiene copolymer (SBR), tensile strength can be improved by a factor of 10 or 12 over the pure gum vulcanizate through the addition of only 50 parts reinforcing carbon black. The magnitude of this enhancement is somewhat unique and typifies the special nature of the forces existing between blacks and elastomers. Consider a noncrystallizing, unfilled elastomer. During elongation of the unfilled vulcanizate, the most highly elongated polymer chains reach their maximum elongation first and then break. When this occurs, these chains no longer carry tension and hence no longer contribute to tensile strength. As elongation continues, the next most highly elongated chains break. This process continues until eventually only a few chains contribute to the final tensile strength. It can therefore be concluded that those materials displaying low tensile strength experience an unequal distribution of loading. This heterogeneous distribution can be improved by wetting or slippage. For example, if the most highly elongated molecular chains do not break down and retain their part in the elongation, a more uniform stress distribution results and the final material is stronger. The

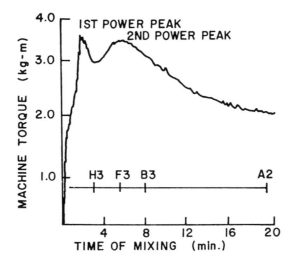

FIGURE 4. Torque-response curve for rubber mixing.

mechanism of mobile adsorption can contribute to the homogeneity of stresses. Elastomer chains that are adsorbed onto the carbon black surface remain in two-dimensional motion on the surface without being separated from it. At excessive elongation, they tend to slip along the surface. In this case the amount of energy per chain segment required for the process is lower than that needed for complete separation from the surface. Adsorptive energies are additive, and hence they can attain considerable values if a large number of successive segments are adsorbed. At the same time a chain slips along the surface during elongation, distributed segments line up sequentially on the surface, building up tension that balances the elastic forces at a given elongation. The greatest contribution is derived from the points of high adsorptive activity.

The phenomenon just described parallels that of crystallization, which occurs naturally in rubbers. Pure gum vulcanizates of crystallizing elastomers display greater tensile strength than elastomers that are noncrystallizing (amorphous). As such, crystalline polymers show only slight reinforcement through carbon black incorporation.

Carbon black vulcanizates display a permanent set after elongation. In contrast, pure gum vulcanizates show little if any permanent set. This permanent set tends to increase with increasing filler loading.

As noted earlier, incorporation refers to the wetting of carbon black with rubber. During this operation, entrapped air is squeezed out between the voids of the rubber and carbon black particles. Early stages of mixing reveal that as the carbon black becomes incorporated, relatively large agglomerates (of the order of 10–100 μm) form. The time required for full carbon black incorporation can be determined by measuring the time required to reach the second power peak during a mixing cycle. One can perform mixing studies in a Banbury mixer, observing from microscopic examination of rubber-carbon black compounds the progression of black dispersion at different times. The rates of carbon black dispersion in such a study can be computed from the maximum torque data. A typical power curve is shown in Figure 4 using an oil-extended EPDM. In this example the rubber was first masticated for about 2 minutes and then the rotors were stopped. Carbon black was charged in the chute, the mixer ram inserted, and the rotors started again. Mixing times were measured from the instant when the rotors were restarted. The carbon black incorporation time was taken to be the time required to attain the second power peak shown in Figure 4.

Bound rubber in such studies can be measured by standard solvent extraction techniques. For example, toluene at room temperature or boiling heptane or hexane extractions can be

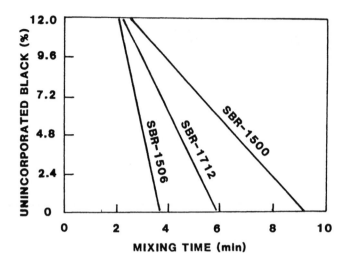

FIGURE 5. Correlation of unicorporated black with mixing time.

used with a known aliquot of rubber sample. From measurement of the residue weight, one can calculate the percentage of insoluble polymer.

The percentage (by volume) of unincorporated (nonwetted) carbon black can be estimated from density measurements. The volume of air in the batch V at time t can be computed from the difference in densities ρ' and ρ at time t and the time at the termination of mixing when carbon black dispersion is better. The formula used for this calculation is

$$V = \frac{B}{\rho'} - \frac{B}{\rho} \qquad (1)$$

where B is the batch weight (phr).

The first power peak corresponds to ingestion of the batch into the mixing chamber. This coincides with the instant that the ram in the loading chute reaches the bottom of its flight and thus removes any additional hydrostatic pressure from the mixing chamber. It can be shown that the fraction of undispersed carbon black decreases linearly with the time of mixing at the point at which the second power peak occurs. At shorter mixing times the compound can appear inhomogeneous and crumbly. Figure 5 illustrates some typical correlations for SBR. The mixing times can be observed to be inversely proportional to the rotor speed. When normalized, data tend to collapse to a single linear correlation. Hence, there is a strong correlation between incorporation time and the rate of incorporation as computed from the slope of regression lines for a change in densities with time from carbon black compounds in a single polymer.

The existence of a double power peak depends in part on the properties of the carbon black itself. Using pelletized carbon black results in the absence of the first power peak because no additional hydrostatic pressure is applied when the mixer chute is lowered (hence the measured torque is no longer affected). The time to reach the second power peak, however, remains the same. It has been noted that with fluffy carbon black, the time to reach the second power peak increases if no pressure is applied. Large agglomerates of carbon black form during the initial mixing stages regardless of the type of carbon black used.

One may also find good correlation between the measured black incorporation time (BIT) and weight-average molecular weight M_w of polymers. Incorporation times increase sharply with increasing polymer molecular weight. This effect tends to mask expected decreases in incorporation times when the Mooney viscosity is decreased through the addition of oil.

Decreasing concentrations of carbon black and increasing oil loading tend to reduce the incorporation time. This is generally thought to be related to a lowering of the polymer viscosity.

Typical filler-reinforced elastomers contain curatives, plasticizers, and stabilizers in the rubber filler. After mixing and curing a masterbatch, both physical and chemical cross-linking processes transform the system into a network structure. The structure is composed essentially of two networks: a stationary and a transient network. The stationary network is comprised of chemical cross-links that are formed by both curatives and fillers. This network also includes permanent crystalline structures from crystallizable polymers. In contrast, the transient network is comprised of trapped entanglements and transient order structures. There are two mechanisms that explain the reinforcing effects of filler in elastomers. The first of these mechanisms can be described as a hydrodynamic interaction in which the filler particles are responsible for the reinforcement. The second mechanism is that of strong chemical interactions between the filler and the matrix. Both these reinforcing interactions contribute to the mechanical properties of filler-reinforced polymers.

The mechanical properties of engineering importance are the tensile modulus, the tensile strength, and the ultimate stretch ratio. The tensile modulus is mathematically defined as the slope of the stress-strain curve at zero strain for a given masterbatch. Tensile strength is defined as the force at break per unit area of the original sample. The ultimate stretch ratio is the uniaxially fractured gage length divided by the original gage length of the filler-reinforced polymer specimen.

A problem in designing polymer structures for a particular reinforced elastomeric application is a lack of quantitative physical relationships that define the reinforcing effects of fillers on mechanical properties. Such constitutive expressions should ideally account both for the effects of fillers on the detailed structure and for the degree of mixing or dispersion of the masterbatch.

A simple relationship for the elastic modulus of filled rubber, in terms of the modulus of the gum E_1 is

$$E = E_1(1 + 2.5\phi) \qquad (2)$$

where ϕ is the volume fraction of filler. This expression is in fact the universal viscosity relation of Enstein. The relation accurately describes the modulus of filled rubber systems for dilute filler concentrations (up to 1%). A more accurate expression based on a polynomial relation is

$$E = E_1(1 + 2.5\phi + 14.1\phi^2) \qquad (3)$$

This relation accounts for the increase in stiffness caused by spherical particles.

For particles that tend to agglomerate into chainlike clusters, a factor that accounts for the asymmetric nature of the cluster must be introduced:

$$E = E_1(1 + 0.67f\phi + 1.62f^2\phi^2) \qquad (4)$$

where f is the geometric factor defined simply as the ratio of cluster length to width.

From the concepts of occluded volume and effective occluded volume a slightly more complex relation can be developed:

$$E + E_1(1 + 2.5\phi'_{aggr} + 14.1\phi'^2_{aggr}) \qquad (5)$$

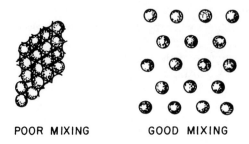

POOR MIXING   GOOD MIXING

FIGURE 6. Hard sphere model showing good and poor dispersion.

where ϕ'_{aggr} is the effective volume fraction of the aggregates.

In the preceding expressions, the values of f and ϕ'_{aggr} can be deduced from a polynomial regression of the modulus data. The literature provides some working correlations for projecting compounding properties. For heat-treated and thermal blacks $1.3 < \phi'_{aggr}/\phi < 2.4$. For the ultimate stretch ratio of filled polymers λ_b, a one-third power-law relation (i.e., $\lambda_b \simeq \phi^{1/3}$) is applicable for some filled plastics.

From thermodynamics, the entropy of mixing defines the tensile modulus and the ultimate stretch ratio. For a given elastomer composition, different mixing histories give rise to different states of entropy. Therefore, a logical basis for defining tensile properties of filled reinforced elastomers is in terms of thermodynamic mixing rules.

As noted earlier, conventional mixing practices of fillers into different rubbers result in the formation of filler-rubber structures (e.g., bound rubber for carbon black-filled elastomers). The specific filler-rubber structures formed affect the thermodynamic state of the elastomer system. This means that the reinforcement properties of the filler change because of the formation of these rubber-filled structures. The entropy of mixing for an ideal system can be computed from the Flory-Huggins equation:

$$\Delta S'_m = -k_B(N, \ln \phi_1 + N_2 \ln \phi_2) \tag{6}$$

where:

k_B = Boltzmann constant
ϕ_1 = volume fraction of gum
ϕ_2 = volume fraction of filler
N_1, N_2 = numbers of effective polymer chain segments and effective filler particles, respectively

The entropy of mixing for a real system is

$$\Delta S^R_m = -k_B (N_1 \ln \phi_1 + N_2 \ln \phi'_2) \tag{7}$$

where ϕ'_2 = volume fraction of filler actually wetted by the polymer matrix.

Figure 6 illustrates schematically the difference between ideal and real mixing in terms of a hard sphere model. For real mixing not all the fillers are incorporated into the polymer-filler wetting process because of the formation of filler agglomerates. The excess entropy due to the heterogeneous distribution of the ith-type filler in the rubber matrix after mixing is

$$\overline{\Delta S}_{p-i} = \overline{\Delta S^R_m} - \overline{\Delta S'_m} \qquad i = 2, 3, \ldots \tag{8}$$

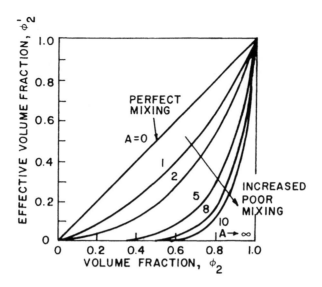

FIGURE 7. Relationship between mixing parameters.

And combining expressions,

$$\overline{\Delta S}_{p-2} = \frac{\Delta S_{p-2}}{N_2} = -k_B \ln \frac{\phi_2'}{\phi_2} \tag{9}$$

For an ith component system, $\phi_i' = \phi_i \, \Delta S_{p-i} = 0$, and hence

$$\overline{\Delta S}_{p-i} = k_B \ln \left(\frac{\phi_i'}{\phi_i} \right) \tag{10}$$

Equation 10 defines the case of perfect mixing.

Using the effective volume fraction of the filler, a semiempirical mixing rule is proposed to describe the relative tensile modulus of filled elastomers:

$$\ln E^{\text{rel}} = \ln \left(\frac{E}{E_1} \right) = \phi' \ln K \tag{11}$$

where:

E = modulus of filled elastomer
E_1 = modulus of gum
ϕ' = effective volume fraction of filler
$\ln k$ = reinforcement effectiveness factor

Equation 11 can also be written as

$$E^{\text{rel}} = \exp\{(\ln K)\phi \exp[A(\phi - 1)]\} \tag{12}$$

where A is an index defining the degree of mixing.

The parameters $\ln K$ and A can be evaluated by a nonlinear regression of E^{rel} and ϕ data. Figure 7 shows the relationship between the effective volume fraction ϕ_2' and the actual volume fraction ϕ_2 of the filler computed for different mixing indices A and zero strain. Equation 9 provides a good fit of the modulus data for different rubbers and carbon blacks. Different rubbers evaluated in this study were chlorosulfonated polyethylene (CPSE), ethylene-propylene-diene rubbers, and polychloroprene (CR).

The effective volume fraction of aggregate ϕ'_{aggr} can be correlated with the effective volume fraction of filler:

$$\frac{\phi'_{aggr}}{\phi} = \left(\frac{\ln K}{2.5}\right) exp[A(\phi - 1)] \tag{13}$$

Using tensile modulus data, the relative mixing index for different elastomers can be determined (e.g., through A). This is a useful approach to assessing the degree of mixing of carbon black in different polymers and hence is practical for the end user in developing compound formulations for different applications, as well as the polymer technologist in developing elastomeric products for customers. The mixing index A, based on tensile modulus data, is found to correlate qualitatively with morphological studies using SEM (scanning electron microscopy). Hence the effective volume fraction of filler ϕ' can be used for characterizing the nonideal mixing behavior of filled elastomers.

Increasing the concentration of carbon black results in a decrease in the ultimate stretch ratio of the compound stock. Also, laboratory tests show that the ultimate stretch ratio of a good mixing system is higher than that of a poorer mixing elatomer system. The reason for this is that the poor degree of mixing of the filler produces weak spots that lead to premature failure. A relationship for the ultimate stretch ratio of filled elastomers based on a power-law expression is

$$\frac{\lambda_b}{\lambda_{b_1}} = \lambda_b^{rel} = (1 - \phi')^{J/3} \tag{14}$$

where:

λ_b = ultimate stretch ratio of filled elastomer
λ_{b_1} = ultimate stretch ratio of gum
J = fracture coefficient
ϕ' = effective volume fraction of filler

The following expression for the relative percentage ultimate strain can be used:

$$\frac{\epsilon_b}{\epsilon_{b_1}} = (1 - \phi^{1/3}) \tag{15}$$

where:

ϵ_b = percentage ultimate strain of filled elastomer
    [i.e., $\epsilon_b = (\lambda_b - 1)100$]
ϵ_{b_1} = percentage ultimate strain of the gum
ϕ = volume fraction of filler

The problem with this correlation is that it does not account for the effect of the degree of mixing on the ultimate stretch ratios.

A power-law relationship between the mixing index A and the fracture coefficient J for conventionally mixed CSPE, CR, and EPDM is

$$J = 3.89 A^{2.87} \qquad \text{for } A \geq 1 \tag{16}$$

This correlation can be combined with Equation 14 to give

$$\frac{\lambda_b}{\lambda_{b_1}} = \{1 - \phi \exp[A(\phi - 1)]\}^{1.30 A^{2.87}} \tag{17}$$

This correlation is applicable for the range $1 \leq A \leq 3$ (i.e., reasonably well-mixed elastomer systems).

VI. ROLL MIXING

Mixing rolls subject pastes and deformable solids to intense shear by passing them between smooth or corrugated metal rolls that revolve at different speeds. These machines are widely used in the rubber and plastics industries. The principal design consists of two horizontal rolls or cylinders arranged side by side and rotating toward each other at different speeds. The ratio of the peripheral speeds of the rolls, known as the friction ratio, ranges from 1 to 2 but is usually around 1.2. The higher friction ratios lead to greater heat generation. Friction, speed, and the size of the rolls influence the cooling of the material mass and the intensity of its treatment. As such, data obtained from a unit are specific to that machine and cannot be applied to a different machine.

Figure 8 illustrates the basic mixing operation. The material enters the mixing rolls in the form of lumps, powder, or friable laminated materials. As a result of rotation, adhesion, and friction, the material is entrained into the gap between the rolls, and upon discharge it sticks to one of the rolls, depending on their temperature difference and velocities. The rolls are temperature controlled. The rolling process is also influenced by the gap between the rolls. In batch mixing rolls the mass, after loading, passes through the gap between the rolls several times and the mixing action is due to the different speeds of the rolls (Figure 8). In continuous mixing rolls the mass enters continuously from one side of the machine and passes between the rolls in rotational and forward motion along the unit's axis. The mix is continuously discharged in narrow strips (refer to Figure 8). Both the shearing action and the entrainment of material into the gap are very important to the mixing process and in transporting the material through the unit.

Shearing strain is required for initial mixing of regular or isolated systems consisting of two components. The objective of this strain is to mix the system as a determined statistical variation of any of the properties of a series of samples to a minimum. According to mixing theory, three basic principles may be formulated.

1. The total component surface contact area should increase during mixing.
2. The elements of surface contact area should be evenly distributed in the mixing mass.
3. Mixture components must be distributed such that for any element of volume the relative content of components is the same as in the total system.

Parametric calculations for mixing rolls make it possible to estimate the basic parameters of the machine: its productivity and the thrust forces between rolls, torque, and horsepower. Calculation of these parameters marks the starting point of the structural design of mixing rolls. The questions of time and quality of mixing and the required number of passes of the mixture through the rolls, as well as other mixing characteristics, must be established from industrial and laboratory experiment.

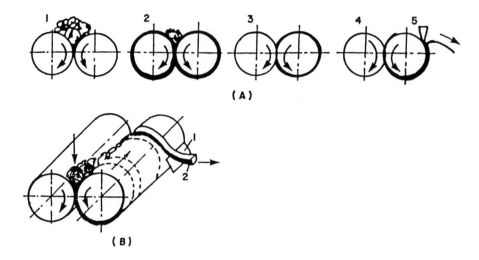

FIGURE 8. Operation of mixing rolls: (A) Batch operation: (1) loading; (2) rolling; (3) end of rolling; (4) mass shear; and (5) knife. (B) Continuous operation: (1) knife and (2) continuous discharge.

The issues outlined following are related to the deformation and flow of polymeric non-Newtonian materials between the gap of the mixer rolls. Examination of the flow behavior in this region provides insight into computing the basic parametric values that characterize the operation of mixing rolls. Figure 9 shows the basic scheme of the rolling process. The rolls act on the material by compressing and pushing it along the machine's horizontal axis. We define the rolls as having radius R, as being located at a distance $2h_0$ from each other, and as rotating at velocity v. The material wedge thickness at the gap entrance is $2h_2$ and at discharge, $2h_1$. The streamline equation can be obtained by integration of the following stream function:

$$\psi = \int v_x \, dy \tag{18}$$

where:

ψ = streamline (constant)
v_x = velocity component in the x-axis direction
v_y = velocity component in the y-axis direction determined from the continuity
     equation

$$\frac{2v_x}{2x} + \frac{2v_y}{2y} = 0 \tag{19}$$

Defining the following dimensionless parameters:

$$\varphi = \frac{x}{\sqrt{2Rh_0}} = \pm \sqrt{\frac{h}{h_0} - 1}$$

and   (20)

$$\eta = \frac{y}{\sqrt{2Rh_0}}$$

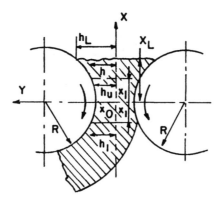

FIGURE 9. Analysis of components of roll mixing.

The velocity in the *x*-axis direction is

$$v_x = \frac{3v}{2} \left\{ \left[\frac{\rho^2 - \rho_1^2}{(1 + \rho^2)^3} \right] \frac{4h^2}{C^2} + \frac{2 - \rho^2 - 3\rho_1^2}{3(1 + d^2)} \right\} \tag{21}$$

where ρ_1 is the value of ρ at the gap outlet and $C = \sqrt{2h_0 R} = $ constant. Substituting the value of v_x from Equation 21 into Equation 18 and integrating using the boundary conditions ($\psi = 0$ at $v_x = y = 0$) we obtain

$$\psi = \frac{vh_0\mu}{C} \left[\frac{4h^2 (\rho^2 - \rho_1^2)}{C^2 (1 + \rho^2)} + \frac{2 + 3\rho_1^{2'} - \rho^2}{1 + \rho^2} \right] \tag{22}$$

The results of this expression are presented in Figure 10 for a series of ρ and η values. As shown in Figure 10, near the roll surface the material flows parallel and the roll surface itself is the streamline. At the central zone located at the wedge entrance, countercurrent flow is observed. For a wedge of a final size, the countercurrent pattern takes the form of two closed eddies. This is the amount of material that may be observed in rolls of equal peripheral velocities. The material passing out of the gap sticks to one of the rolls rotating with it, and again, touching the portion of the wedge where rotational motion takes place to enter in the gap.

The contact area increases very slowly for part of the material. As shown by Figure 10A, the streamline $\psi = 0$ forms a closed contour that passes through the countercurrent zone where the velocity gradient is very small. Beyond the stagnation point the velocity gradient for $\psi = 0$ is equal to zero. Thus, for the case considered, the interface of one part of the material increased considerably slower than the one on the streamlines that pass through the zones of high-velocity gradients. The elements of the interface are not evenly distributed in the mass. Since only the interface is oriented along the streamlines, the material stops to change its orientation because there is no motion across the streamlines. The existence of closed streamlines and the absence of a distinct motion along the z axis parallel to the roll generatrix are the reasons the elements are not distributed evenly in the mixing mass. Even if it is assumed that the components are evenly distributed along the roll axis, there will always be some nonbroken interfaces. Thus, the symmetrical rolling process does not provide satisfactory distribution of the ingredients.

The mixing process may be intensified by imposing different velocities and temperatures on the rolls. The equations describing the flow in the gap between the rolls may be obtained

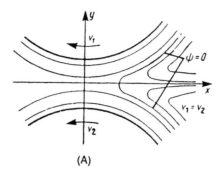

(A)

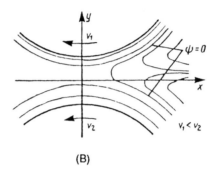

(B)

FIGURE 10.   Streamlines in gap of double-roll mixer: (A) rolls having same peripheral velocities; (B) rolls with different peripheral velocities.

by essentially the same method. In this latter situation, a double integration of Equation 14 is made over the limits of $v_x(h) = v_1 > v_x(-h) = v_2$ where v_1 and v_2 are the peripheral velocities of the front and rear rolls. The equation of streamlines for rolling with friction is

$$\psi = \frac{v_0 h_0 v}{C} \left[\frac{4\eta(\rho^2 - \rho^2)}{C^2(1 + \rho^2)} + \frac{2 + 3\rho_1^2 - \rho^2}{1 + \rho^2} + \frac{2\lambda\eta}{3C(1 + \rho^2)} \right]$$

where   (23)

$$v_0 = \frac{v_1 + v_2}{2} \qquad \lambda = \frac{v_1 - v_0}{2v_0}$$

where

As follows from Equation 23, there is a closed streamline $\psi = 0$ at $\eta = 0$ (Figure 10B). In this case the countercurrent zone is displaced to the roll with a lower peripheral velocity. It is important in this case that beyond the stagnation point the streamlines pass through the zone of definite shear strain, and in comparison to symmetrical rolling the shear of material increases. However, even for this case the streamlines are closed and the rolling with friction does not satisfy the requirements of perfect mixing because the interfaces oriented along the streamlines maintain their positions.

   To eliminate these drawbacks, the operator uses the technique of material undercut and then introduces it again to the gap from the other side of the roll. The closeness of streamlines is violated, and the material is displaced along the axial direction. The material undercut is a necessary part of the process in two-roll mixers, in which for a continuous process a special automatic device is provided.

The methods for evaluating thrust forces are based on the hydrodynamics of rolling and the theories of elastic and plastic deformation, as well as the theory of similarity.

Polymeric materials mixed by rolling undergo both combined and total deformation:

$$\epsilon = \epsilon_{el} + \epsilon_{h,el} + \epsilon_{Pl} \tag{24}$$

where:

ϵ	= total deformation
ϵ_{el}	= elastic deformation
$\epsilon_{h,el}$	= high elastic deformation
ϵ_{pl}	= plastic deformation

The plastic deformation of a polymer involves viscous flow with irreversible mutual displacement of molecules. Plastic deformation is highly sensitive to temperature.

It should be noted that the relationships between the temperatures of different components and total deformation play different roles. From this point of view there are three different amorphous states of polymers: glassy in the zone of lowest temperatures (in this case elastic deformation prevails); high elastic in the zone of medium temperatures (in this case high elastic deformation prevails); and the viscous flowing state in the zone of high temperature (in this case the deciding factor in plastic deformation). The lower regions of elastic and high elastic deformation are best suited for processing polymers by the rolling method.

The law of viscous flow depends on the properties of the materials and may be written as

$$\text{a Newtonian fluid} \qquad \tau = \mu_a \frac{dv}{dx} \tag{25a}$$

$$\text{a Bingham material} \qquad \tau = \tau_0 + \mu_a \frac{dv}{dx} \tag{25b}$$

$$\text{a non-Newtonian fluid} \qquad \tau = k \left(\frac{dv}{dx}\right)^n \tag{25c}$$

where:

τ	= shear stress
μ_a	= apparent viscosity
dv/dx	= velocity gradient
τ_0	= limiting shear stress
K,n	= rheological constants of the flow (n is the tangent of the linearized section of the curve)

To determine the flow curve of a non-Newtonian material, the rheological parameters K and n must be known. Flow curves may appear linear on logarithmic coordinates over a portion of the shear rates. This means that the material may be more viscous over a portion of the shear rates and less viscous over others. These properties are accounted for through the coefficient of apparent viscosity:

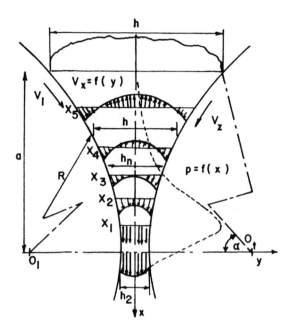

FIGURE 11.   Pressure and velocity distribution between roller gaps.

$$\eta = \mu_a = \frac{\tau}{\dfrac{dv}{dy}} = K \left(\frac{dv}{dy}\right)^{n-1}$$

(26)

We now consider the method based on the hydrodynamic theory of rolling. The behavior of a polymeric material approaches that of a non-Newtonian (quasi-viscous) fluid; therefore, polymers may be considered viscous fluids entrained in the gap by rotating rolls. In deriving design equations the following assumptions are applied:

1.   The flow regime in the gap is laminar.
2.   The material sticks to the roll surface; that is, there is no slip between the relative motion of the polymer and the walls, which limits the flow (the layers of material adjacent to the rolls are in motion at the same velocity as the rolls).
3.   The gravitational and inertial forces are small compared to surface forces and thus may be neglected.
4.   The flow of material is assumed to be one-dimensional; that is, the material passes mostly through the gap ($v_z = 0$, $v_x \geq v_y$ and $\delta v_x/\delta y \leq (\delta v_y/\delta y)$), as illustrated in Figure 11.
5.   Hydrodynamic pressure at the polymer gap entrance and at the outlet from the rolls is zero. The pressure in the plane parallel to the roll axis does not change in the y and z directions; that is, $\delta P/\delta y = 2P/\delta z = 0$.

The Navier-Stokes equations are the starting point in the analysis. Considering the system to consist of a steady plane-parallel flow of viscous liquid in the gap between the rolls and applying the preceding assumptions, we write

$$\frac{\delta^2 vs}{\delta y^2} = \frac{1}{\mu_a} \frac{\delta'' P}{\delta x}$$

(27)

$$\frac{\delta P}{\delta y} = 0$$

where p = pressure.

In the zone of maximum pressure $\delta P/\delta x = 0$ (i.e., $\delta v/\delta y$ = constant). In this zone the flow rate of polymer changes uniformly from v_1 to v_2 (taking into account that material sticks to the roll surface).

Upon integration of Equation 27, we obtain

$$v_x = \frac{1}{\mu_a} \times \frac{\delta P}{\delta x} \times \frac{y^2}{2} + C_1 y + C_2 \tag{28}$$

Substituting the friction value $f = v_2/v_1$ into Equation 28 and applying the boundary conditions $(v_{-h/2} = v_1; \; v_{h/2} = v_2 = fv_1)$, we obtain (after determining and substituting C_1 and C_2)

$$v_x = \frac{1+f}{2} v_1 + \frac{1}{2\mu_a} \times \frac{\delta P}{\delta x} \left(y^2 - \frac{h^2}{4} \right) + (1+f)\frac{v_1}{h} \tag{29}$$

where h is a moving coordinate (the distance between the rolls).

Using Equation 29 and determining the flow rate in the gap, the differential equation of pressure becomes

$$\frac{\delta P}{\delta x} = \pm \frac{1+f}{2} 6\mu_a v_1 \sqrt{2} \, \frac{h - h_1}{h^3 \sqrt{h - h_k}} \tag{30}$$

After integration, we obtain a formula for determining the specific thrust force:

$$P = \frac{1+f}{2} 2.22 \frac{\mu_a v_1 R}{h_k} \qquad \text{kg/cm} \tag{31}$$

where:

μ_a = apparent dynamic viscosity, $\text{kg} - \text{s/cm}^2$
v_1 = peripheral velocity of the roll, cm/s
R = roll radius, cm
h_k = minimum gap between rolls, cm

The thrust force acting between the rolls is

$$P = pL \tag{32}$$

where:

P = specific thrust force from Equation 31
L = length of the roll

The coefficient of apparent viscosity μ_a may be approximately determined from the velocity gradient formula

$$\frac{\delta v}{\delta y} = \frac{2v_1}{h_k} \tag{33}$$

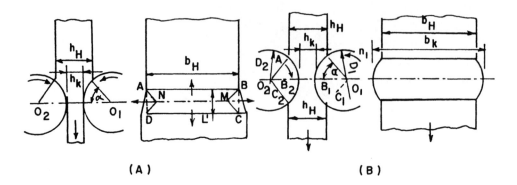

(A)   **(B)**

FIGURE 12.   Deformation of material between rolls: (A) plastic deformation; (B) elastic deformation.

From either literature data or rheological testing a flow curve of $\mu_a = f(\delta v/\delta y, t°C)$ can be constructed and a value for μ_a obtained.

From the laws of plastic deformation of material between the rolls, the average specific pressure on the roll is

$$P_{av} = K\sigma_\tau \frac{2h_{n.l.}}{(\delta - 1)\Delta h} \left[\left(\frac{h_{nl}}{h_k}\right)^\delta - 1 \right] \qquad (34)$$

where:

K = coefficient (≈ 1.15)
σ_t = yield limit of rolling material, kg/cm^2
h_{nl} = neutral layer thickness
δ = $\mu/\tan \alpha/2$, coefficient
Δh = $2R(1 - \cos \alpha)$, linear rolling reduction
α = angle of material seizure

The thrust force is

$$P_{th} = P_{av}B_{in}l_s \qquad kg \qquad (35)$$

where $l_s = \sqrt{R\,\Delta h}$, the arch of seizure (refer to Figure 12).

According to the laws of elastic deformation of materials the thrust force between rolls is

$$P_{th} = \frac{4ERb_H}{h_H} \sin \alpha(1 - \cos \alpha) \qquad kg \qquad (36)$$

where E = elastic modulus of material, kg/cm^2; and b_H, h_H are from Figure 12B.

From similarity theory the following empirical equations, which may be used for evaluating the thrust forces in rubber mixtures are presented:

$$\frac{P}{\gamma D^3} = C_1 B_1^x \left(\frac{h_k}{D}\right)^{y_1} \left(\frac{L}{D}\right)^{z_1} \left(\frac{D_1}{D}\right)^{k_1}$$

$$\frac{P}{\gamma D^3} = C_2 M_2^x \left(\frac{h_k}{D}\right)^{y_2} \left(\frac{L}{D}\right)^{z_2} \left(\frac{D_1}{D}\right)^{k_2} \qquad (37)$$

where:

B,M	= reducibility and softness of mixture
6	= specific gravity of mixture
D	= diameter of the roll
L	= length of material band on rolls
h_k	= minimum gap between rolls

The coefficients C_1, C_2, x_1, x_2, y_1, y_2, z_1, z_2, k_1, and k_2 are determined from experiment. Design equations for some synthetic and natural mixtures are also reported by Lukach et al. (1967). For example, for butadiene-acrylonitrile blends,

$$P = C_1\gamma D^{1.4}L^{0.7}h_k^{0.1}B^{-0.4}$$
$$P = C_2\gamma D^{1.4}L^{0.7}h_k^{0.1}M^{-1.8}$$

(38)

The values for C_1 and C_2 are 260 and 71 for CSN -40 and 180 and 54 for CSN -26.

As in estimating thrust forces, different methods of calculating torque and horsepower are available. The principal methods are those based on the theory of plastic or elastic deformation, on the hydrodynamic theory of rolling, and on the theory of similarity. The power consumed by mixing rolls depends on several factors: the properties and temperatures of materials, the peripheral velocities of the rolls, the rate of friction, the gap size, and others.

The influence of some of these factors on the energy consumption of the mixing rolls can be understood through the theory of plastic deformation. An increase in the peripheral velocity leads to an increase in the volume of material that can be deformed between the rolls. This results in an increase in the work of deformation. Changes in the peripheral velocity also influence the rate of deformation. Elastic deformation, which does not depend on time, does not play an important role. For highly elastic, and particularly for plastic deformation, a decrease in the time of deformation results in an increase in the necessary load on the rolls, which in turn leads to higher horsepower requirements.

Greater friction results in an increase in the volume of material undergoing deformation as well as the rate of deformation and consequently the horsepower consumed.

Enlarging the gap between rolls leads to an increase in horsepower requirements; however, the horsepower increase falls behind the gap increase. This can be explained by the fact that for larger gaps an increase in the volume of material passing through requires more work for deformation. On the other hand, as the gap is enlarged a decrease in the linear rolling occurs, which reduces the work of deformation. The opposite may also occur and the role of relative deformation becomes determinant.

The energy consumption in batch operations may change significantly with time because of variations in the material's plastic properties. The resistance torque to the rotation of the rolls may be assumed to consist of two components:

$$M = M_p + M_f \quad \text{kg} - \text{m}$$

(39)

where M_p is the torque required for overcoming the resistance of the material deformation as determined by thrust forces and M_f is the torque for overcoming friction in the bearings with respect to gravity and thrust forces.

The torque for overcoming thrust forces is

$$M_p = PD \sin \frac{\alpha}{2} \quad \text{kg} - \text{cm}$$

(40)

where:

P = thrust force determined from Equation 32
D = roll diameter
α = angle of seizer

The torque attributed to friction is

$$M_f = \mu(P + G)d \qquad kg - cm \tag{41}$$

where:

μ = friction coefficient in bearings
G = roll gravity force
d = diameter of journal of roll

The horsepower can be computed from the formula

$$N = \frac{(M_p + M_f)}{9700\eta'} \qquad kW \tag{42}$$

where:

n = rotational velocity of the roll, rpm
η' = efficiency of the drive

The electrical power of mixing rolls reaches its maximum value only for a short time (within the first 2 to 3 min). Therefore, the horsepower of an electric motor may be 1.5 to 2.0 times less than the value computed from Equation 42.

Based on the hydrodynamic theory of rolling, the horsepower consumed by the mixing rolls is established by the shear stresses arising in the material. The change in shear stress in the direction of deformation for a roll that rotates slowly is (Bekin and Nemytkov, 1966)

$$\tau_{+h_k/2} = \mu_a \left[\frac{3(1 - f)(\delta - \delta_0)}{\delta^2 h_k} + \frac{(1 - f)}{\delta h_k} \right] \tag{43}$$

For rapidly rotating rolls

$$\tau_{-h_k/2} = -\mu_a \left(\frac{3(1 - f)(\delta - \delta_0)}{\delta^2 h_k} + \frac{(1 - f)}{\delta h_k} \right) \tag{44}$$

where:

μ_a = apparent dynamic viscosity
v_1 = peripheral velocity of a slowly rotating roll
f = friction
δ = $\Delta = h/h_k$, dimensionless parameter
δ_0 = h_1/h_k, dimensionless parameter
h = moving width of the gap
h_k = minimum width of gap

The corresponding section modulus from the forces of viscous friction is

$$M_{+h/2} = \frac{3C_1 K\alpha_{ap}''(1 + f)R\sqrt{RL}v_1''}{\alpha_{ap_1}\sqrt{h_k}}$$

$$M_{-h/2} = \frac{3C_2 K\alpha_{ap}''(1 + f)R\sqrt{RL}v_1''}{\alpha_{ap_2}\sqrt{h_k}} \tag{45}$$

where α_{ap} = coefficient determined from the formulas

$$\alpha_{ap_1} = \frac{(1 + f)\sqrt{RT_1}}{l\sqrt{h_k}}$$

$$\alpha_{ap_2} = \frac{(1 + f)\sqrt{RT_2}}{l\sqrt{h_k}} \tag{46}$$

$$\alpha_{ap} = \frac{3(1 + f)\sqrt{RT}}{2l\sqrt{h_k}}$$

where C_1, C_2, T_1, and T_2 are the experimental constants and l is the length of the roll.
The total torque is

$$M = \frac{6CK\alpha_{ap}''v_1''(1 + f)^2\sqrt{RL}}{\alpha_{ap}\sqrt{h_k}} \tag{47}$$

The total technological horsepower is

$$N = \frac{3EK\alpha_{ap}''v_1''(1 + f)^2\sqrt{RL}}{\alpha_{ap}\sqrt{h_k}} \tag{48}$$

where

E = constant
L = length of the roll

The third method available for establishing horsepower requirements is based on similarity theory. The following equations were obtained for basic marks of butadiene-acrilonitrile caoutchoucs:

$$N = K_1\gamma\omega D^2 L^{0.6}h_k^{0.1}f^{-0.2}B^{-0.7}$$

$$N = K_2\gamma\omega D^2 L^{0.6}h_k^{0.1}f^{-0.2}B^{-2.5} \tag{49}$$

and for styrene-butadiene rubbers:

$$N = K_1\gamma\omega D^{2.3}L^{0.6}h_k^{0.1}f^{-0.2}B^{-0.7}$$

$$N = K_2\gamma\omega D^{2.3}L^{0.6}h_k^{0.1}f^{-0.2}M^{-2.5} \tag{50}$$

Equations 49 and 50 were obtained on the basis of experiments in which parameters were varied over the following limits: h_k = 0.6–2.5 mm; v_1 = 6.28–18 m/min; f = 1–3; L =

150–1050 mm; and D = 200–400 mm; initial caoutchouc plasticity η_a = 0.06–0.15. The temperature of the rolls and caoutchouc was kept constant over all plastification regimes (rolls were $45 \pm 5°C$; caoutchouc $80 \pm 5°C$).

Considering rolls as a double energy contour, the friction gear transfers power that is slightly greater than that obtained by the driving roll. Part of the horsepower may be transferred because of friction from the lower moving roll to the faster roll. As a result, the horsepower required for the rapidly moving roll may exceed that consumed by the drive.

Torques exerted on the rolls are not equal. The total torque may be estimated from the formula

$$M = M_{d_2} \left(1 \pm \frac{K_{ci_2}}{f} \right) + M_f \left(1 + \frac{1}{f} \right) \qquad (51)$$

where

M = total torque
M_{d_2} = torque on driving roll
M_f = torque of friction on the bearing of one roll
K_{ci_2} = coefficient of circulation determined from the graph developed by Lukach et al. (1967)

In addition, the influence of circulation patterns must also be included in the structural design of mixing rolls.

Another area important to parametric evaluations is roll productivity. Rolls may operate in either batch (multiple passing of material through rolls) or continuous processes (single pass of material through the gap of the rolls). The productivity design of mixing rolls should be done correspondingly.

The productivity of batch mixing rolls depends on a single-time volume loading of the machine, the processing cycle, the size of the rolls, and the operational factor. Productivity may be calculated from the formula

$$q = \frac{60 q_1 \rho \eta_e}{\tau} \qquad \text{kg/h} \qquad (52)$$

where:

q_1 = volume of mixture for one-time loading, dm³
ρ = mixture density, kg/dm³ (ρ = 0.9–1.1 kg/dm³)
η_e = operation factor (η_e = 0.8–0.9)
τ = time cycle, min

The one-time volume q_1 is determined from experiments for each material and roll size. Preliminary calculations can be made using the correlation

$$q_1 = (0.0065–0.0085)DL \qquad \text{dm}^3 \qquad (53)$$

where D is the rear roll diameter (cm).

Continuous mixing roll productivity may be determined from

$$Q = 60\pi Dnhb\rho v_e \qquad \text{kg/h} \qquad (54)$$

where:

D	=	rear roll diameter, dm
V_e	=	rear roll velocity, rpm
h	=	thickness of material band leaving the gap
b	=	width of material band leaving the gap

The average velocity of the material is determined from

$$v_m = \frac{Q}{bh\rho} \quad \text{m/min} \tag{55}$$

where:

Q	=	productivity of rolls, kg/min
b,h	=	width and height of band from the gap, dm

Complicated heat processes occur during the treatment of plastic materials in mixing rolls. Depending on the operating regime and material properties, heat flow may either be introduced into or evacuated from the machine. The requirements for cooling or heating are determined by a heat balance over the machine:

$$G_m i_{in} + Q_N + G_c C_c i_{w.in} = G_m i_{out} + G_{w.in} C_w, l_{w.out} + Q_l \tag{56}$$

where:

G_m	=	amount of material to be treated per unit time
i_{in}	=	enthalpy of material at initial temperature
i_{out}	=	enthalpy of material at final temperature
Q_N	=	amount of heat obtained by transformation of mechanical energy into thermal energy
G_c	=	amount of cooling water entering the rolls
$C_c, t_{w,in}$	=	specific heat and temperature of water entering the rolls
$t_{w,out}$	=	temperature of water leaving the rolls
Q_l	=	heat loss of the rolls

From Equation 56, the amount of heat absorbed by the cooling water is established.

$$Q_w = G_c C_c(t_{in} - t_{out}) = Q_N - Q_l - G_m(i_{in} - i_{out}) \tag{57}$$

The amount of heat due to energy dissipation is

$$Q_N = 860N\eta_e \tag{58}$$

where:

N	=	average horsepower consumed by the rolls
η_e	=	drive efficiency

The heat losses consist of losses by convection and radiation:

$$Q_l = (q_c + q_R)F \tag{59}$$

where:

q_c $= h_c(t_w - t_a)$, the specific heat flow by convection
h_c = heat transfer coefficient from a roll to air
t_w = temperature of the wall of the roll (assumed to be equal to the temperature of
         the material to be treated)
t_a = temperature of the ambient air
q_R = specific heat flow by radiation

$$q^R = \epsilon C_0 \left(\frac{T_w}{100}\right)^4 \tag{60}$$

where:

ϵ = emittance; accounts for the fact that one roll is covered with material
T_w = absolute temperature of the wall of the roll
C_0 = radiation constant of an absolute black body

According to calculations, depending on material properties the amount of heat spent in raising the temperature of the mass from t_{in} to t_{out}, including heat losses, can be as high as 10 to 25% of the total heat generation Q_n. The amount of heat absorbed by the cooling water is determined from a heat balance equation

$$Q_w = UL\,\Delta t_{av} \tag{61}$$

where U is the heat transfer coefficient through the cylindrical wall.

$$U = \frac{1}{1/2\pi K \ln\left(d_{out}/d_{in}\right) + (1/h\pi d_{in})} \tag{62}$$

where:

k = conductivity of roll material
d_{out} = outside roll diameter
d_{in} = inside roll diameter (diameter of roll hollow or average diameter of a circle
         passing through the center's peripheral channels)
h = heat transfer coefficient from the internal roll space to the moving water (deter-
         mined by one of the equations of forced convection, depending on the flow
         regime)
L = roll length
Δt_{av} = average temperature difference defined as

$$\Delta t_{av} = t_w - \frac{t_{w.out} - t_{w.in}}{2} \tag{63}$$

The method of heat calculation of the rolls amounts to the following procedure. First, determine Q_1 and G_c (a temperature difference can be assumed; typically $t_{w,out} - t_{w,in} = 5$–$7°C$). Knowing the size of the roll hollow (or the sizes of peripheral channels) it is then possible to determine the velocity of the cooling water to estimate the heat transfer coefficient.

From Q_w, U, and L, determine Δt_{av}, and from Equation 63 calculate $t_{w,out}$ or $t_{w,in}$. For specified $t_{w,out}$ and $t_{w,in}$, apply Equation 61 to establish whether the length is sufficient for evacuation of the given amount of heat Q_w.

SYMBOLS

A	area
A', B'	constants
a	exponent or variable
a'	interfacial area
c	velocity
C	concentration
c_1, C_2	coefficients
D, d	diameter
Eu	Euler number
e	elasticity coefficient
F	area
F'	force
Fr	Froude number
G	modulus of rigidity, also gravitational force
Ga	Galileo number
g	gravitational acceleration
H	depth or head
h, h_e	heat transfer coefficients
J	heat transfer coefficient
k', k	constants
k_n	power number
L	length
L_p	turbulent mixing length
M	torque moments
N_p	mixer power
N_t	mixing number
N_L	modified power number
N	rotational speed
n, n_0	rotational speed
P	pressure, also centrifugal force
Pr	Prandtl number
p	power input
Q	volumetric rate or pumping capacity
q	mass or volumetric throughput
Re	Reynolds number
R, r	radius
s	pitch
T	absolute temperature
t	temperature or time
U	overall heat transfer coefficient
u	velocity
V	velocity or volume
v	velocity
W	dimension
We	Weber number

GREEK LETTERS

α, β	angle or coefficients
$\dot{\gamma}$	shear rate
γ	specific weight or concentration
δ	clearance
ϵ	deformation
λ	friction
λ_1	time constant
λ_v	heat of vaporization
μ	Newtonian viscosity
η_p	plastic viscosity
η_c, η_d	effective viscosities of
η'	continuous and disperse phases, and emulsion, respectively
η_e	drive efficiency
θ	characteristic time
τ_i, τ'	mixing time
τ	shear stress
τ_y	yield stress
ρ	density
ϕ	parameter or angle
ψ	angle, stream function
ψ'	void fraction
ψ_0	angle

REFERENCES

Azbel, D. S. and N. P. Cheremisinoff, *Fluid Mechanics and Unit Operations,* Ann Arbor Science Publishers, Ann Arbor, MI, 1983.

Azbel, D. S., *Two-Phase Flows in Chemical Engineering,* Cambridge University Press, New York, 1981.

Bake, C. K. and G. H. Walter, *Heat Transfer Eng., 1* (2):28, 1979.

Bekin, N. G. and V. A. Nemytkov, *Kauchiek i Resina* (USSR), *10*:31, 1966.

Bernhardt, E., *Pererabotka Termoplasticheskikh Materialov,* M. Goskhinisdat, USSR, 1962.

Bingham, E. C., *Fluidity and Plasticity,* McGraw-Hill, New York, 1922.

Bott, T. R. and S. Azoory, *Chem. Proc. Eng., 50* (1):85–90, 1969.

Calderbank, P. H., in *Mixing,* Vol. 2 (V. H. Uhl and Gray, eds.), Academic Press, New York, 1967.

Calderbank, P. H. and M. B. Moo-Young, *Trans. Inst. Ch. E.* (London), *39*:337–347, 1961.

Carreau, P. J., I. Patterson, and Yap, *Can. J. Chem. Eng., 54:* 135, 1976.

Chavan, V. V. and R. A. Mashelkar, in *Advances in Transport Processes,* Vol. 1 (A. S. Mujudar, ed.), Wiley, New York, 1980, pp. 210–252.

Chavan, V. V. and J. Ulbrecht, *Chem. Eng. J.,* 6:213–223, 1973.

Chavan, V. V., M. Arumugam, and J. Ulbrecht, *AICHE J., 21:*613, 1975.

Cheremisinoff, N. P., *Polymer Mixing and Extrusion Technology,* Marcel Dekker, New York, 1987.

Coyled, C. K., H. E. Hirschland, B. J. Michal, and J. Y. Oldshue, *AICHE J., 16:* 903, 1970.

Ford, D. E. and J. Ulbrecht, *AICHE J., 21:*1239, 1975.

Frohlich, H. and R. Sack, *Proc. Roy. Soc. Lond., A185:* 415, 1946.

Guber, F. B., *Kauchuk i Resina* (USSR), *9:*28, 1966.

Hagedorn, D. and J. J. Salamone, *Ind. Eng. Chem. Proc. Des. Dev., 6:* 469, 1967.

Harriott, P., *Chem. Eng. Prog. Symp. Ser., 55* (29):137 (1959).
Harriott, P., *AICHE J., 8:* 93, 1962.
Houlton, H. G., *Ind. Eng. Chem., 36:* 522, 1944.
Huggins, R. E., *Ind. Eng. Chem., 23:* 749, 1931.
Kapitonov, E. N., *Kauchuk i Resina* (USSR), *4:*20, 1968.
Karpachev, P. S., M. M. Maisel, and N. A. Plevako, *Mashiny i Apparaty Proisvodstviskustvenoy kozhy i Plenochnykh Materialov,* M. Legkay Industria, USSR, 1964.
Kelkar, J. F., R. A. Mahelkar, and J. Ulbrecht, *J. Trans. Inst. Chem. Eng.,* (London), *50:*343, 1972.
Kharkhuta, N. Ya., M. I. Kapustin, and V. P. Semenov, *Dorozhnye Mashiny i Oborudovanie,* M. Mashgiz, USSR, 1968.
Lukach, U. E., D. D. Raybinin, and B. N. Metlov, *Valkovye Mashiny dlia Pererabotki Plastmass i Resinovykh Smesey,* M. Mashinostroenie, USSR, 1967.
Magnusson, K., *IVA, 23* (2):86–99, 1952.
Metzner, A. B., *Non-Newtonian Technology, Fluid Mechanics, Mixing and Heat Transfer,* Academic Press, New York, 1956, pp. 77–153.
Metzner, A. B. and R. E. Otto, *AICHE J., 3:* 3, 1957.
Metzner, A. B. and J. S. Taylor, *AICHE J., 6:* 109, 1960.
Metzner, A. B., R. D. Vaughn, and G. L. Houghton, *AICHE J., 3:* 92, 1957.
Mizushina, J., et al., *Kagaku Kogaku, 31* (12):1208, 1967.
Mohr, W. D., R. L. Saxton, and C. H. Japson, *Ind. Eng. Chem., 49* (11):1857, 1957.
Nagata, S., *Mixing Principles and Application,* Wiley, New York, 1975.
Oldroyd, J. G., *Proc. Roy. Soc. London, A128:* 122, 1953.
Orr, C. and J. M. Dalla Valle, *Chem. Eng. Prog. Symp.,* Service No. 9, *50:*29, 1954.
Ostwald, W., *Kollaidzchr, 38:* 26, 1926.
Pikovsky, Y. M., S. M. Polosin-Nikitin, and N. N. Votshchinia, *Dorozhyne Mashiny i Oborudovanie,* M. Mashgiz, USSR, 1960.
Pryce-Jones, J., *Coll. Zeits., 96:* 129, 1952.
Ramdas, V., V. W. Uhl, M. W. Osborne, and J. R. Ortt, *Heat Transfer Eng., 1* (4):38–46, 1980.
Reiner, M., *Deformation and Flow,* Lewis Pub., London, 1949.
Rushton, J. H., E. W. Costich, and H. J. Everett, *Chem. Eng., Prog., 46:* 395–467, 1950.
Schofield, R. K. and G. W. Scott-Bair, *Proc. Roy. Soc. London, A138:* 707, 1932.
Skelland, A. H. P., *Ind. Eng. Sci., 7:* 166, 1958.
Skelland, A. H. P., Mixing and agitation for non-Newtonia fluids, in *Handbook of Fluids in Motion,* (N. P. Cheremisinoff and R. Gupta, eds.), Ann Arbor Science Publishers, Ann Arbor, MI, 1983, pp. 179–211.
Skelland, A. H. P. and G. R. Dimick, *Ind. Eng. Chem. Proc., Des. Dev., 8:* 267, 1969.
Skelland, A. H. P. and R. Seksaria, *Ind. Eng. Chem. Proc. Des. Dev., 17:* 56, 1978.
Skelland, A. H. P., D. R. Oliver, and E. Tooke, *Brit. Chem. Eng.,* 346, 1962.
Soroka, B. S. and B. A., *Khimichestioe Mashinostroenie,* Kiev, Tekhnika, USSR, *2:*15–19, 1965.
Starov, I. M., et al., *Kauchuk i Resina,* USSR, *6:*19, 1961.
Toms, B. A. and D. J. Strawbridge, *Trans. Faraday Soc., 49:* 1225, 1953.
Treybal, R. E., *Liquid Extraction,* 2nd ed., McGraw-Hill, New York, 1963.
Ulbrecht, J., *Chem. Eng., London,* 367:347–353, 1974.
Wilkinson, W. L., *Non-Newtonian Flow,* Pergamon Press, New York, 1960.
Winding, C. C., F. W. Dittman, and W. L. Kranich, Cornell University Report, Ithaca, New York, 1944.
Zakharkin, O. A., et al., *Kauchuk i Resina,* USSR, *7:*8–11, 1966.

Chapter 22

RUBBER EXTRUSION PRINCIPLES

Nicholas P. Cheremisinoff

TABLE OF CONTENTS

0-8493-4401-8/93/$0.00 + $.50

I. INTRODUCTION

Extruders are used in rubber and plastic processing for such operations as injection and blow molding, for processing thermosetting resins, and for the hybrid process of injection blow molding. The apparatus is fed room-temperature resin in the form of beads, pellets, or powders, or if rubbers are being processed, the feed material may be in the form of particulates or strips. The unit converts the feedstock into a molten polymer at sufficiently high pressure to enable the highly viscous melt to be forced through a nozzle into the mold cavity (injection molding) or through a die (e.g., blow molding or continuous extrusion of articles). In the initial portion of the extruder, the polymer is conveyed along the extruder barrel and is compressed. The material is then heated until soft, eventually reaching a molten state. As fresh feed material enters, heat transfer takes place between the molten fluid and solid polymer. Once in the molten state, the extruder acts as a pump, transferring the molten polymer through the extruder channel and building up pressure before flow through the discharge nozzle or die. The principal components of a single-screw extruder are illustrated in Figure 1. The machine has a motor drive, a gear train, and a screw that is keyed into the gear reducing train. The fluid layers between the screw flights and barrel wall maintain the screw balanced and centered.

Units are equipped with continuous variable speeds, and barrels are usually electrically heated, usually by band heaters with either on-off or proportional control. The barrel can be "zoned" according to the number of controllers on the heater bands. Depending on the application and type of service, the screw may be cored to allow heating or cooling. In operations requiring close tolerances on extruded articles, several pressure transducers and thermocouples are positioned along the barrel to ensure uniform extrusion and to control barrel and stock temperatures. Some designs may include thermocouples on the screw to monitor and control conveying flights.

A die can be attached at the end of the extruder. Die designs range from very simple geometries, such as an annulus for pipe and tubing profiles, to very complex shapes, such as the rubber seals used as glass run channels around the windows of automobiles.

There are numerous types of extruders, the most common of which is screw extrusion, which can be the single or twin type. Another common type is the plunger or ram extruder. Materials like tetrafluoroethylene (TFE) Teflon® and ultrahigh molecular-weight polyethylene are normally handled by ram extrusion. The melting temperatures of these polymers are very high; hence, these materials cannot be pumped readily as in screw extrusion.

II. SINGLE-SCREW DESIGNS

Many commercial extruders plasticate and pump materials in the range of 10–15 lb/h-hp. However, pumping capacity is a relative quantity that depends on the material. In adiabatic mixing, machine capacities can be as low as 3–5 lb/h-hp. Also, a machine that handles a thermoplastic elastomer could show as much as a three to fourfold increase in mass throughput when switched to a low melt viscosity material like nylon.

Screw configuration depends to a large extent on the properties of the material being processed. Figure 2 illustrates several common arrangements. A constant-pitch metering screw is usually employed in applications not requiring intensive mixing. When mixing is important, say for color dispersion, a two-stage screw equipped with a letdown zone in the center of the screw is appropriate. Turbulence promoters can also be included at or near the tip of the screw. In some applications with two-stage screws, venting at the letdown section may be needed.

The screw section immediately ahead of the gear train acts as a solids metering or feed zone. It is characterized by a deep channel between the root of the screw and the barrel

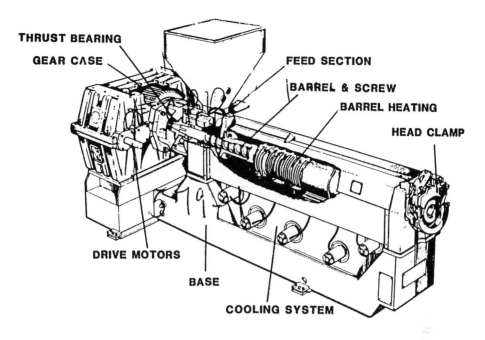

THRUST BEARING

GEAR CASE

FEED SECTION

BARREL & SCREW

BARREL HEATING

HEAD CLAMP

DRIVE MOTORS

BASE

COOLING SYSTEM

FIGURE 1. Single-screw extruder.

wall. The plasticating zone is ahead of the solids metering zone. This is a transition region where the channel narrows. The purpose of this zone is to provide intense friction between solids and a region for melting of the polymer to take place. Near the tip of the screw is the melt metering zone, where pressure builds up. In this region, the polymer melt is essentially homogenized and raised to the proper temperature for extrusion of the article.

The action of an extruder is analogous to that of a positive-displacement pump. The flight depth along the screw (i.e., the ratio of the solids metering channel depth to the melt pump channel depth) is known as the compression ratio. The purpose of screw flights is to enable the screw to transport polymer down the barrel. The pitch angle of the flights again depends on the type of material being handled. Many elastomer applications employ a general-purpose screw of a constant pitch (i.e., the flight equals the diameter). The pitch angle of this single-flighted screw is usually 17.61°. Typical extruder specifications are as follows:

Compression ratio: 2:1–6:1 [for materials ranging from low-density polyethylene (LDPE) to some nylons]
Pitch angle: 12–20°
Length-diameter ratio (L/D): 16:1–36:1 (low 40s typically for easily melting and flowing polymers requiring high mixing and venting)
Extrusion pressures: 10,000–30,000 psi

For extrusion pressures, low ratings (~10,000 psi) are usually sufficient for many thermoplastic materials. The upper limit typifies fluorinated ethylene propylene (FEP) Teflon®.

The *lead* on a screw is defined as the distance between the flights. As an approximation, it is equal to the ID (inner diameter) of the barrel for a single-flighted screw. The radial clearance between the flight tip and the barrel is tight (usually ~0.001 in./in. of barrel ID). The reason for such a tight clearance is that if the gap is too great, the material may flow back along the barrel, resulting in a loss of melt pumping zone capacity.

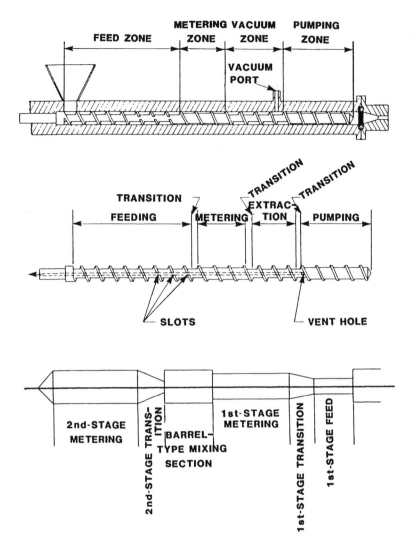

FIGURE 2. Screw profiles and cross section of vented barrel extruder.

The initial region of the extruder plays an important role in the machine's overall operation. It consists of a hopper or feed arrangement and a solid feedstock conveying region. Its purpose is to transfer the cold polymer feed from the feed hopper into the barrel, where it is initially compressed. This compression forces air out between the interstices of resin pellets or rubber chunks (air being expelled back through the hopper) and breaks up lumps and polymer agglomerates. This action creates a more homogeneous feedstock that can be readily melted.

Important features of the screw are shown in Figure 3. The relative motion of the extruder barrel to the screw is at an angle to the flights, where the helix angle of the flight is ϕ. From a vector diagram of the relative motion, the forward velocity of the plug is the velocity down the flights divided by ϕ. The volumetric capacity of the screw is

$$Q_s = U_z \rho \int_{R_s}^{R_b} \frac{2\pi R - pe}{\sin \phi} \, dR \qquad (1a)$$

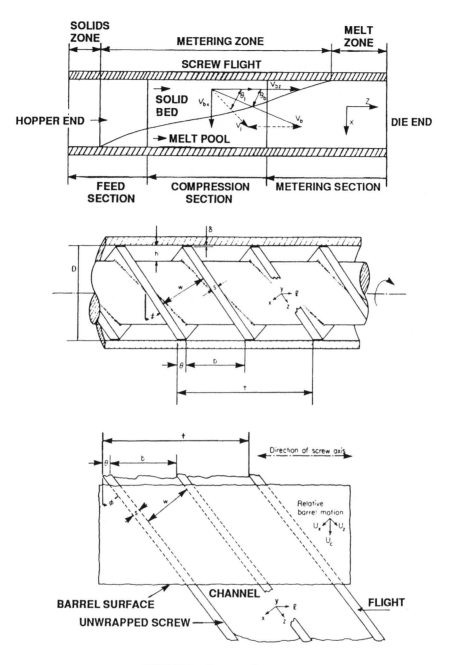

FIGURE 3. Screw configuration.

or

$$Q_s = \rho U_z \left(\frac{\pi(D_b^2(-D_s^2)}{4} - \frac{peh}{\sin \phi} \right) \tag{1b}$$

where

U_z = forward plug velocity
P = number of flights in parallel
e = width of flight
h = channel depth
R_s, R_b = radii of the screw root and inside barrel, respectively

Throne (1979) notes that the plug velocity at the barrel surface is

$$U_{zp} = \frac{U_z}{\sin \theta} \tag{2}$$

where θ is the angle of advance of the plug relative to the screw axis (typically 10–25°). Frictional forces between the plug and barrel depend on angle θ, which in turn is a function of the force and torque exerted on the polymer plug.

This relationship is an oversimplification, however. A more detailed force and torque balance on the plug shows that ϕ is a function of the forces and pressures:

$$Q_s = \rho \pi^2 N h D_b (D_b - h) \frac{W}{W + e} \frac{\tan \phi \, \tan \theta}{\tan \phi + \tan \theta} \tag{3}$$

where W is the average channel width (note that for most designs $W \gg e$). This expression is based on the relationships

$$U_{zp} = \frac{U_b \tan \phi \, \tan \theta}{\tan \theta + \tan \phi} \tag{4}$$

where

U_b = ND_b
N = screw speed, rpm
W = $\left(\dfrac{\pi}{P}\right)(D_b - h) \sin \phi - e$

Following Throne (1979) and Tadmor and Klein (1969), the following values are introduced, which typify a more or less conventional single-screw design: $\phi = 17.61°$ and $\theta = 17.61°$ with negligible pressure drop along solids transport zone. Hence, $\tan \phi = \cot \theta$, and for no friction between the screw and solid plug,

$$Q'_s = \rho \pi^2 N h D_b (D_b - h) \sin \phi \cos \phi \tag{6}$$

and $\sin \phi \cos \phi = 0.2884$.

This provides a basis for defining a screw discharge coefficient in terms of the solids-conveying capacity, as follows:

$$Q^* = \frac{Q_s}{Q'_s} = 0.2884 \frac{\tan \phi \, \tan \theta}{\tan \phi + \tan \theta} \tag{7}$$

This ratio provides a relative rating of the solids flow rate. For practical ranges of θ, Equation 7 indicates that the actual solids throughput can be of the order of 6% less than the maximum theoretical throughput.

Defining ψ_0 as the friction between the plug and screw channels,

$$\psi = F \pi D_b L_s \cos (\phi + \theta) - \psi_0 \tag{8}$$

where

ψ = frictional force or push that a self-containing solid plug can exert on an obstruction

F = frictional force per unit area between the barrel and polymer solids

L_s = channel length filled with polymer

The resistance force that would obstruct the solid plug is

$$\psi_n = 2\pi D_b \xi \tag{9}$$

where ξ is the nonself-conveying friction per unit surface.

For solids to be conveyed, the condition that $\psi > \psi_n$ must exist. This can be stated more explicitly as

$$\frac{F L_s \cos (\phi + \theta)}{2\xi} - \frac{\psi_0}{2\pi D_b \xi} \gg 1 \tag{10}$$

The form of the feed material can have a dramatic effect on the solids-conveying efficiency of the extruder. For example, in plastic extrusion applications when the resin is fed as pellets, the coefficient of friction generally tends to be lower than in powder form; hence pellets can tend to show higher conveying rates. In general, Equation 10 indicates that increasing F (e.g., increasing the coefficient of friction between the material and the barrel) results in more effective conveying.

Mechanical heat is generated in the solids-conveying zone, which can be described in terms of the amount of heat added per pound of polymer conveyed:

$$\frac{H_s}{Q_s} = \frac{F D_b L_s (\tan \phi + \tan \theta)}{\rho h (D_b - h)[W/(W + e)]} \tan \phi \tan \theta \tag{11}$$

This thermal energy is the heat generated as a result of friction. This relationship shows that to generate frictional heat, a large frictional force between the barrel and solids is needed. This can be accomplished in part by using an extruder with a relatively shallow channel height and a long feed section.

Following Tadmor and Klein (1969), the pressure in the solids-conveying zone increases exponentially with distance down the flight and can be described by

$$P = P_1 \, e^{[(B_1 - A_1 k) Z_s / C B_2 + A_2 K)]} \tag{12}$$

where

P_1 = pressure at the extruder inlet

Z_s = distance down the channel

$A_{1,2}, B_{1,2}, K$ = functions of the solids properties and screw geometry

The inlet of the extruder usually has a hopper immediately above it when handling plastic resins. The pressure at the extruder inlet therefore follows the Janssen equation, which shows that the pressure exerted by solids on the floor of a vessel is independent of the height or head of material; i.e.,

$$P_{v\infty} \simeq \frac{D\rho_b}{4\phi'K}$$ (13)

where ϕ' is the friction coefficient between the solid pellets or powder and the wall.

The value of $\phi'K$ can be determined experimentally. More accurate predictions of the pressure exerted at the base of the hopper can be obtained by considering the pressure profiles of the solids through the bin (Bridgwater and Scott, 1983).

An important limitation of the Janssen equation is that it is based on a static bin analysis. Extruder hoppers in fact are usually conical. In addition, they are not operated like static hopper bins or freely discharging hopper bins. Flow rates through extruder hoppers are controlled by the feed rate to the extruder. The operator must ensure that the feed rate from the hopper matches that of the extruder output. Pressure fluctuations due to variations in feed material levels in the hopper can cause surging and uneven extrudate properties.

Once the polymer feed has been compacted, the solid resin particles or rubber chunks must be transformed into a homogeneous melt. Two major requirements are needed for proper operation:

1.   There must be adequate heat conduction from the barrel to the resin or elastomer.
2.   There must be sufficient mixing of the melt with the unmelted resin.

Shear heating between the resin and barrel surface may be necessary for the latter to occur. In general, the larger the amount of viscous dissipation within the melt-solid mixture, the less amount of energy needs to be supplied to the barrel, which translates into lower operating costs. Materials like acrylonitrile butadiene styrene (ABS) can be heated from its resin state to a melt without supplying heat along the barrel. In this manner, the extruder is operated adiabatically. In processing different elastomers, the extruder operation can be isothermal, i.e., the barrel, screw, and rubber are maintained at constant temperature, and we rely on shearing energy to convert the polymer from a solid to a melt state.

Considering the melt as a Newtonian fluid and y as the direction from the top of the solid bed to the barrel surface, the velocity profile in the melt is

$$\frac{d^2 U_z}{dy^2} = 0$$ (14)

Imposing the no-slip condition at the wall, $U_z(0) = 0$, and the velocity component of the barrel in the downchannel direction is $U_z(\delta) = U'_z$ (where δ is the thickness of the melt layer), then

$$U_z = \left(\frac{y}{\delta}\right) U'_z$$ (15)

Equation 15 describes the velocity profile across the melt. Now the heat transfer, which includes viscous dissipation, is

$$k_m \frac{d^2 T}{dy^2} + \eta \left(\frac{dU_z}{dy}\right)^2 = 0$$ (16)

where

k_m = melt thermal viscosity
η = melt viscosity

Conduction is the primary mechanism for heat transfer in laminar flow. The melt temperature is $T(y = 0) = T_m$, and at $y = \delta$, $T = T_b$, which is the barrel temperature. These relations combine to provide the temperature profile through the melt:

$$\frac{T - T_m}{T_b - T_m} = \text{Br}\,\frac{y}{\delta}\,\frac{1 - y}{\delta} + \frac{y}{\delta} \tag{17}$$

where Br is the dimensionless Brinkman number. The Brinkman number provides a measure of the reflective importance of viscous heat generation to heat conduction due to the imposed temperature differential, $T_b - T_m$. It is defined as

$$\text{Br} = \frac{\eta U_z'^2}{2k_m(T_b - T_m)} \tag{18}$$

Differentiating Equation 17 and evaluating the expression $y = 0$ gives the heat flow from the melt into the solid plug of polymer preceding it:

$$-q = k_m(T_b - T_m)\,\delta + \frac{\eta U_z'^2}{2\delta} \tag{19}$$

In the solid bed, the material's surface temperature is $T(0) = T_m$, and the temperature at some sufficiently large distance from the inlet is T_s. If the solids move in a plug-flow fashion, the heat conduction for this material is

$$\rho_s c_s V_{sy}\,\frac{dT}{dy} = k_s\,\frac{d^2T}{dy^2} \tag{20}$$

where

V_{sy} = bed velocity in the y direction
c_s = specific heat
k_s = thermal conductivity

Solving Equation 20 gives

$$\frac{T - T_s}{T_m - T_s} = e(V_{sy}/\alpha_s) \tag{21}$$

where α_s = thermal diffusivity of the bed. Equation 21 states that the temperature profile in the solid exponentially decreases from its melt temperature at the surface to the screw temperature T_s (if the extruder is sufficiently long) or some other reference temperature in the interior of the bed.

The rate at which the melt enters the region above the solid bed is approximately the mass rate of flow leaving the melt pool; hence,

$$Q' = V_{sy}\rho_s x = \frac{V_{bx}\rho_m \delta}{2} \tag{22}$$

where

Q' = rate of melting per unit downchannel distance
ρ_s = solid density
V_{bx} = velocity of barrel in the screw axis direction relative to the screw
ρ_m = melt density
X = average solids bed width

The heat flux into the solid is obtained from the heat conduction Equation 20. From an overall heat balance

$$\frac{k_m(T_b - T_m)}{\delta + \eta U_z'^2/2\delta} - \rho_s c_s V_{sy}(T_m - T_s) = V_{sy}\rho_s H' \tag{23}$$

where H' = heat of fusion solids.

Throne (1979) gives the following two expressions for the thickness of the melt and the velocity of the bed:

$$\delta = \left\{ \frac{[2k_m(T_b - T_m) + \eta U_z'^2]x}{V_{bx}\rho_m c_s(T_m - T_s) + H'} \right\}^{1/2} \tag{24}$$

$$v = \left\{ \frac{[V_{bx}\rho_m k_m (T_b - T_m) + \eta U_z'^2/2]}{2[c_s(T_m - T_s) + H']} \right\}^{1/2} \tag{25}$$

where v is the speed of rotation and thus the velocity of the solid bed.

The rate of melting is proportional to the square root of the width of the solid bed, and the solid's volumetric flow rate is related to the rate of melting:

$$-\rho_s V_{sz} \frac{d(hx)}{dz} = v \tag{26}$$

Throne (1979) imposes the following model assumptions:

1. The melt film thickness is small and constant during the melt phase.
2. The height of the bed h_s is equal to the height in the channel.
3. The channel is tapered, and therefore the height can be described by $h = h_0 - a_z$, where h_0 = channel height at beginning of melt zone and a = slope of the channel.
4. At the beginning of melting (i.e., at $z = 0$), $X = W$, the channel width.

By applying these assumptions, substituting Equation 25 into Equation 26, and simplifying terms, the following relation is obtained:

$$(h_0 - az) \frac{dx}{dz} = ax + \frac{v\phi}{\rho_s V_{sz}} x^{1/2} \tag{27}$$

The solution is

$$\frac{x}{W} = \left\{ \frac{\psi}{a} - \frac{\psi}{a-1} \left[\frac{h_0}{(h_0 - az)} \right]^{1/2} \right\}^2 \tag{28}$$

where ψ is a transformation variable defined in terms of the mass flow rate G:

$$\psi = \frac{\phi W^{1/2}}{G/h_0} \tag{29}$$

and

$$G = V_{sz} h_0 W \rho_s \tag{30}$$

A plot can be prepared of X/W versus z/Z for different a/ψ values [where $Z = h_0/\psi(2 - a/\psi)$, the total melting length], which shows the effect of screw flight tapering on the solid bed profile. The relationship for Z shows that by increasing the extruder channel width, the length of melting required decreases. Further, increasing barrel temperature, melt viscosity, and screw speed decrease for melting length.

After the polymer has been completely melted, it enters the metering or pumping zone of the extruder. This portion of the machine is designed for a certain length so that sufficient pressure builds up for injection or extrusion through dies. The most widely referenced model of this portion of extrusion is that of Squires (1964). The model was later modified by Carley (1971). This is a drag flow model. The relationship between pressure drop in the flow channel and the velocity profile is

$$\frac{dP}{dz} = \eta \frac{d^2 V_z}{dy^2} \tag{31}$$

where

η = melt viscosity (assume Newtonian for immediate discussions)
V_z = velocity component along Z direction
y = distance above the root of the screw

By integration, we obtain

$$V_z = \frac{y V_z}{h} - \frac{y(h - y) \, dP/dz}{2\eta} \tag{32}$$

Integration of the velocity profile across the channel height gives the volumetric flow rate:

$$Q = W \int_0^h V_z \, dy$$
$$Q = \frac{V_z W h}{2} - \frac{W h^3 \, dP/dz}{12\eta} \tag{33}$$

By incorporating terms describing the channel geometry, Equation 33 transforms to

$$Q = F_d \left(\frac{\pi^2}{2} \sin \phi \cos \phi \, D^3 hN \right) - F_p \left(\frac{\pi}{12} \sin^3 \phi h^3 \frac{D}{L} \frac{P}{\eta} \right) \tag{34}$$

where

φ	=	helix angle
D	=	barrel diameter
N	=	screw speed
L	=	length of melt zone of screw in the axial direction
F_d, F_p	=	geometric functions of W/h; drag flow and pressure flow shape factors, respectively

The first term on the right-hand side of Equation 34 represents the drag flow induced by the moving barrel surface. The second term represents the pressure flow induced by the axial pressure gradient in the extruder. Note that for melt flow in very narrow channels, $F_p = F_d = 1$.

III. TWIN-SCREW EXTRUSION

There are a variety of twin-screw extruder designs employed throughout the polymer industry, with each type having distinct operating principles and applications in processing. Designs can be generally categorized as corotating and counterrotating twin-screw extruders. Eise et al. (1981) has classified twin-screw extruders in terms of the mechanisms of operation, and the same terminology is adopted for the present discussion.

Different arrangements of twin-screw systems are illustrated in Figure 3. The figure illustrates differences between corotating and counterrotating as well as between intermeshing and nonintermeshing screw arrangements. Note the difference between fully intermeshing and partially intermeshing systems and between open- and closed-chamber types. For example, a counterrotating screw shown in Figure 4 (type 1) is an axially closed, single-chamber pumping system, whereas a corotating screw (type 4) is an axially open mixing system relying on drag forces.

A screw system open in the axial direction is actually open in the longitudinal direction of the screw channel since it has a passage from the inlet to the outlet of the apparatus. This means that material exchange can take place lengthwise along the channel. In a closed arrangement, the screw flights in the longitudinal direction are at closed intervals. It is important that the cross section of the screw channel be open for material exchange to take place from one flight to the other in a direction normal to the screw channel. There is usually some leakage over the screw crests and through the areas required for mechanical clearances. Whether the screws are open lengthwise or crosswise or have a closed geometry has a direct effect on conveying conditions, mixing action, and the pressure buildup capacity of the system. Nonintermeshing systems, for example, are open lengthwise and crosswise (refer to Figure 4). Fully intermeshing, counterrotating systems can be closed lengthwise and crosswise, and regardless of mechanical clearances, they can develop closed chambers. This is the case with screw pumps. Lengthwise and crosswise closed, fully intermeshing corotating systems (Figure 4, type 2) are theoretically impossible, as is a lengthwise open and crosswise closed counterrotating system.

Fully intermeshing, corotating screws are open lengthwise. When normal screw flights are employed, they are closed crosswise, and when staggered screw discs are used, they are open crosswise. This design is illustrated by types 4 and 6 in Figure 4.

There are several types of partially intermeshing screws, for example, lengthwise open and crosswise closed systems and lengthwise and crosswise open systems. Most commercial intermeshing counterrotating screw extruders are combinations of systems 1, 9a, and 9b (Figure 4). Commercial nonintermeshing, counterrotating screws correspond to system 11 and are used along with a single-screw discharge extruder.

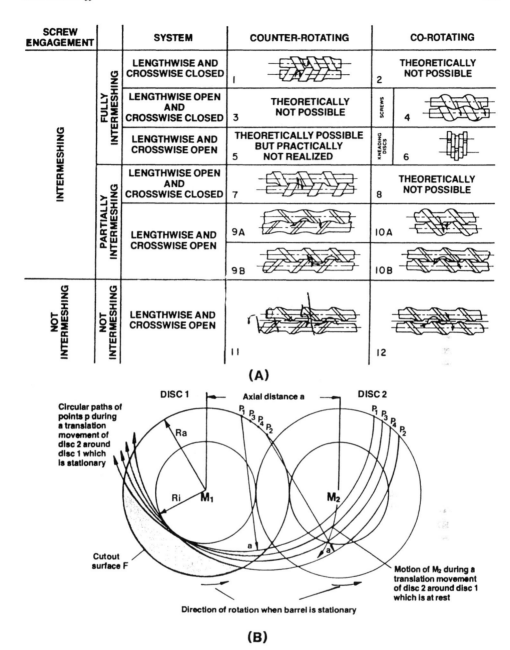

SCREW ENGAGEMENT		SYSTEM	COUNTER-ROTATING	CO-ROTATING
INTERMESHING	FULLY INTERMESHING	LENGTHWISE AND CROSSWISE CLOSED 1		THEORETICALLY NOT POSSIBLE 2
		LENGTHWISE OPEN AND CROSSWISE CLOSED 3	THEORETICALLY NOT POSSIBLE	SCREWS 4
		LENGTHWISE AND CROSSWISE OPEN 5	THEORETICALLY POSSIBLE BUT PRACTICALLY NOT REALIZED	KNEADING DISCS 6
	PARTIALLY INTERMESHING	LENGTHWISE OPEN AND CROSSWISE CLOSED 7		THEORETICALLY NOT POSSIBLE 8
		LENGTHWISE AND CROSSWISE OPEN 9A		10A
		9B		10B
NOT INTERMESHING	NOT INTERMESHING	LENGTHWISE AND CROSSWISE OPEN	11	12

(A)

DISC 1 — Axial distance a — DISC 2

Circular paths of points p during a translation movement of disc 2 around disc 1 which is stationary

Ra

$P_1 P_3 P_4 P_2$ $P_1 P_3 P_4 P_2$

Ri M_1 M_2

Cutout surface F

a a

Motion of M_2 during a translation movement of disc 2 around disc 1 which is at rest

Direction of rotation when barrel is stationary

(B)

FIGURE 4. (A) Twin-screw arrangements and configurations. (B) Corotating screw arrangement.

Counterrotating screws operate with both shafts turning with equal angular velocities in a counteracting rotation. Figure 5B is a vector diagram illustrating the velocity fields and forces of a counterrotating screw operation. In contrast, with corotating screws as shown in Figure 4, when screw 1 is held stationary, screw 2 does not roll off but does make a circular translation motion around screw 1. The center point M_2 moves in this fashion on a circular path around M_1, with the radius M_{1MPV2}; screw 2 itself produces no rotation of its own. Each point describes a circular path with the same radius. If the arc $P_1 - P_2$ on the circumference of disc 2 is a screw crest, the points of this arc describe (as shown with two

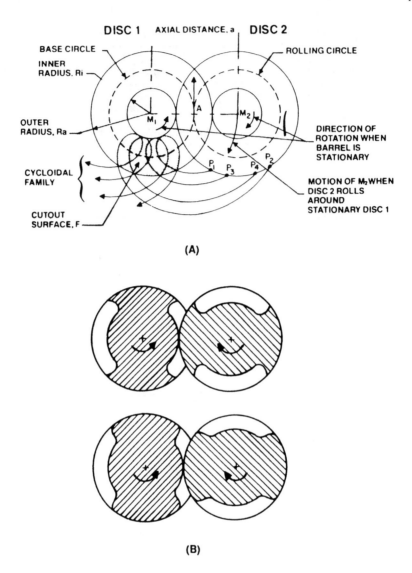

FIGURE 5.   (A) Vector diagram of counterrotating screw. (B) Cross-sectional view.

intermediary points P_3 and P_4; Figure 4) a family of arcs that cut out a half moon-shaped surface area F from screw 1.

Each point along the axial screw surface is tangentially wiped by the other screw with a constant relative velocity. For simplicity, consider the screw simply as a flat disc as shown in Figure 5. If disc 1 is held stationary, disc 2 with rolling circle radius AM_2 rolls off the base circle with the radius M_1A of disc 1. Each point of disc 2 describes an elongated or abbreviated epicycloid depending on its distance from M_2. Let the arc $P_1 - P_2$ of the circumference of disc 2 define a screw crest; the points of this arc, represented with two intermediary points (P_3 and P_4), then describe a family of cycloids that "cut out" the surface area F from disc 1. The shape of the cutout F corresponds to the crosscuts of the screw channel perpendicular to the turning axis. Figure 5B illustrates two examples of these screw crosscuts. The forward rotation of the thin discs in a direction lengthwise to the shaft axis results in screws with positive and negative pitch. Each point of the surface area is wiped by the other screw. The screw flanks are developed in the crosscut from a cycloidal family, and both screws roll off one another in a fashion resembling that of gear wheels. In the wiping points, the relative velocity depends on the distance of these points from the rotating axis and is in the area of $-2\pi n(R_a - R_1) \leq V_{rel} \leq + 2\pi n(R_a - R_1)$. In contrast, with counterrotating arrangements, the screws in the wedge area roll off one another in a rolling motion with locally different gliding velocities. This gliding velocity is small compared to the circumferential speed and is typically $2n(R_a - R_1)$. Corotating screws do not roll off one another at any point but instead have a translation movement in which one crest edge wipes a screw flank tangentially with equally high relative velocity. The relative velocity is high compared to the circumferential speed and is $V_{rel} = 2\pi n(R_a + r_1)$. In the wedge area, the screws with equal pitch and equal pitch direction are intermeshing.

The open area in the wedge part between the two screws is different. In counterrotating designs, there are closed C-shaped chambers with a very small volume such that no material transfer from one screw to the other takes place. In this type of design, the distributive mixing efficient is greatly reduced.

In contrast, corotating screws form V-shaped wedge areas that have four to five times more volume. This enables material to be transferred from one screw to the other, resulting in a renewal of material layers and surfaces. The net result is a higher degree of mixing.

Self-cleaning screws are often employed to prevent material from adhering to the screw root. Material that adheres to the screw root can degrade, because of a broad residence time distribution. They eventually fall off and are carried out with the product, showing up as contamination.

Self-cleaning action is achieved in both counterrotating and corotating screws through an opposite roll-off motion or wiping motion. This self-wiping is achieved in different ways in each system, each having different degrees of effectiveness. With counterrotating screws, the roll-off process between the screw crest and screw root and between the screw flanks simulates the action of a calendar. The necessary shear velocity required to wipe the boundary layers is proportionally lower because of the low relative velocity. Also, the material is drawn into the roller gap and is squeezed onto the surface.

In corotating screws, one crest edge wipes the flanks of the other screw with a tangentially oriented, constant relative velocity. There is a higher relative velocity in this arrangement, and hence a sufficiently high shear velocity is available to wipe the boundary layers. The calendar effect does not occur; however, a more efficient and uniform self-cleaning action is achieved.

The rolling motion of the screws in the counterrotating arrangement causes a calendar effect that results in pressure between the roll nip surfaces. This pressure acts to push apart the shafts. Figure 6 illustrates the pressure and velocity distribution in the roller gap. The resultant parting forces on the two shafts can cause wear of metal surfaces.

Wearing of the screw surface increases in proportion to the relative velocity of the shafts. To minimize this effect, intermeshing, self-wiping, counterrotating machines should be operated at relatively low speeds.

Since corotating screws have no calendering effect between the crest and the flank of the screws, there is considerably less wear. Therefore, corotating machines can be operated at much higher screw speeds with greater throughputs. A typical corotating, self-wiping, twin-screw extruder with 170 mm diameter screws can process 9,000–11,000 lb/h of polypropylene at a screw speed of 300 rpm.

Some attention must be given to the design of the feed intake zone of twin screws. Figure 7 shows the cross section in the feed zone of a counterrotating screw machine. To maximize feed intake, the screws turn outward on top and inward at the bottom. If the screws turned inward at the top, the material would have to be pulled in by the calender gap and there would be little free volume available for the material to be drawn in. The polymer is conveyed to the lower wedge, where it is partially compressed and conveyed as

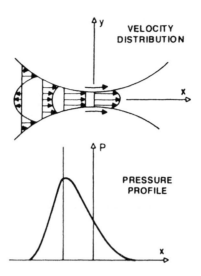

FIGURE 6. Velocity distribution and pressure profile in calender gap of counterrotating screw.

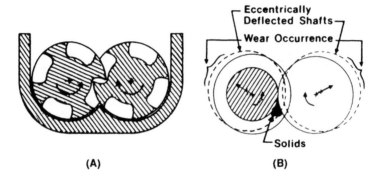

FIGURE 7. (A) Feed intake zone. (B) Solids crushing in wedge region.

a unit volume toward the downstream melting zone. To maximize the intake free volume, multiflighted deep-cut screw elements are sometimes used. The conveying principle of the counterrotating screws is based on closed-chamber unit volumes, which is desirable for feeding such solids as plastic resin pellets. The entire free volume can be utilized, assuming optimum pitch angle. It should be noted, however, that because of the calendering effect, solids are force between the screws and crushed (refer to Figure 7). This action causes the screw elements to be forced apart in an angle at the top, where only the barrel can provide support, which can result in severe wear. The wearing action can be minimized if the gap between the screws is relatively large and if the screw speed is low. However, the conveying volume is proportional to the screw speed, and hence a low screw speed means lower throughput. Larger flank and root clearance measures can reduce the throughput.

In contrast, corotating screw elements convey the material from one screw toward the lower wedge (depending on wedge resistance); the material is compressed and then picked up by the other screw and conveyed farther. This is illustrated in Figure 8. The wedge area provides a twist restraint on the resin, which reduces the tendency of the material to rotate in the channel and increases conveying capacity in the axial direction. The twist restraint in the wedge area varies with the number of flights per unit length. The larger the wedge angle, the larger the twist restraint and the greater the cross-sectional reduction in the wedge

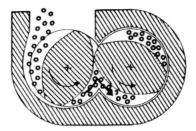

FIGURE 8. Feed intake zone of corotating screw.

area during transfer of the material from one screw to the other. Screw bushings with wide crests can cause significant changes in direction of material flow, leading to larger twist restraints. High conveying angles can be achieved with fully intermeshing, self-wiping corotating screws. The conveying angle is the angle between the vertical line to the screw axis and the true conveying direction of the product. With single-screw machines, the conveying angle can be calculated by analyzing the machine design parameters and the physical properties of the solids. Such relationships do not yet exist for corotating twin-screw extruders since the function of the wedge area has not been mathematically formulated. Therefore, experimental measurements of the conveying angle must be made clear. Figure 9 shows the conveying angle of a fully intermeshing, self-wiping corotating twin screw compared with a single-screw system in the processing of high-density polyethylene powder. Note that each time the pitch approaches $H = D_a$, a maximum is reached. Furthermore, the slope of the pitch in corotating screws is lower than that of single screws. The conveying angle of fully intermeshing, self-wiping corotating screws is generally two to three times greater than that of a single-screw machine, as noted by Eise et al. (1981).

We now direct attention to the melt zone of the machine, which can be divided into two sections:

1. Compacting of the fluffy products to minimize air pockets: the air in most cases escapes through the feed opening. Simultaneously, a melt film is formed on the barrel walls and the resin solids fuse because of heat and pressure.
2. Generation of melt through the shearing section of the product and mixing of the unmolten particles with the already formed melt: as a result, a large heat exchange surface is generated between the solid particles and the melt.

With fully intermeshing, self-wiping counterrotating screws, compaction of the material is achieved by several methods:

1. Continuously reducing the screw channel
2. Using various melt zones with different screw pitches and numbers of flights
3. Continuously increasing the crest widths
4. Continuously decreasing outer and/or inner diameters

In machines having a tight clearance between screws, a compression over several screw flights cannot be obtained, nor will all the air completely escape the individual chambers toward the feed. It is for this reason that counterrotating screws are constructed with loose clearances. This means that the advantages of closed chambers offered by the counterrotating principles are ignored. The melting process is similar to that in single-screw machines, with the following actions:

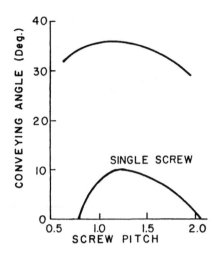

FIGURE 9.   Conveying angles of twin-screw and single-screw machines as functions of screw pitch.

1.   Development of a melt film on the barrel wall
2.   Scraping off the melt film by means of a pushing side of the flank
3.   Collection of melt in front of the pushing flank

The rolling motion of the counterrotating screws produces the following additional effects:

1.   In the wedge intake, a melt whirlpool is created that draws in solid particles and softens them.
2.   The plasticized mass is then taken in by the calendar gap and collected at the trailing flank.

Because of the calendering effect in counterrotating screws, within one rotation the chamber volumes are axially displaced by distance H. One screw can roll off the other one time; therefore only a fraction of the chamber volume passes the calender gap. Despite the high axial velocity, a relatively low shear action of the calender gap is achieved, which is dependent on the low relative velocity $-2\pi n(R_z - R_i) \leq V_{rel} \leq 2\pi n(R_z - R_i)$ and the width of the calendar gap. If the machine has tight clearances, a high shear velocity results but only small amounts of the product are drawn into the gap. High clearance draws in a greater amount of the chamber volume, however, with a low shear effect. Increasing the screw speed results in displacement of the melting zone toward the discharge end, which can result in an inhomogeneous product. Additional mixing devices (e.g., interrupted flights) result in improved mixing and uniform heating of the product. In general, however, the plasticizing process is only marginally improved. Unmolten particles can often be observed in the discharge during such operations.

In corotating screws (fully intermeshing and self-wiping) compression of the solids that are taken in is produced through restriction, which is caused by the reversed screw (conveying in an opposite pitch direction). These reversed screws are referred to as left-handed screws. Figure 10 shows several screw arrangements for the plasticating zone. In the left-handed screw restriction, pressure buildup is generated by the upstream right-handed screws. Plasticizing results from the backup length generated by the left-handed screws. Melting of the resin is similar to that of a single screw, that is, generation of melt film on the barrel wall and sintering of the product; scraping off the film from the barrel wall by pushing flank; and collection of a melt pool in front of the pushing flank.

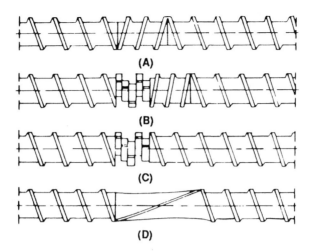

(A)

(B)

(C)

(D)

FIGURE 10. Screw rearrangements in plasticating zone.

Through proper specification of pitch angle and screw speed, it is possible to shift the zero shear strain point of the axial flow components outside the screw channel. This can result in greater shear stress imposed on the suspended solids particles than is possible with counterrotating screws. In the wedge area, the material is picked up by the other screw and does not undergo a calendering effect. The stresses are applied between the flanks by constant relative viscosity $V_{rel} = 2\pi n(R_a + R_i)$.

With the help of special mixing and shear elements, such as kneading discs or self-handed elements with large pitch, plasticating can be performed. These elements improve the self-wiping action. Kneading discs are designed to have the same cross section as the corresponding screws. Discs with various widths and staggered angles can be arranged along the shaft. A kneading block works as follows. When the barrel is rotated relative to the disc, a wedge-shaped flow develops. Figure 10 shows the velocity distribution and pressure profile in such a wedge when it is partially filled.

Counterrotating screws essentially have plug flow caused by their low clearance. To increase longitudinal mixing, the flanks and the root clearances must be enlarged. A system open in length and cross directions is purposely created to produce better homogenization of material.

In contrast, the corotating screw conveys material by means of frictional force. As such, there exists a flow in the partially filled melting zone, in which each particle is subjected to the minimum shear velocity of $\dot{\gamma} = 2nRa/n$. In this situation there is no shear-free dead zone. Varying the screw pitch and speed provides different degrees of fill in the screw channel.

Counterrotating screw systems offer a positive conveying capability. Melt conveying against pressure in closed chambers has disadvantages. The chamber volume is intermittently released, which leads to pressure and throughput fluctuations. To minimize these fluctuations, the passage and conveying capabilities of the screw geometry must be in phase with each other. Overfeeding or underfeeding of the screws increases the magnitude of these fluctuations. Another way of minimizing the fluctuations is to purposely open the clearances to approach an open lengthwise and open cross-direction system.

In corotating screws, the screw crests of corotating screws are connected to form the free wedge volume. Even in a fully intermeshing corotating mechanism, an open system exists. Partial positive conveying is obtained because of the influence of the wedge resistance. Downstream restrictions, such as dies, cause the melt to accumulate along the screw channel upstream, usually over several turns of the screw. Through proper screw configuration and appropriate processing parameters, the development of dead shear zones in the screw channel can be prevented. Optimal processing conditions can be calculated and adjusted, which results in uniform shear stress distribution and minimal backup length.

As noted earlier, the effects of velocity distribution, stress distribution, and strain imposed in the material are distinct in each process. In addition to basic geometric variations possible in corotating and counterrotating mechanisms, an additional degree of process flexibility is possible with corotating mechanisms, such as screw speed and throughput. Fully intermeshing counterrotating screws form lengthwise and crosswise closed C-shaped chambers, resulting in a plug flow similar to that observed in a screw pump. In contrast, fully intermeshing corotating screws do not form closed chambers but, rather, lengthwise open channels and crosswise closed chambers when mechanical clearances are disregarded. Polymer conveyance in this case depends on drag flow. Forced conveyance can be defined as an angle of conveyance based on (opposing) screw intermesh, in which material is conveyed forward by each screw rotation for at least the screw crest width. Because of the presence of an uninterrupted open channel, material exchange can take place lengthwise along the machine, resulting in a mixing system based on the drag flow within the lengthwise open screw channel.

Because of geometric differences, there are basic differences in the velocity and stress distribution of downchannel flow for intermeshing corotating and counterrotating twin-screw mechanisms that have a significant effect on the plasticizing and mixing capabilities of twin-screw systems.

To gain a better perspective, consider a coordinate system superimposed on the rotating screw surface so that the screw surface is stationary relative to the observer while the barrel rotates at a constant velocity.

Positioning the coordinate system on the screw surface of twin-screw intermeshing mechanisms would not be meaningful since the figure eight-shaped barrel opening and corresponding screw would rotate in relation to the stationary screw where the coordinate is placed. Therefore, the coordinate system should be placed on the inner barrel surface parallel to the screw axis. It is moved along in the axial direction of the screw with the velocity $V_1 = H_n$, where H designates screw pitch and n screw rpm. With a constant screw speed and the same screw geometry along the machine length, the boundary conditions become constant. Following Eise et al. (1981), the coordinate system is such that the x-axis is normal to the screw channel, the y-axis perpendicular to the barrel surface, pointing toward the rotating axis, and the z-axis parallel to the screw channel, as shown in Figure 11.

From a force balance, where y_{0z} is the integration constant that denotes the location of zero stress point of downchannel flow, the counterrotating screw rearrangement forms closed C-shaped chambers. Therefore, the flow Q in the z-direction for the selected coordinate system is

$$\frac{Q}{W} = \int_0^h V_z dy = 0 \tag{35}$$

where W = channel width.

Assuming a Newtonian fluid, Equation 35 can be integrated and solved for the velocity and stress distributions:

$$V_z V_0 = \frac{1}{\cos \phi} \left\{ 3 \left[\left(\frac{y}{h} \right)^2 - \frac{y}{h} \right] (\sin^2 \phi + 1) + \frac{y}{h} \cos^2 \phi + \sin^2 \phi) \right\} \tag{36}$$

$$\frac{\tau_{yz}}{n(V_0/h)} = -\frac{1}{\cos \phi} \left[3 \left(2\frac{y}{h} - 1 \right) (\sin^2 \phi + 1) \cos^2 \phi \right] \tag{37}$$

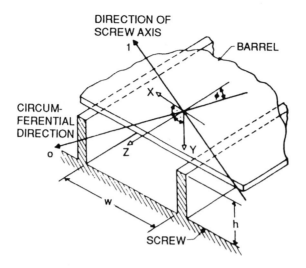

FIGURE 11. System configuration for twin-screw machine.

The velocity and stress distributions for a conventional screw pitch ($\phi = 17.66°$) are shown in Figure 12A for the counterrotating system. At $y/h = 0.36$, there is a zero stress point. The same zero stress point exists for each screw geometry.

In corotating screws, open channels are formed. The material can therefore pass from one screw to the other along the z direction. The flow Q can never reach zero in a selected moving coordinate system when mechanical clearances are neglected. Pressure drop can develop in the z direction, which varies depending on operating conditions between zero (pure drag flow) and a maximum value. The highest pressure is developed when the discharge end is closed. Although of no practical significance, such an analysis establishes the maximum pressure buildup capabilities of the machine. In this limiting case the observer witnesses a throughflow that corresponds to a theoretical plug flow, for which the throughput is

$$Q = W \int_0^h V_z \, dy \tag{38}$$

Upon integration, the maximum pressure gradient is obtained:

$$\left. \frac{\partial P}{\partial z} \right|_{max} = \frac{6\pi DN\eta \cos \phi}{h^2} \tag{39}$$

For this maximum pressure gradient, the following velocity and stress distributions are derived:

$$\frac{V_z}{V_0} = \cos \phi \left[\tan^2 \phi + 4\frac{y}{h} - 3\left(\frac{y}{h}\right)^2 \right] \tag{40}$$

$$\frac{\tau_{yz}}{n(V_0h)} = -2 \cos \phi \left(2 - 3\frac{y}{h} \right) \tag{41}$$

The distributions are shown plotted in Figure 12B, indicating that the zero shear stress point exists in the screw channel at $y_{0z}/h = 2/3$. In the open discharge case, the pure drag flow

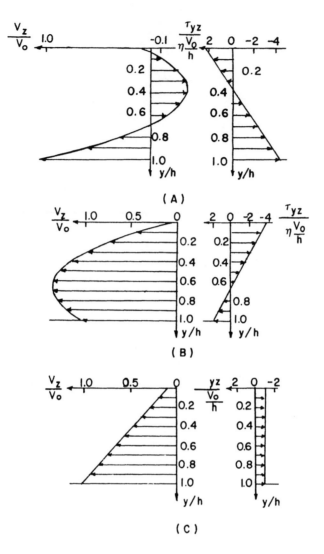

FIGURE 12. (A) Velocity and shear stress distributions of downchannel flow for counterrotating screws. (B) Distributions of downchannel flow for corotating system with closed discharge. (C) Distributions with drag flow.

($\delta P/\delta z = 0$) and the velocity and stress distributions can be determined by solving Equation 38:

$$\frac{V_z}{V_0} = \frac{y}{h} \cos \phi + \frac{\sin^2 \phi}{\cos \phi} \tag{42}$$

$$\frac{\tau_{vz}}{n(V_0/h)} = -\cos \phi \tag{43}$$

Velocity and stress distributions in partially filled channels are given in Figure 12C. As shown, there is a constant stress component for the downchannel flow in the screw channel. The relationship between pressure gradient, velocity, and the location of the zero stress point can be obtained from the following equations:

$$\frac{V_z}{V_0} = \cos \phi \, \frac{1/2 \, y^2/h - y_{0z}/h \, y/h}{1/2 - y_{0z}/h} + \tan^2 \phi \tag{44}$$

and

$$\frac{T_{yz}}{n(y_0/h)} = \cos \phi \, \frac{y/h - y_{0z}/h}{1/2 - y_{0z}/h} \tag{45}$$

In the preceding analyses, the effects of screw clearances and the wedge area were not considered. Also, it was assumed that a Newtonian fluid was used. With viscous materials, parabolas of a higher order for velocity distribution are observed. The shear rate at the wall is increased accordingly as the areas with lower shear stresses become broader. Significant qualitative differences are not obvious; however, the quantitative differences can be significant.

The velocity distribution and the shear stress distribution of the downchannel and cross-flow for counterrotating screws are given by Eise et al. (1981). With both flow components, the apex of the velocity distribution, which is identical to the zero shear stress point, is located in the screw channel. For downchannel flow this is $y_{0z}/h = 0.36$ and for cross-flow it is $y_{0x}/h = 0.66$. The velocity distribution in fully intermeshing, counterrotating screws that convey materials in C-shaped chambers can only be influenced by means of the screw pitch. The boundary layers, particularly at the screw root, are subjected to the greatest stresses, and since adherence to the wall is a prerequisite, they are also subjected to the greatest amount of shear.

The velocity distribution of corotating screws can be influenced by changing the through-put or by changing the screw speed as well as by changing screw pitch. Eise et al. (1981) have shown that, with the same boundary conditions, different velocity distributions and therefore different shear stress distributions result. Equation 84 states that the velocity distribution is dependent not only on the pitch angle ϕ but also on the position of the zero shear stress point y_{0z}. However, this is a function of the pressure gradient, the viscosity, the screw speed, and the screw geometry as shown by the relation

$$\frac{y_{0z}}{h} = \frac{\pi D n \eta}{\cos \phi \, h^2 (\partial p/\partial z)} (\sin^2 \phi - 1) + 1/2 \tag{46}$$

The dependence between y_{0z}/h and the flow Q can be seen from Equations 38 and 49. The velocity distribution can therefore be altered considerably by preselecting the throughput and/or screw speed.

The relationships between the intensity, uniformity, and location of maximum and minimum shear stresses define the basic differences between corotating and counterrotating screws with identical screw geometries. These differences can be summarized as follows:

- The shear stress τ_b at the barrel wall is greater with counterrotating screws than with corotating screws.
- The shear stress τ_s at the screw root is greater with counterrotating screws than with corotating screws.
- The maximum shear stress τ_{max} in the counterrotating screws (not considering the wedge area and mechanical clearances) is identical to the shear stress at the screw root, τ_s. In corotating screws, it is identical to the shear stress at the barrel τ_b.
- In counterrotating screws the minimum shear stress is at $y/h = 0.38$; in corotating screws, however, it is between $0.66 < D(Y/h)A < 0.9$.

- In counterrotating screws, the change in the shear stress $\delta\gamma/\delta z$ in the channel depth is considerably larger than with corotating screws, assuming that in actual production closed discharge does not occur. This means that the shear stress distribution of counterrotating screws is considerably broader than in corotating screws.
- The shear stress distribution is dependent only on the screw pitch H in counterrotating screws. In contrast, in corotating screws not only does the shear stress distribution change with the pitch angle ϕ, but it is also dependent on the throughput and the screw speed.

Because of the first two differences between the designs, as well as the increased tendency toward mechanical wear in counterrotating screws, the operating speeds of counterrotating screws must be relatively low. When considering throughput, the superior conveyance of the counterrotating screw is offset by its higher operational speeds.

In counterrotating screws, the material at the screw root is drawn into the wedge area, where it is exposed to high strain. In counterrotating screws the minimal shear stress is inside the chamber. On the other hand, by properly selecting operating conditions for corotating screws, the shear stress is nearer the screw root.

In corotating screws the shear stress distribution under actual operating conditions is in the area of $1 < y_{0z}/h < \infty$ and is considerably more uniform. In comparison to counterrotating screws, the layers at the screw root are subjected to less shear strain. These layers are wiped by the crest of the oncoming screw into the wedge area and are then recombined near the barrel wall. The material layers from the area of the barrel wall are forced in the wedge area into the bottom of the channel of the oncoming screw. In corotating screws, because of the absence of the low shear stress zone and the layer transfer rearrangement in the wedge area, the shear strain on the product is more uniform. This is a favorable condition for uniform plasticizing and homogeneous mixing. Because external heating during the plasticizing process plays a secondary role, higher screw speeds and large screw diameters can be employed.

IV. COUNTERROTATING, TANGENTIAL TWIN-SCREW EXTRUSION

The counterrotating, intermeshing extruder is most commonly used for the processing of rigid polyvinyl chloride (PVC) for pipe or siding. This type of extruder provides positive conveying characteristics suitable for feeding rigid PVC powder dry blend, and when combined with a low-speed screw (typically 50 rpm or less) provides a low shear environment suitable for the processing of materials with poor thermal stability.

The corotating, intermeshing system does not have positive conveying characteristics like the counterrotating intermeshing system. As noted, conveying takes place by drag flow. Because the channel is axially open, this system, like the counterrotating, tangential type, can be applied to process engineering (e.g., venting).

The counterrotating, tangential type was the first twin-screw extruder to be manufactured in the U.S. The counterrotating, tangential screw arrangement has been used in a broad range of process applications throughout the polymer processing industry over the last 33 years. Nichols (1982) compares the twin-screw extruder with single-screw extruders.

As noted, tangentially opposed screws are not positive conveying devices. However, unlike single-screw extruders, which depend upon frictional forces for solids conveying, in the twin-screw extruder the lower apex applies a "keying" effect analogous to grooved feed bushings in single-screw extruders. This effect continues through all the functional zones and can be readily observed in the partially filled vent zones as a rolling bank of polymer on the lower apex. The outstanding feed acceptance of these extruders allows extremely

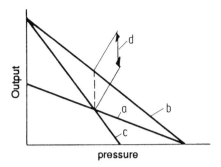

FIGURE 13. Screw characteristics of (a) single, (b) two single screws, and (c) a twin screw.

uniform performance while starve feeding, which is the feeding method most commonly employed with all twin-screw extruders. Although starve feeding has been applied successfully to single-screw extruders, its use has been limited because of feed acceptance problems. The highly advantageous feed acceptance of counterrotating, tangential extruders also provides the possibility of sequential feeding. A benefit of sequential feeding is that the separate ingredients of a compound can be handled with dedicated feeding equipment, thus eliminating the need for preblending. Other advantages of sequential feeding include higher total throughput rates, lower product temperature, and improved product mixing.

Melting in single-screw extruders was described earlier. In the counter-rotating, tangential twin-screw extruder, the melting mechanism varies considerably from that of the single-screw model in that the keying effect of the lower apex and the interaction of the staggered screw flights tend to promote a considerable amount of exchange between the channels of the two screws. Thus, the polymer melt and solid granules become mixed together instead of these being a clear separation of melt pool and solid bed.

In single-screw systems, screw designs to accelerate melting for high efficiency have developed along two opposing concepts. In the first case the melting rate is controlled by providing a barrier between the solid bed and the melt pool to assure that the solid bed does not break up prematurely and become encapsulated in the melt. An opposing concept is the wave screw. In this design concept, conventional feed and melting zones are employed until the point at which about 50% of melting is completed, and then the melt and solids are mixed together. This is accomplished by varying the metering channel depth in a sinusoidal pattern, alternating between very shallow, high-shear zones and rather deep, low-shear zones. The effect of this design is to promote the distributive mixing of the now thoroughly broken up solid bed with the melt pool, thus using the residual melt pool heat to complete the melting of the remaining solids, resulting in a very low average extrudate temperature. This mechanism is similar to that postulated for the counterrotating, tangential extruder except that the mixing effect of twin-screw extruders is generally conceded to be significantly superior to that which can be obtained in a single-screw design.

A comparison of the characteristic curves of the single-screw and the counter-rotating, tangential extruders is given in Figure 13.

The single-screw characteristics at zero pressure or open discharge are defined by the drag flow equation, and the slope of this screw characteristic over a range of pressure is established by subtraction of the pressure flow characteristic for various pressures. The screw characteristic curve for two single screws is the algebraic sum of the result for a single screw. The screw characteristic for the counterrotating, tangential twin-screw machine has a steeper slope than the characteristic of either one or two single screws. The distance between the intersection of the twin-screw characteristic and the single-screw characteristic lines and a point at equal pressure on the characteristic line of two single screws can be defined as the longitudinal mixing flow through the open area at the apexes (Figure 13). This flow component is a pressure flow term, and hence as pressure flow increases, pumping capacity is correspondingly reduced. Several components combine to produce the total mixing capability: the rotation of the material in the screw channel, as would normally be experienced in a single-screw extruder; interchange of material between the two screws; and the longitudinal mixing flow component. These mechanisms combine to homogenize materials going through the apparatus.

In general, all extruders are assumed to be nominally "once-through equipment." This implies that preblending, or bulk mixing, is essential for good mixing results. In fact, for single-screw extruders in which feed acceptance is often poor, preblending is a necessity. In counterrotating, tangential twin-screw extruders, however, since feed acceptance is unusually good as long as sufficient accuracy is provided in the associated feeding devices, uniform product quality can be obtained by feeding separate streams either at the same axial location or sequentially in various axial locations.

Two important types of mixing are distribution and dispersion. Distributive mixing is used for any operation employed to increase the randomness of the spatial distribution of the ultimate particles without reducing their size. In simple mixing, the ultimate particles of the system pass from a less probable to a more probable arrangement. The term "dispersive mixing" applies to those mixing processes that reduce the size of the ultimate particles as well as randomize their positions. Dispersive mixing depends on exposing agglomerates to ultimate particles to sufficiently high shear stresses to cause them to break down. The level of shear stress required for deagglomeration to occur is related to the size of the ultimate particles and to the nature of the bonds holding the agglomerate together. The shear stress in the molten polymer is proportional to the product of viscosity and shear rate. In general, the shear stress in normal melt pumping channels is not sufficiently high to provide adequate dispersive mixing. In counterrotating, tangential extruders, it is often beneficial to introduce all the components requiring dispersion in the initial feed so that they are exposed to the high shear stresses existing in the melting zone, where the viscosity is very high. Consequently, the polymer melt is already prepared for distributive mixing in a lower viscosity zone downstream. Alternatively, high shear stress mixing devices, such as reverse flight compounders or very restrictive cylindrical compounders, may also be used in the fully developed melt region, but this may lead to undesirable temperature rise in the polymer melt, causing degradation. Distribution is somewhat limited in single-screw extruders, and various auxiliary devices must be used to improve distributed mixing in these machines.

One approach is the use of a static mixer, which provides a high degree of radial mixing as a result of the placement of a series of fixed blades that redirect the polymer flow path. This type of device is usually employed between the end of the extruder screw and the extrusion die.

In many applications, there is the additional requirement of removing undesired volatiles from molten polymers. This has resulted in the development of various types of vented extruders. The screw section under the vent cannot be completely full or a path is not provided for the escape of the volatiles, and further, if any positive pressure exists in this region, the polymer flows out of the vent opening. The relationships between the pumping section upstream of a vent section (which pumps against zero pressure) and the downstream pump (which must overcome the die resistance) are well defined if certain simplifying assumptions are made. The key assumptions involved by the designer are as follows.

The flights in the vent region between the two pumping sections are deeper than the pumping sections and are therefore only partially filled.

The upstream pumping (or melt metering) section is assumed to be full and pumping against zero discharge pressure and is pumping at a rate consistent with the drag flow

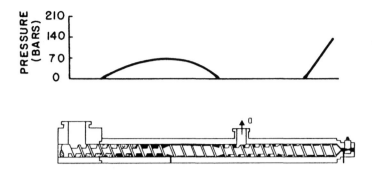

FIGURE 14. Screw design and pressure profile for a two-stage vented screw.

calculation. Figure 14 illustrates a typical screw design (single) along with the pressure profile for a two-stage venting system. The assumption that the first-stage metering section is the rate-controlling section of the entire screw leads to the following relationships. Since the first metering section does not pump against a restriction, its output rate is defined by the drag flow relationship

$$Q_{net} = Q_{d1} \tag{47}$$

Since the second metering section must overcome the resistance of the die assembly, its rate must be defined by the algebraic sum of the drag and pressure flow, as noted earlier:

$$Q_{net} = Q_{d2} - Q_{p2} \tag{48}$$

Combining these expressions provides the mass balance for the screw:

$$Q_{net} = Q_{d1} = Q_{d2} - Q_{p2} \tag{49}$$

There are two imbalances in this expression: vent flow and surging. Vent flow Q_{vent} is

$$Q_{d1} > Q_{d2} - Q_{p2} \tag{50a}$$

or

$$Q_{d1} = Q_{d2} - Q_{p2} + Q_{vent} \tag{50b}$$

Surging is given by the inequality

$$Q_{d1} < Q_{d2} - Q_{p2} \tag{51}$$

At steady-state conditions, Statement 51 cannot exist. It does, however, indicate a real condition in which the pumping capacity of the second metering section is such that until it "fills up", it pumps inefficiently, and when full, it pumps with such great efficiency that it empties before the first-stage pump can replenish it. This in fact is the surging due to a poorly designed two-stage screw.

In the design of a vented extruder screw, several criteria are needed in addition to the requirements of rate, melt quality, and average melt temperature. These criteria are devolatilizing capability, pressure stability, and pressure-generating capability. Several design

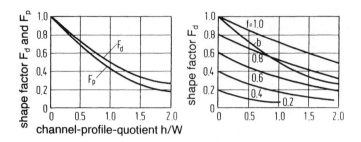

FIGURE 15.   Shape factors for drag F_d and pressure flow F_p depend on channel profile h/W.

factors affect pressure-generating capability. Inspection of the terms in the drag and pressure flow equations, neglecting leakage flow, help to define these.

The drag flow equation is

$$Q_d = F_d \frac{\pi D^2 h(1 - ne/t) \sin \phi \cos \phi}{2} N \tag{52}$$

The pressure flow equation is

$$Q_p = F_p \frac{nDh^3(1 - ne/t) \sin \phi \cos \phi}{2} N \tag{53}$$

where the geometric shape factors F_d and F_p depend on the ratio channel depth h and channel width W. This dependence is shown in Figure 15.

These relationships show that drag flow is directly proportional to the channel depth and the screw speed and is not related to the polymer viscosity, nor is it affected by the metering section length. In contrast, pressure flow is proportional to the third power of the metering section depth, is independent of screw speed, is a function of polymer viscosity, and is inversely proportional to metering section length. This suggests that the second pump depth is the most convenient variable to modify to correct a vent flow problem due to an imbalance as described by Equation 50a. In some cases, this is a correct action to take, but often it can aggravate the problem. Simplifying Equations 52 and 53 and substituting appropriately in Equation 49, we obtain

$$K_1 h_1 = K_1 h_2 - K_2 h_2^3 \Delta P \tag{54}$$

where K_1 represents all the terms in the drag flow equation 52 except the channel depth (h_1 or h_2) and K_2 represents all the terms in the pressure flow equation 53 except the channel depth h^3 and head pressure ΔP.

By rearranging Equation 94, the following is obtained:

$$\Delta P = C \frac{h_2 - h_1}{h_2^3} \tag{55}$$

This relationship can be expressed in a dimensionless form and is shown graphically in Figure 16 as pressure capability vs. pump ratio for the two metering zones. Note that the pressure-developing capability reaches a maximum (for Newtonian fluids) at a 1.5:1 pump ratio. With non-Newtonian polymers, the maximum pressure development can occur at approximately a 1.75:1 pump ratio. Hence, the second-stage pump cannot be greatly deepened.

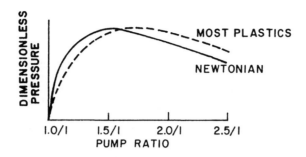

FIGURE 16. Dimensionless pressure capability vs. pump ratio.

Nichols (1982) reports that the practical limits of pump ratio are a minimum of 1.5:1 and a maximum of 2.0:1. Working toward the lower limit tends to enhance stability but limits pressure capability; working toward the upper limit may help maximize pressure capability but may also result in somewhat less stability, depending on how the extruder is operated. A fixed orifice valve with degradable polymers or an adjustable valve with stable polymers is recommended.

Since there is a limitation on the second pump depth, the only two variables that can significantly affect the pumping (pressure-generating) capability are viscosity and the length of the second metering section. The extruder manufacturer does not have the ability to specify the polymer viscosity, and therefore the only variable left to work with is the second pump length.

Attention is now given to devolatilization. The devolatilization rate is generally thought to be limited by diffusion of the solvent molecule through a thick film of polymer. From this viewpoint, the polymer fill can be considered an infinite slab; hence the approach to equilibrium is limited by end-to-end mixing (backmixing) and that surface generation is combined with surface exposure time.

A counterrotating, tangential extruder has a unique devolatilizing capability and is found in widespread use in some of the most demanding large-scale applications at rates of 5000–7500 kg/h dry product, starting with solvent levels as high as 50%. Single-screw extruders, in comparison are rarely used for volatile levels above 1%.

With counterrotating extruders, the additional pressure flow through the open apex area (the longitudinal mixing flow component) is detrimental to high pressure generation. Therefore, counterrotating, tangential extruders often utilize single-screw discharge pump sections. These can be designed as an integral part of the main extruder screw operating at the process screw speed, or they can be incorporated as crosshead extruders of different diameters and speeds. To calculate the final discharge pump depth in a multistage system, one need only determine the apparent metering capability of the first pump (in twin-screw extruders, this is often determined by a feed rate setting) and then the single-screw equations apply directly to design the discharge pump. To calculate pumping in the twin-screw sections, based on an isothermal Newtonian model using drag flow and pressure-flow correction factors to the basic single-screw equations, use the equation

$$q = \tfrac{1}{2} W H V_{br} F_{DTW} - \frac{W H^3}{12\eta} \frac{\Delta P_T}{\Delta Z_T} F_{PTW} \qquad (56)$$

where

$$F_{DTW} = \frac{4f}{1 + 3f} \qquad (57)$$

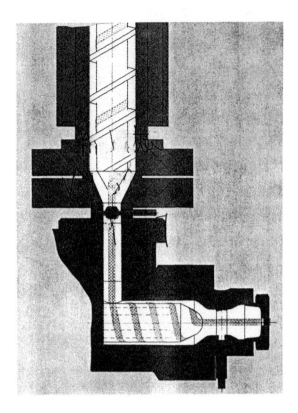

FIGURE 17.   Shear head configuration.

$$F_{PTW} = \frac{4}{1 + 3f} \tag{58}$$

Parameter f is the fraction of helical length occupied by zone 1 and typically has a value of 0.9. The ratio $q_s/q_D \approx 0.5$, and twin-screw extruders typically deliver about 1.73 times the flow rate of a single-screw machine.

Because of good devolatilizing capabilities, typical applications of the counterrotating, tangential twin-screw extruder are desolventizing during polymer manufacture from solvent solutions, latex coagulation during polymer manufacture from latex emulsions, and compounding of plastics. This includes, among other things, coloration, fiberglass addition, and preparation of hot melt adhesives. Cheremisinoff (1987) provides details on parametric scale-up calculations for extruders.

V. SHEAR HEAD EXTRUSION

In recent years, a technique known as shear head extrusion has been developed for rubber extrusion operations. Extrusion lines employing this technique are essentially controlled by the shear head, which consists of a rotating mandrel contained within a stationary barrel wall. The entire unit is positioned between the discharge of the extruder and the nozzle or die (refer to Figure 17). The rubber compound conditioned in the extruder is heated to a predetermined temperature just below the material's scorch range by the shearing action of the shear head mandrel.

The shear head is controlled by a microprocessor and is driven independently of the extruder and autonomously in keeping with the properties of the compounds being processed.

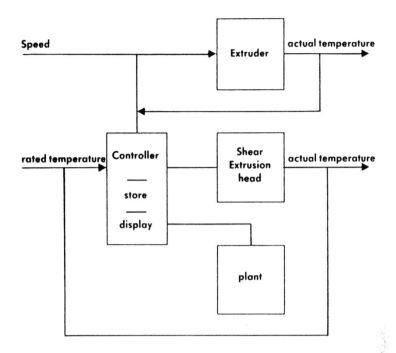

FIGURE 18. Microprocessor control loop for shear head extrusion.

In this manner, the shear head acts in the functional chain as a power-saving warm-up machine. Warm-up is controlled in a manner such that curing actually initiates within the shear head. The low viscosity extrudate flows through the final die while the crosslinking reaction begins. When properly operated this results in a more uniform, stable final shaping of the extruded article while improving dimensional stability by minimizing swell. More advanced systems are equipped with a data acquisition unit that makes it possible to reproduce in the machinery line the production conditions in the shear head, resulting in optimal operation. Downstream from the shear head is a hot air channel in which curing is continued. The final stage in the overall extrusion line is a cooling channel that serves to terminate the curing process from further essential elements of a shear head line. A simplified block diagram of a microprocessor control loop for shear head extrusion is illustrated in Figure 18. Since the key to a stable operation with this arrangement is maintaining the rubber compound in the shear head below scorching (i.e., prevulcanization), the extrudate's melt temperature is the principle parameter by which the shear head speed is controlled. Figure 19 shows one manufacturer's reported data on temperature response vs. shear head mandrel revolution for several different rubbers. Two typical arrangements for shear head extrusion lines are illustrated in Figure 20. Finally, some typical performance data are given in Tables 1 and 2.

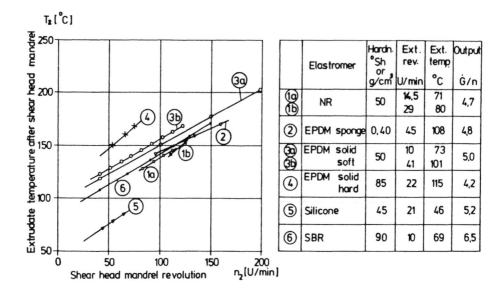

FIGURE 19. Typical data on shear head temperature response for different rubbers. (Courtesy of Krupp Industrietechnik GmbH, Postfach 900880, Hamburg 90, Germany.)

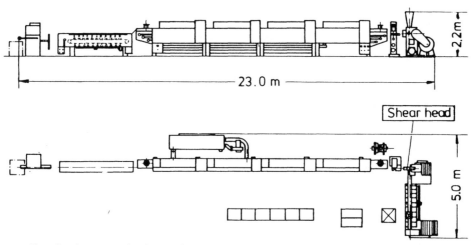

△ Shear head extrusion plant for manufacturing of single compound sealing profiles.

▽ Shear head extrusion plant for manufacturing of dual compound sealing profiles with metal inserts.

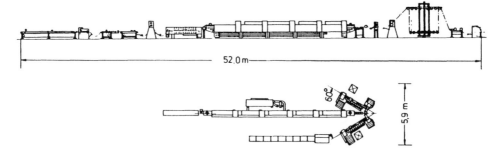

FIGURE 20. Shear head line arrangements.

TABLE 1
Performance Data for Shear Head Extrusion[a]

Cold-feet vented extruder	Screw diameter (mm)	Output[b] (kg/h)	Screw speed (rpm)	Driving power (kW)	Weight (kg)
KGE 90/24	90	40–330	max. 60	70	2300
KGE 120/24	120	60–550	max. 50	114	4100
KGE 150/24	150	80–750	max. 40	150	6000

[a] Motar is dc shunt wound. Speed reduce is two-stage.

[b] The output depends on the obtainable screw speed, type and viscosity of the compound to be processed, and extrusion die.

Courtesy of Krupp Industrietechnik GmbH, Postfach 900880, Hamburg 90, Germany.

TABLE 2
Performance Data for Shear Head Extrusion

Shear head G[a]	3300
Profile cross section, mm^2	110–210
Infinitely variable mass	10:1
temperature control, °C	21
Control range	24
Driving power, kW	35
KGE 90/24	1700
KGE 120/24	
KGE 150/24	
Weight, kg	
Mass temperature control system[b]	±2
Control accuracy, °C	120
Number storable programs	
Profile measurement and control unit[c]	100
Maximum profile	4–40
dimension, mm	3.5
Conveying speed, m/min	0–10 (maximum 50 mA)
Connected load, kW	
Output signal for controlling follow-up equipment, V dc	
Hot air system with heat exchanger[d]	12–15
System length,[e] m	150–250
Conveyor belt width, mm	4–40
Conveying speed, m/min	80–120
Connected load, kW	150–320
Air temperature,[f] °C	
Profile cooling system[g]	3
Cooling length,[e] m	240
Conveyor belt width, mm	4–40
Conveying speed, m/min	3
Connected load, kW	0.7
Water consumption, m^3/h	
Length cutting equation for shapes	100
Without metal reinforcement	45
Cross-sectional width	4–40
maximum, mm	2.2
Cross-sectional height	2–99,999
maximum, mm	180
Conveying speed, m/min	
Connected load, kW	
Advance in stages of 0.1 mm, mm	
Maximum number cuts per minute	

TABLE 2 (continued)
Performance Data for Shear Head Extrusion

With metal reinforcement[h]	10–50
Outer profile diameter, mm	1–30
Conveying speed, m/min	3
Connected load, kW	2.5–99,999
Advance in stages of 0.2 mm, mm	90
Maximum number cuts per minute	

[a]   Drive is dc shunt wound.
[b]   Temperature control system is a function of shear head mandrel speed and extruder screw speed. Control is by microprocessor.
[c]   Processible profile strands: maximum three pieces in parallel.
[d]   Belt control by compressed air and electric motor.
[e]   Triple pass.
[f]   Infinitely variable.
[g]   System type: spray cooling and drying nozzle.
[h]   Cutting system: circular knife.

Courtesy of Krupp Industrietechnik GmbH, Postfach 900880, Hamburg 90, Germany.

VI. SYMBOLS

$A_{1,2}$, $B_{1,2}$	physical and geometrical extrusion constants
a	slope of channel
A_r	Arrhenius shift factor
Br	dimensionless Brinkman number
F	frictional force per unit area
F_d, F_p	extruder geometric function and drag flow and pressure flow shape factors
f	fraction of helical length occupied by zone 1 (Z_1/Z_r)
G	mass flow rate
g	Bagely correction factor
H_s	mechanical heat generation rate
H'	heat of fusion
h	channel depth
IV	intrinsic velocity
K	geometric extrusion parameter
k	rheological coefficient
k_m	melt thermal conductivity
k'	swell index coefficient
L	length
L_s	channel length filled with polymer
MI	melt index
M_n	number-average molecular weight
M_v	viscosity-average molecular weight
M_w	weight-average molecular weight

M_z, M_{z+1}	third and fourth moments of molecular weight distribution
m_0	flow consistency index
c_s	specific heat
D	hopper diameter
D_b	inside barrel diameter
d	die diameter
E	activation energy
e	width of flight
N	screw speed
N_i	number molecular weight of ith species
n	power-law exponent
n'	swell power-law exponent
P	pressure
ΔP_t	pressure rise per turn in twin extruder
p	number of flights in parallel
Q'_s	theoretical volumetric throughput (no friction)
Q^*	screw discharge coefficient
q	heat flow
R	universal gas constant
R_s, R_b	radii of screw root and inside barrel, respectively
T	absolute temperature
T_0	reference temperature
t	temperature
t_c	residence time
U	velocity
u	mean flow velocity
V_{sy}	bed velocity in the y direction
W	average channel width
X	average solids bed width
Y	distance
Z	total melting length
ΔZ_1	helical length per turn ($\Delta Z + \Delta Z_2$)
z	axial distance

VII. GREEK LETTERS

α	solvent property constant in Mark-Houwink viscosity correlation
α_s	thermal diffusivity of bed
$\dot{\eta}$	inherent viscosity
θ	angle of plug advance relative to screw axis
ν	speed of screw rotation nonself-conveying
ξ	friction force per surface area density
Γ	nominal wall shear rate
γ	shear rate
δ	melt layer thickness viscosity
η	apparent viscosity
τ	wall shear stress
ϕ	helix angle of flight
ϕ'	solids coefficient of friction
ψ	friction coefficient or force
Ψ	transformation variable

REFERENCES

Baldwin, F. P. and G. Ver Strate, *Rubber Chem. Technol., 45,* 709–881, 1972.

Bridgwater, J. and A. M. Scott, Flow of solids in bunkers, in *Handbook of Fluids in Motion,* (N. P. Cheremisinoff and R. Gupta, Eds.), Ann Arbor Science Publishers, Ann Arbor, MI, 1983.

Carley, J. F., *Mod. Plast., 60* (2):77, 1971.

Cheremisinoff, N. P., *Polymer Mixing and Extrusion Technology,* Marcel Dekker, New York, 1987.

Chung, C. I., W. J. Hennessey, and M. H. Tusim, *Polym. Eng. Sci., 17:*9, 1977.

Dekker, J., *Kunststoffe, 66:*130, 1976.

Eise, K., H. Werner, H. Herrmann, and U. Burkhardt, An analysis of twin-screw extruder mechanisms, in *Advances in Plastics Technology,* Vol. 1, No. 2, Van Nostrand-Reinhold, New York, April 1981.

Flory, R., *Principles of Polymer Chemistry,* Cornell University Press, Ithaca, 1952.

Johnson, P. S., *Rubber Chem. Technol., 56:*3, July-August 1983.

Kacir, L. and Z. Tadmor, *Polym. Eng. Sci., 12:*387, 1972.

Keller, R. C., paper presented at Rubber Division Meeting, ACS, Los Angeles, CA, April 24, 1985.

Klein, I. and D. I. Marshall, *SPE J., 21:*1376, 1965.

Klein, I., D. I. Marshall, and C. A. Friehe, *SPE J., 21:1299, 1965.*

Kresge, E. N. and C. Cozewith, paper presented at Rubber Division Meeting, ACS, 1984.

Krevelen, D. van, *Properties of Polymers,* Elsevier, New York, 1972.

Marshall, D. I., I. Klein, and R. H. Uhl, *SPE J., 21:*1192, 1965.

Munstedt, H., *Kunststoffe, 68:*92–98, 1978.

Nichols, R. J., *Kunststoffe German Plastics,* Vol. 9, Carls Hanser Verlag, Munich, 1982, pp. 510–605.

Scholte, Th, in *Developments in Polymer Characterization,* Vol. 4 (J. Dawkins, Ed.), Applied Science Publishers, 1983.

Sezna, J. A., *Elastomerics, 16:*8, August 1984.

Slade, P. E., *Polymer Molecular Weights,* Part I, Marcel Dekker, New York, 1975.

Squires, P. H., *Soc. Plast. Eng. Trans., 4* (1):7, 1964.

Tadmor, Z. and E. Broyer, *Polym. Eng. Sci., 12:*378, 1972.

Tadmor, Z. and I. Klein, *Principles of Plasticating Extruders,* Van Nostrand-Reinhold, New York, 1969.

Tadmor, Z. and I. Klein, *Engineering Principles of Plasticating Extrusion,* Van Nostrand-Reinhold, New York, 1970.

Tadmor, Z., D. I. Marshall, and I. Klein, *Polym. Eng. Sci.,* 6:185, 1966.

Throne, J. L., *Plastics Process Engineering,* Marcel Dekker, New York, 1979.

Ver Strate, G., Ethylene-propylene elastomers, in *Encyclopedia of Polymer Science and Engineering,* McGraw-Hill, New York, 1991.

Chapter 23

BLOWN FILM TECHNOLOGY

John D. Culter

TABLE OF CONTENTS

0-8493-4401-8/93/$0.00 + $.50

819

I. INTRODUCTION

Blown film extrusion is a continuous process in which the polymer is melted. The melt is forced through a circular die; the resulting tube is inflated with air and cooled with air impinging on the resulting film. The film is stretched in the longitudinal and circumferential direction simultaneously during the cooling process, which results in properties unique to this process. While this sounds rather simple, a great deal of technology is involved in the process. This chapter will attempt to describe much of that in a very brief manner. The reader will be directed to more extensive texts or articles on the subject.

The principal polymers used in the production of blown films are polyolefins, although other polymers can be used (this will be discussed later). The major markets for blown film tend to be those applications that require biaxial strength. Typical of these markets are bags of all types and agricultural and construction film. The food applications often require coextruded structures of three to five layers or more; the principal food applications are packaging for cereal, meat, snacks, and frozen foods.

The blown film industry in North America is a major part of the plastics processing industry. According to Knittel,[1] the current industry consists of 2300 lines producing 4 billion pounds per year of product. Approximately 80 new lines are added each year, 60% of which replace existing capacity. Obviously, the industry is growing because the new capacity equals about 300 million pounds additional per year.

This review will cover the polymers most used in the production of blown film, the equipment necessary, and some of the technology involved. Since the topic is sufficient for a book unto itself and several do exist on the topic,[2-3] only a summary will be possible in the space alloted here.

II. POLYMERS

The primary polymers used in the production of blown film are the polyolefins and their copolymers. Other important polymers are nylon, ethylene vinyl alcohol, and tie resins. In addition to these, one can blow films of polycarbonate, elastomers, polyester, and others, but they are a minority of the poundage. Because of this fact, these other polymers will not be covered here.

A. POLYETHYLENE

Polyethylene is one of the most important polymers for blown film. It is produced by several processes, and each results in unique polymer properties due to differences in branching and molecular weight distribution. The variations in short chain branching affects the density of the polymer and thereby its physical properties. The various polymerization processes can produce low density (LDPE), medium density (MDPE), or high density polyethylene (HDPE). Table 1 shows the standard definitions for each.

The first process developed for making polyethylene was the autoclave process. This is basically a large stirred pot reactor operating at very high temperatures and high pressures (15,000–40,000 psi and 300–500°F), using either oxygen or a peroxide to initiate the reaction. The resulting polymer is highly branched. Controlling the branching is accomplished by varying the pressure and temperature and the location in the reaction vessel for the introduction of ethylene. These variables also affect the molecular weight and molecular weight distribution (MWD). Therefore, one is not totally free to vary the polymer properties independently. Short chain branching is the prime factor in determining the polymer density, and long chain branching affects the viscoelasticity of the polymer. The elastic properties of the polymer affect its processability.

TABLE 1
Density Ranges for Polyethylene

Polymer	Density range
LDPE	0.910–0.925
MDPE	0.926–0.940
HDPE	>0.941

LDPE can also be made by a tubular reactor, which is also operated at high temperature and high pressures as referenced above. The initiators are the same. Branching is more short chain in nature than in the autoclave, and the same variables control the molecular weight, density, and MWD.

The density of the polymer is defined in the standard physical chemical manner, i.e., mass/unit volume. The molecular weight measure used represents an average molecular weight, since the polymer chains do not grow to a constant length. In the polyolefin polymers, the molecular weight is described in terms of a melt flow rate through a given orifice under a given pressure. The units of measure for polyethylene are g/10 min, and the value is called Melt Index or MI. The MWD is not commonly used in the specification of the polymer; however, manufacturers concerned about the effects of MWD on the processing properties may specify a viscosity ratio to help define the MWD. The MWD affects the high shear viscosity, and therefore, a ratio of a high shear viscosity to the MI can be used as a measure of the MWD. The standard MI is measured at 2054 g, and a high shear value can be measured at 10,000 g. Therefore, the ratio, I_2/I_{10}, can be used to describe the polydispersity of the MWD. A more detailed analysis of the MWD can be done by size exclusion chromatograph, which used to be termed gel permeation chromatography.

LDPE is characterized as having good clarity, flexibility, and ease of processing. All of these properties are the result of the highly amorphous nature of the solid polymer. The side chains limit the amount of crystallinity that can form. The low crystallinity decreases the barrier properties of the polymer, and LDPE has a higher moisture vapor transmission rate than HDPE.

HDPE is made in a low pressure process (300 psi) using a stereospecific catalyst. The catalyst causes the ethylene molecules to be joined in a particular manner, so that there is nearly no branching unless some comonomer is introduced to cause it. A HDPE with a density of 0.955 or greater has only 2 to 3 branches per 1000 carbon atoms. Some larger amount of branching is introduced for certain applications to improve the processing of the polymer. The lack of side chains on the polymer molecule allows HDPE to crystallize to a greater degree, which increases its density, creates a hazy film, reduces the permeation of water vapor through the film, and makes it stiffer and more difficult to heat seal. All of these properties are reasons that HDPE is selected for its various applications. To improve processing characteristics, 1-hexene is often added in small quantities as a comonomer. This reduces the density by making some side chains, but some applications require the change in properties.

The introduction of comonomers in the low pressure process is also used to make linear low density polyethylene (LLDPE). By varying the amount of the comonomer, one can control the density to any desired value. Typical comonomers are 1-butene, 1-hexene, and 1-octane. Since the branching created in this reaction depends on the molecular weight of the comonomer, only short chain branches are formed and the stereochemistry is set. This results in very different physical properties of the films made from these polymers compared to normal LDPE of the same density. LDPE has the characteristics of clarity, flexibility, and ease of extrusion. LLDPE is more hazy because the short regular side chains change the crystallinity slightly, and this also results in a slightly stiffer polymer at the same density. Other improved properties are the amount of elongation before break and the hot tack strength of a heat seal. Both of these properties are the result of the greater mobility of the chain segments due to the shorter side chains. The linearity of the molecules also results in more difficult processing of the polymer. The amount of horsepower to produce the same number of pounds per hour is increased, and the molten film does not have a high enough elongational viscosity to allow the production of high gage films as with LDPE. It is the toughness, or puncture resistance, and the heat seal characteristics that make LLDPE the polymer of choice in many applications, assuming that the equipment can handle the processing.

B. ETHYLENE VINYL ACETATE (EVA)

By incorporating 5 to 50% vinyl acetate (VA) as a comonomer in the ethylene reaction, one can produce a range of polymer properties from thermoplastic to elastomeric. The presence of the vinyl acetate group in the polymer backbone introduces a polar group that can hydrogen bond and is bulky. These aspects make the polymer less crystalline, more flexible, and tougher than LDPE with a comparable molecular weight.

The vinyl acetate content and molecular weight are the parameters controlled in this reaction to vary properties. As the VA content increases, the density increases, but the crystallinity decreases the opposite relationship of density and crystallinity in LDPE. The property changes seen with the increase in VA content are improved clarity, more toughness, more tackiness, increased softness, lower heat seal temperatures, better low temperature performance, and increased stress crack resistance.

To improve processing and give a minor increase in toughness with improved clarity, blown films may be made from 5% EVAs. Heat seal layers in coextrusions are usually in the 15 to 18% VA range. These are the two most commonly used ranges of VA content in the blown film business.

C. IONOMERS AND ACID COPOLYMERS

The copolymerization of ethylene with acrylic acid, methacrylic acid, produces a group of polymers known as acid copolymers. These materials are similar to LDPE, but exhibit higher melt strength at an equivalent molecular weight because of the inter- and intramolecular hydrogen bonding. Similar to EVAs, as the comonomer content increases, the crystallinity decreases; melt strength, density, and adhesion to polar substrates increase at the same time. A major use of these resins is as adhesive laminating layers to bond polyethylene to polar polymer layers; a second major use is as a heat seal layer.

When the acid copolymers are neutralized to a metal salt, it produces polymers known as ionomers because of the ionic bonds in the side chains. It is these ionic bonds that increase the toughness in the solid state and the elongational viscosity in the molten state. Both of these properties give these polymers the unique properties for which they are most used. The ionic bonds produce a toughness and oil resistance characteristic of a much higher molecular weight, while the processability is characteristic of its actual molecular weight. Also, the melting point is low, but the hot tack is high, making the polymer an excellent heat seal material. The rheology of the polymer is excellent for caulking the apexes of fin-sealed packages and sealing through dusty products. These characteristics make packaging films a major market for ionomers.

Both types of polymers are difficult to treat with additives that are expected to bloom to the surface, such as slip or antistat. The ionic nature or the acid groups cause the additives, which are polar, to hydrogen bond to the polymer and not migrate.

D. POLYPROPYLENE

Polypropylene is an important polymer for the blown film industry. It is produced by stereospecific catalysis under condition similar to HDPE. The difference is that a methyl group is pendant from every other carbon atom of the backbone. The similarity is that there are almost no side chains formed in the polymerization. The methyl groups stiffen the backbone of the polymer and give it a higher melting point and tensile strength than HDPE. The stereochemistry can produce atactic, isotactic, or syndiotactic polymer — tacticity referring to the position of alternating methyl groups relative to the backbone. The most common commercial polymer is isotactic where the methyl groups are all on one side of the backbone. This arrangement allows the polymer to readily crystallize. The high rate of crystallization makes the polymer impossible to blow by the normal processes and requires some type of bubble quenching quickly after inflation. More will be discussed later about this.

Polypropylene homopolymer film is not produced by many converters. Those that do convert it typically use a double bubble process with a water quench as the first stage; the second stage reheats and reblows the bubble to orient the material. This increases its clarity, barrier, and stiffness, while reducing its gauge and thereby increasing the yield.

Polypropylene can also be produced as a copolymer with ethylene to increase its toughness, flexibility, and clarity. The ethylene can be introduced into the process as a gas or as an oligomer, which will produce a random and a block copolymer, respectively. Obviously, differing properties result from the two types of polymerization schemes; properties are also affected by the amount of the comonomer added.

The double bubble process is used to produce packaging films and capacitor substrates. In packaging films, a copolymer heat seal layer is usually coextruded on the homopolymer base.

E. ETHYLENE VINYL ALCOHOL

An important polymer for barrier layer is made from saponifying a copolymer of ethylene and vinyl acetate; the resulting polymer is polyethylene vinyl alcohol, or EVOH. One can produce a noncopolymer form; pure polyvinyl alcohol has extremely high barrier to gases, but is highly water soluble. The solubility of water vapor affects the barrier properties by causing the polymer to swell, increasing the segmental mobility and thereby the permeability. Therefore, the addition of ethylene reduces the effects of moisture. It also improves the ability to process the material by extrusion. The polymer is still affected by moisture; therefore, to get the best performance from EVOH, it is used as a buried layer. Several grades of EVOH are now available, depending on the requirements of the finished product. An additional concern about EVOH is that it is not stable at high temperatures and will degrade if abused at high temperatures or allowed to stay too long at normal extrusion temperatures. Therefore, concerns about residence time in the equipment and temperature control must be addressed to use EVOH successfully. Detailed information can be obtained from either EVAL Co. of America or Dupont.

F. NYLON

Nylons, or more correctly, polyamides, are engineering resins, some grades of which can be extruded as blown film. Polyamides are addition polymers made from a dibasic acid and a diamine; the resulting polymer contains an amide group as part of each repeat unit. This polar group has some bulk which stiffens the backbone slightly, but more importantly, it offers sites for hydrogen bonding intra- and interchain. Hydrogen bonding gives the polymer the toughness, and its barrier properties make it useful as a blown film. The fact that the segmental motion is reduced by hydrogen bonds means that the polymer toughness and barrier characteristics are very subject to the moisture content of the material; moisture interrupts the hydrogen bonding.

The most commonly used nylon is nylon 6, polycaproamide. Nylon nomenclature uses numbers to refer to the number of carbons in the dibasic acid and diamine. In the case of nylon 6, only ϵ-caprolactam, an amine containing six carbon atoms, is involved in the

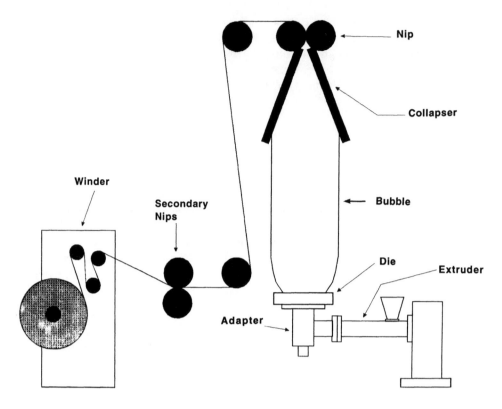

FIGURE 1.   Schematic of blown film line.

polymerization. Because of the moisture sensitivity of nylon 6, it is usually used as a buried layer in a structure, such as HDPE/tie/nylon 6/tie/sealant. Because of the polar nature of the polymer, it does not bond to nonpolar polyolefins; therefore, a tie layer is necessary between the HDPE and the sealant. Some sealants can also be used as tie layers due to the nature of the polymer, e.g., Surlyn. Other examples of tie layers will be discussed below.

G. TIE LAYER POLYMERS

Because polar polymers, such as EVOH or nylon 6, will not bond to polyolefins, it is necessary to use some type of tie, or adhesive, layer between them. An adhesive or tie layer polymer must have part of the molecule that is nonpolar and another part that is polar. Some of the sealant polymers that were discussed above have this characteristic. Other more specialized adhesive resins have been developed by copolymerizing various polyolefins with maleic anhydride. These materials are offered by several polymer suppliers, and their properties are so varied and specialized that the reader is directed to explore their needs directly with the polymer vendor, such as Du Pont or Quantum.

III. EQUIPMENT

A blown film line consists of many pieces of equipment (Figure 1). The polymer must be brought to the hopper of the extruder, which then melts it. The die forms the tube blown into the bubble. The stretching action of the haul-off and the blowing is what gives the film its unique characteristics. The cooling, collapsing, and winding also play an important part in the film quality. The following descriptions discuss each section of the line.

Flight depths exaggerated for clarity

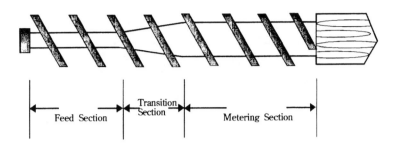

FIGURE 2. Schematic of extruder screw.

A. POLYMER DELIVERY SYSTEMS

Polymer to be melted must obviously be delivered to the extruder. A whole series of equipment options also exist at this point. If one is running only virgin polymer with reground side trim, then the polymer delivery system is a simple hopper loader capable of pulling two streams of material and a hopper. The amount of regrind being used may necessitate the use of a crammer feeder-type hopper to control the bulk density to the extruder.

If colorant or other additives are to be used, then the complexity of the feeding system must escalate. Blending can be done on a volumetric basis, if accuracy is not important, or a weigh feeder may be used. If polymer blends are being done at the extruder, as is often the case in today's business, then several polymer streams must be brought together, metered, and mixed before going into the extruder hopper.

The latest technology in polymer delivery is the use of loss-weight feeders, which control the extruder screw speed depending on the amount of material being pulled from the hopper. This type of system is one approach to layer thickness control in coextrusion.

B. EXTRUDERS

One of the most important pieces of the whole line is the extruder. Its job is to melt the polymer and pump, at high pressure, a specified quantity of molten material to the die. In today's complex film markets, this is not necessarily an easy task. A typical film extruder screw (Figure 2) has three distinct jobs to accomplish. The first section of the screw conveys material from the hopper; the second section compresses the material and melts it. The final section of the screw is only a pump. Because of the wide range of materials that an extruder may be expected to run on the same screw, it may not be able to deliver a melt that is as good as needed, so a mixing section will be added. Extruder lengths vary between 10/1 L/D to 30 + L/D. Most common in the U.S. is 24/1 L/D for film application. Extruder throughput, i.e., lb/h, are a function of the diameter and the amount of horsepower in the motor.

Modern extruder design has added grooves to the feed throat section of the barrel. These were first used in Europe on very short L/D extruders running HMW-HDPE to improve the feeding of the very hard polymer. Feeding of any polymer in the conveying section depends on the friction of the polymer with the wall of the barrel and assumes that the material will slip on the screw root.[5] When the polymer is very stiff and has a low coefficient of friction (COF) on the barrel, the feed rate is reduced. The placement of grooves, either straight or spiral, in the surface of the feed throat section, increases the effective COF of the polymer on the barrel wall. They also guide the movement of the material down the barrel. Grooved feed sections have proven valuable in extruding even LDPE by increasing the output of a given extruder size through better solids conveying.[6]

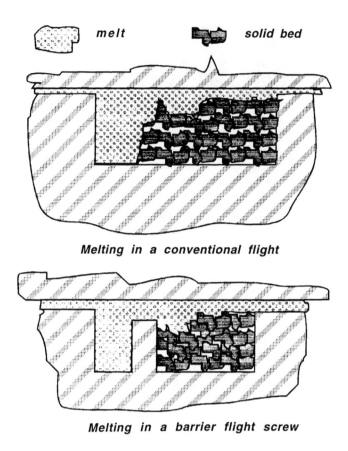

FIGURE 3.   A comparison of melting in a conventional and barrier screw.

The transition, or compression, section of the extruder is where the air is forced out of the solid bed and the melting of the material begins. In the extruder, the melting is accomplished by a combination of the heat being added by the heaters on the barrel and the heat being generated by internal friction, as the polymer is sheared (forced to flow). In smaller diameter extruders (<2 in.), most of the heat addition is from the heaters; while in larger extruders, most of the heat is generated by shear flow. In 3.0 in. and above, it is often necessary to remove heat from the melt through the barrel wall to control the melt temperature. Therefore, the use of water-jacketed heaters is necessary. The melting occurs at the wall of the barrel, where the maximum heat is being produced, and the melt flows into the back of the channel opening (Figure 3). At high production rates, the solid bed may be overwhelmed by the amount of melt, and unmelted material may be delivered at the extruder discharge. To overcome this problem, screws designed with separate melt channels were introduced. These are designed so that there is a partial flight between the melt channel and the solids channel (Figure 3). The melt is scraped off the wall by the trailing flight of the melt channel and conveyed down the barrel separately. The solid bed channel can then be reduced in size and depth at an appropriate rate to keep the melting occurring until it is complete. The solids channel then disappears.

The extrusion of HMW-HDPE for merchandise and grocery bags has resulted in a completely different and specialized design of equipment for this application. In the short L/D extruders used for HMW-HDPE, which are also usually small in diameter, the melting mechanism is different than in most small extruders. The grooved feed throat allows very high pressure to develop in a very short length. These pressures force the polymer to flow, and the melting is through viscous heating, rather than through heat transfer through the wall of the barrel. This type of extruder design tends to run adiabatically, i.e., heat on the barrel is only needed to get the process started. The melt temperature is often determined by the polymer viscosity, rather than by the control system. If one wishes to control the temperature of the melt in a HMW-HDPE application, the extruder must have effective cooling on the barrel.

Extruder screws can be designed to do a very specific job extremely well. However, in the real world, most plants must run a wide variety of materials through an extruder without changing screws. It is therefore necessary to compromise extruder screw design and work around the short comings. One does this through polymer selection and processing condition modification. The extrusion process is a complex topic in its own right, and the reader is directed to several good references for more detail.[2,4]

C. DIES AND ADAPTERS

After the polymer leaves the extruder, it enters a transfer pipe and the adapter section of the die. The transfer piping is not usually a place of great technology. The only concern is to keep it as short as possible with no dead spots. The adapter section of the die changes the polymer direction from horizontal to vertical, and in the case of coextrusion, it keeps the polymer streams separate until they are delivered to the die. If the die rotates, or oscillates, the adapter section must also provide the necessary bearing surfaces for this motion and seals for the moving polymer channels. The seal design technology is a real challenge for the vendors and an area of much proprietary development. When coextruding polymers with a wide variation of melt temperatures and heat stability characteristics, attempts must be made to provide some insulation between the channels through dead-air spaces, etc. These approaches are only partially successful, and more system cleaning is necessary when running such structures. The use of stationary dies requires a subsequent step to spread any gage nonuniformities across the rolls. Such steps, i.e., oscillating nips and rotating winders, will be discussed later.

The difficulty of designing, manufacturing, and maintaining seals for a rotating or oscillating die on the factory floor is one of the reasons manufacturers are going back to stationary dies, particularly for coextrusion systems. The object of rotation is to spread any temperature or thickness variation of the polymer stream across the rolls being formed from the process. Temperature streaks in the melt may not be completely erased by the mixing head of the extruder screw or the die design. These will result in thickness variation in the final film, just as will a misadjusted die gap. In steady rotation, the temperature streak may not be spread across the bubble circumference, since the rotation of the die may only cause the streak to be dragged to some equilibrium position. For this reason, oscillating or stationary adapters can correct this problem more effectively.

While some antique side-entry and spider dies may still be in use in the industry, today's technology is that of the spiral channel dies. The polymer melt is introduced through several ports into spiral channels (Figure 4); it then flows partially up the channel and between the channel and the die wall. This type of smearing flow causes the temperature difference that exists in the melt to be reduced. As the flow moves up the die channel, the spiral disappears, and the melt flows into a relaxation chamber where it is hoped that the melt will loose its memory of the previous shear history. Beyond the relaxation chamber is the die land itself, which shapes the melt to its final dimensions as it leaves the die.

If the process is coextrusion, the spiral section and relaxation chamber of the die are replicated for each polymer layer in the final structure. For this reason, the size of coextrusion dies can be very large. Above the relaxation chamber, the melt streams must be combined. Combining sections can vary in design, depending on the number of layer and the materials expected to be run. In the simplest form, all of the streams are brought together in a short

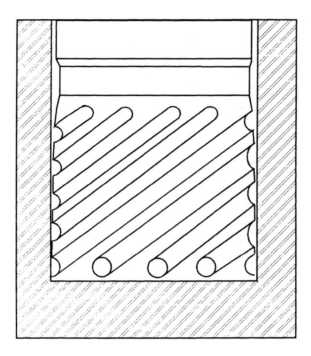

FIGURE 4.   Spiral die.

section just before the die lands. If the die is rather large, or if there are many layers of polymer of widely varying viscosities involved, several sections may be used to make the combining in a more controlled manner. Polymers of widely varying viscosities do not flow together in a stable manner, so the channel length for such a flow situation must be kept as short as possible. The polymer with the lowest viscosity tends to move to the outer perimeter.

Die channel design, like extruder design, is often compromised because of the wide array of materials that must be run through the system.[7] Channels designed for tie layers and barrier polymers are generally small because small amount of these materials are needed. It is not a good idea to switch any polymer used in small flows into a large extruder or die channel because the residence time will likely exceed the polymer's tolerance for heat.[8]

The die land opening, die gap, should consider the types of polymer to be run in the system and the final product thickness. For monolayer films, one uses the rule of thumb that branched polyolefins can only withstand a drawdown of 16:1 because of melt strain hardening due to chain entanglements; a linear polymer such as HDPE or LLDPE can be run at a 160:1 drawdown.[9] The bubble tears off as a result of too much drawdown. In a coextrusion system, the choice of die gap is more complicated because the various polymers are not behaving individually. However, in composite structures, a weighted average of the tensile strengths generally gives a good estimate. Therefore, the melt properties will be dominated by the layer characteristics of the heaviest component.

In addition to the die gap, one must consider the manner in which the die gap is centered. Since the polymer is being extruded in an annulus, centering the external wall is important. One method is to adjust the position of the outer ring with bolts. This allows a slightly easier machining job to make the die, but may result in too many adjustments available to the operators for quality production. The other method is to machine the die parts so that precise alignment can be made during assembly, and sufficient sturdiness is built in to keep the parts from moving once they are assembled.

Because all of the temperature variations that occurred in the extruder and in the rest of the piping and die can not be completely removed from the melt by the die channels, the

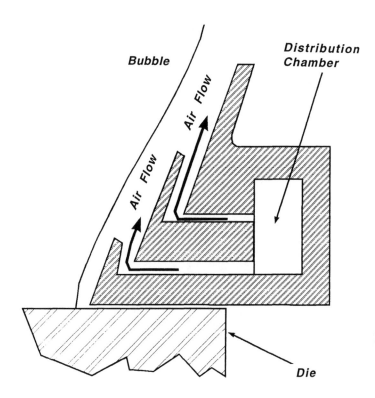

FIGURE 5. Dual lip air ring.

die is rotated or oscillated to spread the nonuniformity over the rolls being made. While this makes flat rolls of film, the point-to-point gage variation can still be $+/-10\%$. Therefore, the latest die design has introduced automatically adjustable die lips.[10] In these new designs, the flexible lip technology used in cast film dies has been translated into blown film dies. A downstream gaging system senses spots in the film that vary from the set point gage and feed a signal back to an adjustment device on the die lip. These types of dies, while allowing better gage control, are very complex, and they introduce additional maintenance problems.

D. AIR RINGS AND INTERNAL BUBBLE COOLING (IBC)

Bubble cooling is the next important step in determining the final properties of the film. The original air ring was a pipe with holes drilled in it that wrapped around the bubble. Modern technology has utilized an advanced understanding of aerodynamics to develop dual lip air rings and IBC packages that greatly increase the rates of production and uniformity of the film.

The dual lip concept of an air ring is shown in Figure 5. The first air channel is designed to lock in the correct bubble shape and to give the bubble stability in the first few diameters above the die. The outer air channel produces the main cooling of the outer surface of the bubble. The older designs of air rings where both functions were done by one air delivery slit could not accomplish the job correctly. If the air volume was high enough to give adequate cooling, the velocity was too high near the neck of the bubble for stable flow. (This usually meant that production rates for low gage films were limited by the bubble instability, and a nonuniformity of gage resulted.)

Another improvement in gage uniformity comes from the application of IBC, so that part of the heat is transferred from both sides of the bubble. For very large layflats, the IBC package is the only solution to high production rates. For any size bubble, the use of IBC

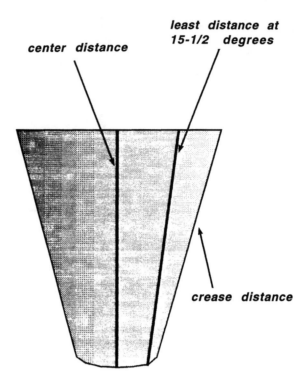

FIGURE 6. Bubble collapsing dimensions.

can increase the maximum output of a line. Also, IBC packages have come a long way in design since their introduction. The latest technology utilizes aerodynamic designs for the application and extraction of air from within the bubble. In addition, the controls on the air flows have been automated and keep the bubble dimensions more accurate than in the past. The use of a sizing cage, or more recently an ultrasonic gaging ring, to control the exhaust air volume flow is now common practice.

The extrusion of HMW-HDPE has resulted in the development of new cooling technology specialized for this application. Because of the linearity of the polymer, it can relax any orientation put into the film by the process, if the orientation is introduced at too high a temperature. Therefore, HMW-HDPE bubbles are run with a long stalk before they are forced to stretch in the circumferential direction. This allows the melt temperature to drop sufficiently so that the orientation in the circumferential direction is introduced just before the frost line, where solidification locks it in place. Air ring design had to be changed to allow for this type of processing. An additional bubble stabilizing iris is also used just above the frost line. Even if IBC is not used in the process, an internal bubble stabilizer is often included in the design. Newer developments have introduced special air flow guides to produce a more stable bubble and to minimize the effect of drafts on the neck region. All of these efforts are important because the gage of the film in most of the applications is 0.5 ml or less, and a +/ − 10% variation is much more significant to the final product at this low thickness.

E. COLLAPSING FRAMES

As the bubble approaches the nip rolls, the bubble shape must be transformed from a round tube to a flatten tube. If one looks at the geometry, the path for particle in the bubble is not the same around the circumference. Knittel[11] presented the data showing that the shortest path up the bubble is at a location 15 1/2° from the crease (Figure 6). These path differences are not too problematic for compliant polymer films, such as LDPE or LLDPE. For HDPE and some stiff coextrusions, this path difference is noticable and results in wrinkling of the film.

The original, and still commonly used, collapser frame is made from maple slats. The film must drag across the maple surface, which may induce wrinkles from stretching the film, or in the case of high gloss films, scratches. The next best approach is to use rollers in the collapsing frame. However, the best collapser now available is the air collapsing frame. This type is an angled air bearing surface that never really touches the film; some additional cooling is also accomplished by the air flow. Air collapser frames definitely reduce the incidence of scratches on the film surface and result generally in less wrinkling and stretching of the film.

However, to overcome the wrinkling of HDPE, or gage variation in some coextrusions, caused by path differences, some other approaches have been used. For HDPE, one approach is to lower the nips to the point where the film is being collapsed before it has cooled completely, thereby making it more compliant. This does not work well for heavier gage films or for films with lower melting sealant layers as part of the structure. Knittel also suggested[11] that one might install an initial bubble shape modifier before the collapser to take the bubble from round to square. This reduced the path differences considerably, but had other problems in production. Therefore, the best solution is to use very tall towers. The longer the path to the nips, the less the path difference is for the two parts of the bubble.

F. NIPS

The nip section of the tower on a conventional blown film tower is not "high tech." It is a pair of rollers, one metal and one rubber, that provides the energy to stretch the melt in the machine direction and to transport the film up the tower. Accurate speed control is necessary at this location as in all of the process motors, but that is about the extent of the technology required. However, in the more demanding applications of today's marketplace, one is finding oscillating and rotating nips being applied more often. In some operations, the rotating or oscillating die can not successfully spread all of the gage variations over the rolls. There may be drafts in the room that cause gage variations, or there may be temperature histories in the melt that do not become randomized. Using an oscillating nip will spread these variations over the rolls. The other more pressing economic reason is that stationary dies that can be used with oscillating nips allow for less expensive construction, particularly in multilayer applications. Stationary adaptors do not need expensive and problematic bearings and seals. Also, when the number of layers reaches five or more, using a stationary die produces a much more reliable design from all process considerations.

Fully rotating nips with the winder attached are offered for HMW-HDPE film because the approach results in better roll formation of the very low gage film, ≤ 0.5 ml. If these are used with upward extrusion, it means dealing with roll handling on the second story. Such systems are offered in a downward extrusion mode, which makes a much easier job of roll handling, but complicates the extrusion process.

G. WINDERS

The winder produces the final product rolls and can affect the quality of the product. The ease of winder use will dictate how much attention is given to the process by the crew. Both are important factors in the final step of producing a quality product. Winder design can be varied depending on the features desired. There are center drive winders, i.e., the core is driven; surface drive winders, where the roll is driven from the outer surface; and surface/center drive winder, the ultimate in control. Most applications in the blown film industry use the center drive design. Beyond the choice of drives, one must choose the method of roll changing. Most winders used in the film industry are turret type, where the

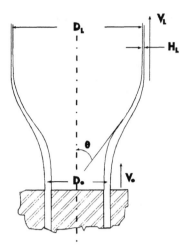

FIGURE 7. Scaling geometry.

empty core is rotated into position by a turret mechanism and an automatic cutoff is made. Many variations on the themes are available from various manufactures, and one should determine the needs of the products to be produced on the line before choosing a design.

There is a secondary nip, either as a part of the winder or just prior to the film entering the winder. This secondary nip provides the tension control point necessary as part of the winder operation. The main functions of a winder are to guide the film evenly into the roll form and to provide tension control on the film to produce evenly wound rolls. The winder must apply enough tension to make a roll that will not telescope when handled, and it must not apply so much tension to the film as to cause crushed cores or center roll bagginess. Tensions on center drive winders are generally tapered from the core to the outer wraps to compensate for the change in diameter of the roll. On a surface winder, the compensation is automatic because of the drive mechanism. When tensions are being adjusted on a winder, it must be kept equal to or less than 1/2 of 1% of the film secant modulus to avoid excessive stress in the film. One reason for the low value of recommended tension is that the film is not totally at equilibrium by the time it is wound into the roll. Dimensional changes occur as the film crystallizes or cools; these dimensional changes also induce stresses. Both Loonsbury[12] and Knittel[9] have produced recent discussions about winders that offer insight into important design parameters. Winding technology is an area of study that has received more attention in the paper industry than in the film industry, but one can learn the principals from paper and apply them to film. One major difference between the two materials is that the elasticity of films is much greater and has to be considered in the winding theory.

H. SCALING OF THE PROCESS AND THE EQUIPMENT

Scaling the blown film process is not easy because the process parameters are generally nonlinear. Simplifying assumptions have to be made to make the problem tractable, and these usually make the scaling procedure inaccurate. The work of Kanai et al.,[13] building on the work of Pearson and Petrie[14] and others, showed a scaling procedure that seems to overcome many of the former limitations, despite the fact that they used a Newtonian fluid model. The geometry of their model is shown in Figure 7. Here are the initial parameters:

At the die		At the frost line	
Die diameter	(D_O)	Bubble diameter	(D_L)
Die gap	(H_O)	Film thickness	(H_L)
Melt velocity	(V_O)	Haul-off velocity	(V_L)

If the scaling is done so that stresses in the bubble, at the frost line, are the same, then the properties of the final film are the same in the small and large scale. The scaling is done so that

- D_L/D_O, V_L/V_O, Z_L/D_O and the melt temperature are constants
- The output, $Q_1 \propto k^2 1 \, Q_2$
- The thicknesses, $H_1 \propto 1 \, H_2$
- The die diameter, $D_1 \propto k \, D_2$.

This scaling works well for HDPE and should be applicable to other polymers. For other types of scaling, the reader is referred to the original paper.[12] The usual caveat about scaling is, of course, always applicable. Experimental verification is necessary to check the results.

To determine the proper size motors and other scaling parameters in the choice of equipment, the reader is referred to the article by Knittle,[8] where he discusses these issues in detail.

REFERENCES

1. **Knittel, R.,** *Mod. Plast.,* mid-October, 245, 1991.
2. **Rauwendaal, C.,** *Polymer Extrusion,* Hanser, Munich, 1990.
3. **Hensen, F.,** *Plastic Extrusion Technology,* Hanser, Munich, 1988.
4. **Tadmor, Z. and Klien, I.,** *Engineering Principles of Plasticating Extrusion,* Van Nostrand Reinhold, New York, 1970.
5. **Darnell, W. H. and Mol, E. A.,** *Soc. Plast. Eng. J.,* 12, 20, 1956.
6. **Bode, W. W.,** *Proc. ANTEC,* 153, 1991.
7. **Perdikoulias, J. and Tzinanakis, C.,** *J. Polym. Film Sheet.,* 7, 118, 1991.
8. **Perdikoulias, J., Vlcek, J., and Vlachopoulos, J.,** *Polym. Film Sheet.,* 5, 18, 1989.
9. **Knittel, R.,** *Proc. ANTEC,* 170, 1991.
10. **Feistkorn, W.,** *J. Polym. Film Sheet.,* 5, 8, 1989.
11. **Knittel, R.,** *J. Polym. Film Sheet.,* 3, 23, 1987.
12. **Lounsbury, D.,** *Proc. ANTEC,* 158, 1988.
13. **Kanai, T., Kimura, M., and Asano, Y.,** *J. Polym. Film Sheet.,* 2, 224, 1986.
14. **Pearson, J. R. A. and Petrie, C. J. S.,** *Plastics Polym. Sci.,* 38, 85, 1970.

Chapter 24

PRINCIPLES OF ADHESIVE RHEOLOGY

Franklin M. C. Chen

TABLE OF CONTENTS

0-8493-4401-8/93/$0.00 + $.50

I. INTRODUCTION

An adhesive is a substance that functions as a "molecular link" at the interfacial zone between two solid surfaces. Sandwiching between them, an adhesive fills the voids and replaces the air pockets at the juncture. This "molecular link" makes intimate contacts with both surfaces. As a result, mechanical force can be transferred across the interface.

The interfacial forces holding two surfaces together in this way may arise from van der Waals forces or specific chemical interactions. The strength of the bond is determined by both the interfacial forces and the mechanical properties of the interfacial zone.

Adhesive performance is, therefore, determined not only by the degree to which the adhesive forms intimate contacts with both surfaces, but also by its ability to resist separation from these surfaces. Adhesives initially must flow and wet both surfaces. If the adherends are made of polymers, molecular diffusion can occur at the interface during bonding. Once the interfacial zone is established, the adhesives must set, or solidify, and provide cohesive strength to resist any process that may cause debonding.

Considering principles of rheology, the adhesives must possess those rheological properties that promote bonding but resist debonding. The adhesive substance must flow and engage in molecular diffusion during bonding yet engage in large energy dissipation and resist delamination during debonding.

To fulfill these criteria, adhesives must have dual sets of rheological properties: properties to facilitate bonding and properties that resist bond delamination when the strengths of the bond are evaluated with various methods. These dual sets of properties can be achieved either by changing states of adhesives from wetting to setting or by manifesting the adhesive viscoelastic nature.

Hot melt and epoxy are examples that the adhesives change states from an initial liquid to a final solid. In the former, the adhesives are initially heated to the liquid melt. The molten materials flow between two surfaces, filling the voids and replacing the air pockets at the interface. As the interfacial zone is established, the adhesive solidifies. In the latter, epoxy adhesives are initially low viscosity fluids consisting of low molecular weight monomers and oligomers. The epoxy flows and wets both surfaces during bonding. Once the interfacial zone is established, these substances are heated to form three dimensional networks.

The coatings on the medical tape are examples that the adhesives exhibit different viscoelastic properties in bonding and debonding applications. These coatings are pressure-sensitive adhesives. The useful property of these coatings are their *tackiness*, which is a manifestation of their viscoelastic nature. Pressure-sensitive adhesives flow during contact and under pressure. During debonding, they exhibit large energy dissipations.

A. OUTLINE

This chapter consists of two parts. The first reviews the fundamental rheological principles and properties involved in both bonding and debonding. The second part uses pressure-sensitive adhesive to illustrate these principles.

II. RHEOLOGICAL PRINCIPLES AND PROPERTIES

Table 1 lists fundamental rheological phenomena involved in adhesion. Flow, creep, and diffusion are necessary for bonding. The ability to participate in plastic flow and to engage in large energy dissipation is important in resisting bond delamination during debonding.*

---

* Adhesive strength is often evaluated by its ability to resist bond separation in simulated debonding process such as peel.

TABLE 1
Rheological Properties of Adhesives

Processes	Adhesive properties
Bonding	Flow
	Creep
	Diffusion
Debonding	Plastic flow
	Energy dissipation

These rheological properties are time and temperature dependent which can be brought together to produce master curve using time-temperature equivalance [or known as William-Landall-Ferry (WLF) equation]. Furthermore, the response of an adhesive to stress or strain is linearly additive. This is known as Boltzmann Super-position principle. The rheological properties can be measured by small oscillatory deformations using the linear viscoelastic functions.

In this section, we will first discuss the rheological properties required for bonding and debonding. Next, we will discuss the Boltzmann principle and the WLF equation to illustrate their relevancy to the adhesive performance. Finally, using dynamic mechanical measurements as examples, we will illustrate how these rheological properties are measured.

A. RHEOLOGICAL PRINCIPLES IN BONDING

The important rheological properties of adhesives in bonding are (1) their ability to flow or creep to displace air pockets at the interfacial zone, and (2) their ability to engage in molecular diffusion to establish molecular mutual entanglement at the interface.

Creep is a property of viscoelastic material while flow is a liquid property. Ordinary liquids flow under applied stresses. Their rheological behavior is weakly dependent on previous stress history. There is no memory effect in the deformations of ordinary liquids. On the other hand, viscoelastic materials creep under applied stresses. The response of viscoelastic materials depends strongly on the stress history. The memory effect is predominant. The creep function of viscoelastic materials depends on their molecular weight, molecular weight distribution, and molecular architecture. Most adhesives are made of polymeric materials that exhibit pronounced viscoelastic responses.

The different responses of ordinary liquids and viscoelastic liquids to flow or creep result from molecular relaxation. In ordinary liquids, structural relaxation of molecules is fast, and the response to the structural deformation is essentially viscous. In viscoelastic liquids, the relaxation is spread over many orders of magnitude in time. Most viscoelastic liquids are polymeric, and rearrangement of molecules over larger chain distances is slow due to presence of entanglements. Relaxation associated with such rearrangement depends strongly on large scale motions of chain architecture. These sluggish processes, referred to as the terminal relaxations, govern the flow and/or creep properties. These properties are manifestations of the polymeric architecture, and depend on chain length distribution and the presence of long chain branching.

Molecular weight and architecture are important for the diffusion of polymeric molecules at the interface. A polymer molecular weight influences both diffusion (measured by the diffusion coefficient) and the effectiveness of molecular mutual entanglement.[1] Diffusion is hastened if adhesives contain many mobile chain ends. This occurs when the moecular weight of polymeric components of an adhesive is low, since the number of chain ends increases as the molecular weight is lowered. The diffusion coefficient is inversely proportional to the square of the molecular weight in the linear polymeric molecules.[2] On the other hand, polymers of higher molecular weights have more effective intermolecular penetrations and molecular mutual entanglements. These molecular penetrations and chain en-

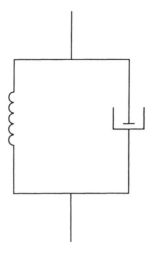

FIGURE 1.   Kelvin arrangement of spring and dash pot element.

tanglements allow the force to transmit across the interfacial zone more effectively for higher molecular weight polymers. This is one of the reasons that a high molecular weight polymer is normally included in pressure sensitive adhesive formulations. We defer discussion of the role of high molecular weight polymers in adhesives until later sections.

The strength of intermolecular coupling also depends on the concentration of high molecular weight polymer chain at the interface,[1] which may be substantially reduced if the adhesive is compounded with plasticizers or other low molecular weight compounds. These species may accumulate at the interface, preventing high molecular weight polymers from engaging in effective chain entanglement at the interface.

Molecular architecture also influences the interfacial molecular diffusion. A short branched polymer is expected to exhibit a higher rate of interdiffusion, compared to a linear polymer, since it contains more chain ends for a given molecular weight. However, when the branches are sufficiently long, the interdiffusion coefficient is strongly suppressed. These aspects of the diffusion behavior of polymeric molecules can be understood from the reptation model of P. G. deGennes and from Doi and Edwards' theory of chain dynamics.[3]

The creep properties of polymeric materials are characterized by the steady state viscosity (at zero shear rate), η_0, and by the steady state recoverable shear compliance, Je^0. They are obtained quite directly from the results of creep experiments. The derivation of fundamental flow and creep properties from creep experiments can be understood using a simple mechanical model due to Kelvin, which involves juxtaposition of the spring and the dash pot elements as shown in Figure 1.[4] Typical results of creep and recovery experiments are shown in Figure 2. The steady state viscosity at zero shear rate is obtained from the creep experiment in the steady state region. The recoverable shear compliance is obtained from the total recoil strain after the flow reaches steady state and after removal of the shear stress.

The flow and/or creep properties of polymeric materials can be controlled by temperature and/or by the experiment's time scale. Figure 3 shows the variation of compliance with time (or temperature) over a very wide time scale (or temperature scale) for a linear amorphous polymer. For short-time (or low temperature) experiments, the observed compliance is $10^{-9}m^2N^{-1}$. This is a typical compliance value for a glassy solid. The compliance for a rubbery solid ranges from $10^{-7}m^2N^{-1}$ to $10^{-5}m^2N^{-1}$.

The creep behavior of polymeric materials depends on the ratio of the experiment's time scale relative to some fundamental time parameter characteristic of the polymer. In creep experiments, this parameter is defined as the ratio of viscous component over elastic component in Kelvin model is called the retardation time τ_0. The distinction between a rubber

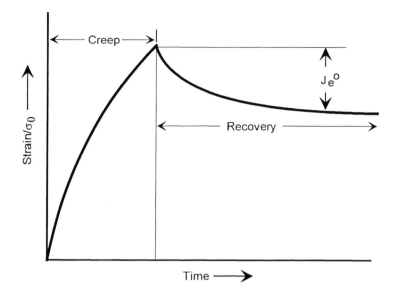

FIGURE 2. Creep and creep recovery experiment with control stress σ_0 applied at $t < t_0$ and released at $t > t_0$.

and a glassy plastic is determined by the values of τ_0 at room temperature. Thus, for a rubber, τ_0 is very small at room temperature, compared with typical experiment times which are greater than 1 s. The opposite is true for a glassy plastic.

These considerations lead immediately to a qualitative understanding of the influence of temperature on polymer properties. With decreasing temperature, the frequency of molecular rearrangements is decreased, increasing the value of τ_0. Thus, at very low temperatures, a rubber will behave like a glassy plastic (high τ_0) and will become rubber-like (low τ_0) at elevated temperatures.

Rubber *alone* does not show *tackiness* property. The rubbery solid does not have sufficient creep compliance (hereafter referred to as compliance),* at less than a value $(10^{-6} m^2 N^{-1})$ as proposed by Dahlquist[5] as a bonding criterion. Additionally, its compliance is time independent on a relatively long experimental time scale (Figure 3). This behavior is unfavorable for both bonding and debonding. To remedy this, additives such as plasticizers and tackifiers are frequently added. Although the elastomers dominate the rheological properties of the adhesives, the rheology can be modified by the presence of the plasticizers and tackifiers. Plasticizers increase the creep compliance on all time scales. Tackifiers, on the other hand, raise the creep compliance only at long-time scale. Tackifiers lower the creep compliance at short-time scale. Additional discussions of the role of tackifiers appear in later sections.

When two surfaces are pressed together, bonds are formed at the interface. The interfacial bond strength depends on both the *number of the bonds* and *the strength of each bond*. The strength of each bond depends on the types of interactions that occur at the interface. Such interaction can be a specific (e.g., an acid-base interaction) or a general friction force that originates from molecular interpenetration. Total strength of interfacial bonds involving frictional force depends on the sum of individual interactions that vary in extent. One must consider both the *degree of chain interdiffusion* and *the number of bonds* (*the extent of contacts*). The diffusion coefficient, D, of a typical industrial elastomer with a molecular weight $200-300 \times 10^3$, is approximately 10^{-13} cm^2/s.[6] One can easily estimate the molecular interdiffusion distance of an elastomer after 1 s of contact to be approximately 45 Å. At

---

* Dahlquist[5] proposed a minimum creep compliance of 10^{-6} m^2 n^{-1} as a bonding criterion for any solid to a substrate. The creep compliances of most rubbery solids are smaller than this value.

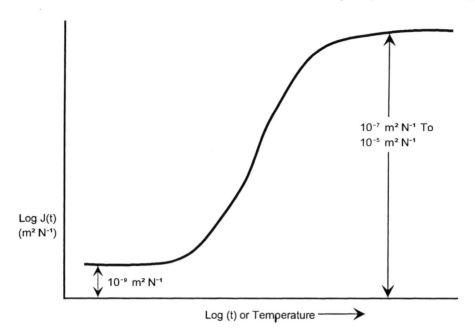

FIGURE 3. The creep compliance J(t) as a function of time or temperature.

this distance, the friction force is generally sufficient to transmit force across the interface at each bond. Thus, for most adhesive systems with a molecular weight $200-300 \times 10^3$, increasing the number of molecular contacts would be the strategy to increase the interfacial bonding strength.

Many experiments have shown that interfacial adhesion strength can be enhanced by increasing the number of molecular contacts at the interface. For example, Hammond[7] observed that tack was increased when pressure was applied during the surface contacts. Since molecular interpenetration is relatively independent of the contact pressure, the increased tack was ascribed to the increase in the extent of surface contacts. In another experiment, Bussemaker and van Beek[8] observed that tack was increased when a small amount of interfacial shear was introduced. Since the interdiffusion distance should not be affected by interfacial shear but the number of molecular contacts should, this experiment convincingly demonstrated that tack is mainly determined by the extent of the contacts between two surfaces. The number of molecular contacts can also be increased by improving the flow properties of the adhesives using plasticizers. However, one must be careful not to improve the flow at the expense of interdiffusion effectiveness.

In summary, the rheological properties involved in bonding are creep, flow, and interdiffusion. Creep and flow in the adhesive improve interfacial molecular contacts. Chain diffusion at the interface secures the ''bonding'' when the bond does not incude specific interactions other than the general friction force. Creep and flow are functions of both the experiment's temperature and its time scale, and their behavior can be determined from simple creep experiments.

B. RHEOLOGICAL PROPERTIES IN DEBONDING

The important rheological properties of adhesives in resisting debonding are their ability to engage in plastic flow and to dissipate energy. Plastic flow truncates local stress at the tip of the crack to the yield value. Energy dissipation necessitates a higher external load to debond the adhesive from the interfacial zone. Plastic flow generally also involves energy dissipation; these two properties are interrelated.

When cracks are created in the adhesives (or at the interface between the adhesive and the adherend), the stresses are normally concentrated at the crack's tips. The stress concentration is related to the curvature; the smaller the curvature, the larger the stress concentration. However, when the adhesive engages in plastic flow, the local flow will blunt the tip of a preexisting crack and truncate the local stress at the value of the yield strength. Thus, the ability of the adhesives to engage in plastic flow helps prevent debonding.

When the crack propagates, it creates new fracture surfaces. The energy expended to create these surfaces comes from the release of potential energy. The potential energy consists of the elastic strain energy and the work performed by the movement of external loads. Griffith[9] has shown that as long as the release of the potential energy per unit area is greater than the fracture energy, crack propagation is spontaneous. The fracture energy specified by Griffith's criterion is the surface-free energy of the newly created surfaces. This criterion applies very well for ideally brittle materials (such as soda-lime glass), but poorly for polymeric materials.

In polymeric materials, the fracture energy is many times greater than the surface-free energy. In these materials, fracture consists of many irreversible dissipative processes, including plastic yielding (or plastic flow) around the crack tip, cavitation, static electrification, and so on. Among the dissipative processes, plastic yielding (or plastic flow) is most important, as energy dissipated during that process is much greater than that in the other processes.

Since fracture energy is the sum of the thermodynamic work and the plastic work, it can be conveniently expressed by a relation proposed by Gent, Andrews, and their co-workers:[10,11]

$$G = G_0 \psi(R) \tag{1}$$

where G_0 is the fracture energy at zero rate (equilibrium fracture energy), R is the rate, and $\psi(R)$ is a rate-dependent viscoelastic function. In a viscoelastic material, rate and temperature effects are equivalent in many rheological experiments. In fracture experiments, the rate and temperature effects are likewise equivalent. Thus, the rate and temperature effects can be superimposed to give a dimensionless rate. Therefore, Equation 1 can be rewritten as

$$G = G_0 \psi(Ra_T) \tag{2}$$

where a_T is a shift factor, and Ra_T is the effective (or reduced) rate. The function ψ is defined such that

$$\psi(Ra_T) = 1 \text{ if } Ra_T = 0 \tag{3}$$

The equivalence of the rate and temperature effects on fracture energy has been amply demonstrated. A typical example was shown by Gent[12,13] in the T-peel test of adhesives. In this example, the adhesive was an uncrosslinked rubber with Tg $\sim -40°C$. The adhesive was supported on a woven cloth backing. Many different polymeric materials were used as adherends, including polyester film, cellophane films, polyethylene films, and films made of polystyrene block. Gent's experiments demonstrated that a general master curve can be constructed irrespective of which adherend was used (Figure 4). The master curve exhibits a high peak at low rates, and a low peak at high rates. It also has an interesting connection to the loss modulus curve as shown in Figure 5. The peak positions in the master curve correspond to the transitions in the modulus curves. The peak at low transition rate corresponds to the liquid-to-rubber transition, and the peak at high rates is associated with the rubber-to-glass transition. The concurrent transitions evident in the peel strength data and the material loss modulus data are quite remarkable.

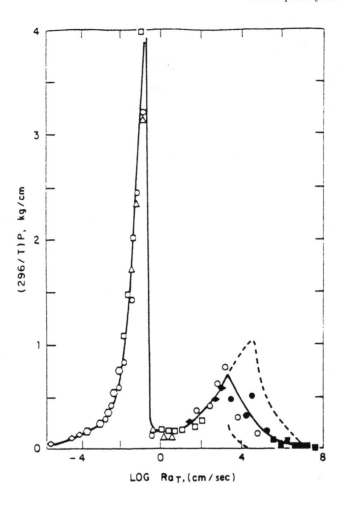

FIGURE 4. Master curve[12] for peel force P vs. peel rate Ra_T reduced to 23°C for an uncrosslinked butadiene-styrene rubber adhering to PET polyester film. Dashed lines are extreme values for stick-slip behavior.

Figure 5 is consistent with the idea that the peel strength of an adhesive-adherend system is directly related to the system's energy dissipation. Associated with each transition (e.g., liquid-to-rubber or rubber-to-glass transition), energy dissipation is a local maximum — so is the peel strength. The difference in the peak heights associated with each energy dissipation mode is due differences in the failure modes. The failure mode corresponding to the large peak height at low rate was a cohesive failure; the mode associated with the smaller peak height at high rate was an interfacial failure.

Each viscoelastic transition (liquid-to-rubber and rubber-to-glass) is associated with changes in the mode of molecular motion. The liquid-to rubber transition (the terminal relaxation) accompanies the onset of the coordinated structural rearrangement of a polymer over large chain distances. The rubber-to-glass transition represents the "freeze in" of those long-range, coordinated molecular motions. Both transitions involve energy absorption, like the energy absorption that occurs with the transitions between the electronic states of the atoms. One can easily demonstrate energy absorption associated with the glass transition by a simple experiment.[4] A rubber ball is initially placed in liquid nitrogen. After sufficient cooling, the ball is removed from the liquid nitrogen container. The percent rebound of the rubber ball from a constant height is then recorded as a function of time. The data follow

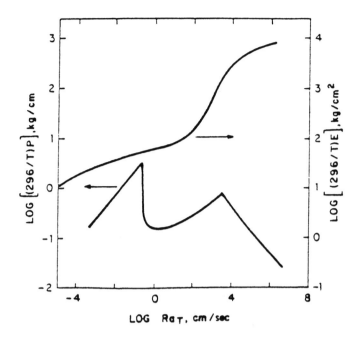

FIGURE 5. Peel force[12] vs. rate for an uncrosslinked butadiene-styrene rubber adhesive on PET polyester film and tensile modulus vs. rate for the bulk rubber adhesive.

the trend as shown in Figure 6. The percent rebound at the glass transition temperature is zero. The energy absorption, and therefore the energy dissipation at the glass transition, is maximum. Further discussions of the relation between energy dissipation and mechanical measurement appear in later sections.

In summary, the viscoelastic properties of plastic flow and energy dissipation are important for an adhesive to resist debonding. Maximum energy dissipation occurs at the point of molecular motion transition.

C. THE BOLTZMANN SUPERPOSITION PRINCIPLE AND THE WLF EQUATION

In Section II.A we showed the importance of creep and flow properties during bonding. In Section II.B, we illustrated the importance of energy dissipation in resisting debonding. The parameters associated with these viscoelastic properties are interrelated by two fundamental rheological principles: the Boltzmann superposition principle and the WLF time-temperature superposition principle.

The Boltzmann superposition principle is at the heart of *linear* viscoelasticity. In the creep experiment, the principle states (1) that creep is a function of the specimen's entire loading history, and (2) that each loading step makes an independent contribution to the final deformation, which can thus be obtained by totaling all contributions. Because of Boltzmann's principle, linear viscoelastic behavior can be described by an infinite set of linear differential equations. Innumerable ways exist to specify these sets, *all of which are equivalent*. Thus, rheological parameters associated with each set are interchangeable by suitable mathematical transformations. Examples of these transformations have been amply illustrated.[14] Table 2[15] exhibits the relationship between the rheological parameters.

Linear viscoelastic behavior might be expected to be valid for substances whose basic structure is not changing with time during the viscoelastic experiment. Typical substances

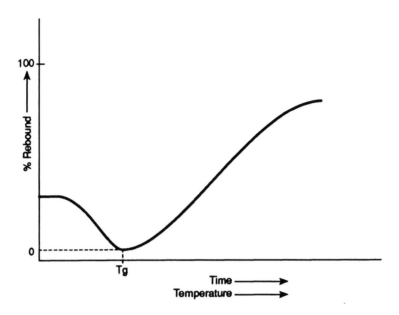

FIGURE 6. Experiment to demonstrate the energy dissipation at the glass transition.[4]

to which the superposition principle applies are glass and amorphous polymers (e.g., polyisobutylene[16]). Oriented crystalline fibers,[17] such as rayon, silk, and nylon 66, show marked deviations. This has been attributed to the fact that when subject to loading, crystalline substances often show orientation and further crystallization. At the end of loading experiments (such as creep or stress relaxation), these crystalline substances are structurally different from their initial states. Thus, in these materials, the Boltzmann principle applies only under special conditions, such as extremely small loads, and at temperatures at which appreciable crystallization does not occur.

Another important rheological principle is the WLF time-temperature superposition principle. Under this principle, rheological properties can be expressed either as a function of the temperature or the time. Master curves can be constructed from these experiments to define a material's viscoelastic function. Examples of such experiments are: creep, stress relaxation, peel tests, and so on. The master curve, and the process of constructing it, provide a framework that can be understood in terms of an equivalence between the temperature and the time scale involved in the experiment. Thus, varying the temperature is equivalent to changing the experiment's time scale.

In the WLF theory, varying the temperature is equivalent to shifting a material's retardation (or the relaxation) time spectrum. Normally, the retardation times are increased as the temperature is decreased. As the retardation times increase, the Deborah number, expressed as the ratio of the experiment's time scale to the material's average retardation time, is decreased. One can also decrease the Deborah number by increasing the rate of the experiment at the same temperature. Thus, decreasing the temperature in any rheological experiment is equivalent to increasing the rate of the same experiment at the original reference temperature. Any viscoelastic property expressed as a function of temperature can also be expressed as a function of time (or of the logarithm of time).

The importance of the WLF theory is made obvious by the fact that different experimental time scales are involved in bonding and debonding. In typical pressure-sensitive adhesive

TABLE 2
Linear Viscoelastic Responses

$\sigma(t)$ = stress at time t (dyn/cm^2)

$\dot{\gamma}(t)$ = strain rate at time t (sec^{-1})

$G(t)$ = stress relaxation modulus (dyn/cm^2)

$J(t)$ = creep compliance (cm^2/dyn)

η_0 = zero shear rate viscosity = $\int_0^\infty G(t)dt$

J_e^0 = recoverable shear compliance = $(1/\eta_0^2)\int_0^\infty tG(t)\,dt$

τ_0 = average relaxation time = $\eta_0 J_e^0$

$\sigma(t)$ = $\int_{-\infty}^t G(s)\dot{\gamma}(t-s)\,ds$

applications, bonding normally takes a few seconds, whereas debonding takes only a few milliseconds. The WLF theory permits us to do rheological measurements at several temperatures and to construct a master curve expressing the rheological property as a function of time at the reference temperature. In this way, the performance of a pressure-sensitive adhesive can be predicted at the reference temperature.

The WLF equation can be derived by considering the requirement for a minimum-free volume to permit rotation of the chain segments, and the hindrance to such rotation which is caused by neighboring molecules. We omit a detailed derivation of the WLF equation, as this derivation can be found in many polymer textbooks. We will present here only a conceptional description of the WLF equation, which follows the arguments given by Sperling.[4]

Let us first consider molecular motions under the constraints of energy barriers per unit time and assign a probability value, P, that motion occurs in unit time. If ΔE_{act} is the activation energy to overcome the barrier, P and ΔE_{act} are related by an Arrhenius-type relationship:

$$P = \exp(-\Delta E_{act}/kT) \tag{4}$$

We shall now consider the time factor. The probability that molecular motions can overcome the barriers will be larger if a longer time, t, is allowed. The onset of molecular motion should depend on the value of tP. We may therefore assume that when the product tP reaches a critical value, rotation of the chain segments is allowed. Hence, we may write:

$$\ln(tP) = \text{constant} = -\Delta E_{act}/kT + \ln(t) \tag{5}$$

and,

$$\ln(t) = \text{constant} + \Delta E_{act}/kT \tag{6}$$

Taking the differential,

$$\Delta\ln(t) = (-\Delta E_{act}/kT^2)\,\Delta T \tag{7}$$

The relation in Equation 7 can be understood in the context of a time-temperature relationship for the onset of a particular cooperative motion. As the temperature is lowered ($\Delta T < 0$), a longer time is needed for the required motion, and the retardation time (or the relaxation time) is increased. Therefore, a longer experimental time is required for the onset of molecular motions.

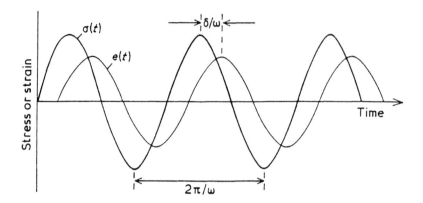

FIGURE 7. Dynamic mechanical measurements of viscoelastic materials.

The time-temperature WLF relationship is universal for all amorphous viscoelastic materials, including adhesives. This relationship allows us to appreciate viscoelastic functions involved in both bonding and debonding. On the other hand, the Boltzmann principle applies only to linear viscoelastic materials. It allows us to convert one viscoelastic function into another by using mathematical transformations. It is by means of this principle that construction of viscoelastic functions based on simple rheological measurements is allowed. In the next section, we will discuss application of such measurements — dynamic mechanical analysis — to study the rheology of the adhesive systems.

D. DYNAMIC MECHANICAL MEASUREMENTS

In Section II.B, we showed that energy absorption (or dissipation) of a material is at a local maximum when the material is engaged in molecular dynamic transitions, such as the glass-to-rubber and liquid-to-rubber transition. The principles underlying the absorption of mechanical energy are not much different from those underlying the absorption of electromagnetic energy such as IR and UV. Hence, the interaction of electromagnetic radiation with matter results in energy absorption during changes in the electronic, vibrational, and rotational energy levels of atoms and molecules. In mechanical spectroscopy, the interaction of mechanical energy with the material results in energy absorption during a change in molecular dynamics. Such an interaction is both time and temperature dependent.

Energy is commonly applied in mechanical spectroscopy by imposing a sinusoidal mechanical strain on the material. In this way, the viscoelastic body can be described as a classical damped harmonic oscillator. Such a model allows for a more penetrating analysis. If the material is elastic, the stress response will be in phase with the sinusoidal mechanical strain. Conversely, if the material is viscous, the stress response will be 90° out of phase with the sinusoidal mechanical strain. For a viscoelastic material, the stress response will exhibit a phase shift with respect to the mechanical strain. Analysis of the phase shift of the mechanical stress with respect to the mechanical strain permits us to calculate the useful rheological properties.

The foregoing arguments permit us to write

$$\text{strain } e = e_0 \sin(\omega t) \tag{8}$$

$$\text{stress } \sigma = \sigma_0 \sin(\omega t + \delta) \tag{9}$$

Expanding $\sin(\omega t + \delta)$, we see that the stress consists of two components: one with a magnitude $\sigma_0\cos(\delta) \sin(\omega t)$ in phase with the strain and a second one with a magnitude $\sigma_0\sin(\delta) \cos(\omega t)$, which is 90° out of phase with the strain.

The stress-strain relationship can, therefore, be defined by a quantity G' in phase with the strain and by a quantity G" which is 90° out of phase with the strain, i.e.,

$$\sigma = e_0 G' \sin(\omega t) + e_0 G'' \cos(\omega t) \tag{10}$$

where

$$G' = (\sigma_0/e_0) \cos(\delta) \text{ and } G'' = (\sigma_0/e_0) \sin(\delta) \tag{11}$$

Figure 7 shows the graphic representations of elastic, viscous, and viscoelastic responses of materials under oscillatory deformation.

If we write

$$e = e_0 \exp(i\omega t), \quad \sigma = \sigma_0 \exp(i(\omega t + \delta)) \tag{12}$$

then

$$(\sigma/e) = G^* = (\sigma_0/e_0) \exp(i\delta)$$

$$G^* = (\sigma_0/e_0)(\cos(\delta) + i \sin(\delta))$$

$$G^* = G' + iG'' \tag{13}$$

The real part of the modulus G', which is in phase with the strain, is often called the *storage modulus*. This term defines the energy stored in the specimen due to the applied strain. The imaginary part of the modulus G", which is 90° out of phase with the strain, is called the *loss modulus*. The loss modulus defines the dissipation of energy of the material. We can easily see this by calculating the energy dissipated per cycle, $\Delta\xi$:

$$\Delta\xi = \int \sigma \, de = \int_0^{2\pi/\omega} (\sigma \, de/dt) dt \tag{14}$$

Substituting for σ and e, we have

$$\Delta\xi = \omega e_0^2 \int_0^{2\pi/\omega} (G' \sin \omega t \cos \omega t + G'' \cos^2(\omega t)) dt = \pi G'' e_0^2 \tag{15}$$

The ratio of the loss modulus to the storage modulus defines a phase angle δ. To a good approximation $\tan\delta = \delta$ when the loss modulus G" is small. Typical values of G', G", and $\tan\delta$ for a polymer in the glassy state would be $10^9 Nm^{-2}$, $10^7 Nm^{-2}$, and 0.01.

In Equation 15, we have shown that G" is a quantitative measure of the mechanical energy absorption (or dissipation). Transitions in molecular dynamics, such as liquid-to-rubber and rubber-to-glass, can therefore be seen in G", which goes through a local frequency maximum at these transitions.

Figure 8 shows the variation of G', G" and $\tan\delta$ with frequency for a polymer. The local maximum of G" in Figure 8 may be associated with either the liquid-to-rubber or the rubber-to-glass transitions. The storage modulus G' is independent of the frequency when the polymer is either in the rubbery state or in the glass state. The G' value ranges from $10^5 \ Nm^{-2}$ to $10^7 Nm^{-2}$ for the rubbery state and $10^9 Nm^{-2}$ for the glassy state.

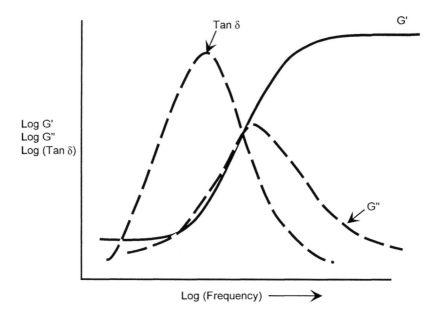

FIGURE 8.   The complex modulus G′, G″, tanδ as a function of frequency ω.

Because of the inverse relationship of frequency and time, variation in G′, G″, and tanδ with *time* (Figure 9) can be constructed from Figure 8 (which shows variation in G′, G″, and tanδ with *frequency*). By comparing both Figure 8 and 9, we can see that these functions are actually mirror images of each other.

An analogous diagram (Figure 10) shows the variation of the compliances J′ and J″ with frequency. The moduli and the compliance functions are interconvertable through the Fourier transformation.

In summary, the viscoelastic functions G′, G″, and tanδ can be measured using dynamic mechanical spectroscopy. The parameter G″ measures the energy dissipation during a mechanical process. It exhibits a local maximum during a transition such as a liquid-to-rubber or a rubber-to-glass transition. The parameter G′ is independent of the frequency (or temperature) when the material is either in the rubbery or glassy state. In viscoelastic materials, which go through a rubber-to-glass transition, G′ changes in value by approximately four orders of magnitude. In viscoelastic materials, J′ also changes value by four orders of magnitude at the glass transition. This combination of G″ and J′ responses of viscoelastic material involving a rubber-to-glass transition make such materials ideal for adhesive applications. We will discuss this in detail in the following section.

III. PRESSURE-SENSITIVE ADHESIVES

In this section, we will use pressure-sensitive adhesives as examples to illustrate the fundamental principles of rheology.

Pressure-sensitive adhesives exhibit tack upon contact under pressure. Typically, a pressure-sensitive adhesive is used in combination with a flexible adherend. The combination of the two is often called a pressure-sensitive adhesive tape. Tape is applied by contacting its adhesive side, under slight pressure to the surface of a substrate. Tack that develops between tape and substrate can be understood in terms of two properties: (1) The tape establishes bonding with the substrate, and (2) the tape resists separation from the substrate. A typical example of a pressure-sensitive tape is commercial bandages.

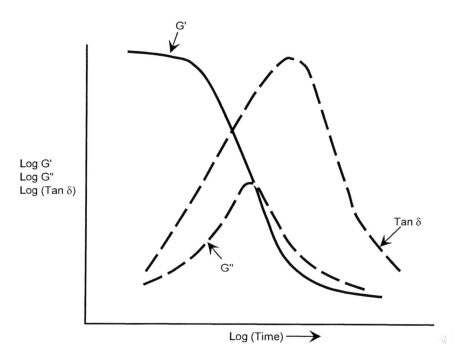

FIGURE 9. The complex modulus G′, G″, tanδ as a function of time.

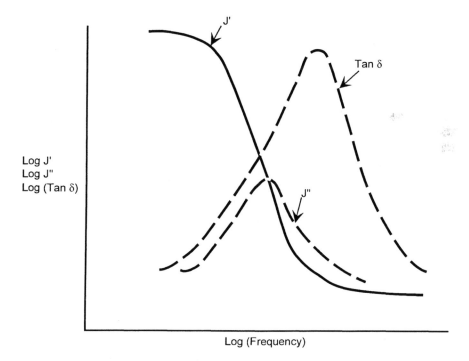

FIGURE 10. The complex compliance J′, J″, tanδ as a function of frequency ω.

It is essential to distinguish between tack and permanent bonding. In using permanent bonding adhesives, such as epoxy, to connect different components into a structure, the

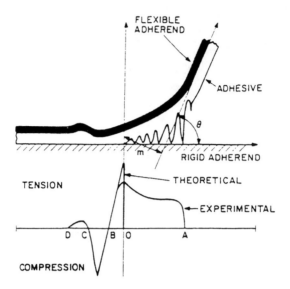

FIGURE 11.   Schematics of peel profile[13] and normal stresses in adhesive layer.

adhesive becomes one of the components of the structure. Separation of the adhesive component generally cannot be achieved without simultaneously destroying other components.

On the other hand, tack is the quality or state of sticking, adhering, adhesiveness, and/or stickiness. According to ASTM D1878-61T, tack is defined as the property of a material which enables it to form a bond of measurable strength immediately upon contact with another surface. This definition implies that the *bond can be separated eventually*, although some external force of measurable strength must be applied. In many pressure-sensitive tape applications, one often desires to cleanly separate the adhesive from the substrate. Any adhesive residue left behind on the substrate's surface is undesirable.

We will present discussions of pressure-sensitive adhesives in two parts. In the first, we will discuss composition; in the second, we will discuss the rheological properties.

A. COMPOSITIONS OF PRESSURE-SENSITIVE ADHESIVES

Pressure-sensitive adhesives generally consist of two major components: (1) an elastomer and (2) a tackifier. In hot melt pressure-sensitive adhesives, a third component such as a plasticizing agent is often added as a processing aid.

In this section, we will use hot melt pressure-sensitive adhesives based on triblock copolymer systems to illustrate the functions of various adhesive components.

1. Rubber

Rubbers are often used to impart cohesive strength to the pressure-sensitive adhesive. Rubbers have two distinct properties that are useful for adhesive applications: extensibility and recoverability. Both properties assure not only adequate strength of the adhesive bond but also *clean* adhesive removal when the tape is separated from the substrate.

The functionality of the rubber component in the adhesives can be appreciated from a stress analysis of the peel test. Figure 11 schematically shows both the peel and normal stress profiles in the adhesive layer.[13] The specimen is placed on the abscissa with its origin O at the point of detachment. The normal tensile stress in the adhesive is greatest at the point of detachment. In the region AO, the normal tensile stress is zero if the adhesive completely detaches from the substrate. However, in most cases, some unbroken ligaments of the adhesive still bridge between the peeling strip and the rigid adherend. Therefore,

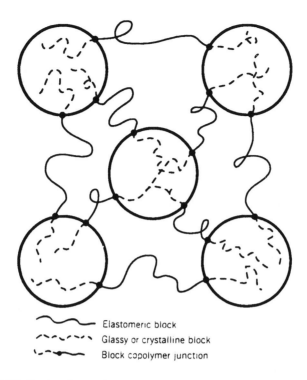

Elastomeric block
Glassy or crystalline block
Block copolymer junction

FIGURE 12.   Idealized triblock copolymer thermoplastic elastomer morphology.

large normal tensile stresses persist in the region AO. Should the adhesive be weak and low in extensibility and recoverability, the ligaments can be broken at any point in region AO. This leaves adhesive residues behind. The premature breakage of the adhesive ligaments also substantially reduces the mechanical strength during separation from the substrate.

Rubber also helps to establish effective intermolecular diffusion friction force at the interface. Adhesive bond formation to the substrate can be assisted by combining the rubber with other components such as tackifiers and plasticizers. These substances help to increase the compliance of this type of adhesive. Pressure-sensitive adhesive creeps under applied stresses. Once this occurs, the rubber component engages in *effective* molecular mutual entanglement at the interface. In Section II.A, we mentioned that molecular entanglement of high molecular weight polymers at the interface provides ideal *molecular link* functionality. Because of such entanglements, the mechanical forces can be effectively transferred across the interface. It is interesting to note that typical rubbers have molecular weights in the range 10^5 to 10^6, which is several orders of magnitude higher than the other components of the adhesive.

Rubbers commonly used in hot melt pressure-sensitive adhesive formulations are block copolymers. These materials contain physical rather than chemical crosslinks. The simplest structure contains two hard blocks (with a Tg above ambient temperature) at the ends and a soft block (with a low Tg) in the middle segment of the molecule (Figure 12). The soft block is amorphous and above Tg under application temperatures. The hard block is glassy. These types of rubbers are called thermoplastic elastomers.

One example of a thermoplastic elastomer is polystyrene-block polybutadiene-block polystyrene triblock polymer (SBS). Another example is polystyrene-block polyethylene butylene-block polystyrene (SEBS). In SEBS, the EB stands for ethylene-butylene, where a combination of 1,2 and 1,4 copolymerization of butadiene on hydrogenation produces a random copolymer of ethylene and butylene. When a linear block copolymer (Figure 12) is

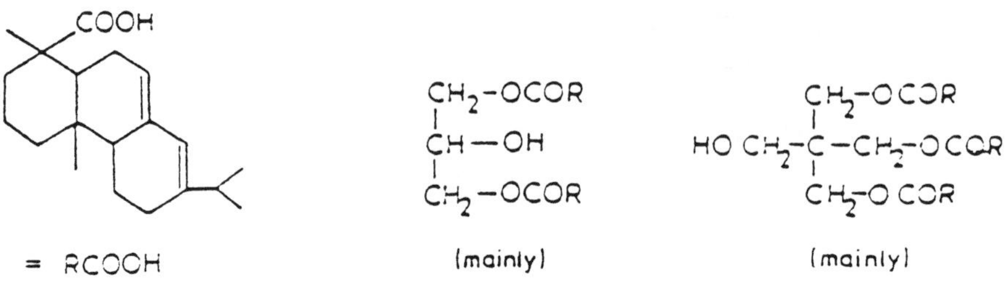

$$= RCOOH$$

I

$$CH_2-OCOR$$
$$CH-OH$$
$$CH_2-OCOR$$

(mainly)

II

$$CH_2-OCOR$$
$$HO\ CH_2-C-CH_2-OCOR$$
$$CH_2-O\ COR$$

(mainly)

III

FIGURE 13. Key components[18] of rosin tackifiers: abietic acid (I) and/or pimaric acid as the monocarboxylic acid. Esterifying alcohols are either glycerol (II) or pentaerythritol (III).

TABLE 3
Number Average Molecular Weight and Tg
for Glycerol Esters

Resins	$\overline{M}_n$	Tg K
Staybelite ester 10 (glycerol ester)	734 ± 14	323
Pentalyn H(pentaerythritol ester)	900 ± 10	344
Piccolyte S70(polyβ pinene)	920 ± 10	335
Piccolyte S115(polyβ pinene)	1200 ± 10	366

of the SBS or SEBS type, it is sold under the trade name Kraton. The block copolymers can also have star-like structures. Several companies, including Firestone and Phillips Petroleum, manufacture this type of block copolymers.

2. Tackifiers

Tackifiers are viscoelastic modifiers present in pressure-sensitive adhesive formulations. They primarily raise Tg so that the adhesives are viscoelastic at use temperatures.

Tackifiers normally are low molecular weight and high Tg materials. They are sometimes characterized as highly condensed alicyclic structures. The most commonly used tackifiers are rosin derivatives, terpene resins (oligomers of α and β pinenes), and petroleum resins (oligomers of unsaturated petroleum fractions). These chemicals are effective tackifiers especially in the SEBS-based adhesive formulations. A tackifier's effectiveness is largely determined by its compatibility with the rubber component and by its ability to modify the viscoelastic property.

Unmodified rosins are complex mixtures of monocarboxylic acids with a smaller amount of neutral components. The monocarboxylic acids are two main types known as the abietic and pimaric types (Figure 13). For use in adhesives, it is desirable to eliminate the carboxylic acid functionality and reduce or eliminate the olefinic unsaturation present. The rosin esters principally used in pressure-sensitive adhesives are esters of glycerol or pentaerythritol. As a result of such esterification, rosin esters are highly branched molecules with a high degree of cyclicity. Esters of hydrogenated rosins, such as those shown in Table 3, may be idealized as structures II and III and their dimer, trimer derivatives with most of the unsaturation eliminated.

Terpene resins are derived from polymerization of either α or β pinenes. Polymerization of β pinenes is shown to go through intramolecular cyclization (Figure 14).[18] The resultant terpenes are cyclic structures with a low degree of saturation.

Petroleum-derived tackifier resins are obtained by the cationic polymerization of complex petroleum fractions, which include diene and cyclic monomers. During polymerization, a

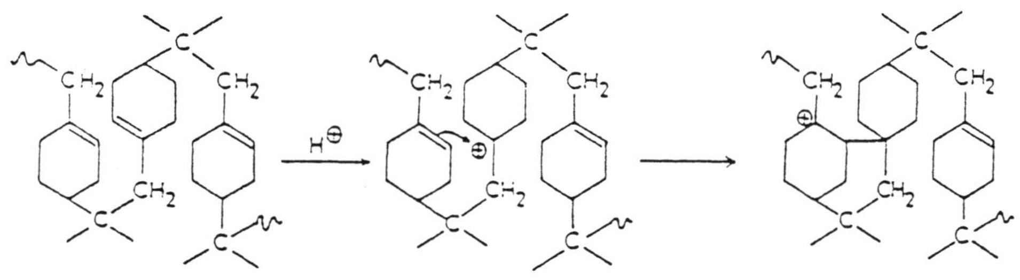

FIGURE 14. Intramolecular cyclization[18] during polymerization of β pinene.

high degree of cyclization and ring condensation likely occurs, producing structures which again give rise to high Tg and low levels of unsaturation. The polymerization chemistry and the structural aspects of the derived tackifiers have been reviewed.[19]

In Section III.B., we will discuss the modification of the viscoelastic behavior of pressure-sensitive adhesives resulting from the addition of tackifying resins.

3. Plasticizers

Plasticizers are used to improve the processability of adhesives. Under processing conditions, plasticizers lower the viscosity of the adhesives so that the adhesives can be easily processed.

In use, however, plasticizers must be compatible with the rubber component. When the rubber component in the adhesives is a SEBS-type block copolymers, commonly used plasticizers are hydrogenation products of petroleum-based aromatics: mixtures of naphthenes, paraffins of low molecular weight compounds, and unreacted aromatic species.

The compatibility of the plasticizers with the rubber component produces an increase of free volume of the mixtures. As a result, plasticizers lower the adhesive's modulus *on all time scales*. In contrast, tackifiers lower the modulus at long-time scales but raise the modulus at short-time scales.

Plasticizers may accumulate at the interface and interfere with interfacial molecular bonding (see Section II.A). Caution must therefore be used in compounding pressure-sensitive adhesives with plasticizers.

B. RHEOLOGICAL PRINCIPLES OF PRESSURE-SENSITIVE ADHESIVES

Applications of pressure-sensitive adhesives place unique demands on the properties of these materials. Pressure-sensitive adhesives are expected to provide tack on contact under slight pressure and to assure clean removal (while offering some resistance) when removing the adhesive. A combination of rubber and tackifying resins will provide these unique properties.

The viscoelastic properties of rubbers are not rate or temperature sensitive at typical application temperatures which are room temperatures in most applications. The rubber modulus or compliance is constant over a wide range of temperatures. When plotted against the logarithm of time, the modulus or the compliance is *flat* over at least five decades of time (Figure 15). Rubber by itself does little to facilitate bonding. Rubbers also offer only minimal energy dissipation during debonding process.

When rubbers are combined with tackifying resins, the viscoelastic properties are greatly modified. First of all, the Tg is increased, from -80 to $-60°C$ for the neat rubbers and from 5 to 30°C for the adhesives. The increase of Tg is ascribed to an increase of monomeric friction coefficient* associated with the blending of the tackifiers with the rubber.

---

* Large monomeric friction coefficients of the tackifiers results from the stiffness of the structures. The coefficient (ζ) is directly related to the activation energy barrier (ΔE_{act}) of the free rotations of the molecules, or $\zeta \sim \exp(\Delta E_{act}/RT)$.

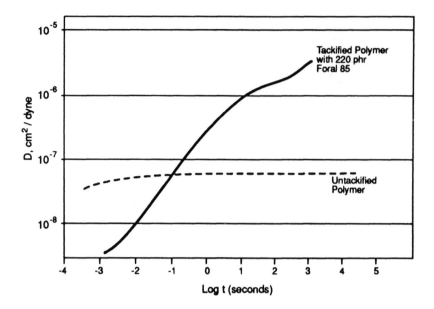

FIGURE 15.   Creep compliance[20] at 22°C for isoprene-styrene block polymer (Solprene 418) with and without rosin ester tackifier.

Accompanying this change, the adhesives become viscoelastic; namely, their modulus or compliance becomes sensitive to the experimental time scales and use temperature. The viscoelastic properties of the rubber-tackifier mixtures are depicted in Figure 15. Overall, a reduction of modulus (or an increase of the compliance) occurs at long times, and an increase of modulus (or a decrease of compliance) occurs at short times.

The reduction in modulus (increase in compliance) at long times shown by the rubber/ tackifier blends of Figure 15 reflects the easier viscous flow that can occur in the bonding stage (typically of the order of a second in duration) in ordinary applications. This assures better contact and hence increases the tack value.

The increase in modulus at short times and the strong dependence of the modulus on the rate implies that a large energy dissipation will occur during debonding. This property assures clean removal while providing substantial separation resistance when removing pressure-sensitive tapes from substrates.

The debonding process generally involves the propagation of a fracture front at the interface. The peel force usually fluctuates between well-defined maxima and minima. This is a phenomenon known as slip-stick fluctuation. A typical slip-stick fluctuation is shown in Figure 16. In slip-stick rupture, failure is initiated when the peel force reaches the maximum. The stored energy is then expended to propagate the crack. As the stored energy is depleted, crack propagation ceases at the minimum peel force. As the peel force is increased again to the maximum, another failure cycle is repeated.

In this slip-stick fluctuation, both the elastic modulus and the energy dissipation make important contributions to adhesive performance. High elastic modulus prevents adhesive-cohesive fracture when one stretches the tape to remove it from the substrate. This assures clean tape removal. On the other hand, high energy dissipation in the adhesive will impart resistance to the propagation of the crack between the adhesive and the substrate.

In summary, blends of rubber and tackifiers offer viscoelastic properties that are advantageous for the applications of pressure-sensitive adhesives. The blends are characterized

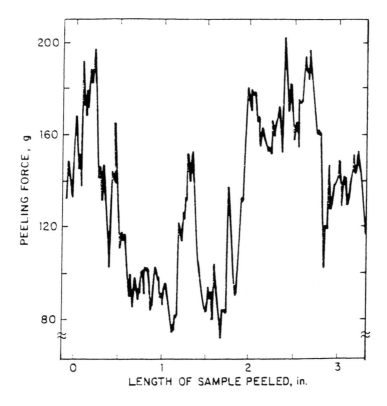

FIGURE 16. A typical slip-stick steady state[21] peel force with continuous failure for T peel.

by a low modulus at long times and a high modulus at short times. The low modulus at long times facilitates bonding. High modulus at short time imparts separation resistance and provides clean adhesive removal.

IV. CONCLUSIONS

The functional requirements of adhesives demand that these substances possess unique rheological properties. In bonding, adhesives must flow and/or creep. In debonding, they must possess toughness and exhibit large energy dissipation. Further rheological criteria are imposed by the fact that different time scales are involved in bonding and debonding of adhesives. Typically, a few seconds are required for bonding, whereas only a fraction of a second is involved in debonding. Knowledge of these rheological principles can help us formulate better adhesives.

REFERENCES

1. **Yoo, J. N., Sperling, L. H., Glinka, C. J., and Klein, A.,** *Macromolecules,* 24, 2868, 1991.
2. **de Gennes, P. G.,** *Scaling Concepts in Polymer Physics,* Cornell University Press, Ithaca, 1979.
3. **Doi, M. and Edwards, S. F.,** *J. Chem. Soc. Faraday Trans. 2,* 74, 1789, 1978.
4. **Sperling, L. H.,** *Introduction to Physical Polymer Science,* John Wiley & Sons, New York, 1985.
5. **Dahlquist, C. A.,** in *Adhesion, Fundamentals and Practice,* Gordon and Breach Science Publishers, New York, 1969, 143.
6. **Skewis, J. D.,** *Rubber Chem. Technol.,* 39, 217, 1966.
7. **Hammond, G. R.,** *Rubber Chem. Technol.,* 54, 403, 1981.
8. **Bussemaker, O. K. F. and van Beek, W. V. C.,** *Rubber Chem. Technol.,* 37, 28, 1964.
9. **Griffith, A. A.,** *Phil. Trans. R. Soc. London,* A221, 163, 1920.
10. **Gent, A. N. and Schultz, J.,** *J. Adhesion,* 3, 281, 1972.
11. **Andrews, E. H. and Kinloch, A. J.,** *Proc. R. Soc. London,* A332, 385, 1973.
12. **Gent, A. N. and Petrich, R. P.,** *Proc. R. Soc. London,* A310, 433, 1969.
13. **Wu, S.,** *Polymer Interface and Adhesion,* Marcel Dekker, New York, 541, 1982.
14. **Tobolsky, A. V.,** *Properties and Structures of Polymers,* John Wiley & Sons, 1960.
15. **Graessley, W. W.,** in *Physical Properties of Polymers,* Mark, J., Eisenberg, A., Graessley, W., Mandelkern, L., and Koneig, J., American Chemical Society, Washington, D.C., 1984, 97.
16. **Tobolsky, A. V. and Andrews, R. D.,** *J. Chem. Phys.,* 13, 3, 1945.
17. **Leaderman, H.,** in *Elastic and Creep Properties of Filamentous Materials and Other High Polymers,* Textile Foundation, Washington, D.C., 1943.
18. **Aubrey, D. W.,** *Rubber Chem. Technol.,* 61, 448, 1988.
19. **Holoham, J. F., Penn, J. Y., and Vredenburgh, W. A.,** in *Encyclopedia of Chemical Technology,* 3rd ed., Vol. 12, Wiley Interscience, New York, p. 852.
20. **Kraus, G., Rollmann, K. W., and Gray, R. A.,** *J. Adhesion,* 10, 221, 1979.
21. **Gardon, J. L.,** *J. Appl. Polym. Sci.,* 7, 643, 1963.

Chapter 25

ADHESION BETWEEN COMPONENTS OF ELASTOMERIC COMPOSITE MATERIALS

Yoichiro Kubo

TABLE OF CONTENTS

0-8493-4401-8/93/$0.00 + $.50

I. INTRODUCTION

Composite materials are those in which one type is incorporated in any way within one or more types of materials, and which do not exist as an individual material in the normal sense. These components then become useful materials for industry. Composite materials are defined as "something that is made up of diverse elements" and thus are a complicated combination of different kinds of materials consciously made in anticipation of the merits of the combination.

Even in the case of fibers dispersed in the composite materials from among the many types available, many kinds of materials, such as plastics, elastomers, metals, and ceramics, are used as the matrix in the composites, and many kinds of fibers are used. However, for actually useful fiber-reinforced composite materials, the number of the well-chosen combinations of raw materials used for these composite materials is limited. This is because one raw material may be unable to combine freely with a different material due to many factors, such as wettability between different materials, adhesion, moldability, and processing.

It is important to develop techniques to combine these different materials in order that the original properties of each raw material are not damaged by the combination of the different materials.

Elastomers are actually used as composite materials in many applications. Elastomer-based composites include an enormous range of types of materials used in a myriad of applications, all of which the author is unable to cover. This chapter will describe only composite materials comprised of elastomers and fibers, and will deal with primarily the adhesion between fibers and elastomers, which is important for these fiber-reinforced elastomers.

II. FIBER-REINFORCED ELASTOMERS

A. APPLICATION OF FIBER-REINFORCED ELASTOMERS

Fiber-reinforced elastomers as a type of composite material is very important for practical applications. They are the major type of product from among many rubber products produced and are useful for tires, conveyor and v-belts, hoses, and rubber-coated fabrics for other products, such as diaphragms, air springs, and membranes for construction.

B. COMPONENT MATERIALS OF FIBER-REINFORCED ELASTOMERS
1. Elastomer

All kinds of elastomers have been utilized to take advantage of their outstanding features for each specific application. Figure 1 shows the relationship between various applications and the combination of elastomers and fibers.[1]

Natural rubber (NR), isoprene rubber (IR), styrene-butadiene rubber (SBR), and butadiene rubber (BR) are used in tires, a fiber-reinforced composite rubber product. The elastomers are the main materials for tire applications because they are economical and have high tensile strength, good abrasion resistance, and low heat build-up.

Chloroprene rubber (CR) is often used for belts in automotive and industrial areas because of its good physical properties, processability, ozone resistance, inflammability, and ashesion to fibers.

Ethylene-propylene-diene rubber (EPDM) is used mainly as heat-resistant conveyor belts and steam hoses because of its good resistance to heat, ozone, and hot water.

Nitrile rubber (NBR), acrylic rubber (ACM), and epichlorohydrin rubber (ECO) are used in rubber parts that are in contact with oils, fuels, and solvents. Applications for these, such as hoses, gaskets, diaphragms, and seals, are used because of their good oil, fuel, and solvent resistance.

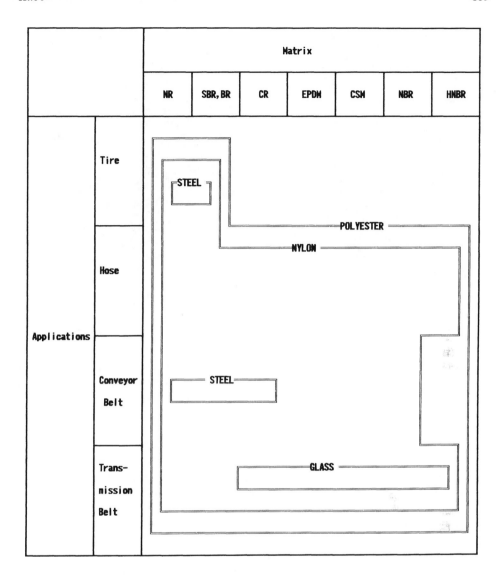

FIGURE 1. Typical applications and types of fiber-reinforced elastomer products.

Hydrogenated nitrile rubber (HNBR) is used widely for automotive fuel and oil hoses, seals, belts, oil fields parts, and so on because it exhibits excellent dynamic and physical properties and environmental resistance with a better balance of properties than other typical oil-resistant elastomers.[2-5]

2. Fibers

Fibers are key components in making the skeleton of fiber-reinforced rubber products, such as tires, belts, and hoses; they provide strength to the products, increase their life, and give dimensional stability to rubber components. The fibers usually are used in the form of twisted fine filaments to form a cord and then woven to form fabrics. Once, cotton and rayon were used as reinforcing fibers, but now, nylon, polyester, glass, steel, and so on, are used as reinforcing members because of their better resistance under severe environments.

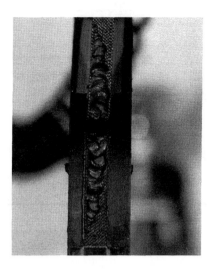

FIGURE 2.   Photographs of peeled sample of PET to NR.

III. ADHESION BETWEEN ELASTOMERS AND FIBERS

Rubber products are being used under severer environments, which create more stringent requirements for the long life of rubber products. These include greater heat and fatigue resistance, which becomes severer year by year. With this background, there have been aggressive programs to improve quality and develop new grades of various types of elastomers and fibers to meet these needs. It is of overriding importance that for long life of rubber products, there must be excellent adhesion between elastomers and fibers.

Fiber-reinforced rubber products fulfill their functions only when the rubbers are united with the fibers by a suitable bonding mechanism. When partial separation occurs between the rubber and fiber of the composite due to inferior adhesive strength, the stress is concentrated at the point of the separation, and the temperature of that part rises. The point of separation becomes larger until the damage to the composite materials results in catastrophic failure.

As an example, Figure 2 shows photographs of the adhesion interface between the polyester cord and natural rubber using an adhesive, after peeling the sample. The left photograph shows the inferior adhesive force between the polyester cord, adhesive layer, and natural rubber compound. The failed portion is the adhesive layer between the polyester cord and the natural rubber, wherein the elastomer does not adhere to the surface of the polyester cord. On the other hand, the right photograph shows a case of excellent adhesion. The adhesion layer did not fail, with failure occurring in the rubber itself. In cases of excellent adhesive strength, the strength of the adhesive layer is stronger than that of the rubber layer, and hence, the relatively weaker rubber layer fails.

In the case of cotton, the naps of the cotton are anchored into the rubber, and they give adhesive strength between cotton and rubber. However, the adhesion between rubber and synthetic fibers, such as rayon and nylon, which have smooth surfaces, created major problems.

Adhesion systems are divided into mechanical and chemical systems. The chemical systems are further divided into the aqueous and solution adhesion systems. The former consists of compounds from resorcinol, formaldehyde resin, and an elastomer latex, and it is called the RFL adhesion system. The latter is composed of dissolved elastomers and compounding ingredients in the solution, often with adhesion promoters.

A. TYPES OF ADHESION SYSTEMS

1. Mechanical Adhesion to Fibers

In the case of cotton, the naps anchor into the rubber, and they give strength to the adhesive between cotton and rubber. The adhesive strength is given by the pull strength of one nap of cotton out of the rubber, and the useful bonding force is derived from the many naps in cotton. However, the adhesive strength of cotton and the strength of the fiber itself is inferior to synthetic fibers; it will not meet the current severe requirements.

2. Chemical Adhesion to Fibers

Chemical system may be divided into the aqueous and solution adhesion systems.

a. Solution Systems

Adhesive Using Isocyanate Compound (e.g., Dichloromethane Solution of Triphenylisocyanate[6]

After mixing an isocyanate solution with a rubber cement, the mixture is coated onto the fiber surface. The isocyanate compound reacts with OH or NH_3 groups to give good adhesive strength. It is difficult to handle this system because of the rapid reaction speed between the isocyanate groups and other functional groups. This leads to gelation caused by crosslinking the rubber chains and to a short shelf-life of the adhesive due to its instability in solution. This system is not used often because of its impracticality.

Rubber Cement

Rubber cement is composed of dissolved elastomers and the compounding ingredients in a suitable solvent. Fibers are dipped into the solution, and the solvent is allowed to escape, after which the fibers are combined with the rubber compounds. Then the rubber compounds and rubber cement on the fibers adhere by crosslinking during vulcanization. This system is less desirable because of the environmental effect during the evaporation of solvent on heating.

b. Aqueous Systems

The aqueous adhesion system is a combination of resins composed of resorcinol, formaldehyde, and an elastomer latex; this is called the RFL adhesion system. This system is handled easily; however, the disadvantage is that it is necessary to dry and heat-treat after coating with the RFL adhesive. This system is described in more detail in Section IV.

Rubber Compounds with Internal Adhesion Promoters

Internal adhesion promoters in a rubber compound function to provide adhesion to the fibers. The adhesion between rubbers and fibers is affected during the curing process, without pretreatment by an adhesive on the fibers. As an example, hexamethylenetetramine and resorcinol are mixed into a rubber compound containing fine particles of silica, and an RF resin is created during the curing process. It is sometimes difficult to disperse finely each reagent in the rubber compound.

B. PRETREATMENT OF FIBERS

In order to enhance the effect of the above mentioned adhesion system, it is usual for the low reactive surface of fibers to be treated by a highly reactive reagent. There are some cases where these pretreatments can hold the fibers in a bundle and protect them from fraying back to the original filaments.

1. Isocyanate Compound Treatment[7]

Polyester does not adhere well enough to elastomers using only an RFL adhesive because hydrogen bonding energy is weaker than that of nylon. In order to enhance the adhesion force of polyester fiber, an isocyanate compound pretreatment is used to strengthen the adhesion between the polyester and elastomer.

Isocyanate compounds cannot be used in a water solution because of their high reactivity with water. The reaction products of an isocyanate and a phenol compound or the dimer of an isocyanate compound are used typically as the isocyanate adhesive pretreatment.

2. Epoxy Resins[7,8]

Because polyester and aramid fibers have low fiber surface reactivity, generally, these fibers are treated with an epoxy resin that has high reactivity.

The solid epoxy resins are used as a water solution or a dispersion in water, and the liquid epoxy resins are used as is.

3. Coupling Reagents[9,10]

With glass fiber, any physical adhesion force or anchoring effect cannot be expected because the glass fiber's cross-section is round in shape and its surface is very smooth.

Coupling reagents provide strong adhesion between glass fiber and matrix, thus enabling it to function as a composite material.

Coupling reagents are mixed with the sizing agent, which protects the glass fiber surface. They also slick the surface in order to prevent damage during the process of spinning the glass filament into fiber. In general, these have hydroxyl reactive groups and reaction groups, and these coupling reagents may be a silane, chromate, or titanate compound. Silane coupling reagent is the most commonly used one.

4. Metal Plating[11]

Steel cord is plated with brass to adhere between elastomer and steel cord. In order to protect the adhesion strength by water, Co or Ni metal is mixed to brass alloy.

IV. ADHESION TECHNIQUES USING RFL ADHESIVES

A. ADHESION MECHANISM OF FIBERS TO ELASTOMERS

Aqueous adhesion systems are typically used in adhering elastomers and fibers for tires and belts. This system consists of compounds from resorcinol, formaldehyde resins, and an elastomeric latex. This system is called the RFL adhesion system.

A microscopic photograph of a cross-section of an RFL-treated polyester cord is shown in Figure 3.[1] The adhesive covers the outside of the filaments with a diameter of approximately 25 μm and penetrates into the fine filaments. The adhesive is about 5 μm thick, and this thin layer gives adhesion between the fibers and the elastomers during crosslinking.

A carbon black dispersion is often added to RFL adhesive to increase adhesive strength.

Many proposals have been made for the adhesion mechanism of elastomer and fiber.[1,11-16] Using our explanation, a three-layer model composed of fibers, adhesive agents, and elastomers is shown in Figure 4.[1,15,16]

The mechanism of bonding forces between the fiber layer and the adhesive agent layer varies with each fiber. For example, the amide groups in nylon assist in hydrogen bonding with the hydroxy groups in the RFL adhesive. Although these bonds have only relatively weak bonding energy with each functional group, compared to a covalent bond, the overall bonding force results from the high level of hydrogen bonds derived from many functional groups in the nylon.

The bonding force between the adhesive agent layer and the elastomer layer comes from the vulcanization of the elastomer layer. The unsaturated bonds in the latex are bonded by a crosslinking reaction with the unsaturated bonds in the elastomer. This crosslinking reaction is similar to that of the conventional elastomer compounds consisting of various compounding

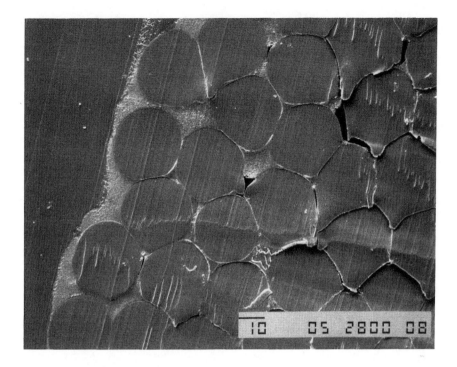

FIGURE 3. Photomicrographs of an outside part of polyester. ($\times$ 750)

ingredients, such as sulfur, accelerators, etc. These compounding ingredients migrate into the unsaturated bonds in the latex by the heat of the vulcanization process, and the crosslinking reaction occurs between the latex and elastomer.[17]

The cohesive strength of the adhesive layer is very important in this adhesion model. When the cohesive strength is inferior, the brittle part is destroyed at the adhesive layer. The latex particles are dispersed as spherical shapes in the RF resin.[1] The RFL resin forms a three-dimensional network that encloses the latex particles (Figure 5).[15,16] This reaction reinforces the cohesive strength of the adhesive layer.

B. ADHESIVE FORMULATIONS

The adhesion strength of fibers, which have been activated on their surface to elastomers, is affected by the relation of the elastomer to the latex. Table 1 shows typical combinations of elastomers and latexes.[18-38] Basically, the compatibility and cocrosslinking between the elastomer and latex is important. The adhesion of general-purpose elastomers, which are nonpolar, to fibers is easier to achieve than with special-purpose elastomers, which are polar. Saturated elastomers, which have few double bonds, such as EPDM and HNBR, are difficult to adhere to fibers. Few studies of the adhesion between glass fiber and elastomers have been reported.[39] Adhesion techniques of typical fibers to elastomers using latexes are described in the following sections.

1. Adhesion of Nylon to Elastomers

Nylon adheres sufficiently to elastomers using an RFL adhesive agent, as per the previously mentioned adhesion mechanism. The adhesive formulations for nylon are shown in Table 2. The suitable type of latex to adhere the elastomer to nylon must be selected.

When NR, SBR, and BR are adhered to nylon, vinylpyridine-butadiene-styrene copolymer latex (VP latex) increases adhesive strength between the elastomers and nylon. The effect of VP latex on the adhesion of the elastomers to nylon may be explained, whereby

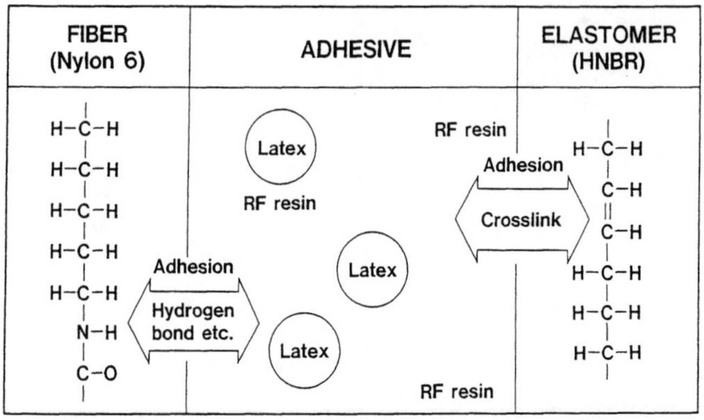

FIBER (Nylon 6)	ADHESIVE	ELASTOMER (HNBR)
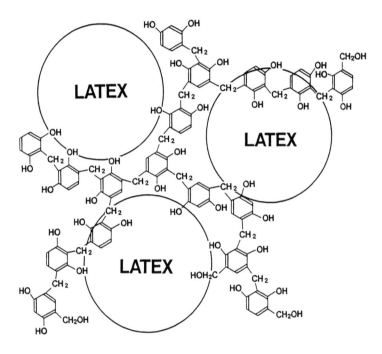		

FIGURE 4. Mechanism of the adhesion of fiber to elastomer.[1,15,16]

FIGURE 5. Presumed structure of RFL.[1,15,16]

the pyridine units in the VP latex bond with the nylon by electrostatic bonding. The adhesion is affected by the crosslinking reaction of the elastomer; however, the function of pyridinyl groups have not been made clear. Moreover, the unsaturated bonds, derived from butadiene units in the VP latex, are bonded by the crosslinking reaction with the unsaturated bonds in the elastomer.

When CR is adhered to nylon, VP latex, a mixture of BP and SBR latex, and CR latex[18] may be suitable for the RFL adhesive. The reason why CR is able to adhere to nylon using the similar latex as with NR is that the function of the chloroprene unit in CR react during the vulcanization process similarly to that of the isoprene units in NR.

If chlorosulfonated polyethylene (CSM) is adhered to nylon, CSM latex, which has the same composition as the CSM elastomer, is inferior to the required minimum standards.

TABLE 1
Latexes for Adhesion of Fiber to Elastomer[18-38]

Elastomer	Latex
NR, SBR	VP, SBR, NR
CR	CR[18], VP
CSM	NBR[19], XNBR[19]
EPM, EPDM	EVA[20], BD (unsaturated ketone[21]), CPE[22], halogenated EPDM[23], CR[24], CSM[25], BR[26,27], VP[26,27], chlorosulfonated EPR[28], acrylamide grafted EPR[29], chlorosulfonated polyolefin[30]
NBR	NBR[31,32], XNBR[32]
HNBR	NBR[33], XNBR[34], CSM[35], HNBR[36,38]

TABLE 2
Adhesive Formulation for Nylon

RF solution	Dry parts	Wet parts
Resorcinol	11.0	11.0
Formaldehyde	6.0	16.2
Sodium hydroxide	0.3	3.0
Soft water	—	235.8
(Maturation: 25°C, 6 h)	17.3	266.0

Final dip solution		
RF solution	17.3	266.0
Latex	100.0	250.0
Ammonium hydroxide	—	11.3
Soft water	—	59.2
(Maturation: 25°C, 20 h)	117.3	586.5

TCS = 20.

TABLE 3
Adhesion of Nylon to CSM[19]

Latex type	F/R (mol)	RF/L (wt)	Pull-out adhesion[a] (N/8 mm)
NBR	0.8/1.0	50/100	131 (85%)
XNBR			133 (75%)
CR			64 (20%)
VP-SBR			69 (25%)
CSM			66 (15%)

[a] Percent rubber coverage shown in parentheses.

The reason for this inferior adhesion has not been made clear. It was reported that NBR latex and carboxylated NBR latex (XNBR latex) are suitable for an RFL adhesive used for CSM to nylon (Table 3).[19]

When EPM and EPDM are adhered to nylon, the adhesion between these elastomers and nylon is very difficult because of the lack of unsaturated bonds. As shown in Table 1, there are many studies of adhesion techniques between elastomers and nylon. Table 4 shows the adhesion between EPDM and nylon, utilizing a chlorosulfonated polyolefin latex.[30] If

TABLE 4
Adhesion of Nylon to EPDM[30]

| Latex type | phr | RFL additives | | | Pull-out adhesion |
		Acceraletor type	phr	ZnO, phr	(N/10 mm)
CPE[a]	100	MBTS	5	—	162
CPE		MBTS	5	1	198
CPP[b]		MBTS	5	1	175
CEPDM[c]		DPTT	5	1	190
—		—	—	5	80

Note: RFL: F/R = 2/1 (mol), RF/L = 20/100 (wt).

[a] Chlorosulfonated polyethylene.
[b] Chlorosulfonated polypropylene.
[c] Chlorosulfonated ethylene-propylene-diene terpolymer.

TABLE 5
Adhesion of Nylon to HNBR

| Latex | Pull-out adhesion (N/8 mm) | |
	Initial adhesion[a]	Heat aged adhesion[b]
HNBR	230 (100%)	215 (100%)
NBR[c]	206 (100%)	162 (60%)
CSM[d]	167 (40%)	—
SBR[e]	70 (0%)	—
VP[f]	88 (0%)	—

Note: Compound recipe: No. 1 of HNBR in Table 5. (%)
        = rubber coverage.

[a] Curing condition: 30 min at 150°C.
[b] Heat aged condition: 168 h at 120°C.
[c] Nipol 1561, Nippon Zeon Co., Ltd.
[d] Hypalon 450, Seitestu Chemical Co., Ltd.
[e] Nipol 110, Nippon Zeon Co., Ltd.
[f] Nipol 2518FS, Nippon Zeon Co., Ltd.

a combination of BR/VP latex is utilized, excellent adhesion is obtained; however, this technique has not been fully developed.[26]

When adhering NBR to nylon, usually a medium or high acrylonitrile content NBR latex is used. It has been reported that if a halogenated phenolic resin is used together with a NBR latex, the adhesion between NBR and nylon is improved.[32]

HNBR adhesion to nylon has been studied utilizing NBR or XNBR latex and reported.[33,34] The composite materials made using these latexes can not achieve their full capabilities utilizing the excellent heat resistance of HNBR because these latexes have inferior heat resistance. The adhesion between HNBR and nylon using various types of latexes in the RFL adhesive system (Table 5) was studied.[16] There was no adhesion between nylon and HNBR when conventional SBR, VP, and CSM latexes were used. The adhesion of HNBR to nylon using these latexes is poor due to the lack of affinity of these latexes to HNBR. The initial adhesive strength using HNBR latex is higher than that of CSM and NBR latex, probably because the crosslinking reaction kinetics of HNBR elastomer during the vulcanization process is equal to that of the HNBR latex. The adhesion strength after heat aging using HNBR latex is considerably higher that of HNBR latex due to the improved heat resistance of HNBR latex (compared to the NBR latex).

2. Adhesion of Polyester to Elastomers

Polyester does not adhere well enough to elastomers using only an RFL adhesive. The following reasons are explained.

In the case of the adhesion between nylon to elastomers, the amide groups in nylon assist in the hydrogen bonding with the hydroxy groups in the RFL adhesive. However, the ester groups in polyester have little or no hydrogen bonding, as there is a large difference in the solubility parameter between RF resin and polyester — thus a weak affinity of polyester to RF resin.

Accordingly, the studies to improve the adhesion between polyester to elastomers were carried out to increase the adhesion between the polyester layer and adhesive layer.[40]

Two methods of improving adhesion between the polyester layer and adhesive layer are proposed:

(1) The low reactivity surface of the polyester can be treated by a highly reactive reagent. After that, an RFL adhesive is applied to the pretreated polyester.

(2) Reagents having a high affinity to polyester may be mixed into the RFL adhesive, and then the polyester is treated by the RFL adhesive.

In the former case, the highly reactive reagents used to pretreat the polyester are iso-cyanate or epoxide compounds (described in Section III.B). In the latter case, reagents with a high affinity to polyester are dispersed in water. These are composed of chlorophenol derivatives, blends with RF resins, and are recommended. Valcabond E® by Vulnax International Ltd. is an available commercial reagent.

These pretreatments for polyester provide a laminate of these reagents on the surface of a polyester during its manufacturing process by means of a dip into these reagents after twisting the filaments into cords or weaving a fabric.[41,44]

In either case, these pretreatments for polyester provide good adhesion between the polyester and NR or SBR. A typical adhesive formulation for polyester are shown in Table 6. In this table, when the double dipping process for the untreated polyester was carried out, the adhesive strength between polyester to elastomers became high. When pretreatment of epoxy resin on polyester is carried out, the second dip of the RFL adhesive formulation gives enough adhesive strength between the polyester and the elastomers without treatment with the first dip.

Polyester has a weakness in that the adhesive strength between polyester and elastomers decreases significantly after being embedded in the elastomer and heated. Iyengar has reported that polyester tended to exhibit hydrolysis of its bonds by amine compounds, such as accelerators or antioxidants compounded into the elastomer.[45] For example, Figure 6 shows the decrease of adhesive strength between a polyester and an elastomer when a sulfenamide compound was used as the vulcanization accelerator. Since the elastomer compounds contained moisture, the adhesive strength decreased very markedly. Figure 7 shows the degradation mechanism of polyester by amine compounds. The amine compounds migrate through the adhesive layer and react with the surface layer of the polyester to decrease cord or fabric strength.[1,16,78]

The following three methods to prevent polyester degradation by amines are proposed:

1. Improve the polyester itself to make it more difficult for degradation by amines.
2. Select effective vulcanization accelerators that generate a minimum amount of reactive amines during decomposition.
3. Use adhesive layer that functions to prevent the penetration of amines from the elastomer to the polyester.

TABLE 6
Adhesive Formulation for Polyester

1st dip

RF solution	Dry parts	Wet parts
Resorcinol	16.6	16.6
Formaldehyde	5.4	14.6
Sodium hydroxide	1.3	13.0
Soft water	—	321.8
(maturation: 25°C, 2 h)	23.3	366.0

RFL solution		
RF solution	23.3	366.0
Latex	100.0	250.0
(maturation: 25°C, 20 h)	123.3	616.0

Final dip solution		
RFL solution	123.3	616.0
Resin[a]	80.0	400.0
	203.3	1016.0 TSC = 20%

[a] Vulcabond E, 2, 6-bis-(2',4'-dihydroxyphenylmethyl)-
4-chlorophenol (Vulnax International Ltd).

2nd dip

RF solution	Dry parts	Wet parts
Resorcinol	11.0	11.0
Formaldehyde	6.0	16.2
Sodium hydroxide	0.3	3.0
Soft water	—	235.8
(maturation: 25°C, 6 h)	17.3	266.0

Final dip solution		
RF solution	17.3	266.0
HNBR latex	100.0	250.0
Ammonium hydroxide	—	11.3
Soft water	—	59.2
(maturation: 25°C, 20 h)	117.3	586.5 TSC = 20%

It was found that a decrease in the number of carboxylic groups combined in the polymer chain ends affects the prevention of polyester degradation; this development has been utilized. This and many methods of improvement have been used, such as finding the optimum balance between the density of polyester and its molecular weight[46] or the degree of crystallization.[47] There are many reports with respect to effective methods of polyester improvement.[7,8]

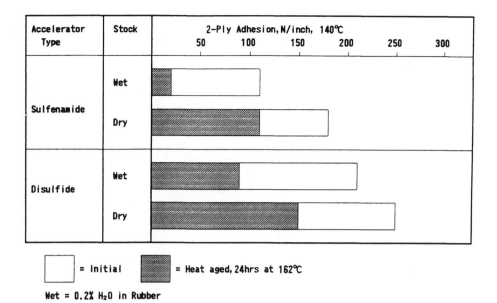

FIGURE 6. Effect of H_2O and amine on adhesion.[45]

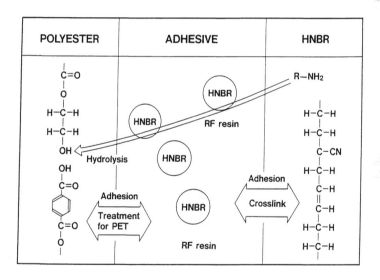

FIGURE 7. Mechanism of the degradation of polyester cord to HNBR by amine compounds.[16]

In the case of adhesion between polyester and NR using thiazole-based compounds, which generate a minimum amount of reactive amines during its decomposition, are recommended as the vulcanization accelerators in place of thiuram-based compounds, which generate a large amount of reactive amines as they decompose. For example, dibenzothiazyl disulfide (MBTS)[48-50] and triazine-based compounds, which combine with the reactive amines during its decomposition, are recommended.[51] The latter improved the adhesive strength between polyester and NR. However, it results in an inferior balance of the vulcanizate properties; hence, this system is not utilized.

The adhesion between polyester and HNBR, when thiazole-based accelerators are used, initially becomes higher (Figure 8). Figure 9 gives the cord strength retained after being

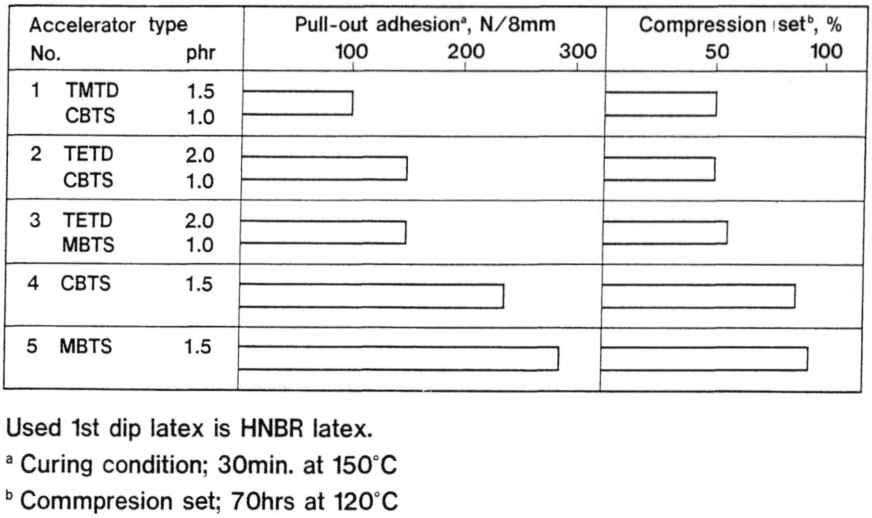

| Accelerator type | | Pull-out adhesion[a], N/8mm | | | Compression set[b], % | |
No.	phr	100	200	300	50	100
1 TMTD	1.5					
CBTS	1.0					
2 TETD	2.0					
CBTS	1.0					
3 TETD	2.0					
MBTS	1.0					
4 CBTS	1.5					
5 MBTS	1.5					

Used 1st dip latex is HNBR latex.

[a] Curing condition; 30min. at 150°C

[b] Commpresion set; 70hrs at 120°C

FIGURE 8. Influence of the accelerator type on adhesion of polyester to HNBR.[16]

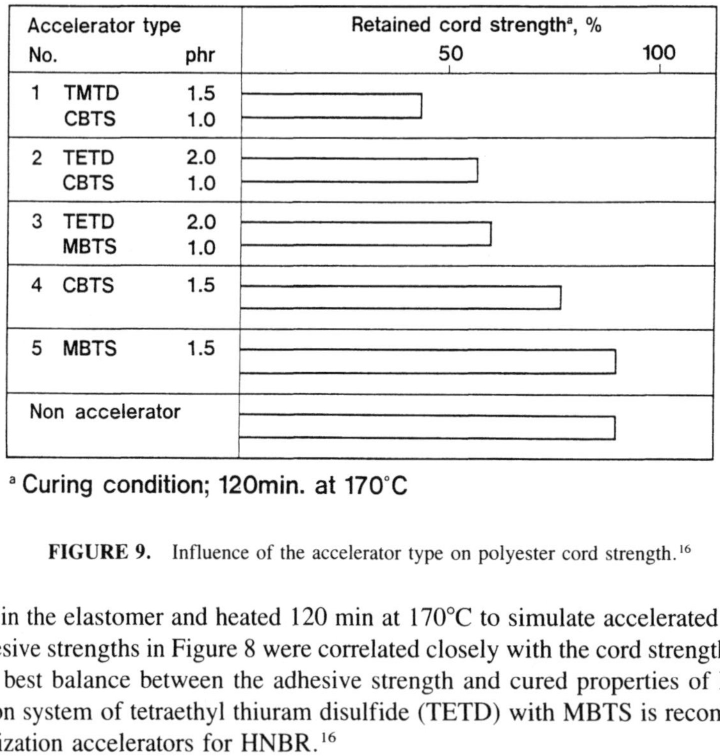

| Accelerator type | | Retained cord strength[a], % | |
No.	phr	50	100
1 TMTD	1.5		
CBTS	1.0		
2 TETD	2.0		
CBTS	1.0		
3 TETD	2.0		
MBTS	1.0		
4 CBTS	1.5		
5 MBTS	1.5		
Non accelerator			

[a] Curing condition; 120min. at 170°C

FIGURE 9. Influence of the accelerator type on polyester cord strength.[16]

embedded in the elastomer and heated 120 min at 170°C to simulate accelerated aging. The initial adhesive strengths in Figure 8 were correlated closely with the cord strengths in Figure 9. For the best balance between the adhesive strength and cured properties of HNBR, the combination system of tetraethyl thiuram disulfide (TETD) with MBTS is recommended as the vulcanization accelerators for HNBR.[16]

Another method is the treatment of a polyester surface with any reagent that prevents the penetration of amines from the elastomer into the polyester. Examples are treatment of polyester by chlorobenzene[52] and by a polymer dispersion containing caboxylated groups.[53] Some studies of additional reagents that function to prevent the penetration of amines to the RFL are reported. For example, anhydrous acid compounds,[54-56] ethylene urea compounds,[57] and blocked isocyanate compounds[58,59] are being utilized.

TABLE 7

**Adhesion of Polyester to NR Using Carboxylated
VP-SBR Latex[73]**

		Pull-out adhesion (N/8 mm)	
1st dip	2nd dip	Initial adhesion[a]	Heat aged adhesion[b]
VP-SBR[c]	VP-SBR	196	108
C-VP[d]	VP-SBR	196	176

[a] Curing condition: 30 min at 150°C.
[b] Curing condition: 90 min at 170°C.
[c] Nipol 2518FS, Nippon Zeon Co., Ltd.
[d] Nipoguard Z-500, Nippon Zeon Co., Ltd.

TABLE 8

Adhesion of Polyester to HNBR

		Pull-out adhesion (N/8 mm)	
1st dip latex	2nd dip latex	Initial adhesion[a]	Heat aged adhesion[b]
C-VP[c]	HNBR	284 (100%)	245 (70%)
HNBR	HNBR	270 (100%)	200 (40%)

Note: Compounding recipe: No. 3 of HNBR in Table 5.

[a] Curing condition: 30 min 15 150°C.
[b] Heat aged condition: 168 h at 120°C.
[c] Nipoguard Z-500, Nippon Zeon Co., LTD.

The latex portion of the RFL adhesive reagent accounts for 50 to 85 wt % of the RFL. To prevent the degradation of polyester by amines, the utilization of a latex that prevents the penetration of amines from an elastomer into a polyester is proposed. Latex containing carboxylic acid groups is preferred, since the carboxylic acid groups react with the amines to inhibit migration of the amines to the latex phase and thus protect the polyester cord or fabric. For example, a caboxylated vinylpyridine-butadiene-styrene latex is recommended as the latex containing caboxylic acid groups.[60-71] This latex exhibits almost complete prevention against degradation of the polyester when adhering it to NR (Table 7).[72-76] By this treatment, the adhesive strength, after heating to simulate accelerated aging, remained higher than without this treatment. For tire applications, this adhesive system using a caboxylated vinylpyridine-butadiene-styrene latex is put to practical use.

For the same reason, utilization of a caboxylated vinylpyridine-butadiene-styrene latex is recommended for adhesion between polyester and HNBR.[16] Unfortunately, it yielded low adhesion of the HNBR to the polyester. In order to provide both sufficient protection against amines and high adhesion levels, the polyester cords are first treated with an RFL adhesive using the caboxylated vinylpyridine-butadiene-styrene latex, followed by treatment with an RFL adhesive using HNBR latex (Table 8).[16] Using these treatments, the adhesive strength before and after heat aging was higher than that obtained when the first and second treatments were with HNBR latex. These results demonstrate that the carboxylic acid groups in the caboxylated vinylpyridine-butadiene-styrene latex inhibit amine penetration. Figure 10 shows the mechanism of polyester adhesion to HNBR using these adhesive systems, which consist the same principle as with the adhesion of polyester to NR.[15]

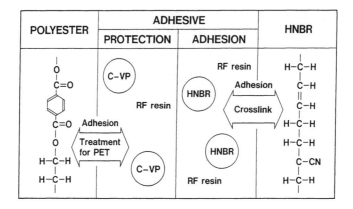

FIGURE 10.   Mechanism of the adhesion of polyester to HNBR.[16]

TABLE 9
Adhesive Formulation for Aramid

Subcoat (IPD-31)	Dry	Wet
Diglycidyl ether of glycerol		2.22
10% NaOH		0.28
5% Aerosol OT		0.56
Soft water		96.94
		100.00

Topcoat (IPD-39)		
Soft water		141.0
Ammonium hydroxide (28%)		6.1
Preformed RF resin[a] (75%)	16.5	22.0
VP-SBR latex[b] (41%)	100.0	244.0
Formaldehyde (37%)	4.1	11.0
Soft water		58.0
HAF carbon black dispersion (25%)	15.1	60.3
	135.7	542.4

Note: Recommended formulation by Du Pont.

[a]   Penacolite R-2170 (Koppers Co.).
[b]   Gen-Tac FS (General Tire).

3. Adhesion of Aramid to Elastomers

Aromatic groups account for more than 80% of the composition of aramid fiber. Bonding between amide groups contained on the main chain of aramid and functional groups of an RFL adhesive is hindered by many aromatic rings having high volume. This yields low adhesive values due to the aromatic groups.

In order to adhere aramid to elastomers, it is necessary to activate the surface of the aramid fiber. Some studies have been reported on proven activation methods for aramid fibers.[77,78]

As sufficient adhesive strength is not generated by conventional adhesion techniques alone, the aramid fibers are first treated, usually with epoxide compounds, and followed by treatment with RFL adhesives.

The adhesive formulations for NR to aramid are shown in Table 9. When the samples in which aramid is adhered to elastomers are peeled apart, the aramid cord was found to be degradated into micro fibrils. In the case of adhesion of aramid to other elastomers, the aramid fibers were observed to degrade in the same way. Many studies for the improvements of adhesion of aramid to elastomers were carried out; however, complete success has not been achieved yet.

REFERENCES

1. **Hisaki, H. and Mori, O.,** *J. Sci. Rubber Ind. Jpn.,* 64, 33, 1991.
2. **Hashimoto, K., Watanabe, N., and Yoshioka, A.,** Paper 43, presented at a meeting of Rubber Div., Am. Chem. Soc., Houston, TX, October 25–28, 1983.
3. **Kubo, Y.,** *Proc. Int. Rubber Conf.,* Org. Comm. IRC, Kyoto, Japan, 1985, 32.
4. **Hashimoto, K., Oyama, M., Watanabe, N., and Todani, Y.,** Paper 5, presented at a meeting of Rubber Div., Am. Chem. Soc., Cleveland, OH, October 1–4, 1985.
5. **Kubo, Y. and Hashimoto, K.,** *Kautsch. Gummi, Kunstst.,* 40, 118, 1987.
6. **Fukuhara, S.,** *Dictionary of Composite Materials,* Asakura Shoten, Tokyo, 1991, 367.
7. **Takahashi, S.,** *Technol. Adhesion,* 10, 1, 45, 1990.
8. **Murayama, S.,** *Technol. Adhesion,* 10, 1, 41, 1990.
9. **Okamura, A.,** *Technol. Adhesion,* 10, 1, 50, 1990.
10. **Ueda, H.,** *J. Adhesion Sci. Jpn.,* 22, 4, 27, 1986.
11. **Fukuhara, S.,** *J. Fiber Sci. Jpn.,* 40, 552, 1984.
12. **Takeyama, T.,** *J. Sci. Rubber Ind. Jpn.,* 45, 958, 1972.
13. **Matsui, J.,** *J. Adhesion Sci. Jpn.,* 8, 26, 1986.
14. **Moult, R. H.,** *Handbook of Adhesive,* Van Nostrand Reinhold, 1962, 495.
15. **Hisaki, H.,** presented at a meeting of Tire Soc., 9th Annu. Meet., Akron, OH, March 20–21, 1990.
16. **Kubo, Y.,** *Rubber Chem. Technol.,* 64, 8, 1991.
17. **Tameyama, T. and Takada, T.,** presented at an annual meeting of Sci. Rubber Ind. Jpn., May 23–24, 1991.
18. Burlington Industries, Inc., U.S. Patent 3,240,649, 1966.
19. Yokohama Tire Co., Ldt., Japan Kokai 61-127738, 1986.
20. Stamicarbon, Belgium Patent 680,492, 1966.
21. Dunlop Rubber, U.S. Patent 3,268,386, 1966.
22. Asahi Chemical Industry Co., Ltd., Japan Kokai 46-23142, 1971.
23. Sumitomo Chemical Co., Ltd., Japan Kokai 46-22358, 1971.
24. Yokohama Tire Co., Ltd., Japan 63-10732, 1988.
25. Du Pont, U.S. Patent 3,276,948, 1966.
26. Toray Co., Ltd., Japan 63-40227, 1988.
27. Teijin Co., Ltd., Japan 64-9349, 1989.
28. Farbwerke Hoechst, Japan Kokai 42-17642, 1967.
29. Monticatini, Japan Kokai 42-23632, 1967.
30. Seitetsu Chemical Co., Ltd. and Sumitomo Chemical Co., Ltd., Japan 62-9631, 1987.
31. Burlington Industries, Inc., U.S. Patent 3,240,659, 1966.
32. Yokohama Tire Co., Ltd., Japan Kokai 1-118671, 1989.
33. Toray Co., Ltd., Japan Kokai 1-174677, 1989.
34. Mitsuboshi Belt Co., Ltd., Japan Kokai 58-45940, 1983.
35. Nippon Zeon Co., Ltd., Japan Kokai 61-207442, 1986.
36. Nippon Zeon Co., Ltd., Japan Kokai 63-248879, 1988.
37. Nippon Glass Fiber Co. and Nippon Zeon Co., Ltd., Japan Kokai 63-248879, 1988.
38. Nippon Zeon Co., Ltd., Japan Kokai 2-91135, 1990.
39. Nippon Glass Fiber Co., Japan Kokai 1-221433, 1989.
40. **Solomon, T. S.,** *Rubber Chem. Technol.,* 58, 561, 1985.
41. **Matsui, J., Toki, M., and Shimizu, M.,** *J. Adhesion Sci. Jpn.,* 7, 303, 1971.
42. **Matsui, J., Toki, M., and Shimizu, M.,** *J. Adhesion Sci. Jpn.,* 8, 26, 1972.

43. **Matsui, J., Toki, M., and Shimizu, M.,** *J. Adhesion Sci. Jpn.,* 8, 329, 1972.
44. **Takada, T.,** *Composite Materials and Interface,* Material Technology Forum, Tokyo, 1987, 161.
45. **Iyenger, Y.,** *J. Appl. Polym. Sci.,* 15, 267, 1971.
46. Toyobo Co., Ltd., Japan Kokai 61-12952, 1986.
47. Teijin Co., Ltd., Japan Kokai 61-132618, 1986.
48. **Iyenger, Y. and Ryder, D.,** *Rubber Chem. Technol.,* 46, 442, 1973.
49. **Bertrand, G.,** *Plast. Rubber Process,* 3, 157, 1978.
50. **Morita, M.,** presented at Symp. Rubber Technol., Chugoku region, May 10, 1988.
51. ICI Co. Ltd., Ger. Offen. 2,051,521, 1971.
52. Toyobo Co., Ltd., Japan Kokai 56-2155, 1981.
53. Toray Co., Ltd., Japan Kokai 55-166235, 1980.
54. Teijin Co., Ltd., Japan Kokai 59-228078, 1984.
55. Teijin Co., Ltd., Japan Kokai 59-228079, 1984.
56. Teijin Co., Ltd., Japan Kokai 59-228080, 1984.
57. Teijin Co., Ltd., Japan Kokai 63-227868, 1988.
58. Toyobo Co., Ltd., Japan Kokai 55-118855, 1980.
59. Toyobo Co., Ltd., Japan Kobai 55-152815, 1980.
60. Nippon Zeon Co., Ltd., Japan Kokai 61-26629, 1986.
61. Nippon Zeon Co., Ltd., Japan Kokai 61-26630, 1986.
62. Nippon Zeon Co., Ltd., Japan Kokai 62-39679, 1987.
63. Nippon Zeon Co., Ltd., Japan Kokai 62-70411, 1987.
64. Nippon Zeon Co., Ltd., Japan Kokai 62-149982, 1987.
65. Nippon Zeon Co., Ltd., Japan Kokai 62-230807, 1987.
66. Nippon Zeon Co., Ltd., Japan Kokai 63-48313, 1988.
67. Nippon Zeon Co., Ltd., Japan Kokai 63-112629, 1988.
68. Nippon Zeon Co., Ltd., Japan Kokai 63-130641, 1988.
69. Nippon Zeon Co., Ltd., Japan Kokai 63-234036, 1988.
70. Nippon Zeon Co., Ltd., Japan Kokai 1-9622, 1989.
71. Nippon Zeon Co., Ltd., Japan Kokai 1-118672, 1989.
72. **Hisaki, H., Kataoka, Y., and Sekiya, M.,** *J. Sci. Rubber Ind. Jpn.,* 61, 728, 1988.
73. **Hisaki, H., Takinami, S., and Suziki, S.,** *J. Sci. Rubber Ind. Jpn.,* 62, 515, 1989.
74. **Hisaki, H., Kataoka, Y., and Sekiya, M.,** *1st Elastomer Symp.,* Sci. Rubber Ind. Jpn., 1987.
75. **Hisaki, H., Takinami, S., and Suziki, S.,** *2nd Elastomer Symp.,* Sci. Rubber Ind. Jpn., 1988.
76. **Hisaki, H., Nakano, Y., and Suziki, S.,** *3rd Elastomer Symp.,* Sci. Rubber Ind. Jpn., 1989.
77. Bando Chemical Ind. Ltd. Japan Kokai 2-175974, 1990.
78. **Takada, T.,** *Polym. Appl.,* 36, 390, 1987.

Chapter 26

POLYMER APPLICATIONS IN THE CONSTRUCTION INDUSTRY

R. Hussein and Nicholas P. Cheremisinoff

TABLE OF CONTENTS

0-8493-4401-8/93/$0.00 + $.50
© 1993 by CRC Press, Inc.

I. INTRODUCTION

It is probably true to say that the building industry, with one major exception, is still largely employing conventional materials. At the same time, it cannot but be aware of the impact that polymers are making in a number of departments. Some of these will be immediately obvious, but not others although they warrant careful consideration.

The main exception referred to is polyvinyl chloride (PVC) flooring. Although it is not possible to give statistics, there can be little doubt that a large proportion of houses and public and industrial buildings being built today are provided with PVC floors of one type or another. Undoubtedly, this development will continue to grow.

Some other developments, which will also be fairly well known, include PVC rainwater goods, polythene film for damp courses and other purposes, plastic pipes, plastic foams for insulation, various lavatory goods, etc. These are examples of a wide range that is continually increasing.

Amidst all this activity in these new materials, the architect and builder will no doubt be puzzled by the conflicting stories they hear about "plastics." In some cases, claims are made that plastics will outlast conventional materials by many years or that they are unbreakable. On the other hand, there will be many stories of plastic goods (particularly lavatory seats) breaking very quickly. Both accounts are true — on two main reasons. In the first place, there is a general lack of knowledge about polymers since there are many kinds. Each one must be employed for its correct sphere of duty or trouble may be experienced. The second reason is that many polymers were developed rapidly in the immediate post-war years, and some of the materials and products made at that time undoubtedly gave trouble. However, there has been considerable development over the years, although it cannot be claimed that there are no troubles. It can be said that in many cases, the products will give trouble-free service for the life of the building.

A new branch of engineering materials has developed rapidly over the last 35 years — the science and technology of high polymers.

In general, these materials consist of the elements, C, H, N, and O, and are therefore conventionally classified as organic polymers. However, in some instances, other elements such as B, Si, P, S, F, and Cl are present in certain proportions and influence the ultimate properties of the products. Nevertheless, this group of compounds is customarily referred to as organic polymers or organic macromolecules. Together with the large family of metallic compounds and ceramic systems, these organic polymers represent the essential engineering materials in the construction of buildings, vehicles, engines, appliances, textiles, packaging and writing sheets, plastics, rubber goods, and household articles of all kinds.

The rapid growth of these relatively new engineering materials in the recent past is due to several factors.

The basic raw materials for their production are readily available in large quantities and, in general, are inexpensive. Natural organic polymers such as celullose (paper, textiles), proteins (wool, silk, leather), starch (food adhesives), and rubber (tires and many other goods) are mainly available as products of farming and forestry activities, whereas the raw materials for the production of synthetic polymers come essentially from the industries of coal and oil. The simplest building units are called *monomers*, some of which (ethylene, propylene, isobutylene, butadiene, and styrene) are by-products of gasoline and luboils, low in cost (5¢/lb and less), and available in very large quantities. Many others are simple derivatives of ethylene, benzene, formaldehyde, phenol, urea, and other basic organic chemicals. They range in cost from 5 to 20¢ and at the most, 25¢/lb. They are also, in general, large scale industrial products. Thus, the basic building units for organic polymers, as defined above, represent a large variety of compounds, are readily available, and are of low or moderate cost.

During the last 30 years, intense research activities in many laboratories have succeeded in elucidating the mechanism of the reactions by which long chain molecules are formed from the above-mentioned basic units. They are called *polymerization reactions* and represent either typical chain reactions of highly exothermic character or step reactions in the course of which chains of systematically repeating units are formed. The results of this research have provided a sound fundamental background for the understanding of these reactions. During the same period, systematic engineering efforts developed a number of relatively simple unit-type processes that permit us to translate polymerization and polycondensation reactions into large scale industrial operations. Today, several types of them (such as polymerization in the gas phase at high pressures, in solution, suspension, emulsion, and even in the solid state) are well-developed standard procedures that allow us to convert monomers into polymers rapidly, conveniently, and inexpensively. In fact, in many cases, the actual conversion cost from monomer to polymer can be as low as 5¢/lb, a condition that has greatly contributed to the rapid expansion of this field.

Thirty years ago, there existed only a few well-developed processes and machines to convert organic polymers from the state of a latex, a sheet of molding power, into the ultimate commercial product such as fibers, bristles, films, plates, rods, tubes, bottles, cups, combs, and other salable commodities. Today, there are many continuous automatic, rapid, and inexpensive methods for spinning, casting, blow molding, injection and compression molding, stamping, and vacuum forming, which give each polymer with attractive properties an almost immediate chance of being converted into useful and salable consumer goods. Obviously, the existence of manifold applications and the availability of standardized and automated methods of developing a new polymer into many different channels simulate the synthesis and development of such new members of the organic polymer family.

The large number of available monomers and the even larger number of polymers and copolymers made from them have provided us with an almost continuous spectrum of composition and structure or organic macromolecules. On the other hand, the systematic exploration of their mechanical, optical, electrical, and thermal behavior has provided us with an equally dense spectrum of characteristic practical properties, and the study of their correspondence has led to a relatively profound and dependable understanding of structure property relationships. This has the great advantage that new polymers or copolymers with desired and prescribed properties need not be looked for by the empirical, more or less random system of synthetic efforts, but can be designed on paper — a successful elimination of many possibilities can be effected before work is actually started in the laboratory. This approach has been so successful that, in many instances, one can speak of a *molecular engineering* approach in the synthesis and development of new polymeric materials.

All fundamental and applied efforts of monomer syntheses, polymerization techniques, and manufacturing processes eventually can be condensed into a few guiding principles that represent, so to speak, the essence of our present understanding and know-how. Such principles are, of course, only qualitative generalizations and, in each individual case, have to be supplemented by quantitative considerations and numerical refinements, but they give a convenient and clarifying "helicopter view" of the present state of our knowledge and its practical applicability. As a consequence, they are good working hypotheses or guideposts, if they are used with caution and with the realization that they are approximations and illustrations.

Building has been with us ever since man moved out of the cave. The building industry is a huge one, utilizing a quarter of a trillion pounds of materials in the U.S. alone.

By contrast, the synthetic polymers that go into plastics, coatings, adhesives, and fibers are products of the 20th century. At present, they comprise less than 1% of the total tonnage of construction materials in the U.S., but their usage and influence are growing rapidly. Not only are they supplanting older materials for many conventional applications, but they are also stimulating the architect and designer to innovate new types of structures and structural components.

The raw material capacity of Canadian polymer manufacturers are shown in Table 1.

The Canadian polymers industry has an ample supply of cheap plastics and raw materials and offers great opportunities for future development, particularly in synthesis, processing, and processing machinery.

TABLE 1
Raw Material Capacity of Canadian
Polymer Manufacturers

Polymer	No. of producers	1980 anticipated capacity × 1000 t
Low density polyethylene	4	446
High density polyethylene	3	256
Polypropylene	2	136
ABS	2	60
Polystyrene	5	156
PVC	3	308

A. ADVANTAGES OF POLYMERS

Polymers offer many advantages over other materials. Of course, one must select the most appropriate material for the particular end use, since every material does not possess all of the following virtues:

- High light transmission, for glazing and lighting fixtures.
- Colorability and aesthetic appeal.
- Infinite texture possibilities — smooth surface if desired.
- Easy maintenance; permanence of color without painting.
- Infinite design possibilities. Sheets can be cast into either simple or compound curvatures and can be corrugated. Thermoplastics can be extruded in continuous lengths with intricate profiles. Complex shapes can be molded.
- Resistance to water, corrosion, and weathering, making them suitable for facades, gutters, pipes, bathroom fixtures, waterproofing, etc.
- Light weight and high strength-to-weight ratio, an especially important feature in high-rise construction. Prebuilt components are easily transported, thus encouraging the use of prefabricated construction, with great savings in labor. The speeding up of on-the-site erection lowers capital investment requirements.
- High impact resistance. In the glazing of schools and public buildings, the replacement of glass by plastics has greatly reduced damage by vandalism.
- Excellent dielectric characteristics, making them suitable for electrical insulation.
- Low thermal conductivity, consequently warm to the touch and pleasant to handle. Plastic foams are outstanding for thermal insulation, especially against cold.
- Excellent adhesion (while still in liquid form) to wood, aluminum, concrete, glass, and other building materials.

B. DISADVANTAGES OF POLYMERS

Several factors have impeded the utilization of synthetic polymers in building. Fortunately, most of these can be overcome, at least by some polymers for some purposes:

- Cost per kilogram is high. However, because of their low density, the cost per cubic inch is comparable to or lower than that of other materials. Furthermore, the cost of synthetic polymers has been coming down consistently, in opposition to the price trends of other materials.
- Plastics are not as strong as some other building materials. Consequently, while they make excellent facades, it has been found expedient to use them in conjunction with structural members of steel or reinforced concrete. But sandwich construction, with faces of fiberglass-reinforced plastics, results in panels that show promise as structural members.

- The scratch resistance of plastics is not as great as that of glass or porcelain.
- Some plastics are subject to deterioration from ultraviolet light, or combinations of light and heat, making them unsuitable in hot climates. Improved formulation can mitigate environmental effects.
- Most polymers burn, in common with wood and other organic materials. However, many plastic materials are not available which are self-extinguishing or nonburning.
- Building codes have in the past been based on specific materials; consequently, they have tended to preclude the use of plastics simply because of their newness. Now, performance codes are becoming more widely accepted, and polymers are thus placed on a fairer competitive basis.
- Some building trade unions, e.g., plumbers and carpenters, have a vested interest in older materials and therefore resist the introduction of plastics. The current trend toward greater prefabrication of building components has lessened the impact of these artisans' opposition.
- The architect and builder are wary of plastics because of their newness and the sparcity of information about them. More engineering data must be published by both the raw materials suppliers and the fabricators of plastic components. Accelerated aging tests must be developed further to compensate for the youth of the synthetic organic materials. Warranties of satisfactory performance for periods of 15 years or longer will do much to encourage the architect to specify plastics more freely.

In Table 2, we may follow the amounts of polymers aged in the U.S. building industry in 1978 and 1979.

II. STRUCTURAL FIBERGLASS-REINFORCED PLASTICS

Fiberglass-reinforced plastics* are structural composites consisting of components that, individually, have no particular attributes as structural materials. Bundles of fine, almost invisible, very strong glass fibers are bound in various controllable patterns within a matrix of moldable, fluid resin that can be hardened to a rigid state. The glass fibers, independently, have limited structural form except perhaps as fabrics. The resins, independently, offer little in the way of structural properties. While they are basically moldable, when cast in large sections, they warp, shrink, and crack. However, when the components are combined, the resin imparts form and structural continuity to the filaments, and the filaments in turn provide strength and dimensional stability to the resin. The composite is crudely analogous to reinforced concrete, but because of the scale of the components, it more closely compares to the Ferro-Cementé — a mixture of fine wire and cement grout introduced by Nervi in the early 1940s.

Fiberglass-reinforced plastics are well-known and respected engineering materials. Indeed, the building industry has had some extensive exposure to FRP in interior decorative panels, skylights and corrugated roofing panels, translucent sidewalls, and concrete formwork. Exciting structures such as the early Monsanto House of the Future, the American Pavilion in Moscow, the American Pavilion at Brussels, and a host of other imaginative structures at the Seattle and New York World's Fairs serve to demonstrate that FRP can be engineered to serve major structural tasks.

An immense amount of data has been generated, particularly in the last decade, on the behavior of reinforced plastics systems. The architect or building engineer, faced for the first time with the task of designing a reinforced plastic structure, not only must master the basic aspects of working with this unique material and its behavior, but is also faced with a formidable amount of literature from which to draw needed details. Thus, the purpose of

---

* Fiberglass-reinforced plastics are abbreviated as RP and FRP by the industry. FRP will be used in this text.

TABLE 2
Polymers in Buildings

Application/material	1000 tons	
	1978	1979
Decorative laminates		
Phenolic	23	24
Urea and melamine	18	18
Flooring[a]		
Epoxy (including paving)	8	8
PVC	152	168
Urethane foam (rug underlay)	60	61
Glazing and skylights		
Acrylic	32	33
Reinforced polyester[b]	17	18
Polycarbonate	33	40
Insulation		
Phenolic (binder)	124	130
Polystyrene foam	83	93
Urethane foam (rigid)	110	133
Lighting fixtures		
Acrylic	10	10
Cellulosics	2	2
Polycarbonate	5	5
Polystyrene	12	11
PVC	7	8
Panels and siding		
Acrylic	6	6
Butyrate	2	2
PVC	77	91
Reinforced polyester	70	70
Pipe, fittings, conduit		
ABS	162	165
Epoxy (coatings)	4	4
HDPE	227	239
LDPE	12	15
Polypropylene	9	11
Polystyrene	6	5
PVC	1007	1101
Reinforced polyester[b]	98	108
Profile extrusions[c]		
PVC (including foam)	77	85
Polyethylene	3	3
Plumbing		
Acrylic	11	13
Polyacetal	10	10
Polyester, thermoplastic	3	3
Polystyrene	2	2
Reinforced polyester[b]	59	63
Resin-bonded woods		
Phenolic	215	218
Urea and melamine	395	390
Vapor barriers		
LDPE	80	78

TABLE 2 (continued)
Polymers in Buildings

Application/material	1000 tons	
	1978	1979
PVC[d]	18	19
Wall coverings		
Polystyrene	5	6
PVC	57	64
Total	3311	3533

[a]   Excluding bonding or adhesive materials.
[b]   Including reinforcements.
[c]   Including windows, rainwater systems, etc.
[d]   Including swimming pool liners.

this chapter is to set forth the important considerations that should be recognized and understood by the designer of fiberglass-reinforced plastic-based building structures. Even when the scope is narrowed to building applications, the degree to which components and hence properties can be varied is still so wide as to preclude detailed discussion of all possible parameters and their inteactions.

A distinction should be made between the kinds of components that can be considered for building applications. First, there are the proprietary standard items of manufacture, such as flat and corrugated sheet, some sandwich panels, and the like. Second, there are the "custom" structures designed for a given application by the architect or engineer. Structural capabilities and manufacturing processes for the proprietary items have usually been established in fair detail; hence, little difficulty should be encountered in incorporating them into a design. But if the building designer is to custom-tailor his own original structure, much more should be known of the basic characteristics and behavior of the material and the types of manufacturing processes that are available to him.

A. RADOMES

Spherical radomes based on either the geodesic or sandwich shell structural framing system are fabricated in many cases exclusively of fiberglass-reinforced plastics. The domes range up to 150 ft in diameter and are designed to resist hurricane wind forces. Fiberglass-reinforced plastics are uniquely qualified for this application because they offer good structural properties and yet they permit passage of electromagnetic radar signals with little attenuation. In some geodesic designs, the moldability of the material allows integral rib and skin construction. In other sandwich shell arrangements, moldability allows the generation of doubly warped spherical segments.

B. PIPES AND TANKS FOR CHEMICALS

The chemical industry has found that fiberglass-reinforced pipe, tanks, and stacks are superior to stainless steel in many applications. While the critical consideration is corrosion resistance under extreme environments, the ease of fabrication of complex tank shapes and simplicity of connection details enhance the competitive cost picture.

C. M.I.T. SCHOOL, CAMBRIDGE, MA

The M.I.T. School represents a prototype study of a totally prefabricated, self-sustaining, modular classroom unit. The concept here is that of installing standard package units with a minimum of site preparation. The roof of this building is a hyperbolic paraboloid sandwich,

8 ft^2 in plan, comprised of fiberglass-reinforced plastic skins separated by a low density plastic core. Obviously, the need here is for a panel system that is light in weight and easily erected. FRP is considered as one material that answers this need.

D. HOLIDAY INN JR.

Motel units represent another application where large scale prefabrication using FRP is contemplated. Units of four rooms are fully prefabricated including utilities and plumbing. FRP is used as the outer skin of the wall sandwich panels. Other areas where FRP has been employed are in an arched roof having a 6-ft span, and in cornice and still bands. To date, a prototype unit has been erected, and several others are under construction. If full production is undertaken, this will probably represent the largest integrated building application of FRP.

E. PROPRIETARY ITEMS

There exists a number of proprietary FRP items that are regularly used in the building field. Translucent panels are used as skylights, glazing panels, roofing panels, or decorative screening. These materials are provided in the forms of flat sheet, corrugated sheet, or as sandwich panels.

F. CONCRETE FORMS

The moldability, strength, durability, and surface finish of fiberglass-reinforced plastics are leading to increased use of the material in concrete formwork. Early uses were in pans for grid-slab construction. Larger custom forms of complex shape have been used in such major structures as the Marina City Towers, Chicago, IL, the University Apartments, Hyde Park, IL, and the multistory buildings of the Technology Square complex, Cambridge, MA. FRP is also extensively used in forms for precast concrete panels and as a durable surface for plywood forms.

III. CHARACTERISTICS OF FIBERGLASS-REINFORCED PLASTICS

The applications discussed in the preceding sections should provide an idea of the characteristics of fiberglass-reinforced plastics that can contribute to successful structural usage. This section will serve to summarize and extend the attributes of the material as they may affect building applications. Limitations will also be discussed in order to provide an objective, balanced, total picture of FRP systems. Only through knowledge of both can the material be engineered with purpose and confidence to the objective of any good design, i.e., to produce a system that embraces every possible advantage while the limitations are held to a tolerable level.

A. ATTRIBUTES
1. Fabrication

Perhaps the greatest advantages accompanying the use of fiberglass-reinforced plastics relate to the ease with which the material can be fabricated and the variety of fabrication methods that are available.

2. Strength and Stiffness

Short-term ultimate strength of reinforced plastics varies from 10,000 to over 200,000 psi, while modulus of elasticity varies from about 0.5 to over 7×10^6 psi. Strength and stiffness are strongly dependent upon the type and relative quantity of resin and reinforcement, the manufacturer, and the manufacturing process used.

3. Low Weight

Density of fiberglass-reinforced plastics is on the order of 100 lb/ft³. The low density of this strong, stiff material allows the use of light sections. In addition to the structural advantages offered by such lightweight sections, there are potential benefits in shipping, handling, and erection.

4. Light Transmission

The light transmission characteristics can be tailored from nearly transparent, although never as clear as glass, to totally opaque. The translucency is, of course, a unique property offered by no other truly structural material.

5. Color and Color Variations

Fiberglass-reinforced plastics can be dyed or pigmented to virtually any color or color intensity. In addition, random or regular colors or patterns can be obtained through spattering, mottling, or incorporating printed paper sheets into the laminate.

6. Corrosion and Chemical Resistance

While corrosion resistance depends to some degree on the resin formulation used, reinforced plastics can usually be considered inert to normal industrial atmosphere.

7. Insulation

Fiberglass-reinforced plastics are good thermal and electrical insulators.

B. LIMITATIONS

1. Fire and Temperature Resistance

The resinous component of fiberglass-reinforced plastics composites, being organic in chemical structure, will burn. While totally fireproof resins have not been forthcoming, the industry has been successful in reducing the degree to which resins will contribute to fire spread. This is done by molecular tailoring of the chemical structure of the resin molecule or by incorporating fire-inhibiting additives into the resin mixture.

The glass-in-plastic systems will lose strength when exposed to high temperatures. Both the glass and the resinous components are responsible for this. Most systems will not suffer significantly under ambient temperatures normally found in buildings, but when heated to a few hundred degrees, the properties of the material degrade rapidly. Because of this, temperature rise due to fire, even though the material may not be ignited, may reduce the strength and stiffness of the structure.

2. Effects of Weather and Sunlight

Reinforced plastics systems vary considerably in their resistance to the effects of outdoor environment. There are tendencies to dull, discolor, lose translucency, or erode. These effects can be reduced substantially by proper choice and careful design of the components of the system.

3. Stress History

Many claims are made regarding the strength capabilities of FRP laminates which are based on conventional test values. But working stresses must be kept well below short-time test values if long-term or cyclic loads must be carried.

4. Cost

A number of factors such as type and quality of resin and reinforcement, and number and complexity of units, will obviously influence the cost of components.

In comparing the cost of FRP systems with other materials, the density of the material must be taken into account. On a volume basis, the cost of the material falls into the price range of some forms of aluminum.

5. Acoustics

Fiberglass-reinforced plastics can be tailored into very efficient, lightweight, thin-skinned systems that are highly elastic. These factors combine to yield a structure that efficiently transmits and radiates sound. Such lightweight elastic members provide little barrier to noise; hence, rain impinging on thin skins or wind fluttering a light membrane can create noise problems within a structure. The architect should consider acoustic treatment if reinforced plastics are to house sound-sensitive areas.

6. Compatibility of Attributes

The preceding sections deal with the often-quoted characteristics of FRP composites with particular emphasis on those that apply to building applications. Seldom, however, can a given component possess all of the attributes (or limitations) that have been mentioned. There are important compromises that must be struck when materials capabilities, aesthetics, choice of process, and economics are considered.

IV. FOAMS IN THE BUILDING INDUSTRY

The Western European polymeric foam market is dominated in tonnage terms by polyurethane and polystyrene. In 1977, it was reported as being around 167,000 t/annum. Structural engineering is said to account for 50,000 t/annum. The estimated spectacular growth for structural engineering foams is quoted as being around 32.5%/annum up to 1980 and around 20% for the following decade. Thus, an annual total of 333,000 t of both structural and nonstructural rigid foam is envisaged for 1980 with a figure double that by 1985. In 1990, the figure was around 1.4 million t. These figures must be set alongside the fact that in Western Europe, the use of all plastics materials in the building and construction industry has increased over the last decade. Indeed, there have been reports of surveys that indicate by the middle of the same period, construction uses accounted for 20 to 35% of total polymer consumption with a figure of 40% being approached in some countries.

The same reports indicate that every product and application was averaging an annual growth rate of 10 to 15% with 30 to 35% being reached in some individual cases. Within this broad area, the general range of structural and insulative polymeric foams coming onto the market has increased markedly. Many of these products may be considered as lightly crosslinked thermosetting materials. In this chapter, both thermosetting and thermoplastic foam materials are considered side by side to give a true sense of perspective; the application areas have been subdivided into roofs and ceilings, walls cavities and panels, floors pavements and roads, pipe fittings and accessories, and complete systems. The intention is to give an insight into the breadth of applications without attempting to list every single item. The production of polyurethane foams in U.S. is shown in Table 3.

V. THE RANGE AND SCOPE OF FOAMED PLASTIC MATERIALS

A. ROOFS AND CEILINGS

The good thermal and insulative qualities of cellular materials are important factors in influencing their commercial exploitation. For this reason, roofs and ceilings containing polystyrene foam are quite common, including one spectacular example — 378 m × 177 m situated above a factory hall on the Continent. Very often they are used in the form of

TABLE 3
Polyurethane Foams: Pattern of Consumption

Market	1000 tons 1978	1000 tons 1979
Flexible foam		
Bedding	85	87
Furniture	227	230
Rug underlay	60	61
Transportation	180	175
Other	56	55
Total	608	608
Rigid foam		
Building insulation	110	133
Furniture	13	12
Household and commercial refrigeration	45	48
Industrial insulation	21	21
Transportation	21	21
Other	22	22
Total	232	258
Grand total	840	866

simple slabs and panels, but in addition, there are a number of interesting reports of polystyrene and polyurethane foams combined with other materials to produce composites that can function as water/vapor barriers as well as thermal insulants. There are examples of polystyrene foam/metal foil/bitumen laminates, polyurethane foam/mastic Hypalon laminates, and a sprayed-on polyurethane foam capped with epoxy resin to replace a circular aluminum theater roof at 6% of the original cost. Similarly, there are examples of polyurethane used beneath pitch and tar and also on the top of waterproof bitumen to reduce heat losses from within the building. In the latter example, the top surface is additionally protected from excessive radiant solar heat and cracking due to loss of low molecular weight fractions. Polyurethane foam can be applied in analogous fashion to paint by airless spray and other methods to concrete slabs and related surfaces. Rigid polystyrene foam is also said to be used as a dampening and stiffening material for steel sheet roofs with the two components bonded by bituminous based adhesives. In another context, polyurethane, steel, and aluminum form the basis of a housing system developed in France where triangular panels can be bolted together to form floors and ceilings. Also, there are a number of examples of imaginative uses of foams in roofing to cope with local environments and availability (or lack of availability) of materials. Included here are bamboo-reinforced composite foams in the tropics and rural U.S. and simple plastic domes produced by foaming over an earth mound and raising to the required height.

B. WALLS, CAVITIES, AND PANELS

Broadly speaking, three main areas can be defined in this section where plastic foams are becoming increasingly important. First, there is cavity insulation where the space between exterior and inner walls, found in most post-1930 houses, is filled with foam to reduce heat losses. Second, there is cladding where plastic foam composites can be hung onto an existing wall or framework for decorative or insulative purposes. Third, there are foam composite materials that actually bear a significant proportion of the building loads and thus may be more accurately termed "walls."

The most commonly encountered plastic foams for cavity insulation purposes are polyurea formaldehydes (which can be injected through holes in the wall during or after construction), polystyrene blocks, panels and slabs (built into the wall during construction), and polystyrene foam granules tipped into the top cavity wall after completion. Polyurethane is also used as an on-site foaming system, but being more expensive and toxic (before reaction), it is usually advertised for use when its structural and reinforcing qualities are required (e.g., to replace corroded tie bars). Similarly, phenolic foams are finding increasing application as nonflammable substitutes for flammable foams such as polystyrene. Technically speaking, according to current building regulations, "bridging" of wall cavities is not allowed, and it is within the power of local authorities to refuse permission for cavity wall filling operations whenever they see fit.

Considerable interest has been aroused recently in the manufacture of sandwich (composite) structures which in some cases actually act as load bearing members. As reported for roof members, reinforced plastic foam is being produced combined with metal and timber and other materials to create partially prefabricated wall elements. These come in many interesting combinations, including steel or aluminum/reinforced plastic/polyurethane foam (in France and Germany) and polyurethane and polyester foamed wall panels filled with clay (in Holland and France). A common combination involves a polyurethane-foamed core lined with reinforced plastic. These are said to pass Germany's strict fire regulation, and there are reports of commercially advantageous continuous production processes. Similarly, steel/foamed polyurethane/steel panels are capable of being made by continuous production. There are also reports of many other types of laminate, including polyurethane/crushed stone/ polyester/plasterboard interior, polyurethane/wood, reinforced plastic/expanded polystyrene foam/cement exterior, polyurethane foam/bitumen, polyurethane foam/asbestos cement sheeting, polyurethane foam/hardboard/plastic sheet lining, expanded polystyrene foam bonded to rigid skin, and glass/foam/melamine. Other plastics used for foam cores worth mentioning in this context include phenolics, foamed acrylics from Japan, and specialist pyrrone foams from the U.S. (NASA).

Alternative types of wall and panel construction involving plastic foams that are increasingly popular are reinforced foam panels (as opposed to sandwich foam laminates). Here, either reinforcing agents and fillers are added to the foam mix to strengthen and/or reduce fire risk or the foam itself is used as a filler for other building materials, thus reducing weight, cost, and thermal losses in the final product. Examples include filled polyester foams marketed by Bayer AG in France and Holland, polyurethane foam filled with expanded clay and glass spheres, foams containing Saran microspheres, and polyurethane foam with glass fibers. There are many examples of expanded polystyrene-filled cements, including those being used in Germany, Austria, Australia, Chile, Venezuela, Peru, etc., and also combinations of expanded polystyrene in clay bricks and building blocks. It is also possible that we shall eventually see building products being produced from so-called foamed polycarbonate, providing prices can be made more competitive.

As an associated activity, it must be mentioned that foamed plastics, such as expanded polystyrene, are playing an important role in the fabrication of concrete walls and bridges, both for decoration and molds for producing voids.

C. FLOORS, PAVEMENTS, AND ROADS

Plastic foams are finding increased usage as insultants beneath roads and pavements to control frost damage and heat flow. Examples include a PVC foam/screen/concrete/DPM/ asphalt layer structured to a shop service road in Derby, England. PU foam beneath pavements in Germany and PS foams beneath concrete roads in Switzerland incidentally are said to reduce construction costs by 15% in addition to providing "antifreeze" surfaces. Further potential applications are to be found in the context of vibration control where a closed cell

PE foam can be incorporated into factory floors, structural support layers in pavements and foundations using PU and PS and, as mentioned in the previous section, light concrete roads that incorporate expanded polystyrene foam sheet being used as mold lining release agents for the casting of concrete floor panels.

D. PIPES, FITTINGS, AND ACCESSORIES

For pipework, the main use for foam seems to be for insulation purposes or corrosion protection. In the former case, this is usually in the form of a foam core sandwich structure such as used for district heating schemes with polyurethane foam laminated between glass-reinforced polyester inner and outer sheaths or external polyurethane or isocyanurate-based (low flammability) cellular insulation lagging. In the latter case, the demand is said to have risen 150% in 1973. There are also examples of complete foamed PVC and PE nonpressure pipes. Some firms have forecast big markets for equipment associated with plumbing (such as hot water tanks) which are supplied as complete systems with an insulating and stiffening core sandwich structure made from polyurethane, isocyanurate, or phenolic foam.

Numerous examples can be quoted where plastic foamed profiles of different types are being promoted as replacements for wood in the fittings and accessories associated with the building trade. Extruded PVC foam and polystyrene foam sections are now finding markets in skirting board, cladding, coving, doors, window sills, window and door frames, pallets, sliding door tracks, ladder sections, etc. Similarly, there are reports of fiber-reinforced foam frames and polyurethane foam door frames. One building group have for some time offered complete foamed polymethylmethacrylate doors said to meet BS 476 Part 5 as "not easily ignitable." Open louvred shutters are also on the market made of polystyrene foam and coated with acrylic paint for UV protection. For structural foam moldings and furniture, the markets are said to be booming with many examples of bedroom cupboards, table units, filing systems, etc., in a variety of foamed materials.

E. COMPLETE SYSTEMS

The concepts of "prefabrication" and "building systems" applied to plastics and foams can mean different things to different people according to their background. Thus, it is important to distinguish between the so-called "all-plastics" house and the perhaps more pragmatic notion of composite prefabricated components playing their part within existing types of structure. The former concept is said to be traceable back to the Chicago Fair in 1933 when extensive and remarkable use was made of PVC leading later to complete plastic houses being exhibited in New York in 1955. There was also the one shown in the ideal Home Exhibition in London in 1956 and perhaps the most significant and heavily publicized of all — the "Monsanto house of the future" in U.S. Disneyland in 1957. It was pointed out in 1969, however, that only 2 out of the 70 or so examples of these types of houses then existing were really all plastic. Architects designing modern day equivalents of these experimental dwellings have been quick to make use of polymeric foams (mostly in the form of sandwich structures), and numerous examples and surveys can be quoted. An interesting by-product of this is that it has acted as a stimulus to research into the "systems" use of more traditional building materials such as wood, thus diversifying and increasing the competition.

In the immediate future, it would appear that interest is likely to grow increasingly in practical balanced combinations of materials rather than "all" concrete, plastic, wood, etc. In this context, there are a number of published surveys that deal with the more sensible part-use of plastic materials and give examples of how traditionally built houses contain many foam applications. Similarly, one can cite the plastics house in the 1969 Toronto Fair where exterior wall construction was based on a system of steam expanded polystyrene blocks designed with tongue and groove jambing. The blocks served as forms into which concrete was poured, providing structural support for the house. The foamed forms stayed in place as insulation pieces. Also, work has been carried out at Cambridge on a so-called "Autonomous House" which incorporates stabilized earth panels and blocks blended together with resin. Perhaps, the most likely manifestation of so-called fully commercial all-plastic structures in the future will be along the lines of smaller integral constructions such as the recently announced one-piece bathroom or portable collapsible shelters made from polyurethane foam/glass-reinforced plastic sandwiches, or disaster area igloos that can be quickly made *in situ* and which, incidentally, are said to show promise as permanent dwellings. Fabrications larger than foam laminated leisure homes still meet considerable scepticism unless, like the famous German "fg 2000" house, it can be demonstrated to pass all relevant performance standards by safety factors (in some cases) several times greater than those expected of traditionally built dwellings.

In conclusion, it is useful to comment on some of the latest trends involving expanded plastics in buildings. In the U.S., the 1978 consumption figures indicate a definite move towards isocyanurate (trimer) fire retardant foams. There is also increasing interest in polyurethane/phenol formaldehyde hybrid foams (made possible by a phenolic based polyol).

Undoubtedly, there will be a general increase in the use of phenolic based materials, especially as large companies are already announcing PF foams with low flame spread, low smoke emission, and mechanical properties similar to the same density polyurethane versions.

VI. FLOORING

A. INTRODUCTION

One of the most active fields of polymers is flooring in buildings of all kinds. It would be a virtually impossible exercise to estimate how many vinyl tiles have been used in the building trade, but it must now be many millions.

The following are the types of floors, which will be discussed:

- Vinyl floor tiles
- Rubber floor tiles
- Epoxy (jointless) floors
- Polyester (jointless) floors
- Electric floor heating: (1) methods and (2) thermal insulation with foam

Perhaps, one of the most interesting developments in the field of plastic flooring is the use of polyester and epoxy resins in concrete flooring. This technique is being adopted rapidly and gives excellent results.

An indication of the extensive use of plastics in flooring was given by the Timber Development Association Ltd. This organization and its members must view with increasing concern the adoption of plastics in flooring, particularly on top of cement bases. They have, however, adopted the sensible view about the situation. The Association has worked on the development of a satisfactory method of laying plastic tiles, etc. on a timber subfloor. Actually, this is of considerable value, since wood enables very good bonds to be achieved.

Vinyl tiles are the most important products in the laying of floors, certainly in the application of plastic and rubber fields, if not in the flooring field as a whole.

The most obvious feature of the tiles, a part from technical properties, is the extensive color range now available. All manufacturers have taken considerable trouble in this respect, employing industrial designers to produce a modern range of colors and patterns.

Most of the companies producing vinyl tiles also offer similar products in sheet form.

Although plastic flooring now takes first place, the merits of rubber flooring should not be forgotten. This particularly applies to some of the new synthetic materials that can give attractive, hard-wearing flooring. There are a number of possible combinations of various materials.

British Technical Cork Products illustrate an example of this combination of materials with their "Britcork," available in seven colors. It is a cork and synthetic rubber combination, giving the warmth of cork and the resilience of rubber. The Bulgomme-Silence flooring gives a combination in a different way, utilizing a hard-wearing rubber surface with a fabric interliner and a cellular rubber backing. The flooring, of course, has great resilience and cushioning powers.

A wide range of colors can be achieved with rubber as well as PVC, some plain and some marbled. Rubber, unlike PVC, lends itself to reinforcement with suitable fillers, and high strength can be achieved. In addition, the synthetic rubber provides resistance to domestic cooking fats (and other chemicals), and the marbling goes right through the thickness of the tile. Hypalon synthetic rubber, in particular, offers considerable advantages in chemical resistance.

In terms of plastics, seamless flooring may be taken to mean compositions including resins which are laid on as whole. These usually employ polyester and epoxy resins mixed with more conventional materials, particularly asbestos. A good example of this type is the "Plastik" plastic flooring (of undisclosed composition) from Vigers Brothers, Ltd. The floor is laid by them, without seams (except for expansion joints), to give a nonslip, oil- and water-resistant surface. It is trowelled on to cement, terrazzo, granolithic, and other surfaces where it sets to a hard finish.

Epoxy resins are very popular for this type of flooring. A good example is that of Tretol Ltd. with their "Epiflor," which will adhere strongly to concrete and cement, giving a hard surface with exceptional chemical resistance. In the case of "Epiflor," the composition is supplied to contractors for laying. It may be laid on to almost any type of flooring, including concrete, timber, mature asphalt floors, metal, etc. to a minimum thickness of 5 mm.

Cementex (U.K.) Co. offers a remarkable range of products for preparing and repairing concrete structures. Their "Epoweld" can be used for applying to floors followed by sprinkling on silicon carbide or sand to give a hard-wearing surface.

Although many companies use epoxy resins, Plastics and Resins Ltd. use polyester resins in similar flooring compositions. The final properties of the two types of resin have much in common, and both set in a short time to hard, tough products. "Certite" resins are for jointing present concrete units. They give excellent adhesion. There is no cracking due to shrinkage, and they provide thin, noncorrosive, waterproof joints.

Roads are, of course, a specialized form of "flooring." The main interest from the rubber and plastics point of view is electric heating and the use of epoxy resins. The latter is particularly prominent in giving hard-wearing surfaces.

B. VINYL FLOOR TILES

These tiles are by far the best known and widely used. It is therefore proposed to consider them in some detail.

1. General Considerations

It is probably true to say that the majority of floors laid today in offices, laboratories, schools, and similar buildings are made of rubber or plastic. All architects must be aware of the revolution in flooring materials that has taken place during the past few years, particularly with plastics. An indication of this is given by the search which was made for a suitable vinyl flooring for the NATO Headquarters building in Paris. Over 200 types of floor covering manufactured by 15 NATO countries were examined before a choice was made.

Why has such a change come about? There is no doubt that parquet and stone flooring are very attractive and reasonably durable. They are expensive, however, and the former, in particular, needs polishing regularly to maintain the attractive finish. Plastic floors, on the other hand, need little polishing; cleaning with soap and water is often sufficient. They are also relatively cheap, just as durable, considerably more so in laboratories, and available in a variety of colors and designs that cannot be approached in natural materials.

These and other properties have assured the ready acceptance of plastic and, to a lesser extent, rubber floors.

Virtually all so-called plastic tiles are based on PVC. This material is blended with a variety of fillers, mainly a special grade of asbestos, which gives greater rigidity and increases resistance to damage.

One of the advantages of these tiles is that they have great resistance to fire. They will not burn at all unless held continuously in a naked flame, and when the flame is removed, they extinguish immediately. At the same time, the floors should not have very hot objects (above about 80 to 90°C.) placed on them, or they will be damaged.

The chemical resistance of the tiles is very good. The following materials, after 24 h on the floors, have a negligible effect on them: water, soap and detergents, paraffin, petrol, white spirit, methylated spirit, washing soda, bleaching powders, hydrogen peroxide, concentrated ammonia, caustic soda, beer, milk, cooking fats, oils, blood, fruit juices, and vinegar. Dilute acids will cause slight surface marking, but serious damage is caused by concentrated acids and such solvents as ketones and chlorinated types.

Other interesting properties may be mentioned. Although PVC floors will not resist high temperatures, the action of a cigarette on such floors is very small. Electrical properties of such floors are good, and this makes them suitable in electrical installations. Scratching resistance is also good.

There are several types of flooring available. The most common is perhaps relatively thin vinyl tiles of varying area and thickness. Usually they are square but some companies additionally manufacture various matching shapes. They are secured by adhesives.

To avoid the labor of laying so many tiles, continuous lengths of PVC flooring are available. The joints can be welded in certain cases to produce seamless flooring. Similar designs are available, but obviously they must be continuous.

Similar products are available in rubber and described later. They include tiles in much the same sizes, thicknesses, and continuous flooring. Rubber floors can be very attractive and are more resilient than vinyl floors, but require more attention to keep them in first-class condition.

In addition to these standard floors, there are other types. The most common are hessian-backed flexible vinyl flooring which gives a feeling of great resilience, something like a deep pile carpet. Foamed products are also used in sandwich constructions to give exceptional resilience. One of the most attractive features about vinyl floors is the tremendous range of colors and designs available. A word of caution is required in choosing colors from catalogues. Printed patterns cannot well match tiles, and colors should therefore be chosen from tile samples. In addition, it is not easy during manufacturing to guarantee exact color matching from one batch to another so care must be exercised in making a choice. As far as possible, one consignment should be used for one room.

2. Composition of PVC

A PVC compound used for the manufacture of vinyl floor tiles contains a number of ingredients. It is important to know a little about these to understand the variations which can be made to affect the properties that have been described.

A PVC compound consists essentially of the pure PVC polymer itself, which is an extremely hard material with plasticizer, stabilizer, filler, lubricant, and pigment. The lubrication is purely for processing purposes and need not be considered further.

The plasticizer modifies the properties of the hard resin to give it the typical flexibility of PVC compounds. In flexible vinyl toys, there may be 30 to 50% of plasticizer for high flexibility, but in vinyl tiles, the proportion is usually very low, perhaps less than 5%. Thus, PVC tiles are not very flexible.

The filler is one of the most important ingredients of the mix, and large quantities are added to give a suitable product. The filler reduces cost but also confers stiffness. One of the most common fillers is asbestos.

Another ingredient, the stabilizer, is not of such great concern as it is mainly required for processing. However, if the correct stabilizer is not chosen, it is liable to react with impurities in the asbestos to cause discoloration. A choice of a suitable stabilizer is therefore important.

The only other ingredient is the color. Suitable pigments must, of course, be chosen to give color fastness and, where resistance to chemicals is involved, there should be no attack on the colors by the chemicals. A typical mix used for tiles is as follows:

Ingredient	Parts by weight
Vinyl polymer (PVC)	100
DIOP (plasticizer)	15
Lankroflex ED.3 (stabilizer)	12
Lankro Mark 225 (stabilizer)	6
Asbestos (filler)	100
Calcium carbonate (cheap filler)	150

There are, of course, many variations on this theme, but this may be taken as an example. A wax may also be added to supply bloom to the surface and give the tiles a high polish.

3. Electrical Conductivity

A consideration of the electrical conductivity of PVC is a good point at which to start a study of technical characteristics. This is because a direct relationship between the composition of the mix and conductivity properties can be established.

Normally, PVC compounds are reasonably good electrical insulators and are used as such. Asbestos is not a good filler from the electrical point of view as it tends to lower conductivity on any floor tiles. This, of course, is of no importance. In general, the idea is to reduce conductivity so as to avoid the accumulation of static charges. Such an accumulation usually occurs at relative humidities below 80% and where some type of rubbing of the surface occurs.

The elimination of static electricity can frequently be important. Obviously, in hospitals or rooms where inflammable solvents are kept, a discharge causing a spark could have disastrous results. Special flooring should therefore be used in such circumstances. Another point is that a static charge will attract dust and cause an unsightly appearance or the need for cleaning far more frequently than should be necessary.

To be antistatic, a compound must have a volume resistivity (specific resistance) of less than 10^{10} Ω/cm, or preferably, less than 10^{8} Ω/cm.

4. Fire and Heat Resistance

These properties, which are often confused are important in a number of instances. Fire resistance is primarily concerned with safety, but in the case of heat, it is the resistance to hot objects placed on the floors or incorporated floor heating.

Fire resistance is the ability to resist a flame. PVC compounds, quite rightly, are normally claimed to be flame resistant. However, the term is only relative and calls for further comment.

If a PVC compound is placed in a flame, it will begin to burn. Once the flame is withdrawn, however, it should quickly extinguish itself. By compounding fillers, it is possible to make the PVC extinguish itself immediately upon removal of the flame, and this is most desirable. As it happens, most vinyl flooring compounds contain a large proportion of asbestos, and this, coupled with the inherent fire resistance of PVC itself, makes the tiles fire resistant in the sense already described.

Heat resistance is rather a different matter. PVC is a thermoplastic which means that it softens when warmed and returns to its original state when cooled. Any hot object placed on it, therefore, depending on its weight, will tend to mark the PVC tiles. Up to temperatures of about 70°C and for relatively light loads, there will virtually be no marking, but with increasing temperature and pressure, the indentation is likely to become marked. Some manufacturers do take this into account and issue suitable warnings.

Although PVC compounds are thermoplastic, they may safely be employed with heated floors. Fortunately, the temperature of heating is usually quite low and well below the PVC danger point. In more specific terms, the heated floor should not reach a temperature greater than 30°C, although this does depend on the compound.

Associated with fire and heat resistance is the question of resistance to cigarette burns. Some manufacturers appear to make exaggerated claims in this connection to the effect that their tiles do, in fact, resist cigarette burns. Others, however, are practical about the matter and give a warning that a lighted cigarette applied to the surface will cause local softening and a scar. The severity of such burns will undoubtedly vary with the composition of the mix and may be less with higher filler content. Nevertheless, all vinyl floors will tend to suffer from this defect, and a suitable warning should be given. If burning does occur, then the scar should be removed with some abrasive and repolished. Although a householder could not achieve a repair by welding, a deep scar should be repaired by a contractor using a stick of PVC and a suitable ironing instrument.

Finally, the associated feature of thermal conductivity may be mentioned. In general terms, normal PVC compounds for floor purposes are poor conductors of heat.

5. Low Temperature Properties

In general terms, all PVC compounds gradually stiffen as the temperature is lowered. As the temperature approaches 0°C, they may become so brittle that they will shatter from a blow.

The low temperature properties can be improved in a number of ways by compounding. However, this involves the use of special plasticizers which are usually required in greater quantities than normally employed in PVC flooring compounds. It is unusual, therefore, to cater for these properties, but in really cold climates, it would be advisable to devise a special compound.

Generally speaking, the main problem in the northern countries arises during laying. If a roll of PVC flooring is exposed at very low temperatures, it may easily crack when unrolled. It is, therefore, advisable to allow such flooring to be at a moderate temperature before attempting to lay. One of the few companies who mention the point in their literature suggest a temperature of +7°C, which is reasonable for most compounds.

Once the PVC is installed, the temperature which it will withstand is not quite so critical. However, in cold rooms and in exposed buildings during construction in the winter, the floors might well crack if heavy loads are dropped on them. The point is worth keeping in mind.

There is one certain way of improving flex resistance: reduce the filler content considerably. This obviously is undesirable from the cost point of view so that the method can rarely be used. Nevertheless, there is a compromise solution which some manufacturers have adopted. This is to make a laminate flooring with a pure vinyl resin on the top. This excellent practice confers other advantages as will be indicated later.

6. Abrasion Resistance

It is difficult to be precise about abrasion resistance with regard to PVC. Few figures are available on the subject, one of the reasons being that the tests to measure it are unsatisfactory. They do not necessarily correlate with what is obtained in practice.

In theory, the use of large quantities of filler in flooring should lower its resistance to abrasion. Unlike natural rubber, fillers, with minor exceptions, do not reinforce PVC in any way; they merely act as diluants. The abrasion resistance of PVC flooring should therefore be very low.

7. Deformation

One of the most difficult problems to solve with PVC floors is that of deformation under a load. The loads are usually industrial trucks and equipment, furniture, or stiletto heels.

Fortunately, the deformation sustained by PVC is reduced with increasing filler content. The penetration can readily be measured in the laboratory by penetration tests. In one of these, a 5-mm ball is pressed into a sample under a load of 2 kg for 1 min and the deformation measured. In the case of a filler called Omya BSH, the following results were obtained:

Temperature of test (°C)	Parts of Omya BSH			
	0	10	30	60
30	86	83	72	67
40	97	100	87	78
50	122	119	108	93
60	148	141	132	120

It will be seen from these results that the deformation (in microns) is considerably reduced with increasing filler content. Considering that the PVC compound contained 250 parts of filler compared with the 60 above, the improvement expected will be obvious.

The results given in the table also show that the deformation increases at elevated temperatures. Even here, however, the deformation is reduced with increasing filler content.

Unfortunately, deformation as such is not the only consideration. Permanent set is also vitally important and means the degree by which any deformation is retained after removal of the load. With minor exceptions, unfilled compounds have the best resistance to permanent set, and this is rapidly lowered as filler content is increased. The minor exception is with the use of Omya BSH, where the addition of 10 parts of filler actually reduced permanent set. After the addition of this filler, however, the permanent set starts to rise rapidly as with other materials.

Once again, therefore, the use of an unfilled compound in a laminate may offer advantages. It is problematical, however, whether increased deformation with this type of material or the reduced permanent set will show up to most advantage. This can be determined only by experience and experiment.

8. Chemical Resistance

The claims by manufacturers for chemical resistance are normally in general terms. One company claims, for example, that their tiles are "grease-proof, stain-proof, resistant to alcohol, detergents, soaps and most acids and solvents." Others talk of "grease and oil" and yet others of "general kitchen chemicals." PVC compounds are, in fact, highly resistant to chemicals, but there are some that attack the material seriously. These include aliphatic and aromatic ketones, aromatic amino compounds, acetic anhydride, organic compounds containing nitro and chlorine groups, and lacquer solvents. However, the compounds resist such materials as nitric, sulphuric, hydrofluoric, and chromic acids and strong alkalis, all at reasonable concentrations.

It should be stressed that the resistance of PVC to chemicals varies with concentration and temperature.

One feature of the chemical resistance of PVC should be mentioned. PVC is seriously attacked or stained by certain bitumen compounds. It is impossible to be specific about this since there are hundreds of types, some of which are satisfactory and others not. For this reason, care should be taken when laying PVC floors on freshly laid bitumen bases, and these are best avoided unless the supplier gives specific information to the contrary.

The compounding of PVC does not affect the chemical resistance very much. Certain changes can be effected, of course, but they are usually unnecessary. A special class of plasticizer (polymeric), when used, reduces maring and the extraction of plasticizer by chemicals.

Another source of possible damage to vinyl floors are certain types of cleaners. Manufacturers, again, must give instructions here so that cleaners and polishes containing solvents which attack PVC may be avoided.

9. Other Features

It is often necessary to lay tiles in situations that are likely to be damp. Such conditions usually allow the growth of various bacteria and molds. Generally, PVC compounds are highly resistant to such organisms, but resistance can be improved by adding insecticides to the compounds. In spite of statements to the contrary, PVC is attacked by termites and, in general, there is little that can be done about it.

The resistance of PVC floors to marking by rubber is a formidable problem. Many rubber compounds contain what are called "staining antioxidants." The antioxidants will be absorbed by the vinyl floors, and the action of light will cause discoloration. If this is seen to occur, the only real remedy, unless there is some control over the composition of the rubber, is to protect the floor from contact with rubber.

C. RUBBER FLOOR TILES

Rubber flooring as such has been used for so long and is so well known that it warrants little description. Rubber flooring, available in a range of attractive colors, has been known to last for 20 or 30 years, and there is much to recommend it.

There is probably little or no true natural rubber flooring today. It is all at least blends with general-purpose synthetic rubbers, the properties of which are much the same.

1. Hypalon Floors

The outstanding success and ready acceptance of vinyl floors has been so overwhelming that the consideration of yet another flooring material may seem to be superfluous. This position, however, should not prevent the flooring industry from examining other products since vinyl floors are certainly not perfect and have limitations in certain applications.

Two of the inherent defects of PVC are its thermoplasticity and the relatively high permanent set which it takes. Both of these limitations can be offset to a certain extent, but they cannot be eliminated.

Hypalon is superior to PVC in both properties mentioned and must clearly be considered as an alternative in certain cases. It is highly unlikely that it will ever become a serious competitor of PVC for general purposes because of cost consideration alone, but for specialist applications, it might well find a useful place in the field.

Hypalon is a synthetic rubber that is flame and weather resistant and highly resistant to a large variety of chemicals. It thus has a number of valuable characteristics so vital in a material to be used for flooring.

Chemically, Hypalon is a chlorosulphonated polyethylene made by reacting polyethylene (the plastic polythene) with chlorine and sulphur which converts the thermoplastic into a synthetic rubber. This means that in vulcanization, it becomes a thermosetting material that cannot be softened by heat. Similarly, it acquires a good resiliency (i.e., low permanent set) so that the material at once has two advantages over vinyl flooring. These properties are considered in detail later.

Hypalon is manufactured by E. I. Du Pont de Nemours and Co., Inc., in the U.S.

2. Summary of Properties

The claimed properties of Hypalon floors may be summarized as follows:

- Super abrasion resistance
- Easy to clean
- Available in a wide range of stable colors
- Extreme resistance to aging and can be used for outdoor applications
- Does not shrink or spread during service
- Resistant to indentation
- Deadens sound
- Will not burn and is not damaged by buring cigarette ends
- Resistant to chemicals; animal, vegetable, and mineral oils and greases
- Resistant to heat
- Resilient and flexible

These properties are not just a list designed for publicity purposes, but genuinely represent the characteristics of the material. They indicate, without doubt, how valuable Hypalon can be under certain conditions. They might even be a good reason for using the tiles everywhere, a practice largely prevented by the relatively high cost.

3. Mechanical Properties

One of the most important properties of Hypalon is its resistance to indentation with subsequent high recovery. The indentation resistance is several times greater than any similar floor covering in use at the present time. Tests conducted by Du Pont indicate that Hypalon has residual indentation 6 times lower than vinyl, 8 times lower than asphalt, 14 times lower than rubber, and 16 times lower than linoleum. Field trials and installations have confirmed these tests.

As far as abrasion resistance is concerned, Hypalon tiles compare very favorably with vinyl and rubber tiles and linoleum. This again has been confirmed in installations. For scratching and scuff resistance, Hypalon appears to be superior to vinyl and equal to rubber and other resilient floor coverings. The hardness is Shore 92° to 95°.

One difficulty with vinyl flooring is the reduced flexibility at low temperatures. At very low temperatures, brittleness is likely to be encountered. Asphalt behaves in a similar way. In the case of Hypalon, however, the resilience remains constant at lowered temperatures.

The material also has good dimensional stability. The rubber is compounded with fillers and other ingredients to resist shrinkage or extension during service on the floor.

Traditionally, rubber and cork have been the better types of floor coverings to reduce the noise of foot traffic. Hypalon tiles have even greater sound-deadening characteristics.

4. Other Properties Affecting Use

Quite apart from actual chemical resistance, to be discussed in detail later, a desirable characteristic of any flooring material is to resist staining from foods, inks, cleaning agents, and a number of common solvents. The way of achieving this resistance is to have a completely nonporous surface. This, Hypalon possesses, and it thus has excellent staining resistance.

A corollary of this property is the ease with which the floors may be cleaned. The smooth, hard surface and the resiliency prevent impregnation of dirt which facilitates cleaning. At the same time, a fine luster may be obtained with a minimum amount of wax.

In certain applications, the weathering resistance cannot only be important but extend the fields of application. In weathering, resistance may include resistance to oxygen, ozone, and water. Hypalon is excellent in this respect, so much so that it may safely be employed out-of-doors. Terraces and patios are obvious instances where this property will be of use.

This durability is irrespective of the color. In this connection, the range of colors that may be obtained is virtually unlimited. Light pastel shades resistant to sunlight and discoloration are created without difficulty.

Some figures for tests carried out in a weatherometer, simulating many months of exposure and water resistance, may be quoted:

Weatherometer Discoloration

After:

1 d	None
2 d	None
4 d	None
8 d	Faded

Water absorption (70 h at 70°C)

Volume increase	2.3%
Weight increase	1.1%

A further important feature of Hypalon floors is their flame resistance. This is achieved, not just by the addition of flame resistant fillers, but by the inherent characteristics of the material. In this respect, it is similar to vinyl floors, but considerably superior to rubber floors.

5. Chemical Properties of Hypalon

Although the chemical resistance can vary according to the compounding ingredients used in the tiles, the resultant products all have basic chemical resistance. It should always be stressed, however, that the results given are for general guidance, and if any doubt exists, then the manufacturers should be consulted.

Some time ago, various interesting tests were undertaken by the University of Ottawa on Hypalon tiles made by Canadian General-Tower, Ltd. In these tests, a comparison was made between resilient vinyl, vinyl-asbestos, and Hypalon tiles by subjecting them all to the following chemicals for 5 min and 20 min: concentrated sulphuric, hydrochloric, nitric, hydrofluoric, and acetic acids; concentrated sodium hydroxide, acetone, xylene, chloroform, carbon tetrachloride, ethyl acetate, and benzene.

The Hypalon tiles withstood the effects of all these chemicals, whereas the vinyl tiles were affected by some of them. Some of the chemicals removed the original gloss off the Hypalon tiles, but this was restored by polishing with steel wool. No visible mark was left after the test. The worst offender of these chemicals was chloroform, but even in this case, the original gloss could be restored by polishing.

The chemicals stated were chosen as being those most commonly used in the laboratories of the University. As a result of the tests, Dr. Pierre R. Gendon, Dean of the Faculty of Pure and Applied Science, thought that the tiles would be ideal for their building.

This is clearly a case of the specific application of Hypalon because of its superior properties to other materials.

In view of the possible importance of the use of Hypalon tiles in special applications, it has been thought helpful to give detailed chemical resistance properties of the material.

6. Chemicals Having Little or No Effect upon Hypalon

Chemical	Concentration by weight (%)	Temperature (°F)
Ammonia	Liquid anhydrous	Room temperature (R.T.)
Chlorine dioxide	14	150
Chrome plating sol		158
Chromic acid	50	200
Chromic acid	Concentrated	R.T.
Cottonseed oil		R.T.
Diethyl sebacate		R.T.
Dimethyl ether		R.T.
Ethylene glycol		158
Ferric chloride	15	200
Ferric chloride	Saturated	R.T.
Formaldehyde	37	R.T.
Freon 12		R.T.
Hydrochloric acid	37	122
Hydrofluoric acid	48	158
Hydrogen peroxide	50	212
Hydrogen peroxide	88.5	R.T.
Methyl alcohol		R.T.
Mineral oil		R.T.
Motor oil (SAE 10)		R.T.
Nitric acid	Up to 20	158
Nitric acid	70	R.T.
Phosphoric acid	85	200
Pickling solution	20% nitric, 4 HF%	158
Potassium hydroxide	Concentrated	R.T.
Sodium dichromate	20	R.T.
Sodium hydroxide	20	200
Sodium hydroxide	50	158
Sodium hypochlorite	20	200
Stannous chloride	15	200
Sulphur dioxide	Liquid	R.T.
Sulphuric acid	Up to 50	200
Sulphuric acid	Up to 80	158
Sulphuric acid	Up to 95.5	R.T.
Tetrabutyl titanate		R.T.
Tributyl phosphate		R.T.
Water		200

7. Chemicals Having a Moderate Effect upon Hypalon

Chemical	Concentration by weight (%)	Temperature (°F)
Acetic acid	Glacial	R.T.
Acetone		R.T.
Aniline		R.T.
Chlorine (dry)	Liquid	R.T.
Cottonseed oil		158
Diethyl ether		R.T.
Hydrochloric acid	37	158
Mineral oil		212
Motor oil (SAE 10)		158
Nitric acid	70	122
Sulphuric acid	95.5	122
Turbo oil No. 15 (jet engine lubricant)		R.T.

8. Chemicals Having a Severe Effect upon Hypalon

Chemical	Concentration by weight (%)	Temperature (°F)
Acetic acid	Glacial	158
Carbon tetrachloride		R.T.
Citrus oils		R.T.
Dichlorobutene		R.T.
Formaldehyde	37	158
Gasoline		R.T.
Hydrochloric acid	37	200
Jet engine fuel (JP-4)		R.T.
Nitric acid	30	158
Nitric acid	Fuming	R.T.
Nitrobenzene		R.T.
Perchlorethylene		R.T.
Turbo oil No. 15		350
Xylene		R.T.

As can be seen, extensive testing by Du Pont has shown that Hypalon possesses excellent resistence to most chemicals. In application where contact with strong oxidizing chemicals is involved, the material has shown its superiority to all other elastomers, which means natural rubber, styrene-butadiene (SBR), neoprene, etc.

The following is a list of chemical properties expressed in a different manner to the list given previously. (Key: A = little or no adverse effect; B = minor to moderate effect; C = severe effect.)

	Rating	Temperature (°F)
Acetic acid, glacial	B	R.T.
Acetone	B	R.T.
Ammonium hydroxide	A	200
Amyl acetate	C	R.T.
Amyl alcohol	C	R.T.
Aniline	B	R.T.
Asphalt	C	
Beer	A	R.T.
Beer juice	A	R.T.
Benzene	C	R.T.
Butter and buttermilk	A	158 and 100
Calcium hydroxide (lime)	A	200
Calcium hypochlorite (bleach)	A	200
Carbolic acid (phenol)	B-C	R.T.
Carbon disulphide	C	R.T.
Carbon tetrachloride	C	R.T.
Castor oil	A	158
Chlorine water	B	R.T.
Chloroform	C	R.T.
Chrome plating solutions	C	158
Citric acid	A	R.T.
Copper plating solution	C	190
Creosote	B-C	R.T.
Cyclohexanone	C	R.T.
Diethyl ether	B	R.T.
Ethyl alcohol	A	200
Ethylene glycol	A	200

	Rating	Temperature (°F)
Ferric chloride	A	200
Formaldehyde	A	R.T.
Freon 12	A	R.T.
Fuel oil	B	158
Glycerin	A	200
Hydrochloric acid	A	R.T.
do.	C	158
Lubricating oil	B	158
Methyl ethyl ketone	C	R.T.
Milk	A	R.T.
Molasses	A	200
Nickel plating solution	A	140
Nitric acid (up to 30%)	A	R.T.
do.	B	122
do.	C	158
Olive oil	B	R.T.
Phosphoric acid	A	200
Potassium dichromate	A	200
Sea water (and sodium chloride solutions)	A	158
Soap solutions	A	200
Sodium hydroxide	A	200
Sulphuric acid (up to 50%)	A	250
Tin salts	A	200
Toluene	C	R.T.
Tricresyl phosphate	B-C	R.T.
Turpentine	C	R.T.
Urine	A	R.T.
Wines and spirits	A	R.T.
Zinc salts	A	200

9. Applications of Hypalon Flooring

It is useful to consider some of the circumstances in which Hypalon tiles are used since it shows why the material was chosen for the application concerned.

As already indicated, one of the most important applications is in chemical and similar laboratories. The Du Pont laboratories at Hemel Hempstead have an area covered with Hypalon floor which is open to inspection at any time.

Another obvious application is in canteens and cafeterias. Reasons include ease of cleaning and resistance to food spillage and cigarette burns. Banks and similar buildings will also use Hypalon for long life, ease of cleaning, minimum maintenance, and lasting attractive appearance. In addition, there is the usual reason of resistance to stiletto heels and furniture and nonslip characteristics. The latter is also important on ships.

D. EPOXY (JOINTLESS) FLOORS

Many materials have been used for floors, some of them with only mixed success. In the decorative field, parquet floors are very attractive, but are expensive and require much attention to maintain their beautiful finish. Vinyl floors are much cheaper, are attractive, and easy to maintain, but are still not as resistant to stiletto heels as they might be.

In the industrial and utilitarian fields, asphalt and concrete are very common. The former is not particularly abrasion resistant and is restricted to dark colors. The latter soon sustains abrasive damage, forms a surface dust, and is not particularly attractive.

It is clear, therefore, that there is still much room for new types of floors. These have become available in the last year or two in the form of compositions based on epoxy resins, members of the plastics family. These are about 10 times more resistant to abrasion than concrete, can be laid as jointless floors, can be colored in any way, are very light, are resistant to chemicals, and sustain no damage from stiletto heels. They can also be used for restoring concrete and as varnishes or paints for protecting woodwork and giving it a high gloss. The epoxy resins are obviously of the greatest value and potential.

1. What are Epoxy Resins?

The epoxy resins are synthetic materials of the polymers family. They are also often called "epoxide" resins and chemically are usually glycidyl ethers of bisphenol-A. The chemical composition, however, need be of no concern.

As far as industry is concerned, the important points are the form in which the materials are available and the products which can be obtained from them. They are normally available as two liquids: the basic resin itself and the hardener. As long as these two are kept apart, they may be stored for long periods, but as soon as they are mixed, they will set hard, at room temperature, in a matter of hours. They may thus be applied by all the conventional means and yet, without further treatment, give all the properties previously described.

2. Some Outstanding Properties

The materials are resistant to a wide variety of chemicals; this makes them suitable for many purposes. A typical range of chemicals is as follows:

- Hydrochloric acid (below 36%)
- Sulphuric acid (below 50%)
- Nitric acid (below 10%)
- Phosphoric acid
- Acetic acid (and vinegar dilute)
- Citric acid
- Tartaric acid
- Carbon tetrachloride
- Lactic acid (and milk and other fats)
- Sodium hydroxide
- Hydrocarbon solvents
- Sugar solutions
- Fruit Juices
- Ethyl alcohol
- White spirit
- Calcium hypochlorite

The floors are very resistant to the chemicals given, although the stronger acids will cause rapid deterioration. Ketones cause softening in about a week, but the materials harden up on the removal of the solvent.

A number of colors are available with proprietary epoxy floors. For example, in the case of products of Evode Ltd., the standard colors are light gray, Venetian red, pale green, and light blue. There is obviously no limit here to colors, and others will be made to special order.

The mechanical properties of such floors are very good. In general terms, the compressive, flexural, tensile, and impact strengths are roughly double the value for the average concrete. Resistance to high pressure is good, which means resistance to, e.g., stiletto heels. The abrasion resistance is also good, which means the absence of dust, a valuable characteristic in hospitals and kitchens.

In addition to the merits of epoxy floors, the snags must also be pointed out. The main one is the much higher cost as compared to concrete, so the materials may well be retained for topping, repairs, and finishes. In addition, the materials can be damaged by excessive heat and flame and have poor resistance above about 110°C. They can be made relatively flame resistant, however. As compensation for this, the products are unaffected by the free-thaw cycle that is disastrous for concrete.

3. Application Properties

Some properties of the resins are particularly applicable to the installations where they are used. The first one of these is weight. The specific gravity is low but, in addition, only 1/8 to 1/4 in. of material need be applied to give floors equivalent to much thicker layers of other substances.

These low weight floors are obviously of the greatest value. In particular, architects will be able to make considerable weight savings in multistorey buildings. In this connection, most of the usual grades of epoxy flooring are rather more flexible than other *in situ* and are therefore invaluable for use on suspended steel floors. If architects are not satisfied with the normal degree of flexibility, then additional flexibility may be readily provided.

Another important point with the floors is that they can be applied to all existing floors whether they are concrete, metal, or wood. It is only essential that the floors be clean and dry. Another valuable feature is that the slip properties of the floors can be controlled and any degree of nonslip characteristics imparted. Even when wet and with operators wearing rubber boots, there is no danger of slipping.

4. Some Typical Applications

The fact that epoxy floors are chemical resistant, hard wearing and nonslip has meant application in a number of specific fields, such as in buildings that are likely to have chemicals spilt on the floor. Typical examples include laboratories, plating shops, battery rooms, photographic darkrooms, oil refineries, garage lubricating bays, and chemical process areas.

Other normal applications are places where food is processed. Examples include hospitals, kitchens, sugar refineries, dairies, confectionery factories, bakeries, breweries, and similar establishments.

The other important series of buildings are those which have to withstand much wear and tear. There are many of these, including warehouse gangways, factory gangways, blacksmith shops, garage forecourts, loading bays and ramps, and any industrial establishments where heavy wheeled trucks are employed.

It should not be thought that all applications are necessarily semi or fully industrial. Many of the products are used in offices and the home. The epoxy paints are applied even to parquet flooring. This enhances the already high gloss and gives a really hard surface. Car showrooms have been treated in this way and the surfaces have been unaffected for years.

E. POLYESTER (JOINTLESS) FLOORS

In certain fields, the conventional flooring materials are, at best, just good enough because it has been thought that no suitable alternatives were available. Obvious examples include any factory shops where corrosive chemicals are being used.

Again, for hygienic reasons, jointless floors may be necessary to prevent the trapping of dust and germs such as in hospitals and food manufacturing plants. In some cases, exceptional abrasion resistance may be necessary, in areas where heavy trucks are moved about. At the same time, it may be necessary to avoid the noise which would be expected should steel floors be employed.

Some of the advantages of this type of flooring may be summarized as follows:

1. They are self-colored with the color running right through the material.
2. They are exceptionally resistant to chemicals and water.

3. Thermal insulation is good which means that heat will be saved compared with conventional floors.
4. Abrasion resistance is exceptional.
5. Mechanical properties are excellent.
6. They are silent.
7. They have low specific gravity and therefore are light in weight.

Other properties are given under a specific example of polyester flooring described later.

For floors, spraying with chopped glass strand is undoubtedly an important way of proceeding. It is also possible to make exceptionally strong laminates on site.

Some flooring manufacturers merely use the polyester resins with colored chippings of wood or granite, for example, to give a terrazzo-like effect. The ingredients are not fillers in the usual sense, but the flooring material is merely bound by the resin. Such floors are very strong.

1. The Resins

The gel coats may be applied by brush, roller, or spray equipment. In the latter case, great care is necessary since one of the components (styrene) can evaporate quickly and cause undercure resulting in a most unsatisfactory product. A very humid atmosphere can cause the same trouble as, in fact, can moisture of any nature. The presence of moisture will mean a longer curing time (which can be speeded up by the application of heat). In any case, the cure should be made as rapid as possible.

Various faults are possible in laying floors with any of the foregoing products. One of the most common is crazing. This can be caused by a poor formulation, insufficient mixing, addition of extra styrene (to make up losses), and use of the wrong resin. Using plasticizing resin as well as attending to the matters mentioned will help overcome the problem.

A suitable gel coat formulation is as follows:

	Parts by weight
Polymer	70
Plasticizing resin	30
Catalyst	4
Accelerator	4

It is essential to mix thoroughly because if this is not done, an inconsistent mix will result in which some selections will have gelled prematurely and other sections will be incompletely cured. Mechanical stirring is recommended for all production mixing. However, excessive stirring should be avoided. Otherwise, air bubbles will be introduced, and there will be loss of styrene.

Since the compounded mix has a short pot life, only sufficient should be mixed for the work in hand. The temperature of the mixing shop should be kept constant and as low as is consistent with comfortable working.

2. Spraying Techniques

Obviously, a convenient spray method can save much time and money, and suitable equipment is now available. But the quality of the molding can still rely on the operator's skill.

3. Indentation Resistance

Loads of 200 kg, 400 kg, and 1 ton were applied to the material surface through a cylindrical steel bar of 1 in.^{2} section. The loads were each held for 5 min and then removed.

The surface was inspected. After subjection to these respective loads, there was no detectable indentation of the surface.

4. Hardness

The impression made by a 1 in. diameter steel ball falling onto the surface from a height of 6 ft was measured, using the standard "carbon paper" method. Diameter of the impression was 3.5 mm.

Corresponding figures for other materials were:

Material	Diameter of impression (mm)
Magnesite	3.5–6.8
Concrete	4.4
Quarry tiles	4.0–4.5
Wood blocks	8.9

5. Abrasion

A sample was scrubbed 100 times with an abrasive washing powder, and the effect on the surface noted. Small pinholes in the surface were rendered more easily visible due to the accumulation of the white abrasive powder therein. Most of the surface was not matte. There was no real damage or color change.

6. Wear Resistance

The specimens were dried to constant weight and then placed in a rectangular container with their wearing surfaces inwards. Placed in the container, were 1000 hard steel balls of $1/2$ in. diameter. The whole unit revolved along its horizontal axis at 60 rpm for 22 h. The specimens were removed for examination after 22 h revolving in one direction only.

The dust was brushed off, and the specimens reweighed with the following results:

Color	Weight before test (lb)	Weight after test (lb)	Loss in weight (lb)
Black	42.47	42.44	0.03
Green	41.89	41.53	0.36
Red	41.31	41.06	0.25

F. ELECTRIC FLOOR HEATING

1. Methods

This section is not meant to give detailed information on floor heating but only an indication of where rubber and plastics are employed in the various systems involved.

The main points to be considered are: (1) the type of insulation to be used for the heating elements, (2) whether to bed direct into the screed or use a conduit, and (3) the permitted surface temperature of the floor.

The heating element is always a cable of some type, the most common being the conductor insulated with a high temperature PVC. This is a satisfactory type and perhaps the safest if any wet conditions are likely to occur. It does, of course, suffer from the disadvantages of all thermoplastics and, in addition, if there is a fault followed by excessive increase in temperature, embrittlement of the PVC can occur followed by cracking and failure. On the whole, however, such failures are rare. Most trouble occurs at the joints (element to cold tails), a problem outside the scope of this book.

For those who prefer it, there are cables insulated with butyl rubber that does not embrittle. It is sheathed usually with neoprene or nitrile rubber. Whatever the material, the end result is the same except that the butyl-neoprene or nitrile is perhaps more robust and suitable for roads, football fields, and the like.

To place a cable in conduit appears to offer a big advantage. If a fault occurs, it is a simple matter to withdraw the cable and replace it. Unfortunately, the air gap between cable and replace it. Unfortunately, the air gap between cable and conduit (which can be metal but is often PVC) means that the cable has to be run at a higher temperature to give the same floor temperature; this causes many more faults to develop.

Most manufacturers of PVC floors recommend that the temperature of the floor should not exceed 80 to 85°F. Since most floor heating systems recommend these operating temperatures, there should not be much trouble. Although PVC and Hypalon tiles could operate at higher temperatures than this, the limiting factor may well be the adhesives used for bonding the tiles to the screeds.

If a higher temperature is required, then it is best to employ the jointless plastic floors. Epoxy and polyester floors could easily operate at 100 to 120°F.

A development in floor heating is to underlay the screed (and elements) with polystyrene foam.

A typical set-up may be illustrated by an example from Heating Investments Ltd. What they claim is as follows:

1. Even heat distribution — no heat ridging or convected currents
2. Utilization of the structure of the building
3. Fully automatic controls — no supervision required
4. Attendant labor costs completely eliminated
5. Completely free floor and wall areas
6. No heat flare or damage to furniture or decoration
7. Complies with smokeless zone air requirements
8. Architecturally, simple integrated
9. Low capital installation costs

Running costs are extremely reasonable, being at least comparable to more conventional forms of central heating.

A competent design with full heat loss assessment is essential. Average loadings of 15 W/ft^2 can be misleading, particularly with the modern light forms of construction, if a full temperature rise is to be provided. In this respect, Heating Investments Ltd. or any other member of the Electrical Floor Warming Association would be pleased to give advice.

As far as the element is concerned, development has taken place over many years, and basically (for Heating Investments Ltd.), a medium voltage heating element is used with PVC insulation. Experience has proved that the best element is of stranded semihigh tensile steel. Each strand is about 0.015 in. in diameter. This is laid up to provide an element of giving a loading of between 4 and 5 W/linear foot. The element is insulated with PVC.

Each made-up element is then fitted with copper conductor double insulated PVC cold tails; the length is suited to the requirements of the building (12 ft minimum).

The joint between the cold tail and the element is manufactured with extreme care and rigidly tested to ensure that no consequential trouble arises after this has been buried in the floor. The joint is crimped and soldered for electrical continuity and mechanical strength, and a PVC sleeve is molded overall for a completely waterproof joint.

2. Thermal Insulation with Foam

The loss of heat through floors, walls, and roofs is accepted with resignation, although the cost of this loss is incalculable. Even now, there is little thought of saving energy where heated floors are not to be employed.

Fortunately, the advent of floor heating has drawn our attention to the limitations of the existing systems, and efforts are being made to improve the situation. This development has been made possible, to a large degree, by the development of plastic foams. They have extremely low "K" and "U" values and are ideal for floor insulation.

3. Floors in General

Slabs of polystyrene foam are laid on concrete floors to form an insulated foundation for flooring cements or composition floors. The thickness of the slabs usually varies between $1/2$ and 1 in. If the expanded slabs are laid before any part of the floor, then the ground should be covered with a layer of fine ashes. This is an advantage since it assists in excluding moisture.

On the other hand, an advantage of applying the polystyrene slabs over the concrete is that floors can be laid immediately as there is no seepage of moisture from the concrete. However, joints between slabs should be sealed with a filler or tape and covered with bitumen paper. It is also important that the cement or concrete foundation be even. Where joints are specified, the spaces between them can be filled with expanded beads or slabs.

In the case of uneven concrete floors, these should be leveled with sand to avoid local overloading and consequent cracking of the cement screed.

Floors over basements and above archways and terraces require a thicker layer of foam for perfect insulation. It is also advisable to use a layer of bitumen paper in such cases.

Polystyrene insulated floors have excellent sound and heat insulation properties. So far as heat is concerned, polystyrene is ideal for incorporation into floor heating systems where the cables, or conduits carrying them, are laid directly on the blocks. So far as sound insulation is concerned, by employing 1 in. of polystyrene over a bare concrete floor, the improvement is approximately 15 phons. To improve the elimination of sound transmission, polystyrene edge strips are laid along the wall corners before placing the main slabs into position and casting the cement floors.

4. Piggery Floor Insulation

As a specific example of the use of foam in floors, a piggery installation may be considered. A large increase in the food conversion factor is obtained when a piggery floor is properly insulated. This is quite well known but what is not so well known, however, is how to obtain proper insulation of a reasonable and economical cost.

Experiments with a number of trial floors under all kinds of conditions has shown that foamed polystyrene (e.g., Poron) used in the following form of floor construction gives first-class insulation properties. It provides a good hard-wearing floor, resulting in a minimum loss of body heat by the pigs.

The floor must be flat and of good surface (new construction should be of well-rammed hardfilling, e.g., 3 to 4 in. thick with 3 to 4 in. concrete of 6:3:1 mix with a screeded finish). Existing floors should be broken up, rammed, surfaced with at least 2 to 3 in. of 3:1 cement-sand mix, and screeded, or if otherwise mechanically sound, screeded to level with 3:1 mix of cement-sand (minimum 1 in. thick). It is recommended that the floor should be hacked to give a key for the screed or a coat of grout or slurry well brushed in and allowed to dry to form a key.

This surface screed must then be coated with a bituminous solution at the rate of 15 to 20 yd/gal (Synthaprufe is one such suitable material). It is recommended that one coat be applied, allowed to dry, and followed by a second coat.

While the bituminous solution is still tacky, slide into place $3/4$ in. slabs of foam so that a small quantity of solution is forced up between the joints as a seal. It is recommended that Poron grade P.30 should be used as this has a higher compressive strength than the standard P.15 grade.

A top screed of cement-sand in the ratio of 3:1, incorporating a wire mesh or expanded metal reinforcement, should be laid over the Poron to a thickness of 2 in.

5. Use of Beads

The use of expanded polystyrene beads to "dilute" concrete is a valuable development in concrete flooring technology. A number of advantages result from this development. First, the "K" value of the concrete is considerably reduced. Second, moisture resistance and absorption (the cells for this purpose are noncommunicating) are reduced. Finally, there is a considerable weight reduction. Expanded polystyrene beads at 2 lb/ft³ may be compared with a stone aggregate at 140 to 150 lb/ft³. The importance of this in structural steelwork and load-bearing wall considerations is obvious.

The following figures, based on information given by Dr. W. D. Brown of Monsanto Chemicals Ltd., illustrate the change in strength with varying polystyrene bead content.

6. Strength of Polystyrene-Concrete Mixtures

| Parts by volume | | | Bulk density | Crushing strength |
Granules	Cement	Sand	(lb/ft³)	(lb/in.²)
4	1	—	21	23
3	1	—	28	70
2	1	—	52	350
1	1	—	76	600
12	1	3	28	20
8	1	3	42	46
4	1	3	63	150
2	1	3	88	380
3	1	2	79	260

In mixing the screed, the water ratio should be sufficient only to give a workable mix because an excess of water will give a tendency for the granules to "float" to the surface. Concrete must be "placed" and not "dumped" into position. By the use of reinforcement, extremely lightweight slabs capable of high load bearing can be obtained.

7. Advantages of Foams

An outline of foam technology has been given to enable those concerned with floors to reconsider their designs and views thereon. The use of expanded plastics is not extensive at the moment, but there can be little doubt that their application will be extended when the properties and advantages of the materials have become better understood.

The biggest advantage of the use of foams is to improve heat insulation, and the improvement is most marked. Sound insulation can also be improved and moisture ingress reduced. A weight reduction can also be obtained when expanded polystyrene is mixed with cement.

One or another of these advantages can be achieved in most installations. The cost, of course, is increased, at least as far as raw materials are concerned, but the saving, in terms of heat conservation, quickly offsets this. All flooring manufacturers would do well to consider these new materials.

Chapter 27

POLYMERS IN SANDWICH CONSTRUCTION

R. Hussein

TABLE OF CONTENTS

I. INTRODUCTION

There is growing interest in the development and applications of composite construction. Sandwich construction as a special form of composite materials is based on the concept of combining dissimilar materials to form a composite, fulfilling specific design requirements.

Sandwich construction is characterized by the use of two thin layers of strong material, denoted as faces, between which a thick layer of lightweight and comparatively weak core is sandwiched (Figure 1). In a structural sandwich panel, the faces resist bending moments and in plane compressive or shear forces. The shear forces normal to the plane of the panel are resisted by the core, which also stabilizes the faces against buckling (Figure 2). This type of construction is efficient structurally due to the large stiffness achieved by spacing apart the most highly stressed elements, namely, the faces. The basic principle is much the same as that of an I-beam.

Sandwich construction has found many applications in the aerospace, naval vessels, railway engineering, and building industry.

II. SANDWICH PANELS IN THE BUILDING INDUSTRY

Answers to many problems related to buildings may be found in sandwich panels which offer several advantages. Panel virtues related to environmental requirements may include structural integrity, durable finishes, weather tightness, dimensional stability, and sound or microwave absorption. From the point of view of structural performance, sandwich construction is an efficient structural design due to the large stiffness achieved. Other advantages of sandwich-type construction may include high strength-to-weight ratios, increased fatigue life, endurance, low moisture permeability, electrical insulation, color processability, reduced shop labor, and potentially lower costs of house manufacturing and transportation.

As any other component, there are some factors to be considered. When the core or facing materials may lead to corrosion problems, special pretreatments are required. No foam is fireproof, but many of them can be made nonflammable. Polyester sandwich panels were tested, both painted and unpainted, by the Forest Products Laboratory. It was found that the unpainted panels deteriorated the most in three years of weathering, the edgewise compressive strength reduced by 40%, and the flexural strength reduced by 30%. Painting the other polyester panels reduced the loss in edgewise compressive strength to 22%, but the reduction in flexural strength was still 30%. Similar effects on sandwich panel strengths were observed with moisture contents. Although several advantages can be attributed to most plastic foams, their resistance to chemical agents should be considered in the manufacturing process.

A. MATERIAL REQUIREMENTS

A sandwich panel in building systems should combine the features stated previously. Sandwich constituents contribute to achieve these features as well as fulfilling the basic principle of this type of construction. A list of the required properties of each layer in a sandwich panel is presented in Table 1.

B. PANEL MATERIALS

Sandwich panel developments provide the opportunity of not favoring any one material, but rather employing most of the available materials. Therefore, a desired design requirement can be achieved with many more alternatives of material combinations.

Some of the materials that have been or might be used are listed in Table 2.

C. POLYMERS IN SANDWICH PANELS

Combining several materials into one composite unit is not a new idea. Ancient Egyptians understood by intuition and experience the principles of a laminated system. Spaced facings

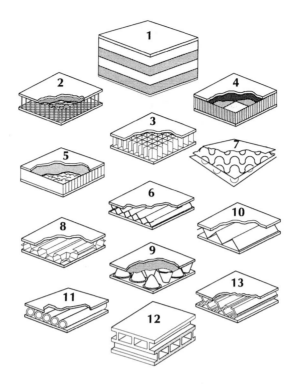

FIGURE 1. Various types of sandwich panels.

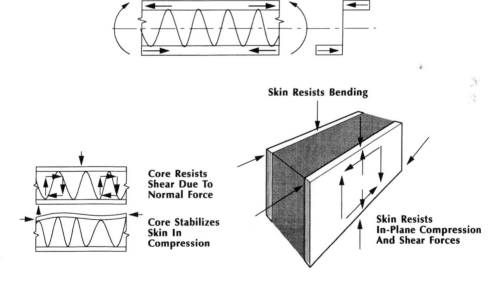

FIGURE 2. Structural action of a true sandwich panel.

were developed by a Frenchman named Dulcan in 1820. During World War II, demands for efficient use of labor and materials imposed by the aircraft industry resulted in promoting the use of sandwich panels with plywood faces and honeycomb cores. However, with today's advanced technologies, new materials, most of which are polymers, are developed. This resulted in an impact on the area of sandwich construction.

TABLE 1
Required Properties for Sandwich Constituents

Panel	Skin	Core	Bonding	Frame	Connections
Strength	Strength	Strength	Strength	Strength	Cheap
Rigidity	Resistance to	Rigidity	High peel	Rigidity	Simple
Repairability	local damage	Impermeability	strength	Insulation	Structural
Durability	Weather resis-	Nonrotting	Durability	Repairability	Weathertight
No loss of	tance	Fire resistance	Low creep	Weather resis-	
strength up to	Nonrotting	Sound absorp-	No loss of	tance	
high tempera-	Fire resistance	tion	strength up to	Sound absorp-	
ture	Impermeability	Insulation	high tempera-	tion	
Minimum	Durability	Durability	ture		
weight	Reflective	Minimum	Fatigue		
Fatigue	Sound Absorp-	weight	Long life		
Vibration	tion	Incombustible	Suitable for		
		Can accept	varied mate-		
		adhesion	rials		
			Adhesion		
			maintenance		
			during contact		
			with moisture		
			or water		
			Chemical sta-		
			bility regard-		
			ing both faces		
			and core		

TABLE 2
Materials for Sandwich Panels

Faces	Core	Bonding	Frame	Connections
Plywood	Kraft paper honey-	Latex	Timber	Nails
Gypsum board	comb	Neoprene base con-	Steel	Screws
High pressure lami-	Balsa wood	tact cement	Aluminum	Bolts
nates	Lightweight con-	Epoxy resins		Locking devices
Painted steel sheet	crete			Coverstrip
Stainless steel sheet				
Asbestos				

Sandwich panels may be made entirely of plastics or a combination of plastic with other materials. For examples, the core may be made of rigid foams and skins of aluminum. A list of other plastics for the use in sandwich panels is given in Table 3.

The behavior of a sandwich panel depends on its constituents, its geometry, and the applied technology in the manufacturing process. The virtues of some polymers for use in sandwich panels will be discussed next.

D. FACES

Among the synthetic materials often used for the faces of sandwich panels are:

- Polyvinyl chloride (PVC) and vinyl chloride copolymers
- Polymethyl methacrylate (PMMA)
- Laminated wood-plastic
- Fiber-reinforced plastics (FRP)

TABLE 3
Polymers for Sandwich Panels

Faces	Core	Bonding
Phenolic-asbestos laminates	Fiberglass	Latex
Epoxy-glass cloth laminates	Expanded polystyrene	Neoprene base contact cement
Polyester-fiberglass fiber laminates	Expanded polyurethane	Alkyd
PVC	Expanded PVC	Acrylate
PMMA	Acrylonitrile styrene	Casein
Laminated wood-plastic	Cellulose acetate	Cellulose-nitrate
FRP	Epoxy	Cellulose-vinyl
Polyamide (nylon)	Methylmethacrylate styrene	Epoxy
Plastic clad plywood	Phenolic	Epoxy-novalac
	Polyethylene	Epoxy-phenolic
	Polypropylene	Epoxy-polyamide
	Polystyrene	Epoxy-polysulfide
	Polyvinyl chloride	Epoxy-silicone
	Silicone	Melamine formaldehyde
	Urea formaldehyde	Phenolic
	Urethane	Phenolic-butadiene acrylonitrile
		Phenolic-neoprene
		Phenolic-neoprene
		Phenolic-vinyl
		Polacrylonitrile
		Polamide
		Polyethylene
		Polyimide
		Polyvinyl acetate
		Polyvinyl butyral
		Resorcinol-phenol formaldehyde
		Silicone
		Vinyl butyral-phenolic
		Vinyl copolymers
		Urea formaldehyde
		Urethane

E. POLYVINYL CHLORIDE (PVC)

Unplasticized PVC is a hard horny material, insoluble in most solvents and not easily softened. If it is mixed while hot with certain plasticizers, it forms a rubbery material.

PVC has a tendency to liberate hydrochloric acid, particularly at temperatures approaching the upper limit of serviceability. To prevent this reaction, stabilization with acid neutralizers or absorbers is necessary. PVC is nonflammable, moisture-resistant, odorless, and tasteless. It offers exceptional resistance against a great number of corrosive media.

Advances in rigid PVC formulation and in processing equipment design have made available a new dimension in building panels to the building industry. PVC elements lend themselves admirably for use in roofing, siding, various shelter and enclosure structures, and interior paneling. In fact, they have already caused some revolutionary changes in building concepts.

Rigid PVC panels have some outstanding advantages as follows: attractive appearance, weatherability, light weight, corrosion and fire resistance, easy to clean, and easy installation. New rigid PVC, resistant to higher temperature, contains chlorinated PVC; these plastics may be used at temperatures up to 100°C.

Copolymerization of vinyl chloride with other vinyl monomers leads to some products with improved properties.

TABLE 4
Fracture Energy

Material	Energy to propagate a crack (J/cm²)
Glass	0.061
Epoxy resin	0.61
Metal	183
Glass-reinforced epoxy	
Parallel fiber	3.1
Perpendicular	430

F. POLYMETHYL METHACRYLATE (PMMA)

This is a clear colorless transparent plastic with a high softening point, good impact strength, and weatherability. Sheets of PMMA are commonly made by extrusion.

The resistance of PMMA to outdoor exposure is outstanding. In this respect, this polymer is markedly superior to other thermoplastics. It has a low water absorption, and the abrasion resistance is roughly comparable with that of aluminium. Like other polymers, PMMA is thermally insulating.

In sheet form, PMMA and other acrylic polymers can be softened at high temperatures and formed (usually vacuum formed) into practically any desired three-dimensional shape.

These polymers have excellent transparency (total luminous transmission up to 93%) while polystyrenes in their initial stage rarely exceed 88 to 90%. Due to the latter characteristics, PMMA have been used for many lighting and glazing applications.

G. PLASTIC CLAD PLYWOOD

The properties of this well-known and proven product are essentially those of solid wood. But the cross-plies greatly increase strength and stability in the cross-grain direction at the expense of some reduction in the grain direction.

Plastic clad plywood touches almost all of us. This laminate is the familiar work surface of kitchens and many restaurant tables. Again, the advantages of the plywood and light weight are enhanced for the specific use — hardness, decorative quality, and wear resistance of the plastic outer layer.

H. COMPOSITES

Composite materials are being applied in increasing amounts for essentially the same reasons as for other applications. Namely, they provide properties and behavior not attainable in single-phase materials, or they provide these features more efficiently, at lower cost, or both.

The types of composite materials or composite structures principally found in building are fibrous: fibers embedded in a continuous matrix laminar; layers of materials bonded together and possibly interpenetrated by a building material particulate; particles embedded in a continuous matrix.

An important subcategory of the laminar class of composites is the structural sandwich.

Composites have three outstanding advantages: (1) ease of fabrication, (2) high fracture energy, and (3) potential low cost.

The latter is particularly true for glass-reinforced resins, glass being by far the most common fiber used in resin-matrix composites. Added advantages of the resin-matrix composites often include low density, low electrical and thermal conductivity, transluscence aesthetic, color effects, and corrosion resistance. The resin-matrix composites are highly formable, and their fracture energy is enormous. Table 4 illustrates the unusual fracture energy possibilities of resin-matrix composites.

TABLE 5
Properties of the More Common Fibers Used in
Reinforced Plastics

Material	Tensile strength (N/cm² × 10³)	Modulus (N/cm² × 10⁶)	Density (g/cm³)
E glass	345	7.2	2.55
S glass	450	8.6	2.5
PRD-49-III	275	13	1.45
Boron	275–310	38–41	2.4
Carbon	103–310	69–62	1.4–1.9
Steel wire	206–512	20	7.7–7.8

TABLE 6
Formulation of E Glass

Constituent	Content (%)
Silicon dioxide	52–56
Calcium oxide	16–25
Aluminium oxide	12–16
Boron oxide	8–13
Sodium and potassium oxides	0–1
Magnesium oxide	0–6

Mechanical properties of a composite depend primarily on the type of fibers, their quantity in the laminate, and their direction. The type of fiber, i.e., its tension strength and modulus, determine these properties in the finished laminate. The percentage value of these fibers is directly proportional to the tensile strength and modulus. These are short fiber reinforcements such as wollastonite, asbestos, or continuous filament reinforcements, such as glassfiber, basalt fibers, high modulus organic fibers, aluminium oxide, and other ceramic and metal filaments.

The more important fibrous materials for advanced composites are compared in Table 5.

There are several glass formulations that have been used as fibers in polymeric composites. These include A, C, D, E, M, and S glass. In terms of advanced composites, E and S glass are the most important. E glass, of low alkali oxide content (composition given in Table 6), has become the most widely used formulation for both textile and industrial applications.

Between fiber and matrix (resin), there is an interfacial bond (a coupling agent) whose function seems to prevent the complete adhesion of the fiber to resin so that relative motion is possible.

The cracks, instead of penetrating through the matrix and directly through the fibers, are deflected along the length of the fibers.

The most common coupling agents for glass are organosilanes and chrome complexes; the latter is cheaper and preferred in a great number of applications.

The availability of a wide variety of manufacturing techniques has undoubtedly been an important factor in the steady growth of the reinforced plastics industry, and very likely, it will exert considerable influence on the progress of advanced fibrous composites and sandwich panels.

Table 7 shows the relative efficiencies of various forms of fibrous reinforcement; the superiority demonstrated by the continuous-oriented fiber structures is apparent.

TABLE 7
Typical Composite Efficiencies Attained in Reinforced Plastics

Fiber configuration	Fiber length	Total fiber content (by volume) V_f	F_{long}[a] ksi (N/cm^2 × 10^3) F_{theor}[b]	F_{test}[c]	Composite[d] efficiency (%)
Filament-wound (unidirectional)	Continuous	0.77	310(214)	180(124)	58.0
Cross-laminated fibers	Continuous	0.48	197(136)	72.5(50.0)	36.8
Cloth-laminated fibers	Continuous	0.48	197(136)	43.0(29.6)	21.8
Mat-laminted fibers	Continuous	0.48	197(136)	57.2(39.4)	29.0
Chopped fiber systems (random)	Noncontinuous	0.13	60.7(41.8)	15.0(10.3)	24.7
Glass flake composites	Noncontinuous	0.70	165.5(114.1)	20.0(13.8)	12.1

[a] F_{long} = Ultimate tensile strength in direction of greatest fiber content (longitudinal), if there is one.
[b] Theoretical strength based on "Rule of Mixtures": $F_{theor} = V_f S_f + (1 - V_f)S_m$, where $S_f = 400/(275.8)$ksi(N/cm^2 × 10^3) = typical boron or carbon fiber strength, and $S_m = 10/(6.9)$ksi(N/cm^2 × 10^3) = typical resin strength.
[c] F_{test} = typical experimental strength values.
[d] Composite efficiency = (F_{test}/F_{theor}) × 100.

It has already been pointed out that the properties of a fibrous composite depend upon the characteristics of fibers, matrix, and coupling agent. Several polymeric matrices are used and may be divided into thermosets and thermoplastics. Sometimes, polyblends are also used.

Cross-linked polyesters, phenol formaldehyde, melamine formaldehyde resins, epoxide, and silicone polymers are among the most common in the thermosetting field. The characteristics of these polymers may be seen in Table 8.

The second group, the thermoplastics, is poor in one or more of the following engineering properties: creep resistance rigidity and tensile strength, dimensional stability, impact strength, maximum service temperatures, and hardness. The use of reinforcement is one way of overcoming some of these difficulties. Table 9 shows some of the relative strength-to-weight ratios of various materials and illustrates well the advantages to be gained by using thermoplastics in composite form. Polyamide nylon 66 is a good example of a thermoplastic amendable to modification. A lot of properties can be improved by incorporating 20 to 40% fiber glass.

In many cases, the new composites have to demonstrate their superiority over timber and its derived products, such as hardboard, chipboard, and plywood. Polymer composites have the advantages that they are generally immediately ready for use and do not require finishing processes. Moreover, they usually need very little maintenance after exposure to weather.

I. CORE

Different core materials are used: foamed or cellular plastics, honeycomb, and others.

Cellular structural materials are produced by making "solid" foams from various polymers, such as phenolic resins, cellulose acetate, polystyrene, polyvinyl chloride, polyurethane, and others. Depending on manufacturing methods and composition, these structures may have different properties (Table 10).

Polystyrene and polyurethane are used to make a wide variety of foamed products. Most polystyrene foams are based on foamed in-place beads, made for suspension polymerization

TABLE 8
Characteristics and Use of Reinforced Thermosetting resins

Resins	Characteristics	Uses	Limitations
Diallyl phthalate polymer	Good electrical properties, dimensional stability, chemical and heat resistance	Prepegs, ducting, radomes, aircraft, missiles	
Epoxy	Good electrical properties, chemical resistance, high strength	Printed circuit board tooling, filament winding	Require heat curing for maximum performance
Melamine formaldehyde	Good electric properties, chemical and heat resistance	Decorative, electrical (arc and track resistance), circuit breakers	
Phenolic	Low cost, chemical resistance, good electrical properties, heat resistance, nonflammable, can be used to 350–400°F (177–204°C)	General diverse mechanical and electrical applications	Dissolve in caustic unless specially treated
Polyester	Good all-round properties, ease of fabrication, low cost, versatile	Corrugated sheeting, seating, boats, automotive, tanks and piping, aircraft, tote boxes	Degraded by strong oxidizers, aromatic solvents, concentrated caustic
Silicone (qv)	Heat resistance, good electrical properties	Electrical, aerospace	

TABLE 9
Relative Strength-to-Weight Ratios of Various Types of Materials

Material	Strength-to-weight ratio relative to polycarbonate
Polycarbonate	1.00
Polystyrene	1.09
Nylon	1.24
Styrene-acrylonitrile copolymers	1.48
Brass (yellow cast)	1.52
Zinc alloys (cast)	1.67
Glass/polystyrene	1.71
Glass/polycarbonate	1.76
Glass/styrene-acrylonitrile copolymers	1.95
Magnesium	2.19
Aluminium	2.52
Glass/nylon	2.62

in the presence of a foaming agent. Subsequent heating softens the resin and volatizes the foaming agent.

Polyurethane foams are expanded by a somewhat different chemical reaction. These expanded plastics may have separate interconnected or partially interconnected cells. Rigid polyurethane foams are resistant to compression and may be used to reinforce hollow structural units with a minimum of weight. In addition, they consist of closed cells and have low rates of heat transmission. They develop excellent adhesion when they are formed in voids or between sheets of material. Finally, they are resistant to oils and do not absorb appreciable amounts of water. These properties make the rigid foams valuable for prefabricated sandwich structures used in the building industry for thermal insulation.

TABLE 10
Important Properties for Core Materials

1. Relative weight
2. Flexibility or rigidity
3. Preparation from thermosetting or thermoplastic polymer
4. Presence or absence of plasticizers
5. Method of cells formation
6. Nature of cells in the solid foam
7. Whether foamed in mold in place or freely expanded

Polymers were used in an experimental study of sandwich foam coextrusion using a sheet-forming die with a feedblock. The polymers used in the experiment were low density polyethylene for the outer layers and ethylene-vinyl-acetate copolymer for the foamed core component. The study has shown that cell size and its distribution in the foamed core and the mechanical properties of the sandwiched foam product can be controlled by a judicious choice of the thickness ratio of the core to the skin components, melt extrusion temperature, and concentration of chemical blowing agent.

The production of panels by foaming is placed between two skins, takes advantage of the capability of polyurethanes to enter the molds as a liquid or semiexpanded froth, and subsequently expands to fill every section of the mold cavity. Obviously, this provides a degree of freedom in panel design that is not available with lamination from slab stock. Complex shaped panels which would require an impractical amount of cutting and fitting with slab are produced rather simply by foaming in place.

Honeycomb core consists of thin foils in the form of hexagonal cells perpendicular to the faces.

Treatment of the draft paper with phenolic resin assures adequate durability if the panels are made and assembled in place to remain dry, or at least drain freely at edges and joints. Adhesives should be chosen to ''wet'' the honeycomb and skin in order to form a fillet to distribute the shear stresses on the thin paper edges. Metal skins can be fully stabilized by the high strength of these hexagonal paper structures, but the open cells provide little bending resistance. For many uses, paper-phenolic honeycomb remains the standard for structural cores.

III. BONDING

These are the bonding elements between the faces and core of a sandwich structure. The bonded core-face allows two thin faces to be used structurally together with the core, making the panel tough.

There are a lot of synthetic polymers that may be used to manufacture sandwich panels. They may be thermosets or thermoplastics as follows:

Urea-formaldehyde polymers — They are the cheapest, but have inferior properties because of their sensitivity to moisture.

Phenol-formaldehyde resins and their derivatives — They are cheap, but need heating or acid catalysts, which may attack faces and/or core.

Resorcinol-formaldehyde resins — They may be used at room temperature with neutral catalysts that do not influence the panel elements.

Unsaturated polyesters — They form the basis of glass-reinforced plastics used for constructional purposes, but are also marketed as adhesives.

Epoxy polymers — They have good adhesion with metals and good mechanical resistance, but are expensive. They also have the same drawbacks with the unsaturated polyesters, i.e., to be made of two different components.

Acrylic copolymers — They may be used as adhesives which cure through solvent evaporation or mixture of monomers.

Polychloroprene — It is the most used elastomer adhesive; it is cheap and has the advantage of no time restriction to assemblage.

Polyblends — Like mixture of epoxy polymers and elastomers (polychloroprenes), they appear good, are resistant to water, but are expensive. Their use eliminates the necessity of pressing after gluing.

Adhesives for sandwich panels are prepared as solutions, latices, or films. With the last one, a constant thickness will be needed on the glued surface to ensure that the strength is uniform. However, films are more expensive than liquid adhesives and are useful only for honeycomb cores.

Recommended curing conditions will vary depending on the facing and core material. In many cases, modification of a given curing procedure can be brought about if advantageous to the producer. The adhesive manufacturer should be consulted for specific recommendations on his adhesive system, and these recommendations should be followed exactly.

IV. STRUCTURAL DESIGN OF SANDWICH PANELS

The structural design of thin face sandwich panels will be discussed in this section. The faces are considered thin, which means that the flexural rigidity of each face about its own middle plane is negligible. The core resists only vertical shear stresses. The sandwich panel is made of two faces of equal thickness and of the same material. Materials are homogeneous, isotropic, and linearly elastic.

Criteria for thin faces sandwich panels are

$$\frac{h}{t_f} > 5.75 \tag{1}$$

where t_f = face thickness, t_c = core thickness, and $h = t_f + t_c$.

The flexural rigidity of the core is negligible in comparison with that of faces if

$$\frac{E_f}{E_c} \frac{t_f}{t_c} \left(\frac{h}{t_c}\right)^2 > 16.70 \tag{2}$$

where E_f = Young's modulus of the faces material; and E_c = Young's modulus of the core material.

The shear stress in the core is uniform over the depth of it if

$$\frac{E_f}{E_c} \frac{t_f}{t_c} \frac{h}{t_c} > 25 \tag{3}$$

The flexural stiffness of a sandwich panel, D_f, is given by

$$D_f = \frac{E_f t_f h^2}{2(1 - v^2)} \tag{4}$$

where v = Poisson's ratio of the face material.

The twisting stiffness, D_t, is given by

$$D_t = \frac{D_f}{2} (1 - v) \tag{5}$$

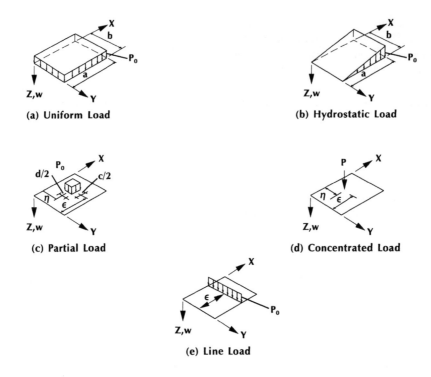

(a) Uniform Load

(b) Hydrostatic Load

(c) Partial Load

(d) Concentrated Load

(e) Line Load

FIGURE 3. Load types on simply supported sandwich plate.

The shear stiffness of the panel, S, may be calculated from the following formula

$$S = G \frac{h^2}{t_c} \tag{6}$$

where G = Shear modulus of the core material.

A. BENDING OF SIMPLY SUPPORTED SANDWICH PANELS

Consider a sandwich panel simply supported along the four edges and subjected to a transverse load. The deflection at the center of this panel may be calculated from the following general formula:

$$w = \frac{Po^{a^4}}{D_f} \quad K_{wb} + \frac{Po^{a^2}}{S} \quad K_{ws} \tag{7}$$

where a = sandwich panel length (Figure 3), Po = load intensity, and K_{wb}, K_{ws} = deflection factors. It can be seen from this equation that two terms are to be added: the bending and shear components of the deflection, and each involves two factors. The first contains parameters reflecting the material properties and panel geometry, and the second the aspect ratio of the plate. To facilitate the use of Equation 7, numerical results are calculated for K_{wb} and K_{ws} and presented in the second and third columns of Tables 11 to 35. Five types of transverse load are considered: uniformly distributed, hydrostatic pressure, partial load on rectangular area, concentrated load, and line load uniformly distributed across the plate width. In the case of concentrated and line loads, the following equations should be used instead of Equation 7.

TABLE 11
Numerical Values for a Simply Supported Sandwich
Plate Under Uniform Distribution Load

R	K_{wb}	K_{ws}	K_{qx}	K_{qy}	K_{mx}	K_{my}	K_{mxy}
1.0000	.0041	.0737	.3357	.3357	.0479	.0479	− .0325
1.1000	.0033	.0666	.3129	.2958	.0408	.0459	− .0267
1.2000	.0027	.0602	.2921	.2621	.0348	.0435	− .0220
1.3000	.0022	.0545	.2732	.2333	.0298	.0411	− .0182
1.4000	.0018	.0494	.2561	.2087	.0256	.0385	− .0152
1.5000	.0015	.0448	.2407	.1875	.0222	.0361	− .0127
1.6000	.0013	.0407	.2267	.1690	.0193	.0337	− .0107
1.7000	.0011	.0371	.2141	.1530	.0168	.0314	− .0091
1.8000	.0009	.0339	.2027	.1389	.0148	.0293	− .0078
1.9000	.0007	.0310	.1923	.1266	.0131	.0273	− .0067
2.0000	.0006	.0285	.1828	.1158	.0116	.0254	− .0058
3.0000	.0002	.0136	.1217	.0545	.0045	.0132	− .0018
4.0000	.0001	.0078	.0908	.0310	.0024	.0077	− .0007
5.0000	.0000	.0050	.0722	.0199	.0015	.0050	− .0004

Note: R = a/b
v = .30

TABLE 12
Numerical Values for a Simply Supported Sandwich
Plate Under Uniform Hydrostatic Pressure

R	K_{wb}	K_{ws}	K_{qx}	K_{qy}	K_{mx}	K_{my}	K_{mxy}
1.0000	.0020	.0368	.1678	.1678	.0239	.0239	− .0162
1.1000	.0017	.0333	.1565	.1479	.0204	.0229	− .0133
1.2000	.0014	.0301	.1461	.1310	.0174	.0218	− .0110
1.3000	.0011	.0273	.1366	.1167	.0149	.0205	− .0091
1.4000	.0009	.0247	.1281	.1044	.0128	.0193	− .0076
1.5000	.0008	.0224	.1203	.0937	.0111	.0180	− .0064
1.6000	.0006	.0204	.1134	.0845	.0096	.0168	− .0054
1.7000	.0005	.0186	.1070	.0765	.0084	.0157	− .0046
1.8000	.0004	.0169	.1013	.0695	.0074	.0146	− .0039
1.9000	.0004	.0155	.0961	.0633	.0065	.0136	− .0033
2.0000	.0003	.0142	.0914	.0579	.0058	.0127	− .0029
3.0000	.0001	.0068	.0609	.0273	.0023	.0066	− .0009
4.0000	.0000	.0039	.0454	.0155	.0012	.0039	− .0004
5.0000	.0000	.0025	.0361	.0100	.0008	.0025	− .0002

Note: R = a/b
v = .30

$$w = \frac{Pa^2R}{Df} K_{wb} + \frac{PR}{S} K_{ws} \qquad \text{concentrated load} \qquad (8)$$

$$w = \frac{2Po a^3}{Df} K_{wb} + \frac{2Po a}{S} K_{ws} \qquad \text{line load} \qquad (9)$$

where P = concentrated load, and R = aspect ratio of the plate.

Concerning the bending moments at the center of the plate and the twisting moment at its corners, for each case of the loading types, the following equations should be applied.

TABLE 13
Influence Coefficients for Simply Supported Sandwich Plates Under Partial Load

R	K_{wb}	K_{ws}	K_{qx}	K_{qy}	K_{mx}	K_{my}	K_{mxy}
1.0000	.0001	.0019	.0160	.0160	.0012	.0012	− .0054
1.1000	.0001	.0017	.0164	.0128	.0010	.0012	− .0044
1.2000	.0001	.0015	.0167	.0104	.0008	.0012	− .0037
1.3000	.0001	.0014	.0168	.0084	.0006	.0012	− .0031
1.4000	.0001	.0012	.0168	.0069	.0005	.0011	− .0026
1.5000	.0000	.0011	.0167	.0058	.0004	.0010	− .0022
1.6000	.0000	.0009	.0165	.0048	.0003	.0009	− .0019
1.7000	.0000	.0008	.0163	.0040	.0002	.0009	− .0017
1.8000	.0000	.0007	.0160	.0034	.0001	.0008	− .0015
1.9000	.0000	.0006	.0157	.0029	.0001	.0007	− .0013
2.0000	.0000	.0006	.0154	.0025	.0001	.0007	− .0011
3.0000	.0000	.0002	.0154	.0006	− .0000	.0003	− .0004
4.0000	.0000	.0001	.0090	.0002	− .0000	.0001	− .0002
5.0000	.0000	.0000	.0068	.0001	− .0000	.0000	− .0001

Note: R = a/b
$\xi' = .17 \; (\xi' = \xi/a)$
$\eta' = .17 \; (\eta' = \eta/b)$
$\upsilon = .30$

TABLE 14
Influence Coefficients for Simply Supported Sandwich Plate Under Partial Load

R	K_{wb}	K_{ws}	K_{qx}	K_{qy}	K_{mx}	K_{my}	K_{mxy}
1.0000	.0002	.0037	.0166	.0551	.0028	.0021	− .0044
1.1000	.0002	.0035	.0158	.0465	.0024	.0021	− .0034
1.2000	.0002	.0032	.0149	.0396	.0021	.0021	− .0027
1.3000	.0001	.0029	.0139	.0340	.0018	.0020	− .0022
1.4000	.0001	.0027	.0130	.0294	.0016	.0020	− .0017
1.5000	.0001	.0025	.0120	.0256	.0013	.0019	− .0014
1.6000	.0001	.0023	.0111	.0224	.0012	.0018	− .0012
1.7000	.0001	.0021	.0102	.0197	.0010	.0017	− .0010
1.8000	.0001	.0019	.0094	.0174	.0009	.0016	− .0008
1.9000	.0000	.0018	.0086	.0155	.0008	.0015	− .0007
2.0000	.0000	.0016	.0079	.0138	.0007	.0014	− .0006
3.0000	.0000	.0008	.0032	.0052	.0002	.0008	− .0001
4.0000	.0000	.0004	.0013	.0024	.0001	.0005	− .0000
5.0000	.0000	.0002	.0006	.0013	.0000	.0003	− .0000

Note: R = a/b
$\xi' = .33 \; (\xi' = \xi/a)$
$\eta' = .17 \; (\eta' = \eta/b)$
$\upsilon = .30$

$$M_x = P_o a^2 K_{mx} \qquad \text{uniform distributed, hydrostatic}$$

$$M_y = p_o a^2 K_{my} \qquad \text{pressure, and partial load on} \qquad (10)$$

$$M_{xy} = -P_o a^2 R K_{mxy} \qquad \text{rectangular area}$$

TABLE 15
Influence Coefficients for Simply Supported Sandwich Plate Under Partial Load

R	K_{wb}	K_{ws}	K_{qx}	K_{qy}	K_{mx}	K_{my}	K_{mxy}
1.0000	.0003	.0046	.0114	.1064	.0037	.0023	$-.0029$
1.1000	.0002	.0044	.0101	.0937	.0033	.0023	$-.0022$
1.2000	.0002	.0041	.0089	.0832	.0030	.0023	$-.0017$
1.3000	.0002	.0038	.0078	.0743	.0027	.0023	$-.0013$
1.4000	.0001	.0036	.0068	.0668	.0024	.0022	$-.0010$
1.5000	.0001	.0034	.0059	.0603	.0022	.0022	$-.0008$
1.6000	.0001	.0032	.0051	.0548	.0020	.0021	$-.0006$
1.7000	.0001	.0030	.0045	.0499	.0019	.0020	$-.0005$
1.8000	.0001	.0028	.0039	.0457	.0017	.0019	$-.0004$
1.9000	.0001	.0026	.0033	.0420	.0016	.0018	$-.0003$
2.0000	.0001	.0025	.0029	.0387	.0015	.0018	$-.0003$
3.0000	.0000	.0015	.0007	.0197	.0008	.0012	$-.0000$
4.0000	.0000	.0010	.0002	.0118	.0005	.0008	$-.0000$
5.0000	.0000	.0007	.0000	.0078	.0003	.0006	$-.0000$

Note: R = a/b
$\xi' = .50$ $(\xi' = \xi/a)$
$\eta' = .17$ $(\eta' = \eta/b)$
$\upsilon = .30$

TABLE 16
Influence Coefficients for Simply Supported Sandwich Plate Under Partial Load

R	K_{wb}	K_{ws}	K_{qx}	K_{qy}	K_{mx}	K_{my}	K_{mxy}
1.0000	.0002	.0037	.0551	.0166	.0021	.0028	$-.0044$
1.1000	.0002	.0033	.0536	.0142	.0016	.0026	$-.0038$
1.2000	.0002	.0029	.0520	.0122	.0013	.0025	$-.0033$
1.3000	.0001	.0025	.0503	.0104	.0010	.0023	$-.0029$
1.4000	.0001	.0022	.0486	.0089	.0008	.0021	$-.0025$
1.5000	.0001	.0019	.0469	.0076	.0006	.0019	$-.0022$
1.6000	.0001	.0017	.0452	.0066	.0004	.0018	$-.0020$
1.7000	.0001	.0015	.0435	.0057	.0003	.0016	$-.0017$
1.8000	.0000	.0013	.0419	.0049	.0002	.0015	$-.0015$
1.9000	.0000	.0011	.0404	.0042	.0002	.0013	$-.0014$
2.0000	.0000	.0010	.0389	.0037	.0001	.0012	$-.0012$
3.0000	.0000	.0003	.0267	.0010	$-.0001$	.0004	$-.0005$
4.0000	.0000	.0001	.0188	.0003	$-.0000$	.0002	$-.0002$
5.0000	.0000	.0000	.0136	.0001	$-.0000$	.0001	$-.0001$

Note: R = a/b
$\xi' = .17$ $(\xi' = \xi/a)$
$\eta' = .33$ $(\eta' = \eta/b)$
$\upsilon = .30$

$$M_x = PR\,K_{mx} \qquad \text{concentrated load}$$

$$M_y = PR\,K_{my} \tag{11}$$

$$M_{xy} = -PR^2 K_{mxy}$$

TABLE 17
Influence Coefficients for Simply Supported Sandwich
Plate Under Partial Load

R	K_{wb}	K_{ws}	K_{qx}	K_{qy}	K_{mx}	K_{my}	K_{mxy}
1.0000	.0005	.0083	.0370	.0370	.0054	.0054	−.0051
1.1000	.0004	.0075	.0337	.0331	.0046	.0052	−.0042
1.2000	.0003	.0066	.0307	.0296	.0039	.0049	−.0035
1.3000	.0003	.0062	.0280	.0266	.0033	.0047	−.0029
1.4000	.0002	.0056	.0254	.0239	.0029	.0044	−.0024
1.5000	.0002	.0050	.0231	.0215	.0025	.0041	−.0020
1.6000	.0001	.0046	.0210	.0194	.0021	.0038	−.0017
1.7000	.0001	.0042	.0191	.0175	.0018	.0036	−.0014
1.8000	.0001	.0038	.0174	.0159	.0016	.0033	−.0012
1.9000	.0001	.0035	.0158	.0144	.0014	.0031	−.0010
2.0000	.0001	.0032	.0144	.0131	.0012	.0029	−.0009
3.0000	.0000	.0014	.0057	.0058	.0004	.0015	−.0002
4.0000	.0000	.0008	.0024	.0029	.0001	.0009	−.0001
5.0000	.0000	.0004	.0010	.0017	.0000	.0005	−.0000

Note: R = a/b
$\xi' = .33$ ($\xi' = \xi/a$)
$\eta' = .33$ ($\eta' = \eta/b$)
$\upsilon = .30$

TABLE 18
Influence Coefficients for Simply Supported Sandwich
Plate Under Partial Load

R	K_{wb}	K_{ws}	K_{qx}	K_{qy}	K_{mx}	K_{my}	K_{mxy}
1.0000	.0005	.0115	.0215	.0493	.0083	.0066	−.0041
1.1000	.0004	.0106	.0187	.0454	.0074	.0064	−.0032
1.2000	.0004	.0098	.0162	.0418	.0065	.0062	−.0025
1.3000	.0003	.0091	.0141	.0386	.0059	.0059	−.0020
1.4000	.0003	.0084	.0121	.0357	.0053	.0057	−.0016
1.5000	.0002	.0078	.0105	.0331	.0048	.0054	−.0013
1.6000	.0002	.0072	.0091	.0307	.0043	.0051	−.0010
1.7000	.0002	.0067	.0078	.0286	.0039	.0048	−.0008
1.8000	.0001	.0063	.0068	.0267	.0036	.0046	−.0007
1.9000	.0001	.0059	.0058	.0249	.0033	.0043	−.0005
2.0000	.0001	.0055	.0050	.0234	.0030	.0041	−.0004
3.0000	.0000	.0032	.0012	.0133	.0016	.0026	−.0001
4.0000	.0000	.0020	.0003	.0085	.0009	.0017	−.0000
5.0000	.0000	.0014	.0001	.0059	.0006	.0013	−.0000

Note: R = a/b
$\xi' = .50$ ($\xi' = \xi/a$)
$\eta' = .33$ ($\eta' = \eta/b$)
$\upsilon = .30$

$$M_x = 2\,P_o a k_{mx}$$

$$My = 2\,P_o a K_{my} \qquad\qquad \text{line load} \qquad\qquad (12)$$

$$M_{xy} = -2\,P_o Ra K_{mxy}$$

TABLE 19
Influence Coefficients for Simply Supported Sandwich
Plate Under Partial Load

R	K_{wb}	K_{ws}	K_{qx}	K_{qy}	K_{mx}	K_{my}	K_{mxy}
1.0000	.0003	.0046	.1064	.0114	.0023	.0037	− .0029
1.1000	.0002	.0040	.0994	.0103	.0018	.0034	− .0026
1.2000	.0002	.0035	.0931	.0093	.0014	.0031	− .0023
1.3000	.0002	.0030	.0874	.0083	.0011	.0028	− .0021
1.4000	.0001	.0026	.0822	.0074	.0008	.0026	− .0018
1.5000	.0001	.0023	.0774	.0066	.0006	.0023	− .0016
1.6000	.0001	.0020	.0730	.0058	.0005	.0021	− .0015
1.7000	.0001	.0017	.0690	.0051	.0003	.0019	− .0013
1.8000	.0001	.0015	.0653	.0045	.0002	.0017	− .0012
1.9000	.0000	.0013	.0619	.0040	.0002	.0016	− .0011
2.0000	.0000	.0012	.0587	.0035	.0001	.0014	− .0010
3.0000	.0000	.0003	.0366	.0010	− .0001	.0005	− .0004
4.0000	.0000	.0001	.0244	.0003	− .0001	.0002	− .0002
5.0000	.0000	.0000	.0171	.0001	− .0000	.0001	− .0001

Note: R = a/b
$\xi' = .17\ (\xi' = \xi/a)$
$\eta' = .50\ (\eta' = \eta/b)$
$\upsilon = .30$

TABLE 20
Influence Coefficients for Simply Supported Sandwich
Plate Under Partial Load

R	K_{wb}	K_{ws}	K_{qx}	K_{qy}	K_{mx}	K_{my}	K_{mxy}
1.0000	.0005	.0115	.0493	.0215	.0066	.0083	− .0041
1.1000	.0004	.0102	.0439	.0201	.0056	.0077	− .0035
1.2000	.0004	.0091	.0392	.0187	.0047	.0071	− .0030
1.3000	.0003	.0081	.0352	.0173	.0040	.0066	− .0025
1.4000	.0002	.0073	.0316	.0159	.0034	.0061	− .0021
1.5000	.0002	.0065	.0284	.0147	.0029	.0056	− .0018
1.6000	.0002	.0059	.0256	.0135	.0025	.0052	− .0016
1.7000	.0001	.0053	.0231	.0124	.0021	.0048	− .0013
1.8000	.0001	.0048	.0209	.0114	.0019	.0044	− .0012
1.9000	.0001	.0044	.0189	.0105	.0016	.0041	− .0010
2.0000	.0001	.0040	.0172	.0097	.0014	.0038	− .0009
3.0000	.0000	.0018	.0068	.0047	.0004	.0019	− .0002
4.0000	.0000	.0009	.0029	.0025	.0001	.0010	− .0001
5.0000	.0000	.0005	.0013	.0015	.0000	.0006	− .0000

Note: R = a/b
$\xi' = .33\ (\xi' = \xi/a)$
$\eta' = .50\ (\eta' = \eta/b)$
$\upsilon = .30$

Values for the factors K_{mx}, K_{my} and K_{mxy} are obtained and presented in Tables 11 to 35.

Stresses are sometimes of interest in many practical problems. The inplane stresses in the faces may be obtained from Equations 10 to 12. For example, the normal stress σ_x and σ_y are calculated by

TABLE 21
Influence Coefficients for Simply Supported Sandwich Plate Under Partial Load

R	K_{wb}	K_{ws}	K_{qx}	K_{qy}	K_{mx}	K_{my}	K_{mxy}
1.0000	.0007	.0182	.0261	.0261	.0118	.0118	−.0037
1.1000	.0005	.0165	.0224	.0248	.0104	.0111	−.0030
1.2000	.0004	.0150	.0193	.0234	.0091	.0104	−.0024
1.3000	.0004	.0137	.0166	.0220	.0081	.0097	−.0020
1.4000	.0003	.0125	.0143	.0207	.0072	.0091	−.0016
1.5000	.0003	.0115	.0123	.0194	.0065	.0085	−.0013
1.6000	.0002	.0106	.0106	.0182	.0059	.0079	−.0011
1.7000	.0002	.0098	.0092	.0171	.0053	.0074	−.0009
1.8000	.0002	.0091	.0079	.0161	.0049	.0070	−.0007
1.9000	.0001	.0085	.0068	.0152	.0044	.0066	−.0006
2.0000	.0001	.0079	.0059	.0144	.0041	.0062	−.0005
3.0000	.0000	.0044	.0014	.0087	.0020	.0037	−.0001
4.0000	.0000	.0028	.0004	.0058	.0012	.0024	−.0000
5.0000	.0000	.0019	.0001	.0041	.0008	.0017	−.0000

Note: R = a/b $\eta' = .50$ ($\eta' = \eta/b$)
$\xi' = .50$ ($\xi' = \xi/a$) $\upsilon = .30$

TABLE 22
Influence Coefficients for Simply Supported Sandwich Plate Under Concentrated Load

R	K_{wb}	K_{ws}	K_{qx}	K_{qy}	K_{mx}	K_{my}	K_{mxy}
1.0000	.0023	.0306	.2556	.2556	.0199	.0199	−.1052
1.1000	.0019	.0275	.2675	.1997	.0157	.0201	−.0864
1.2000	.0015	.0245	.2762	.1561	.0122	.0197	−.0716
1.3000	.0012	.0216	.2821	.1265	.0093	.0189	−.0597
1.4000	.0010	.0190	.2855	.1022	.0070	.0178	−.0501
1.5000	.0008	.0167	.2867	.0833	.0051	.0166	−.0424
1.6000	.0007	.0146	.2859	.0684	.0036	.0154	−.0360
1.7000	.0005	.0128	.2836	.0566	.0025	.0141	−.0308
1.8000	.0004	.0112	.2799	.0471	.0016	.0129	−.0265
1.9000	.0004	.0097	.2750	.0394	.0009	.0118	−.0229
2.0000	.0003	.0085	.2691	.0331	.0004	.0107	−.0199
3.0000	.0000	.0022	.1897	.0068	−.0009	.0037	−.0059
4.0000	.0000	.0006	.1191	.0016	−.0005	.0012	−.0022
5.0000	.0000	.0002	.0720	.0005	−.0002	.0004	−.0010

Note: R = a/b $\eta' = .17$ ($\eta' = \eta/b$)
$\xi' = .17$ ($\xi' = \xi/a$) $\upsilon = .30$

$$\sigma_x \quad \frac{M_x}{ht_f}$$

$$\sigma_y = \frac{M_y}{ht_f} \tag{13}$$

$$\tau_{xy} = \frac{M_{xy}}{ht_f}$$

TABLE 23
Influence Coefficients for Simply Supported Sandwich Plate Under Concentrated Load

R	K_{wb}	K_{ws}	K_{qx}	K_{qy}	K_{mx}	K_{my}	K_{mxy}
1.0000	.0041	.0599	.2712	.8535	.0457	.0321	−.0768
1.1000	.0034	.0557	.2585	.7061	.0393	.0331	−.0592
1.2000	.0028	.0515	.2434	.5878	.0337	.0332	−.0462
1.3000	.0023	.0473	.2269	.4924	.0288	.0327	−.0365
1.4000	.0019	.0434	.2101	.4150	.0246	.0318	−.0292
1.5000	.0016	.0397	.1934	.3518	.0210	.0306	−.0235
1.6000	.0013	.0362	.1773	.2999	.0179	.0292	−.0191
1.7000	.0011	.0331	.1619	.2570	.0153	.0278	−.0157
1.8000	.0009	.0302	.1475	.2213	.0130	.0263	−.0129
1.9000	.0008	.0276	.1341	.1915	.0111	.0248	−.0107
2.0000	.0007	.0253	.1217	.1664	.0094	.0234	−.0089
3.0000	.0002	.0107	.0442	.0495	−.0015	.0124	−.0018
4.0000	.0000	.0049	.0158	.0186	−.0003	.0066	−.0005
5.0000	.0000	.0023	.0058	.0081	−.0006	.0036	−.0001

Note: R = a/b
$\xi' = .33 \ (\xi' = \xi/a)$
$\eta' = .17 \ (\eta' = \eta/b)$
$\upsilon = .30$

TABLE 24
Influence Coefficients for Simply Supported Sandwich Plate Under Concentrated Load

R	K_{wb}	K_{ws}	K_{qx}	K_{qy}	K_{mx}	K_{my}	K_{mxy}
1.0000	.0049	.0759	.1849	1.5505	.0628	.0358	−.0488
1.1000	.0041	.0720	.1641	1.4348	.0568	.0368	−.0365
1.2000	.0034	.0681	.1441	1.3343	.0517	.0368	−.0276
1.3000	.0028	.0642	.1257	1.2465	.0472	.0363	−.0212
1.4000	.0024	.0606	.1090	1.1692	.0433	.0354	−.0164
1.5000	.0020	.0572	.0942	1.1007	.0400	.0343	−.0128
1.6000	.0017	.0541	.0812	1.0396	.0372	.0331	−.0101
1.7000	.0014	.0512	.0699	.9848	.0347	.0318	−.0080
1.8000	.0012	.0485	.0601	.9355	.0326	.0305	−.0063
1.9000	.0011	.0461	.0516	.8908	.0307	.0292	−.0051
2.0000	.0009	.0439	.0443	.8502	.0290	.0280	−.0041
3.0000	.0003	.0294	.0096	.5837	.0190	.0191	−.0005
4.0000	.0001	.0220	.0022	.4440	.0142	.0143	−.0001
5.0000	.0001	.0176	.0006	.3571	.0114	.0114	−.0000

Note: R = a/b
$\xi' = .50 \ (\xi' = \xi/a)$
$\eta' = .17 \ (\eta' = \eta/b)$
$\upsilon = .30$

It is of interest to observe that the expressions for the stresses do not include G, nor any other term which refers to the shear stiffness of the plate. Indeed, not only are the stresses in the faces independent of the shear, but also the results in Equation 13 are identical to those determined by the classical theory of homogeneous plate. This is true due to the

TABLE 25
Influence Coefficients for Simply Supported Sandwich
Plate Under Concentrated Load

R	K_{wb}	K_{ws}	K_{qx}	K_{qy}	K_{mx}	K_{my}	K_{mxy}
1.0000	.0041	.0599	.8535	.2712	.0321	.0457	− .0768
1.1000	.0033	.0524	.8428	.2308	.0249	.0431	− .0674
1.2000	.0027	.0457	.8245	.1960	.0191	.0402	− .0592
1.3000	.0022	.0397	.8012	.1663	.0145	.0372	− .0521
1.4000	.0018	.0345	.7744	.1413	.0108	.0341	− .0460
1.5000	.0014	.0300	.7455	.1202	.0078	.0312	− .0407
1.6000	.0012	.0261	.7155	.1024	.0055	.0284	− .0362
1.7000	.0009	.0226	.6851	.0875	.0037	.0257	− .0322
1.8000	.0008	.0197	.6547	.0749	.0023	.0233	− .0287
1.9000	.0006	.0171	.6248	.0642	.0012	.0210	− .0256
2.0000	.0005	.0149	.5955	.0552	.0004	.0190	− .0230
3.0000	.0001	.0038	.3584	.0136	− .0015	.0065	− .0084
4.0000	.0000	.0010	.2124	.0039	− .0008	.0022	− .0035
5.0000	.0000	.0003	.1252	.0013	− .0003	.0007	− .0016

Note: R = a/b
$\xi' = .17 \ (\xi' = \xi/a)$
$\eta' = .33 \ (\eta' = \eta/b)$
$v = .30$

TABLE 26
Influence Coefficients for Simply Supported Sandwich
Plate Under Concentrated Load

R	K_{wb}	K_{ws}	K_{qx}	K_{qy}	K_{mx}	K_{my}	K_{mxy}
1.0000	.0076	.1324	.5988	.5988	.0861	.0861	− .0874
1.1000	.0062	.1196	.5450	.5365	.0721	.0834	− .0716
1.2000	.0051	.1078	.4942	.4797	.0604	.0797	− .0588
1.3000	.0042	.0970	.4471	.4285	.0506	.0756	− .0485
1.4000	.0035	.0874	.4039	.3827	.0424	.0711	− .0401
1.5000	.0029	.0787	.3645	.3420	.0356	.0667	− .0333
1.6000	.0024	.0709	.3287	.3059	.0299	.0623	− .0278
1.7000	.0020	.0640	.2964	.2738	.0251	.0581	− .0232
1.8000	.0017	.0578	.2671	.2454	.0211	.0540	− .0195
1.9000	.0014	.0523	.2407	.2202	.0178	.0502	− .0165
2.0000	.0012	.0474	.2169	.1979	.0150	.0467	− .0139
3.0000	.0003	.0191	.0768	.0736	.0020	.0228	− .0030
4.0000	.0001	.0085	.0276	.0310	− .0007	.0117	− .0008
5.0000	.0000	.0040	.0108	.0143	− .0011	.0063	− .0002

Note: R = a/b
$\xi' = .33 \ (\xi' = \xi/a)$
$\eta' = .33 \ (\eta' = \eta/b)$
$v = .30$

manner in which a simply supported sandwich plate deforms. Under the applied transverse load, the thin faces of a simply supported sandwich plate undergo uniform extension or contractions as they bend about the middle plane of the whole sandwich plate. In this way, the inplane stresses are uniformly distributed across the thicknesses of the faces. In addition, the sandwich plate undergoes shear strains which correspond to an additional transverse

TABLE 27
Influence Coefficients for Simply Supported Sandwich
Plate Under Concentrated Load

R	K_{wb}	K_{ws}	K_{qx}	K_{qy}	K_{mx}	K_{my}	K_{mxy}
1.0000	.0092	.1894	.3499	1.2182	.1466	.0997	− .0678
1.1000	.0076	.1775	.3029	1.1021	.1328	.0979	− .0529
1.2000	.0063	.1663	.2614	1.0028	.1209	.0952	− .0415
1.3000	.0052	.1559	.2250	.9175	.1108	.0919	− .0326
1.4000	.0044	.1464	.1934	.8437	.1021	.0883	− .0258
1.5000	.0037	.1378	.1660	.7795	.0945	.0846	− .0204
1.6000	.0031	.1299	.1424	.7234	.0881	.0809	− .0163
1.7000	.0026	.1228	.1221	.6740	.0824	.0773	− .0130
1.8000	.0023	.1163	.1046	.6303	.0774	.0738	− .0104
1.9000	.0019	.1104	.0897	.5915	.0731	.0705	− .0084
2.0000	.0017	.1051	.0768	.5569	.0692	.0674	− .0068
3.0000	.0005	.0702	.0166	.3466	.0456	.0457	− .0009
4.0000	.0002	.0526	.0040	.2496	.0341	.0343	− .0001
5.0000	.0001	.0420	.0016	.1945	.0272	.0274	− .0000

Note: R = a/b
$\xi' = .50 \ (\xi' = \xi/a)$
$\eta' = .33 \ (\eta' = \eta/b)$
$\upsilon = .30$

TABLE 28
Influence Coefficients for Simply Supported Sandwich
Plate Under Concentrated Load

R	K_{wb}	K_{ws}	K_{qx}	K_{qy}	K_{mx}	K_{my}	K_{mxy}
1.0000	.0049	.0759	1.5505	.1849	.0358	.0628	− .0488
1.1000	.0039	.0653	1.3815	.1682	.0280	.0569	− .0440
1.2000	.0031	.0564	1.2405	.1512	.0217	.0516	− .0397
1.3000	.0025	.0488	1.1210	.1347	.0167	.0467	− .0358
1.4000	.0020	.0422	1.0184	.1192	.0126	.0423	− .0323
1.5000	.0017	.0366	.9293	.1051	.0094	.0382	− .0292
1.6000	.0013	.0318	.8512	.0924	.0069	.0345	− .0265
1.7000	.0011	.0277	.7821	.0810	.0049	.0311	− .0241
1.8000	.0009	.0241	.7206	.0709	.0033	.0281	− .0219
1.9000	.0007	.0210	.6654	.0620	.0021	.0253	− .0200
2.0000	.0006	.0184	.6155	.0543	.0011	.0228	− .0182
3.0000	.0001	.0052	.2976	.0144	− .0011	.0060	− .0077
4.0000	.0000	.0019	.1419	.0041	− .0005	.0029	− .0035
5.0000	.0000	.0009	.0574	.0013	− .0000	.0012	− .0017

Note: R = a/b
$\xi' = .17 \ (\xi' = \xi/a)$
$\eta' = .50 \ (\eta' = \eta/b)$
$\upsilon = .30$

deflection. The faces accomodate this extra deflection by bending about their own middle planes, as well as by displacing vertically only. This implies that while shear deformations are taking place, the thin faces of a simple supported sandwich plate do not undergo stretching or contraction; thus, they retain their bending stresses unaltered.

The shear forces in a simply supported sandwich plate under each type of the loads considered can be obtained from the following formulas:

TABLE 29
Influence Coefficients for Simply Supported Sandwich
Plate Under Concentrated Load

R	K_{wb}	K_{ws}	K_{qx}	K_{qy}	K_{mx}	K_{my}	K_{mxy}
1.0000	.0092	.1894	1.2182	.3499	.0997	.1466	− .0678
1.1000	.0075	.1659	1.1087	.3283	.0826	.1330	− .0581
1.2000	.0061	.1459	1.0136	.3054	.0687	.1210	− .0496
1.3000	.0050	.1289	.9303	.2825	.0573	.1103	− .0423
1.4000	.0041	.1143	.8567	.2603	.0479	.1007	− .0360
1.5000	.0034	.1017	.7914	.2393	.0402	.0920	− .0307
1.6000	.0028	.0907	.7330	.2197	.0338	.0842	− .0262
1.7000	.0023	.0812	.6807	.2015	.0285	.0771	− .0224
1.8000	.0019	.0729	.6336	.1848	.0241	.0708	− .0192
1.9000	.0016	.0657	.5912	.1695	.0204	.0650	− .0165
2.0000	.0014	.0593	.5528	.1554	.0172	.0598	− .0141
3.0000	.0003	.0238	.3149	.0674	.0030	.0279	− .0034
4.0000	.0001	.0110	.2123	.0310	− .0001	.0144	− .0009
5.0000	.0000	.0056	.1615	.0149	− .0006	.0079	− .0002

Note: R = a/b $\eta' = .50 \ (\eta' = \eta/b)$
$\xi' = .33 \ (\xi' = \xi/a)$ $v = .30$

TABLE 30
Influence Coefficients for Simply Supported Sandwich
Plate Under Concentrated Load

R	K_{wb}	K_{ws}	K_{qx}	K_{qy}	K_{mx}	K_{my}	K_{mxy}
1.0000	.0116	.7057	.6679	.6679	.4587	.4587	− .0610
1.1000	.0095	.6407	.5977	.6130	.4104	.4225	− .0497
1.2000	.0078	.5853	.5369	.5639	.3700	.3909	− .0403
1.3000	.0065	.5376	.4843	.5201	.3359	.3630	− .0326
1.4000	.0054	.4962	.4385	.4811	.3069	.3382	− .0264
1.5000	.0045	.4600	.3987	.4465	.2819	.3161	− .0213
1.6000	.0038	.4281	.3639	.4157	.2604	.2962	− .0173
1.7000	.0033	.3999	.3335	.3883	.2416	.2783	− .0140
1.8000	.0028	.3747	.3069	.3638	.2251	.2621	− .0113
1.9000	.0024	.3522	.2836	.3418	.2105	.2475	− .0092
2.0000	.0021	.3320	.2631	.3221	.1975	.2341	− .0075
3.0000	.0006	.2055	.1508	.2008	.1189	.1483	− .0010
4.0000	.0003	.1448	.1095	.1445	.0823	.1058	− .0002
5.0000	.0001	.1097	.0884	.1126	.0616	.0810	− .0000

Note: R = a/b $\eta' = .50 \ (\eta' = \eta/b)$
$\xi' = .50 \ (\xi' = \xi/a)$ $v = .30$

$$Q_x = P_o a K_{Qx} \qquad \text{uniform distributed, hydrostatic pressure,}$$
$$Q_y = P_o a R K_{Qy} \qquad \text{and partial load on rectangular area} \tag{14}$$

$$Q_x = \frac{PR}{a} K_{Qx} \qquad \text{concentrated load}$$
$$Q_y = \frac{PR^2}{a} K_{Qy} \tag{15}$$

TABLE 31
Influence Coefficients for Simply Supported Sandwich Plate Subjected to Line Load

R	K_{wb}	K_{ws}	K_{qx}	K_{qy}	K_{mx}	K_{my}	K_{mxy}
1.0000	.0009	.0127	.4210	.0524	.0067	.0098	.0288
1.1000	.0007	.0110	.4118	.0436	.0051	.0093	.0247
1.2000	.0006	.0096	.4026	.0365	.0038	.0086	.0214
1.3000	.0005	.0083	.3934	.0306	.0028	.0080	.0187
1.4000	.0004	.0072	.3843	.0258	.0020	.0073	.0165
1.5000	.0003	.0062	.3753	.0218	.0014	.0067	.0146
1.6000	.0003	.0053	.3664	.0185	.0009	.0060	.0130
1.7000	.0002	.0046	.3576	.0157	.0005	.0055	.0116
1.8000	.0002	.0040	.3489	.0133	.0002	.0049	.0104
1.9000	.0001	.0034	.3404	.0114	.0000	.0044	.0094
2.0000	.0001	.0029	.3320	.0097	−.0001	.0039	.0085
3.0000	.0000	.0007	.2565	.0021	−.0004	.0012	.0037
4.0000	.0000	.0002	.1965	.0005	−.0002	.0004	.0019
5.0000	.0000	.0000	.1504	.0001	−.0001	.0001	.0011

Note: $R = a/b$
$\xi' = .10 \ (\xi' = \xi/a)$
$\upsilon = .30$

TABLE 32
Influence Coefficients for Simply Supported Sandwich Plate Subjected to Line Load

R	K_{wb}	K_{ws}	K_{qx}	K_{qy}	K_{mx}	K_{my}	K_{mxy}
1.0000	.0018	.0265	.3006	.1161	.0148	.0196	.0330
1.1000	.0014	.0233	.2849	.0975	.0117	.0186	.0276
1.2000	.0012	.0204	.2696	.0823	.0091	.0174	.0233
1.3000	.0010	.0179	.2547	.0698	.0071	.0162	.0198
1.4000	.0008	.0157	.2404	.0595	.0054	.0150	.0170
1.5000	.0006	.0137	.2266	.0508	.0041	.0138	.0146
1.6000	.0005	.0120	.2135	.0436	.0030	.0126	.0126
1.7000	.0004	.0105	.2009	.0375	.0021	.0115	.0110
1.8000	.0003	.0092	.1889	.0324	.0015	.0105	.0096
1.9000	.0003	.0080	.1775	.0280	.0009	.0095	.0084
2.0000	.0002	.0070	.1667	.0243	.0005	.0086	.0074
3.0000	.0000	.0020	.0861	.0064	−.0006	.0031	.0024
4.0000	.0000	.0006	.0417	.0019	−.0004	.0011	.0009
5.0000	.0000	.0002	.0177	.0006	−.0002	.0004	.0004

Note: $R = a/b$
$\xi' = .20 \ (\xi' = \xi/a)$
$\upsilon = .30$

$$Q_x = 2P_o K_{Qx} \qquad \text{line load}$$
$$Q_y = 2P_o R K_{Qy} \qquad\qquad\qquad\qquad\qquad (16)$$

Numerical values for the coefficient K_{Qx} and K_{Qy} are evaluated and shown in Tables 11 to 35. While the transverse shear force Q_x is greatest at the middle of the sides of length b, the transverse shear force Q_y is greatest at the middle of the side of length a.

TABLE 33
Influence Coefficients for Simply Supported Sandwich
Plate Subjected to Line Load

R	K_{wb}	K_{ws}	K_{qx}	K_{qy}	K_{mx}	K_{my}	K_{mxy}
1.0000	.0026	.0424	.2403	.2095	.0260	.0291	.0316
1.1000	.0021	.0379	.2217	.1782	.0215	.0278	.0258
1.2000	.0017	.0339	.2041	.1526	.0178	.0263	.0213
1.3000	.0014	.0302	.1876	.1314	.0146	.0247	.0177
1.4000	.0012	.0270	.1722	.1138	.0120	.0231	.0147
1.5000	.0009	.0241	.1580	.0990	.0099	.0215	.0124
1.6000	.0008	.0216	.1449	.0864	.0081	.0199	.0104
1.7000	.0007	.0193	.1328	.0758	.0066	.0185	.0088
1.8000	.0005	.0173	.1217	.0666	.0054	.0171	.0075
1.9000	.0005	.0155	.1115	.0588	.0044	.0158	.0064
2.0000	.0004	.0139	.1022	.0520	.0035	.0146	.0055
3.0000	.0001	.0051	.0439	.0173	.0000	.0066	.0014
4.0000	.0000	.0020	.0209	.0066	− .0005	.0031	.0004
5.0000	.0000	.0009	.0120	.0028	− .0004	.0015	.0001

Note: R = a/b
$\xi' = .30 \ (\xi' = \xi/a)$
$\upsilon = .30$

TABLE 34
Influence Coefficients for Simply Supported Sandwich
Plate Subjected to Line Load

R	K_{wb}	K_{ws}	K_{qx}	K_{qy}	K_{mx}	K_{my}	K_{mxy}
1.0000	.0031	.0615	.1683	.3899	.0419	.0380	.0278
1.1000	.0026	.0561	.1498	.3379	.0364	.0366	.0222
1.2000	.0021	.0512	.1328	.2951	.0317	.0349	.0179
1.3000	.0018	.0468	.1175	.2595	.0277	.0332	.0145
1.4000	.0015	.0428	.1036	.2296	.0243	.0314	.0118
1.5000	.0012	.0392	.0912	.2042	.0214	.0296	.0097
1.6000	.0010	.0360	.0802	.1825	.0189	.0279	.0079
1.7000	.0009	.0331	.0703	.1639	.0168	.0263	.0065
1.8000	.0007	.0305	.0616	.1477	.0149	.0247	.0054
1.9000	.0006	.0282	.0539	.1336	.0133	.0233	.0045
2.0000	.0005	.0260	.0470	.1212	.0119	.0219	.0037
3.0000	.0001	.0130	.0103	.0528	.0043	.0125	.0007
4.0000	.0001	.0072	− .0002	.0269	.0016	.0077	.0001
5.0000	.0000	.0042	− .0032	.0151	.0004	.0050	.0000

Note: R = a/b
$\xi' = .40 \ (\xi' = \xi/a)$
$\upsilon = .30$

The shear stresses in the core can be obtained from

$$\tau_{xz} = \frac{Q_x}{h}$$

$$\tau_{yz} = \frac{Q_y}{h}$$

(17)

TABLE 35
Influence Coefficients for Simply Supported Sandwich
Plate Subjected to Line Load

R	K_{wb}	K_{ws}	K_{qx}	K_{qy}	K_{mx}	K_{my}	K_{mxy}
1.0000	.0034	.0834	.1282	1.2201	.0627	.0457	.0232
1.1000	.0028	.0777	.1117	1.1016	.0568	.0443	.0182
1.2000	.0023	.0725	.0970	1.0017	.0517	.0426	.0144
1.3000	.0019	.0678	.0840	.9166	.0473	.0408	.0114
1.4000	.0016	.0635	.0727	.8432	.0436	.0390	.0090
1.5000	.0014	.0597	.0628	.7795	.0404	.0372	.0072
1.6000	.0011	.0562	.0543	.7237	.0376	.0354	.0058
1.7000	.0010	.0530	.0470	.6744	.0352	.0338	.0046
1.8000	.0008	.0502	.0407	.6307	.0330	.0322	.0037
1.9000	.0007	.0476	.0353	.5917	.0311	.0307	.0030
2.0000	.0006	.0452	.0306	.5567	.0294	.0293	.0025
3.0000	.0002	.0299	.0089	.3398	.0191	.0198	.0003
4.0000	.0001	.0222	.0044	.2365	.0141	.0148	.0001
5.0000	.0000	.0176	.0034	.1774	.0111	.0118	.0000

Note: R = a/b
$\xi' = .50 \; (\xi' = \xi/a)$
$\upsilon = .30$

B. CONTINUOUS SANDWICH PLATES

A rectangular sandwich plate of width a and length $(b^l + b^r)$, supported along the edges and intermediate line bb (Figure 4), forms a simply supported continuous sandwich plate over two spans. Values for the deflections, shears, and moments can be obtained by combining those for laterally loaded, simply supported plates with those for plates under distributed moments along their edges. Practical formulas for the structural design of continuous sandwich panels are given in Report No. CBS-103 at the Center for Building Studies, Concordia University, Toronto, Canada.

C. SANDWICH PANELS WITH BONDING HAVING FINITE STIFFNESS

As with any other material, bonding in sandwich panels have finite stiffness in resisting interlayer shear stresses. This stiffness may change with time due to creep, cyclic loads, or effect of temperature changes.

It is a very common assumption in practice that bonding is rigid in shear. This assumption does not represent the real behavior of bonding materials and their effects on the responses of sandwich components. The deflection of a panel shows greater sensitivity to variation of the bonding stiffness when the latter is in its lower range. Beyond a certain level of stiffness, the bonding can be practically considered as rigid. An increase in the bonding stiffness is accompanied by a decrease in the normal stress of the core. The resulting loss in the resisting moment is compensated by a slight increase in the face normal stresses. As a consequence, the interlayer shear flux is practically independent of the bonding stiffness. Also, by increasing the bonding stiffness, the shear stresses in the core change from nonlinear to linear distributions with constant values.

The quantities of primary interest in structural design of sandwich panels in this case are: the maximum normal forces or stresses in facings (which also imply moments), maximum interlayer shear fluxes, and maximum deflection. Practical formulas are given here to calculate these quantities for sandwich beams under three types of loads: (1) uniform distribution load, (2) mid-span concentrated load, and (3) axial compression.

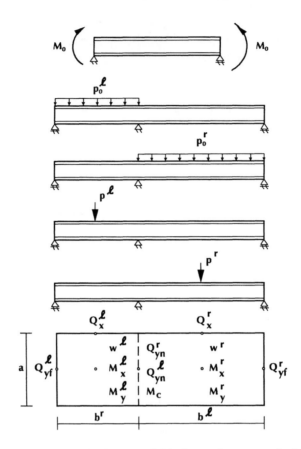

FIGURE 4. Continuous sandwich plates under transverse loads.

$$N = .5\, P_o\, t_c\, F_{nu}$$

$$q = P_o\, F_{qu} \qquad \text{uniform distributed load} \tag{18}$$

$$W = \frac{P_o t_c}{2 E_c\, b}\, F_{wu}$$

$$N = Q\, F_{nc}$$

$$q = \frac{Q}{a}\, F_{qc} \qquad \text{mid-span concentrated load} \tag{19}$$

$$w = \frac{Q}{E_{cb}}\, F_{wc}$$

$$P_{cr} = \frac{1}{2}\, E_c\, t_c\, b\, F_p \qquad \text{axial compression} \tag{20}$$

where

N = maximum normal force in facings
q = maximum interlayer shear flux
w = maximum deflection

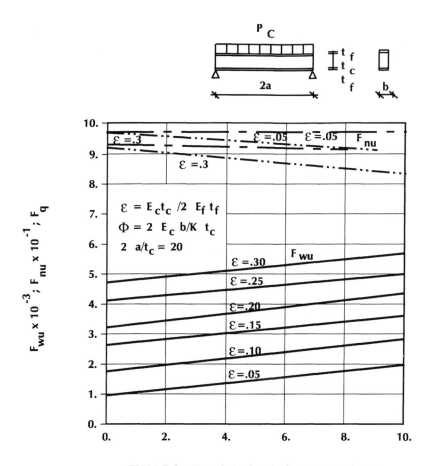

FIGURE 5. Numerical values for factors in Equation 18.

P_{cr} = overall buckling load
P_O = load intensity
Q = concentrated load
b = beam width

F_{nu}, F_{qu}, F_{wu}, F_{nc}, F_{qc}, F_{wc}, and F_p are factors to be obtained from Figures 5 to 14.

In the case of sandwich plates, these quantities can also be obtained from the equations given in the Report CBS-103 at the Center for Building Studies, Concordia University.

D. INSTABILITY OF SANDWICH PANELS

A sandwich panel subjected to inplane loads may fail in different modes. The two major failure modes, in which instability may manifest itself are overall buckling and local instability.

Overall buckling occurs when a sandwich panel becomes elastically unstable under the application of inplane loads with the buckling mode characterized by long waves. Local instability of sandwich panels may be classified into three categories: (1) dimpling, (2) wrinkling, and (3) crimping. Dimpling occurs in sandwich construction with a honeycomb core, where in the region above the cell, the face buckles in a plate-like fashion with the cell walls acting as edge supports. Failure by wrinkling occurs in the form of short-wave lengths in the facings and involves straining of the core material. Such failure mode could

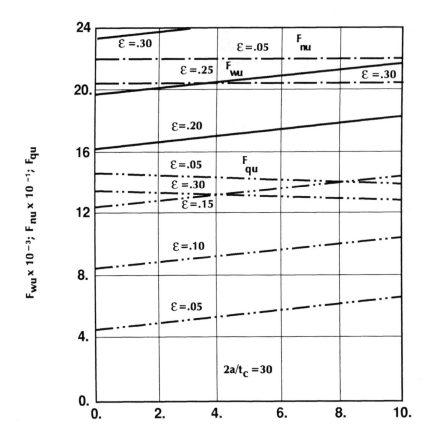

FIGURE 6.   Numerical values for factors in Equation 18.

be symmetric or antisymmetric with respect to the middle plane of the sandwich panel. Finally, crimping is a shear failure mode.

V. OVERALL BUCKLING OF SIMPLY SUPPORTED SANDWICH PANELS

The critical load in the x-direction of a simply supported sandwich panel may be calculated from

$$P_{cr} = \frac{\pi^2 D_f}{b^2} F_{cr} \tag{21}$$

where

P_{cr} = critical buckling edge load per unit length in the x-direction.
b = panel dimension in the y-direction (normal to x-direction)
F_{cr} = dimensionless coefficient to be obtained from Figure 15 in terms of ration a/b and p, where a is the panel dimension in x-direction and p is defined by p = π^2/b^2 times ratio of flexural rigidity to the shear rigidity.

In this figure, m represents the number of waves in the direction of the applied load.

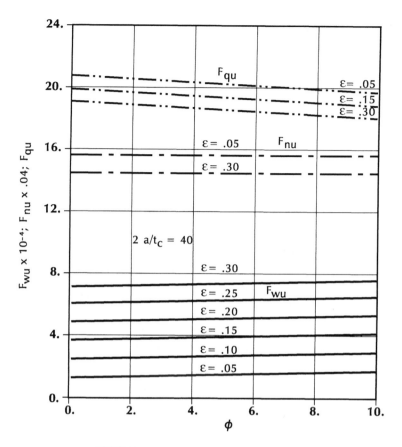

FIGURE 7. Numerical values for factors in Equation 18.

A. WRINKLING STRESSES OF A PERFECTLY FLAT SANDWICH PANEL

When the core is neither thin enough to cause overall buckling nor thick enough to cause symmetric wrinkling, the possible wrinkling mode is a skew ripple. The antisymmetrical wrinkling stress can be calculated from

$$\sigma_{cr} = 0.59 \ \frac{(E_f E_c t_f)^{.5}}{t_c} + 0.387 \ \frac{G t_c}{t_f} \ \text{when} \ t_m = \frac{t_c}{2} \tag{22}$$

$$\sigma_{cr} = 0.51 \ (E_f E_c G)^{1/3} + 0.33 \ \frac{G t_c}{t_f} \ \text{when} \ t_m < \frac{t_c}{2} \tag{23}$$

where

t_m = width of a marginal zone
   = $2.38 \ t_f \ (E_f/E_c)^{1/3}$
E_c = Young's modulus for core material

When the core is thick enough, the possible wrinkling mode is symmetrical with respect to the middle plane of the panel. The critical stress in this case may be obtained from

$$\sigma_{cr} = 0.50 \ (E_f E_c G)^{1/3} \tag{24}$$

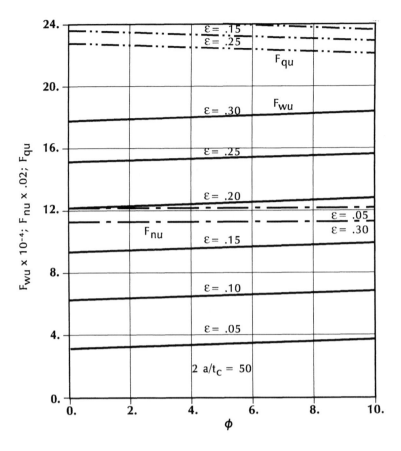

FIGURE 8.   Numerical values for factors in Equation 18.

B. FAILING STRESSES OF IMPERFECT SANDWICH PANELS

In the case of a sandwich panel with an initial imperfection in the form of a sine shape, and after applying the compressive load, the deflection amplitude will be amplified by the factor

$$\text{Amplification factor} = 1\Big/\!\left(1 - \frac{PL^2}{\pi^2 E_f I_f}\right) \tag{25}$$

where

P   =   applied axial load
L   =   wavelength
I_f = $t_f^3/12$

As an increase in the applied compressive load will give rise to the core deformations, loading to critical situation where the ultimate strength of the core material is reached before the critical wrinkling can occur. Thus, this results in an internal core failure. The failing stresses may be calculated from

$$\sigma_{fn} = \frac{\sigma_{cr}}{1 + \dfrac{E_c K_o^w}{T}} \tag{26}$$

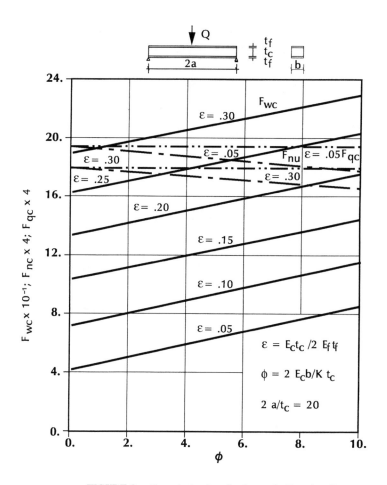

FIGURE 9. Numerical values for factors in Equation 19.

$$\sigma_{fs} = \frac{\sigma_{cr}}{L + \dfrac{\pi Gwo}{L_{cr}S_t}} \qquad (27)$$

where

σ_{fn} = failing stress of the core material based on its strength in compression or tension
σ_{fs} = same meaning as σ_{fn} but for shear strength
T = core compressive or tensile strength
S_t = core shear strength
w_O = initial amplitude of face irregularity
K = $\pi/L \, (G/E_c)^{1/2}$
L_{cr} = $(2 \, \pi^3 \, D/\sqrt{E_c G})^{1/3}$
D = $1/12 \, E_f t_f^3$

By means of Equations 26 and 27 for any given initial imperfection and core strength, the failing stress can be determined. Or conversely, for any given initial imperfection, the core strength required to sustain a specific axial load can be computed.

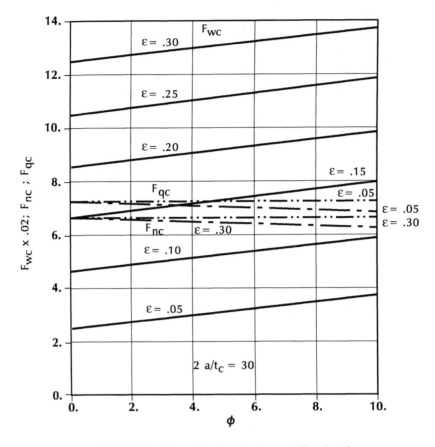

FIGURE 10.   Numerical values for factors in Equation 19.

C. CRIMPING LOAD OF SANDWICH PANELS

The crimping load of a sandwich panel can be calculated from the following equation

$$P_{cr} = S \tag{28}$$

where S is the shear stiffness as defined previously.

D. THERMAL STRESSES IN SANDWICH PANELS

Sandwich plates are called upon to experience thermal stresses when they are subjected to a temperature change. Whether the temperature change is uniform or not and whether the boundaries are free or not, the panel will undergo thermal stresses. This is because each layer in a sandwich panel is unable to deform freely in a manner compatible with the temperature distribution through it.

Consider a sandwich plate subjected to a temperature change T_t at its upper surface and T_b at its lower surface. This linear distribution through the thickness may be divided into two superimposed distributions:

$$\text{Thermal gradient} \pm T_g = \frac{T_t + T_b}{2}$$

$$\text{Uniform temperature change } T_u = \frac{T_t - T_b}{2}$$

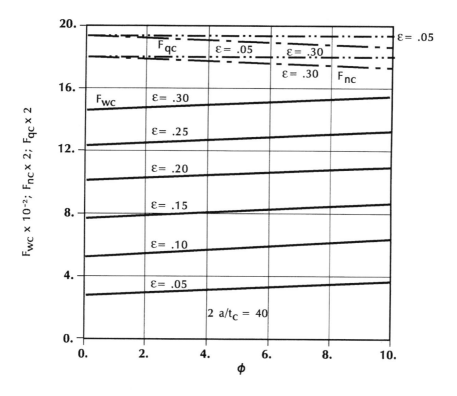

FIGURE 11. Numerical values for factors in Equation 19.

Practical formulas are presented here to calculate the maximum normal stress in the facings and the maximum deflection in sandwich panel under each of these two distributions.

E. SIMPLY SUPPORTED SANDWICH PLATES SUBJECTED TO THERMAL GRADIENTS

The maximum normal stress in the faces can be determined from

$$\sigma_f = \frac{1}{2}\,\alpha_f T_g E_f \tag{29}$$

where α_f = thermal expansion coefficient of the faces material, and T_g = temperature change at the plate surfaces.

The maximum deflection in this case may be determined by the following equation:

$$w = 4\alpha_f T_g a^2 \frac{(1 + v_f)}{\pi^3 h} K_w \tag{30}$$

where

a = panel length

v_f = Poisson's ratio of the faces material

K_w = a factor to be obtained from Figure 16 in terms of the aspect ratio b/a

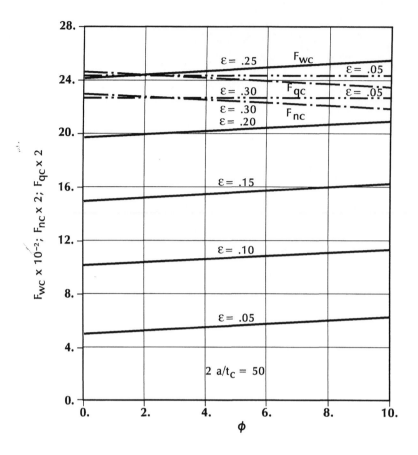

FIGURE 12.   Numerical values for factors in Equation 19.

F. FREE EDGES SANDWICH PANEL SUBJECTED TO A THERMAL GRADIENT

The maximum normal stress in the faces in this case may be obtained from

$$\sigma_f = \frac{E_f \, \Delta_m}{1 - v_f + \dfrac{6 \, E_f t_f h}{E_c t_c^2}} \tag{31}$$

where Δ_m = relative free strain at the interface between facings and core due to a uniform temperature change.

G. FREE EDGES SANDWICH PANEL SUBJECTED TO A UNIFORM TEMPERATURE CHANGE

The maximum normal stress in the faces in this case may be obtained from

$$\sigma_f = \frac{E_f \Delta m}{1 - v_f + \dfrac{2 \, E_f t_f}{E_c t_c}} \tag{32}$$

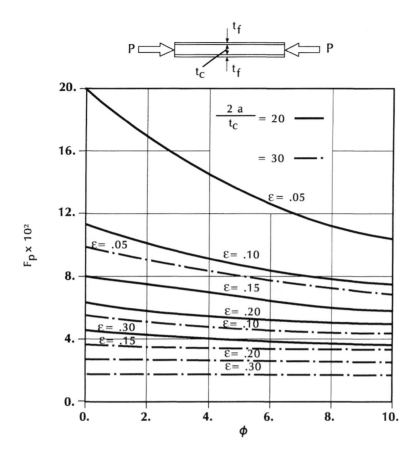

FIGURE 13. Numerical values for factors in Equation 20.

VI. SUMMARY

Demands on new developments of building systems imposed by today's advanced technologies have become so diverse and severe that they are often difficult to be fulfilled by a single compound material acting alone. It is sometimes necessary to combine dissimilar materials into a composite to provide performance unattainable by single constituent.

Sandwich panels, as a special form of composite materials, are being used in building industry. In the foregoing sections, several features of sandwich panels were discussed. The use of polymers to meet with the construction requirements was elaborated through several examples. For structural design of sandwich panels, simple formulas are given. Numerical values for the coefficients in these formulas may be obtained from the tables and graphs presented.

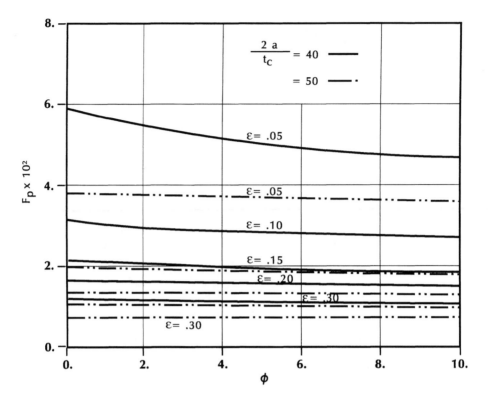

FIGURE 14.   Numerical values for factors in Equation 20.

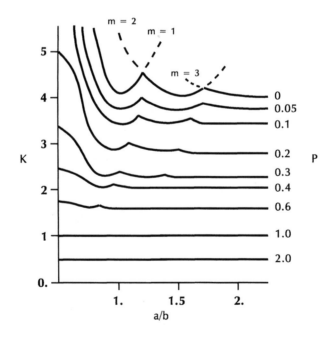

FIGURE 15.   Relationship between K and a/b for known values of p.

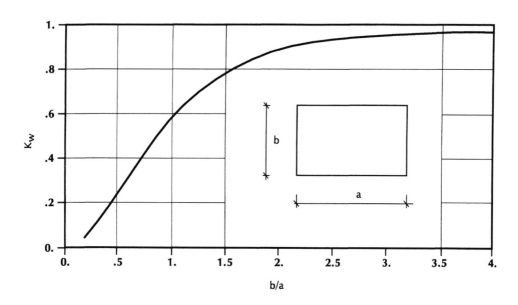

FIGURE 16. Values for K_w in Equation 30.

Chapter 28

POLYMER-CONCRETE COMPOSITES

R. Hussein

TABLE OF CONTENTS

0-8493-4401-8/93/$0.00 + $.50

I. INTRODUCTION

As the world's needs for housing, transportation, and industry increase, the consumption of concrete products is expected to increase correspondingly. At the same time, product management of energy and natural resources demands ever higher levels of performance. Although Portland cement concrete is one of the most remarkable and versatile construction materials, a clear need is perceived for the improvement of properties such as strength, toughness, ductility, and durability. One valid approach is to improve the concrete itself; another is to combine technologies in order to make new composites based on concrete.

Polymers containing large amounts of filler, such as polymer mortars, without cement and polymer concretes, are increasingly being used in buildings and other structures. Polymer mortars are mainly used as protective coatings in concrete, reinforced concrete, and more rarely, steel, while polymer concretes represent a new type of structural material capable of withstanding highly corrosive environments.

The mechanical properties, chemical stability, and some other useful properties are the reasons for continuous interest shown in these materials by various research, design, and production organizations. However, polymer mortars and polymer concretes have been introduced only recently, and many of their properties are still imperfectly known.

Polymers are relatively cheap and completely resistant to alkali attack by cement paste. They offer hope in overcoming one of the main problems of fiber-reinforced concrete — lack of ductility. The material tends to crack rather than bend under relatively modest loads. Apart from polymers, which are only now being developed, interest is being shown in natural and synthetic fibers, mainly to provide asbestos cement substitutes.

II. POLYMERS IN CONCRETE

The development of concrete-polymer composite materials is directed at both improved and new concrete materials by combining the well-known technology of hydraulic cement concrete formation with the modern technology of polymers.

A wide range of concrete-polymer composites is being investigated. The most important are the following:

- Polymer-impregnated concrete (PIC)
- Polymer cement concrete (PCC)
- Polymer concrete (PC)
- Fiber-reinforced concrete
- Polymer and fibers-reinforced concrete

Polymer-impregnated concrete is a precase and cured hydrated cement concrete that has been impregnated with a monomer which is polymerized inside. This type of product is the more developed of the composites.

PCC is a premixture of cement paste and aggregate to which a monomer is added prior to setting and curing.

PC is an aggregate bound with a polymer binder. This product may be obtained in the field. It is called a concrete because by general definition, concrete consists of any aggregate bound with a binder.

The following two composite products are natural and especially synthetic fibers.

A. POLYMER-IMPREGNATED CONCRETE

This composite system has obtained the largest improvement in structural and durability properties. In the presence of polymer, the compressive strength can be increased four times,

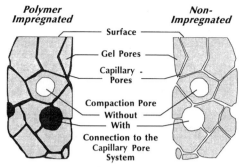

Pore Size not Drawn to the Same Scale

FIGURE 1. Scheme of pores in hardened cement paste.

water absorption is reduced by 99%, freeze-thaw resistance is enormously improved, and in contrast to conventional concrete, polymer-impregnated concrete exhibits essentially zero creep properties.

The ability to vary the shape of the stress-strain curve presents some interesting possibilities for tailoring desired properties of concrete for particular structural applications.

Normal concrete consists of particles of fine and coarse aggregates dispersed in a matrix of hardened cement paste. In common concretes, the aggregate volume is around 70%. The remaining volume is the hardened cement matrix. Since stone aggregates and sand have very low porosity, the bulk of the porosity of concrete is in the cement phase. Therefore, the macromolecular compound of polymer-impregnated concrete must be concentrated in the cement phase.

The system becomes more complex when polymer is added. This one fills the pores of the hardened cement paste phase, creating still another composite, the hardened cement paste-polymer system. The system may be considered to be a composite of coarse and fine aggregates in a matrix of impregnated cement.

The mechanism of polymer-concrete system formation and the physicochemical phenomena are not yet very well known; they are considered very complex.

The processes that accompany the formation of such types of composites are known only in general lines, based on the observations made on exterior phenomena. The decrease in porosity is one factor in controlling the properties of the composite, but this cannot account for the different performance recorded with different polymers.

For sometime, it has been recognized that the pore structure of hardened cement paste plays a decisive role with respect to many important characteristics of concrete. It is expected, therefore, that property improvement due to impregnation is brought about mostly due to pore structure modification and should be a function of this. (Figure 1).

Considerable importance has been attached to the role of interphase bonding in determining the role of polymer-impregnated composites, though this improvement has not been measured directly.

The chemicals required for producing polymer-concrete systems are relatively new to the construction industry.

It is important to have a good understanding of their properties, their hazards, and the safety precautions required. Safety procedures should be carefully followed whenever different chemicals for polymer-concrete composites are used.

Polymer-impregnated concrete is generally prepared by impregnating dry precast concrete with a liquid monomer and polymerizing the resin *in situ* by thermal catalytic or by radiation method.

A monomer may be defined as a small molecule from which polymerization or poly-condensation, macromolecular compounds can be made. Some of the most widely used monomers for polymer-concrete system include:

- Methyl methacrylate (MMA)
- Styrene (S)
- Butyl acrylate (BA)
- Unsaturated polyester-styrene
- Vinyl acetate (VA)
- Acrylonitrile (AN)
- Methylacrylate (MA)
- Trimethylolpropane trimethacrylate (TMPTMA)

These monomers may be used alone or in mixtures. TMPTMA also serves as a cross-linking agent and is used primarily to increase the rate of polymerization; this means to decrease the time of curing.

Generally, monomers are volatile, combustible, and toxic liquids. However, practice has shown that prolonged stability and safety can be achieved by the following recommended practices in storage and handling.

The principal factors that influence stability in these chemical compounds and determine the methods to be used for safe handling are

- Level and effectiveness of stabilizers (inhibitors)
- Storage conditions (temperature, light, humidity, oxygen, etc.)
- Toxicity
- Flammability
- Effect on construction materials with which they are used

The role of stabilizers is to prevent premature polymerization; they can act as antioxidants and prevent polymerization by reacting with oxidation products that may be formed in the storage vessel through contamination or through the formation of peroxides from air oxygen.

The time required for initial stabilizer concentrations to fall to a critical level varies greatly with storage and handling conditions.

Factors affecting the depletion of stabilizer are water, air, and heat, with last one being the most important.

The manufacturing process is generally divided into

- Fabrication of precast concrete
- Drying the precast concrete
- Impregnation (saturation with monomer)
- Polymerization *in situ*

In Figure 2, we may follow a schematic of PIC technological process.

Processing requirements for each of these steps are being developed. The degree of dryness of concrete strongly affects the amount of monomer absorbed during impregnation. From their results, it appears that a drying temperature of 150°C is required to remove most of the free water without seriously affecting the mechanical properties of polymer-concrete composite.

Several methods of concrete curing, including fog, low pressure steam, and high pressure steam, have been studied to determine the effects on the properties of polymer-impregnated

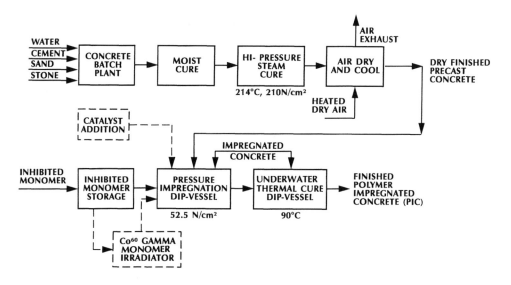

FIGURE 2. Illustrates thermal process (with optional monomer irradiation treatment or catalyst addition) for production of polymer-impregnated concrete (PIC).

concrete. The data indicate that high pressure steam curing results in specimens with higher polymer loadings and generally in higher strengths than for comparable fog-cured specimens.

Concrete mix variable such as water/cement ratio, entrained air content, and aggregate size and quality do not greatly affect the properties of such kind of composites. Some results indicate that the impregnation of low strength concrete produced polymer-impregnated concrete with essentially the same properties as PIC produced from high strength mixes.

Property improvement due to impregnation is mostly due to pore structure modification and shall be a formation of the pore structure.

Pore size distributors can be obtained by two independent means: (1) analysis of the mercury intrusion data, and (2) analysis of the absorption data. The mercury intrusion data can be analyzed by assuming cylindrical pores and relating the intrusion pressure P to the pore diameter. They can also be analyzed by Equation 1

$$P = \frac{4\delta \cos \theta}{d} \tag{1}$$

where δ is the air/mercury interfacial tension, and θ is the angle of contact between mercury and the pore wall.

Impregnation of the concrete pore structure with a liquid monomer occurs by a viscous flow mechanism, and consequently monomer viscosity is one existing factor. In the laboratory, this process can be assisted by applying a vacuum to the concrete to remove entrapped air. After initial vacuum soaking, positive pressure can be applied to the monomer for additional driving force for flow into the pore structure.

However, these approaches are more difficult for field impregnation of large surface areas.

Porosity can be determined before and after impregnation by measuring solid volume by helium comparison pycnometry. The apparent volume is determined by weighing in methanol — samples saturated with methanol.

For hydraulic setting materials, like Portland cement and plaster of Paris, a large volume of porosity is inherent in the set structure. This porosity derives mainly from the excess water required to lubricate the powders so that the material may be placed. The porosity that derives from this excess water is called *capillary porosity*. The amount of porosity present at any time is dependent upon the original water-cement ratio, and the maturity, or degree of hydration, of the cement paste.

The additional porosity in cement that forms within the hydration products has been called *gel porosity*; it is finer than the first one and may be present in mature pastes, to the extent of about 0.1 cm^3/g. So the total cement porosity can be about 0.35 cm^3/g.

For a cement paste of density 1.4 g/cm^3, the total pore content is 49%. For the cement with 30 to 35% cement paste, the pore volume of the concrete is 15 to 17%. If all these pores become filled upon monomer impregnation, a weight loading of 6 to 7% is calculated. Because of compositional variations, most normal concretes indicate polymer weight loadings of between 5 to 8%.

The advantage of polymers over other systems of impregnation is that the particle size of the molecules is very small (1 to 10 nm), whereas in cement grouts, the particle sizes are much greater (0.1 to 10 μm). Therefore, complete penetration of fine silts is now possible by dispersed mineral grouts. These may set too slowly to give the rapidly required strength, impermeability, and other properties. Another factor important in some workings is the need for chemical resistance to sulphates, chlorides, and alkalis. Several grouts may fail after long exposure to these agents, but macromolecular systems, being based on a different type of chemical compounds, can be superior.

The costs of such grouts is higher than the inorganic types, but their special characteristics of penetratability, cure rate, and corrosion resistance ensure that they have a significant place amongst polymer composites.

Moreover, diffusion into existing concretes is being carefully studied because special concretes with improved environmental resistance and tensile properties are required.

Impregnation is deeper when the 3% water content of a typical concrete is removed, and an air evacuation stage in the production cycle raises the polymer content by 50%.

Fundamental problems arise in thick sections, which require very low viscosity monomers, but these are inevitably volatile products. When polymerization is induced, the system must be heated. The process, being exothermic in large quantities of heat, are valued, so that large losses of monomer take place from the outer layers.

The volatility, toxicity, and flammability of most monomers restrict their use in composite making, but in situations where penetration of porous structures is called for, they have obvious advantages. They leave no residue nor require evaporation of solvents, water, etc. It will however, often be found that the viscosity of monomers themselves is too low for convenience, and then perhaps oligomers are used instead.

Evaporation of the monomer during impregnation or polymerization is an important technical and economical factor. This problem may be dealt with by selecting low vapor pressure monomers or by employing evaporation barriers or encapsulation methods.

One of the steps in the procedure for fully impregnating precast concrete has been to wrap the saturated with monomer, concrete, in aluminum, or polyethylene sheets in order to reduce monomer evaporation and drainage losses. In an attempt to further reduce these losses and to eliminate the necessity of wrapping, the following encapsulation techniques have been evaluated:

- Encapsulation in a form during processing
- Underwater polymerization
- Application of olygomer technique

Maximum polymer loadings are obtained with the second method, which consists of a standard vacuum soak impregnation, placing the specimens in water and radiopolymerizing under water.

The drying temperature, degree of dryness, and some impregnation parameters, such as vacuum, pressure, soaking time, method of encapsulation, etc., have an important effect on the mechanical properties (strength) of polymer-impregnated concrete.

Concrete composition has a significant effect on the monomer filling rate. Samples of concrete having high air and water contents may be saturated with monomer in shorter period of time than Portland cement-type concrete.

For the polymerization of monomer introduced in concrete, the following methods are usually applied: (1) irradiation with γ rays and (2) application of heating.

The use of radiopolymerization involves some difficult problems with regard to the potential safety hazard on the field.

Attention was paid especially to methods of heating application, such as solar energy, microwave heating, steam, heating blankets, hot water.

From the standpoint of economy, solar energy is a very appealing heat source, but the use of sunlight involves too many variables which cannot be fully controlled. The use of microwave heating for monomer polymerization on field, so far, does not appear feasible.

At present, the use of hot water appears to be the most feasible method of heat application for surface treatments.

For heat catalytic polymerization of monomers initiators and promoters must be employed.

Initiators are compounds which initiate polymer draining formation by decomposing into free radicals that actually start the chain growth. When free radicals are generated in the presence of a vinyl monomer, the radical adds to the double bond with the regeneration of another radical. Evidence for the radical mechanism of addition polymerization comes not only from the capability of radicals to accelerate vinyl polymerization, but also from the demonstration that the polymers so formed contain fragments of the initiator.

Promoters are used together with the initiators to form redox initiator systems which are applied when it is desirable to lead polymerization at ambient temperature.

Initiators most commonly used producing concrete-polymer composites are: 2,2'-azo *bis* (iso butyro-nitrile)-(AIBN); 1 2,2'-azo *bis* (2,4-dimethyl valero nitrile) (AMVN), benzoyl peroxide (BP), laruoyl peroxide (LP); and methyl ethyl ketone peroxide (MEKP). The quantity of initiator used as percentage of monomer weight is 0.5 to 4%.

AIBN is the most used initiator for polymer impregnation of cast in place initiator.

The role of porosity is to increase the decomposition rate of peroxide initiators. They are used in small quantities compared to the monomer. The promoter-initiator system can be designed to cause polymerization over a wide range of times and temperatures.

Some of the most commonly used initiator-promoter systems are:

- BP with dimethyl-toluidine (DMT) to polymerize MAA
- BP with dimethyl-aniline (DMA) to polymerize MMA
- LP with DMA to polymerize styrene and TMPTMA
- MEKP with cobalt napthtenate for rapid curing of polyester-styrene

To obtain polymer-impregnated paste and mortar, various monomers such as methyl methacrylate, styrene, vinyl acetate methylacrylate, and anylonitrale have been used. In particular, methylacrylate was chosen because its polymer glass transition temperature is about 3°C, and in the bulk state, PMA is a rubber-like material at room temperature.

In all polymerization, AIBN can be used as an initiator. Its concentration is typically 0.5% in MMA, MA, and S, 1.0% in VA, and 1.5% in styrene. Polymerizations can be done at 80 ± 5°C.

Polymer loading depends on the temperature of thermal polymerization in water with the following formula:

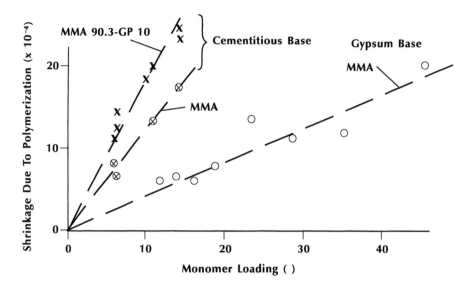

FIGURE 3. Shrinkage due to polymerization.

$$\text{Polymer loading } \% = \frac{\text{Weight of PIC specimens (g)} - \text{weight of dried concrete specimen (g)}}{\text{Weight of dried concrete specimen (g)}}$$

It can be argued that shrinkage occurs through two stages of impregnation treatment namely through initial drying and polymerization. Shrinkage through polymerization is peculiar to PIC and could be several times greater than the normal drying shrinkage (Figure 3) where such shrinkage is plotted against polymer loading. The measurements were made by the method prescribed in ASTM C-490-65 T. Samples were all impregnated completely, and polymer loadings were controlled by varying the amount of water and aggregates. Polymerization was accomplished by heating at 60°C for 16 h and using 0.5% of AIBN.

Concrete specimens can be impregnated with methyl methacrylate, and after polymerization *in situ*, the extraction of polymer can be carried out according to the steps displayed in Figure 4.

The molecular weight of PMMA can be determined from the intrinsic viscosity values at $30 \pm 0.05°C$ in benzene solutions using the following relationship:

$$[\eta] = 8.69 \times 10^{-3} \overline{M}_v 0.76 \tag{2}$$

where $[\eta]$ = intrinsic viscosity, and $\overline{M}_v$ = average viscometric molecular weight.

The influence of polymer on the properties of composite is determined primarily by the change in properties of the cement paste phase and not by changes at the paste-aggregate interface. The polymer increases the amount of solid per unit volume in the cement paste and simultaneously reduces the effect of stress concentrations from pores and microcracks. The properties of the polymer in the pores of the cement paste is likely to be the most important factor in determining the effectiveness of any particular polymer. These properties are not always a reflection of the bulk polymer properties.

B. PROPERTIES OF POLYMER-IMPREGNATED CONCRETE

Incorporating a relatively small volume of a high polymer in the pore structure of concrete results in an important increase of the elastic and mechanical properties of the material. The

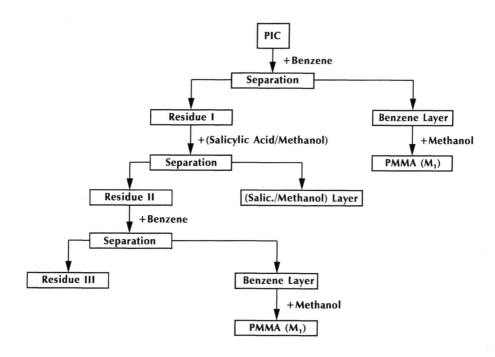

FIGURE 4. Polymerization scheme.

main effects of the presence of polymer in such type of composites are (1) to improve the strength and increase the modulus of the cement phase and (2) to improve the bonding between the cement and aggregate phases.

A number of workers have developed composite models based on the structure of concrete. One such concrete model is

$$\frac{E}{Ea} = 1 + \frac{2 \, Va \, (r - 1)}{(r - 1) - Va \, (r - 1)} \tag{3}$$

where E is the modulus of the composite:

E_m = modulus of the matrix
V_a = volume fraction of aggregate
r = ratio Ea/E_m, where Ea is modulus of the aggregate phase.

This equation can also be applied to mortars. The aggregate phase now consists of sand particles of modulus in 70×10^3 MN/m^2, and the matrix is either normal cement of modulus 7×10^3 MN/m^2 or polymer-filled cement of modulus 19.6×10^3 MN/m^2. The moduli of the two mortars can be calculated for various aggregate loadings. This is illustrated in Figure 5, where theoretical curves predicted from the equation are presented along with Young's modulus for a series of normal and polymer-filled mortars with sand contents ranging from 40 to 65% volume. The calculated results are in fairly good agreement with the experimental results, although the agreement appears better for polymer-filled series. The results indicate that for mortar, the only effect of polymer is to increase the modulus of cement phase.

Some concretes were also investigated by adding aggregate to mortar in concentrations of 10 to 50% by volume. The Young's moduli of the concretes were calculated with Equation 3 using 56×10^3 MN/m^2 as the modulus for the limestone aggregate and the observed modulus of the 40% sand mortar. These results are shown in Figure 6.

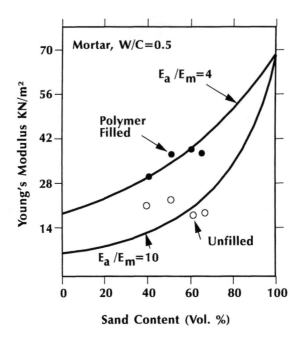

FIGURE 5.   Predicted and experimental variation of Young's modulus of mortar with sand content.

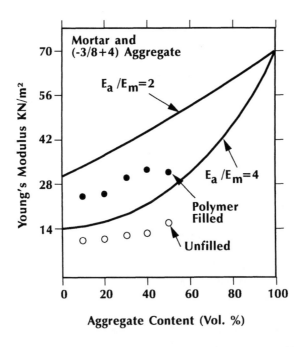

FIGURE 6.   Predicted and experimental variation of Young's modulus of concrete with $-3/8 + 4$ mesh aggregate content.

In this case, the experimental results lie below the calculated curves and indicate an extrapolated mortar modulus (0% aggregate) lower than that used for the calculation.

In contrast to conventional concrete, PIC exhibits essentially zero creep properties. By polymer impregnating concrete, conventional concrete is transformed from a plastic material to essentially an elastic one with an increase of at least two times in the modulus of elasticity.

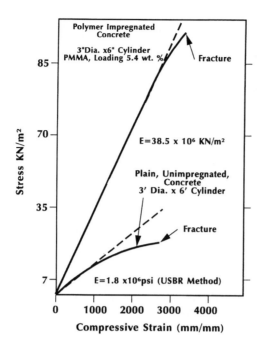

FIGURE 7. Compressive stress-strain curve for PMMA — impregnated concrete. Impregnated shows elastic behavior. Unimpregnated shows plastic behavior.

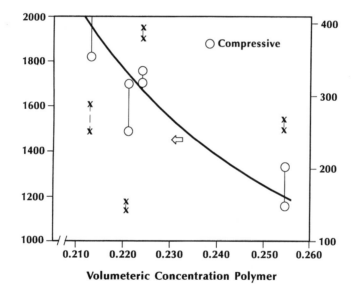

FIGURE 8. Relationship between volumetric concentration of polymer and mechanical strength.

This is indicated by the linearity of the stress-strain curve for PIC in Figure 7. The ability to vary the shape of the stress-strain curve presents, as we told before, some very interesting possibilities for tailoring desired properties of concrete for particular structural applications.

The bending strength is not simply related to polymer loading or water-cement ratio, while the compressive strength showed an inverse relation with polymer loading or volumetric concentration of polymer (Figure 8).

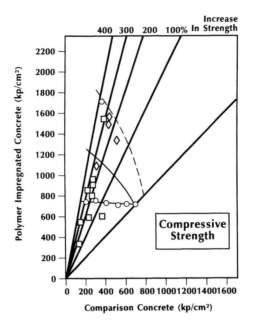

FIGURE 9.   Variations of compressive strength values after transformation into polymer-impregnated concrete.

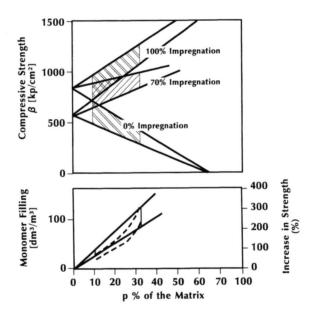

FIGURE 10.   Example of strength developments by transformation of cement concrete into polymer-impregnated concrete.

Variations of compressive strength values after impregnation may be seen also in Figures 9 and 10.

In some of their investigations, MMA has been replaced by different amounts of DOP. Bending strength and compressive strength measured with these samples are shown in Figures 11 and 12. Both of these properties were affected by the quantity of DOP, with the optimum replacement being about 5% by weight.

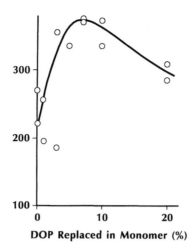

FIGURE 11. Effect of DOP dosage on bending strength.

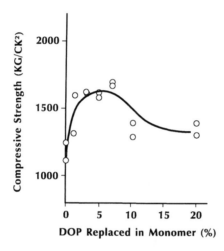

FIGURE 12. Effect of DOP dosage on compressive strength.

Polystyrene-impregnated concrete specimens for compressive strength are well known in the literature. Test results are shown in Figure 13. Here the strength improvements factor α is expressed as follows:

$$\alpha = \frac{\sigma_i}{\sigma_0} \tag{4}$$

where σ_i = strength of polymer-impregnated concrete; and σ_0 = strength of impregnated concrete.

Authors have argued that the composite strength of polystyrene-impregnated and un-impregnated autoclaved concretes tends to increase (1) with an increase in the cement and silica content and (2) with a decrease in the water-cement ratio in the mixture proportions of concrete.

At the same cement content, the strength improvement factors (α) of polystyrene-impregnated autoclaved concrete are much the same, regardless of water-cement ratio and silica content.

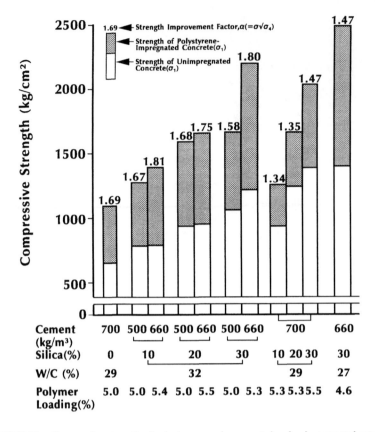

FIGURE 13. Compressive strength of polystyrene — impregnated and unimpregnated autoclaved concretes vs. mix proportions of concrete and polymer loading.

The compressive strength of this type of composite is hardly influenced by polymer loading.

The influence of AN concentration on the mechanical properties of PIC in the case of impregnation with styrene-acrylonitrile copolymer has been studied by Zeldin and Cowark.

Figures 14 and 15 shown some of these results.

The authors have concluded that the concentration of AN is critical and dependent on the concentration of other monomers. Thus, at 40 to 45 mol% concentration of AN, 5 to 5.5 acrylamide, and 1.15 to 1.25 mol% TMPTMA, a true homogenous copolymer is formed exclusively, which leads to the formation of highly stable, chemically resistant composites.

The monomers and technological process used in the fabrication of PIC may be changed to produce a material having special requirements. For instance, special monomer systems may be formulated to support high temperatures, the impregnation depth may be controlled to give a product having good durability rather than high strength, or the impregnated concrete may be given a seal coat to improve durability.

PIC can be divided into four major categories as follows:

1.   PIC for normal temperature applications
2.   PIC for desalting plant applications
3.   Partially impregnated concrete
4.   Surface impregnated concrete

Durability tests have shown highly significant increases of PIC as compared with conventional concrete. The testing program includes freeze-thaw, sulfate attack, and resistance

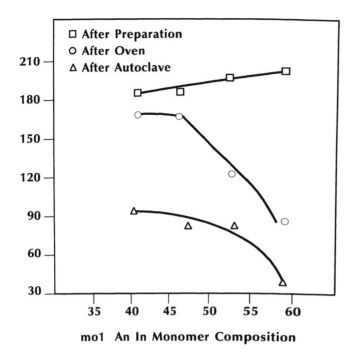

FIGURE 14. Compressive strength of polymer concrete vs. AN concentration. Monomer load, 13 wt%; sand/cement ratio, 7:3; crosslinking agent, 1.7–1.8 vol%; Aa, 5.75 vol%.

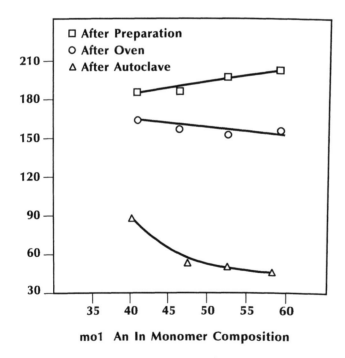

FIGURE 15. Compressive strength of polymer 13 wt%; sand/cement ratio, 7:3; crosslinking agent, 2.4–2.5 vol%; Aa, 6.0 vol%.

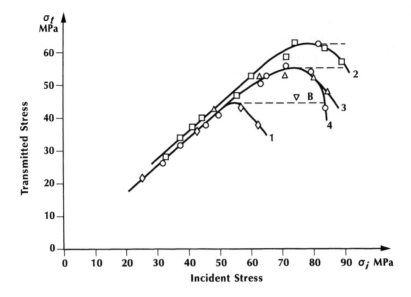

FIGURE 16.   Relation between incident and transmitted pulses. (1) Plain concrete, (2) polymer-cement concrete, (3) fiber-reinforced concrete, and (4) fiber-reinforced polymer concrete.

to acids and bases for PIC designed for normal temperature applications, and exposure to brine and demineralized water at elevated temperatures for PIC designed for desalting plant applications.

Between the most important applications of PIC, we may mention

- Tunnel support — lining systems
- Concrete pipe (hydrostatic tests showed that PIC pipe supported about twice as much hydrostatic pressure as the unimpregnated pipe)
- Desalting plants
- Bridge decking
- Beams (ordinary reinforced beams, posttensioned beams)
- Underwater habitats

After PIC has been proven as a new composite material, more research is needed to determine structural behavior. The 300% or more increase in compressive and tensile strengths and up to 100% increase in stiffness indicate excellent potential for PIC structural beams, columns, floor and roof joists, and piling, especially in severe environments where the excellent durability properties can be utilized.

PIC must be considered a new complex material with specific characteristics, which place it in a position from the viewpoint of quality and cost, between traditional concrete and metallic and ceramic materials.

C. POLYMER CEMENT CONCRETE

The results obtained by introduction of various organic compounds to a concrete mix are relatively modest in improvement of strength and durability.

Under the best conditions, compressive strength improvement over conventional concrete of about 50% are obtained with relatively high polymer concentrations of 30%.

Production of PCC have been used with limited success in vinylidene chloride, furans unsaturated polyester-styrene, and epoxy-styrene. Organic materials are incompatible with

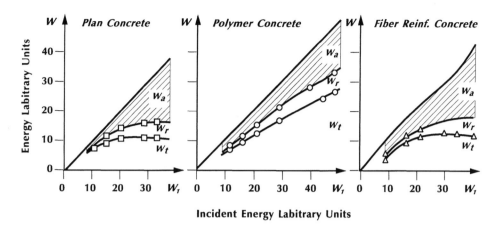

FIGURE 17. Energy transmission and attenuation for different concretes.

aqueous systems in many cases and interfere with the alkaline cement hydration process. The incentive to attain improved premix concrete materials is that it can be cast in place for field applications, whereas PIC requires a precast structure.

From the amplitude of incident and transmitted pulses, the stresses in concrete shown in Figure 16 were calculated. As the amplitude of loading pulse was gradually increased, the transmitted stresses reached an optimum value and then decreased. This optimum value was considered as the dynamic strength of the material.

The value of the stress due to transmitted wave — (σ_t) max-obtained when the specimens were subjected to maximum amplitude pulse directly (marked by ∇ in Figure 16) was approximately the same as that obtained by the increasing amplitude method. Thus, the complete $\sigma_t - \sigma_i$ curve AB (σ_i = the stress due to the incident wave), including the horizontal extension, gives the stress amplitude envelope for the particular concrete quality.

The fact that the slope of σ_t/σ_i curve is the same for all the concretes indicated that the viscoelastic character of concrete was not changed by polymer and fiber admixtures. The authors considered that the increase in impact strength was rather due to the improvement in deformation capacity.

Figure 17 shows the relative magnitudes of W_t, W_n, and W_a (transmitted and reflected wave, and energy attenuated in the specimen) with increasing incident energy levels for different concretes.

From this figure, the following features can be observed:

1. Even at early stages of loading, about 10 to 20% of the energy was attenuated in the specimens, which resulted in some stable crack formation.
2. Polymer cement concrete showed a much higher impact endurance (lower energy attenuation) and improved energy transmission capacity than the corresponding plain concrete.
3. Reflex was rather large even in polymer concrete. This was probably due to the large amount entrapped during mixing.

D. POLYMER CONCRETE

PC may be considered as an aggregate filled with a high polymer.

The main technique in producing PC is to minimize void volume in the aggregate mass so as to reduce the quantity of the relatively impressive polymer necessary for binding the aggregate.

Monomer which diffuses up through the aggregated and polymerization process is initiated by either radiation or thermochemical means.

Some of the drawbacks of conventional concrete are

- Formation of integral; voids when alkaline Portland hydraulic cement is used.
- Because of water entrapping, on freezing, it can readily crack.
- Alkaline cement can be chemically attacked by acidic materials and concrete rapidly deteriorates under this condition.

With polymer as a binder, most of these difficulties are overcome.

A great number of macromolecular compounds are hydrophobic, resistant to chemical attack, and can be made compact with a low amount of voids. To improve the bond strength between the macromolecular compound and the aggregate, a silane coupling agent is added to the monomer before polymerization.

It can be shown that the main problems arise from the viscoelastic properties of polymer. Usually, the macromolecular compounds have a low modulus of elasticity, which means they are flexible and manifest creep properties. These explain why high polymers cannot be used alone in structural units. Using polymer as a binder with aggregates, some of these difficulties are overcome.

Polymer concretes generally consist of a mixture of aggregate with phenol formaldehyde, furan, epoxy, or polyester resins. These mixes give composite materials high tensile and bending strengths. The addition of fibrous materials such as asbestos or glass wool confers further strengthening effects. To obtain the best chemical resistance, complete curing of the polymer is necessary. This usually may be obtained by using a crosslinking agent or by heating.

Polymer concretes are viscoelastic in nature and are more efficient as structural materials for short-term loadings. In the composite beam, the polymer concrete cap would be effective during short-term loadings, even though it might relax during long-term loadings. Thus, the composite beam would have a high, short-term capacity and a lower, long-term capacity.

III. SYNTHETIC FIBERS — CEMENT COMPOSITES

A solution to the problem of reinforcing a brittle cement matrix with fibers was obtained in the late 19th century by the inclusion of asbestos fibers. The success of this material is probably unparalleled in the field of fiber composites. The world production of asbestos cement in recent years has exceeded 20×10^6 t/annum.

However, this market is likely to decline because it has been forecast that the supply of asbestos may be exhausted by the year 2000.

Organic fibers have, without exceptions, a lower young modulus than cement or concrete matrix in which they must be incorporated.

Between the synthetic fibers polypropylene is one of the cheapest; propylene fibers are readily available in a great variety of prices lower than those of polyesters, polyamides, polyacrylonitriles, etc.

The potential of polypropylene as a substitute for mineral asbestos fiber in cement matrix has not yet been fully explored, although it seems to be very promising. This may be due to the differences in properties between asbestos and polypropylene fibers. For instance, asbestos fibers have a high tensile strength, a high modulus of elasticity, and a good compatibility with cement, this means a high affinity for cement particles. On the other hand, polypropylene fibers have a low young modulus, are hydrophobic, and are incompatible with cement particles. A good bond between fiber and cement matrix is a prime requisite for any fiber concrete. The hydrophobic polypropilene cannot ensure the same type of bonding as exists between cement and mineral fibers such as asbestos, glass fibers, etc.

In the early 1960s, stretched films with a longitudinal fibrillation were developed for twines used in agricultural and rope making applications. This fibrillated polypropylene film applied in cement technology allows the fluid water cement paste to penetrate between the fibrils and to achieve a mechanical bond within the hardened matrix, in place of the physical adhesion.

The stretching of polypropylene films in the extrusion process induces orientation and crystallization, and provides a strong film in the processing direction, which prones to split over short distances.

Near good mechanical properties polypropylene has a relatively high melting point (165°C), low density, and good stability to chemical agents. In the case of a chemical attack, the cement matrix will be destroyed long before polypropylene is affected.

Tables 1 and 2 show typical properties of fibers and matrices used in fiber-cement composites.

In the past, polypropylene fibers have not been regarded as potentially satisfactory for reinforcing materials in direct tension or flexure because of their low elastic mondulus, high Poisson's ratio, and poor bonding with cement.

The use of continuous networks of fibrillated polypropylene film to produce adequate bonding and thus to utilize the full strength of the polymer is an important consideration. The inclusion of polypropylene mesh in a cement mortar mix more than doubles the load capacity of beams and can produce closely spaced multiple cracking at bond stresses well within the measured range of polypropylene and cement. The material has the additional advantage of high energy absorption under impact.

Polypropylene films were chosen for their cheapness, strength, and chemical inertness, to the extent that the cement matrix would be the first to deteriorate should there be contact with aggressive chemicals.

Continuous polypropylene fibrillated films, when aligned approximately parallel to the direction of tensile stress, can yield a composite with high flexural strength combined with fine multiple cracking. This finding results in a cement-based material with high ductility and opens the way to production of economic thin sheet products suitable for replacing asbestos/cement in some applications.

Figure 18 represents stress-strain curve in tension for a composite containing 2.3% by volume of opened polypropylene film networks.

The main shortcomings of polypropylene fibers are combustibility, low modulus of elasticity, poor bond between monofilaments and matrix, and sensitivity to sunlight and oxygen.

The low modulus of elasticity means that the inclusion of fibers reduces the cracking strength of the composite system and results in a very large strain before multiple cracking is complete.

With respect to photo and oxygen sensitivity, the surrounding concrete in the composite protects the fibers so well that this drawback is removed altogether.

Polypropylene fibers have been used with good results to increase the toughness of concrete subject to impact loading. However, the fibers have a high Poisson's ratio, a poor bonding with cement paste, a low elastic modulus, and therefore have rarely been considered as promising fibers for reinforcing materials in direct tension or flexure or for the achievement spaced multiple cracking in the relatively brittle matrix of cements and mortars.

The high impact strength of polypropylene fibers-reinforced cement is partly due to the large amount of energy absorbed in debonding, stretching, and pulling out the fibers; this occurs after the matrix has cracked.

TABLE 1
Typical Fiber Properties

Fiber	Diameter (μm)	Length (mm)	Density kg/m³/10³	Young's modulus GN/m²	Poisson's ratio	Tensile strength MN/m²	Elongation at break %	Typical volume in composite %[a]
Chrysotile (white)	0.02–30	<40	2.55	164	0.3	200–1800 (fiber bundles)	2–3	10
Asbestos								
Crocidolite(blue)	0.1–20		3.37	196	—	3500	2–3	—
Type 1 (high modulus)	8		1.90	380	0.35	1800	0.5	
Carbon		10-cont.						2–12
Type 2 (high strength)	9		1.90	230		2600	1.0	
Cellulose			1.2	10		300–500		10–20
E	8–10		2.54	72	0.25	3500	4.8	
Glass Cem-Fil filament	12.5	10–50	2.7	80	0.22	2500	3.6	2–8
204 filament strand	110×650			70		1250	—	
PRD 49	10		1.45	133	0.32	2900	2.1	
Kevlar		6–65						<2
PRD 29	12		1.44	69		2900	4.0	
Nylon (type 242)	>4	5–50	1.14	Rate dependent up to 4	0.40	750–900	13.5	0.1–6
Monofilament	100–200	5–50	0.9	Rate dependent up to 5		400	18	0.1–6
Polypropylene					0.29			
Fibrillated	500–4000	20–75	0.9	up to 8	0.46	400	8	0.2–1.2
High tensile	100–600			200		700–2000	3.5	
Steel		10–60	7.86					0.5–2
Stainless	10–330			160	0.28	2100	3	

[a] 1% elongation — 10,000 × 10^{-6} strain.

TABLE 2
Typical Properties of Matrix

Matrix	Density kg/m	Young's modulus GN/m²	Tensile strength MN/m²	Strain at failure × 10⁻⁶
Ordinary Portland cement paste	2000–2200	10–25	3–6	100–500
High alumina cement paste	2100–2300	10–25	3–7	100–500
O.P.C. Mortar	2200–2300	25–35	2–4	50–150
O.P.C. Concrete	2300–2450	30–40	1–4	50–150

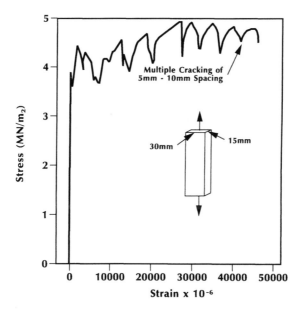

FIGURE 18. Stress-strain curve in tension for a composite containing 2.3 vol% of opened polypropylene film networks.

Comparative impact tests between polypropylene and steel fibers using a modified Charpy machine have indicated that polypropylene can absorb as much energy as some steel fibers for the same fiber volume.

The area under the load-deflection curve in slow flexure is another measure of the ability to absorb flexural impact energy. Figure 19 shows such a curve for short chopped fibers at 1.2% by volume.

It can be seen that compared with the plain matrix, the work to fracture is greatly increased.

With regard to the durability, there was little change in modulus of rupture or impact strength with time after a severe test (accelerated curing under water at 60°C for 1 year). Results indicated that the high impact strength derived from polypropylene will remain stable over very long periods of time in normal use.

Compared with concrete without fibers, the presence of fiber had made no difference in the behavior of concrete under fine conditions.

The polypropylene fibers had melted during the test, and when the temperature had been high enough, the polymer had volatilized, leaving fine channels and an additional porosity in the panel, which was otherwise fully intact.

Among other organic fibers used in cement composites, we have to mention carbon fibers, polyamide fibers, and cellulose fibers.

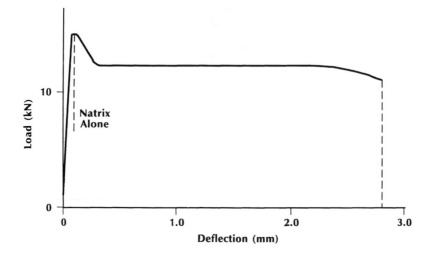

FIGURE 19. Load-deflection curve for 100 mm × 100 mm × 500 mm beam containing polypropylene chopped fibers (1.2 vol% of fibrillated polypropylene, 700 m/kg, length 75 mm).

TABLE 3
Mechanical Properties of High Modulus Carbon
Fiber-Cement Composites

Fiber volume (%)	Fiber orientation	Young's modulus GN/m²	Ultimate tensile properties		
			Stress MN/m²	Strain × 10⁻⁶	Impact strength kj/m²
0	—	13.8	5.52	300–400	2
3.0	Random in plane chopped fiber mat	18.2	9.6	570	1.4–1.8
3.7	Continuous aligned	26.1	26.6	2160	3.6–4.5

Table 3 shows some results obtained with carbon fiber cement using different fiber orientation. From this table, it can be seen that 3% by volume of random fiber mat produced about a 70% increase in tensile properties, but the impact strength is decreased. However, with 3.7% by volume of continuous aligned fibers, a fivefold increase in tensile strength can be achieved.

It can be seen from Figure 20 that the random mat composite has very little postcracking ductility.

The authors have established the good durability of carbon fiber-cement composites by measuring the strength retention of the samples which were kept under water at 18°C and 50°C up to 1 year.

Polyamide fibers, known as Kevlar 49, nylon, and Perlon, have been studied.

With 2 vol% fiber addition, typical mechanical properties measured at ambient temperature were ultimate tensile strength, 16 MN/m²; modulus of rupture, 44 MN/m²; and impact strength, 47 K_j/m^2. These values are not likely to be affected by age. The composite showed excellent fatigue resistance at stresses well above the elastic limit. No failures were recorded for samples stressed below the elastic limit of 15 MN/m² after 10^6 cycles.

The tensile stress-strain diagram of the Kevlar-cement composite is shown in Figure 21 with some other fiber composites.

The matrix cracking begins in the case of Kevlar-cement composites at more or less the same stress as in glass-reinforced composite. But the ultimate failure strain is much higher

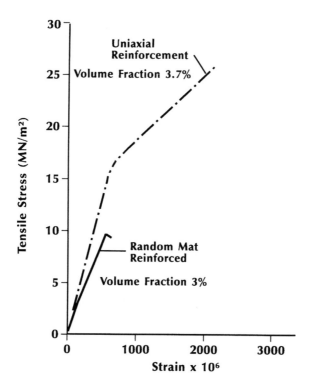

FIGURE 20. Stress-strain curves for carbon fiber — reinforced cement composites in tension.

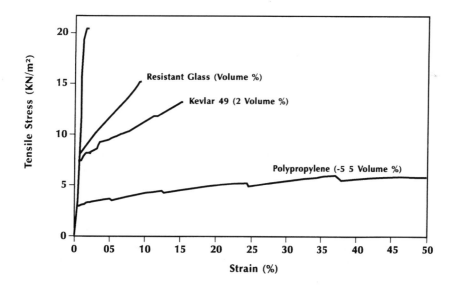

FIGURE 21. Tensile properties of some typical fiber cement composites.

in the composite containing Kevlar, reflecting perhaps higher effective tensile strength of this fiber and relatively poor interfacial bond strength.

Because Kevlar fibers decompose at a temperature much higher than can be sustained by polypropylene on nylon-type polyamide containing cement composites, the fire resistance of Kevlar cement is expected to be appreciably better than the former.

Nylon was one of the first of the synthetic fibers to be included in cement and concrete, but because of its relatively high cost compared with that of polypropylene, its commercial potential may be limited.

A. APPLICATIONS

The existing and potential applications for fiber-reinforced (FR) concrete can be conveniently placed into two main categories: (1) mass concrete applications, in which the final product is produced at the construction or building site, and (2) precase concrete applications, in which the product is produced in a plant and shipped elsewhere for use.

In the U.S., it is in the mass concrete area that fiber concrete has had the greatest success in attracting the interest and active participation of potential users of the material. In England and Western Europe, the interest may be about equally divided, and the overall activity is less than that in the U.S. Refer to the list of applications in Tables 4 and 5.

IV. GLASS FIBER-REINFORCED CEMENT (GRC)

GRC is a composite in which a strong brittle fiber is used to strengthen and toughen a weak brittle matrix. The basic mechanics of this material is no different from that of other ceramic and brittle matrix composites or reinforced plastics and metals. But when examining both its microscopic structure and characteristics, it is one of the most complex composites ever made. Some of the special characteristics are given below.

The purpose of reinforcing the cement matrix is not quite the same as reinforcing a plastic or metal. For example, in the case of plastics, fibers are added to strengthen and stiffen the matrix. While cement is relatively stiff but brittle, adding fibers inhibit crack propagation, increase tensile strength and fracture energy, and provide a degree of ductility in the behavior of the material and a local stress relieving mechanism. The localized cracking of the cement gives the required energy absorbing mechanism, but the catastrophic propagation of these cracks is prevented by the presence of unbroken bridging fibers that hold the matrix together and confer a load-bearing function on the locally cracked material. For this type of function, neither very small nor very high fiber content is required. Glass fiber which is 3 times stiffer and 300 times stronger than the cement matrix, is an ideal reinforcing material and 5 to 10% of glass by volume is adequate for this function.

The microscopic characteristics of cement are also different from those of plastics and metals. Hardened cement has a very low tensile strength (1.4 to 7 N/mm^2) and failure strain (200 to 600 10^{-6}). Microscopically, it has a porous structure which that changes itself with age due to continuing hydration of cement. The size and extent of the porosity depends on the water-cement ratio and fabrication and curing methods. The Young's modulus, shear modulus, and compressive strength of cement paste vary with the porosity.

Like concrete and cement mortar, plain cement paste contains, albeit to a lesser extent, randomly oriented and distributed microcracks. Differential shrinkage of cement paste and the presence of unhydrated cement particles or additives like sand determine the size and density of the microcracks. Wetting and drying shrinkage, alternate freezing and thawing, together with the constraining effects of fibers, could increase the microcracking of cement paste within the composite. The failure of cement paste is considered to be, by the slow growth and merging of such microcracks to form effect of all the voids, inclusions. Microcracks is to make the stress-strain relation somewhat nonlinear and the strength and failure strain erratic. In other words, unreinforced cement paste possesses neither a stable stress-strain curve nor a well-defined tensile strength, and therefore cannot be used to bear tensile, bending, or cyclic loads.

TABLE 4
Mass Concrete Application Areas for FR Concrete for Which Field Trials have been Performed

Application area	Fiber types used in concretes evaluated in the field	Countries in which significant field work has been done
Bridge decks, overlays, and construction	Steel	U.S.
Highway, street, and airfield pavement overlays and construction	Steel, glass	U.S., England, Canada
New pavement construction	Steel	U.S.
Mass concrete maintenance and repairs (dams, slabs, pavements, bridges, culverts, etc.)	Steel	U.S., England, Western Europe
Mining and tunneling	Steel	U.S., England
Rock slope stabilization	Steel	U.S.
Industrial floors	Steel, glass	U.S., Italy, England
Refractory applications	Steel	U.S.

Note: In the U.S., it is estimated that at least 70% of the total field work effort has involved the construction, overlay, or repair of bridge decks and highway, street, and airfield pavements and the repair of other mass concrete members such as dams and culverts.

TABLE 5
Precast Concrete Product Applications for FR Concrete

Product application	Type of fiber evaluated	Countries in which significant field work has been done
Car park deck slabs	Steel	England
Concrete pipe	Steel, glass	U.S., England
Concrete piling	Polypropylene	England
Ceramic tooling	Steel	U.S.
Floating pontoon units	Steel, glass	England
Dolosse (break waters)	Steel	U.S.
Boat hulls	Steel, glass	U.S., England
Burial vaults	Steel	U.S.
Concrete steps	Glass	U.S.
Decorative garden units	Glass	U.S.
Utility poles	Steel	
Decorative building panels	Polypropylene	Canada
Structural units	Steel	U.S.
Manhole assembly	Steel	England
Weight coatings for undersea gas and oil transmission line	Steel, polypropylene	U.S., England
Pile tips	Steel, glass	U.S.
Machine pads	Steel	U.S.
Machine frames	Steel	U.S.
Precast refractory shapes	Steel (stainless and carbon)	U.S.
Underground utility vaults	Steel	U.S.

Glass fiber, unlike cement paste, has a well-defined Young's modulus, and the stress-strain behavior is linear up to the point of failure. The spray method of fabrication inflicts relatively little damage on the fiber.

The bonding characteristics of glass-cement interface play a significant role in determining the composite behavior. The interface bond in a 28-d-old material is predominantly frictional, but with age, it may remain frictional or change to an adhesive bond followed by frictional slip after debonding. There are indications that the bond strength increases with the age of the composite, however, the relative increase in the magnitude of the two forms of bond has not yet been conclusively determined.

A. MICROSTRUCTURE OF GRC

The major difference between GRC and most other composites is in its special microstructure. Because the fibers are sprayed as strands of 204 filaments and because the cement particles with their 20 μm average size are unable to penetrate into the strand, the strands remain integral as short bundles or fiber-rich layers separated by cement (illustrated in Figure 2). The low glass content and lack of filament dispersion make the proportion of such fiber-rich layers small compared to the cement paste. The thickness of cement paste between two strands stacked one above the other is about 560 μm compared to 40 μm between fibers that are uniformly dispersed.

The fiber-rich regions are oval or nearly rectangular in cross-section and are randomly packed in two dimensions. However, in many spray-molded GRC products, a preferential orientation of strands in directions has been observed.

The continuous change of microstructure with age is another feature that is of some significance. The fiber bundle of 28-d-old material has practically no cement hydrates within it and therefore tends to act as a single reinforcing element. As chemical reaction continues in the cement and to some limited extent at the interface, the reaction products penetrate into the strand. Those filaments in the strand covered by cement products this way tend to act as independent reinforcements. In 6 to 12 months, the entire strands of a water-stored GRC are filled by reaction products; the fiber bundle may be effectively modified from a single reinforcement to a "miniature composite." The bonded surface area increases from the initial surface area of the bundle to the total surface area of all the filaments in the bundle.

A lightweight material can be produced by incorporating in glass-reinforced cement 20 to 60 wt% pulverized fuel (also known as "cenospheres" or "floats"). Composite sheets having initial dry densities in the range of 700 to 1500 kg/m^3 have been produced by spray dewatering method. (Asbestos wallboards have densities in the range of 900 to 1450 kg/m^3, and the values for insulating boards are somewhat lower).

The authors have determined the mechanical and thermal properties of lightweight glass-reinforced cement as well as its durability. Typical property values are flexural strength, 12 MN/m^2; limit of proportionality in flexure, 6 MN/m^2; tensile strength, 4.5 MN/m^2; impact strength, 13 kg/m^2; Young's modulus, 99 N/m^2; shear modulus, 39 N/m^2; and thermal conductivity, 0.2 W/mc/°C.

On completion of a 1-h long fire test, the material was found to retain its integrity and showed little change in appearance.

The composition of lightweight glass-reinforced boards and some mechanical properties may be seen in Tables 6 to 8.

The new material is considered to be suitable as replacement for some asbestos wall and/or insulation boards in fire protection.

TABLE 6
Composition of Lightweight GRC Boards (wt%)

Dewatered board at time of fabrication (28 d old)

Board number	OPC (%)	Cenospheres	Glass fiber (%)	Water (%)	Water/solid	Water/cement	Porosity[a] (%)	Bulk density[b] (kg/m³)
1	74	0	4.9	21	0.28	0.28	—	2150
2	57	18	6.3	20	0.27	0.35	—	1400
3	48	26	8.3	18	0.25	0.38	18	1080
4	46	30	7.9	16	0.21	0.35	12	1060
6	40	33	7.6	19	0.26	0.47	15	950
9	37	36	8.4	18	0.25	0.49	18	900
10	24	42	8.5	26	0.39	1.1	7	750

[a] Porosity values refer to open pores only.
[b] From 7 to 28 d, stored in air at 20°C and 60% RH.

TABLE 7
Modulus of Rupture of Lightweight GRC in Different Environments (MN/m²)

Board number	Air days 7	Air days 28	Air days 90	Air days 180	Air days 365	Water days 90	Water days 180	Weathering days 365	Weathering days 365
1	37	40	39	—	31	—	—	19	30
2	19	25	—	23	24	—	—	19	23
3	9	15	17	—	17	—	15	16	19
4	13	14	14	—	14	12	—	15	19
6	6	12	—	12	13	—	11	12	14
9	4	10	—	10	12	—	9	11	12
10	2	5	—	5	5	—	4	5	6

TABLE 8
Specific Tensile Strength and Failure Strain of Lightweight GRC (kNm/kg, μ-strain)[a]

Board number	Air 28 days	Air 1 year	Water 1 year	Weathering 1 year
1	—	6	4	4.5
		9190	320	3880
2	7	—	—	—
	10,070			
3	6.5	6.5	4.5	6
	9030	8530	6870	7600
4	5	6	5	7
	8220	8040	7390	8470
6	—	5	3.5	5.5
	—	8610	5780	8620
9	4.5	4.5	3.5	5
	7220	8280	7060	7940
10	2.5	—	—	—
	3870	—	—	—

[a] Top figure is specific strength; failure strain is below

Chapter 29

PROPERTIES OF POLYMERS SUITABLE FOR SOLAR ENERGY APPLICATIONS

R. Hussein

TABLE OF CONTENTS

0-8493-4401-8/93/$0.00 + $.50

975

I. INTRODUCTION

Various polymers already play a prominent role in the solar industry. This role is likely to increase steadily, for an unexpected reason. Even though the raw material for most of our plastics is oil, the energy content of the finished product is much less than most of the alternatives. Not much energy is used in processing most polymers, while enormous amounts are used in transporting and transforming the raw materials for glass, aluminium, steel, etc. As conventional energy sources diminish, solar manufacturers and builders will shift to the use of polymers wherever possible. In addition, many polymers can be produced from renewable raw materials like alcohol and methane.

Anyone who has returned to a closed, unshaded automobile after a sunny afternoon at the beach understands the basic principles of primary solar-thermal converters or collectors. Any dark material or coating converts light to heat energy by degrading the energy of excitation of electrons into vibrational energy of molecules or atoms. If we slow down the loss of this heat by putting a glazing over the absorber and insulation behind it, some of the heat can be removed and put to use before it returns to the environment. Thus, the main elements of a collector or a passive solar heater are an absorber, a glazing, some insulation, plus some arrangement to facilitate transfer of the heat to the end use or storage. The latter usually consists of a heat exchanger attached to the absorber and conduits or pipes to contain a moving heat exchange fluid, most often air or water. In some cases, the absorber is placed on a solid or liquid storage, and conduction is used to remove the available heat. Some air heating collector designs are shown in Figure 1.

Polymer products such as fiberglass, polycarbonates (PCO), polyvinyl fluorides (PVF), and Teflon® are already used extensively as glazings because of their light weight, freedom from breakage, and good transmission for the solar input. Foamed plastic insulation is often used in spite of serious long-term degradation problems. The largest profit center of the solar manufacturing industry, swimming pool heaters, is served almost exclusively by extruded integral collectors of black plastic with no glazing. Because of their low price ($2 or $3/ft^2), these collectors have found a large market. Flexible combined heat exchangers and absorbers of ethylene-propylene-diene rubber (EPDM) are finding a rapidly growing market because of their ease of installation, durability, and ability to stretch when ice forms in the conduits. Any solar collector must be carefully sealed to prevent heat leaks. If the circulating pump or fan fails, temperatures as high as 250°C can occur, causing unusually high thermal expansions. High quality (i.e., high temperature) polymer sealants, gaskets, and caulking, such as the silicones and EPDM, are used almost exclusively to seal the glazing and openings for the heat transfer fluid pipes or ducts. In addition, most solar heating systems require piping or ducting from the collector to the points of use, and an assortment of hardware such as pressure release valves, back flow prevention valves or dampers, routing valves, etc. Long and reliable service life is essential so corrosion-free polymers are often specified here. Water storage containers are often made from fiberglass. Storages and many of the pipes or ducts require insulation to conserve the hard-won solar heat. A market is also developing for Fresnel lens concentrators made of polymethylmethacrylate (PMMA) for use with solar cells.

As always, the choice of materials is dictated by the stresses encountered in use, as well as the design parameters. For collectors, the stresses can be much more severe than those on materials used for the rest of the building. Ultraviolet intensity is necessarily high on the glazing. Thermal and the consequent mechanical cycling is necessarily high, since most systems are designed for at least 60°C absorber temperature under normal noontime operation, and can go much higher if the heat transfer fluid flow is reduced by a power failure or component malfunction. Condensation will occur inside the collector at night, so

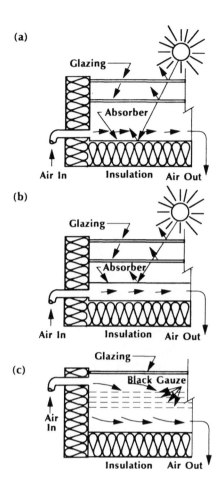

(a)

Glazing

Absorber

Air In   Insulation   Air Out

(b)

Glazing

Absorber

Air In   Insulation   Air Out

Glazing

Black Gauze

(c)

Air In

Insulation   Air Out

FIGURE 1. Cross-sections of three air heating flat plate collectors: (a) front pass collector, (b) back pass collector, and (c) collector with mesh absorber.

it must be weather proof inside as well as out. To minimize radiation heat losses, the glazing should be as opaque as possible in the middle infrared (I.R.). As shown in Figure 2, the peak of the radiation from the absorber lies at about 9 μm. No plastic is as effective as glass in this respect, but the polycarbonates and Teflon® fluorinated ethylene-propylene (FEP) are quite good (Table 1 and Figure 3). Similarly, absorbers with low emissivity in the middle infrared have a thermal advantage, as shown in Figures 4 and 5. It may be that the graphite epoxy composites now being used for airplane skins will serve well as absorbers since the I.R. emissivity of graphite is only 45%. Of course, as one designs for lower losses, the operating temperatures inside the collector rise (the range of velocities for the heat transfer fluid is limited by mechanical and corrosion considerations), and the choice of polymers is reduced.

One very useful trick is to reduce convection and radiation losses simultaneously, without much loss in visible input by using a transparent honeycomb inside the outer glazing. As shown in Figures 6 and 7, the honeycomb material must have good transparency in the visible and good absorption in the middle I.R. It must also withstand high temperatures and be light weight. Thus far, the only material used for this purpose in a commercial collector has been Teflon® FEP film. Teflon® is expensive and difficult to fabricate. A cheaper, extrudable material would be most welcome here. Typical collector efficiency curves for different systems are shown in Figure 8.

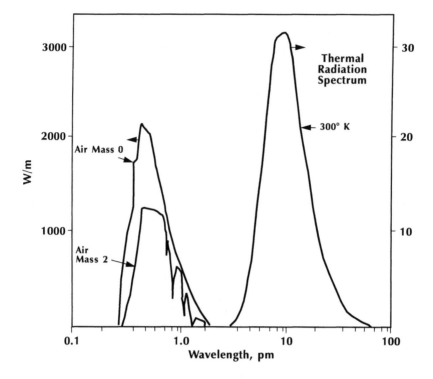

FIGURE 2. Shows peak of radiation from absorber.

II. GLAZINGS

Polymer films can have a higher visible transmittance than glass, both because of their strength, which allows thinner films, and their lower index of refraction, which gives lower reflectance of the solar input. Polymers containing strong polar groups, such as oxygen or halogen atoms, have the strongest middle I.R. absorption. In order of decreasing absorption, they are: PMMA, PCO, PVF, Teflon® FEP, and polyester (PETP).

PETP, PMMA, and PVF soften above 90°C. Fiberglass can be used to somewhat higher temperatures than PETP without reinforcing fibers. None of these should be used as inner collector glazings. PCOs have an upper limit of 130°C and can be used fairly close to the absorber in air heating collectors. The maximum temperature for Teflon® FEP is 200°C. If allowance is made for the sag that occurs at high temperatures, FEP can be used as an inner glazing.

PMMA and the various fluorocarbons weather well without surface coatings or additives. They are completely transparent to UV. If used as an outer glazing, Teflon® should have a surface treatment to increase wettability, so rain will wash down dust accumulations which would reduce transmission. Many simpler polymers can probably be stabilized with the addition of U.V. absorbers, either in bulk or as a surface layer, but long-term test results are not yet available. Pure PCO and PETP films are subject to discoloration and cracking on weathering. However, a surface film of PVF or FEP can slow this attack considerably. To avoid scratching and abrasion, a layer of polymethyl siloxane can be applied to PMMA or PCO. Tables 2 and 3 provide typical properties and comparisons of glazing materials.

TABLE 1
Glazing Materials: Optical Properties

Cover material identification	Trade names	Formula	Light transmittance	Cost $/m² (1977)	Longwave transmission %[a]	Daily shortwave transmission	Solar energy transmission[b]			
							Type[c]	Angle of incidence (degrees)		
								0	45	67
Fluorinated ethylene-propylene copolymer (FEP)	Suntek Teflon		95 96 (1 mm film)	3.8 5.3	13					
Polyvinyl fluoride (PVF) 0.07 mm	Tedlar	$[CH_2 - CHF]_m$		1.83	43	91.0	Total Dir.	93 85	90 78	65 48
Polymethyl methacrylate (PMMA)	Plexi-glass	$[-CH_2-C-\ ]_m$ CH_3 / $COOCH_3$	91 (after 10 years)			91.0				
Polycarbonate (PCO) 1.6 mm	Tuffak (Lexan)	$[-O-C_6H_4-C-C_6H_4-O-C]_m$ CH_3 $=O$	82–89 depending on thickness	26.91	6	84.4	Tot. Dir.	86	83	72
Polyethylene-terephthalate (PETP) 0.12 mm	Mylar	$[-O-(CH)_2-O-C-C_6H_4-C-]_m$ $=O$ $=O$		1.94	32	86.5	Tot. Dir.	88 87	86 84	72 72
Fiberglass reinforced polyester (FRP) 1 mm	Sunlite premium		80 (after 20 years)	4.52	6	83.1	Tot. Dir.	81 65	80 60	57 34

TABLE 1 (continued)
Glazing Materials: Optical Properties

Cover material identification	Trade names	Formula	Light transmittance	Cost $/m² (1977)	Longwave transmission %[a]	Solar energy transmission[b]				
						Daily shortwave transmission	Type[c]	Angle of incidence (degrees)		
								0	45	67
Corrugated fiberglass 1 mm[d]				5.27	7–8	78.1–	Tot.	81	75	43
				5.92		79.2	Dir.	52	45	21
Glass, double strength, 3 mm				5.38	3	87.8	Tot.	89	85	74
							Dir.	88	85	70

Note: Tot. = Total solar energy; Dir = Direct beam component of the total solar energy.

[a] Longwave > 2.8 m

[b] Shortwave < 2.8 m

[c] Type of solar energy flux transmission measurement (January)

[d] Coated with PVF

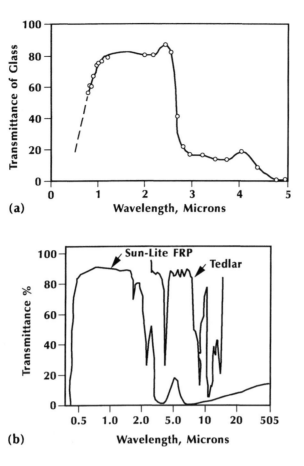

FIGURE 3. (a) The transmittance curve for ordinary glass shows that solar radiation (see Figure 6) passes through, but thermal infrared (>5 μm) is almost completely absorbed. (b) The transmittance curve for Sun-Lite® (a glass fiber-reinforced polyester) and Tedlar®.

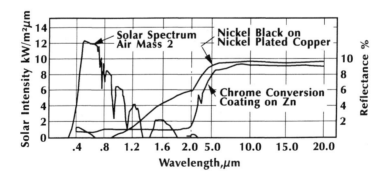

FIGURE 4. Reflectivity of selected surfaces. The solar spectrum is superimposed for reference.

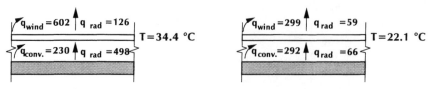

One Cover, Plate Emittance = 0.95; $\mu\tau$ =8.1 W/M²° C   One Cover, Plate Emittance = 0.10; $\mu\tau$ =4.0 W/M²° C

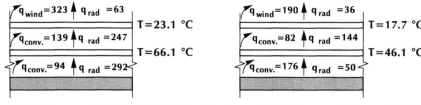

Two Covers, Plate Emittance = 0.95; $\mu\tau$ =4.3 W/M²° C   Two Covers, Plate Emittance = 0.10; $\mu\tau$ =2.5 W/M²° C

FIGURE 5.   Temperature distribution and upward heat loss terms for flat-rate collectors operating at 100°C with ambient and sky temperatures at 10°C, plate spacing of 2.5 cm, tilt of 45°, and wind speed of 5 m/s (all heat flux terms in W/m²).

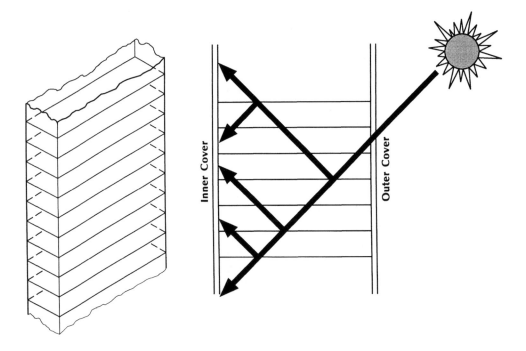

FIGURE 6.   Light transmission in honey comb cells (all reflections reach the inner cover).

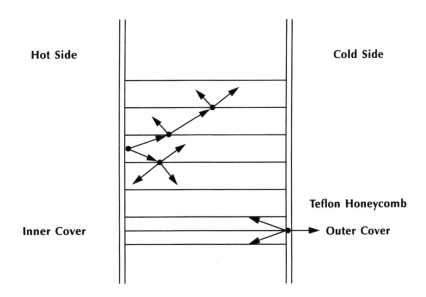

FIGURE 7. Infrared reradiation in honeycomb cells.

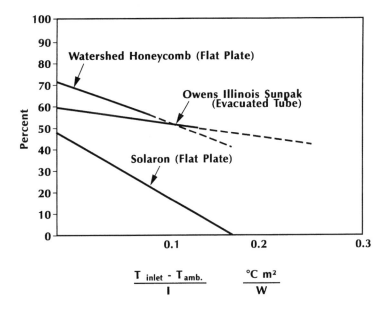

FIGURE 8. Tested collector efficiency curves.

TABLE 2
Glazing Materials; Mechanical, Thermal and Other Properties

Cover Material	Mechanical properties				Thermal properties		Flammability	Chemical resistance	Surface weathering	Maximum Operating Temperature (°C)	Heat shrink capability and temperature (°C)	Outdoor longevity (years)
	Tensile strength (MPa)	Impact strength (J)	Elongation (%)	Tensile modulus KPa 10^5	Conductivity $J/mm^2/s/°C$	coefficient of thermal Expansion $°C^{-1}/10^{-5}$						
Suntek					0.028	10.8				204		
FEP										148		30
PVF	20–100	—	100–190	20	0.25	4.8	Nonflammable	Highly resistant	Undamaged by 20 years	208	Yes 148	Data not available
PMMA	55–75	11	207	24–35	0.19	7.3	Combustible	Alkalis, weak acids, oils	Little or none	82–93	0.002–0.008% at 154–261	10–20
PCO	55–65	22	60–100	24	0.19	6.84	Combustible self-extinguishing	Weak acids aromatic solvents	Resistant	121–132	0.005–0.007% at 261–315	5–7
PETP — film	3.4–7		50–100	0.06								
Sunwall (two layers 1 mm with 12 mm air gap)	75–79	7	—	0.066–0.079	0.06–0.08	0.72–2.23	Less than 3.81 mm/min ASTM D-635	Most airborne chemicals	20 years expectancy	93		20; minimum maintenance
Flexigard (PET and PMMA) composite film						5–9	No specification		Highly resistant	121	30% unsupported 0% framed	7–10
FRP	75–79	14			0.13	2.52	Less than 3.81 mm/min	Most acids and alkalis	Good color stability	93	Does not shrink	20

TABLE 3
Comparing Glazings

	Thickness (in.)	Cost($/ft)	Transmittance	Weight/area (lb/ft²)	Thermal expansion (F⁻¹ × 10⁻¹)	Ease in handling	Strength	Sheet size (ft)	Remarks
Water white glass "Solatex" (ASG)	0.125	0.99	0.90	1.60	0.47	Poor	Good (tempered)	2, 3 or 4 × 8	Very durable —no degradation
Float glass	0.125	2.35	0.84	1.60	0.47	Poor	Good (tempered)	4 × 8	Very durable —no degradation
Window glass (ASG SS Lustraglass)	0.090	1.80	0.91	1.20	0.47	Poor	Poor (nontempered)	4 × 7	Fragile
Sunlite Premium II (Kalwall)	0.040	0.60	0.88	0.29	2.00	Excellent	Very good	4 or 5 width rolls	Maximum temperature 300°F
Filon with Tedlar (Vistron Corp.)	—	1.00	0.86	0.25	2.30	Very good	Very good	4.25 × 16	Maximum temperature 300°F
Flexiguard 7410 (3M)	7 ml	0.38	0.89	0.053	—	Fair	Good	4 × 150 roll	Maximum temperature 275°F
Tedlar (Dupont)	4 ml	0.05	0.95	0.029	2.80	Fair	Good, some embrittlement	up to 5.33 width roll (64 in.)	4–5 year lifetime at 150°F
Teflon FEP 100A (Dupont)	1 ml	0.58	0.96	0.02	5.85	Poor	Fair, not for exterior glazing	4.83 width roll (58 in.)	Maximum temperature 300°F

TABLE 3 (continued)
Comparing Glazings

	Thickness (in.)	Cost($/ft)	Transmittance	Weight/area (lb/ft^2)	Thermal expansion ($F^{-1} \times 10^{-1}$)	Ease in handling	Strength	Sheet size (ft)	Remarks
Swedcast 300 Acrylic (Swed-low Inc.)	0.125	0.81	0.93	0.77	4	Excellent	Very good	9 wide	Maximum service temperature 200°F
Lucite acrylic (Dupont)	0.125	1.14	0.92	0.73	4	Very good	Very good	4 × 8	Maximum temperature 200°F
Tuffak-Twinwall (Rhorm & Haas)	—	1.25 (2 layers)	Equiv. to 0.89 for 1 layer	0.25	3.3	Very good	High impact strength fatigue cracking	4 × 8	5% reduction in transmittance over 5 years
Acrylite SDP (Cyro)	—	2.15 (2 layers)	Equiv. to 0.93 for 1 layer	1.00	4	Very good	Good	6 × 8	Maximum temperature 230°F
Sunlite Insulated Panels (Kalwall)	—	2.50 (2 layers)	Equiv. to 0.88 for 1 layer	0.7	—	Good	Good	4 × 8 4 × 10 4 × 12 4 × 14	Maximum temperature 300°F
Solar glass panels (ASG)	—	2.99 (2 layers)	Equiv. to 0.90 for 1 layer	4.5	0.47	Poor	Good	3 or 4 × 6 3 or 4 × 8	Very durable

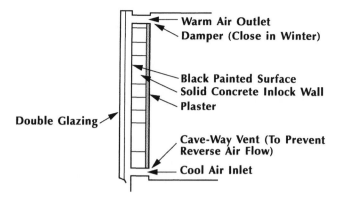

FIGURE 9. Trombe wall section and retrofit.

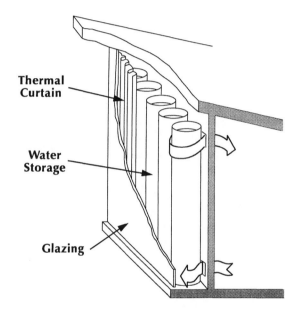

FIGURE 10. Thermal curtain with water tube storage.

III. INSULATION

Most of the plastic foam insulations will shrink under the cyclic internal and external mechanical stresses resulting from the daily temperature swings inside the collector. Gaps as large as half an inch have been reported in 7-ft long collectors. The most suitable of the common foams are the polyurethanes and polystyrenes, but even these will show some shrinkage, especially in the presence of moisture. They should be laminated or foamed in place, so that the inevitable gaps do as little thermal damage as possible. The polystyrenes especially should be protected from melting with an inner layer of higher temperature insulation such as fiberglass. Polyisocyanurate foams have shown good stability, but are not yet easily available. Figures 9 through 13 illustrate different applications in construction.

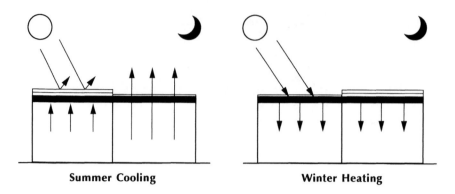

FIGURE 11.   Thermal storage roofs: summer and winter operation.

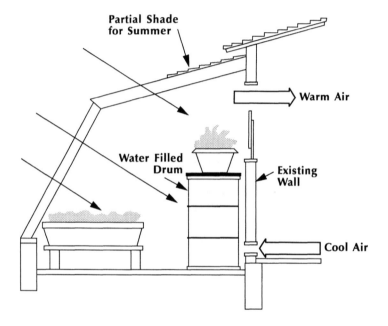

FIGURE 12.   Cross-section of solar sustenance greenhouse project.

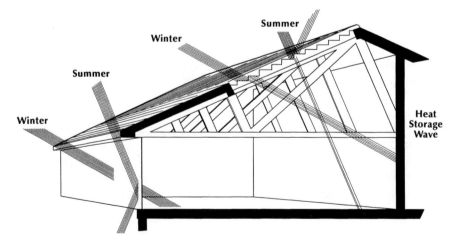

FIGURE 13.   Solar staircase (patented by N. Saunders).

GENERAL PROPERTY DESCRIPTIONS OF PRINCIPLE TYPES OF ELASTOMERS

This appendix contains a brief description of widely used rubber and elastomers. For an introductory background on the chemistry of elastomers, please read the following paragraphs.

Elastomer Chemistry — The molecules of chemicals are called monomers. Synthesized rubbers are made from monomers and are polymerizable. The term polymerizable refers to the fact that monomers can join with themselves and/or other monomers to form links in a chain of molecules. As this process occurs, the material changes from a liquid or gaseous state to a rubbery state.

These chains are called polymers. The chains of two monomers are called copolymers. The term terpolymer refers to a chain containing three monomer types.

The process of polymerization takes place under the influence of heat and in the presence of catalysts. The constituents are emulsified in an aqueous soap solution in a large, closed vessel called a polymerizer. The material formed in the polymerizer is a latex that is used for some products or it is coagulated by acids or salts. The solid rubber that comes from the vessel is usually in the form of lumps or crumbs. It is then washed, dried, and pressed into bales, or bagged as pieces/crumbs or rolled into sheets.

There are many variations to this process. Oil can be added to the rubber producing an oil-extended rubber. Carbon black and oil can be added to produce carbon black masterbatches. The physical and chemical properties of the final product are established by the type and portion of the monomers and other chemicals in the polymerization formula. Other ingredients can include catalysts/co-catalysts, antioxidants, modifiers, and catalyst activators. Additionally, temperature, pressure, and residence time affect the properties of the end product.

Solution polymerization is conducted in a solvent rather than water. This is more costly than emulsion polymerization as previously described; however the control of molecular weights and polymer structure are many times greater. Tailor-made or specialty polymers are more readily made by solution polymerization methods.

Description of Elastomers:

Natural rubber - NR

Also known as: Natural polyisoprene, tree rubber, *Hevea*, plantation or estate rubber, crude. Many grades and varieties.

Available as: Bales, sheets, crumb, ground, liquid, oil-extended, black masterbatch.

Specific gravity	0.92 - 0.93
Tensile strength	E (4500 psi)
Resilience	E
Compression set	E
Heat aging at 212°F	G
Low temp. stiffening, °F	−20 to −50
Low temp brittle point, °F	−80
Resistance to: abrasion	E
impact	E
tear	E
flame	P
weather & sunlight	P
oxidation	F
ozone	P
HC solvents & oils	P
acid, dilute/conc.	E/F to G
alkali, dilute/conc.	E/F to G

Acrylate-butadiene rubber - ABR

And/or copolymers of ethyl or other acrylates with a small amount of a monomer to facilitate vulcanization, ACM. A family of copolymers also called acrylic rubber, polyacrylic rubber, polyacrylate.

Specific gravity	1.09
Tensile strength	F (2000 psi)
Resilience	G
Compression set	G
Heat aging at 212°F	E
Low temp. stiffening, °F	+35 to +10
Low temp. brittle point, °F	−15
Resistance to: abrasion	F
impact	P
tear	P to F
flame	P
weather & sunlight	E
oxidation	E
ozone	E
HC solvents & oils	G to E
acid, dilute/conc.	F/F
alkali, dilute/conc.	F/F

End uses:
All high performance rubber products not requiring resistance to solvents, oils, ozone, flame or direct sunlight. End uses include tires (especially aircraft, heavy-duty, off-the-road), gaskets, seals, belting, rolls, hose, tubing, vibration dampers, motor and engine mounts, bearings, electrical goods, bumpers, drive wheels, drug sundries, surgical goods, sheeting, rubber bands, toys, athletic and sports goods, adhesives, mastics, caulks, O-rings, flooring, shoe products, hard rubber (Ebonite) goods, cellular products, latex foam mattresses and pillows.

End Uses:
Hydraulic hose, transmission gaskets and lip seals, packings, conveyor belts, O-rings, electric light gaskets, valve stem seals, pinion seals, spark plug boots, ignition cable jackets, fabric coatings and tank linings.

E = excellent
G = good
F = fair
P = poor

Butadiene rubber - BR
Solution polymerized *cis*-polybutadiene.
Available as bales, slabs, non-pigmented, oil-extended, black masterbatches, other varieties.

Specific gravity	0.91
Tensile strength	G (3000 psi)
Resilience	E+
Compression set	G
Heat aging at 212°F	F

ozone	G to E
HC solvents & oils	F to G
acid, dilute/conc.	E/E
alkali, dilute/conc.	E/E

End Uses:
Bridge bearing pads, adhesives, caulks, sealants, paint, conveyor and V-belts, hose, packings, diaphragms, wire and cable jacketing, motor mounts, bellows, bushings, seals, rolls, flooring, coated fabrics, miscellaneous extruded, calendered and molded goods, dipped goods, tank linings, automotive parts, cellular products, railroad crossing blocks, expansion joints and strips, footwear and underground wire insulation.

Chlorosulfonylpolyethylene - CSM

Also called chlorosulfonated polyethylene and by its tradename, Hypalon. There are several varieties.

Specific gravity	1.18 ave.
Tensile strength	F to G (3500 psi)
Resilience	F to G
Compression set	F to G
Heat aging at 212°F	G to E
Low temp. stiffening, °F	-30 to -50
Low temp. brittle point, °F	-70
Resistance to: abrasion	G
impact	G
tear	G
flame	G
weather & sunlight	E
oxidation	E
ozone	E+
HC solvents & oils	P to E
acid, dilute/conc.	E/E
alkali, dilute/conc.	E/E

Low temp. stiffening, °F	-30 to -60
Low temp. brittle point, °F	-100
Resistannce to: abrasion	E+
impact	G
tear	G
flame	P
weather & sunlight	F
oxidation	G
ozone	P
HC solvents & oils	P
acid, dilute/conc.	F/F
alkali, dilute/conc.	F/F

End Uses:
Blended with NR, IR, SBR for improving abrasion and groove cracking resistance of all types of tires. For improving resilience and low temperature properties in blends with other rubbers. Used in mechanical goods; adhesives; shoe products; solid golf balls; hose; conveyor belts; motor mounts; grommets; tubing and high-resilience play balls.

Chloro-isobutene-isoprene rubber - CIIR

Also referred to as chlorobutyl. Faster cure rates than butyl IIR with conventional cure systems. Compatible with other rubbers such as IIR, CR, NR, SBR; improved adhesion to other rubbers.

Specific gravity	0.92
Compression set	G
Heat aging to 400°F	E
Resistance to: tear	G
oxidation	G
ozone	G
gas permeability	G

End Uses:
See supplier literature for additional properties and for end use applications.

Chloro-polyethylene - CM

Also called chlorinated polyethylene elastomers and/or rubbers.
Chlorine content: 36 percent to 42 percent. Available in bags.

Specific gravity	1.16 - 1.25
Compression set	G
Resistance to: heat	G
low temperature	G
weather	G
ozone	G
flame	G
oil & chemicals	G

End Uses:
For good resistance to oxidizing chemicals and oils. Coated fabrics; solution coatings; sundries; paints; caulks; protective coatings; extruded, molded and calendered mechanical goods; rolls; wire and cable insulation; fluid handling hose; tank linings; roof sheeting; film; ditch liners; colored products and tire white sidewalls.

End Uses:
Molded and extruded goods, hydraulic hose and linings, gaskets, O-rings, cable jacketing. For blends with NR, EPDM, SBR and other rubbers.
See supplier literature for additional properties and other end use applications.

Chloroprene rubber - CR

Polychloroprene, neoprene, many varieties. Available as bagged nuggets, liquids, latex.

Specific gravity	1.23 - 1.25
Tensile strength	E (4000 psi)
Resilience	E
Compression set	F to G
Heat aging at 212°F	G to E
Low temp. stiffening, °F	+10 to −40
Low temp. brittle point, °F	−45 to −75
Resistance to: abrasion	G
impact	G
tear	G
flame	G
weather & sunlight	E
oxidation	G to E

Epichlorohydrin rubbers. Polychloromethyl oxirane (a homopolymer) - CO. Ethylene oxide (oxirane) and chloromethyl oxirane (a copolymer) - ECO.

These specialty polyether rubbers are also referred to as poly-alkylene oxide polymers. Several varieties.

Specific gravity	1.25 - 1.36
Tensile strength	NA*
Resilience	G
Compression set	NA*
Heat aging at 212°F	E
Low temp. stiffening, °F	−40
Low temp. brittle point, °F	−75
Resistance to: abrasion	NA*
impact	NA*
tear	NA*
flame	NA*
weather & sunlight	E
oxidation	G to E
ozone	E
HC solvents & oils	E
acid, dilute/conc.	NA*
alkali, dilute/conc.	NA*

*Not available; check supplier literature for specific type and application.

End Uses:
For applications requiring very low gas permeability and outstanding ozone and heat resistance: gaskets, oil field specialists, jar seals, fuel pump diaphragms, pump and valve parts, adhesives,

hose and tubing, sheet goods, oil seals, rollers, power and conveyor belting, coated fabrics, air conditioning hose. Items requiring resistance to chemicals, solvents, aromatic gasoline, gasohol and other exotic fuels.

Ethylene-acrylic rubbers

Available as black and non-black masterbatches with 81 percent polymer content.

Specific gravity	1.04 (raw polymer)
Resistance to: high temperature	G
low temperature	G
ozone	G
"chemicals"	G

End Uses:
See supplier literature for additional properties and end use applications.

Ethylene-propylene rubbers - EPDM, EPM

EPDM - Terpolymer of ethylene, propylene and a diene with the residual unsaturated portion of the diene in the side chain. EPM- Ethylene-propylene copolymer. Available in bales and crumb.

Specific gravity	0.86
Tensile strength	F to G (3000 psi)
Resilience	G
Compression set	G to E
Heat aging at 212°F	E
Low temp. stiffening, °F	−20 to −50
Low temp. brittle point, °F	−90
Resistance to: abrasion	F to G
impact	G
tear	P to F
flame	P
weather & sunlight	E
oxidation	E
ozone	E
HC solvents & oils	P
acid, dilute/conc.	E/E
alkali, dilute/conc.	E/E

acid, dilute/conc.	NA*
alkali, dilute/conc.	NA*

*Not available. Check supplier for additional properties.

End Uses:
Injection and blow molding, extruded goods, cellular products, weatherstripping, flexible packaging, shoe soles, protective padding, hot-melt adhesives and adhesive coatings, colored products, cable and ignition insulation, textile coatings, coolant and fuel hose, O-rings, belting, roll covers, machinery mounts and grommets.

Fluorosilicone rubbers

Also referred to as fluoro-vinyl silane.

Specific gravity	1.4
Tensile strength	F (1500 psi)
Resilience	F
Compression set	F to G
Heat aging at 212°F	E
Low temp. stiffening, °F	−70
Low temp. brittle point, °F	−85
Resistance to: abrasion	P
impact	P
tear	P
flame	E
weather & sunlight	E
oxidation	E
ozone	E
HC solvents & oils	G to E
acid, dilute/conc.	E/G
alkali, dilute/conc.	E/G

End Uses:
Fuel line seals, gaskets, sealants; shaft and access door seals; carburetor parts; hose; ducts and diaphragms.

End Uses:
Used as a polyolefin modifier; to blend with CR, IR and IIR to improve weatherability; gas and air impermeability equal to or better than butyl; excellent electrical properties for wire and cable insulations and connectors. Other uses include O-rings, brake cups and components, molded and extruded parts, tire flaps, steam and other hose tubes and covers, matting, rolls, brake wheel cylinder boots, master cylinder cover diaphragms, marine applications, roofing, window and other seals and gaskets, bumpers, tarp liner coating, cellular products, white and light-colored items, tire white sidewalls, curing bladders and bags, conveyor belts, soil pipe gaskets, coated fabrics, rug underlay, boot uppers and bindings, tie-down and tarp straps, and dock bumpers.

Ethylene vinyl acetate - EAM

A rubbery, flexible copolymer of ethylene and vinyl acetate. Several different varieties. Available in boxes.

Specific gravity	0.98 - 1.02
Tensile strength	G
Resilience	G
Compression set	G
Heat aging at 212°F	G
Low temp. stiffening, °F	-75
Low temp. brittle point, °F	NA*
Resistance to: abrasion	NA*
impact	G
tear	NA*
flame	P
weather & sunlight	G
oxidation	G
ozone	G
HC solvents & oils	P to F

End Uses:

Hexafluoropropylene-vinylidene fluoride copolymers - FKM and hexafluoropropylene-vinylidene fluoride-tetrafluoroethylene terpolymers

These fluoroelastomers are also commonly referred to as fluorinated rubbers and fluorinated hydrocarbons. They contain no chlorine in the polymer chain. Two commercial tradenames are Fluorel and Viton. Viton A and E are copolymers with about 66% fluorine; Viton B types, terpolymers with up to 68% fluorine and Viton G types, terpolymers similar to B with the addition of a peroxide curable crosslink monomer. Viton GF, the latest G type terpolymer, contains around 69% fluorine and is the most fluid resistant in the series. See also, "Fluoroelastomers - CFM."

Specific gravity	1.4 - 1.95
Tensile strength	F to G (1725 - 2550 psi)
Resilience	F
Compression set	G to E (in fluids)
Heat aging at 212°F to 600°F	E
Low temp. stiffening, °F	+7 to +23
Low temp. brittle point, °F	+10 to -60
Resistance to: abrasion	G
impact	P to G
tear	P to G
flame	E
weather & sunlight	E
oxidation	E+
ozone	E+
HC solvents & oils	E
acid, dilute/conc.	E/G
alkali, dilute/conc.	E/G

End Uses:
Specialized polymers useful where chemical resistance, low gas permeability and continuous service at temperature extremes are required. Typical applications include helicopter rotor seals, military and aerospace elastomeric components, universal seals, radiator seals, oil field (downhole) packer seals, V-cup packings,

heat exchanger gaskets, pipeline balls, O-rings, automotive and heavy-duty equipment, auto fuel delivery systems, fuel injector seals, locomotive radiator seals, carburetor valves, brake components, in-tank fuel lines and connectors, fuel filler cap pressure/vacuum relief seals, flue duct expansion joints, chimney coatings, blow-out preventors, agrichemical pump and tank seals.

Isobutene-isoprene rubber - IIR

Also called polyisobutylene-isoprene and more commonly referred to as butyl. Best known for outstanding impermeability to air and other gases. Available in bales and liquid (depolymerized).

Property	
Specific gravity	0.92
Tensile strength	F (2500 psi)
Resilience	P to F
Compression set	F
Heat aging at 212°F	E
Low temp. stiffening, °F	-10 to -40
Low temp. brittle point, °F	-80
Resistance to: abrasion	G
impact	G
tear	G
flame	P
weather & sunlight	E
oxidation	E
ozone	E
HC solvents & oils	P
acid, dilute/conc.	E/E
alkali, dilute/conc.	E/E

End Uses:
Inner tubes, tubeless tire liners, curing bags and bladders, conveyor belts, dairy and other hose, vibration dampers, electrical insulations, tank and reservoir linings, automotive and industrial molded parts, caulks, sealants and sealant tapes, pipe wrappings, appliance parts, freezer gaskets, light colored products and cellular goods.

End Uses:
For very high oil and solvent resistance: gaskets; fuel lines & hose; adhesives; oil well parts; fuel cell liners; rolls; packings; O-rings; molded, calendered, extruded goods; sheeting; shoe soles and heels; kitchen mats; coated fabrics; paper binder; diaphragms; implement handles; tubing; seals; conveyor belts and parts; printing blankets; caster wheels; wire and cable jackets; grinding and abrasion wheel binder and miscellaneous cellular products.

Polychlorotrifluoroethylene - CFM

Properties listed pertain primarily to copolymers of chlorotrifluoroethylene and vinylidine fluoride commonly known as fluoroelastomers, fluorinated hydrocarbons or by their tradenames, Fluorel and Kel-F. There are many other varieties of specialty fluorine type polymers. See also "Fluoroelastomers - FKM."

Property	
Specific gravity	1.4 - 1.9
Tensile strength	F to G (3000 psi)
Resilience	F
Compression set	G to E
Heat aging at 212°F	E
Low temp. stiffening, °F	+20 to -30
Low temp. brittle point, °F	+10 to -60
Resistance to: abrasion	G
impact	P to G
tear	P to G
flame	G to E
weather & sunlight	E
oxidation	E+
ozone	E+
HC solvents & oils	E
acid, dilute/conc.	E/G
alkali, dilute/conc.	G/P

Isoprene-acrylonitrile rubber
A low-plasticity copolymer with around 34 percent ACN, available in bales.

Specific gravity	0.96
Tensile strength	G
Resistance to oil	G

End Uses:
Thread, tape, pharmaceutical closures, adhesives, milking inflations, fabric frictions, cork and asbestos gaskets.
See supplier literature for additional properties and end use applications.

Nitrile-butadiene rubber - NBR
A wide variety of acrylonitrile-butadiene copolymers available as bales, slabs, crumb, latex, liquid, emulsions, also, blends with PVC and other resins.

Specific gravity	1.0
Tensile strength	E (4000 psi)
Resilience	G
Compression set	G
Heat aging at 212°F	G
Low temp. stiffening, °F	+30 to −20
Low temp. brittle point, °F	−40 to −65
Resistance to: abrasion	G
impact	F
tear	G
flame	P
weather & sunlight	F
oxidation	G
ozone	F
HC solvents & oils	E
acid, dilute/conc.	G/G
alkali, dilute/conc.	G/G

End Uses:
Specialized polymers useful where chemical resistance, low gas permeability and continuous service at high temperatures are required. Applications include O-rings, oil seals, gaskets, diaphragms, pump and valve linings, other molded products; hose and tubing; coated fabrics and sheet goods; solution coatings and rollers.

Polyisobutene - IM
Also called polyisobutylene, commonly referred to by the tradename, Vistanex. Available in bales, semi-solids, viscous liquids and aqueous dispersions.

Specific gravity	0.92 - 0.97
Resistance to ozone	G
Resistance to "chemicals"	G
Electrically inert	

End Uses:
Inks, adhesives, caulks, binders, hot-melt adhesives and tackifiers for grease and some rubbers.
See supplier literature for additional properties and other end use applications.

Polyisoprene rubber (synthetic polyisoprene) - IR
Available as bales, black masterbatches, oil-extended.

Specific gravity	0.91 - 0.93
Tensile strength	E (4500 psi)
Resilience	E
Compression set	E
Heat aging at 212°F	G
Low temp. stiffening, °F	−20 to −50
Low temp. brittle point, °F	−80
Resistance to: abrasion	E
impact	E
tear	E
flame	P

oxidation	F
ozone	P
HC solvents & oils	P
acid, dilute/conc.	E/F to G
Alkali, dilute/conc.	E/F to G

End Uses:
Essentially the same as those listed for natural rubber.

Polynorbornene

Available as a white powder.
Can absorb large quantities of oil, pigments, fillers to produce soft (20 Duro.) compounds with good tensile, abrasion resistance and excellent damping properties.

Compression set	G
Resistance to: hot tear	G
low temperatures	G
ozone	G
gas permeability	G
Electrical properties	G

End Uses:
Engine and body mounts. bridge bearing pads, gaskets, seals, printing blankets, implement handles, rug underlay, electrical insulation, footwear, roll covers and conduit seals.
See supplier literature for additional properties and other end use applications.

Polysulfide rubbers - T

These are rubbers having sulfur in the polymer chain, commonly referred to under the trademark. Thiokol. Available as millable gums and reactive liquid polymers.

Resistance to: abrasion	F
impact	P
tear	F
flame	F to E
weather & sunlight	E
oxidation	E
ozone	E
HC solvents & oils	P
acid, dilute/conc.	G/F
alkali, dilute/conc.	E/E

End Uses:
Coatings, pastes, adhesives, caulks. Miscellaneous molded, calendered and extruded goods; surgical and medical specialties; wire and cable insulation; appliance and TV lead wires; sealing, encapsulating and potting compounds; flexible molds; marine sealants. Roofing; flange gaskets; lighting seal gaskets; automotive seals; navy shipboard cable; nuclear plant power cable insulation; spark plug boots; O-rings; anode cups; vibration dampers; release agents; fabric coating; oxygen masks and oven gaskets.

Styrene-butadiene rubber - SBR

Solution and emulsion polymerized; hot and "cold" polymerized; non-pigmented; oil extended; black masterbatches; oil-black masterbatches; resin masterbatches; block polymers; many polymer and composition variations.
Available as bales, slabs, crumb, latex.

Specific gravity	0.94
Tensile strength	G (3500 psi)
Resilience	F-G
Compression set	G
Heat aging at 212°F	G
Low temp. stiffening, °F	0 to −50
Low temp. brittle point, °F	−80

Specific gravity	1.35
Tensile strength	P to F (1500 psi)
Resilience	P to F
Compression set	P to F
Heat aging at 212°F	F to G
Low temp. stiffening, °F	−10 to −45
Low temp. brittle point, °F	−60
Resistance to: abrasion	P
impact	P
tear	P
flame	P
weather & sunlight	G to E
oxidation	G to E
ozone	E
HC solvents & oils	G to E
acid, dilute/conc.	F/P
alkali, dilute/conc.	G/G

End Uses:
Paint spray, lacquer and fuel hose; printing rollers; molded and extruded goods; gaskets, putties, cements. Sealants for aircraft wing tanks and fuselages; insulating glass sealants; marine, building and machine sealers and adhesives; potting and sealing electrical parts; joint sealers and adhesives; miscellaneous coating, casting and potting compounds and flexible molds.

Silicone rubbers - Q

Also referred to as polysiloxane, these are rubbers having silicone in the polymer chain. Available as basic gums; solid, uncured compounds; reactive liquid polymers. There are many varieties and grades.

Specific gravity	0.95 - 1.6 (type dependent)
Tensile strength	P to F (1500 psi)
Resilience	P to E (type dependent)
Compression set	G to E
Heat aging at 212°F to 600°F	E
Low temp. stiffening, °F	−60 to −90

Resistance to: abrasion	E
impact	E
tear	F
flame	P
weather & sunlight	P
oxidation	F
ozone	P
HC solvents & oils	P
acid, dilute/conc.	F/F
alkali, dilute/conc.	F/F

End Uses:
Passenger car, implement and small truck tires; tread rubber; miscellaneous molded, extruded and calendered goods; hose; belts; rolls; shoe products; adhesives; coatings; caulks; carpet backing; wire and cable insulation; sporting goods; balls; toys; tool handles; brake linings; proofed goods; sheet packings; flooring; milking inflations; hard rubber products; chute linings; additive to asphalt for roads; latex-base paints; cellular products; latex foam mattresses and pillows.

Styrene-isoprene rubbers

Available in bags.

Specific gravity	0.93
Tensile strength	G (3000 psi)
Resistance to low temperature	G

End Uses:
Adhesives, caulks, coatings, films, hot-melt and solvent pressure-sensitive adhesives.
See supplier literature for additional properties and end use applications.

Tetrafluoroethylene propylene copolymer

A new class of synthetic rubber commercialized as Aflas elastomer.

Specific gravity	1.55

Tensile strength, psi	F to G (1800 to 3200)
Resilience	G
Compression set	G to E
Heat aging at 212°F to 600°F	E
Low temp. stiffening, °F	+30
Low temp. brittle point, °F	-55
Resistance to: abrasion	G
impact	G
tear	P to G
flame	N/A
weather & sunlight	E
oxidation	E
ozone	E
HC solvents & oils	F to E
acid, dilute/conc.	E/E
alkali, dilute/conc.	E/E

End Uses:
Useful where resistance to a wide range of chemicals and/or high heat is required. Also provides exceptional electrical resistance properties. Typical applications are in industries such as oil field (packing, seal, packer elements and seals, electrical boots and connectors, BOP's); chemical and agrichemical (seals, flexible couplings, heat exchangers and pipe gaskets, hose, wire and cable); aerospace (O-ring, shaft seal, wire and cable); truck/automotive (cooling and brake systems, engines, axles, transmissions, lubricating systems).

Thermoplastic-copolyester (multiblock copolymers) - Y
Structure: hard component — polyester
soft component — polyol segments

Specific gravity	1.17 - 1.25
Hardness range	53-74 Shore D; 89-94 Shore A
Tensile strength	G
Resilience	G
Set	G
Elastic recovery	G

Set	G
Elastic recovery	G
Service temperature range, °F	-60 to +300
Electrical properties	F to G (Some grades approved for wire insulation)
Other properties	Ultimate elongation in 200-300% for molded items — higher for extruded items
Resistance to: oxidation, ozone, weather	G
aliphatic HC solvents, fuels, oils	P to G depending on grade
low MW polar organic solvents	G
high MW nonpolar organic solvents	P to F

End Uses:
New applications for this synthetic are growing rapidly. Any attempt on our part to list specific end uses would be incomplete, and therefore misleading.

Thermoplastic polyurethane (multiblock copolymers) - YAU, YEU
Structure: alternating hard and soft blocks:
hard component — chain extenders and polyisocyanate
soft component — polyol segments

Specific gravity	1.10 - 1.35
Hardness range	80 to 93 Shore A to 65 Shore D
Tensile strength, psi	3000 to 8000
Resilience	G

Service temperature range, °F — −20 to +300 and above
Electrical properties — P to F (limited to low-voltage insulation)
Other properties — Good flex properties; resistant to hydraulic fluids.
Resistance to: tear — G
   heat aging — E
   oxidation, ozone, weather — G
   aliphatic HC solvents, fuels, oils — G
   aromaticc hydrocarbons — G
   low MW polar organic solvents — G (to some solvents)

End Uses:
New applications for this synthetic are growing rapidly. Any attempt on our part to list specific end uses would be incomplete, and therefore misleading.

Thermoplastic polyolefins (blends) - YEPM
Structure: hard component — crystalline polyolefin such as PP or PE
   soft component — ethylene-propylene rubber

Specific gravity — 0.88 - 1.30 (most varieties); 2.0 (some compounds)
Hardness range — 55 to 95 + Shore A or 50 Shore D
Tensile strength, psi — 500 to 2000, low hardness to 3000, higher hardness
Resilience — G

Set — G
Elastic recovery — G
Service temperature range, °F — −60 to +300
Electrical properties — F
Other properties — Polyether types have better water resistance than polyester types
Resistance to: abrasion — E
   tear — E (500 to 900 lb./linear inch, Die C)
   heat aging — Polyesters better than polyethers
   oxidation, ozone, weather — G
   aliphatic HC solvents, fuels, oils — G
   aromatic hydrocarbons — P
   low MW polar organic solvents — P

End Uses:
New applications for this synthetic are growing rapidly. Any attempt on our part to list specific end uses would be incomplete and therefore misleading.

Thermoplastic styrene-butadiene-styrene (block copolymers) - YSBR, YSIR
Structure: hard component — styrene blocks
   soft, mid-block component — butadiene, ethylene-butene or isoprene

Specific gravity — NA
Hardness range — 35 to 95 Shore A

Tensile strength, psi	500 to 2000
Resilience	G
Set	G
Elastic recovery	G
Service temperature range, °F	−60 to +300 (limited)
Electrical properties	F to G (Some grades approved for wire insulation)
Other properties	Bound styrene can range from 14% to 50% by weight. Some varieties are oil-extended
Resistance to: oxidation, ozone, weather	P to F*
aliphatic HC solvents, fuels, oils	P*
aromatic hydrocarbons	P*
low MW polar organic solvents	P*

*As saturation or hydrogenation of the styrene blocks increases, resistance improves.

End Uses:
New applications for this synthetic are growing rapidly. Any attempt on our part to list specific end uses would be incomplete, and therefore misleading.

Urethane rubber (polyurethane). Polyester urethanes - AU. Polyether urethanes - EU.
Many varieties: millable gum, reactive prepolymers, liquids.

Specific gravity	1.05 - 1.25
Tensile strength	E + (8000 psi)
Resilience	F to E
Compression set	P to E (type dependent)

Heat aging at 212°F   G to E
Low temp. stiffening, °F   −10 to −30
Low temp. brittle point, °F   −60 to −20
Resistance to: abrasion   E
         impact   G to E
         tear   E
         flame   P to F
         weather & sunlight   E
         oxidation   G
         ozone   E
         HC solvents & oils   F to G
         acid, dilute/conc.   F/P
         alkali, dilute/conc.   F/P

End Uses:
Binders, coated fabrics, tapes, adhesives; leather finishes for coats, ski boots, purses and luggage. Hot-melt sealants; life jackets; tarpaulins; paper and foil coatings; automotive parts; appliance gears; O-rings; seals; film; sheeting; tubing; wire jacketing; caster wheels; rollerskate wheels; extrusions; laminates; potting compounds; foamed-in-place insulation; foam mattresses and pillows; shock abosorbing pads; vibration dampers; paints and varnishes.

Vinyl pyridine polymers
These terpolymers of butadiene, styrene and vinyl pyridine are available as latex.
May be compounded with SBR latex and resorcinol-formaldehyde resins for high adhesion to rubber compounds.

End Uses:
Threatment of tire cord and other textiles to promote adhesion to rubber compounds.
See supplier literature for technical information and guidance.

APPENDIX B

MISCELLANEOUS DATA AND INFORMATION

THIRD MONOMERS IN EP TERPOLYMERIZATIONS

Termonomer Name	Structure	Availability	Extent of Investigations	1	2	3	4	5	6
1. Propadiene (Allene)	C=C=C	Comm.	Slight	—	—	—			—
2. Pentadiene-1,4	C=C-C=C	Exper.	Slight	(–)	(–)	(+)	‡	+	‡
3. Hexadiene-1,4	C=C-C=C-C	Comm.	Moderate	+	‡	–	‡	‡	‡
4. Hexadiene-1,5	C=C-C-C=C	Comm.	Slight	(–)	(–)	(+)	(±)	(±)	
5. Heptadiene-1,5	C=C-C-C=C-C	Exper.	Moderate	+	‡	–	‡	‡	‡
6. 2-Me-hexadiene-1,5	C=C-C-C-C=C (2-Me)	Exper.	Slight	‡	–	–	‡	‡	
7. 2,5-Me₂-hexadiene-1,5	C=C-C-C-C=C (2,5-Me₂)	Exper.	Slight	—	–	(– – –)			
8. 2,5-Me₂-hexadiene-1,4	C=C-C-C=C-C (2,5-Me₂)	Exper.	Slight	—	–	(‡‡)			
9. 2,4-Me₂-octadiene-2,7	C=C-C-C-C=C-C (2,4-Me₂)		Slight	+	‡	(‡‡)	‡	–	‡
10. 11-Et-Tridecadiene-1,11	C=C-C-C-C-C-C-C-C=C-C (11-Et)		Slight	+	‡	(‡‡)	–	–	+
11. Squalene	$C_{30}H_{50}$	Exper.	Slight	–?		—			
12. 4,8-Me₂-decatriene-1,4,9	C=C-C=C-C-C-C=C-C-C (4,8-Me₂)		Moderate	+	+	–	—	+	‡

No.	Compound		Availability
13.	1,2-Vinylcyclopropane	Slight	
14.	2,2-Br₂-Vinylcyclopropane	Slight	
15.	3-Allyl-cyclopentene	Slight	
16.	4-Vinyl-cyclohexene	Moderate	Comm.
17.	3-Butenyl-cyclopentene	Slight	
18.	Limonene	Slight	Comm.?
19.	Cyclohexadiene-1,4	Slight	Exper.
20.	Cyclododecatriene-1,5,9	Slight	Comm.?
21.	Cyclooctadiene-1,5	Slight	Comm.

THIRD MONOMERS IN EP TERPOLYMERIZATIONS

	Termonomer Name	Structure	Availability	Extent Of Investigations	Properties As Termonomer					
					1	2	3	4	5	6
22.	1-Me-cyclohexadiene-1,4	CH₃ (ring structure)	Exper.	Slight	(—)		(—→)			
23.	Heptadiyne-1,6	C≡C-C-C-C-C≡C		Slight				‡		‡
24.	[4.3.0] Bicyclononadiene-3.7 = Tetrahydroindene		Pot. Comm.	Moderate	‡	-	+	‡	↓	‡
25.	Me-Tetrahydroindene			Moderate	+	+	+		-	‡
26.	β-Pinene		Comm.	Slight	|		‡			|
27.	Norbornene (henceforth abbrev. N.)		Comm.	Moderate	‡		↓	‡	|	|
28.	5-Me-N.	CH₃	Pot. Comm.	Moderate	‡		‡	‡	|	|
29.	5-Cl-N.	Cl		Slight	‡	No unsaturation	+	+	↓	'

Compound							Extent	Availability
30. 5-ClCH₂-N.	ˌ	↓	↓	+		‡	Slight	
31. Norbornadiene	(˙)		+	↓	¦	‡	Slight	Comm.
32. 2-Me-norbornadiene	‡	↓	‡	‡	‡	‡	Slight	
33. 5-Methylene-N.	‡	‡	‡	‡	+	‡	Thorough	Comm.
34. 5-Ethylidene-N.	+	+	‡	‡	+	‡	Thorough	Comm.
35. 5-Vinyl-N.	+	+	‡	↓	|	↓	Moderate	(Pot.) Comm.
36. 5-Propenyl-N.	‡	+	‡	‡	‡	‡	Moderate	

THIRD MONOMERS IN EP TERPOLYMERIZATIONS

				Properties As Termonomer					
Name (Termonomer)	Structure	Availability	Extent Of Investigations	1	2	3	4	5	6
37. 5-1-Propenyl-N.		Pot. Comm.	Slight	±	±	+	±	+	±
38. 5-Crotyl-N.		Pot. Comm.	Slight	±	±	±	±	±	±
39. 5-Methallyl-N.		Pot. Comm.	Slight	±	+	+	±	+	±
40. 5-(2-Butenyl-2)-N.		Pot. Comm.	Slight	±	±	±	±	+	±
41. 5-(3-Butenyl-1)-N.			Slight	±	−	+	±	+	[±]
42. 5-Me-5-Vinyl-N.			Slight	±	±	±	±	±	−
43. 5-Me-5-Propenyl-N.			Slight	±	±	±	±	+	+

No.	Monomer	Availability	Properties as termonomers	1	2	3	4	5	6
44.	4e-5-1-Propenyl-N.	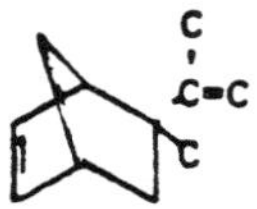	Slight	‡	‡	‡	‡	‡	‡
45.	5-(β-Styrenyl)-N.		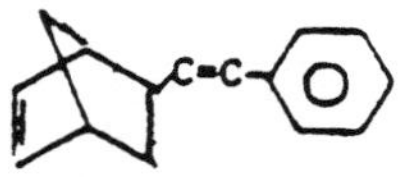 Slight	‡	‡	+	-	↓	?
46.	Tricyclo [5.2.1.0^{2,6}] decadiene-3,8 = Dicyclopentadiene	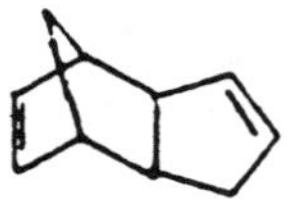 Comm.	Moderate	‡	+	‡	+	+	‡
47.	Me$_2$-dicyclopentadiene	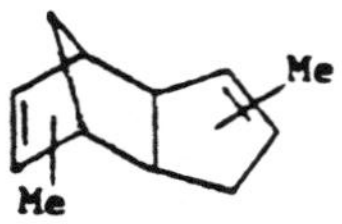	Slight	+	↓	↓	↓	↓	+
48.	5,6-Bismethylene norbornene	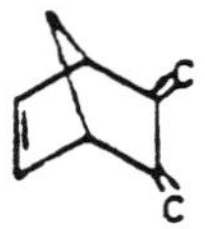	Moderate	‡	+	‡	‡	‡	+
49.	Acenaphthylene	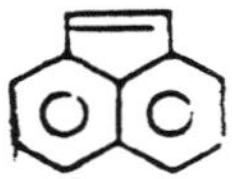 Comm.	Slight	‡	-	↓	↓	‡	(--)

Code for the column labelled "properties as termonomers":

1. Incorporability (reactivity of 1st double bond)
2. Gelation (reactivity of 2nd double bond)
3. Effect on polymerization rate, MW, etc.
4. Recovery and separation of unreacted material
5. Analysis in polymer
6. (Sulfur) curability

Ratings In Order Of Decreasing Desirability:

‡ + ± - -- ---

Availability: Comm. = commercially in large quantities
               Exper. = commercially in small quantities

TABLE 1 STRUCTURE: CHEMICAL COMPOSITION

Characterization technique[a]	Variables measured	Symbol (and unit)	Principle of operation
Elemental analysis	Fraction of particular element in a polymer	Number of atoms/ base unit (weight fraction, or percent)	Varied. usually involves destruction of sample into fragments which are analyzed
Absorption spectroscopy			
Infrared (ir), 1–16μm[b]	Wave number / Intensity of absorption	n (cm^{-1})[b] or ν / A (none)	Measures characteristic energies of bond deformations by their location in the spectrum
Raman	Wave number / Intensity of emission	n (cm^{-1})	Same as infrared
Nuclear magnetic resonance H (NMR) 5×10^7 μm ^{13}C	Chemical shift, coupling constant / Intensity of absorption / Chemical shift / Intensity of absorption	δ (ppm), J (Hz) (Arbitrary)	Measures characteristic energies of magnetic moment transitions / Same as H except ^{13}C nucleus transitions
Ultraviolet (uv) 0.2–0.35μm	Wave number / Intensity of absorption	n (cm^{-1}) / A	Measures characteristic energies of electronic transitions
Visible 0.35–0.8 μm	Wave number / Intensity of absorption	n (cm^{-1}) / A	Same as ultraviolet
Functional group analysis	Acid, hydroxyl, olefins, amine, isocyanate, etc. concentration	Fraction, percent, or concentration	Depends on known reactions of the particular type of functionality
Pyrolysis			
Gas chromatography Mass spectroscopy	Small molecules and oxidation products indicative of macro-molecular composition		Sample is pyrolyzed with chromatographic or mass spectrographic identification of characteristic fragments.
Electron paramagnetic resonance (EPR) 10^5 μm	Position of resonance / Intensity of absorption / Hyperfine splitting constants	g / A / a_s (G)	Measures characteristic energies of unpaired electron spin transitions which are a function of the chemical environment

[a] Other techniques discussed in general references [9, 10] and Table 1 [32].

[b] Absorption location can be wavelength λ (μm), wave number $n = 1/\lambda$ (cm^{-1}), or frequency Hz $= (c\lambda)$.   [c] Range of wavelength applicable.

TABLE IIA STRUCTURE: COMPOSITIONAL DISTRIBUTION

Characterization technique	Variables measured	Symbol (and unit)	Principle of operation
Fractionation: Precipitation	Composition, molecular weight of fractions		Different composition and molecular weight polymer components have different solubility.
Solution	Composition, molecular weight of fractions	Composition of components must be measured after separation by these techniques.	Different composition and molecular weight polymer components have different solubility.
Gel permeation chromatography (GPC)	Composition, molecular weight of fractions		Different size molecules are separated on the basis of exclusion from pores; uv, ir, etc. detectors on column show compositional distribution.
Thermal diffusion	Composition, molecular weight of fractions		A gradient in concentration and molecular weight is set up by a thermal gradient.
Isothermal diffusion	Diffusion coefficients of components	D (cm^2·sec)	Diffusion coefficients depend on molecular weight and composition.
Sedimentation	Sedimentation and diffusion coefficients, density gradient equilibrium	S (sec) D (cm^2·sec)	Sedimentation and diffusion coefficients depend on molecular weight and composition.
Thin layer chromatography (TLC)	Ratio of distance moved by polymer component relative to solvent front, must be calibrated.	R_f (0 to 1)	Separations based on differences in diffusion and adsorption through a particular matrix
Light scattering (LS)	Apparent molecular weights Heterogeneity parameter	$\overline{M}_w$ Q	Apparent $\overline{M}_w$ depends on the solvent refractive index, which is related to compositional heterogeneity of the polymer.

TABLE IIB

STRUCTURE: MONOMER SEQUENCE DISTRIBUTION, TACTICITY, AND CONFORMATIONS

Characterization technique	Variables measured	Symbol (and unit)	Principle of operation
Infrared (ir)	Intensity of absorption	A	Arrangement and geometry of units affects spectra.
	Frequency	n (cm^{-1})	
Nuclear magnetic resonance (NMR)	Chemical shift	∂ (ppm)	Arrangement and geometry of units affects spectra.
	Intensity of absorption	Area (arbitrary units)	
	Coupling constants	J (Hz)	
Pyrolysis–gas chromatography (GC)	Size of fragments	g/mole	Arrangement and chemical structure of units affects types of fragments produced.
Infrared	Structure of fragments		
Mass spectroscopy (MS)	Size of fragments		
Dipole moments	Mean square dipole moment	$\langle u^2 \rangle$ (D^2)	Geometry and sequence distribution affects dipole moments.
Light scattering	Mean square radius of gyration	$\langle S^2 \rangle$ (cm^2)	Geometry and sequence distribution affects polymer dimensions.
	Characteristic ratio	C,	
Viscometry	Characteristic ratio	C,	Geometry and sequence distribution affects polymer dimensions.

TABLE IIIA

STRUCTURE: MOLECULAR WEIGHT AND ITS DISTRIBUTION

Characterization technique	Variables measured	Symbol (and unit)	Principle of operation
Membrane osmometry	Number average molecular weight	$\overline{M}_n$ (g/mole)	Thermodynamic potential for mixing is measured by separating solvent and solution by membrane "impermeable" to polymer, a colligative property.
	Virial coefficient	A_i (mole cc/g^2)	
Vapor pressure osmometry (VPO)	Number average molecular weight	$\overline{M}_n$ (g/mole)	Vapor pressure lowering measured by a dynamic evaporation technique, needs calibration
Cryoscopy	Number average molecular weight	$\overline{M}_n$ (g/mole)	Freezing point depression colligative property
Ebullometry	Number average molecular weight	$\overline{M}_n$ (g/mole)	Boiling point elevation, colligative property

Method	Quantities measured	Symbol (units)	Principle
Light scattering (LS)	Weight average molecular weight	$\bar{M}_w$ (g/mole)	Light scattered by a polymer solution is related to the molecular weight of the solute.
	Virial coefficient	A_i (mole cc/g^2)^n	Angular dependence of scattering is related to particle size (0.05–0.2 μm).
	Radius of gyration	$\langle S^2 \rangle_z$ (cm^2), $\langle R_G \rangle_z$ (cm^2)	
Intrinsic viscosity	Intrinsic viscosity	$[\eta]$ (dl/g)	Ability of a polymer to increase the viscosity of a solution depends on polymer molecular weight.
	Viscosity average molecular weight	$\bar{M}_v$ (g/mole)	
	Estimates of MWD breadth	$\bar{M}_z$ (g/mole)	
Ultracentrifugation — Equilibrium	$\bar{M}_w$ and other molecular weight averages	$\bar{M}_w, \bar{M}_z$ (g/mole)	Concentration gradient in a large gravitational field is related to molecular weights.
Equilibrium in density gradient	General heterogeneity; Molecular weight averages	$\bar{M}_w, \bar{M}_z$; $\bar{M}_v, \bar{M}_z$	Concentration gradient in a large gravitational field is related to molecular weights.
Sedimentation	Molecular weights; Sedimentation coefficient	S (sec)	Rates of diffusion in a large gravitational field are related to molecular weights.
Terrestrial sedimentation	Molecular weight		Polymer settles under the influence of gravity near the critical point, or if of large enough size in normal solutions.
Gel permeation chromatography	Average molecular weights; Shape of distribution	$\bar{M}_v, \bar{M}_w, \bar{M}_n$ (g/mole)	Permeation of polymer into a porous structure depends on molecular weight.
Fractionation by solubility	Molecular weight averages if measurements are made on fractions; Shape of distribution	$\bar{M}_v, \bar{M}_w$ (g/mole)	Solubility of polymer depends on molecular weight and composition.
Turbidimetry or cloud point measurement	Cloud point temperature; Spinodals, concentration of precipitated material	T_c (C)	Phase separation temperatures depend on molecular weight.
Diffusion	Molecular weight; Diffusion coefficients	$\bar{M}_w$; D (cm^2/sec)	Diffusion coefficients depend on molecular weight.
Gel–Sol analysis of cross-linked sample	Gel fraction; Sol fraction	W_g gel fraction; S sol fraction	Gel point and partition between gel and sol depends on molecular weight and MWD.

Bulk viscosity	Viscosity	η (poise)	Empirical correlation of viscosity with molecular weight
Filtration	Gel fraction	W_s	
Comparisons of methods			

* Changes depending on form of equation: this unit corresponds to the equation in the text.

TABLE IIIB

STRUCTURE: BRANCHING

Characterization technique	Variables measured	Symbol (and unit)	Principle of operation
Fractionation with [η] and $\overline{M}_n$ measurements or D and $\overline{M}_n$ measurement	$\overline{M}_n$ and [η] on fractions	$g = \dfrac{[\eta]_{br}}{[\eta]_{lin}}$	Branching reduces intrinsic viscosity by amounts which can be correlated with theoretical calculations or model polymers. Need [η] = KM^a relation for linear polymer.
GPC with [η] measurement on fractions, or on line	GPC of fractions for which [η] is known, or [η] of GPC effluent	g	Same as above except the molecular weight of the polymer is determined by postulating $(M[\eta])_{lin} = (M[\eta])_{br}$ at a given elution volume. $\overline{M}_v$ appears to be the correct MW average.
GPC with [η] on whole polymer	GPC, [η] on whole polymer	g	Assumptions are made about branching distribution and the degree of branching is chosen to make GPC and whole polymer [η] agree, or [η] measured as GPC cuts elute.

GPC—sedimentation and [η]	GPC Sedimentation coefficients [η]	S, (g)	Branching increases sedimentation coefficient.
Huggins constant	Concentration dependence of intrinsic viscosity	k	Branching causes "systematic" variations in these quantities which have been recommended as measures of branching.
Virial coefficients	Concentration dependence of osmotic pressure	A_2 (mole cm^3/g^2)	
Relaxation times	Theta temperatures Time dependence of mechanical properties	Θ (°C) τ, (sec)	Either by comparison with model polymers or through use of theoretical calculations on models such as the Zimm theory; changes in these quantities at given molecular weights in dilute solution are related to degrees of branching.
Compliances	Strain response to imposed stress is measured	J', J_e' (area/force)	
Moduli	Stress response to imposed strain is measured	G', G'' (force/area)	
Concentrated polymer viscosity	Viscosity is measured as a function of shear rate and temperature.	η (poise)	Changes in this quantity are related to branching empirically by comparison with model compounds; least rigorous of the methods, because no theory equivalent to that for dilute solutions is involved.

SHORT NOTES SECTION ON BUTYL RUBBER POLYMERIZATION

POLYMERIZATION REACTIONS

POLYMERIZATIONS ARE CLASSIFIED IN VARIOUS WAYS

I. ACCORDING TO MECHANISM
STEP POLYMERIZATIONS
ADDITION POLYMERIZATIONS

II. ACCORDING TO SPECIES PARTICIPATING IN REACTION

FREE-RADICAL POLYMERIZATIONS
ANIONIC POLYMERIZATIONS
CATIONIC POLYMERIZATIONS
ZIEGLER-NATTA POLYMERIZATIONS

III. ACCORDING TO METHOD OF POLYMERIZATION
BULK POLYMERIZATIONS
SOLUTION POLYMERIZATIONS
SUSPENSION POLYMERIZATIONS
EMULSION POLYMERIZATIONS
GAS POLYMERIZATIONS

- UNREACTED ISOBUTYLENE CONCENTRATION IN REACTOR SLURRY DEPENDS ON CONVERSION AND IS TYPICALLY 4-6 %. "POISON" CONCENTRATION DICTATES MOONEY VISCOSITY (MW) - CONVERSION RATIO, OR REACTOR INDEX.

- POLYMER MOLECULES EXIST FOR A TIME AS INDIVIDUAL COLLAPSED CHAINS OF ABOUT 50 ANGSTROMS DIAMETER. SOON AFTER POLYMERIZATION COMMENCES AN ABRUPT AGGREGATION (NUCLEATION) OF CHAINS OCCURS AND A SLURRY IS FORMED. SLURRY PARTICLE SIZE IS 5-30 MICRONS DIAMETER.

- IT IS THE NATURE OF THE BUTYL SLURRY PARTICLES - SIZE, SHAPE, CONCENTRATION, STATE OF ASSOCIATION - THAT ESTABLISHES THE VISCOSITY OF THE SLURRY. THIS VISCOSITY HAS A PRIMARY INFLUENCE ON REACTOR REACTOR HEAT TRANSFER.

- A REACTOR WARMS-UP SLOWLY DURING ITS PRODUCTION LIFE, IT IS NORMALLY STOPPED WHEN SLURRY TEMPERATURE REACHES -87 TO -90°C, IT IS THEN FLUSHED, EMPTIED AND WASHED WITH NAPHTA. A CLEAN REACTOR IS STARTED. AS SLURRY TEMPERATURE INCREASES, MW DECREASES, "CATALYST" FLOW RATE IS REDUCED.

● THE REACTION HAS FIVE BASIC STEPS :

    + INITIATION TO PRODUCE A CARBENIUM ION/ANION ION PAIR
    + PROPAGATION OF A T-BUTYL CARBENIUM ION ADDING ISOBUTYLENE
    + CROSS-PROPAGATION REACTIONS INVOLVING ISOPRENE
    + CHAIN TRANSFER WHICH STOPS THE GROWTH OF ONE CHAIN AND STARTS ANOTHER
      (ACTIVE ION PAIR IS PRESERVED).
    + TERMINATION WHICH STOPS A KINETIC CHAIN AND DESTROYS THE ACTIVE ION
      PAIR.

SEVERAL THOUSAND PROPAGATION STEPS NEED TO OCCUR BEFORE ANOTHER REACTION
INTERVENES. DEGREE OF POLYMERIZATION IS 2700-3600 FOR BUTYL 268, IT IS UP TO
15,000-20,000 FOR POLYISOBUTYLENE (VISTANEX).

● PROPAGATION RATE CONSTANTS ARE VERY HIGH :

    T-BUTYL C$^+$ AS CHAIN END : 10^7 TO 10^8 L/MOL-SEC
    ISOPRENYL C$^+$ AS CHAIN END : 10^5 L/MOL-SEC

TOTAL CHAIN LIFETIME IS MEASURED IN MICROSECONDS.

CONCEPT OF LOW FUNCTIONALITY

<u>BUTYL</u> :   ● 0.5 TO 2.5 MOLES ISOPRENE PER 100 MOLES
          ● MOLECULAR WEIGHT (BUTYL 268) : M_V = 350,000 TO 450,000 AND
            M_N = 150,000 - 200,000.

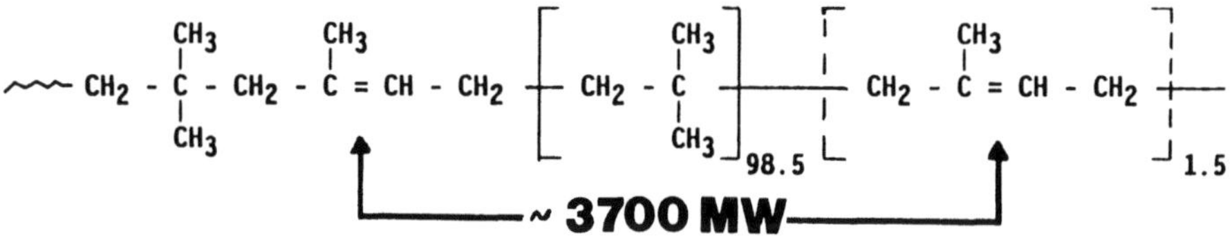

A BUTYL MOLECULE OF BUTYL 268 OF M_N 200,000 TYPICALLY CONTAINS
    ● 3570 MOLES OF ISOBUTYLENE
    ●   53 MOLES OF ISOPRENE

<u>POLYISOPRENE</u> : DISTANCE BETWEEN $\overset{\backslash}{\underset{/}{C}} = \overset{/}{\underset{\backslash}{C}}$: 68 MW.

POLYISOBUTYLENE (VISTANEX[R]) AND BUTYL RUBBER ARE PRODUCED BY | CATIONIC POLYMERIZATION |

- CATIONIC POLYMERIZATION PROCEEDS BY ATTACK ON THE MONOMER BY AN ACIDIC SPECIES RESULTING IN HETEROLYTIC SPLITTING OF THE C = C DOUBLE BOND TO PRODUCE A TRIVALENT TERT-BUTYL CARBOCATION :

$$H - A \ + \ H_2C = \underset{\underset{CH_3}{|}}{\overset{\overset{CH_3}{|}}{C}} \quad\longrightarrow\quad H_3C - \underset{\underset{CH_3}{|}}{\overset{\overset{CH_3}{|}}{C^{\oplus}}} \ \cdots \ A^{\ominus}$$

- THE TERT-BUTYL CARBOCATION/ANION PAIR WILL BE THE CHAIN CARRIER DURING CHAIN PROPAGATION.

- ANY DISRUPTION OF THE ION PAIR WILL CAUSE CHAIN TRANSFER OR CHAIN TERMINATION.

| INITIATION |

- PRODUCTION OF A T-BUTYL CARBENIUM ION/ANION ION PAIR :

$$AlCl_3 + H\text{-}X + \underset{\delta^-}{CH_2} = \overset{\delta^+}{\underset{\underset{CH_3}{|}}{\overset{\overset{CH_3}{|}}{C}}} \quad\longrightarrow\quad CH_3 - \underset{\underset{CH_3}{|}}{\overset{\overset{CH_3}{|}}{C^{\oplus}}} \ [\ AlCl_3X\ ]^{\ominus}$$

LEWIS ACIDS ALSO EFFECTIVE ARE :

BF_3, $SnCl_4$, $TiCl_4$, $AlRCl_2$, AlR_2Cl, $AlBr_3$, $FeCl_3$, SbF_5, WCl_6, TaF_5 ...

● FROM THE SCIENTIFIC AND TECHNOLOGICAL POINT OF VIEW, THE MOST IMPORTANT
  CARBOCATIONIC INITIATING SYSTEMS ARE COMBINATIONS OF CATIONOGENS AND A
  FRIEDEL-CRAFTS ACID.

● FRIEDEL-CRAFTS ACID ARE LEWIS ACID OF GENERAL FORMULA :

MR_NX_M M : METAL
                  R : ALKYL OR HYDROGEN
                  X : HALOGEN
                  N + M : METAL VALENCE

● PUREST FRIEDEL-CRAFTS HALIDES ARE, MORE OFTEN THAN NOT, UNABLE TO INITIATE BUTYL
  POLYMERIZATION, THEY USUALLY REQUIRE OR WILL IN FACT BE ASSOCIATED WITH A CATION
  SOURCE SUCH AS A BRONSTED ACID.   TYPICAL FRIEDEL-CRAFTS ACIDS ARE $AlCl_3$ AND BF_3.

CATALYST OR INITIATOR

● IN OUR BUTYL PLANTS, $AlCl_3$ IS CALLED THE <u>CATALYST</u>, H_2O OR HCL IS GENERALLY CALLED
  THE <u>ACTIVATOR</u> OR <u>COCATALYST</u>.

● A MORE LOGICAL NOMENCLATURE, WHICH BETTER REFLECTS THE CARBOCATION GENERATION IS :

<u>INITIATOR</u> :   THE CATION SOURCE (CATIONOGEN) SUCH AS A BRONSTED ACID (HCL), OR
                        H_2O, OR ORGANIC HALIDE (T-BUT.CL), SOMETIMES ALSO CALLED
                        PROMOTERS.

<u>COINITIATOR</u> :   THE LEWIS ACID $AlCl_3$, WHICH ASSISTS IN GENERATING THE
                        INITIATING ENTITY.

$$H_2O \; + \; AlCl_3 \; + \; CH_2 = C(CH_3)_2 \; \longrightarrow \; CH_3 - \overset{CH_3}{\underset{CH_3}{\overset{|}{\underset{|}{C^{\oplus}}}}} \; [\,AlCl_3OH\,]^{\ominus}$$

$$(CH_3)_3C\text{-}Cl \; + \; AlCl_3 \; \longrightarrow \; CH_3 - \overset{CH_3}{\underset{CH_3}{\overset{|}{\underset{|}{C^{\oplus}}}}} \; [\,AlCl_4\,]^{\ominus}$$

LEWIS ACID - ALCL₃

- ALCL₃ IS A TYPICAL FRIEDEL-CRAFTS ACID, THE AL ATOM IS SURROUNDED BY ONLY 6 ELECTRONS, IT HAS A STRONG TENDENCY TO COMPLETE ITS OCTET AND SEEK TO FILL ITS VACANT 3P ORBITAL BY COORDINATING WITH AVAILABLE ELECTRON DONORS :

COORDINATION WITH UNSHARED ELECTRONS ON CL.

COORDINATION WITH 2 PI ELECTRONS OF OLEFIN LEADING TO POLARIZATION.

COORDINATION WITH UNSHARED ELECTRONS ON OXYGEN ATOM OF H₂O, ROH, R-O-R

COORDINATION WITH UNSHARED ELECTRONS ON CL ATOM OF HCL

- COMPLEXATION OCCURS PREFERENTIALLY WITH THE STRONGEST BASES AVAILABLE, THE UNSHARED ELECTRON ON THE OXYGEN ATOM OF H₂O OR ON THE CHLORINE ATOM OF HCL ARE MORE AVAILABLE THAN ARE THE CHLORINE ATOMS OF ALCL₃, SO THE DIMER AL₂CL₆ OPENS THE COORDINATE WITH H₂O OR HCL.

EXAMPLES OF SOME POSSIBLE CARBOCATION FORMATION

- **WITH HCL**

$$Cl_3Al : ClH + (CH_3)_2C = CH_2 \longrightarrow CH_3 - \overset{\overset{\displaystyle CH_3}{|}}{\underset{\underset{\displaystyle CH_3}{|}}{C^{\oplus}}} [\ AlCl_4\ ]^{\ominus}$$

- **WITH T-BUTYL CHLORIDE :**

$$Cl_3Al : Cl - C(CH_3)_3 \longrightarrow CH_3 - \overset{\overset{\displaystyle CH_3}{|}}{\underset{\underset{\displaystyle CH_3}{|}}{C^{\oplus}}} \quad [\ AlCl_4\ ]^{\ominus}$$

- **WITH AN ALCOHOL :**

$$Cl_6Al_2 : O\overset{\diagup CH_3}{\diagdown H} + (CH_3)_2C = CH_2 \longrightarrow CH_3 - \overset{\overset{\displaystyle CH_3}{|}}{\underset{\underset{\displaystyle CH_3}{|}}{C^{\oplus}}} [\ Al_2Cl_6\ OCH_3\ ]^{\ominus}$$

- **WITH H₂O :** $Cl_3Al : O\overset{\diagup H}{\diagdown H} + (CH_3)_2C = CH_2 \longrightarrow CH_3 - \overset{\overset{\displaystyle CH_3}{|}}{\underset{\underset{\displaystyle CH_3}{|}}{C^{\oplus}}} [\ AlCl_3OH\ ]^{\ominus}$

H₂O WHEN ADDED TO WARM AlCl₃ SOLUTION REACTS TO GENERATE HCL :

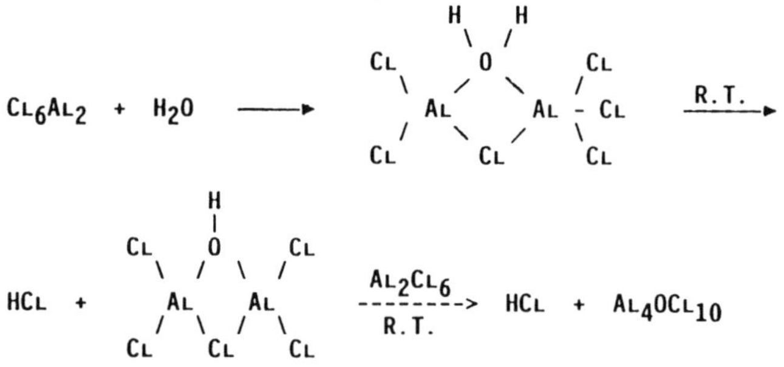

$$Cl_6Al_2 + H_2O \longrightarrow
\begin{array}{c}
\quad\quad \overset{H}{\diagdown}\ O\ \overset{H}{\diagup} \\
Cl \diagdown \quad \diagup \quad Cl \\
\quad Al \quad\quad Al - Cl \quad \xrightarrow{R.T.} \\
Cl \diagup \quad Cl \diagdown \quad Cl
\end{array}$$

$$HCL +
\begin{array}{c}
\quad \overset{H}{\underset{\displaystyle |}{}} \\
Cl \diagdown \ O\ \diagup \ Cl \\
\quad Al \quad Al \\
Cl \diagup \ Cl \diagup \ Cl
\end{array}
\xrightarrow[R.T.]{Al_2Cl_6} HCL + Al_4OCl_{10}$$

1 MOLE H₂O $\longrightarrow$ 2 MOLES HCL

Al₂Cl₆ IS CONVERTED TO OXY CONTAINING SPECIES OF LOWER LEWIS ACIDITY WITH LITTLE OR NO "CATALYTIC" ACTIVITY.

TOO MUCH H₂O DESTROYS Al₂Cl₆ ABILITY TO POLYMERIZE ISOBUTYLENE, HIGH MOLAR RATIO HCL/Al₂Cl₆ DOES NOT BUT HCL IS GOOD TRANSFER AGENT.

● BUTYL REACTORS OPERATE IN AN INITIATOR ("ACTIVATOR") LIMITED CONDITION, IN THE
  ABSENCE OF ADDED "ACTIVATOR".
  THERE IS AN EXCESS OF UNACTIVATED $AlCl_3$ IN THE "CAT" FLOW AND POSSIBLY IN THE
  REACTOR SLURRY.

-----> IF AN "ACTIVATOR" ENTERS THE REACTOR IN THE FEED, CONVERSION AND MOONEY
       VISCOSITY CHANGE, MOONEY CONTROL WILL BE VERY DIFFICULT, AND MAY NOT BE
       VERY RESPONSIVE TO NORMAL "$AlCl_3$" FLOW RATES CHANGES.

-----> WHEN CHANGES IN $AlCl_3$ EFFICIENCY ARE REQUIRED IT CAN BEST BE ACCOMPLISHED
       BY MODIFYING THE $AlCl_3$/"ACTIVATOR" MOLE RATIO (OR COMPLEX "POOL") IN THE
       $AlCl_3$ STREAM.

-----> OPTIMUM CONTROL OF ALL INITIATORS, WHETHER IN FEED OR IN $AlCl_3$, IS OF
       PARAMOUNT IMPORTANCE FOR REACTOR CONTROL AND OPERATION.

PROPAGATION

● A MONOMER ADDS TO THE CARBENIUM ION TO GROW THE CHAIN, WITH THE ANION REMAINING
  ASSOCIATED WITH THE $C^\oplus$ CHAIN END.

● IT IS AN EXTREMELY FAST AND ENERGETICALLY FAVORABLE REACTION. FOR A T-BUTYL
  CARBENIUM ION CHAIN END, κ_p IS OF THE ORDER OF 10^7 - 10^8 L/MOLE.SEC., THE
  TOTAL CHAIN LIFETIME IS MEASURED IN MICROSECONDS.

$$\sim CH_2 - \underset{\underset{CH_3}{|}}{\overset{\overset{CH_3}{|}}{C}} - CH_2 - \underset{\underset{CH_3}{|}}{\overset{\overset{CH_3}{|}}{C^{\oplus}}} \ldots A^{\ominus}$$

● THERE IS A VARYING DEGREE OF IONIC CHARACTER IN THE BOND BETWEEN C^+ ... AND ... A^-.

$$M - A \rightleftharpoons M^+A^- \rightleftharpoons M^+/A^- \rightleftharpoons M^+ + A^-$$

| COVALENT | CONTACT ION PAIR | SOLVENT SEPARATED ION PAIR | FREE SOLVATED IONS |

● THE RATE CONSTANT k_P WILL BE MARKEDLY DEPENDENT ON THE DEGREE OF ASSOCIATION WHICH IS A FUNCTION OF :
    NATURE OF A^-, SOLVENT POLARITY, TEMPERATURE

● MeCl IS A LOW DIELECTRIC CONSTANT MEDIUM, IN WHICH C^+ AND A^- EXIST AS SOLVATED ION PAIR.

HOMO- AND CROSS-PROPAGATION

● THERE ARE <u>FOUR PROPAGATION REACTIONS</u>, EACH WITH ITS OWN RATE CONSTANT :

+ <u>ADDITION TO T-BUTYL CARBENIUM ION</u>

$$\underset{\underset{C}{|}}{\overset{\overset{C}{|}}{C^{\oplus}}} [A]^{\ominus} \quad + \quad CH_2 = C(CH_3)_2 \quad \xrightarrow{K_{11}} \quad \sim C - \underset{\underset{C}{|}}{\overset{\overset{C}{|}}{C}} - C - \underset{\underset{C}{|}}{\overset{\overset{C}{|}}{C^{\oplus}}} [A]^{\ominus}$$

OR

$$\underset{\underset{C}{|}}{\overset{\overset{C}{|}}{C^{\oplus}}} [A]^{\ominus} \quad + \quad C = \overset{\overset{C}{|}}{C} - C = C \quad \xrightarrow{K_{12}} \quad \sim C - \underset{\underset{C}{|}}{\overset{\overset{C}{|}}{C}} - \underset{\delta_+ \quad [A]^{\ominus} \quad \delta_+}{C - C - C}$$

+ <u>ADDITION TO ISOPRENYL CARBENIUM ION</u>

$$C - \underset{\delta_+}{\overset{\overset{C}{|}}{C}} - \underset{\delta_+}{C} - C \quad + \quad CH_2 = C(CH_3)_2 \quad \xrightarrow{K_{21}} \quad \sim C - \overset{\overset{C}{|}}{C} = C - C - \underset{\underset{C}{|}}{\overset{\overset{C}{|}}{C^{\oplus}}} [A]^{\ominus}$$

OR

$$C - \underset{\delta_+}{\overset{\overset{C}{|}}{C}} - \underset{\delta_+}{C} - C \quad + \quad C = \overset{\overset{C}{|}}{C} - C = C \quad \xrightarrow{K_{22}} \quad \sim C - \overset{\overset{C}{|}}{C} = C - C - C - \underset{\delta_+}{\overset{\overset{C}{|}}{C}} - \underset{\delta_+}{C} - C + [A]^{\ominus}$$

- LET $CH_2 = C(CH_3)_2$ BE <u>A</u>   AND $C = C - C = C$ (with C substituent on second carbon) BE <u>B</u>.

$$A^{\oplus} + A \xrightarrow{\kappa_{11}} A - A^{\oplus}$$

$$A^{\oplus} + B \xrightarrow{\kappa_{12}} A - B^{\oplus}$$

$$B^{\oplus} + A \xrightarrow{\kappa_{21}} B - A^{\oplus}$$

$$B^{\oplus} + B \xrightarrow{\kappa_{22}} B - B^{\oplus}$$

$$\kappa_{11} > \kappa_{12} \gg \kappa_{21} > \kappa_{22}$$

- $R_1 = \kappa_{11}/\kappa_{12} \;\; = \;\; 3.5 \; (1) \; - \; 2.5 \pm 0.5 \; (2)$
- $R_2 = \kappa_{22}/\kappa_{21} \;\; = \;\; 0.2 \; (1) \; - \; 0.4 \pm 0.1 \; (2)$

THOMAS AND SPARKS (1944) : $ALCL_3/MECL/-103^oC$

- THE RELATIVE RATES OF HOMOPROPAGATION AND CROSS-PROPAGATION REACTIONS TOGETHER WITH STEADY-STATE MONOMER CONCENTRATIONS IN THE REACTOR DETERMINE BUTYL UNSATURATION OR ISOPRENE CONTENT.

- ADDITION OF ISOBUTYLENE IS ALWAYS FAVORED, THE POLYMER IS ALWAYS LOWER IN ISOPRENE CONTENT THAN IN THE FEED, BUT GRADUALLY INCREASES IN ISOPRENE CONTENT WITH INCREASED CONVERSION.

- <u>ISOPRENE DOMINATES THE COPOLYMERIZATION</u>, κ_{21} IS $10^2 - 10^3$ TIMES LOWER THAN κ_{11} OR κ_{12}, THE GROWING CHAINS ARE NEARLY ALL ISOPRENYL CARBENIUM ION ENDED AND <u>MOST TRANSFERS AND TERMINATIONS OCCUR FROM ISOPRENYL ENDED CHAINS</u>.

CHAIN TRANSFER REACTIONS

- CHAIN TRANSFER REACTIONS YIELD A "DEAD" CHAIN BUT A NEW CARBENIUM ION ASSOCIATED WITH THE SAME ANION STARTS A NEW CHAIN. PURE CHAIN TRANSFER REDUCES MOLECULAR WEIGHT AND SHOULD NOT AFFECT CONVERSION (POLYMER YIELD) NOR "REAL CATALYST EFFICIENCY".

- HOWEVER, AS WE RUN THE REACTORS TO PRODUCE MOONEY VISCOSITY, TRANSFER AGENTS CAUSE TO REDUCE LBS OF $AlCl_3$ TO RESTORE VISCOSITY, HENCE RAISE UNREACTED ISOBUTYLENE CONCENTRATION (LOWER CONVERSION), AND INCREASE LBS OF BUTYL/LB $AlCl_3$.

- USUALLY TRANSFER REACTIONS INVOLVE REACTION OF THE CHAIN END CARBENIUM ION WITH SOME OTHER SPECIES WHICH CAN ACCEPT A PROTON OR DONATE A HYDRIDE ION OR ANOTHER NEGATIVE FRAGMENT TO THE PROPAGATING CARBENIUM ION.

EXAMPLES ARE :

- <u>WITH T-BUTYL CHLORIDE</u> (DONATE Cl^-)

$$\sim \overset{\displaystyle C}{\underset{\displaystyle C}{\overset{|}{\underset{|}{C^{\oplus}}}}}\ [A]^{\ominus}\ +\ (CH_3)_3CCl\ \longrightarrow\ \sim \overset{\displaystyle C}{\underset{\displaystyle C}{\overset{|}{\underset{|}{C}}}} - Cl\ +\ C - \overset{\displaystyle C}{\underset{\displaystyle C}{\overset{|}{\underset{|}{C^{\oplus}}}}}\ [A]^{\ominus}$$

- <u>WITH ISOPRENE</u> (ACCEPTS A PROTON)

$$\sim \overset{\displaystyle CH_3}{\underset{\displaystyle CH_3}{\overset{|}{\underset{|}{C^{\oplus}}}}} + [A]^{\ominus}\ +\ C = C - \overset{\displaystyle C}{\overset{|}{C}} = C\ \longrightarrow\ \sim \overset{\displaystyle }{\underset{\displaystyle C}{\overset{}{\underset{|}{C}}}} = CH_2\ +\ C - C\overset{C\ \ C}{\oplus}\ +\ [A]^{\ominus}$$

- <u>WITH 2 METHYL BUTENE 1</u> (DONATE H^-, AN HYDRIDE ION)

$$\sim \overset{\displaystyle C}{\underset{\displaystyle C}{\overset{|}{\underset{|}{C^{\oplus}}}}} + [A]^{\ominus}\ +\ H_2C = \overset{\displaystyle CH_3}{\overset{|}{C}} - CH_2 - CH_3\ \longrightarrow\ \sim \overset{\displaystyle C}{\underset{\displaystyle C}{\overset{|}{\underset{|}{C}}}} - H\ +\ H_2C = \overset{\displaystyle CH_3}{\overset{|}{C}} - \underset{\oplus}{CH} - CH_3$$

$$\longrightarrow\ [A]^{\ominus}\quad \overset{H_2C}{\underset{H_2C}{\oplus C}} - CH_2 - CH_3 \qquad \text{MORE STABLE ALLYLIC CATION.}$$

- <u>WITH ISOBUTYLENE</u> (PROTON β-ELIMINATION)

$$\sim CH_2 - \overset{\overset{\displaystyle CH_3}{|}}{\underset{\underset{\displaystyle CH_3}{|}}{C}}\oplus \longrightarrow \sim CH_2 - \overset{\overset{\displaystyle CH_2}{\|}}{\underset{\underset{\displaystyle CH_3}{|}}{C}} + H^\oplus \xrightarrow{+M} CH_3 - \overset{\overset{\displaystyle CH_3}{|}}{\underset{\underset{\displaystyle CH_3}{|}}{C}}\oplus$$

- <u>ISOBUTYLENE DIMER</u> (2,4,4-TRI ME -1- (AND 2) - PENTENE

    CLASSICAL EXAMPLE OF STERIC FACTOR AND FACILE TRANSFER

$$\sim \overset{\overset{\displaystyle C}{|}}{\underset{\underset{\displaystyle C}{|}}{C}} - \overset{}{C}\oplus \;+\; \overset{\overset{\displaystyle C}{|}}{C} = \overset{}{C} - C - \overset{\overset{\displaystyle C}{|}}{\underset{\underset{\displaystyle C}{|}}{C}} - C \longrightarrow \sim \overset{\overset{\displaystyle C}{|}}{\underset{\underset{\displaystyle C}{|}}{C}} - C - \underset{\oplus}{C} - C - \overset{\overset{\displaystyle C}{|}}{\underset{\underset{\displaystyle C}{|}}{C}} - C$$

    RESULTING TERTIARY $C^\oplus$ IS TOO STERICALLY CROWDED TO PROPAGATE (A BURIED CARBENIUM ION), PROTON ELIMINATION OCCURS INSTEAD :

$$\sim\sim \overset{\overset{\displaystyle CH_3}{|}}{\underset{\underset{\displaystyle CH_3}{|}}{C}} - CH_2 - \overset{\overset{\displaystyle CH_3}{|}}{C} = CH - \overset{\overset{\displaystyle CH_3}{|}}{\underset{\underset{\displaystyle CH_3}{|}}{C}} - CH_3 \quad\text{AND}\quad H^\oplus + C = \overset{\overset{\displaystyle C}{|}}{\underset{\underset{\displaystyle C}{|}}{C}} \longrightarrow CH_3 - \overset{\overset{\displaystyle CH_3}{|}}{\underset{\underset{\displaystyle CH_3}{|}}{C}}\oplus$$

    OR FAVORABLE BETA-SCISSION :

$$\sim\sim \overset{\overset{\displaystyle C}{|}}{\underset{\underset{\displaystyle C}{|}}{C}} - C - \underset{\oplus}{\overset{\overset{\displaystyle C}{|}}{C}} - C - \overset{\overset{\displaystyle C}{|}}{\underset{\underset{\displaystyle C}{|}}{C}} - C \longrightarrow \sim\sim\sim \overset{\overset{\displaystyle C}{|}}{\underset{\underset{\displaystyle C}{|}}{C}} - C - \overset{\overset{\displaystyle C}{|}}{C} = C \;+\; C - \overset{\overset{\displaystyle C}{|}}{\underset{\underset{\displaystyle C}{|}}{C}}\oplus$$

    AN IDEAL T-BUTYL CARBENIUM ION FOR ISOBUTYLENE PROPAGATION IS GENERATED.

- TRANSFER REACTIONS HAVE DIFFERENT RATES AND DIFFERENT ACTIVATION ENERGIES THAN PROPAGATION, THEY AFFECT POLYMERIZATION RATE, AND THEIR EFFECT ON MOLECULAR WEIGHT WILL BE A FUNCTION OF TEMPERATURE.

$$P_N^\oplus + M \xrightarrow{K_P} P_{N+1}^\oplus \qquad \text{RATE} = K_P \,(P_N^\oplus)\,(M)$$

$$P_N^\oplus + T \xrightarrow{K_{TR}} P_N + T^\oplus \qquad \text{RATE} = K_{TR}\,(P_N^\oplus)\,(T)$$

$$T^\oplus + M \xrightarrow{K_A} P_N^\oplus \qquad \text{RATE} = K_A\,(T^\oplus)\,(M)$$

TERMINATION REACTIONS

- TERMINATION RESULTS IN THE DESTRUCTION OF THE PROPAGATION ION PAIR. THE KINETIC CHAIN STARTED BY AN INITIATION IS INTERRUPTED.

  SOME SPECIES FORMED IN THE TERMINATION REACTION MAY HOWEVER SIMPLY BE DORMANT AND CAPABLE OF REINITIATION AT SOME LATER TIME OR UNDER OTHER CONDITIONS (E.G. AT HIGHER TEMPERATURE).
  IN SOME SENSE THERE IS A CONTINUUM OF REACTIONS FROM RAPID CHAIN TRANSFER TO TOTAL TERMINATION WITH PERMANENT DESTRUCTION OF ACTIVE INITIATING SPECIES.

- TERMINATION AGENTS ARE TRUE REACTION POISONS, BUT A LOT OF COMPOUNDS WILL HAVE SOME TRANSFER ACTIVITY AS WELL AS SOME TERMINATION ACTIVITY. SOME CAN EVEN BE "ACTIVATORS" AS WELL !

EXAMPLES OF TERMINATION REACTIONS ARE :

- SPONTANEOUS COLLAPSE, REACTION WITH ANION

$$\sim C - \underset{\underset{C}{|}}{\overset{\overset{C}{|}}{C}} \oplus \ [AlCl_3D]^{\ominus} \longrightarrow \sim \underset{\underset{C}{|}}{\overset{\overset{C}{|}}{C}} - D \ + \ AlCl_3 \ (\text{"DORMANT"})$$

- WITH "OXY" POISON

$$\sim C - \underset{\underset{C}{|}}{\overset{\overset{C}{|}}{C}} \oplus \ [AlCl_4]^{\ominus} \ + \ R - OH \longrightarrow \sim \underset{\underset{C}{|}}{\overset{\overset{C}{|}}{C}} - \underset{\underset{R}{\backslash}}{\overset{\overset{H}{/}}{O}} \oplus \ [AlCl_4]^{\ominus}$$

"STABLE" OXONIUM ION

$$\longrightarrow \sim \underset{\underset{C}{|}}{\overset{\overset{C}{|}}{C}} - OH \ + \ R^{\oplus} \ [AlCl_4]^{\ominus} \quad \begin{array}{l}(\text{IF } R = CH_3 \text{ OR } C_2H_5 \\ \text{WILL NOT REINITIATE})\end{array}$$

- **BY HYDRIDE TRANSFER (ALLYLIC TERMINATION)**

$$- \overset{\displaystyle C}{\underset{\displaystyle C}{\overset{|}{\underset{|}{C}}}}^{\oplus} \; [AlCl_4]^{\ominus} \; + \; C = \overset{\displaystyle C}{\overset{|}{C}} - C = C \; \longrightarrow \; \sim \overset{\displaystyle C}{\underset{\displaystyle C}{\overset{|}{\underset{|}{C}}}} - H \; + \; C - C - C \; [AlCl_4]^{\ominus}$$

ISOPRENE

DORMANT ALLYLICALLY
STABILIZED

$$\sim \overset{\displaystyle CH_3}{\underset{\displaystyle CH_3}{\overset{|}{\underset{|}{C}}}}^{\oplus} \; [AlCl_4]^{\ominus} \; + \; CH_2 = CH - CH_3 \; \longrightarrow \; \sim \overset{\displaystyle CH_3}{\underset{\displaystyle CH_3}{\overset{|}{\underset{|}{C}}}} - H \; + \; CH_2 - \overset{\oplus}{CH} - CH_2 \; [AlCl_4]^{\ominus}$$

OR CL TRANSFER :
$$+ \; C = C - C - C \; \longrightarrow \; \sim \overset{\displaystyle C}{\underset{\displaystyle C}{\overset{|}{\underset{|}{C}}}} - C - \overset{\oplus}{C} - C - C \; [AlCl_4]^{\ominus}$$

UNSTABLE

$$\longrightarrow \; AlCl_3 \; + \; \sim \overset{\displaystyle C}{\underset{\displaystyle C}{\overset{|}{\underset{|}{C}}}} - C - \underset{\displaystyle CL}{\overset{|}{C}} - C - C$$

(DORMANT)

EFFECT OF POISON AND TRANSFER AGENT ON KINETIC CHAIN

├─────**MW**──────►┤	POLYMER YIELD
├──────────────────┤	"PURE" PROPAGATING MONOMER
├──────┤├──────┤├──────┤├──────┤	WEAK TRANSFER AGENT
├──┤├──┤├──┤├──┤├──┤├──┤├──┤├──┤	STRONG TRANSFER AGENT
├──────────────────┤ M M M M M	WEAK POISON
├──────┤ M M M M M M	STRONG POISON

- TRANSFER DOES NOT INTERRUPT KINETICS CHAIN, IT REDUCES MW
- POISON STOPS KINETIC CHAIN, IT REDUCES CONVERSION.

  MORE INITIATION IS REQUIRED TO SUSTAIN CONVERSION, THEN MOLECULAR WEIGHT DROPS.

MOST COMPOUNDS HAVE BOTH TERMINATION AND TRANSFER CAPABILITIES

EMPIRICAL POISON COEFFICIENTS AND TRANSFER COEFFICIENTS

Material	P.C.	T.C.
Propylene	4.9	0
1-Butene	6.2	0
1-Pentene	9.7	0
1-Hexene	11.9	0
2-Octene	20.8	12.1
Butadiene	7.6	2.0
2,5-Dimethyl-2,4-hexadiene	26.7	15.2
2,3-Dimethyl-1,3-butadiene	107.0	3.6
Cyclohexadiene	103.0	48.0
Isoprene	140.0	60.0
Piperylene	170.0	327.0
Cyclopentadiene	900.0	
2-Methylcyclopentadiene	685.0	
2-Methyl-1-pentene	3.5	53.8
2-Ethyl-1-hexene	24.7	248.0
2,4,4-Trimethyl-1-pentene	66.7	700.0
2,4,4-Trimethyl-2-pentene	66.7	34.6
3-Methyl-1-butene	0	0
4-Methyl-1-pentene	2.9	0
3,3-Dimethyl-1-butene	0	0
Vinylcyclohexane	3.7	6.6

MOST COMPOUNDS HAVE BOTH TERMINATION AND TRANSFER CAPABILITIES

EMPIRICAL POISON COEFFICIENTS AND TRANSFER COEFFICIENTS

Material	P.C.	T.C.
Propylene	4.9	0
1-Butene	6.2	0
1-Pentene	9.7	0
1-Hexene	11.9	0
2-Octene	20.8	12.1
Butadiene	7.6	2.0
2,5-Dimethyl-2,4-hexadiene	26.7	15.2
2,3-Dimethyl-1,3-butadiene	107.0	3.6
Cyclohexadiene	103.0	48.0
Isoprene	140.0	60.0
Piperylene	170.0	327.0
Cyclopentadiene	900.0	
2-Methylcyclopentadiene	685.0	
2-Methyl-1-pentene	3.5	53.8
2-Ethyl-1-hexene	24.7	248.0
2,4,4-Trimethyl-1-pentene	66.7	700.0
2,4,4-Trimethyl-2-pentene	66.7	34.6
3-Methyl-1-butene	0	0
4-Methyl-1-pentene	2.9	0
3,3-Dimethyl-1-butene	0	0
Vinylcyclohexane	3.7	6.6

- TERMINATORS DECREASES MW, CONVERSION AND "REAL" AlCl₃ EFFICIENCY, BUT BECAUSE MW IS REDUCED THEY FORCE US TO DROP CONVERSION FURTHER BY REDUCING AlCl₃ FLOW WHICH MAY OVERRIDE THE "REAL" EFFICIENCY LOSS AND CAUSE TO SEE AN APPARENT AlCl₃ EFFICIENCY INCREASE.

-----> FOR GOOD MOONEY CONTROL, IT IS DESIRABLE FOR THE LEVEL OF FORTUITOUS "ACTIVATORS" OR "POISONS" TO BE LOW <u>AND STEADY</u>, WITH AN INITIATOR INTENTIONALLY ADDED AND METERED INTO AlCl₃ STREAM OR IN AlCl₃ SOLVENT.

MOONEY VISCOSITY IS RELATIVELY INDEPENDENT OF AlCl₃ CONCENTRATION IN THE AlCl₃ FLOW RATE. INCREASING AlCl₃ CONCENTRATION INCREASES THE AlCl₃/INITIATOR RATIO, INCREASING THE AMOUNT OF UNEFFICIENT AlCl₃ AND DECREASING "AlCl₃ EFFICIENCY".

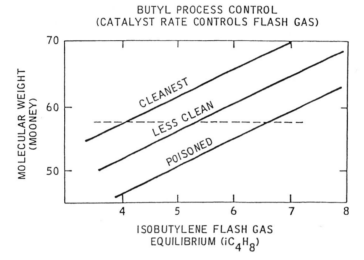

REACTOR INDEX, MOONEY/FLASH GAS, IS A MEASURE OF WHICH CURVE THE PLANT IS OPERATING ON

BUTYL POLYMERS

HISTORICAL : • 1873 GORIANOV AND BUTLEROV, OILY LOW MW HOMOPOLYMERS OF ISOBUTYLENE ($H_2C=C(CH_3)_2$) WITH H_2SO_4.

• 1928-29 : I.G. FARBEN : HIGH MW PIB (M.OTTO) BF_3 AT $-75^{\circ}C$ - RUBBER-LIKE PROPERTIES.

• 1937 : W.J. SPARKS AND R.M. THOMAS (STANDARD OIL DEVELOPMENT CO) : FIRST VULCANIZABLE ISOBUTYLENE BASED ELASTOMER BY INCORPORATION OF SMALL AMOUNT OF 1.4 BUTADIENE THEN LATER ISOPRENE.

- FIRST CONCEPT OF LOW FUNCTIONALITY FOR VULCANIZATION (LATER IN EPDM AND HYPALON)

• 1942-43 COMMERCIAL PRODUCTION IN BATON ROUGE
• 1955   PLANTS PURCHASED FROM U.S. GOVERNMENT
• 1954   BROMINATED BUTYL COMMERCIALIZED BY B.F. GOODRICH
• 1961   CHLOROBUTYL COMMERCIALIZED (F.P. BALDWIN)
• 1974   BROMOBUTYL COMMERCIALIZED BY POLYSAR AND LATER BY EXXON.

END USES FOR BUTYL PRODUCTS

BUTYL	VISTANEX	CHLOROBUTYL	BROMOBUTYL
INNERTUBES	PLASTICS BLENDING	INNERLINERS	INNERLINERS
SHEETING	SHEETING	PHARMACEUTICALS	PHARMACEUTICALS
INSULATION	ADHESIVES	INNERTUBES	
AUTO BODY MOUNTS		ADHESIVES	ADHESIVES
MECH GOODS		STEAM HOSE	STEAM HOSE
		MECH GOODS	MECH GOODS

BUTYL CHARACTERISTICS

COLD, RAPID REACTION

- REACTION SENSITIVE TO POISONS AT PPM LEVEL

POISON IMPACT

MILD ————————————————————————→ SEVERE

n-Olefins   Isoprene   Ethers   t-Chlorides H_2O Alcohols

- LOWEST TEMP COMMERCIAL POLYMERIZATION

    + -100ºC FAVORS CHAIN PROPAGATION

- REACTION PROCESS EXTREMELY RAPID (<0.001 SEC)

- REACTOR SIZE BASED ON HEAT TRANSFER, NOT RESIDENCE TIME

BUTYL CHARACTERISTICS

PRECIPITATION POLYMERIZATION

- MONOMERS IN SOLUTION OF DILUENT

- MONOMER REACTS TO FORM POLYMER IN SOLUTION

- DILUENT IS POOL SOLVENT FOR POLYMER

- POLYMER CHAINS PRECIPITATE FROM SOLUTION

- 1000'S OF CHAIN COLLECT TO FORM PARTICLES

- PARTICLES REMAIN IN COLLOIDAL SUSPENSION

- SUSPENSION APPEAR AS SKIM MILK ——→ HEAVY CREAM

- SLURRY VISCOSITY REMAINS LOW (1-20CP)

BUTYL CHARACTERISTICS

OTHER REACTOR HIGHLIGHTS

- PARTIAL MONOMER CONVERSION AND RECYCLE

    + ISOBUTYLENE RECYCLES REQUIRES FRACTIONATION
    + ISOPRENE RECYCLE REQUIRES ADDITIONAL PURIFICATION TO
      REMOVE t-BUTYL CHLORIDE

- METHYL CHLORIDE (25 PPM OEL) USED AS DILUENT

    + GOOD SOLVENT FOR MONOMERS AND CAT
    + NON-SOLVENT FOR POLYMER

- POLYMER PRECIPITATES ON WALLS AS WELL

    + FOR HEAT TRANSFER, WALLS MUST BE CLEANED

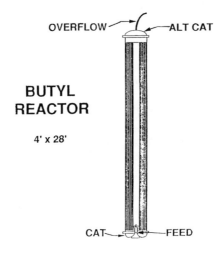

OVERFLOW — ALT CAT

BUTYL REACTOR

4' x 28'

CAT — FEED

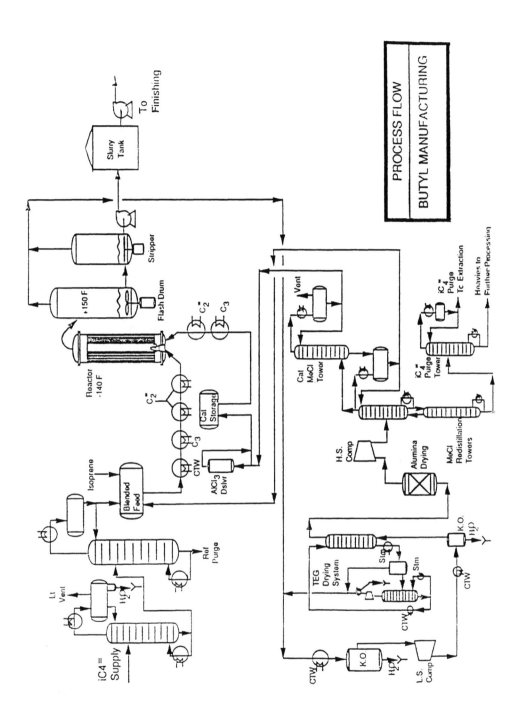

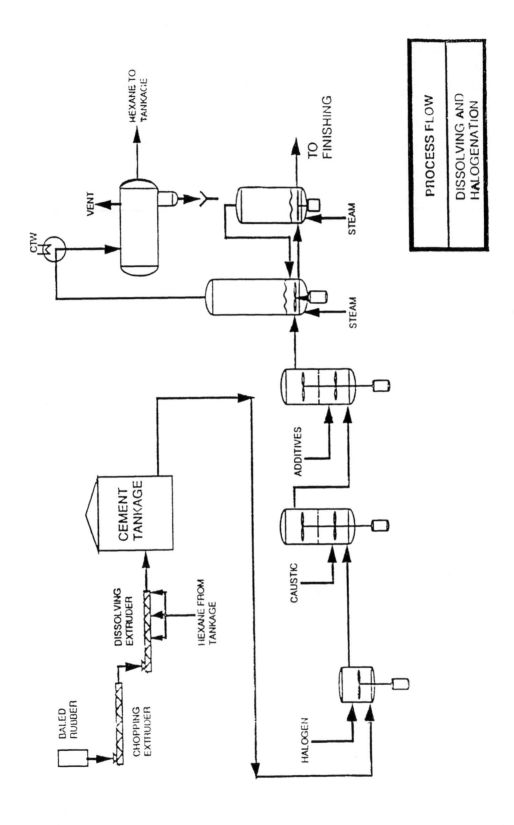

PROCESS FLOW

DISSOLVING AND HALOGENATION

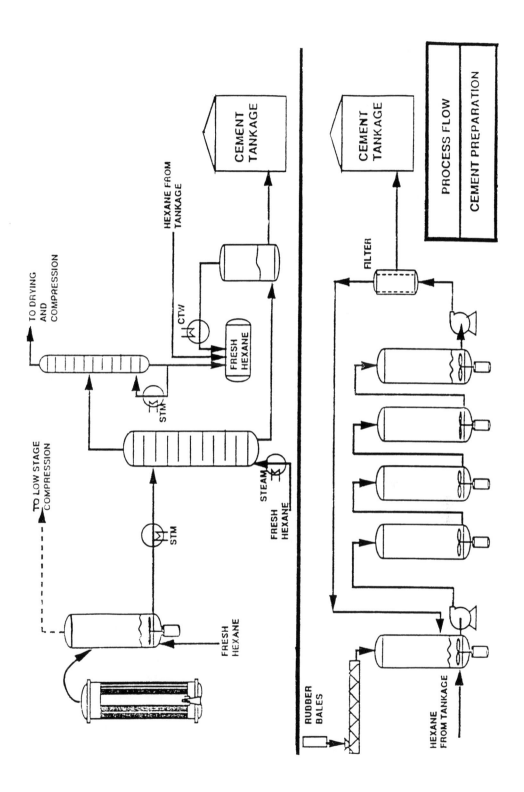

APPENDIX D

SHORT NOTES SECTION ON POLYMERIZATION BASICS

POLYMERIZATION REACTIONS

POLYMERIZATIONS ARE CLASSIFIED IN VARIOUS WAYS

I. ACCORDING TO MECHANISM
   STEP POLYMERIZATIONS
   ADDITION POLYMERIZATIONS

II. ACCORDING TO SPECIES PARTICIPATING IN REACTION

   FREE-RADICAL POLYMERIZATIONS
   ANIONIC POLYMERIZATIONS
   CATIONIC POLYMERIZATIONS
   ZIEGLER-NATTA POLYMERIZATIONS

III. ACCORDING TO METHOD OF POLYMERIZATION
   BULK POLYMERIZATIONS
   SOLUTION POLYMERIZATIONS
   SUSPENSION POLYMERIZATIONS
   EMULSION POLYMERIZATIONS
   GAS POLYMERIZATIONS

MONOMER ——————————▶ POLYMER

IT IS SELF-EVIDENT THAT THE MONOMER MOLECULES WHICH ARE TO BE
MADE INTO POLYMER CHAINS MUST BE <u>BIFUNCTIONAL</u>.

MONOMER		POLYMER
⋆ <u>MONOFUNCTIONAL</u> ——————▶		<u>DIMER</u>
R-NCO + HO-R'		R-NH-CO-O-R'
⋆ <u>BIFUNCTIONAL</u> ——————▶		<u>POLYMERIC CHAIN</u>

HOOC-R-NH$_2$ HO-(-OC-R-NH-CO-R-NH-)-H

HETEROCYCLIC ——————▶RING OPENING ——————▶LINEAR CHAIN
COMPOUNDS

⋆ <u>TRI-OR TETRA-
FUNCTIONAL</u> ——————▶ <u>BRANCHED POLYMER
AND NETWORK</u>

GENERAL CONSIDERATIONS OF POLYMERIZABILITY

1. POLYMERIZATION MUST BE <u>THERMODYNAMICALLY POSSIBLE,</u>
   THE FREE ENERGY DIFFERENCE ΔG BETWEEN MONOMER AND
   POLYMER MUST BE NEGATIVE :

$$\Delta G = \Delta H - T\Delta S \quad < 0$$

   WITH ΔS USUALLY < 0 IT FOLLOWS THAT ΔH MUST BE < 0 ;
   AND HEAT MUST BE REMOVED.

2. POLYMERIZATION MUST BE <u>KINETICALLY FEASIBLE</u>
   NATURE OF THE <u>INTERMEDIATE</u> IS VERY IMPORTANT,
   VERY SPECIFIC REACTION CONDITIONS MAY BE REQUIRED
   TO ACTUALLY ACCOMPLISH A PARTICULAR POLYMERIZATION
   I. E. NEED OF SPECIFIC CATALYSTS.

FORMATION OF MACROMOLECULES

* <u>STEP - GROWTH REACTION</u> (POLYCONDENSATION).

CLASSIC ORGANIC REACTIONS BETWEEN ANTAGONIST FUNCTIONS A AND B

$$A \longrightarrow A \quad + \quad B \longrightarrow B \longrightarrow A \longrightarrow C \longrightarrow B$$

EX. $HOOC-R-COOH + H_2N-R'-NH_2 \longrightarrow HOOC-R-CO-NH-R'-NH_2 + H_2O$

* <u>POLYADDITION</u> : <u>CHAIN REACTION MECHANISM</u>

- OPENING OF $C = C$

$$R^* + C = C \longrightarrow R - C - C^* + C = C \longrightarrow R -(-C - C)-C - C^* \qquad ETC...$$

- RING OPENING OF HETEROCYCLIC COMPOUND

$$H^+A^- + CH_2 - CH_2 \longrightarrow CH_2 - CH \quad + \quad CH_2 - CH_2 \longrightarrow HO - CH_2 - CH_2 - O$$

STEP - GROWTH POLYMERIZATION

REACTION PROCEEDS VIA FUNCTIONAL GROUPS :

$$N \quad A \text{———} A \quad + \quad N \quad B \text{———} B \quad \longrightarrow A -(- C -)- B_N$$

EXAMPLES :

DIACID + DIOL $\longrightarrow$ POLYESTER ($- CO - O -$)

DIAMINE + DIACIDE $\longrightarrow$ POLYAMIDE ($- CO - NH -$)

DIOL + DIISOCYANATE $\longrightarrow$ POLYURETHANE ($- NH - CO - O -$)

DIAMINE + DIISOCYANTE $\longrightarrow$ POLYUREA ($- NH - CO - NH -$)

BISPHENOL + COCL$_2$ $\longrightarrow$ POLYCARBONATE ($- O - CO - O -$)

TRANSESTERIFICATION
OF BISPHENOL AND $\longrightarrow$ POLYCARBONATE
DIPHENYLCARBONATE

DIANHYDRIDE + DIAMINE $\longrightarrow$ POLYIMIDE ($- CO - N - CO -$)

PHENOL + FORMALDEHYDE $\longrightarrow$ NOVOLAK RESINS

DIEPOXY + DIAMINE $\longrightarrow$ EPOXY RESINS ($- CHOH - CH_2 - NR -$)

SILANOL CONDENSATION
OR DICHLOROSILANE + H$_2$O $\longrightarrow$ POLYSILICONE ($- \overset{|}{\underset{|}{Si}} - O - \overset{|}{\underset{|}{Si}} -$)

* <u>STEP-GROWTH POLYMERIZATION PROCEEDS WITH SLOW INCREASE OF MOLECULAR WEIGHT</u>

  - DISAPPEARENCE OF 'MONOMER' OCCURS VERY EARLY

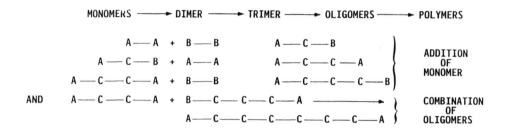

MONOMERS ⟶ DIMER ⟶ TRIMER ⟶ OLIGOMERS ⟶ POLYMERS

* <u>INTRINSIC REACTIVITY OF ALL FUNCTIONAL GROUPS IS CONSTANT AND INDEPENDANT OF MOLECULAR SIZE</u> .

* <u>MOLECULAR WEIGHT INCREASES WITH TIME</u>.

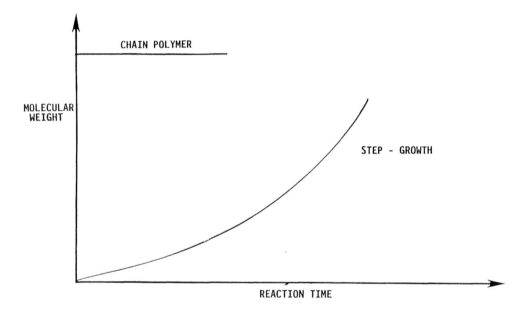

MOLECULAR WEIGHT CONTROL

$$A \text{———} A \quad + \quad B \text{———} B$$

LET : N_A AND N_B = NUMBER OF A AND B GROUPS

$R = N_A / N_B$ = EXTENT OF STOICHIOMETRIC IMBALANCE

P = EXTENT OF REACTION

⋆ NUMBER AVERAGE DEGREE OF POLYMERIZATION AT TIME T :

$$(\overline{X_N})_T = \frac{\text{NUMBER OF MONOMER MOLECULES AT T = 0}}{\text{TOTAL NUMBER OF MOLECULES AT T}}$$

$$= (1 + R) / (1 + R - 2 RP)$$

FOR $\underline{R = 1}$: $(\overline{X_N})_T = \underline{1 / 1 - P}$

EFFECT OF STOICHIOMETRIC RATIO ON MOLECULAR WEIGHT

$$A - A + B - B$$

$R = N_A/N_B$ AND P = EXTENT OF REACTION

	P :	0	0.5	0.95	0.98	0.99	0.999	1
● $\underline{R = 1}$	X_N	1	2	20	50	100	1000	∞
● $\underline{R = 100/100.1}$ (0.1% IMBALANCE)	X_N			19.8	49	96	667	2001
● $\underline{R = 100/101}$ (1% IMBALANCE)	X_N			18	40	67	167	201

SOME PROCESS CHARACTERISTICS

- HIGH PURITY REACTANTS

- HIGH M.W. PRODUCTS PRODUCES ONLY AT VERY LAST STAGES OF REACTION
  VERY HIGH M.W. ACHIEVED ONLY AT HIGH CONVERSION

- M.W. CONTROL THROUGH CONVERSION AND SETTING OF STOICHIOMETRIC
  IMBALANCE

- REMOVAL OF WATER TO PROCEED WITH REACTION

- PRODUCT SEPARATION IS RELATIVELY SIMPLE

ADDITION POLYMERIZATION

NATURE OF POLYADDITION

REACTIVE INITIATING SPECIES R* MUST BE CAPABLE OF ADDING TO MONOMER

R* CAN BE :   - A FREE RADICAL $R^\bullet$ $R^\bullet + C = C \longrightarrow R - C - C^\bullet$

               - AN ELECTROPHILIC R^+ $R^+ + C = C \longrightarrow R - C - C^+$

                                                   (CARBOCATION)

               - A NUCLEOPHILIC R^- $R^- + C = C \longrightarrow R - C - C^-$

                                                    (CARBANION)

               - COORDINATION BOND WITH A TRANSITION METAL
                            (ZIEGLER - NATTA)

- NOT EVERY MONOMER IS SUSCEPTIBLE OF R^+ OF R^- POLYMERIZATION,
  SUSTITUENT EFFECT (INDUCTIVE AND RESONANCE) IS DETERMINING.

- **FOR R•, R⁺ AND R⁻ :**

  IT IS THE NATURE OF THE <u>REACTIVE CHAIN END</u> THAT
  DRIVES THE POLYMERIZATION.

- **IN ZIEGLER-NATTA CATALYZED POLYMERIZATION, POLYMER GROWTH**

  PROCEEDS THROUGH INSERTION OF MONOMER AT <u>CATALYST SITE</u>

$$\boxed{\textbf{FREE RADICAL POLYADDITION}}$$

<u>MONOMERS CAPABLE OF FREE RADICAL POLYMERIZATION</u>

- MOST OF THE VINYLIC, ACRYLIC AND DIENIC MONOMERS
- ETHYLENE ONLY AT ELEVATED TEMPERATURE AND PRESSURE

<u>MONOMERS THAT DO NOT POLYMERIZE UNDER FREE RADICAL ATTACK</u>

- PROPYLENE YIELDS ONLY LOW MW OLIGOMERS
- SYMMETRICALLY SUBSTITUTED MONOMERS (2,3-DIMEBUTENE, ISOBUTENE)
- MONOMERS WITH BULKY SUBSTITUENTS (1, 1-DIPHENYL ETHYLENE)
- ALLYL MONOMERS (RESONANCE STABILIZED RADICALS EXHIBIT LOW REACTIVITY)
- RING OPENING POLYMERIZATIONS RARELY INVOLVE FREE RADICAL MECHANISMS

MECHANISM OF RADICAL CHAIN POLYMERIZATION

INITIATOR :
- HEAT (THERMAL POLYMERIZATION)
- UV LIGHT AND PHOTOSENSITIVE AGENT
- HIGH ENERGY IRRADIATION (GAMMA - ,ELECTRON - ,XRAYS)
- HOMOLYTIC SCISSION OF WEAK COVALENT BONDS
  (25- 40 KCAL/MOLE) CONTAINING COMPOUNDS BY HEAT

$$R - O - O - R \longrightarrow 2\ RO^{\bullet}$$
$$R - O - O - H \longrightarrow RO^{\bullet} + {}^{\bullet}OH$$

- REDOX SYSTEMS (LOW ACTIVATION ENERGY)

$$Fe^{++} + RO - OH \longrightarrow Fe^{3+} + RO^{-} + {}^{\bullet}OH$$
$$Fe^{++} + S_2O_8^{--} \longrightarrow Fe^{3+} + SO_4^{=} + SO_4^{-\bullet}$$

AND

$$Fe^{3+} + HSO_3^{-} \longrightarrow HSO_3^{\bullet} + Fe^{++}$$

FOUR EVENTS ...

1. **INITIATION** : $RO - OR \longrightarrow 2\ RO^{\bullet}$ (K_D)
   $R^{\bullet} + M \longrightarrow R - M^{\bullet}$

$$R_D = 2\ F\ K_D\ [I_N]$$

2. **PROPAGATION** : $R - M^{\bullet} + xM \longrightarrow R - (M)_X - M^{\bullet}$ (K_P)

$$R_{PR} = K_P\ [M]\ [M^{\bullet}]$$

3. **TERMINATION** :

COUPLING : $\sim\!\sim\!\sim M^{\bullet}{}_M + {}^{\bullet}M_N\sim\!\sim\!\sim \longrightarrow \sim\!\sim M_{M+N}$ (K_T)

DISPROPORTIONATION : $\sim\!\sim\!\sim M^{\bullet}{}_M + {}^{\bullet}M_N\sim\!\sim\!\sim \longrightarrow \sim\!\sim\!\sim M_M + M_N \sim\!\sim\!\sim$

$$R_T = 2\ K_T\ [\overset{\bullet}{M}]^2$$

4. <u>TRANSFER</u> :

$$\sim\sim M^\bullet + TX \longrightarrow \sim\sim MX + T^\bullet$$

AND $T^\bullet + M \longrightarrow TM^\bullet$

$$\underline{R_{TR} = K_{TR}\ [M^\bullet]\ [TX]}$$

TRANSFER DOES NOT INTERRUPT KINETICS CHAIN (CONVERSION OF M CONTINUES)

BUT AFFECTS MOLECULAR WEIGHT (CHAIN LENGTH)

<u>TX CAN BE</u> :

MONOMER	POLYMER
SOLVENT	CHAIN TRANSFER AGENT
INITIATOR	

(ALSO : $\sim\sim CH_2 - CH_2 - CH_2 - CH_2 - \overset{\bullet}{C}H_2 \longrightarrow \sim\sim\overset{\bullet}{C}H - [CH_2]_3 - CH_3$)

$$\boxed{\text{MOLECULAR WEIGHT } (\overline{M}_N = \overline{X}_N M_O)}$$

• AT STATIONARY-STATE CONDITIONS :

$$\overline{X}_N = \frac{\text{RATE OF PROPAGATION}}{\text{SUM OF RATES OF ALL REACITONS LEADING TO DEAD POLYMER}}$$

$$= \frac{K_P\ [M]\ [M^\bullet]}{K_T\ [M^\bullet]^2 + K_{TR}\ [M^\bullet]\ [S] + K_{TR}\ [M^\bullet]\ [M] + K_{TR}\ [M^\bullet]\ [I] + K_{TR}\ [M^\bullet]\ [P]}$$

EFFECT OF POISON AND TRANSFER AGENT ON KINETIC CHAIN

MW → POLYMER YIELD

"PURE" PROPAGATING MONOMER

WEAK TRANSFER AGENT

STRONG TRANSFER AGENT

M M M M WEAK POISON

M M M M M STRONG POISON

- TRANSFER DOES NOT INTERRUPT KINETICS CHAIN, IT REDUCES MW
- POISON STOPS KINETIC CHAIN, IT REDUCES CONVERSION.

  MORE INITIATION IS REQUIRED TO SUSTAIN CONVERSION, THEN MOLECULAR WEIGHT DROPS.

POLYMERIZATION RATE EXPRESSION

$$R_P = -\frac{D[M]}{DT} = K_I \, [R^{\bullet}] \, [M] + K_P \, [M] \, [M^{\bullet}] = K_P \, [M] \, [M^{\bullet}]$$

STEADY-STATE REQUIRES THAT RATE OF RADICAL PRODUCTION = RATE OF RADICAL DISAPPEARANCE

$$2 F K_D \, [I] = 2 \, K_T \, [M^{\bullet}]^2 \longrightarrow [M^{\bullet}] = \left(F \frac{K_D}{K_T} \, [I]\right)^{0.5}$$

$$\Longrightarrow \quad R_P = K_P \, [M] \, \left(F \frac{K_D}{K_T} \, [I]\right)^{0.5}$$

- ALL κ FOLLOWS ARRHENIUS EQUATION : $\kappa = A E^{-E_A/RT}$

  $E_D = 30 - 40$ KCAL/MOLE $- E_P : 5 - 10$ KCAL/MOLE $- E_T : 2 - 5$ KCAL/MOLE

EFFECT OF TEMPERATURE ON R_P

$$R_P = K_P \ [M] \ (F \frac{K_D}{K_T} \ [IN] \ )^{0.5}$$

$$E_A = E_P + E_D/2 - E_T/2$$

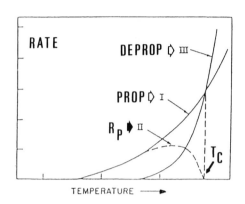

T_C : <u>CEILING TEMPERATURE</u>

  DUE TO REVERSIBILITY

  OF PROPAGATION STEP

$$RM^{\bullet}_x + M \ \underset{K_{DP}}{\overset{K_P}{\rightleftarrows}} \ R - M^{\bullet}_{x+1}$$

T_C : α-ME-STYRENE $61^{\circ}C$

  ME-METHACRYLATE $220^{\circ}C$

  STYRENE $310^{\circ}C$

INHIBITION - RETARDATION

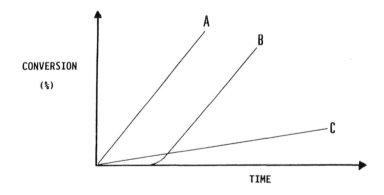

A. NO ADDITIVES PRESENT

B. WITH INHIBITOR (EX. BENZOQUINONE)

C. WITH RETARDER (EX. NITROBENBENE)

COPOLYMERIZATION

- LET US CONSIDER TWO MONOMERS A AND B

- THERE ARE FOUR POSSIBLE PROPAGATION REACTIONS :

    1. $\sim\sim A^\bullet + A \longrightarrow \sim\sim AA^\bullet$ (κ_{AA})
    2. $\sim\sim A^\bullet + B \longrightarrow \sim\sim AB^\bullet$ (κ_{AB})
    3. $\sim\sim B^\bullet + B \longrightarrow \sim\sim BB^\bullet$ (κ_{BB})
    4. $\sim\sim B^\bullet + A \longrightarrow \sim\sim BA^\bullet$ (κ_{BA})

- THE RATES OF CONSUMPTION OF A AND B ARE EXPRESSED AS :

$$- \frac{D[A]}{DT} = \kappa_{AA}\,[A^\bullet]\,[A] + \kappa_{BA}\,[B^\bullet]\,[A] \qquad (1)$$

$$- \frac{D[B]}{DT} = \kappa_{BB}\,[B^\bullet]\,[B] + \kappa_{AB}\,[A^\bullet]\,[B] \qquad (2)$$

- AT STEADY-STATE :

$$\kappa_{AB}\,[A^\bullet][B] = \kappa_{BA}\,[B^\bullet][A]$$

- <u>RADICAL REACTIVITY RATIOS</u> ARE DEFINED AS :

$$R_A = \frac{\kappa_{AA}}{\kappa_{AB}} \qquad \text{AND} \quad R_B = \frac{\kappa_{BB}}{\kappa_{BA}}$$

R_A : REACTIVITY OF A TO REACT WITH A RATHER THAN B

R_B : REACTIVITY OF B TO REACT WITH B RATHER THAN A

● LET $\quad F_A = \dfrac{[A]}{[A] + [B]}$ $\qquad$ MOLE FRACTION OF A IN FEED
$\qquad\qquad\qquad\qquad\qquad\qquad\qquad$ ($F_B = 1 - F_A$)

● LET $\quad F_A = \dfrac{D[A]}{D[A] + D[B]}$ $\qquad$ MOLE FRACTION OF A IN POLYMER
$\qquad\qquad\qquad\qquad\qquad\qquad\qquad\qquad$ <u>AT TIME T</u> (INSTANTANEOUS COMPOSITION)

ON CAN SHOW THAT :

$$F_A = \frac{R_1 F_A{}^2 + F_A F_B}{R_1 F_A{}^2 + 2 F_A F_B + R_2 F_B{}^2}$$

F_A IS A FUNCTION OF F_A AND OF R_1 AND R_2

PHYSICAL SIGNIFICANCE OF REACTIVITY RATIOS

1. $R_A = R_B = 0$: NEITHER RADICAL CAN REGENERATE ITSELF :

$\qquad\qquad\qquad\qquad$ A $\quad$ + A $\quad$ —#►
$\qquad\qquad\qquad\qquad$ B $\quad$ + B $\quad$ —#►

$\qquad\qquad\qquad\qquad$ PERFECT <u>ALTERNATING</u> COPOLYMER

2. $R_A > 1$ AND $R_B > 1$: EACH REACTING SPECIES PREFERS ITS OWN MONOMER :

   $\qquad\qquad\qquad\qquad$ TENDENCY TO FORM BLOCK COPOLYMERS
   $\qquad\qquad\qquad\qquad$ <u>DOES NOT EXIST</u>.

3. $R_A = R_B = 1$: A AND B CANNOT DISTINGUISH BETWEEN A AND B.

   $\qquad\qquad\qquad\qquad$ ADDITION IS <u>COMPLETELY RANDOM</u>   (EX. E - VA)
   $\qquad\qquad\qquad\qquad$ NO FLUCTUATIONS IN COMPOSITION WITH CONVERSION

4. $R_A R_B = 1$: $R_A = 1/R_2$ OR $\kappa_{AA}/\kappa_{AB} = \kappa_{AB}/\kappa_{BB}$

          BOTH RADICALS A   AND B   EXHIBIT THE SAME RELATIVE

          REACTIVITIES FOR A AND B

          IDEAL COPOLYMERIZATION

5. $R_A R_B \neq 1$:

    A. $R_A R_B > 1$ AND $R_A > 1$, $R_B > 1$ SEE CASE 2

    B. $R_A R_B < 1$ AND $R_A < 1$, $R_B < 1$

          CROSS-PROPAGATION IS FAVORED

    C. $R_A R_B < 1$ AND $R_A > 1$, $R_B < 1$

          VERY FREQUENT CASE

          IF R_A VERY DIFFERENT FROM R_2 : POOR COPOLYMER YIELD

DISTILLATION   **COPOLYMERIZATION**

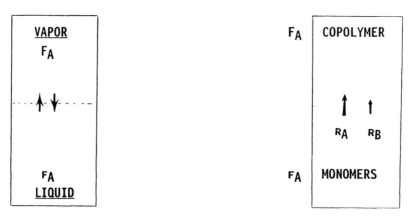

RELATIVE VOLATILITY   **RELATIVE REACTIVITY**

INSTANTANEOUS COMPOSITIONS

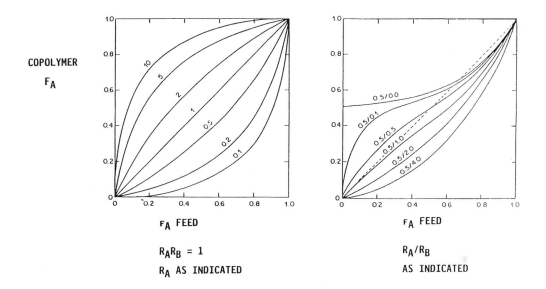

COPOLYMER F_A

F_A FEED

$R_A R_B = 1$
R_A AS INDICATED

F_A FEED

R_A / R_B
AS INDICATED

CATIONIC POLYMERIZATION

- CATIONIC POLYMERIZATION PROCEEDS BY ATTACK ON THE MONOMER OF AN ACIDIC SPECIES RESULTING IN SPLITTING OF THE DOUBLE BOND OR IN RING OPENING, TO PRODUCE A TRIVALENT CARBOCATION ($-\overset{|}{\underset{|}{C}}{}^{+}$) OR AN OXONIUM ION ($-\overset{|}{O}-$).

- <u>INITIATORS</u> : - SOME PROTONIC ACIDS (H_2SO_4, $HClO_4$)
    - LEWIS ACIDS ($AlCl_3$, BF_3, $SnCl_4$, $TiBr_4$, ...)
    - HALOGENOALKYL / ALUMINUM ALKYL

$$HA + H_2C = \overset{\displaystyle X}{\underset{\displaystyle Y}{\overset{|}{\underset{|}{C}}}} \longrightarrow H_3C - \overset{\displaystyle X}{\underset{\displaystyle Y}{\overset{|}{\underset{|}{C}}}}{}^{\oplus} \ldots A^{\ominus}$$

INITIATION BY LEWIS ACIDS $\left(\begin{array}{c} F \\ F \end{array} \!\! \diagdown \!\! B - F \right)$

COCATALYST USUALLY REQUIRED (H_2O, ROH, RCOOH, RCL)

$$BF_3 + H_2O \rightleftharpoons [H^+ \ BF_3OH^-] \ \text{COMPLEX}$$

$$[H^+ \ BF_3OH^-] + \overset{}{\underset{}{C}} = \overset{}{\underset{}{C}} \longrightarrow \overset{}{\underset{}{C}} - \overset{}{\underset{}{C}} \oplus \quad \dots \quad BF_3OH \ominus$$

- CARBOCATIONS ARE INHERENTLY UNSTABLE, REACTIVE SPECIES (ENTHALPY OF ALKYL-X -> ALKYL $^+$ X $^-$ IS HIGH, 140-215 KCAL/MOLE).

- CARBOCATION STABILITY IS DETERMINED BY ITS MICROARCHITECTURE AND NATURE OF ATOMS IN THE VICINITY OF THE $C \oplus$

  - ELECTRON - DONATING GROUPS STABILIZE CATIONS
  - CONJUGATION TENDS TO STABILIZE $C \oplus$

PROPAGATION : CARBENIUM ION + NEGATIVE GEGENION PROCEED TO GROW

$$CH_3 - \overset{X}{\underset{Y}{C}} \oplus \dots A \ominus + CH_2 = \overset{X}{\underset{Y}{C}} \xrightarrow{\ K_P\ } CH_3 - \overset{X}{\underset{Y}{C}} - CH_2 - \overset{X}{\underset{Y}{C}} \oplus \dots A \ominus$$

THERE IS A VARYING DEGREE OF IONIC CHARACTER IN THE BOND BETWEEN $C^+ \dots$ AND $\dots A^-$

- THE RATE CONSTANT K_P WILL BE MARKEDLY DEPENDENT ON THE DEGREE OF ASSOCIATION (= F (NATURE OF A^-, SOLVENT POLARITY, TEMPERATURE).

- R_P CATIONIC IS MUCH LARGER THAN R_P FREE RADICAL

  $[C^+]$ IS 10^{-3} TO 10^{-4} MOL/LITER VERSUS $[C^\bullet] = 10^{-8}$ MOL/LITER

  STEADY - STATE ASSUMPTION NOT VALID

TERMINATION

- BY <u>COMBINATION</u> : $\sim CH_2 - \overset{\oplus}{\underset{X}{CH}} \ldots \overset{\ominus}{A} \longrightarrow \sim CH_2 - \underset{X}{CH} - A$
  (COLLAPSE)

$$\sim CH_2 - \underset{|}{\overset{|}{C}} \oplus \; .. \; BF_3OH^{\ominus} \longrightarrow \sim CH_2 - \underset{|}{\overset{|}{C}} - OH + BF_3$$

- BY <u>TRANSFER</u> : $\sim CH_2 - \underset{|}{\overset{|}{C}} \oplus \; .. \; A^{\ominus} + H_2O \longrightarrow \sim CH_2 - \underset{|}{\overset{|}{C}} - OH + H^{\oplus}A^{\ominus}$

- BY <u>HYDRIDE TRANSFER</u> (ALLYLIC TERMINATION) :

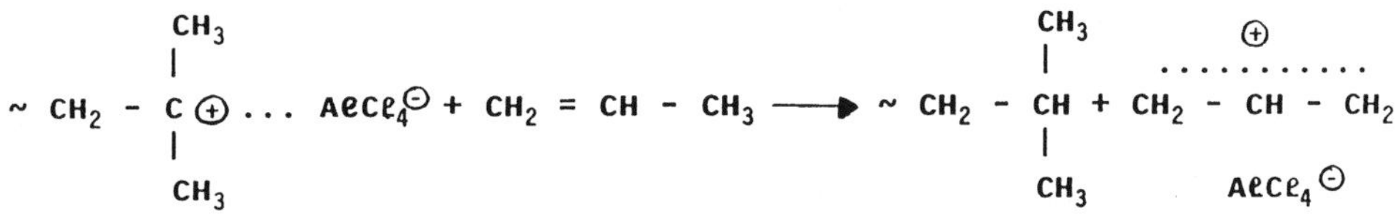

$$\sim CH_2 - \underset{\underset{CH_3}{|}}{\overset{\overset{CH_3}{|}}{C}} \oplus \; \ldots \; A\ell C\ell_4^{\ominus} + CH_2 = CH - CH_3 \longrightarrow \sim CH_2 - \underset{\underset{CH_3}{|}}{\overset{\overset{CH_3}{|}}{CH}} + \underset{A\ell C\ell_4^{\ominus}}{CH_2 - CH - CH_2 \overset{\oplus}{\cdots\cdots\cdots}}$$

TRANSFER : ESSENTIALLY ON MONOMER

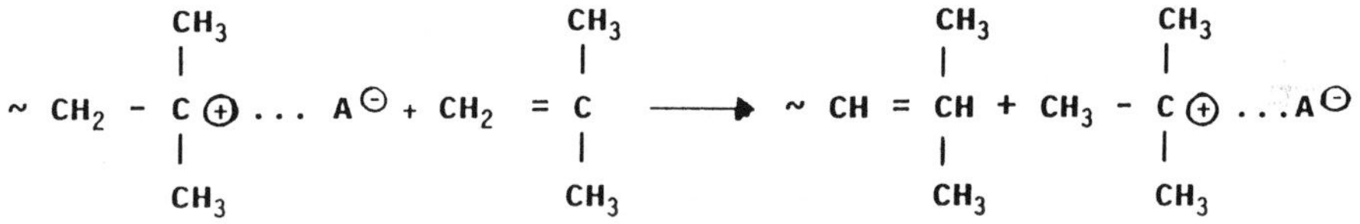

$$\sim CH_2 - \underset{\underset{CH_3}{|}}{\overset{\overset{CH_3}{|}}{C}} \oplus \; \ldots \; A^{\ominus} + CH_2 = \underset{\underset{CH_3}{|}}{\overset{\overset{CH_3}{|}}{C}} \longrightarrow \sim CH = CH + CH_3 - \underset{\underset{CH_3}{|}}{\overset{\overset{CH_3}{|}}{C}} \oplus \; \ldots A^{\ominus}$$

ANIONIC ADDITION POLYMERIZATION

- ANIONIC POLYMERIZATION PROCEEDS BY ATTACK ON MONOMER OF A BASIC
  (NUCLEOHILIC) SPECIES TO   PRODUCE A CARBANION $-C^{\ominus}$ FOLLOWED BY
  PROPAGATION OF THIS ION.

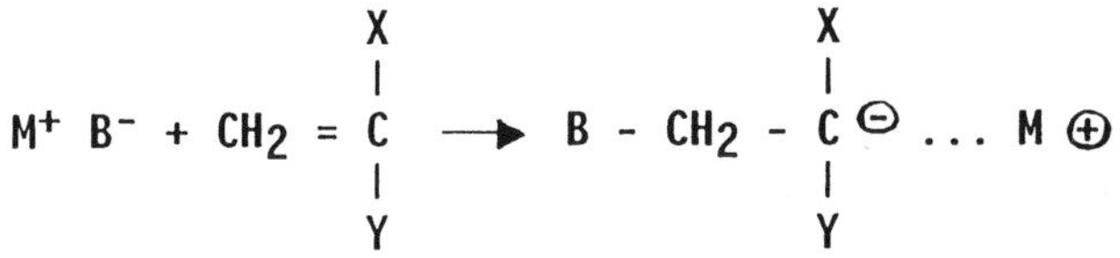

$$M^+ B^- + CH_2 = \underset{\underset{Y}{|}}{\overset{\overset{X}{|}}{C}} \longrightarrow B - CH_2 - \underset{\underset{Y}{|}}{\overset{\overset{X}{|}}{C}}{}^{\ominus} \ldots M^{\oplus}$$

- <u>INITIATORS</u> : ALKYL AND ARYL DERIVATIVES OF ALKALI METALS
        n BULI, NA NAPHTALENE, NaNH2, CH3ONa

- <u>STABILITY OF ANION</u> $-C^{\ominus}$

  ENHANCED BY ELECTRON-WITHDRAWING SUBSTITUENTS SUCH AS
  $-CN > -COOR > -C_6H_5 > -CH = CH_2$ THEY STABILIZE THE ANION
  BY RESONANCE, THEREBY FACILITATING POLYMERIZATION.

- <u>MONOMERS</u> : ACRYLONITRILE ($CH_2 = CH-CN$)

    ACRYLATES ($CH_2 = CH - COOR$)

    CYCLIC OXIDES (R-CH —CH2)
    $$\underset{O}{\diagdown \diagup}$$

    DIENES

    STYRENE

- BEING IONIC IN NATURE IT IS VERY FAST, USUALLY COMPLETED IN 10-30 SECONDS.

- PROPAGATION RATE CONSTANT IS VERY MUCH EFFECTED BY :
    - NATURE OF COUNTERION
    - SOLVENT POLARITY

- SOLVENT MUST BE DRY AND CONTAIN NO ACIDIC COMPOUND "LIVING" POLYMER CAN THEN BE OBTAINED.

PARTICULAR FEATURES

- THE REACTION STOPS WHEN THERMODYNAMIC EQUILIBRIUM IS REACHED, <u>TERMINA-TION REACTIONS</u> MAY BE VIRTUALLY ABSENT AND THE ANIONIC CENTERS REMAIN INTACT SINCE TRANSFER OF H^+ FROM SOLVENT USED (THF, DIOXANE...) DOES NOT OCCUR.

    POLYMER CHAIN IS "<u>LIVING</u>" : $\sim\!\!\sim\!\!\sim\!\!\sim C^- ... M^+$

- ALL $M^+ B^-$ SPECIES ARE AVAILABLE AT THE START OF THE REACTION, <u>ALL CHAINS WILL GROW IMMEDIATELY AND SIMULTANEOUSLY</u>.

- COMBINATION OF SIMULTANEOUS INITIATION AND NO TERMINATION YIELDS <u>VERY NARROW MOLECULAR WEIGHT DISTRIBUTION</u>.

    USUALLY $\bar{X}_W/\bar{X}_N = 1.05 - 1.10$

BLOCK COPOLYMERIZATION

- CONTAIN LONG SEQUENCES OF A AND B WITHIN THE MAIN CHAIN

$$\sim AAA \sim\sim ABBB \text{---} BAAA \sim$$

- ANIONIC POLYMERIZATION (LIVING POLYMER TECHNIQUE) IS PARTICULARLY SUITED :

$$M^+ B^- + A \; CH_2 = CXY \longrightarrow B \left[CH_2 - CXY \right]_{A-1} CH_2 - \overset{\ominus}{CXY} \; .... \; M^{\oplus}$$

$$\xrightarrow{ N \; CH_2 = CHR } \qquad B \left[CH_2 - CXY \right]_A \left[CH_2 - CHR \right]_N$$

E.G. KRATON :

ABA WHERE A IS POLYSTYRENE AND B POLYBUTADIENE OR POLY-ISOPRENE. THEY EXHIBIT ELASTOMERIC PROPERTIES WITHOUT THE NEED FOR VULCANIZATION UP TO THE SOFTENING TEMPERATURE OF POLYSTYRENE.

THEY ARE <u>THERMOPLASTIC ELASTOMERS</u>.

TRIBLOCK COPOLYMER MICROSTRUCTURE

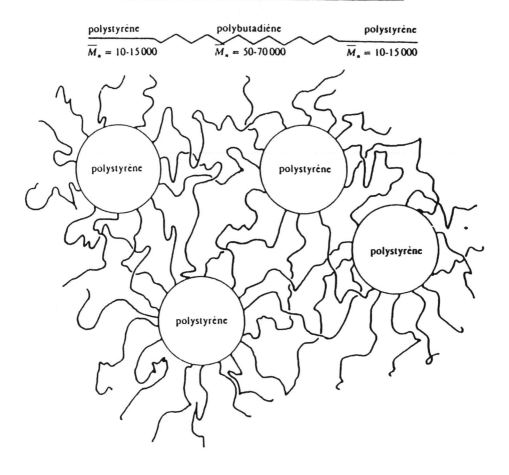

polystyrène	polybutadiène	polystyrène
$\overline{M}_n = 10\text{-}15\,000$	$\overline{M}_n = 50\text{-}70\,000$	$\overline{M}_n = 10\text{-}15\,000$

COORDINATION POLYMERIZATION
(ZIEGLER - NATTA)

A NEW CLASS OF POLYMERIZATION CATALYSTS WAS DISCOVERED IN 1954 TO POSSESS POWERFUL STEREO REGULATING PROPERTIES. THEY FUNCTION BY A COORDINATION COMPLEX BETWEEN THE CATALYST, GROWING CHAIN, AND INCOMING MONOMER ; THE PROCESS IS THUS REFERRED TO AS <u>COORDINATION ADDITION POLYMERIZATION</u>.

Z.N. CATALYSTS ARE USED FOR :

- STEREOSPECIFIC POLYMERIZATION OF α-OLEFINS SUCH AS PROPYLENE
- SYNTHESIS OF LINEAR POLYETHYLENE
- COPOLYMERIZATION OF ETHYLENE AND PROPYLENE
- POLYMERIZATION OF POLYDIENES WITH CONTROLLED MICROSTRUCTURE

COORDINATION CATALYSTS

<u>CATALYSTS</u> : THE CATALYST IS BASED ON VARIOUS COMPONENTS INCLUDING :

1. A GROUP IV - VII TRANSITION METAL COMPOUND
2. A GROUP I - III ORGANOMETALLIC COMPOUND
3. SOMETIMES AN ELECTRON - DONATING LIGAND (AMINE, ETHER..)

COMBINATIONS THAT SHOW ACTIVITY ARE COUNTLESS, IN PRACTICE METAL ALKYLS ARE USUALLY LIMITED TO THOSE OF AL, AND THE TRANSITION METAL COMPONENT TO IONIC SPECIES CONTAINING BETWEEN 0 AND 3 D-ELECTRONS

- TRANSITION METALS OF INTEREST IN ZIEGLER CATALYSIS ARE CHARACTERIZED BY THE INCOMPLETENESS OF THE 3D ORBITAL.

- ANOTHER CHARACTERISTIC OF THE TRANSITION METAL IS THAT IT CAN BE EASILY REDUCED BY THE ORGANOMETALLIC COMPOUND.

Ex. $TiCl_4 \xrightarrow[Al\ (Et)_3]{} TiCl_3 \xrightarrow[Al(Et)_3]{} TiCl_2$

TRANSITION METAL COMPOUNDS

THE ELECTRONIC CONFIGURATION OF THE TRANSITION METALS IS :

Ti [Ar]	$4s^2\ 3d^2$
V	$4s^2\ 3d^3$
Cr	$4s^1\ 3d^5$
Mn	$4s^2\ 3d^5$
Fe	$4s^2\ 3d^6$
Co	$4s^2\ 3d^7$
Ni	$4s^2\ 3d^8$
(Cu	$4s^1\ 3d^{10}$)
(Zn	$4s^2\ 3d^{10}$)

ORBITAL ENERGY LEVELS

- VANADIUM HAS A MAXIMUM VALENCY OF 5 (5 UNPAIRED ELECTRONS), ONE ELECTRON CAN EASILY JUMP FROM 4S TO 3D, VERY LITTLE ENERGY IS REQUIRED, AND IS AT LEAST COMPENSATED BY THE ENERGY RELEASED WHEN TWO EXTRA BONDS ARE FORMED.

- THE NATURE OF THE TRANSITION METAL HAS A LARGE EFFECT ON THE STEREOSPECIFICITY OF THE POLYMER :

  TI   IS MOST USED IN <u>HETEROGENEOUS</u> CATALYSIS FOR ISOTACTIC POLYOLEFINS MANUFACTURE

  V   IS USED FOR <u>SOLUBLE</u>, SYNDIOSPECIFIC CATALYSTS

  IN <u>DIENES</u> POLYMERIZATION :

  CO, NI, FE, RH   FAVOR CIS 1-4 STRUCTURES
  TI, V   LEAD TO 1-4 CIS OR TRANS
  CR   GIVES 1-2 STRUCTURES

- THE NATURE OF THE METAL (BE, AL, MG, ZN) OF THE ORGANOMETALLIC COMPOUND (M. Et_2 OR M. Et_3) ALSO EFFECTS THE DEGREE OF ISOTACTICITY.

MECHANISM OF INITIATION AND PROPAGATION

THE DETAILED CHEMICAL NATURE OF THE ACTIVE CENTER IS VERY COMPLEX AND HAS BEEN DEBATED SINCE THE FIRST DISCOVERIES.

IT IS GENERALLY ACCEPTED THAT A FIRST KEY STEP IS : FORMATION OF AN ALKYL-METAL BOND (M - C) ILLUSTRATED FOR $TiCl_3$ / $AlEt_3$ (TI IN OCTAHEDRAL CONFIGURATION).

REPRESENTS A VACANT D ORBITAL CAPABLE OF COMPLEXING AN OLELINIC MONOMER MOLECULE

STEREOREGULATION

HETEROGENEOUS MODEL

o HEAD-TO-TAIL ENCHAINMENT IS THE ONLY ONE PERMITTED BY THE GEOMETRY OF
THE CHLORINE VACANCY AT THE CRYSTAL SURFACE.

o DRIVING FORCE FOR ISOTACTIC PLACEMENT ORIGINATES FORM STERIC
INTERACTIONS BETWEEN THE ALPHA-SUBSTITUENT AND THE LIGANDS.
THE OLEFIN CAN APPROACH THE M-C BOND IN ONLY ONE CONFIGURATION.

OCTAHEDRIC
CATALYTIC
SITE

A SECOND KEY STEP IS THE COORDINATION OF THE OLEFIN WITH THE TRANSITION
METAL AT THE VACANT OCTAHEDRAL POSITION THROUGH η-BONDING AND INSERTION
OF A C-C BOND BETWEEN THE METAL AND THE ALKYL GROUP :

GROWTH STEP OCCURS AT <u>CATALYTIC SITE</u> NOT AT CHAIN END.

CONFIGURATION OF ALKYLATED Tɪ³⁺ SITES

A

B

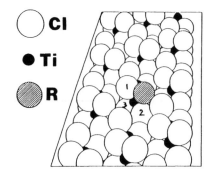

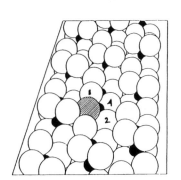

○ **Cl**

● **Ti**

◍ **R**

BEFORE MONOMER ADDITION
OR
AFTER INVERSION OF
VACANT SITE

AFTER MONOMER
INSERTION

STEREOREGULATION

HOMOGENEOUS MODEL :

- MONOMER COMPLEXING MODE LEADING TO HEAD-TO TAIL ENCHAINMENT IS ENERGETICALLY FAVORED (METHYLENE FACING THE VACANCY) BUT SOME HEAD-TO-HEAD ENCHAINMENT IS ALSO OBTAINED.

- STERIC METHYL-METHYL INTERACTIONS FORCE THE INCOMING PROPENE TO BE COMPLEXED IN OPPOSITE CONFIGURATIONS AT EACH CONSECUTIVE GROWTH STEP LEADING TO SYNDIOTACTIC PROPAGATION.

- ROTATION ABOUT THE M-C BOND IS MINIMIZED OR PREVENTED DUE TO LIGAND ENVIRONMENT. WHENEVER THIS ROTATIONAL BARRIER IS OVERCOME, E.G. AT HIGHER TEMPERATURES OR WHEN SMALLER LIGANDS ARE ATTACHED TO M, THE STERIC RESTRICTIONS PLACED ON THE INCOMING MONOMER ARE THE SAME FOR BOTH SYNDIO- AND ISOTACTIC PLACEMENTS RESULTING IN AN ATACTIC POLYMER.

- METHYL-METHYL INTERACTIONS ARE MINIMIZED BY ROTATIONS ABOUT THE M-C GROWTH BOND, MOVING THE SUBSTITUENT OF THE LAST ADDED MONOMER AWAY FROM THE VACANCY, THUS THE SYNDIOTACTIC DRIVING FORCE IS ABSENT.

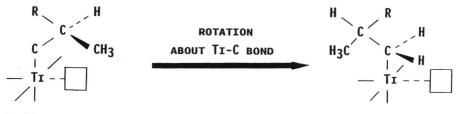

H, CH$_3$ FACE VACANCY

CH$_3$/CH$_3$ INTERACTIONS

H, H FACE VACANCY

No CH$_3$/CH$_3$ INTERACTIONS

POLYMERIZATION SCHEME

INITIATION : [ME] $-$ C$_2$H$_5$ + CH$_2$ = CHR $\longrightarrow$ [ME] $-$ CH$_2$ $-$ CHR $-$ C$_2$H$_5$
∂_+ $\quad$ ∂_- $\qquad\qquad$ ∂_+ $\quad$ ∂_-

PROPAGATION : [ME] $-$ CH$_2$ $-$ CHR $-$ C$_2$H$_5$ + N CH$_2$ = CHR
∂_+ $\quad$ ∂_-

$\longrightarrow$ [ME] $-$ CH$_2$ $-$ CHR $-$(CH$_2$ $-$ CHR)$_N$ $-$ C$_2$H$_5$
∂_+ $\quad$ ∂_-

TRANSFER : A) SPONTANEOUS : [ME] $-$ H + CH$_2$ = CR $-$(CH$_2$ $-$ CHR)$_N$ $-$ C$_2$H$_5$
∂_+ $\quad$ ∂_-

B) ON AL (C$_2$H$_5$)$_3$: [ME] $-$ C$_2$H$_5$ + (C$_2$H$_5$)$_2$ AL $-$ CH$_2$ $-$ CHR $\sim\sim\sim$
∂_+ $\quad$ ∂_-

C) WITH H$_2$: [ME] $-$ H + H $-$ CH$_2$ $-$ CHR $\sim\sim\sim$
∂_+ $\quad$ ∂_-

TERMINATION :
- WHEN CATALYST GETS CLOGGED WITH POLYMER.
- WITH A POISON SUCH AS ROH.

INDEX

A

O

S

For Product Safety Concerns and Information please contact our EU
representative GPSR@taylorandfrancis.com Taylor & Francis Verlag GmbH,
Kaufingerstraße 24, 80331 München, Germany

Printed and bound by CPI Group (UK) Ltd, Croydon, CR0 4YY

01/05/2025
01858619-0001